T0317445

ADAPTIVE FILTERS

ADAPTIVE FILTERS
THEORY AND APPLICATIONS

Second Edition

Behrouz Farhang-Boroujeny
University of Utah
USA

A John Wiley & Sons, Ltd., Publication

This edition first published 2013

First Edition published in 1998

Registered office
John Wiley & Sons Ltd, The Atrium, Southern Gate, Chichester, West Sussex, PO19 8SQ, United Kingdom

For details of our global editorial offices, for customer services and for information about how to apply for permission to reuse the copyright material in this book please see our website at www.wiley.com.

Library of Congress Cataloguing-in-Publication Data has been applied for.

A catalogue record for this book is available from the British Library.

ISBN: 978-1-119-97954-8

Set in 10/12 Times by Laserwords Private Limited, Chennai, India

To Diana for her continuous support, understanding
and love throughout my career

Contents

Preface

This book has grown out of the author's research work and teaching experience in the field of adaptive signal processing as well as signal processing applications to a variety of communication systems. The second edition of this book, while preserving the presentation of the basic theory of adaptive filters as in the first edition, expands significantly on a broad range of applications of adaptive filters. Six new chapters are added that look into various applications of adaptive filters.

This book is designed to be used as a text to teach graduate-level courses in adaptive filters at different levels. It is also intended to serve as a technical reference for practicing engineers.

A typical one-semester introductory course on adaptive filters may cover Chapters 1, 3–6, and 12, and the first half of Chapter 11, in depth. Chapter 2, which contains a short review of the basic concepts of the discrete-time signals and systems, and some related concepts from random signal analyses, may be left as self-study material for students. Selected parts of the rest of this book may also be taught in the same semester, or, broader range of chapters may be used for a second semester course on advanced topics and applications.

In the study of adaptive filters, computer simulations constitute an important supplemental component to theoretical analyses and deductions. Often, theoretical developments and analyses involve a number of approximations and/or assumptions. Hence, computer simulations become necessary to confirm the theoretical results. Apart from this, computer simulation turns out to be a necessity in the study of adaptive filters for gaining an in-depth understanding of the behavior and properties of the various adaptive algorithms. MATLAB® from MathWorks Inc. appears to be the most commonly used software simulation package. Throughout this book, MATLAB® is used to present a number of simulation results to clarify and/or confirm the theoretical developments. The programs as well as data files used for generating these results can be downloaded from the accompanying website of this book at www.wiley.com/go/adaptive_filters

Another integral part of this text is exercise problems at the end of chapters. With the exception of the first few chapters, two kinds of exercise problems are provided in each chapter:

1. **The usual problem exercises**. These problems are designed to sharpen the readers' skill in theoretical development. They are designed to extend results developed in the text and illustrate applications to practical problems. Solutions to these

problems are available to instructors on the companion website to the book: www.wiley.com/go/adaptive_filters

2. **Simulation-oriented problems**. These involve computer simulations and are designed to enhance the readers' understanding on the behavior of the different adaptive algorithms that are introduced in the text. Most of these problems are based on the MATLAB® programs, which are provided on the accompanying website. In addition, there are also other (open-ended) simulation-oriented problems, which are designed to help the readers to develop their own programs and prepare them to experiment with practical problems.

The book assumes that the reader has some background of discrete-time signals and systems (including an introduction to linear system theory and random signal analysis), complex variable theory, and matrix algebra. However, a review of these topics is provided in Chapters 2 and 4.

This book starts with a general overview of adaptive filters in Chapter 1. Many examples of applications such as system modeling, channel equalization, echo cancellation, and antenna arrays are reviewed in this chapter. This follows with a brief review of discrete-time signals and systems in Chapter 2, which puts the related concepts in a framework appropriate for the rest of this book.

In Chapter 3, we introduce a class of optimum linear systems collectively known as *Wiener filters*. Wiener filters are fundamental to the implementation of adaptive filters. We note that the cost function used to formulate the Wiener filters is an elegant choice, leading to a mathematically tractable problem. We also discuss the unconstrained Wiener filters with respect to causality and duration of the filter impulse response. This study reveals many interesting aspects of Wiener filters and establishes a good foundation for the study of adaptive filters for the rest of this book. In particular, we find that, in the limit, when the filter length tends to infinity, a Wiener filter treats different frequency components of underlying processes separately. Numerical examples reveal that when the filter length is limited, separation of frequency components may be replaced by separation of frequency bands within a good approximation. This treatment of adaptive filters, which is pursued throughout this book, turns out to be an enlightening engineering approach for the study of adaptive filters.

Eigenanalysis is an essential mathematical tool for the study of adaptive filters. A thorough treatment of this topic is covered in the first half of Chapter 4. The second half of this chapter gives an analysis of the performance surface of transversal Wiener filters. This is followed by search methods, which are introduced in Chapter 5. The search methods discussed in this chapter are idealized versions of the statistical search methods that are used in practice for actual implementation of adaptive filters. They are idealized in the sense that the statistics of the underlying processes are assumed to be known *a priori*.

The celebrated least-mean-square (LMS) algorithm is introduced in Chapter 6 and extensively studied in Chapters 7–11. The LMS algorithm, which was first proposed by Widrow and Hoff in 1960's, is the most widely used adaptive filtering algorithm, in practice, owing to its simplicity and robustness to signal statistics.

Chapters 12 and 13 are devoted to the method of least-squares. This discussion, although brief, gives the basic concept of the method of least-squares and highlights its advantages and disadvantages compared to the LMS-based algorithms. In Chapter 13, the reader is introduced to the fast versions of least-squares algorithms. Overall, these two chapters lay a good foundation for the reader to continue his/her study of this subject with reference to more advanced books and/or papers.

The problem of tracking is discussed in Chapter 14. In the context of a system modeling problem, we present a generalized formulation of the LMS algorithm, which covers most of the algorithms that are discussed in the previous chapters of this book, thus bringing a common platform for comparison of different algorithms. We also discuss how the step-size parameter(s) of the LMS algorithm and the forgetting factor of the RLS algorithm may be optimized for achieving good tracking behavior.

Chapters 15–20 cover a range of applications where the theoretical results of the previous chapters are applied to a wide range of practical problems. Chapter 15 presents a number of practical problems related to echo cancelers. The chapter emphasis is on acoustic echo cancellation that is encountered in teleconferencing applications. In such applications, one has to deal with specific problems that do not fall into the domain of the traditional theory of adaptive filters. For instance, when both parties at the two sides of the conferencing line talk simultaneously, their respective signals interfere with one another and hence, the adaptation of both echo cancelers on the two sides of the line may be disrupted. Therefore, specific double-talk detection methods should be designed. The stereophonic acoustic echo cancelers that have gained some momentums in recent years are also discussed in detail.

Chapter 16 presents and discusses the underlying problems related to active noise cancellation control, which are also somewhat different from the traditional adaptive filtering problems.

Chapters 17 is devoted to the issues related to synchronization and channel equalization in communication systems. Although many fundamentals of the classical adaptive filters theory have been developed in the context of channel equalization, there are a number of specific issues in the domain of communication systems that can be only presented as new concepts, which may be thought of as extensions of the classical theory of adaptive filters. Many such extensions are presented in this chapter.

Sensor array processing and code division multiple access (CDMA) are two areas where adaptive filters have been used extensively. Although these seem to be two very different applications, there are a number of similarities that if understood allows one to use results of one application for the other as well. As sensor array processing has been developed well ahead of CDMA, we follow this historical development and present sensor array processing techniques in Chapter 18, followed by a presentation of CDMA theory and the relevant algorithms in Chapter 19.

The recent advancement in adaptive filters as applied to the design and implementation multicarrier systems (respectively, orthogonal frequency division multiplexing–OFDM) and communication systems with multiple antennas at both transmitter and receiver sides (known as multi-input multi-output–MIMO) are discussed in Chapter 20. This chapter explains some practical issues related to these modern signal processing techniques and presents a few solutions that have been adopted in the current standards, e.g., WiFi, WiMax, and LTE.

The following notations are adopted in this book. We use nonbold lowercase letters for scalar quantities, bold lowercase for vectors, and bold uppercase for matrices. Nonbold uppercase letters are used for functions of variables, such as $H(z)$, and lengths/dimensions of vectors/matrices. The lowercase letter "n" is used for the time index. In the case of block processing algorithms, such as those discussed in Chapters 8 and 9, we reserve the lowercase letter "k" as the block index. The time and block indices are put in brackets, while subscripts are used to refer to elements of vectors and matrices. For example, the ith element of the time-varying tap-weight vector $\mathbf{w}(n)$ is denoted as $\omega_i(n)$. The superscripts "T" and "H" denote vector or matrix transposition and Hermitian transposition, respectively. We keep all vectors in column form. More specific notations are explained in the text as and when found necessary.

Behrouz Farhang-Boroujeny

Acknowledgments

Development of the first edition of this book as well as the present edition has relied on a variety of contributions from my students in the past 20 years. I wish to thank all of them for their support, enthusiasm, and encouragements.

I am grateful to Professor Stephen Elliott, University of Southampton, UK, for reviewing Chapter 16 and making many valuable suggestions. I am also thankful to Professor Vellenki Umapathi Reddy for reviewing Chapters 2–7, providing valuable suggestions, and giving me moral support. My special thanks go to Dr. George Mathew of Hitachi Global Storage Technologies for critically reviewing the entire manuscript of the first edition of this book.

Many colleagues who adopted the first edition of this book for their classes have been the source of encouragement for me to take the additional task of preparing the second edition of this book. I hope they find this edition more useful in broadening the knowledge of our graduate students by giving them many examples of advanced applications of adaptive filters.

1

Introduction

As we begin our study of "adaptive filters," it may be worth trying to understand the meaning of the terms *adaptive* and *filters* in a very general sense. The adjective "adaptive" can be understood by considering a system that is trying to adjust itself so as to respond to some phenomenon that is taking place in its surroundings. In other words, the system tries to adjust its parameters with the aim of meeting some well-defined goal or target that depends on the state of the system as well as its surrounding. This is what "adaptation" means. Moreover, there is a need to have a set of steps or certain procedure by which this process of "adaptation" is carried out. And finally, the "system" that carries out and undergoes the process of "adaptation" is called by the more technical, yet general enough, name "filter" – a term that is very familiar to and a favorite of any engineer. Clearly, depending on the time required to meet the final target of the adaptation process, which we call convergence time, and the complexity/resources that are available to carry out the adaptation, we can have a variety of adaptation algorithms and filter structures. From this point of view, we may summarize the contents/contribution of this book as "the study of some selected adaptive algorithms and their implementations along with the associated filter structures from the points of view of their convergence and complexity performance."

1.1 Linear Filters

The term *filter* is commonly used to refer to any device or system that takes a mixture of particles/elements from its input and processes them according to some specific rules to generate a corresponding set of particles/elements at its output. In the context of signals and systems, *particles/elements* are the *frequency components* of the underlying signals and, traditionally, filters are used to retain all the frequency components that belong to a particular band of frequencies, while rejecting the rest of them, as much as possible. In a more general sense, the term *filter* may be used to refer to a system that *reshapes* the frequency components of the input to generate an output signal with some desirable features, and this is how we view the concept of filtering throughout the chapters which follow.

Filters (or systems, in general) may be either *linear* or *nonlinear*. In this book, we consider only linear filters and our emphasis will also be on *discrete-time* signals and systems. Thus, all the signals will be represented by sequences, such as $x(n)$. The most basic feature of linear systems is that their behavior is governed by *the principle of superposition*. This means that if the responses of a linear discrete-time system to input sequences

Adaptive Filters: Theory and Applications, Second Edition. Behrouz Farhang-Boroujeny.

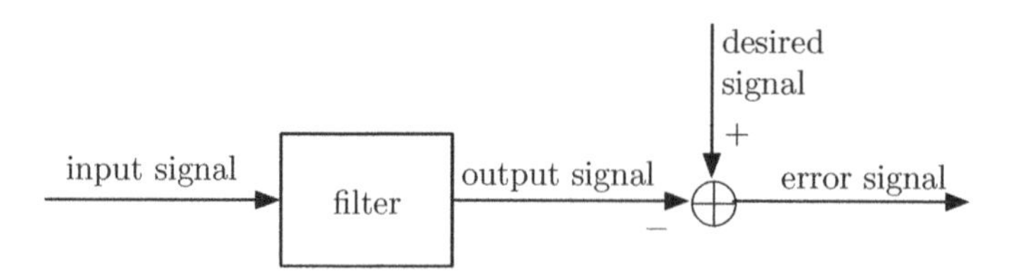

Figure 1.1 Schematic diagram of a filter emphasizing its role of reshaping the input signal to match the desired signal.

$x_1(n)$ and $x_2(n)$ are $y_1(n)$ and $y_2(n)$, respectively, the response of the same system to the input sequence $x(n) = ax_1(n) + bx_2(n)$, where a and b are arbitrary constants, will be $y(n) = ay_1(n) + by_2(n)$. This property leads to many interesting results in "linear system theory." In particular, a linear system is completely characterized by its impulse response or the Fourier transform of its impulse response known as *transfer function*. The transfer function of a system at any frequency is equal to its gain at that frequency. In other words, in the context of our discussion above, we may say that the transfer function of a system determines how the various frequency components of its input are reshaped by the system.

Figure 1.1 depicts a general schematic diagram of a filter emphasizing the purpose for which it is used in different problems addressed/discussed in this book. In particular, the filter is used to reshape a certain *input signal* in such a way that its output is a good estimate of the given desired signal. The process of selecting the filter parameters (coefficients) so as to achieve the best match between the desired signal and the filter output is often done by optimizing an appropriately defined *performance function*. The performance function can be defined in a *statistical* or *deterministic* framework. In the statistical approach, the most commonly used performance function is the *mean-squared value* of the *error signal*, that is, difference between the desired signal and the filter output. For stationary input and desired signals, minimizing the mean squared error (MSE) results in the well-known *Wiener filter*, which is said to be *optimum in the mean-square sense*. The subject of Wiener filters is extensively covered in Chapter 3. Most of the adaptive algorithms that are studied in this book are practical solutions to Wiener filters. In the deterministic approach, the usual choice of performance function is *a weighted sum of the squared error signal*. Minimizing this function results in a filter that is optimum for the given set of data. However, under some assumptions on certain statistical properties of the data, the deterministic solution will approach the statistical solution, that is, the Wiener filter, for large data lengths. Chapters 12 and 13 deal with the deterministic approach in detail. We refer the reader to Section 1.4 for a brief overview of the adaptive formulations under the stochastic (i.e., statistical) and deterministic frameworks.

1.2 Adaptive Filters

As we mentioned in the previous section, the filter required for estimating the given desired signal can be designed using either the stochastic or the deterministic formulations. In the deterministic formulation, the filter design requires the computation of certain average quantities using the given set of data that the filter should process. On the other hand, the design of Wiener filter (i.e., in the stochastic approach) requires *a priori* knowledge of the statistics of the underlying signals. Strictly speaking, a large number of realizations of the underlying signal sequences are required for reliably estimating these statistics. This procedure is practically not feasible because we usually have only one realization for each of the signal sequences. To resolve this problem, it is assumed that the underlying signal sequences are *ergodic*, which means that they are stationary and their statistical and time averages are identical. Thus, using the time averages, Wiener filters can be designed, even though there is only one realization for each of the signal sequences.

Although, direct measurement of the signal averages to obtain the necessary information for the design of Wiener or other optimum filters is possible, in most of the applications, the signal averages (statistics) are used in an indirect manner. All the algorithms that are covered in this book take the output error of the filter, correlate that with the samples of filter input in some way, and use the result in a recursive equation to adjust the filter coefficients iteratively. The reasons for solving the problem of adaptive filtering in an iterative manner are as follows:

1. Direct computation of the necessary averages and their application for computing the filter coefficients requires accumulation of a large amount of signal samples. Iterative solutions, on the other hand, do not require accumulation of signal samples, thereby *resulting in a significant amount of saving in memory.*
2. Accumulation of signal samples and their postprocessing to generate the filter output, as required in noniterative solutions, introduces a large delay in the filter output. This is unacceptable in many applications. *Iterative solutions*, on the contrary, *do not introduce any significant delay in the filter output.*
3. The use of iterations results in adaptive solutions with some *tracking* capability. That is, if the signal statistics are changing with time, the solution provided by an iterative adjustment of the filter coefficients will be able to adapt to the new statistics.
4. Iterative solutions, in general, are much *simpler* to code in software or implement in hardware than their noniterative counterparts.

1.3 Adaptive Filter Structures

The most commonly used structure in the implementation of adaptive filters is the *transversal structure*, depicted in Figure 1.2. Here, the adaptive filter has a single input, $x(n)$, and an output, $y(n)$. The sequence $d(n)$ is the desired signal. The output, $y(n)$, is generated as a linear combination of the delayed samples of the input sequence, $x(n)$, according to Equation (1.1)

$$y(n) = \sum_{i=0}^{N-1} w_i(n)x(n-i) \tag{1.1}$$

where $w_i(n)$'s are the filter *tap weights* (coefficients) and N is the filter length. We refer to the input samples, $x(n-i)$, for $i = 0, 1, \ldots, N-1$, as the filter *tap inputs*. The tap weights, $w_i(n)$'s, which may vary with time, are controlled by the adaptation algorithm.

In some applications, such as beamforming (Section 1.6.4), the filter tap inputs are not the delayed samples of a single input. In such cases, the structure of the adaptive filter

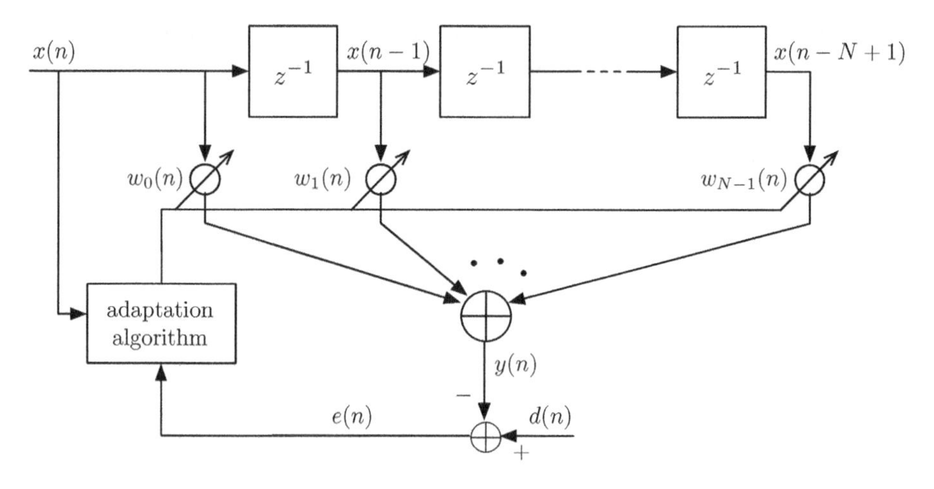

Figure 1.2 Adaptive transversal filter.

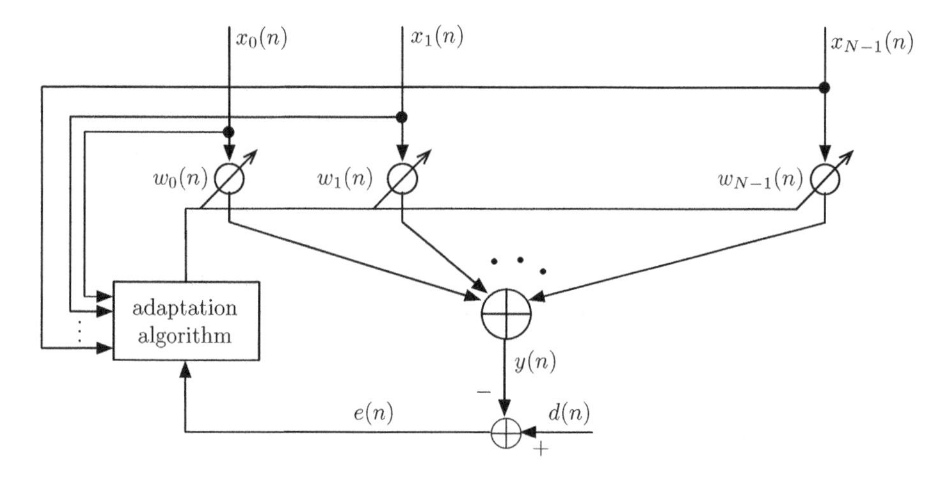

Figure 1.3 Adaptive linear combiner.

assumes the form shown in Figure 1.3. This is called *a linear combiner* as its output is a linear combination of the different signals received at its tap inputs:

$$y(n) = \sum_{i=0}^{N-1} w_i(n) x_i(n) \tag{1.2}$$

Note that the linear combiner structure is more general than the transversal. The latter, as a special case of the former, can be obtained by choosing $x_i(n) = x(n-i)$.

The structures of Figures 1.2 and 1.3 are those of the nonrecursive filters, that is, computation of filter output does not involve any feedback mechanism. We also refer to Figure 1.2 as a finite-impulse response (FIR) filter as its impulse response is of finite duration in time. An infinite-impulse response (IIR) filter is governed by recursive equations such as (Figure 1.4)

$$y(n) = \sum_{i=0}^{N-1} a_i(n)x(n-i) + \sum_{i=1}^{M-1} b_i(n)y(n-i) \tag{1.3}$$

where $a_i(n)$ and $b_i(n)$ are the forward and feedback tap weights, respectively. IIR filters have been used in many applications. However, as we shall see in the later chapters, because of the many difficulties involved in the adaptation of IIR filters, their application in the area of adaptive filters is rather limited. In particular, they can easily become unstable because their poles may get shifted out of the unit circle (i.e., $|z| = 1$, in the z-plane, Chapter 2) by the adaptation process. Moreover, the performance function (e.g., MSE as a function of filter coefficients) of an IIR filter usually has many local minima points. This may result in convergence of the filter to one of the local minima and not to the desired global minimum point of the performance function. On the contrary, the MSE functions of FIR filter and linear combiner are well-behaved quadratic functions with a single minimum point, which can easily be found through various adaptive algorithms. Because of these points, the nonrecursive filters are the sole candidates in most of the applications of adaptive filters. Hence, most of our discussions in the subsequent chapters are limited to the nonrecursive filters. The IIR-adaptive filters with two specific examples of their applications are discussed in Chapter 10.

The FIR and IIR structures shown in Figures 1.2 and 1.4 are obtained by direct realization of the respective difference equations (1.1) and (1.3). These filters may alternatively be implemented using the *lattice structures*. The lattice structures, in general, are more complicated than the direct implementations. However, in certain applications, they have some advantages which make them better candidates than the direct forms. For instance, in the application of linear prediction for speech processing where we need to realize all-pole (IIR) filters, the lattice structure can be more easily controlled to prevent possible instability of the filter. Derivation of lattice structures for both FIR and IIR filters are presented in Chapter 11. Also, in the implementation of the method of least-squares (Section 1.4.2), the use of lattice structure leads to a computationally efficient algorithm known as *recursive least-squares (RLS) lattice*. A derivation of this algorithm is presented in Chapter 13.

The FIR and IIR filters which were discussed above are classified as linear filters because their outputs are obtained as linear combinations of the present and past samples of input and, in the case of IIR filter, the past samples of the output also. Although most applications are restricted to the use of linear filters, nonlinear adaptive filters become necessary in some applications where the underlying physical phenomena to be modeled are far from being linear. A typical example is magnetic recording where the recording channel becomes nonlinear at high densities because of the interaction among the magnetization transitions written on the medium. The Volterra series representation of systems is usually used in such applications. The output, $y(n)$, of a Volterra system is related to

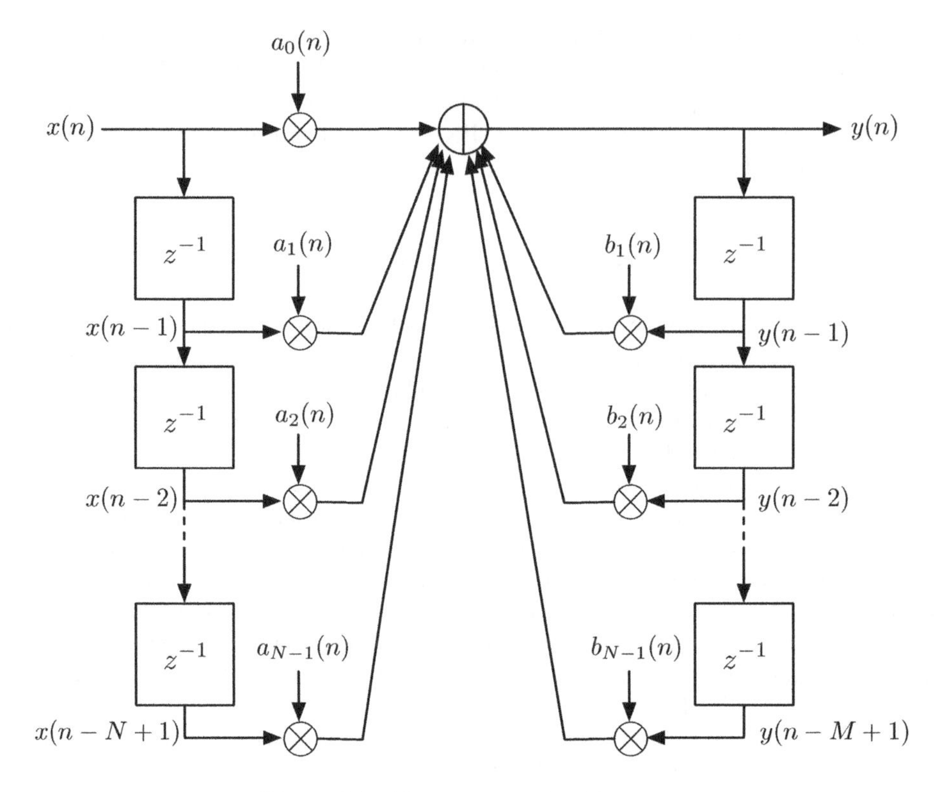

Figure 1.4 The structure of an IIR filter.

its input, $x(n)$, according to the equation

$$y(n) = w_{0,0}(n) + \sum_i w_{1,i}(n)x(n-i) + \sum_{i,j} w_{2,i,j}(n)x(n-i)x(n-j) + \sum_{i,j,k} w_{3,i,j,k}(n)x(n-i)x(n-j)x(n-k) + \ldots \tag{1.4}$$

where $w_{0,0}(n)$, $w_{1,i}(n)$'s, $w_{2,i,j}(n)$'s, $w_{1,i,j,k}(n)$'s, ... are filter coefficients. In this book, we do not discuss the Volterra filters any further. However, we note that all the summations in Eq. (1.4) may be put together and the Volterra filter may be thought of as a linear combiner whose inputs are determined by the delayed samples of $x(n)$ and their cross-multiplications. Noting this, we find that the extension of most of the adaptive filtering algorithms to the Volterra filters is straightforward.

1.4 Adaptation Approaches

As introduced in Sections 1.1 and 1.2, there are two distinct approaches that have been widely used in the development of various adaptive algorithms, viz. stochastic and deterministic. Both these approaches have many variations in their implementations leading to a rich variety of algorithms; each of which offers desirable features of its own. In this section, we present a review of these two approaches and highlight the main features of the related algorithms.

1.4.1 Approach Based on Wiener Filter Theory

According to the Wiener filter theory, which comes from the stochastic framework, the optimum coefficients of a linear filter is obtained by minimization of its MSE. As was noted before, strictly speaking, the minimization of MSE requires certain statistics obtained through ensemble averaging, which may not be possible in practical applications. The problem is resolved using ergodicity so as to use time averages instead of ensemble averages. Furthermore, to come up with simple recursive algorithms, very rough estimates of the required statistics are used. In fact, the celebrated *least-mean square* (LMS) algorithm, which is the most basic and widely used algorithm in various adaptive filtering applications, uses the *instantaneous* value of the square of the error signal as an estimate of the MSE. It turns out that this very rough estimate of the MSE, when used with a *small* step-size parameter in searching for the optimum coefficients of the Wiener filter, leads to a very simple and yet reliable adaptive algorithm.

The main disadvantage of the LMS algorithm is that its convergence behavior is highly dependent on the power spectral density of the filter input. When the filter input is white, that is, its power spectrum is flat across the whole range of frequencies, the LMS algorithm converges very fast. However, when certain bands of frequencies are not well excited (i.e., the signal energy in those bands is relatively low), some slow modes of convergence appear, thus resulting in very slow convergence compared to the case of white input. In other words, to converge fast, the LMS algorithm requires equal excitation over the whole range of frequencies. Noting this, over the years, researchers have developed many algorithms that effectively divide the frequency band of the input signal into a number of subbands and achieve some degree of signal whitening using some power normalization mechanism before applying the adaptive algorithm. These algorithms which appear in different forms are presented in Chapters 7, 9, and 11.

In some applications, we need to use adaptive filters whose length exceeds a few hundreds or even a few thousands of taps. Clearly, such filters are computationally expensive to implement. An effective way of implementing such filters at a much lower computational complexity is to use the fast Fourier transform (FFT) algorithm to implement the time domain convolutions in the frequency domain, as is commonly done in the implementation of long digital filters (Oppenheim and Schafer, 1975, 1989). Adaptive algorithms which use FFT for reducing computational complexity are presented in Chapter 8.

1.4.2 Method of Least-Squares

The adaptive filtering algorithms whose derivations are based on the Wiener filter theory have their origin in a statistical formulation of the problem. In contrast to this, the *method of least-squares* approaches the problem of filter optimization from a deterministic point of view. As mentioned before, in the Wiener filter theory, the desired filter is obtained by minimizing the MSE, that is, a statistical quantity. In the method of least-squares, on the other hand, the performance index is the *sum of weighted error squares* for the given data, that is, a deterministic quantity. A consequence of this deterministic approach (that will become clear as we go through its derivation in Chapter 12) is that the least-squares-based algorithms, in general, converge much faster than the LMS-based algorithms. They are also insensitive to the power spectral density of the input signal. The price that is paid for achieving this improved convergence performance is higher computational complexity and poorer numerical stability.

Direct formulation of the least-squares problem results in a matrix formulation of its solution which can be applied on block-by-block basis to the incoming signals. This, which is referred to as *block estimation of the least-squares method*, has some useful applications in areas such as linear predictive coding (LPC) of speech signals. However, in the context of adaptive filters, recursive formulations of the least-squares method that update the filter coefficients after the arrival of every sample of input are preferred because of the reasons that were given in Section 1.2. There are three major classes of RLS adaptive filtering algorithms and are as follows:

Standard RLS algorithm. The derivation of this algorithm involves the use of a well-known result from linear algebra known as the *matrix inversion lemma*. Consequently, the implementation of the standard RLS algorithm involves matrix manipulations that result in a computational complexity proportional to the square of the filter length.

QR-decomposition-based RLS (QRD-RLS) algorithm. This formulation of RLS algorithm also involves matrix manipulations, which leads to a computational complexity that grows with the square of the filter length. However, the operations involved here are such that they can be put into some regular structures known as *systolic arrays*. Another important feature of the QRD-RLS algorithm is its robustness to numerical errors compared to other types of RLS algorithms (Haykin, 1991, 1996)

Fast RLS algorithms. In the case of transversal filters, the tap inputs are successive samples of input signal, $x(n)$ (Figure 1.1). The fast RLS algorithms use this property of the filter input and solve the problem of least-squares with a computational complexity, which is proportional to the length of the filter, thus the name *fast RLS*. Two types of fast RLS algorithms may be recognized:

RLS lattice algorithms. These lattice algorithms involve the use of order-update as well as the time-update equations. A consequence of this feature is that it results in modular structures, which are suitable for hardware implementations using the pipelining technique. Another desirable feature of these algorithms is that certain variants of them are very robust against numerical errors arising from the use of finite word lengths in computations.

Fast transversal RLS algorithm. In terms of number of operations per iteration, the fast transversal RLS algorithm is less complex than the lattice RLS algorithms. However, it suffers from numerical instability problems that require careful attention to prevent undesirable behavior in practice.

In this book, we present a complete treatment of the various LMS-based algorithms in seven chapters. However, our discussion on RLS algorithms is rather limited. We present a comprehensive treatment of the properties of the method of least-squares and a derivation of the standard RLS algorithm in Chapter 12. The basic results related to the development of fast RLS algorithms and some examples of such algorithms are presented in Chapter 13. A study of the *tracking behavior* of selected adaptive filtering algorithms is presented in Chapter 14 of this book. The use of these algorithms to various applications are discussed in Chapters 15 through 20.

1.5 Real and Complex Forms of Adaptive Filters

There are some practical applications in which the filter input and its desired signal are complex-valued. A good example of this situation appears in digital data transmission, where the most widely used signaling techniques are phase shift keying (PSK) and quadrature-amplitude modulation (QAM). In this application, the baseband signal consists of two separate components, which are the real and imaginary parts of a complex-valued signal. Moreover, in the case of frequency domain implementation of adaptive filters (Chapter 8) and subband adaptive filters (Chapter 9), we will be dealing with complex-valued signals, even though the original signals may be real-valued. Thus, we find cases where the formulation of the adaptive filtering algorithms must be given in terms of complex-valued variables.

In this book, to keep our presentation as simple as possible, most of the derivations are given for real-valued signals. However, wherever we find it necessary, the extensions to complex forms will also be followed.

1.6 Applications

Adaptive filters by their very nature are self-designing systems that can adjust themselves to different environments. As a result, adaptive filters find applications in such diverse fields as control, communications, radar and sonar signal processing, interference cancellation, active noise control (ANC), biomedical engineering, and so on. The common feature of these applications that brings them under the same basic formulation of adaptive filtering is that they all involve a process of filtering some input signal to match a desired response. The filter parameters are updated by making a set of measurements of the underlying signals and applying that to the adaptive filtering algorithm such that the difference between the filter output and the desired response is minimized in either statistical or deterministic sense. In this context, four basic classes of adaptive filtering applications are recognized. Namely, modeling, inverse modeling, linear prediction, and interference cancellation. In the rest of this chapter, we present an overview of these applications.

1.6.1 Modeling

Figure 1.5 depicts the problem of modeling in the context of adaptive filters. The aim is to estimate the parameters of the model, $W(z)$, of a plant, $G(z)$. On the basis of some

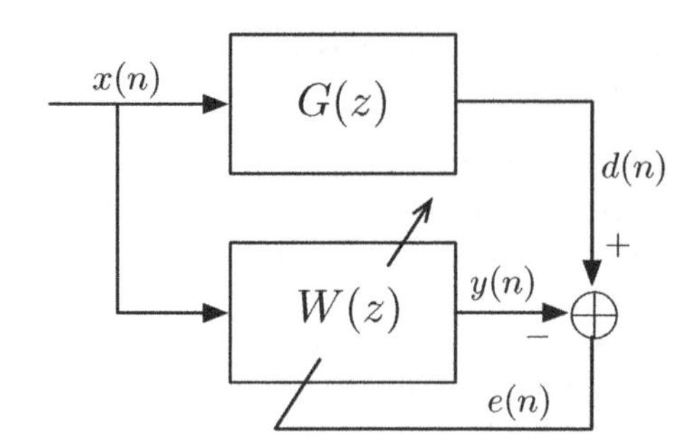

Figure 1.5 Adaptive system modeling.

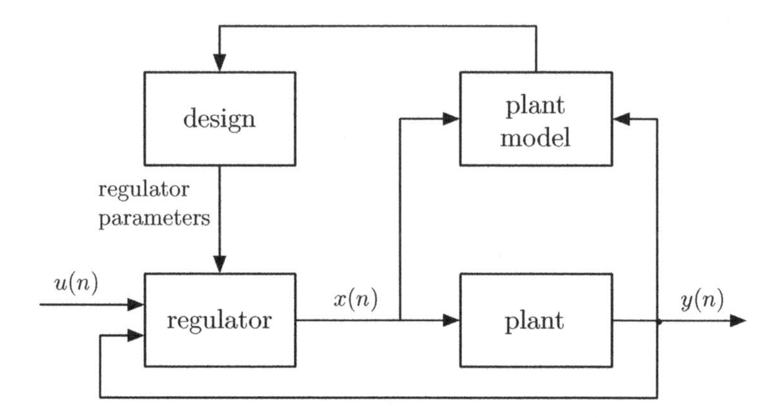

Figure 1.6 Block diagram of a self-tuning regulator.

a priori knowledge of the plant, $G(z)$, a transfer function, $W(z)$, with certain number of adjustable parameters is selected first. The parameters of $W(z)$ are then chosen by an adaptive filtering algorithm such that the difference between the plant output, $d(n)$, and the adaptive filter output, $y(n)$, is minimized.

An application of modeling, which may be readily thought of, is *system identification*. In most modern control systems, the plant under control is identified on-line and the result is used in a *self-tuning regulator* (STR) loop, as depicted in Figure 1.6 (see e.g., Astrom and Wittenmark (1980)).

Another application of modeling is *echo cancellation*. In this application, an adaptive filter is used to identify the impulse response of the path between the source from which the echo originates and the point where the echo appears. The output of the adaptive filter, which is an estimate of the echo signal, can then be used to cancel the undesirable echo. The subject of echo cancellation is discussed further under the topic of interference cancellation.

Nonideal characteristics of communication channels often result in some distortion in the received signals. To mitigate such distortion, channel equalizers are usually used. This technique, which is equivalent to implementing the inverse of the channel response,

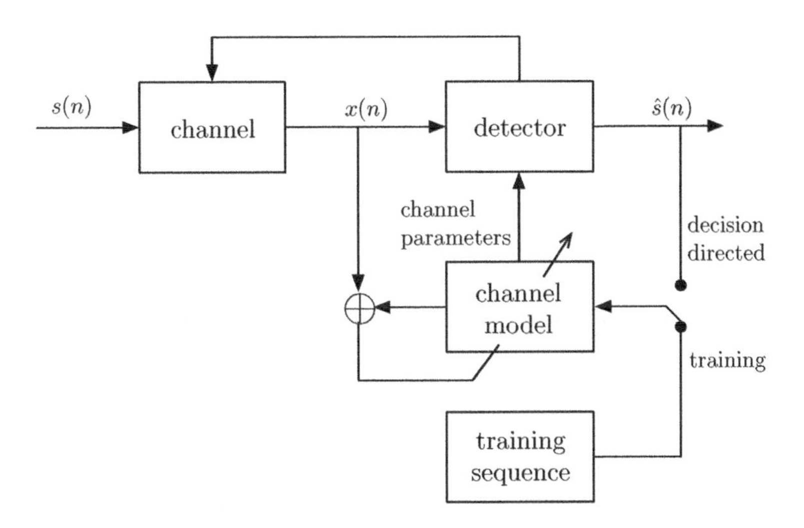

Figure 1.7 An adaptive data receiver using channel identification.

is discussed in the following under the topic of inverse modeling. Direct modeling of the channel, however, has also been found useful in some implementations of data receivers. For instance, data receivers equipped with maximum-likelihood detectors require an estimate of the channel response (Proakis, 1995). Furthermore, computation of equalizer coefficients from channel response has been proposed by some researchers because this technique has been found to result in better tracking of time-varying channels (Fechtel and Meyr (1991) and Farhang-Boroujeny and Wang (1995)). In such applications, a *training pattern* is transmitted in the beginning of every connection. The received signal, which acts as the desired signal to an adaptive filter, is used in a setup, as shown in Figure 1.7 to identify the channel. Once the channel is identified and the normal mode of transmission begins, the detected data symbols, $\hat{s}(n)$, are used as input to the channel model and the adaptation process continues for tracking possible variations of the channel. This is known as *decision-directed mode* and is also shown in Figure 1.7.

1.6.2 Inverse Modeling

Inverse modeling, also known as *deconvolution*, is another application of adaptive filters that has found extensive use in various engineering disciplines. The most widely used application of inverse modeling is in communications where an inverse model (also called *equalizer*) is used to mitigate the channel distortion. The concept of inverse modeling has also been applied to adaptive control systems where a controller is to be designed and cascaded with a plant so that the overall response of this cascade matches a desired (target) response (Widrow and Stearns, 1985). The process of prediction, which is explained later, may also be viewed as an inverse modeling scheme (Section 1.6.3). In this section, we concentrate on the application of inverse modeling in channel equalization. The full treatment of the subject of channel equalization is presented in Chapter 17.

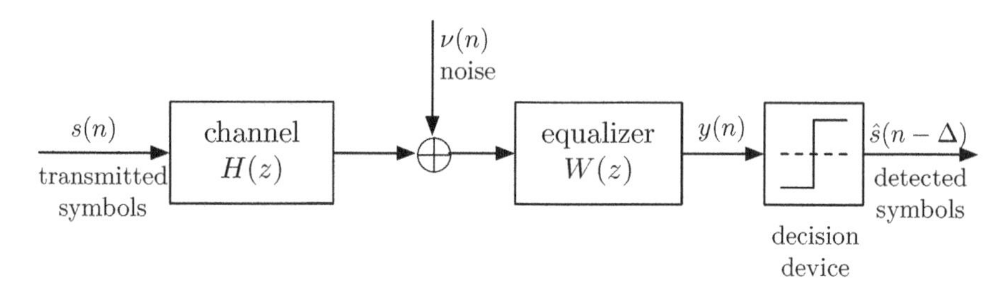

Figure 1.8 A baseband data transmission system with channel equalizer.

Channel Equalization

Figure 1.8 depicts the block diagram of a baseband transmission system equipped with a channel equalizer. Here, the channel represents the combined response of the transmitter filter, the actual channel, and the receiver front-end filter. The additive noise sequence, $\nu(n)$, arises from thermal noise in the electronic circuits and possible crosstalks from neighboring channels. The transmitted data symbols, $s(n)$, which appear in the form of amplitude/phase modulated pulses, are distorted by the channel. The most significant among the different distortions is the *pulse-spreading effect,* which results because the channel impulse response is not equal to an ideal impulse function, and instead a response which is nonzero over many symbol periods. This distortion results in interference of neighboring data symbols with one another, thereby making the detection process through a simple threshold detector unreliable. The phenomenon of interference among neighboring data symbols is known as *intersymbol interference* (ISI). The presence of the additive noise samples, $\nu(n)$, further deteriorates the performance of data receivers. The role of the equalizer, as a filter, is to resolve the distortion introduced by the channel (i.e., rejection or minimization of ISI), while minimizing the effect of additive noise at the threshold detector input (equalizer output) as much as possible. If the additive noise could be ignored, the task of equalizer would be rather straightforward. For a channel $H(z)$, an equalizer with transfer function $W(z) = 1/H(z)$ could do the job perfectly as this results in an overall channel equalizer transfer function $H(z)W(z) = 1$, which implies that the transmitted data sequence, $s(n)$, will appear at the detector input without any distortion. Unfortunately, this is an ideal situation which cannot be used in most of the practical applications.

We note that the inverse of the channel transfer function, that is, $1/H(z)$, may be noncausal if $H(z)$ happens to have a zero outside the unit circle, thus making it unrealizable in practice. This problem is solved by selecting the equalizer so that $H(z)W(z) \approx z^{-\Delta}$, where Δ is an appropriate integer delay. This is equivalent to saying that a delayed replica of the transmitted symbols appears at the equalizer output. Example 3.4 of Chapter 3 clarifies the concept of noncausality of $1/H(z)$ and also the way the problem is (approximately) solved by introducing a delay, Δ. Greater details appear in Chapter 17.

We also note that the choice of $W(z) = 1/H(z)$ (or $W(z) \approx z^{-\Delta}/H(z)$) may lead to a significant enhancement of the additive noise, $\nu(n)$, in those frequency bands where the magnitude of $H(z)$ is small (i.e., $1/H(z)$ is large). Hence, in choosing an equalizer, $W(z)$, one should keep a balance between residual ISI and noise enhancement at the equalizer output. Wiener filter is a solution with such a balance (Chapter 3, Section 3.6.4).

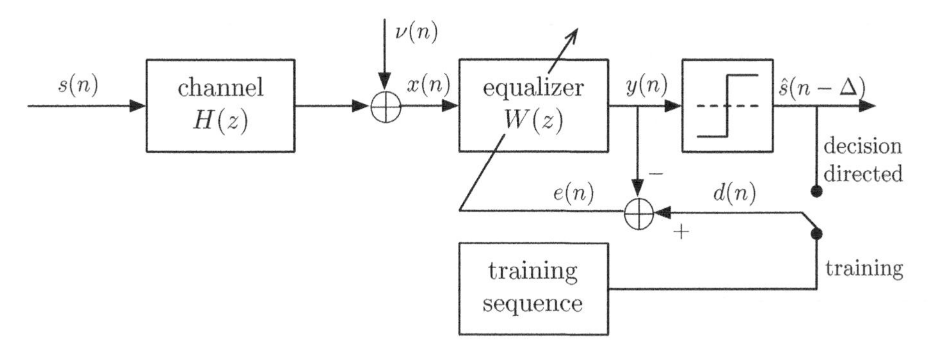

Figure 1.9 Details of a baseband data transmission system equipped with an adaptive channel equalizer.

Figure 1.9 presents the details of a baseband transmission system, equipped with an adaptive equalizer. The equalizer is usually implemented in the form of a transversal filter. Initial training of the equalizer requires knowledge of the transmitted data symbols as they (to be more accurate, a delayed replica of them) should be used as the desired signal samples for adaptation of the equalizer tap weights. This follows from the fact that the equalizer output should ideally be the same as the transmitted data symbols. We thus require an initialization period during which the transmitter sends a sequence of training symbols that are known to the receiver. This is called *the training mode*. Training symbols are usually specified as part of the standards, and the manufacturers of data modems[1] should comply with these so that the modems of different manufacturers can communicate with one another.

At the end of the training mode, the tap weights of the equalizer would have converged close to their optimal values. The detected symbols would then be similar to the transmitted symbols with a probability close to 1. Hence, then onward, the detected symbols can be treated as the desired signal for further adaptation of the equalizer so that possible variations of the channel can be tracked. This mode of operation of the equalizer is called *the decision-directed mode*. The decision-directed mode successfully works as long as the channel variation is slow enough so that the adaptation algorithm is able to follow the channel variations satisfactorily. This is necessary for the purpose of ensuring low-symbol error rates in detection so that these symbols can still be used as the desired signal.

The inverse modeling discussed previously defines the equalizer as an approximation of $z^{-\Delta}/H(z)$, that is, the target/desired response of the cascade of channel and equalizer is $z^{-\Delta}$, a pure delay. This can be generalized by replacing the target response $z^{-\Delta}$ by a general target response, say $\Gamma(z)$. In fact, to achieve higher efficiency in the usage of the available bandwidth, some special choices of $\Gamma(z) \neq z^{-\Delta}$ are usually considered in communication systems. Systems which incorporate such nontrivial target responses are referred to as *partial-response signaling systems*. The detector in such systems is no more the simple threshold detector, but one which can exploit the information that the overall channel is now $\Gamma(z)$, instead of the trivial memoryless channel $z^{-\Delta}$. The Viterbi detector

[1] The term *modem* which is the abbreviation for "modulator and demodulator" is commonly used to refer data transceivers (transmitter and receiver).

(Proakis, 1995) is an example for such a detector. The target response, $\Gamma(z)$, is selected so that its magnitude response approximately matches the channel response, that is, $|\Gamma(e^{j\omega})| \approx |H(e^{j\omega})|$, over the range of frequencies of interest. The impact of this choice is that the equalizer, which is now $W(z) \approx \Gamma(z)/H(z)$, has a magnitude response that is approximately equal to 1, thereby minimizing the noise enhancement. To clarify further on this and also to mention another application of inverse modeling, we discuss the problem of magnetic recording next.

Magnetic Recording

The process of writing data bits on a magnetic medium (tape or disk) and reading them back later is similar to sending data bits over a communication channel from one end of a transmission line and receiving them at the other side of the line. The data bits, which are converted to signal pulses before recording, undergo some distortion because of nonperfect behavior of the head and medium, as it happens in communication channels because of the nonideal response of the channel. Additive thermal noise and interference from neighboring recording tracks (just like neighboring channels in communications) are also present in the magnetic recording channels (Bergmans, 1996).

Magnetic recording channels are usually characterized by their response to an isolated pulse of width 1-bit interval, T. This is known as *dibit response* and in the case of hard-disk channels, it is usually modeled by the superposition of a positive and negative *Lorentzian* pulses, separated by 1-bit interval, T. In other words, the Lorentzian pulse models the step response of the channel. The Lorentzian pulse is defined as

$$g_a(t) = \frac{1}{1 + \left(\frac{2t}{t_{50}}\right)^2} \tag{1.5}$$

where t_{50} is the pulse width measured at 50% of its maximum amplitude. The subscript "a" in $g_a(t)$ and other functions that appear in the rest of this subsection are to emphasize that they are analog (nonsampled) signals. The ratio $D = t_{50}/T$ is known as the *recording density*. Typical values of D are in the range of 1 to 3. A higher density means more bits are contained in one t_{50} interval, that is, more ISI. We may also note that t_{50} is a temporal measure of the recording density. When measured spatially, we obtain another parameter $pw_{50} = t_{50}/v$, where v is the velocity of the medium with respect to head. Accordingly, for a given speed, v, the value of D specifies the actual number of bits written on a length pw_{50} along the track on the magnetic medium.

Using Eq. (1.5), the dibit response of a hard-disk channel is obtained as

$$h_a(t) = g_a(t) - g_a(t - T) \tag{1.6}$$

The response of the channel to a sequence $s(n)$ of data bits is then given by the convolution sum

$$u_a(t) = \sum_n s(n) h_a(t - nT) \tag{1.7}$$

Thus, the dibit response, $h_a(t)$, is nothing but the impulse response of the recording channel.

Figure 1.10a and b shows the dibit (time domain) and magnitude (frequency domain) responses, respectively, of the magnetic channels (based on the Lorentzian model) for densities $D = 1$, 2, and 3. From Figure 1.10b, we note that most of the energy in the read-back signals is concentrated in a midband range between zero and an upper limit around $1/2T$. Clearly, the bandwidth increases with increase in density. In the light of our previous discussions, we may thus choose the target response, $\Gamma(z)$, of the equalizer so that it resembles a bandpass filter whose bandwidth and magnitude response are close to those of the Lorentzian dibit responses. In magnetic recording, the most commonly used partial responses (i.e., target responses) are given by the class-IV response

$$\Gamma(z) = z^{-\Delta}(1 + z^{-1})^K(1 - z^{-1}) \tag{1.8}$$

where Δ, as before, is an integer delay and K is an integer greater than or equal to 1. As the recording density increases, higher values of K will be required to match the channel characteristics. But, as K increases, the channel length also increases, implying higher complexity in the detector. In Chapter 10, we elaborate on these aspects of partial-response systems.

1.6.3 Linear Prediction

Prediction is a spectral estimation technique that is used for modeling correlated random processes for the purpose of finding a parametric representation of these processes. In general, different parametric representations could be used to model the processes. In the context of linear prediction, the model used is shown in Figure 1.11. Here, the random process, $\tilde{x}(n)$, is assumed to be generated by exciting the filter $G(z)$ with the input $u(n)$. As $G(z)$ is an all-pole filter, this is known as *autoregressive* (AR) modeling. The choice/type of the excitation signal, $u(n)$, is application dependent and may vary depending on the nature of the process being modeled. However, it is usually chosen to be a white process.

Other models used for parametric representation are *moving average* (MA) models, where $G(z)$ is an all-zero (transversal) filter, and *autoregressive-moving average* (ARMA) models, where $G(z)$ has both poles and zeros. However, the use of AR model is more popular than other two.

The rationale behind the use of AR modeling may be explained as follows. As the samples of any given nonwhite random signal, $x(n)$, are correlated with one another, these correlations could be used to make a prediction of the present sample of the process, $x(n)$, in terms of its past samples, $x(n-1)$, $x(n-2)$, ..., $x(n-N)$, as shown in Figure 1.12. Intuitively, such prediction improves as the predictor length increases. However, the improvement obtained may become negligible once the predictor length, N, exceeds certain value, which depends on the extent of correlation in the given process. The prediction error, $e(n)$, will then be approximately white. We now note that the transfer function between the input process, $x(n)$, and the prediction error, $e(n)$, is

$$H(z) = 1 - \sum_{i=1}^{N} a_i z^{-i} \tag{1.9}$$

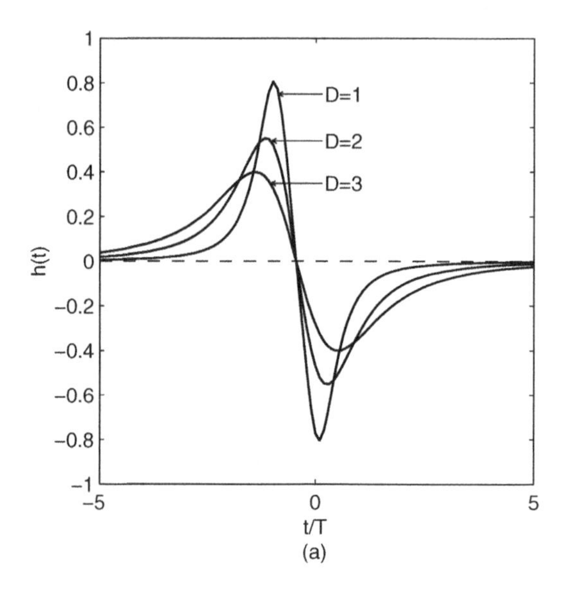

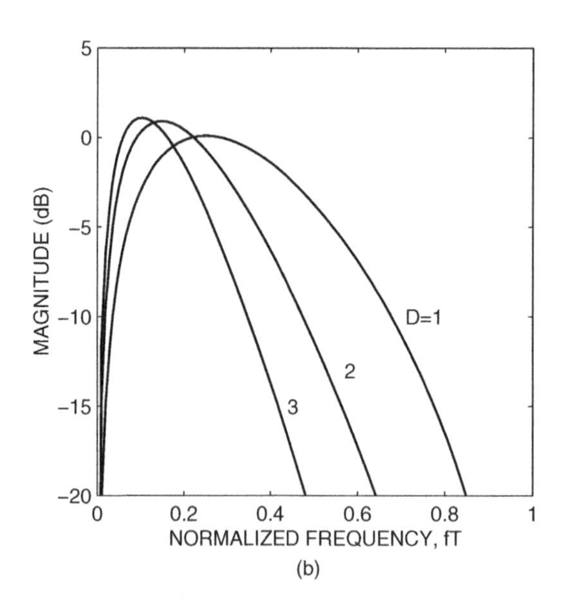

Figure 1.10 Time and frequency domain responses of magnetic recording channels for densities $D = 1$, 2, and 3 modeled using the Lorentzian pulse: (a) dibit response; (b) magnitude response of dibit response.

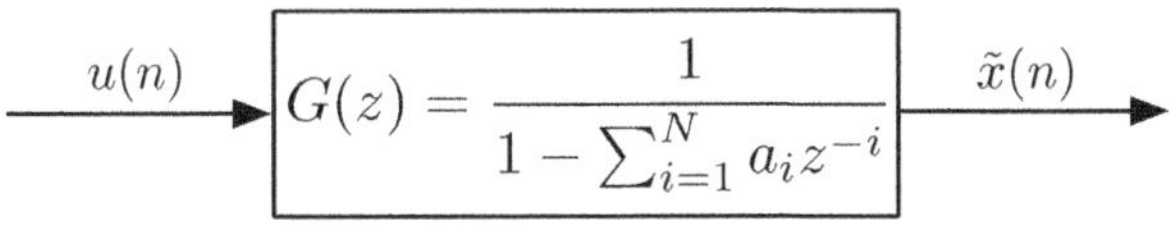

$u(n)$ → $G(z) = \frac{1}{1 - \sum_{i=1}^{N} a_i z^{-i}}$ → $\tilde{x}(n)$

Figure 1.11 Autoregressive modeling of a random process.

where a_i's are the predictor coefficients. Now, if a white process, $u(n)$, with similar statistics as $e(n)$ is passed through an all-pole filter with the transfer function

$$G(z) = \frac{1}{1 - \sum_{i=1}^{N} a_i z^{-i}} \tag{1.10}$$

as shown in Figure 1.11, the generated output, $\tilde{x}(n)$, will clearly be a process with the same statistics as $x(n)$.

With the background developed above, we are now ready to discuss a few applications of adaptive prediction.

Autoregressive Spectral Analysis

In certain applications, we need to estimate the power spectrum of a random process. A trivial way of obtaining such estimate is to take the Fourier transform (discrete Fourier transform (DFT) in the case of discrete-time processes) and use some averaging (smoothing) technique to improve the estimate. This comes under the class of *nonparametric spectral estimation techniques* (Kay, 1988). When the number of samples of the input is limited, the estimates provided by nonparametric spectral estimation techniques will become unreliable. In such cases, the *parametric spectral estimation*, as explained above, may give more reliable estimates.

As mentioned already, parametric spectral estimation could be done using either AR, MA, or ARMA models (Kay, 1988). In the case of AR modeling, we proceed as follows. We first choose a proper order, N, for the model. The observed sequence, $x(n)$, is then applied to a predictor structure similar to Figure 1.12 whose coefficients, a_i's, are optimized by minimizing the prediction error, $e(n)$. Once the predictor coefficients have

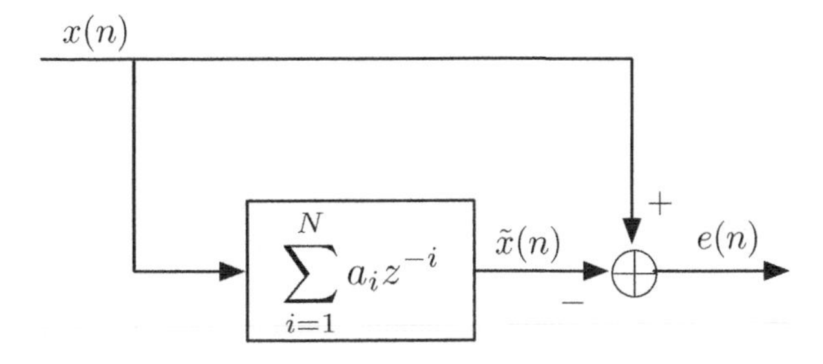

Figure 1.12 Linear predictor.

converged, an estimate of the power spectral density of $x(n)$ is obtained according to the following equation:

$$\Phi_{xx}(e^{j\omega}) = N_o \left| \frac{1}{1 - \sum_{i=1}^{N} a_i e^{-j\omega i}} \right|^2 \tag{1.11}$$

where N_o is an estimate of the power of the prediction error, $e(n)$. This follows from the model of Figure 1.11 and the fact that after the convergence of the predictor, $e(n)$ is approximately white. For further explanation on the derivation of Eq. (1.11) from the signal model of Figure 1.11, refer to Chapter 2 (Section 2.4.4).

Adaptive Line Enhancement

Adaptive line enhancement refers to the situation where a narrow-band signal embedded in a wide-band signal (usually, white) needs to be extracted. Depending on the application, the extracted signal may be the signal of interest, or an unwanted interference that should be removed. Examples of the latter case are a spread spectrum signal that has been corrupted by a narrow-band signal and biomedical measurement signals that have been corrupted by the 50/60 Hz power-line interference.

The idea of using prediction to extract a narrow-band signal when mixed with a wide-band signal follows from the following fundamental result of signal analysis: successive samples of a narrow-band signal are highly correlated with one another, whereas there is almost no correlation between successive samples of a wide-band process. Because of this, if a process $x(n)$ consisting of the sum of a narrow-band and wide-band processes is applied to a predictor, the predictor output, $\hat{x}(n)$, will be a good estimate of the narrow-band portion of $x(n)$. In other words, the predictor will act as a narrow-band filter, which rejects most of the wide-band portion of $x(n)$ and keeps (enhances) the narrow-band portion, thus the name *line enhancer*. Examples of line enhancers can be found in Chapters 6 and 10. In particular, in Chapter 10, we find that line enhancers can be best implemented using IIR filters.

We also note that in the applications where the narrow-band portion of $x(n)$ has to be rejected (such as the examples mentioned above), the difference between $x(n)$ and $\hat{x}(n)$, that is, the estimation error, $e(n)$, is taken as the system output. In this case, the transfer function between the input, $x(n)$, and the output, $e(n)$, will be that of a notch filter.

Speech Coding

Since the advent of digital signal processing, speech processing has always been one of the focused research areas. Among various processing techniques that have been applied to speech signals, linear prediction has been found to be the most promising technique leading to many useful algorithms. In fact, most of the theory of prediction was developed in the context of speech processing.

There are two major speech coding techniques that involve linear prediction (Jayant and Noll, 1984). Both these techniques aim at reducing the number of bits used for every second of speech to achieve saving in storage and/or transmission bandwidth. The first

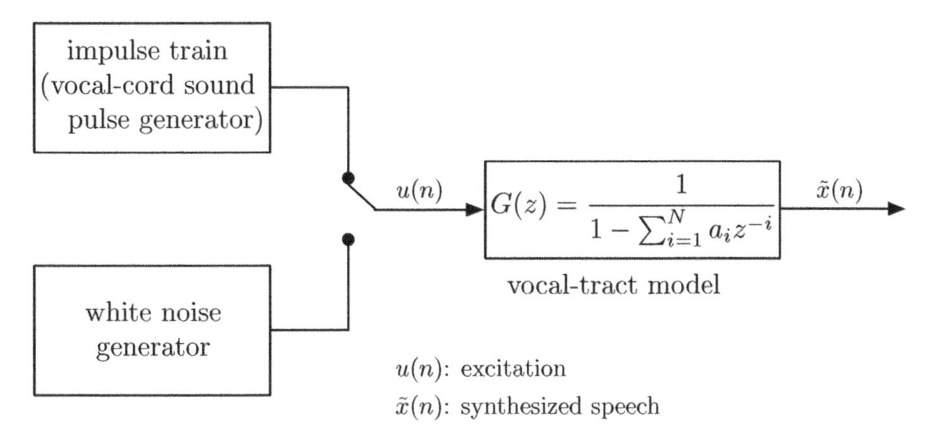

Figure 1.13 Speech-production model.

technique, which is categorized under the class of *source coders*, strives to produce digitized voice data at low bit rates in the range of 2 to 10 kb/s. The synthesized speech, however, is not of a high quality. It sounds more synthetic, lacking naturalism. Hence, it becomes difficult to recognize the speaker. The second technique, which comes under the class of *waveform coders*, gives much better quality at the cost of a much higher bit rate (typically, 32 kb/s).

The main reason for linear prediction being widely used in speech coding is that speech signals can be accurately modeled as shown in Figure 1.13. Here, the all-pole filter is the vocal-tract model. The excitation to this model, $u(n)$, is either a white noise in the case of unvoiced sounds (fricatives such as /s/ and /f/), or an impulse train in the case of voiced sounds (vowels such as /i/). The period of the impulse train, known as *pitch period*, and the power of the white noise, known as *excitation level*, are parameters of the speech model which are to be identified in the coding process.

Linear Predictive Coding (LPC)

Speech signal is a highly nonstationary process. The vocal-tract shape undergoes variations to generate different sounds in uttering each word. Accordingly, in LPC, to code a speech signal, it is first partitioned into segments of 10-30 ms long. These segments are short enough for the vocal-tract shape to be nearly stationary, so that the parameters of the speech-production model of Figure 1.13 could be assumed fixed. Then, the following steps are used to obtain the parameters of each segment:

1. Using the predictor structure shown in Figure 1.12, the predictor coefficients, a_i's, are obtained by minimizing the prediction error $e(n)$ in the least-squares sense, for the given segment.
2. The energy of the prediction error $e(n)$ is measured. This specifies the level of excitation required for synthesizing this segment.
3. The segment is classified as voiced or unvoiced.
4. In the case of voiced speech, the pitch period of the segment is measured.

The following parameters are then stored or transmitted for every segment, as the coded speech: (i) predictor coefficients, (ii) energy of excitation signal, (iii) voiced/unvoiced classification, and (iv) pitch period in the case of voiced speech. These parameters can then (when necessary) be used in a model similar to Figure 1.13 to synthesize the speech signal.

Waveform Coding

The most direct way of waveform coding is the *standard pulse-code modulation* (PCM) technique, where the speech signal samples are directly digitized into a prescribed number of bits to generate the information bits associated with the coded speech. Direct quantization of speech samples requires relatively large number of bits (usually, 8 bits per sample) in order to be able to reconstruct the original speech with an acceptable quality.

A modification of the standard PCM, known as *differential pulse-code modulation* (DPCM), employs a linear predictor such as Figure 1.12 and uses the bits associated with the quantized samples of the prediction error, $e(n)$, as the coded speech. The rationale here is that the prediction error, $e(n)$, has much smaller variance than the input, $x(n)$. Thus, for a given quantization level, $e(n)$ may be quantized with less number of bits compared to $x(n)$. Moreover, as the number of information bits per every second of the coded speech is directly proportional to the number of bits used per sample, bit rate of the DPCM will be less compared to the standard PCM.

The prediction filter used in DPCM can be fixed or be made adaptive. A DPCM system with an adaptive predictor is called *adaptive DPCM* (ADPCM). In the case of speech signals, use of ADPCM results in superior performance compared to the case where a nonadaptive DPCM is used. In fact, the ADPCM has been standardized and widely used in practice (ITU Recommendation G.726).

Figure 1.14 depicts a simplified diagram of the ADPCM system, as proposed in ITU Recommendation G.726.[2]. Here, the predictor is a six-zero, two-pole adaptive IIR filter. The coefficients of this filter are adjusted adaptively so that the quantized error $\tilde{e}(n)$ is minimized in mean-square sense. The predictor input $\tilde{x}(n)$ is same as the original input $x(n)$ except for the quantization error in $\tilde{e}(n)$. To understand the joint operation of the encoder and decoder shown in Figure 1.14, note that the same signal, $\tilde{e}(n)$, is used as inputs to the predictor structures at the encoder and decoder. Hence, if the stability of the loop consisting of the predictor and adaptation algorithm could be guaranteed, then the steady-state value of the reconstructed speech at the decoder, that is, $\tilde{x}'(n)$, will be equal to that at the encoder, that is, $\tilde{x}(n)$, as nonequal initial conditions of the encoder and decoder loops will die away after their transient phase.

1.6.4 Interference Cancellation

Interference cancellation refers to situations where it is required to cancel an interfering signal/noise from the given signal which is a mixture of the desired signal and the interference. The principle of interference cancellation is to obtain an estimate of interfering

[2] ITU stands for International Telecommunication Union

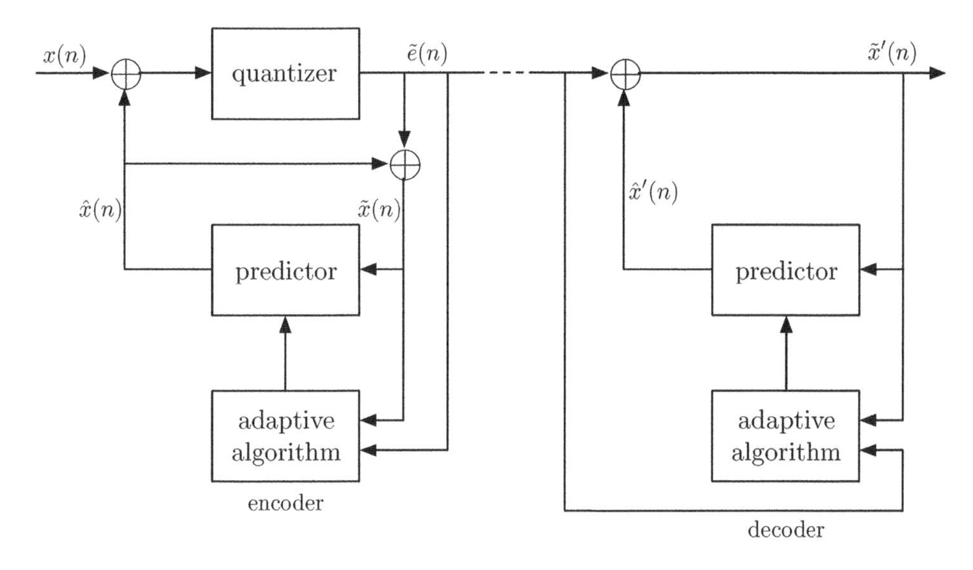

Figure 1.14 ADPCM encoder–decoder.

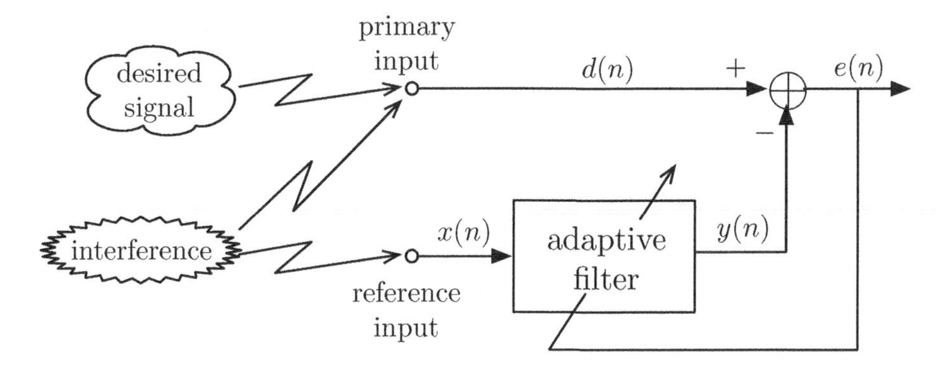

Figure 1.15 Interference cancellation.

signal and subtract that from the corrupted signal. Feasibility of this idea relies on the availability of a reference source from which the interfering signal originates.

Figure 1.15 depicts the concept of interference cancellation, in its simplest form. There are two inputs to the canceler: *primary* and *reference*. The primary input is the corrupted signal, that is, the desired signal plus interference. The reference input, on the other hand, originates from the interference source only.[3] The adaptive filter is adjusted so that a replica of the interference signal that is present in the primary signal appears at its output, $y(n)$. Subtracting this from the primary input results in an output which is cleared from interference, thus the name interference cancellation.

[3] In some applications of interference cancellation, there might also be some leakage of the desired signal to the reference input. Here, we have ignored this situation for simplicity.

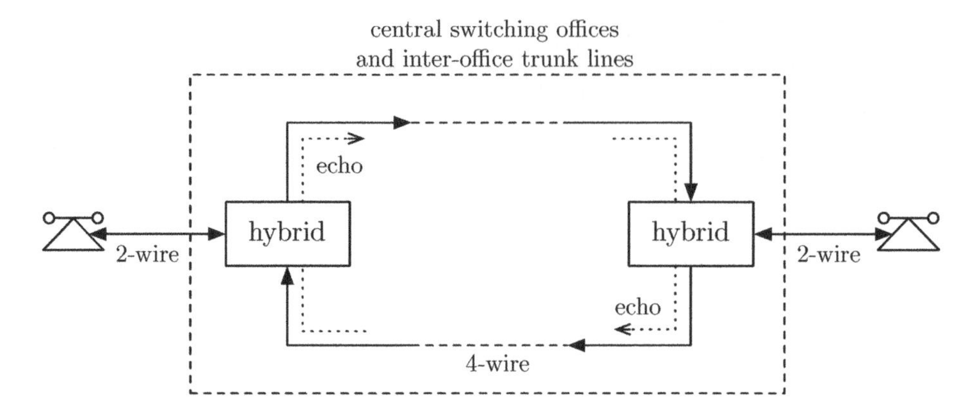

Figure 1.16 Simplified diagram of a telephone network.

We note that the interference cancellation configuration of Figure 1.15 is different from the previous cases of adaptive filters, in the sense that the residual error (which was discarded in other cases) is the cleaned-up signal, here. The desired signal in the previous cases has been replaced here by a noisy (corrupted) version of the actual desired signal. Moreover, the use of the term *reference* to refer the adaptive filter input is clearly related to the role of this input in the canceler.

In the rest of this section, we present some specific applications of interference canceling.

Echo Cancellation in Telephone Lines

Echoes in telephone lines mostly occur at points where hybrid circuits are used to convert four-wire networks to two-wire ones. Figure 1.16 presents a simplified diagram of a telephone connection network, highlighting the points where echoes occur. The two wires at the ends are subscriber loops connecting customers' telephones to central offices. It may also include some portions of the local network. The four wires, on the other hand, are carrier systems (trunk lines) for medium-to-long-haul transmission. The distinction is that the two-wire segments carry signals in both directions on the same lines, while in the four-wire segment signals in the two directions are transmitted on two separate lines. Accordingly, the role of hybrid circuit is to separate the signals in the two directions. Perfect operation of the hybrid circuit requires that the incoming signal from the trunk lines should be directed to the subscriber line and that there be no leakage (echo) of that to the return line. In practice, however, such ideal behavior cannot be expected from hybrid circuits. There would always be some echo on the return path. In the case of voice communications (i.e., ordinary conversation on telephone lines), effect of the echoes becomes more obvious (and annoying to the speaker) in long-distance calls, where the delay with which the echo returns to the speaker may be in the range of a few hundred milliseconds. In digital data transmission, both short- and long-delay echoes are serious.

As was noted before and also can clearly be seen from Figure 1.17, the problem of echo cancellation may be viewed as one of system modeling. An adaptive filter is put

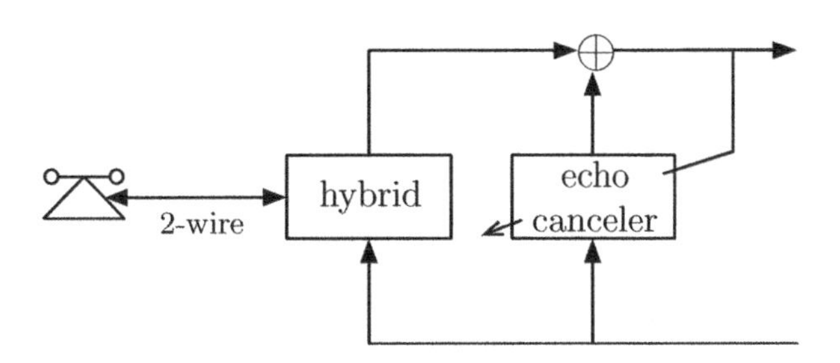

Figure 1.17 Adaptive echo canceler.

between the incoming and outgoing lines of the hybrid. By adapting the filter to realize an approximation of the echo path, a replica of the echo is obtained at its output. This is then subtracted from the outgoing signal to clear that from the undesirable echo.

Echo cancelers are usually implemented in transversal form. The time spread of echoes in a typical hybrid circuit is in the range of 20–30 ms. If we assume a sampling rate of 8 kHz for the operation of the echo canceler, an echo spread of 30 ms requires an adaptive filter with at least 240 taps (30 ms×8 kHz). This is a relatively long filter, requiring a high-speed digital signal processor for its realization. Frequency domain processing is often used to reduce the high computational complexity of long filters. The subject of frequency domain adaptive filters is covered in Chapter 8.

The echo cancelers described previously are applicable to both voice and data transmission. However, more stringent conditions need to be satisfied in the case of data transmission. To maximize the usage of the available bandwidth, full-duplex data transmission is often used. This requires the use of a hybrid circuit for connecting the data modem to the two-wire subscriber loop, as shown in Figure 1.18. The leakage of the transmitted data back to the receiver input is thus inevitable and an echo canceler has to be added, as indicated in Figure 1.18. However, we note that the data echo cancelers are different from the voice echo cancelers used in central switching offices in many ways. For instance, because the input to the data echo canceler are data symbols, it can operate at the data symbol rate that is in the range of 2.4–3 kHz (about three times smaller than the 8 kHz sampling frequency used in voice echo cancelers). For a given echo spread, a lower sampling frequency implies less number of taps for the echo canceler. Clearly, this simplifies the implementation of the echo canceler, greatly. On the other hand, the data echo cancelers require to achieve a much higher level of echo cancellation to ensure reliable transmission of data at higher bit rates. In addition, the echoes returned from the other side of the trunk lines should also be taken care of. Detailed discussions on these issues can be found in Lee and Messerschmitt (1994) and Gitlin, Hayes, and Weinstein (1992).

Acoustic Echo Cancellation

The problem of acoustic echo cancellation can be best explained by referring to Figure 1.19, which depicts the scenario that arises in teleconferencing applications. The speech signal from a far-end speaker, received through a communication channel, is broadcast by a loudspeaker in a room and its echo is picked up by a microphone.

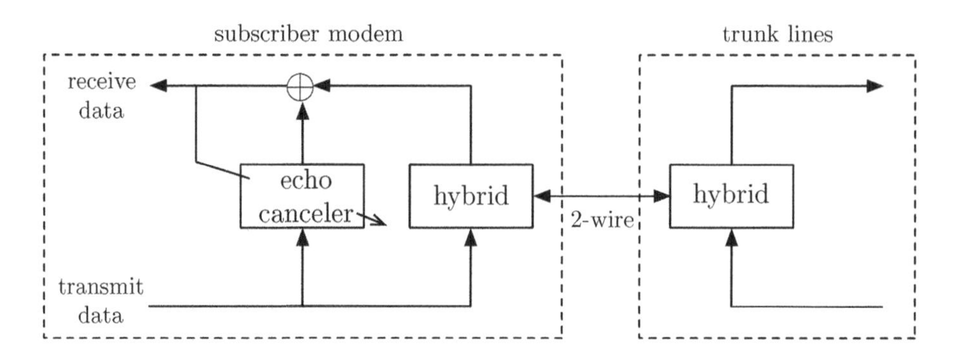

Figure 1.18 Data echo canceler.

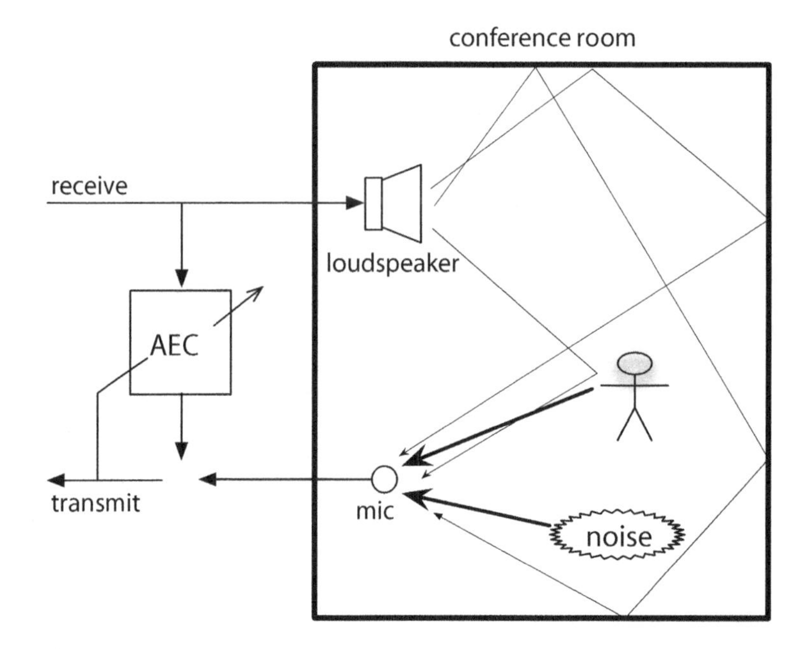

Figure 1.19 Acoustic echo cancellation.

This echo must be canceled to prevent its feedback to the far-end speaker. The microphone also picks up the near-end speaker(s) speech and possible background noise, which may exist in the room. An adaptive transversal filter with sufficient length is used to model the acoustics of the room. A replica of the loudspeaker echo is then obtained and subtracted from the microphone signal before the transmission.

Clearly, the problem of acoustic echo cancellation can also be posed as one of system modeling. The main challenge here is that the echo paths spread over a relatively long length in time. For typical office rooms, echoes in the range of 100–250 ms spread is quite common. For a sampling rate of 8 kHz, this would mean 800–2000 taps! Thus, the main problem of acoustic echo cancellation is that of realizing very long adaptive filters. In addition, as speech is a lowpass signal, it becomes necessary to use special algorithms

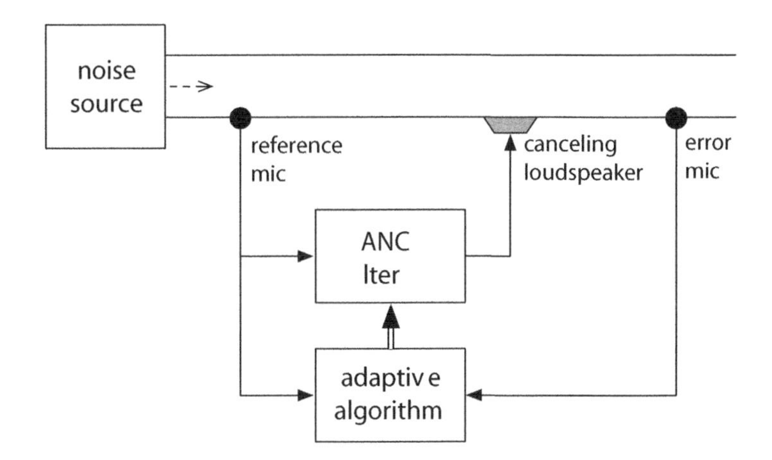

Figure 1.20 Active noise cancellation in a narrow duct.

to ensure fast adaptation of the echo canceler. The algorithms discussed in Chapters 8 and 9 have been widely used to overcome these difficulties in the implementation of acoustic echo cancelers. The topic of echo cancelers, with particular emphasis on acoustic echo cancelers, is covered in Chapter 15.

Active Noise Control

ANC refers to situations where acoustic *antinoise* waves are generated from electronic circuits (Kuo and Morgan, 1996). The ANC can be best explained by the following example.

A well-examined application of ANC is cancellation of noise in narrow ducts, such as exhaust pipes and ventilation systems, as illustrated in Figure 1.20. The acoustic noise traveling along the duct is picked up by a microphone at position A. This is used as reference input to an ANC filter whose parameters are adapted so that its output after conversion to an acoustic wave (through the canceling loudspeaker), is equal to the negative value of the duct noise at position B, thereby canceling that. The residual noise, picked up by the error microphone at position C, is the error signal used for adaptation of the ANC filter.

Comparing this ANC setup with the interference cancellation setup shown in Figure 1.15, we may note the following. The source of interference here is the duct noise, reference input is the noise picked up by the reference microphone, desired output (i.e., what we wish to see after canceling the duct noise) is zero, and primary input is the duct noise reaching position B. Accordingly, the role of ANC filter is to model the response of the duct from position A to B.

The above description of ANC assumes that the duct is narrow and the acoustic noise waves are traveling along the duct, which is like a one-dimensional model. The acoustical models of wider ducts and large enclosures, such as cars and aircrafts, are usually more complicated. Multiple microphones/loudspeakers are needed for successful implementation of ANCs in such enclosures. The adaptive filtering problem is then that of a multiple-input multiple-output system (Kuo and Morgan, 1996). Nevertheless, the basic principle remains the same, that is, generation of antinoise to cancel the actual noise. The subject of active noise control is covered in detail in Chapter 16.

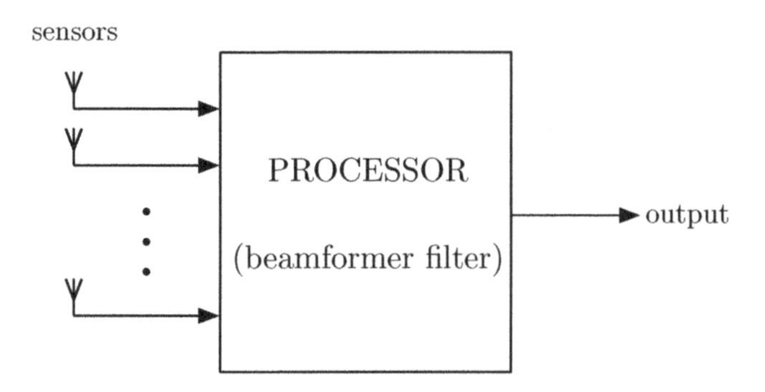

Figure 1.21 Spatial filtering (beamforming).

Beamforming

In the applications that have been discussed so far, the filters/predictors are used to combine together samples of the input signal(s) at different time instants to generate the output. Hence, these are classified as *temporal filtering*. Beamforming, however, is different from these in the sense that the inputs to a beamformer are samples of incoming signals at different positions in space. This is called *spatial filtering*. Beamforming finds applications in communications, radar, and sonar (Johnson and Dudgeon, 1993), and also imaging in radar and medical engineering (Soumekh, 1994).

In spatial filtering, a number of independent sensors are placed at different points in space to pick up signals coming from various sources (Figure 1.21). In radar and communications, the signals are usually electromagnetic waves and the sensors are thus antenna elements. Accordingly, the term *antenna arrays* is often used to refer to these applications of beamformers. In sonar applications, the sensors are hydrophones designed to respond to acoustic waves.

In a beamformer, the samples of the signals picked up by the sensors at a particular instant of time constitutes a *snapshot*. The samples of snapshot (spatial samples) play the same role as the successive (temporal) samples of input in a transversal filter. The beamformer filter linearly combines the sensor signals so that signals arriving from some particular direction are amplified, while signals from other directions are attenuated. Thus, in analogy with the frequency response of temporal filters, spatial filters have responses that vary according to the direction-of-arrival of the incoming signal(s). This is given in the form of a polar plot (gain versus angle) and is referred to as *beam pattern*.

In many applications of beamformers, the signals picked up by sensors are narrow bands having the same carrier (center) frequency. These signals differ in their direction-of-arrival, which are related to the location of their sources. The operation of beamformers in such applications can be best explained by the following example.

Consider an antenna array consisting of two omnidirectional elements A and B, as presented in Figure 1.22. The tone (as approximation to narrow-band) signals $s(n) = \alpha \cos \omega_o n$ and $\nu(n) = \beta \cos \omega_o n$ arriving at angles 0 and θ_o (with respect to the line perpendicular to the line connecting A and B), respectively, are the inputs to the array (beamformer) filter, which consists of a phase-shifter and a subtracter. The signal $s(n)$ arrives at elements A and B at the same time, whereas the arrival times of signal $\nu(n)$ at

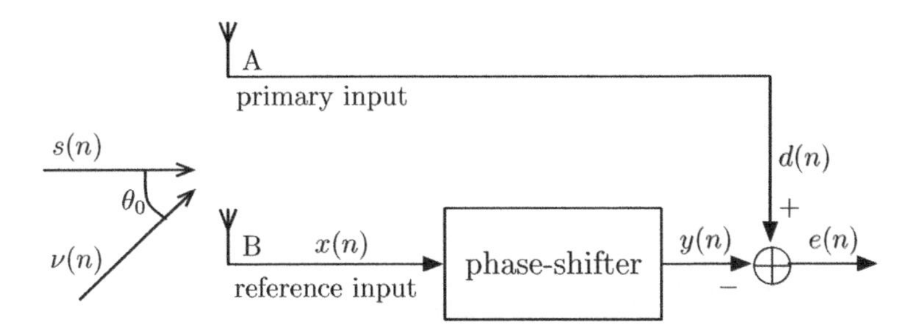

Figure 1.22 A two-element beamformer.

A and B are different. We may thus write

$$s_A(n) = s_B(n) = \alpha \cos \omega_o n$$

$$\nu_B(n) = \beta \cos \omega_o n$$

and

$$\nu_A(n) = \beta \cos(\omega_o n - \varphi)$$

where subscripts A and B are used to denote the signals picked up by elements A and B, respectively, φ is the phase shift arising from the time delay of arrival of $\nu(n)$ at element A with respect to its arrival at element B.

Now, if we assume that $s(n)$ is the desired signal and $\nu(n)$ is an interference, by inspection, one can see that if the phase-shifter phase is chosen equal to φ, then the interference, $\nu(n)$, will be completely canceled by the beamformer. The desired signal, on the other hand, reaches the beamformer output as $\alpha(\cos \omega_o n - \cos(\omega_o n - \varphi))$, which is nonzero (and still holding the information contained in its envelope, α) when $\varphi \neq 0$, that is, when the interference direction is different from the direction of the desired signal. This shows that one can tune a beamformer so as to allow the desired signal arriving from a direction to pass through it, while rejecting the unwanted signals (interferences) arriving from other directions.

The idea of using a phase-shifter to adjust the beam pattern of two sensors is easily extendible to the general case of more than two sensors. In general, by introducing appropriate phase shifts and also gains at the output of the various sensors and summing up these outputs, one can realize any arbitrary beam pattern. This is similar to the selection of tap weights of a transversal filter so that the filter frequency response becomes a good approximation to the desired response. Clearly, by increasing the number of elements in the array, better approximations to the desired beam pattern can be achieved.

The last point that we wish to add here is that in cases where the input signals to the beamformer are not narrow band, a combination of spatial and temporal filtering needs to be used. In such cases, spatial information is obtained by having sensors at different positions in space, as was discussed previously. The temporal information is obtained using a transversal filter at the output of each sensor. The output of the broadband beamformer is the summation of the outputs of these transversal filters. Detailed discussions on these points and the relevant mathematical backgrounds are presented in Chapter 18.

2

Discrete-Time Signals and Systems

Most of the adaptive algorithms have been developed for discrete-time (sampled) signals. Hence, discrete-time systems are used for implementation of adaptive filters. In this chapter, we present a short review of discrete-time signals and systems. Our assumption is that the reader is familiar with the basic concepts of discrete-time systems, such as the Nyquist sampling theorem, z-transform and system function, and also with the theory of random variables and stochastic processes. Our goal, in this chapter, is to review these concepts and put them in a framework appropriate for the rest of the book.

2.1 Sequences and z-Transform

In discrete-time systems, we are concerned with processing signals that are represented by sequences. Such sequences may be samples of a continuous-time analog signal or may be discrete in nature. As an example, in the channel equalizer structure presented in Figure 1.9, the input sequence to the equalizer, $x(n)$, consists of the samples of the channel output which is an analog signal, but the original data sequence, $s(n)$, is discrete in nature.

A discrete-time sequence, $x(n)$, may be equivalently represented by its z-transform defined as

$$X(z) = \sum_{n=-\infty}^{\infty} x(n)z^{-n} \tag{2.1}$$

where z is a complex variable. The range of values of z for which the above summation converges is called the *region of convergence of* $X(z)$. The following two examples illustrate this.

Example 2.1

Consider the sequence

$$x_1(n) = \begin{cases} a^n, & n \geq 0 \\ 0, & n < 0 \end{cases} \tag{2.2}$$

Adaptive Filters: Theory and Applications, Second Edition. Behrouz Farhang-Boroujeny.

The z-transform of $x_1(n)$ is

$$X_1(z) = \sum_{n=0}^{\infty} a^n z^{-n} = \sum_{n=0}^{\infty} (az^{-1})^n$$

which converges to

$$X_1(z) = \frac{1}{1 - az^{-1}} \tag{2.3}$$

for $|az^{-1}| < 1$, that is, $|z| > |a|$. We may also write

$$X_1(z) = \frac{z}{z - a} \tag{2.4}$$

for $|z| > |a|$.

Example 2.2

Consider the sequence

$$x_2(n) = \begin{cases} 0, & n \geq 0 \\ b^n, & n < 0 \end{cases} \tag{2.5}$$

The z-transform of $x_2(n)$ is

$$X_2(z) = \sum_{n=-\infty}^{-1} b^n z^{-n} = \sum_{n=1}^{\infty} (b^{-1}z)^n$$

which converges to

$$X_2(x) = \frac{b^{-1}z}{1 - b^{-1}z} = \frac{z}{b - z} \tag{2.6}$$

for $|z| < |b|$.

The two sequences presented in the above examples are different in many respects. The sequence $x_1(n)$ in Example 2.1 is called *right-sided*, since its non-zero elements start at a finite $n = n_1$ (here, $n_1 = 0$) and extend up to $n = +\infty$. On the other hand, the sequence $x_2(n)$ in Example 2.2 is a *left-sided* one. Its nonzero elements start at a finite $n = n_2$ (here, $n_2 = -1$) and extend up to $n = -\infty$. This definition of right-sided and left-sided sequences also implies that the region of convergence of a right-sided sequence is always the exterior of a circle ($|z| > |a|$, in Example 2.1), while that of a left-sided sequence is always the interior of a circle ($|z| < |b|$, in Example 2.2).

We thus note that the specification of the z-transform, $X(z)$, of a sequence is complete only when its region of convergence is also specified. In other words, the inverse

z-transform of $X(z)$ can be uniquely found, only if its region of convergence is also specified. For example, one may note that $X_1(z)$ and $X_2(z)$ in the above examples have exactly the same form, except a sign reversal. Hence, if their regions of convergence are not specified, both may be interpreted as the z-transforms of either left-sided or right-sided sequences.

Two-sided sequences may also exist. A two-sided sequence is one that extends from $n = -\infty$ to $n = +\infty$. The following example shows how to deal with two-sided sequences.

Example 2.3

Consider the sequence

$$x_3(n) = \begin{cases} a^n, & n \geq 0 \\ b^n, & n < 0 \end{cases} \tag{2.7}$$

where $|a| < |b|$. The condition $|a| < |b|$, as we shall see, is necessary to make the convergence of the z-transform of $x_3(n)$ possible.

The z-transform of $x_3(n)$ is

$$X_3(z) = \sum_{n=-\infty}^{-1} b^n z^{-n} + \sum_{n=0}^{\infty} a^n z^{-n} \tag{2.8}$$

Clearly, the first sum converges when $|z| < |b|$, and the second sum converges when $|z| > |a|$. Thus, we obtain

$$X_3(z) = \frac{z}{b-z} + \frac{z}{z-a} = \frac{z(a-b)}{(z-a)(z-b)} \tag{2.9}$$

for $|a| < |z| < |b|$.

We may note that the region of convergence of $X_3(z)$ is the area in between two concentric circles. This is true, in general, for all two-sided sequences. For a sequence with a rational z-transform, the radii of the two circles are determined by two of the poles of the z-transform of sequence. The right-sided part of the sequence is determined by the poles which are surrounded by the region of convergence, and the poles surrounding the region of convergence determine the left-sided part of the sequence. The following example which also shows one way of calculating inverse z-transform clarifies the above points.

Example 2.4

Consider a two-sided sequence, $x(n)$, with the z-transform

$$X(z) = \frac{-0.1z^{-1} + 3.05z^{-2}}{(1-0.5z^{-1})(1+0.7z^{-1})(1+2z^{-1})} \tag{2.10}$$

and the region of convergence $0.7 < |z| < 2$.

To find $x(n)$, that is, the inverse z-transform of $X(z)$, we use the method of partial fraction and expand $X(z)$ as

$$X(z) = \frac{A}{1-0.5z^{-1}} + \frac{B}{1+0.7z^{-1}} + \frac{C}{1+2z^{-1}},$$

where A, B, and C are constants that can be determined as follows:

$$A = (1 - 0.5z^{-1})X(z)|_{z=0.5} = 1$$

$$B = (1 + 0.7z^{-1})X(z)|_{z=-0.7} = -2$$

$$C = (1 + 2z^{-1})X(z)|_{z=-2} = 1$$

This gives

$$X(z) = \frac{1}{1 - 0.5z^{-1}} - \frac{2}{1 + 0.7z^{-1}} + \frac{1}{1 + 2z^{-1}} \tag{2.11}$$

We treat each of the terms in the above equation separately. To expand these terms and, from there, extract their corresponding sequences, we use the following identity, which holds for $|a| < 1$,

$$\frac{1}{1 - a} = 1 + a + a^2 + \cdots$$

We note that within the region of convergence of $X(z)$, $|0.5z^{-1}|$ and $|0.7z^{-1}|$ are both less than 1, and thus

$$\frac{1}{1 - 0.5z^{-1}} = 1 + 0.5z^{-1} + 0.5^2z^{-2} + \cdots \tag{2.12}$$

and

$$-\frac{2}{1 + 0.7z^{-1}} = -\frac{2}{1 - (-0.7)z^{-1}}$$
$$= -2(1 + (-0.7)z^{-1} + (-0.7)^2z^{-2} + \cdots) \tag{2.13}$$

However, for the third term on the right-hand side of Eq. (2.11), $|2z^{-1}| > 1$, and, thus, an expansion similar to the last two is not applicable. A similar expansion will be possible, if we rearrange this term as

$$\frac{1}{1 + 2z^{-1}} = \frac{0.5z}{1 + 0.5z}$$

Here, within the region of convergence of $X(z)$, $|0.5z| < 1$, and, thus, we may write

$$\frac{0.5z}{1 + 0.5z} = 0.5z(1 + (-0.5)z + (-0.5)^2z^2 + \cdots)$$
$$= -(-2)^{-1}z - (-2)^{-2}z^2 - (-2)^{-3}z^3 - \cdots \tag{2.14}$$

Substituting Eqs. (2.12), (2.13), and (2.14) in Eq. (2.11) and recalling Eq. (2.1), we obtain

$$x(n) = \begin{cases} -(-2)^n, & n < 0 \\ 0.5^n - 2(-0.7)^n, & n \geq 0 \end{cases} \tag{2.15}$$

An alternative way of performing inverse z-transform can be derived using the *Cauchy integral theorem*, which is stated as follows:

$$\frac{1}{2\pi j}\oint_C z^{k-1}\mathrm{d}z = \begin{cases} 1, & k = 0 \\ 0, & k \neq 0 \end{cases} \tag{2.16}$$

where C is a counterclockwise contour that encircles the origin.

The z-transform relation, reproduced here for convenience, is given by

$$X(z) = \sum_{n=-\infty}^{\infty} x(n)z^{-n} \tag{2.17}$$

Multiplying both sides of Eq. (2.17) by z^{k-1} and integrating, we obtain

$$\frac{1}{2\pi j}\oint_C X(z)z^{k-1}\mathrm{d}z = \frac{1}{2\pi j}\oint_C \sum_{n=-\infty}^{\infty} x(n)z^{-n+k-1}\mathrm{d}z \tag{2.18}$$

where C is a contour within the region of convergence of $X(z)$ and encircling the origin. Interchanging the order of integration and summation on the right-hand side of Eq. (2.18), we obtain

$$\frac{1}{2\pi j}\oint_C X(z)z^{k-1}\mathrm{d}z = \sum_{n=-\infty}^{\infty} x(n)\frac{1}{2\pi j}\oint_C z^{-n+k-1}\mathrm{d}z \tag{2.19}$$

Application of the Cauchy integral theorem in Eq. (2.19) gives the inverse z-transform relation

$$x(n) = \frac{1}{2\pi j}\oint_C X(z)z^{n-1}\mathrm{d}z \tag{2.20}$$

where C is a counterclockwise closed contour in the region of convergence of $X(z)$ and encircling the origin of the z-plane.

For rational z-transforms, contour integrals are often conveniently evaluated using the residue theorem, that is,

$$\begin{aligned} x(n) &= \frac{1}{2\pi j}\oint_C X(z)z^{n-1}\mathrm{d}z \\ &= \sum[\text{residues of } X(z)z^{n-1} \text{ at the poles inside } C] \end{aligned} \tag{2.21}$$

In general, if $X(z)z^{n-1}$ is a rational function of z, and z_p is a pole of $X(z)z^{n-1}$, repeated m times

$$\text{residue of } X(z)z^{n-1} \text{ at } z_\mathrm{p} = \frac{1}{(m-1)!}\left[\frac{d^{m-1}\psi(z)}{dz^{m-1}}\right]_{z=z_\mathrm{p}} \tag{2.22}$$

where $\psi(z) = (z - z_\mathrm{p})^m X(z)z^{n-1}$. In particular, if there is a first-order pole at $z = z_\mathrm{p}$, that is, $m = 1$, then

$$\text{residue of } X(z)z^{n-1} \text{ at } z_\mathrm{p} = \psi(z_\mathrm{p}) \tag{2.23}$$

2.2 Parseval's Relation

Among various important results and properties of z-transform, in this book, we are in particular interested in *Parseval's relation* which states that for any pair of sequences $x(n)$ and $y(n)$,

$$\sum_{n=-\infty}^{\infty} x(n)y^*(n) = \frac{1}{2\pi j}\oint X(z)Y^*(1/z^*)z^{-1}\mathrm{d}z \tag{2.24}$$

where the superscript $*$ denotes complex conjugation and the contour of integration is taken in the overlap of the regions of convergence of $X(z)$ and $Y^*(1/z^*)$. If $X(z)$ and $Y(z)$ converge on the unit circle, we can choose $z = e^{j\omega}$, and Eq. (2.24) becomes

$$\sum_{n=-\infty}^{\infty} x(n)y^*(n) = \frac{1}{2\pi}\int_{-\pi}^{\pi} X(e^{j\omega})Y^*(e^{j\omega})d\omega \tag{2.25}$$

Furthermore, if $y(n) = x(n)$, for all n, Eq. (2.25) becomes

$$\sum_{n=-\infty}^{\infty} |x(n)|^2 = \frac{1}{2\pi}\int_{-\pi}^{\pi} |X(e^{j\omega})|^2 d\omega \tag{2.26}$$

Equation (2.26) has the following interpretation. *The total energy in a sequence* $x(n)$, *that is,* $\sum_{n=-\infty}^{\infty} |x(n)|^2$, *may be equivalently obtained by averaging* $|X(e^{j\omega})|^2$ *over one cycle of that.*

2.3 System Function

Consider a discrete-time linear time-invariant system with the impulse response $h(n)$. With $x(n)$ and $y(n)$ denoting, respectively, the input and output of the system,

$$y(n) = x(n) \star h(n) \tag{2.27}$$

where $\star$ denotes convolution and is defined as

$$x(n) \star h(n) = \sum_{k=-\infty}^{\infty} h(k)x(n-k) \tag{2.28}$$

Equation (2.27) suggests that any linear time-invariant system is completely characterized by its impulse response, $h(n)$. Taking z-transform from both sides of Eq. (2.27), we obtain

$$Y(z) = X(z)H(z) \tag{2.29}$$

This shows that the input–output relation for a linear time-invariant system corresponds to a multiplication of the z-transforms of the input and the impulse response of the system.

The z-transform of the impulse response of a linear time-invariant system is referred to as its *system function*. The system function evaluated over the unit circle, $|z| = 1$, is the frequency response of the system, $H(e^{j\omega})$. For any particular frequency ω, $H(e^{j\omega})$ is the gain (complex-valued, in general) of the system, when its input is the complex sinusoid $e^{j\omega n}$.

Any stable linear time-invariant system has a finite frequency response for all values of ω. This means that the region of convergence of $H(z)$ has to include the unit circle. This fact can be used to uniquely determine the region of convergence of any rational system function, once its poles are known. As an example, if we consider a sequence with the z-transform

$$H(z) = \frac{1}{(1-0.5z^{-1})(1-2z^{-1})} \tag{2.30}$$

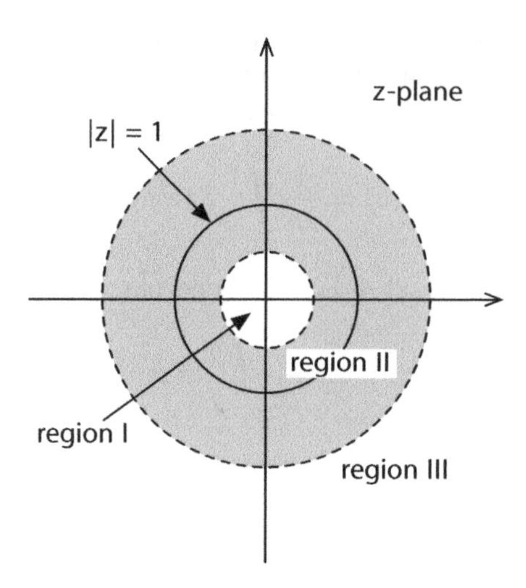

Figure 2.1 Possible regions of convergence for a two pole z-transform.

we find that there are three possible regions of convergence for $H(z)$ as specified in Figure 2.1. These are regions I, II, and III, each giving a different time sequence. However, if we assume that $H(z)$ is the system function of a stable time-invariant system, the only acceptable region of convergence will be region II. Noting this, we obtain (Problem 2.1)

$$h(n) = \begin{cases} -\frac{4}{3} \times 2^n, & n < 0 \\ -\frac{1}{3} \times 0.5^n, & n \geq 0 \end{cases} \tag{2.31}$$

We note that the impulse response $h(n)$, obtained above, extends from $n = -\infty$ to $n = +\infty$. This means that, although the input, $\delta(n)$, is applied at time $n = 0$, the system output takes nonzero values even before that. Such a system is called *noncausal*. In contrast to this, a system is said to be *causal* if its impulse response is nonzero only for nonnegative values of n. Noncausal systems, although not realistic, may be encountered in some theoretical developments. It is important that we find a practical solution for handling such cases. The following example considers such a case and gives a solution to that.

Example 2.5

Figure 2.2 shows a communication system. It consists of a communication channel which is characterized by the system function

$$C(z) = 1 - 2.5z^{-1} + z^{-2} = (1 - 0.5z^{-1})(1 - 2z^{-1}) \tag{2.32}$$

The equalizer, $H(z)$, should be selected so that the original transmitted signal, $s(n)$, can be recovered from the equalizer output without any distortion.

For this, we shall select $H(z)$ so that we get $y(n) = s(n)$. This can be achieved if $H(z)$ is selected so that $C(z)H(z) = 1$. This gives

$$H(z) = \frac{1}{C(z)} = \frac{1}{(1 - 0.5z^{-1})(1 - 2z^{-1})} \tag{2.33}$$

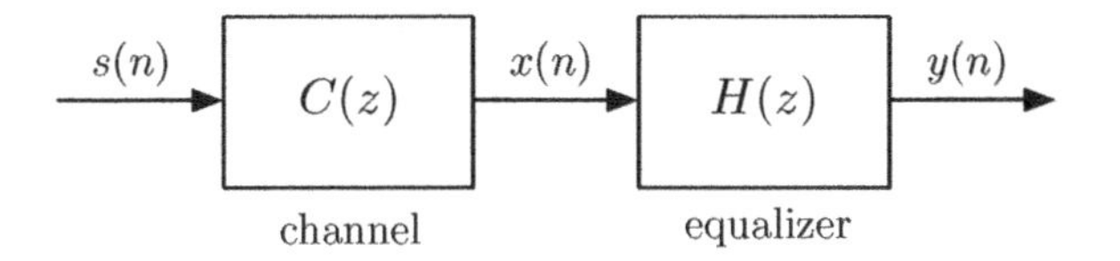

Figure 2.2 A communication system.

Noting that this is similar to $H(z)$ in Eq. (2.30), we find that the equalizer impulse response is the one given in Eq. (2.31). This, of course, is noncausal and, therefore, not realizable. The problem can be easily solved by shifting the noncausal response of the equalizer to the right by sufficient number of samples so that the remaining noncausal samples are sufficiently small and can be ignored. Mathematically, we say

$$H(z) \approx \frac{z^{-\Delta}}{C(z)} \tag{2.34}$$

where Δ is the number of sample delays introduced to achieve a realizable causal system. We use the approximation sign, $\approx$, in Eq. (2.34), since we ignore the noncausal samples of $z^{-\Delta}/C(z)$. The equalizer output is then $s(n-\Delta)$.

2.4 Stochastic Processes

Input signal to an adaptive filter and its desired output are, in general, random, that is, they are not known *a priori*. However, they exhibit some statistical characteristics which have to be utilized for optimum adjustment of the filter coefficients. Such random signals are called *stochastic processes*. Adaptive algorithms are designed to extract these characteristics and use them for adjusting the filter coefficients.

A discrete-time stochastic process is an indexed set of random variables $\{x(n); n = \ldots, -2, -1, 0, 1, 2, \ldots\}$. As a random signal, the index n is associated with time or possibly some other physical dimension. In this book, for convenience, we frequently refer to n as time index. So far, we have used the notation $x(n)$ to refer to a particular sequence $x(n)$ that extends from $n = -\infty$ to $n = +\infty$. We use the notation $\{x(n)\}$ for a stochastic process which a particular sequence $x(n)$ may be a single realization of that.

The elements of a stochastic process, $\{x(n)\}$, for different values of n, are in general complex-valued random variables that are characterized by their probability distribution functions. The inter-relationships between different elements of $\{x(n)\}$ is determined by their joint distribution functions. Such distribution functions, in general, may change with the time index n. A stochastic process is called *stationary in the strict sense*, if all of its (single and joint) distribution functions are independent of a shift in the time origin.

2.4.1 Stochastic Averages

It is often useful to characterize stochastic processes by statistical averages of their elements. These averages are called ensemble averages and, in general, are time dependent. For example, the mean of the nth element of a stochastic process $\{x(n)\}$, is defined as

$$m_x(n) = E[x(n)] \tag{2.35}$$

where $E[\cdot]$ denotes statistical expectation. It shall be noted that since $m_x(n)$ is in general a function of n, it may not be possible to obtain $m_x(n)$ by time averaging of a single realization of the stochastic process $\{x(n)\}$, unless the process possesses certain special properties indicated at the end of this chapter. Instead, n has to be fixed and averaging has to be done over the nth element of the stochastic process, as a single random variable.

In our later developments, we are heavily dependent on the following averages:

1. *Autocorrelation Function:* For a stochastic process $\{x(n)\}$, it is defined as

$$\phi_{xx}(n,m) = E[x(n)x^*(m)] \tag{2.36}$$

where the superscript $*$ denotes complex conjugation.

2. *Cross-Correlation Function:* It is defined for two stochastic processes $\{x(n)\}$ and $\{y(n)\}$ as

$$\phi_{xy}(n,m) = E[x(n)y^*(m)] \tag{2.37}$$

A stochastic process $\{x(n)\}$ is said to be *stationary in the wide sense*, if $m_x(n)$ and $\phi_{xx}(n,m)$ are independent of a shift of time origin. That is, for any k, m, and n,

$$m_x(n) = m_x(n+k)$$

and

$$\phi_{xx}(n,m) = \phi_{xx}(n+k,m+k)$$

These imply that $m_x(n)$ is a constant for all n and $\phi_{xx}(n,m)$ depends on the difference $n-m$ only. Then, it would be more appropriate to define the autocorrelation function of $\{x(n)\}$ as

$$\phi_{xx}(k) = E[x(n)x^*(n-k)] \tag{2.38}$$

Similarly, the processes $\{x(n)\}$ and $\{y(n)\}$ are said to be jointly stationary in the wide sense, if their means are independent of n and $\phi_{xy}(n,m)$ depends on $n-m$ only. We may, then, define the cross-correlation function of $\{x(n)\}$ and $\{y(n)\}$ as

$$\phi_{xy}(k) = E[x(n)y^*(n-k)] \tag{2.39}$$

Besides the autocorrelation and cross-correlation functions, autocovariance and cross-covariance functions are also defined. For stationary processes, these are defined as

$$\gamma_{xx}(k) = E[(x(n)-m_x)(x(n-k)-m_x)^*] \tag{2.40}$$

and

$$\gamma_{xy}(k) = E[(x(n)-m_x)(y(n-k)-m_y)^*] \tag{2.41}$$

respectively. By expanding the right-hand sides of Eqs. (2.40) and (2.41), we obtain

$$\gamma_{xx}(k) = \phi_{xx}(k) - |m_x|^2 \tag{2.42}$$

and

$$\gamma_{xy}(k) = \phi_{xy}(k) - m_x m_y^* \tag{2.43}$$

respectively. This shows that the correlation and covariance functions differ by some bias which are determined by the means of the corresponding processes.

We may also note that for many random signals, the signal samples become less correlated as they become more separated in time. Thus, we may write

$$\lim_{k\to\infty} \phi_{xx}(k) = |m_x|^2 \tag{2.44}$$

$$\lim_{k\to\infty} \gamma_{xx}(k) = 0 \tag{2.45}$$

$$\lim_{k\to\infty} \phi_{xy}(k) = m_x m_y^* \tag{2.46}$$

$$\lim_{k\to\infty} \gamma_{xy}(k) = 0 \tag{2.47}$$

Other important properties of the correlation and covariance functions that should be noted here are their symmetry properties which are summarized below:

$$\phi_{xx}(k) = \phi_{xx}^*(-k) \tag{2.48}$$

$$\gamma_{xx}(k) = \gamma_{xx}^*(-k) \tag{2.49}$$

$$\phi_{xy}(k) = \phi_{yx}^*(-k) \tag{2.50}$$

$$\gamma_{xy}(k) = \gamma_{yx}^*(-k) \tag{2.51}$$

We may also note that

$$\phi_{xx}(0) = E[|x(n)|^2] = \text{mean-square of } x(n) \tag{2.52}$$

$$\gamma_{xx}(0) = \sigma_x^2 = \text{variance of } x(n) \tag{2.53}$$

2.4.2 *z-Transform Representations*

The z-transform of $\phi_{xx}(k)$ is given by

$$\Phi_{xx}(z) = \sum_{k=-\infty}^{\infty} \phi_{xx}(k) z^{-k} \tag{2.54}$$

We note that a necessary condition for $\Phi_{xx}(z)$ to be convergent is that m_x should be zero (Problem P2.4). We assume this for the random processes that are considered in the rest of this chapter, and also the following chapters. Exceptional cases will be mentioned explicitly.

From Eq. (2.48), we note that

$$\Phi_{xx}(z) = \Phi_{xx}^*(1/z^*) \tag{2.55}$$

Similarly, if $\Phi_{xy}(z)$ denotes the z-transform of $\phi_{xy}(k)$, then

$$\Phi_{xy}(z) = \Phi_{yx}^*(1/z^*) \tag{2.56}$$

Equation (2.55) implies that if $\Phi_{xx}(z)$ is a rational function of z, its poles and zeros must occur in complex-conjugate reciprocal pairs, as depicted in Figure 2.3.

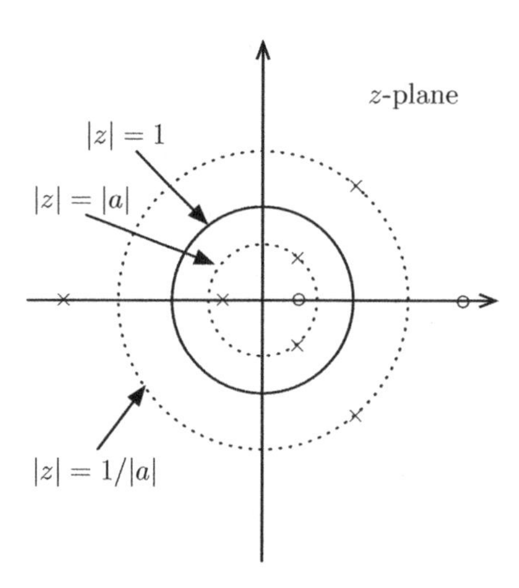

Figure 2.3 Poles and zeros of a typical z-transform of an autocorrelation function.

Moreover, Eq. (2.55) implies that the points which belong to the region of convergence of $\Phi_{xx}(z)$ also occur in complex-conjugate reciprocal pairs. This in turn suggests that the region of convergence of $\Phi_{xx}(z)$ must be of the form (Figure 2.3)

$$|a| < |z| < \frac{1}{|a|} \tag{2.57}$$

It is important that we note this covers the unit circle, $|z| = 1$.

The inverse z-transform relation (Eq. 2.20) may be used to evaluate $\phi_{xx}(0)$ as

$$\phi_{xx}(0) = \frac{1}{2\pi j} \oint_C \Phi_{xx}(z) z^{-1} \mathrm{d}z \tag{2.58}$$

We assume that $\Phi_{xx}(z)$ is convergent on the unit circle and select the unit circle as the contour of integration. For this, we substitute z by $\mathrm{e}^{j\omega}$. Then, ω changes from $-\pi$ to $+\pi$ as we traverse the unit circle once. Noting that $z^{-1}\mathrm{d}z = jd\omega$, Eq. (2.58) becomes

$$\phi_{xx}(0) = \frac{1}{2\pi} \int_{-\pi}^{\pi} \Phi_{xx}(\mathrm{e}^{j\omega}) \mathrm{d}\omega \tag{2.59}$$

Since $m_x = 0$, we can combine Eqs. (2.52) and (2.53) with Eq. (2.59) to obtain

$$\sigma_x^2 = E[|x(n)|^2] = \frac{1}{2\pi} \int_{-\pi}^{\pi} \Phi_{xx}(\mathrm{e}^{j\omega}) \mathrm{d}\omega \tag{2.60}$$

2.4.3 The Power Spectral Density

The function $\Phi_{xx}(z)$, when evaluated on the unit circle is the Fourier transform of the autocorrelation sequence $\phi_{xx}(k)$. It is called *power spectral density* since it reflects the spectral content of the underlying process as a function of frequency. It is also called *power spectrum* or simply *spectrum*. Convergence performance of an adaptive filter is directly related to the spectrum of its input process. Next, we present a direct development of the power spectral density of a wide-sense stationary discrete-time stochastic process which reveals its important properties.

Consider a sequence $x(n)$ which represents a single realization of a zero-mean wide-sense stationary stochastic process $\{x(n)\}$. We consider a window of $2N+1$ elements of $x(n)$ as

$$x_N(n) = \begin{cases} x(n), & -N \le n \le N \\ 0, & \text{otherwise} \end{cases} \tag{2.61}$$

By definition, the discrete-time Fourier transform of $x_N(n)$ is

$$X_N(e^{j\omega}) = \sum_{n=-\infty}^{\infty} x_N(n)e^{-j\omega n} = \sum_{n=-N}^{N} x(n)e^{-j\omega n} \tag{2.62}$$

Conjugating both sides of Eq. (2.62), and replacing n by m, we obtain

$$X_N^*(e^{j\omega}) = \sum_{m=-N}^{N} x^*(m)e^{j\omega m} \tag{2.63}$$

Next, we multiply Eqs. (2.62) and (2.63) to obtain

$$|X_N(e^{j\omega})|^2 = \sum_{n=-N}^{N} \sum_{m=-N}^{N} x(n)x^*(m)e^{-j\omega(n-m)} \tag{2.64}$$

Taking the expectation on both sides of Eq. (2.64), and interchanging the order of expectation and double summation, we get

$$E[|X_N(e^{j\omega})|^2] = \sum_{n=-N}^{N} \sum_{m=-N}^{N} E[x(n)x^*(m)]e^{-j\omega(n-m)} \tag{2.65}$$

Noting that $E[x(n)x^*(m)] = \phi_{xx}(n-m)$, and letting $k = n-m$, we may rearrange the terms in Eq. (2.65) to obtain

$$\frac{1}{2N+1} E[|X_N(e^{j\omega})|^2] = \sum_{k=-2N}^{2N} \left(1 - \frac{|k|}{2N+1}\right) \phi_{xx}(k)e^{-j\omega k} \tag{2.66}$$

To simplify Eq. (2.66), we assume that for k greater than an arbitrary large constant, but less than infinity, $\phi_{xx}(k)$ is identically equal to zero. This, in general, is a fair assumption, unless the $\{x(n)\}$ contains sinusoidal components, in which case the summation on the right-hand side of Eq. (2.66) will not be convergent. With this assumption, we get, from Eq. (2.66)

$$\lim_{N\to\infty} \frac{1}{2N+1} E[|X_N(e^{j\omega})|^2] = \sum_{k=-\infty}^{\infty} \phi_{xx}(k)e^{-j\omega k} \tag{2.67}$$

which is nothing but the Fourier transform of the autocorrelation function, $\phi_{xx}(k)$. We may, thus, write

$$\Phi_{xx}(e^{j\omega}) = \lim_{N\to\infty} \frac{1}{2N+1} E[|X_N(e^{j\omega})|^2] \tag{2.68}$$

The function $\Phi_{xx}(e^{j\omega})$ is called the power spectral density of the stochastic wide-sense stationary process $\{x(n)\}$. It is defined as in Eq. (2.68) or more conveniently as the Fourier transform of the autocorrelation function of $\{x(n)\}$,

$$\Phi_{xx}(e^{j\omega}) = \sum_{k=-\infty}^{\infty} \phi_{xx}(k) e^{-j\omega k} \tag{2.69}$$

The power spectral density possesses certain special properties. These are indicated below for our later reference.

Property 1: *When the limit in* Eq. (2.68) *exists,* $\Phi_{xx}(e^{j\omega})$ *has the following interpretation*:

$$\begin{aligned}\frac{1}{2\pi}\Phi_{xx}(e^{j\omega})d\omega = {} & \text{average contribution of the} \\ & \text{frequency components of } \{x(n)\} \\ & \text{located between } \omega \text{ and } \omega + d\omega \end{aligned} \tag{2.70}$$

This interpretation matches Eq. (2.60), if both sides of Eq. (2.70) are integrated over ω from $-\pi$ to $+\pi$. We will elaborate more on this later, once we introduce response of linear systems to random signals; see Example 2.6.

Property 2: *The power spectral density* $\Phi_{xx}(e^{j\omega})$ *is always real and nonnegative.*

This property is obvious from the definition (Eq. 2.68), as $|X_N(e^{j\omega})|^2$ is always real and nonnegative.

Property 3: *The power spectral density of a real-valued stationary stochastic process is even, that is, symmetric with respect to the origin* $\omega = 0$. In other words,

$$\Phi_{xx}(e^{j\omega}) = \Phi_{xx}(e^{-j\omega}) \tag{2.71}$$

However, this may not be true when the process is complex-valued.

This follows from Eq. (2.69), by replacing k with $-k$ and noting that for a real-valued stationary process $\phi_{xx}(k) = \phi_{xx}(-k)$.

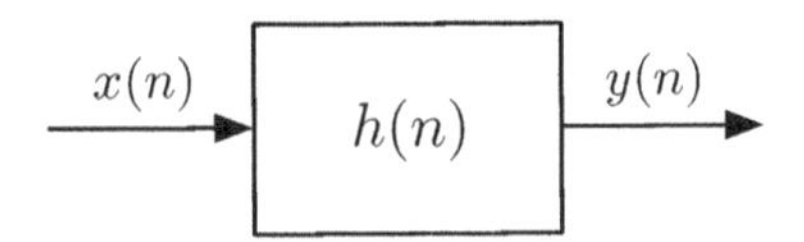

Figure 2.4 A linear time-invariant system.

2.4.4 Response of Linear Systems to Stochastic Processes

We consider a linear time-invariant discrete-time system, with input $\{x(n)\}$, output $\{y(n)\}$ and impulse response $h(n)$, as depicted in Figure 2.4. The input and output sequences are stochastic processes, but the system impulse response, $h(n)$, is a deterministic sequence. Since $\{x(n)\}$ and $\{y(n)\}$ are stochastic processes, we are interested in finding how they are statistically related together. We assume that $\phi_{xx}(k)$ is known, and find the relationships which relate this with $\phi_{xy}(k)$ and $\phi_{yy}(k)$. These relationships can be conveniently established through the z-transforms of the sequences.

We note that

$$\begin{aligned}\Phi_{xy}(z) &= \sum_{k=-\infty}^{\infty} \phi_{xy}(k)z^{-k} \\ &= \sum_{k=-\infty}^{\infty} E[x(n)y^*(n-k)]z^{-k} \\ &= \sum_{k=-\infty}^{\infty} E\left[x(n)\sum_{l=-\infty}^{\infty} h^*(l)x^*(n-k-l)\right]z^{-k} \end{aligned} \tag{2.72}$$

Since both summation and expectation are linear operators, their orders can be interchanged. Using this in Eq. (2.72), we get

$$\begin{aligned}\Phi_{xy}(z) &= \sum_{k=-\infty}^{\infty}\sum_{l=-\infty}^{\infty} h^*(l)E[x(n)x^*(n-k-l)]z^{-k} \\ &= \sum_{l=-\infty}^{\infty} h^*(l)\sum_{k=-\infty}^{\infty} \phi_{xx}(k+l)z^{-k} \end{aligned} \tag{2.73}$$

If we substitute $k+l$ by m, we get

$$\begin{aligned}\Phi_{xy}(z) &= \sum_{l=-\infty}^{\infty} h^*(l)\sum_{m=-\infty}^{\infty} \phi_{xx}(m)z^{-(m-l)} \\ &= \sum_{l=-\infty}^{\infty} h^*(l)z^{l}\sum_{m=-\infty}^{\infty} \phi_{xx}(m)z^{-m} \end{aligned} \tag{2.74}$$

This gives

$$\Phi_{xy}(z) = H^*(1/z^*)\Phi_{xx}(z) \tag{2.75}$$

where $H(z)$ follows the conventional definition

$$H(z) = \sum_{n=-\infty}^{\infty} h(n)z^{-n} \tag{2.76}$$

Furthermore, using Eqs. (2.55), (2.56), and (2.75), we can also get

$$\Phi_{yx}(z) = H(z)\Phi_{xx}(z) \tag{2.77}$$

The autocorrelation of the output process $\{y(n)\}$ is obtained as follows:

$$\begin{aligned}
\Phi_{yy}(z) &= \sum_{k=-\infty}^{\infty} \phi_{yy}(k)z^{-k} \\
&= \sum_{k=-\infty}^{\infty} E[y(n)y^*(n-k)]z^{-k} \\
&= \sum_{k=-\infty}^{\infty} E\left[\sum_{l=-\infty}^{\infty} h(l)x(n-l) \sum_{m=-\infty}^{\infty} h^*(m)x^*(n-k-m)\right]z^{-k} \\
&= \sum_{l=-\infty}^{\infty} h(l) \sum_{m=-\infty}^{\infty} h^*(m) \sum_{k=-\infty}^{\infty} E[x(n-l)x^*(n-k-m)]z^{-k} \\
&= \sum_{l=-\infty}^{\infty} h(l) \sum_{m=-\infty}^{\infty} h^*(m) \sum_{k=-\infty}^{\infty} \phi_{xx}(k+m-l)z^{-k}
\end{aligned} \tag{2.78}$$

Substituting $k+m-l$ by p, we get

$$\Phi_{yy}(z) = \sum_{l=-\infty}^{\infty} h(l)z^{-l} \sum_{m=-\infty}^{\infty} h^*(m)z^{m} \sum_{p=-\infty}^{\infty} \phi_{xx}(p)z^{-p} \tag{2.79}$$

or

$$\Phi_{yy}(z) = H(z)H^*(1/z^*)\Phi_{xx}(z) \tag{2.80}$$

It would be also convenient if we assume that z varies only over the unit circle, that is, $|z| = 1$. In that case $1/z^* = z$ and Eqs. (2.75) and (2.80) simplify to

$$\Phi_{xy}(z) = H^*(z)\Phi_{xx}(z) \tag{2.81}$$

and

$$\Phi_{yy}(z) = H(z)H^*(z)\Phi_{xx}(z) = |H(z)|^2\Phi_{xx}(z) \tag{2.82}$$

respectively. Also, by replacing z with $\mathrm{e}^{j\omega}$, we obtain

$$\Phi_{xy}(\mathrm{e}^{j\omega}) = H^*(\mathrm{e}^{j\omega})\Phi_{xx}(\mathrm{e}^{j\omega}) \tag{2.83}$$

$$\Phi_{yx}(\mathrm{e}^{j\omega}) = H(\mathrm{e}^{j\omega})\Phi_{xx}(\mathrm{e}^{j\omega}) \tag{2.84}$$

$$\Phi_{yy}(\mathrm{e}^{j\omega}) = |H(\mathrm{e}^{j\omega})|^2\Phi_{xx}(\mathrm{e}^{j\omega}) \tag{2.85}$$

These equations show how the cross power spectral densities, $\Phi_{xy}(\mathrm{e}^{j\omega})$ and $\Phi_{yx}(\mathrm{e}^{j\omega})$, and also the output power spectral density, $\Phi_{yy}(\mathrm{e}^{j\omega})$, are related with the input power spectral density, $\Phi_{xx}(\mathrm{e}^{j\omega})$, and the system transfer function, $H(\mathrm{e}^{j\omega})$. As a useful application of the above results, we consider the following example.

Example 2.6

Consider a band-pass filter with a magnitude response as in Figure 2.5. The input process to the filter is a zero-mean wide-sense stationary stochastic process $\{x(n)\}$. We are interested in finding the variance of the filter output, $\{y(n)\}$.

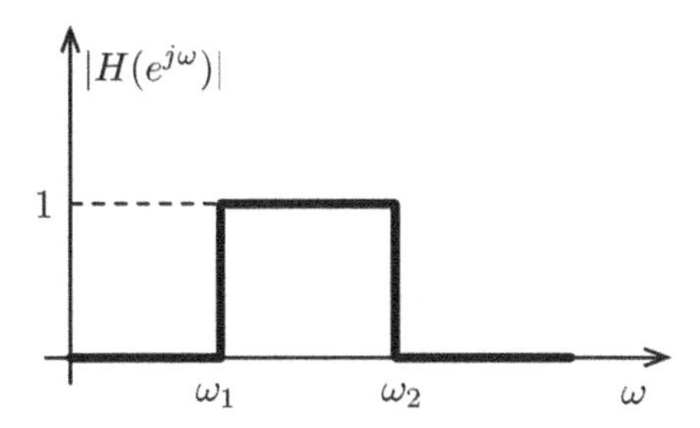

Figure 2.5 Magnitude response of a band-pass filter.

We note that

$$|H(e^{j\omega})| = \begin{cases} 1, & \omega_1 \leq \omega \leq \omega_2 \\ 0, & \text{otherwise} \end{cases} \tag{2.86}$$

Substituting this in Eq. (2.85) and using an equation similar to Eq. (2.60) for $\{y(n)\}$, we obtain

$$\sigma_y^2 = \frac{1}{2\pi}\int_{-\pi}^{\pi} |H(e^{j\omega})|^2 \Phi_{xx}(e^{j\omega})d\omega = \frac{1}{2\pi}\int_{\omega_1}^{\omega_2} \Phi_{xx}(e^{j\omega})d\omega \tag{2.87}$$

If ω_2 approaches ω_1, then we may write $\omega_2 - \omega_1 = d\omega$, where $d\omega$ is a variable approaching zero. In that case, we may write

$$\sigma_y^2 = \frac{1}{2\pi}\Phi_{xx}(e^{j\omega_1})d\omega.$$

This proves the interpretation of the power spectral density given by Property 1 in Section 2.4.3, that is, Eq. (2.70).

Consider the case where there is a third process, $\{d(n)\}$, whose cross-correlation with the input process, $\{x(n)\}$, of Figure 2.4 is known. We are interested in finding the cross-correlation of $\{d(n)\}$ and $\{y(n)\}$.

In terms of z-transforms, we have

$$\begin{aligned}
\Phi_{dy}(z) &= \sum_{k=-\infty}^{\infty} \phi_{dy}(k) z^{-k} \\
&= \sum_{k=-\infty}^{\infty} E[d(n)y^*(n-k)]z^{-k} \\
&= \sum_{k=-\infty}^{\infty} E\left[d(n)\sum_{l=-\infty}^{\infty} h^*(l)x^*(n-k-l)\right]z^{-k} \\
&= \sum_{l=-\infty}^{\infty} h^*(l) \sum_{k=-\infty}^{\infty} \phi_{dx}(k+l)z^{-k}
\end{aligned} \tag{2.88}$$

Substituting $k+l$ by m, we obtain

$$\Phi_{\mathrm{d}y}(z) = \sum_{l=-\infty}^{\infty} h^*(l)z^l \sum_{m=-\infty}^{\infty} \phi_{\mathrm{d}x}(m)z^{-m} \tag{2.89}$$

or

$$\Phi_{\mathrm{d}y}(z) = H^*(1/z^*)\Phi_{\mathrm{d}x}(z) \tag{2.90}$$

Also, using Eq. (2.56), we can get, from Eq. (2.90)

$$\Phi_{y\mathrm{d}}(z) = H(z)\Phi_{x\mathrm{d}}(z) \tag{2.91}$$

2.4.5 *Ergodicity and Time Averages*

Estimation of stochastic averages, as was suggested above, requires a large number of realizations (sample sequences) of the underlying stochastic processes, even under the condition that the processes are stationary. This is not feasible, in practice, where usually only a single realization of each stochastic process is available. In that case, we have no choice, but to use time averages to estimate the desired ensemble averages. Then, a fundamental question that arises is the following. Under what condition(s) do time averages become equal to ensemble averages? As one may intuitively understand, it turns out that under rather mild conditions, the only requirement for the time and ensemble averages to be the same is that the corresponding stochastic process be stationary (Papoulis, 1991).

A stationary stochastic process $\{x(n)\}$ is said to be ergodic if its ensemble averages are equal to time averages. Ergodicity is usually defined for specific averages. For example, we may come across the terms such as mean-ergodic or correlation-ergodic. In adaptive filters theory, it is always assumed that all the underlying processes are *ergodic in the strict sense*. This means, all averages can be obtained by time averages. We make such assumption throughout this book, whenever necessary.

Problems

P2.1 Find the z-transform and its region of convergence of the following sequences

(i)

$$x(n) = \begin{cases} 3^n, & n < 0 \\ 0.7^n, & n \geq 0. \end{cases}$$

(ii)

$$x(n) = \begin{cases} 2^n, & n < 0 \\ 0.7^n, & 0 \leq n < 5 \\ 0.5^n & n \geq 5 \end{cases}$$

P2.2 Find the inverse z-transform of Eq. (2.30) when

(i) its region of convergence is region I of Figure 2.1.
(ii) its region of convergence is region II of Figure 2.1.
(iii) its region of convergence is region III of Figure 2.1.

P2.3 Use the basic definitions of the correlation and covariance functions to prove the symmetry properties (Eq. 2.48) to (Eq. 2.51).

P2.4 Consider a stationary stochastic process,

$$x(n) = \nu(n) + \sin(\omega_o n + \theta)$$

where $\{\nu(n)\}$ is a stationary white noise, ω_o is a fixed angular frequency, and θ is a random phase which is uniformly distributed in the interval $-\pi \leq \theta \leq \pi$, but constant for each realization of $\{x(n)\}$. Find the autocorrelation function of $\{x(n)\}$ and show that $\Phi_{xx}(z)$ has no region of convergence in the z-plane.

P2.5 Consider a stationary stochastic stationary process, $\{x(n)\}$, with mean $m_x \neq 0$.

(i) Show that $\Phi_{xx}(z)$ contains a summation which is not convergent for any value of the complex variable z.
(ii) Consider the case where $|z| = 1$, that is, $z = e^{j\omega}$, for $0 \leq \omega \leq 2\pi$. For this case, argue that $X(e^{j\omega})$ at the vicinity of $\omega = 0$ is an impulse with the magnitude of m_x.
(iii) What is the power spectral density $\Phi_{xx}(e^{j\omega})$ at the vicinity of $\omega = 0$
Hint: To answer this part, you may evaluate $\phi_{xx}(k)$ first.

P2.6 In Problem P2.5, the nonzero mean of $x(n)$ may be interpreted as a presence of a tone at $\omega = 0$. Repeat Problem P2.5 when $x(n)$ contains a tone at an arbitrary frequency $\omega = \omega_0$.

P2.7 Prove the symmetry equations (2.55) and (2.56).

P2.8 A stationary unit-variance white noise process, $\{\nu(n)\}$, is passed through a linear time-invariant system with the system function

$$H(z) = \frac{1}{1 - az^{-1}}, \quad |a| < 1.$$

If the system output is referred to as $\{u(n)\}$, find the following:

(i) $\Phi_{\nu u}(z)$ and $\Phi_{uu}(z)$.
(ii) The cross-correlation and autocorrelation functions $\phi_{\nu u}(k)$ and $\phi_{uu}(k)$.
(iii) Variance of $\{u(n)\}$.

P2.9 Repeat P2.8 when

$$H(z) = \frac{1}{(1 - az^{-1})(1 - bz^{-1})}, \quad |a| \text{ and } |b| < 1.$$

Find the answers for the two cases when $a = b$ and $a \neq b$.

P2.10 Repeat P2.8 when $H(z)$ is a finite-impulse response system with

$$H(z) = \sum_{n=0}^{N-1} h(n) z^{-n}.$$

P2.11 Work out the details of derivation of Eq. (2.66) from Eq. (2.65).

P2.12 Show that for a linear time-invariant system with input $\{x(n)\}$, output $\{y(n)\}$, and system function $H(z)$

$$\Phi_{yx}(z) = H(z)\Phi_{xx}(z).$$

Also, if $\{d(n)\}$ is a third process

$$\Phi_{y\mathrm{d}}(z) = H(z)\Phi_{x\mathrm{d}}(z).$$

P2.13 Write the following z-transform relations in terms of the time series $h(n)$ and the correlation functions:

(i) $\Phi_{yx}(z) = H(z)\Phi_{xx}(z)$.
(ii) $\Phi_{xy}(z) = H^*(1/z^*)\Phi_{xx}(z)$.
(iii) $\Phi_{y\mathrm{d}}(z) = H(z)\Phi_{x\mathrm{d}}(z)$.
(iv) $\Phi_{\mathrm{d}y}(z) = H^*(1/z^*)\Phi_{\mathrm{d}x}(z)$.

P2.14 Consider the system shown in Figure P2.14. The input processes, $\{u(n)\}$ and $\{v(n)\}$ are zero-mean and uncorrelated with each other. Derive the relationships which relate $\Phi_{uu}(z)$, $\Phi_{vv}(z)$, $H(z)$, and $G(z)$ with the following functions:

(i) $\Phi_{uy}(z)$.
(ii) $\Phi_{vy}(z)$.
(iii) $\Phi_{yy}(z)$.

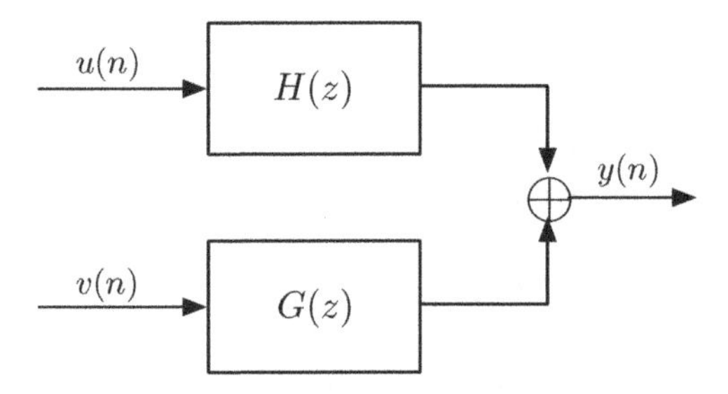

Figure P2.14

P2.15 Consider the system shown in Figure P2.15. The input, $\{\nu(n)\}$, is a stationary zero-mean unit-variance white noise process. Show that

(i) $\Phi_{xx}(z) = \frac{1}{(1-0.5z^{-1})(1-0.5z)}$.
(ii) $\Phi_{yy}(z) = 4$.
(iii) $\Phi_{\nu y}(z) = \frac{1-2z}{1-0.5z}$.

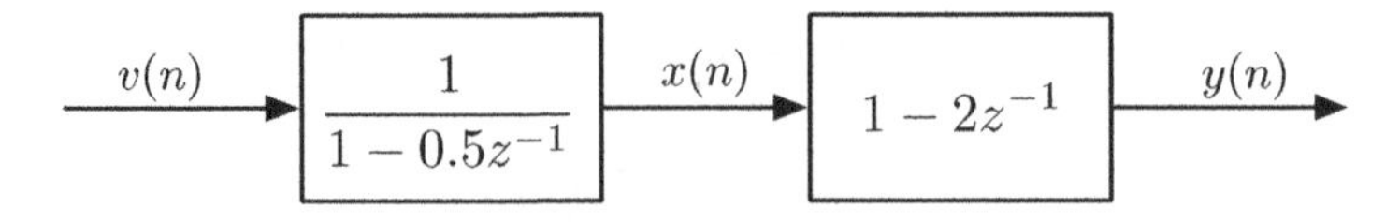

Figure P2.15

P2.16 Consider the system shown in Figure P2.16. The input, $\{v(n)\}$, is a stationary zero-mean unit-variance white noise process. Show that

(i) $\phi_{xy}(m) = \sum_l h(l+m)g^*(l)$.
(ii) $\Phi_{xy}(z) = H(z)G^*(1/z^*)$.

where $h(n)$ and $g(n)$ are the impulse responses of the subsystems $H(z)$ and $G(z)$, respectively.

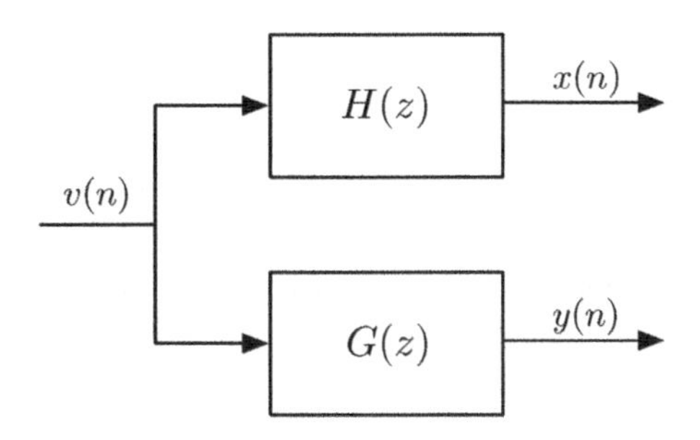

Figure P2.16

P2.17 Consider the random process $\{u(n)\}$, where $u(n) = x(n) + y(n)$, and $\{x(n)\}$ and $\{y(n)\}$ are the random processes that were introduced in Problem P2.16. Derive an expression for $\Phi_{uu}(z)$ in terms of $\Phi_{xx}(z)$, $\Phi_{yy}(z)$ and $\Phi_{xy}(z)$,

(i) through direct use of the equation $u(n) = x(n) + y(n)$.
(ii) by noting that $\{u(n)\}$ is the output of a system with input $\{v(n)\}$ and transfer function $H(z) + G(z)$.
(iii) Confirm that the results obtained in Parts (i) and (ii) are similar.

3

Wiener Filters

In this chapter, we study a class of optimum linear filters known as *Wiener filters*. As we will see in later chapters, the concept of Wiener filters is essential as well as helpful to understand and appreciate adaptive filters. Furthermore, Wiener filtering is general and applicable to any application that involves linear estimation of a desired signal sequence from another related sequence. Applications such as prediction, smoothing, joint process estimation, and channel equalization (deconvolution) are all covered by Wiener filters.

We study Wiener filters by looking at them from different angles. We first develop the theory of causal transversal Wiener filters for the case of discrete-time real-valued signals. This will then be extended to the case of complex-valued signals. Our discussion follows with a study of *unconstrained Wiener filters*. The term *unconstrained* signifies that the filter impulse response is allowed to be noncausal and infinite in duration. The study of unconstrained Wiener filters is very instructive as it reveals many important aspects of Wiener filters, which otherwise would be difficult to see.

In the theory of Wiener filters, the underlying signals are assumed to be random processes, and the filter design is done using the statistics obtained by ensemble averaging. We follow this approach while doing the theoretical development and analysis of Wiener filters. However, from the implementation point of view and, in particular, while developing adaptive algorithms in later chapters, we have to consider the use of time averages instead of ensemble averages. Adoption of this approach in the development of Wiener filters is also possible, once we assume all the underlying processes are ergodic; that is, their time and ensemble averages are the same (Section 2.4.5).

3.1 Mean-Squared Error Criterion

Figure 3.1 shows the block schematic of a linear discrete-time filter $W(z)$ in the context of estimating a desired signal $d(n)$ based on an excitation $x(n)$. Here, we assume that both $x(n)$ and $d(n)$ are samples of infinite length random processes. The filter output is $y(n)$, and $e(n)$ is the estimation error. Clearly, the smaller the estimation error, the better the filter performance. As the error approaches zero, the output of the filter approaches the desired signal, $d(n)$. Hence, the question that arises is the following: what is the most appropriate choice for the parameters of the filter, which would result in the smallest possible estimation error? To a certain extent, the statement of this question itself gives

---

Adaptive Filters: Theory and Applications, Second Edition. Behrouz Farhang-Boroujeny.

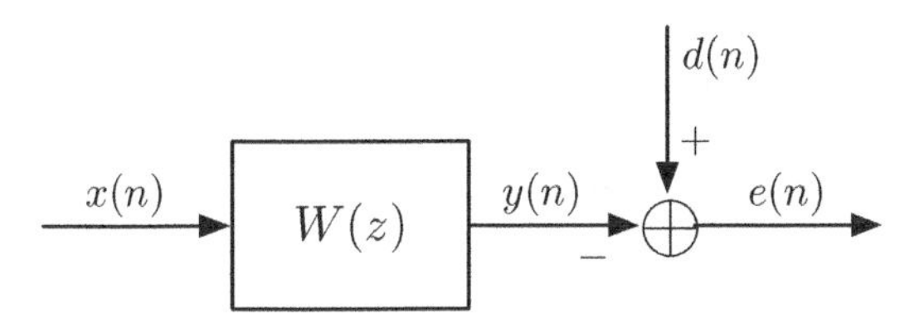

Figure 3.1 Block diagram of a filtering problem.

us some hints on the choice of the filter parameters. As we want the estimation error to be as small as possible, a straightforward approach to the design of the filter parameters appears to be "to choose an *appropriate* function of this estimation error as a *cost function* and select that set of the filter parameters which *optimizes* this cost function *in some sense*." This is indeed the philosophy that underlies almost all filter design approaches. The various details of this design principle will become clear as we go along. Commonly used synonyms for cost function are *performance function* and *performance surface*.

In choosing a performance function, the following points have to be considered:

1. The performance function must be mathematically tractable.
2. The performance function should preferably have a single minimum (or maximum) point, so that the optimum set of filter parameters could be selected unambiguously.

The tractability of the performance function is essential as it permits the analysis of the filter and also greatly simplifies the development of adaptive algorithms for adjustment of the filter parameters. The number of minima (or maxima) points for a performance function is closely related to the filter structure. The *recursive* (infinite-impulse response – IIR) filters, in general, result in performance functions that may have many minima (or maxima) points, whereas the *non-recursive* (finite-impulse response – FIR) filters are guaranteed to have a single global minimum (or maximum) point if a proper performance function is used. Because of this, application of the IIR filters in adaptive filtering has been very limited. In this book, also, with the exception of a few cases, our discussion is limited to the FIR-adaptive filters.

In Wiener filters, the performance function is chosen to be

$$\xi = E[|e(n)|^2] \tag{3.1}$$

where $E[\cdot]$ denotes the statistical expectation. In fact, the performance function ξ, which is also called *mean-squared error criterion*, turns out to be the simplest possible function, which satisfies the two requirements noted above. It can easily be handled mathematically, and in many cases of interest, it has a single global minimum. In particular, in the case of FIR filters, the performance function ξ is a hyperparaboloid (bowl shaped) with a single minimum point, which can easily be calculated using the second-order statistics of the underlying random processes.

It is instructive to note that a possible generalization of the mean-squared error criterion (3.1) is

$$\xi_p = E[|e(n)|^p] \tag{3.2}$$

where p takes the integer values 1, 2, 3,.... Clearly, the case of $p = 2$ leads to the Wiener filter performance function, which is defined above. Cases where $p > 2$, with p being even, may result in more than one minimum and/or maximum point. Furthermore, the case of odd p turns out to be difficult to handle mathematically because of the modulus sign on $e(n)$.

3.2 Wiener Filter – Transversal, Real-Valued Case

Consider a transversal filter which is shown in Figure 3.2. The filter input, $x(n)$, and its desired output, $d(n)$, are assumed to be real-valued stationary processes. The filter tap weights, $w_0, w_1, \ldots, w_{N-1}$, are also assumed to be real-valued. The filter tap-weight and input vectors are defined, respectively, as the column vectors

$$\mathbf{w} = [w_0 \; w_1 \; \cdots \; w_{N-1}]^{\mathrm{T}} \tag{3.3}$$

and

$$\mathbf{x}(n) = [x(n) \; x(n-1) \; \cdots \; x(n-N+1)]^{\mathrm{T}} \tag{3.4}$$

where superscript T stands for transpose.

The filter output is

$$y(n) = \sum_{i=0}^{N-1} w_i x(n-i) = \mathbf{w}^{\mathrm{T}}\mathbf{x}(n) \tag{3.5}$$

which can also be written as

$$y(n) = \mathbf{x}^{\mathrm{T}}(n)\mathbf{w} \tag{3.6}$$

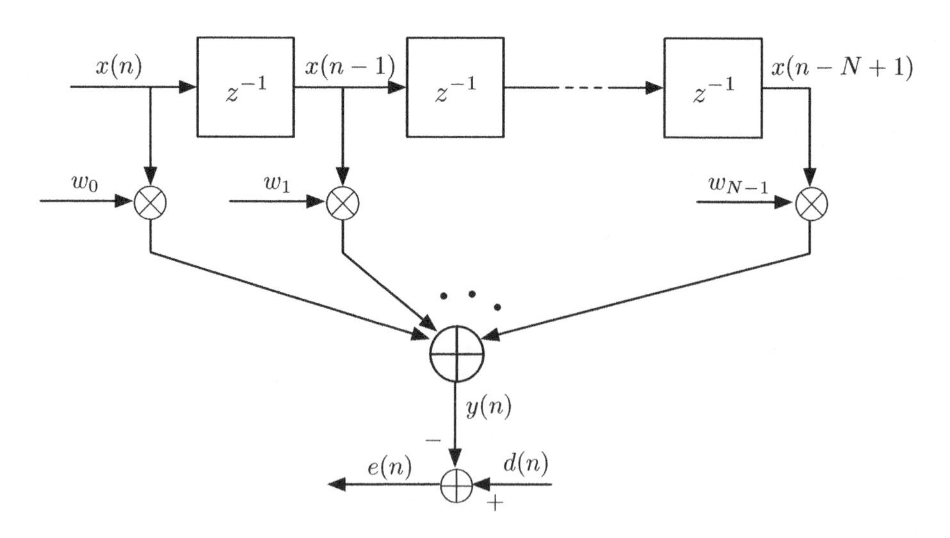

Figure 3.2 A transversal filter.

because $\mathbf{w}^{\mathrm{T}}\mathbf{x}(n)$ is a scalar and thus it is equal to its transpose, that is, $\mathbf{w}^{\mathrm{T}}\mathbf{x}(n) = (\mathbf{w}^{\mathrm{T}}\mathbf{x}(n))^{\mathrm{T}} = \mathbf{x}^{\mathrm{T}}(n)\mathbf{w}$. Thus, we may write

$$\begin{aligned} e(n) &= d(n) - y(n) \\ &= d(n) - \mathbf{w}^{\mathrm{T}}\mathbf{x}(n) \\ &= d(n) - \mathbf{x}^{\mathrm{T}}(n)\mathbf{w} \end{aligned} \tag{3.7}$$

Using Eq. (3.7) in Eq. (3.1), we get

$$\xi = E[e^2(n)] = E[(d(n) - \mathbf{w}^{\mathrm{T}}\mathbf{x}(n))(d(n) - \mathbf{x}^{\mathrm{T}}(n)\mathbf{w})] \tag{3.8}$$

Expanding the right-hand side of Eq. (3.8) and noting that $\mathbf{w}$ can be shifted out of the expectation operator, $E[\cdot]$, because it is not a statistical variable, we obtain

$$\begin{aligned} \xi = E[d^2(n)] &- \mathbf{w}^{\mathrm{T}}E[\mathbf{x}(n)d(n)] \\ &- E[d(n)\mathbf{x}^{\mathrm{T}}(n)]\mathbf{w} + \mathbf{w}^{\mathrm{T}}E[\mathbf{x}(n)\mathbf{x}^{\mathrm{T}}(n)]\mathbf{w} \end{aligned} \tag{3.9}$$

Next, if we define the N-by-1 cross-correlation vector

$$\mathbf{p} = E[\mathbf{x}(n)d(n)] = [p_0, p_1, \cdots p_{N-1}]^{\mathrm{T}} \tag{3.10}$$

and the N-by-N autocorrelation matrix

$$\mathbf{R} = E[\mathbf{x}(n)\mathbf{x}^{\mathrm{T}}(n)] = \begin{bmatrix} r_{00} & r_{01} & r_{02} & \cdots & r_{0,N-1} \\ r_{10} & r_{11} & r_{12} & \cdots & r_{1,N-1} \\ r_{20} & r_{21} & r_{22} & \cdots & r_{2,N-1} \\ \vdots & \vdots & \vdots & \ddots & \vdots \\ r_{N-1,0} & r_{N-1,1} & r_{N-1,2} & \cdots & r_{N-1,N-1} \end{bmatrix} \tag{3.11}$$

and note that $E[d(n)\mathbf{x}^{\mathrm{T}}(n)] = \mathbf{p}^{\mathrm{T}}$, and also $\mathbf{w}^{\mathrm{T}}\mathbf{p} = \mathbf{p}^{\mathrm{T}}\mathbf{w}$, we obtain

$$\xi = E[d^2(n)] - 2\mathbf{w}^{\mathrm{T}}\mathbf{p} + \mathbf{w}^{\mathrm{T}}\mathbf{R}\mathbf{w} \tag{3.12}$$

This is a quadratic function of the tap-weight vector $\mathbf{w}$ with a single global minimum.[1] We will give the full details of this function in Chapter 4.

To obtain the set of tap weights, which minimizes the performance function ξ, we need to solve the system of equations that results from setting the partial derivatives of ξ with respect to every tap weight to zero. That is,

$$\frac{\partial \xi}{\partial w_i} = 0, \quad \text{for } i = 0, 1, \ldots, N-1 \tag{3.13}$$

These equations may collectively be written as

$$\nabla \xi = \mathbf{0} \tag{3.14}$$

[1] It may be noted that for Eq. (3.12) to correspond to a convex quadratic surface, so that it has a unique minimum point, and not a saddle point, $\mathbf{R}$ has to be a positive definite matrix. This point, which is missed out here, will be examined in detail in Chapter 4.

where ∇ is the gradient operator defined as the column vector

$$\nabla = \left[\frac{\partial}{\partial w_0} \; \frac{\partial}{\partial w_1} \; \cdots \; \frac{\partial}{\partial w_{N-1}}\right]^{\mathrm{T}} \tag{3.15}$$

and $\mathbf{0}$ on the right-hand side of Eq. (3.14) denotes the column vector consisting of N zeros.

To find the partial derivatives of ξ with respect to the filter tap weights, we first expand Eq. (3.12) as

$$\xi = E[d^2(n)] - 2\sum_{l=0}^{N-1} p_l w_l + \sum_{l=0}^{N-1}\sum_{m=0}^{N-1} w_l w_m r_{lm} \tag{3.16}$$

Also, we note that the double summation on the right-hand side of Eq. (3.16) may be expanded as

$$\begin{aligned}\sum_{l=0}^{N-1}\sum_{m=0}^{N-1} w_l w_m r_{lm} &= \sum_{\substack{l=0\\ l\neq i}}^{N-1}\sum_{\substack{m=0\\ m\neq i}}^{N-1} w_l w_m r_{lm} + w_i \sum_{\substack{l=0\\ l\neq i}}^{N-1} w_l r_{li} \\ &\quad + w_i \sum_{\substack{m=0\\ m\neq i}}^{N-1} w_m r_{im} + w_i^2 r_{ii}\end{aligned} \tag{3.17}$$

Substituting Eq. (3.17) in Eq. (3.16), taking partial derivative of ξ with respect to w_i and replacing m by l, we obtain

$$\frac{\partial \xi}{\partial w_i} = -2p_i + \sum_{l=0}^{N-1} w_l (r_{li} + r_{il}), \quad \text{for } i = 0, 1, \ldots, N-1 \tag{3.18}$$

To simplify this, we note that

$$r_{li} = E[x(n-l)x(n-i)] = \phi_{xx}(i-l) \tag{3.19}$$

where $\phi_{xx}(i-l)$ is the autocorrelation function of $x(n)$ for lag $i-l$. Similarly,

$$r_{il} = \phi_{xx}(l-i) \tag{3.20}$$

Considering the symmetry property of the autocorrelation function, that is, $\phi_{xx}(k) = \phi_{xx}(-k)$, we get

$$r_{li} = r_{il} \tag{3.21}$$

Substituting Eq. (3.21) in Eq. (3.18), we obtain

$$\frac{\partial \xi}{\partial w_i} = 2\sum_{l=0}^{N-1} r_{il} w_l - 2p_i, \quad \text{for } i = 0, 1, \ldots, N-1 \tag{3.22}$$

which can be expressed using matrix notation as

$$\nabla \xi = 2\mathbf{R}\mathbf{w} - 2\mathbf{p} \tag{3.23}$$

Letting $\nabla\xi = 0$ gives the following equation from which the optimum set of the Wiener filter tap weights can be obtained.

$$\mathbf{R}\mathbf{w}_o = \mathbf{p} \tag{3.24}$$

Note that we have added the subscript "o" to $\mathbf{w}$ to emphasize that it is the optimum tap-weight vector. Equation (3.24), which is known as the *Wiener–Hopf* equation, has the following solution:

$$\mathbf{w}_o = \mathbf{R}^{-1}\mathbf{p} \tag{3.25}$$

assuming that $\mathbf{R}$ has an inverse.

Replacing $\mathbf{w}$ by $\mathbf{w}_o$ and $\mathbf{R}\mathbf{w}_o$ by $\mathbf{p}$ in Eq. (3.12), we obtain

$$\begin{aligned}\xi_{\min} &= E[d^2(n)] - \mathbf{w}_o^{\mathrm{T}}\mathbf{p} \\ &= E[d^2(n)] - \mathbf{w}_o^{\mathrm{T}}\mathbf{R}\mathbf{w}_o.\end{aligned} \tag{3.26}$$

This is the *minimum mean-squared error* that can be achieved by the transversal Wiener filter $W(z)$ and is obtained when its tap weights are chosen according to the optimum solution given by Eq. (3.25).

For our later reference, we may also note that by substituting Eq. (3.25) in Eq. (3.26), we obtain

$$\xi_{\min} = E[d^2(n)] - \mathbf{p}^{\mathrm{T}}\mathbf{R}^{-1}\mathbf{p} \tag{3.27}$$

Example 3.1

Consider the modeling problem shown in Figure 3.3. The plant is a two-tap filter with an additive noise, $\nu(n)$, added to its output. A two-tap Wiener filter with tap weights w_0 and w_1 is used to model the plant parameters. The same input is applied to both the plant and Wiener filter. The input, $x(n)$, is a stationary white process with variance of unity. The additive noise, $\nu(n)$, is zero-mean and uncorrelated with $x(n)$, and its variance is $\sigma_\nu^2 = 0.1$. We want to compute the optimum values of w_0 and w_1, which minimize $E[e^2(n)]$.

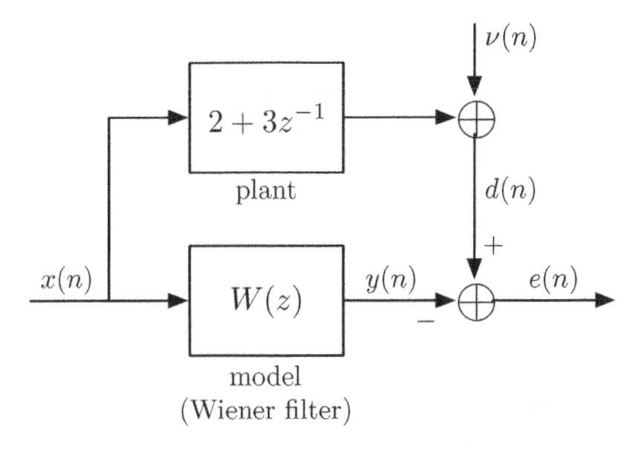

Figure 3.3 A modeling problem.

We need to compute $\mathbf{R}$ and $\mathbf{p}$ to obtain the optimum values of w_0 and w_1, which minimize $E[e^2(n)]$. For this example, we get

$$\mathbf{R} = \begin{bmatrix} E[x^2(n)] & E[x(n)x(n-1)] \\ E[x(n-1)x(n)] & E[x^2(n-1)] \end{bmatrix} = \begin{bmatrix} 1 & 0 \\ 0 & 1 \end{bmatrix} \tag{3.28}$$

This follows, as $x(n)$ is white, thus, $E[x(n)x(n-1)] = E[x(n-1)x(n)] = 0$, and also it has a variance of unity. The latter implies that $E[x^2(n)] = E[x^2(n-1)] = 1$.

Also, we note that $d(n) = 2x(n) + 3x(n-1) + \nu(n)$, and, thus,

$$\begin{aligned} \mathbf{p} &= \begin{bmatrix} E[x(n)d(n)] \\ E[x(n-1)d(n)] \end{bmatrix} \\ &= \begin{bmatrix} E[x(n)(2x(n) + 3x(n-1) + \nu(n))] \\ E[x(n-1)(2x(n) + 3x(n-1) + \nu(n))] \end{bmatrix} \end{aligned} \tag{3.29}$$

Expanding the terms under the expectation operators, and noting that $E[x^2(n)] = E[x^2(n-1)] = 1$ and $E[x(n)x(n-1)] = E[x(n)\nu(n)] = E[x(n-1)\nu(n)] = 0$, we get

$$\mathbf{p} = \begin{bmatrix} 2 \\ 3 \end{bmatrix} \tag{3.30}$$

Similarly, we obtain

$$\begin{aligned} E[d^2(n)] &= E[(2x(n) + 3x(n-1) + \nu(n))^2] \\ &= 4E[x^2(n)] + 9E[x^2(n-1)] + \sigma_\nu^2 = 13.1 \end{aligned} \tag{3.31}$$

Substituting Eqs. (3.28), (3.30), and (3.31) in Eq. (3.12), we get

$$\xi = 13.1 - 4w_0 - 6w_1 + w_0^2 + w_1^2 \tag{3.32}$$

This is a paraboloid in the three-dimensional space with the axes w_0, w_1 and ξ. Figure 3.4 shows this paraboloid. We may note that the optimum tap weights of the Wiener filter are given by Eq. (3.25), which for the present example may be written as

$$\begin{bmatrix} w_{o,0} \\ w_{o,1} \end{bmatrix} = \begin{bmatrix} 1 & 0 \\ 0 & 1 \end{bmatrix}^{-1} \begin{bmatrix} 2 \\ 3 \end{bmatrix} = \begin{bmatrix} 2 \\ 3 \end{bmatrix} \tag{3.33}$$

Also, from Eq. (3.26),

$$\xi_{\min} = 13.1 - [2 \quad 3] \begin{bmatrix} 2 \\ 3 \end{bmatrix} = 0.1 \tag{3.34}$$

Clearly, the values of $w_{o,0}$, $w_{o,1}$, and $\xi_{\min}$ coincide with the minimum point shown in Figure 3.4.

The features of interest on the performance surface in Figure 3.4 and the results obtained in Eqs. (3.33) and (3.34) and also Figure 3.4 may be understood better, if we note that the right-hand side of Eq. (3.32) may also be expressed as

$$\xi = 0.1 + (w_0 - 2)^2 + (w_1 - 3)^2 \tag{3.35}$$

Clearly, the minimum value of ξ is achieved when the last two terms on the right-hand side of Eq. (3.35) are forced to zero. This coincides with the results in Eqs. (3.33) and (3.34).

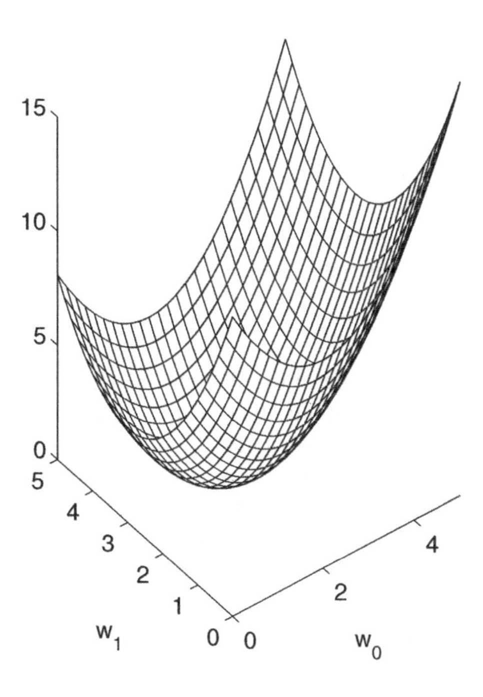

Figure 3.4 The performance surface of the modeling problem of Figure 3.3.

3.3 Principle of Orthogonality

In this section, we present an alternative approach for the design of Wiener filters. This presentation is a complement to the derivations in the last section in the sense that the approach presented below can be considered as a simplified/shortened version of the approach in the last section. More importantly, it leads to more insight into the concept of Wiener filtering problem.

We start with the cost function equation (3.1), which in the case of real-valued data may be written as

$$\xi = E[e^2(n)] \tag{3.36}$$

Taking partial derivatives of ξ with respect to the filter tap weights, $\{w_i; i = 0, 1, \ldots, N-1\}$, and interchanging the derivative and expectation operators (since these are linear operators), we obtain

$$\frac{\partial \xi}{\partial w_i} = E\left[2e(n)\frac{\partial e(n)}{\partial w_i}\right], \quad \text{for } i = 0, 1, \ldots, N-1 \tag{3.37}$$

where $e(n) = d(n) - y(n)$. As $d(n)$ is independent of the filter tap weights, we get

$$\frac{\partial e(n)}{\partial w_i} = -\frac{\partial y(n)}{\partial w_i} = -x(n-i) \tag{3.38}$$

where the last result is obtained by replacing for $y(n)$ from Eq. (3.5). Using this result in Eq. (3.37), we obtain

$$\frac{\partial \xi}{\partial w_i} = -2E[e(n)x(n-i)], \quad \text{for } i = 0, 1, \ldots, N-1 \tag{3.39}$$

From our discussion in the last section, we know that when the Wiener filter tap weights are set to their optimal values, the partial derivatives of the cost function, ξ, with respect to the filter tap weights are all zero. Hence, if $e_o(n)$ is the estimation error when the filter tap weights are set equal to their optimal values, Eq. (3.39) becomes

$$E[e_o(n)x(n-i)] = 0, \quad \text{for } i = 0, 1, \ldots, N-1 \tag{3.40}$$

This shows that at the optimal setting of the Wiener filter tap weights, the estimation error is uncorrelated with the filter tap inputs, that is, the input samples used for estimation. This is known as the *principle of orthogonality*.

The principle of orthogonality is an elegant result of the Wiener filtering that is frequently used for simple derivations of results which otherwise would seem far more difficult to derive. We will use the principle of orthogonality throughout this book for many of our derivations.

As a useful corollary to the principle of orthogonality, we note that the filter output is also uncorrelated with the estimation error when its tap weights are set to their optimal values. This may be shown as follows:

$$\begin{aligned} E[e_o(n)y_o(n)] &= E\left[e_o(n)\sum_{i=0}^{N-1} w_{o,i}x(n-i)\right] \\ &= \sum_{i=0}^{N-1} w_{o,i}E[e_o(n)x(n-i)] \end{aligned} \tag{3.41}$$

where $y_o(n)$ is the Wiener filter output when its tap weights are set to their optimal values. Then, using Eq. (3.40) in Eq. (3.41), we obtain

$$E[e_o(n)y_o(n)] = 0 \tag{3.42}$$

We may also refer to the above result by saying that *the optimized Wiener filter output and the estimation error are orthogonal.*

The words *orthogonality* and *orthogonal* are commonly used for referring to pairs of random variables that are uncorrelated with each other. This originates from the fact that the set of all random variables with finite second-order moments constitutes a linear space with an inner product. The inner product in this space is defined to be the correlation between its elements. In particular, if x and y are two elements of the linear space of random variables, the inner product of x and y is defined as $E[xy]$, when x and y are real-valued, or $E[xy^*]$, in the more general case of complex-valued random variables. Then, in analogy with the Euclidean space in which the elements are vectors, the geometrical concepts such as orthogonality, projection, and subspaces may also be defined for the space of random variables. Interested readers may refer to Honig and Messerschmitt (1984) for an excellent, yet simple, discussion on this topic.

Next, we use the principle of orthogonality to give an alternative derivation of the Wiener–Hopf equation (3.24) and also the minimum mean-squared error of Eq. (3.26). We note that

$$e_o(n) = d(n) - \sum_{l=0}^{N-1} w_{o,l}x(n-l) \tag{3.43}$$

where $w_{o,l}$'s are the optimum values of the Wiener filter tap weights. Substituting Eq. (3.43) in Eq. (3.40) and rearranging the results, we get

$$\sum_{l=0}^{N-1} E[x(n-i)x(n-l)]w_{o,l} = E[d(n)x(n-i)], \quad \text{for } i = 0, 1, \ldots, N-1 \tag{3.44}$$

We also note that $E[x(n-i)x(n-l)] = r_{il}$ and $E[d(n)x(n-i)] = p_i$. Using these in Eq. (3.44), we obtain

$$\sum_{l=0}^{N-1} r_{il}w_{o,l} = p_i, \quad \text{for } i = 0, 1, \ldots, N-1 \tag{3.45}$$

which is nothing but Eq. (3.24) in expanded form.

Also, we note that

$$\begin{aligned}
\xi_{\min} &= E[e_o^2(n)] \\
&= E[e_o(n)(d(n) - y_o(n))] \\
&= E[e_o(n)d(n)] - E[e_o(n)y_o(n)] \\
&= E[e_o(n)d(n)]
\end{aligned} \tag{3.46}$$

where Eq. (3.42) has been used to obtain the last equality. Now, substituting Eq. (3.43) in Eq. (3.46), we obtain

$$\begin{aligned}
\xi_{\min} &= E\left[\left(d(n) - \sum_{i=0}^{N-1} w_{o,i}x(n-i)\right)d(n)\right] \\
&= E[d^2(n)] - \sum_{i=0}^{N-1} w_{o,i}E[d(n)x(n-i)] \\
&= E[d^2(n)] - \sum_{i=0}^{N-1} w_{o,i}p_i
\end{aligned} \tag{3.47}$$

which is nothing but Eq. (3.26) in expanded form.

3.4 Normalized Performance Function

Equation (3.43) can be written as

$$d(n) = e_o(n) + y_o(n) \tag{3.48}$$

Squaring both sides of Eq. (3.48) and taking expectation, we get

$$E[d^2(n)] = E[e_o^2(n)] + E[y_o^2(n)] + 2E[e_o(n)y_o(n)] \tag{3.49}$$

We may note that $E[e_o^2(n)] = \xi_{\min}$, and the last term in Eq. (3.49) is zero because of Eq. (3.42). Thus, we obtain

$$\xi_{\min} = E[d^2(n)] - E[y_o^2(n)] \tag{3.50}$$

which suggests that *the minimum mean-squared error at the Wiener filter output is the difference between the mean-squared of the desired output and the mean-squared of the best estimate of that at the filter output.*

It is appropriate if we define the ratio

$$\zeta = \frac{\xi}{E[d^2(n)]} \tag{3.51}$$

as the normalized performance function. We may note that $\zeta = 1$ when $y(n)$ is forced to zero; that is, when no estimation of $d(n)$ has been made. It reaches its minimum value, $\zeta_{\min}$, when the filter tap weights are chosen to achieve the minimum mean-squared error. This is given by

$$\zeta_{\min} = 1 - \frac{E[y_o^2(n)]}{E[d^2(n)]} \tag{3.52}$$

Noting that $\zeta_{\min}$ cannot be negative, we find that its value remains between 0 and 1. *The value of $\zeta_{\min}$ is an indication of the ability of the filter in estimating the desired output. A value of $\zeta_{\min}$ close to zero is an indication of good performance of the filter, and a value of $\zeta_{\min}$ close to 1 indicates poor performance of the filter.*

3.5 Extension to Complex-Valued Case

There are some practical applications in which the underlying random processes are complex-valued. For instance, in data transmission (Chapter 17), the most frequently used signaling techniques are phase shift keying (PSK) and quadrature-amplitude modulation (QAM) in which the baseband signal consists of two separate components which are the real and imaginary parts of a complex-valued signal. Moreover, in the case of frequency domain implementation of adaptive filters (Chapter 8) and subband-adaptive filters (Chapter 9), we will be dealing with complex-valued signals, even though the original signals may be real-valued.

In this section, we extend the results of the last two sections to the case of complex-valued signals. We assume a transversal filter as shown in Figure 3.2. The input, $x(n)$, the desired output, $d(n)$, and the filter tap weights are all assumed to be complex-valued. Then, the estimation error, $e(n)$, is also complex-valued and we may write

$$\xi = E[|e(n)|^2] = E[e(n)e^*(n)] \tag{3.53}$$

where the asterisk denotes complex conjugation.

As in the real-valued case, the performance function, ξ, in the complex-valued case is also a quadratic function of the filter tap weights. Similarly, to find the optimum set of the filter tap weights, we have to solve the system of equations, which results from setting the partial derivatives of ξ with respect to every tap weight to zero. However, noting that the filter tap weights are complex variables, the conventional definition of derivative with respect to an independent variable is not applicable to the present case. In fact, one may note that each tap weight, in the present case, consists of two independent variables that make the real and imaginary parts of that. Thus, the partial derivatives with respect to these two independent variables have to be performed separately and the results have to be set to zero to obtain the optimum tap weights of the Wiener filter. In particular, to obtain the optimum set of the filter tap weights, the following set of equations has to be solved, simultaneously.

$$\frac{\partial \xi}{\partial w_{i,R}} = 0 \quad \text{and} \quad \frac{\partial \xi}{\partial w_{i,I}} = 0, \quad \text{for } i = 0, 1, \ldots, N-1 \tag{3.54}$$

where $w_{i,R}$ and $w_{i,I}$ denote the real and imaginary parts of w_i, respectively. To write Eq. (3.54) in a more compact form, we note that ξ, $w_{i,R}$, and $w_{i,I}$ are all real. This implies that the partial derivatives in Eq. (3.54) are also all real and thus the pairs of equations in (3.54) may be combined together to obtain

$$\frac{\partial \xi}{\partial w_{i,R}} + j\frac{\partial \xi}{\partial w_{i,I}} = 0, \quad \text{for } i = 0, 1, \ldots, N-1 \tag{3.55}$$

where $j = \sqrt{-1}$. This, in turn, suggests the following definition of gradient of a function with respect to a complex variable $w = w_R + jw_I$.

$$\nabla_w^C \triangleq \frac{\partial}{\partial w_R} + j\frac{\partial}{\partial w_I} \tag{3.56}$$

We note that when ξ is a real function of w_R and w_I, the real and imaginary parts of $\nabla_w^C \xi$ are, respectively, equal to $\partial\xi/\partial w_R$ and $\partial\xi/\partial w_I$, and in that case $\nabla_w^C \xi = 0$ implies that $\partial\xi/\partial w_R = \partial\xi/\partial w_I = 0$. It is in this context that we can say Eqs. (3.54) and (3.55) are equivalent. This would not be true, in general, if ξ was complex-valued (Problem P3.9).

With the above background, we may now continue with the derivation of the principle of orthogonality and its subsequent results, for the case of complex-valued signals. From Eq. (3.53), we note that

$$\nabla_{w_i}^C \xi = E[e(n)\nabla_{w_i}^C e^*(n) + e^*(n)\nabla_{w_i}^C e(n)] \tag{3.57}$$

Noting that

$$e(n) = d(n) - \sum_{k=0}^{N-1} w_k x(n-k) \tag{3.58}$$

we obtain

$$\nabla_{w_i}^C e(n) = -x(n-i)\nabla_{w_i}^C w_i \tag{3.59}$$

and

$$\nabla_{w_i}^C e^*(n) = -x^*(n-i)\nabla_{w_i}^C w_i^* \tag{3.60}$$

Applying definition (3.56), we obtain

$$\nabla^C_{w_i} w_i = \frac{\partial w_i}{\partial w_{i,R}} + j\frac{\partial w_i}{\partial w_{i,I}} = 1 + j(j) = 1 - 1 = 0 \tag{3.61}$$

and

$$\nabla^C_{w_i} w_i^* = \frac{\partial w_i^*}{\partial w_{i,R}} + j\frac{\partial w_i^*}{\partial w_{i,I}} = 1 + j(-j) = 1 + 1 = 2 \tag{3.62}$$

Substituting Eqs. (3.61) and (3.62) in Eqs. (3.59) and (3.60), respectively, and the results in Eq. (3.57), we obtain

$$\nabla^C_{w_i} \xi = -2E[e(n)x^*(n-i)] \tag{3.63}$$

When the Wiener filter tap weights are set to their optimal values, $\nabla^C_{w_i} \xi = 0$. This gives

$$E[e_o(n)x^*(n-i)] = 0, \quad \text{for } i = 0, 1, \ldots, N-1 \tag{3.64}$$

where $e_o(n)$ is the optimum estimation error. The set of equations (3.64) represent *the principle of orthogonality for the case of complex-valued signals*.

To proceed with the derivation of the Wiener–Hopf equation, we define the input and tap-weight vectors of the filter as

$$\mathbf{x}(n) \triangleq [x(n)\ x(n-1)\ \ldots\ x(n-N+1)]^{\mathrm{T}} \tag{3.65}$$

and

$$\mathbf{w} \triangleq [w_0^*\ w_1^*\ \ldots\ w_{N-1}^*]^{\mathrm{T}} \tag{3.66}$$

respectively, where the asterisk and T denote complex conjugation and transpose, respectively. Note that the elements of the column vector $\mathbf{w}$ are complex conjugates of the actual tap weights of the filter, while conjugation is not applied to the samples of input in $\mathbf{x}(n)$. Also, for our further reference, later, we may write

$$\mathbf{x}(n) = [x^*(n)\ x^*(n-1)\ \cdots\ x^*(n-N+1)]^{\mathrm{H}} \tag{3.67}$$

and

$$\mathbf{w} = [w_0\ w_1\ \cdots\ w_{N-1}]^{\mathrm{H}} \tag{3.68}$$

where the superscript H denotes complex-conjugate transpose or Hermitian.

The set of equations (3.64) may also be written as

$$E[e_o^*(n)x(n-i)] = 0, \quad \text{for } i = 0, 1, \ldots, N-1 \tag{3.69}$$

Using the definition (3.65), these may be packed together as

$$E[e_o^*(n)\mathbf{x}(n)] = 0 \tag{3.70}$$

Also, we note that

$$e_o(n) = d(n) - \mathbf{w}_o^{\mathrm{H}}\mathbf{x}(n) \tag{3.71}$$

where $\mathbf{w}_o$ is the optimum tap-weight vector of the Wiener filter.

Replacing Eq. (3.71) in Eq. (3.70), we obtain

$$E[\mathbf{x}(n)(d^*(n) - \mathbf{x}^{\mathrm{H}}(n)\mathbf{w}_{\mathrm{o}})] = 0 \quad (3.72)$$

Rearranging Eq. (3.72), we get

$$\mathbf{R}\mathbf{w}_{\mathrm{o}} = \mathbf{p} \quad (3.73)$$

where $\mathbf{R} = E[\mathbf{x}(n)\mathbf{x}^{\mathrm{H}}(n)]$ and $\mathbf{p} = E[\mathbf{x}(n)d^*(n)]$. This is *the Wiener–Hopf equation for the case of complex-valued signals.*

Also, following the same derivations as Eqs. (3.46) through (3.47), for the present case, we obtain

$$\begin{aligned}\xi_{\min} &= E[|d(n)|^2] - \mathbf{w}_{\mathrm{o}}^{\mathrm{H}}\mathbf{p} \\ &= E[|d(n)|^2] - \mathbf{w}_{\mathrm{o}}^{\mathrm{H}}\mathbf{R}\mathbf{w}_{\mathrm{o}} \quad (3.74)\end{aligned}$$

3.6 Unconstrained Wiener Filters

The developments in the last three sections put some constraints on the Wiener filter by assuming that it is causal and the duration of its impulse response is limited. In this section, we remove such constraints and let the Wiener filter impulse response, w_i, to extend from $i = -\infty$ to $i = +\infty$ and derive equations for the filter performance function and its optimal system function. Such developments are very instructive for understanding many of the important aspects of the Wiener filter, which otherwise could not be easily understood.

Consider the Wiener filter shown in Figure 3.1, and repeated here in Figure 3.5, for convenience. We assume that the filter $W(z)$ may be noncausal and/or IIR. In order to keep the derivations in this section as simple as possible and also to emphasize more on the concepts, we consider only the case in which the underlying signals and system parameters are real-valued. Moreover, we assume that the complex variable z remains on the unit circle, that is, $|z| = 1$. This implies that $z^* = z^{-1}$. Also, for our later reference, we note that when the coefficients of a system function, such as $W(z)$, are real-valued, $W^*(1/z^*) = W(z^{-1})$, for all values of z, and $W(z^{-1}) = W^*(z)$, when $|z| = 1$.

The derivations that follow in this section highly depend on the results developed in Section 2.4.4 of Chapter 2. The reader is encouraged to review the latter section before continuing with the rest of this section.

3.6.1 Performance Function

Recall that the Wiener filter performance function is defined as

$$\xi = E[e^2(n)]$$

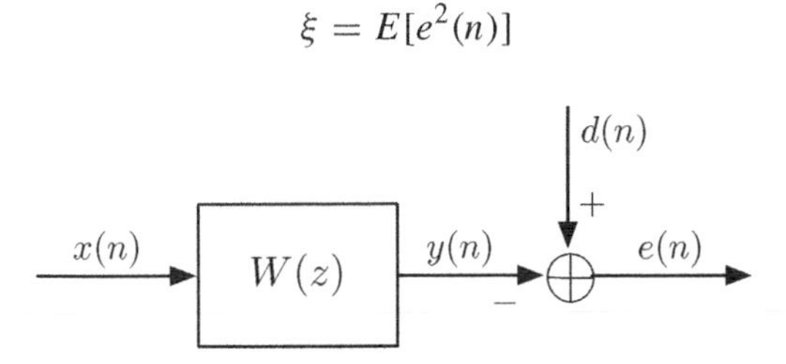

Figure 3.5 Block diagram of a Wiener filter.

Substituting $e(n)$ by $d(n) - y(n)$ and expanding, we get

$$\xi = E[d^2(n)] + E[y^2(n)] - 2E[y(n)d(n)] \tag{3.75}$$

In terms of autocorrelation and cross-correlation functions (Chapter 2), we may write

$$\xi = \phi_{dd}(0) + \phi_{yy}(0) - 2\phi_{yd}(0) \tag{3.76}$$

Replacing the last two terms on the right-hand side of Eq. (3.76) with their corresponding inverse z-transform relations, we obtain

$$\xi = \phi_{dd}(0) + \frac{1}{2\pi j}\oint_C \Phi_{yy}(z)\frac{\mathrm{d}z}{z} - 2 \times \frac{1}{2\pi j}\oint_C \Phi_{yd}(z)\frac{\mathrm{d}z}{z} \tag{3.77}$$

Also, from our discussion in Chapter 2, Section 2.4.4, we recall that when $x(n)$ and $y(n)$ are related as shown in Figure 3.5, for an arbitrary sequence $d(n)$, $\Phi_{yd}(z) = W(z)\Phi_{xd}(z)$. Also, if z is selected to be on the unit circle in the z-plane, $\Phi_{yy}(z) = |W(z)|^2\Phi_{xx}(z)$, $|W(z)|^2 = W(z)W^*(z)$, and $W^*(z) = W(z^{-1})$. Using these in Eq. (3.77), we obtain

$$\begin{aligned}\xi &= \phi_{dd}(0) + \frac{1}{2\pi j}\oint_C |W(z)|^2\Phi_{xx}(z)\frac{\mathrm{d}z}{z} - 2 \times \frac{1}{2\pi j}\oint_C W(z)\Phi_{xd}(z)\frac{\mathrm{d}z}{z} \\ &= \phi_{dd}(0) + \frac{1}{2\pi j}\oint_C [W^*(z)\Phi_{xx}(z) - 2\Phi_{xd}(z)]W(z)\frac{\mathrm{d}z}{z}\end{aligned} \tag{3.78}$$

where the contour of integration, C, is the unit circle. *This is the performance function for a Wiener filter with the system function* $W(z)$, *in its most general form. It covers IIR and FIR, as well as causal and noncausal filters.* The following examples show some of the flexibilities of Eq. (3.78).

Example 3.2

Consider the case where the Wiener filter is an N-tap FIR filter with the system function

$$W(z) = \sum_{l=0}^{N-1} w_l z^{-l} \tag{3.79}$$

This is the case that we studied in Sections 3.1 and 3.2.

Using Eq. (3.79) in the first line of Eq. (3.78), we obtain

$$\begin{aligned}\xi = \phi_{dd}(0) &+ \frac{1}{2\pi j}\oint_C \left(\sum_{l=0}^{N-1} w_l z^{-l}\right)\left(\sum_{m=0}^{N-1} w_m z^{m}\right)\Phi_{xx}(z)\frac{\mathrm{d}z}{z} \\ &-2 \times \frac{1}{2\pi j}\oint_C \left(\sum_{l=0}^{N-1} w_l z^{-l}\right)\Phi_{xd}(z)\frac{\mathrm{d}z}{z}\end{aligned} \tag{3.80}$$

Interchanging the order of the integrations and summations, Eq. (3.80) is simplified to

$$\xi = \phi_{dd}(0) + \sum_{l=0}^{N-1}\sum_{m=0}^{N-1} w_l w_m \frac{1}{2\pi j}\oint_C \Phi_{xx}(z) z^{m-l-1} dz$$
$$-2\sum_{l=0}^{N-1} w_l \frac{1}{2\pi j}\oint_C \Phi_{xd}(z) z^{-l-1} dz \tag{3.81}$$

Using the inverse z-transform relation, this gives

$$\xi = \phi_{dd}(0) + \sum_{l=0}^{N-1}\sum_{m=0}^{N-1} w_l w_m \phi_{xx}(m-l) - 2\sum_{l=0}^{N-1} w_l \phi_{xd}(-l) \tag{3.82}$$

Now, using the notations $\phi_{dd}(0) = E[d^2(n)]$, $\phi_{xd}(-l) = p_l$, and $\phi_{xx}(m-l) = \phi_{xx}(l-m) = r_{lm}$, we see that the performance function given by Eq. (3.82) is the same as what we had derived earlier in Eq. (3.16).

Example 3.3

Consider the modeling problem depicted in Figure 3.6, where a plant $G(z)$ is being modeled by a single-pole single-zero Wiener filter

$$W(z) = \frac{1 - w_0 z^{-1}}{1 - w_1 z^{-1}} \tag{3.83}$$

To keep our discussion simple, we assume that all the involved signals and system parameters are real-valued. The input sequence, $x(n)$, is assumed to be a white process with zero-mean and variance of unity, and uncorrelated with the additive noise $\nu(n)$. This implies that

$$\Phi_{xx}(z) = 1 \quad \text{and} \quad \Phi_{\nu x}(z) = 0 \tag{3.84}$$

We note that $d(n)$ is the noise corrupted output of the plant, $G(z)$, when it is excited with the input $x(n)$. Then, using the relationship (2.75) given in Chapter 2, and noting that all

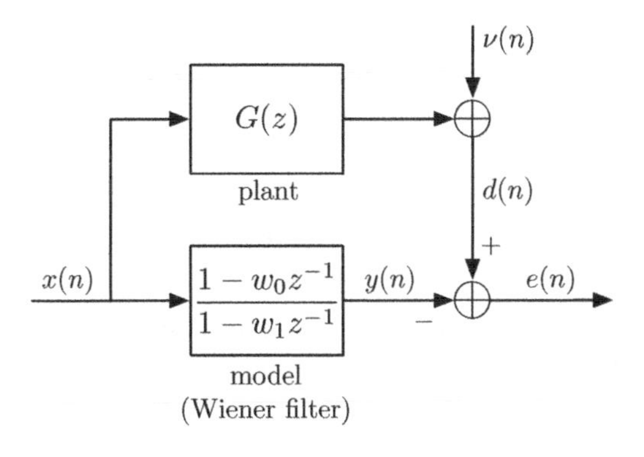

Figure 3.6 A modeling problem with an IIR model.

the signals here are real-valued, we get

$$\Phi_{xd}(z) = G(z^{-1})\Phi_{xx}(z) \tag{3.85}$$

Using this in Eq. (3.78), we obtain

$$\begin{aligned}\xi = \phi_{dd}(0) &+ \frac{1}{2\pi j}\oint_C \frac{1 - w_0 z^{-1}}{1 - w_1 z^{-1}} \cdot \frac{1 - w_0 z}{1 - w_1 z} \cdot \frac{\mathrm{d}z}{z} \\ &-2 \times \frac{1}{2\pi j}\oint_C \frac{1 - w_0 z^{-1}}{1 - w_1 z^{-1}} G(z^{-1})\frac{\mathrm{d}z}{z}\end{aligned} \tag{3.86}$$

Using the residue theorem to calculate the above-mentioned integrals and assuming that $G(z^{-1})$ has no pole inside the unit circle, we get

$$\begin{aligned}\xi = \phi_{dd}(0) &+ \frac{w_1 - w_0}{w_1} \cdot \frac{1 - w_0 w_1}{1 - w_1^2} + \frac{w_0}{w_1} \\ &-2\left[\frac{w_1 - w_0}{w_1} G(w_1^{-1}) + \frac{w_0}{w_1} G(\infty)\right]\end{aligned} \tag{3.87}$$

This is the performance function of the IIR filter shown in Figure 3.6. We note that, although we have selected a very simple example, the resulting performance function is a complicated one. It is clear that a performance function such as Eq. (3.87) or more complicated ones that would result for higher order filters is difficult to be handled. In particular, one may find that there can be many local minima and searching for the global minimum of the performance function may not be a trivial task. This, when compared with the nicely shaped quadratic performance function of FIR filters, makes it clear why most of the attention in adaptive filters have been devoted to the transversal structure.

3.6.2 *Optimum Transfer Function*

We now derive an equation for the optimum transfer function of unconstrained Wiener filters, that is, when the filter impulse response is allowed to extend from time $n = -\infty$ to $n = +\infty$. We use the principle of orthogonality for this purpose. As the filter impulse response stretches from time $n = -\infty$ to $n = +\infty$, the principle of orthogonality for real-valued signals suggests

$$E[e_o(n)x(n-i)] = 0, \quad \text{for } i = \ldots, -2, -1, 0, 1, 2, \ldots \tag{3.88}$$

where $e_o(n)$ is the optimum estimation error and is given by

$$e_o(n) = d(n) - \sum_{l=-\infty}^{\infty} w_{o,l}\, x(n-l) \tag{3.89}$$

Here, $w_{o,l}$'s are the samples of the optimized Wiener filter impulse response.

Substituting Eq. (3.89) in Eq. (3.88) and rearranging the result, we obtain

$$\sum_{l=-\infty}^{\infty} w_{o,l} E[x(n-l)x(n-i)] = E[d(n)x(n-i)] \tag{3.90}$$

We may also note that $E[x(n-l)x(n-i)] = \phi_{xx}(i-l)$ and $E[d(n)x(n-i)] = \phi_{dx}(i)$. Using these in Eq. (3.90), we get

$$\sum_{l=-\infty}^{\infty} w_{o,l}\phi_{xx}(i-l) = \phi_{dx}(i), \quad \text{for } i = \ldots, -2, -1, 0, 1, 2, \ldots \tag{3.91}$$

Noting that Eq. (3.91) holds for all values of i, we may take z-transforms on both sides to obtain

$$\Phi_{xx}(z)W_o(z) = \Phi_{dx}(z) \tag{3.92}$$

This is referred to as *the Wiener–Hopf equation for unconstrained Wiener filtering problem.* The optimum unconstrained Wiener filter is given by

$$W_o(z) = \frac{\Phi_{dx}(z)}{\Phi_{xx}(z)} \tag{3.93}$$

Replacing z by $e^{j\omega}$ in Eq. (3.93), we obtain

$$W_o(e^{j\omega}) = \frac{\Phi_{dx}(e^{j\omega})}{\Phi_{xx}(e^{j\omega})} \tag{3.94}$$

This result has an interesting interpretation. It shows that *the frequency response of the optimal Wiener filter, for a particular frequency, say $\omega = \omega_i$, is determined by the ratio of the crosspower spectral density of $d(n)$ and $x(n)$, to the power spectral density of $x(n)$, at $\omega = \omega_i$*. This, in turn, may be obtained through a sequence of filtering and averaging steps, as depicted in Figure 3.7. The sequences $x(n)$ and $d(n)$ are first filtered by two identical narrow-band filters, centered at $\omega = \omega_i$. To keep the phase information of the underlying signals, these filters are designed to pick-up signals from the positive side

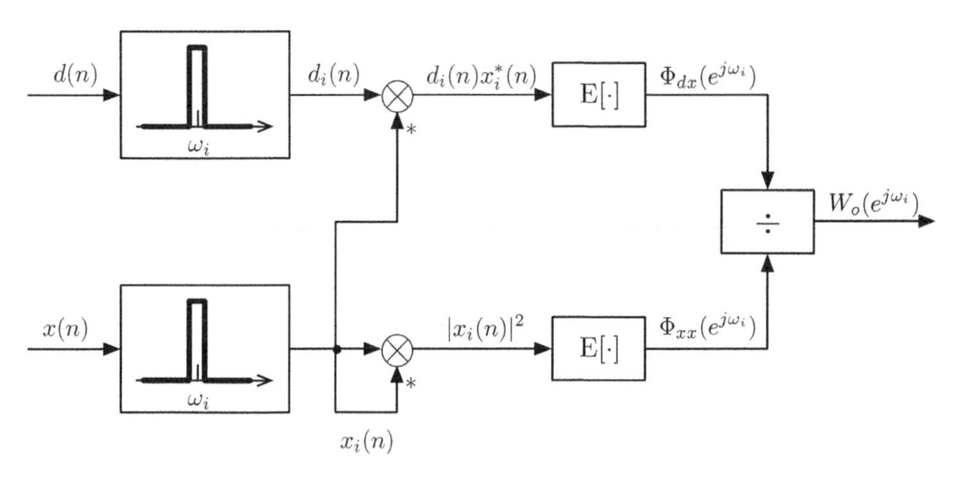

Figure 3.7 Procedure for calculating the transfer function of a Wiener filter through a sequence of filtering and averaging.

of the frequency axis, only. The signal spectrums belonging to negative frequencies are completely rejected. As a result, the filtered signals, $d_i(n)$ and $x_i(n)$, are both complex-valued. The cross-correlation of $d_i(n)$ and $x_i(n)$ with zero lag, that is, $E[d_i(n)x_i^*(n)]$, gives a quantity proportional to $\Phi_{dx}(e^{j\omega_i})$, and the average energy of $x_i(n)$ gives a quantity proportional to $\Phi_{xx}(e^{j\omega_i})$ – see Papoulis (1991). The ratio of these two quantities gives $W_o(e^{j\omega_i})$. This interpretation becomes more interesting, if we note that $W_o(e^{j\omega_i})$ is also the optimum tap weight of a single-tap Wiener filter whose input and desired output are the complex-valued random processes $x_i(n)$ and $d_i(n)$, respectively; see Problem P3.21.

The minimum mean-squared estimation error for the unconstrained Wiener filtering case can be obtained by substituting Eq. (3.93) in Eq. (3.78). For this, we first note that when $|z| = 1$,

$$\begin{aligned}|W_o(z)|^2 &= W_o(z)W_o^*(z)\\ &= W_o(z)\frac{\Phi_{dx}^*(z)}{\Phi_{xx}^*(z)}\\ &= W_o(z)\frac{\Phi_{xd}(z)}{\Phi_{xx}(z)}\end{aligned} \tag{3.95}$$

as, on the unit circle, $\Phi_{dx}^*(z) = \Phi_{xd}(z)$ and $\Phi_{xx}^*(z) = \Phi_{xx}(z)$. Using this result in Eq. (3.78), we get

$$\xi_{\min} = \phi_{dd}(0) - \frac{1}{2\pi j}\oint_C W_o(z)\Phi_{xd}(z)\frac{dz}{z} \tag{3.96}$$

This may be considered as a dual of the previous derivations in Eqs. (3.26), (3.47), and (3.74); see Problem P3.18.

Replacing z by $e^{j\omega}$ in Eq. (3.96), we obtain

$$\xi_{\min} = \phi_{dd}(0) - \frac{1}{2\pi}\int_{-\pi}^{\pi} W_o(e^{j\omega})\Phi_{xd}(e^{j\omega})d\omega \tag{3.97}$$

3.6.3 Modeling

In this and the subsequent two subsections, we discuss three specific applications of Wiener filters, namely, modeling, inverse modeling, and noise cancellation. These cover most of the cases that we encounter in adaptive filtering. Our aim, in these presentations, is to highlight some of the important features of Wiener filters when applied to various applications of adaptive signal processing.

Consider the modeling problem depicted in Figure 3.8. An estimate of the model of a plant $G(z)$ is to be obtained by the Wiener filter $W(z)$. The plant input $u(n)$, contaminated with an additive noise $\nu_i(n)$, is available as the Wiener filter input. The noise sequence $\nu_i(n)$ may be thought of as introduced by a transducer that is used to get samples of the plant input. There is also an additive noise $\nu_o(n)$ at the plant output. The sequences $u(n)$, $\nu_i(n)$, and $\nu_o(n)$ are assumed to be stationary, zero-mean, and uncorrelated with one another.

We note that, for the present problem, the Wiener filter input and its desired output are, respectively,

$$x(n) = u(n) + \nu_i(n) \tag{3.98}$$

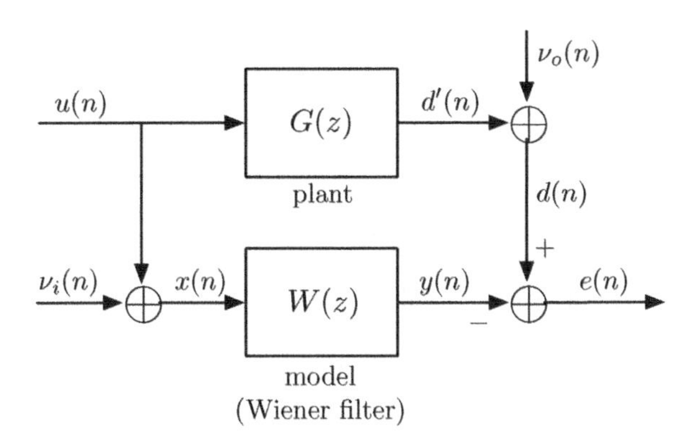

Figure 3.8 Block diagram of a modeling problem.

and

$$d(n) = g_n \star u(n) + v_o(n) \tag{3.99}$$

where g_n's are the samples of the plant impulse response, and $\star$ denotes convolution.

We use Eq. (3.93) to obtain the optimum transfer function of the Wiener filter, $W_o(z)$. For this, we should first find $\Phi_{xx}(z)$ and $\Phi_{dx}(z)$. We note that

$$\begin{aligned}\phi_{xx}(k) &= E[x(n)x(n-k)] \\ &= E[(u(n) + \nu_i(n))(u(n-k) + \nu_i(n-k))] \\ &= E[u(n)u(n-k)] + E[u(n)\nu_i(n-k)] \\ &\quad + E[\nu_i(n)u(n-k)] + E[\nu_i(n)\nu_i(n-k)]\end{aligned} \tag{3.100}$$

As $u(n)$ and $\nu_i(n)$ are uncorrelated with each other, the second and third terms on the right-hand side of Eq. (3.100) are zero. Thus, we obtain

$$\phi_{xx}(k) = \phi_{uu}(k) + \phi_{\nu_i\nu_i}(k) \tag{3.101}$$

Taking z-transform on both sides of Eq. (3.101), we get

$$\Phi_{xx}(z) = \Phi_{uu}(z) + \Phi_{\nu_i\nu_i}(z) \tag{3.102}$$

To find $\Phi_{dx}(z)$, we note that only $u(n)$ is common to $x(n)$ and $d(n)$, and the signals $u(n)$, $\nu_i(n)$, and $\nu_o(n)$ are uncorrelated with one another. Considering these and following a procedure similar to the one used to arrive at Eq. (3.102), one can show that

$$\Phi_{dx}(z) = \Phi_{d'u}(z) \tag{3.103}$$

where $d'(n)$ is the plant output when the additive noise $\nu_o(n)$ is excluded from that. Moreover, from our discussions in Chapter 2, we have

$$\Phi_{d'u}(z) = G(z)\Phi_{uu}(z) \tag{3.104}$$

Thus,

$$\Phi_{dx}(z) = G(z)\Phi_{uu}(z) \tag{3.105}$$

Using Eqs. (3.105) and (3.102) in Eq. (3.93), we obtain

$$W_o(z) = \frac{\Phi_{uu}(z)}{\Phi_{uu}(z) + \Phi_{\nu_i\nu_i}(z)} G(z) \tag{3.106}$$

We note that $W_o(z)$ is equal to $G(z)$, only when $\Phi_{\nu_i\nu_i}(z)$ is equal to zero. That is, when $\nu_i(n)$ is zero for all values of n.

It is also instructive to replace z by $e^{j\omega}$ in Eq. (3.106). This gives

$$W_o(e^{j\omega}) = \frac{\Phi_{uu}(e^{j\omega})}{\Phi_{uu}(e^{j\omega}) + \Phi_{\nu_i\nu_i}(e^{j\omega})} G(e^{j\omega}) \tag{3.107}$$

This result has the following interpretation. *Matching between the unconstrained Wiener filter and the plant frequency response at any particular frequency, ω, depends on the signal-to-noise power spectral density ratio $\Phi_{uu}(e^{j\omega})/\Phi_{\nu_i\nu_i}(e^{j\omega})$. Perfect matching is achieved when this ratio is infinity (i.e., when $\Phi_{\nu_i\nu_i}(e^{j\omega}) = 0$), and the mismatch between the plant and its model increases as $\Phi_{uu}(e^{j\omega})/\Phi_{\nu_i\nu_i}(e^{j\omega})$ decreases.* Note that $\Phi_{uu}(e^{j\omega})$ and $\Phi_{\nu_i\nu_i}(e^{j\omega})$ are power spectral density functions, and, thus, are real and nonnegative.

We may also define

$$K(e^{j\omega}) \triangleq \frac{\Phi_{uu}(e^{j\omega})}{\Phi_{uu}(e^{j\omega}) + \Phi_{\nu_i\nu_i}(e^{j\omega})} \tag{3.108}$$

and note that $K(e^{j\omega})$ is real and varies in the range of 0 to 1 as the power spectral density functions $\Phi_{uu}(e^{j\omega})$ and $\Phi_{\nu_i\nu_i}(e^{j\omega})$ are both real and nonnegative. Furthermore, to prevent ambiguity of the above ratio, we assume that for all values of ω, $\Phi_{uu}(e^{j\omega})$ and $\Phi_{\nu_i\nu_i}(e^{j\omega})$ are never equal to zero simultaneously. Using this, we obtain

$$W_o(e^{j\omega}) = K(e^{j\omega})G(e^{j\omega}) \tag{3.109}$$

An expression for the minimum mean-squared error of the modeling problem is obtained by replacing Eqs. (3.105) and (3.109) in Eq. (3.97). This gives

$$\xi_{\min} = \phi_{dd}(0) - \frac{1}{2\pi}\int_{-\pi}^{\pi} K(e^{j\omega})\Phi_{uu}(e^{j\omega})|G(e^{j\omega})|^2 d\omega \tag{3.110}$$

We may also note that $d'(n)$ and $\nu_o(n)$ are uncorrelated, and, thus,

$$\phi_{dd}(0) = \phi_{\nu_o\nu_o}(0) + \phi_{d'd'}(0) \tag{3.111}$$

Also,

$$\phi_{d'd'}(0) = \frac{1}{2\pi}\int_{-\pi}^{\pi} \Phi_{uu}(e^{j\omega})|G(e^{j\omega})|^2 d\omega \tag{3.112}$$

Substituting Eqs. (3.111) and (3.112) in Eq. (3.110), we obtain

$$\xi_{\min} = \phi_{\nu_o\nu_o}(0) + \frac{1}{2\pi}\int_{-\pi}^{\pi} (1 - K(e^{j\omega}))\Phi_{uu}(e^{j\omega})|G(e^{j\omega})|^2 d\omega \tag{3.113}$$

We note that the minimum mean-squared of the estimation error consists of two distinct components. The first one comes directly from the additive noise, $\nu_o(n)$, at the plant output. The Wiener filter will not be able to reduce this component as $\nu_o(n)$ is uncorrelated with its input $x(n)$. The second component arises because of the input noise, $\nu_i(n)$, which, in turn, results in some mismatch between $G(z)$ and $W_o(z)$. Thus, the best performance that one can expect from the optimum unconstrained Wiener filter is $\xi_{\min} = \phi_{\nu_o\nu_o}(0)$ and this happens when the input noise $\nu_i(n)$ is absent.

Another very important and useful concept that can be understood based on the above theoretical exercise is *the principle of correlation cancellation*. We remarked previously that the Wiener filter cannot do anything to reduce the contribution $\phi_{\nu_o\nu_o}(0)$ from the total mean-squared error. This is because the input $x(n)$ of the Wiener filter is uncorrelated with the output noise $\nu_o(n)$ and hence, the filter tries to match its output $y(n)$ with the plant output $d'(n)$ without bothering about $\nu_o(n)$. In other words, the Wiener filter attempts to estimate that part of the target signal $d(n)$, which is correlated with its own input $x(n)$ (i.e., $d'(n)$) and leave the remaining part of $d(n)$ (i.e., $\nu_o(n)$) unaffected. This is known as *the principle of correlation cancellation*. However, as noted above, perfect cancellation of the correlated part $d'(n)$ from $d(n)$ will only be possible when the input noise $\nu_i(n)$ is absent.

3.6.4 Inverse Modeling

Inverse modeling has applications both in communications and control. However, most of the theory of inverse modeling has been developed in the context of channel equalization. We also emphasize on the latter. Figure 3.9 depicts a channel equalization scenario. The data samples, $s(n)$, are transmitted through a communication channel with the system function $H(z)$. The received signal at the channel output is contaminated with an additive noise $\nu(n)$, which is assumed to be uncorrelated with the data samples, $s(n)$. An equalizer, $W(z)$, is used to process the received noisy signal samples, $x(n)$, to recover back the original data samples, $s(n)$.

When the additive noise at the channel output is absent, the equalizer has the following trivial solution:

$$W_o(z) = \frac{1}{H(z)} \tag{3.114}$$

In the absence of channel noise, this results in perfect recovery of the original data samples, as $W_o(z)H(z) = 1$. This implies that $y(n) = s(n)$, and thus $e(n) = 0$, for all n. This, clearly, is the optimum solution, as it results in zero mean-squared error which of

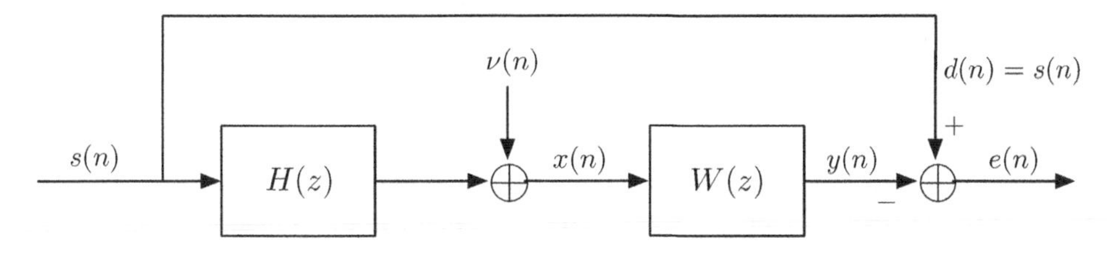

Figure 3.9 Channel equalization.

course is the minimum as mean-squared error is a nonnegative quantity. The following example gives a better view of the problem.

Example 3.4

Consider a channel with

$$H(z) = -0.4 + z^{-1} - 0.4z^{-2} \tag{3.115}$$

Also, assume that the channel noise, $\nu(n)$, is zero, for all n. The channel output then is obtained by convolving the input data sequence, $s(n)$, with the channel impulse response, h_n, which consists of three nonzero samples $h_0 = -0.4$, $h_1 = 1$, and $h_2 = -0.4$. This gives

$$x(n) = -0.4s(n) + s(n-1) - 0.4s(n-2) \tag{3.116}$$

in the absence of channel noise.

We note that each sample of $x(n)$ is made of a mixture of three successive samples of the original data. This is called *intersymbol interference (ISI)*, and it should be compensated or canceled for correct detection of transmitted data. For this purpose, we may use an equalizer with the system function (3.114)

$$W_o(z) = \frac{1}{H(z)} = \frac{1}{-0.4 + z^{-1} - 0.4z^{-2}} \tag{3.117}$$

Factorizing the denominator of $W_o(z)$ and rearranging, we get

$$W_o(z) = \frac{-2.5}{(1 - 0.5z^{-1})(1 - 2z^{-1})} \tag{3.118}$$

This is a system function with one pole inside and one pole outside the unit circle. With reference to our discussions in Chapter 2, we recall that Eq. (3.118) will correspond to a stable time-invariant system, if the region of convergence of $W_o(z)$ includes the unit circle. Considering this and finding the inverse z-transform of $W_o(z)$, we obtain (Chapter 2, Section 2.2)

$$w_{o,i} = \begin{cases} \frac{10}{3} \times 2^i, & i < 0 \\ \frac{2.5}{3} \times 0.5^i, & i \geq 0 \end{cases} \tag{3.119}$$

To obtain this result, we have noted that $W_o(z)$ of Eq. (3.118) is similar to $H(z)$ of Eq. (2.30), except for the factor -2.5 in Eq. (3.118). Figure 3.10a–c shows the samples of the impulse responses of the channel, equalizer, and their convolution, respectively. Existence of ISI at the channel output, as noted above, is because of more than one (here, three) nonzero samples in the channel impulse response. This is observed in Figure 3.10a. Figure 3.10c shows that the ISI is completely removed after passing the received signal through the equalizer.

When the channel noise, $\nu(n)$, is nonzero, the solution provided by Eq. (3.114) may not be optimal. The channel noise also passes through the equalizer and may be greatly enhanced in the frequency bands where $H(e^{j\omega})$ is small. In this situation, a compromise has to be made between cancellation of ISI and noise enhancement. As we show in the following, the optimal Wiener filter achieves this trade-off in an effective way.

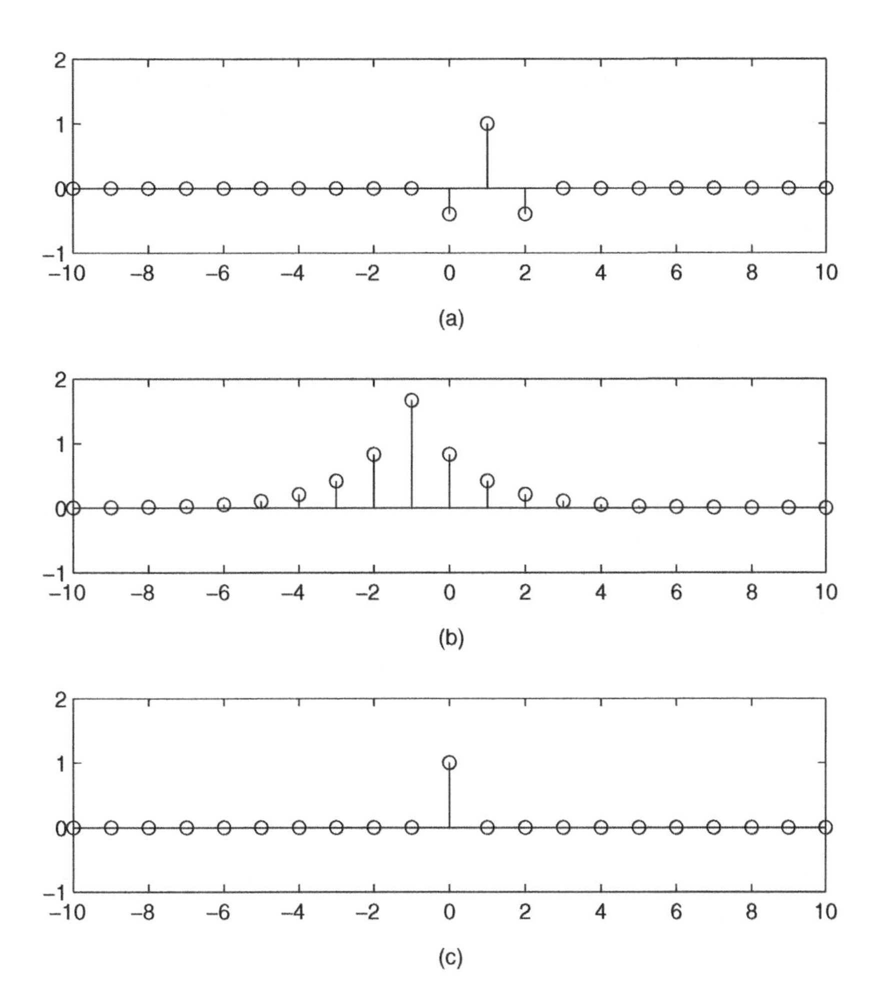

Figure 3.10 Impulse response of (a) channel response, (b) equalizer response, and (c) cascade of channel and equalizer.

To derive an equation for $W_o(z)$ when the channel noise is nonzero, we use Eq. (3.93). We note that

$$x(n) = h_n \star s(n) + \nu(n) \tag{3.120}$$

and

$$d(n) = s(n) \tag{3.121}$$

where h_n is the impulse response of the channel, $H(z)$.

Noting that $s(n)$ and $\nu(n)$ are uncorrelated and using the results of Section 2.4.4, we obtain, from Eq. (3.120)

$$\Phi_{xx}(z) = \Phi_{ss}(z)|H(z)|^2 + \Phi_{\nu\nu}(z) \tag{3.122}$$

Also, from Eqs. (3.120) and (3.121), we may note that $x(n)$ is the output of a system with input $s(n)$ and impulse response h_n, plus an uncorrelated noise, $\nu(n)$. Noting these and the fact that all the processes and system parameters are real-valued, we obtain

$$\Phi_{dx}(z) = \Phi_{sx}(z) = H(z^{-1})\Phi_{ss}(z) \tag{3.123}$$

Note that the above result is independent of $\nu(n)$. Also, with $|z| = 1$, we may write

$$\Phi_{dx}(z) = H^*(z)\Phi_{ss}(z) \tag{3.124}$$

Using Eqs. (3.122) and (3.124) in Eq. (3.93), we obtain

$$W_o(z) = \frac{H^*(z)\Phi_{ss}(z)}{\Phi_{ss}(z)|H(z)|^2 + \Phi_{\nu\nu}(z)} \tag{3.125}$$

This is the general solution to the equalization problem, when there is no constraint on the equalizer length and, also, it may be let to be noncausal. Equation (3.125) includes the effects of autocorrelation function of the data, $s(n)$, and the noise, $\nu(n)$.

To give an interpretation of Eq. (3.125), we divide the numerator and denominator by the first term in the denominator to obtain

$$W_o(z) = \frac{1}{1 + \frac{\Phi_{\nu\nu}(z)}{\Phi_{ss}(z)|H(z)|^2}} \cdot \frac{1}{H(z)} \tag{3.126}$$

Next, we replace z by $e^{j\omega}$, and define the parameter

$$\rho(e^{j\omega}) \triangleq \frac{\Phi_{ss}(e^{j\omega})|H(e^{j\omega})|^2}{\Phi_{\nu\nu}(e^{j\omega})} \tag{3.127}$$

We may note that this is the signal-to-noise power spectral density ratio at the channel output. $\Phi_{ss}(e^{j\omega})|H(e^{j\omega})|^2$ and $\Phi_{\nu\nu}(e^{j\omega})$ are the signal power spectral density and noise power spectral density, respectively, at the channel output. Substituting Eq. (3.127) in Eq. (3.126) and rearranging, we obtain

$$W_o(e^{j\omega}) = \frac{\rho(e^{j\omega})}{1 + \rho(e^{j\omega})} \cdot \frac{1}{H(e^{j\omega})} \tag{3.128}$$

We note that the frequency response of the optimized equalizer is proportional to the inverse of the channel frequency response, with a proportionality constant that is frequency dependent. Furthermore, $\rho(e^{j\omega})$ is a nonnegative real quantity, for power spectra are nonnegative real functions. Hence,

$$0 \leq \frac{\rho(e^{j\omega})}{1 + \rho(e^{j\omega})} \leq 1 \tag{3.129}$$

This brings us to the following interpretation of Eq. (3.128). *The frequency response of the optimum equalizer resembles the channel inverse within a real-valued constant in the range of 0 to 1. This constant, which is frequency dependent, depends on the signal-to-noise power spectral density ratio,* $\rho(e^{j\omega})$, *at the equalizer input. It approaches 1 when* $\rho(e^{j\omega})$ *is large, and reduces with* $\rho(e^{j\omega})$.

Once again, it is important to note that different frequencies are treated independent of one another by the equalizer. In particular, at a given frequency $\omega = \omega_i$, $W_o(e^{j\omega_i})$ depends only on the values of $H(e^{j\omega})$ and $\rho(e^{j\omega})$ at $\omega = \omega_i$. With this background, we shall now examine Eq. (3.128) closely to see how the equalizer is able to make a good trade-off between cancellation of ISI and noise enhancement. In the frequency regions where the noise is almost absent, the value of $\rho(e^{j\omega})$ is very large and hence the equalizer approximates the inverse of the channel closely, without any significant enhancement of noise. On the other hand, in the frequency regions where the noise level is high (relative to the signal level), the value of $\rho(e^{j\omega})$ is not large and hence the equalizer does not approximate the channel inverse well. This, of course, is to prevent noise enhancement.

Example 3.5

Consider the channel $H(z)$ of Example 3.4. We assume that the data sequence, $s(n)$, is binary (taking values of $+1$ and -1) and white. We also assume that $\nu(n)$ is a white noise process with variance of 0.04. With these, we obtain

$$\Phi_{ss}(z) = 1 \quad \text{and} \quad \Phi_{\nu\nu}(z) = 0.04$$

Using these in Eq. (3.125), we get

$$W_o(z) = \frac{-0.4 + z - 0.4z^2}{(-0.4 + z^{-1} - 0.4z^{-2})(-0.4 + z - 0.4z^2) + 0.04} \tag{3.130}$$

Figure 3.11 presents the plots of $1/|H(e^{j\omega})|$ and $|W_o(e^{j\omega})|$. We note that at those frequencies where $1/|H(e^{j\omega})|$ is small, a near-perfect match between $1/|H(e^{j\omega})|$ and $|W_o(e^{j\omega})|$ is observed. On the other hand, at those frequencies where $1/|H(e^{j\omega})|$ is large, the deviation

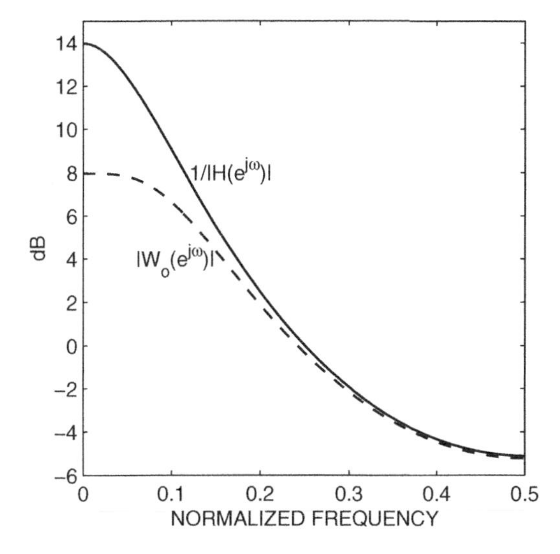

Figure 3.11 Plots of $1/|H(e^{j\omega})|$ and $|W_o(e^{j\omega})|$.

between the two increases. We may also note that $|W_o(e^{j\omega})|$ remains less than $1/|H(e^{j\omega})|$, for all values of ω.

This is consistent with the conclusion drawn previously because a small value of $1/|H(e^{j\omega})|$ implies that $|H(e^{j\omega})|$ is large and thus, according to Eq. (3.127), $\rho(e^{j\omega})$ is also large. This, in turn, implies that the ratio $\rho(e^{j\omega})/1 + \rho(e^{j\omega})$ is close to 1; hence, from Eq. (3.128), we get

$$W_o(e^{j\omega}) \approx \frac{1}{H(e^{j\omega})}$$

Similarly, the same argument may be used to explain why $W_o(e^{j\omega})$ is significantly smaller than $1/|H(e^{j\omega})|$ when the latter is large. Furthermore, the fact that $|W_o(e^{j\omega})|$ remains less than $1/|H(e^{j\omega})|$, for all values of ω, is predicted by Eq. (3.128).

3.6.5 Noise Cancellation

Figure 3.12 depicts a typical noise canceler setup. There are two inputs to this setup: a signal source, $s(n)$, and a noise source, $\nu(n)$. These two signals, which are assumed to be uncorrelated with each other, are mixed together through the system functions $H(z)$ and $G(z)$, and result in the *primary input*, $d(n)$, and *reference input*, $x(n)$, as shown in Figure 3.12. The reference input is passed through a Wiener filter $W(z)$, which is designed so that the difference between the primary input and the filter output is minimized in the mean-square sense. The noise canceler output is the error sequence $e(n)$. The aim of a noise canceler setup, as explained above, is to extract the signal $s(n)$ from the primary input $d(n)$.

We note that

$$x(n) = \nu(n) + h_n \star s(n) \tag{3.131}$$

and

$$d(n) = s(n) + g_n \star \nu(n) \tag{3.132}$$

where h_n and g_n are the impulse responses of the filters $H(z)$ and $G(z)$, respectively.

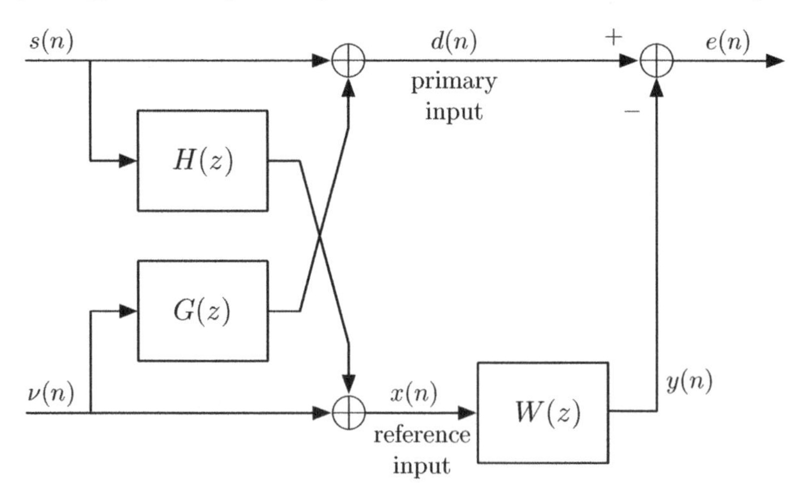

Figure 3.12 Noise canceler setup.

Noting that $s(n)$ and $\nu(n)$ are uncorrelated with each other and recalling the results of Section 2.4.4, we obtain, from Eq. (3.131)

$$\Phi_{xx}(z) = \Phi_{\nu\nu}(z) + \Phi_{ss}(z)|H(z)|^2 \tag{3.133}$$

To find $\Phi_{dx}(z)$, we note that $d(n)$ and $x(n)$ are related with each other through the signal sequences $s(n)$ and $\nu(n)$ and the filters $H(z)$ and $G(z)$. As $s(n)$ and $\nu(n)$ are uncorrelated with each other, their contribution in $\Phi_{dx}(z)$ may be considered separately. In particular, we may write

$$\Phi_{dx}(z) = \Phi^{s}_{dx}(z) + \Phi^{\nu}_{dx}(z) \tag{3.134}$$

where $\Phi^{s}_{dx}(z)$ is $\Phi_{dx}(z)$ when $\nu(n) = 0$, for all values of n, and $\Phi^{\nu}_{dx}(z)$ is $\Phi_{dx}(z)$ when $s(n) = 0$, for all values of n. Thus, we obtain

$$\Phi^{s}_{dx}(z) = H^*(z)\Phi_{ss}(z) \tag{3.135}$$

and

$$\Phi^{\nu}_{dx}(z) = G(z)\Phi_{\nu\nu}(z) \tag{3.136}$$

Substituting Eqs. (3.135) and (3.136) in Eq. (3.134), we get

$$\Phi_{dx}(z) = H^*(z)\Phi_{ss}(z) + G(z)\Phi_{\nu\nu}(z) \tag{3.137}$$

Using Eqs. (3.133) and (3.137) in Eq. (3.93), we obtain

$$W_{\text{o}}(z) = \frac{H^*(z)\Phi_{ss}(z) + G(z)\Phi_{\nu\nu}(z)}{\Phi_{\nu\nu}(z) + \Phi_{ss}(z)|H(z)|^2} \tag{3.138}$$

A comparison of Eq. (3.138) with Eqs. (3.106) and (3.125) reveals that Eq. (3.138) may be thought as a generalization of the results we obtained in the last two sections for the modeling and inverse modeling scenarios. In fact, if we refer to Figure 3.12, we can easily find that the modeling and inverse modeling scenarios are embedded in the noise canceler setup. While trying to minimize the mean-squared value of the output error, one must strike a balance between noise cancellation and signal cancellation at the output of the noise canceler. Cancellation of the noise $\nu(n)$ occurs when the Wiener filter $W(z)$ is chosen to be close to $G(z)$, and cancellation of the signal $s(n)$ occurs when $W(z)$ is close to the inverse of $H(z)$. In this sense, we may note that the noise canceler treats $s(n)$ and $\nu(n)$ without making any distinction between them and tries to cancel both of them as much as possible so as to achieve the minimum mean-squared error in $e(n)$. This seems contrary to the main goal of the noise canceler, which is meant to cancel only the noise. The following discussion aims at revealing some of the peculiar characteristics of the noise canceler setup and show under which condition an acceptable cancellation occurs.

To proceed with our discussion, we define $\rho_{\text{pri}}(\mathrm{e}^{j\omega})$, $\rho_{\text{ref}}(\mathrm{e}^{j\omega})$, and $\rho_{\text{out}}(\mathrm{e}^{j\omega})$ as the signal-to-noise power spectral density ratios at the primary input, reference input, and output, respectively. By direct inspection of Figure 3.12 and application of Eq. (2.85), we obtain

$$\rho_{\text{pri}}(\mathrm{e}^{j\omega}) = \frac{\Phi_{ss}(\mathrm{e}^{j\omega})}{|G(\mathrm{e}^{j\omega})|^2\Phi_{\nu\nu}(\mathrm{e}^{j\omega})} \tag{3.139}$$

and

$$\rho_{\rm ref}(e^{j\omega}) = \frac{|H(e^{j\omega})|^2\Phi_{ss}(e^{j\omega})}{\Phi_{\nu\nu}(e^{j\omega})} \tag{3.140}$$

To derive a similar equation for $\rho_{out}(e^{j\omega})$, we note that $s(n)$ reaches the canceler output through two routes: one direct and one through the cascade of $H(z)$ and $W(z)$. This gives

$$\Phi_{ee}^{s}(e^{j\omega}) = |1 - H(e^{j\omega})W(e^{j\omega})|^2\Phi_{ss}(e^{j\omega}) \tag{3.141}$$

where the superscript s refers to the portion of $\Phi_{ee}(e^{j\omega})$ which comes from $s(n)$. Similarly, $\nu(n)$ reaches the output through the routes $G(z)$ and $W(z)$. Thus,

$$\Phi_{ee}^{\nu}(e^{j\omega}) = |G(e^{j\omega}) - W(e^{j\omega})|^2\Phi_{\nu\nu}(e^{j\omega}) \tag{3.142}$$

Replacing $W(e^{j\omega})$ by $W_o(e^{j\omega})$ and using Eq. (3.138) in Eqs. (3.141) and (3.142), we obtain

$$\Phi_{ee}^{s}(e^{j\omega}) = \frac{|1 - G(e^{j\omega})H(e^{j\omega})|^2\Phi_{\nu\nu}^2(e^{j\omega})}{[\Phi_{\nu\nu}(e^{j\omega}) + |H(e^{j\omega})|^2\Phi_{ss}(e^{j\omega})]^2}\Phi_{ss}(e^{j\omega}) \tag{3.143}$$

and

$$\Phi_{ee}^{\nu}(e^{j\omega}) = \frac{|H(e^{j\omega})|^2|1 - G(e^{j\omega})H(e^{j\omega})|^2\Phi_{ss}^2(e^{j\omega})}{[\Phi_{\nu\nu}(e^{j\omega}) + |H(e^{j\omega})|^2\Phi_{ss}(e^{j\omega})]^2}\Phi_{\nu\nu}(e^{j\omega}) \tag{3.144}$$

respectively. Hence, $\rho_{\rm out}(e^{j\omega})$ can now be obtained as

$$\rho_{\rm out}(e^{j\omega}) = \frac{\Phi_{ee}^{s}(e^{j\omega})}{\Phi_{ee}^{\nu}(e^{j\omega})} = \frac{\Phi_{\nu\nu}(e^{j\omega})}{|H(e^{j\omega})|^2\Phi_{ss}(e^{j\omega})} \tag{3.145}$$

Comparing Eq. (3.145) with Eq. (3.140), we find that

$$\rho_{\rm out}(e^{j\omega}) = \frac{1}{\rho_{\rm ref}(e^{j\omega})} \tag{3.146}$$

This is known as *power inversion* (Widrow, McCool, and Ball, 1975). It shows that *the signal-to-noise power spectral density ratio at the noise canceler output is equal to the inverse of the signal-to-noise power spectral density ratio at the reference input.* This means that if the signal-to-noise power spectral density ratio at the reference input is low, we should expect a good cancellation of the noise at the output. On the other hand, we should expect a poor performance from the canceler when the signal-to-noise power spectral density ratio at the reference input is high. This surprising result suggests that *the noise canceler works better in situations when noise level is high and signal level is low.* The following example gives a clear picture of this general result.

Example 3.6

To demonstrate how the power inversion property of the noise canceler may be utilized in practice, we consider a receiver with two omnidirectional (equally sensitive to all directions) antennas, A and B, as shown in Figure 3.13.

A desired signal $s(n) = \alpha(n)\cos n\omega_o$ arrives in the direction perpendicular to the line connecting A and B. An interferer (jammer) signal $\nu(n) = \beta(n)\cos n\omega_o$ arrives at an angle

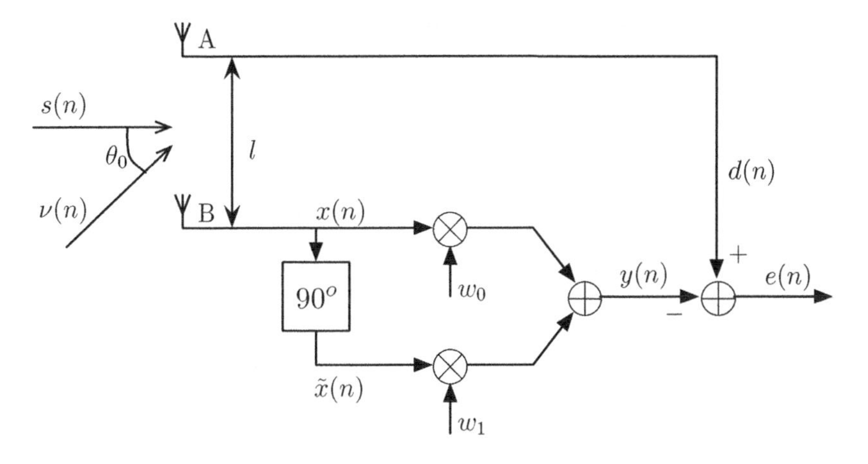

Figure 3.13 A receiver with two omnidirectional antennas.

θ_o with respect to the direction of $s(n)$. The amplitudes $\alpha(n)$ and $\beta(n)$ are narrow-band baseband signals. This implies that $s(n)$ and $\nu(n)$ are narrow-band signals concentrated around $\omega = \omega_o$. Such signals may be treated as single tones, and, thus a filter with two degrees of freedom is sufficient for any linear filtering that may have to be performed on them. This is why only a two-tap linear combiner is considered in Figure 3.13. This is expected to perform almost as good as any other unconstrained linear (Wiener) filter. We also assume that $\alpha(n)$ and $\beta(n)$ are zero-mean and uncorrelated with each other. The two omnis are separated by a distance of l meters. The linear combiner coefficients are adjusted so that the output error, $e(n)$, is minimized in the mean-square sense.

The desired signal, $s(n)$, arrives at the same time at both omnis. However, $\nu(n)$ arrives at B first, and arrives at A with a delay

$$\delta = \frac{l \sin \theta_o}{c} \tag{3.147}$$

where c is the propagation speed. To add this to the time index n, it has to be normalized by the time step T, which corresponds to one increment of n. This gives

$$\delta_o = \frac{l \sin \theta_o}{cT} \tag{3.148}$$

Noting these, in Figure 3.13, we have

$$d(n) = \alpha(n) \cos n\omega_o + \beta(n) \cos[(n - \delta_o)\omega_o] \tag{3.149}$$

$$x(n) = \alpha(n) \cos n\omega_o + \beta(n) \cos n\omega_o \tag{3.150}$$

$$\tilde{x}(n) = \alpha(n) \sin n\omega_o + \beta(n) \sin n\omega_o \tag{3.151}$$

It may be noted that in Eq. (3.149), we have used $\beta(n)$ instead of $\beta(n - \delta_o)$. This, which has been done to simplify the following equations, in practice is valid with a very good approximation because of the narrow bandwidth of $\beta(n)$, which implies that its variation in time is slow, and the small size of δ_o.

To find the optimum coefficients of the linear combiner, we shall derive and solve the Wiener–Hopf equation governing the linear combiner. We note that, here,

$$\mathbf{R} = \begin{bmatrix} E[x^2(n)] & E[x(n)\tilde{x}(n)] \\ E[x(n)\tilde{x}(n)] & E[\tilde{x}^2(n)] \end{bmatrix} \tag{3.152}$$

Also,

$$E[x^2(n)] = E[(\alpha(n)\cos n\omega_o + \beta(n)\cos n\omega_o)^2] \tag{3.153}$$

Expanding Eq. (3.153) and recalling that $\alpha(n)$ and $\beta(n)$ are uncorrelated with each other, we obtain

$$E[x^2(n)] = \frac{\sigma_\alpha^2 + \sigma_\beta^2}{2} + \frac{1}{2}(\sigma_\alpha^2 E[\cos 2n\omega_o] + \sigma_\beta^2 E[\cos 2n\omega_o]) \tag{3.154}$$

where σ_α^2 and σ_β^2 are variances of $\alpha(n)$ and $\beta(n)$, respectively. Also, $E[\cos 2n\omega_o]$ is replaced by its time average.[2] This is assumed to be zero. Thus, we obtain

$$E[x^2(n)] = \frac{\sigma_\alpha^2 + \sigma_\beta^2}{2} \tag{3.155}$$

Similarly, one can get

$$E[\tilde{x}^2(n)] = \frac{\sigma_\alpha^2 + \sigma_\beta^2}{2} \tag{3.156}$$

and

$$E[x(n)\tilde{x}(n)] = 0 \tag{3.157}$$

Substituting these in Eq. (3.152), we have

$$\mathbf{R} = \frac{\sigma_\alpha^2 + \sigma_\beta^2}{2} \begin{bmatrix} 1 & 0 \\ 0 & 1 \end{bmatrix} \tag{3.158}$$

It is also straightforward to show that

$$\mathbf{p} = \begin{bmatrix} E[d(n)x(n)] \\ E[d(n)\tilde{x}(n)] \end{bmatrix} = \frac{1}{2}\begin{bmatrix} \sigma_\alpha^2 + \sigma_\beta^2 \cos\delta_o\omega_o \\ \sigma_\beta^2 \sin\delta_o\omega_o \end{bmatrix} \tag{3.159}$$

Using Eqs. (3.158) and (3.159) in the Wiener–Hopf equation $\mathbf{R}\mathbf{w}_o = \mathbf{p}$, we get

$$\mathbf{w}_o = \begin{bmatrix} \dfrac{\sigma_\alpha^2 + \sigma_\beta^2 \cos\delta_o\omega_o}{\sigma_\alpha^2 + \sigma_\beta^2} \\ \dfrac{\sigma_\beta^2 \sin\delta_o\omega_o}{\sigma_\alpha^2 + \sigma_\beta^2} \end{bmatrix} \tag{3.160}$$

---

[2] Strictly speaking the replacement of the time average of the periodic sequence $\cos 2n\omega_o$ as $E[\cos 2n\omega_o]$ does not fit into the conventional definitions of stochastic processes. The sequence $\cos 2n\omega_o$ is deterministic and thus it does not really make sense to talk about its expectation, which is conventionally defined as an ensemble average. On the other hand, this is a reality that in many occasions in adaptive filters (such as our example here), the involved signals are deterministic and the time averages are used to evaluate the performance of the filters and/or calculate their parameters. This is in this context that we replace statistical expectations by time averages. We may note that the problem stated in Example 3.6 could also be put in a more statistical form to prevent the above arguments; see Problem P3.30. Here, we have decided not to do this, in order to emphasize on the fact that in practice, time averages are used instead of statistical expectations.

The optimized output of the receiver is

$$e_o(n) = d(n) - \mathbf{w}_o^T \mathbf{x}(n) \tag{3.161}$$

where $\mathbf{x}(n) = [x(n)\ \tilde{x}(n)]^T$. Using Eqs. (3.149), (3.150), (3.151), and (3.160) in Eq. (3.161), we get, after some manipulations,

$$e_o(n) = \frac{\cos n\omega_o - \cos[(n - \delta_o)\omega_o]}{\sigma_\alpha^2 + \sigma_\beta^2}(\sigma_\beta^2 \alpha(n) - \sigma_\alpha^2 \beta(n)) \tag{3.162}$$

Now, by inspection of Eqs. (3.150) and (3.162), we find that

$$\text{signal-to-noise ratio at the reference input} = \frac{\sigma_\alpha^2}{\sigma_\beta^2} \tag{3.163}$$

and

$$\text{signal-to-noise ratio at the output} = \frac{(\sigma_\beta^2)^2 \sigma_\alpha^2}{(\sigma_\alpha^2)^2 \sigma_\beta^2} = \frac{\sigma_\beta^2}{\sigma_\alpha^2} \tag{3.164}$$

which match the power inversion equation (3.146).

3.7 Summary and Discussion

In this chapter, we reviewed a class of optimum linear systems collectively known as *Wiener filters*. We noted that the performance function used in formulating the Wiener filters is an elegant choice, which leads to a mathematically tractable problem. We discussed the Wiener filters in the context of discrete-time signals and systems, and presented different formulations of the Wiener filtering problem. We started with the Wiener filtering problem for a FIR filter. The case of real-valued signals was dealt with first, and the formulation was then extended to the case of complex-valued signals.

The unconstrained Wiener filters were also discussed in detail. By unconstrained, we mean there is no constraint on the duration of the impulse response of the filter. It may extend from time $n = -\infty$ to $n = +\infty$. This study, although nonrealistic in actual implementation, turned out to be very instructive in revealing many aspects of the Wiener filters, which could not be easily perceived when the duration of the filter impulse response is limited.

The eminent features of the Wiener filters that were observed are as follows:

- For a transversal Wiener filter, the performance function is a quadratic function of its tap weights with a single global minimum. The set of tap weights, which minimizes the Wiener filter cost function, can be obtained analytically by solving a set of simultaneous linear equations known as *Wiener–Hopf equation*.
- When the optimum Wiener filter is used, the estimation error is uncorrelated with the input samples of the filter. This property of Wiener filters, which is referred to as *the principle of orthogonality*, is useful and handy for many related derivations.
- Wiener filter can also be viewed as a correlation canceler in the sense that the optimum Wiener filter cancels that part of the desired output, which is correlated with its input, while generating the estimation error.

- In the case of unconstrained Wiener filters, the Wiener filter treats different frequency components of the underlying processes separately. In particular, the Wiener filter transfer function at any particular frequency depends only on the power spectral density of the filter input and crosspower spectral density between the filter input and its desired output at that frequency.

The last property, although could only be derived in the case of unconstrained Wiener filters, is also approximately valid when the filter length is constrained. The concept of power spectra and their influence on the performance of Wiener filters is fundamental for understanding the behavior of adaptive filters. We note that the adaptive filters, as commonly implemented, are aimed at implementing Wiener filters. In this chapter, we observed that the optimum coefficients of the Wiener filter are a function of the autocorrelation function of the filter input and the crosscorrelation function between the filter input and its desired output. As correlation functions and power spectra are uniquely related, we also observed that the optimum coefficients can be expressed in terms of the corresponding power spectra instead of correlation functions. In the next few chapters, we will show that the convergence behavior of adaptive filters is closely related to the power spectrum of their inputs. In the rest of this book, we will be making frequent references to the results derived in this chapter.

Problems

P3.1 Consider a two-tap Wiener filter with the following statistics:

$$E[d^2(n)] = 2, \quad \mathbf{R} = \begin{bmatrix} 1 & 0.7 \\ 0.7 & 1 \end{bmatrix}, \quad \mathbf{p} = \begin{bmatrix} 1 \\ 0.5 \end{bmatrix}$$

(i) Use the above information to obtain an expression for the performance function of the filter.

(ii) By letting the derivatives of the performance function with respect to the involved variables equal to zero, obtain the optimum values of the filter tap weights.

(iii) Insert the result obtained in (ii) in the performance function expression to obtain the minimum mean-squared error of the filter.

(iv) Find the optimum tap weights of the filter and its minimum mean-squared error using the equations derived in this chapter to confirm the results obtained in (ii) and (iii).

P3.2 Consider a three-tap Wiener filter with the following statistics:

$$E[d^2(n)] = 10, \quad \mathbf{R} = \begin{bmatrix} 1 & 0.5 & 0.25 \\ 0.5 & 1 & 0.5 \\ 0.25 & 0.5 & 1 \end{bmatrix}, \quad \mathbf{p} = \begin{bmatrix} 3 \\ 1 \\ 0 \end{bmatrix}$$

Repeat Steps (i) through (iv) of Problem P3.1.

P3.3 In Section 3.2, the Wiener–Hopf and minimum mean-squared error equations were derived for a transversal filter. Derive similar equations when the transversal filter is replaced by a linear combiner similar to the one presented in Figure 1.3

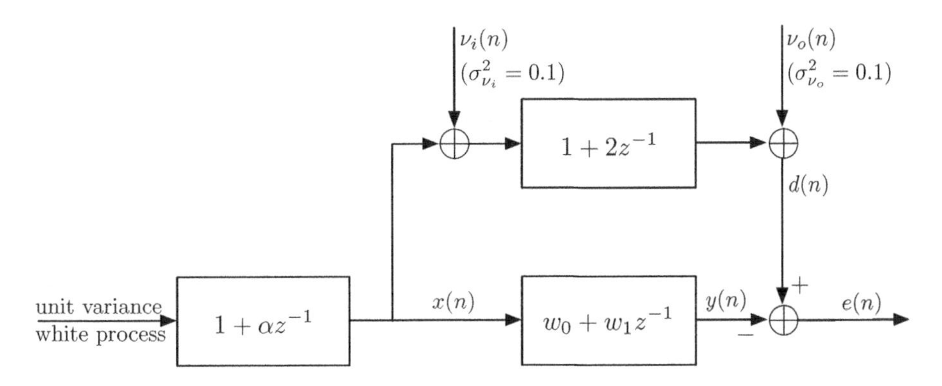

Figure P3.4

P3.4 Consider the modeling problem shown in Figure P3.4. For $\alpha = 0$,

(i) find the correlation matrix $\mathbf{R}$ of the filter tap inputs and the crosscorrelation vector $\mathbf{p}$ between the filter tap inputs and its desired output.
(ii) find the optimum tap weights of the Wiener filter and the minimum mean-squared error at the filter output.

P3.5 Repeat Problem P3.4 when $\alpha = 1$.

P3.6 Repeat Problems P3.4 and P3.5 when Figure P3.4 is replaced by Figure P3.6.

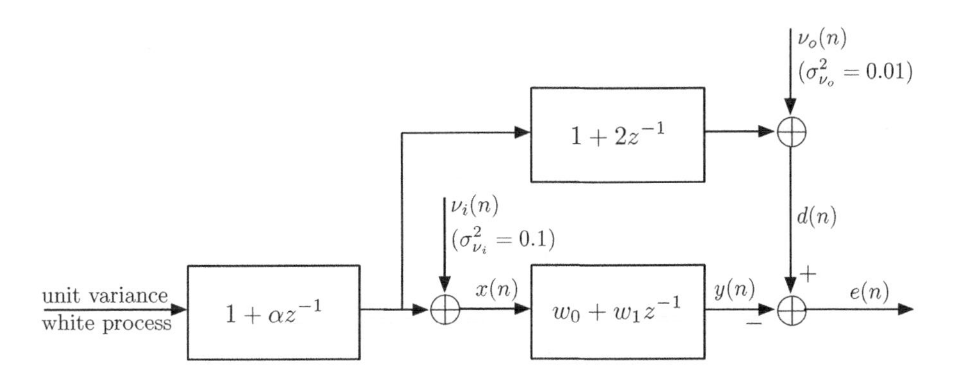

Figure P3.6

P3.7 Consider the modeling problem shown in Figure P3.7.

(i) Find the correlation matrix $\mathbf{R}$ of the filter tap inputs and the crosscorrelation vector $\mathbf{p}$ between the filter tap inputs and its desired output.
(ii) Find the optimum tap weights of the Wiener filter.
(iii) What is the minimum mean-squared error? Obtain this analytically as well as by direct inspection of Figure P3.7.

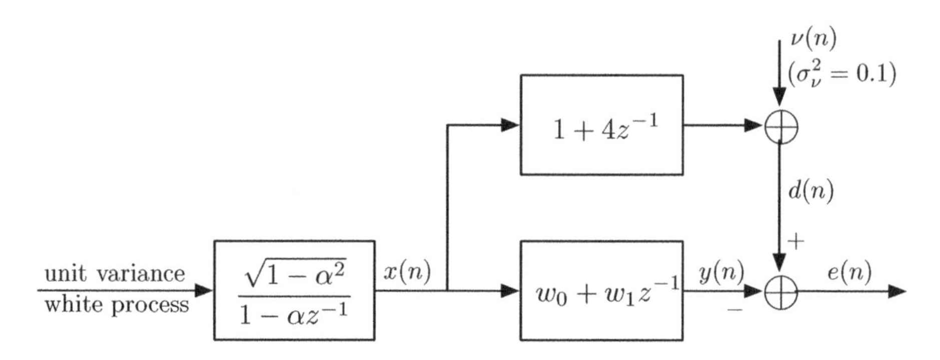

Figure P3.7

P3.8 Consider the channel equalization problem shown in Figure P3.8. The data symbols, $s(n)$, are assumed to be samples of a stationary white process.

(i) Find the correlation matrix $\mathbf{R}$ of the equalizer tap inputs and the crosscorrelation vector $\mathbf{p}$ between the equalizer tap inputs and the desired output.
(ii) Find the optimum tap weights of the equalizer.
(iii) What is the minimum mean-squared error at the equalizer output.
(iv) Could you guess the results obtained in (ii) and (iii) without going through the derivations? How and why?

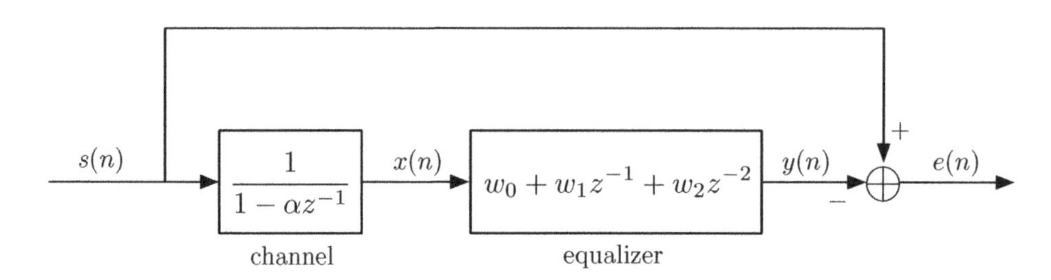

Figure P3.8

P3.9 In Section 3.5, we emphasized that for a complex variable w

$$\nabla_w^C f(w) = 0 \tag{P3.9.1}$$

does not imply that

$$\frac{\partial f(w)}{\partial w_R} = \frac{\partial f(w)}{\partial w_I} = 0 \tag{P3.9.2}$$

in general. In this problem, we want to elaborate on this further.

(i) Assume that $f(w) = w^m$ and show that for this function Eq. (P3.9.1) is true, but Eq. (P3.9.2) is false.

Can you extend this result to the case when

$$f(w) = \sum_{i=0}^{L} a_i w^i$$

and a_i's are fixed real or complex coefficients?

(ii) What is your answer to the case when $f(w) = w^* w^m$? The asterisk denotes complex conjugation.

P3.10 Workout the details of the derivation of Eq. (3.74).

P3.11 In Section 3.5, for the complex-valued signals, we used the principle of orthogonality to derive the Wiener–Hopf equation and the minimum mean-squared error (3.74). Starting with the definition of the performance function, derive an equation similar to Eq. (3.12) for the case of complex-valued signals. Use this equation to give a direct derivation for the Wiener–Hopf equation, in the present case. Also, confirm the minimum mean-squared error equation (3.74).

P3.12 Show that for a Wiener filter with complex-valued tap-input vector $\mathbf{x}(n)$ and optimum tap-weight vector $\mathbf{w}_o$

$$\mathbf{w}_o^H \mathbf{p} = E[|\mathbf{w}_o^H \mathbf{x}(n)|^2]$$

where $\mathbf{p} = E[d(n)\mathbf{x}^*(n)]$ and $d(n)$ is the desired output of the filter. Use this result to argue that $\mathbf{w}_o^H \mathbf{p}$ is always positive. Also, use the above result to derive an equation similar to Eq. (3.50), for the general case of Wiener filters with complex-valued signals.

P3.13 Consider the channel equalization problem depicted in Figure P3.13. Assume that the underlying processes are real-valued with

$$E[s(n)s(n-k)] = \begin{cases} 1, & k = 0 \\ 0, & k \neq 0 \end{cases}$$

and

$$E[\nu(n)\nu(n-k)] = \begin{cases} \sigma_\nu^2, & k = 0 \\ 0, & k \neq 0 \end{cases}$$

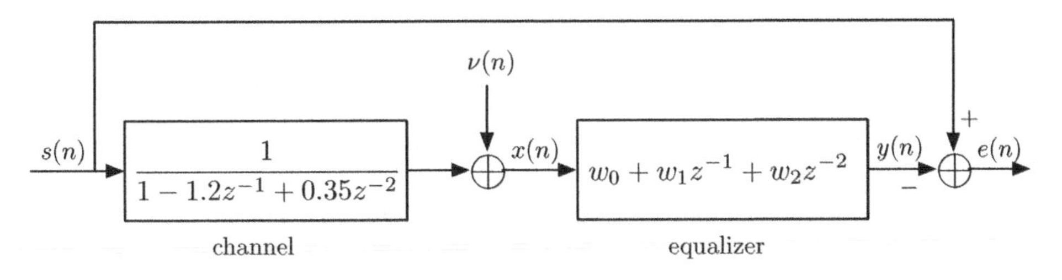

Figure P3.13

(i) For $\sigma_\nu^2 = 0$, obtain the equalizer tap weights by direct solution of the Wiener–Hopf equation. To be sure of your results, you may also guess the equalizer tap weights and compare them with the calculated ones.

(ii) Find the equalizer tap weights when $\sigma_\nu^2 = 0.1$ and compare the results with what you have obtained in (i).

(iii) Plot the magnitude and phase responses of the two designs obtained above and compare the results.

P3.14 Consider the equalization setup in Figure P3.14. Assume that the input data $s(n)$ is binary and takes equally probable values of ± 1. Also, assume that the data symbols, $s(m)$ and $s(n)$, for $m \neq n$ are uncorrelated with one another. Δ is a delay that should be chosen to optimize the performance of the equalizer.
Find the equalizer tap weights w_0 through w_7 and the minimum mean-squared error, $\xi_{\min}$, at the equalizer output for the choices of $\Delta = 0, 1, 2, \ldots, 7$. Discuss your observation on how $\xi_{\min}$ varies with Δ.
Note: You may use MATLAB or any other numerical software to complete your answer to this problem.

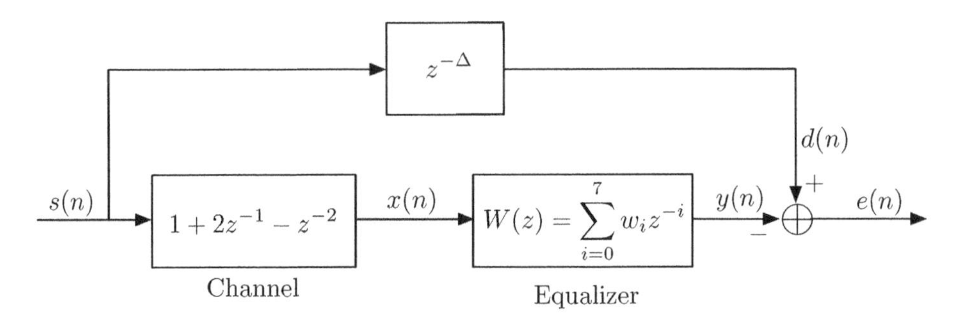

Figure P3.14

P3.15 Repeat Problem P3.14 when the channel transfer function is replaced by $1 + 2z^{-1} + z^{-2}$.

P3.16 Consider the system modeling problem of Figure P3.16. Assume that $x(n)$ and $\nu_o(n)$ are both white processes and $\sigma_x^2 = 1$ and $\sigma_{\nu_o}^2 = 0.01$.

(i) Find an expression for the performance function ξ.

(ii) Present a plot of ξ versus w for values of w in the range of -1 to $+1$ and show that it has a minimum at $w = 0.5$.

P3.17 Consider the system modeling problem of Figure P3.17. Assume that $x(n)$ and $\nu_o(n)$ are both white processes and $\sigma_x^2 = 1$ and $\sigma_{\nu_o}^2 = 0.001$.

(i) Find an expression for the performance function ξ in terms of the parameters w_0 and w_1.

(ii) Present a mesh plot of ξ versus the parameters w_0 and w_1.

(iii) Show that at the point where $(w_0, w_1) = (0.5, -0.5)$, $\xi(w_0, w_1) = \sigma_\nu^2$. Is there any other choice of (w_0, w_1) for which also $\xi(w_0, w_1) = \sigma_\nu^2$. Explain your answer.

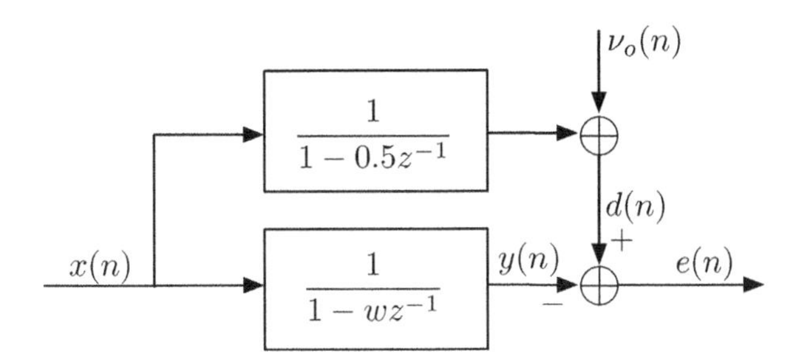

Figure P3.16

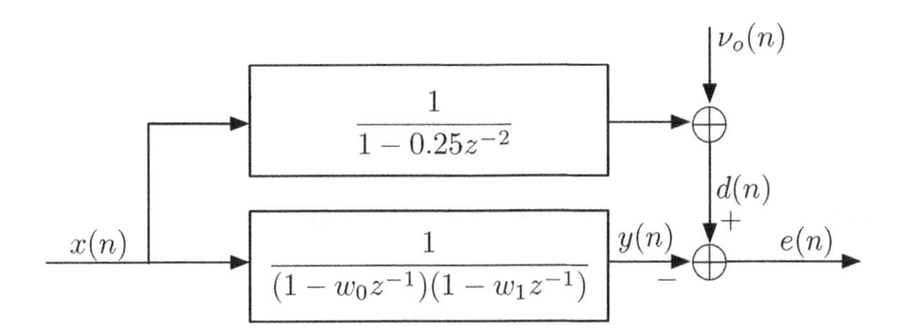

Figure P3.17

P3.18 Recalling that in any Wiener filter

$$\xi_{\min} = E[|d(n)|^2] - E[|y_o(n)|^2]$$

prove Eq. (3.97) by first deriving an expression for $E[|y_o(n)|^2]$ and then replacing the result in the above equation.

P3.19 By following a procedure similar to the one given in Section 3.6.1, show that when the involved processes and system parameters are complex-valued

$$\xi = \phi_{dd}(0) + \phi_{yy}(0) - 2\Re\{\phi_{yd}(0)\}$$

where $\Re\{x\}$ denotes the real part of x. Proceed with this result to develop the dual of equation (3.78).

P3.20 Show that Eq. (3.93) is a valid result even when the involved processes are complex-valued.

P3.21 Consider Figure P3.21, in which $x_i(n)$ and $d_i(n)$ are the outputs of two similar narrow-band filters centered at $\omega = \omega_i$, as in Figure 3.7. Show that if $w_{i,o}$ is the optimum value of w_i which minimizes the mean-squared of the output error, $e_i(n)$, then

$$w_{i,o} \approx \frac{\Phi_{dx}(e^{j\omega_i})}{\Phi_{xx}(e^{j\omega_i})}$$

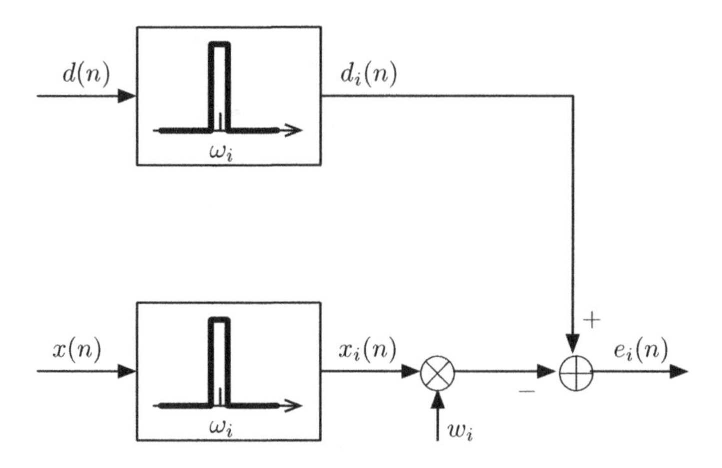

Figure P3.21

P3.22 Assuming that $W_o(z)$ is the optimum system function of a FIR filter, show that Eq. (3.96) can be converted to Eq. (3.74) and vice versa.

P3.23 Give a detailed derivation of Eq. (3.122) from Eq. (3.120).

P3.24 Give a detailed derivation of Eq. (3.123).

P3.25 For the noise canceler setup shown in Figure 3.12, consider the case when $\Phi_{ss}(z)|H(z)|^2 \ll \Phi_{\nu\nu}(z)$. Recall that the term *signal* is used to refer to $s(n)$ and the term *noise* is used to refer to $\nu(n)$.

(i) Show that, in this case,

$$W_o(z) \approx G(z) + \frac{\Phi_{ss}(z)}{\Phi_{\nu\nu}(z)} H^*(z)$$

(ii) Show that the power spectral density of the noise reaching the noise canceler output is

$$\Phi_{\text{output noise}}(z) = \Phi_{\nu\nu}(z)\rho_{\text{ref}}(z)\rho_{\text{pri}}(z)|G(z)|^2$$

(iii) Define the signal distortion at the canceler output, $D(z)$, as the ratio of the power spectral density of the signal propagating through $W_o(z)$ to the output to the power spectral density of the signal at the primary input. Show that

$$D(z) \approx |H(z)G(z) + \rho_{\text{ref}}(z)|^2$$

(iv) Show that the result obtained in (iii) may be written as

$$D(z) \approx \frac{\rho_{\text{ref}}(z)}{\rho_{\text{pri}}(z)}$$

when $\rho_{\text{ref}}(z) \ll |H(z)| \cdot |G(z)|$.

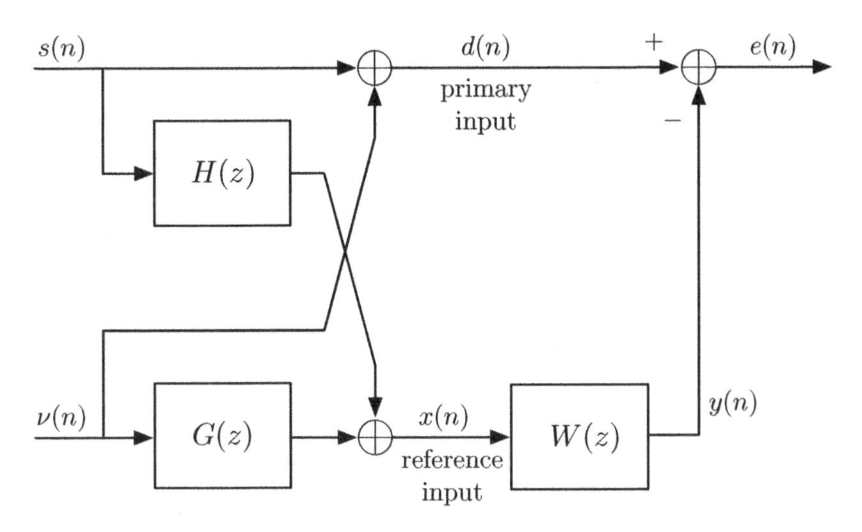

Figure P3.26

P3.26 Consider the noise canceler setup shown in Figure P3.26.

(i) Derive an unconstrained Wiener filter $W_o(z)$.
(ii) Show that the power inversion formula (3.146) is also valid for this setup.

P3.27 Consider an array of three omnidirectional antennas, as shown in Figure P3.27. The signal, $s(n)$, and jammer, $\nu(n)$, are narrow-band processes, as in Example 3.6. To cancel the jammer, we use a two-tap filter, similar to the one used in Figure 3.13, at either of the points 1 or 2, in Figure P3.27.

(i) To maximize the cancellation of the jammer, where will you place the two-tap filter?

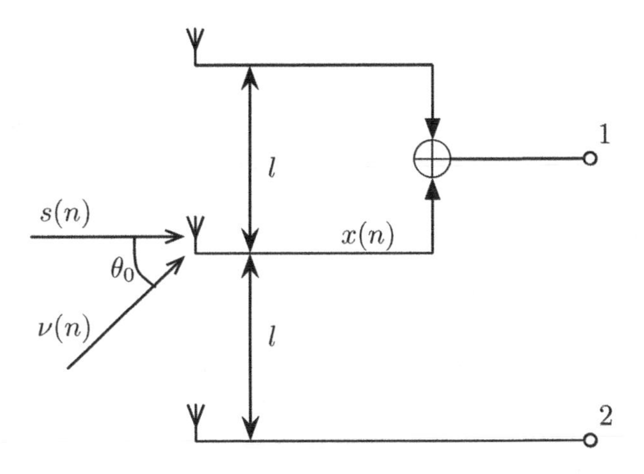

Figure P3.27

(ii) For your choice in (i), find the optimum values of the filter tap weights.
(iii) Find an expression for the signal and jammer components reaching the canceler output, and confirm the power inversion formula.

P3.28 Consider an array of three omnidirectional antennas as shown in Figure P3.28. The signal, $s(n)$, and jammer, $\nu(n)$, are narrow-band processes, as in Example 3.6.

(i) Find the optimum values of the filter tap weights, which minimize the mean-squared of the output error, $e(n)$.
(ii) Find an expression for the canceler output, and confirm the power inversion formula.

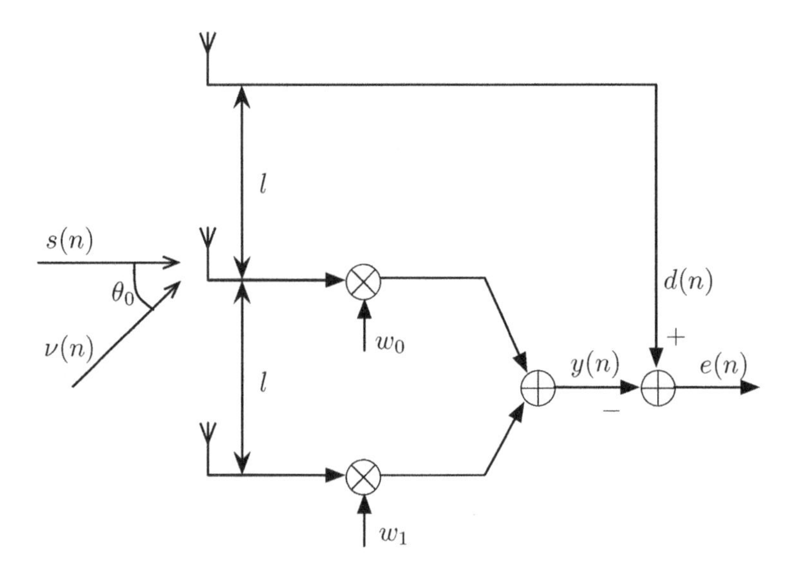

Figure P3.28

P3.29 Repeat P3.28 for the array shown in Figure P3.29, and compare the results obtained with those of P3.28.

P3.30 To prevent time averages and derive the results presented in Example 3.6 through ensemble averages, the desired signal and jammer may be redefined as $s(n) = \alpha(n)\cos(n\omega_o + \varphi_1)$ and $\nu(n) = \beta(n)\cos(n\omega_o + \varphi_2)$, respectively, where φ_1 and φ_2 are random initial phases of the carrier, and assumed to be uniformly distributed in the interval $-\pi$ to $+\pi$. The amplitudes $\alpha(n)$ and $\beta(n)$, as in Example 3.6, are uncorrelated narrow-band baseband signals. Furthermore, the random phases φ_1 and φ_2 are assumed to be independent among themselves as well as with respect to $\alpha(n)$ and $\beta(n)$.

(i) Using the new definitions of $s(n)$ and $\nu(n)$, show that the same result as in Eq. (3.160) is also obtained through ensemble averages.

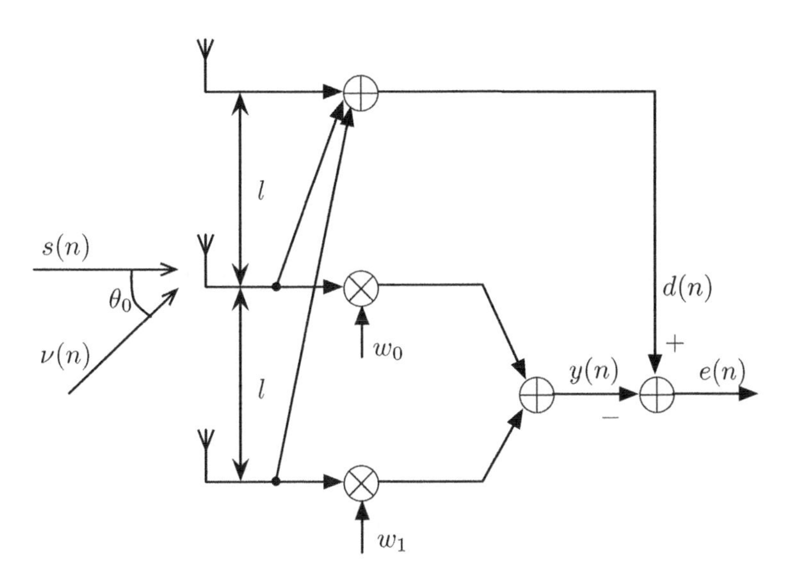

Figure P3.29

(ii) Show that, for the present case

$$e_o(n) = \frac{\sigma_\beta^2 \alpha(n)}{\sigma_\alpha^2 + \sigma_\beta^2}[\cos(n\omega_o + \varphi_1) - \cos((n - \delta_o)\omega_o + \varphi_1)]$$

$$-\frac{\sigma_\alpha^2 \beta(n)}{\sigma_\alpha^2 + \sigma_\beta^2}[\cos(n\omega_o + \varphi_2) - \cos((n - \delta_o)\omega_o + \varphi_2)]$$

(iii) Use the result in (ii) to verify the power inversion formula, in the present case.

4

Eigenanalysis and Performance Surface

The transversal Wiener filter was introduced in the last chapter as a powerful signal processing structure with a unique performance function, which has many desirable features for adaptive filtering applications. In particular, it was noted that the performance function of the transversal Wiener filter has a unique global minimum point, which can be easily obtained using the second-order moments of the underlying processes. This is a consequence of the fact that the performance function of the transversal Wiener filter is a convex quadratic function of its tap weights.

Our goal in this chapter is to analyze the quadratic performance function of the transversal Wiener filter in detail. We get a clear picture of the shape of the performance function when it is visualized as a surface in the $(N+1)$-dimensional space of variables consisting of the filter tap weights, as the first N axes, and the performance function, as the $(N+1)$th axis. This is called *performance surface*.

The shape of the performance surface of a transversal Wiener filter is closely related to the eigenvalues of the correlation matrix $\mathbf{R}$ of the filter tap inputs. Hence, we start with a thorough discussion on the eigenvalues and eigenvectors of the correlation matrix $\mathbf{R}$.

4.1 Eigenvalues and Eigenvectors

Let

$$\mathbf{R} = E[\mathbf{x}(n)\mathbf{x}^{\mathrm{H}}(n)] \tag{4.1}$$

be the N-by-N correlation matrix of a complex-valued wide-sense stationary stochastic process represented by the N-by-1 observation vector $\mathbf{x}(n) = [x(n)\ x(n-1) \cdots x(n-N+1)]^{\mathrm{T}}$, where the superscripts H and T denote Hermitian and transpose, respectively.

A nonzero N-by-1 vector $\mathbf{q}$ is said to be an *eigenvector* of $\mathbf{R}$, if it satisfies the equation

$$\mathbf{R}\mathbf{q} = \lambda\mathbf{q} \tag{4.2}$$

for some scalar constant λ. The scalar λ is called the *eigenvalue* of $\mathbf{R}$ associated with the eigenvector $\mathbf{q}$. We note that if $\mathbf{q}$ is an eigenvector of $\mathbf{R}$, then for any nonzero scalar a,

---

Adaptive Filters: Theory and Applications, Second Edition. Behrouz Farhang-Boroujeny.

$a\mathbf{q}$ is also an eigenvector of $\mathbf{R}$, corresponding to the same eigenvalue, λ. This is easily verified by multiplying Eq. (4.2) through by a.

To find the eigenvalues and eigenvectors of $\mathbf{R}$, we note that Eq. (4.2) may be rearranged as

$$(\mathbf{R} - \lambda\mathbf{I})\mathbf{q} = \mathbf{0} \tag{4.3}$$

where $\mathbf{I}$ is the N-by-N identity matrix, and $\mathbf{0}$ is the N-by-1 null vector. To prevent the trivial solution $\mathbf{q} = \mathbf{0}$, the matrix $\mathbf{R} - \lambda\mathbf{I}$ has to be singular. This implies

$$\det(\mathbf{R} - \lambda\mathbf{I}) = 0 \tag{4.4}$$

where $\det(\cdot)$ denotes determinant. Equation (4.4) is called the *characteristic equation* of the matrix $\mathbf{R}$. The characteristic equation (4.4), when expanded, is an Nth order equation in the unknown parameter λ. The roots of this equation, which may be called $\lambda_0, \lambda_1, \ldots, \lambda_{N-1}$, are the eigenvalues of $\mathbf{R}$. When λ_is are distinct, $\mathbf{R} - \lambda_i\mathbf{I}$, for $i = 0, 1, \ldots, N-1$ will be of rank $N-1$. This leads to N eigenvectors $\mathbf{q}_0, \mathbf{q}_1, \ldots, \mathbf{q}_{N-1}$, for the matrix $\mathbf{R}$, which are unique up to a scale factor. On the other hand, when the characteristic equation (4.4) has repeated roots, the matrix $\mathbf{R}$ is said to have *degenerate* eigenvalues. In that case, the eigenvectors of $\mathbf{R}$ will not be unique. For example, if λ_m is an eigenvalue of $\mathbf{R}$ repeated p times, then the rank of $\mathbf{R} - \lambda_m\mathbf{I}$ is $N - p$, and thus the solution of the equation $(\mathbf{R} - \lambda_m\mathbf{I})\mathbf{q}_m = \mathbf{0}$ can be any vector in a p-dimensional subspace of the N-dimensional complex vector space. This, in general, creates some confusion in eigenanalysis of matrices, which should be handled carefully. To prevent such confusions, in the discussion that follows, wherever necessary, we start with the case that the eigenvalues of $\mathbf{R}$ are distinct. The results will then be extended to the case of repeated eigenvalues.

4.2 Properties of Eigenvalues and Eigenvectors

We discuss the various properties of the eigenvalues and eigenvectors of the correlation matrix $\mathbf{R}$. Some of the properties derived here are directly related to the fact that the correlation matrix $\mathbf{R}$ is Hermitian and nonnegative definite. A matrix $\mathbf{A}$, in general, is said to be Hermitian if $\mathbf{A} = \mathbf{A}^{\mathrm{H}}$. This for the correlation matrix $\mathbf{R}$ is observed by direct inspection of Eq. (4.1). The N-by-N Hermitian matrix $\mathbf{A}$ is said to be nonnegative definite or *positive semidefinite,* if

$$\mathbf{v}^{\mathrm{H}}\mathbf{A}\mathbf{v} \geq 0 \tag{4.5}$$

for any N-by-1 vector $\mathbf{v}$. The fact that $\mathbf{A}$ is Hermitian implies that $\mathbf{v}^{\mathrm{H}}\mathbf{A}\mathbf{v}$ is real-valued. This can be seen easily, if we note that with the dimensions specified above $\mathbf{v}^{\mathrm{H}}\mathbf{A}\mathbf{v}$ is a scalar and $(\mathbf{v}^{\mathrm{H}}\mathbf{A}\mathbf{v})^* = (\mathbf{v}^{\mathrm{H}}\mathbf{A}\mathbf{v})^{\mathrm{H}} = \mathbf{v}^{\mathrm{H}}\mathbf{A}\mathbf{v}$. For the correlation matrix $\mathbf{R}$, to show that $\mathbf{v}^{\mathrm{H}}\mathbf{R}\mathbf{v}$ can never be negative, we replace for $\mathbf{R}$ from Eq. (4.1) to obtain

$$\begin{aligned}\mathbf{v}^{\mathrm{H}}\mathbf{R}\mathbf{v} &= \mathbf{v}^{\mathrm{H}}E[\mathbf{x}(n)\mathbf{x}^{\mathrm{H}}(n)]\mathbf{v}\\ &= E[\mathbf{v}^{\mathrm{H}}\mathbf{x}(n)\mathbf{x}^{\mathrm{H}}(n)\mathbf{v}]\end{aligned} \tag{4.6}$$

We note that $\mathbf{v}^{\mathrm{H}}\mathbf{x}(n)$ and $\mathbf{x}^{\mathrm{H}}(n)\mathbf{v}$ constitute a pair of complex-conjugate scalars. This, when used in Eq. (4.6), gives

$$\mathbf{v}^{\mathrm{H}}\mathbf{R}\mathbf{v} = E[|\mathbf{v}^{\mathrm{H}}\mathbf{x}(n)|^2] \tag{4.7}$$

which is nonnegative for any vector $\mathbf{v}$.

From Eq. (4.7), we note that when $\mathbf{v}$ is nonzero, the Hermitian form $\mathbf{v}^{\mathrm{H}}\mathbf{R}\mathbf{v}$ may be zero only when there is a consistent dependency between the elements of the observation vector $\mathbf{x}(n)$, so that $\mathbf{v}^{\mathrm{H}}\mathbf{x}(n) = 0$, for all observations of $\mathbf{x}(n)$. For a random process $\{x(n)\}$, this can only happen when $\{x(n)\}$ consists of a sum of L sinusoids with $L < N$. In practice, we find that this situation is very rare and thus for any nonzero $\mathbf{v}$, $\mathbf{v}^{\mathrm{H}}\mathbf{R}\mathbf{v}$ is almost always positive. We thus say that the correlation matrix $\mathbf{R}$ is almost always *positive definite*.

With this background, we are now prepared to discuss the properties of the eigenvalues and eigenvectors of the correlation matrix $\mathbf{R}$.

Property 1: *The eigenvalues of the correlation matrix* $\mathbf{R}$ *are all real and nonnegative.*

Consider an eigenvector $\mathbf{q}_i$ of $\mathbf{R}$ and its corresponding eigenvalue λ_i. These two are related together according to the equation

$$\mathbf{R}\mathbf{q}_i = \lambda_i \mathbf{q}_i \tag{4.8}$$

Premultiplying Eq. (4.8) by $\mathbf{q}_i^{\mathrm{H}}$ and noting that λ_i is a scalar, we get

$$\mathbf{q}_i^{\mathrm{H}}\mathbf{R}\mathbf{q}_i = \lambda_i \mathbf{q}_i^{\mathrm{H}}\mathbf{q}_i \tag{4.9}$$

The quantity $\mathbf{q}_i^{\mathrm{H}}\mathbf{q}_i$ on the right-hand side is always real and positive, because it is the squared length of the vector $\mathbf{q}_i$. Furthermore, the Hermitian form $\mathbf{q}_i^{\mathrm{H}}\mathbf{R}\mathbf{q}_i$ on the left-hand side of Eq. (4.9) is always real and nonnegative, because the correlation matrix $\mathbf{R}$ is nonnegative definite. Noting these, it follows from Eq. (4.9) that

$$\lambda_i \geq 0, \quad \text{for } i = 0, 1, \ldots, N-1 \tag{4.10}$$

Property 2: *If* $\mathbf{q}_i$ *and* $\mathbf{q}_j$ *are two eigenvectors of the correlation matrix* $\mathbf{R}$ *that correspond to two of its distinct eigenvalues, then*

$$\mathbf{q}_i^{\mathrm{H}}\mathbf{q}_j = 0 \tag{4.11}$$

In other words, eigenvectors associated with the distinct eigenvalues of the correlation matrix $\mathbf{R}$ *are mutually orthogonal.*

Let λ_i and λ_j be the distinct eigenvalues corresponding to the eigenvectors $\mathbf{q}_i$ and $\mathbf{q}_j$, respectively. We have

$$\mathbf{R}\mathbf{q}_i = \lambda_i \mathbf{q}_i \tag{4.12}$$

and

$$\mathbf{R}\mathbf{q}_j = \lambda_j \mathbf{q}_j \tag{4.13}$$

Applying conjugate transpose on both sides of Eq. (4.12) and noting that λ_i is a real scalar and for the Hermitian matrix $\mathbf{R}$, $\mathbf{R}^{\mathrm{H}} = \mathbf{R}$, we obtain

$$\mathbf{q}_i^{\mathrm{H}}\mathbf{R} = \lambda_i \mathbf{q}_i^{\mathrm{H}} \tag{4.14}$$

Premultiplying Eq. (4.13) by $\mathbf{q}_i^{\mathrm{H}}$, postmultiplying Eq. (4.14) by $\mathbf{q}_j$, and subtracting the two resulting equations, gives

$$(\lambda_j - \lambda_i)\mathbf{q}_i^{\mathrm{H}}\mathbf{q}_j = 0 \tag{4.15}$$

Noting that λ_i and λ_j are distinct, this gives Eq. (4.11).

Property 3: *Let $\mathbf{q}_0, \mathbf{q}_1, \ldots, \mathbf{q}_{N-1}$ be the eigenvectors associated with the distinct eigenvalues $\lambda_0, \lambda_1, \ldots, \lambda_{N-1}$ of the N-by-N correlation matrix* **R**, *respectively. Assume the eigenvectors $\mathbf{q}_0, \mathbf{q}_1, \ldots, \mathbf{q}_{N-1}$ are all normalized to have a length of unity, and define the N-by-N matrix*

$$\mathbf{Q} = [\mathbf{q}_0\ \mathbf{q}_1 \cdots \mathbf{q}_{N-1}]. \tag{4.16}$$

Q *is then a unitary matrix, i.e.,*

$$\mathbf{Q}^{\mathrm{H}}\mathbf{Q} = \mathbf{I} \tag{4.17}$$

This implies that the matrices **Q** *and* $\mathbf{Q}^{\mathrm{H}}$ *are the inverse of each other.*

To show this property, we note that the ijth element of the N-by-N matrix $\mathbf{Q}^{\mathrm{H}}\mathbf{Q}$ is the product of the ith row of $\mathbf{Q}^{\mathrm{H}}$, which is $\mathbf{q}_i^{\mathrm{H}}$, and the jth column of **Q**, which is $\mathbf{q}_j$. That is,

$$\text{the } ij\text{th element of } \mathbf{Q}^{\mathrm{H}}\mathbf{Q} = \mathbf{q}_i^{\mathrm{H}}\mathbf{q}_j. \tag{4.18}$$

Noting this, Eq. (4.17) follows immediately from Property 2.
In cases where the correlation matrix **R** has one or more repeated eigenvalues, as was noted earlier, attached to each of these repeated eigenvalues there is a subspace of the same dimension as the multiplicity of the eigenvalue in which any vector is an eigenvector of **R**. From Property 2, we can say that the subspaces that belong to distinct eigenvalues are orthogonal. Moreover, within each subspace one can always find a set of orthogonal basis vectors that span the whole subspace. Clearly, such a set is not unique, but can always be chosen. This means, for any repeated eigenvalue with multiplicity p, one can always find a set of p orthogonal eigenvectors. Noting this, we can say, in general, that for any N-by-N correlation matrix **R**, one can always make a unitary matrix **Q** whose columns are made up of a set of eigenvectors of **R**.

Property 4: *For any N-by-N correlation matrix* **R**, *one can always find a set of mutually orthogonal eigenvectors. Such a set may be used as a basis to express any vector in the N-dimensional space of complex vectors.*

This property follows from the earlier discussion.

Property 5: Unitary Similarity Transformation. *The correlation matrix* **R** *can always be decomposed as*

$$\mathbf{R} = \mathbf{Q}\boldsymbol{\Lambda}\mathbf{Q}^{\mathrm{H}} \tag{4.19}$$

where the matrix **Q** *is made up of a set of unit-length orthogonal eigenvectors of* **R** *as specified in* Eqs. (4.16) *and* (4.17),

$$\boldsymbol{\Lambda} \triangleq \begin{bmatrix} \lambda_0 & 0 & \cdots & 0 \\ 0 & \lambda_1 & \cdots & 0 \\ \vdots & \vdots & \ddots & \vdots \\ 0 & 0 & \cdots & \lambda_{N-1} \end{bmatrix} \tag{4.20}$$

and the order of the eigenvalues $\lambda_0, \lambda_1, \ldots, \lambda_{N-1}$ *matches that of the corresponding eigenvectors in the columns of* $\mathbf{Q}$.

To prove this property, we note that the set of equations

$$\mathbf{R}\mathbf{q}_i = \lambda_i \mathbf{q}_i, \quad \text{for } i = 0, 1, \ldots, N-1 \tag{4.21}$$

may be packed together as a single matrix equation

$$\mathbf{R}\mathbf{Q} = \mathbf{Q}\mathbf{\Lambda}. \tag{4.22}$$

Then, postmultiplying Eq. (4.22) by $\mathbf{Q}^{\mathrm{H}}$ and noting that $\mathbf{Q}\mathbf{Q}^{\mathrm{H}} = \mathbf{I}$, we can get Eq. (4.19). The right-hand side of Eq. (4.19) may be expanded as

$$\mathbf{R} = \sum_{i=0}^{N-1} \lambda_i \mathbf{q}_i \mathbf{q}_i^{\mathrm{H}} \tag{4.23}$$

Property 6: *Let* $\lambda_0, \lambda_1, \ldots, \lambda_{N-1}$ *be the eigenvalues of the correlation matrix* $\mathbf{R}$. *Then,*

$$\mathrm{tr}[\mathbf{R}] = \sum_{i=0}^{N-1} \lambda_i \tag{4.24}$$

where $\mathrm{tr}[\mathbf{R}]$ *denotes trace of* $\mathbf{R}$ *and is defined as the sum of the diagonal elements of* $\mathbf{R}$.

Taking the trace on both sides of Eq. (4.19), we get

$$\mathrm{tr}[\mathbf{R}] = \mathrm{tr}[\mathbf{Q}\mathbf{\Lambda}\mathbf{Q}^{\mathrm{H}}] \tag{4.25}$$

To proceed, we may use the following result of matrix algebra. If $\mathbf{A}$ and $\mathbf{B}$ are N-by-M and M-by-N matrices, respectively, then,

$$\mathrm{tr}[\mathbf{A}\mathbf{B}] = \mathrm{tr}[\mathbf{B}\mathbf{A}] \tag{4.26}$$

Using this result, we may swap $\mathbf{Q}\mathbf{\Lambda}$ and $\mathbf{Q}^{\mathrm{H}}$ on the right-hand side of Eq. (4.25). Then, noting that $\mathbf{Q}^{\mathrm{H}}\mathbf{Q} = \mathbf{I}$, Eq. (4.25) is simplified as

$$\mathrm{tr}[\mathbf{R}] = \mathrm{tr}[\,\mathbf{\Lambda}\,] \tag{4.27}$$

Using the definition (4.20) in Eq. (4.27) completes the proof.
An alternative way of proving the above result is by direct expansion of Eq. (4.4); see Problem P4.8. This proof shows that the identity (4.24) is not limited to the Hermitian matrices. It applies to any square matrix.

Property 7: Minimax Theorem.[1] *The distinct eigenvalues* $\lambda_0 > \lambda_1 > \cdots > \lambda_{N-1}$ *of the correlation matrix* $\mathbf{R}$ *of an observation vector* $\mathbf{x}(n)$, *and their corresponding eigenvectors,* $\mathbf{q}_0, \mathbf{q}_1, \ldots \mathbf{q}_{N-1}$, *may be obtained through the following optimization procedure*:

$$\lambda_{\max} = \lambda_0 = \max_{\|\mathbf{q}_0\|=1} E[|\mathbf{q}_0^{\mathrm{H}}\mathbf{x}(n)|^2] \tag{4.28}$$

[1] In matrix algebra literature, the minimax theorem is usually stated using the Hermitian form $\mathbf{q}_i^{\mathrm{H}}\mathbf{R}\mathbf{q}_i$ instead of $E[|\mathbf{q}^{\mathrm{H}}\mathbf{x}(n)|^2]$ (Haykin, 1991). The method that we have adopted here is to simplify some of our discussions in the following chapters. This method has been adopted from Farhang-Boroujeny and Gazor (1992).

and for $i = 1, 2, \ldots, N-1$

$$\lambda_i = \max_{\|\mathbf{q}_i\|=1} E[|\mathbf{q}_i^{\mathrm{H}}\mathbf{x}(n)|^2] \tag{4.29}$$

with

$$\mathbf{q}_i^{H}\mathbf{q}_j = 0, \quad \textit{for} \ \ 0 \leq j < i \tag{4.30}$$

where $\|\mathbf{q}_i\| \triangleq \sqrt{\mathbf{q}_i^{\mathrm{H}}\mathbf{q}_i}$ *denotes the length or norm of the complex vector* $\mathbf{q}_i$*. Alternatively, the following procedure may also be used to obtain the eigenvalues of the correlation matrix* $\mathbf{R}$*, in the ascending order:*

$$\lambda_{\min} = \lambda_{N-1} = \min_{\|\mathbf{q}_{N-1}\|=1} E[|\mathbf{q}_{N-1}^{\mathrm{H}}\mathbf{x}(n)|^2] \tag{4.31}$$

and for $i = N-2, \ldots, 1, 0$

$$\lambda_i = \min_{\|\mathbf{q}_i\|=1} E[|\mathbf{q}_i^{\mathrm{H}}\mathbf{x}(n)|^2] \tag{4.32}$$

with

$$\mathbf{q}_i^{\mathrm{H}}\mathbf{q}_j = 0, \quad \textit{for} \ \ i < j \leq N-1. \tag{4.33}$$

Let us assume that the set of vectors that satisfy the minimax optimization procedure are the unit-length vectors $\mathbf{p}_0, \mathbf{p}_1, \ldots, \mathbf{p}_{N-1}$. From Property 4, we recall that the eigenvectors $\mathbf{q}_0, \mathbf{q}_1, \ldots, \mathbf{q}_{N-1}$ are a set of basis vectors for the N-dimensional complex vector space. This implies that, we may write

$$\mathbf{p}_i = \sum_{j=0}^{N-1} \alpha_{ij}\mathbf{q}_j, \quad \text{for } i = 0, 1, \ldots, N-1 \tag{4.34}$$

where the complex-valued coefficients α_{ij}s are the coordinates of the complex vectors $\mathbf{p}_0, \mathbf{p}_1, \ldots, \mathbf{p}_{N-1}$ in the N-dimensional space spanned by the basis vectors $\mathbf{q}_0, \mathbf{q}_1, \ldots, \mathbf{q}_{N-1}$. Let $\mathbf{p}_0$ be the unit-length complex vector, which maximizes $E[|\mathbf{p}_0^{\mathrm{H}}\mathbf{x}(n)|^2]$. We note that

$$\begin{aligned} E[|\mathbf{p}_0^{\mathrm{H}}\mathbf{x}(n)|^2] &= E[\mathbf{p}_0^{\mathrm{H}}\mathbf{x}(n)\mathbf{x}^{\mathrm{H}}(n)\mathbf{p}_0] \\ &= \mathbf{p}_0^{\mathrm{H}}E[\mathbf{x}(n)\mathbf{x}^{\mathrm{H}}(n)]\mathbf{p}_0 \\ &= \mathbf{p}_0^{\mathrm{H}}\mathbf{R}\mathbf{p}_0 \end{aligned} \tag{4.35}$$

Substituting Eq. (4.23) in Eq. (4.35), we obtain

$$E[|\mathbf{p}_0^{\mathrm{H}}\mathbf{x}(n)|^2] = \sum_{i=0}^{N-1} \lambda_i \mathbf{p}_0^{\mathrm{H}}\mathbf{q}_i\mathbf{q}_i^{\mathrm{H}}\mathbf{p}_0 \tag{4.36}$$

Using Property 3, we get

$$\mathbf{p}_0^{\mathrm{H}}\mathbf{q}_i = \alpha_{0i}^{*} \tag{4.37}$$

and

$$\mathbf{q}_i^{\mathrm{H}}\mathbf{p}_0 = \alpha_{0i} \tag{4.38}$$

Substituting these in Eq. (4.36), we obtain

$$E[|\mathbf{p}_0^{\mathrm{H}}\mathbf{x}(n)|^2] = \sum_{i=0}^{N-1} \lambda_i |\alpha_{0i}|^2 \tag{4.39}$$

On the other hand, we may note that, because $\lambda_0 > \lambda_1, \lambda_2, \ldots, \lambda_{N-1}$,

$$\sum_{i=0}^{N-1} \lambda_i |\alpha_{0i}|^2 \leq \lambda_0 \sum_{i=0}^{N-1} |\alpha_{0i}|^2 \tag{4.40}$$

where the equality holds (i.e., $\mathbf{p}_0$ maximizes $E[|\mathbf{p}_0^{\mathrm{H}}\mathbf{x}(n)|^2]$) only when $\alpha_{0i} = 0$, for $i = 1, 2, \ldots, N-1$. Furthermore, the fact that the $\mathbf{p}_0$ is constrained to the length of unity implies that

$$\sum_{i=0}^{N-1} |\alpha_{0i}|^2 = 1 \tag{4.41}$$

Application of Eqs. (4.39) and (4.41) in Eq. (4.40) gives

$$\max_{\|\mathbf{p}_0\|=1} E[|\mathbf{p}_0^{\mathrm{H}}\mathbf{x}(n)|^2] = \lambda_0 \tag{4.42}$$

and this is achieved when

$$\mathbf{p}_0 = \alpha_{00}\mathbf{q}_0 \quad \text{with} \quad |\alpha_{00}| = 1 \tag{4.43}$$

We may note that the factor α_{00} is arbitrary and has no significance as it does not affect the maximum in Eq. (4.42) because of the constraint $|\alpha_{00}| = 1$ in Eq. (4.43). Hence, without any loss of generality, we assume $\alpha_{00} = 1$. This gives

$$\mathbf{p}_0 = \mathbf{q}_0 \tag{4.44}$$

as a solution to the maximization problem

$$\max_{\|\mathbf{p}_0\|=1} E[|\mathbf{p}_0^{\mathrm{H}}\mathbf{x}(n)|^2]. \tag{4.45}$$

The fact that the solution obtained here is not unique follows from the more general fact that the eigenvector corresponding to an eigenvalue is always arbitrary to the extent of a scalar multiplier factor. Here, the scalar multiplier is constrained to have a modulus of unity to satisfy the condition that both $\mathbf{p}_i$ and $\mathbf{q}_i$ vectors are constrained to the length of unity.

In proceeding to find $\mathbf{p}_1$, we note that the constraint (4.30), for $i = 1$, implies that

$$\mathbf{p}_1^{\mathrm{H}}\mathbf{q}_0 = 0 \tag{4.46}$$

This, in turn, requires $\mathbf{p}_1$ to be limited to a linear combination of $\mathbf{q}_1, \mathbf{q}_2, \ldots, \mathbf{q}_{N-1}$, only. That is,

$$\mathbf{p}_1 = \sum_{j=1}^{N-1} \alpha_{1,j}\mathbf{q}_j \tag{4.47}$$

Noting this and following a procedure similar to the one used to find $\mathbf{p}_0$, we get

$$\max_{\|\mathbf{p}_1\|=1} E[|\mathbf{p}_1^{\mathrm{H}}\mathbf{x}(n)|^2] = \lambda_1 \tag{4.48}$$

and

$$\mathbf{p}_1 = \mathbf{q}_1 \tag{4.49}$$

Following the same procedure for the rest of the eigenvalues and eigenvectors of $\mathbf{R}$ completes the proof of the first procedure of the minimax theorem.
The alternative procedure of the minimax theorem, suggested by Eqs. (4.31)–(4.33), can also be proved in a similar way.

Property 8: *The eigenvalues of the correlation matrix* $\mathbf{R}$ *of a discrete-time stationary stochastic process* $\{x(n)\}$ *are bounded by the minimum and maximum values of the power spectral density,* $\Phi_{xx}(\mathrm{e}^{j\omega})$, *of the process.*

The minimax theorem, as introduced in Property 7, views the eigenvectors of the correlation matrix of a discrete-time stochastic process as the conjugate of a set of tap-weight vectors corresponding to a set of FIR filters that are optimized in the minimax sense introduced there. Such filters are conveniently called *eigenfilters*. The minimax optimization procedure suggests that the eigenfilters may be obtained through a maximization or a minimization procedure, which looks at the output powers of the eigenfilters. In particular, the maximum and minimum eigenvalues of $\mathbf{R}$ may be obtained by solving the following two independent problems, respectively:

$$\lambda_{\max} = \max_{\|\mathbf{q}_0\|=1} E[|\mathbf{q}_0^{\mathrm{H}}\mathbf{x}(n)|^2] \tag{4.50}$$

and

$$\lambda_{\min} = \min_{\|\mathbf{q}_{N-1}\|=1} E[|\mathbf{q}_{N-1}^{\mathrm{H}}\mathbf{x}(n)|^2] \tag{4.51}$$

Let $Q_i(z)$ denote the system function of the ith eigenfilter of the discrete-time stochastic process $\{x(n)\}$. Using the Parseval's relation (Eq. (2.26) of Chapter 2), we obtain

$$\|\mathbf{q}_i\|^2 = \mathbf{q}_i^{\mathrm{H}}\mathbf{q}_i = \frac{1}{2\pi}\int_{-\pi}^{\pi} |Q_i(\mathrm{e}^{j\omega})|^2 d\omega \tag{4.52}$$

With the constraint $\|\mathbf{q}_i\| = 1$, this gives

$$\frac{1}{2\pi}\int_{-\pi}^{\pi} |Q_i(\mathrm{e}^{j\omega})|^2 d\omega = 1 \tag{4.53}$$

On the other hand, if we define $x_i'(n)$ as the output of the ith eigenfilter of $\mathbf{R}$, that is,

$$x_i'(n) = \mathbf{q}_i^{\mathrm{H}}\mathbf{x}(n) \tag{4.54}$$

then, using the power spectral density relationships provided in Chapter 2, we obtain

$$\Phi_{x_i'x_i'}(\mathrm{e}^{j\omega}) = |Q_i(\mathrm{e}^{j\omega})|^2\Phi_{xx}(\mathrm{e}^{j\omega}). \tag{4.55}$$

We may also recall from the results presented in Chapter 2 that

$$E[|x_i'(n)|^2] = \frac{1}{2\pi}\int_{-\pi}^{\pi} \Phi_{x_i'x_i'}(e^{j\omega})d\omega. \tag{4.56}$$

Substituting Eq. (4.55) in Eq. (4.56), we obtain

$$E[|x_i'(n)|^2] = \frac{1}{2\pi}\int_{-\pi}^{\pi} |Q_i(e^{j\omega})|^2\Phi_{xx}(e^{j\omega})d\omega. \tag{4.57}$$

This result has the following interpretation. *The signal power at the output of the ith eigenfilter of the correlation matrix* **R** *of a stochastic process* $\{x(n)\}$ *is given by a weighted average of the power spectral density of* $\{x(n)\}$. *The weighting function used for averaging is the squared magnitude response of the corresponding eigenfilter.*
Using the above results, Eq. (4.50) may be written as

$$\lambda_{\max} = \max \frac{1}{2\pi}\int_{-\pi}^{\pi} |Q_0(e^{j\omega})|^2\Phi_{xx}(e^{j\omega})d\omega \tag{4.58}$$

subject to the constraint

$$\frac{1}{2\pi}\int_{-\pi}^{\pi} |Q_0(e^{j\omega})|^2 = 1 \tag{4.59}$$

We may also note that

$$\frac{1}{2\pi}\int_{-\pi}^{\pi} |Q_0(e^{j\omega})|^2\Phi_{xx}(e^{j\omega})d\omega \le \Phi_{xx}^{\max}\frac{1}{2\pi}\int_{-\pi}^{\pi} |Q_0(e^{j\omega})|^2 d\omega \tag{4.60}$$

where

$$\Phi_{xx}^{\max} \triangleq \max_{-\pi\le\omega\le\pi} \Phi_{xx}(e^{j\omega}) \tag{4.61}$$

With the constraint (4.59), Eq. (4.60) simplifies to

$$\frac{1}{2\pi}\int_{-\pi}^{\pi} |Q_0(e^{j\omega})|^2\Phi_{xx}(e^{j\omega})d\omega \le \Phi_{xx}^{\max} \tag{4.62}$$

Using Eq. (4.62) in Eq. (4.58), we obtain

$$\lambda_{\max} \le \Phi_{xx}^{\max} \tag{4.63}$$

Following a similar procedure, we may also find that

$$\lambda_{\min} \ge \Phi_{xx}^{\min} \tag{4.64}$$

where

$$\Phi_{xx}^{\min} \triangleq \min_{-\pi\le\omega\le\pi} \Phi_{xx}(e^{j\omega}) \tag{4.65}$$

Property 9: *Let* $\mathbf{x}(n)$ *be an observation vector with the correlation matrix* **R**. *Assume that* $\mathbf{q}_0, \mathbf{q}_1, \ldots, \mathbf{q}_{N-1}$ *are a set of orthogonal eigenvectors of* **R** *and the matrix* **Q** *is defined as in* Eq. (4.16). *Then, the elements of the vector*

$$\mathbf{x}'(n) = \mathbf{Q}^{\mathrm{H}}\mathbf{x}(n) \tag{4.66}$$

constitute a set of uncorrelated random variables. The transformation defined by Eq. (4.66) *is called Karhunen–Loéve transform.*

Using Eq. (4.66), we obtain

$$E[\mathbf{x}'(n)\mathbf{x}'^{\mathrm{H}}(n)] = \mathbf{Q}^{\mathrm{H}}E[\mathbf{x}(n)\mathbf{x}^{\mathrm{H}}(n)]\mathbf{Q} = \mathbf{Q}^{\mathrm{H}}\mathbf{R}\mathbf{Q}. \tag{4.67}$$

Substituting for $\mathbf{R}$ from Eq. (4.19) and assuming that the eigenvectors $\mathbf{q}_0, \mathbf{q}_1, \ldots, \mathbf{q}_{N-1}$ are normalized to the length of unity[2], so that $\mathbf{Q}^{\mathrm{H}}\mathbf{Q} = \mathbf{I}$, we obtain

$$E[\mathbf{x}'(n)\mathbf{x}'^{\mathrm{H}}(n)] = \mathbf{\Lambda} \tag{4.68}$$

Noting that $\mathbf{\Lambda}$ is a diagonal matrix, this clearly shows that the elements of $\mathbf{x}'(n)$ are uncorrelated with one another.

It is worth noting that the ith element of $\mathbf{x}'(n)$ is the output of the ith eigenfilter of the correlation matrix of the process $\{\mathrm{x}(n)\}$, that is, the variable $\mathrm{x}'_i(n)$ as defined by Eq. (4.54). Thus, an alternative way of stating Property 9 is to say that *the eigenfilters associated with a process* $\mathrm{x}(n)$ *may be selected so that their output samples, at any time instant* n, *constitute a set of mutually orthogonal random variables.*

It may also be noted that by premultiplying Eq. (4.66) with $\mathbf{Q}$ and using $\mathbf{Q}\mathbf{Q}^{\mathrm{H}} = \mathbf{I}$, we obtain

$$\mathbf{x}(n) = \mathbf{Q}\mathbf{x}'(n) \tag{4.69}$$

Replacing $\mathbf{x}'(n)$ by the column vector $[\mathbf{x}'_0(n)\ \mathbf{x}'_1(n)\ \cdots\ \mathbf{x}'_{N-1}(n)]^{\mathrm{T}}$, and expanding Eq. (4.69) in terms of the elements of $\mathbf{x}'(n)$ and columns of $\mathbf{Q}$, we get

$$\mathbf{x}(n) = \sum_{i=0}^{N-1} \mathbf{x}'_i(n)\mathbf{q}_i. \tag{4.70}$$

This is known as *Karhunen–Loéve expansion.*

Example 4.1

Consider a stationary random process $\{x(n)\}$ that is generated by passing a real-valued stationary zero-mean unit-variance white noise process $\{\nu(n)\}$ through a system with the system function

$$H(z) = \frac{\sqrt{1-\alpha^2}}{1-\alpha z^{-1}} \tag{4.71}$$

where α is a real-valued constant in the range of -1 to $+1$. We want to verify some of the results developed above for the process $\{x(n)\}$.

We note that for the unit-variance white noise process $\{\nu(n)\}$

$$\Phi_{\nu\nu}(z) = 1.$$

[2] This is not necessary for the above property to hold. However, it is a useful assumption as it simplifies our discussion.

Also, using Eq. (2.80) of Chapter 2 and noting that α is real-valued, we obtain

$$\Phi_{xx}(z) = H(z)H(z^{-1})\Phi_{\nu\nu}(z) = \frac{1-\alpha^2}{(1-\alpha z^{-1})(1-\alpha z)}. \quad (4.72)$$

Taking inverse z-transform, we get

$$\phi_{xx}(k) = \alpha^{|k|}, \quad \text{for } k = \ldots -2, -1, 0, 1, 2, \ldots \quad (4.73)$$

Using this result, we find that the correlation matrix of an N-tap transversal filter with input $\{x(n)\}$ is

$$\mathbf{R} = \begin{bmatrix} 1 & \alpha & \alpha^2 & \cdots & \alpha^{N-1} \\ \alpha & 1 & \alpha & \cdots & \alpha^{N-2} \\ \vdots & \vdots & \vdots & \ddots & \vdots \\ \alpha^{N-1} & \alpha^{N-2} & \alpha^{N-3} & \cdots & 1 \end{bmatrix}. \quad (4.74)$$

Next, we present some numerical results, which demonstrate the relationships between the power spectral density of the process $\{x(n)\}$, $\Phi_{xx}(e^{j\omega})$, and its corresponding correlation matrix.

Figure 4.1 shows a set of the plots of $\Phi_{xx}(e^{j\omega})$ for values of $\alpha = 0$, 0.5, and 0.75. We note that $\alpha = 0$ corresponds to the case where $\{x(n)\}$ is white and, therefore, its power spectral density is flat. As α increases from 0 to 1, $\{x(n)\}$ becomes more colored and for values of α close to 1, most of its energy is concentrated around $\omega = 0$.

From Property 8, we recall that the eigenvalues of the correlation matrix $\mathbf{R}$ are bounded by the minimum and maximum values of $\Phi_{xx}(e^{j\omega})$. To illustrate this, in Figure 4.2a, b, and c, we have plotted the minimum and maximum eigenvalues of $\mathbf{R}$ for values of $\alpha = 0.5$, 0.75, and 0.9, as N varies from 2 to 20. It may be noted that the limits predicted by the minimum and maximum values of $\Phi_{xx}(e^{j\omega})$ are achieved asymptotically as N increases.

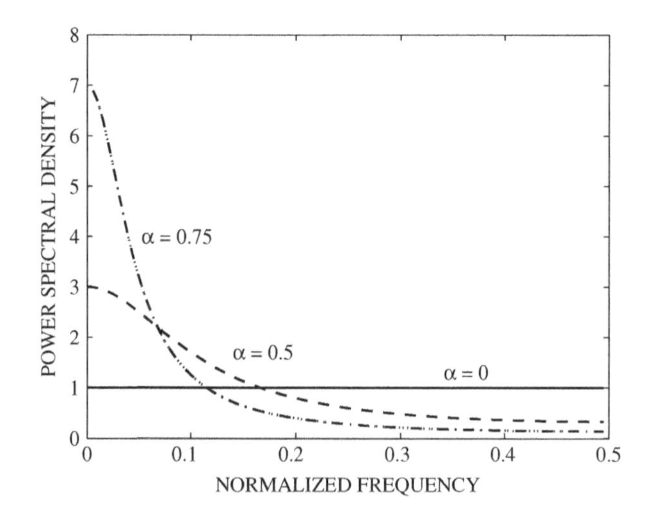

Figure 4.1 Power spectral density of $\{x(n)\}$ for different values of the parameter α.

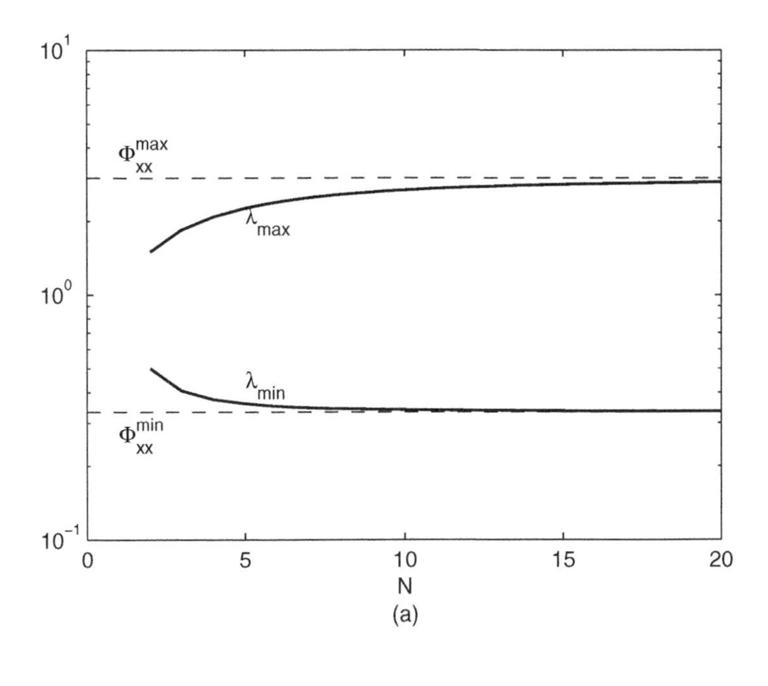

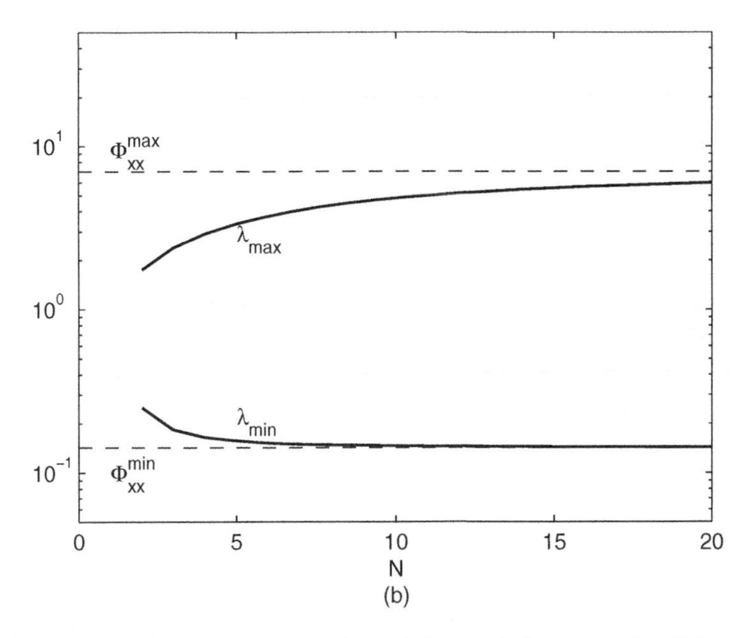

Figure 4.2 Minimum and maximum eigenvalues of the correlation matrix for different values of the parameter α: (a) $\alpha = 0.5$, (b) $\alpha = 0.75$, and (c) $\alpha = 0.9$.

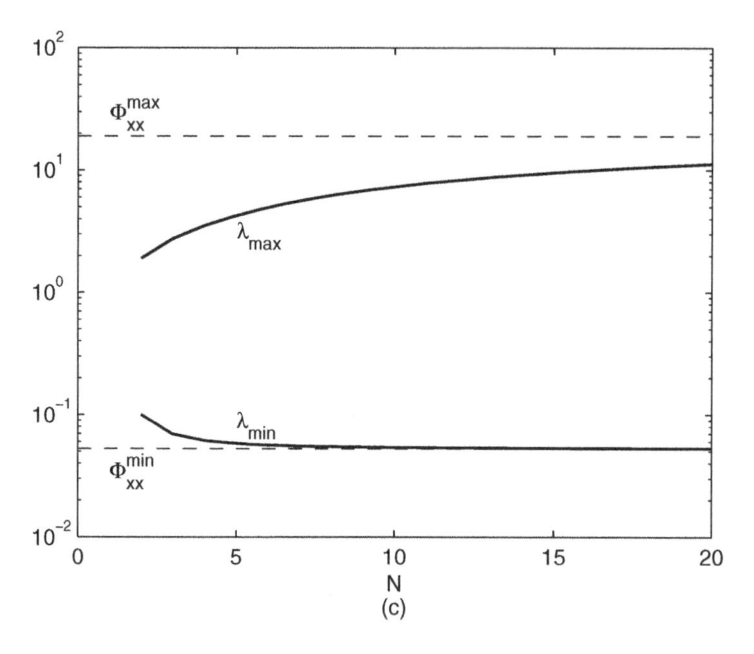

Figure 4.2 Continued.

However, for values of α close to 1, such limits are approached only when N is very large. This may be explained using the concept of eigenfilters. We note that when α is close to 1, the peak of the power spectral density function $\Phi_{xx}(e^{j\omega})$ is very narrow; see the case of $\alpha = 0.75$ in Figure 4.1. To pick up this peak accurately, an eigenfilter with a very narrow pass-band (i.e., high selectivity) is required. On the other hand, a narrow band filter can be realized only if the filter length, N, is selected long enough.

Example 4.2

Consider the case where the input process, $\{x(n)\}$, to an N-tap transversal filter consists of the summation of a zero-mean white noise process, $\{\nu(n)\}$, and a complex sinusoid, $\{e^{j(\omega_o n+\theta)}\}$, where θ is an initial random phase, which varies for different realizations of the process. The correlation matrix of $\{x(n)\}$ is

$$\mathbf{R} = \sigma_\nu^2 \mathbf{I} + \begin{bmatrix} 1 & e^{j\omega_o} & \cdots & e^{j(N-1)\omega_o} \\ e^{-j\omega_o} & 1 & \cdots & e^{j(N-2)\omega_o} \\ \vdots & \vdots & \ddots & \vdots \\ e^{-j(N-1)\omega_o} & e^{-j(N-2)\omega_o} & \cdots & 1 \end{bmatrix} \tag{4.75}$$

where the first term on the right-hand side is the correlation matrix of the white noise process and the second term is that of the sinusoidal process. We are interested in finding the eigenvalues and eigenvectors of $\mathbf{R}$. These are conveniently obtained through the minimax theorem and the concept of eigenfilters.

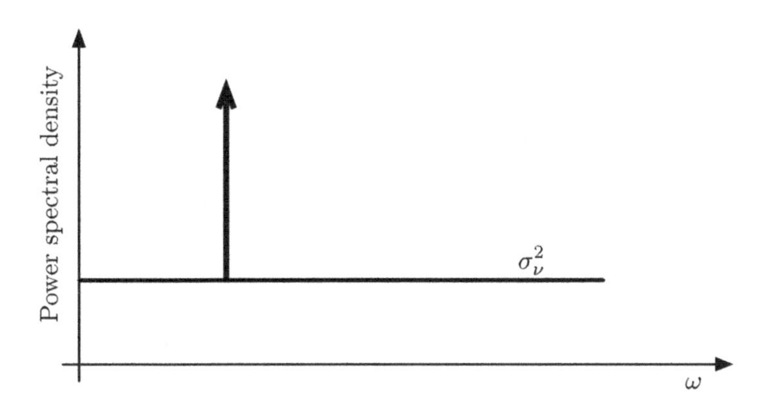

Figure 4.3 Power spectral density of the process $\{x(n)\}$ consisting of a white noise plus a single tone sinusoidal signal.

Figure 4.3 shows the power spectral density of the process $\{x(n)\}$. It consists of a flat level, which is contributed by $\{\nu(n)\}$ and an impulse at $\omega = \omega_o$ due to the sinusoidal part of $\{x(n)\}$. The eigenfilter, which picks up maximum energy of the input, is the one that is matched to the sinusoidal part of the input. The coefficients of this filter are the elements of the eigenfilter

$$\mathbf{q}_0 = \frac{1}{\sqrt{N}}[1 \ \ e^{-j\omega_o} \ldots e^{-j(N-1)\omega_o}]^{\mathrm{T}} \tag{4.76}$$

The factor $1 \,/\, \sqrt{N}$ in Eq. (4.76) is to normalize $\mathbf{q}_0$ to the length of unity. The vector $\mathbf{q}_0$ can easily be confirmed to be an eigenvector of $\mathbf{R}$ by evaluating $\mathbf{Rq}_0$ and noting that this gives

$$\mathbf{Rq}_0 = (\sigma_\nu^2 + N)\mathbf{q}_0 \tag{4.77}$$

This also shows that the eigenvalue corresponding to the eigenvector $\mathbf{q}_0$ is

$$\lambda_0 = \sigma_\nu^2 + N \tag{4.78}$$

Also, from the minimax theorem, we note that the rest of eigenvectors of $\mathbf{R}$ have to be orthogonal to $\mathbf{q}_0$, that is,

$$\mathbf{q}_i^{\mathrm{H}}\mathbf{q}_0 = 0, \quad \text{for } i = 1, 2, \ldots, N-1 \tag{4.79}$$

Using this, it is not difficult (Problem P4.7) to show that

$$\mathbf{Rq}_i = \sigma_\nu^2\mathbf{q}_i, \quad \text{for } i = 1, 2, \ldots, N-1 \tag{4.80}$$

This result shows that as long as Eq. (4.79) holds, the eigenvectors $\mathbf{q}_1, \mathbf{q}_2, \ldots, \mathbf{q}_{N-1}$ of $\mathbf{R}$ are arbitrary. In other words, any set of vectors, which belongs to the subspace orthogonal to the eigenvector $\mathbf{q}_0$, makes an acceptable set for the rest of the eigenvectors of $\mathbf{R}$. Furthermore, the eigenvalues corresponding to these eigenvectors are all equal to σ_ν^2.

4.3 Performance Surface

With the background developed so far, we are now ready to proceed with exploring the performance surface of transversal Wiener filters. We start with the case where the filter coefficients, input, and desired output are real-valued. The results will then be extended to the complex-valued case.

We recall from Chapter 3 that the performance function of a transversal Wiener filter with a real-valued input sequence $x(n)$ and a desired output sequence $d(n)$ is

$$\xi = \mathbf{w}^{\mathrm{T}}\mathbf{R}\mathbf{w} - 2\mathbf{p}^{\mathrm{T}}\mathbf{w} + E[d^2(n)] \tag{4.81}$$

where the superscript T denotes vector or matrix transpose, $\mathbf{w} = [w_0\ w_1 \cdots\ w_{N-1}]^{\mathrm{T}}$ is the filter tap-weight vector, $\mathbf{R} = E[\mathbf{x}(n)\mathbf{x}^{\mathrm{T}}(n)]$ is the correlation matrix of the filter tap-input vector $\mathbf{x}(n) = [x(n)\ \ x(n-1) \cdots x(n-N+1)]^{\mathrm{T}}$, and $\mathbf{p} = E[d(n)\mathbf{x}(n)]$ is the cross-correlation vector between $d(n)$ and $\mathbf{x}(n)$. We want to study the shape of the performance function ξ when it is viewed as a surface in the $(N+1)$-dimensional Euclidean space constituted by the filter tap weights w_i, $i = 0, 1, \ldots, N-1$, and the performance function, ξ.

Also, we recall that the optimum value of the Wiener filter tap-weight vector is obtained from the Wiener-Hopf equation

$$\mathbf{R}\mathbf{w}_{\mathrm{o}} = \mathbf{p} \tag{4.82}$$

The performance function ξ may be rearranged as follows:

$$\xi = \mathbf{w}^{\mathrm{T}}\mathbf{R}\mathbf{w} - \mathbf{w}^{\mathrm{T}}\mathbf{p} - \mathbf{p}^{\mathrm{T}}\mathbf{w} + E[d^2(n)] \tag{4.83}$$

where we have noted that $\mathbf{w}^{\mathrm{T}}\mathbf{p} = \mathbf{p}^{\mathrm{T}}\mathbf{w}$. Next, we substitute for $\mathbf{p}$ in Eq. (4.83) from Eq. (4.82) and add and subtract the term $\mathbf{w}_{\mathrm{o}}^{\mathrm{T}}\mathbf{R}\mathbf{w}_{\mathrm{o}}$ to obtain

$$\xi = \mathbf{w}^{\mathrm{T}}\mathbf{R}\mathbf{w} - \mathbf{w}^{\mathrm{T}}\mathbf{R}\mathbf{w}_{\mathrm{o}} - \mathbf{w}_{\mathrm{o}}^{\mathrm{T}}\mathbf{R}^{\mathrm{T}}\mathbf{w} + \mathbf{w}_{\mathrm{o}}^{\mathrm{T}}\mathbf{R}\mathbf{w}_{\mathrm{o}} + E[d^2(n)] - \mathbf{w}_{\mathrm{o}}^{\mathrm{T}}\mathbf{R}\mathbf{w}_{\mathrm{o}} \tag{4.84}$$

As $\mathbf{R}^{\mathrm{T}} = \mathbf{R}$, the first four terms on the right-hand side of Eq. (4.84) can be combined to obtain

$$\xi = (\mathbf{w} - \mathbf{w}_{\mathrm{o}})^{\mathrm{T}}\mathbf{R}(\mathbf{w} - \mathbf{w}_{\mathrm{o}}) + E[d^2(n)] - \mathbf{w}_{\mathrm{o}}^{\mathrm{T}}\mathbf{R}\mathbf{w}_{\mathrm{o}} \tag{4.85}$$

We may also recall from Chapter 3 that

$$\xi_{\min} = E[d^2(n)] - \mathbf{w}_{\mathrm{o}}^{\mathrm{T}}\mathbf{R}\mathbf{w}_{\mathrm{o}} \tag{4.86}$$

where $\xi_{\min}$ is the minimum value of ξ, which is obtained when $\mathbf{w} = \mathbf{w}_{\mathrm{o}}$. Substituting Eq. (4.86) in Eq. (4.85), we get

$$\xi = \xi_{\min} + (\mathbf{w} - \mathbf{w}_{\mathrm{o}})^{\mathrm{T}}\mathbf{R}(\mathbf{w} - \mathbf{w}_{\mathrm{o}}) \tag{4.87}$$

This result has the following interpretation. The nonnegative definiteness of the correlation matrix $\mathbf{R}$ implies that the second term on the right-hand side of Eq. (4.87) is nonnegative. When $\mathbf{R}$ is positive definite (a case very likely to happen in practice), the second term on the right-hand side of Eq. (4.87) is zero only when $\mathbf{w} = \mathbf{w}_{\mathrm{o}}$, and in that case ξ coincides with its minimum value. This is depicted in Figure 4.4 where a typical performance

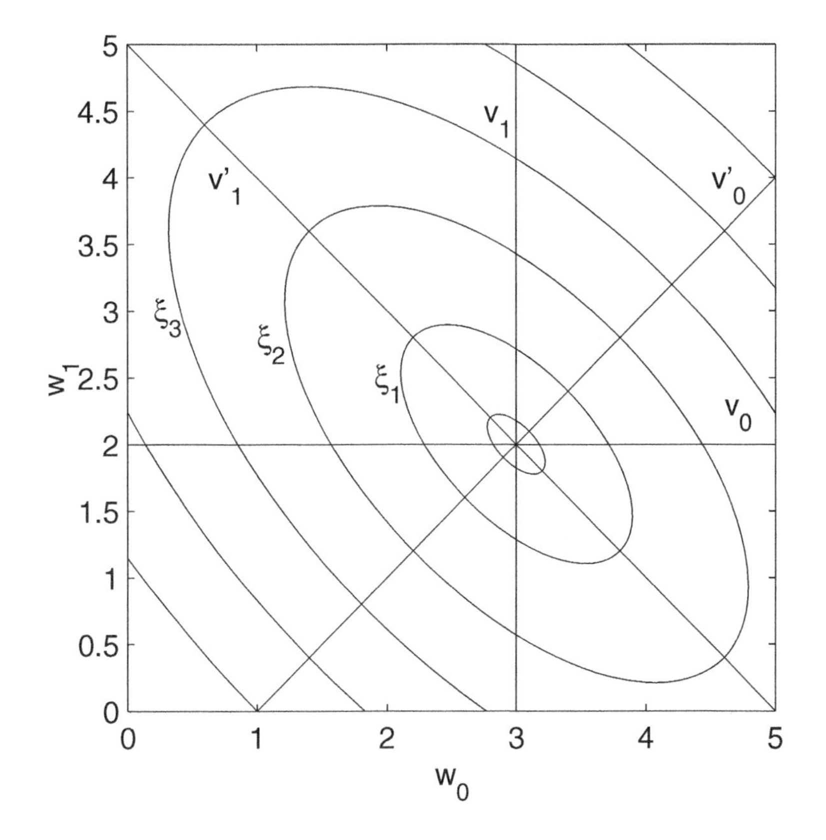

Figure 4.4 A typical performance surface of a two-tap transversal filter.

surface of a two-tap Wiener filter is presented by a set of contours, which correspond to different levels of ξ, and

$$\xi_{\min} < \xi_1 < \xi_2 < \cdots$$

To proceed further, we define the vector

$$\mathbf{v} \triangleq \mathbf{w} - \mathbf{w}_o \tag{4.88}$$

and substitute it in Eq. (4.87) to obtain

$$\xi = \xi_{\min} + \mathbf{v}^{\mathrm{T}}\mathbf{R}\mathbf{v} \tag{4.89}$$

This simpler form of the performance function in effect is equivalent to shifting the origin of the N-dimensional Euclidean space defined by the elements of $\mathbf{w}$ to the point $\mathbf{w} = \mathbf{w}_o$. The new Euclidean space has a new set of axes given by $v_0, v_1, \ldots, v_{N-1}$ (Figure 4.4). These are in parallel with the original axes $w_0, w_1, \ldots, w_{N-1}$. Obviously, the shape of the performance surface is not affected by the shift in the origin.

To simplify Eq. (4.89) further, we use the unitary similarity transformation, that is, Eq. (4.19) of the last section, which, for real-valued signals, is written as

$$\mathbf{R} = \mathbf{Q}\boldsymbol{\Lambda}\mathbf{Q}^{\mathrm{T}} \tag{4.90}$$

Substituting Eq. (4.90) in Eq. (4.89), we obtain

$$\xi = \xi_{\min} + \mathbf{v}^{\mathrm{T}}\mathbf{Q}\boldsymbol{\Lambda}\mathbf{Q}^{\mathrm{T}}\mathbf{v} \tag{4.91}$$

We define

$$\mathbf{v}' = \mathbf{Q}^{\mathrm{T}}\mathbf{v} \tag{4.92}$$

and note that multiplication of the vector $\mathbf{v}$ by the unitary matrix $\mathbf{Q}^{\mathrm{T}}$ is equivalent to rotating the v-axes to a new set of axes given by $v'_0, v'_1, \ldots, v'_{N-1}$, as depicted in Figure 4.4. The new axes are in the directions specified by the rows of the transformation matrix $\mathbf{Q}^{\mathrm{T}}$. We may further note that the rows of $\mathbf{Q}^{\mathrm{T}}$ are the eigenvectors of the correlation matrix $\mathbf{R}$. This means that the v'-axes, defined by Eq. (4.92), are in the directions of the basis vectors specified by the eigenvectors of $\mathbf{R}$.

Substituting Eq. (4.92) in Eq. (4.91), we obtain

$$\xi = \xi_{\min} + \mathbf{v}'^{\mathrm{T}}\boldsymbol{\Lambda}\mathbf{v}' \tag{4.93}$$

This is known as the *canonical form* of the performance function. Expanding Eq. (4.93) in terms of the elements of the vector $\mathbf{v}'$ and the diagonal elements of the matrix $\boldsymbol{\Lambda}$, we get

$$\xi = \xi_{\min} + \sum_{i=0}^{N-1} \lambda_i v_i'^2 \tag{4.94}$$

This, when compared with the previous forms of the performance function in Eqs. (4.81) and (4.89), is a much easier function to visualize. In particular, if all the variables v'_0, $v'_1, \ldots, v'_{N-1}$, except v'_k, are set to zero,

$$\xi = \xi_{\min} + \lambda_k v_k'^2 \tag{4.95}$$

This is a parabola whose minimum occur at $v'_k = 0$. The parameter λ_k determines the shape of the parabola, in the sense that for smaller values of λ_k the resulting parabolas are wider (flatter in shape) when compared with those obtained for larger values of λ_k. This is demonstrated in Figure 4.5 where ξ, as a function of v'_k, is plotted for a few values of λ_k.

When all variables $v'_0, v'_1, \ldots, v'_{N-1}$ are varied simultaneously, the performance function ξ, in the $(N+1)$-dimensional Euclidean space, is a hyperparabola. The path traced by ξ as one moves along any of the axes $v'_0, v'_1, \ldots, v'_{N-1}$ is a parabola whose shape is determined by the corresponding eigenvalue.

The hyperparabola shape of the performance surface can be best understood in the case of a two-tap filter when the performance surface can easily be visualized in the three-dimensional Euclidean space whose axes are the two independent taps of the filter and the function ξ; see Figure 3.4 as an example. Alternatively, the contour plots, such as those presented in Figure 4.4, may be used to visualize the performance surface in a very convenient way.

For $N = 2$, the canonical form of the performance function is

$$\xi = \xi_{\min} + \lambda_0 v_0'^2 + \lambda_1 v_1'^2 \tag{4.96}$$

This may be rearranged as

$$\left(\frac{v'_0}{a_0}\right)^2 + \left(\frac{v'_1}{a_1}\right)^2 = 1 \tag{4.97}$$

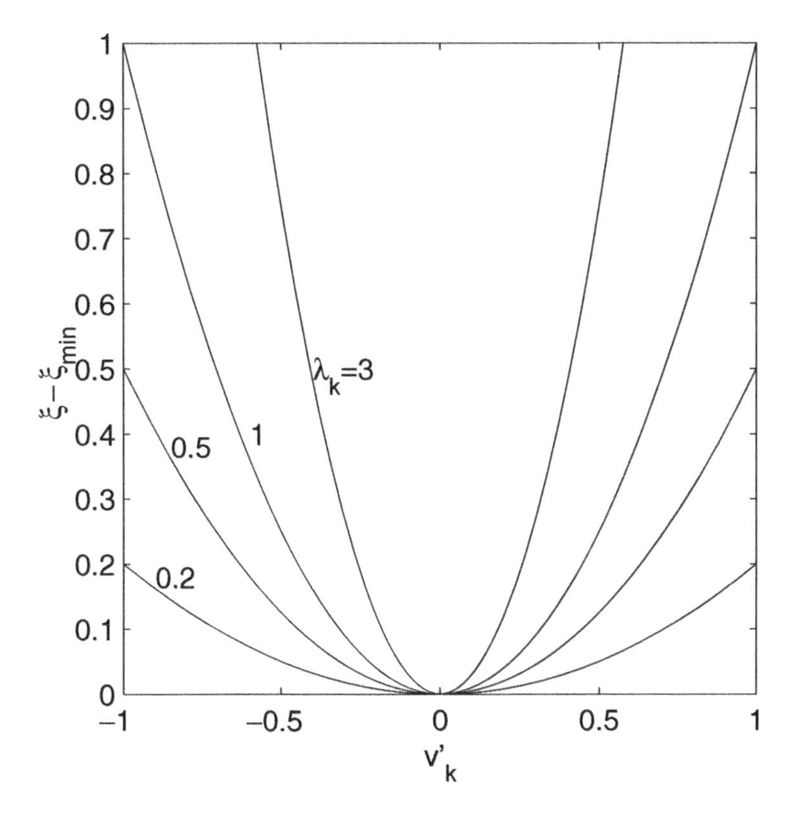

Figure 4.5 The effect of eigenvalues on the shape of the performance function when only one of the filter tap weights is varied.

where

$$a_0 = \sqrt{\frac{\xi - \xi_{\min}}{\lambda_0}} \tag{4.98}$$

and

$$a_1 = \sqrt{\frac{\xi - \xi_{\min}}{\lambda_1}} \tag{4.99}$$

Equation (4.97) represents an ellipse whose principal axes are along v_0' and v_1' axes, and for $a_1 > a_0$, the lengths of its major and minor principal axes are $2a_1$ and $2a_0$, respectively. These are highlighted in Figure 4.6, where a typical plot of the ellipse defined by Eq. (4.97) is presented. We may also note that $a_1 \ / \ a_0 = \sqrt{\lambda_0 \ / \ \lambda_1}$. This implies that for a particular performance surface, the aspect ratio of the contour ellipses is fixed and is equal to the square root of the ratio of its eigenvalues. In other words, the eccentricity of the contour ellipses of a performance surface is determined by the ratio of the eigenvalues of the corresponding correlation matrix. A larger ratio of the eigenvalues results in more eccentric ellipses and, thus, a narrower bowl-shape performance surface.

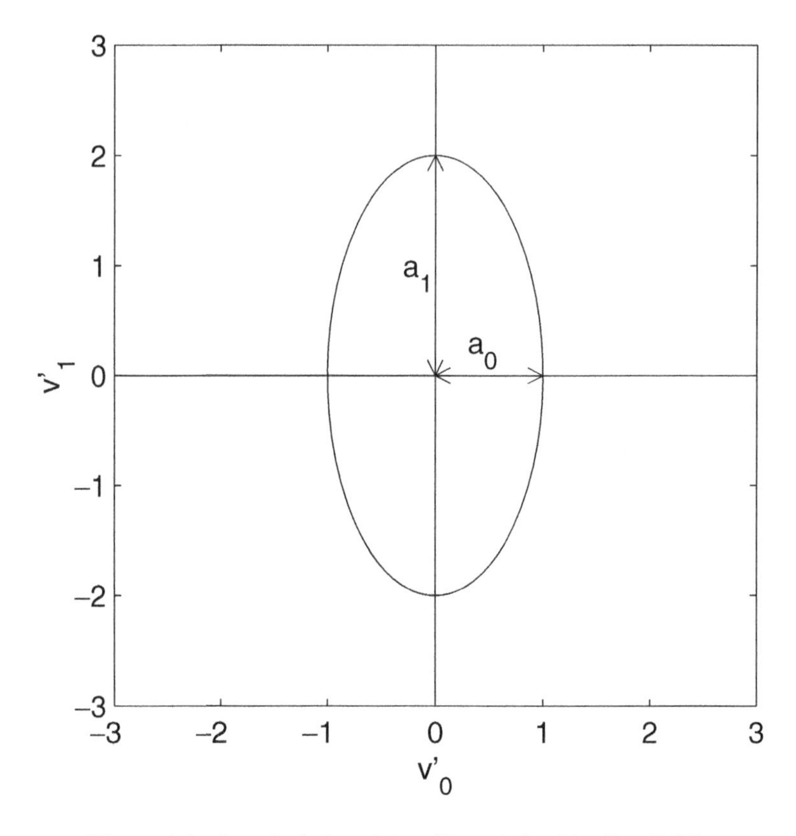

Figure 4.6 A typical plot of the ellipse defined by Eq. (4.97).

Example 4.3

Consider the case where a two-tap transversal Wiener filter is characterized by the following parameters:

$$\mathbf{R} = \begin{bmatrix} 1 & \alpha \\ \alpha & 1 \end{bmatrix}, \quad \mathbf{p} = \begin{bmatrix} 1 \\ 1 \end{bmatrix}, \quad \text{and } E[d^2(n)] = 2$$

We want to explore the performance surface of this filter for values of α ranging from 0 to 1.

The performance function of the filter is obtained by substituting the above parameters in Eq. (4.81). This gives

$$\xi = [w_0 \ \ w_1]\begin{bmatrix} 1 & \alpha \\ \alpha & 1 \end{bmatrix}\begin{bmatrix} w_0 \\ w_1 \end{bmatrix} - 2[1 \ \ 1]\begin{bmatrix} w_0 \\ w_1 \end{bmatrix} + 2 \tag{4.100}$$

Solving the Wiener–Hopf equation to obtain the optimum tap weights of the filter, we obtain

$$\begin{bmatrix} w_{o,0} \\ w_{o,1} \end{bmatrix} = \mathbf{R}^{-1}\mathbf{p} = \begin{bmatrix} 1 & \alpha \\ \alpha & 1 \end{bmatrix}^{-1}\begin{bmatrix} 1 \\ 1 \end{bmatrix} = \begin{bmatrix} \frac{1}{1+\alpha} \\ \frac{1}{1+\alpha} \end{bmatrix} \tag{4.101}$$

Using this result, we get

$$\begin{aligned}\xi_{\min} &= E[d^2(n)] - \mathbf{w}_o^T \mathbf{p} \\ &= 2 - \begin{bmatrix} \frac{1}{1+\alpha} & \frac{1}{1+\alpha} \end{bmatrix} \begin{bmatrix} 1 \\ 1 \end{bmatrix} = \frac{2\alpha}{1+\alpha}\end{aligned} \tag{4.102}$$

Also,

$$\begin{aligned}\xi &= \xi_{\min} + (\mathbf{w} - \mathbf{w}_o)^T \mathbf{R} (\mathbf{w} - \mathbf{w}_o) \\ &= \frac{2\alpha}{1+\alpha} + [v_0 \; v_1] \begin{bmatrix} 1 & \alpha \\ \alpha & 1 \end{bmatrix} \begin{bmatrix} v_0 \\ v_1 \end{bmatrix}\end{aligned} \tag{4.103}$$

To convert this to its canonical form, we should first find the eigenvalues and eigenvectors of $\mathbf{R}$. To find the eigenvalues of $\mathbf{R}$, we should solve the characteristic equation

$$\det(\lambda \mathbf{I} - \mathbf{R}) = \begin{vmatrix} \lambda - 1 & -\alpha \\ -\alpha & \lambda - 1 \end{vmatrix} = 0 \tag{4.104}$$

Expanding Eq. (4.104), we obtain

$$(\lambda - 1)^2 - \alpha^2 = 0$$

which gives

$$\lambda_0 = 1 + \alpha \tag{4.105}$$

and

$$\lambda_1 = 1 - \alpha \tag{4.106}$$

The eigenvectors $\mathbf{q}_0 = [q_{00} \; q_{01}]^T$ and $\mathbf{q}_1 = [q_{10} \; q_{11}]^T$ of $\mathbf{R}$ are obtained by solving the equations

$$\begin{bmatrix} \lambda_0 - 1 & -\alpha \\ -\alpha & \lambda_0 - 1 \end{bmatrix} \begin{bmatrix} q_{00} \\ q_{01} \end{bmatrix} = 0 \tag{4.107}$$

and

$$\begin{bmatrix} \lambda_1 - 1 & -\alpha \\ -\alpha & \lambda_1 - 1 \end{bmatrix} \begin{bmatrix} q_{10} \\ q_{11} \end{bmatrix} = 0 \tag{4.108}$$

Substituting Eqs. (4.105) and (4.106) in Eqs. (4.107) and (4.108), respectively, we obtain

$$q_{00} = q_{01} \quad \text{and} \quad q_{10} = -q_{11}.$$

Using these results and normalizing $\mathbf{q}_0$ and $\mathbf{q}_1$ to have lengths of unity, we obtain

$$\mathbf{q}_0 = \frac{1}{\sqrt{2}} \begin{bmatrix} 1 \\ 1 \end{bmatrix} \quad \text{and} \quad \mathbf{q}_1 = \frac{1}{\sqrt{2}} \begin{bmatrix} 1 \\ -1 \end{bmatrix}$$

It may be noted that the eigenvectors $\mathbf{q}_0$ and $\mathbf{q}_1$ of $\mathbf{R}$ are independent of the parameter α. This is an interesting property of the correlation matrices of two-tap transversal filters, which implies that the v'-axes are always obtained by a 45° rotation of v-axes. The eigenvectors associated with the correlation matrices of three-tap transversal filters also have some special form. This is discussed in Problem P4.5.

With the above results, we get

$$\begin{bmatrix} v_0' \\ v_1' \end{bmatrix} = \frac{1}{\sqrt{2}} \begin{bmatrix} 1 & 1 \\ 1 & -1 \end{bmatrix} \begin{bmatrix} v_0 \\ v_1 \end{bmatrix} = \frac{1}{\sqrt{2}} \begin{bmatrix} v_0 + v_1 \\ v_0 - v_1 \end{bmatrix} \tag{4.109}$$

and

$$\xi = \frac{2\alpha}{1+\alpha} + (1+\alpha)v_0'^2 + (1-\alpha)v_1'^2 \tag{4.110}$$

Figures 4.7a, b, and c show the contour plots of the performance surface of the two-tap transversal filter for $\alpha = 0.5$, 0.8, and 0.95, which correspond to the eigenvalue ratios of 3, 9, and 39, respectively. These plots clearly show how the eccentricity of the performance surface changes as the eigenvalue ratio of the correlation matrix **R** increases.

The above results may be generalized as follows. The performance surface of an N-tap transversal filter with real-valued data is a hyperparaboloid in the $(N+1)$-dimensional Euclidean space whose axes are the N tap-weight variables of the filter and the performance function ξ. The performance function may also be represented by a set of hyperellipses in the N-dimensional Euclidean space of the filter tap-weight variables. Each hyperellipse corresponds to a fixed value of ξ. The directions of the principal axes of the hyperellipses are determined by the eigenvectors of the correlation matrix **R**. The size of the various principal axes of each hyperellipse are proportional to the square root of the inverse of the corresponding eigenvalues. Thus, the eccentricity of the hyperellipses is determined by the spread of the eigenvalues of the correlation matrix **R**. This shows that the shape of the performance surface of a Wiener FIR filter is directly related to the spread of the eigenvalues of **R**. In addition, from Property 8 of the eigenvalues and eigenvectors, we recall that the spread of the eigenvalues of the correlation matrix of a stochastic process $\{x(n)\}$ is directly linked to the variation of the power spectral density function $\Phi_{xx}(e^{j\omega})$ of the process. This, in turn, means there is a close relationship between the power spectral density of a random process and the shape of the performance surface of an FIR Wiener filter for which the latter is used as input.

The above results can easily be extended to the case where the filter coefficients, input, and desired output are complex-valued. One should only remember that the elements of all the involved vectors and matrices are complex-valued and replace all the transpose operators in the developed equations by Hermitian operators. Doing this, Eq. (4.93) becomes

$$\xi = \xi_{\min} + \mathbf{v}'^{\mathrm{H}} \boldsymbol{\Lambda} \mathbf{v}' \tag{4.111}$$

This can be expanded as

$$\xi = \xi_{\min} + \sum_{i=0}^{N-1} \lambda_i |v_i'|^2 \tag{4.112}$$

The difference between this result and its dual (for the real-valued case) in Eq. (4.94) is an additional modulus sign on v_i''s, in Eq. (4.112). This, of course, is due to the fact that here v_i''s are complex-valued.

The performance function ξ of Eq. (4.112) may be thought of as a hyperparabola in the $(N+1)$-dimensional space whose first N-axes are defined by the complex-valued variables v_i''s and its $(N+1)$th axis is the real-valued performance function ξ. To prevent

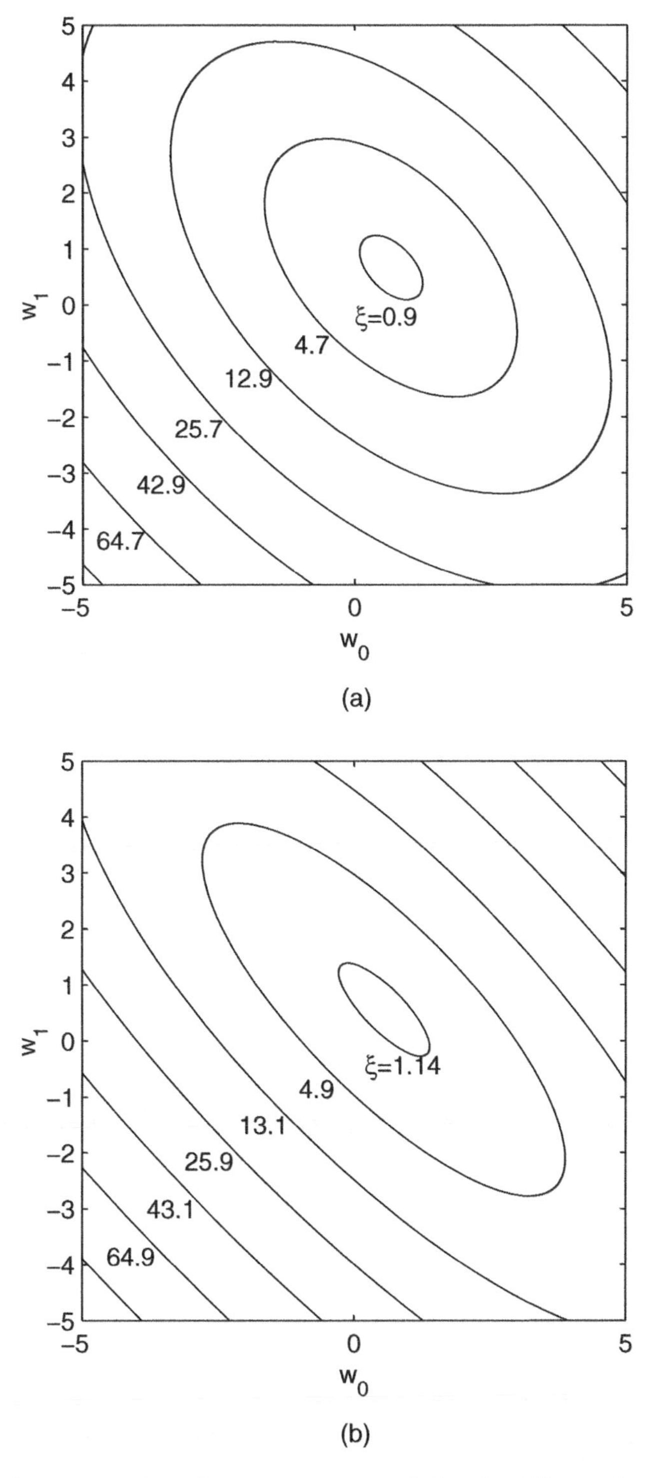

Figure 4.7 Performance surface of a two-tap transversal filter for different eigenvalue spread of **R**: (a) $\lambda_0/\lambda_1 = 3$, (b) $\lambda_0/\lambda_1 = 9$, and (c) $\lambda_0/\lambda_1 = 39$.

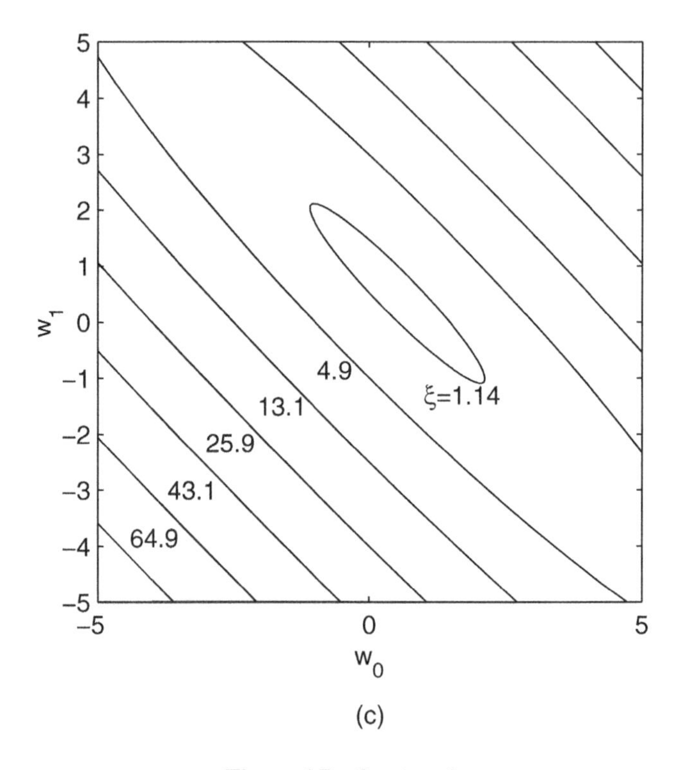

Figure 4.7 Continued.

such a mixed domain and have a clearer picture of the performance surface in the case of complex signals, we may expand Eq. (4.112) further by replacing v'_i with $v'_{i,R} + jv'_{i,I}$, where $v'_{i,R}$ and $v'_{i,I}$ are the real and imaginary parts of v'_i. With this, we obtain

$$\xi = \xi_{\min} + \sum_{i=0}^{N-1} \lambda_i (v'^2_{i,R} + v'^2_{i,I}) \tag{4.113}$$

Here, $v'_{i,R}$ and $v'_{i,I}$ are both real-valued variables. Equation (4.113) shows that *the performance surface of an N-tap transversal Wiener filter with complex-valued coefficients is a hyperparabola in the* $(2N + 1)$*-dimensional Euclidean space of the variables consisting of the real and imaginary parts of the filter coefficients and the performance function.*

Problems

P4.1 Consider the performance function

$$\xi = w_0^2 + w_1^2 + w_0 w_1 - w_0 + w_1 + 1.$$

(i) Convert this to its canonical form.

(ii) Plot the set of contour ellipses of the performance surface of ξ for values of $\xi = 1$, 2, 3, and 4.

P4.2 $\mathbf{R}$ is a correlation matrix.

(i) Using the unitary similarity transformation, show that for any integer n

$$\mathbf{R}^n = \mathbf{Q}\boldsymbol{\Lambda}^n\mathbf{Q}^{\mathrm{H}}.$$

(ii) The matrix $\mathbf{R}^{1/2}$ with the property $\mathbf{R}^{1/2}\mathbf{R}^{1/2} = \mathbf{R}$ is defined as the square root of $\mathbf{R}$. Show that

$$\mathbf{R}^{1/2} = \mathbf{Q}\boldsymbol{\Lambda}^{1/2}\mathbf{Q}^{\mathrm{H}}.$$

(iii) Show that the identity

$$\mathbf{R}^a = \mathbf{Q}\boldsymbol{\Lambda}^a\mathbf{Q}^{\mathrm{H}}$$

is valid for any rational number a.

P4.3 Consider the correlation matrix $\mathbf{R}$ of an N-by-1 observation vector $\mathbf{x}(n)$, and an arbitrary N-by-N unitary transformation matrix $\mathbf{U}$. Define the vector

$$\mathbf{x}_U(n) = \mathbf{U}\mathbf{x}(n)$$

and its corresponding correlation matrix $\mathbf{R}_U = E[\mathbf{x}_U(n)\mathbf{x}_U^{\mathrm{H}}(n)]$.

(i) Show that $\mathbf{R}$ and $\mathbf{R}_U$ share the same set of eigenvalues.
(ii) Find an expression for the eigenvectors of $\mathbf{R}_U$ in terms of the eigenvectors of $\mathbf{R}$ and the transformation matrix $\mathbf{U}$.

P4.4 In Example 4.3, we noted that the eigenvectors of the correlation matrix of any two-tap transversal filter with real-valued input are fixed and are

$$\mathbf{q}_0 = \frac{1}{\sqrt{2}}\begin{bmatrix}1\\1\end{bmatrix}$$

and

$$\mathbf{q}_1 = \frac{1}{\sqrt{2}}\begin{bmatrix}1\\-1\end{bmatrix}$$

Plot the magnitude responses of the eigenfilters defined by $\mathbf{q}_0$ and $\mathbf{q}_1$ and verify that $\mathbf{q}_0$ corresponds to a low-pass filter and $\mathbf{q}_1$ corresponds to a high-pass one. How do you relate this observation with the minimax theorem?

P4.5 Consider the correlation matrix $\mathbf{R}$ of a three-tap transversal filter with a real-valued input $x(n)$.

(i) Show that when $E[x^2(n)] = 1$, $\mathbf{R}$ has the form

$$\mathbf{R} = \begin{bmatrix}1 & \rho_1 & \rho_2\\ \rho_1 & 1 & \rho_1\\ \rho_2 & \rho_1 & 1\end{bmatrix}.$$

(ii) Show that

$$\mathbf{q}_0 = \frac{1}{\sqrt{2}}\begin{bmatrix}1\\0\\-1\end{bmatrix}$$

is an eigenvector of $\mathbf{R}$ and find its corresponding eigenvalue.

(iii) Show that the other eigenvectors of **R** are

$$\mathbf{q}_i = \frac{1}{\sqrt{2+\alpha_i^2}}\begin{bmatrix} 1 \\ \alpha_i \\ 1 \end{bmatrix}, \qquad \text{for } i = 1, 2$$

where

$$\alpha_1 = \frac{-\rho_2/\rho_1 + \sqrt{(\rho_2/\rho_1)^2 + 8}}{2} \quad \text{and} \quad \alpha_2 = \frac{-\rho_2/\rho_1 - \sqrt{(\rho_2/\rho_1)^2 + 8}}{2}.$$

Find the eigenvalues that correspond to $\mathbf{q}_1$ and $\mathbf{q}_2$.

(iv) For the following numerical values, plot the magnitude responses of the eigenfilters defined by $\mathbf{q}_0$, $\mathbf{q}_1$, and $\mathbf{q}_2$, and find that in all cases these correspond to a band-pass, a low-pass, and a high-pass filter, respectively.

(a) $\rho_1 = 0.5$, $\rho_2 = 0.25$.
(b) $\rho_1 = 0.8$, $\rho_2 = 0.3$.
(c) $\rho_1 = 0.9$, $\rho_2 = -0.4$.

How do you relate this observation to the minimax theorem?

P4.6 Consider the correlation matrix **R** of an observation vector $\mathbf{x}(n)$. Define the vector

$$\tilde{\mathbf{x}}(n) = \mathbf{R}^{-1/2}\mathbf{x}(n)$$

where $\mathbf{R}^{-1/2}$ is the inverse of $\mathbf{R}^{1/2}$, and $\mathbf{R}^{1/2}$ is defined as in Problem P4.2. Show that the correlation matrix of $\tilde{\mathbf{x}}(n)$ is the identity matrix.

P4.7 Consider the case discussed in Example 4.2, and the eigenvector $\mathbf{q}_0$ as defined by Eq. (4.76). Show that any vector $\mathbf{q}_i$, which is orthogonal to $\mathbf{q}_0$ (i.e., $\mathbf{q}_i^{\mathrm{H}}\mathbf{q}_0 = 0$), is a solution to the equation

$$\mathbf{R}\mathbf{q}_i = \sigma_\nu^2 \mathbf{q}_i.$$

P4.8 The determinant of an N-by-N matrix **A** can be obtained by iterating the equation

$$\det(\mathbf{A}) = \sum_{j=0}^{N-1} (-1)^j a_{0j} \mathrm{cof}_{ij}(\mathbf{A})$$

where a_{ij} is the ijth element of **A**, and $\mathrm{cof}_{ij}(\mathbf{A})$ denotes the ijth cofactor of **A**, which is defined as

$$\mathrm{cof}_{ij}(\mathbf{A}) = (-1)^{i+j}\det(\mathbf{A}_{ij})$$

where $\mathbf{A}_{ij}$ is the $(N-1)$-by-$(N-1)$ matrix obtained by deleting the ith row and jth column of **A**. This procedure is general and applicable to all square matrices. Use this procedure to show that

(i) Equation (4.24) is a valid result for any arbitrary square matrix **A**.
(ii) For any square matrix **A**

$$\det(\mathbf{A}) = \prod_{i=0}^{N-1} \lambda_i$$

where λ_is are the eigenvalues of **A**.

P4.9 Give a proof for the minimax procedure suggested by Eqs. (4.31)–(4.33).

P4.10 Consider a filter whose input is the vector $\tilde{\mathbf{x}}(n)$, as defined in Problem P4.6, and its output is $y(n) = \tilde{\mathbf{w}}^{\mathrm{T}}\tilde{\mathbf{x}}(n)$, where $\tilde{\mathbf{w}}$ is N-by-1 tap-weight vector of the filter. Discuss on the shape of the performance surface of this filter.

P4.11 Workout the details of the derivation of Eq. (4.70).

P4.12 Give a detailed derivation of Eq. (4.111).

P4.13 The input process to an N-tap transversal filter is

$$x(n) = a_1 \mathrm{e}^{j\omega_1 n} + a_2 \mathrm{e}^{j\omega_2 n} + \nu(n)$$

where a_1 and a_2 are uncorrelated complex-valued zero-mean random variables with variances σ_1^2 and σ_2^2, respectively, and $\{\nu(n)\}$ is a white noise process with variance unity.

(i) Derive an equation for the correlation matrix, $\mathbf{R}$, of the observation vector at the filter input.

(ii) Following an argument similar to the one in Example 4.2, show that the smallest $N-2$ eigenvalues of $\mathbf{R}$ are all equal to σ_ν^2.

(iii) Let

$$\mathbf{u}_0 = \frac{1}{\sqrt{N}}[1 \;\; \mathrm{e}^{-j\omega_1} \;\; \mathrm{e}^{-j2\omega_1} \;\; \cdots \;\; \mathrm{e}^{-j(N-1)\omega_1}]^{\mathrm{T}}$$

and

$$\mathbf{u}_1 = \frac{1}{\sqrt{N}}[1 \;\; \mathrm{e}^{-j\omega_2} \;\; \mathrm{e}^{-j2\omega_2} \;\; \cdots \;\; \mathrm{e}^{-j(N-1)\omega_2}]^{\mathrm{T}}.$$

Show that the eigenvectors corresponding to the largest two eigenvalues of $\mathbf{R}$ are

$$\mathbf{q}_0 = \alpha_{00}\mathbf{u}_0 + \alpha_{01}\mathbf{u}_1$$

and

$$\mathbf{q}_1 = \alpha_{10}\mathbf{u}_0 + \alpha_{11}\mathbf{u}_1$$

where α_{00}, α_{01}, α_{10}, and α_{11} are a set coefficients to be found. Propose a minimax procedure for finding these coefficients.

(iv) Find the coefficients α_{00}, α_{01}, α_{10}, and α_{11} of Part (iii) in the case where $\mathbf{u}_0^{\mathrm{H}}\mathbf{u}_1 = 0$. Discuss on the uniqueness of the answer in the cases where $\sigma_1^2 \neq \sigma_2^2$ and $\sigma_1^2 = \sigma_2^2$.

P4.14 Equation (4.113) suggests that the performance surface of an N-tap FIR Wiener filter with complex-valued input is equivalent to the performance surface of a $2N$-tap filter with real-valued input. Furthermore, the eigenvalues corresponding to the latter surface appear with multiplicity of at least two. This problem suggests an alternative procedure, which also leads to the same results.

(i) Show that the Hermitian form $\mathbf{w}^{\mathrm{H}}\mathbf{R}\mathbf{w}$ may be expanded as

$$\mathbf{w}^{\mathrm{H}}\mathbf{R}\mathbf{w} = [\mathbf{w}_R^{\mathrm{T}} \;\; \mathbf{w}_I^{\mathrm{T}}]\begin{bmatrix} \mathbf{R}_R & -\mathbf{R}_I \\ \mathbf{R}_I & \mathbf{R}_R \end{bmatrix}\begin{bmatrix} \mathbf{w}_R \\ \mathbf{w}_I \end{bmatrix}$$

where the subscripts R and I refer to real and imaginary parts.

Hint: Note that $\mathbf{R}_I^{\mathrm{T}} = -\mathbf{R}_I$ and this implies that for any arbitrary vector $\mathbf{v}$, $\mathbf{v}^{\mathrm{T}}\mathbf{R}_I\mathbf{v} = 0$.

(ii) Show that equation

$$\mathbf{R}\mathbf{q}_i = \lambda_i \mathbf{q}_i \tag{P4.14.1}$$

implies

$$\begin{bmatrix} \mathbf{R}_R & -\mathbf{R}_I \\ \mathbf{R}_I & \mathbf{R}_R \end{bmatrix} \begin{bmatrix} \mathbf{q}_{i,R} \\ \mathbf{q}_{i,I} \end{bmatrix} = \lambda_i \begin{bmatrix} \mathbf{q}_{i,R} \\ \mathbf{q}_{i,I} \end{bmatrix}.$$

Also, multiplying Eq. (P4.14.1) through by $j = \sqrt{-1}$, we get $\mathbf{R}(j\mathbf{q}_i) = \lambda_i(j\mathbf{q}_i)$. Show that this implies

$$\begin{bmatrix} \mathbf{R}_R & -\mathbf{R}_I \\ \mathbf{R}_I & \mathbf{R}_R \end{bmatrix} \begin{bmatrix} -\mathbf{q}_{i,I} \\ \mathbf{q}_{i,R} \end{bmatrix} = \lambda_i \begin{bmatrix} -\mathbf{q}_{i,I} \\ \mathbf{q}_{i,R} \end{bmatrix}.$$

Relate these with Eq. (4.113).

Computer-Oriented Problems

The following problems involve numerical evaluation/analysis of large matrices. MATLAB will work best for completing the solutions to these problems.

P4.15 Consider a random process $x(n) = \nu(n) + \cos(0.3\pi n + \theta)$, where $\nu(n)$ is a white process with power spectral density $\Phi_{\nu\nu}(e^{j\omega}) = \sigma_\nu^2$, and θ is a random variable uniformly distributed in the interval 0 to 2π.

(i) Find an expression for the autocorrelation coefficients $\phi_{xx}(k)$.

(ii) Using the result of (i), find an expression for the power spectral density $\Phi_{xx}(e^{j\omega})$.

(iii) Present the autocorrelation matrix $\mathbf{R}$ of $\mathbf{x}(n) = [x(n)\ x(n-1) \cdots x(n-N+1)]^{\mathrm{T}}$.

(iv) For $N = 10$ and $\sigma_\nu^2 = 0.01$, find the numerical results for the eigenvectors and eigenvalues of $\mathbf{R}$.

(v) Repeat (iv), for $N = 10$ and the choices of $\sigma_\nu^2 = 0$ and 0.1 and compare your results of the three choices of σ_ν^2. Explain any relationship that you may find and explain your findings with the theoretical results in this chapter.

(vi) Present a plot of $\Phi_{xx}(e^{j\omega})$, in a form similar to Figure 4.3, within the normalized frequency range 0 to 1.

(vii) For the numerical results evaluated in (iv) and (v), add the plots of the magnitude responses of the associated eigenfilters to the results of (vi). Make the observation that all eigenfilters, except one, have zeros at the normalized frequencies 0.15 and 0.85. Following the argument made in Example 4.2, explain this observation.

P4.16 Repeat Problem P4.15 for the case where $x(n) = \nu(n) + e^{j0.3\pi n + \theta}$ and explain differences that you observe in the results compared to those in Problem P4.15.

P4.17 Consider a random process $x(n) = \nu(n) + \cos(0.3\pi n + \theta_1) + \cos(0.5\pi n + \theta_2)$, where $\nu(n)$ is a white process with power spectral density $\Phi_{\nu\nu}(e^{j\omega}) = \sigma_\nu^2$, and θ_1 and θ_2 are two independent random variables both uniformly distributed in the interval 0 to 2π.

(i) Find an expression for the autocorrelation coefficients $\phi_{xx}(k)$.
(ii) Using the result of (i), find an expression for the power spectral density $\Phi_{xx}(e^{j\omega})$.
(iii) Present the autocorrelation matrix $\mathbf{R}$ of $\mathbf{x}(n) = [x(n)\ x(n-1) \cdots x(n-N+1)]^{\mathrm{T}}$.
(iv) For $N = 10$ and $\sigma_\nu^2 = 0.01$, find the numerical results for the eigenvectors and eigenvalues of $\mathbf{R}$.
(v) Repeat (iv), for $N = 10$ and the choices of $\sigma_\nu^2 = 0$ and 0.1 and compare your results of the three choices of σ_ν^2. Explain any relationship that you may find and explain your findings with the theoretical results in this chapter.
(vi) Present a plot of $\Phi_{xx}(e^{j\omega})$, in a form similar to Figure 4.3, within the normalized frequency range 0 to 1.
(vii) For the numerical results evaluated in (iv) and (v), add the plots of the magnitude responses of the associated eigenfilters to the results of (vi). Make the observation that all eigenfilters, except two, have zeros at the normalized frequencies 0.15, 0.25, 0.75, and 0.85. Expand the argument made in Example 4.2 to explain this observation.

P4.18 Consider the case where a random process $\{x(n)\}$ is generated as in Figure P4.18. Let $\mathbf{x}(n) = [x(n)\ x(n-1)\ \cdots\ x(n-9)]^{\mathrm{T}}$.

(i) Find and present the correlation matrix $\mathbf{R} = E[\mathbf{x}(n)\mathbf{x}^{\mathrm{H}}(n)]$.
(ii) Find and present the eigenvalues, λ_i, and eigenvectors, $\mathbf{q}_i$, of $\mathbf{R}$.
(iii) Present a plot of the power spectral density $\Phi_{xx}(e^{j\omega})$.
(iv) Add the plots of the magnitude responses of the eigenfilters of $\mathbf{R}$ to the power spectral density plot in Part (c).
(v) Make an attempt to relate the plots in Part (d) to the eigenvalues in Part (b).

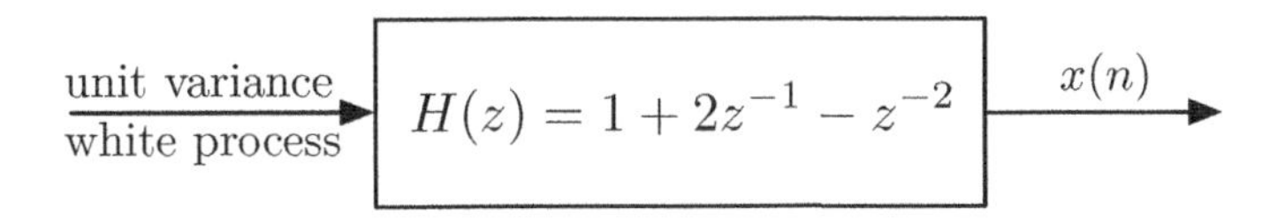

Figure P4.18

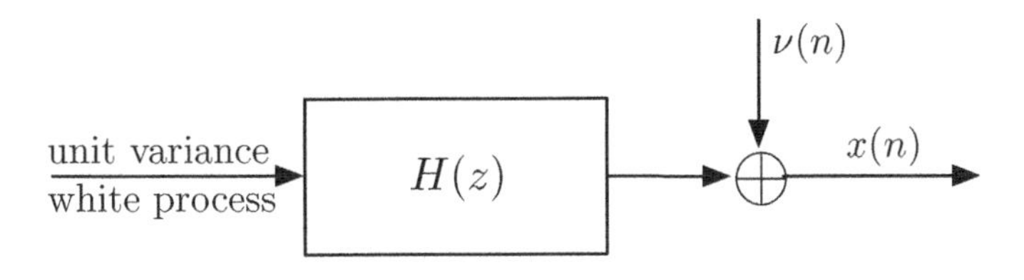

Figure P4.20

P4.19 Repeat Problem P4.18 for the case where $H(z) = 1 + 2z^{-1} + z^{-2}$.

P4.20 Repeat Problems P4.18 and P4.19 for the case where $\{x(n)\}$ is generated as in Figure P4.20. Here, $\nu(n)$ is a white random process with variance of $\sigma_\nu^2 = 0.1$. Compare the eigenvalues and eigenvectors obtained here with those in Problems P4.18 and P4.19.

5

Search Methods

In the last two chapters, we established that the optimum tap weights of a transversal Wiener filter can be obtained by solving the Wiener–Hopf equation, provided the required statistics of the underlying signals are available. We arrived at this solution by minimizing a cost function, which is a quadratic function of the filter tap-weight vector. An alternative way of finding the optimum tap weights of a transversal filter is to use an iterative search algorithm that starts at some arbitrary initial point in the tap-weight vector space and progressively moves toward the optimum tap-weight vector in steps. Each step is chosen so that the underlying cost function is reduced. If the cost function is convex (which is so for the transversal filter problem), such an iterative search procedure is guaranteed to converge to the optimum solution. The principle of finding the optimum tap-weight vector by progressive minimization of the underlying cost function by means of an iterative algorithm is central to the development of adaptive algorithms, which will be extensively discussed in the forthcoming chapters of this book. Using a highly simplified language, we might state at this point that adaptive algorithms are nothing but iterative search algorithms derived for minimizing the underlying cost function with the true statistics replaced by their estimates obtained in some manner. Hence, a very thorough understanding of the iterative algorithms from the point of view of their development and convergence property is an essential prerequisite for the study of adaptive algorithms, and this is the subject of this chapter.

In this chapter, we discuss two *gradient-based* iterative methods for searching the performance surface of a transversal Wiener filter to find the tap weights that correspond to its minimum point. These methods are idealized versions of the class of practical algorithms, which will be presented in the next few chapters. We assume that the correlation matrix of the input samples to the filter and the cross-correlation vector between the desired output and filter input are known *a priori*.

The first method that we discuss is known as the *method of steepest descent*. The basic concept behind this method is simple. Assuming that the cost function to be minimized is convex, one may start with an arbitrary point on the performance surface and take a small step in the direction in which the cost function decreases fastest. This corresponds to a step along the steepest-descent slope of the performance surface at that point. Repeating this successively, convergence toward the bottom of the performance surface, at which point the set of parameters that minimize the cost function assume their optimum values, is guaranteed. For the transversal Wiener filters, we find that this method may suffer from

Adaptive Filters: Theory and Applications, Second Edition. Behrouz Farhang-Boroujeny.

slow convergence. The second method that we introduce can overcome this problem at the cost of additional complexity. This, which is known as the *Newton's method,* takes steps that are in the direction pointing toward the bottom of the performance surface.

Our discussion in this chapter is limited to the case where the filter tap weights, input, and desired output are real-valued. The extension of the results to the case of complex-valued signals is straightforward and deferred to a problem at the end of the chapter.

5.1 Method of Steepest Descent

Consider a transversal Wiener filter, as in Figure 5.1. The filter input, $x(n)$, and its desired output, $d(n)$, are assumed to be real-valued sequences. The filter tap weights, $w_0, w_1, \ldots, w_{N-1}$, are also assumed to be real-valued. The filter tap-weight and input are defined, respectively, by the column vectors

$$\mathbf{w} = [w_0 \; w_1 \; \cdots \; w_{N-1}]^{\mathrm{T}} \tag{5.1}$$

and

$$\mathbf{x}(n) = [x(n) \; x(n-1) \; \cdots \; x(n-N+1)]^{\mathrm{T}} \tag{5.2}$$

where superscript T stands for transpose. The filter output is

$$y(n) = \mathbf{w}^{\mathrm{T}}\mathbf{x}(n) \tag{5.3}$$

We recall from Chapter 3 that the optimum tap-weight vector $\mathbf{w}_{\mathrm{o}}$ is the one that minimizes the performance function

$$\xi = E[e^2(n)] \tag{5.4}$$

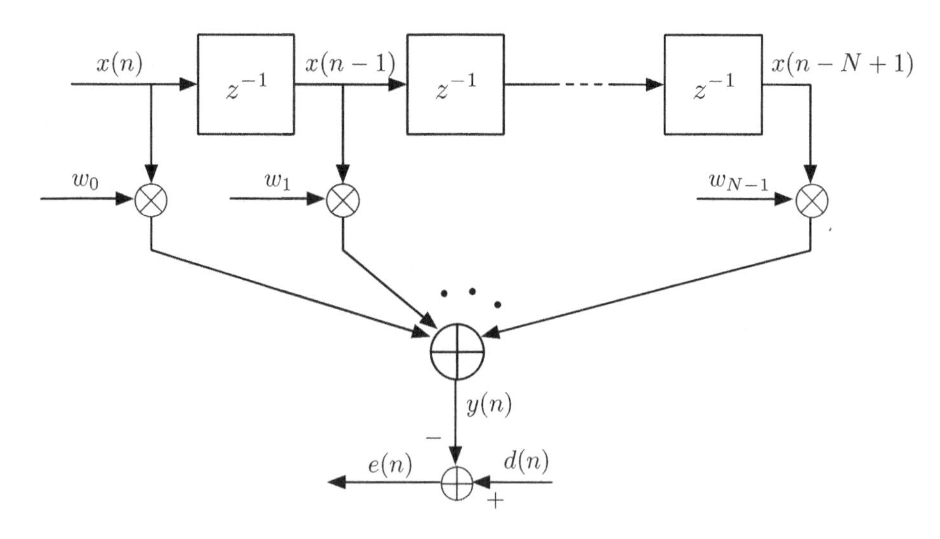

Figure 5.1 A transversal filter.

where $e(n) = d(n) - y(n)$ is the *estimation error* of the Wiener filter. Also, we recall that the performance function ξ can be expanded as

$$\xi = E[d^2(n)] - 2\mathbf{w}^{\mathrm{T}}\mathbf{p} + \mathbf{w}^{\mathrm{T}}\mathbf{R}\mathbf{w} \tag{5.5}$$

where $\mathbf{R} = E[\mathbf{x}(n)\mathbf{x}^{\mathrm{T}}(n)]$ is the autocorrelation matrix of the filter input and $\mathbf{p} = E[\mathbf{x}(n)d(n)]$ is the cross-correlation vector between the filter input and its desired output. The function ξ (whose details were given in the last chapter) is a quadratic function of the filter tap-weight vector $\mathbf{w}$. It has a single global minimum, which can be obtained by solving the Wiener–Hopf equation

$$\mathbf{R}\mathbf{w}_{\mathrm{o}} = \mathbf{p} \tag{5.6}$$

if $\mathbf{R}$ and $\mathbf{p}$ are available. Here, we assume that $\mathbf{R}$ and $\mathbf{p}$ are available, but resort to a different approach to find $\mathbf{w}_{\mathrm{o}}$. Instead of trying to solve Eq. (5.6) directly, we choose an *iterative search method* in which starting with an initial guess for $\mathbf{w}_{\mathrm{o}}$, say $\mathbf{w}(0)$, a recursive search method that may require many iterations (steps) to converge to $\mathbf{w}_{\mathrm{o}}$ is used. Understanding of this method is basic to the development of the *iterative algorithms*, which are commonly used in the implementation of adaptive filters in practice.

The method of steepest descent is a general scheme that uses the following steps to search for the minimum point of any convex function of a set of parameters:

1. Start with an initial guess of the parameters whose optimum values are to be found for minimizing the function.
2. Find the gradient of the function with respect to these parameters at the present point.
3. Update the parameters by taking a step in the opposite direction of the gradient vector obtained in Step 2. This corresponds to a step in the direction of steepest descent in the cost function at the present point. Furthermore, the size of the step taken is chosen proportional to the size of the gradient vector.
4. Repeat Steps 2 and 3 until no further significant change is observed in the parameters.

To implement this procedure in the case of the transversal filter shown in Figure 5.1, we recall from Chapter 3 that

$$\nabla\xi = 2\mathbf{R}\mathbf{w} - 2\mathbf{p} \tag{5.7}$$

where ∇ is the gradient operator defined as the column vector

$$\nabla = \left[\frac{\partial}{\partial w_0}\ \frac{\partial}{\partial w_1}\ \cdots\ \frac{\partial}{\partial w_{N-1}}\right]^{\mathrm{T}} \tag{5.8}$$

According to the above procedure, if $\mathbf{w}(k)$ is the tap-weight vector at the kth iteration, the following recursive equation may be used to update $\mathbf{w}(k)$.

$$\mathbf{w}(k+1) = \mathbf{w}(k) - \mu\nabla_k\xi \tag{5.9}$$

where μ is a positive scalar called *step-size*, and $\nabla_k\xi$ denotes the gradient vector $\nabla\xi$ evaluated at the point $\mathbf{w} = \mathbf{w}(k)$. Substituting Eq. (5.7) in Eq. (5.9), we get

$$\mathbf{w}(k+1) = \mathbf{w}(k) - 2\mu(\mathbf{R}\mathbf{w}(k) - \mathbf{p}) \tag{5.10}$$

As we shall soon show, the convergence of $\mathbf{w}(k)$ to the optimum solution $\mathbf{w}_\mathrm{o}$ and the speed at which this convergence takes place are dependent on the size of the step-size parameter μ. A large step-size may result in divergence of this recursive equation.

To see how the recursive update $\mathbf{w}(k)$ converges toward $\mathbf{w}_o$, we rearrange Eq. (5.10) as

$$\mathbf{w}(k+1) = (\mathbf{I} - 2\mu\mathbf{R})\mathbf{w}(k) + 2\mu\mathbf{p} \tag{5.11}$$

where $\mathbf{I}$ is the N-by-N identity matrix. Next, we substitute for $\mathbf{p}$ from Eq. (5.6). Also, we subtract $\mathbf{w}_\mathrm{o}$ from both sides of Eq. (5.11) and rearrange the result to obtain

$$\mathbf{w}(k+1) - \mathbf{w}_\mathrm{o} = (\mathbf{I} - 2\mu\mathbf{R})(\mathbf{w}(k) - \mathbf{w}_\mathrm{o}) \tag{5.12}$$

Defining the vector $\mathbf{v}(k)$ as

$$\mathbf{v}(k) = \mathbf{w}(k) - \mathbf{w}_\mathrm{o} \tag{5.13}$$

and substituting this in Eq. (5.12), we obtain

$$\mathbf{v}(k+1) = (\mathbf{I} - 2\mu\mathbf{R})\mathbf{v}(k) \tag{5.14}$$

This is the tap-weight update equation in terms of the v-axes (see Chapter 4 for further discussion on the v-axes). This result can be simplified further if we transform these to the v'-axes (see Eq. (4.92) of Chapter 4 for the definition of v'-axes). Recall from Chapter 4 that $\mathbf{R}$ has the following unitary similarity decomposition

$$\mathbf{R} = \mathbf{Q}\mathbf{\Lambda}\mathbf{Q}^\mathrm{T} \tag{5.15}$$

where $\mathbf{\Lambda}$ is a diagonal matrix consisting of the eigenvalues $\lambda_0, \lambda_1, \ldots, \lambda_{N-1}$ of $\mathbf{R}$ and the columns of $\mathbf{Q}$ contain the corresponding orthonormal eigenvectors. Substituting Eq. (5.15) in Eq. (5.14) and replacing $\mathbf{I}$ with $\mathbf{Q}\mathbf{Q}^\mathrm{T}$, we get

$$\begin{aligned}\mathbf{v}(k+1) &= (\mathbf{Q}\mathbf{Q}^\mathrm{T} - 2\mu\mathbf{Q}\mathbf{\Lambda}\mathbf{Q}^\mathrm{T})\mathbf{v}(k) \\ &= \mathbf{Q}(\mathbf{I} - 2\mu\mathbf{\Lambda})\mathbf{Q}^\mathrm{T}\mathbf{v}(k)\end{aligned} \tag{5.16}$$

Premultiplying Eq. (5.16) by $\mathbf{Q}^\mathrm{T}$ and recalling the transformation

$$\mathbf{v}'(k) = \mathbf{Q}^\mathrm{T}\mathbf{v}(k) \tag{5.17}$$

we obtain the recursive equation in terms of v'-axes as

$$\mathbf{v}'(k+1) = (\mathbf{I} - 2\mu\mathbf{\Lambda})\mathbf{v}'(k) \tag{5.18}$$

The vector recursive Eq. (5.18) may be separated into the scalar recursive equations

$$v'_i(k+1) = (1 - 2\mu\lambda_i)v'_i(k), \quad \text{for } i = 0, 1, \ldots, N-1 \tag{5.19}$$

where $v'_i(k)$ is the ith element of the vector $\mathbf{v}'(k)$.

Starting with a set of initial values $v'_0(0)$, $v'_1(0)$, ..., $v'_{N-1}(0)$ and iterating Eq. (5.19) k times, we get

$$v'_i(k) = (1 - 2\mu\lambda_i)^k v'_i(0), \quad \text{for } i = 0, 1, \ldots, N-1 \tag{5.20}$$

From Eqs. (5.13) and (5.17), we see that $\mathbf{w}(k)$ converges to $\mathbf{w}_o$ if and only if $\mathbf{v}'(k)$ converges to the zero vector. But Eq. (5.20) implies that $\mathbf{v}'(k)$ can converge to zero if and only if the step-size parameter μ is selected so that

$$|1 - 2\mu\lambda_i| < 1, \quad \text{for } i = 0, 1, \ldots, N-1 \tag{5.21}$$

When Eq. (5.21) is satisfied, the scalars $v_i'(k)$, for $i = 0, 1, \ldots, N-1$, exponentially decay toward zero as the number of iterations, k, increases. Furthermore, Eq. (5.21) provides the condition for the recursive equations (5.20) and, hence, the steepest-descent algorithm to be stable. The inequalities (5.21) may be expanded as

$$-1 < 1 - 2\mu\lambda_i < 1$$

or

$$0 < \mu < \frac{1}{\lambda_i}, \quad \text{for } i = 0, 1, \ldots, N-1 \tag{5.22}$$

Noting that the step-size parameter μ is common for all values of i, convergence (stability) of the steepest-descent algorithm is guaranteed only when

$$0 < \mu < \frac{1}{\lambda_{\max}} \tag{5.23}$$

where $\lambda_{\max}$ is the maximum of the eigenvalues $\lambda_0, \lambda_1, \ldots, \lambda_{N-1}$. The left limit in Eq. (5.23) refers to the fact that the tap-weight correction must be in the opposite direction of the gradient vector. The right limit is to ensure that all the scalar tap-weight parameters in the recursive equations (5.19) decay exponentially as k increases.

Figure 5.2 depicts a set of plots that show how a particular tap-weight parameter $v_i'(k)$ varies as a function of the iteration index k and for different values of the step-size parameter μ. The cases considered here correspond to the typical distinct ranges of μ, referred to as *overdamped* $\left(0 < \mu < \frac{1}{2\lambda_i}\right)$, *underdamped* $\left(\frac{1}{2\lambda_i} < \mu < \frac{1}{\lambda_i}\right)$, and *unstable* $\left(\mu < 0 \text{ or } \mu > \frac{1}{\lambda_i}\right)$.

We may now derive a more explicit formulation for the transient behavior of the steepest-descent algorithm in terms of the original tap-weight vector $\mathbf{w}(k)$. We note that

$$\begin{aligned}
\mathbf{w}(k) &= \mathbf{w}_o + \mathbf{v}(k) \\
&= \mathbf{w}_o + \mathbf{Q}\mathbf{v}'(k) \\
&= \mathbf{w}_o + [\mathbf{q}_0 \ \mathbf{q}_1 \ \cdots \ \mathbf{q}_{N-1}] \begin{bmatrix} v_0'(k) \\ v_1'(k) \\ \vdots \\ v_{N-1}'(k) \end{bmatrix} \\
&= \mathbf{w}_o + \sum_{i=0}^{N-1} \mathbf{q}_i v_i'(k)
\end{aligned} \tag{5.24}$$

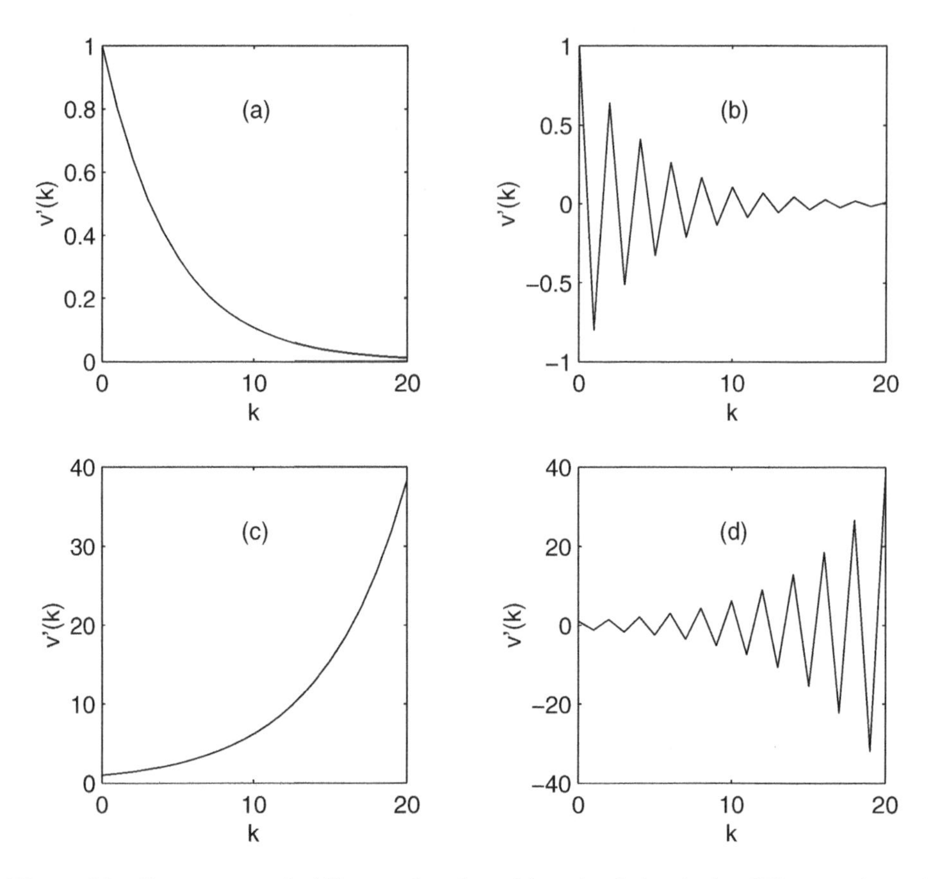

Figure 5.2 Convergence of $v_i'(k)$ as a function of iteration index k, for different values of the step-size parameter μ. (a) Overdamped case: $0 < \mu < 1/2\lambda_i$, (b) Underdamped case: $1/2\lambda_i < \mu < 1/\lambda_i$, (c) Unstable: $\mu < 0$, and (d) Unstable: $\mu > 1/\lambda_i$.

where $\mathbf{q}_0, \mathbf{q}_1, \ldots, \mathbf{q}_{N-1}$ are the eigenvectors associated with the eigenvalues $\lambda_0, \lambda_1, \ldots, \lambda_{N-1}$ of the correlation matrix $\mathbf{R}$. Substituting Eq. (5.20) in Eq. (5.24), we obtain

$$\mathbf{w}(k) = \mathbf{w}_o + \sum_{i=0}^{N-1} v_i'(0)(1 - 2\mu\lambda_i)^k \mathbf{q}_i \tag{5.25}$$

This result shows that the transient behavior of the steepest-descent algorithm for an N-tap transversal filter is determined by a sum of N exponential terms, each of which is controlled by one of the eigenvalues of the correlation matrix $\mathbf{R}$. Each eigenvalue λ_i determines a particular mode of convergence in the direction defined by its associated eigenvector $\mathbf{q}_i$. The various modes work independent of one another. For a selected value of the step-size parameter μ, the geometrical ratio factor $1 - 2\mu\lambda_i$, which determines how fast the ith mode converges, is determined by the value of λ_i.

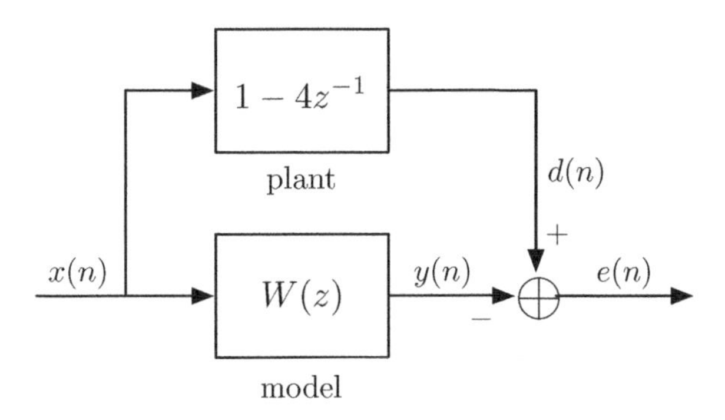

Figure 5.3 A modeling problem.

Example 5.1

Consider the modeling problem depicted in Figure 5.3. The input signal, $x(n)$, is generated by passing a white noise signal, $\nu(n)$, through a coloring filter with the system function

$$H(z) = \frac{\sqrt{1-\alpha^2}}{1-\alpha z^{-1}} \tag{5.26}$$

where α is a real-valued constant in the range of -1 to $+1$. The plant is a two-tap FIR system with the system function

$$P(z) = 1 - 4z^{-1}$$

An adaptive filter with the system function

$$W(z) = w_0 + w_1 z^{-1}$$

is used to identify the plant system function. The steepest-descent algorithm is used to find the optimum values of the tap weights w_0 and w_1. We want to see, as the iteration number increases, how the tap weights w_0 and w_1 converge toward the plant coefficients 1 and -4, respectively. We examine this for different values of the parameter α.

From the results derived in Example 4.1 of Chapter 4, we note that

$$E[x^2(n)] = 1 \qquad \text{and} \qquad E[x(n)x(n-1)] = \alpha$$

These give

$$\mathbf{R} = E[\mathbf{x}(n)\mathbf{x}^{\mathrm{T}}(n)] = \begin{bmatrix} 1 & \alpha \\ \alpha & 1 \end{bmatrix} \tag{5.27}$$

where $\mathbf{x}(n) = [x(n)\ x(n-1)]^{\mathrm{T}}$. Furthermore, the elements of the cross-correlation vector $\mathbf{p} = E[\mathbf{x}(n)d(n)]$ are obtained as follows:

$$\begin{aligned} p_0 &= E[x(n)d(n)] = E[x(n)(x(n) - 4x(n-1))] \\ &= E[x^2(n)] - 4E[x(n)x(n-1)] = 1 - 4\alpha \end{aligned}$$

and

$$p_1 = E[x(n-1)d(n)] = E[x(n-1)(x(n) - 4x(n-1))]$$
$$= E[x(n-1)x(n)] - E[x^2(n-1)] = \alpha - 4$$

These give

$$\mathbf{p} = \begin{bmatrix} 1 - 4\alpha \\ \alpha - 4 \end{bmatrix} \tag{5.28}$$

Substituting Eqs. (5.27) and (5.28) in Eq. (5.11), we get

$$\begin{bmatrix} w_0(k+1) \\ w_1(k+1) \end{bmatrix} = \begin{bmatrix} 1 - 2\mu & -2\mu\alpha \\ -2\mu\alpha & 1 - 2\mu \end{bmatrix} \begin{bmatrix} w_0(k) \\ w_1(k) \end{bmatrix} + 2\mu \begin{bmatrix} 1 - 4\alpha \\ \alpha - 4 \end{bmatrix} \tag{5.29}$$

Starting with an initial value $\mathbf{w}(0) = [w_0(0)\ w_1(0)]^{\mathrm{T}}$ and letting the recursive equation (5.29) to run, we get two sequences of the tap-weight variables $w_0(k)$ and $w_1(k)$. We may then plot $w_1(k)$ versus $w_0(k)$ to get the *trajectory* (path) that the steepest-descent algorithm follows. Figure 5.4a, b, c, and d show four of such trajectories that we have obtained for values of $\alpha = 0, 0.5, 0.75$, and 0.9, respectively. Also shown in the figures are the contour plots, which highlight the performance surface of the filter. The convergence of the algorithm along the steepest-descent slope of the performance surface can be clearly seen. The results presented are for $\mu = 0.05$ and 30 iterations, for all cases. It is interesting to note that in the case $\alpha = 0$, which corresponds to a white input sequence, $x(n)$, the convergence is almost complete within 30 iterations. However, the other three cases require some more iterations before they converge to the minimum point of the performance surface. This can be understood if one notes that the eigenvalues of $\mathbf{R}$ are $\lambda_0 = 1 + \alpha$ and $\lambda_1 = 1 - \alpha$, and for α close to 1, the geometrical ratio factor $1 - 2\mu\lambda_1$ may be very close to 1. This introduces a slow mode of convergence along v'_1-axis (i.e., in the direction defined by the eigenvector $\mathbf{q}_1$).

5.2 Learning Curve

Although the recursive equations (5.19) and (5.24) provide detailed information about the transient behavior of the steepest-descent algorithm, the multiparameter nature of the equations makes it difficult to visualize such behavior graphically. Instead, it is more convenient to consider the variation of the mean squared error (MSE), that is, the performance function ξ, versus the number of iterations.

We define $\xi(k)$ as the value of the performance function ξ when $\mathbf{w} = \mathbf{w}(k)$. Then, using Eq. (4.94) of Chapter 4, we get

$$\xi(k) = \xi_{\min} + \sum_{i=0}^{N-1} \lambda_i v_i'^2(k) \tag{5.30}$$

where $\xi_{\min}$ is the minimum MSE. Substituting Eq. (5.20) in Eq. (5.30), we obtain

$$\xi(k) = \xi_{\min} + \sum_{i=0}^{N-1} \lambda_i (1 - 2\mu\lambda_i)^{2k} v_i'^2(0) \tag{5.31}$$

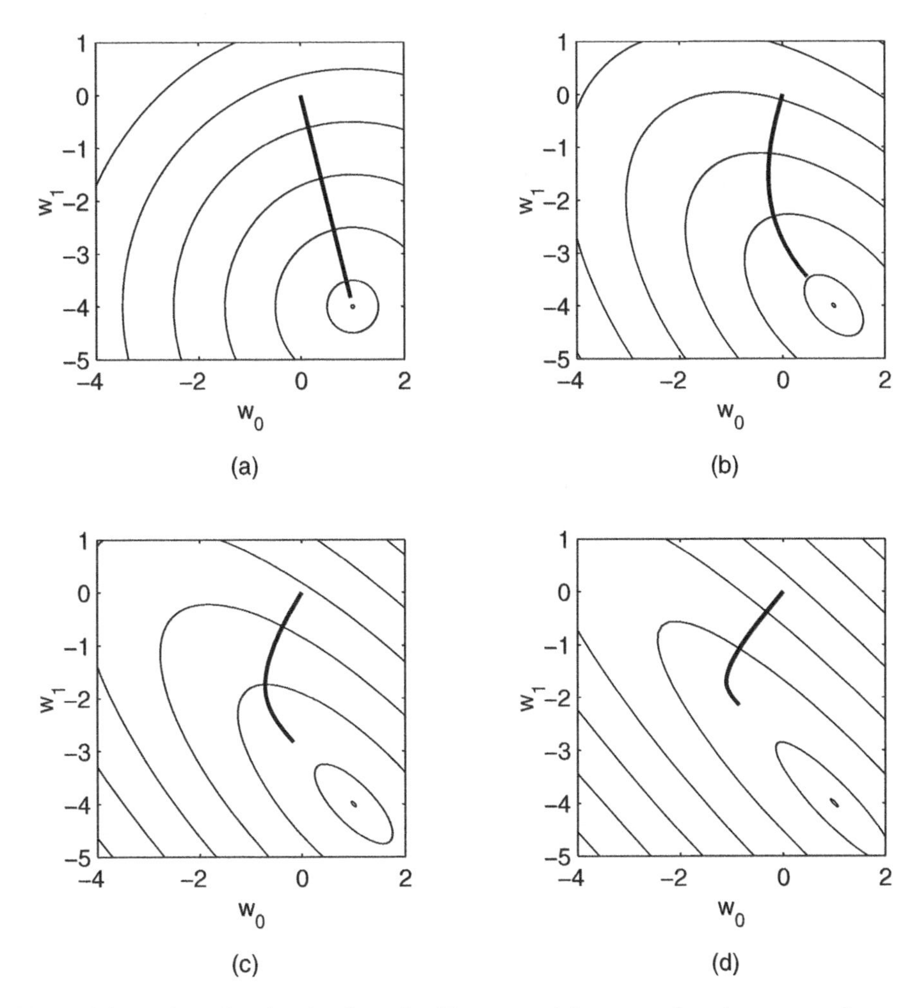

Figure 5.4 Trajectories showing how the filter tap weights vary when the steepest-descent algorithm is used: (a) $\alpha = 0$, (b) $\alpha = 0.5$, (c) $\alpha = 0.75$, and (d) $\alpha = 0.9$. Each plot is based on 30 iterations and $\mu = 0.05$.

When μ is selected within the bounds defined by Eq. (5.23), the terms under the summation in Eq. (5.31) converge to zero as k increases. As a result, the minimum MSE is achieved after a sufficient number of iterations.

The curve obtained by plotting $\xi(k)$ as a function of the iteration index, k, is called *learning curve*. A learning curve of the steepest-descent algorithm, as can be seen from Eq. (5.31), consists of a sum of N exponentially decaying terms, each of which corresponds to one of the modes of convergence of the algorithm. Each exponential term may be characterized by a time constant, which is obtained as follows.

Let

$$(1 - 2\mu\lambda_i)^{2k} = e^{-k/\tau_i} \tag{5.32}$$

and define τ_i as the time constant associated with the exponential term $(1-2\mu\lambda_i)^{2k}$. Solving Eq. (5.32) for τ_i, we get

$$\tau_i \approx \frac{-1}{2\ln(1-2\mu\lambda_i)} \tag{5.33}$$

For small values of the step-size parameter μ, when $2\mu\lambda_i \ll 1$, we note that

$$\ln(1-2\mu\lambda_i) \approx -2\mu\lambda_i \tag{5.34}$$

Substituting this in Eq. (5.33), we obtain

$$\tau_i = \frac{1}{4\mu\lambda_i} \tag{5.35}$$

This result, which is true for all values of $i = 0, 1, \ldots, N-1$, shows that, in general, the number of time constants that characterize a learning curve are equal to the number of filter taps. Furthermore, the time constants that are associated with the smaller eigenvalues are larger than those associated with the larger eigenvalues.

Example 5.2

Consider the modeling arrangement that was discussed in Example 5.1. The correlation matrix $\mathbf{R}$ of the filter input is given by Eq. (5.27). The eigenvalues of $\mathbf{R}$ are

$$\lambda_0 = 1+\alpha \qquad \text{and} \qquad \lambda_1 = 1-\alpha$$

Using these in Eq. (5.35), we obtain

$$\tau_0 \approx \frac{1}{4\mu(1+\alpha)} \tag{5.36}$$

and

$$\tau_1 \approx \frac{1}{4\mu(1-\alpha)} \tag{5.37}$$

These are the time constants that characterize the learning curve of the modeling problem. Figure 5.5 shows a learning curve of the modeling problem when $\mathbf{w}(0) = [2\ 2]^{\mathrm{T}}$, $\alpha = 0.75$, and $\mu = 0.05$. For these values, we obtain

$$\tau_0 \approx 2.85 \qquad \text{and} \qquad \tau_1 \approx 20 \tag{5.38}$$

The existence of two distinct time constants on the learning curve in Figure 5.5 is clearly observed.

The two time constants could be observed more clearly if the ξ axis is scaled logarithmically. To see this, the learning curve of the modeling problem is plotted in Figure 5.6 with the ξ axis scaled logarithmically. The two exponentials appear as two straight lines on this plot. The first part of the plot, with a steep slope, is dominantly controlled by τ_0. The remaining part of the learning curve shows the contribution of the second exponential, which is characterized by τ_1. Estimates of the time constants may be obtained by finding the number of iterations required for ξ to drop 2.73 (i.e., the Napier number)

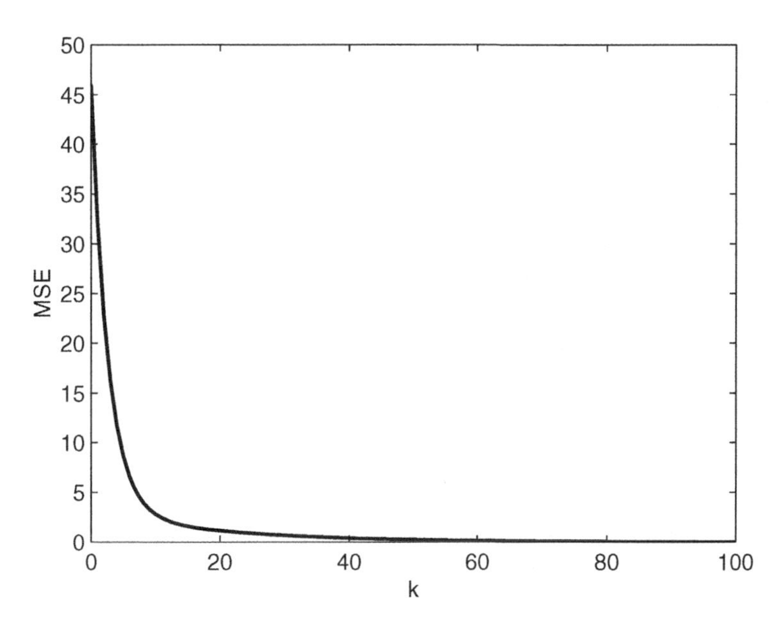

Figure 5.5 A learning curve of the modeling problem. The ξ (MSE) axis is scaled linearly.

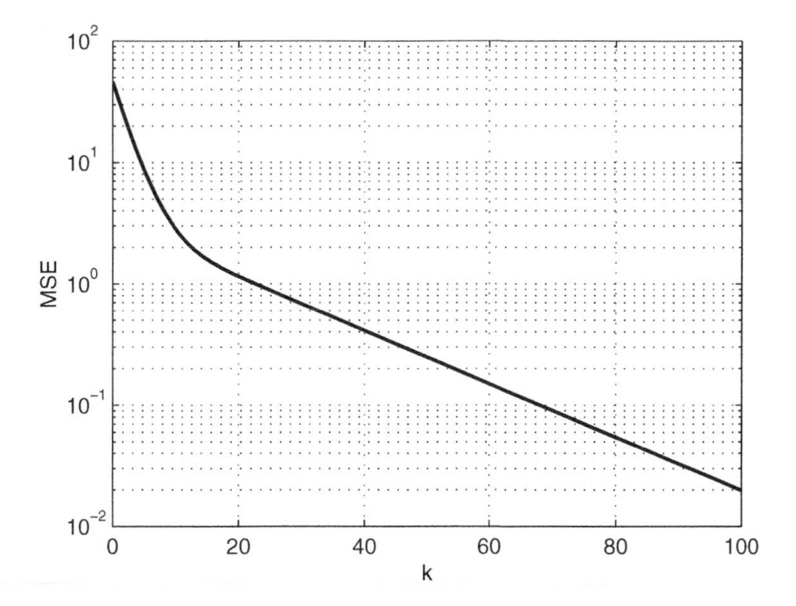

Figure 5.6 A learning curve of the modeling problem. The ξ (MSE) axis is scaled logarithmically.

times along each of the slopes. This gives

$$\tau_0 \approx 3 \qquad \text{and} \qquad \tau_1 \approx 20$$

which match well with those in Eq. (5.38).

5.3 Effect of Eigenvalue Spread

Our study in the last two sections shows that the performance of the steepest-descent algorithm is highly dependent on the eigenvalues of the correlation matrix **R**. In general, a wider spread of the eigenvalues results in a poorer performance of the steepest-descent algorithm. To gain further insight into this property of the steepest-descent algorithm, we find the optimum value of the step-size parameter μ, which results in the fastest possible convergence of the steepest-descent algorithm.

We note that the speeds at which various modes of the steepest-descent algorithm converge are determined by the size (absolute value) of the geometrical ratio factors $1 - 2\mu\lambda_i$, for $i = 0, 1, \ldots, N-1$. For a given value of μ, the transient time of the steepest-descent algorithm is determined by the largest element in the set $\{|1 - 2\mu\lambda_i|, i = 0, 1, \ldots, N-1\}$. The optimum value of μ, which minimizes the largest element in the latter set, is obtained by looking at the two extreme cases that correspond to $\lambda_{\max}$ and $\lambda_{\min}$, that is, the maximum and minimum eigenvalues of **R**. Figure 5.7 shows the plots of $|1 - 2\mu\lambda_{\min}|$ and $|1 - 2\mu\lambda_{\max}|$ as functions of μ. The plots for the other eigenvalues lie in between these two plots. From these plots, one can clearly see that the optimum value of the step-size parameter μ corresponds to the point where the two plots meet. This is the point highlighted as μ_{opt} in Figure 5.7. It corresponds to the case where

$$1 - 2\mu_{\text{opt}}\lambda_{\min} = -(1 - 2\mu_{\text{opt}}\lambda_{\max}) \tag{5.39}$$

Solving this for μ_{opt}, we obtain

$$\mu_{\text{opt}} = \frac{1}{\lambda_{\min} + \lambda_{\max}} \tag{5.40}$$

For this choice of the step-size parameter $1 - 2\mu_{\text{opt}}\lambda_{\min}$ is positive and $1 - 2\mu_{\text{opt}}\lambda_{\max}$ is negative. These correspond to overdamped and underdamped cases presented in Figure 5.2a and b, respectively. However, the two modes converge at the same speed. For $\mu = \mu_{\text{opt}}$, the speed of convergence of the steepest-descent algorithm is determined by the geometrical ratio factor

$$\beta = 1 - 2\mu_{\text{opt}}\lambda_{\min} \tag{5.41}$$

Substituting Eq. (5.40) in Eq. (5.41), we obtain

$$\beta = \frac{\frac{\lambda_{\max}}{\lambda_{\min}} - 1}{\frac{\lambda_{\max}}{\lambda_{\min}} + 1} \tag{5.42}$$

This has a value that remains between 0 and 1. When $\lambda_{\max} = \lambda_{\min}$, $\beta = 0$ and the steepest-descent algorithm can converge in one step. As the ratio $\lambda_{\max}/\lambda_{\min}$ increases, β also increases and becomes close to 1 when $\lambda_{\max}/\lambda_{\min}$ is large. Clearly, a value of β close

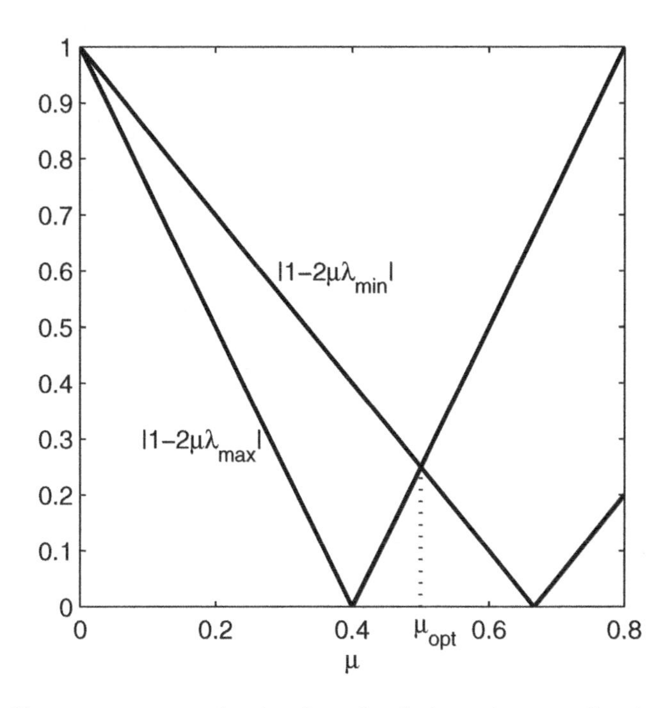

Figure 5.7 The extreme cases showing how $1 - 2\mu\lambda_i$ varies as a function of the step-size parameter μ.

to 1 corresponds to a slow mode of convergence. Thus, we note that the ratio $\lambda_{\max}/\lambda_{\min}$ plays a fundamental role in limiting the convergence performance of the steepest-descent algorithm. This ratio is called *eigenvalue spread.*

We may also recall from the last chapter that the values of $\lambda_{\max}$ and $\lambda_{\min}$ are closely related to the maximum and minimum values of the power spectral density of the underlying process. Noting this, we may say that *the performance of the steepest-descent algorithm is closely related to the shape of the power spectral density of the underlying input process.* A wide distribution of the energy of the underlying process within different frequency bands introduces slow modes of convergence, which result in a poor performance of the steepest-descent algorithm. When the underlying process contains very little energy in a band of frequencies, we say the filter is *weakly excited* in that band. *Weak excitation,* as we see, degrades the performance of the steepest-descent algorithm.

5.4 Newton's Method

Our discussions in the last few sections show that the steepest-descent algorithm may suffer from slow modes of convergence, which arise due to the spread in the eigenvalues of the correlation matrix **R**. This means that if we can somehow get rid of the eigenvalue spread, we can get much better convergence performance. This is exactly what Newton's method does. To derive Newton's method for the quadratic case, we start from the steepest-descent algorithm given in Eq. (5.10). Using $\mathbf{p} = \mathbf{R}\mathbf{w}_o$, Eq. (5.10) becomes

$$\mathbf{w}(k+1) = \mathbf{w}(k) - 2\mu\mathbf{R}(\mathbf{w}(k) - \mathbf{w}_o) \tag{5.43}$$

We may note that it is the presence of $\mathbf{R}$ in Eq. (5.43), which causes the eigenvalue spread problem in the steepest-descent algorithm. Newton's method overcomes this problem by replacing the scalar step-size parameter μ with a matrix step-size given by $\mu\mathbf{R}^{-1}$. The resulting algorithm is

$$\mathbf{w}(k+1) = \mathbf{w}(k) - \mu\mathbf{R}^{-1}\nabla_k\xi \tag{5.44}$$

Figure 5.8 demonstrates the effect of the addition of $\mathbf{R}^{-1}$ in front of the gradient vector in Newton's update Eq. (5.44). This has the effect of rotating the gradient vector to the direction pointing toward the minimum point of the performance surface.

Substituting Eq. (5.7) in Eq. (5.44), we obtain

$$\begin{aligned}\mathbf{w}(k+1) &= \mathbf{w}(k) - 2\mu\mathbf{R}^{-1}(\mathbf{R}\mathbf{w}(k) - \mathbf{p}) \\ &= (1 - 2\mu)\mathbf{w}(k) + 2\mu\mathbf{R}^{-1}\mathbf{p}\end{aligned} \tag{5.45}$$

We also note that $\mathbf{R}^{-1}\mathbf{p}$ is equal to the optimum tap-weight vector $\mathbf{w}_o$. Using this in Eq. (5.45), we obtain

$$\mathbf{w}(k+1) = (1 - 2\mu)\mathbf{w}(k) + 2\mu\mathbf{w}_o \tag{5.46}$$

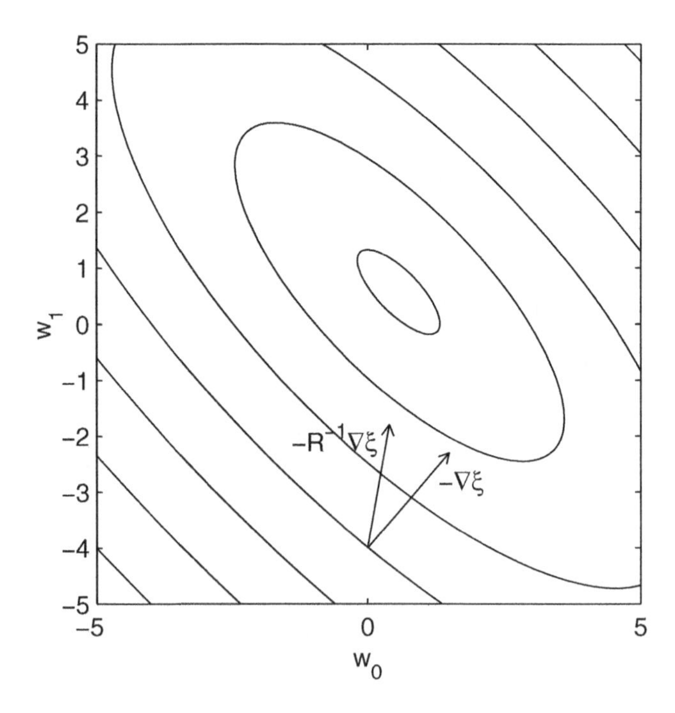

Figure 5.8 The negative gradient vector and its correction by Newton's method.

Subtracting $\mathbf{w}_o$ from both sides of Eq. (5.46), we get

$$\mathbf{w}(k+1) - \mathbf{w}_o = (1 - 2\mu)(\mathbf{w}(k) - \mathbf{w}_o) \tag{5.47}$$

Starting with an initial value $\mathbf{w}(0)$ and iterating Eq. (5.47), we obtain

$$\mathbf{w}(k) - \mathbf{w}_o = (1 - 2\mu)^k(\mathbf{w}(0) - \mathbf{w}_o) \tag{5.48}$$

The original Newton's method selects the step-size parameter μ equal to 0.5. This leads to convergence of $\mathbf{w}(k)$ to its optimum value, $\mathbf{w}_o$, in one iteration. In particular, we note that setting $\mu = 0.5$ and $k = 1$ in Eq. (5.48), we obtain $\mathbf{w}(1) = \mathbf{w}_o$. However, in actual implementation of adaptive filters where the exact values of $\nabla_k\xi$ and $\mathbf{R}^{-1}$ are not available and they have to be estimated, one needs to use a step-size parameter much smaller than 0.5. Thus, an evaluation of Newton's recursion (5.44) for values of $\mu \neq 0.5$ is instructive for our further study in the later chapters.

Using Eq. (5.48) and following the same line of derivations as in the case of the steepest-descent method, it is straightforward to show that (Problem P5.5)

$$\xi(k) = \xi_{\min} + (1 - 2\mu)^{2k}\xi(0) \tag{5.49}$$

where $\xi(k)$ is the value of the performance function, ξ, when $\mathbf{w} = \mathbf{w}(k)$.

From Eq. (5.49), we note that the stability of Newton's algorithm is guaranteed when $|1 - 2\mu| < 1$ or, equivalently,

$$0 < \mu < 1 \tag{5.50}$$

With reference to Eq. (5.49), we make the following observations: *The transient behavior of Newton's algorithm is characterized by a single exponential whose corresponding time constant is obtained by solving the equation*

$$(1 - 2\mu)^{2k} = e^{-k/\tau}. \tag{5.51}$$

When $2\mu \ll 1$, this gives

$$\tau \approx \frac{1}{4\mu} \tag{5.52}$$

This result shows that Newton's method has only one mode of convergence and that is solely determined by its step-size parameter μ.

5.5 An Alternative Interpretation of Newton's Algorithm

Further insight into the operation of Newton's algorithm is developed by giving an alternative derivation of that. This derivation uses the Karhunen-Loéve transform (KLT), which was introduced in the last chapter.

For an observation vector $\mathbf{x}(n)$ with real-valued elements, the KLT is defined by the equation

$$\mathbf{x}'(n) = \mathbf{Q}^{\mathrm{T}}\mathbf{x}(n) \tag{5.53}$$

where $\mathbf{Q}$ is the N-by-N matrix whose columns are the eigenvectors $\mathbf{q}_0, \mathbf{q}_1, \ldots, \mathbf{q}_{N-1}$ of the correlation matrix $\mathbf{R} = E[\mathbf{x}(n)\mathbf{x}^{\mathrm{T}}(n)]$. We recall from Chapter 4 that the elements of the transformed vector $\mathbf{x}'(n)$, denoted by $x'_0(n)$, $x'_1(n)$, ..., $x'_{N-1}(n)$ constitute a set of mutually uncorrelated random variables. Furthermore, Eq. (4.68) implies that

$$E[x_i'^2(n)] = \lambda_i, \quad \text{for } i = 0, 1, \ldots, N-1 \tag{5.54}$$

where λ_is are the eigenvalues of the correlation matrix $\mathbf{R}$.

We define the vector $x'^{\mathrm{n}}(n)$ whose elements are

$$x_i'^{\mathrm{n}}(n) = \lambda_i^{-1/2} x_i'(n), \quad \text{for } i = 0, 1, \ldots, N-1 \tag{5.55}$$

where the superscript n signifies the fact that $x_i'^{\mathrm{n}}(n)$ is normalized to the power of unity (see Eq. (5.57), below). These equations may collectively be written as

$$\mathbf{x}'^{\mathrm{n}}(n) = \mathbf{\Lambda}^{-1/2}\mathbf{x}'(n) \tag{5.56}$$

where $\mathbf{\Lambda}$ is a diagonal matrix consisting of the eigenvalues λ_0, λ_1, ..., λ_{N-1}. It is straightforward to show that

$$\mathbf{R}'^{\mathrm{n}} = E[\mathbf{x}'^{\mathrm{n}}(n)\mathbf{x}'^{\mathrm{nT}}(n)] = \mathbf{I} \tag{5.57}$$

where $\mathbf{I}$ is the N-by-N identity matrix.

We also define

$$\mathbf{w}'^{\mathrm{n}} = \mathbf{\Lambda}^{1/2}\mathbf{Q}^{\mathrm{T}}\mathbf{w} \tag{5.58}$$

and note that

$$\mathbf{w}'^{\mathrm{nT}}\mathbf{x}'^{\mathrm{n}}(n) = \mathbf{w}^{\mathrm{T}}\mathbf{Q}\mathbf{\Lambda}^{1/2}\mathbf{\Lambda}^{-1/2}\mathbf{Q}^{\mathrm{T}}\mathbf{x}(n) = \mathbf{w}^{\mathrm{T}}\mathbf{x}(n) \tag{5.59}$$

This result shows that a filter with an input vector $\mathbf{x}(n)$ and output $y(n) = \mathbf{w}^{\mathrm{T}}\mathbf{x}(n)$ may alternatively be realized by $\mathbf{x}'^{\mathrm{n}}(n)$ and $\mathbf{w}'^{\mathrm{n}}$ as the filter input and tap-weight vectors, respectively. The steepest-descent algorithm for this realization may be written as Eq. (5.11)

$$\mathbf{w}'^{\mathrm{n}}(k+1) = (\mathbf{I} - 2\mu\mathbf{R}'^{\mathrm{n}})\mathbf{w}'^{\mathrm{n}}(k) + 2\mu\mathbf{p}'^{\mathrm{n}} \tag{5.60}$$

where

$$\mathbf{p}'^{\mathrm{n}} = E[\mathbf{x}'^{\mathrm{n}}(n)d(n)]. \tag{5.61}$$

As $\mathbf{R}'^{\mathrm{n}} = \mathbf{I}$, Eq. (5.60) simplifies to

$$\mathbf{w}'^{\mathrm{n}}(k+1) = (1 - 2\mu)\mathbf{w}'^{\mathrm{n}}(k) + 2\mu\mathbf{w}'^{\mathrm{n}}_{\mathrm{o}} \tag{5.62}$$

where $\mathbf{w}'^{\mathrm{n}}_{\mathrm{o}} = \left(\mathbf{R}'^{\mathrm{n}}\right)^{-1}\mathbf{p}'^{\mathrm{n}} = \mathbf{p}'^{\mathrm{n}}$ is the optimum value of the tap-weight vector $\mathbf{w}'^{\mathrm{n}}$. Comparing this with Newton's algorithm (5.46), we find that the steepest-descent algorithm in this case works just similar to Newton's algorithm.

Next, we show that the recursive equation (5.60) is nothing but Newton's recursive equation (5.44) written in a slightly different form. For this, we use Eq. (5.58) in Eq. (5.62) to obtain

$$\mathbf{\Lambda}^{1/2}\mathbf{Q}^{\mathrm{T}}\mathbf{w}(k+1) = (1 - 2\mu)\mathbf{\Lambda}^{1/2}\mathbf{Q}^{\mathrm{T}}\mathbf{w}(k) + 2\mu\mathbf{\Lambda}^{1/2}\mathbf{Q}^{\mathrm{T}}\mathbf{w}_{\mathrm{o}} \tag{5.63}$$

Premultiplying both sides of this equation by $(\mathbf{\Lambda}^{1/2}\mathbf{Q}^{\mathrm{T}})^{-1} = \mathbf{Q}\mathbf{\Lambda}^{-1/2}$ (as $(\mathbf{Q}^{\mathrm{T}})^{-1} = \mathbf{Q}$), we get Eq. (5.46), which can easily be converted to Eq. (5.44).

The above development shows that Newton's algorithm may be viewed as *a steepest-descent algorithm for the transformed input signal.* The eigenvalue spread problem associated with the steepest-descent algorithm is resolved by decorrelating the filter input samples (through their corresponding KLT) followed by a power-normalization procedure. *This is a whitening process, viz., the input samples are decorrelated and then normalized to the unit power before the filtering process.*

Problems

P5.1 Use the method of steepest descent to solve the equation $\mathbf{Rw} = \mathbf{p}$ for the following choices of $\mathbf{R}$ and $\mathbf{p}$. For each case, find the range of the step-size parameter μ for which the steepest-descent algorithm is convergent. Also, for each case, find the value of μ that results in the fastest convergence of the steepest-descent algorithm. You may write a MATLAB code for finding the solutions.

(i)

$$\mathbf{R} = \begin{bmatrix} 2 & 1 \\ 1 & 2 \end{bmatrix}, \qquad \mathbf{p} = \begin{bmatrix} 1 \\ 1 \end{bmatrix}$$

(ii)

$$\mathbf{R} = \begin{bmatrix} 2 & 1 & 0.5 \\ 1 & 2 & 1 \\ 0.5 & 1 & 2 \end{bmatrix}, \qquad \mathbf{p} = \begin{bmatrix} 1 \\ -1 \\ 0 \end{bmatrix}$$

P5.2 By applying the method of steepest descent to the canonical form of the performance function, that is, Eq. (4.93), suggest an alternative derivation of Eq. (5.25).

P5.3 Show that when the steepest-descent algorithm (5.10) is used, the time constants that control the variation of the tap weights of a transversal filter are

$$\tau_i' = \frac{1}{2\mu\lambda_i}, \qquad \text{for } i = 0, 1, \ldots, N-1$$

P5.4 Give a detailed derivation of Eq. (5.25) in the case where underlying signals are complex-valued.

P5.5 Give a detailed derivation of Eq. (5.49).

P5.6 Show that if in the steepest-descent algorithm, the tap-weight vector is initialized to zero,

$$\mathbf{w}(k) = [\mathbf{I} - (\mathbf{I} - 2\mu\mathbf{R})^k]\mathbf{w}_\text{o}$$

where $\mathbf{w}_\text{o}$ is the optimum tap-weight vector.

P5.7 $\mathbf{R}$ is a correlation matrix with the eigenvalues λ_i, $i = 0, 1, \ldots, N-1$. Find the eigenvalues of the matrix $\mathbf{G} = \mathbf{I} + \mathbf{R} + \mathbf{R}^2$.

P5.8 $\mathbf{R}$ is a correlation matrix with the eigenvalues λ_i, $i = 0, 1, \ldots, N-1$. Prove that if $0 < \lambda_i < 1$, for $i = 0, 1, \ldots, N-1$,

(i)

$$\lim_{n\to\infty} (\mathbf{I} + \mathbf{R} + \mathbf{R}^2 + \cdots + \mathbf{R}^n) = (\mathbf{I} - \mathbf{R})^{-1}$$

(ii)

$$\lim_{n\to\infty}(\mathbf{I}+(\mathbf{I}-\mathbf{R})+(\mathbf{I}-\mathbf{R})^2+\cdots+(\mathbf{I}-\mathbf{R})^n)=\mathbf{R}^{-1}$$

P5.9 Consider the modeling problem depicted in Figure P5.6. Note that the input to the model is a noisy version of the plant input. The additive noise at the model input, $\nu_{\mathrm{i}}(n)$, is white and its variance is σ_i^2. The sequence $\nu_{\mathrm{o}}(n)$ is the plant noise. It is uncorrelated with $u(n)$ and $\nu_{\mathrm{i}}(n)$. The correlation matrix of the plant input, $u(n)$, is denoted by $\mathbf{R}$. The model has to be selected so that the MSE at the model output is minimized.

(i) Find the correlation matrix of the model input and show that it shares the same set of eigenvectors with $\mathbf{R}$.
(ii) Derive the corresponding Wiener–Hopf equation.
(iii) Show that the difference between the plant tap-weight vector, $\mathbf{w}_{\mathrm{o}}$, and its estimate, $\hat{\mathbf{w}}_{\mathrm{o}}$, which is obtained through the Wiener–Hopf equation derived in (ii), is

$$\mathbf{w}_{\mathrm{o}}-\hat{\mathbf{w}}_{\mathrm{o}}=\sigma_i^2\sum_{l=0}^{N-1}\frac{\mathbf{q}_l^{\mathrm{T}}\mathbf{p}}{\lambda_l(\lambda_l+\sigma_i^2)}\mathbf{q}_l$$

where $\mathbf{q}_l$s are the eigenvectors of $\mathbf{R}$ and $\mathbf{p}$ is the cross-correlation between the model input and the desired output.
(iv) Show that

$$\mathrm{MMSE}=\sigma_o^2+\sigma_i^4\sum_{l=0}^{N-1}\frac{(\mathbf{q}_l^{\mathrm{T}}\mathbf{p})^2}{\lambda_l(\lambda_l+\sigma_i^2)^2}$$

(v) If the steepest-descent algorithm is used to find $\hat{\mathbf{w}}_{\mathrm{o}}$, find the time constants of the resulting learning curve. How do these time constants vary with σ_i^2? Discuss on the eigenvalue spread problem as σ_i^2 varies.

P5.10 Consider a transversal filter with the input and tap-weight vectors $\mathbf{x}(n)$ and $\mathbf{w}$, respectively, and output

$$y(n)=\mathbf{w}^{\mathrm{T}}\mathbf{x}(n)$$

Define the vector

$$\breve{\mathbf{x}}(n)=\mathbf{R}^{-1/2}\mathbf{x}(n)$$

where $\mathbf{R}=E[\mathbf{x}(n)\mathbf{x}^{\mathrm{T}}(n)]$. Let $\breve{\mathbf{x}}(n)$ be the input to a filter whose output is obtained through the equation

$$\breve{y}(n)=\breve{\mathbf{w}}^{\mathrm{T}}\breve{\mathbf{x}}(n)$$

where $\breve{\mathbf{w}}$ is the filter tap-weight vector.

(i) Derive an equation for $\breve{\mathbf{w}}$ so that the two outputs $y(n)$ and $\breve{y}(n)$ be the same.
(ii) Derive a steepest-descent update equation for the tap-weight vector $\breve{\mathbf{w}}$.
(iii) Derive an equation that demonstrates the variation of the tap weights of the filter as the steepest-descent algorithm derived in Part (ii) is running.
(iv) Find the time constants of the learning curve of the algorithm.
(v) Show that the update equation derived in (ii) is equivalent to Newton's algorithm.

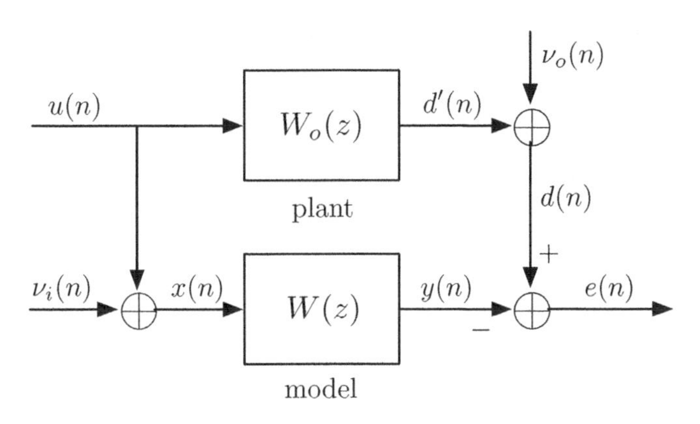

Figure P5.10

P5.11 Consider a two-tap Wiener filter, which is characterized by the following parameters

$$\mathbf{R} = \begin{bmatrix} 1 & 0.8 \\ 0.8 & 1 \end{bmatrix} \quad \text{and} \quad \mathbf{p} = \begin{bmatrix} 2 \\ 2 \end{bmatrix}$$

where $\mathbf{R}$ is the correlation matrix of the filter tap-input vector, $\mathbf{x}(n)$, and $\mathbf{p}$ is the cross-correlation between $\mathbf{x}(n)$ and the desired output, $d(n)$.

(i) Find the range of the step-size parameter μ, which ensures convergence of the steepest-descent algorithm. Does this result depend on the cross-correlation vector $\mathbf{p}$?

(ii) Run the steepest-descent algorithm for $\mu = 0.05$, 0.1, 0.5, and 1 and plot the corresponding trajectories in the (w_0, w_1)-plane.

(iii) For $\mu = 0.05$, plot $w_0(k)$ and $w_1(k)$, separately, as functions of the iteration index, k.

(iv) On the plots obtained in (iii), you should find that the variation of each tap weight is signified by two distinct time constants. This implies that the variation of each tap weight may be decomposed into a summation of two distinct exponential series. Explain this observation.

P5.12 For the modeling problem discussed in Examples 5.1 and 5.2, develop a MATLAB program to present the contour plots of the performance surface and plot the trajectories of the steepest-descent and Newton's algorithm on the same plane for $\mu = 0.05$ and $\alpha = 0$, 0.5, 0.75, and 0.9. Comment on your observations.
Hint: To generate the contour plots, follow the procedure discussed in Section 4.3.

P5.13 Consider the modeling problem depicted in Figure 5.3. Let $x(n) = 1$, for all values of n.

(i) Derive the steepest-descent algorithm that may be used to find the model parameters.

(ii) Derive an equation for the performance function of the present problem, and plot the contours that show its performance surface.

(iii) Run the algorithm that you have derived in (i) and find the model parameters that it converges to.

(iv) On the performance surface obtained in (ii), plot the trajectory showing the variation of the model parameters. Comment on your observation.

P5.14 Repeat Problem P5.13 for the case where $x(n) = (-1)^n$.

P5.15 All the derivations in this chapter were for the case that all the underlying processes were real-valued. In this problem, you are guided to repeat some of the results for the case where the underlying processes are complex-valued.

(i) Starting with the definition $\xi = E[|e(n)|^2] = E[e(n)e^*(n)]$, show that in the case where the underlying processes are complex-valued

$$\xi = E[|d(n)|^2] - \mathbf{w}^{\mathrm{H}}\mathbf{p} - \mathbf{p}^{\mathrm{H}}\mathbf{w} + \mathbf{w}^{\mathrm{H}}\mathbf{R}\mathbf{w}.$$

Here, the definitions for $\mathbf{w}$, $\mathbf{p}$, and $\mathbf{R}$ follow those in Section 3.5.

(ii) Show that

$$\nabla_{\mathbf{w}}^{C}\xi = 2(\mathbf{R}\mathbf{w} - \mathbf{p}).$$

(iii) Using the result of (ii), present a steepest-descent update equation for the case of this problem and compare it with Eq. (5.10).

(iv) Continuing with the result in (iii), can we say Eqs. (5.18), (5.23), and (5.25) are also valid in the case where the underlying processes are complex-valued? Why? Explain.

6

LMS Algorithm

The celebrated least-mean square (LMS) algorithm is introduced in this chapter. The LMS algorithm, which was first proposed by Widrow and Hoff in 1960, is the most widely used adaptive filtering algorithm, in practice. This wide spectrum of applications of the LMS algorithm can be attributed to its simplicity and robustness to signal statistics. The LMS algorithm has also been cited and worked upon by many researchers and over the years many modifications to that have been proposed. In this and the subsequent few chapters, we introduce and study several of such modifications.

6.1 Derivation of LMS Algorithm

Figure 6.1 depicts an N-tap transversal adaptive filter. The filter input, $x(n)$, desired output, $d(n)$, and the filter output

$$y(n) = \sum_{i=0}^{N-1} w_i(n)x(n-i) \tag{6.1}$$

are assumed to be real-valued sequences. The tap weights $w_0(n)$, $w_1(n)$, ..., $w_{N-1}(n)$ are selected, so that the difference (error)

$$e(n) = d(n) - y(n) \tag{6.2}$$

is minimized in some sense. It may be noted that the filter tap weights are explicitly indicated to be functions of the time index n. This signifies the fact that in an adaptive filter, in general, tap weights are time varying, as they are continuously being adapted, so that any variations in the signals statistics could be tracked. The LMS algorithm changes (adapts) the filter tap weights, so that $e(n)$ is minimized in the mean-square sense, thus the name LMS. When the processes $x(n)$ and $d(n)$ are jointly stationary, this algorithm converges to a set of tap weights, which, on average, are equal to the Wiener–Hopf solution discussed in Chapter 3. In other words, the LMS algorithm is a practical scheme for realizing Wiener filters, without explicitly solving the Wiener–Hopf equation. It is a sequential algorithm that can be used to adapt the tap weights of a filter by continuous observation of its input, $x(n)$, and desired output, $d(n)$.

The conventional LMS algorithm is a stochastic implementation of the steepest-descent algorithm. It simply replaces the cost function $\xi = E[e^2(n)]$ by its instantaneous coarse

Adaptive Filters: Theory and Applications, Second Edition. Behrouz Farhang-Boroujeny.

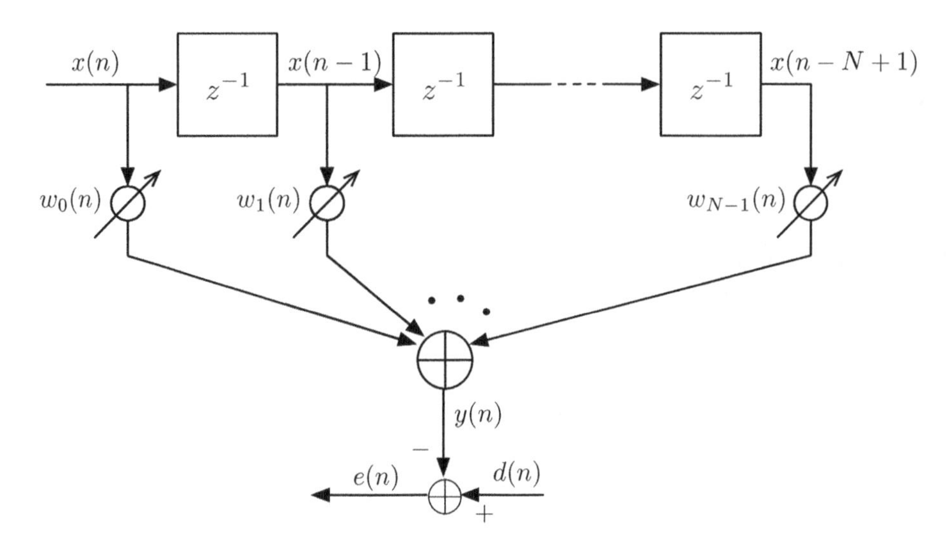

Figure 6.1 An N-tap transversal adaptive filter.

estimate $\hat{\xi}(n) = e^2(n)$. Substituting $\hat{\xi}(n) = e^2(n)$ for ξ in the steepest-descent recursion (5.9), of Chapter 5, and replacing the iteration index k by the time index n, we obtain

$$\mathbf{w}(n+1) = \mathbf{w}(n) - \mu \nabla e^2(n) \tag{6.3}$$

where $\mathbf{w}(n) = [w_0(n)\, w_1(n) \cdots w_{N-1}(n)]^{\mathrm{T}}$, μ is the algorithm step-size parameter and ∇ is the gradient operator defined as the column vector

$$\nabla = \left[\frac{\partial}{\partial w_0} \; \frac{\partial}{\partial w_1} \; \cdots \; \frac{\partial}{\partial w_{N-1}} \right]^{\mathrm{T}} \tag{6.4}$$

We note that the ith element of the gradient vector $\nabla e^2(n)$ is

$$\frac{\partial e^2(n)}{\partial w_i} = 2e(n) \frac{\partial e(n)}{\partial w_i} \tag{6.5}$$

Substituting Eq. (6.2) in the last factor on the right-hand side of Eq. (6.5) and noting that $d(n)$ is independent of w_i, we obtain

$$\frac{\partial e^2(n)}{\partial w_i} = -2e(n) \frac{\partial y(n)}{\partial w_i} \tag{6.6}$$

Substituting for $y(n)$ from Eq. (6.1), we get

$$\frac{\partial e^2(n)}{\partial w_i} = -2e(n)x(n-i) \tag{6.7}$$

Using Eqs. (6.4) and (6.7), we obtain

$$\nabla e^2(n) = -2e(n)\mathbf{x}(n) \tag{6.8}$$

Table 6.1 Summary of the LMS algorithm.

Input:	Tap-weight vector, $\mathbf{w}(n)$,
	Input vector, $\mathbf{x}(n)$,
	and desired output, $d(n)$
Output:	Filter output, $y(n)$,
	Tap-weight vector update, $\mathbf{w}(n+1)$

1. Filtering:
$$y(n) = \mathbf{w}^{\mathrm{T}}(n)\mathbf{x}(n)$$
2. Error estimation:
$$e(n) = d(n) - y(n)$$
3. Tap-weight vector adaptation:
$$\mathbf{w}(n+1) = \mathbf{w}(n) + 2\mu e(n)\mathbf{x}(n)$$

where $\mathbf{x}(n) = [x(n)x(n-1)\cdots x(n-N+1)]^{\mathrm{T}}$. Substituting this result in Eq. (6.3), we get

$$\mathbf{w}(n+1) = \mathbf{w}(n) + 2\mu e(n)\mathbf{x}(n) \tag{6.9}$$

This is referred to as the *LMS recursion*. It suggests a simple procedure for recursive adaptation of the filter coefficients after arrival of every new input sample, $x(n)$, and its corresponding desired output sample, $d(n)$. Equations (6.1), (6.2), and (6.9), in this order, specify the three steps required to complete each iteration of the LMS algorithm. Equation (6.1) is referred to as *filtering*. It is performed to obtain the filter output. Equation (6.2) is used to calculate the estimation error. Equation (6.9) is tap-weight adaptation recursion. Table 6.1 gives a summary of the LMS algorithm.

The eminent feature of the LMS algorithm, which has made it the most popular adaptive filtering scheme, is its simplicity. Its implementation requires, $2N+1$ multiplications (N multiplications for calculating the output $y(n)$, one to obtain $(2\mu) \times e(n)$ and N for scalar by vector multiplication $(2\mu e(n)) \times \mathbf{x}(n)$) and $2N$ additions. Another important feature of the LMS algorithm, which is equally important from implementation point of view, is its stable and robust performance against different signal conditions. This aspect of the LMS algorithm will be studied in the later chapters when it is compared with other alternative adaptive filtering algorithms. The major problem of the LMS recursion (6.9) is its slow convergence when the underlying input process is highly colored. This aspect of the LMS algorithm is discussed in the next section and solutions to that will be given in the later chapters.

6.2 Average Tap-Weight Behavior of the LMS Algorithm

Consider the case where the filter input, $x(n)$, and its desired output, $d(n)$, are stationary. In that case, the optimum tap-weight vector, $\mathbf{w}_{\mathrm{o}}$, of the transversal Wiener filter is fixed and can be obtained according to the Wiener–Hopf equation (3.24). Subtracting $\mathbf{w}_{\mathrm{o}}$ from both sides of Eq. (6.9), we obtain

$$\mathbf{v}(n+1) = \mathbf{v}(n) + 2\mu e(n)\mathbf{x}(n) \tag{6.10}$$

where $\mathbf{v}(n) = \mathbf{w}(n) - \mathbf{w}_o$ is the weight-error vector. We also note that

$$\begin{aligned} e(n) &= d(n) - \mathbf{w}^T(n)\mathbf{x}(n) \\ &= d(n) - \mathbf{x}^T(n)\mathbf{w}(n) \\ &= d(n) - \mathbf{x}^T(n)\mathbf{w}_o - \mathbf{x}^T(n)(\mathbf{w}(n) - \mathbf{w}_o) \\ &= e_o(n) - \mathbf{x}^T(n)\mathbf{v}(n) \end{aligned} \tag{6.11}$$

where

$$e_o(n) = d(n) - \mathbf{x}^T(n)\mathbf{w}_o \tag{6.12}$$

is the estimation error when the filter tap weights are optimum. Substituting Eq. (6.11) in Eq. (6.10) and rearranging, we obtain

$$\mathbf{v}(n+1) = (\mathbf{I} - 2\mu\mathbf{x}(n)\mathbf{x}^T(n))\mathbf{v}(n) + 2\mu e_o(n)\mathbf{x}(n) \tag{6.13}$$

where $\mathbf{I}$ is the identity matrix. Taking expectation on both sides of Eq. (6.13), we get

$$\begin{aligned} E[\mathbf{v}(n+1)] &= E[(\mathbf{I} - 2\mu\mathbf{x}(n)\mathbf{x}^T(n))\mathbf{v}(n)] + 2\mu E[e_o(n)\mathbf{x}(n)] \\ &= E[(\mathbf{I} - 2\mu\mathbf{x}(n)\mathbf{x}^T(n))\mathbf{v}(n)] \end{aligned} \tag{6.14}$$

where the last equality follows from the fact that $E[e_o(n)\mathbf{x}(n)] = \mathbf{0}$, according to the principle of orthogonality.

The main difficulty with any further analysis of the right-hand side of Eq. (6.14) is that it involves evaluation of the third- order moment vector $E[\mathbf{x}(n)\mathbf{x}^T(n)\mathbf{v}(n)]$, which, in general, is a difficult mathematical task. Different approaches have been adopted by researchers to overcome this mathematical hurdle. The most widely used analysis assumes that the present observation data samples $(\mathbf{x}(n), d(n))$ are independent of the past observations $(\mathbf{x}(n-1), d(n-1))$, $(\mathbf{x}(n-2), d(n-2))$, ... – see, for example, Widrow *et al.* (1976) and Feuer and Weinstein (1985). This is referred to as *the independence assumption*. Using the independence assumption, one can argue that as $\mathbf{v}(n)$ depends only on the past observations $(\mathbf{x}(n-1), d(n-1))$, $(\mathbf{x}(n-2), d(n-2))$, ..., it is independent of $\mathbf{x}(n)$, and thus

$$E[\mathbf{x}(n)\mathbf{x}^T(n)\mathbf{v}(n)] = E[\mathbf{x}(n)\mathbf{x}^T(n)]E[\mathbf{v}(n)] \tag{6.15}$$

We may note that in most of the practical cases, the independence assumption is questionable. For example, in the case of a length N transversal filter, the input vectors

$$\mathbf{x}(n) = [x(n)x(n-1)\cdots x(n-N+1)]^T$$

and

$$\mathbf{x}(n-1) = [x(n-1)x(n-2)\cdots x(n-N)]^T$$

have $(N-1)$ terms in common, out of N. Nevertheless, experience with the LMS algorithm has shown that the predictions made by the independence assumption match the computer simulations and the actual performance of the LMS algorithm, in practice.

This may be explained as follows. The tap-weight vector $\mathbf{w}(n)$ at any given time has been affected by the whole past history of the observation data samples $(\mathbf{x}(n-1), d(n-1))$, $(\mathbf{x}(n-2), d(n-2))$, ... When the step-size parameter μ is small, the share of the last N observations in the present value of $\mathbf{w}(n)$ is small, and thus we may say $\mathbf{x}(n)$ and $\mathbf{w}(n)$ are weakly dependent. This clearly leads to Eq. (6.15), with some degree of approximation, if we can assume that the observation samples, which are apart from each other at a distance of N or greater, are weakly dependent. This reasoning seems to be more appealing than the independence assumption. In any case, we use Eq. (6.15) and other similar equations (approximations), which will be introduced later to proceed with our analysis in this book.

Substituting Eq. (6.15) in Eq. (6.14), we obtain

$$E[\mathbf{v}(n+1)] = (\mathbf{I} - 2\mu\mathbf{R})E[\mathbf{v}(n)] \tag{6.16}$$

where $\mathbf{R} = E[\mathbf{x}(n)\mathbf{x}^{\mathrm{T}}(n)]$ is the correlation matrix of the input vector $\mathbf{x}(n)$.

Comparing the recursions (6.16) and (5.14), we find that they are of exactly the same mathematical form. The deterministic weight-error vector $\mathbf{v}(k)$ in Eq. (5.14) of the steepest-descent algorithm is replaced by the averaged weight-error vector $E[\mathbf{v}(n)]$ of the LMS algorithm. This suggests that, on average, the LMS algorithm behaves just like the steepest-descent algorithm. In particular, similar to the steepest-descent algorithm, the LMS algorithm is controlled by N modes of convergence, which are characterized by the eigenvalues of the correlation matrix $\mathbf{R}$. Consequently, the convergence behavior of the LMS algorithm is directly linked to the eigenvalue spread of the correlation matrix $\mathbf{R}$. Furthermore, recalling the relationship between the eigenvalue spread of $\mathbf{R}$ and the power spectrum of $x(n)$, we can say that the convergence of the LMS algorithm is directly related to the flatness in the spectral content of the underlying input process.

Following a similar procedure as in Chapter 5, by manipulating Eq. (6.16), one can show that $E[\mathbf{v}(n)]$ converges to zero when μ remains within the range

$$0 < \mu < \frac{1}{\lambda_{\max}} \tag{6.17}$$

where $\lambda_{\max}$ is the maximum eigenvalue of $\mathbf{R}$. However, we should point out here that the above range does not necessarily guarantee the stability of the LMS algorithm. *The convergence of the LMS algorithm requires convergence of the mean of* $\mathbf{w}(n)$ *toward* $\mathbf{w}_{\mathrm{o}}$ *and also convergence of the variance of elements of* $\mathbf{w}(n)$ *to some limited values.* As we shall show later, to guarantee the stability of the LMS algorithm, the latter requirement imposes a much stringent condition on the size of μ. Furthermore, we may note that the independence assumption used to obtain Eq. (6.16) was based on the assumption that μ was very small. The upper limit of μ in Eq. (6.17) may badly violate this assumption. Thus, the validity of Eq. (6.17), even for the convergence of $E[\mathbf{w}(n)]$, is questionable.

Example 6.1

Consider the modeling problem of Example 5.1 which is repeated in Figure 6.2, for convenience. As in Example 5.1, the input signal, $x(n)$, is generated by passing a white noise signal, $\nu(n)$, through a coloring filter with the system function

$$H(z) = \frac{\sqrt{1-\alpha^2}}{1-\alpha z^{-1}} \tag{6.18}$$

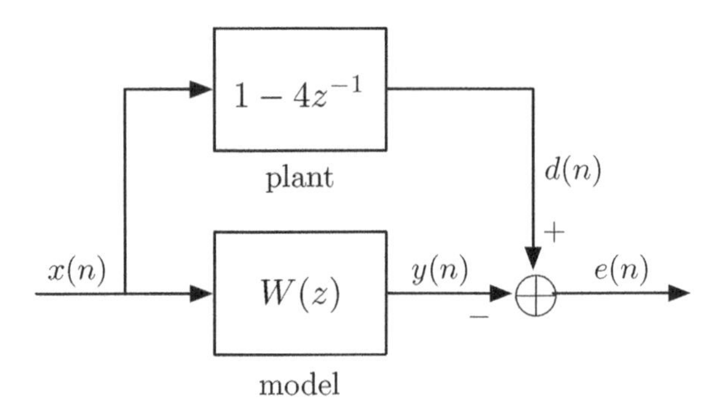

Figure 6.2 A modeling problem.

where α is a real-valued constant in the range of -1 to $+1$. The plant is a two-tap FIR system with the system function $P(z) = 1 - 4z^{-1}$. An adaptive filter with the system function $W(z) = w_0 + w_1 z^{-1}$ is used to identify the plant. Here, the LMS algorithm is used to find the optimum values of the tap weights w_0 and w_1. We want to see, as the iteration number increases, how the tap weights w_0 and w_1 converge toward the plant coefficients, 1 and -4, respectively. We examine this for different values of the parameter α. We recall from Example 5.1 that the parameter α controls the eigenvalue spread of the correlation matrix $\mathbf{R}$ of the input samples to the filter $W(z)$.

Figure 6.3a–d presents four plots showing typical trajectories of the LMS algorithm, which have been obtained for the values of $\alpha = 0$, 0.5, 0.75, and 0.9, respectively. Also shown in the figures are the contour plots that highlight the performance surface of the filter. The results presented are for $\mu = 0.01$ and 150 iterations, for all cases. In comparison with the parameters used in Figure 5.4 of Example 5.1, here μ is selected five times smaller, while the number of iterations is chosen five times larger. Comparing the results here with those of Figure 5.4, we can clearly see that, as predicted above, the LMS algorithm, on average, follows the same trajectories as the steepest-descent algorithm. In particular, the convergence of the LMS algorithm along the steepest-descent slope of the performance surface is clearly observed. Also, we note that in the case $\alpha = 0$, which corresponds to a white input sequence, the convergence of the LMS algorithm is almost complete within 150 iterations. However, the other three cases require some more iterations before they converge to the vicinity of the minimum point of the performance surface. This, as was noted in Example 5.1, can be understood if one notes that the eigenvalues of the correlation matrix $\mathbf{R}$ of the input samples to the adaptive filter are $\lambda_0 = 1 + \alpha$ and $\lambda_1 = 1 - \alpha$, and for α close to 1, the time constant $\tau_1 = (1/4\mu\lambda_1)$ may be very large.

6.3 MSE Behavior of the LMS Algorithm

In this section, the variation of $\xi(n) = E[e^2(n)]$ as LMS algorithm is being iterated is studied.[1] This study is directly related to the convergence of LMS algorithm.

[1] The derivations provided in this section follow the work of Feuer and Weinstein (1985). Prior to Feuer and Weinstein (1985), Horowitz and Senne (1981) have also arrived at similar results, using a different approach.

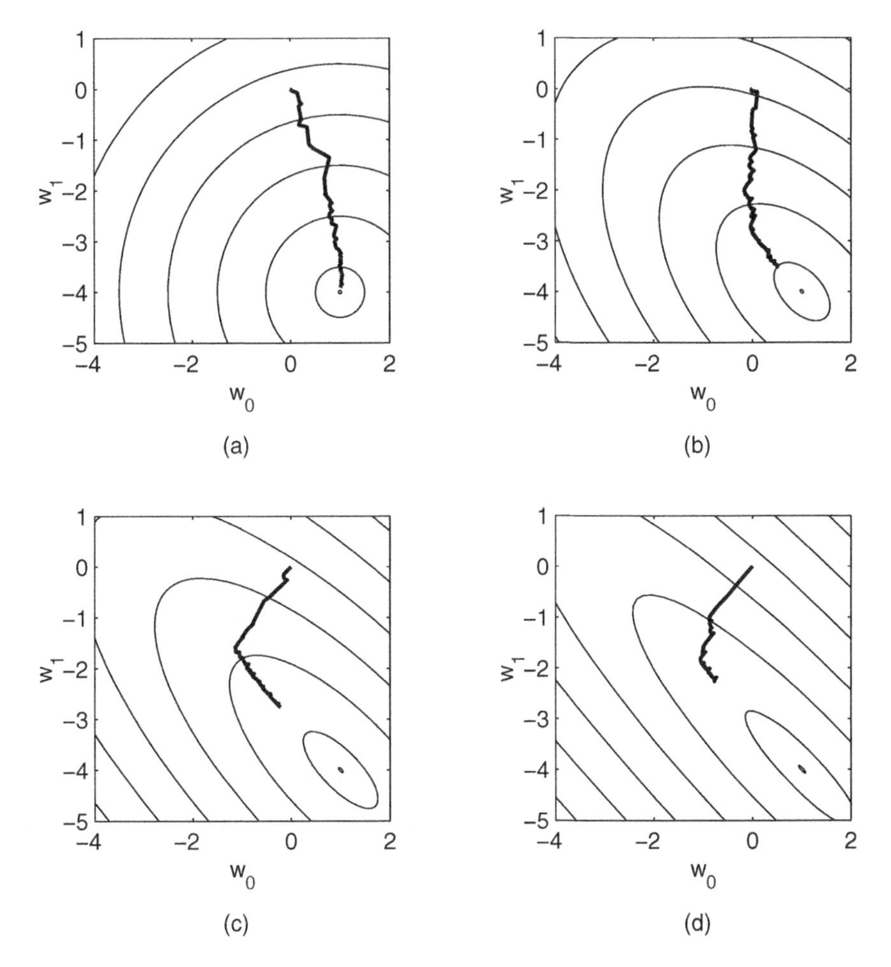

Figure 6.3 Trajectories showing how the filter tap weights vary when the LMS algorithm is used: (a) $\alpha = 0$, (b) $\alpha = 0.5$, (c) $\alpha = 0.75$, and (d) $\alpha = 0.9$. Each plot is based on 150 iterations and $\mu = 0.01$.

In the derivations that follow, it is assumed that

1. the input, $x(n)$, and desired output, $d(n)$, are zero-mean stationary processes;
2. $x(n)$ and $d(n)$ are jointly Gaussian-distributed random variables, for all n; and
3. at time n, the tap-weight vector $\mathbf{w}(n)$ is independent of the input vector $\mathbf{x}(n)$ and the desired output $d(n)$.

The validity of the last assumption is justified for small values of the step-size parameter μ, as was discussed in the previous section. This, as was noted before, is referred to as the *independence assumption*. Assumption 1 greatly simplifies the analysis. Assumption 2 results in some simplification in the final results, as the third- and higher order moments, which appear in the derivations, can be expressed in terms of the second-order moments when the underlying random variables are jointly Gaussian.

6.3.1 Learning Curve

We note from Eq. (6.11) that the estimation error, $e(n)$, can be expressed as

$$e(n) = e_o(n) - \mathbf{v}^T(n)\mathbf{x}(n) \quad (6.19)$$

Squaring both sides of Eq. (6.19) and taking the expectation on both sides, we obtain

$$E[e^2(n)] = E[e_o^2(n)] + E[(\mathbf{v}^T(n)\mathbf{x}(n))^2] - 2E[e_o(n)\mathbf{v}^T(n)\mathbf{x}(n)] \quad (6.20)$$

Noting that $\mathbf{v}^T(n)\mathbf{x}(n) = \mathbf{x}^T(n)\mathbf{v}(n)$ and using the independence assumption, the second term on the right-hand side of Eq. (6.20) can be expanded as[2]

$$\begin{aligned} E[(\mathbf{v}^T(n)\mathbf{x}(n))^2] &= E[\mathbf{v}^T(n)\mathbf{x}(n)\mathbf{x}^T(n)\mathbf{v}(n)] \\ &= E[\mathbf{v}^T(n)E[\mathbf{x}(n)\mathbf{x}^T(n)]\mathbf{v}(n)] \\ &= E[\mathbf{v}^T(n)\mathbf{R}\mathbf{v}(n)] \end{aligned} \quad (6.21)$$

Noting that $E[(\mathbf{v}^T(n)\mathbf{x}(n))^2]$ is a scalar and using Eq. (6.21), we may also write

$$\begin{aligned} E[(\mathbf{v}^T(n)\mathbf{x}(n))^2] &= \mathrm{tr}[E[(\mathbf{v}^T(n)\mathbf{x}(n))^2]] \\ &= \mathrm{tr}[E[\mathbf{v}^T(n)\mathbf{R}\mathbf{v}(n)]] \\ &= E[\mathrm{tr}[\mathbf{v}^T(n)\mathbf{R}\mathbf{v}(n)]] \end{aligned} \quad (6.22)$$

where $\mathrm{tr}[\cdot]$ denotes the trace of a matrix, and in writing the last identity we have noted that "trace" and "expectation" are linear operators and, thus, could be exchanged. This result can be further simplified using the following result from matrix algebra. For any pair of N-by-M and M-by-N matrices $\mathbf{A}$ and $\mathbf{B}$,

$$\mathrm{tr}[\mathbf{AB}] = \mathrm{tr}[\mathbf{BA}] \quad (6.23)$$

Using this identity, we obtain

$$\begin{aligned} E[\mathrm{tr}[\mathbf{v}^T(n)\mathbf{R}\mathbf{v}(n)]] &= E[\mathrm{tr}[\mathbf{v}(n)\mathbf{v}^T(n)\mathbf{R}]] \\ &= \mathrm{tr}[E[\mathbf{v}(n)\mathbf{v}^T(n)]\mathbf{R}] \end{aligned} \quad (6.24)$$

Defining the correlation matrix of the weight-error vector $\mathbf{v}(n)$ as

$$\mathbf{K}(n) = E[\mathbf{v}(n)\mathbf{v}^T(n)] \quad (6.25)$$

[2] We note that when x and y are two *independent* random variables

$$E[xy] = E[x]E[y] = E[xE[y]]$$

Also,

$$E[x^2y^2] = E[x^2]E[y^2] = E[x^2E[y^2]] = E[xE[y^2]x]$$

Similar procedure is used to arrive at Eq. (6.21) and in other similar derivations that appear in the rest of this book.

the above result reduces to

$$E[(\mathbf{v}^{\mathrm{T}}(n)\mathbf{x}(n))^2] = \mathrm{tr}[\mathbf{K}(n)\mathbf{R}] \tag{6.26}$$

Using the independence assumption and noting that $e_o(n)$ is a scalar, the last term on the right-hand side of Eq. (6.20) can be written as

$$\begin{aligned} E[e_o(n)\mathbf{v}^{\mathrm{T}}(n)\mathbf{x}(n)] &= E[\mathbf{v}^{\mathrm{T}}(n)\mathbf{x}(n)e_o(n)] \\ &= E[\mathbf{v}^{\mathrm{T}}(n)]E[\mathbf{x}(n)e_o(n)] \\ &= 0 \end{aligned} \tag{6.27}$$

where the last step follows from the principle of orthogonality, which states that the optimal estimation error and the input data samples to a Wiener filter are orthogonal (uncorrelated), that is, $E[e_o(n)\mathbf{x}(n)] = 0$.

Using Eqs. (6.26) and (6.27) in Eq. (6.20), we obtain

$$\xi(n) = E[e^2(n)] = \xi_{\min} + \mathrm{tr}[\mathbf{K}(n)\mathbf{R}] \tag{6.28}$$

where $\xi_{\min} = E[e_o^2(n)]$, that is, the minimum mean-squared error (MSE) at the filter output.

This result may be written in a more convenient form for our further analysis later, if we recall from Chapter 4 that the correlation matrix $\mathbf{R}$ may be decomposed as

$$\mathbf{R} = \mathbf{Q}\mathbf{\Lambda}\mathbf{Q}^{\mathrm{T}} \tag{6.29}$$

where $\mathbf{Q}$ is the N-by-N matrix whose columns are the eigenvectors of $\mathbf{R}$, and $\mathbf{\Lambda}$ is the diagonal matrix consisting of the eigenvalues $\lambda_0, \lambda_1, \ldots, \lambda_{N-1}$ of $\mathbf{R}$. Substituting Eq. (6.29) in Eq. (6.28) and using the identity (6.23), we obtain

$$\xi(n) = \xi_{\min} + \mathrm{tr}[\mathbf{K}'(n)\mathbf{\Lambda}] \tag{6.30}$$

where $\mathbf{K}'(n) = \mathbf{Q}^{\mathrm{T}}\mathbf{K}(n)\mathbf{Q}$. Furthermore, using Eq. (6.25), and recalling the definition $\mathbf{v}'(n) = \mathbf{Q}^{\mathrm{T}}\mathbf{v}(n)$, from Chapter 4, we find that

$$\mathbf{K}'(n) = E[\mathbf{v}'(n)\mathbf{v}'^{\mathrm{T}}(n)] \tag{6.31}$$

Also, we recall that $\mathbf{v}'(n)$ is the weight-error vector in the coordinates defined by the basis vectors specified by the eigenvectors of $\mathbf{R}$.

Noting that $\mathbf{\Lambda}$ is a diagonal matrix, Eq. (6.30) can be expanded as

$$\xi(n) = \xi_{\min} + \sum_{i=0}^{N-1} \lambda_i k'_{ii}(n) \tag{6.32}$$

where $k'_{ij}(n)$ is the ijth element of the matrix $\mathbf{K}'(n)$.

The plot $\xi(n)$ versus the time index n, defined by Eq. (6.28) or its alternative forms in Eq. (6.30) or Eq. (6.32), is called the *learning curve of the LMS algorithm.* It is very similar to the learning curve of the steepest-descent algorithm because according to the derivations in the previous section, the LMS algorithm on average follows the same trajectory as the steepest-descent algorithm. The noisy variations of the filter tap weights in the case of LMS algorithm introduce some additional error and push up its learning curve compared to that of the steepest-descent algorithm. However, when the step-size parameter, μ, is small (which is usually the case in practice), one finds that the difference between the two curves is noticeable only when they have converged and approached their steady state. The following example shows this.

Example 6.2

Figure 6.4 shows the learning curves of the LMS algorithm and the steepest-descent algorithm for the modeling problem discussed in Examples 5.1 and 6.1, when $\alpha = 0.75$ and $\mu = 0.01$. For both cases, the filter tap weights have been initialized with $w_0(0) = w_1(0) = 0$. The learning curve of the steepest-descent algorithm has been obtained by inserting the numerical values of the parameters in Eq. (5.31). The learning curve of the LMS algorithm is obtained by an ensemble average of the sequence $e^2(n)$ over 1000 independent runs. We note that the two curves match closely. The learning curve of the LMS algorithm remains slightly above the learning curve of the steepest-descent algorithm. This is because of the use of noisy estimates of the gradient vector in the LMS algorithm.

We shall emphasize that, despite the noisy variation of the filter tap weights, the learning curve of the LMS algorithm matches closely with the theoretical results of the steepest-descent algorithm. In particular, Eq. (5.31) is applicable and the time constant equation

$$\tau_i = \frac{1}{4\mu\lambda_i} \tag{6.33}$$

can be used for predicting the transient behavior of the LMS algorithm.

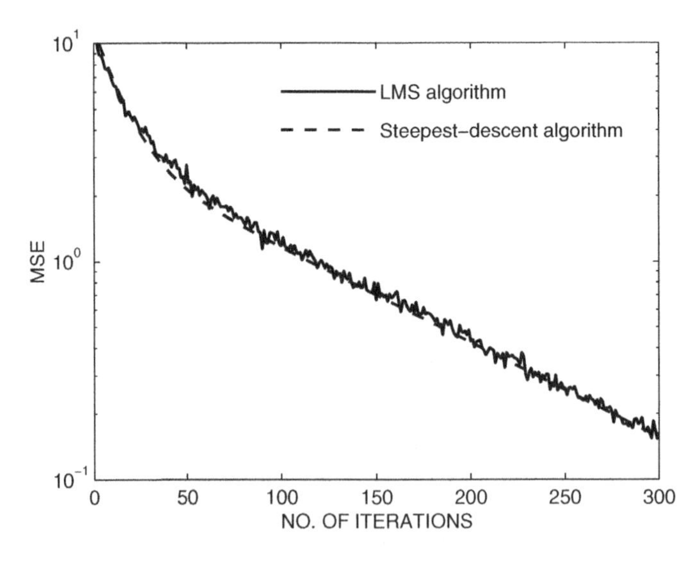

Figure 6.4 Learning curves of the steepest-descent algorithm and LMS algorithm for the modeling problem of Figure 6.2 and the parameter values of $\alpha = 0.75$ and $\mu = 0.01$.

6.3.2 Weight-Error Correlation Matrix

The weight-error correlation matrix $\mathbf{K}(n)$ plays an important role in the study of the LMS algorithm. From Eq. (6.28), we note that the value of $\xi(n)$ is directly related to $\mathbf{K}(n)$. Equation (6.28) implies that the stability of the LMS algorithm is guaranteed if, as n increases, the elements of $\mathbf{K}(n)$ remain bounded. Also, from Eqs. (6.30) and (6.32), we note that $\mathbf{K}'(n)$ may equivalently be used in the study of the convergence of LMS algorithm. Here, we develop a time-update equation for $\mathbf{K}'(n)$.

Multiplying both sides of Eq. (6.13) from the left by $\mathbf{Q}^{\mathrm{T}}$, using the definitions $\mathbf{v}'(n) = \mathbf{Q}^{\mathrm{T}}\mathbf{v}(n)$ and $\mathbf{x}'(n) = \mathbf{Q}^{\mathrm{T}}\mathbf{x}(n)$, and rearranging the result, we obtain

$$\mathbf{v}'(n+1) = (\mathbf{I} - 2\mu\mathbf{x}'(n)\mathbf{x}'^{\mathrm{T}}(n))\mathbf{v}'(n) + 2\mu e_{\mathrm{o}}(n)\mathbf{x}'(n) \tag{6.34}$$

Next, we multiply both sides of Eq. (6.34) from the right by their respective transposes, take statistical expectation of the result and expand to obtain

$$\begin{aligned}
\mathbf{K}'(n+1) = \mathbf{K}'(n) &- 2\mu E[\mathbf{x}'(n)\mathbf{x}'^{\mathrm{T}}(n)\mathbf{v}'(n)\mathbf{v}'^{\mathrm{T}}(n)] \\
&-2\mu E[\mathbf{v}'(n)\mathbf{v}'^{\mathrm{T}}(n)\mathbf{x}'(n)\mathbf{x}'^{\mathrm{T}}(n)] \\
&+4\mu^2 E[\mathbf{x}'(n)\mathbf{x}'^{\mathrm{T}}(n)\mathbf{v}'(n)\mathbf{v}'^{\mathrm{T}}(n)\mathbf{x}'(n)\mathbf{x}'^{\mathrm{T}}(n)] \\
&+2\mu E[e_{\mathrm{o}}(n)\mathbf{x}'(n)\mathbf{v}'^{\mathrm{T}}(n)] \\
&+2\mu E[e_{\mathrm{o}}(n)\mathbf{v}'(n)\mathbf{x}'^{\mathrm{T}}(n)] \\
&-4\mu^2 E[e_{\mathrm{o}}(n)\mathbf{x}'(n)\mathbf{v}'^{\mathrm{T}}(n)\mathbf{x}'(n)\mathbf{x}'^{\mathrm{T}}(n)] \\
&-4\mu^2 E[e_{\mathrm{o}}(n)\mathbf{x}'(n)\mathbf{x}'^{\mathrm{T}}(n)\mathbf{v}'(n)\mathbf{x}'^{\mathrm{T}}(n)] \\
&+4\mu^2 E[e_{\mathrm{o}}^2(n)\mathbf{x}'(n)\mathbf{x}'^{\mathrm{T}}(n)].
\end{aligned} \tag{6.35}$$

We note that the independence assumption (which states that $\mathbf{v}(n)$ is independent of $\mathbf{x}(n)$ and $d(n)$) is also applicable to the transformed (prime) variables in Eq. (6.35). That is, the random vector $\mathbf{v}'(n)$ is independent of $\mathbf{x}'(n)$ and $d(n)$. This is immediately observed if we note that $\mathbf{x}'(n)$ and $\mathbf{v}'(n)$ are independently obtained from $\mathbf{x}(n)$ and $\mathbf{v}(n)$, respectively. Also, the assumption that $d(n)$ and $\mathbf{x}(n)$ are zero-mean and mutually Gaussian-distributed implies that $d(n)$ and $\mathbf{x}'(n)$ are also zero-mean and jointly Gaussian. Furthermore, using the definition $\mathbf{x}'(n) = \mathbf{Q}^{\mathrm{T}}\mathbf{x}(n)$, we note that the principle of orthogonality, that is, $E[e_{\mathrm{o}}(n)\mathbf{x}(n)] = \mathbf{0}$, may also be written as

$$E[e_{\mathrm{o}}(n)\mathbf{x}'(n)] = \mathbf{0} \tag{6.36}$$

Noting this which shows that $e_{\mathrm{o}}(n)$ and $\mathbf{x}'(n)$ are uncorrelated, and the fact that $d(n)$ and $\mathbf{x}'(n)$ and, thus, $e_{\mathrm{o}}(n)$ and $\mathbf{x}'(n)$ are jointly Gaussian, one can say that the random variables $e_{\mathrm{o}}(n)$ and $\mathbf{x}'(n)$ are independent of each other.[3] Also, the independence of $\mathbf{v}'(n)$ from $d(n)$ and $\mathbf{x}(n)$ implies that $\mathbf{v}'(n)$ and $e_{\mathrm{o}}(n)$ are independent, since $e_{\mathrm{o}}(n)$ depends only on $d(n)$ and $\mathbf{x}(n)$. With these points in mind, the expectations on the right-hand side

[3] We recall that when random variables x and y are Gaussian and uncorrelated, they are also independent (Papoulis, 1991).

of Eq. (6.35) can be simplified as follows:

$$E[\mathbf{x}'(n)\mathbf{x}'^{\mathrm{T}}(n)\mathbf{v}'(n)\mathbf{v}'^{\mathrm{T}}(n)] = E[\mathbf{x}'(n)\mathbf{x}'^{\mathrm{T}}(n)]E[\mathbf{v}'(n)\mathbf{v}'^{\mathrm{T}}(n)]$$

$$= \mathbf{\Lambda}\mathbf{K}'(n) \tag{6.37}$$

where we have noted that $E[\mathbf{x}'(n)\mathbf{x}'^{\mathrm{T}}(n)] = \mathbf{\Lambda}$. Similarly

$$E[\mathbf{v}'(n)\mathbf{v}'^{\mathrm{T}}(n)\mathbf{x}'(n)\mathbf{x}'^{\mathrm{T}}(n)] = \mathbf{K}'(n)\mathbf{\Lambda} \tag{6.38}$$

Simplification of the third expectation requires some algebraic manipulations. These are provided in Appendix 6A. The result is

$$E[\mathbf{x}'(n)\mathbf{x}'^{\mathrm{T}}(n)\mathbf{v}'(n)\mathbf{v}'^{\mathrm{T}}(n)\mathbf{x}'(n)\mathbf{x}'^{\mathrm{T}}(n)] = 2\mathbf{\Lambda}\mathbf{K}'(n)\mathbf{\Lambda} + \mathrm{tr}[\mathbf{\Lambda}\mathbf{K}'(n)]\mathbf{\Lambda} \tag{6.39}$$

Using the independence of $e_{\mathrm{o}}(n)$, $\mathbf{x}'(n)$, and $\mathbf{v}'(n)$ and noting that $e_{\mathrm{o}}(n)$ has zero mean, we get

$$E[e_{\mathrm{o}}(n)\mathbf{x}'(n)\mathbf{v}'^{\mathrm{T}}(n)] = E[e_{\mathrm{o}}(n)]E[\mathbf{x}'(n)\mathbf{v}'^{\mathrm{T}}(n)] = \mathbf{0} \tag{6.40}$$

where $\mathbf{0}$ denotes the N-by-N zero matrix. Similarly,

$$E[e_{\mathrm{o}}(n)\mathbf{v}'(n)\mathbf{x}'^{\mathrm{T}}(n)] = \mathbf{0} \tag{6.41}$$

$$E[e_{\mathrm{o}}(n)\mathbf{x}'(n)\mathbf{v}'^{\mathrm{T}}(n)\mathbf{x}'(n)\mathbf{x}'^{\mathrm{T}}(n)] = \mathbf{0} \tag{6.42}$$

$$E[e_{\mathrm{o}}(n)\mathbf{x}'(n)\mathbf{x}'^{\mathrm{T}}(n)\mathbf{v}'(n)\mathbf{x}'^{\mathrm{T}}(n)] = \mathbf{0} \tag{6.43}$$

and

$$E[e_{\mathrm{o}}^2(n)\mathbf{x}'(n)\mathbf{x}'^{\mathrm{T}}(n)] = E[e_{\mathrm{o}}^2(n)]E[\mathbf{x}'(n)\mathbf{x}'^{\mathrm{T}}(n)]$$

$$= \xi_{\min}\mathbf{\Lambda} \tag{6.44}$$

Substituting Eqs. (6.37)–(6.44) in Eq. (6.35), we obtain

$$\mathbf{K}'(n+1) = \mathbf{K}'(n) - 2\mu(\mathbf{\Lambda}\mathbf{K}'(n) + \mathbf{K}'(n)\mathbf{\Lambda})$$

$$+8\mu^2\mathbf{\Lambda}\mathbf{K}'(n)\mathbf{\Lambda} + 4\mu^2\mathrm{tr}[\mathbf{\Lambda}\mathbf{K}'(n)]\mathbf{\Lambda} + 4\mu^2\xi_{\min}\mathbf{\Lambda} \tag{6.45}$$

The difference equation Eq. (6.45) is difficult to be handled. However, the fact that $\mathbf{\Lambda}$ is a diagonal matrix can be used to simplify the analysis. Consider the ith diagonal element of $\mathbf{K}'(n)$ and note that its corresponding time-update equation, obtained from Eq. (6.45), is

$$k'_{ii}(n+1) = \rho_i k'_{ii}(n) + 4\mu^2\lambda_i \sum_{j=0}^{N-1} \lambda_j k'_{jj}(n) + 4\mu^2\xi_{\min}\lambda_i \tag{6.46}$$

where

$$\rho_i = 1 - 4\mu\lambda_i + 8\mu^2\lambda_i^2 \tag{6.47}$$

and we have noted that

$$\mathrm{tr}[\Lambda\mathbf{K}'(n)] = \sum_{j=0}^{N-1} \lambda_j k'_{jj}(n) \tag{6.48}$$

The important feature of Eq. (6.46) to be noted is that the update of $k'_{ii}(n)$ is independent of the off-diagonal elements of $\mathbf{K}'(n)$. Furthermore, we note that as $\mathbf{K}'(n)$ is a correlation matrix, $k'^2_{ij}(n) \leq k'_{ii}(n)k'_{jj}(n)$, for all values of i and j. This suggests that the convergence of the diagonal elements of $\mathbf{K}'(n)$ is sufficient to ensure the convergence of all elements of that, which, in turn, are required to guarantee the stability of the LMS algorithm. Thus, we concentrate on Eq. (6.46), for $i = 0, 1, \ldots, N-1$.

Let us define the column vectors

$$\mathbf{k}'(n) = [k'_{00}(n)\ k'_{11}(n) \cdots k'_{N-1,N-1}(n)]^{\mathrm{T}} \tag{6.49}$$

and

$$\boldsymbol{\lambda} = [\lambda_0 \lambda_1 \cdots \lambda_{N-1}]^{\mathrm{T}} \tag{6.50}$$

and the matrix

$$\mathbf{F} = \mathrm{diag}[\rho_0, \rho_1, \ldots, \rho_{N-1}] + 4\mu^2 \boldsymbol{\lambda}\boldsymbol{\lambda}^{\mathrm{T}} \tag{6.51}$$

where diag$[\cdots]$ refers to a diagonal matrix consisting of the indicated elements. Considering these definitions and the time-update equation (6.46), for $i = 0, 1, \ldots, N-1$, we get

$$\mathbf{k}'(n+1) = \mathbf{F}\mathbf{k}'(n) + 4\mu^2 \xi_{\min} \boldsymbol{\lambda} \tag{6.52}$$

The difference equation (6.52) can be used to study the stability of the LMS algorithm. As was noted before, the stability of the LMS algorithm is guaranteed if the elements of $\mathbf{K}(n)$ (or, equivalently, the elements of $\mathbf{k}'(n)$) remain bounded, as n increases. The necessary and sufficient condition for this to happen is that all the eigenvalues of the coefficient matrix $\mathbf{F}$ of Eq. (6.52) be less than 1, in magnitude. Feuer and Weinstein (1985) have discussed on the eigenvalues of $\mathbf{F}$ and given the condition required to keep the LMS algorithm stable. Here, we will comment on the stability of the LMS algorithm in an indirect way. This is done after we find an expression for the excess MSE of the LMS algorithm, which is defined in the following.

6.3.3 *Excess MSE and Misadjustment*

We note that even when the filter tap-weight vector $\mathbf{w}(n)$ approaches its optimal value, $\mathbf{w}_o$, and the mean of the stochastic gradient vector $\nabla e^2(n)$ tends to zero, the instantaneous value of this gradient may not be zero. This results in a perturbation of the tap-weight vector $\mathbf{w}(n)$ around its optimal value, $\mathbf{w}_o$, even after convergence of the algorithm. This, in turn, increases the MSE of the LMS algorithm to a level above the minimum MSE, which would be obtained if the filter tap weights were fixed at their optimal values. This additional error is called *excess MSE*. In other words, the excess MSE of an adaptive filter is defined as the difference between its steady-state MSE and its minimum MSE.

The steady-state MSE of the LMS algorithm can be found from Eq. (6.28) or, equivalently, Eq. (6.30) or Eq. (6.32) by letting the time index n to tend to infinity. Thus, subtracting $\xi_{\min}$ from both sides of Eq. (6.28), we obtain

$$\xi_{\mathrm{excess}} = \mathrm{tr}[\mathbf{K}(\infty)\mathbf{R}] \tag{6.53}$$

where ξ_{excess} denotes the excess MSE. Alternatively, if Eq. (6.32) is used, we get

$$\xi_{\text{excess}} = \sum_{i=0}^{N-1} \lambda_i k'_{ii}(\infty) = \boldsymbol{\lambda}^{\mathrm{T}} \mathbf{k}'(\infty) \tag{6.54}$$

When the LMS algorithm is convergent, $\mathbf{k}'(n)$ converges to a bounded steady-state value and we can say $\mathbf{k}'(n+1) = \mathbf{k}'(n)$, when $n \to \infty$. Noting this, from Eq. (6.52), we obtain

$$\mathbf{k}'(\infty) = 4\mu^2 \xi_{\min} (\mathbf{I} - \mathbf{F})^{-1} \boldsymbol{\lambda} \tag{6.55}$$

Substituting this in Eq. (6.54), we get

$$\xi_{\text{excess}} = 4\mu^2 \xi_{\min} \boldsymbol{\lambda}^{\mathrm{T}} (\mathbf{I} - \mathbf{F})^{-1} \boldsymbol{\lambda} \tag{6.56}$$

We note that ξ_{excess} is proportional to $\xi_{\min}$. This is intuitively understandable, if we note that when $\mathbf{w}(n)$ has converged to a vicinity of $\mathbf{w}_{\mathrm{o}}$, the variance of the elements of the stochastic gradient vector $\nabla e^2(n)$ is proportional to $\xi_{\min}$ (Problem P6.1). We also note that similar to $\xi_{\min}$, ξ_{excess} also has the units of power. It is convenient to normalize ξ_{excess} to $\xi_{\min}$, so that a dimension-free degradation measure is obtained. The result is called *misadjustment* and denoted as $\mathcal{M}$. For the LMS algorithm, from Eq. (6.56), we obtain

$$\mathcal{M} = \frac{\xi_{\text{excess}}}{\xi_{\min}} = 4\mu^2 \boldsymbol{\lambda}^{\mathrm{T}} (\mathbf{I} - \mathbf{F})^{-1} \boldsymbol{\lambda} \tag{6.57}$$

The special structure of the matrix $(\mathbf{I} - \mathbf{F})$ can be used to find its inverse.

We note from Eq. (6.51) that

$$\mathbf{I} - \mathbf{F} = \operatorname{diag}[1 - \rho_0, 1 - \rho_1, \ldots, 1 - \rho_{N-1}] - 4\mu^2 \boldsymbol{\lambda}\boldsymbol{\lambda}^{\mathrm{T}} \tag{6.58}$$

On the other hand, we note that according to the matrix inversion lemma, for an arbitrary positive-definite N-by-N matrix $\mathbf{A}$, any N-by-1 vector $\mathbf{a}$ and a scalar α,

$$(\mathbf{A} + \alpha \mathbf{a}\mathbf{a}^{\mathrm{T}})^{-1} = \mathbf{A}^{-1} - \frac{\alpha \mathbf{A}^{-1} \mathbf{a}\mathbf{a}^{\mathrm{T}} \mathbf{A}^{-1}}{1 + \alpha \mathbf{a}^{\mathrm{T}} \mathbf{A}^{-1} \mathbf{a}} \tag{6.59}$$

Moreover, for our further reference later, in this book, we also recall that the general form of the matrix inversion lemma states: if $\mathbf{A}$ and $\mathbf{B}$ are positive-definite N-by-N and M-by-M matrices, respectively, and $\mathbf{C}$ is an arbitrary N-by-M matrix, then

$$(\mathbf{A} + \mathbf{C}\mathbf{B}\mathbf{C}^{\mathrm{T}})^{-1} = \mathbf{A}^{-1} - \mathbf{A}^{-1}\mathbf{C}(\mathbf{B}^{-1} + \mathbf{C}^{\mathrm{T}}\mathbf{A}^{-1}\mathbf{C})^{-1}\mathbf{C}^{\mathrm{T}}\mathbf{A}^{-1} \tag{6.60}$$

Letting $\mathbf{A} = \operatorname{diag}[1 - \rho_0, 1 - \rho_1, \ldots, 1 - \rho_{N-1}]$, $\mathbf{a} = \boldsymbol{\lambda}$, and $\alpha = -4\mu^2$ in Eq. (6.59) to obtain the inverse of $(\mathbf{I} - \mathbf{F})$, substituting the result in Eq. (6.57), and after some straightforward manipulations, we get

$$\mathcal{M} = \frac{\sum_{i=0}^{N-1} \mu\lambda_i/(1 - 2\mu\lambda_i)}{1 - \sum_{i=0}^{N-1} \mu\lambda_i/(1 - 2\mu\lambda_i)} \tag{6.61}$$

It is useful to simplify this result, by making some appropriate approximations, so that it can conveniently be used for the selection of the step-size parameter, μ. In practice, one usually selects μ, so that a misadjustment of 10% ($\mathcal{M} = 0.1$) or less is achieved. In that case, we may find that

$$\sum_{i=0}^{N-1} \frac{\mu\lambda_i}{1 - 2\mu\lambda_i} \approx \mu \sum_{i=0}^{N-1} \lambda_i = \mu\text{tr}[\mathbf{R}] \tag{6.62}$$

where the last equality is obtained from Eq. (4.24). This approximation is understood if we note that when $\mathcal{M}$ is small, the summation on the left-hand side of Eq. (6.62) is also small. Moreover, when the latter summation is small, $\mu\lambda_i \ll 1$, for $i = 0, 1, \ldots, N-1$, and, thus, these may be deleted from the denominators of the terms under the summation on the right-hand side of Eq. (6.62). Thus, we obtain

$$\mathcal{M} = \frac{\mu\text{tr}[\mathbf{R}]}{1 - \mu\text{tr}[\mathbf{R}]} \tag{6.63}$$

Furthermore, we note that when $\mathcal{M}$ is small, say $\mathcal{M} \leq 0.1$, $\mu\text{tr}[\mathbf{R}]$ is also small and, thus, it may be ignored in the denominator of Eq. (6.63), to obtain

$$\mathcal{M} \approx \mu\text{tr}[\mathbf{R}] \tag{6.64}$$

This is a very convenient equation, as $\text{tr}[\mathbf{R}]$ is equal to the sum of the powers of the signal samples at the filter tap inputs. This can be easily measured and used for the selection of the step-size parameter, μ, for achieving a certain level of misadjustment. Furthermore, when the input process to the filter is nonstationary, $\text{tr}[\mathbf{R}]$ may be updated recursively and the step-size parameter, μ, chosen accordingly to keep a certain level of misadjustment.

6.3.4 Stability

In Chapter 5, we noted that the steepest-descent algorithm remains stable only when its corresponding step-size parameter, μ, takes a value between zero and an upper bound value, which was found to be dependent on the statistics of the filter input. The same is true for the LMS algorithm. However, the use of stochastic gradient in the LMS algorithm makes it more sensitive to the value of its step-size parameter, μ, and, as a result, the upper bound of μ, which can ensure a stable behavior of the LMS algorithm, is much lower than the corresponding bound in the case of the steepest-descent algorithm. To find the upper bound of μ which guarantees the stability of the LMS algorithm, we elaborate on the misadjustment equation (6.61).

We define

$$\mathcal{J} = \sum_{i=0}^{N-1} \frac{\mu\lambda_i}{1 - 2\mu\lambda_i} \tag{6.65}$$

and note that

$$\mathcal{M} = \frac{\mathcal{J}}{1 - \mathcal{J}} \tag{6.66}$$

We also note that

$$\frac{\partial \mathcal{J}}{\partial \mu} = \sum_{i=0}^{N-1} \frac{\lambda_i}{(1 - 2\mu\lambda_i)^2} \tag{6.67}$$

From Eq. (6.67), we note that $\mathcal{J}$ is an increasing function of μ, since its derivative with respect to μ is always positive. In a similar way, one can show that $\mathcal{M}$ is an increasing function of $\mathcal{J}$. This, in turn, implies that *the misadjustment $\mathcal{M}$ of* Eq. (6.61) *is an increasing function of the step-size parameter*, μ. Thus, starting with $\mu = 0$ (i.e., the lower bound of μ) and increasing μ, we find that $\mathcal{J}$ and $\mathcal{M}$ also start from zero and increase with μ. We also note that when $\mathcal{J}$ approaches unity, $\mathcal{M}$ tends to infinity. This clearly coincides with the upper bound of the step-size parameter, say μ_{max}, below which μ has to remain to ensure a stable behavior of the LMS algorithm. Thus, the value of μ_{max} is obtained by finding the first positive root of the equation

$$\sum_{i=0}^{N-1} \frac{\mu\lambda_i}{1 - 2\mu\lambda_i} = 1 \tag{6.68}$$

Finding the exact solution of this problem, in general, turns out to be a difficult mathematical task. Furthermore, from a practical point of view, such solution is not rewarding as it depends on the statistics of the filter input in a complicated way. Here, we give an upper bound of μ, which depends only on $\sum_{i=0}^{N-1} \lambda_i = \text{tr}[\mathbf{R}]$. This results in a smaller (more stringent) value as the upper bound of μ, but a value that can easily be measured in practice. For this, we note that when

$$0 < \mu < \frac{1}{2\sum_{i=0}^{N-1} \lambda_i} \tag{6.69}$$

the following inequality always holds:

$$\sum_{i=0}^{N-1} \frac{\mu\lambda_i}{1 - 2\mu\lambda_i} \leq \frac{\mu \sum_{i=0}^{N-1} \lambda_i}{1 - 2\mu \sum_{i=0}^{N-1} \lambda_i} \tag{6.70}$$

The proof of this inequality is discussed in Problem P6.4. From Eq. (6.70), we find that the value of μ, which satisfies the equation

$$\frac{\mu \sum_{i=0}^{N-1} \lambda_i}{1 - 2\mu \sum_{i=0}^{N-1} \lambda_i} = 1 \tag{6.71}$$

satisfies the inequality

$$\sum_{i=0}^{N-1} \frac{\mu\lambda_i}{1 - 2\mu\lambda_i} \leq 1 \tag{6.72}$$

Furthermore, any value of μ that remains between zero and the solution of Eq. (6.71) satisfies Eq. (6.72). This means that Eq. (6.71) gives an upper bound for μ, which is sufficient for the stability of the LMS algorithm, but is not necessary, in general. If we call the solution of Eq. (6.71) μ'_{max}, we obtain

$$\mu'_{max} = \frac{1}{3\sum_{i=0}^{N-1} \lambda_i} = \frac{1}{3\text{tr}[\mathbf{R}]} \tag{6.73}$$

To summarize, we found that, under the assumptions made at the beginning of this section, the LMS algorithm remains stable when

$$0 < \mu < \frac{1}{3\mathrm{tr}[\mathbf{R}]} \tag{6.74}$$

The significance of the upper bound of μ, which is provided by Eq. (6.74), is that it can easily be measured from the filter input samples. We also note that the range of μ, which is provided by Eq. (6.74), is sufficient for the stability of the LMS algorithm but is not necessary. The first positive root of Eq. (6.68) gives a more accurate upper bound of μ. However, this depends on the filter input statistics in a very complicated way, which prohibits its applicability in actual practice.

6.3.5 The Effect of Initial Values of Tap Weights on the Transient Behavior of the LMS Algorithm

As was noted before, the LMS algorithm on average follows the same trajectory as the steepest-descent algorithm. As a result, the learning curves of the two algorithms are found to be similar when the same step-size parameter is used for both. In particular, the learning curve equation (5.31) is also (approximately) applicable to the LMS algorithm. Thus, we may write

$$\xi(n) \approx \xi_{\min} + \sum_{i=0}^{N-1} \lambda_i (1 - 2\mu\lambda_i)^{2n} v_i'^2(0) \tag{6.75}$$

In most applications, the filter tap weights are all initialized to zero. In that case,

$$\mathbf{v}(0) = \mathbf{w}(0) - \mathbf{w}_{\mathrm{o}} = -\mathbf{w}_{\mathrm{o}} \tag{6.76}$$

Using this result and recalling the definition $\mathbf{v}'(0) = \mathbf{Q}^{\mathrm{T}}\mathbf{v}(0)$, we get

$$\mathbf{v}'(0) = -\mathbf{w}'_{\mathrm{o}} \tag{6.77}$$

where $\mathbf{w}'_{\mathrm{o}} = \mathbf{Q}^{\mathrm{T}}\mathbf{w}_{\mathrm{o}}$. Using Eq. (6.77) in Eq. (6.75), we obtain

$$\xi(n) \approx \xi_{\min} + \sum_{i=0}^{N-1} \lambda_i (1 - 2\mu\lambda_i)^{2n} w_{\mathrm{o},i}'^2 \tag{6.78}$$

where $w'_{\mathrm{o},i}$ is the ith element of $\mathbf{w}'_{\mathrm{o}}$.

The contribution of various modes of convergence of the LMS algorithm (i.e., the terms under the summation on the right-hand side of Eq. (6.75)) on its learning curve depends on the coefficients $\lambda_i w_{\mathrm{o},i}'^2$'s. As a result, one finds that even for a similar eigenvalue distribution, the convergence behavior of the LMS algorithm is application dependent. For instance, if the $w'_{\mathrm{o},i}$'s corresponding to the smaller eigenvalues of $\mathbf{R}$ are all close to zero, the transient behavior of the LMS algorithm is determined by the larger eigenvalues of $\mathbf{R}$ whose associated time constants are small; thus a fast convergence is observed. On the contrary, if the $w'_{\mathrm{o},i}$'s corresponding to the smaller eigenvalues of $\mathbf{R}$ are significantly large, one finds that the slower modes of the LMS algorithm are prominent on its learning curve. Examples given in the next section show that these two extreme cases can happen in practice.

6.4 Computer Simulations

In the study of adaptive filters, computer simulation plays a major role. In the analysis that was presented in the previous section, we had to consider a number of assumptions to make the problem mathematically tractable. The validity of these assumptions and the matching between mathematical results and the actual performance of adaptive filters are usually verified through computer simulations.

In this section, we present a few examples of computer simulations. We present examples of four different applications of adaptive filters:

- System modeling
- Channel equalization
- Adaptive line enhancement (this is an example of prediction)
- Beamforming.

Our objectives in this presentation are:

1. To help the novice readers to have a fast start in doing computer simulations.
2. To check the accuracy of the developed theoretical results.
3. To enhance the understanding of the theoretical results by careful observation and interpretation of simulation results.

All the results, which are given in the following have been generated by using the MATLAB numerical package. The MATLAB programs used to generate the results presented in this section and other parts of this book are available on an accompanying website. A list of these programs (m-files as they are called in MATLAB) is given at the end of the book and also in the `read.me` file on the accompanying website. We encourage all the novice readers to try to run these programs, as this, we believe, is essential for a better understanding of the adaptive filtering concepts.

6.4.1 System Modeling

Consider a system modeling problem, as depicted in Figure 6.5. The filter input is obtained by passing a unit variance white Gaussian sequence, $\nu(n)$, through a filter with the system function $H(z)$. The plant, $W_{\rm o}(z)$, is assumed to be a FIR system with the impulse response duration of N samples. The plant output is contaminated with an additive white Gaussian noise sequence, $e_{\rm o}(n)$, with variance $\sigma_{\rm o}^2$. An N-tap adaptive filter, $W(z)$, is used to estimate the plant parameters.

For simulations, in this section, we select $N = 15$, $\sigma_o^2 = 0.001$ and

$$W_o(z) = \sum_{i=0}^{7} z^{-i} - \sum_{i=8}^{14} z^{-i} \tag{6.79}$$

We present results of simulations for two choices of input, which are characterized by

$$H(z) = H_1(z) = 0.35 + z^{-1} - 0.35z^{-2} \tag{6.80}$$

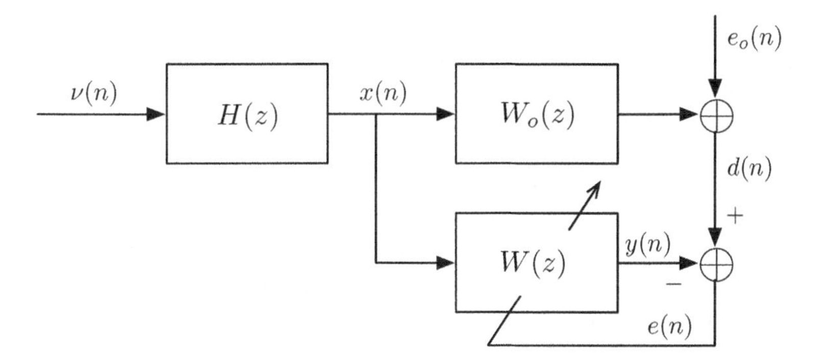

Figure 6.5 Adaptive modeling of an FIR plant.

and

$$H(z) = H_2(z) = 0.35 + z^{-1} + 0.35z^{-2} \tag{6.81}$$

The first choice results in an input, $x(n)$, whose corresponding correlation matrix has an eigenvalue spread of 1.45. This is close to a white input. On the contrary, the second choice of $H(z)$ results is a highly colored input with an associated eigenvalue spread of 28.7. From the results of Chapter 4, we recall that the eigenvalue spread figures can approximately be obtained from the underlying power spectral densities. Figure 6.6 shows the power spectral densities of the two inputs generated using the filters $H_1(z)$ and $H_2(z)$. These plots are obtained by noting that

$$\Phi_{xx}(e^{j\omega}) = \Phi_{\nu\nu}(e^{j\omega})|H(e^{j\omega})|^2 \tag{6.82}$$

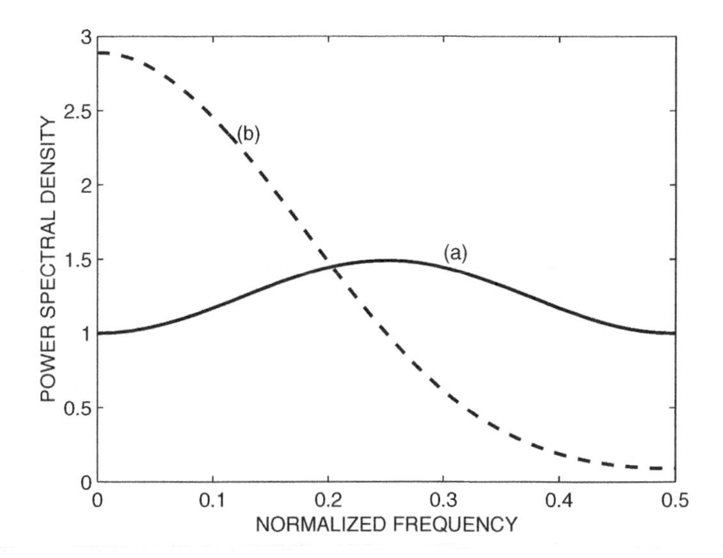

Figure 6.6 Power spectral densities of the two input processes used for the simulation of the modeling problem: (a) $H(z) = H_1(z)$ and (b) $H(z) = H_2(z)$.

and $\Phi_{\nu\nu}(e^{j\omega}) = 1$, since $\nu(n)$ is a unit variance white noise process. The fact that $H_2(z)$ generates a process that is highly colored, while the process generated by $H_1(z)$ is relatively flat, is clearly seen.

Figure 6.7a and b shows the learning curves of the LMS algorithm for the two choices of $H(z)$. The step-size parameter, μ, is selected according to the simplified misadjustment equation (6.64) for the misadjustment values 10%, 20%, and 30%. The filter tap weights are initialized to zero. Each plot is obtained by an ensemble average of 100 independent simulation runs. We note that $\xi_{\min} = \sigma_o^2 = 0.001$, and this is achieved when the model and plant coefficients match. Careful examination of the results presented in Figure 6.7a and b reveals that the predictions made by Eq. (6.64) are accurate for the cases where μ is set for a misadjustment of 10% (or less). For larger values of μ, one finds that a more accurate theoretical estimate of the misadjustment is obtained using Eq. (6.61). Such estimate, of course, requires calculation of the eigenvalues of the correlation matrix $\mathbf{R}$. The MATLAB program "`modeling.m`" on the accompanying website contains instructions, which generate matrix $\mathbf{R}$ and the other parameters required for these calculations. The reader is encouraged to use this program and experiment with that to examine the effect of various parameters, such as the step-size, μ, the plant model, $W_o(z)$, and the input sequence to the adaptive filter. Such experiments will greatly enhance the reader's understanding of the concepts of convergence and misadjustment.

Experiments with the LMS algorithm show that the accuracy of the misadjustment equations developed above varies with the statistics of the filter input and the step-size parameter. For example, one finds that all of the three plots in Figure 6.7a and two of the plots in Figure 6.7b match the theoretical predictions made by Eq. (6.61), but the third plot in Figure 6.7b (i.e., the case $\mathcal{M} = 30\%$) does not match Eq. (6.61). In the latter case, the LMS algorithm experiences some instability problem. The mismatch between the theory and experiments here is attributed to the fact that the independence assumption made in the development of the theoretical results is badly violated for larger values of μ.

6.4.2 Channel Equalization

Figure 6.8 depicts a channel equalization problem. The input sequence to the channel is assumed to be binary (taking values of +1 and −1) and white. The channel system function is denoted by $H(z)$. The channel noise, $\nu_c(n)$, is modeled as an additive white Gaussian process with variance $\sigma_{\nu_c}^2$. The equalizer is implemented as an N-tap transversal filter. The desired output of the equalizer is assumed to be $s(n-\Delta)$, that is, a delayed replica of the transmitted data symbols. For the training of the equalizer, it is assumed that the transmitted data symbols are available at the receiver. This is called *training mode*. Once the equalizer is trained and switched to the data mode, its output, after passing through a slicer, gives the transmitted symbols. A discussion on the training and data modes of equalizers can be found in Chapter 1. More detailed explanations and adaptation algorithms are presented in Chapter 17.

Two choices of the channel response, $H(z)$, are considered for our study, here. These are purposefully selected to be the same as the two choices of $H(z)$ in the modeling problem, above, where $H(z)$ was used to shape the power spectral density of the input process to the plant and model. This facilitates a comparison of the results in the two

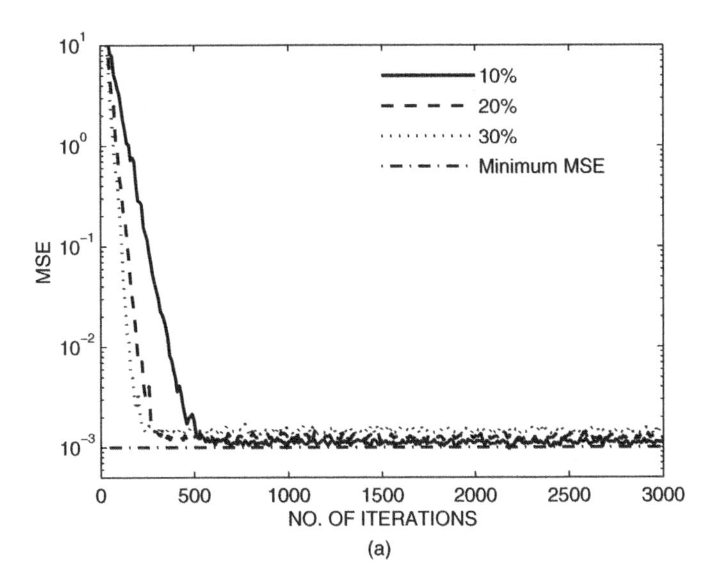

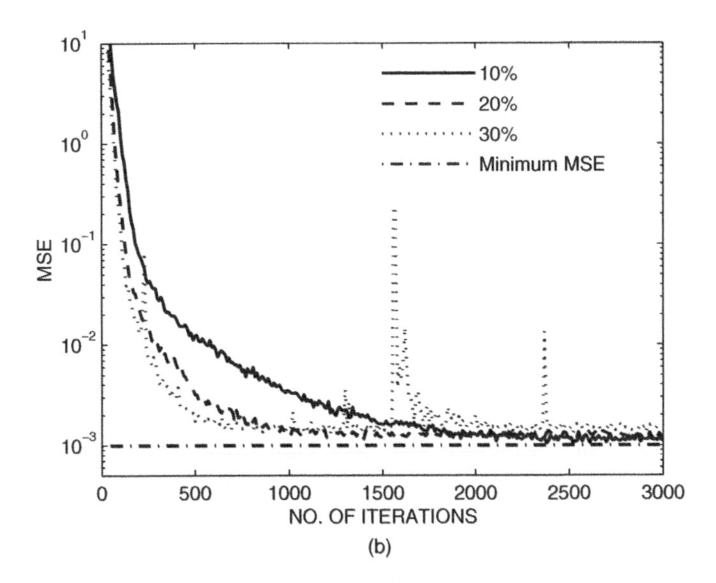

Figure 6.7 Learning curves of the LMS algorithm for the modeling problem of Figure 6.5, for the two input processes discussed in the text: (a) $H(z) = H_1(z)$ and (b) $H(z) = H_2(z)$. The step-size parameter, μ, is selected for the misadjustment values 10%, 20%, and 30%, according to the simplified equation (6.64).

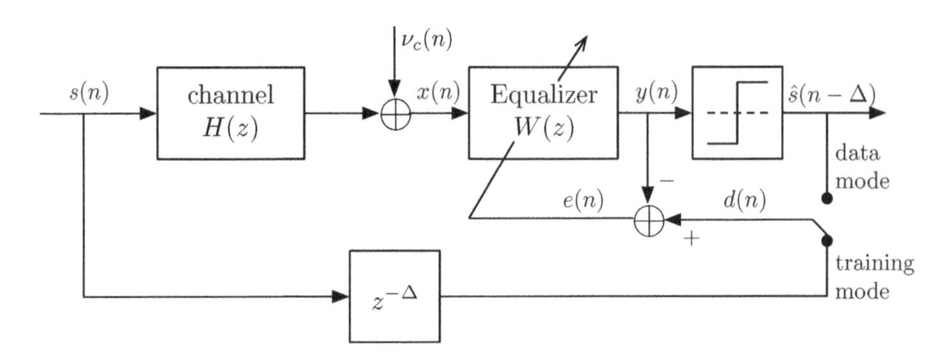

Figure 6.8 Adaptive channel equalization.

cases. In particular, we note that, in the present problem,

$$\begin{aligned}\Phi_{xx}(e^{j\omega}) &= \Phi_{ss}(e^{j\omega})|H(e^{j\omega})|^2 + \Phi_{\nu_c\nu_c}(e^{j\omega})\\ &= |H(e^{j\omega})|^2 + \sigma_{\nu_c}^2\end{aligned} \tag{6.83}$$

Comparing Eqs. (6.82) and (6.83), we note that when a similar $H(z)$ is used for both cases and the signal-to-noise ratio at the channel output is high (i.e., $\sigma_{\nu_c}^2$ is small), the power spectral densities of the input samples to the two adaptive filters are almost the same. This, in turn, implies that the convergence of both the filters is controlled by the same set of eigenvalues. As a result, on average, one may expect to see similar learning curves for both cases.

Figure 6.9a and b presents the learning curves of the equalizer for the two choices of the channel response, that is, $H_1(z)$ and $H_2(z)$ of Eqs. (6.80) and (6.81), respectively. The equalizer length, N, and the delay, Δ, are set equal to 15 and 9, respectively. The step-size parameter, μ, is chosen according to the simplified equation (6.64) for the three misadjustment values 10%, 20%, and 30%. The equalizer tap weights are initialized to zero. Each plot is based on an ensemble average of 100 independent simulation runs. The MATLAB program used to obtain these results is available on the accompanied website. It is called "`equalizer.m`." Careful study of Figure 6.9a and b and further numerical tests (using the "`equalizer.m`" or any similar simulation program) reveal that similar to the modeling case, the theoretical and simulation results match well when the step-size parameter, μ, is small. However, the accuracy of the theoretical results is lost for larger values of μ. The latter effect is more noticeable when the eigenvalue spread of the correlation matrix $\mathbf{R}$ is large.

Comparing the results presented in Figures 6.7a and 6.9a, we find that the performance of the adaptive filters in both cases are about the same. Moreover, these results compare very well with the predictions made by theory. We recall that these correspond to the case where the eigenvalue spread of the correlation matrix $\mathbf{R}$ is small. Some differences between the results of the two cases are observed as the eigenvalue spread of $\mathbf{R}$ increases. In particular, a comparison of Figures 6.7b and 6.9b shows that the learning curve of the channel equalizer is dominantly controlled by its slower modes of convergence, while in the modeling case, a balance of slow and fast modes of convergence is observed.

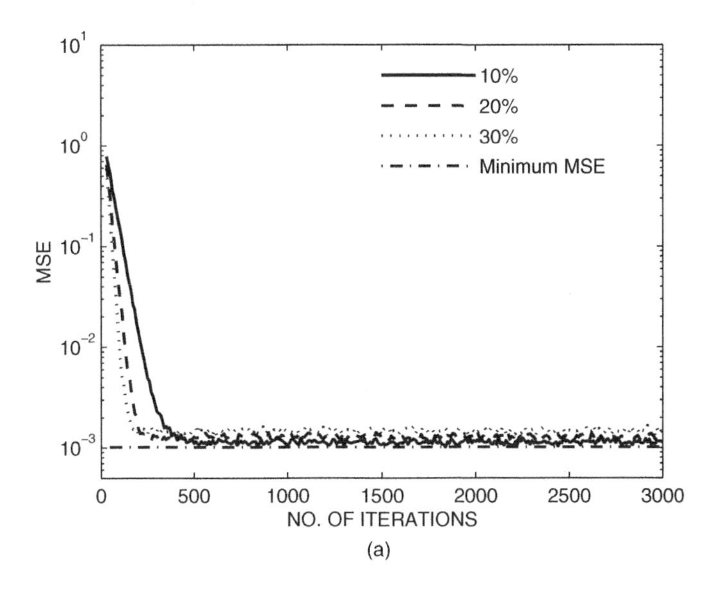

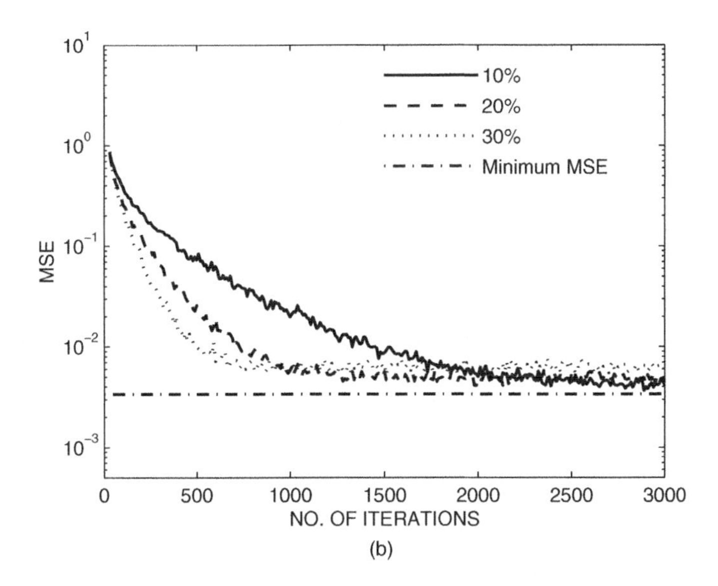

Figure 6.9 Learning curves of the LMS algorithm for the channel equalizer, for the two choices of channel responses discussed in the text: (a) $H(z) = H_1(z)$ and (b) $H(z) = H_2(z)$. The step-size parameter, μ, is selected for the misadjustment values 10%, 20%, and 30%, according to the simplified equation (6.64).

In the latter case, a drop of MSE from 10 to 0.1 within the first 100 iterations of the LMS algorithm is observed. The slower modes of the algorithm are observed after the filter output MSE has dropped to a relatively low level. As a result, the existence of slow and fast modes of convergence on the learning curve is clearly visible. On the contrary, in the case of channel equalization, we find that the convergence of the LMS algorithm is dominantly determined by its slower modes. We can hardly see any fast mode of convergence on the learning curves presented in Figure 6.9b. An explanation of this phenomenon, which is usually observed when the LMS algorithm is used to adapt channel equalizers, is instructive.

As was noted before, besides the eigenvalue spread of **R**, the transient behavior of the LMS algorithm is also affected by the initial offset of the filter tap weights from their optimal values; see Eq. (6.75). We also noted that when the filter tap weights are initialized to zero, the transient behavior of the LMS algorithm is affected by the optimum tap weights of the filter; see Eq. (6.78). To be more precise, the contribution of various modes of convergence of the LMS algorithm in shaping its learning curve is determined by the values of $\lambda_i w_{o,i}'^2$, for $i = 0, 1, \ldots, N-1$.

For a modeling problem, the statistics of the filter input and its optimum tap weights, $\mathbf{w}_o$, (i.e., the plant response) are, in general, independent of each other. In this situation, it is hard to make any comment on the values of $\lambda_i w_{o,i}'^2$ terms. The only comment which may be made is that if one assumes the statistics of the filter input are fixed and the plant response is arbitrary, the elements of $\mathbf{w}_o'$, that is, $w_{o,i}'$'s, may be thought of as a set of zero-mean random variables whose values change from one plant to another, and they all have the same variance, say $\sigma_{w'}^2$. Using this in Eq. (6.78), we obtain, for the modeling problem,

$$E[\xi(n)] \approx \xi_{\min} + \sigma_{w'}^2 \sum_{i=0}^{N-1} \lambda_i (1 - 2\mu\lambda_i)^{2n} \tag{6.84}$$

where the statistical expectation on $\xi(n)$ is with respect to the variations of $w_{o,i}'$'s, that is, the plant response.

On the contrary, in the case of channel equalization, there is a close relationship between the filter (equalizer) input statistics and the optimum setting of its tap weights. The equalizer is adapted to implement the inverse of the channel response, that is,

$$W_o(z) \approx \frac{z^{-\Delta}}{H(z)} \tag{6.85}$$

This result, which may be referred to as *spectral inversion property of channel equalizer*, can be used to evaluate $\lambda_i w_{o,i}'^2$ terms, when the equalizer length is relatively long. A procedure for approximation of $\lambda_i w_{o,i}'^2$ is discussed in Problem P6.14. The result there is that when the equalizer length N is relatively long

$$\lambda_i w_{o,i}'^2 \approx \frac{1}{N}, \qquad \text{for} \quad i = 0, 1, \ldots, N-1 \tag{6.86}$$

Substituting this in Eq. (6.78), we get, for an N-tap channel equalizer,

$$\xi(n) \approx \xi_{\min} + \frac{1}{N} \sum_{i=0}^{N-1} (1 - 2\mu\lambda_i)^{2n} \tag{6.87}$$

The difference between the learning curves of the modeling and channel equalization problems may now be explained by comparing Eqs. (6.84) and (6.87). When the eigenvalues λ_0, λ_1, ..., λ_{N-1} are widely spread and n is small (i.e., the adaptation has just started), the summation on the right-hand side of Eq. (6.84) is dominantly determined by the larger λ_i's. However, noting that the geometrical regressor factors, $(1 - 2\mu\lambda_i)^{2n}$'s, corresponding to the larger λ_i's converge to zero at a relatively fast rate, the summation on the right-hand side of Eq. (6.84) experiences a fast drop to a level significantly below its initial value, when $n = 0$. The slower modes of the LMS algorithm are observed after this initial fast drop of the MSE. This, of course, is what we observe in Figure 6.7b. In the case of channel equalizer, we note that when n is small, all the terms under the summation on the right-hand side of Eq. (6.87) are about the same. This means there is no dominant term in the latter summation and as a result, unlike the modeling problem case, the convergence of the faster modes of the LMS algorithm may not reduce $\xi(n)$ significantly. A significant reduction of $\xi(n)$ after convergence of the faster modes of the LMS algorithm may only be observed when the filter length, N, is large and only a few of the eigenvalues of $\mathbf{R}$ are small.

6.4.3 Adaptive Line Enhancement

Adaptive line enhancement refers to the case where a noisy signal consisting of a few sinusoidal components is available and the aim is to filter out the noise part of the signal. The filtering solution to this problem is trivial. The noisy signal is passed through a filter, which is tuned to the sinusoidal components. When the frequency of the sine-waves present in the noisy signal is known, of course, a fixed filter will suffice. However, when the sine-wave frequencies are unknown or may be time-varying, an adaptive solution has to be adopted.

Figure 6.10 depicts the block schematic of an adaptive line enhancer. It is basically an M-step-ahead predictor. The assumption is that the noise samples, which are more than M samples apart, are uncorrelated with one another. As a result, the predictor can only make a prediction of the sinusoidal components of the input signal and when adapted to minimize the output MSE, the line enhancer will be a filter tuned to the sinusoidal components. The maximum possible rejection of the noise will also be achieved as any portion of the noise, which passes through the prediction filter, will enhance the output MSE whose minimization is the criterion in adapting the filter tap weights.

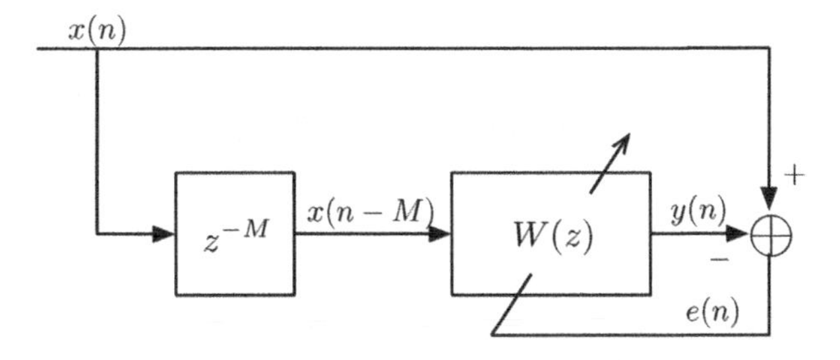

Figure 6.10 Adaptive line enhancer.

Here, to simplify our discussion, we assume that the enhancer input consists of a single sinusoidal component and the additive noise is white. More specifically, we assume that

$$x(n) = a\sin(\omega_o n + \theta) + \nu(n) \tag{6.88}$$

where $\nu(n)$ is a white noise sequence. The delay parameter M is set to 1 as $\nu(n)$ is white.

Figure 6.11 shows the learning curves of the adaptive line enhancer when $x(n)$ is chosen as in Eq. (6.88). The following parameters are used to obtain these results: $N = 30$, $M = 1$, $a = 1$, $\omega_o = 0.1$, and θ is chosen to be a random variable with constant distribution in the range of 0 to 2π, for different simulation runs. The variance of $\nu(n)$ is chosen 10 dB below the sinusoidal signal energy. The learning curves are given for three choices of the step-size parameter, μ, which result in 1%, 5%, and 10% misadjustment. The predictor tap weights are initialized to zero. The program used to obtain these results is available on the accompanying website. It is called "`lenhncr.m`."

From the results presented in Figure 6.11, it appears that the convergence of the line enhancer is governed by only one mode. Examination of the eigenvalues of the underlying process and the resulting time constants of the various modes of the line enhancer reveals that the mode which is observed in Figure 6.11 coincides with the fastest convergence mode of the LMS algorithm in the present case. An explanation of this phenomenon is instructive.

We note that the optimized predictor of the line enhancer is a filter tuned to the peak of the spectrum of $x(n)$. Furthermore, from the minimax theorem (of Chapter 4), we may say that the latter is the eigenfilter associated with the maximum eigenvalue of the correlation matrix $\mathbf{R}$ of the underlying process. This implies that the optimum tap-weight vector of the line enhancer coincides with the eigenvector associated with the largest eigenvalue of its

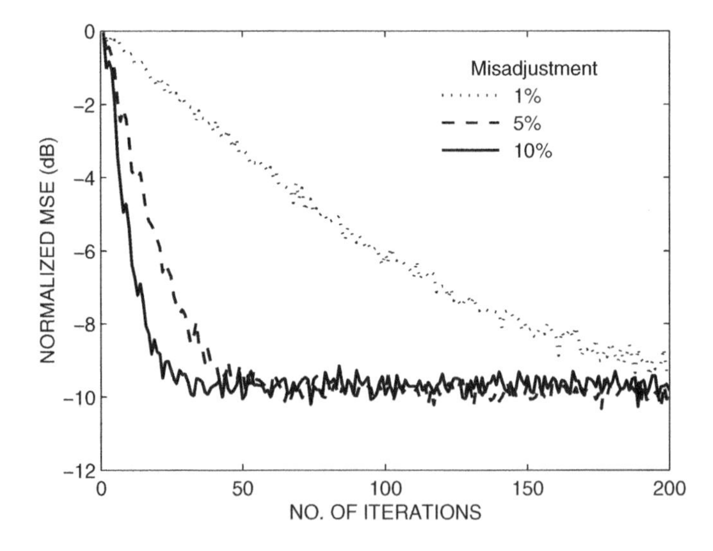

Figure 6.11 Learning curves of the adaptive line enhancer. The line enhancer MSE is normalized to the input signal power.

corresponding correlation matrix. In other words, in the Euclidean space associated with the tap weights of the line enhancer, the line connecting the origin to the point defined by the optimized tap weights is along the eigenvector associated with largest eigenvalue of its corresponding correlation matrix. This clearly explains why the learning curves of the line enhancer presented in Figure 6.11 are dominantly controlled by only one mode and this coincides with the fastest mode of convergence of the corresponding LMS algorithm.

6.4.4 Beamforming

Consider a two-element antenna array similar to the one discussed in Example 3.6. The array consists of two omnidirectional (equally sensitive to all directions) antennas A and B, as shown in Figure 6.12. A desired signal $s(n) = \alpha(n)\cos(n\omega_o + \phi_1)$ arrives in the direction perpendicular to the line connecting A and B. An interferer (jammer) signal $\nu(n) = \beta(n)\cos(n\omega_o + \phi_2)$ arrives at an angle θ_o relative to $s(n)$. The signal sequences $s(n)$ and $\nu(n)$ are assumed to be narrow-band processes with random phases ϕ_1 and ϕ_2, respectively. It is also assumed that the random amplitudes $\alpha(n)$ and $\beta(n)$ are zero-mean and uncorrelated with each other. The two omnis are separated by a distance of $l = \lambda_c/2$ meters, where λ_c is the wavelength associated with the continuous time carrier frequency

$$\omega_c = \frac{\omega_o}{T} \tag{6.89}$$

with T being the sampling period. The coefficients, w_0 and w_1, of the beamformer are adjusted, so that the output error, $e(n)$, is minimized in the mean-square sense.

As in Example 3.6, the adaptive beamformer of Figure 6.12 is characterized by the following signal sequences[4]:

1. Primary input

$$d(n) = \alpha(n)\cos(n\omega_o + \phi_1) + \beta(n)\cos(n\omega_o + \phi_2 - \phi_o) \tag{6.90}$$

2. Reference tap-input vector

$$\begin{aligned}\mathbf{x}(n) &= \begin{bmatrix} x(n) \\ \tilde{x}(n) \end{bmatrix} \\ &= \begin{bmatrix} \alpha(n)\cos(n\omega_o + \phi_1) + \beta(n)\cos(n\omega_o + \phi_2) \\ \alpha(n)\sin(n\omega_o + \phi_1) + \beta(n)\sin(n\omega_o + \phi_2) \end{bmatrix}\end{aligned} \tag{6.91}$$

The phase shift ϕ_o is introduced because of the difference between the arrival time of the jammer at A and B. It is given by

$$\phi_o = \frac{l \sin\,\theta_o}{c}\omega_c \tag{6.92}$$

where c is the propagation speed. Replacing l with $\lambda_c/2$ in Eq. (6.92) and noting that $\omega_c/c = 2\pi/\lambda_c$, we obtain

$$\phi_o = \pi \sin\,\theta_o \tag{6.93}$$

We note that, as expected, ϕ_o is independent of the sampling period T. It depends only on the angle of arrival of the jammer signal, θ_o.

[4] In Example 3.6, to simplify the derivations, ϕ_1 and ϕ_2 were assumed to be zero.

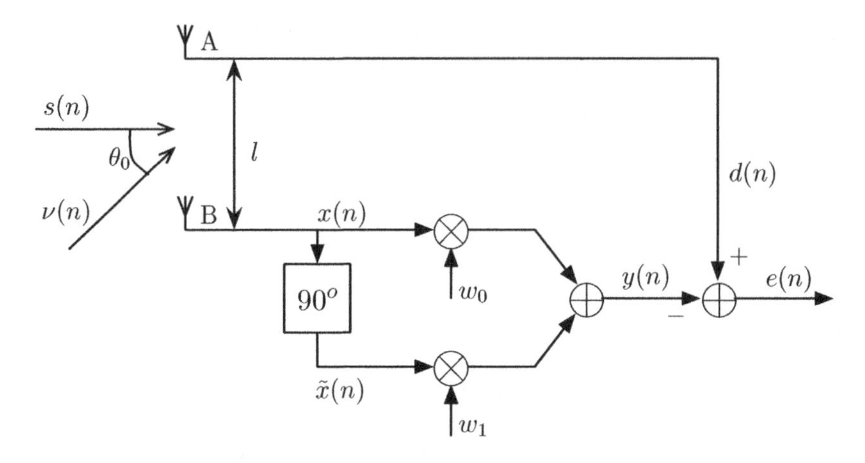

Figure 6.12 A two-element antenna array.

The beamformer coefficients, w_0 and w_1, are selected (adapted), so that the difference

$$e(n) = d(n) - \mathbf{w}^{\mathrm{T}}\mathbf{x}(n)$$

where $\mathbf{w} = [w_0\ w_1]^{\mathrm{T}}$, is minimized in the mean-square sense. The error signal $e(n)$ is the beamformer output.

For a given set of the beamformer coefficients w_0 and w_1 and a signal arriving at an angle θ, the array power gain, $\mathcal{G}(\theta)$, is defined as the ratio of the signal power in the output $e(n)$ to the signal power at one of the omnis. Assuming that a narrow-band signal $\gamma(n)\cos n\omega_o$ is arriving at an angle θ,

$$\begin{aligned} e(n) &= \gamma(n)[\cos(n\omega_o - \pi\sin\theta) - w_0\cos\ n\omega_o - w_1\sin\ n\omega_o] \\ &= \gamma(n)[(\cos(\pi\sin\theta) - w_0)\cos\ n\omega_o + (\sin(\pi\sin\theta) - w_1)\sin\ n\omega_0] \\ &= a(\theta)\gamma(n)\sin(n\omega_o + \varphi(\theta)) \end{aligned} \tag{6.94}$$

where

$$a(\theta) = \sqrt{(\cos(\pi\sin\theta) - w_0)^2 + (\sin(\pi\sin\theta) - w_1)^2}$$

and

$$\varphi(\theta) = \tan^{-1}\left(\frac{\cos(\pi\sin\theta) - w_0}{\sin(\pi\sin\theta) - w_1}\right)$$

Using these, we get

$$\mathcal{G}(\theta) = a^2(\theta) = (\cos(\pi\sin\theta) - w_0)^2 + (\sin(\pi\sin\theta) - w_1)^2 \tag{6.95}$$

$\mathcal{G}(\theta)$ when plotted against the angle of arrival of the received signal is called *directivity pattern* of the array (beamformer). The names *beam pattern*, *array pattern*, and *spatial response* are also used to refer to $\mathcal{G}(\theta)$. The directivity patterns are usually plotted in polar coordinates.

Figure 6.13 shows the directivity pattern of the two-element beamformer of Figure 6.12 when its coefficients have been adjusted near their optimal values using the LMS

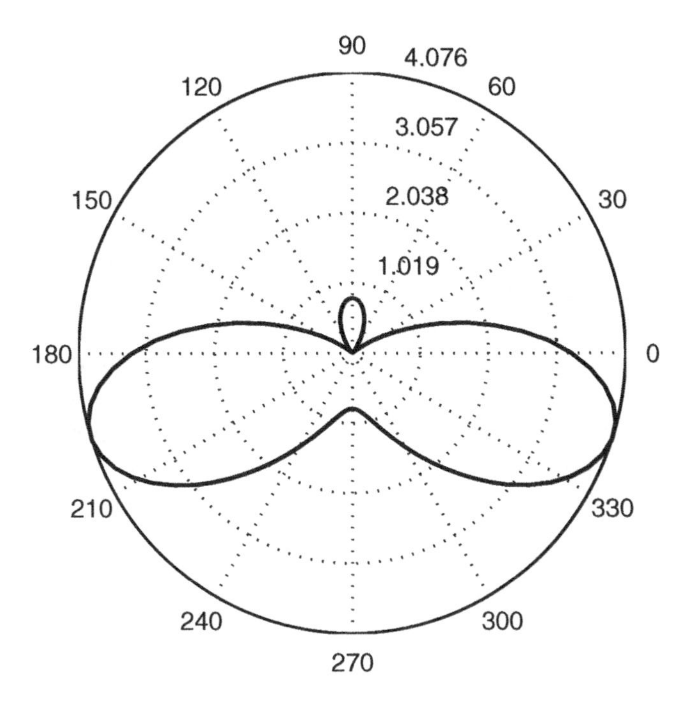

Figure 6.13 The directivity pattern of the two-element antenna array when a jammer arrives from the direction 45° with respect to the desired signal, as defined in Figure 6.12.

algorithm. The following parameters have been used to obtain these results:

$$\theta_o = 45^\circ, \quad \sigma_\alpha^2 = 0.01, \quad \sigma_\beta^2 = 1$$

where σ_α^2 and σ_β^2 are the variances of $\alpha(n)$ and $\beta(n)$, respectively. The results, as could be predicted from the theory, show a clear deep null in the direction that the jammer arrives ($\theta = \theta_o$) and a reasonably good gain in the direction of the desired signal ($\theta = 0$). The array pattern is symmetrical with respect to the line connecting A to B because of the omnidirectional properties of the antennas. The MATLAB program used to obtain this result is available on the accompanying website. It is called "`bformer.m`." We encourage the readers to try this program for different values of θ_o, σ_α^2, and σ_β^2. An interesting observation that can be made is that a null is always produced in the direction of arrival of the desired signal or jammer, whichever is stronger. The theoretical results related to these observations can be found in Chapter 3, Section 3.6.5. The subject of beamforming is presented in great details in Chapter 18, under the more generic name *sensor array processing*.

6.5 Simplified LMS Algorithms

Over the years, a number of modifications, which simplify hardware implementation of the LMS algorithm, have been proposed (Hirsch and Wolf (1970); Claasen and

Mecklenbräuker (1981), and Duttweiler (1982)). These simplifications are discussed in this section. The most important members of this class of algorithms are:

Sign algorithm. This algorithm is obtained from the conventional LMS recursion (6.9) by replacing $e(n)$ with its sign. This leads to the following recursion

$$\mathbf{w}(n+1) = \mathbf{w}(n) + 2\mu \mathrm{sign}(e(n))\mathbf{x}(n) \tag{6.96}$$

Because of replacement of $e(n)$ by its sign, implementation of this recursion may be cheaper than the conventional LMS recursion, especially in high-speed applications where a hardware implementation of the adaptation recursion may be necessary. Furthermore, the step-size parameter is usually selected to be a power-of-two, so that no multiplication would be required for implementing the recursion (6.96). A set of shift and add/subtract operations would suffice to update the filter tap weights.

Signed-Regressor algorithm. The signed-regressor algorithm is obtained from the conventional LMS recursion (6.9) by replacing the tap-input vector $\mathbf{x}(n)$ with the vector $\mathrm{sign}(\mathbf{x}(n))$, where the sign function is applied to the vector $\mathbf{x}(n)$ on element-by-element basis. The signed-regressor recursion is then

$$\mathbf{w}(n+1) = \mathbf{w}(n) + 2\mu e(n)\mathrm{sign}(\mathbf{x}(n)) \tag{6.97}$$

Although, quite similar in form, the signed-regressor algorithm performs much better than the sign algorithm. This will be shown later through a simulation example.

Sign–Sign algorithm. The sign–sign algorithm, as may be understood from its name, combines the sign and signed-regressor recursions together, resulting in the following recursion:

$$\mathbf{w}(n+1) = \mathbf{w}(n) + 2\mu \mathrm{sign}(e(n))\mathrm{sign}(\mathbf{x}(n)) \tag{6.98}$$

It may be noted that even though in many practical cases, all of the above algorithms are likely to converge to the optimum Wiener–Hopf solution, this may not be true in general. For example, the sign–sign algorithm converges toward a set of tap weights, which satisfy the equation

$$E[\mathrm{sign}(e(n)\mathbf{x}(n))] = 0 \tag{6.99}$$

which in general may not be equivalent to the principle of orthogonality

$$E[e(n)\mathbf{x}(n)] = 0 \tag{6.100}$$

which leads to the Wiener–Hopf equation. For instance, when the elements of the vector $\mathbf{x}(n)$ are zero-mean but have a nonsymmetrical distribution around zero, the elements of $e(n)\mathbf{x}(n)$ may also have a nonsymmetrical distribution around zero. In that case, it is likely that the solutions to Eqs. (6.99) and (6.100) lead to two different set of tap weights. Nevertheless, we shall emphasize that in most of the practical applications, the scenario that was just mentioned is unlikely to happen. Even if it happens, the solutions obtained from Eqs. (6.99) and (6.100) are usually about the same.

To compare the performance of these algorithms, with the conventional LMS algorithm and among themselves, we run the system modeling problem that was introduced in Section 6.4.1. Figure 6.14 shows the convergence behavior of the algorithms when

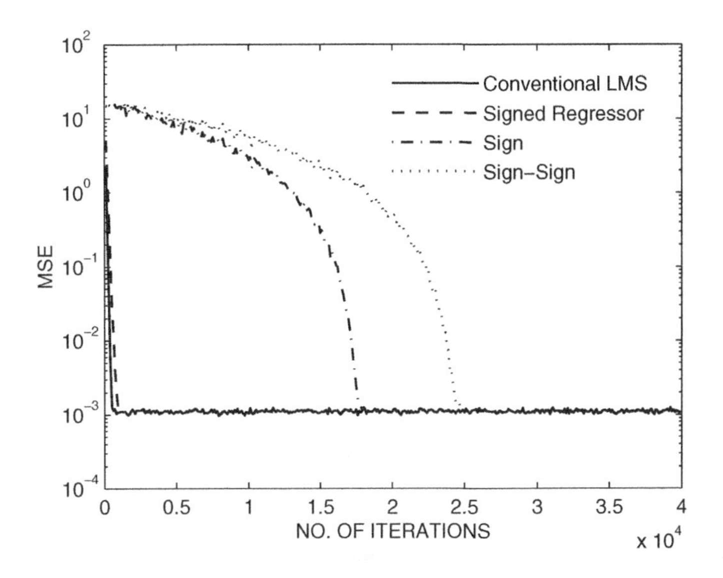

Figure 6.14 Learning curves of the conventional LMS algorithm and its simplified versions. Different step-size parameters are used. These have been selected experimentally, so that all algorithms approach the same steady-state MSE.

the input coloring filter $H(z) = H_1(z)$ is used and the step-size parameters for different algorithms are selected experimentally, so that they all reach the same steady-state MSE.

From the results presented in Figure 6.14, we see that the performance of the signed-regressor algorithm is only slightly worse than the conventional LMS algorithm. However, the sign and sign–sign algorithms are both much slower than the conventional LMS algorithm. The convergence behavior of them is also rather peculiar. They converge very slowly at the beginning but speed up as the MSE level drops. This can be explained as follows.

Consider the sign algorithm recursion and note that it may be written as

$$\mathbf{w}(n+1) = \mathbf{w}(n) + 2\mu \frac{e(n)}{|e(n)|}\mathbf{x}(n) \tag{6.101}$$

as $\text{sign}(e(n)) = e(n)/|e(n)|$. This may be rearranged as

$$\mathbf{w}(n+1) = \mathbf{w}(n) + 2\frac{\mu}{|e(n)|}e(n)\mathbf{x}(n) \tag{6.102}$$

Inspection of Eq. (6.102) reveals that the sign algorithm may be thought as an LMS algorithm with a variable step-size parameter $\mu'(n) = \mu/|e(n)|$. The step-size parameter $\mu'(n)$ increases, on an average, as the sign algorithm converges as $e(n)$ decreases in magnitude. Thus, to keep the sign algorithm stable, with a small steady-state error, a very small step-size parameter μ has to be used. Choosing a very small μ leads to an equally small value (on an average) for $\mu'(n)$ in the initial portion of the sign algorithm. This clearly explains why the sign algorithm initially converges very slowly. However, as the algorithm converges and $e(n)$ becomes smaller in magnitude, the step-size parameter $\mu'(n)$ becomes larger, on an average, and this, of course, leads to a faster convergence of the algorithm. A rigorous analysis of the sign algorithm for a nonstationary case can be found in Eweda (1990b).

The same procedure may be followed to explain the behavior of the signed-regressor algorithm. In this case, each tap of the filter is controlled by a separate variable step-size parameter. In particular, the step-size parameter of the ith tap of the filter at nth iteration is $\mu'_i(n) = \mu/|x(n-i)|$, where μ is a common parameter to all taps. The fundamental difference between the variable step-size parameters, $\mu'_i(n)$'s, here and what was observed above for the sign algorithm is that in the present case, the variations of $\mu'_i(n)$'s are independent of the filter convergence. The selection of the common parameter μ is based on the average size of $|x(n)|$. This leads to a more homogeneous convergence of the signed-regressor algorithm when compared with the sign algorithm. In fact, the analysis of the signed-regressor algorithm given by Eweda (1990a) shows that for Gaussian signals, the convergence behavior of the signed-regressor algorithm is very similar to the conventional LMS algorithm. The replacement of $x(n-i)$ terms by their signs leads to an increase of the time constants of the algorithm learning curve by a fixed factor of $\pi/2$. This, clearly, increases the convergence time of the signed-regressor algorithm by the same factor when it is compared with the conventional LMS algorithm. Problem P6.16 contains the necessary theoretical elements, which lead to this result.

Another interesting proposal, which also leads to some simplification of the LMS algorithm, was suggested by Duttweiler (1982). He suggested that in calculating the gradient vector $e(n)\mathbf{x}(n)$, $e(n)$, and/or $\mathbf{x}(n)$ may be quantized to their respective nearest power-of-two. This leads to an algorithm that performs very similar to the conventional LMS algorithm.

6.6 Normalized LMS Algorithm

Normalized LMS (NLMS) algorithm may be viewed as a special implementation of the LMS algorithm that takes into account the variation of the signal level at the filter input and selects a normalized step-size parameter, which results in a stable as well as fast converging adaptation algorithm. The NLMS algorithm may be developed from different viewpoints. Goodwin and Sin (1984) formulated the NLMS algorithm as a constrained optimization problem; see also Haykin (1991). Nitzberg (1985) obtained the NLMS recursion by running the conventional LMS algorithm many times, for every new sample of the input. Here, we start with a rather straightforward derivation of the NLMS recursion and later show that the recursion obtained satisfies the constrained optimization criterion of Goodwin and Sin and also that it matches the result of Nitzberg.

We consider the LMS recursion

$$\mathbf{w}(n+1) = \mathbf{w}(n) + 2\mu(n)e(n)\mathbf{x}(n) \tag{6.103}$$

where the step-size parameter $\mu(n)$ is time-varying. We select $\mu(n)$, so that the *a posteriori* error

$$e^{+}(n) = d(n) - \mathbf{w}^{\mathrm{T}}(n+1)\mathbf{x}(n) \tag{6.104}$$

is minimized in magnitude. Substituting Eq. (6.103) in Eq. (6.104) and rearranging, we obtain

$$e^{+}(n) = (1 - 2\mu(n)\mathbf{x}^{\mathrm{T}}(n)\mathbf{x}(n))e(n) \tag{6.105}$$

Minimizing $(e^{+}(n))^2$ with respect to $\mu(n)$ results in the following:

$$\mu(n) = \frac{1}{2\mathbf{x}^{\mathrm{T}}(n)\mathbf{x}(n)} \tag{6.106}$$

which forces $e^{+}(n)$ to zero. Substituting Eq. (6.106) in Eq. (6.103), we obtain

$$\mathbf{w}(n+1) = \mathbf{w}(n) + \frac{1}{\mathbf{x}^{\mathrm{T}}(n)\mathbf{x}(n)}e(n)\mathbf{x}(n) \tag{6.107}$$

This is the NLMS recursion. When this is combined with the filtering equation (6.1) and the error estimation equation (6.2), we obtain the NLMS algorithm.

There have been a variety of interpretations to the NLMS algorithm. We review some of these in the following as it can help in enhancing our understanding of this algorithm.

1. The use of $\mu(n)$ as in Eq. (6.106) is appealing as it selects a step-size parameter proportional to the inverse of the instantaneous signal samples energy at the adaptive filter input. This matches the misadjustment equation (6.64), which suggests that the step-size parameter of the LMS algorithm should be selected proportional to the inverse of the average total energy at the filter tap inputs. Note that

$$\mathrm{tr}[\mathbf{R}] = \sum_{i=0}^{N-1} E[x^2(n-i)] = E\left[\sum_{i=0}^{N-1} x^2(n-i)\right]$$

   and $\sum_{i=0}^{N-1} x^2(n-i)$ is the total instantaneous signal energy at the filter tap inputs.
2. The NLMS recursion (6.107) is equivalent to running the LMS recursion for every new sample of input many iterations until it converges (Nitzberg, 1985); see Problem P6.17.
3. The NLMS recursion may also be derived by solving the following constrained optimization problem (Goodwin and Sin, 1984):

   Given the tap-input vector $\mathbf{x}(n)$ and the desired output sample $d(n)$, choose the updated tap-weight vector $\mathbf{w}(n+1)$ so as to minimize the squared Euclidean norm of the difference

$$\boldsymbol{\eta}(n) = \mathbf{w}(n+1) - \mathbf{w}(n) \tag{6.108}$$

   subject to the constraint

$$\mathbf{w}^{\mathrm{T}}(n+1)\mathbf{x}(n) = d(n) \tag{6.109}$$

   Observe that the solution given by Eq. (6.107) satisfies the constraint (6.109). Hence, define $\boldsymbol{\eta}_{\mathrm{NLMS}}(n)$ as

$$\boldsymbol{\eta}_{\mathrm{NLMS}}(n) = \mathbf{w}(n+1) - \mathbf{w}(n) = \frac{1}{\mathbf{x}^{\mathrm{T}}(n)\mathbf{x}(n)}e(n)\mathbf{x}(n) \tag{6.110}$$

   We will now show that $\boldsymbol{\eta}_{\mathrm{NLMS}}(n)$ is indeed the solution to the problem posed above.

Let the optimum $\boldsymbol{\eta}(n)$ be given by

$$\boldsymbol{\eta}_{\text{o}}(n) = \boldsymbol{\eta}_{\text{NLMS}}(n) + \boldsymbol{\eta}_1(n) \tag{6.111}$$

where $\boldsymbol{\eta}_1(n)$ indicates any difference that may exist between $\boldsymbol{\eta}_{\text{o}}(n)$ and $\boldsymbol{\eta}_{\text{NLMS}}(n)$. As the updated vector $\mathbf{w}(n+1) = \mathbf{w}(n) + \boldsymbol{\eta}_{\text{NLMS}}(n)$ satisfies the constraint (6.109), we get

$$(\mathbf{w}(n) + \boldsymbol{\eta}_{\text{NLMS}}(n))^{\text{T}}\mathbf{x}(n) = d(n) \tag{6.112}$$

The tap-weight vector $\mathbf{w}(n+1) = \mathbf{w}(n) + \boldsymbol{\eta}_{\text{o}}(n)$ also satisfies the constraint (6.109), since $\boldsymbol{\eta}_o(n)$ is the optimum solution. Thus,

$$(\mathbf{w}(n) + \boldsymbol{\eta}_{\text{o}}(n))^{\text{T}}\mathbf{x}(n) = d(n) \tag{6.113}$$

Subtracting Eqs. (6.112) from (6.113) and using Eq. (6.111), we get

$$\boldsymbol{\eta}_1^{\text{T}}(n)\mathbf{x}(n) = 0 \tag{6.114}$$

Multiplying the left- and right-hand sides of Eq. (6.111) by their respective transposes, from left, we obtain

$$\begin{aligned}\boldsymbol{\eta}_{\text{o}}^{\text{T}}(n)\boldsymbol{\eta}_{\text{o}}(n) &= (\boldsymbol{\eta}_{\text{NLMS}}(n) + \boldsymbol{\eta}_1(n))^{\text{T}}(\boldsymbol{\eta}_{\text{NLMS}}(n) + \boldsymbol{\eta}_1(n)) \\ &= \boldsymbol{\eta}_{\text{NLMS}}^{\text{T}}(n)\boldsymbol{\eta}_{\text{NLMS}}(n) + \boldsymbol{\eta}_1^{\text{T}}(n)\boldsymbol{\eta}_1(n) \\ &\quad + 2\boldsymbol{\eta}_{\text{NLMS}}^{\text{T}}(n)\boldsymbol{\eta}_1(n)\end{aligned} \tag{6.115}$$

Premultiplying Eq. (6.110) by $\boldsymbol{\eta}_1^{\text{T}}(n)$ and using Eq. (6.114), we obtain

$$\boldsymbol{\eta}_1^{\text{T}}(n)\boldsymbol{\eta}_{\text{NLMS}}(n) = 0 \tag{6.116}$$

Substituting Eq. (6.116) in Eq. (6.115), we obtain

$$\boldsymbol{\eta}_{\text{o}}^{\text{T}}(n)\boldsymbol{\eta}_{\text{o}}(n) = \boldsymbol{\eta}_{\text{NLMS}}^{\text{T}}(n)\boldsymbol{\eta}_{\text{NLMS}}(n) + \boldsymbol{\eta}_1^{\text{T}}(n)\boldsymbol{\eta}_1(n) \tag{6.117}$$

This suggests that the squared Euclidean norm of the vector $\boldsymbol{\eta}_{\text{o}}(n)$, that is, $\boldsymbol{\eta}_{\text{o}}^{\text{T}}(n)\boldsymbol{\eta}_{\text{o}}(n)$, attains its minimum when the squared Euclidean norm of the vector $\boldsymbol{\eta}_1(n)$ is minimum. This, of course, is achieved when $\boldsymbol{\eta}_1(n) = \mathbf{0}$. Thus, we obtain

$$\boldsymbol{\eta}_{\text{o}}(n) = \boldsymbol{\eta}_{\text{NLMS}}(n) \tag{6.118}$$

This completes our proof.[5]

Figure 6.15 gives a geometrical interpretation of the above result. The tap-weight vector $\mathbf{w}(n)$ is represented by a point. The constraint $\mathbf{w}^{\text{T}}(n+1)\mathbf{x}(n) = d(n)$ limits $\mathbf{w}(n+1)$ to the points in a subspace whose dimension is one less than the filter length, N, that is, $N-1$. This is represented as a plane in Figure 6.15. The vector $\boldsymbol{\eta}_{\text{NLMS}}(n)$ is orthogonal to this subspace. It is also the vector connecting the point associated with $\mathbf{w}(n)$ to its projection on the subspace. This, clearly, shows that $\boldsymbol{\eta}_{\text{NLMS}}(n)$ is the minimum length vector,

[5] The above results could also be derived by the application of the method of Lagrange multipliers; see Section 6.10.1 for an example of the use of Lagrange multipliers. Here, we have selected to give a direct derivation of the results from the first principles of vector calculus. This derivation is also instructive because its application leads to the geometrical interpretation of the NLMS recursion depicted in Figure 6.15.

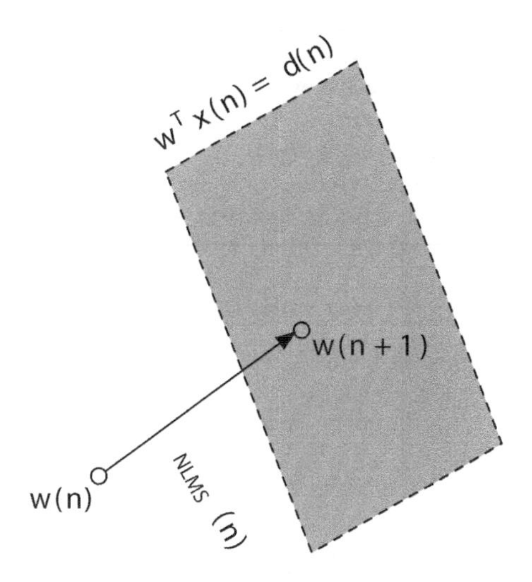

Figure 6.15 Geometrical interpretation of the NLMS recursion.

which results in the updated tap-weight vector $\mathbf{w}(n+1) = \mathbf{w}(n) + \eta_{\text{NLMS}}(n)$ subject to the constraint (6.109).

Despite its appealing interpretations, the NLMS recursion Eq. (6.107) is seldom used in actual applications. Instead, it is often observed that the following relaxed recursion results in a more reliable implementation of adaptive filters:

$$\mathbf{w}(n+1) = \mathbf{w}(n) + \frac{\tilde{\mu}}{\mathbf{x}^{\mathrm{T}}(n)\mathbf{x}(n) + \psi} e(n)\mathbf{x}(n) \tag{6.119}$$

In this recursion, $\tilde{\mu}$ and ψ are positive constants, which should be selected appropriately. The rationale for the introduction of the constant ψ is to prevent division by a small value when the squared Euclidean norm $\mathbf{x}^{\mathrm{T}}(n)\mathbf{x}(n)$ is small. This results in a more stable implementation of the NLMS algorithm. The constant $\tilde{\mu}$ may be thought of as a step-size parameter, which controls the rate of convergence of the algorithm and also its misadjustment. We also note that the recursion (6.119) reduces to Eq. (6.107) when $\tilde{\mu} = 1$ and $\psi = 0$. Table 6.2 gives a summary of the NLMS algorithm.

6.7 Affine Projection LMS Algorithm

Affine projection LMS (APLMS) algorithm, also called generalized NLMS algorithm, is in fact a generalization of the NLMS algorithm. Following the Goodwin and Sin's (1984) formulation of the NLMS, APLMS algorithm is obtained by solving the following constrained optimization problem:

> Given the set of tap-input vectors $\mathbf{x}(n), \mathbf{x}(n-1), \ldots, \mathbf{x}(n-M+1)$ and the set of desired output samples $d(n), d(n-1), \ldots, d(n-M+1)$, choose the updated tap-weight vector $\mathbf{w}(n+1)$ so as to minimize the squared Euclidean norm of the difference
>
> $$\eta(n) = \mathbf{w}(n+1) - \mathbf{w}(n) \tag{6.120}$$

Table 6.2 Summary of the normalized LMS algorithm.

Input:	Tap-weight vector, $\mathbf{w}(n)$, input vector, $\mathbf{x}(n)$, and desired output, $d(n)$
Output:	Filter output, $y(n)$, tap-weight vector update, $\mathbf{w}(n+1)$

1. Filtering:

$$y(n) = \mathbf{w}^{\mathrm{T}}(n)\mathbf{x}(n)$$

2. Error estimation:

$$e(n) = d(n) - y(n)$$

3. Tap-weight vector adaptation:

$$\mathbf{w}(n+1) = \mathbf{w}(n) + \frac{\tilde{\mu}}{\mathbf{x}^{\mathrm{T}}(n)\mathbf{x}(n) + \psi} e(n)\mathbf{x}(n)$$

subject to the set of constraints

$$\mathbf{w}^{\mathrm{T}}(n+1)\mathbf{x}(n-k) = d(n-k), \qquad k = 0, 1, \ldots, M-1 \tag{6.121}$$

This problem can be best solved using the method of Lagrange multipliers. To this end, we define the $N \times M$ matrix

$$\mathbf{X}(n) = [\mathbf{x}(n)\mathbf{x}(n-1)\ldots\mathbf{x}(n-M+1)] \tag{6.122}$$

and the length M column vector

$$\mathbf{d}(n) = [d(n)d(n-1)\cdots d(n-M+1)]^{\mathrm{T}} \tag{6.123}$$

and note that the set of constraints (6.121) can be written as

$$\mathbf{X}^{\mathrm{T}}(n)\mathbf{w}(n+1) = \mathbf{d}(n) \tag{6.124}$$

Then, following the method of Lagrange multipliers, we define

$$\xi^{\mathrm{c}} = \|\mathbf{w} - \mathbf{w}(n)\|^2 + (\mathbf{X}^{\mathrm{T}}(n)\mathbf{w} - \mathbf{d}(n))^{\mathrm{T}}\boldsymbol{\lambda} \tag{6.125}$$

where $\boldsymbol{\lambda}$ is a column vector of the Lagrange multipliers $\lambda_0, \lambda_1, \ldots, \lambda_{M-1}$.

The solution to the above-constrained optimization problem is obtained by forming and solving the system of equations

$$\nabla_{\mathbf{w}}\xi^{\mathrm{c}} = 0 \tag{6.126}$$

and

$$\nabla_{\Lambda}\xi^{\mathrm{c}} = 0 \tag{6.127}$$

We note that Eq. (6.127) reduces to the constraints (6.124), and Eq. (6.126) leads to the minimization of the norm $\|\mathbf{w}(n+1) - \mathbf{w}(n)\|^2$.

Noting that $\|\mathbf{w}-\mathbf{w}(n)\|^2 = (\mathbf{w}-\mathbf{w}(n))^{\mathrm{T}}(\mathbf{w}-\mathbf{w}(n))$ and substituting Eq. (6.125) in Eqs. (6.126) and (6.127), we obtain

$$2(\mathbf{w}(n+1)-\mathbf{w}(n)) + \mathbf{X}(n)\boldsymbol{\lambda} = 0 \tag{6.128}$$

and

$$\mathbf{X}^{\mathrm{T}}(n)\mathbf{w}(n+1) - \mathbf{d}(n) = 0 \tag{6.129}$$

respectively. Note that here $\mathbf{w}$ is replaced by $\mathbf{w}(n+1)$ because the solution to Eqs. (6.128) and (6.129) is indeed $\mathbf{w}(n+1)$.

Multiplying Eq. (6.128) from left by $\mathbf{X}^{\mathrm{T}}(n)$, noting that according to Eq. (6.129), $\mathbf{X}^{\mathrm{T}}(n)\mathbf{w}(n+1) = \mathbf{d}(n)$, and defining

$$\mathbf{e}(n) = \mathbf{d}(n) - \mathbf{X}^{\mathrm{T}}(n)\mathbf{w}(n) \tag{6.130}$$

we obtain

$$\boldsymbol{\lambda} = -2(\mathbf{X}^{\mathrm{T}}(n)\mathbf{X}(n))^{-1}\mathbf{e}(n) \tag{6.131}$$

Substituting Eq. (6.131) in Eq. (6.128) and rearranging the result, we obtain

$$\mathbf{w}(n+1) = \mathbf{w}(n) + \mathbf{X}(n)(\mathbf{X}^{\mathrm{T}}(n)\mathbf{X}(n))^{-1}\mathbf{e}(n) \tag{6.132}$$

This is the APLMS update equation.

Because of the same reasons as in the case of NLMS algorithm, in practice, Eq. (6.132) is often replaced by its relaxed form

$$\mathbf{w}(n+1) = \mathbf{w}(n) + \tilde{\mu}\mathbf{X}(n)(\mathbf{X}^{\mathrm{T}}(n)\mathbf{X}(n) + \psi\mathbf{I})^{-1}\mathbf{e}(n) \tag{6.133}$$

where $\tilde{\mu}$ is an step-size parameter, $\mathbf{I}$ is the $M \times M$ identity matrix, and ψ is a small positive constant that ensures the numerical stability of the algorithm when $\mathbf{X}^{\mathrm{T}}(n)\mathbf{X}(n)$ is a near singular matrix. Table 6.3 presents a summary of the APLMS algorithm.

Table 6.3 Summary of the affine projection LMS algorithm.

Input:	Tap-weight vector, $\mathbf{w}(n)$, Input matrix, $\mathbf{X}(n) = [\mathbf{x}(n)\mathbf{x}(n-1) \quad \cdots \mathbf{x}(n-M+1)]$, and Desired output vector, $\mathbf{d}(n) = [d(n) \quad d(n-1)\cdots d(n-M+1)]^{\mathrm{T}}$
Output:	Filter output, $y(n)$, Tap-weight vector update, $\mathbf{w}(n+1)$

1. Filtering:

$$\mathbf{y}(n) = \mathbf{X}^{\mathrm{T}}(n)\mathbf{w}(n),\ y(n) = \text{the first element of } \mathbf{y}(n)$$

2. Error estimation:

$$\mathbf{e}(n) = \mathbf{d}(n) - \mathbf{y}(n)$$

3. Tap-weight vector adaptation:

$$\mathbf{w}(n+1) = \mathbf{w}(n) + \tilde{\mu}\mathbf{X}(n)(\mathbf{X}^{\mathrm{T}}(n)\mathbf{X}(n) + \psi\mathbf{I})^{-1}\mathbf{e}(n)$$

For $M = 1$, the APLMS algorithm reduces to the NLMS algorithm. However, the APLMS algorithm offers a significant convergence improvement over the NLMS algorithm as M increases. Clearly, this improvement is at a cost of additional computational complexity. A comparison of the number of operations in Tables 6.2 and 6.3 reveal that the APLMS algorithm is at least M times more complex than the NLMS algorithm. This does not include the computation and inversion of the $M \times M$ matrix $\mathbf{X}^{\mathrm{T}}(n)\mathbf{X}(n) + \psi\mathbf{I}$. Nevertheless, further studies reveal that the tap-weight vector adaptation (6.133) may be executed once after every M sample, without any significant loss in the convergence behavior, hence, brings down its complexity to a level comparable to that of the NLMS algorithm. A summary of various versions of the APLMS algorithm can be found in Morgan and Kratzer (1996).

A convergence analysis of the APLMS algorithm is presented in Shin and Sayed (2004). The following observations are made from the results presented in this work. (i) The convergence rate of APLMS algorithm improves as M increases. (ii) The misadjustment of APLMS algorithm, on the other hand, increases (i.e., degrades) as M increases. Here, without getting involved into the details of mathematical derivations, we make an attempt to explain these observations through some intuitions.

Using the matrix inversion lemma formula (6.60), one can show that

$$\mathbf{X}(n)(\mathbf{X}^{\mathrm{T}}(n)\mathbf{X}(n) + \psi\mathbf{I})^{-1} = (\mathbf{X}(n)\mathbf{X}^{\mathrm{T}}(n) + \psi\mathbf{I})^{-1}\mathbf{X}(n) \tag{6.134}$$

Substituting Eq. (6.134) in Eq. (6.133), we obtain

$$\mathbf{w}(n+1) = \mathbf{w}(n) + \tilde{\mu}(\mathbf{X}(n)\mathbf{X}^{\mathrm{T}}(n) + \psi\mathbf{I})^{-1}\mathbf{X}(n)\mathbf{e}(n) \tag{6.135}$$

Examining Eq. (6.135), one finds that $\mathbf{X}(n)\mathbf{e}(n) = \sum_{i=0}^{M-1} \mathbf{x}(n-i)e(n-i)$ and this in turn implies that $\mathbf{X}(n)\mathbf{e}(n)$ is a random vector whose mean is equal to $-\frac{M}{2}\nabla_{\mathbf{w}}\xi$. Also, as M increases, the mean of the $N \times N$ matrix $\mathbf{X}(n)\mathbf{X}^{\mathrm{T}}(n)$ approaches $M\mathbf{R}$. Hence, one may think of $(\mathbf{X}(n)\mathbf{X}^{\mathrm{T}}(n) + \psi\mathbf{I})^{-1}\mathbf{X}(n)\mathbf{e}(n)$ as a noisy (and regularized) sample of $-\frac{1}{2}\mathbf{R}^{-1}\nabla_{\mathbf{w}}\xi$, and, thus, argue that the update equation of APLMS algorithm attempts to implement a stochastic version of the Newton's method. In other words, one may argue that the premultiplication of the stochastic gradient vector $\mathbf{X}(n)\mathbf{e}(n)$ by $(\mathbf{X}(n)\mathbf{X}^{\mathrm{T}}(n) + \psi\mathbf{I})^{-1}$ results in a vector that, on average, points toward the minimum of the performance surface of the adaptive filter, hence, avoiding the slow modes of convergence of the LMS algorithm.

To explain why the misadjustment of APLMS algorithm increases with M, we resort to a generalization of the geometrical interpretation of the NLMS algorithm that was presented earlier in Figure 6.15. Figure 6.16 presents a diagram that expands Figure 6.15 to the case where $M = 2$. Here, to satisfy the pair of constraints $\mathbf{w}^{\mathrm{T}}\mathbf{x}(n) = d(n)$ and $\mathbf{w}^{\mathrm{T}}\mathbf{x}(n-1) = d(n-1)$, while minimizing $\|\boldsymbol{\eta}_{\mathrm{APLMS}}(n)\|$, the error vector $\boldsymbol{\eta}_{\mathrm{APLMS}}(n)$ must be orthogonal to the intersection of the subspaces of the two constraints. Hence, one may note that $\boldsymbol{\eta}_{\mathrm{APLMS}}(n)$ is not necessarily orthogonal to the subspace of the constraint $\mathbf{w}^{\mathrm{T}}\mathbf{x}(n) = d(n)$ and thus $\|\boldsymbol{\eta}_{\mathrm{APLMS}}(n)\| \geq \|\boldsymbol{\eta}_{\mathrm{NLMS}}(n)\|$. Obviously, $\|\boldsymbol{\eta}_{\mathrm{APLMS}}(n)\|$ increases further as M is given larger values. On the other hand, we recall that $\boldsymbol{\eta}_{\mathrm{NLMS}}(n)$ and, similarly, $\boldsymbol{\eta}_{\mathrm{APLMS}}(n)$ may be thought as a perturbation that may be imposed on the filter tap weights as the NLMS and APLMS algorithms proceed. In the steady state, a larger perturbation of the tap weights, clearly, results in a larger misadjustment. A few problems

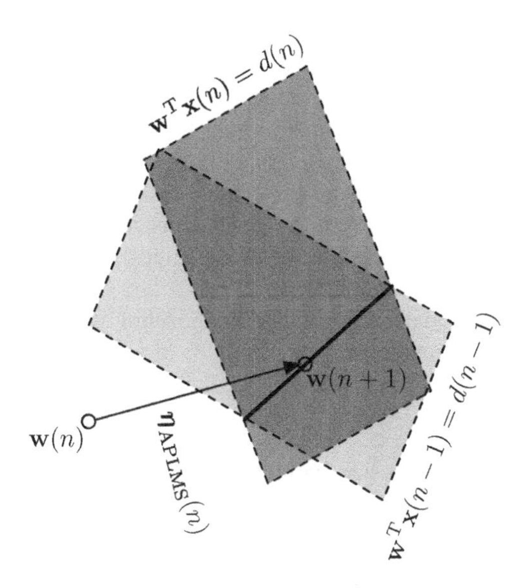

Figure 6.16 Geometrical interpretation of the affine projection LMS recursion.

at the end of this chapter guide the reader to develop a better understanding of the NLMS and APLMS algorithms through a sequence of computer simulations.

6.8 Variable Step-Size LMS Algorithm

The analysis presented in Section 6.3 shows that the step-size parameter, μ, plays a significant role in controlling the performance of the LMS algorithm. On the one hand, the speed of convergence of the LMS algorithm changes proportional to its step-size parameter. As a result, a large step-size parameter may be required to minimize the transient time of the LMS algorithm. On the other hand, to achieve a small misadjustment, a small step-size parameter has to be used. These are conflicting requirements and, thus, a compromise solution has to be adopted. The variable step-size LMS (VSLMS) algorithm, which is introduced in this section, is an effective solution to this problem (Shin and Lee, 1985; Harris, Chabries, and Bishop, 1986).

The VSLMS algorithm works based on a simple heuristic that comes from the mechanism of the LMS algorithm. Each tap of the adaptive filter is given a separate time-varying step-size parameter and the LMS recursion is written as

$$w_i(n+1) = w_i(n) + 2\mu_i(n)e(n)x(n-i), \qquad \text{for} \quad i = 0, 1, \ldots, N-1 \tag{6.136}$$

where $w_i(n)$ is the ith element of the tap-weight vector $\mathbf{w}(n)$ and $\mu_i(n)$ is its associated step-size parameter at iteration n. The adjustment of the step-size parameter $\mu_i(n)$ is done as follows. The corresponding stochastic gradient term $g_i(n) = e(n)x(n-i)$ is monitored over the successive iterations of the algorithm, and $\mu_i(n)$ is increased if the latter term consistently shows positive or negative direction. This happens when the adaptive filter has not yet converged. As the adaptive filter tap weights converge to some vicinity of their optimum values, the averages of the stochastic gradient terms approach zero and hence they change signs more frequently. This is detected by the algorithm and the corresponding step-size parameters are gradually reduced to some minimum values. If the situation changes and the algorithm begins to hunt for a new optimum point, the gradient terms will indicate consistent (positive or negative) directions, resulting in increase of the corresponding step-size parameters. To ensure that the step-size parameters do not become too large (which may result in the system instability) or too small (which may result in a slow reaction of the system to sudden changes), upper and lower limits should be specified for each step-size parameter.

Following the above argument, the VSLMS algorithm step-size parameters, $\mu_i(n)$'s, may be adjusted using the recursions

$$\mu_i(n) = \mu_i(n-1) + \rho\text{sign}[g_i(n)]\text{sign}[g_i(n-1)] \tag{6.137}$$

where ρ is a small positive step-size parameter. The "sign" functions may be dropped from Eq. (6.137). This results in the following alternative step-size parameter update equation:

$$\mu_i(n) = \mu_i(n-1) + \rho g_i(n) g_i(n-1) \tag{6.138}$$

Both update equations (6.137) and (6.138) work well in practice. Which of the two choices works better is application dependent. The choice of one over the other may also be decided based on the available hardware/software platform on which the algorithm is to be implemented. For instance, if a digital signal processor is being used, the recursion (6.138) may be much easier to implement. On the other hand, if a custom chip is to be designed, the update equation (6.137) may be preferred.

Derivation of an inequality similar to Eq. (6.74) to determine the range of the step-size parameters that ensure the stability of the VSLMS algorithm is rather difficult because of the time variation of the step-size parameters. Here, we adopt a simple approach by assuming that the step-size parameters vary slowly, so that for the stability analysis, they may be assumed fixed and use the analogy between the resulting VSLMS algorithm equations and the conventional LMS algorithm to arrive at a result, which through computer simulations has been found to be reasonable. Further results on the VSLMS algorithm misadjustment and its tracking behavior, along with computer simulation results, can be found in Chapter 14.

The set of update equations (6.136) may be written in vector form as

$$\mathbf{w}(n+1) = \mathbf{w}(n) + 2\boldsymbol{\mu}(n)e(n)\mathbf{x}(n) \tag{6.139}$$

where $\boldsymbol{\mu}(n)$ is a diagonal matrix consisting of the step-size parameters $\mu_0(n)$, $\mu_1(n)$, ..., $\mu_{N-1}(n)$. Equation (6.139) may further be rearranged as

$$\mathbf{v}(n+1) = (\mathbf{I} - 2\boldsymbol{\mu}(n)\mathbf{x}(n)\mathbf{x}^{\mathrm{T}}(n))\mathbf{v}(n) + 2\boldsymbol{\mu}(n)e_{\mathrm{o}}(n)\mathbf{x}(n) \tag{6.140}$$

where notations follow those of Section 6.2. Comparing Eq. (6.140) with Eq. (6.13) and the subsequent discussions on the stability of the conventional LMS algorithm in Section 6.3, we may argue that to ensure the stability of the VSLMS algorithm, the scalar step-size parameter μ in Eq. (6.74) should be replaced by the diagonal matrix $\boldsymbol{\mu}(n)$.

Table 6.4 Summary of an implementation of variable step-size LMS algorithm.

Input:	Tap-weight vector, $\mathbf{w}(n)$, input vector, $\mathbf{x}(n)$, Gradient terms $g_0(n-1), g_1(n-1), \ldots, g_{N-1}(n-1)$, Step-size parameters, $\mu_0(n-1), \mu_1(n-1), \ldots, \mu_{N-1}(n-1)$, and desired output, $d(n)$
Output:	Filter output, $y(n)$, tap-weight vector update, $\mathbf{w}(n+1)$, gradient terms $g_0(n), g_1(n), \ldots, g_{N-1}(n)$, and updated step-size parameters $\mu_0(n), \mu_1(n), \ldots, \mu_{N-1}(n)$

1. Filtering:

$$y(n) = \mathbf{w}^{\mathrm{T}}(n)\mathbf{x}(n)$$

2. Error estimation: $e(n) = d(n) - y(n)$
3. Tap weights and step-size parameters adaptation:

$$\begin{aligned}
&\text{For} \quad i = 0, 1, \ldots, N-1 \\
&\qquad g_i(n) = e(n)x(n-i) \\
&\qquad \mu_i(n) = \mu_i(n-1) + \rho\,\mathrm{sign}[g_i(n)]\mathrm{sign}[g_i(n-1)] \\
&\qquad\quad \text{if } \mu_i(n) > \mu_{\max}, \mu_i(n) = \mu_{\max} \\
&\qquad\quad \text{if } \mu_i(n) < \mu_{\min}, \mu_i(n) = \mu_{\min} \\
&\qquad w_i(n+1) = w_i(n) + 2\mu_i(n)g_i(n) \\
&\text{end}
\end{aligned}$$

This leads to the inequality[6]

$$\mathrm{tr}[\boldsymbol{\mu}(n)\mathbf{R}] < \frac{1}{3} \tag{6.141}$$

as a sufficient condition, which ensures the stability of the VSLMS algorithm. Although the inequality (6.141) may be used to impose some bounds on the step-size parameters $\mu_i(n)$'s dynamically as the adaptation of the filter proceeds, this leads to a rather complicated process. Instead, in practice, one usually prefers to use Eq. (6.74) to limit all $\mu_i(n)$'s to the same maximum value, say $\mu_{\max}$.

The minimum bound, which may be imposed on the variable step-size parameters, $\mu_i(n)$'s, can be as low as zero. However, in actual practice, a positive bound is usually used, so that the adaptation process will be on all the time and possible variations in the adaptive filter optimum tap weights can always be tracked. Here, we use the notation $\mu_{\min}$ to refer to this lower bound. Table 6.4 gives the summary of an implementation of the VSLMS algorithm.

6.9 LMS Algorithm for Complex-Valued Signals

In applications such as data transmission with quadrature-amplitude modulation (QAM) signaling and beamforming with baseband processing of signals, the underlying data

[6] See Chapter 14 for a formal derivation of Eq. (6.141).

signals and filter coefficients are complex-valued. To modify the LMS recursion for such applications, we use the definition of gradient of real-valued functions of complex-valued variables, as was defined in Section 3.5. We consider an adaptive filter with complex-valued tap-input vector $\mathbf{x}(n)$, tap-weight vector $\mathbf{w}(n) = [w_0^*(n) w_1^*(n) \cdots w_{N-1}^*(n)]^{\mathrm{T}}$, output $y(n) = \mathbf{w}^{\mathrm{H}}(n)\mathbf{x}(n)$, and desired output $d(n)$.

The LMS algorithm in this case works based on the update equation

$$\mathbf{w}(n+1) = \mathbf{w}(n) - \mu \nabla_{\mathbf{w}}^{C} |e(n)|^2 \tag{6.142}$$

where $\nabla_{\mathbf{w}}^{C}$ denotes complex gradient operator with respect to the variable vector $\mathbf{w}$. This is defined as

$$\nabla_{\mathbf{w}}^{C} = \begin{bmatrix} \nabla_{w_0^*}^{C} \\ \nabla_{w_1^*}^{C} \\ \vdots \\ \nabla_{w_{N-1}^*}^{C} \end{bmatrix} \tag{6.143}$$

where ∇_{w}^{C}, as was defined in Section 3.5, is complex gradient with respect to the complex variable w. We recall that

$$\nabla_{w}^{C} = \frac{\partial}{\partial w_R} + j\frac{\partial}{\partial w_I} \tag{6.144}$$

where w_{R} and w_{I} are real and imaginary parts of w, respectively, and $j = \sqrt{-1}$. We also note that in Eq. (6.143), the elements of the gradient vector $\nabla_{\mathbf{w}}^{C}$ are complex gradient with respect to the elements of $\mathbf{w}$ and these elements are the conjugates of the actual tap weights, that is, $w_0^*, w_1^*, \cdots, w_{N-1}^*$. Furthermore, we note that a direct substitution in Eq. (6.144) gives

$$\nabla_{w_i^*}^{C} = \frac{\partial}{\partial w_{i,R}} - j\frac{\partial}{\partial w_{i,I}} \tag{6.145}$$

Replacing $|e(n)|^2$ by $e(n)e^*(n)$, using Eq. (6.145), and following a derivation similar to the one which has led to Eq. (3.63), we obtain

$$\nabla_{w_i^*}^{C} |e(n)|^2 = -2e^*(n)x(n-i), \qquad \text{for} \quad i = 0, 1, \ldots, N-1 \tag{6.146}$$

where the asterisk denotes complex conjugation. Substituting Eq. (6.146) and definition (6.143) in Eq. (6.142), we obtain

$$\mathbf{w}(n+1) = \mathbf{w}(n) + 2\mu e^*(n)\mathbf{x}(n) \tag{6.147}$$

This is the desired LMS recursion for the case where the underlying processes are complex-valued. Table 6.5 gives a summary of implementation of the LMS algorithm for complex-valued signals.

The convergence properties of the LMS algorithm for complex-valued signals are very similar to those of the real-valued signals. These properties are summarized as follows for reference:

- The time constant equations (6.33) is also applicable to adaptive filters with complex-valued signals.

Table 6.5 Summary of the complex LMS algorithm.

Input:	Tap-weight vector, $\mathbf{w}(n)$, input vector, $\mathbf{x}(n)$, and desired output, $d(n)$
Output:	Filter output, $y(n)$, Tap-weight vector update, $\mathbf{w}(n+1)$

1. Filtering:
$$y(n) = \mathbf{w}^{\mathrm{H}}(n)\mathbf{x}(n)$$
2. Error estimation:
$$e(n) = d(n) - y(n)$$
3. Tap-weight vector adaptation:
$$\mathbf{w}(n+1) = \mathbf{w}(n) + 2\mu e^*(n)\mathbf{x}(n)$$

- The misadjustment equation (6.61) has to be slightly modified. This modification is the result of the fact that for complex-valued jointly Gaussian random variables, the equality (6A.6) has to be replaced by

$$E[x_1 x_2^* x_3 x_4^*] = E[x_1 x_2^*]E[x_3 x_4^*] + E[x_1 x_4^*]E[x_2^* x_3] \tag{6.148}$$

Taking note of this and following a similar derivation as in Section 6.3, we obtain[7]

$$\mathcal{M} = \frac{\sum_{i=0}^{N-1} \mu\lambda_i/(1-\mu\lambda_i)}{1 - \sum_{i=0}^{N-1} \mu\lambda_i/(1-\mu\lambda_i)} \tag{6.149}$$

- When the step-size parameter, μ, is small, so that $\mu\lambda_i \ll 1$, for $i = 0, 1, \ldots, N-1$, Eq. (6.149) reduces to Eq. (6.64). Thus, the approximation (6.64) is also applicable to the case where the underlying signals are complex-valued.
- Using Eq. (6.149) and following the same arguments as given in Section 6.3.4, we find that in the case of complex-valued signals, the LMS algorithm remains stable when

$$0 < \mu < \frac{1}{2\mathrm{tr}[\mathbf{R}]} \tag{6.150}$$

Comparing this result with Eq. (6.74), we find that in the case of complex-valued signals, the upper bound of μ is more relaxed when compared with the corresponding bound for real-valued signals.

[7] A detailed derivation of this result can be found in Haykin (1991).

6.10 Beamforming (Revisited)

The beamforming structure presented in Example 3.6 as well as earlier in this chapter works with signals at their associated radio frequency (RF) or an intermediate frequency (IF). We also recall that the carrier phase plays a major role in implementing a desired beam pattern. To extract and use the carrier phase angles of the signals picked up by the array elements, it was previously proposed that modulated carrier signals and their associated 90° phase-shifted version be processed, simultaneously. The amplitude and phase angle of an amplitude-modulated signal, such as $u(t) = \alpha(t)\cos(\omega_c t + \phi)$ (where t denotes continuous time), are preserved if it is converted to an equivalent complex-valued baseband signal using a phase-quadrature demodulator structure, as depicted in Figure 6.17. This structure suggests that the baseband equivalent of an RF (or IF) signal $u(t) = \alpha(t)\cos(\omega_c t + \phi)$, which preserves both phase and amplitude of $u(t)$, is the sampled signal

$$\underline{u}(n) = \alpha(n)\mathrm{e}^{j\phi}$$

This is known as *phasor*. Note that we have used underline notation to indicate that $\underline{u}(n)$ is a phasor.

In the implementation of beamformers, working with phasor signals is more convenient than RF (or IF) signals. In particular, from implementation point of view, digital processing of RF (or IF) signals requires a very high sampling rate to prevent aliasing and allow any postprocessing of the sampled signals, while the required sampling (Nyquist) rate for equivalent baseband signals is much lower.

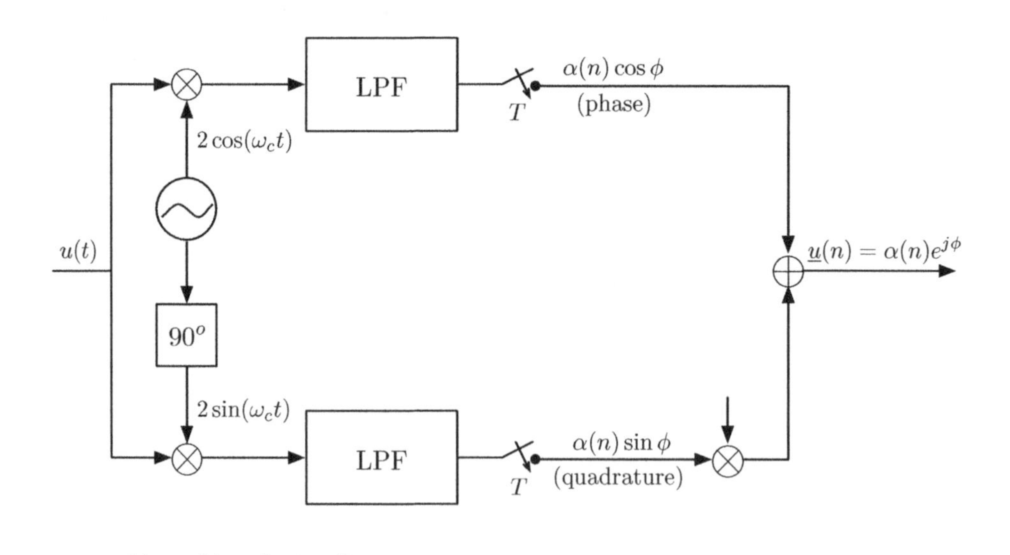

Figure 6.17 Conversion of an amplitude-modulated signal to its equivalent phase-quadrature baseband (phasor) signal.

Example 6.3

In this example, we discuss the implementation of the beamformer of Figure 6.12 at baseband using phasor signals. Figure 6.18 shows an equivalent implementation of Figure 6.12 when all signals are converted to their equivalent phasors. This implementation, as shown, involves an adaptive filter with only one complex tap weight, w, whose optimum value is obtained by minimizing

$$\xi = E[|\underline{d}(n) - w\underline{x}(n)|^2] \tag{6.151}$$

Using the definition (6.144) to obtain the gradient of ξ with respect to w and setting the result equal to zero, we obtain

$$w_{\text{o}} = \frac{E[\underline{d}(n)\underline{x}^*(n)]}{E[|\underline{x}(n)|^2]} \tag{6.152}$$

where w_{o} is the optimum value of w.

Converting $d(n)$ of Eq. (6.90) to its equivalent phasor, we get

$$\underline{d}(n) = \alpha(n)\text{e}^{j\phi_1} + \beta(n)\text{e}^{j(\phi_2 - \phi_o)} \tag{6.153}$$

Similarly,

$$\underline{x}(n) = \alpha(n)\text{e}^{j\phi_1} + \beta(n)\text{e}^{j\phi_2} \tag{6.154}$$

Substituting Eqs. (6.153) and (6.154) in Eq. (6.152) and recalling that $\alpha(n)$ and $\beta(n)$ are zero-mean, real-valued and uncorrelated random variables, we obtain

$$w_{\text{o}} = \frac{\sigma_\alpha^2 + \sigma_\beta^2 \text{e}^{-j\phi_o}}{\sigma_\alpha^2 + \sigma_\beta^2} \tag{6.155}$$

With this value of w_o, the array power gain, $\mathcal{G}(\theta)$, for a narrow-band signal arriving at an angle θ, is obtained as follows.

Assuming that the signal arriving at the angle of θ is $\gamma(t)\cos\omega_{\text{c}}t$ and using Eq. (6.93), with θ_o replaced by θ, we get $\underline{d}(n) = \gamma(n)e^{-j\pi\sin\theta}$. We also note that $\underline{x}(n) = \gamma(n)$. Thus,

$$\underline{e}(n) = \gamma(n)\text{e}^{-j\pi\sin\theta} - w_o\gamma(n) = \gamma(n)(\text{e}^{-j\pi\sin\theta} - w_o)$$

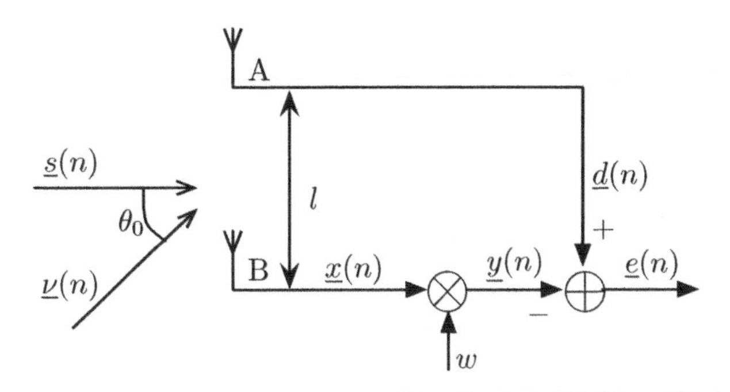

Figure 6.18 Baseband implementation of a two-element beamformer.

and

$$\mathcal{G}(\theta) = \frac{E[|\underline{e}(n)|^2]}{E[|\underline{x}(n)|^2]} = |e^{-j\pi \sin\theta} - w_o|^2 \tag{6.156}$$

Careful examination of Eqs. (6.95) and (6.156) reveals that, as one may expect, both implementations of the beamformer (i.e., Figures 6.12 and 6.18) result in the same optimized power gain. The beamformer tap weights w_0 and w_1 in Eq. (6.95) correspond to the real and negative of the imaginary parts, respectively, of the complex tap weight w_o in Eq. (6.156). This, in turn, confirms that the two implementations are equivalent.

To adjust w adaptively, we may use the complex LMS algorithm of Table 6.5 with the following substitutions:

$$\mathbf{x}(n) = \underline{x}(n), \quad d(n) = \underline{d}(n), \quad e(n) = \underline{e}(n)$$

and

$$\mathbf{w}(n) = w^*(n)$$

If we run the resulting algorithm for sufficient number of iterations and then use the converged tap weight in Eq. (6.156), we will obtain the same directivity pattern as the one presented in Figure 6.13 as the two implementations are equivalent.

So far, we have introduced beamformers that are limited to only two antennas. Such beamformers are capable of canceling only one jammer. Use of more elements, as shown in Figure 6.19, allows cancellation of more than one jammer. In general, to cancel M jammers, one requires at least $M + 1$ antennas. We may also recall that the implementation proposed in Example 6.3 and also those that were discussed previously do not differentiate between the jammer(s) and the desired signal. They simply adapt, so that the stronger signal(s) is (are) canceled, leaving behind the weaker signal(s). In cases where no jammer is present or the desired signal is strong, the latter is deleted by the beamformer. This problem can be prevented using an amended version of the LMS algorithm, which imposes a linear constraint on the tap weights of the adaptive filter. This, which is known as *linearly constrained LMS algorithm*, is introduced in the next section. To be able to apply the latter algorithm, the beamformer structure has to be modified as shown in Figure 6.20, where $M + 1$ antennas are used for cancellation of up to M jammers arriving from different directions.

The fundamental difference between the two structures shown in Figures 6.19 and 6.20 is that in the latter, there is no primary input. The tap weights of the beamformer of Figure 6.20 are optimized, so that its output, $y(n)$, is minimized in the mean-square sense. To prevent the trivial solution of $w_i = 0$, for all i, a linear constraint, which ensures a nonzero gain in the desired direction, is imposed on the beamformer tap weights before their optimization. The discussions provided in the next section and, especially, Example 6.4 will clarify this concept.

We may also recall that the beamformer structures depicted in Figures 6.19 and 6.20 assume that the signals picked up by array elements (antennas) are narrow-band. When the underlying signals are wideband, the output of each element has to go through a transversal filter, so that there would be some control over different frequency bins (Widrow and Stearns, 1985; Johnson and Dudgeon, 1993).

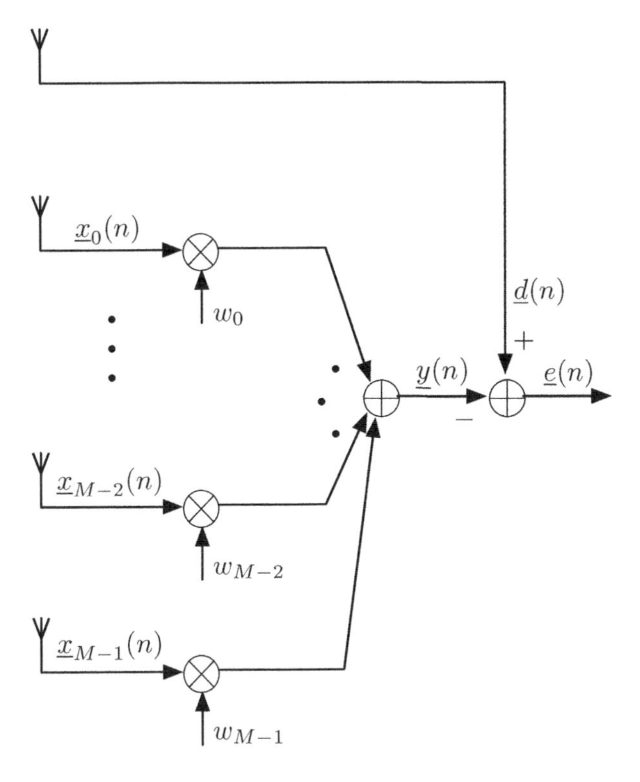

Figure 6.19 Baseband implementation of an $(M+1)$-element beamformer.

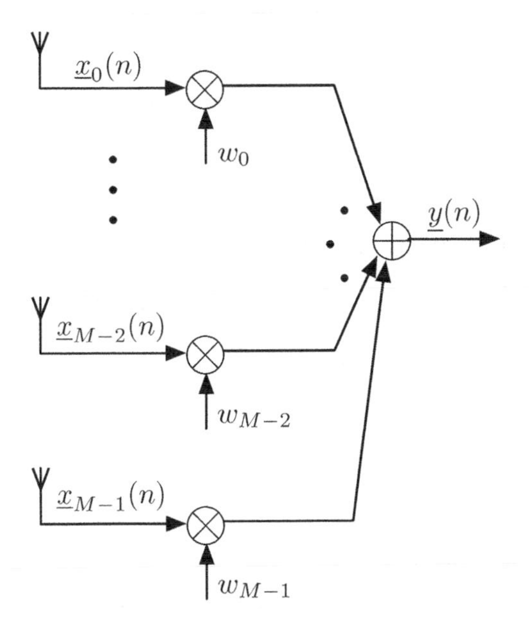

Figure 6.20 Alternative implementation of the $(M+1)$-element baseband beamformer.

6.11 Linearly Constrained LMS Algorithm

In this section, we discuss the problem of Wiener filtering with a linear constraint imposed on the filter tap weights. We also present an LMS algorithm for adaptive adjustment of the filter tap weights subject to the required constraint. For the sake of simplicity, all derivations are given for the case of real-valued signals. However, we also give a summary of the final results for the case of complex-valued signals. Application of the proposed algorithm to narrow-band beamforming is then discussed as an example.

6.11.1 Statement of the Problem and Its Optimal Solution

Given an observation vector $\mathbf{x}(n)$ and a desired response $d(n)$, we wish to find a tap-weight vector $\mathbf{w}$, so that

$$e(n) = d(n) - \mathbf{w}^{\mathrm{T}}\mathbf{x}(n) \tag{6.157}$$

is minimized in the mean-square sense, subject to the constraint

$$\mathbf{c}^{\mathrm{T}}\mathbf{w} = a \tag{6.158}$$

where a is a scalar and $\mathbf{c}$ is a fixed column vector.

This problem can be solved using the method of Lagrange multipliers. According to the method of Lagrange multipliers, we define (the superscript c stands for constraint)

$$\xi^{\mathrm{c}} = E[e^2(n)] + \lambda(\mathbf{c}^{\mathrm{T}}\mathbf{w} - a) \tag{6.159}$$

where λ is the Lagrange multiplier, and solve the equations

$$\nabla_{\mathbf{w}}\xi^{\mathrm{c}} = \mathbf{0} \quad \text{and} \quad \frac{\partial \xi^{\mathrm{c}}}{\partial \lambda} = 0 \tag{6.160}$$

simultaneously. We note that $\partial\xi^{\mathrm{c}}/\partial\lambda = 0$ results in the constraint (6.158).

Substituting Eq. (6.157) in Eq. (6.159) and going through some manipulations similar to those in Chapter 4 (Section 4.3), we obtain

$$\xi^{\mathrm{c}} = \xi_{\min} + \mathbf{v}^{\mathrm{T}}\mathbf{R}\mathbf{v} + \lambda(\mathbf{c}^{\mathrm{T}}\mathbf{v} - a') \tag{6.161}$$

where $\mathbf{v} = \mathbf{w} - \mathbf{w}_{\mathrm{o}}$, $\mathbf{w}_{\mathrm{o}} = \mathbf{R}^{-1}\mathbf{p}$, $\mathbf{R} = E[\mathbf{x}(n)\mathbf{x}^{\mathrm{T}}(n)]$, $\mathbf{p} = E[d(n)\mathbf{x}(n)]$, and $a' = a - \mathbf{c}^{\mathrm{T}}\mathbf{w}_{\mathrm{o}}$. With this, the above problem is reduced to the minimization of $\mathbf{v}^{\mathrm{T}}\mathbf{R}\mathbf{v}$, subject to the constraint $\mathbf{c}^{\mathrm{T}}\mathbf{v} = a'$. The solution to this problem is obtained by simultaneous solution of

$$\nabla_{\mathbf{v}}\xi^{\mathrm{c}} = 2\mathbf{R}\mathbf{v}_{\mathrm{o}}^{\mathrm{c}} + \lambda\mathbf{c} = 0 \tag{6.162}$$

and

$$\frac{\partial \xi^{\mathrm{c}}}{\partial \lambda} = \mathbf{c}^{\mathrm{T}}\mathbf{v}_{\mathrm{o}}^{\mathrm{c}} - a' = 0 \tag{6.163}$$

where $\mathbf{v}_{\mathrm{o}}^{\mathrm{c}}$ is the constrained optimum value of $\mathbf{v}$.

From Eq. (6.162), we obtain

$$\mathbf{v}_{\mathrm{o}}^{\mathrm{c}} = -\frac{\lambda}{2}\mathbf{R}^{-1}\mathbf{c} \tag{6.164}$$

Substituting Eq. (6.164) in Eq. (6.163), we get

$$-\frac{\lambda}{2}\mathbf{c}^{\mathrm{T}}\mathbf{R}^{-1}\mathbf{c} - a' = 0$$

or

$$\lambda = -\frac{2a'}{\mathbf{c}^{\mathrm{T}}\mathbf{R}^{-1}\mathbf{c}} \tag{6.165}$$

Finally, substituting Eq. (6.165) in Eq. (6.164), we obtain

$$\mathbf{v}_{\mathrm{o}}^{\mathrm{c}} = \frac{a'\mathbf{R}^{-1}\mathbf{c}}{\mathbf{c}^{\mathrm{T}}\mathbf{R}^{-1}\mathbf{c}} \tag{6.166}$$

The minimum value of ξ^{c} is obtained by substituting Eq. (6.166) in Eq. (6.161). This gives

$$\xi_{\mathrm{min}}^{\mathrm{c}} = \xi_{\mathrm{min}} + \frac{a'^2}{\mathbf{c}^{\mathrm{T}}\mathbf{R}^{-1}\mathbf{c}} \tag{6.167}$$

We note that the second term on the right-hand side of Eq. (6.167) is the excess MSE, which is introduced as a result of the imposed constraint.

Also, noting that $\mathbf{w} = \mathbf{v} + \mathbf{w}_o$, and using Eq. (6.166), we obtain

$$\mathbf{w}_{\mathrm{o}}^{\mathrm{c}} = \mathbf{w}_{\mathrm{o}} + \frac{a'\mathbf{R}^{-1}\mathbf{c}}{\mathbf{c}^{\mathrm{T}}\mathbf{R}^{-1}\mathbf{c}} \tag{6.168}$$

6.11.2 Update Equations

The adaptation of the tap-weight vector $\mathbf{w}$, while the constraint (6.158) holds, may be done in two steps as follows:

Step 1.

$$\mathbf{w}^{+}(n) = \mathbf{w}(n) + 2\mu e(n)\mathbf{x}(n) \tag{6.169}$$

Step 2.

$$\mathbf{w}(n+1) = \mathbf{w}^{+}(n) + \boldsymbol{\vartheta}(n) \tag{6.170}$$

where $\boldsymbol{\vartheta}(n)$ is chosen, so that $\mathbf{c}^{\mathrm{T}}\mathbf{w}(n+1) = a$, while $\boldsymbol{\vartheta}^{\mathrm{T}}(n)\boldsymbol{\vartheta}(n)$ is minimized. That is, *we choose* $\boldsymbol{\vartheta}(n)$*, so that the constraint* (6.137) *holds after Step 2, while the perturbation introduced by* $\boldsymbol{\vartheta}(n)$ *is minimized.*

The latter problem can also be solved using the method of Lagrange multipliers and following a procedure similar to the one used above to obtain $\mathbf{v}_{\mathrm{o}}^{\mathrm{c}}$. This gives

$$\boldsymbol{\vartheta}(n) = \frac{a - \mathbf{c}^{\mathrm{T}}\mathbf{w}^{+}(n)}{\mathbf{c}^{\mathrm{T}}\mathbf{c}}\mathbf{c} \tag{6.171}$$

Substituting this result in Eq. (6.170), we obtain

$$\mathbf{w}(n+1) = \mathbf{w}^{+}(n) + \frac{a - \mathbf{c}^{\mathrm{T}}\mathbf{w}^{+}(n)}{\mathbf{c}^{\mathrm{T}}\mathbf{c}}\mathbf{c} \tag{6.172}$$

The above derivations are summarized in Table 6.6.

Table 6.6 Summary of the linearly constrained LMS algorithm.

Input:	Tap-weight vector, $\mathbf{w}(n)$, input vector, $\mathbf{x}(n)$, and desired output, $d(n)$
Output:	Filter output, $y(n)$, Tap-weight vector update, $\mathbf{w}(n+1)$

1. Filtering:

$$y(n) = \mathbf{w}^{\mathrm{T}}(n)\mathbf{x}(n)$$

2. Error estimation:

$$e(n) = d(n) - y(n)$$

3. Tap-weight vector adaptation:

$$\mathbf{w}^{+}(n) = \mathbf{w}(n) + \mu e(n)\mathbf{x}(n)$$

$$\mathbf{w}(n+1) = \mathbf{w}^{+}(n) + \frac{a - \mathbf{c}^{\mathrm{T}}\mathbf{w}^{+}(n)}{\mathbf{c}^{\mathrm{T}}\mathbf{c}}\mathbf{c}$$

6.11.3 Extension to the Complex-Valued Case

When the underlying signal/variables are complex-valued, the following amendments have to be made to the previous results:

- The constraint equation (6.158) is written as

$$\mathbf{w}^{\mathrm{H}}\mathbf{c} = a \tag{6.173}$$

  where the vector $\mathbf{c}$ and the scalar a are both complex-valued.
- The constrained optimum tap-weight vector of the filter is obtained according to the equation

$$\mathbf{w}_{\mathrm{o}}^{\mathrm{c}} = \mathbf{w}_{\mathrm{o}} + \frac{a'^{*}\mathbf{R}^{-1}\mathbf{c}}{\mathbf{c}^{\mathrm{H}}\mathbf{R}^{-1}\mathbf{c}} \tag{6.174}$$

  where $a' = a - \mathbf{w}_{\mathrm{o}}^{\mathrm{H}}\mathbf{c}$.
- The adaptation of the filter tap weights is made according to the following equations:

$$e(n) = d(n) - \mathbf{w}^{\mathrm{H}}(n)\mathbf{x}(n) \tag{6.175}$$

$$\mathbf{w}^{+}(n) = \mathbf{w}(n) + 2\mu e^{*}(n)\mathbf{x}(n) \tag{6.176}$$

  and

$$\mathbf{w}(n+1) = \mathbf{w}^{+}(n) + \frac{a^{*} - \mathbf{c}^{\mathrm{H}}\mathbf{w}^{+}(n)}{\mathbf{c}^{\mathrm{H}}\mathbf{c}}\mathbf{c} \tag{6.177}$$

Example 6.4

As an example of the linearly constrained LMS algorithm, we consider the two-element narrow-band beamformer of Figure 6.21. Here, we consider the processing of signals in

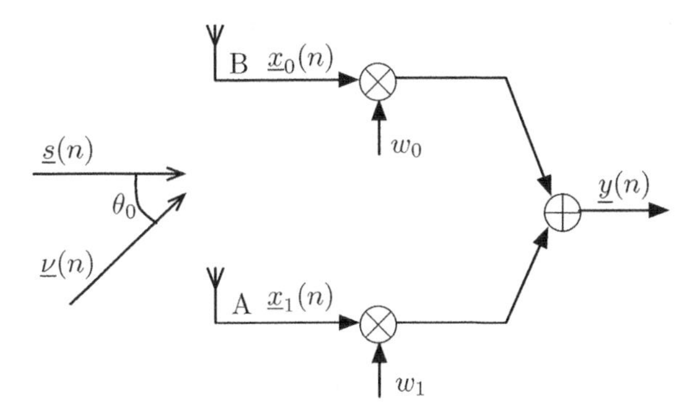

Figure 6.21 Baseband implementation of a two-element beamformer with a linearly constrained beam pattern.

baseband, that is, phasor (complex-valued) signals are considered. We note that with $s(n)$ and $\nu(n)$, as defined in Section 6.4.4,

$$\underline{x}_0(n) = \alpha(n)e^{j\phi_1} + \beta(n)e^{j(\phi_2-\phi_o)} \tag{6.178}$$

$$\underline{x}_1(n) = \alpha(n)e^{j\phi_1} + \beta(n)e^{j\phi_2} \tag{6.179}$$

The beamformer tap weights, w_0 and w_1, are adjusted, so that its output, $y(n)$, is minimized in the mean-square sense. This is equivalent of saying $\underline{d}(n) = 0$. It is clear that if there is no constraint on the tap weights and they are adjusted to minimize $E[|y(n)|^2]$, we obtain the undesirable result of $w_{0,o} = w_{1,o} = 0$, which cancels both the jammer and the desired signal. To ensure that the desired signal $s(n)$, arriving at the direction perpendicular to the line connecting A to B, passes through the beamformer with no distortion, the following constraint must hold:

$$w_0 + w_1 = 1$$

Using vector notations, this may be written as

$$\mathbf{w}^{\mathrm{H}}\mathbf{c} = 1 \tag{6.180}$$

where $\mathbf{c} = [1 \ \ 1]^{\mathrm{T}}$ and $\mathbf{w} = [w_0^* \ \ w_1^*]^{\mathrm{T}}$. We note that, in general, the value of $\mathbf{c}$ will depend on the angle of arrival of the desired signal $s(n)$ with respect to the perpendicular to the line connecting A to B; see Problem P6.29.

Letting $\mathbf{x}(n) = [\underline{x}_0(n) \ \ \underline{x}_1(n)]^{\mathrm{T}}$ and noting that $\alpha(n)$ and $\beta(n)$ are uncorrelated with each other, we get

$$\mathbf{R} = \begin{bmatrix} \sigma_\alpha^2 + \sigma_\beta^2 & \sigma_\alpha^2 + \sigma_\beta^2 e^{-j\phi_o} \\ \sigma_\alpha^2 + \sigma_\beta^2 e^{j\phi_o} & \sigma_\alpha^2 + \sigma_\beta^2 \end{bmatrix} \tag{6.181}$$

Using this result and noting that in the present case $\mathbf{c} = [1 \;\; 1]^{\mathrm{T}}$, $a = 1$, and $\mathbf{w}_{\mathrm{o}} = 0$, we obtain, from Eq. (6.174),

$$\mathbf{w}_{\mathrm{o}}^{\mathrm{c}} = \frac{1}{2(1 - \cos\,\phi_o)} \begin{bmatrix} 1 - \mathrm{e}^{-j\phi_o} \\ 1 - \mathrm{e}^{j\phi_o} \end{bmatrix} \tag{6.182}$$

Using this, we get

$$y_{\mathrm{o}}(n) = (\mathbf{w}_{\mathrm{o}}^{\mathrm{c}})^{\mathrm{H}}\mathbf{x}(n) = \alpha(n)\mathrm{e}^{j\phi_1}$$

which means that the desired signal, $\underline{s}(n)$, passes through the beamformer with no distortion, while the jammer, $\underline{v}(n)$, is completely canceled.

Problems

P6.1 Show that when an adaptive filter has converged and $\mathbf{w}(n) \approx \mathbf{w}_{\mathrm{o}}$

$$\text{Variance of } \nabla e^2(n) \approx 4\xi_{\min} E[x^2(n)].$$

P6.2 Formulate the LMS algorithm for a one-step ahead N-tap linear predictor, that is, a filter that predicts $x(n)$ based on a linear combination of its past samples, $x(n-1), x(n-2), \ldots, x(n-N)$.

P6.3 By multiplying $\mathbf{A} + \alpha\mathbf{a}\mathbf{a}^{\mathrm{T}}$ with the right-hand side of Eq. (6.59), confirm the equality (6.59).

P6.4 Prove that if a and b are two positive values and $a + b < 1$,

$$\frac{a}{1-a} + \frac{b}{1-b} < \frac{a+b}{1-(a+b)}$$

Use this result to establish the inequality (6.70).

P6.5 A 10-tap transversal adaptive filter is adapted using LMS algorithm. Consider five cases of the filter input, which are characterized by the following eigenvalues:

Case	1	2	3	4	5
λ_0	1.0000	1.8182	5.2632	1.8182	1.0989
λ_1	1.0000	1.6364	0.5263	1.8182	1.0989
λ_2	1.0000	1.4545	0.5263	1.8182	1.0989
λ_3	1.0000	1.2727	0.5263	1.8182	1.0989
λ_4	1.0000	1.0909	0.5263	1.8182	1.0989
λ_5	1.0000	0.9091	0.5263	0.1818	1.0989
λ_6	1.0000	0.7273	0.5263	0.1818	1.0989
λ_7	1.0000	0.5455	0.5263	0.1818	1.0989
λ_8	1.0000	0.3636	0.5263	0.1818	1.0989
λ_9	1.0000	0.1818	0.5263	0.1818	0.1099

Note that for all cases, the eigenvalues are normalized such that $\sum_i \lambda_i = \text{tr}[\mathbf{R}] = 10$. This implies that the tight stability bound Eq. (6.74) for all cases is the same. The goal of this problem is to show the impact of distribution of eigenvalues of the correlation matrix of an adaptive on the variation of the true stability bound of the LMS algorithm, that is, the range of μ that guarantees a stable LMS algorithm.

(i) Make a plot of $\mathcal{J}$ (as defined in Eq. (6.65)) for each case when μ varies from 0 to 0.1.
(ii) Find the range of μ in each case, which results in a stable LMS algorithm.
(iii) Discuss the various ranges that you have obtained in (ii) and compare them with the tight-bound $1/(3\text{tr}[\mathbf{R}])$ and a softer bound that may be defined as $1/\text{tr}[\mathbf{R}]$. Discuss in which cases one is closer to the former bound or to the latter bound.

P6.6 Equations (6.84) and (6.87) provide approximate expressions for expected learning curves of the LMS algorithm in the two cases of system modeling and channel equalization. For the five cases noted in Problem P6.5, plot the expected learning curves of the LMS algorithm for system modeling and channel equalization and discuss your observation.

P6.7 The input process to a system modeling problem, using a 10-tap FIR adaptive filter, has the power spectral density shown in Figure P6.7. Assume that the MSE at zeroth iteration is equal to 1 and $\xi_{\text{min}} = 0.0001$, and the step-size parameter μ has been chosen for a 10% misadjustment. Present a typical learning curve of an LMS algorithm in this setup. Indicate the time constants of the various modes of convergence on the presented curve and the mean squared error (MSE) that the LMS algorithm converges to.

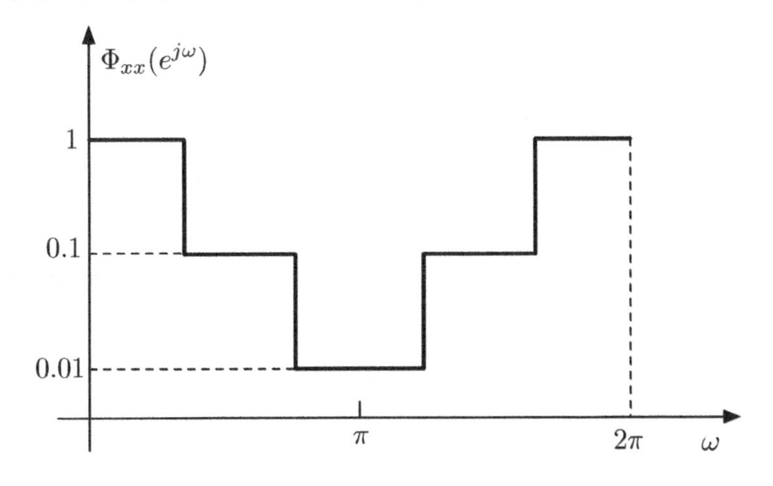

Figure P6.7

P6.8 Consider a channel equalization problem similar to the one depicted in Figure 6.8. The magnitude response of the channel, $|H(z)|$, is as shown in Figure P6.8.

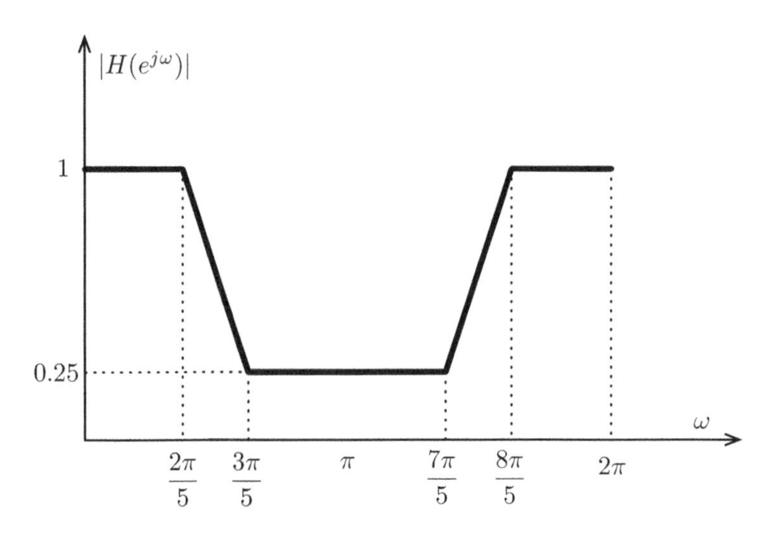

Figure P6.8

The additive noise at the channel output has a variance of $\sigma_\nu^2 = 0.04$. The transmitted data symbols, $s(n)$'s, take the values of $+1$ and -1 and are samples of a white noise process.

(i) Draw the power spectral density of the sequence $x(n)$ and obtain an estimate of $E[|x(n)|^2]$.
(ii) Give estimates of the maximum and minimum eigenvalues of the correlation matrix of the input process to the equalizer.
(iii) When the conventional LMS algorithm is used to adjust the equalizer tap weights and the equalizer has 20 taps, what is the value of the step-size parameter μ which results in 10% misadjustment?
(iv) Obtain the range of time constants of the LMS algorithm in the present case and plot a typical learning curve for that.

P6.9 It has been noted that when the tap weights of a line enhancer are initialized to zero, it converges very fast. However, this will not be the case when tap weights are randomized to some nonzero values. Explain why this is true.

P6.10 The sequence $u(n) = \cos(n\omega_o + \phi(n))$ is a narrow-band phase-modulated sampled signal. The phase angle $\phi(n)$ is random, but varies slowly in time, so that $\phi(n) \approx \phi(n-1) \approx \phi(n-2)$. The aim is to detect the carrier frequency ω_o of $u(n)$. It is proposed that the setup shown in Figure P6.10 be used. The coefficient w has to be adjusted so as to maximize the output, $y(n)$, in the mean-square sense.

(i) Show that the optimized value of w is

$$w_o \approx 2\cos\,\omega_o$$

(ii) Formulate the LMS algorithm for the present problem. In particular, specify the filter tap-weight vector, $\mathbf{w}(n)$, input vector, $\mathbf{x}(n)$, the desired output, $d(n)$, and how the output error is defined in present case.

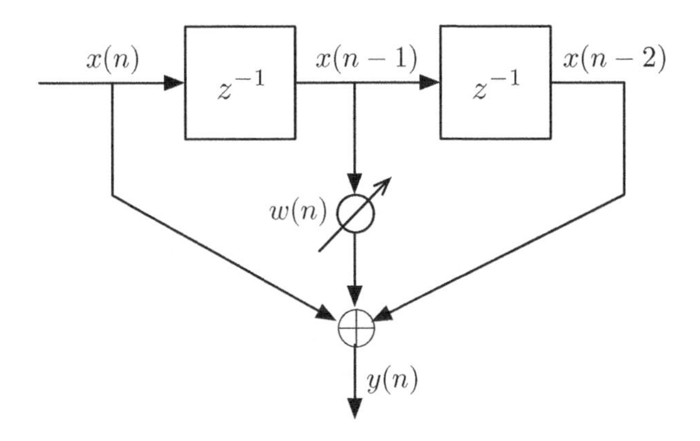

Figure P6.10

P6.11 The LMS algorithm is used to adapt an adaptive filter with tap-weight vector $\mathbf{w}(n)$. Define $\bar{\mathbf{v}}(n) = E[\mathbf{w}(n) - \mathbf{w}_o]$, where $E[\cdot]$ denotes statistical expectation and $\mathbf{w}_o$ is the optimum value of the filter tap-weight vector.

(i) Show that if the step-size parameter, μ, is properly selected, $|\bar{\mathbf{v}}(n)|^2 = \bar{\mathbf{v}}^{\mathrm{T}}(n)\bar{\mathbf{v}}(n)$ will approach zero, as n increases.
(ii) Find the range of μ that guarantees convergence of $|\bar{\mathbf{v}}(n)|^2$. Does this range guarantee the convergence of the LMS algorithm?
(iii) Find the time constants that govern the convergence of $|\bar{\mathbf{v}}(n)|^2$.

P6.12 A communication channel with a FIR shorter than or equal to M-bit interval is to be identified using the setup shown in Figure P6.12. The transmitted data bits, $s(n)$, which take values of $+1$ and -1, are passed through the channel, $H_o(z)$. The same data bits are passed through an adaptive filter, $H(z)$, which is adapted through the LMS algorithm, so that its output matches the output of the channel in the mean-square sense. The channel noise is modeled as an additive noise sequence $\nu(n)$ with variance σ_ν^2. The sequences $s(n)$ and $\nu(n)$ are independent of each other. Define the length M column vector $\mathbf{g}(n) = \mathbf{h}(n) - \mathbf{h}_o$, where $\mathbf{h}(n)$ is the channel model tap-weight vector at iteration n and the elements of the vector $\mathbf{h}_o$ are the samples of channel response.

(i) Show that

$$\mathbf{g}(n+1) = (\mathbf{I} - 2\mu\mathbf{s}(n)\mathbf{s}^{\mathrm{T}}(n))\mathbf{g}(n) + 2\mu\nu(n)\mathbf{s}(n)$$

where $\mathbf{I}$ is the identity matrix, $\mathbf{s}(n) = [s(n)\, s(n-1)\, \cdots\, s(n-M+1)]^{\mathrm{T}}$, and μ is the LMS algorithm step-size parameter.
(ii) Use the independence assumption to show that

$$\|\mathbf{g}(n+1)\|^2 = E[\mathbf{g}^{\mathrm{T}}(n)(\mathbf{I} - (4\mu - 4M\mu^2)\mathbf{R}_{ss})\mathbf{g}(n)] + 4M\mu^2\sigma_\nu^2$$

where $\|\mathbf{g}(n)\|^2 = E[\mathbf{g}^{\mathrm{T}}(n)\mathbf{g}(n)]$ and $\mathbf{R}_{ss} = E[\mathbf{s}(n)\mathbf{s}^{\mathrm{T}}(n)]$.

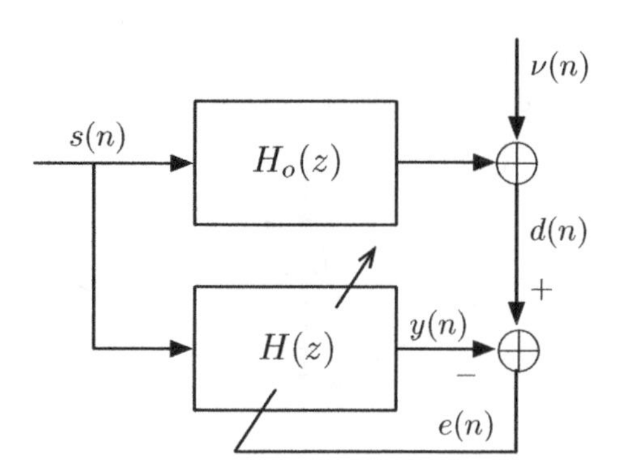

Figure P6.12

(iii) Use the result of Part (ii) to find the range of μ, which guarantees the convergence of $\|\mathbf{g}(n)\|^2$. Does this also guarantees the convergence of the LMS algorithm?

(iv) Compare the range obtained in Part (iii) with the range of μ given in Eq. (6.74).

P6.13 In this problem, we discuss the effect of the power level of the input process to an adaptive filter and its variation on the convergence of the LMS algorithm.

(i) Consider the LMS recursion (6.9) and assume that the time constants of its different modes of convergence are $\tau_0, \tau_1, \ldots, \tau_{N-1}$. Keep μ fixed, replace $\mathbf{x}(n)$ by $\mathbf{x}'(n) = \alpha\mathbf{x}(n)$, where α is a constant, and obtain the corresponding time constants of the resulting recursion, in terms of τ_i's, under the condition that the step-size parameter μ is small enough to guarantee the convergence of the algorithm.

(ii) Under the condition that the power levels of the elements of $\mathbf{x}(n)$ are time varying and fluctuate slowly between high and low levels, what is the shortcoming of the LMS algorithm (discuss)? Can you suggest any solution to this?

P6.14 This problem attempts to show the validity of the approximation (6.86) in a nonrigorous manner.
Consider a random process $x(n)$ and its associated $(2M+1)$-by-$(2M+1)$ correlation matrix $\mathbf{R}$. Let

$$\mathbf{q}_i = [q_{i,-M} \cdots q_0 \cdots q_{i,M}]^{\mathrm{T}}, \qquad \text{with} \quad \mathbf{q}_i^{\mathrm{H}}\mathbf{q}_i = 1$$

be the ith eigenvector of $\mathbf{R}$ and λ_i be its corresponding eigenvalue.

(i) Show that the expansion of the relationship $\mathbf{R}\mathbf{q}_i = \lambda_i\mathbf{q}_i$ leads to

$$\sum_{k=-M}^{M} \phi_{xx}(k-l)q_{i,k} = \lambda_i q_{i,l}, \qquad \text{for } -M \le l \le M \tag{P6.11.1}$$

where $\phi_{xx}(k-l)$ is the autocorrelation function of $x(n)$ for lag $k-l$.

(ii) Let $M \to \infty$ and take Fourier transform on both sides of equation (P6.11.1). Show that this leads to the identity

$$\Phi_{xx}(e^{j\omega})Q_i(e^{j\omega}) = \lambda_i Q_i(e^{j\omega}) \tag{P6.11.2}$$

where $\Phi_{xx}(e^{j\omega})$ is the power spectral density of $x(n)$ and

$$Q_i(e^{j\omega}) = \sum_{k=-\infty}^{\infty} q_{i,k} e^{-j\omega k}$$

(iii) Consider the case when $\Phi_{xx}(e^{j\omega})$ is a single-valued function of the angular frequency ω. Using Eq. (P6.11.2), show that

$$Q_i(e^{j\omega}) = \begin{cases} \text{a nonzero value, for } \omega = \omega_i \\ 0, \text{ otherwise.} \end{cases}$$

Thus, argue that when M is large, the set of vectors

$$\mathbf{q}_i = \frac{1}{\sqrt{2M+1}}[e^{-j\frac{2\pi iM}{2M+1}} \quad e^{-j\frac{2\pi i(M-1)}{2M+1}} \cdots e^{j\frac{2\pi iM}{2M+1}}]^{\mathrm{T}}$$

for $-M \leq i \leq M$, may be considered as an approximation to the eigenvectors of $\mathbf{R}$.
Also, from the Parseval relation (Chapter 2), recall that

$$\mathbf{q}_i^{\mathrm{H}}\mathbf{q}_i = \frac{1}{2\pi}\int_{-\pi}^{\pi} |Q_i(e^{j\omega})|^2 d\omega$$

Thus, conclude that

$$|Q_i(e^{j\omega})|^2 = 2\pi\delta(\omega - \omega_i)$$

where $\delta(\cdot)$ is the Kronecker delta function and $\omega = \omega_i$ is the solution of the equation $\Phi_{xx}(e^{j\omega}) = \lambda_i$.

(iv) Extend the above result and argue that the latter approximation is also valid in the cases where $\Phi_{xx}(e^{j\omega})$ is not necessarily single-valued.

Now consider the case where $x(n)$ is the output of a channel with system function $H(z)$ and also the input to a $(2M+1)$-tap equalizer $W(z)$, as in Section 6.4.2. Ignore the channel noise and recall that, if the system delay Δ is assumed to be zero, the transmitted data symbols are assumed to be uncorrelated, and the equalizer is allowed to be noncausal,

$$W_{\mathrm{o}}(z) \approx \frac{1}{H(z)}$$

where $W_{\mathrm{o}}(z)$ is the optimum setting of $W(z)$.

(v) Using the approximation derived in Parts (iii) and (iv), show that when M is large

$$w'_{o,i} = \mathbf{q}_i^{\mathrm{H}}\mathbf{w} \approx \frac{1}{\sqrt{2M+1}} \cdot \frac{1}{H(\mathrm{e}^{j\frac{2\pi i}{2M+1}})} \qquad \text{for} - M \leq i \leq M$$

where $\mathbf{w}$ is the equalizer tap-weight vector.

(vi) Using this result, show that

$$\lambda_i |w'_{o,i}|^2 \approx \frac{1}{2M+1}.$$

P6.15 Power-line-induced noise is common in many equipments/instruments. Examples are humming noise in electric guitars and in electrocardiograms. A noise-canceling setup similar to the one presented in Figure P6.15a may be used to cancel such humming noise. Here, to simplify the derivation, the reference input (the source of humming noise) is chosen to be the complex-valued sign-wave $\mathrm{e}^{j\omega_0 n}$. It is assumed that the frequency ω_0 is known. The primary input $d(n)$ consists of desired signal, which has been contaminated by a humming noise of the same frequency as the reference input, but with an unknown complex-valued gain w_o. The single-tap adaptive filter, $W(z)$, will ideally adjust $w(n) = w_o$ to cancel the humming noise perfectly.

These problems that follow a similar procedure to the noise canceler of Widrow *et al.* (1975) show that the system presented in Figure P6.15a is effectively a notch filter whose bandwidth is controlled by the step-size parameter μ of the LMS algorithm.

(i) Present a recursive equation for the adaptation of the tap-weight $w(n)$.

(ii) Using the result of (i), show that Figure P6.15a can be redrawn as in Figure P6.15b.

(iii) By developing the difference equation that relates $y(n)$ and $e(n)$, show that Figure P6.15b can be simplified to Figure P6.15c.

(iv) Using the result presented in Figure P6.15c, show that the signal sequences $d(n)$ and $e(n)$ are related by the transfer function

$$\frac{E(z)}{D(z)} = \frac{1 - \mathrm{e}^{j\omega_0}z^{-1}}{1 - (1 - 2\mu)\mathrm{e}^{j\omega_0}z^{-1}}$$

(v) By presenting the pole and zero of the transfer function obtained in (iv) and proper argument, show that this is a notch filter, with a notch frequency at $\omega = \omega_0$, whose bandwidth decreases with μ.

Note that even though we started with an adaptive filter, which is naturally a nonlinear time-varying system, we ended up showing that this effectively is a linear time-invariant system!

P6.16 This problem whose aim is to study the signed-regressor algorithm in some details is based on the derivations of Eweda (1990a).

Using the Price's theorem (Papoulis, 1991), one can show that if x and y are a pair of zero-mean jointly Gaussian random variables

$$E[x \cdot \mathrm{sign}(y)] = \frac{1}{\sigma_y}\sqrt{\frac{2}{\pi}}E[xy]$$

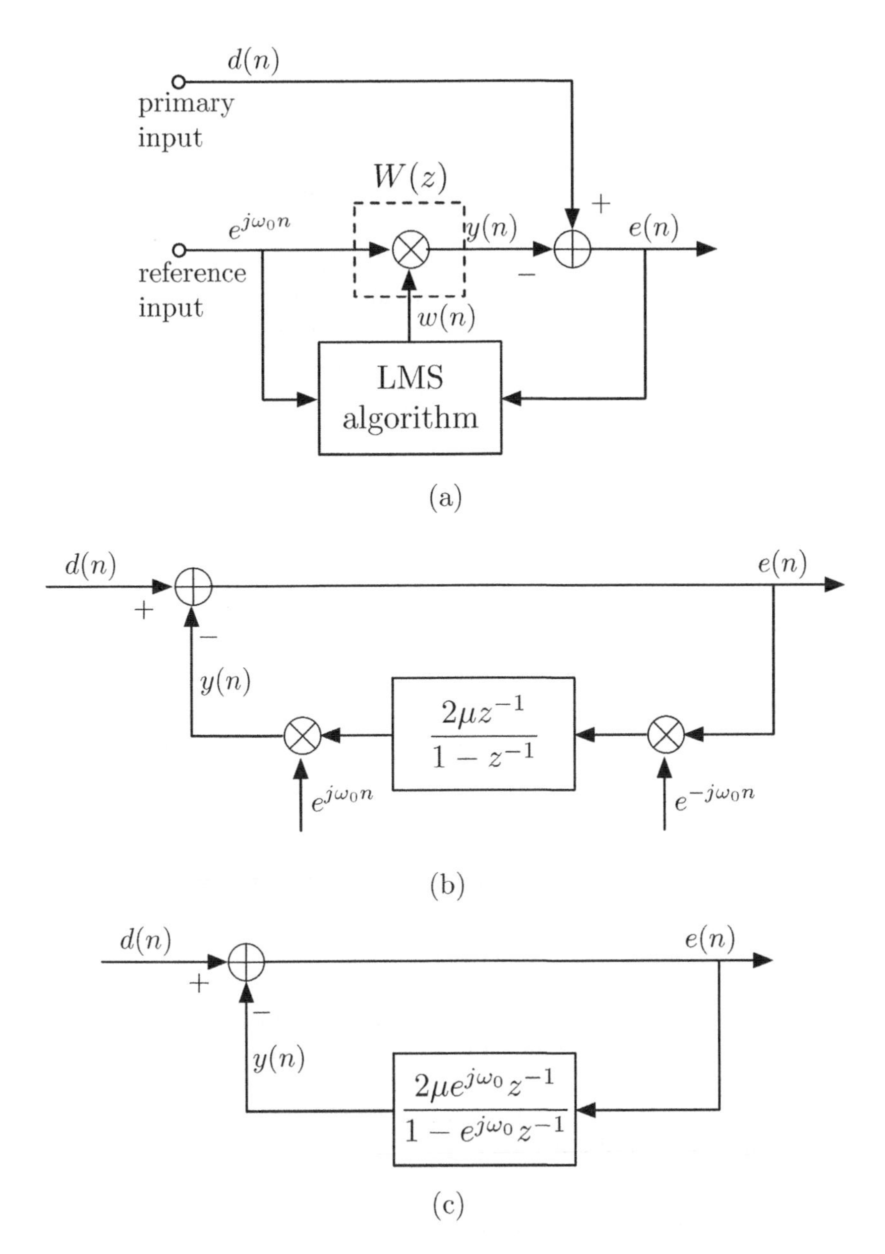

Figure P6.15

Consider the signed-regressor algorithm introduced in Section 6.5, and let the assumptions made at the beginning of Section 6.3 apply. Show that:

(i)

$$E[\text{sign}(\mathbf{x}(n))\mathbf{x}^{\mathrm{T}}(n)] = E[\mathbf{x}(n)\text{sign}(\mathbf{x}^{\mathrm{T}}(n))] = \frac{1}{\sigma_x}\sqrt{\frac{2}{\pi}}\mathbf{R}$$

(ii)

$$\mathbf{v}(n+1) = (\mathbf{I} - 2\mu \text{sign}(\mathbf{x}(n))\mathbf{x}^{\mathrm{T}}(n))\mathbf{v}(n) + 2\mu e_{\mathrm{o}}(n)\text{sign}(\mathbf{x}(n))$$

(iii)

$$E[\mathbf{v}(n+1)] = \left(\mathbf{I} - 2\mu\frac{1}{\sigma_x}\sqrt{\frac{2}{\pi}}\mathbf{R}\right)E[\mathbf{v}(n)] \tag{P6.16.1}$$

and from there argue that the signed-regressor algorithm follows the same trajectory as the conventional LMS algorithm.

(iv) Define $\|\mathbf{v}(n)\|^2 = E[\mathbf{v}^{\mathrm{T}}(n)\mathbf{v}(n)]$ and show that

$$\|\mathbf{v}(n+1)\|^2 = \|\mathbf{v}(n)\|^2 - 4\left(\frac{1}{\sigma_x}\sqrt{\frac{2}{\pi}}\mu - \mu^2 N\right)E[\mathbf{v}^{\mathrm{T}}(n)\mathbf{R}\mathbf{v}(n)] + 4\mu^2 N\xi_{\min}$$

(v) Assuming that the signed-regressor algorithm is convergent, show that its misadjustment is given by

$$\mathcal{M} = \frac{\mu N}{\frac{1}{\sigma_x}\sqrt{\frac{2}{\pi}} - \mu N} \tag{P6.16.2}$$

(vi) From this result and following a line of argument similar to the one in Section 6.3.5, show that the signed-regressor algorithm remains stable when

$$0 < \mu < \frac{1}{N\sigma_x}\sqrt{\frac{2}{\pi}}$$

(vii) When the step-size parameter, μ, is small, Eq. (P6.16.2) reduces to

$$\mathcal{M} \approx \mu N\sigma_x\sqrt{\frac{\pi}{2}}$$

Using this result and Eq. (P6.16.1) and comparing these against their counterparts in the conventional LMS algorithm, show that when the step-size parameters of the two algorithms are chosen, so that both result in the same misadjustment, the signed-regressor algorithm is $2/\pi$ times slower than the conventional LMS algorithm.

P6.17 Consider a case where the input vector, $\mathbf{x}(n)$, to an adaptive filter and its desired output, $d(n)$, are fixed for all values of n. Assuming an initial value, $\mathbf{w}(0)$, for the filter tap-weight and running the LMS algorithm with a small step-size parameter (which guarantees the stability of that), find the final setting of the filter tap weights after the convergence of the LMS algorithm. Using the result obtained, confirm Nitzberg's interpretation (Section 6.6) of the NLMS algorithm.

P6.18 In the derivation of NLMS recursion (6.107), we searched for the step-size parameter, $\mu(n)$, which would minimize $(e^+(n))^2$, where $e^+(n)$ is as defined in Eq. (6.104). We may also note that $e^+(n) = e(n) - \mathbf{x}^{\mathrm{T}}(n)\boldsymbol{\eta}(n)$, where $\boldsymbol{\eta}(n) = \mathbf{w}(n+1) - \mathbf{w}(n)$, as defined in Eq. (6.108). Furthermore, we note that to have an

LMS algorithm with variable step-size parameter, the increment $\boldsymbol{\eta}(n)$ has to be in the direction of $\mathbf{x}(n)$, that is, we may write $\boldsymbol{\eta}(n) = \alpha(n)\mathbf{x}(n)$, where $\alpha(n)$ is scalar.

(i) Give an alternative derivation of the NLMS algorithm by optimizing $\alpha(n)$, so that $(e^+(n))^2$ is minimized.

(ii) To limit the perturbation introduced by the vector $\boldsymbol{\eta}(n)$, it is proposed that $(e^+(n))^2 + \psi\boldsymbol{\eta}^{\mathrm{T}}(n)\boldsymbol{\eta}(n)$ be minimized. Show that this leads to the recursion

$$\mathbf{w}(n+1) = \mathbf{w}(n) + \frac{1}{\mathbf{x}^{\mathrm{T}}(n)\mathbf{x}(n) + \psi} e(n)\mathbf{x}(n)$$

P6.19 Consider the case where the input $x(n)$ to an N-tap adaptive filter is generated by passing a white noise $\nu(n)$ through an L-tap transversal filter. Let $L < N$ and recall the definition (6.122). Show that for $L \geq M$, $E[\mathbf{X}(n)\mathbf{X}^{\mathrm{T}}(n)] = L\mathbf{R}$. Does this observation have any implication on the performance of the APLMS algorithm? Explain.

P6.20 In the APLMS algorithm, consider the case where $M = N$. Show that in this case, the APLMS recursion (6.132), irrespective of the value of $\mathbf{w}(n)$ always converges to the solution $\mathbf{w}(n+1) = (\mathbf{X}^{\mathrm{T}}(n))^{-1}\mathbf{d}(n)$, provided that $\mathbf{X}^{\mathrm{T}}(n)$ is invertible.

P6.21 Develop and present a version of the APLMS algorithm for the case where the underlying processes are complex-valued.

P6.22 The following recursions have been proposed for implementation of a VSLMS algorithm:

$$\boldsymbol{\mu}(n) = \boldsymbol{\mu}(n-1) - \rho\nabla_{\boldsymbol{\mu}} e^2(n)$$

and

$$\mathbf{w}(n+1) = \mathbf{w}(n) - \boldsymbol{\mu}(n)\nabla_{\mathbf{w}} e^2(n)$$

where $\boldsymbol{\mu}(n)$ is a diagonal matrix consisting of N separate variable step-size parameters, $\mu_0(n)$, $\mu_1(n)$, ..., $\mu_{N-1}(n)$, ρ is a small positive step-size parameter, the gradient $\nabla_{\boldsymbol{\mu}} e^2(n)$ is a diagonal matrix compatible with $\boldsymbol{\mu}(n)$, and is evaluated at $\boldsymbol{\mu} = \boldsymbol{\mu}(n-1)$, the gradient $\nabla_{\mathbf{w}} e^2(n)$ is a column vector, as usual, and is evaluated at $\mathbf{w} = \mathbf{w}(n)$.

Show that the proposal given above leads to the VSLMS algorithm, which was introduced in Section 6.7 and summarized in Table 6.4.

P6.23 Assuming that a scalar variable step-size parameter, $\mu(n)$, is used for all taps of a transversal adaptive filter, show that a derivation similar to the one discussed in Problem P6.22 leads to the following recursion:

$$\mu(n) = \mu(n-1) - \rho e(n)e(n-1)\mathbf{x}^{\mathrm{T}}(n)\mathbf{x}(n-1)$$

P6.24 Give details of the derivation of Eq. (6.61) from Eq. (6.57).

P6.25 This problem looks at a variation of the LMS algorithm called *leaky LMS algorithm*. The leaky LMS algorithm works based on the recursion

$$\mathbf{w}(n+1) = \beta\mathbf{w}(n) + 2\mu e(n)\mathbf{x}(n)$$

where β is a constant slightly smaller than 1.

(i) Define $\bar{\mathbf{w}}(n) = E[\mathbf{w}(n)]$, where $E[\cdot]$ denotes statistical expectation, and use the *independence assumption* of Section 6.3 to show that the following recursive equation holds:

$$\bar{\mathbf{w}}(n+1) = (\mathbf{I} - 2\mu\mathbf{R}')\bar{\mathbf{w}}(n) + 2\mu\mathbf{p}$$

Specify $\mathbf{R}'$ and $\mathbf{p}$ and obtain the time constants of the learning curve of the leaky LMS algorithm in terms of the eigenvalues of the correlation matrix $\mathbf{R} = E\{\mathbf{x}(n)\mathbf{x}^{\mathrm{T}}(n)\}$ and the parameters β and μ.

(ii) Assuming that the step-size parameter μ is small enough to guarantee the convergence of the leaky LMS algorithm, derive an equation for $\bar{\mathbf{w}}(\infty)$ in terms of $\mathbf{R}'$ and $\mathbf{p}$.

(iii) Show that the difference between $\bar{\mathbf{w}}(\infty)$ and the optimum tap-weight vector of the adaptive filter is given by the following equation.

$$\bar{\mathbf{w}}(\infty) - \mathbf{w}_{\mathrm{o}} = -\gamma \sum_{i=0}^{N-1} \frac{\mathbf{q}_i^{\mathrm{T}}\mathbf{p}}{\lambda_i(\lambda_i + \gamma)}\mathbf{q}_i$$

where $\gamma = \frac{1-\beta}{2\mu}$, and λ_i's and $\mathbf{q}_i$'s are the eigenvalues and eigenvectors of $\mathbf{R}$, respectively.

P6.26 Define the scalar value $\|\mathbf{v}(n)\|^2 = E[\mathbf{v}^{\mathrm{T}}(n)\mathbf{v}(n)]$ as the misalignment of an adaptive filter tap weights.

(i) Show that

$$\|\mathbf{v}(n)\|^2 = \mathrm{tr}[\mathbf{K}'(n)]$$

where the correlation matrix $\mathbf{K}'(n)$ is defined as in Eq. (6.31).

(ii) Use Eq. (6.52) to show that

$$\|\mathbf{v}(\infty)\|^2 = \frac{\sum_{i=0}^{N-1} \mu/(1-2\mu\lambda_i)}{1 - \sum_{i=0}^{N-1} \mu\lambda_i/(1-2\mu\lambda_i)}$$

(iii) Show that when μ is small, the above result reduces to

$$\|\mathbf{v}(\infty)\|^2 = \mu N$$

P6.27 A complex-valued random process $x(n) = u(n) + \nu(n)$ is available. The process $u(n) = ae^{j(\omega n+\phi)}$, where a and ϕ are random, but fixed for every realization of $x(n)$. The process $\nu(n)$ is a complex-valued noise which may not be white. Assuming that the frequency, ω, of $u(n)$ is known, propose an adaptive filter and its associated adaptation algorithm to filter out $\nu(n)$ from $x(n)$ and enhance $u(n)$ in the minimum MSE sense, preserving its phase, ϕ, and its amplitude, a.

P6.28 Repeat Problem P6.27 when $u(n) = a\cos(\omega n + \phi)$ and $\nu(n)$ is a real-valued noise sequence.

P6.29 Give the details of the linearly constrained LMS algorithm required for adaptation of the tap-weight vector, $\mathbf{w}$, of the beamformer of Example 6.4.

P6.30 The beamformer discussed in Example 6.4 assumes that the desired signal is arriving at the direction perpendicular to the line connecting A to B. What constraint had to be imposed on the tap weights w_0 and w_1, if the desired signal was arriving at an angle $\theta = \theta_d$?

P6.31 Griffiths and Jim (1982) have proposed a structure for beamforming whose performance is similar to that of the Frost algorithm; however, it does not need any constraint to applied. Example 6.4 is an example of Frost algorithm. The equivalent implementation of Figure 6.21, which follows the idea of Griffiths and Jim, is shown in Figure P6.31. Note that here the beamformer has only one tap weight, as opposed to Figure 6.21, which has two tap weights. Also, adaptation of the tap weight w of Figure P6.31 is based on the conventional LMS algorithm, as opposed to the Frost implementation (Figure 6.21), which requires the use of the linearly constrained LMS algorithm.

(i) Explore the validity of the Griffiths and Jim algorithm in the present case.
(ii) Figure P6.31 assumes that the desired signal is arriving in the direction perpendicular to the line connecting A to B. Modify this structure for the case when the desired signal is arriving at an angle $\theta = \theta_d$.

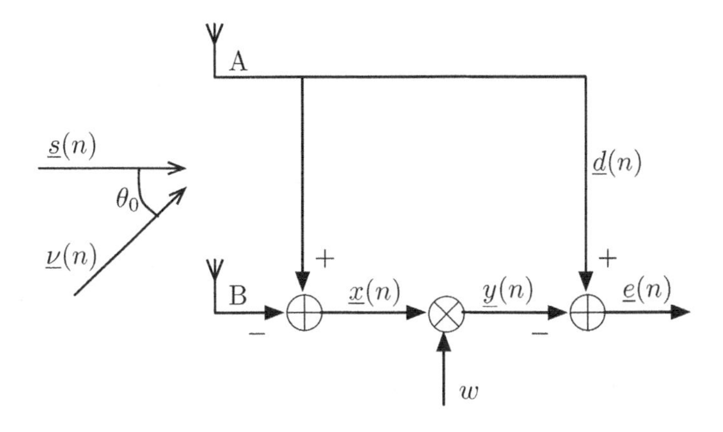

Figure P6.31

P6.32 The antenna array setup shown in Figure P6.32 has to be adopted to see a signal coming from a look angle θ. The spacing between the antennas is $\lambda_c/2$, where λ_c is the wavelength of the carrier of the incoming signal. What constraint should be applied to the tap weights w_0 through w_{M-1}:

(i) when $\theta = 0$.
(ii) when $\theta = \pi/4$.
(iii) For the cases in (i) and (ii), write down a constrained LMS algorithm for adaptation of the tap weights.

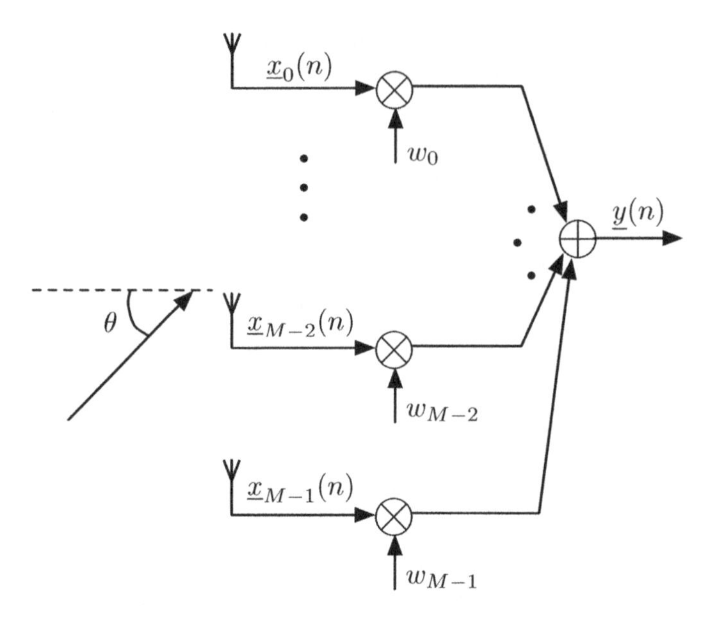

Figure P6.32

Computer-Oriented Problems

All the programs that have been used to generate the results of the various examples of this chapter are written in MATLAB software package and are available in an accompanying website. The reader is encouraged to run these programs and confirm the results of this chapter. It would also be useful and enlightening to the reader if he tries other variations of the simulation parameters, such as coloring filter in modeling, channel response in channel equalization, and noise and sinusoidal signal powers in adaptive line enhancer. The following problems (case studies) are designed to guide the reader to many other interesting results.

P6.33 Consider the transfer function $H_1(z)$, of Eq. (6.80), as the channel response in the equalizer setup of Figure 6.8. Set the equalizer length, N, equal to 15. Find the minimum MSE of the equalizer for values of the delay, Δ, in the range 3–15. Repeat this experiment for the transfer function $H_2(z)$, of Eq. (6.81), as well. For each case, find the optimum value of the delay, Δ, which results in the minimum MSE. From these observations, arrive at a rule-of-thumb for selection of Δ in terms of the duration of channel response and equalizer length.

P6.34 In the line enhancer problem that was studied in Section 6.4.3, we noted that no slow mode appears on its learning curve when the tap weights are initialized to zero. To observe the slow modes of the line enhancer, perform the following experiment. Run the line enhancer program "`lenhncr.m`" (available on the accompanying website), starting with zero tap weights and using an input similar to the one used to obtain the result of Figure 6.11. After convergence of the line enhancer, run it again, starting with the latest values of the tap weights obtained in the first run as initial tap weights, but change the frequency of the sinusoid in the input to $\omega = \pi/2$. Compare the two learning curves that you have obtained and explain your observation.

P6.35 For the modeling problem that was discussed in Section 6.4.2, develop a program to study the convergence behavior of the NLMS algorithm. Compare your results with those of the conventional LMS algorithm for both choices of $H(z) = H_1(z)$ and $H(z) = H_2(z)$, presented in Eqs. (6.80) and (6.81), respectively.

P6.36 Repeat P6.35 when the VSLMS algorithm is used for adaptation of the filter.

P6.37 Repeat P6.35 and P6.36 for the case where the adaptive filter of interest is the channel equalizer discussed in Section 6.4.2.

P6.38 For the modeling problem that was discussed in Section 6.4.1, develop a program to study the convergence behavior of the APLMS algorithm. Compare your results with those of the NLMS algorithm for both choices of $H(z) = H_1(z)$ and $H(z) = H_2(z)$, presented in Eqs. (6.80) and (6.81), respectively. Perform your study for the following cases of parameters and in each case find the misadjustment of both algorithms.

(i) $\tilde{\mu} = 0.5$, $\psi = 0.0001$, and, for the APLMS algorithm, select $M = 2$, 3, 5, and 7.
(ii) $\tilde{\mu} = 1$, $\psi = 0.0001$, and, for the APLMS algorithm, select $M = 2$, 3, 5, and 7.

P6.39 Write a program to study the convergence behavior of the line enhancer of Section 6.4.3 when the input, $x(n)$, is given by

$$x(n) = a\sin(\omega_1 n + \theta_1) + b\sin(\omega_2 n + \theta_2) + v(n)$$

where θ_1 and θ_2 are random phases that are uniformly distributed in the range 0 to 2π. Obtain the learning curves of the line enhancer for the following choices of the signal parameters:

(i) $\omega_1 = \pi/6$, $\omega_2 = 5\pi/8$, $a = 1$, $b = 1$, $\sigma_v^2 = 0.1$
(ii) $\omega_1 = \pi/6$, $\omega_2 = 5\pi/8$, $a = 5$, $b = 1$, $\sigma_v^2 = 0.1$
(iii) $\omega_1 = \pi/6$, $\omega_2 = \pi/4$, $a = 5$, $b = 1$, $\sigma_v^2 = 0.1$

Run your program for the cases when the filter tap weights are initialized to zero and also when they are initialized to some random values. Study the results that you obtain and explain your observations.

P6.40 Write your own program to confirm the results of Figure 6.13.

P6.41 By adding proper lines to the MATLAB program "`equalizer.m`," study the variation of the magnitude response of the equalizer as the LMS adaptation proceeds. Perform your study for both choices of $H(z) = H_1(z)$ and $H(z) = H_2(z)$, presented in Eqs. (6.72) and (6.73), respectively. Run the program for a misadjustment of 10% and present the magnitude response of the equalizer after every 100 iterations.

P6.42 Repeat Problem P6.41 when the LMS algorithm is replaced by the APLMS algorithm. Perform your study for the following choices of the algorithm parameters.

(i) $\tilde{\mu} = 0.5$, $\psi = 0.0001$, and $M = 2$, 3, 5, and 7
(ii) $\tilde{\mu} = 1$, $\psi = 0.0001$, and $M = 2$, 3, 5, and 7

P6.43 Consider a communication channel with a complex-valued impulse response consisting of the following samples:

$$-0.1 + j0.2, \quad 0.15 - j0.4, \quad 1, \quad 0.5 - j0.2, \quad -0.2 + j0.1$$

where $j = \sqrt{-1}$. The input data symbols are randomly selected from the alphabets

$$1 + j, \quad -1 + j, \quad -1 - j, \text{ and } 1 - j$$

with equal probability. The channel noise is white and at 30 dB below the signal level at the equalizer input. Develop a program to simulate this scenario and study the convergence behavior of the LMS algorithm in this case.

P6.44 Consider the scenario that is discussed in Example 6.4. Develop a program for adaptive adjustment of the coefficients w_0 and w_1. By running your program for different choices of σ_α^2, σ_β^2, ϕ_1, and ϕ_2, study the behavior of the constrained LMS algorithm in this case.

Appendix 6A: Derivation of Eq. (6.39)

Using the independence of $\mathbf{x}'(n)$ and $\mathbf{v}'(n)$, we get

$$\begin{aligned} &E[\mathbf{x}'(n)\mathbf{x}'^{\mathrm{T}}(n)\mathbf{v}'(n)\mathbf{v}'^{\mathrm{T}}(n)\mathbf{x}'(n)\mathbf{x}'^{\mathrm{T}}(n)] \\ &\quad = E[\mathbf{x}'(n)\mathbf{x}'^{\mathrm{T}}(n)E[\mathbf{v}'(n)\mathbf{v}'^{\mathrm{T}}(n)]\mathbf{x}'(n)\mathbf{x}'^{\mathrm{T}}(n)] \\ &\quad = E[\mathbf{x}'(n)\mathbf{x}'^{\mathrm{T}}(n)\mathbf{K}'(n)\mathbf{x}'(n)\mathbf{x}'^{\mathrm{T}}(n)] \end{aligned} \tag{6A.1}$$

To expand the right-hand side of Eq. (6A-1), we first note that

$$\mathbf{x}'^{\mathrm{T}}(n)\mathbf{K}'(n)\mathbf{x}'(n) = \sum_{i=0}^{N-1}\sum_{j=0}^{N-1} x_i'(n)x_j'(n)k_{ij}'(n) \tag{6A.2}$$

where $x_i'(n)$ is the ith element of the vector $\mathbf{x}'(n)$. We also define

$$\mathbf{C}(n) = \mathbf{x}'(n)\mathbf{x}'^{\mathrm{T}}(n)\mathbf{K}'(n)\mathbf{x}'(n)\mathbf{x}'^{\mathrm{T}}(n) \tag{6A.3}$$

and note that it is an N-by-N matrix. The lmth element of $\mathbf{C}(n)$ is

$$c_{lm}(n) = x_l'(n)x_m'(n)\sum_{i=0}^{N-1}\sum_{j=0}^{N-1} x_i'(n)x_j'(n)k_{ij}'(n) \tag{6A.4}$$

Taking statistical expectation on both sides of Eq. (6A.4), we obtain

$$E[c_{lm}(n)] = \sum_{i=0}^{N-1}\sum_{j=0}^{N-1} E[x_l'(n)x_m'(n)x_i'(n)x_j'(n)]k_{ij}'(n) \tag{6A.5}$$

Next, if we consider the assumption that the input samples $x(n)$ (and, thus, $x_i'(n)$'s) are a set of mutually Gaussian random variables, and note that for any set of real-valued mutually Gaussian random variables x_1, x_2, x_3, and x_4

$$\begin{aligned} E[x_1x_2x_3x_4] &= E[x_1x_2]E[x_3x_4] + E[x_1x_3]E[x_2x_4] \\ &\quad + E[x_1x_4]E[x_2x_3] \end{aligned} \tag{6A.6}$$

and, also,

$$E[x_i'(n)x_j'(n)] = \lambda_i\delta(i-j) \tag{6A.7}$$

where $\delta(\cdot)$ is the Kronecker delta function, we obtain

$$\begin{aligned} E[x_l'(n)x_m'(n)x_i'(n)x_j'(n)] &= \lambda_l\lambda_i\delta(l-m)\delta(i-j) + \lambda_l\lambda_m\delta(l-i)\delta(m-j) \\ &\quad + \lambda_l\lambda_m\delta(l-j)\delta(m-i) \end{aligned} \tag{6A.8}$$

Substituting Eq. (6A.8) in Eq. (6A.5), we obtain

$$\begin{aligned} E[c_{lm}(n)] &= \sum_{i=0}^{N-1}\sum_{j=0}^{N-1} \lambda_l \lambda_i \delta(l-m)\delta(i-j)k'_{ij}(n) \\ &\quad + \sum_{i=0}^{N-1}\sum_{j=0}^{N-1} \lambda_l \lambda_m \delta(l-i)\delta(m-j)k'_{ij}(n) \\ &\quad + \sum_{i=0}^{N-1}\sum_{j=0}^{N-1} \lambda_l \lambda_m \delta(l-j)\delta(m-i)k'_{ij}(n) \\ &= \lambda_l \delta(l-m)\sum_{i=0}^{N-1} \lambda_i k'_{ii}(n) + \lambda_l \lambda_m k'_{lm}(n) + \lambda_m \lambda_l k'_{ml}(n) \end{aligned} \tag{6A.9}$$

for $l = 0, 1, \ldots, N-1$ and $m = 0, 1, \ldots, N-1$. Noting that $k'_{lm}(n) = k'_{ml}(n)$, $\sum_{i=0}^{N-1} \lambda_i k'_{ii}(n) = \mathrm{tr}[\mathbf{\Lambda K}'(n)]$, and using the result of Eq. (6A.9) to construct the matrix

$$E[\mathbf{C}(n)] = E[\mathbf{x}'(n)\mathbf{x}'^{\mathrm{T}}(n)\mathbf{v}'(n)\mathbf{v}'^{\mathrm{T}}(n)\mathbf{x}'(n)\mathbf{x}'^{\mathrm{T}}(n)]$$

we obtain Eq. (6.39).

7

Transform Domain Adaptive Filters

In Chapter 6, we noted that the convergence behavior of LMS algorithm depends on the eigenvalues of the correlation matrix, **R**, of the adaptive filter input process. Furthermore, we also saw that the eigenvalues of **R** are directly related to the power spectral density of the underlying process. Hence, we may say that the convergence behavior of LMS algorithm is frequency dependent in the sense that for an adaptive filter with the transfer function $W(e^{j\omega})$, the rate of convergence of $W(e^{j\omega})$ toward its optimum value, $W_0(e^{j\omega})$, at a given frequency $\omega = \omega_o$, depends on the relative value of the power spectral density of the underlying input signal at $\omega = \omega_o$, that is, $\Phi_{xx}(e^{j\omega_o})$. A large value of $\Phi_{xx}(e^{j\omega_o})$ (relative to the values of $\Phi_{xx}(e^{j\omega})$ at other frequencies) indicates that the adaptive filter is well excited at $\omega = \omega_o$. This results in fast convergence around $\omega = \omega_o$. On the other hand, the LMS algorithm converges very slowly over those frequency bands in which the adaptive filter is poorly excited. This concept which is intuitively understandable may also be confirmed through computer simulations (see simulation exercise P7.20 at the end of this chapter).

A solution that one might intuitively consider for solving the above-mentioned problem of slow convergence of the LMS algorithm may be to employ a set of bandpass filters to partition the adaptive filter input into a few subbands and use a normalization process to equalize the energy content in each of the subbands. The equalized subband signals can then be used for adaptation of the filter tap weights. The content of this chapter is an elaboration of this principle for developing adaptive algorithms with better convergence behavior than the conventional LMS algorithm.

In this chapter, we present an adaptive filtering scheme that uses an orthogonal transform for partitioning the filter input into subbands. This is called *transform domain adaptive filter (TDAF)* for obvious reasons. We present a thorough study of TDAF that includes not only its convergence behavior but also its efficient implementation.

Adaptive Filters: Theory and Applications, Second Edition. Behrouz Farhang-Boroujeny.

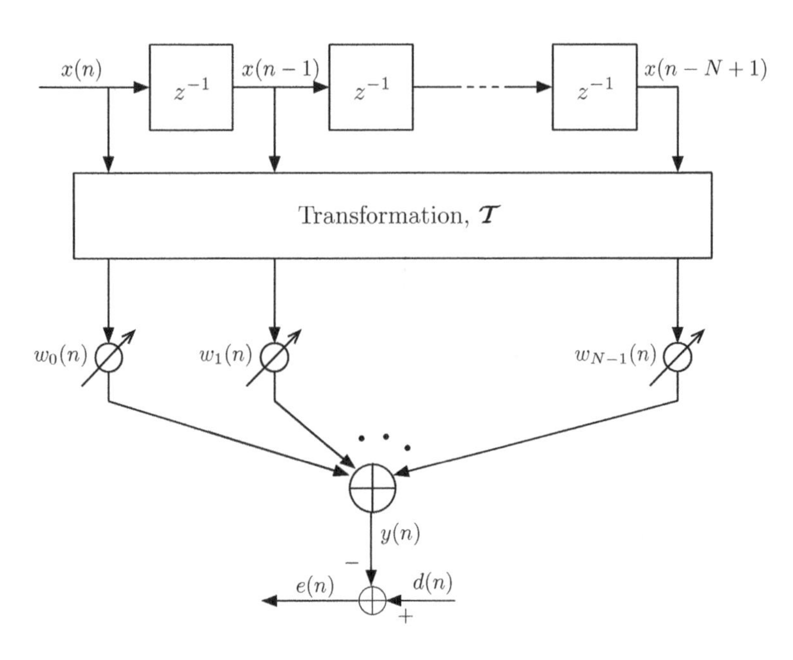

Figure 7.1 Transform domain adaptive filter.

7.1 Overview of Transform Domain Adaptive Filters

Figure 7.1 depicts a block schematic of a TDAF.[1] A set of input samples $x(n)$, $x(n-1),\ldots,$ $x(n-N+1)$ to the filter are transformed to a new set of samples, $x_{\mathcal{T},0}(n)$, $x_{\mathcal{T},1}(n), \ldots, x_{\mathcal{T},N-1}(n)$, through an orthogonal transform ($\mathcal{T}$), before the filtering process. The tap weights $w_{\mathcal{T},0}, w_{\mathcal{T},1}, \ldots, w_{\mathcal{T},N-1}$ are optimized so that the output error, $e(n)$, is minimized in the mean-square sense.

The orthogonal transform ($\mathcal{T}$) is implemented according to the following equation

$$\mathbf{x}_{\mathcal{T}}(n) = \mathcal{T}\mathbf{x}(n) \tag{7.1}$$

where $\mathbf{x}(n) = [x(n)\; x(n-1)\; \cdots\; x(n-N+1)]^{\mathrm{T}}$ is the filter tap-input vector in the time domain, $\mathbf{x}_{\mathcal{T}}(n) = [x_{\mathcal{T},0}(n)\; x_{\mathcal{T},1}(n)\; \cdots\; x_{\mathcal{T},N-1}(n)]^{\mathrm{T}}$ is the filter tap-input vector in the transform domain, and $\mathcal{T}$ is the transformation matrix, which is selected to be a unitary matrix[2], that is,

$$\mathcal{T}^{\mathrm{T}}\mathcal{T} = \mathcal{T}\mathcal{T}^{\mathrm{T}} = \mathbf{I} \tag{7.2}$$

Here, we assume that the elements of $\mathcal{T}$ are real-valued. When the elements of $\mathcal{T}$ are complex-valued, the superscript T in Eq. (7.2), indicating transposition, has to be replaced by H, that is, Hermitian transposition.

The filter output is obtained according to the equation

$$y(n) = \mathbf{w}_{\mathcal{T}}^{\mathrm{T}}\mathbf{x}_{\mathcal{T}}(n) \tag{7.3}$$

[1] The TDAF was first proposed by Narayan *et al.* (1981, 1983).

[2] Throughout this chapter, the symbol $\mathcal{T}$ will be used to represent a unitary matrix satisfying Eq. (7.2).

where $\mathbf{w}_{\mathcal{T}} = [w_{\mathcal{T},0}\ w_{\mathcal{T},1}\ \cdots\ w_{\mathcal{T},N-1}]^{\mathrm{T}}$. We may note that although $\mathbf{x}_{\mathcal{T}}(n)$ is in the transform domain, the filter output, $y(n)$, is in the time domain. The estimation error

$$e(n) = d(n) - y(n) \tag{7.4}$$

is also in the time domain.

The cost function used to optimize the filter tap weights is

$$\xi = E[e^2(n)] \tag{7.5}$$

Substituting Eqs. (7.3) and (7.4) in Eq. (7.5), we obtain

$$\xi = \mathbf{w}_{\mathcal{T}}^{\mathrm{T}}\,\mathbf{R}_{\mathcal{T}}\mathbf{w}_{\mathcal{T}} - 2\mathbf{w}_{\mathcal{T}}^{\mathrm{T}}\,\mathbf{p}_{\mathcal{T}} + E[d^2(n)] \tag{7.6}$$

where $\mathbf{R}_{\mathcal{T}} = E[\mathbf{x}_{\mathcal{T}}(n)\,\mathbf{x}_{\mathcal{T}}^{\mathrm{T}}(n)]$ and $\mathbf{p}_{\mathcal{T}} = E[d(n)\mathbf{x}_{\mathcal{T}}(n)]$. Setting the gradient of ξ with respect to $\mathbf{w}_{\mathcal{T}}$ equal to zero, we obtain the corresponding Wiener–Hopf equation whose solution gives the optimum tap-weight vector of the TDAF as

$$\mathbf{w}_{\mathcal{T},o} = \mathbf{R}_{\mathcal{T}}^{-1}\,\mathbf{p}_{\mathcal{T}} \tag{7.7}$$

Substituting this result in Eq. (7.6), the minimum mean-squared error (MSE) of the TDAF is obtained as

$$\xi_{\min} = E[d^2(n)] - \mathbf{p}_{\mathcal{T}}^{\mathrm{T}}\,\mathbf{R}_{\mathcal{T}}^{-1}\mathbf{p}_{\mathcal{T}} \tag{7.8}$$

To compare this with the minimum MSE associated with the conventional transversal structure case (i.e., without the orthogonal transformation, $\mathcal{T}$), we note that

$$\begin{aligned}\mathbf{R}_{\mathcal{T}} &= E[\mathbf{x}_{\mathcal{T}}(n)\mathbf{x}_{\mathcal{T}}^{\mathrm{T}}(n)]\\ &= \mathcal{T}E[\mathbf{x}(n)\mathbf{x}^{\mathrm{T}}(n)]\mathcal{T}^{\mathrm{T}}\\ &= \mathcal{T}\mathbf{R}\mathcal{T}^{\mathrm{T}}\end{aligned} \tag{7.9}$$

and

$$\begin{aligned}\mathbf{p}_{\mathcal{T}} &= E[d(n)\mathbf{x}_{\mathcal{T}}(n)]\\ &= \mathcal{T}E[d(n)\mathbf{x}(n)]\\ &= \mathcal{T}\mathbf{p}\end{aligned} \tag{7.10}$$

Substituting Eqs. (7.9) and (7.10) in Eq. (7.8) and using Eq. (7.2), after some straightforward manipulations, we get

$$\xi_{\min} = E[d^2(n)] - \mathbf{p}^{\mathrm{T}}\mathbf{R}^{-1}\mathbf{p} \tag{7.11}$$

Comparing this result with Eq. (3.27), we find that the minimum MSE associated with a conventional transversal filter and its corresponding TDAF is the same. This could also be understood, intuitively, if we note that the transformation $\mathbf{x}_{\mathcal{T}}(n) = \mathcal{T}\mathbf{x}(n)$ is reversible (i.e., $\mathbf{x}(n) = \mathcal{T}^{\mathrm{T}}\mathbf{x}_{\mathcal{T}}(n)$) and, thus, any output $y(n) = \mathbf{w}^{\mathrm{T}}\mathbf{x}(n)$ can also be obtained from $\mathbf{x}_{\mathcal{T}}(n)$ using an appropriate tap-weight vector $\mathbf{w}_{\mathcal{T}}$. To find the relationship between $\mathbf{w}$ and

$\mathbf{w}_{\mathcal{T}}$, we simply let $\mathbf{w}^{\mathrm{T}}\mathbf{x}(n) = \mathbf{w}_{\mathcal{T}}^{\mathrm{T}}\mathbf{x}_{\mathcal{T}}(n)$ and use Eq. (7.1) to obtain

$$\mathbf{w}_{\mathcal{T}} = \mathcal{T}\mathbf{w} \tag{7.12}$$

Before going into further details of TDAFs, in the next two sections, we study a very specific feature of orthogonal transforms, which makes them suitable for adaptive filtering algorithms.

7.2 Band-Partitioning Property of Orthogonal Transforms

We explore the discrete cosine transform (DCT) as an example of orthogonal transforms. The DCT of a sequence $\{x(n),\ x(n-1),\ \ldots,\ x(n-N+1)\}$ is defined as

$$x_{DCT,k}(n) = \sum_{l=0}^{N-1} c_{kl}x(n-l), \qquad \text{for } k = 0, 1, \ldots, N-1 \tag{7.13}$$

where

$$c_{kl} = \begin{cases} \frac{1}{\sqrt{N}}, & k = 0 \quad \text{and} \quad l = 0, 1, \ldots, N-1 \\ \sqrt{\frac{2}{N}}\cos\frac{\pi(2l+1)k}{2N}, & k = 1, 2, \ldots, N-1 \\ & \text{and} \quad l = 0, 1, \ldots, N-1 \end{cases} \tag{7.14}$$

are the DCT coefficients. It is also worth noting that Eq. (7.13) may be written as

$$\mathbf{x}_{\mathrm{DCT}}(n) = \mathcal{T}_{\mathrm{DCT}}\mathbf{x}(n) \tag{7.15}$$

where $\mathcal{T}_{\mathrm{DCT}}$ is the N-by-N DCT matrix. The klth element of $\mathcal{T}_{\mathrm{DCT}}$ is c_{kl}, as defined in Eq. (7.14) and

$$\mathbf{x}_{\mathrm{DCT}}(n) = [x_{\mathrm{DCT},0}(n)\ x_{\mathrm{DCT},1}(n)\ \cdots\ x_{\mathrm{DCT},N-1}(n)]^{\mathrm{T}}$$

Besides being a linear transformation, the process defined by Eq. (7.13) (or Eq. (7.15)) may also be viewed as an implementation of a bank of finite-impulse response (FIR) filters whose coefficients are c_{kl}'s. Here, these are referred to as *DCT filters*. The transfer function of the kth DCT filter is

$$C_k(z) = \sum_{l=0}^{N-1} c_{kl}z^{-l} \tag{7.16}$$

Figure 7.2 shows the magnitude responses of the DCT filters when $N = 8$. The plots clearly show the band-partitioning property of the DCT filters. Each response has a large main lobe that may be identified as its passband, and a number of side lobes that correspond to its stop-band. Similar plots (with some variations in the shapes) are also obtained for other commonly used orthogonal transforms, for example, discrete Fourier transform (DFT).

Before we elaborate on how or why this band-partitioning property of orthogonal transforms is important to us, we shall look at this property from a different angle in the next section.

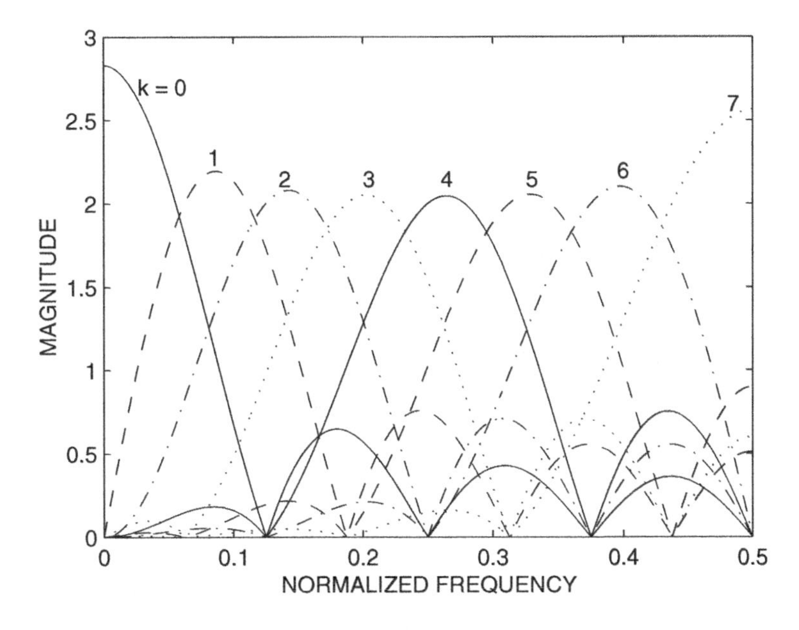

Figure 7.2 Magnitude responses of the DCT filters for $N = 8$.

7.3 Orthogonalization Property of Orthogonal Transforms

The band-partitioning property of orthogonal transforms gives a frequency domain view of them. The dual of this in the time domain is the orthogonalization property of such transforms. This property can be deduced intuitively from the band-partitioning property observed in Section 7.2. We recall that processes with mutually exclusive spectral bands are uncorrelated with one another (Papoulis, 1991). On the other hand, from the band-partitioning property, we note that the elements of the transformed tap-input vector, $\mathbf{x}_{\mathcal{T}}(n)$, constitute a set of random processes with approximately mutually exclusive spectral bands. This implies that the elements of $\mathbf{x}_{\mathcal{T}}(n)$ are (at least) approximately uncorrelated with one another. This, in turn, implies that the correlation matrix $\mathbf{R}_{\mathcal{T}} = E[\mathbf{x}_{\mathcal{T}}(n)\mathbf{x}_{\mathcal{T}}^{\mathrm{T}}(n)]$ is closer to a diagonal matrix than $\mathbf{R}$ is. An appropriate normalization can convert $\mathbf{R}_{\mathcal{T}}$ to a normalized matrix $\mathbf{R}_{\mathcal{T}}^{\mathrm{n}}$ whose eigenvalue spread will be much smaller than that of $\mathbf{R}$, thereby improving the convergence behavior of the LMS algorithm in the transform domain. This can be best explained through a numerical example.

Consider the case where $x(n)$ is a first-order autoregressive process generated by passing a white noise process through the system function[3]

$$H(z) = \frac{\sqrt{1-\alpha^2}}{1-\alpha z^{-1}} \tag{7.17}$$

[3] The reader may recall that we used the same process in many examples in the previous chapters.

where α is a constant in the range of -1 to $+1$. For $\alpha = 0.9$, we obtain

$$\mathbf{R} = \begin{bmatrix} 1.0000 & 0.9000 & 0.8100 & 0.7290 \\ 0.9000 & 1.0000 & 0.9000 & 0.8100 \\ 0.8100 & 0.9000 & 1.0000 & 0.9000 \\ 0.7290 & 0.8100 & 0.9000 & 1.0000 \end{bmatrix} \tag{7.18}$$

For a derivation of **R**, see Example 4.1. Using the DCT as the transformation, we get

$$\mathbf{R}_{\mathcal{T}} = \begin{bmatrix} 3.5245 & 0.0000 & -0.0855 & 0.0000 \\ 0.0000 & 0.3096 & 0.0000 & -0.0032 \\ -0.0855 & 0.0000 & 0.1045 & 0.0000 \\ 0.0000 & -0.0032 & 0.0000 & 0.0614 \end{bmatrix} \tag{7.19}$$

This clearly is much closer to diagonal (i.e., its off-diagonal elements are relatively closer to zero) when compared to **R**.

The normalization performed in the implementation of the LMS algorithm in transform domain (as we shall see later), in effect, is equivalent to normalization of the elements of $\mathbf{x}_{\mathcal{T}}(n)$ to the power of unity. This is done by premultiplying $\mathbf{x}_{\mathcal{T}}(n)$ with a diagonal matrix, $\mathbf{D}^{-1/2}$, before the filtering and adaptation process, where $\mathbf{D}^{-1/2}$ is the inverse of the square root of the diagonal matrix

$$\mathbf{D} = \begin{bmatrix} E[x^2_{\mathcal{T},0}(n)] & 0 & \cdots & 0 \\ 0 & E[x^2_{\mathcal{T},1}(n)] & \cdots & 0 \\ \vdots & \vdots & \ddots & \vdots \\ 0 & 0 & \cdots & E[x^2_{\mathcal{T},N-1}(n)] \end{bmatrix} \tag{7.20}$$

Thus, we get

$$\mathbf{x}^{\mathrm{n}}_{\mathcal{T}}(n) = \mathbf{D}^{-1/2}\mathbf{x}_{\mathcal{T}}(n) \tag{7.21}$$

where $\mathbf{x}^{\mathrm{n}}_{\mathcal{T}}(n)$ is the normalized tap-input vector. The correlation matrix associated with $\mathbf{x}^{\mathrm{n}}_{\mathcal{T}}(n)$ is

$$\mathbf{R}^{\mathrm{n}}_{\mathcal{T}} = \mathbf{D}^{-1/2}\mathbf{R}_{\mathcal{T}}\mathbf{D}^{-1/2} \tag{7.22}$$

Furthermore, we note that

$$\mathbf{D} = \mathrm{diag}[\mathbf{R}_{\mathcal{T}}] \tag{7.23}$$

where $\mathrm{diag}[\mathbf{R}_{\mathcal{T}}]$ denotes the diagonal matrix consisting of the diagonal elements of $\mathbf{R}_{\mathcal{T}}$. The reader may easily verify that the mean-squared values of the elements of $\mathbf{x}^{\mathrm{n}}_{\mathcal{T}}(n)$ as well as the diagonal elements of $\mathbf{R}^{\mathrm{n}}_{\mathcal{T}}$ are all equal to unity as a result of this normalization.

For the above example, we get

$$\mathbf{D}^{-1/2} = \begin{bmatrix} 0.5327 & 0 & 0 & 0 \\ 0 & 1.7972 & 0 & 0 \\ 0 & 0 & 3.0934 & 0 \\ 0 & 0 & 0 & 4.0357 \end{bmatrix} \tag{7.24}$$

and

$$\mathbf{R}_T^{\mathrm{n}} = \begin{bmatrix} 1.0000 & 0.0000 & -0.1409 & 0.0000 \\ 0.0000 & 1.0000 & 0.0000 & -0.0231 \\ -0.1409 & 0.0000 & 1.0000 & 0.0000 \\ 0.0000 & -0.0231 & 0.0000 & 1.0000 \end{bmatrix} \tag{7.25}$$

To compare the performance of the conventional LMS algorithm (in the time domain) with its associated implementation in a transform domain (explained in the following section), the eigenvalue spreads of $\mathbf{R}$ and $\mathbf{R}_T^{\mathrm{n}}$ have to be examined. For the example mentioned previously, we obtain

$$\text{eigenvalue spread of } \mathbf{R} = 57.5$$

and

$$\text{eigenvalue spread of } \mathbf{R}_T^{\mathrm{n}} = 1.33.$$

For the present example, these results predict a much superior performance of the LMS algorithm in the transform domain compared to its conventional implementation in the time domain. This, clearly, is a direct consequence of the orthogonalization property of the DCT, as was demonstrated previously. This argument justifies the application of orthogonal transforms for improving the performance of the LMS algorithm.

In our study in this chapter, we find that for a given transform, the degree of improvement achieved by replacing the conventional LMS algorithm with its transform domain counterpart depends on the power spectral density of the underlying input process. We emphasize on the band-partitioning property of orthogonal transforms and present some theoretical results that explain this phenomenon. We also find that a rough estimate of the power spectral density of the underlying input process is sufficient for the purpose of selecting an appropriate transform.

7.4 Transform Domain LMS Algorithm

In the implementation of transform domain LMS (TDLMS) algorithm, the filter tap weights are updated according to the following recursion:

$$\mathbf{w}_T(n+1) = \mathbf{w}_T(n) + 2\mu\hat{D}^{-1}e(n)\mathbf{x}_T(n) \tag{7.26}$$

where $\hat{\mathbf{D}}$ is an estimate of the diagonal matrix $\mathbf{D}$. This vector recursion can be decomposed into the following N scalar recursions:

$$w_{T,i}(n+1) = w_{T,i}(n) + 2\frac{\mu}{\hat{\sigma}^2_{x_{T,i}}(n)}e(n)x_{T,i}(n), \qquad i = 0, 1, \ldots, N-1 \tag{7.27}$$

where $\hat{\sigma}^2_{x_{T,i}}(n)$ is an estimate of $E[x^2_{T,i}(n)]$. This shows that the presence of $\hat{D}^{-1}$ in Eq. (7.26) is equivalent to using different step-size parameters at various taps of the TDAF. Each step-size parameter is chosen proportional to the inverse of the power of its associated input signal. Noting this, we refer to Eq. (7.26) as a step-normalized LMS recursion. In the present literature, the term *normalized LMS algorithm* has often been used to refer to Eq. (7.26) (Narayan *et al.* 1983; Marshall *et al.*, 1989). In this book, we use the term *step-normalized* (when necessary) to prevent any confusion between the normalization applied to TDAFs and the normalized LMS algorithm, which was introduced in Chapter 6 (Section 6.6).

In the implementation of Eq. (7.26), one needs to obtain the estimates of the signal powers at various taps of the filter, that is, $\hat{\sigma}^2_{x_{T,i}}(n)$'s. The following recursions are usually used for this purpose:

$$\hat{\sigma}^2_{x_{T,i}}(n) = \beta\hat{\sigma}^2_{x_{T,i}}(n-1) + (1-\beta)x^2_{T,i}(n), \qquad i = 0, 1, \ldots, N-1 \tag{7.28}$$

where β is a positive constant close to but less than 1. This recursion estimates the power by calculating a weighted average of the present and past samples of $x^2_{T,i}(n)$'s using an exponential weighting function given by 1, β, β^2, ... (Problem P7.2). The TDLMS algorithm, including this signal power estimation, is summarized in Table 7.1.

The step-normalization, as applied in Eq. (7.26), is equivalent to the normalization of the elements of the transformed tap-input vector, $\mathbf{x}_T(n)$, to the power of unity. To show this, we multiply Eq. (7.26) on both sides by $\hat{D}^{1/2}$ (the diagonal matrix consisting of the square roots of the diagonal elements of $\hat{D}$) and define $\mathbf{w}^{\mathrm{n}}_T(n) = \hat{D}^{1/2}\mathbf{w}_T(n)$ and $\mathbf{x}^{\mathrm{n}}_T(n) = \hat{D}^{-1/2}\mathbf{x}_T(n)$, to obtain

$$\mathbf{w}^{\mathrm{n}}_T(n+1) = \mathbf{w}^{\mathrm{n}}_T(n) + 2\mu e(n)\mathbf{x}^{\mathrm{n}}_T(n) \tag{7.29}$$

Table 7.1 Summary of the TDLMS algorithm.

Input:	Tap-weight vector, $\mathbf{w}_T(n)$, input vector, $\mathbf{x}(n)$, past tap-input power estimates, $\hat{\sigma}^2_{x_{T,i}}(n-1)$, and desired output, $d(n)$
Output:	Filter output, $y(n)$, tap-input power estimate updates, $\hat{\sigma}^2_{x_{T,i}}(n)$, and tap-weight vector update, $\mathbf{w}_T(n+1)$

1. Transformation:
$$\mathbf{x}_T = \mathcal{T}\mathbf{x}(n)$$

1. Filtering:
$$y(n) = \mathbf{w}^{\mathrm{T}}_T(n)\mathbf{x}_T(n)$$

2. Error estimation:
$$e(n) = d(n) - y(n)$$

3. Tap-input power estimate update:
for $i = 0$ to $N-1$
$$\hat{\sigma}^2_{x_{T,i}}(n) = \beta\hat{\sigma}^2_{x_{T,i}}(n-1) + (1-\beta)x^2_{T,i}(n)$$

4. Tap-weight vector adaptation:
$$\mathbf{w}_T(n+1) = \mathbf{w}_T(n) + 2\mu\hat{D}^{-1}e(n)\mathbf{x}_T(n)$$
where $\hat{D} = \mathrm{diag}[\hat{\sigma}^2_{x_{T,0}}(n), \hat{\sigma}^2_{x_{T,1}}(n), \cdots, \hat{\sigma}^2_{x_{T,N-1}}(n)]$

We may also note that

$$\begin{aligned} e(n) &= d(n) - y(n) \\ &= d(n) - \mathbf{w}_{\mathcal{T}}^{\mathrm{T}}(n)\mathbf{x}_{\mathcal{T}}(n) \\ &= d(n) - \mathbf{w}_{\mathcal{T}}^{\mathrm{nT}}(n)\mathbf{x}_{\mathcal{T}}^{\mathrm{n}}(n) \end{aligned} \tag{7.30}$$

Equations (7.29) and (7.30) suggest that *the TDLMS algorithm, in effect, is equivalent to a conventional LMS algorithm with the normalized tap-input vector* $\mathbf{x}_{\mathcal{T}}^{\mathrm{n}}(n)$.

The significance of this result is that as Eq. (7.29) is a conventional LMS recursion, the analytical results of the last chapter can immediately be applied to evaluate the performance of the TDLMS algorithm. In particular, we note that the various modes of convergence of the TDLMS algorithm are determined by the eigenvalues of the correlation matrix $\mathbf{R}_{\mathcal{T}}^{\mathrm{n}} = E[\mathbf{x}_{\mathcal{T}}^{\mathrm{n}}(n)\mathbf{x}_{\mathcal{T}}^{\mathrm{nT}}(n)]$. This matches our conjecture in the last section. Also, by substituting the eigenvalues of $\mathbf{R}_{\mathcal{T}}^{\mathrm{n}}$ for $\lambda_0, \lambda_1, \ldots, \lambda_{N-1}$, in Eqs. (6.61), (6.63), or (6.64), misadjustment of the TDLMS algorithm can be evaluated. In particular, we note that $\mathrm{tr}[\mathbf{R}_{\mathcal{T}}^{\mathrm{n}}] = N$ as the diagonal elements of $\mathbf{R}_{\mathcal{T}}^{\mathrm{n}}$ are all normalized to unity. Thus, using Eq. (6.64), misadjustment of the TDLMS algorithm is obtained as

$$\mathcal{M} \approx \mu N \tag{7.31}$$

7.5 Ideal LMS-Newton Algorithm and Its Relationship with TDLMS

In Chapter 5, we introduced two search methods: the method of steepest-descent and the Newton's algorithm. The LMS algorithm was introduced in Chapter 6 as a stochastic implementation of the method of steepest-descent. In the LMS algorithm, the gradient vector $\nabla_{\mathbf{w}}\xi$ is replaced by its instantaneous estimate, $\nabla_{\mathbf{w}}e^2(n)$. A similar substitution of the gradient vector in the Newton method, given by Eq. (5.44), results in the recursion

$$\mathbf{w}(n+1) = \mathbf{w}(n) + 2\mu\mathbf{R}^{-1}e(n)\mathbf{x}(n) \tag{7.32}$$

which may be called *ideal LMS-Newton algorithm*. The term *ideal* refers to the fact that the knowledge of true $\mathbf{R}^{-1}$ is assumed here. In actual practice, of course, this cannot be true. One can only obtain an estimate of $\mathbf{R}^{-1}$. Methods for obtaining such estimates are available in the literature (see e.g., Widrow and Stearns (1985); Marshall and Jenkins (1992); Farhang-Boroujeny (1993)). Our aim in this section is to show that there is a close relationship between the LMS–Newton and TDLMS algorithms. We show that when the transformation matrix $\mathcal{T}$ is selected to be the Karhunen-Loéve transform (KLT) of the filter input, the TDLMS and LMS–Newton are two different formulations of the same algorithm. Thus, we conclude that when a proper transformation is used, the TDLMS algorithm may be considered as an efficient implementation of the LMS–Newton algorithm.

We recall that the correlation matrix $\mathbf{R}$ can be decomposed as $\mathbf{R} = \mathbf{Q}\boldsymbol{\Lambda}\mathbf{Q}^{\mathrm{T}}$, where $\mathbf{Q}$ is the N-by-N matrix whose columns are the eigenvectors of $\mathbf{R}$, and $\boldsymbol{\Lambda}$ is the diagonal matrix consisting of the associated eigenvalues of $\mathbf{R}$. This, in turn, implies that

$$\mathbf{R}^{-1} = \mathbf{Q}\boldsymbol{\Lambda}^{-1}\mathbf{Q}^{\mathrm{T}} \tag{7.33}$$

because $\mathbf{QQ}^{\mathrm{T}} = \mathbf{I}$.

Substituting Eq. (7.33) in Eq. (7.32) and premultiplying the result by $\mathbf{Q}^{\mathrm{T}}$, we obtain

$$\mathbf{w}'(n+1) = \mathbf{w}'(n) + 2\mu\mathbf{\Lambda}^{-1}e(n)\mathbf{x}'(n) \tag{7.34}$$

where $\mathbf{w}'(n) = \mathbf{Q}^{\mathrm{T}}\mathbf{w}(n)$ and $\mathbf{x}'(n) = \mathbf{Q}^{\mathrm{T}}\mathbf{x}(n)$. Also, from our discussions in Chapter 4, we recall that the eigenvalues of $\mathbf{R}$, that is, the diagonal elements of $\mathbf{\Lambda}$, are equal to the powers (mean-squared values) of the elements of the vector $\mathbf{x}'(n)$. Combining this with our discussion in the previous section, we see that the LMS–Newton algorithm is an alternative formulation of the TDLMS algorithm when $\mathcal{T} = \mathbf{Q}^{\mathrm{T}}$. Furthermore, we note that for a given input process, $x(n)$, with correlation matrix $\mathbf{R}$, the transform $\mathcal{T} = \mathbf{Q}^{\mathrm{T}}$ is the ideal one in the sense that it results in a diagonal $\mathbf{R}_{\mathcal{T}} = \mathbf{\Lambda}$. When $\mathcal{T} \neq \mathbf{Q}^{\mathrm{T}}$, but results in an approximately diagonal $\mathbf{R}_{\mathcal{T}}$, we may say that the TDLMS algorithm is equivalent to a *quasi* LMS–Newton algorithm.

7.6 Selection of the Transform $\mathcal{T}$

It turns out that for a given process, $x(n)$, the performance of the TDLMS algorithm may vary significantly depending on the selection of the transformation matrix, $\mathcal{T}$. A transform that may perform well for a given input process may perform poorly once the statistics of the input changes. This happens to be more prominent when the filter length is short. For long filters, we find that most of the commonly used transforms perform well and result in a significant performance improvement compared with the conventional LMS algorithm.

In this section, we present some theoretical results that explain these observations. This presentation includes a geometrical interpretation of the TDLMS process, which will be given for a two-tap filter. This interpretation will then be generalized using a special performance index, which is also introduced in this section. This leads to a very instructive view of the band-partitioning property of orthogonal transforms, which helps one to select a proper transform once a rough estimate of the power spectral density of the underlying input process is known.

7.6.1 A Geometrical Interpretation

The geometrical interpretation presented here has been adopted from Marshall *et al.* (1989). We recall that the performance surface of a transversal filter with input correlation matrix $\mathbf{R}$ may be written as

$$\xi(\mathbf{v}) = \xi_{\min} + \mathbf{v}^{\mathrm{T}}\mathbf{R}\mathbf{v} \tag{7.35}$$

where the vector $\mathbf{v}$ is the difference between the filter tap-weight vector, $\mathbf{w}$, and its optimum value, $\mathbf{w}_{\mathrm{o}}$.

For the sake of illustrating the principles, let us consider a two-tap filter problem with $\mathbf{R}$ given by

$$\mathbf{R} = \begin{bmatrix} 1.0 & 0.9 \\ 0.9 & 1.0 \end{bmatrix} \tag{7.36}$$

With this $\mathbf{R}$, Eq. (7.35) becomes

$$\xi(v_0, v_1) = \xi_{\min} + v_0^2 + v_1^2 + 1.8v_0v_1 \tag{7.37}$$

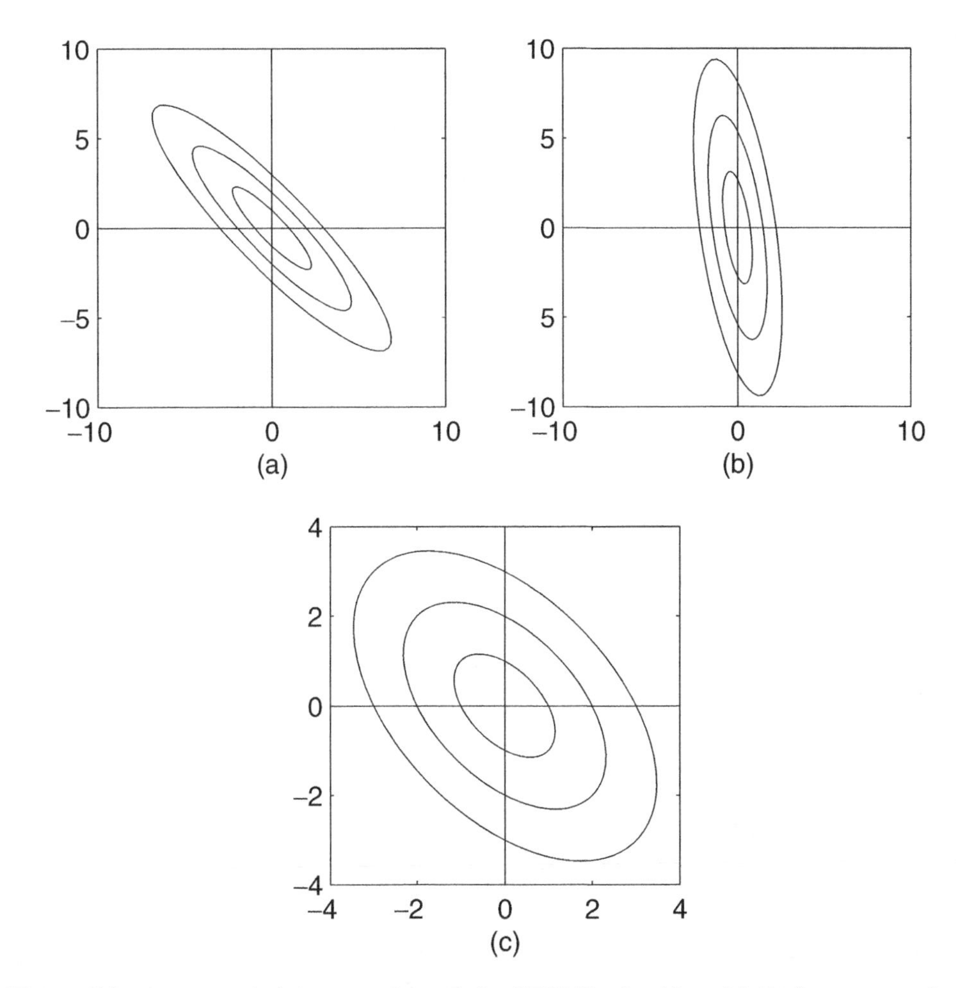

Figure 7.3 A geometrical interpretation of the TDLMS algorithm. (a) Performance surface before transformation; (b) performance surface after transformation, but without normalization; and (c) performance surface after transformation and normalization (adopted from Marshall *et al.* (1989)).

Figure 7.3a shows the contour plot associated with this performance surface. As we may recall from our discussions in the previous chapters, the eccentricity of the contour ellipses in Figure 7.3a, is related to the eigenvalue spread of the correlation matrix $\mathbf{R}$. A large eccentricity is due to a large eigenvalue spread and that, in turn, results in certain slow mode(s) of convergence when the conventional LMS algorithm is used to adjust the filter tap weights.

Application of an orthogonal transform, $\mathcal{T}$, converts the tap-input vector $\mathbf{x}(n)$ to $\mathbf{x}_{\mathcal{T}}(n) = \mathcal{T}\mathbf{x}(n)$, whose associated correlation matrix, $\mathbf{R}_{\mathcal{T}}$, is related to $\mathbf{R}$ according to Eq. (7.9). As a numerical example, let us choose

$$\mathcal{T} = \begin{bmatrix} 0.8 & 0.6 \\ -0.6 & 0.8 \end{bmatrix} \tag{7.38}$$

This, with $\mathbf{R}$ as given in Eq. (7.36), results in

$$\mathbf{R}_{\mathcal{T}} = \begin{bmatrix} 1.864 & 0.252 \\ 0.252 & 0.136 \end{bmatrix} \tag{7.39}$$

If no normalization is applied to the transformed samples, the performance surface associated with the TDAF will be

$$\xi_{\mathcal{T}}(\mathbf{v}_{\mathcal{T}}) = \xi_{\min} + \mathbf{v}_{\mathcal{T}}^T \mathbf{R}_{\mathcal{T}} \mathbf{v}_{\mathcal{T}} \tag{7.40}$$

which for the present numerical example can be expanded as

$$\xi_{\mathcal{T}}(v_{\mathcal{T},0}, v_{\mathcal{T},1}) = \xi_{\min} + 1.864 v_{\mathcal{T},0}^2 + 0.136 v_{\mathcal{T},1}^2 + 0.504 v_{\mathcal{T},0} v_{\mathcal{T},1} \tag{7.41}$$

Figure 7.3b shows the contours associated with the performance surface defined by Eq. (7.41). Note that the effect of the transformation is only to rotate the performance surface with respect to the coordinate axes. The shape of the performance surface, that is, the eccentricity of the contour ellipses, has not changed. This can be mathematically explained by noting that, as $\mathcal{T}\mathcal{T}^{\mathrm{T}} = \mathcal{T}^{\mathrm{T}}\mathcal{T} = \mathbf{I}$,

$$\begin{aligned}
\xi(\mathbf{v}) &= \xi_{\min} + \mathbf{v}^{\mathrm{T}} \mathbf{R} \mathbf{v} \\
&= \xi_{\min} + \mathbf{v}^{\mathrm{T}} \mathcal{T}^{\mathrm{T}} \mathcal{T} \mathbf{R} \mathcal{T}^{\mathrm{T}} \mathcal{T} \mathbf{v} \\
&= \xi_{\min} + \mathbf{v}_{\mathcal{T}}^{\mathrm{T}} \mathbf{R}_{\mathcal{T}} \mathbf{v}_{\mathcal{T}} = \xi_{\mathcal{T}}(\mathbf{v}_{\mathcal{T}})
\end{aligned} \tag{7.42}$$

where $\mathbf{v}$ and $\mathbf{v}_{\mathcal{T}}$ are related according to the equation $\mathbf{v}_{\mathcal{T}} = \mathcal{T}\mathbf{v}$. This result, which can also be written as $\xi_{\mathcal{T}}(\mathbf{v}_{\mathcal{T}}) = \xi(\mathcal{T}^{\mathrm{T}}\mathbf{v}_{\mathcal{T}})$, shows that the performance surface defined by Eq. (7.40) is obtained from the one defined by Eq. (7.35) by a rotation of the coordinate axes according to the relationship $\mathbf{v}_{\mathcal{T}} = \mathcal{T}\mathbf{v}$, or, equivalently, by keeping the coordinate axes fixed and rotating the performance surface in the opposite direction. This observation shows that *transformation without normalization has no effect on the convergence behavior of the steepest-descent method and, thus, the LMS algorithm.* Thus, we emphasize that normalization has to be considered as an integrated part of any transform domain adaptive algorithm (as introduced in the case of TDLMS algorithm, in Section 7.4), otherwise transformation adds up to the filter complexity without any gain in convergence.

When the elements of $\mathbf{x}_{\mathcal{T}}(n)$ are normalized to the power of unity[4], the corresponding correlation matrix is given by Eq. (7.22) and its associated performance surface is defined as

$$\xi_{\mathcal{T}}^{\mathrm{n}}(\mathbf{v}_{\mathcal{T}}^{\mathrm{n}}) = \xi_{\min} + \mathbf{v}_{\mathcal{T}}^{\mathrm{n}T} \mathbf{R}_{\mathcal{T}}^{\mathrm{n}} \mathbf{v}_{\mathcal{T}}^{\mathrm{n}} \tag{7.43}$$

For the present example, we obtain

$$R_{\mathcal{T}}^{\mathrm{n}} = \begin{bmatrix} 1.0000 & 0.5005 \\ 0.5005 & 1.0000 \end{bmatrix} \tag{7.44}$$

and

$$\xi_{\mathcal{T}}^{\mathrm{n}}(v_{\mathcal{T},0}^{\mathrm{n}}, v_{\mathcal{T},1}^{\mathrm{n}}) = \xi_{\min} + (v_{\mathcal{T},0}^{\mathrm{n}})^2 + (v_{\mathcal{T},1}^{\mathrm{n}})^2 + 1.001 v_{\mathcal{T},0}^{\mathrm{n}} v_{\mathcal{T},1}^{\mathrm{n}} \tag{7.45}$$

[4] We recall that the step-normalization, as applied in Eq. (7.26), and normalization of the elements of $\mathbf{x}_{\mathcal{T}}(n)$ to the power of unity are equivalent.

Figure 7.3c shows the contours associated with Eq. (7.45). We note that the normalization reduces the eccentricity of the performance surface. This, of course, will result in a faster convergence of the TDLMS algorithm compared to the conventional LMS algorithm.

A better insight into the effect of normalization is obtained by making the following observations. *The hyperellipses associated with the performance surface of a transversal filter are hyperspherical at the points of their intersection with the* v*-axes,* that is, the intersection points of each contour (hyperellipse) with the v-axes are at equal distance from the origin. This which is clearly observed in Figure 7.3a can be shown to be true, in general, if we note that, for any i

$$\xi_i(v_i) = \xi_{\min} + r_{ii} v_i^2$$

where r_{ii} is the ith diagonal element of $\mathbf{R}$, and $\xi_i(v_i)$ is the performance function $\xi(\mathbf{v})$ when all elements of the vector $\mathbf{v}$, except its ith element, v_i, have been set equal to zero. We also note that for a transversal filter, r_{ii} is the same for all values of i. Thus, the identity

$$\xi_i(v_i) = \xi_j(v_j), \qquad \text{for all } i \text{ and } j$$

implies that

$$|v_i| = |v_j|$$

which, in turn, shows that the hyperellipses associated with the performance surface of a transversal filter are hyperspherical at the points of their intersection with the v-axes.

Following the same argument, we find that as the diagonal elements of $\mathbf{R}_{\mathcal{T}}$ are likely to be unequal (unless the underlying input process, $x(n)$, is white, i.e., when $\mathbf{R}$ is a multiple of the identity matrix), the contour ellipses associated with $\xi_{\mathcal{T}}(\mathbf{v}_{\mathcal{T}})$ are most likely nonhyperspherical at the points of their intersection with the $v_{\mathcal{T}}$-axes. This is clearly observed in Figure 7.3b.

On the other hand, normalization of the transformed samples to the power of unity equalizes the diagonal elements of $\mathbf{R}_{\mathcal{T}}^{\mathrm{n}}$. Thus, the hyperellipses associated with the performance surface $\xi_{\mathcal{T}}^{\mathrm{n}}(\mathbf{v}_{\mathcal{T}}^{\mathrm{n}})$ are hyperspherical at the points of their intersection with the corresponding coordinate axes. To get a better insight, we may also note that

$$\begin{aligned}
\xi_{\mathcal{T}}(\mathbf{v}_{\mathcal{T}}) &= \xi_{\min} + \mathbf{v}_{\mathcal{T}}^{\mathrm{T}} \mathbf{R}_{\mathcal{T}} \mathbf{v}_{\mathcal{T}} \\
&= \xi_{\min} + \mathbf{v}_{\mathcal{T}}^{\mathrm{T}} \mathbf{D}^{1/2} \mathbf{D}^{-1/2} \mathbf{R}_{\mathcal{T}} \mathbf{D}^{-1/2} \mathbf{D}^{1/2} \mathbf{v}_{\mathcal{T}} \\
&= \xi_{\min} + (\mathbf{D}^{1/2}\mathbf{v}_{\mathcal{T}})^{\mathrm{T}} \mathbf{R}_{\mathcal{T}}^{\mathrm{n}} (\mathbf{D}^{1/2}\mathbf{v}_{\mathcal{T}}) = \xi_{\mathcal{T}}^{\mathrm{n}}(\mathbf{v}_{\mathcal{T}}^{\mathrm{n}})
\end{aligned} \tag{7.46}$$

where $\mathbf{v}_{\mathcal{T}}^{\mathrm{n}} = \mathbf{D}^{1/2}\mathbf{v}_{\mathcal{T}}$, and we have noted that $(\mathbf{D}^{1/2})^{\mathrm{T}} = \mathbf{D}^{1/2}$ as $\mathbf{D}$ is a diagonal matrix. This result, which can also be written as $\xi_{\mathcal{T}}^{\mathrm{n}}(\mathbf{v}_{\mathcal{T}}^{\mathrm{n}}) = \xi_{\mathcal{T}}(\mathbf{D}^{-1/2}\mathbf{v}_{\mathcal{T}}^{\mathrm{n}})$, shows that the performance surface defined by Eq. (7.43) is obtained from the one defined by Eq. (7.40) by scaling its coordinate axes according to the relationship $\mathbf{v}_{\mathcal{T}}^{\mathrm{n}} = \mathbf{D}^{1/2}\mathbf{v}_{\mathcal{T}}$. For the example shown in Figure 7.3, this is equivalent to stretching the contour ellipses of Figure 7.3b along $v_{\mathcal{T},0}$-axis and shrinking them along $v_{\mathcal{T},1}$-axis. This clearly reduces the eccentricity of the ellipses. Furthermore, we note that the ellipses in Figure 7.3c would become circles resulting in maximum improvement in convergence if the ellipses in Figure 7.3b had been rotated so that their principal axes would be along the $v_{\mathcal{T}}$-axes. It is interesting to note that this corresponds to the case where $\mathcal{T}$ is the Karhunen-Loéve transform (KLT) associated with the correlation matrix $\mathbf{R}$. We may also recall from Chapter 4 that in the latter case $\mathbf{R}_{\mathcal{T}} = \mathbf{\Lambda}$, that is, the diagonal matrix consisting of the eigenvalues of $\mathbf{R}$. Furthermore, from the minimax theorem (of Chapter 4), we find that this corresponds to the case where the diagonal elements of $\mathbf{R}_{\mathcal{T}}$ are maximally spread. Moreover, a closer look at the present example reveals that the effect of normalization in reducing the eccentricity of the contour ellipses depends on the relative size (i.e., spread) of the diagonal elements of $\mathbf{R}_{\mathcal{T}}$. In other words, the spread of the signal power at the filter taps after transformation appears to be the key factor that determines the success of a TDAF. The discussion that follows in the rest of this section aims at exploring this aspect of TDAFs further.

7.6.2 A Useful Performance Index

In the study of the LMS algorithm, eigenvalue spread, that is, $\lambda_{\mathrm{max}}/\lambda_{\mathrm{min}}$, of the correlation matrix, $\mathbf{R}$, of the underlying input process is the most widely used performance index. In this book also, so far, we have emphasized on the significance of eigenvalue spread. Unfortunately, there is no way of getting closed-form (explicit) equations for the maximum, λ_{max}, and minimum, λ_{min}, eigenvalues of a matrix $\mathbf{R}$, in general. As a result, application of this index for any further study of the TDLMS algorithm and its comparison with the conventional LMS algorithm is not possible. Hence, we shall look for other possible performance indices that may be mathematically tractable.

Farhang-Boroujeny and Gazor (1991, 1992) proposed an index that is mathematically tractable and able to give some further insight into the effect of orthogonal transforms in improving the performance of the LMS algorithm. The proposed index is

$$\rho(\mathbf{R}) = \left(\frac{\lambda_{\mathrm{a}}}{\lambda_{\mathrm{g}}}\right)^{N} \tag{7.47}$$

where λ_{a} and λ_{g} are arithmetic and geometric averages, respectively, of the eigenvalues of $\mathbf{R}$. Namely,

$$\lambda_{\mathrm{a}} = \frac{\sum_i \lambda_i}{N} \tag{7.48}$$

and

$$\lambda_{\mathrm{g}} = \sqrt[N]{\prod_i \lambda_i} \tag{7.49}$$

We note that the value of $\rho(\mathbf{R})$ depends on the distribution of the eigenvalues of $\mathbf{R}$. It is always greater than or equal to 1. It approaches 1 when all the eigenvalues of $\mathbf{R}$ assume about the same values and increases as the eigenvalues of $\mathbf{R}$ spread apart. Furthermore, the lower bound $\rho(\mathbf{R}) = 1$ is reached when the eigenvalues of $\mathbf{R}$ are all equal. Using the identities (Chapter 4)

$$\sum_i \lambda_i = \mathrm{tr}[\mathbf{R}]$$

where $\mathrm{tr}[\mathbf{R}]$ is the trace of $\mathbf{R}$, and

$$\prod_i \lambda_i = \det[\mathbf{R}]$$

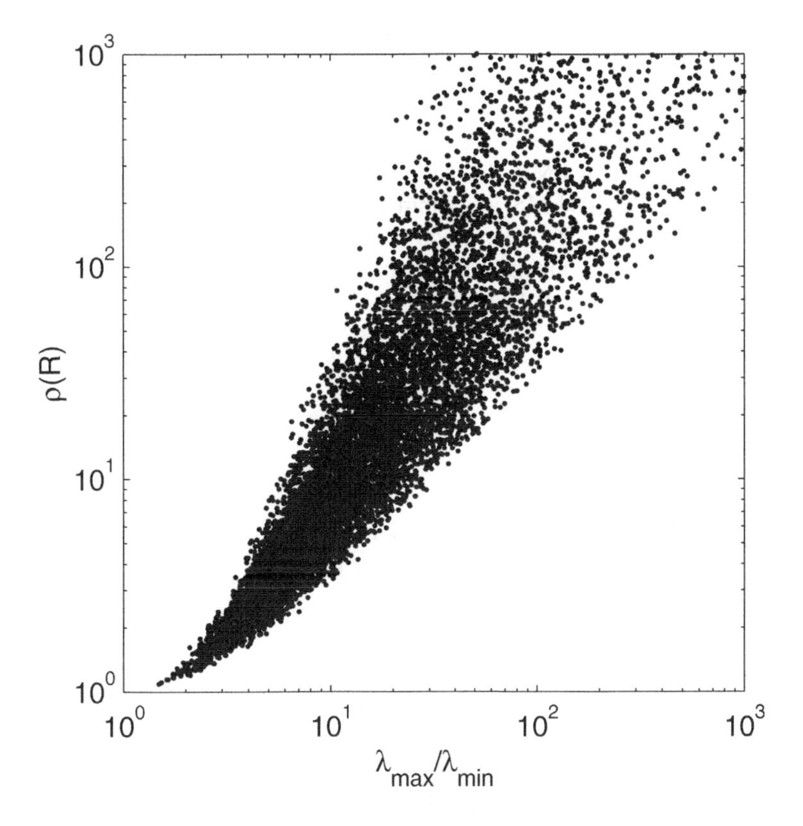

Figure 7.4 Variation of $\rho(\mathbf{R})$ versus eigenvalue spread of $\mathbf{R}$.

where det[$\mathbf{R}$] is the determinant of $\mathbf{R}$, we obtain

$$\rho(\mathbf{R}) = \frac{(\text{tr}[\mathbf{R}]/N)^N}{\det[\mathbf{R}]} \tag{7.50}$$

Now, one may appreciate the index $\rho(\mathbf{R})$ because of its closed-form nature in terms of elements of $\mathbf{R}$.

Before we proceed with the application of the performance index $\rho(\mathbf{R})$ to further study the TDLMS algorithm, we may remark that the relationship between $\rho(\mathbf{R})$ and the eigenvalue spread of $\mathbf{R}$, that is, $\lambda_{max}/\lambda_{min}$, is rather complicated. The index $\rho(\mathbf{R})$ depends on not only $\lambda_{max}/\lambda_{min}$, but also the distribution of the rest of eigenvalues of $\mathbf{R}$ in the range λ_{min} to λ_{max}. However, the general trend is that a large eigenvalue spread of $\mathbf{R}$ implies a large $\rho(\mathbf{R})$ and vice versa. Similarly, a $\rho(\mathbf{R})$ close to 1 implies that the eigenvalue spread of $\mathbf{R}$ is small. Figure 7.4 shows how $\rho(\mathbf{R})$ varies as a function of $\lambda_{max}/\lambda_{min}$ when $N = 10$ and the eigenvalues of $\mathbf{R}$ are assumed to be a set of random numbers distributed in the range 0 to 1.

7.6.3 Improvement Factor and Comparisons

To compare a pair of LMS-based algorithms, say LMS_1 and LMS_2, we define an improvement factor, I_ρ, as the natural logarithm of the ratio of the performance index $\rho(\cdot)$ in the two cases. In particular, when LMS_1 is compared with LMS_2, we define

$$I_\rho = \ln\ \rho(\mathbf{R}_1) - \ln\ \rho(\mathbf{R}_2) \tag{7.51}$$

where $\mathbf{R}_1$ and $\mathbf{R}_2$ are the associated correlation matrices in LMS_1 and LMS_2, respectively. A positive I_ρ indicates that LMS_2 is superior, and a negative I_ρ indicates that LMS_2 is inferior. In comparing a TDLMS algorithm and its conventional LMS counterpart, we shall let $\mathbf{R}_1 = \mathbf{R}$ and $\mathbf{R}_2 = \mathbf{R}_{\mathcal{T}}^{\text{n}}$. On the other hand, if no normalization is applied in the implementation of TDLMS algorithm, we shall let $\mathbf{R}_2 = \mathbf{R}_{\mathcal{T}}$. Thus, for the latter case, the corresponding improvement factor is

$$I_{\rho,\mathcal{T}} = \ln \rho(\mathbf{R}) - \ln \rho(\mathbf{R}_{\mathcal{T}}) \tag{7.52}$$

We note that

$$\begin{aligned}\rho(\mathbf{R}_{\mathcal{T}}) &= \frac{(\text{tr}[\mathbf{R}_{\mathcal{T}}]/N)^N}{\det[\mathbf{R}_{\mathcal{T}}]} \\ &= \frac{(\text{tr}[\mathcal{T}\mathbf{R}\mathcal{T}^{\text{T}}]/N)^N}{\det[\mathcal{T}\mathbf{R}\mathcal{T}^{\text{T}}]}\end{aligned} \tag{7.53}$$

To simplify this, we recall the following results of matrix algebra. If $\mathbf{A}$ and $\mathbf{B}$ are N-by-M and M-by-N matrices, respectively, then

$$\text{tr}[\mathbf{AB}] = \text{tr}[\mathbf{BA}] \tag{7.54}$$

Also, when $\mathbf{A}$ and $\mathbf{B}$ are square matrices

$$\det[\mathbf{AB}] = \det[\mathbf{BA}] = \det[\mathbf{A}]\cdot\det[\mathbf{B}] \tag{7.55}$$

Using Eqs. (7.54) and (7.55) in Eq. (7.53), we obtain

$$\rho(\mathbf{R}_{\mathcal{T}}) = \frac{(\text{tr}[\mathcal{T}^{\text{T}}\mathcal{T}\mathbf{R}]/N)^N}{\det[\mathcal{T}^{\text{T}}\mathcal{T}\mathbf{R}]} = \rho(\mathbf{R}) \tag{7.56}$$

where the last equality follows as $\mathcal{T}^{\text{T}}\mathcal{T} = \mathbf{I}$. Substituting Eq. (7.56) in Eq. (7.52), we get

$$I_{\rho,\mathcal{T}} = 0$$

This shows that transformation without normalization has no effect in improving the performance of the LMS algorithm. This result, which was also predicted by the geometrical interpretation of the TDLMS algorithm before, (Figure 7.3), can also be understood if we recall the definition of $\rho(\mathbf{R})$ (i.e., Eq. (7.47)) while noting that for an arbitrary orthogonal transformation $\mathcal{T}$, with $\mathcal{T}\mathcal{T}^{\text{T}} = \mathbf{I}$, the eigenvalues of $\mathbf{R}$ and $\mathbf{R}_{\mathcal{T}} = \mathcal{T}\mathbf{R}\mathcal{T}^{\text{T}}$ are the same (Problem P7.7).

Another case of interest to be noted here is the comparison of the conventional LMS algorithm and the ideal LMS–Newton algorithm. In this case, $\mathbf{R}_1 = \mathbf{R}$ and $\mathbf{R}_2 = \mathbf{I}$. Thus, the improvement achieved using an ideal LMS–Newton algorithm instead of its conventional LMS counterpart is

$$\begin{aligned}I_{\rho,\max} &= \ln \rho(\mathbf{R}) - \ln \rho(\mathbf{I}) \\ &= \ln \rho(\mathbf{R})\end{aligned} \tag{7.57}$$

as $\rho(\mathbf{I}) = 1$. The notation $I_{\rho,\max}$ reflects the fact that the ideal LMS–Newton algorithm results in the maximum possible improvement that one can achieve by modifying the conventional LMS algorithm. Furthermore, Eq. (7.57) indicates that $\ln \rho(\mathbf{R})$ can be considered as a measure of the *distance* of the LMS algorithm from the ideal LMS–Newton algorithm. Similarly, in evaluating a particular implementation of the TDLMS algorithm, the value of $\ln \rho(\mathbf{R}_{\mathcal{T}}^{\mathrm{n}})$ shows the distance of the TDLMS algorithm from the ideal LMS–Newton algorithm and, thus, it may be considered as a parameter indicating the extent of decorrelation that is achieved by the transformation.

The following theorem shows an easy way to compare a transform-domain-normalized LMS (TDNLMS) algorithm with its conventional LMS counterpart.

Theorem 7.1 *When the conventional LMS algorithm is replaced by its TDLMS counterpart, the resulting improvement factor is*

$$I_{\rho,\mathcal{T}}^{\mathrm{n}} = \ln \rho(\mathrm{diag}[\mathbf{R}_{\mathcal{T}}]) \tag{7.58}$$

where $\mathbf{R}_{\mathcal{T}} = \mathcal{T}\mathbf{R}\mathcal{T}^{\mathrm{T}}$, $\mathbf{R}$ *is the correlation matrix of the underlying input process,* $\mathcal{T}$ *is the transformation matrix, and* $\mathrm{diag}[\mathbf{R}_{\mathcal{T}}]$ *denotes the diagonal matrix consisting of the diagonal elements of* $\mathbf{R}_{\mathcal{T}}$.

Proof. According to Eq. (7.51), the improvement factor is

$$I_{\rho,\mathcal{T}}^{\mathrm{n}} = \ln \rho(\mathbf{R}) - \ln \rho(\mathbf{R}_{\mathcal{T}}^{\mathrm{n}}) \tag{7.59}$$

We recall that the diagonal elements of the normalized N-by-N matrix $\mathbf{R}_{\mathcal{T}}^{\mathrm{n}}$ are all equal to 1. This implies that

$$\mathrm{tr}[\mathbf{R}_{\mathcal{T}}^{\mathrm{n}}] = N \tag{7.60}$$

Noting this and using Eqs. (7.22) and (7.55), we may proceed as follows:

$$\begin{aligned}
\rho(\mathbf{R}_{\mathcal{T}}^{\mathrm{n}}) &= \frac{(\mathrm{tr}[\mathbf{R}_{\mathcal{T}}^{\mathrm{n}}]/N)^N}{\det[\mathbf{R}_{\mathcal{T}}^{\mathrm{n}}]} \\
&= \frac{1}{\det[\mathbf{D}^{-1/2}\mathbf{R}_{\mathcal{T}}\mathbf{D}^{-1/2}]} \\
&= \frac{1}{\det[\mathbf{D}^{-1}\mathbf{R}_{\mathcal{T}}]} \\
&= \frac{1}{\det[\mathbf{D}^{-1}]\cdot\det[\mathbf{R}_{\mathcal{T}}]} \\
&= \frac{\det[\mathbf{D}]}{\det[\mathbf{R}_{\mathcal{T}}]}
\end{aligned} \tag{7.61}$$

where the last equality follows from the identity $\det[\mathbf{D}^{-1}] = (\det[\mathbf{D}])^{-1}$. Next, substituting for $\mathbf{D}$ from Eq. (7.23) and noting that $\mathrm{tr}[\mathbf{R}_{\mathcal{T}}] = \mathrm{tr}[\mathrm{diag}[\mathbf{R}_{\mathcal{T}}]]$, we get

$$\rho(\mathbf{R}_{\mathcal{T}}^{\mathrm{n}}) = \frac{\det[\mathrm{diag}[\mathbf{R}_{\mathcal{T}}]]}{\det[\mathbf{R}_{\mathcal{T}}]}$$

$$= \frac{\det[\text{diag}[\mathbf{R}_{\mathcal{T}}]]}{(\text{tr}[\text{diag}[\mathbf{R}_{\mathcal{T}}]]/N)^N} \bullet \frac{(\text{tr}[\mathbf{R}_{\mathcal{T}}]/N)^N}{\det[\mathbf{R}_{\mathcal{T}}]}$$

$$= \frac{\rho(\mathbf{R}_{\mathcal{T}})}{\rho(\text{diag}[\mathbf{R}_{\mathcal{T}}])} \tag{7.62}$$

Substituting Eq. (7.62) in Eq. (7.59) and noting that $\rho(\mathbf{R}) = \rho(\mathbf{R}_{\mathcal{T}})$, completes the proof.

The corollary is as follows:

Corollary 7.1 *As* $\ln \rho(\text{diag}[\mathbf{R}_{\mathcal{T}}])$ *is always nonnegative, the performance of a TDLMS algorithm can never be worse than its conventional LMS counterpart.*

The following remark may also be made. When comparing a TDLMS algorithm with its conventional LMS counterpart, the degree of improvement achieved depends on the distribution of the signal power at various outputs of the transformation, that is, the tap inputs $x_{\mathcal{T},i}(n)$. A wide spread of signal power at the taps indicates a significant improvement. Similarly, a small spread in signal powers indicates that the improvement achievable is very less.

7.6.4 Filtering View

The quantitative result of the above theorem suggests that for a given input process, a transformation matrix will effectively decorrelate the samples of input if it implements a set of parallel FIR filters whose output powers are close to maximally spread. The maximally spread signal powers, here, is quantified by the *minimax theorem*, which was introduced in Chapter 4. When the correlation matrix of the underlying input process is known, the minimax theorem suggests a procedure for the optimal selection of a set of filters, which achieve maximum power spreading. It starts with the design of a set of filters (with orthogonal coefficient vectors) whose output powers are maximized. Instead, it may also start with the design of another set of filters whose output powers are minimized. We also note that these two optimization procedures are implemented independent of each other, but both result in the same set of eigenvectors. This gives an intuitive feeling of how the minimax theorem (procedure) finds a transformation with a maximum spread of signal powers at its outputs.

We note that while the minimax theorem suggests a procedure for the design of the optimal transform for a given input process, the above theorem gives a measure of effectiveness of a transformation matrix in decorrelating the samples of an underlying input process. We note that for a given input process with correlation matrix $\mathbf{R}$, the maximum attainable improvement factor is $I_{\rho,\max} = \ln \rho(\mathbf{R})$, and this is achieved when $\mathcal{T}$ is the KLT of the underlying input process. On the other hand, for a given transformation, $\mathcal{T}$, $I^{\text{n}}_{\rho,\mathcal{T}} = \ln \rho(\text{diag}[\mathbf{R}_{\mathcal{T}}])$. Thus, the difference $I_{\rho,\max} - I^{\text{n}}_{\rho,\mathcal{T}}$ gives a measure of the success of $\mathcal{T}$ in decorrelating the input samples. A small value of $I_{\rho,\max} - I^{\text{n}}_{\rho,\mathcal{T}}$ indicates that the transformation used is close to optimal and vice versa. Furthermore, as explained in Section 7.6.3, $I_{\rho,\max} - I^{\text{n}}_{\rho,\mathcal{T}} = \ln \rho(\mathbf{R}^{\text{n}}_{\mathcal{T}})$ is also the distance of the TDLMS from the ideal LMS–Newton algorithm.

It is instructive to elaborate more on the power spreading effect of a transformation $\mathcal{T}$ and relate that to the above findings. We recall that the output power of a filter with the transfer function $F(e^{j\omega})$, input $x(n)$, and output $y(n)$ is given by (Chapter 2)

$$E[y^2(n)] = \frac{1}{2\pi}\int_0^{2\pi} \Phi_{xx}(e^{j\omega})|F(e^{j\omega})|^2 d\omega \tag{7.63}$$

where $\Phi_{xx}(e^{j\omega})$ is the power spectral density of $x(n)$. Now, if $F(e^{j\omega})$ is the transfer function of a filter whose coefficients constitute the elements of a row of a transformation matrix $\mathcal{T}$, with $\mathcal{T}\mathcal{T}^T = \mathbf{I}$, then $F(e^{j\omega})$ is constrained to satisfy the following identity

$$\frac{1}{2\pi}\int_0^{2\pi} |F(e^{j\omega})|^2 d\omega = 1 \tag{7.64}$$

This follows from the Parseval's relation (Chapter 2, Section 2.2). Noting this, we may say that the diagonal elements of $\mathbf{R}_{\mathcal{T}}$ (i.e., the signal powers at the outputs of the FIR filters defined by the rows of $\mathcal{T}$) are a set of averaged values of the power spectral density function, $\Phi_{xx}(e^{j\omega})$, of the underlying input process. The weighting functions used to obtain these averages are the squared magnitude responses of the FIR filters associated with the various rows of $\mathcal{T}$.

The numerical example that was given in Section 7.3 shows that the DCT is very effective in decorrelating the samples of the input process, $x(n)$, which was considered there. A closer look at this particular example is very instructive. Figure 7.5 shows the power spectral density, $\Phi_{xx}(e^{j\omega})$, of the underlying input process, $x(n)$. The main characteristic of this process to be noted here is that it is of lowpass nature, that is, most of

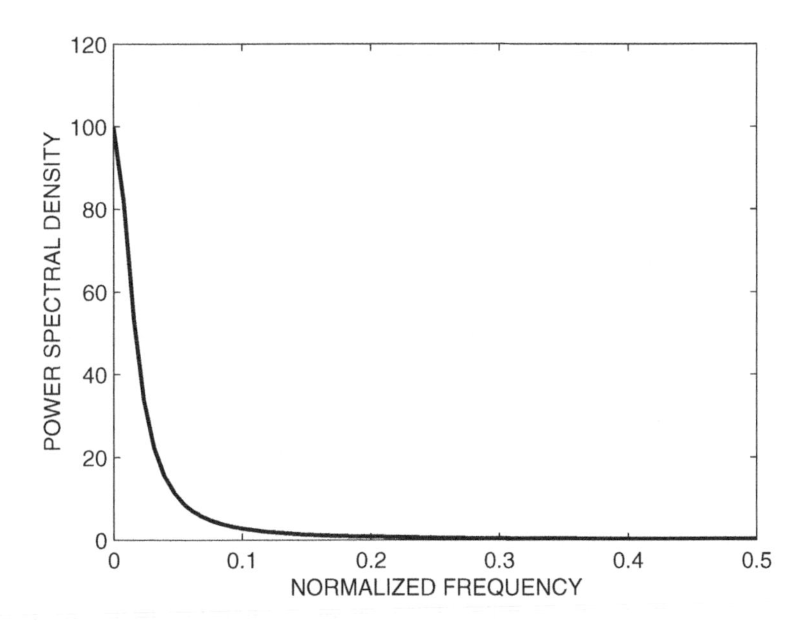

Figure 7.5 Power spectral density of the process $x(n)$ that is generated by the coloring filter Eq.(7.17).

its spectral energy is concentrated over low frequencies. We also refer to Figure 7.2 where the magnitude responses of the DCT filters are shown, for $N = 8$, and note the following features. The side lobes of the filters, whose passbands are over higher frequencies (closer to 0.5), are smaller than the side lobes of the filters whose passbands are over lower frequencies (close to zero). This, as we show next, is a very special characteristic of the DCT, which makes it an effective transform when it is applied for decorrelating the samples of a process that is dominantly lowpass in nature. To see this, we refer to Eq. (7.63) and note that when the main (passband) lobe of $F(e^{j\omega})$ lies in the frequency bands where $\Phi_{xx}(e^{j\omega})$ is large (relative to its values in other frequency bands), the value of $E[y^2(n)]$ is not much affected by the size of the side lobes of $F(e^{j\omega})$. On the other hand, when the main lobe of $F(e^{j\omega})$ lies in frequency bands where $\Phi_{xx}(e^{j\omega})$ is relatively small, the value of $E[y^2(n)]$ may be significantly affected by the side lobes of $F(e^{j\omega})$ as these side lobes, although small, are multiplied by some large values of $\Phi_{xx}(e^{j\omega})$ before integration. In the context of orthogonal transforms and signal power spreading, the minimization of the side lobes of $F(e^{j\omega})$ in the latter case to reduce the value of $E[y^2(n)]$ is very critical. Referring back to the DCT filters and the size of their associated side lobes, we find that the DCT has the necessary properties to be effective in achieving a close to maximum signal power spreading when applied to any lowpass signal.

To get further insight on the above results, we consider two more examples. We consider two choices of the inputs, $x_1(n)$ and $x_2(n)$, that are generated by passing a unit variance white noise process through two coloring filters, which are specified by the system functions

$$H_1(z) = 0.1 + 0.2z^{-1} + 0.3z^{-2} + 0.4z^{-3} + 0.4z^{-4} + 0.2z^{-5} + 0.1z^{-6}$$

and

$$H_2(z) = 0.1 - 0.2z^{-1} - 0.3z^{-2} + 0.4z^{-3} + 0.4z^{-4} - 0.2z^{-5} - 0.1z^{-6}$$

respectively. Figures 7.6 and 7.7 show the power spectral densities of $x_1(n)$ and $x_2(n)$. We note that $x_1(n)$ and $x_2(n)$ are low- and band-pass processes, respectively.

We also consider two choices of $\mathcal{T}$:

1. The DCT matrix whose coefficients are specified by Eq. (7.14).
2. The discrete sine transform (DST) that is specified by the coefficients

$$s_{kl} = \left(\frac{2}{N+1}\right)^{1/2} \sin \frac{kl\pi}{N+1} \qquad k, l = 1, 2, \ldots, N \tag{7.65}$$

We expect the DCT to perform well when applied to $x_1(n)$ as this is a lowpass process.

Figure 7.8 shows the magnitude responses of the DST filters for $N = 8$. For the DST, we observe that the side lobes of the filters whose passbands belong to high or low frequencies are relatively smaller than the side lobes of the filters whose passbands are within the midband frequencies. Thus, according to our discussion above, we expect DST to perform well when applied to $x_2(n)$.

Table 7.2 shows the results of some numerical calculations that have been performed to observe the effect of the two transformations in decorrelating the samples of $x_1(n)$ and $x_2(n)$. These results compare the eigenvalue spread of $\mathbf{R}$ and $\mathbf{R}_{\mathcal{T}}^{\mathrm{n}}$ of the respective

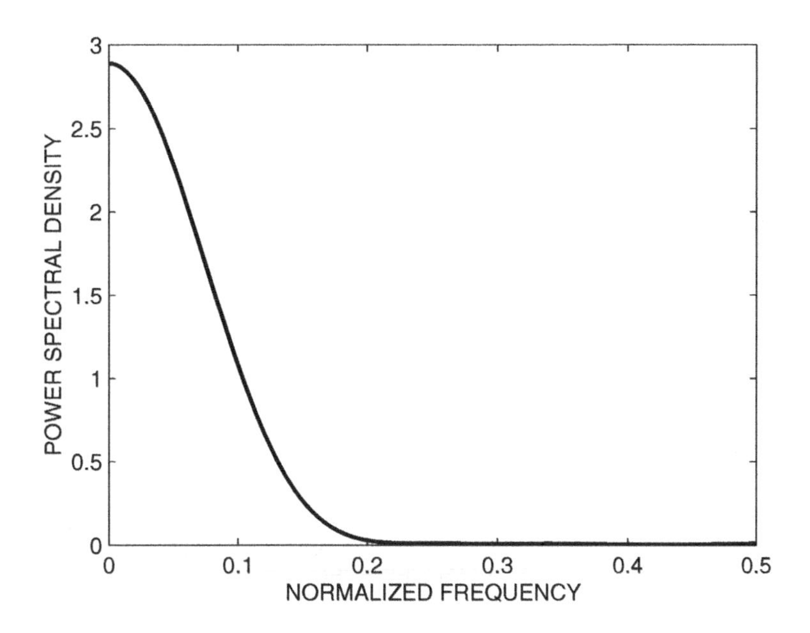

Figure 7.6 Power spectral density of the process $x_1(n)$.

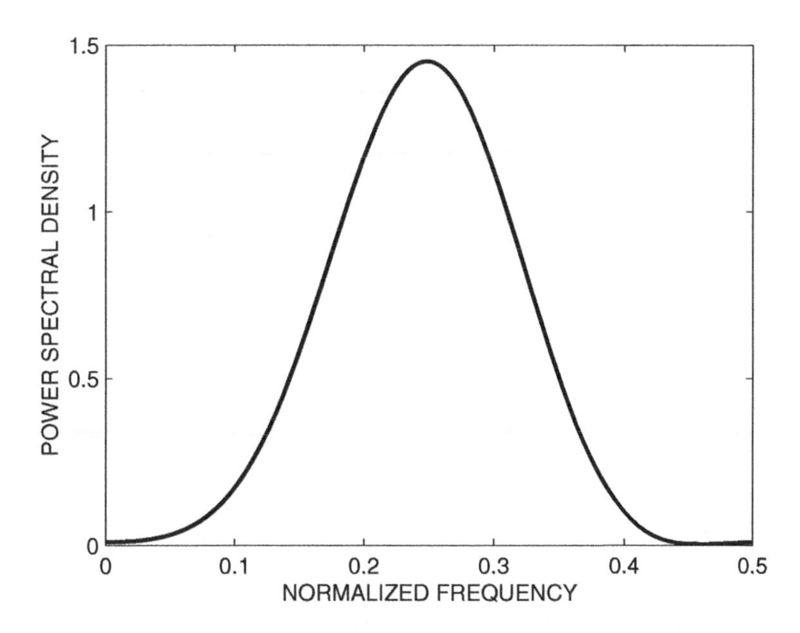

Figure 7.7 Power spectral density of the process $x_2(n)$.

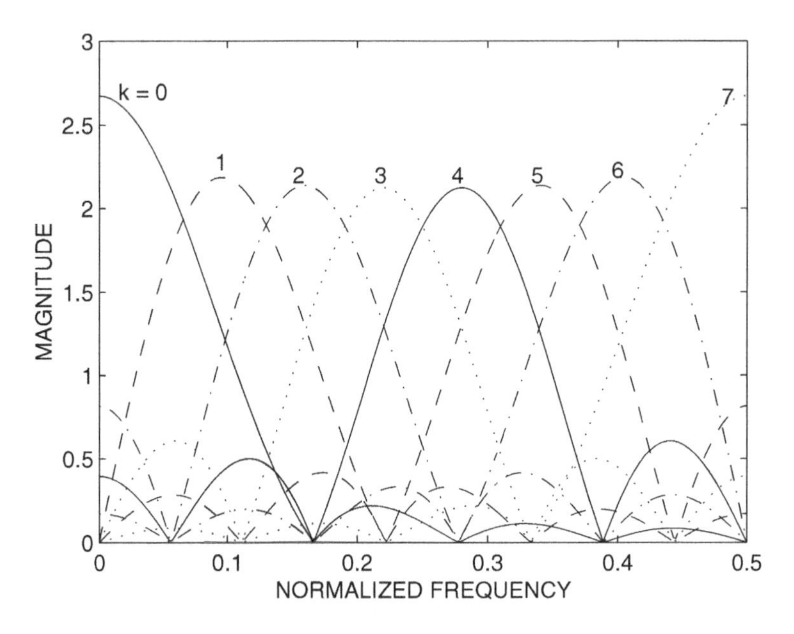

Figure 7.8 The magnitude responses of the DST filters for $N = 8$.

Table 7.2 Comparison of the DST and DCT transformations when applied to lowpass process $x_1(n)$ and bandpass process $x_2(n)$.

Process	N	$\lambda_{max}/\lambda_{min}$			I_ρ		
		$\mathbf{R}$	$\mathbf{R}^n_{DCT}$	$\mathbf{R}^n_{DST}$	DCT	DST	max
$x_1(n)$	8	375.35	3.01	14.19	15.12	12.10	15.75
	20	781.62	3.52	18.15	43.47	37.55	44.66
	30	945.38	3.81	18.18	67.41	60.06	68.79
$x_2(n)$	8	50.69	5.97	2.93	3.86	4.75	5.28
	20	184.74	11.78	3.49	15.44	17.86	18.84
	30	253.42	12.41	3.82	25.82	29.08	30.32

processes for three values of filter length, N. Also, to illustrate that the improvement factor, I_ρ, and variation of eigenvalue spread of the respective matrices are tightly related, values of I_ρ are also presented in Table 7.2. As it was predicted, the DCT performs better for $x_1(n)$, and the DST performs better for $x_2(n)$.

Reviewing the above observations, the following guidelines may be drawn for the selection of the transformation $\mathcal{T}$:

In general, transforms whose associated band-partitioning filters have smaller side lobes are expected to perform better than those with larger side lobes. When an estimate of the power *spectral density, $\Phi_{xx}(e^{j\omega})$, of the underlying input process, $x(n)$, is known, selection of those transforms whose associated filters have smaller side lobes within the frequency bands where $\Phi_{xx}(e^{j\omega})$ is large, leads to a more significant performance improvement.*

7.7 Transforms

Although, in general, there are infinite possible choices of the transformation matrix $\mathcal{T}$, only a few transforms have been widely used in practice. The main feature of such transforms is that there are many fast algorithms for their efficient implementation. They also exhibit a good signal separation, that is, from the band-partitioning point of view, they all offer well-behaved sets of parallel FIR filters with approximately mutually exclusive passbands. In the application of TDAFs, the most commonly used transforms are:

1. DFT. The DFT is the most widely used transform in various applications of signal processing. The klth element of the DFT transformation matrix, $\mathcal{T}_{\text{DFT}}$, is

$$f_{kl} = \frac{1}{\sqrt{N}} e^{-j2\pi kl/N}, \qquad \text{for } 0 \leq k, l \leq N-1 \tag{7.66}$$

The factor $\frac{1}{\sqrt{N}}$ on the right-hand side of Eq. (7.66) is to normalize the DFT coefficients so that $\mathcal{T}_{\text{DFT}} \mathcal{T}^{\text{T}}_{\text{DFT}} = \mathbf{I}$.

The distinct feature of DFT, compared to other transforms, is that it distinguishes between positive and negative frequencies. This, among all the widely used transforms, makes DFT the most effective transform in cases where the underlying input process has a nonsymmetrical power spectral density with respect to $\omega = 0$, that is, for complex-valued inputs. If the input is real-valued, then DFT has no advantage over the other transforms, In fact, its complex-valued coefficients add some unnecessary redundancy to the transformed signal samples, which increases the complexity of the system.

2. Real DFT (RDFT). When N is even, the coefficients of RDFT are given by

$$f^{R}_{kl} = \begin{cases} \frac{1}{\sqrt{N}}, & k = 0,\ 0 \leq l \leq N-1 \\ \sqrt{\frac{2}{N}} \cos \frac{2\pi kl}{N}, & 1 \leq k \leq \frac{1}{2}N - 1,\ 0 \leq l \leq N-1 \\ \frac{1}{\sqrt{N}}(-1)^{l}, & k = \frac{1}{2}N,\ 0 \leq l \leq N-1 \\ \sqrt{\frac{2}{N}} \sin \frac{2\pi kl}{N}, & \frac{1}{2}N + 1 \leq k \leq N-1,\ 0 \leq l \leq N-1 \end{cases} \tag{7.67}$$

3. Discrete Hartley transform (DHT). The DHT coefficients are defined as

$$h_{kl} = \frac{1}{\sqrt{N}} \left(\cos \frac{2\pi kl}{N} + \sin \frac{2\pi kl}{N} \right), \qquad \text{for } 0 \leq k, l \leq N-1 \tag{7.68}$$

Both RDFT and DHT may be viewed as derivatives of the DFT, which for real-valued signals exploit the redundancy of the transformed samples and suggest a lower complexity implementation of TDAFs. Experiments on TDAFs with DFT, RDFT, and DHT show that they all perform the same when the underlying input process is real-valued.

4. DCT. There are a few variations of DCT (Ersoy, 1997). However, the most widely used DCT is the one defined in Eq. (7.14).
5. DST. Similar to DCT, there are also a few variations of DST (Ersoy, 1997). However, the most widely used DST is the one defined in Eq. (7.65).
6. Walsh–Hadamard transform (WHT). The WHT is defined when the transformation length, N, is a power of 2. The WHT coefficients are

$$w_{kl} = \frac{1}{\sqrt{N}} \prod_{p=0}^{m-1} (-1)^{b_p(k) b_{m-1-p}(l)}, \qquad \text{for } 0 \le k, l \le N-1 \tag{7.69}$$

where $m = \log_2 N$, and $b_p(k)$ is the pth bit (with $p = 0$ referring to the least significant bit) of the binary representation of k.

The main characteristic of the WHT is its simplicity as all of its coefficients are $+1$ or -1 and, as a result, its implementation does not involve any multiplication. We note that in the implementation of TDLMS algorithm, the common coefficient $1/\sqrt{N}$, which is just a normalization factor, can be dropped as the step-size normalization of the TDLMS algorithm takes care of signal normalization. The price paid for this simplicity of the WHT is its higher side lobes compared to other transforms. This, of course, results in poorer performance of the WHT when applied to TDAFs, in general.

7.8 Sliding Transforms

The conventional fast algorithms available for the implementation of the transforms introduced in the previous section require $O(N \log N)$ operations (additions, subtractions, or multiplications), where $O(\cdot)$ denotes *order of* and the term *order of* x means a value proportional to x with a fixed proportionality constant. In the context of transversal filters and their corresponding transform domain implementation, there is an important property of the filter tap-input vector, $\mathbf{x}(n)$, that can be used to reduce the complexity of the latter transforms further. Namely, when $\mathbf{x}(n) = [x(n)\ x(n-1)\ \cdots\ x(n-N+1)]^{\mathrm{T}}$, $\mathbf{x}(n)$ and $\mathbf{x}(n+1)$ have $N-1$ elements in common. $\mathbf{x}(n+1)$ is obtained from $\mathbf{x}(n)$ by shifting (sliding) the elements of $\mathbf{x}(n)$ one element down, dropping out the last element of $\mathbf{x}(n)$, and adding the new sample of input, $x(n+1)$, as the first element of $\mathbf{x}(n+1)$. In this section, we exploit this data redundancy in the successive tap-input vectors $\mathbf{x}(n)$ and $\mathbf{x}(n+1)$ and introduce two $O(N)$ complexity schemes for efficient implementation of the transformation part of TDAFs. These are called *sliding transforms*.

7.8.1 Frequency Sampling Filters

A useful common property of the transforms which were introduced in the last section (with the exception of the WHT) is that the transfer functions of their corresponding FIR filters can be written in a compact recursive form. These transfer functions can then be used for efficient implementation of the respective transforms. To clarify this, we consider the DFT filters as an example.

The transfer function of the kth DFT filter is

$$H_{\text{DFT}}^{kn}(z) = \sum_{l=0}^{N-1} f_{kl} z^{-l} \tag{7.70}$$

The superscript n in Eq. (7.70) emphasizes that the coefficients f_{kl}'s have been normalized so that $\sum_{l=0}^{N-1} |f_{kl}|^2 = 1$.

Substituting Eq. (7.66) in Eq. (7.70), we obtain

$$\begin{aligned} H_{\text{DFT}}^{kn}(z) &= \frac{1}{\sqrt{N}} \sum_{l=0}^{N-1} \mathrm{e}^{-j2\pi kl/N} z^{-l} \\ &= \frac{1}{\sqrt{N}} \frac{1 - (\mathrm{e}^{-j2\pi k/N} z^{-1})^N}{1 - \mathrm{e}^{-j2\pi k/N} z^{-1}} \\ &= \frac{1}{\sqrt{N}} \frac{1 - z^{-N}}{1 - \mathrm{e}^{-j2\pi k/N} z^{-1}} \end{aligned} \tag{7.71}$$

When the TDLMS algorithm is used to adapt a DFT-based TDAF, the constant factor $1/\sqrt{N}$ may be dropped from the right-hand side of Eq. (7.71) as signal normalization is taken care of by the step-normalization in TDLMS algorithm, as discussed in the earlier sections. Thus, the (unnormalized) transfer function of the kth DFT filter may be defined as

$$H_{\text{DFT}}^{k}(z) = \frac{1 - z^{-N}}{1 - \mathrm{e}^{-j2k\pi/N} z^{-1}} \tag{7.72}$$

The transfer functions associated with other transforms can also be derived in a similar way. For transforms with real-valued coefficients, one has to start with expanding the sine and cosine coefficients in terms of their associated complex exponents, then proceed as in the case of DFT filters and pack the results. At the end, any fixed scale factor in front of the final results is dropped.

Table 7.3 gives a summary of the transfer functions that are associated with various transforms. We have not included WHT here as its transfer functions do not have any closed-form equivalent. Hence, a different approach has to be adopted to arrive at an efficient implementation of the WHT. This is discussed in Problem P7.19.

The term *frequency sampling filter* is used to refer to the filters defined by the transfer functions given in Table 7.3. This is because each transfer function corresponds to a narrow-band filter which samples a small band of the spectrum of the underlying input process.

Table 7.3 provides all the necessary information for the development of the two realizations of the sliding transforms which are categorized as recursive and nonrecursive structures, and are discussed below.

7.8.2 *Recursive Realization of Sliding Transforms*

A direct realization of the transfer functions given in Table 7.3 suggests a simple recursive scheme for the implementation of the associated transforms. As an example, we present

Table 7.3 Transfer functions associated with the various transforms (frequency sampling filters).

$$H_{\mathrm{DFT}}^{k}(z)=\frac{1-z^{-N}}{1-\mathrm{e}^{-j2\pi k/N}z^{-1}}$$

$$H_{\mathrm{RDFT}}^{k}(z)=\begin{cases}\frac{1-z^{-N}}{1-z^{-1}}, & \text{for } k=0\\ \frac{\left(1-\cos\frac{2\pi k}{N}z^{-1}\right)(1-z^{-N})}{1-2\cos\frac{2\pi k}{N}z^{-1}+z^{-2}}, & \text{for } 1\le k\le \frac{1}{2}N-1\\ \frac{1-z^{-N}}{1+z^{-1}}, & \text{for } k=\frac{1}{2}N\\ \frac{z^{-1}(1-z^{-N})}{1-2\cos\frac{2\pi k}{N}z^{-1}+z^{-2}}, & \text{for } \frac{1}{2}N+1\le k\le N-1\end{cases}$$

$$H_{\mathrm{DHT}}^{k}(z)=\frac{\left(1-\left(\cos\frac{2\pi k}{N}-\sin\frac{2\pi k}{N}\right)z^{-1}\right)(1-z^{-N})}{1-2\cos\frac{2\pi k}{N}z^{-1}+z^{-2}}$$

$$H_{\mathrm{DCT}}^{k}(z)=\frac{(1-z^{-1})(1-(-1)^{k}z^{-N})}{1-2\cos\frac{\pi k}{N}z^{-1}+z^{-2}}$$

$$H_{\mathrm{DST}}^{k}(z)=\frac{1+(-1)^{k}z^{-(N+1)}}{1-2\cos\frac{\pi(k+1)}{N+1}z^{-1}+z^{-2}}$$

here a recursive realization of the DCT filters. Recursive realization of the other transforms which follow the same concept is then straightforward.

From Table 7.3, we have

$$H_{\mathrm{DCT}}^{k}(z)=\frac{(1-z^{-1})(1-(-1)^{k}z^{-N})}{1-2\cos\frac{\pi k}{N}z^{-1}+z^{-2}} \tag{7.73}$$

This is the transfer function of the kth DCT filter. Figure 7.9 depicts a detailed realization of Eq. (7.73). In this realization, we have purposefully divided the transfer function of $H_{\mathrm{DCT}}^{k}(z)$ into three separate parts. Namely, the forward parts, $1-z^{-1}$ and $1-(-1)^{k}z^{-N}$, and the feedback part, $\frac{1}{1-2\cos\pi k/Nz^{-1}+z^{-2}}$. This separation facilitates the integration of the DCT filters (for $k=0, 1, \ldots, N-1$) in a parallel structure.

Figure 7.10 depicts a block diagram of the DCT frequency sampling filters when they are put together in a parallel structure. Points to be noted here are:

1. For $k=0$,

$$\frac{1}{1-2\cos\frac{\pi k}{N}z^{-1}+z^{-2}}=\frac{1}{(1-z^{-1})^{2}}$$

Substituting this result in Eq. (7.73), we obtain

$$H_{\mathrm{DCT}}^{0}(z)=\frac{1-z^{-N}}{1-z^{-1}} \tag{7.74}$$

This has been considered in the block diagram of Figure 7.10 and, thus, the case $k=0$ has been treated separately.

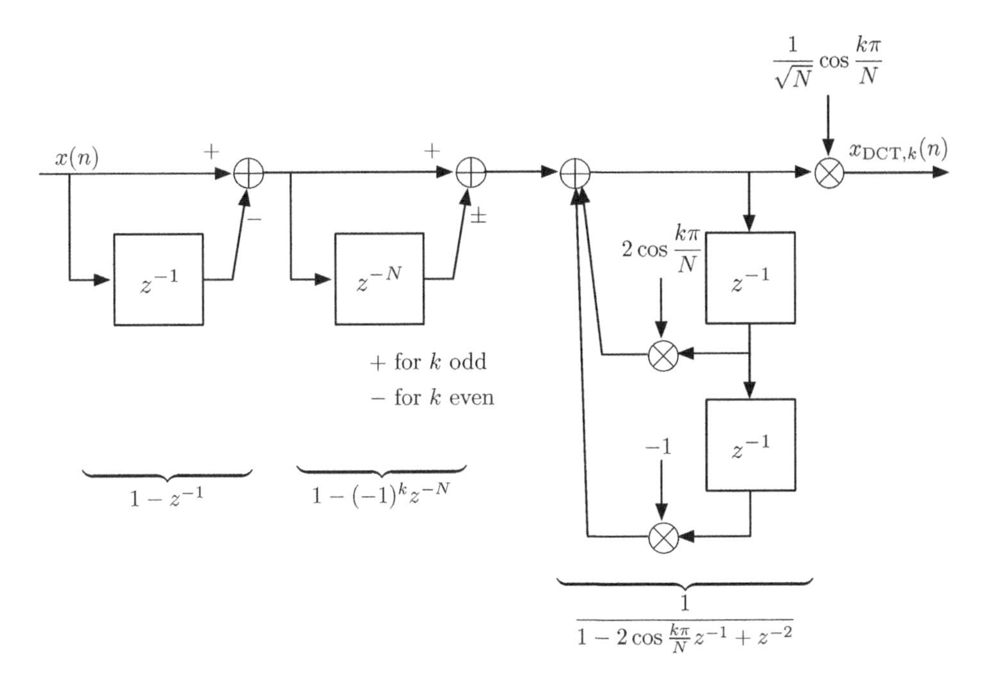

Figure 7.9 A realization of $H^k_{\text{DCT}}(z)$.

2. We also note that

$$1 - (-1)^k z^{-N} = \begin{cases} 1 - z^{-N}, & \text{for } k \text{ even} \\ 1 + z^{-N}, & \text{for } k \text{ odd} \end{cases}$$

Thus, the cases of k even and odd are separated at the first stage of Figure 7.10. However, when implementing the structure of Figure 7.10, one should note that the blocks $1 - z^{-N}$ and $1 + z^{-N}$ have the same common input and, thus, can share the same delay line to hold the past samples of the input. This reduces the memory requirement of the system.

A common problem with the recursive realization of the frequency sampling filters that needs careful attention is that these filters are only marginally stable. They can easily run into instability problems, unless some special care is taken to ensure stability. This is because the poles of the frequency sampling filters are all on the unit circle and, as a result, any round-off error will accumulate and grow unbounded. Furthermore, quantization of the filter coefficients may result in poles outside the unit circle and thus result in unstable filters.

The above problem can be alleviated by replacing z^{-1} with βz^{-1}, where β is a constant smaller than, but close to, 1. This shifts all the poles and zeros of the frequency sampling filters, which are ideally on the unit circle to a circle with radius $\beta < 1$. This stabilizes the filters at the cost of some additional complexity in their realization as addition of β changes some of the filter coefficients which otherwise would have been unity.

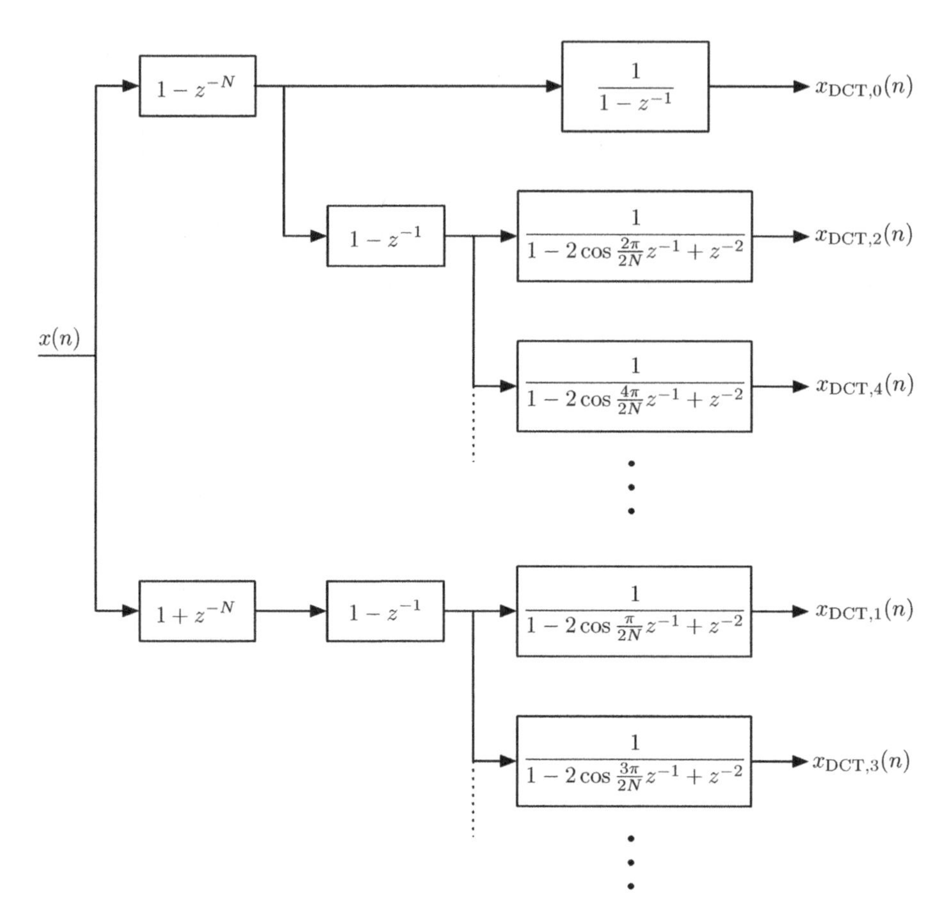

Figure 7.10 A parallel realization of the recursive DCT frequency sampling filters.

7.8.3 Nonrecursive Realization of Sliding Transforms

The nonrecursive sliding transforms, which are introduced in this section, use the following common property of the frequency sampling filters:

> *The frequency sampling filters associated with each transform have a common set of zeros out of which each filter selects* $N - 1$

Bruun (1978) noted the significance of the above property in the case of DFT and used that to develop a fast Fourier transform (FFT) structure. Farhang-Boroujeny *et al.* (1996) noted that a rearrangement of the Bruun's algorithm leads to a sliding DFT structure and extended the concept to the other transforms. In the rest of this section, we present the sliding transforms that have been proposed in (Farhang-Boroujeny *et al.* 1996) and demonstrate their efficiency in the implementation of TDAFs.

Bruun's Algorithm as Sliding DFT

The transfer functions of the DFT frequency sampling filters are (from Table 7.3):

$$H_{\mathrm{DFT}}^{k}(z) = \frac{1 - z^{-N}}{1 - \mathrm{e}^{-j2\pi k/N}z^{-1}}, \quad \text{for } k = 0, 1, \ldots, N-1 \tag{7.75}$$

We note that the zeros of these filters are all taken from the set of Nth roots of unity, that is, $\mathrm{e}^{-j2\pi k/N}$, for $k = 0, 1, \ldots, N-1$. We also note that each DFT filter has one pole that belongs to the same set. As a result, we find that a pole-zero cancellation occurs and, thus, each DFT filter has effectively $N-1$ zeros out of the set of Nth roots of unity and no pole.

Bruun used this simple concept and suggested an elegant factorization of $1 - z^{-N}$ and used these results to form a tree structure, as shown in Figure 7.11 (for $N = 16$), to realize the various FIR frequency sampling filters of DFT. The following identities are used for the factorization of $1 - z^{-N}$:

$$1 - z^{-2M} = (1 - z^{-M})(1 + z^{M}) \tag{7.76}$$

and

$$1 + az^{-2M} + z^{-4M} = (1 + \sqrt{2-a}z^{-M} + z^{-2M})(1 - \sqrt{2-a}z^{-M} + z^{-2M}) \tag{7.77}$$

These factorizations, which are used until the last stage of the tree structure, have the following two features:

1. Each factor consists of either two or three sparse taps.
2. There is at most one nontrivial real-valued coefficient in each factor.

To see how the above identities could be used to develop the tree structure of Figure 7.11, we note that the factors which appear in the first stage are those of $1 - z^{-16} = (1 - z^{-8})(1 + z^{-8})$. The branches that follow after the factor $1 + z^{-8}$ are made of the factors of the other branch of the first stage, that is, $1 - z^{-8} = (1 - z^{-4})(1 + z^{-4})$. Similarly, the branches that follow the factor $1 - z^{-8}$ are made of the factors of $1 + z^{-8} = (1 + \sqrt{2}z^{-2} + z^{-4})(1 - \sqrt{2}z^{-2} + z^{-4})$. The same procedure is used to determine the other branches of the structure. At the end of the third stage (in our particular example), each path of the tree covers 14 out of the 16 zeros of $1 - z^{-16}$. The remaining two zeros that have not been covered by each path are complex conjugates, except for the top path whose corresponding missing zeros are $z = \pm 1$. One out of the two missing zeros is, then, added at the last stage. The same procedure can be used to develop the same structure for any value of N (the transform length), which is a power of 2.

Bruun (1978) elaborated on the tree structure of Figure 7.11 and proposed his FFT structure. In the context of the TDLMS algorithm, we are interested in an efficient implementation of the DFT frequency sampling filters and updating their outputs after the arrival of every new data sample. The tree structure of Figure 7.11 is exactly what we are looking for. Thus, we hold on to this structure as an efficient way of implementing the nonrecursive sliding DFT filters.

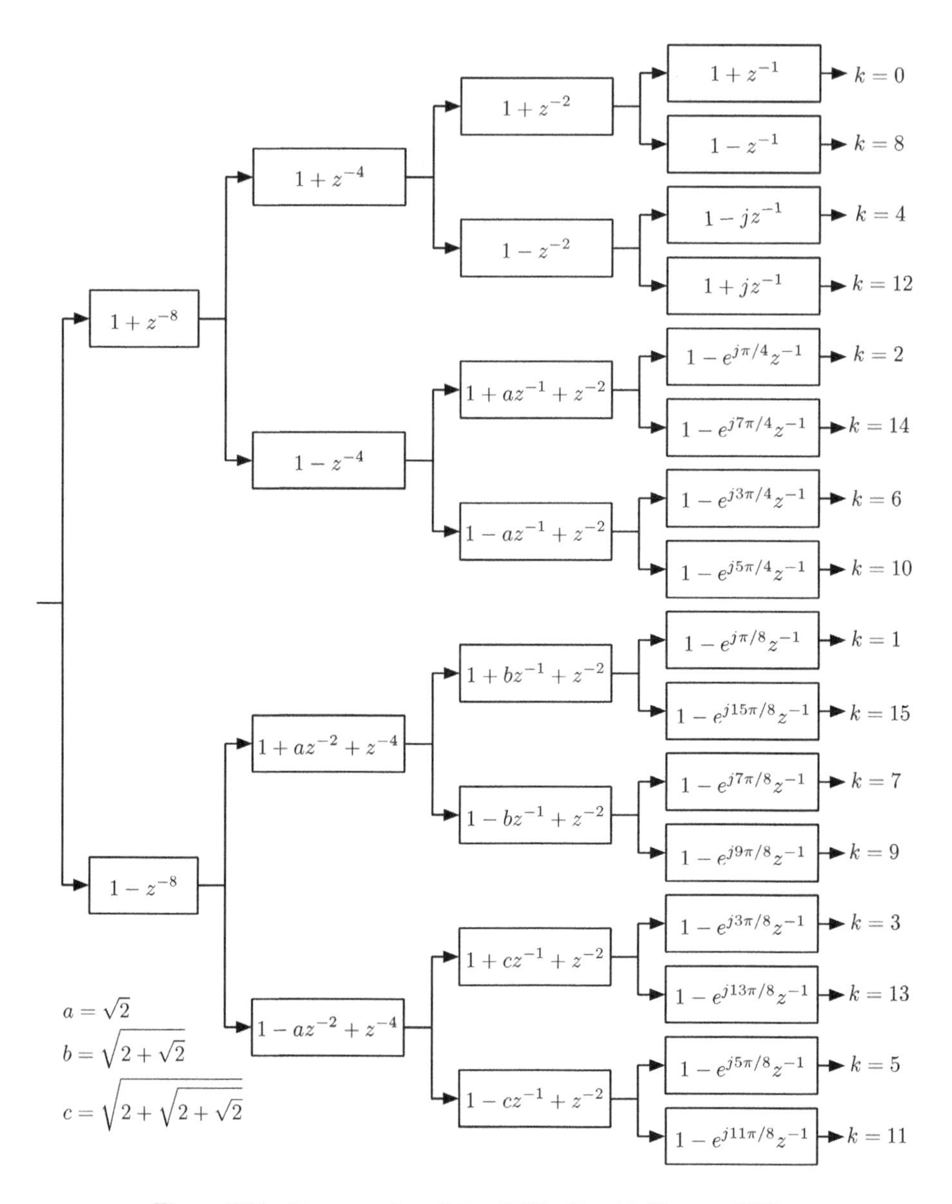

Figure 7.11 Nonrecursive sliding DFT: $N = 16$ (Bruun, 1978).

To appreciate the efficiency of the structure given in Figure 7.11, we shall elaborate on it further. We note that the pair of filters that originate from a common node at any stage share the same coefficients and, thus, they can be implemented jointly, as depicted in Figure 7.12. For a real-valued sequence, this implementation requires only one multiplication and three additions. For a complex-valued input, the number of operations is twice this figure. We may also note that each filter pair at the output stage in Figure 7.11 uses a pair of complex conjugate coefficients, and therefore, the corresponding multiplications

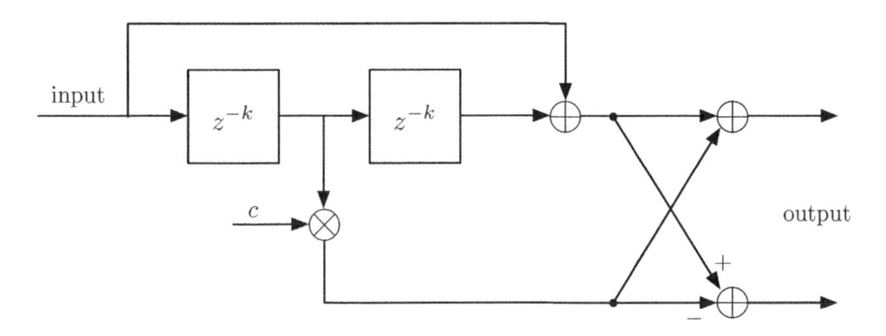

Figure 7.12 An implementation of the filter pair $1 \pm cz^{-k} + z^{-2k}$, when they share a common input.

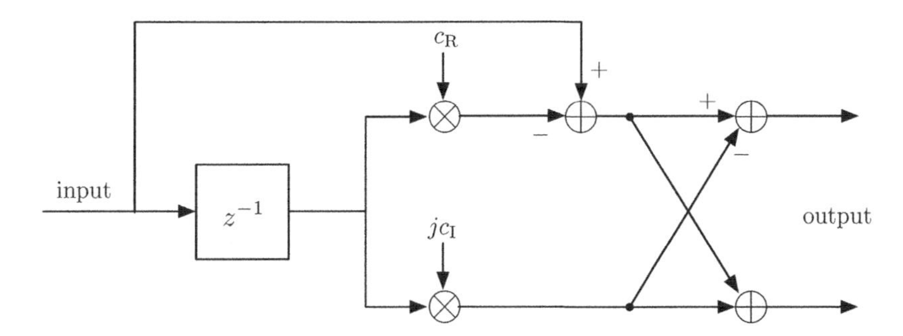

Figure 7.13 An implementation of the filter pair $(1 - cz^{-1}, 1 - c^*z^{-1})$, when they share a common input.

can be shared. Figure 7.13 depicts a joint implementation of a filter pair of the output stage of Figure 7.11. In this implementation, c_R and c_I denote the real and imaginary parts of c, respectively, where c and c^* are the pair of filter coefficients. For a complex-valued input, this implementation requires four real multiplications and six real additions.

Real-Valued Transforms

As was noted before, when the filter input is real-valued, about 50% of the DFT outputs are redundant as they appear in complex conjugate pairs. In such situations, transforms with real-valued coefficients are preferred. Following Bruun's factorization technique, it is not difficult to come up with tree structures similar to ones in Figure 7.11 for other transforms. Figures 7.14–7.17 show a set of such tree structures for nonrecursive sliding RDFT, DHT, DCT, and DST, respectively. Note that for the examples shown, value of N is 16 for RDFT, DHT, and DCT, and 15 in the case of DST. Further details on these structures, along with some efficient programming techniques for their software implementations, can be found in (Farhang-Boroujeny *et al.* 1996).

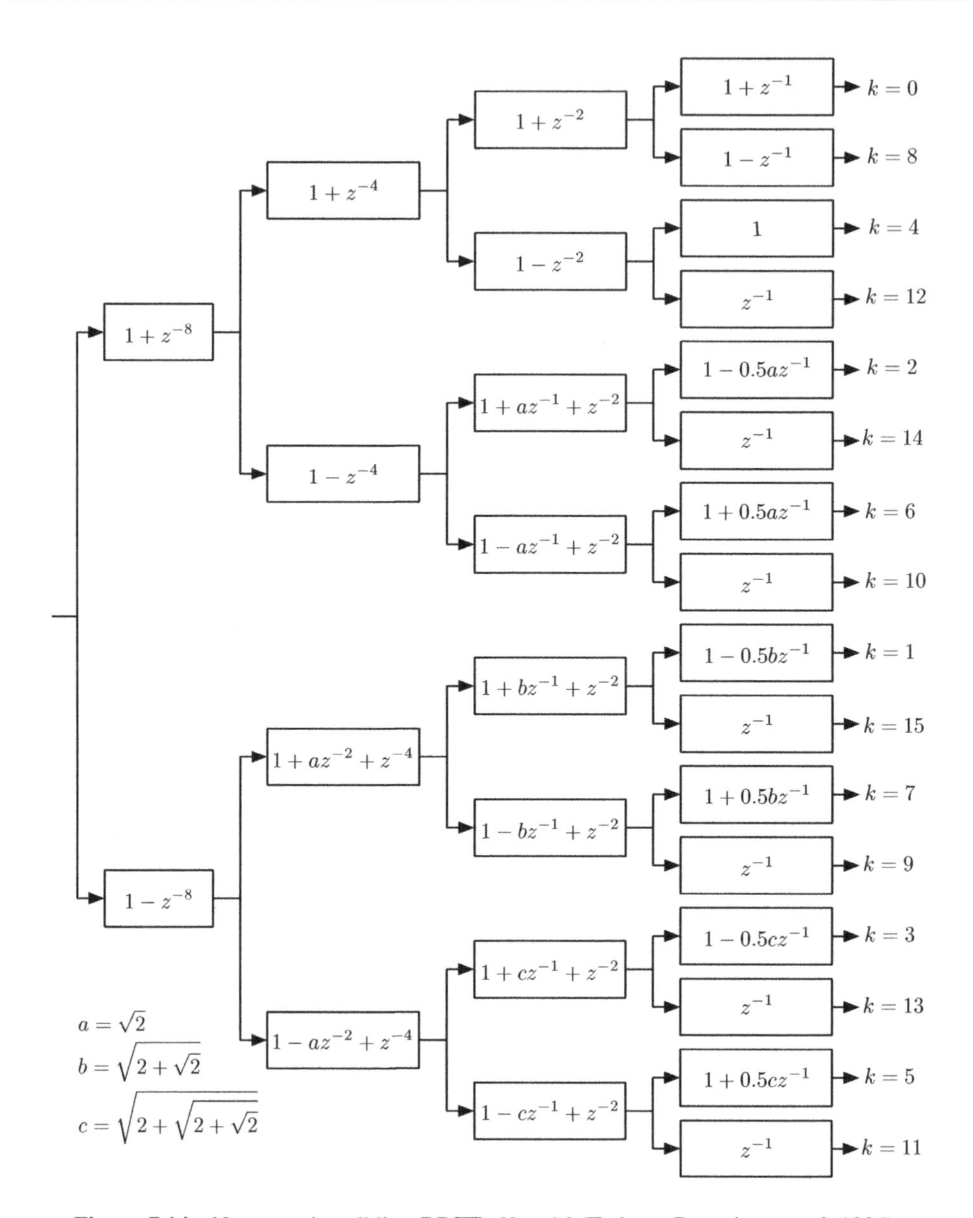

Figure 7.14 Nonrecursive sliding RDFT: $N = 16$ (Farhang-Boroujeny *et al.* 1996).

7.8.4 Comparison of Recursive and Nonrecursive Sliding Transforms

In terms of robustness to numerical round-off errors, the nonrecursive sliding transforms are superior to their recursive counterparts. A simple inspection of the nonrecursive sliding structures shows that each output in these structures is calculated based on a very limited number of multiplications and additions. Furthermore, there is no feedback of numerical errors, thereby avoiding error accumulation. This property, which is inherent to all

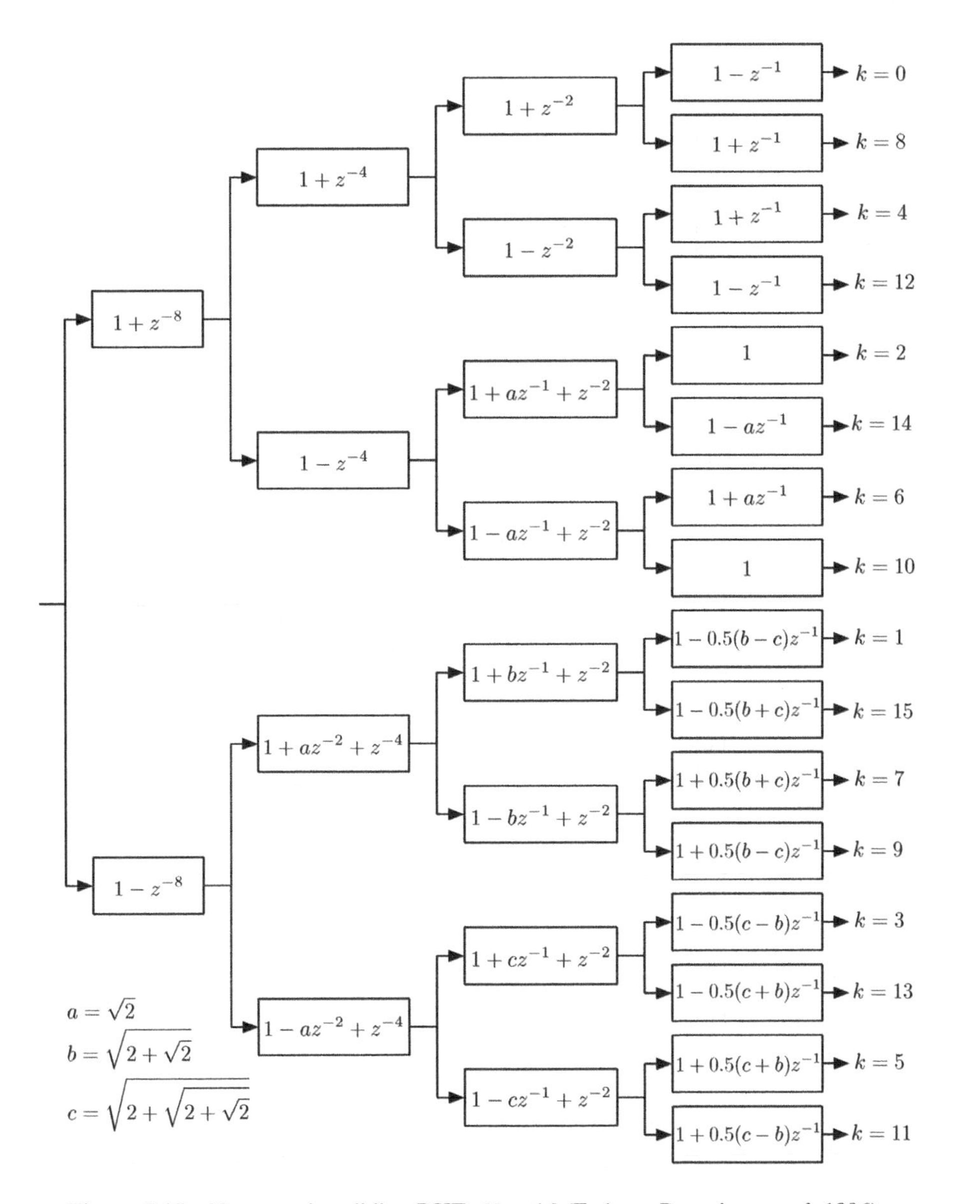

Figure 7.15 Nonrecursive sliding DHT: $N = 16$ (Farhang-Boroujeny *et al.* 1996).

FFT-like structures, results in very low sensitivity to finite wordlength effects (Rabiner and Gold, 1975; Oppenheim and Schafer, 1975). On the contrary, the recursive sliding transforms are highly sensitive to numerical error accumulation, because of the feedback. The variances of such errors are proportional to $\frac{1}{1-\beta^2}$, where β is the stabilizing factor as defined before. Noting that β has to be selected close to 1 so that the deviation of the realized filters from the ideal frequency sampling filters would be minimum, these variances can be excessively large.

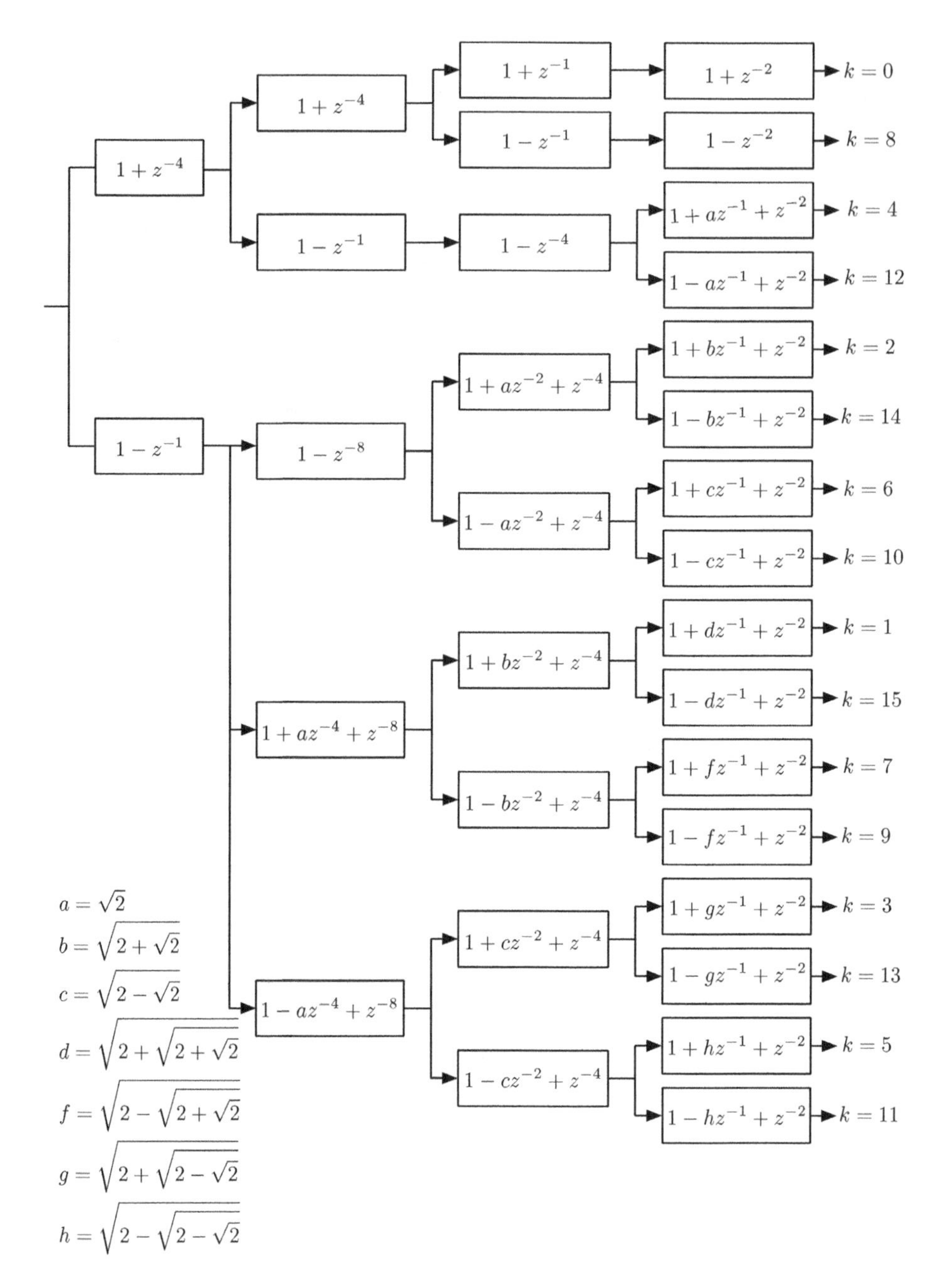

Figure 7.16 Nonrecursive sliding DCT: $N = 16$ (Farhang-Boroujeny *et al.* 1996).

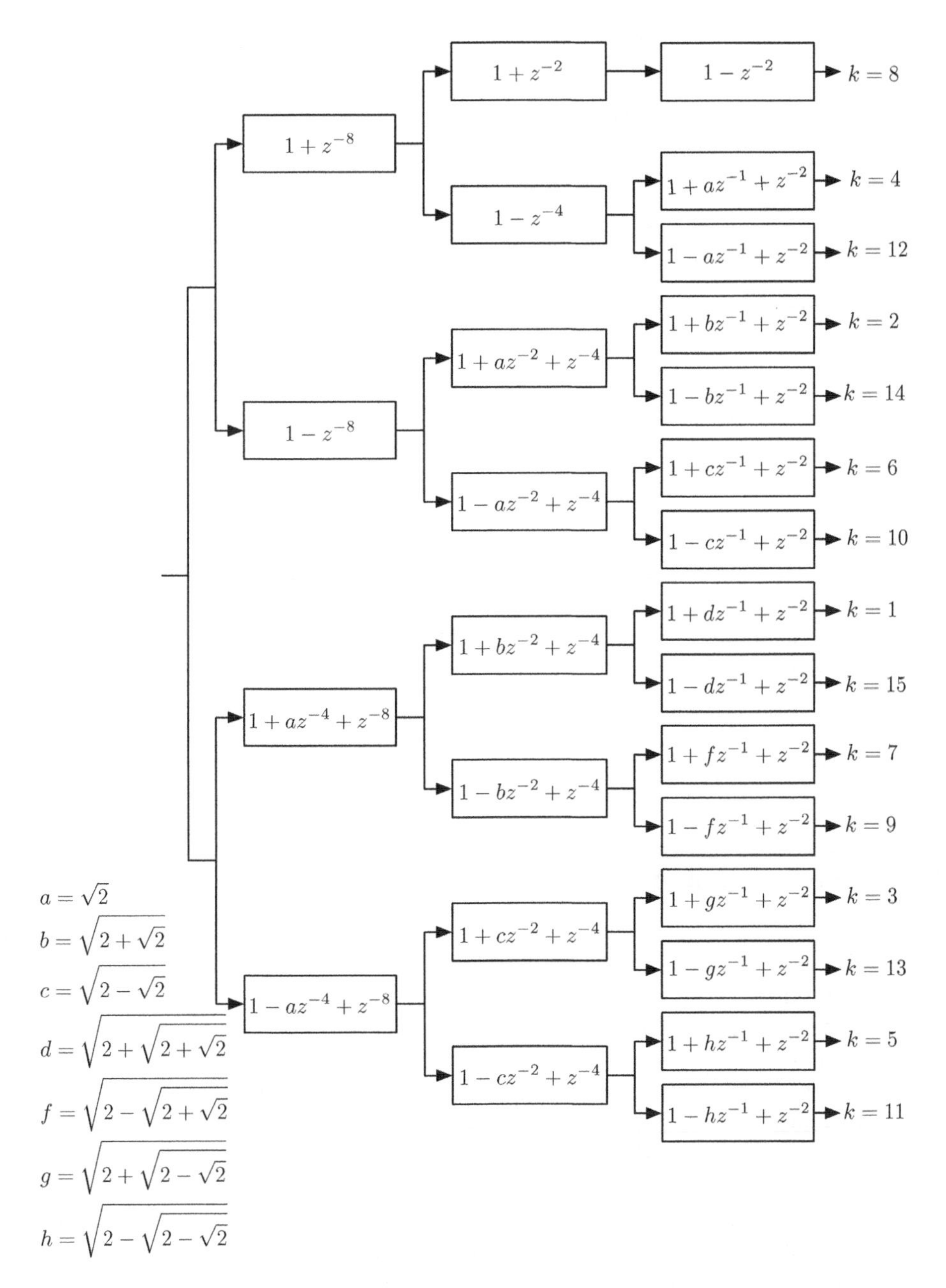

Figure 7.17 Nonrecursive sliding DST: $N = 15$ (Farhang-Boroujeny *et al.*, 1996).

Table 7.4 Computation counts of the nonrecursive and recursive sliding transforms

	Nonrecursive		Recursive	
	Mults	Adds/Subs	Mults	Adds/Subs
DFT	$3N - 2m - 8$	$6N - 2m - 8$	$4N - 6$	$4N - 6$
RDFT	$N - m - 2$	$2N - m - 2$	$\frac{5N}{2} - 5$	$\frac{5N}{2} - 5$
DHT	$\frac{3N}{2} - m - 6$	$\frac{5N}{2} - m - 4$	$3N - 7$	$3N - 7$
DST	$N - m$	$3N - m - 3$	$2N$	$2N + 1$
DCT	$N - m - 1$	$3N - 5$	$2N + 1$	$2N + 2$

$m = \log_2 N$ for DFT, RDFT, DHT, and DCT
$m = \log_2(N + 1)$ for DST

In terms of the number of operations per input sample, also the nonrecursive sliding transforms are found to be superior to their recursive counterparts. Table 7.4 gives the details of the operation counts of the two schemes. For the case of recursive implementations, the figures given in Table 7.4 have taken into account the effect of the stabilizing factor β.

The major drawback of the nonrecursive sliding transforms is that they are limited to the cases where the filter length, N, (filter length plus one in the case of DST) is a power of 2. On the contrary, the recursive sliding transforms can be used for any value of N.

7.9 Summary and Discussion

In this chapter, we reviewed a class of adaptive filters known as *TDAFs*. We gave a filtering interpretation of orthogonal transforms and demonstrated that a transformation may be viewed as a bank of bandpass filters, which are used to separate different parts of the spectrum of the underlying input process. This led to a band-partitioning view of orthogonal transforms. It was thus concluded that the outputs from an orthogonal transformation constitute a set of partially decorrelated processes as they belong to (partially) mutually exclusive bands.

Implementation of the LMS algorithm in transform domain was then presented. This was called *transform domain LMS (TDLMS)* algorithm. It was shown that significant improvement in convergence behavior of the TDLMS algorithm can be achieved if a proper set of normalized step-size parameters is used. This, which was called *step-normalization*, is assumed to be part of the TDLMS algorithm.

We showed that the TDLMS algorithm could equivalently be reformulated by normalizing the transformed samples of the underlying input process to the power of unity and then using the conventional LMS algorithm (with a single step-size parameter for all taps) to adapt the filter tap weights. This formulation is theoretically of interest as it allows one to use the results of the conventional LMS algorithm in evaluating the performance of the TDLMS algorithm.

The ideal LMS–Newton algorithm was introduced as a stochastic implementation of the Newton search method of Chapter 5. The relationship between the TDLMS and ideal LMS–Newton algorithms was also established. We found that the TDLMS algorithm is in fact an approximation to the ideal LMS–Newton algorithm.

We noted that for a given input process, the success of different transforms in decorrelating the samples of an input process varies. We presented a theory that relates the signal decorrelation property of orthogonal transforms to the distribution of signal powers after transformation. We demonstrated how this concept is related to the Karhunen-Loéve transform and drew some general guidelines for the selection of an appropriate transform when a rough estimate of the power spectral density of the underlying input process is known.

We also introduced various standard transforms that can be implemented efficiently using fast transforms. The siding fast implementation of these transforms was then presented. We found that in the application of TDAFs, the commonly used transforms can all be implemented with an order of N computational complexity, where N is the filter length.

Problems

P7.1 Figure P7.1 shows the power spectral densities of four processes and the magnitude responses of their associated eigenfilters for $N = 5$, in some arbitrary order. Considering the maximum signal-power-spreading property of the KLT, identify the magnitude response associated with each power spectral density.

P7.2 By substituting for past values of $\hat{\sigma}^2_{x_{T,i}}(n)$ in Eq. (7.28), show that $\hat{\sigma}^2_{x_{T,i}}(n)$ is an exponentially weighted average of the present and past samples of $x^2_{T,i}(n)$'s using the weighting function characterized by the coefficients $1, \beta, \beta^2, \ldots$, that is,

$$\sigma^2_{x_{T,i}}(n) = \frac{\sum_{k=0}^{\infty} \beta^k x^2_{T,i}(n-k)}{\sum_{k=0}^{\infty} \beta^k}$$

P7.3 Assume a noisy sinusoidal sequence $s(n) = a \sin(\omega n + \phi) + \nu(n)$, where $\nu(n)$ is an uncorrelated noise sequence. The angular frequency ω is known *a priori*. However, the magnitude "a" and phase "ϕ" are unknown. To obtain an estimate of these parameters, a two-tap transversal filter whose input is chosen to be $u(n) = \sin \omega n$ is set up and its tap weights, $w_0(n)$ and $w_1(n)$, are adapted so that the difference between $s(n)$ and the filter output, $y(n)$, is minimized in the mean-square sense. The filter output, $y(n)$, is then a noise-free estimate of the sinusoidal sequence. The LMS algorithm is used for this purpose.

(i) Using time averages, find the correlation matrix $\mathbf{R}$ of the filter tap inputs.
(ii) Find the step-size parameter, μ, of the LMS algorithm that results in 5% misadjustment.
(iii) For the step-size parameter obtained in (ii), find the time constants of the learning curve of the filter and show that the convergence of the LMS algorithm becomes slower as ω decreases.

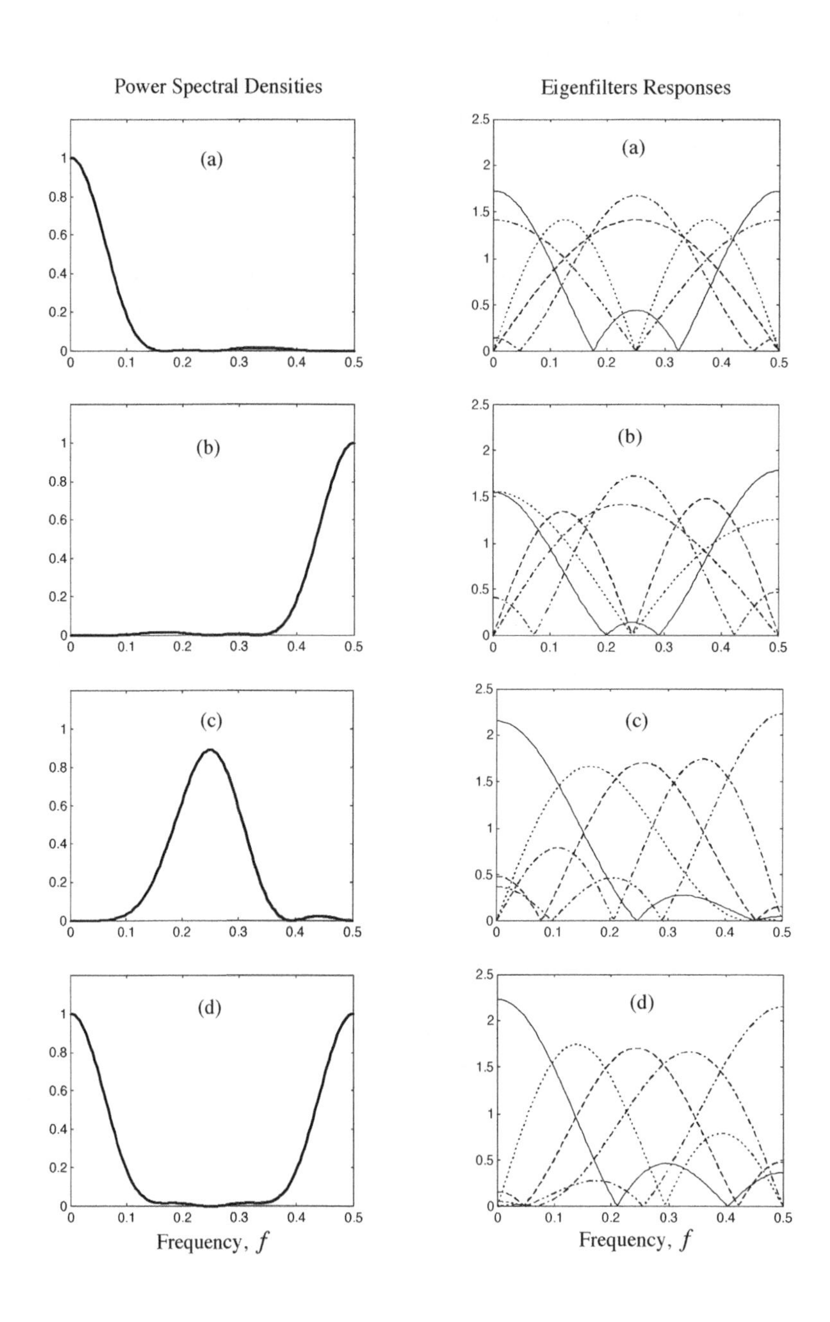
Power Spectral Densities
Eigenfilters Responses
(a)
(b)
(c)
(d)
Frequency, f
0
0.1
0.2
0.3
0.4
0.5
0.8
0.6
1
1.5
2
2.5

Figure P7.1

(iv) Show that the problem of slow convergence of the LMS algorithm can be solved if a TDLMS algorithm with the transformation matrix

$$\mathcal{T} = \frac{1}{\sqrt{2}} \begin{bmatrix} 1 & -1 \\ 1 & 1 \end{bmatrix}$$

is used.

P7.4 An adaptive transversal filter is excited by two different inputs, $u(n)$ and $v(n)$, whose power spectral densities are presented in Figure P7.4a and b.

(i) If the LMS algorithm is used in both cases and its step-size parameter is selected accordingly for a fixed level of misadjustment (say, 10%), which of the two inputs will result in the shortest transient time for the algorithm? Explain.

(ii) What will be your answer to (i), if a DCT-based transform domain implementation of the adaptive filter is employed?

(iii) Will your answer to (ii) change, if the DCT is replaced by DST?

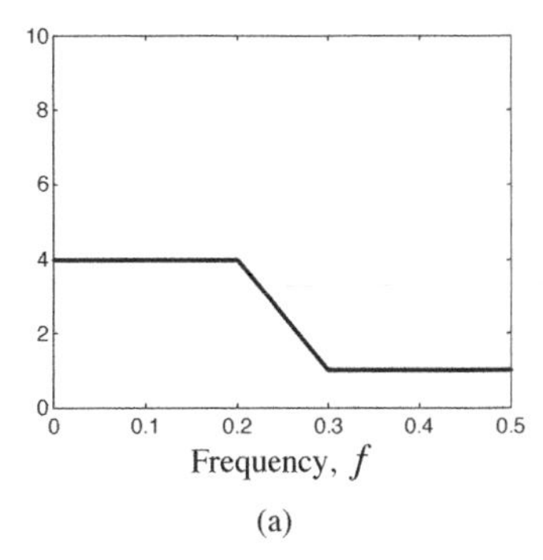

(a)

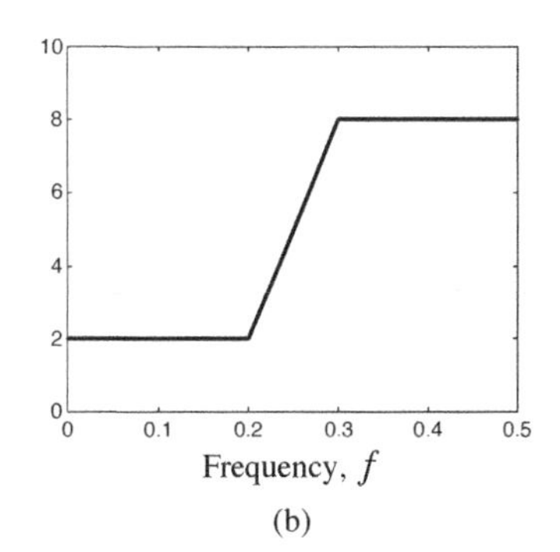

(b)

Figure P7.4

P7.5 Figure P7.5 shows the structure of a special adaptive filter, whose tap inputs are the samples of the processes $u(n)$ and $v(n)$, which are generated from a stationary input process $x(n)$ as shown. Assume that the filter length, N, is an even number.

(i) Define the length N column vector

$$\tilde{\mathbf{x}}(n) = [u(n)\ v(n)\ u(n-2)\ v(n-2)\ \cdots\ u(n-N+2)\ v(n-N+2)]^{\mathrm{T}}$$

and show that

$$\tilde{\mathbf{x}}(n) = \mathcal{T}_2 \mathbf{x}(n)$$

where

$$\mathcal{T}_2 = \begin{bmatrix} 1 & 1 & 0 & 0 & 0 & \cdots & 0 & 0 \\ 1 & -1 & 0 & 0 & 0 & \cdots & 0 & 0 \\ 0 & 0 & 1 & 1 & 0 & \cdots & 0 & 0 \\ 0 & 0 & 1 & -1 & 0 & \cdots & 0 & 0 \\ \vdots & \vdots & \vdots & \vdots & \vdots & \ddots & \vdots & \vdots \\ 0 & 0 & 0 & 0 & 0 & \cdots & 1 & 1 \\ 0 & 0 & 0 & 0 & 0 & \cdots & 1 & -1 \end{bmatrix}$$

(ii) Show that $\mathcal{T}_2$ is an orthogonal matrix and, thus, conclude that the structure presented in Figure P7.5 corresponds to a TDAF with $\mathcal{T} = \mathcal{T}_2$.

(iii) You may note that $\mathcal{T}_2\mathcal{T}_2^{\mathrm{T}} = 2\mathbf{I}$. This is different from the unitary condition $\mathcal{T}\mathcal{T}^{\mathrm{T}} = \mathbf{I}$, which is usually assumed for the transformation matrix $\mathcal{T}$. Does this deviation affect the performance of the TDLMS algorithm?

(iv) If the TDLMS algorithm (with the step-normalization) is to be used for fast adaptation of this structure, give the details of equations required for such implementation.

(v) Compare the structure of Figure P7.5 with that of a conventional LMS-based transversal adaptive filter both in terms of computational complexity and memory requirement.

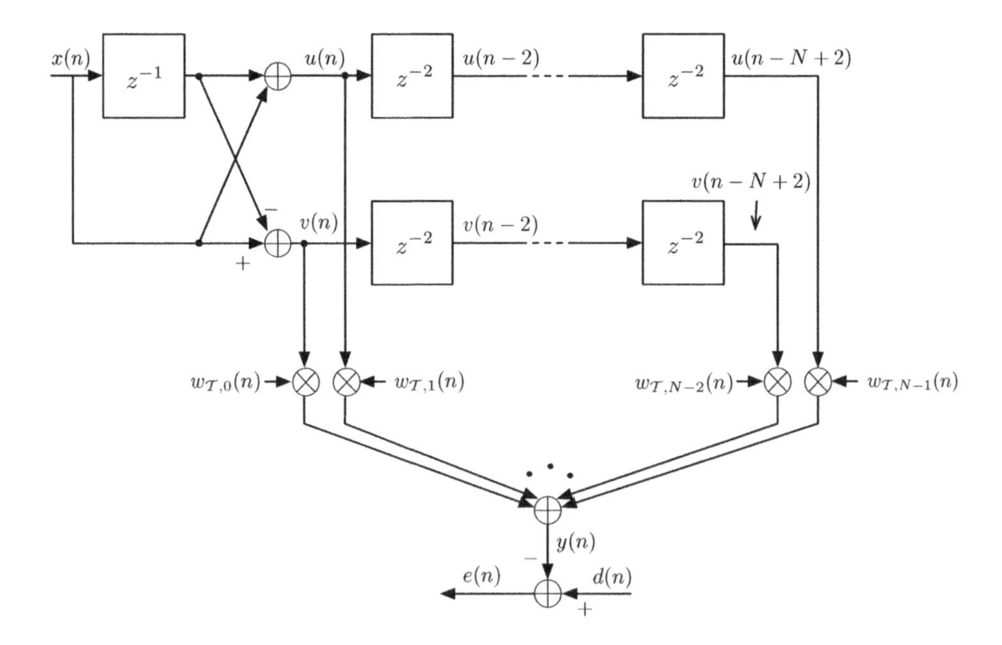

Figure P7.5

P7.6 Generalization of the adaptive filter structure given in Problem P7.5 may be done as follows.[5] Define the N-by-N matrix

$$\mathcal{T} = \begin{bmatrix} \mathcal{T}_{\text{sub}} & \mathbf{0} & \mathbf{0} & \cdots & \mathbf{0} \\ \mathbf{0} & \mathcal{T}_{\text{sub}} & \mathbf{0} & \cdots & \mathbf{0} \\ \vdots & \vdots & \vdots & \ddots & \vdots \\ \mathbf{0} & \mathbf{0} & \mathbf{0} & \cdots & \mathcal{T}_{\text{sub}} \end{bmatrix}$$

where $\mathcal{T}_{\text{sub}}$ is an orthogonal square matrix, and $\mathbf{0}$'s are zero matrices of appropriate dimensions.

(i) Show that $\mathcal{T}$ is an orthogonal matrix.
(ii) Considering the analogy between the transformation matrix $\mathcal{T}$ here and the one in Problem P7.5, construct a generalized version of Figure P7.5.
(iii) Noting that, in general, larger matrices achieve a higher degree of signal decorrelation (orthogonalization), discuss on the convergence behavior of the proposed structure as the size of $\mathcal{T}_{\text{sub}}$ increases.
(iv) Discuss on the memory requirement and computational complexity of the proposed structure as the size of $\mathcal{T}_{\text{sub}}$ increases.

P7.7 Show that the identity $\mathcal{T}\mathcal{T}^{\text{T}} = \mathbf{I}$ implies that the eigenvalues of $\mathbf{R}$ and $\mathbf{R}_{\mathcal{T}} = \mathcal{T}\mathbf{R}\mathcal{T}^{\text{T}}$ are the same. Thus, conclude that $\rho(\mathbf{R}) = \rho(\mathbf{R}_{\mathcal{T}})$.

P7.8 With reference to the notations in Section 7.6, show that

$$I_{\rho,\max} - I_{\rho,\mathcal{T}}^{\text{n}} = \ln \rho(\mathbf{R}_{\mathcal{T}}^{\text{n}}).$$

P7.9 Consider a two-tap transversal filter that is characterized by the performance function

$$\xi(v_0, v_1) = 0.1 + [v_0\ v_1]\mathbf{R}\begin{bmatrix} v_0 \\ v_1 \end{bmatrix}$$

where

$$\mathbf{R} = \begin{bmatrix} 1 & \alpha \\ \alpha & 1 \end{bmatrix}$$

Assume that the filter input is a real-valued random process.

(i) Find the points $(a, 0)$ and $(0, b)$ where the contour ellipse, given by $\xi(v_0, v_1) = 1.1$, meets v_0 and v_1 axes and show that $a = b$. Show that this result is directly related to the fact that the diagonal elements of $\mathbf{R}$ are the same which, in turn, implies that the signal energies at various taps of the filter are equal.
(ii) By sketching an arbitrary ellipse that passes through the points $(a, 0)$ and $(0, b)$ of (i), verify that the principal axes of the sketched ellipse are always in the directions obtained by 45° rotation of the coordinate axes v_0 and v_1.

[5] This problem has been designed based on the work of Petraglia and Mitra (1993).

(iii) Define an orthogonal transformation matrix

$$\mathcal{T} = \begin{bmatrix} \cos\theta & -\sin\theta \\ \sin\theta & \cos\theta \end{bmatrix}$$

and show that the transformation $\mathbf{v}_{\mathcal{T}} = \mathcal{T}\mathbf{v}$, where $\mathbf{v} = [v_0 \ v_1]^{\mathrm{T}}$, is equivalent to rotating the coordinate axes v_0 and v_1 by θ radian counterclockwise.

(iv) Find the rotation angle θ that maximizes the ratio of the diagonal elements of $\mathbf{R}_{\mathcal{T}} = \mathcal{T}\mathbf{R}\mathcal{T}^{\mathrm{T}}$ and show that it is independent of α.

(v) Noting that the diagonal elements of $\mathbf{R}_{\mathcal{T}}$ are the input signal energies after transformation, comment on your results in (iv) and show that for two-tap transversal filters with real-valued input processes, the optimum transformation matrix, $\mathcal{T}_{\mathrm{opt}}$, is fixed and independent of the statistics of the underlying input process. What is $\mathcal{T}_{\mathrm{opt}}$?

P7.10 The autocorrelation matrix $\mathbf{R}$ of the input process to an adaptive filter is known. To use this information to speed up the adaptation of the filter, the following algorithm is proposed.[6]

$$\begin{aligned} \mathbf{x}_{\mathcal{T}}(n) &= \mathcal{T}\mathbf{x}(n) \\ y(n) &= \mathbf{w}_{\mathcal{T}}^{\mathrm{T}}(n)\mathbf{x}_{\mathcal{T}}(n) \\ e(n) &= d(n) - y(n) \\ \mathbf{w}_{\mathcal{T}}(n+1) &= \mathbf{w}_{\mathcal{T}}(n) + 2\mu e(n)\mathbf{x}_{\mathcal{T}}(n) \end{aligned}$$

where $\mathcal{T} = \mathbf{R}^{-1/2}$, which is the inverse of the square root of $\mathbf{R}$ (as defined in Chapter 4), and μ is a scalar step-size parameter. Note that the matrix $\mathcal{T}$ here is not an orthogonal matrix, and, thus, the proposed algorithm is different from the TDLMS algorithm introduced in this chapter. In particular, we may note that the proposed algorithm does not have any step-normalization.

(i) Obtain the correlation matrix of the transformed samples, $\mathbf{x}_{\mathcal{T}}(n)$, and discuss on the significance of $\mathcal{T} = \mathbf{R}^{-1/2}$ in increasing the speed of convergence of the adaptive filter.

(ii) Give an approximate equation for the misadjustment of the proposed algorithm.

(iii) Define $\mathbf{w}(n) = \mathbf{R}^{-1/2}\mathbf{w}_{\mathcal{T}}(n)$ and use that to show that the proposed algorithm is equivalent to the ideal LMS–Newton algorithm.

P7.11 In Section 7.8.1, a derivation of the DFT frequency sampling filters was given. Following the procedure used there, derive the rest of the system functions listed in Table 7.3.

P7.12 Derive a sliding DFT structure for the case where $N = 8$.

P7.13 Derive the sliding RDFT structure presented in Figure 7.14.

P7.14 Derive the sliding DHT structure presented in Figure 7.15.

[6] This problem has been designed based on the work of Widrow and Walach (1984).

P7.15 Derive the sliding DCT structure presented in Figure 7.16.

P7.16 Derive the sliding DST structure presented in Figure 7.17.

P7.17 Derive a sliding DCT structure for the case where $N = 8$.

P7.18 Derive a sliding DHT structure for the case where $N = 7$.

P7.19 The transfer functions associated with the WHT cannot be written in a recursive form such as those given in Table 7.3 for the other transforms. However, we still find that each WHT filter may be implemented as a cascade of $\log_2 N$ nonrecursive sparse coefficients filters similar to the other transforms. In this problem, we clarify this by exploring the WHT for the transformation length $N = 8$. The generalization of the results to any value of N, which is a power of 2, is then obvious.

(i) Use Eq. (7.69) to find the coefficients of the WHT when $N = 8$.
(ii) Use the results of (i) to write down the transfer functions associated with various rows of the WHT when $N = 8$.
(iii) Show that the transfer functions obtained in (ii) can be factorized as

$$\frac{1}{\sqrt{8}}(1 \pm z^{-4})(1 \pm z^{-2})(1 \pm z^{-1})$$

where the various combinations of the $\pm$ signs cover all the eight filter transfer functions.
(iv) Using the latter factorization, propose a tree structure, similar to the nonrecursive sliding transforms introduced in Section 7.8.3, for an $O(N)$ implementation of the WHT.

Computer-Oriented Problems

P7.20 Consider a modeling problem where a plant

$$W_o(z) = 0.4 + z^{-1} - 0.3z^{-2}$$

is modeled using a 15-tap transversal adaptive filter. The plant is assumed to be noise free. The input to the plant and adaptive filter is generated by passing a unit-variance white process through the coloring filter

$$H(z) = 0.1 - 0.3z^{-1} - 0.5z^{-2} + z^{-3} + z^{-4} - 0.5z^{-5} - 0.3z^{-6} + 0.1z^{-7}$$

(i) Write a program to simulate this scenario. In your program after every 10 iterations, plot the magnitude response of the adaptive filter and observe how it converges toward the magnitude response of the plant.
(ii) Obtain and plot the power spectral density of the adaptive filter input and try to relate that to your observation in (i). You should find that the convergence of the magnitude response of the adaptive filter toward the plant response is frequency dependent. Over the frequency bands where the filter input has

higher power, convergence is faster. On the other hand, the slow modes of the adaptive filter correspond to the bands where the filter input is poorly excited, that is, having low power spectral density.

P7.21 Repeat Problem P7.20 when the LMS algorithm is replaced by a DCT-based TDLMS algorithm. Study the performance of the algorithm with and without step-normalization.

P7.22 Repeat Problem P7.20 when the LMS algorithm is replaced by a DST-based TDLMS algorithm. Compare your results here with those of Problem P7.21.

P7.23 Repeat Problem P7.20 when the LMS algorithm is replaced by a DCT-based TDLMS algorithm and

$$H(z) = 0.1 + 0.3z^{-1} + 0.5z^{-2} + z^{-3} + z^{-4} - 0.5z^{-5} + 0.3z^{-6} + 0.1z^{-7}.$$

P7.24 Repeat Problem P7.20 when the LMS algorithm is replaced by a DST-based TDLMS algorithm and

$$H(z) = 0.1 + 0.3z^{-1} + 0.5z^{-2} + z^{-3} + z^{-4} - 0.5z^{-5} + 0.3z^{-6} + 0.1z^{-7}$$

Compare your results here with those of Problem P7.23.

P7.25 Develop and run your own program(s) to confirm the results of Table 7.2.

P7.26 Consider a modeling problem where the plant is a 16-tap transversal filter. The plant output is contaminated with an additive white noise, $e_o(n)$, with variance $\sigma_o^2 = 10^{-4}$. The plant input is generated by passing a unit variance white process through a coloring filter. Here, we consider the following choices of the noise coloring filter:

$$H_1(z) = 0.1 + 0.2z^{-1} + 0.3z^{-2} + 0.4z^{-3} + 0.4z^{-4} + 0.2z^{-5} + 0.1z^{-6},$$

$$H_2(z) = 0.1 - 0.2z^{-1} - 0.3z^{-2} + 0.4z^{-3} + 0.4z^{-4} - 0.2z^{-5} - 0.1z^{-6},$$

and

$$H_3(z) = 0.1 - 0.2z^{-1} + 0.3z^{-2} - 0.4z^{-3} + 0.4z^{-4} - 0.2z^{-5} + 0.1z^{-6}$$

Note that the first two filters are those which were used in Section 7.6.4 to obtain the results of Table 7.2. We also note that the outputs of $H_1(z)$ and $H_2(z)$ are lowpass and bandpass processes, respectively (Figures 7.6 and 7.7). The coloring filter $H_3(z)$ generates a highpass process.

Develop a program (or a set of programs) to study the convergence behavior of the TDLMS algorithm for these choices of input and various choices of transforms. Examine your results and see how consistent are these with the general conclusions of Section 7.6.

8

Block Implementation of Adaptive Filters

There are certain applications of signal processing that require adaptive filters whose length exceeds a few hundreds or even a few thousands of taps. For instance, to prevent the return of speaker echo to the far-end side of the telephone line, in the application of hand-free telephony, the use of an acoustic echo canceler whose length exceeds a few thousand taps is not uncommon. Other applications, such as active noise control and equalization of some communication channels, may also require adaptive filters with exceedingly long lengths. In such applications, one finds that even the conventional LMS algorithm, which is known for its simplicity, is computationally expensive to implement.

In this chapter, we show how block processing of the data samples can significantly reduce the computational complexity of adaptive filters. In *block processing* (or *block implementation*), a block of samples of the filter input and desired output are collected and then processed together to obtain a block of output samples. Thus, the process involves serial-to-parallel conversion of the input data, parallel processing of the collected data, and parallel-to-serial conversion of the generated output data. This is illustrated in Figure 8.1. The computational complexity of the adaptive filter can then be reduced significantly through elegant parallel processing of the data samples. We note that the parallel processing involved in Figure 8.1 is repeated only after collection of every block of data samples. Thus, a good measure of the computational complexity in a block processing system is given by the number of operations required to process one block of data divided by the block length. We may then note that the sharing of the processing time among the samples in each block is the key to achieve high computational efficiency.

In this chapter, we discuss an efficient technique for block processing of data samples in the adaptive filtering context. This involves a special implementation of the LMS algorithm, which is called *block* LMS (BLMS). We introduce a computationally efficient implementation of the BLMS algorithm in the frequency domain. This is called *fast* BLMS (FBLMS) algorithm. The high computational efficiency of the FBLMS algorithm is achieved by employing the following result from the theory of digital signal processing (DSP). Linear convolution of time domain sequences can be efficiently implemented using frequency domain processing. In particular, the linear convolution of an indefinite length sequence, $x(n)$, with a finite length sequence, h_n (which may be that of the impulse response of a FIR filter) is obtained by partitioning $x(n)$ into a set of overlapping finite

Adaptive Filters: Theory and Applications, Second Edition. Behrouz Farhang-Boroujeny.

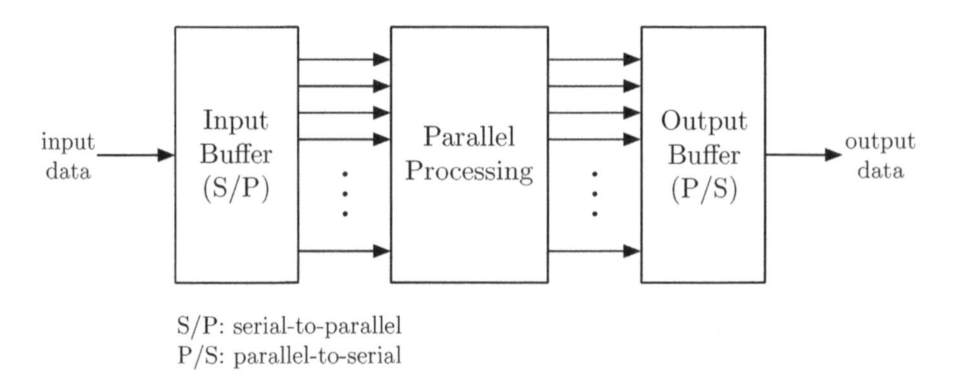

Figure 8.1 Schematic of a block processing system.

duration blocks, finding the circular convolution of h_n (appended with some extra zeros) with these blocks, and then choosing the portions of the circular convolutions, which match the desired linear convolution samples. The circular convolutions can be very efficiently performed in the frequency domain, using the properties of the discrete Fourier transform (DFT).

Throughout this chapter, we adopt the following notations. As in the previous chapters, bold lowercase letters represent vectors, bold uppercase letters denote matrices, and non-bold lowercase letters represent scalars. As before, we use "n" as the time (sample) index. The letter "k" is reserved for block index. The subscript $\mathcal{F}$ is used to refer the frequency domain signals, for example, DFT of the time domain vector $\mathbf{x}$ is denoted as $\mathbf{x}_{\mathcal{F}}$. In the derivations that follow, we frequently need to extend the dimensions of vectors and matrices to some certain dimensions by appending zeros. We use $\mathbf{0}$ (in bold) to refer to zero vectors and zero matrices and the dimensions of these zero vectors and/or matrices will be clear from the context.

Our discussion in this chapter is limited to the case were the filter input, $x(n)$, and the desired output, $d(n)$, are real-valued processes. However, we note that the frequency domain equivalent of these processes are complex-valued and hence, the LMS recursion that is used is the complex LMS algorithm.

8.1 Block LMS Algorithm

The conventional LMS algorithm, which was introduced in Chapter 6, uses the following recursion to adjust the tap weights of an adaptive filter:

$$\mathbf{w}(n+1) = \mathbf{w}(n) + 2\mu e(n)\mathbf{x}(n) \tag{8.1}$$

where $\mathbf{x}(n) = [x(n)x(n-1)\cdots x(n-N+1)]^{\mathrm{T}}$ and $\mathbf{w}(n) = [w_0(n)w_1(n)\cdots\ w_{N-1}(n)]^{\mathrm{T}}$ are the column vectors consisting of the filter tap inputs and tap weights, respectively, $e(n) = d(n) - y(n)$ is the output error, $d(n)$ and $y(n) = \mathbf{w}^{\mathrm{T}}(n)\mathbf{x}(n)$ are the desired and actual outputs of the filter, respectively, and μ is the step-size parameter. We also recall that the conventional LMS algorithm is a stochastic implementation of the steepest-descent method using the instantaneous gradient vector

$$\nabla_{\mathbf{w}} e^2(n) = -2e(n)\mathbf{x}(n) \tag{8.2}$$

The BLMS algorithm works based on the following strategy. The filter tap weights are updated once after collection of every block of data samples. The gradient vector used to update the filter tap weights is an average of the instantaneous gradient vectors of the form (8.2), which are calculated during the current block. Using k to denote the block index, the BLMS recursion is obtained as

$$\mathbf{w}(k+1) = \mathbf{w}(k) + 2\mu_{\mathrm{B}} \frac{\sum_{i=0}^{L-1} e(kL+i)\mathbf{x}(kL+i)}{L} \tag{8.3}$$

where L is the block length and μ_{B} is the algorithm step-size parameter. We also note that for the computation of the output error samples $e(kL+i) = d(kL+i) - y(kL+i)$, for $i = 0, 1, \ldots, L-1$, the output samples $y(kL+i) = \mathbf{w}^{\mathrm{T}}(k)\mathbf{x}(kL+i)$ are calculated using the update of the filter tap-weight vector, $\mathbf{w}(k)$, from the previous block.

The derivations presented in the following sections make use of, to a large extent, the vector formulation of the BLMS algorithm. Hence, we now present this formulation.

Define the matrix

$$\mathbf{X}(k) = [\mathbf{x}(kL) \;\; \mathbf{x}(kL+1) \;\; \cdots \;\; \mathbf{x}(kL+L-1)]^{\mathrm{T}} \tag{8.4}$$

and the column vectors

$$\mathbf{d}(k) = [d(kL) \;\; d(kL+1) \;\; \cdots \;\; d(kL+L-1)]^{\mathrm{T}} \tag{8.5}$$

$$\mathbf{y}(k) = [y(kL) \;\; y(kL+1) \;\; \cdots \;\; y(kL+L-1)]^{\mathrm{T}} \tag{8.6}$$

$$\mathbf{e}(k) = [e(kL) \;\; e(kL+1) \;\; \cdots \;\; e(kL+L-1)]^{\mathrm{T}} \tag{8.7}$$

and note that

$$\mathbf{y}(k) = \mathbf{X}(k)\mathbf{w}(k) \tag{8.8}$$

and

$$\mathbf{e}(k) = \mathbf{d}(k) - \mathbf{y}(k) \tag{8.9}$$

We also note that

$$\sum_{i=0}^{L-1} e(kL+i)\mathbf{x}(kL+i) = \mathbf{X}^{\mathrm{T}}(k)\mathbf{e}(k) \tag{8.10}$$

Substituting Eq. (8.10) in Eq. (8.3), we obtain

$$\mathbf{w}(k+1) = \mathbf{w}(k) + 2\frac{\mu_{\mathrm{B}}}{L}\mathbf{X}^{\mathrm{T}}(k)\mathbf{e}(k) \tag{8.11}$$

Equations (8.8), (8.9), and (8.11), which correspond to filtering, error estimation, and tap-weight vector updating, respectively, define one iteration of the BLMS algorithm.

On the basis of our background from the method of steepest-descent and, also, the conventional LMS algorithm, the following comments may be made, intuitively:

1. Convergence behavior of the BLMS algorithm is governed by the eigenvalues of the correlation matrix $\mathbf{R} = E[\mathbf{x}(n)\mathbf{x}^{\mathrm{T}}(n)]$. This follows from the fact that similar to the

conventional LMS algorithm, the BLMS algorithm is also a stochastic implementation of the steepest-descent method.

2. The BLMS algorithm has N modes of convergence, which are characterized by the time constants:

$$\tau_{\mathrm{B},i} = \frac{1}{4\mu_{\mathrm{B}}\lambda_i}, \quad \text{for} \quad i = 0, 1, \ldots, N-1 \tag{8.12}$$

where λ_i's are the eigenvalues of the correlation matrix $\mathbf{R}$. These time constants are in the unit of iteration (block) interval.

3. Averaging the instantaneous (stochastic) gradient vectors, as done in the BLMS algorithm, results in gradient vectors with a lower variance compared to those in the conventional LMS algorithm. This allows the use of a larger step-size parameter for the BLMS algorithm compared to the conventional LMS algorithm. For block lengths, L, comparable or less than the filter length, N, and small misadjustments, in the range of 10% or less, misadjustment, $\mathcal{M}_{\mathrm{B}}$, of the BLMS algorithm can be approximated by the following expression:

$$\mathcal{M}_{\mathrm{B}} \approx \frac{\mu_{\mathrm{B}}}{L}\mathrm{tr}[\mathbf{R}] \tag{8.13}$$

This result is derived in Appendix 8A.
Comparing Eq. (8.13) with Eq. (6.64), and letting $\mathcal{M}_{\mathrm{B}} = \mathcal{M}$, where $\mathcal{M}$ denotes the misadjustment of the conventional LMS algorithm, we obtain

$$\mu_{\mathrm{B}} = L\mu \tag{8.14}$$

where μ is the step-size parameter of the conventional LMS algorithm. Substituting Eq. (8.14) in Eq. (8.12), we get

$$\tau_{\mathrm{B},i} = \frac{1}{4L\mu\lambda_i} \quad \text{block interval} \tag{8.15}$$

$$= \frac{1}{4\mu\lambda_i} \quad \text{sample interval} \tag{8.16}$$

Comparing this result with Eq. (6.33) and recalling that the time constants associated with the conventional LMS algorithm are in sample intervals, we conclude that the convergence behavior of BLMS and conventional LMS algorithms are the same.

The following example illustrates the above remarks.

Example 8.1

Let us consider the modeling problem discussed in Section 6.4.1 and use the signal coloring filter $H_1(z)$ of Eq. (6.80) to generate the input process, $x(n)$. Figure 8.2 shows the results of simulations that compare the conventional LMS algorithm and the BLMS algorithm for different choices of the block length, L. The results presented here are based on an ensemble average of 100 independent runs for each plot. The step-size parameters μ and μ_{B} have been selected according to Eqs. (6.64) and (8.13), respectively, for 10% misadjustment. We note that the difference between the various learning curves

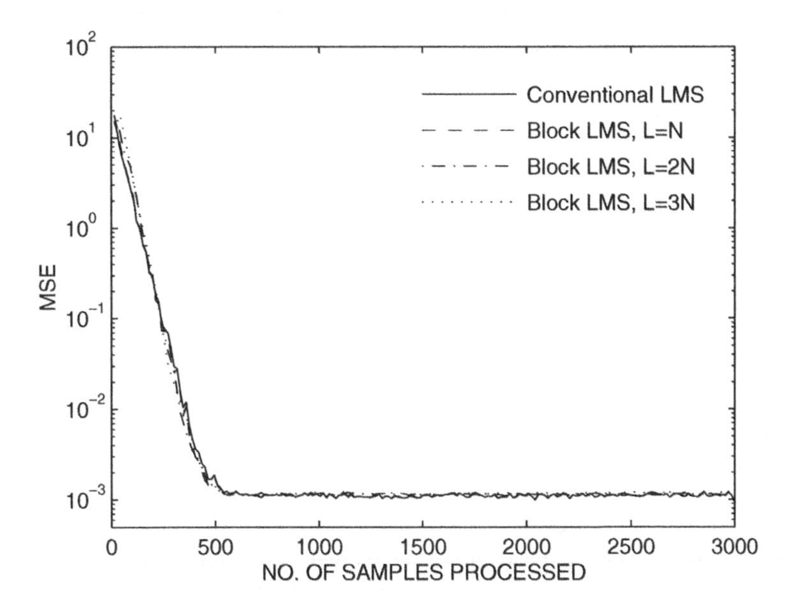

Figure 8.2 Convergence behavior of the BLMS algorithm for various values of the block length, L. Results of the conventional LMS algorithm are also shown for comparison. The step-size parameters μ and μ_B are selected based on Eqs. (6.64) and (8.13), respectively, for 10% misadjustment.

in Figure 8.2 is negligible. This confirms the theoretical predictions made above, which suggest that the BLMS and conventional LMS algorithms perform the same.

The program used to generate the results of Figure 8.2 is available on the accompanying website. It is called `blk_mdlg.m`. The reader is encouraged to try this program for different choices of misadjustment and block length to study the effect of variations of these parameters on the behavior of the BLMS algorithm.

8.2 Mathematical Background

The mathematical and signal processing tools required for the rest of this chapter are briefly reviewed in this section. In particular, we discuss how time domain linear convolutions can be efficiently performed using DFT (Oppenheim and Schafer, 1975, 1989). We also introduce circular matrices and review some of their properties that are relevant to our study of BLMS algorithms.

8.2.1 Linear Convolution Using the Discrete Fourier Transform

We consider the filtering of a sequence $x(n)$ through a FIR filter with coefficients $w_0, w_1, \ldots, w_{N-1}$. This involves computation of the linear convolution

$$y(n) = \sum_{i=0}^{N-1} w_i x(n-i) \tag{8.17}$$

This process requires N multiplications and $N-1$ additions for computing every sample of the output, $y(n)$. When N is large, the samples of $y(n)$ can be obtained with a reduced number of multiplications and additions, as discussed below.

Let us define the column vector $\tilde{\mathbf{x}}(k)$ of length $N' = N + L - 1$ as

$$\tilde{\mathbf{x}}(k) = [x(kL-N+1)\ x(kL-N+2)\ \cdots\ x(kL+L-1)]^{\mathrm{T}} \tag{8.18}$$

and $\tilde{\mathbf{w}}(k)$ of length N' as

$$\tilde{\mathbf{w}}(k) = \begin{bmatrix} \mathbf{w}(k) \\ \mathbf{0} \end{bmatrix} \tag{8.19}$$

where $\mathbf{w}(k) = [w_0(k)\ w_1(k)\ \cdots\ w_{N-1}(k)]^{\mathrm{T}}$ is the filter tap-weight vector, and $\mathbf{0}$ refers to a column vector consisting of $L-1$ zeros. In order to maintain uniformity in the derivations of the subsequent sections, the block index k has been added to the filter tap weights, indicating that the weights vary only from block to block, as it happens in the implementation of the BLMS algorithm.

From the properties of the DFT, we know that the circular convolution of $\tilde{\mathbf{w}}(k)$ and $\tilde{\mathbf{x}}(k)$ can be obtained by transforming both vectors to their respective frequency domain equivalents (using the DFT), performing an element-wise multiplication on the transformed samples and transforming the result back to the time domain (using the inverse DFT (IDFT)). This process can be efficiently implemented using the fast Fourier transform (FFT) and inverse FFT (IFFT) algorithms. Examining the circular convolution of $\tilde{\mathbf{w}}(k)$ and $\tilde{\mathbf{x}}(k)$ reveals that only the last L elements of the result coincide with the corresponding elements of the linear convolution (8.17); see Oppenheim and Schafer (1975) for example.[1] The rest of the elements of the circular convolution do not provide any useful result as the elements of $\tilde{\mathbf{x}}(k)$ are wrapped around and are not in the right order, as required by the linear convolution (8.17). The computation of the circular convolution of $\tilde{\mathbf{w}}(k)$ and $\tilde{\mathbf{x}}(k)$ and the wraparound phenomenon are summarized as

$$\begin{bmatrix} * \\ * \\ \vdots \\ * \\ y(kL) \\ y(kL+1) \\ \vdots \\ y(kL+L-1) \end{bmatrix} = \begin{bmatrix} x(kL-N+1) & x(kL+L-1) & x(kL+L-2) & \cdots & x(kL-N+2) \\ x(kL-N+2) & x(kL-N+1) & x(kL+L-1) & \cdots & x(kL-N+3) \\ \vdots & \vdots & \vdots & \ddots & \vdots \\ x(kL-1) & x(kL-2) & x(kL-3) & \cdots & x(kL) \\ x(kL) & x(kL-1) & x(kL-2) & \cdots & x(kL+1) \\ x(kL+1) & x(kL) & x(kL-1) & \cdots & x(kL+2) \\ \vdots & \vdots & \vdots & \ddots & \vdots \\ x(kL+L-1) & x(kL+L-2) & x(kL+L-3) & \cdots & x(kL-N+1) \end{bmatrix} \begin{bmatrix} w_0(k) \\ w_1(k) \\ \vdots \\ w_{N-2}(k) \\ w_{N-1}(k) \\ 0 \\ \vdots \\ 0 \end{bmatrix} \tag{8.20}$$

In Eq. (8.20), the elements represented by asterisks correspond to circular convolution results, which do not coincide with linear convolution samples, as required by Eq. (8.17). Careful examination of the summations related to these elements reveals that the input samples experience some discontinuity in their order. For example, a jump from

[1] In the original derivation of the FBLMS algorithm by Ferrara and Widrow (1981), and most of the subsequent publications on this, the block length, L, is chosen equal to the filter length, N. Also, the column vector $\mathbf{0}$ in Eq. (8.19) has been assumed to be of length $L = N$, and not $L-1$, as we assume here.

$x(kL - N + 1)$ to $x(kL + L - 1)$ is observed in the first row of the data matrix on the right-hand side of Eq. (8.20). When such discontinuities overlap with the nonzero portion of $\tilde{\mathbf{w}}(k)$, then the corresponding output samples will not correspond to valid linear convolution samples.

The procedure explained by Eq. (8.20) is commonly known as the *overlap-save method*. This name reflects the fact that in each block of input, $\tilde{\mathbf{x}}(k)$ consists of L new samples and $N - 1$ overlapped samples from the previous block(s). Another equally efficient method for computation of linear convolutions using DFT is the *overlap-add method*. However, the overlap-add method has been found to be computationally less efficient than the overlap-save method when applied to the implementation of the BLMS algorithm. Noting this, we do not discuss the overlap-add method in this book.

8.2.2 Circular Matrices

Circular matrices are used extensively in the derivation and analysis of the FBLMS algorithm. Hence, it is very useful as well as necessary to have a good understanding of the properties of these matrices before we start our discussion on the FBLMS algorithm.

Consider the M-by-M circular matrix

$$\mathbf{A}_c = \begin{bmatrix} a_0 & a_{M-1} & a_{M-2} & \cdots & a_1 \\ a_1 & a_0 & a_{M-1} & \cdots & a_2 \\ \vdots & \vdots & \vdots & \ddots & \vdots \\ a_{M-2} & a_{M-3} & a_{M-4} & \cdots & a_{M-1} \\ a_{M-1} & a_{M-2} & a_{M-3} & \cdots & a_0 \end{bmatrix} \tag{8.21}$$

Clearly, the name "circular" refers to the fact that each row (column) of $\mathbf{A}_c$ is obtained by circularly shifting the previous row (column) by one element. A special property of circular matrices, which is extensively used in the following sections, is that such matrices are diagonalized by DFT matrices. That is, if $\mathcal{F}$ is the M-by-M DFT matrix defined as

$$\mathcal{F} = \begin{bmatrix} 1 & 1 & 1 & \cdots & 1 \\ 1 & e^{-j\frac{2\pi}{M}} & e^{-j\frac{4\pi}{M}} & \cdots & e^{-j\frac{2\pi(M-1)}{M}} \\ 1 & e^{-j\frac{4\pi}{M}} & e^{-j\frac{8\pi}{M}} & \cdots & e^{-j\frac{4\pi(M-1)}{M}} \\ \vdots & \vdots & \vdots & \ddots & \vdots \\ 1 & e^{-j\frac{2\pi(M-1)}{M}} & e^{-j\frac{4\pi(M-1)}{M}} & \cdots & e^{-j\frac{2\pi(M-1)^2}{M}} \end{bmatrix} \tag{8.22}$$

then

$$\mathcal{A}_{\mathcal{F}} = \mathcal{F}\mathbf{A}_c\mathcal{F}^{-1} \tag{8.23}$$

is a diagonal matrix. Furthermore, the diagonal elements of $\mathcal{A}_{\mathcal{F}}$ correspond to the DFT of the first column of $\mathbf{A}_c$. In matrix notation, this may be written as

$$\mathcal{A}_{\mathcal{F}} = \text{diag}[\mathbf{a}_{\mathcal{F}}] \tag{8.24}$$

where $\mathbf{a}_{\mathcal{F}} = \mathcal{F}\mathbf{a}$, $\mathbf{a} = [a_0\, a_1\, \cdots\, a_{M-1}]^{\mathrm{T}}$ is the first column of $\mathbf{A}_c$, and $\text{diag}[\mathbf{a}_{\mathcal{F}}]$ denotes the diagonal matrix consisting of the elements of $\mathbf{a}_{\mathcal{F}}$. This can be proved as follows.

Since $\mathcal{F}$ is a DFT matrix, recall that

$$\mathcal{F}^{-1} = \frac{1}{M}\mathcal{F}^* \tag{8.25}$$

where M is the length of DFT and asterisk denotes complex conjugation. In other words, the lth column of $\mathcal{F}^{-1}$ can be given as

$$\mathbf{g}_l = \frac{1}{M}\mathbf{f}_l^* \tag{8.26}$$

where $\mathbf{f}_l$ is the lth column of the DFT matrix $\mathcal{F}$ given by

$$\mathbf{f}_l = [1\,\mathrm{e}^{-j\frac{2\pi l}{M}}\,\mathrm{e}^{-j\frac{4\pi l}{M}}\,\cdots\,\mathrm{e}^{-j\frac{2(M-1)\pi l}{M}}]^{\mathrm{T}} \tag{8.27}$$

Next, by direct insertion, one can easily show that (Problem P8.2)

$$\mathbf{A}_c\mathbf{g}_l = a_{\mathcal{F},l}\mathbf{g}_l, \quad \text{for} \quad l = 0, 1, \ldots, M-1 \tag{8.28}$$

where $a_{\mathcal{F},l} = \sum_{i=0}^{M-1} a_i \mathrm{e}^{-j\frac{2\pi li}{M}}$ is the lth element of $\mathbf{a}_{\mathcal{F}}$. Using Eq. (8.24), the M equations in (8.28) may be put together to obtain

$$\mathbf{A}_c\mathcal{F}^{-1} = \mathcal{F}^{-1}\boldsymbol{\mathcal{A}}_{\mathcal{F}} \tag{8.29}$$

Premultiplying Eq. (8.29) on both sides by $\mathcal{F}$ gives Eq. (8.23).

Another important result of the circular matrices that will be useful for our later application is derived next. Applying Hermitian transposition on both sides of Eq. (8.23), we obtain

$$\boldsymbol{\mathcal{A}}_{\mathcal{F}}^{\mathrm{H}} = \mathcal{F}^{-\mathrm{H}}\mathbf{A}_{\mathrm{c}}^{\mathrm{H}}\mathcal{F}^{\mathrm{H}} \tag{8.30}$$

where $\mathcal{F}^{-H}$ is the shorthand notation for $(\mathcal{F}^{-1})^{\mathrm{H}}$. Since $\boldsymbol{\mathcal{A}}_{\mathcal{F}}$ is diagonal, $\boldsymbol{\mathcal{A}}_{\mathcal{F}}^{\mathrm{H}} = \boldsymbol{\mathcal{A}}_{\mathcal{F}}^*$. Furthermore, from Eqs. (8.22) and (8.25), $\mathcal{F}^{-\mathrm{H}} = \frac{1}{M}\mathcal{F}$ and $\mathcal{F}^{\mathrm{H}} = M\mathcal{F}^{-1}$ as $\mathcal{F}^{\mathrm{T}} = \mathcal{F}$. Using these in Eq. (8.30), we get

$$\boldsymbol{\mathcal{A}}_{\mathcal{F}}^* = \mathcal{F}\mathbf{A}_{\mathrm{c}}^{\mathrm{H}}\mathcal{F}^{-1} \tag{8.31}$$

When elements of $\mathbf{A}_c$ are real-valued, $\mathbf{A}_{\mathrm{c}}^{\mathrm{H}} = \mathbf{A}_{\mathrm{c}}^{\mathrm{T}}$ and, thus, Eq. (8.31) may be written as

$$\boldsymbol{\mathcal{A}}_{\mathcal{F}}^* = \mathcal{F}\,\mathbf{A}_{\mathrm{c}}^{\mathrm{T}}\mathcal{F}^{-1} \tag{8.32}$$

8.2.3 Window Matrices and Matrix Formulation of the Overlap-Save Method

Let us define the N'-by-N' circular matrix, for $N' = L + N - 1$, as

$$\mathbf{X}_{\mathrm{c}}(k) = \begin{bmatrix} x(kL-N+1) & x(kL+L-1) & x(kL+L-2) & \cdots & x(kL-N+2) \\ x(kL-N+2) & x(kL-N+1) & x(kL+L-1) & \cdots & x(kL-N+3) \\ \vdots & \vdots & \vdots & \ddots & \vdots \\ x(kL+L-1) & x(kL+L-2) & x(kL+L-3) & \cdots & x(kL-N+1) \end{bmatrix} \tag{8.33}$$

We note that this is nothing but the data matrix on the right side of Eq. (8.20).

We also define the length N' column vector

$$\tilde{\mathbf{y}}(k) = \begin{bmatrix} \mathbf{0} \\ \mathbf{y}(k) \end{bmatrix} \tag{8.34}$$

where $\mathbf{y}(k)$, as defined in Eq. (8.6), is the column vector consisting of the output samples of kth block, and $\mathbf{0}$ is the length $N-1$ zero vector. Let us denote the column vector by $\mathbf{y}_c(k)$, which appears on the left-hand side of Eq. (8.20), and note that $\tilde{\mathbf{y}}(k)$ can be obtained from $\mathbf{y}_c(k)$ by substituting all the $*$ elements in the latter with zeros. This substitution can be written in the form of a matrix–vector product as

$$\tilde{\mathbf{y}}(k) = \mathbf{P}_{0,L}\mathbf{y}_c(k) \tag{8.35}$$

where $\mathbf{P}_{0,L}$ is the N'-by-N' *windowing matrix* defined as

$$\mathbf{P}_{0,L} = \begin{bmatrix} \mathbf{0} & \mathbf{0} \\ \mathbf{0} & \mathbf{I}_L \end{bmatrix} \tag{8.36}$$

with $\mathbf{I}_L$ being the L-by-L identity matrix, and $\mathbf{0}$'s are zero matrices with appropriate dimensions. Using Eqs. (8.20), (8.19), and the above definitions, we obtain

$$\tilde{\mathbf{y}}(k) = \mathbf{P}_{0,L}\mathbf{X}_c(k)\tilde{\mathbf{w}}(k) \tag{8.37}$$

Implementation of Eq. (8.37) in the frequency domain can now be obtained by simply noting that Eq. (8.37) may be written as

$$\tilde{\mathbf{y}}(k) = \mathbf{P}_{0,L}\mathcal{F}^{-1}\mathcal{F}\mathbf{X}_c(k)\mathcal{F}^{-1}\mathcal{F}\tilde{\mathbf{w}}(k) \tag{8.38}$$

where $\mathcal{F}$ is the N'-by-N' DFT matrix. Next, define

$$\mathbf{w}_{\mathcal{F}}(k) = \mathcal{F}\tilde{\mathbf{w}}(k) \tag{8.39}$$

and

$$\mathcal{X}_{\mathcal{F}}(k) = \mathcal{F}\mathbf{X}_c(k)\mathcal{F}^{-1} \tag{8.40}$$

and note that $\mathcal{X}_{\mathcal{F}}(k)$ is the diagonal matrix consisting of the elements of the DFT of the first column of $\mathbf{X}_c(k)$ as the latter is a circular matrix. We also note that the first column of $\mathbf{X}_c(k)$ is the input vector $\tilde{\mathbf{x}}(k)$, as defined in Eq. (8.18). Using Eqs. (8.39) and (8.40) in Eq. (8.38), we obtain

$$\tilde{\mathbf{y}}(k) = \mathbf{P}_{0,L}\mathcal{F}^{-1}\mathcal{X}_{\mathcal{F}}(k)\mathbf{w}_{\mathcal{F}}(k) \tag{8.41}$$

This equation has the following interpretation. Since $\mathcal{X}_{\mathcal{F}}(k)$ is diagonal, $\mathcal{X}_{\mathcal{F}}(k)\mathbf{w}_{\mathcal{F}}(k)$ is nothing but the element-wise multiplication of the filter input and its coefficients in the frequency domain. This gives the output samples of the filter in the frequency domain. Premultiplication of this result by $\mathcal{F}^{-1}$ converts the frequency domain samples of the output to the time domain. Furthermore, premultiplying the result by the windowing matrix $\mathbf{P}_{0,L}$ results in selecting only those samples that coincide with the required linear convolution samples.

With the background developed in this section, we are now ready to proceed with the derivation and analysis of the FBLMS algorithm.

8.3 The FBLMS Algorithm

The FBLMS algorithm, as mentioned in the introduction, is nothing but a fast (numerically efficient) implementation of the BLMS algorithm in the frequency domain.

Equation (8.41) corresponds to the filtering part of the FBLMS algorithm. Element-by-element multiplication of the frequency domain samples of the input and filter coefficients is followed by an IDFT and a proper windowing of the result to obtain the output vector $\tilde{\mathbf{y}}(k)$, in the extended form, as defined by Eq. (8.34). The vector of desired outputs, in the extended form, is defined as

$$\tilde{\mathbf{d}}(k) = \begin{bmatrix} \mathbf{0} \\ \mathbf{d}(k) \end{bmatrix} \tag{8.42}$$

where $\mathbf{d}(k)$ is defined by Eq. (8.5), and $\mathbf{0}$ is the $N-1$ element zero column vector. We also define the extended error vector

$$\tilde{\mathbf{e}}(k) = \tilde{\mathbf{d}}(k) - \tilde{\mathbf{y}}(k) \tag{8.43}$$

To obtain the frequency domain equivalent of the recursion (8.11), we replace $\mathbf{w}(k)$ and $\mathbf{e}(k)$ by their extended versions and note that Eq. (8.11) may also be written as

$$\tilde{\mathbf{w}}(k+1) = \tilde{\mathbf{w}}(k) + 2\mu\mathbf{P}_{N,0}\mathbf{X}_c^{\mathrm{T}}(k)\tilde{\mathbf{e}}(k) \tag{8.44}$$

where $\mathbf{X}_{\mathrm{c}}(k)$ is the circular matrix of samples of the filter input as defined by Eq. (8.33), $\mu = \mu_{\mathrm{B}}/L$, and

$$\mathbf{P}_{N,0} = \begin{bmatrix} \mathbf{I}_N & \mathbf{0} \\ \mathbf{0} & \mathbf{0} \end{bmatrix} \tag{8.45}$$

is a N'-by-N' windowing matrix that ensures that the last $L-1$ elements of the updated weight vector $\tilde{\mathbf{w}}(k+1)$ remain equal to zero after each iteration of Eq. (8.44). The fact that Eqs. (8.11) and (8.44) are equivalent can easily be shown by substituting for the vectors and matrices in Eq. (8.44) and expanding the result (Problem P8.4).

Conversion of the recursion (8.44) to its frequency domain equivalent can be done by premultiplying that on both sides by the DFT matrix $\mathcal{F}$ and using the identity $\mathcal{F}^{-1}\mathcal{F} = \mathbf{I}$ to obtain

$$\mathbf{w}_{\mathcal{F}}(k+1) = \mathbf{w}_{\mathcal{F}}(k) + 2\mu\mathcal{F}\mathbf{P}_{N,0}\mathcal{F}^{-1}\mathcal{F}\mathbf{X}_c^{\mathrm{T}}(k)\mathcal{F}^{-1}\mathcal{F}\tilde{\mathbf{e}}(k) \tag{8.46}$$

Using Eq. (8.40) and the identity (8.32), Eq. (8.46) can be written as

$$\mathbf{w}_{\mathcal{F}}(k+1) = \mathbf{w}_{\mathcal{F}}(k) + 2\mu\mathcal{P}_{N,0}\mathcal{X}_{\mathcal{F}}^{*}(k)\mathbf{e}_{\mathcal{F}}(k) \tag{8.47}$$

where $\mathbf{e}_{\mathcal{F}}(k) = \mathcal{F}\tilde{\mathbf{e}}(k)$ and

$$\mathcal{P}_{N,0} = \mathcal{F}\mathbf{P}_{N,0}\mathcal{F}^{-1} \tag{8.48}$$

Equations (8.41), (8.43), and (8.47) are the three steps required to complete each iteration of the FBLMS algorithm, namely, filtering, error estimation, and tap-weight adaptation, respectively. Figure 8.3 depicts a block diagram of the FBLMS algorithm, which shows how these steps are realized efficiently. The input samples are collected in an input buffer whose output is the vector $\tilde{\mathbf{x}}(k)$, consisting of L new samples and $N-1$ samples from the previous block(s). The vector $\tilde{\mathbf{x}}(k)$ is converted to the frequency domain

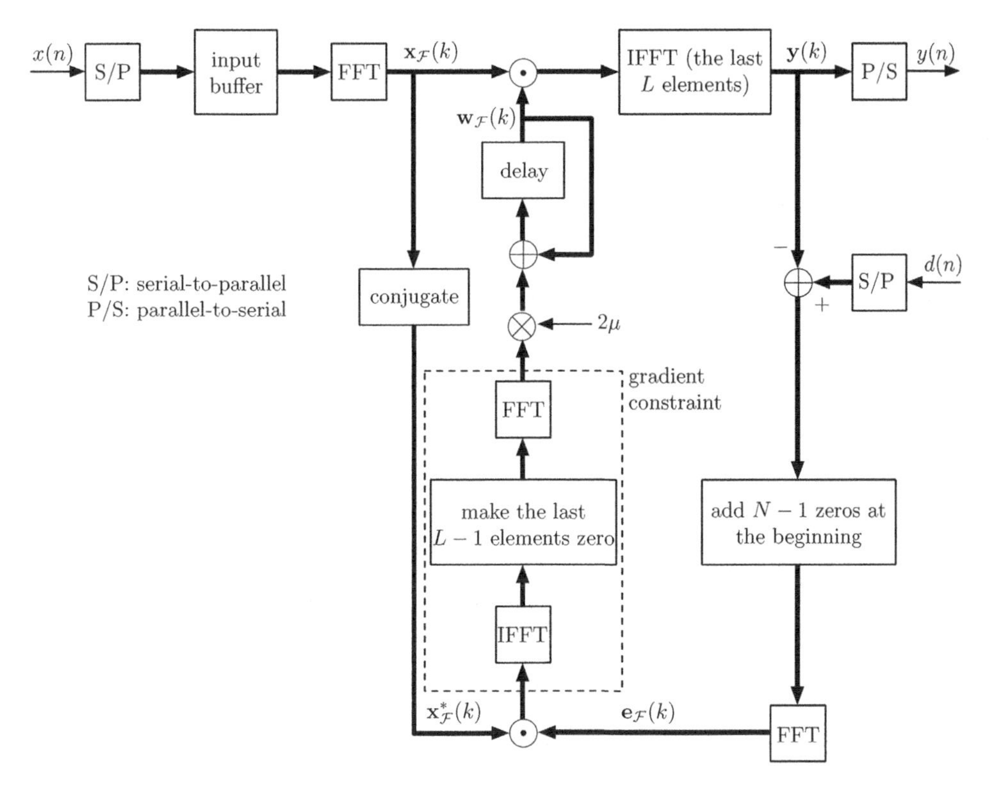

Figure 8.3 Implementation of the FBLMS algorithm.

and multiplied by the associated tap-weight vector, $\mathbf{w}_\mathcal{F}(k)$, on an element-wise basis. This gives the samples of the filter output in the frequency domain, which are subsequently converted to the time domain using an IFFT. The last L samples of this result correspond to the output samples of the current block and are sent to the output buffer as well as the error estimation section. The error vector, $\mathbf{e}(n)$, which consists of L elements, is extended to the length of $N + L - 1$ by appending $N - 1$ zeros at its beginning and converted to the frequency domain using a FFT algorithm. An element-wise multiplication of the error and conjugate of the input samples is performed in the frequency domain and the result is used to update the filter tap weights. Premultiplication of gradient vector $\mathcal{X}^*_\mathcal{F}(k)\mathbf{e}_\mathcal{F}(k)$ by $\mathcal{P}_{N,0}$ is necessary to ensure that the last $L - 1$ elements of the time domain equivalent of the tap-weight vector $\mathbf{w}_\mathcal{F}(k)$ are constrained to zero (Eq. 8.19). This constraining operation is implemented by converting the gradient vector $\mathcal{X}^*_\mathcal{F}(k)\mathbf{e}_\mathcal{F}(k)$ to time domain, making the last $L - 1$ elements zero, and converting back to the frequency domain, as shown in Figure 8.3.

8.3.1 Constrained and Unconstrained FBLMS Algorithms

Mansour and Gray (1982) have shown that under fairly mild conditions, the FBLMS algorithm can work well even when the tap-weight constraining matrix $\mathcal{P}_{N,0}$ is dropped from Eq. (8.47). They have shown that when the filter length, N, is chosen sufficiently large, and the input process, $x(n)$, does not satisfy some specific (unlikely to happen in practice) conditions, the update equation (8.47) and the recursion

$$\mathbf{w}_{\mathcal{F}}(k+1) = \mathbf{w}_{\mathcal{F}}(k) + 2\mu \boldsymbol{\mathcal{X}}_{\mathcal{F}}^{*}(k)\mathbf{e}_{\mathcal{F}}(k) \tag{8.49}$$

converge to the same set of tap weights. To differentiate between the two cases, Eq. (8.49) is called the *unconstrained* FBLMS recursion, while Eq. (8.47) is referred to as the *constrained* FBLMS recursion.

The block diagram given in Figure 8.3 is that of the constrained FBLMS algorithm. However, it is easily converted to the unconstrained FBLMS algorithm if the gradient-constraining operation, enclosed by the dotted line box, is dropped. We may thus note that the unconstrained FBLMS algorithm is much simpler to implement as two of the five FFTs and IFFTs are deleted from Figure 8.3. As we show in the next section, this simplification is at the cost of a higher misadjustment.

8.3.2 Convergence Behavior of the FBLMS Algorithm

In this section, we present a convergence analysis of the FBLMS algorithm. We start with the unconstrained recursion (8.49). Substituting Eq. (8.41) in Eq. (8.43), we get

$$\tilde{\mathbf{e}}(k) = \tilde{\mathbf{d}}(k) - \mathbf{P}_{0,L}\mathcal{F}^{-1}\boldsymbol{\mathcal{X}}_{\mathcal{F}}(k)\mathbf{w}_{\mathcal{F}}(k) \tag{8.50}$$

The fact that the first $N-1$ elements of $\tilde{\mathbf{d}}(k)$ are all-zero implies that $\tilde{\mathbf{d}}(k) = \mathbf{P}_{0,L}\tilde{\mathbf{d}}(k)$. Using this in Eq. (8.50), we obtain

$$\begin{aligned}\tilde{\mathbf{e}}(k) &= \mathbf{P}_{0,L}(\tilde{\mathbf{d}}(k) - \mathcal{F}^{-1}\boldsymbol{\mathcal{X}}_{\mathcal{F}}(k)\mathbf{w}_{\mathcal{F}}(k)) \\ &= \mathbf{P}_{0,L}\mathcal{F}^{-1}(\mathcal{F}\tilde{\mathbf{d}}(k) - \boldsymbol{\mathcal{X}}_{\mathcal{F}}(k)\mathbf{w}_{\mathcal{F}}(k))\end{aligned} \tag{8.51}$$

Premultiplying Eq. (8.51) on both sides by $\mathcal{F}$, we get

$$\mathbf{e}_{\mathcal{F}}(k) = \boldsymbol{\mathcal{P}}_{0,L}(\mathbf{d}_{\mathcal{F}}(k) - \boldsymbol{\mathcal{X}}_{\mathcal{F}}(k)\mathbf{w}_{\mathcal{F}}(k)) \tag{8.52}$$

where $\mathbf{d}_{\mathcal{F}}(k) = \mathcal{F}\tilde{\mathbf{d}}(k)$ and

$$\boldsymbol{\mathcal{P}}_{0,L} = \mathcal{F}\mathbf{P}_{0,L}\mathcal{F}^{-1} \tag{8.53}$$

Substituting Eq. (8.52) in the unconstrained update equation (8.49), we obtain

$$\mathbf{w}_{\mathcal{F}}(k+1) = \mathbf{w}(k) + 2\mu\boldsymbol{\mathcal{X}}_{F}^{*}(k)[\boldsymbol{\mathcal{P}}_{0,L}(\mathbf{d}_{\mathcal{F}}(k) - \boldsymbol{\mathcal{X}}_{\mathcal{F}}(k)\mathbf{w}_{\mathcal{F}}(k))] \tag{8.54}$$

Next, we define the tap-weight error vector

$$\mathbf{v}_{\mathcal{F}}(k) = \mathbf{w}_{\mathcal{F}}(k) - \mathbf{w}_{o,\mathcal{F}} \tag{8.55}$$

where $\mathbf{w}_{0,\mathcal{F}}$ is the optimum value of the filter tap-weight vector in the frequency domain. Using Eq. (8.55) in Eq. (8.54), we obtain, after some simple manipulation

$$\mathbf{v}_{\mathcal{F}}(k+1) = (\mathbf{I} - 2\mu\boldsymbol{\mathcal{X}}_{\mathcal{F}}^{*}(k)\boldsymbol{\mathcal{P}}_{0,L}\boldsymbol{\mathcal{X}}_{\mathcal{F}}(k))\mathbf{v}_{\mathcal{F}}(k) + 2\mu\boldsymbol{\mathcal{X}}_{\mathcal{F}}^{*}(k)\mathbf{e}_{0,\mathcal{F}}(k) \tag{8.56}$$

where $\mathbf{e}_{0,\mathcal{F}}(k)$ is the optimum error vector obtained when $\mathbf{w}_{\mathcal{F}}(k)$ is replaced by $\mathbf{w}_{0,\mathcal{F}}$.

Now, if we use the *independence assumption* and follow the same procedure as in Section 6.2, we will find that the convergence of the unconstrained FBLMS algorithm is controlled by the eigenvalues of the matrix

$$\mathcal{R}_{xx}^{u} = E[\mathcal{X}_{\mathcal{F}}^{*}(k)\mathcal{P}_{0,L}\mathcal{X}_{\mathcal{F}}(k)] \tag{8.57}$$

The matrix $\mathcal{R}_{xx}^{u}$ may be evaluated as follows. Substituting Eq. (8.40) and Eq. (8.53) in Eq. (8.57), we obtain

$$\mathcal{R}_{xx}^{u} = \mathcal{F}\mathbf{R}_{xx}^{u}\mathcal{F}^{-1} \tag{8.58}$$

where

$$\mathbf{R}_{xx}^{u} = E[\mathbf{X}_{c}^{\mathrm{T}}(k)\mathbf{P}_{0,L}\mathbf{X}_{c}(k)] \tag{8.59}$$

A careful examination of $\mathbf{R}_{xx}^{u}$ reveals that when L and N are large and the autocorrelation function of the input process, $x(n)$, that is, $\phi_{xx}(l)$, approaches zero for the lag values l much smaller than L and N, $\mathbf{R}_{xx}^{u}$ can be approximated by the N'-by-N' circular matrix whose first column is (Lee and Un, 1989)

$$\mathbf{r}_{xx}^{u} = L \times [\phi_{xx}(0)\,\phi_{xx}(1)\,\cdots\,\phi_{xx}(l)\,0\,\cdots\,0\,\phi_{xx}(l)\,\phi_{xx}(l-1)\,\cdots\,\phi_{xx}(1)]^{\mathrm{T}} \tag{8.60}$$

Using the properties of the circular matrices, this implies that[2]

$$\mathcal{R}_{xx}^{u} \approx L \times \mathrm{diag}\left(\Phi_{xx}\left(\mathrm{e}^{j\frac{2\pi\times 0}{N'}}\right),\ \Phi_{xx}\left(\mathrm{e}^{j\frac{2\pi}{N'}}\right), \ldots,\ \Phi_{xx}\left(\mathrm{e}^{j\frac{2\pi(N'-1)}{N'}}\right)\right) \tag{8.61}$$

where $\Phi_{xx}(\mathrm{e}^{j\omega})$ is the power spectral density of the input process, $x(n)$. The samples of $\Phi_{xx}(\mathrm{e}^{j\omega})$, on the right-hand side of Eq. (8.61), are obtained by taking the DFT of the vector $\mathbf{r}_{xx}^{u}/L$.

The fact that $\mathcal{R}_{xx}^{u}$ is a diagonal matrix implies that its eigenvalues are equal to its diagonal elements. The diagonal elements of $\mathcal{R}_{xx}^{u}$, as specified in Eq. (8.61), in turn, are proportional to the samples of the power spectral density of the underlying input process. Thus, for colored inputs, as it happens with the conventional LMS algorithm, the unconstrained FBLMS algorithm will also perform poorly. The same is also true for the constrained FBLMS algorithm as it is nothing but a fast implementation of the BLMS algorithm whose convergence behavior was studied in Section 8.1 and found performing very similar to the conventional LMS algorithm.

8.3.3 Step-Normalization

Convergence performance of the FBLMS algorithm can be greatly improved using individually normalized step-size parameters for each element of the tap-weight vector $\mathbf{w}_{\mathcal{F}}(k)$ rather than a common step-size parameter. This technique, known as *step-normalization* is similar to the one that was described in Chapter 7 for improving the convergence of the transform domain LMS (TDLMS) algorithm. It is implemented by replacing the scalar step-size parameter μ by the diagonal matrix

$$\boldsymbol{\mu}(k) = \mathrm{diag}[\mu_0(k),\ \mu_1(k),\ \ldots\ \mu_{N'-1}(k)] \tag{8.62}$$

[2] see also Problem P8.11 for an alternative derivation of Eq. (8.61).

where $\mu_i(k)$ is the normalized step-size parameters for the ith tap. These are obtained according to the equations

$$\mu_i(k) = \frac{\mu_o}{\hat{\sigma}^2_{x_{\mathcal{F},i}}(k)}, \quad \text{for} \quad i = 0, 1, \ldots, N' - 1 \tag{8.63}$$

where μ_o is a common unnormalized step-size parameter and $\hat{\sigma}^2_{x_{\mathcal{F},i}}(k)$'s are the power estimates of the samples of the filter input in the frequency domain, $x_{\mathcal{F},i}(k)$'s. These estimates may be obtained using the following recursion:

$$\hat{\sigma}^2_{x_{\mathcal{F},i}}(k) = \beta\hat{\sigma}^2_{x_{\mathcal{F},i}}(k-1) + (1-\beta)|x_{\mathcal{F},i}(k)|^2 \tag{8.64}$$

for $i = 0, 1, \ldots, N' - 1$, where β is a constant close to, but smaller than, 1.

8.3.4 Summary of the FBLMS Algorithm

Using the results developed in the previous sections, Table 8.1 gives a summary of the FBLMS algorithm. This table is in a form that can be readily converted to an efficient program code for implementing the FBLMS algorithm. In particular, the diagonal matrix $\boldsymbol{\mathcal{X}}_{\mathcal{F}}(k)$ is replaced by the vector $\mathbf{x}_{\mathcal{F}}(k)$ consisting of the diagonal elements of $\boldsymbol{\mathcal{X}}_{\mathcal{F}}(k)$. Also, $\boldsymbol{\mu}(k)$ is redefined as a column vector. Furthermore, the constraining/windowing operations defined by the matrices $\boldsymbol{\mathcal{P}}_{0,L}$ and $\boldsymbol{\mathcal{P}}_{N,0}$ are reexpressed more explicitly by replacing the unwanted elements of the corresponding vectors with zeros. We also use the terms FFT and IFFT to refer the DFT and IDFT operations. This is to emphasize that, in practice, fast Fourier transform algorithms are used to perform these operations efficiently.

In the derivations given in Section 8.3, it is assumed that the frequency domain tap-weight vector $\mathbf{w}_{\mathcal{F}}(k)$ satisfies the required time domain constraint, *viz.*, the last $L - 1$ elements of the IDFT of $\mathbf{w}_{\mathcal{F}}(k)$ are all zero. Thus, the constraint needs to be imposed only on the stochastic gradient vector $-2\boldsymbol{\mathcal{X}}^*_{\mathcal{F}}(k)\mathbf{e}_{\mathcal{F}}(k)$; see Eq. (8.47). This assumption, although theoretically correct if $\mathbf{w}_{\mathcal{F}}(0)$ is initialized to a constraint satisfying vector, may not continue to be true as the algorithm progresses. This is because the roundoff noise that is added to the elements of $\mathbf{w}_{\mathcal{F}}(k+1)$ will accumulate and result in a vector that may seriously violate the constraint after some iterations. In the case of unconstrained FBLMS algorithm, these errors are compensated by the adaptation process, as they propagate back to themselves through the unconstrained gradient vector $-2\boldsymbol{\mathcal{X}}^*_{\mathcal{F}}(k)\mathbf{e}_{\mathcal{F}}(k)$. However, this does not happen in the case of constrained FBLMS algorithm, because the gradient vector $-2\boldsymbol{\mathcal{X}}^*_{\mathcal{F}}(k)\mathbf{e}_{\mathcal{F}}(k)$ is constrained before being used for updating the tap weights. To resolve this problem, the tap-weight vector $\mathbf{w}_{\mathcal{F}}(k)$ should be regularly checked and constrained, as explained below.

Assume that $\mathbf{w}_{\mathcal{F}}(k)$ satisfies the required time-domain constraint. That is, the last $L - 1$ elements of IDFT of $\mathbf{w}_{\mathcal{F}}(k)$ are all zero. This implies that

$$\mathbf{w}_{\mathcal{F}}(k) = \boldsymbol{\mathcal{P}}_{N,0}\mathbf{w}_{\mathcal{F}}(k) \tag{8.65}$$

Using this in the constrained FBLMS recursion (8.47), we obtain

$$\mathbf{w}_{\mathcal{F}}(k+1) = \boldsymbol{\mathcal{P}}_{N,0}[\mathbf{w}_{\mathcal{F}}(k) + 2\mu\boldsymbol{\mathcal{X}}^*_{\mathcal{F}}(k)\mathbf{e}_{\mathcal{F}}(k)] \tag{8.66}$$

Table 8.1 Summary of the FBLMS algorithm.

Input: Tap-weight vector, $\mathbf{w}_{\mathcal{F}}(k)$,
Signal power estimates, $\hat{\sigma}^2_{x_{\mathcal{F},i}}(k-1)$'s,
Extended input vector,
$\tilde{\mathbf{x}}(k) = [x(kL-N+1)\, x(kL-N+2)\, \cdots\, x(kL+L-1)]^{\mathrm{T}}$,
and Desired output vector,
$\mathbf{d}(k) = [d(kL)\, d(kL+1)\, \cdots\, d(kL+L-1)]^{\mathrm{T}}$

Output: Filter output,
$\mathbf{y}(k) = [y(kL)\, y(kL+1)\, \cdots\, y(kL+L-1)]^{\mathrm{T}}$,
Tap-weight vector update, $\mathbf{w}(k+1)$

1. Filtering:

$$\mathbf{x}_{\mathcal{F}}(k) = \mathrm{FFT}(\tilde{\mathbf{x}}(k))$$
$$\mathbf{y}(k) = \text{the last } L \text{ elements of } \mathrm{IFFT}(\mathbf{x}_{\mathcal{F}}(k) \odot \mathbf{w}_{\mathcal{F}}(k))$$

2. Error estimation:

$$\mathbf{e}(k) = \mathbf{d}(k) - \mathbf{y}(k)$$

3. Step-normalization:

$$\text{for} \quad i = 0 \quad \text{to} \quad N'-1$$
$$\hat{\sigma}^2_{x_{\mathcal{F},i}}(k) = \beta\hat{\sigma}^2_{x_{\mathcal{F},i}}(k-1) + (1-\beta)|x_{\mathcal{F},i}(k)|^2$$
$$\mu_i(k) = \mu_o/\hat{\sigma}^2_{x_{\mathcal{F},i}}(k)$$
$$\boldsymbol{\mu}(k) = [\mu_0(k)\ \ \mu_1(k)\ \cdots\ \mu_{N'-1}(k)]^{\mathrm{T}}$$

4. Tap-weight adaptation:

$$\mathbf{e}_{\mathcal{F}}(k) = \mathrm{FFT}\left(\begin{bmatrix}\mathbf{0}\\ \mathbf{e}(k)\end{bmatrix}\right)$$
$$\mathbf{w}_{\mathcal{F}}(k+1) = \mathbf{w}_{\mathcal{F}}(k) + 2\boldsymbol{\mu}(k) \odot \mathbf{x}^*_{\mathcal{F}}(k) \odot \mathbf{e}_{\mathcal{F}}(k)$$

5. Tap-weight constraint:

$$\mathbf{w}_{\mathcal{F}}(k+1) = \mathrm{FFT}\left(\begin{bmatrix}\text{first } M \text{ elements of } \mathrm{IFFT}(\mathbf{w}_{\mathcal{F},l}(k+1))\\ \mathbf{0}\end{bmatrix}\right)$$

Notes:

- N: filter length; L: block length; $N' = N + L - 1$.
- $\mathbf{0}$ denotes the column zero vectors with appropriate length to extend vectors to the length of N'.
- $\odot$ denotes the element-wise multiplication of vectors.
- Here, $\boldsymbol{\mu}(k)$ is defined as a column vector. This is different from the definition of $\boldsymbol{\mu}(k)$ in the text where it is defined as a diagonal matrix.
- Step 5 is applicable only for the constrained FBLMS algorithm.

This recursion constrains $\mathbf{w}_{\mathcal{F}}(k+1)$ after every iteration and thus prevents any accumulation of roundoff noise errors. Implementation of the constrained FBLMS algorithm given in Table 8.1 is based on this recursion.

Before ending this section, some remarks on real- and complex-valued signal cases would be instructive. Although, all the derivations in this chapter are given for real-valued signals in order to prevent some unnecessary confusions, the final algorithm presented in Table 8.1 is applicable to both real- and complex-valued signals. Another point to be noted in the case of real-valued signals is that all the frequency domain vectors will be conjugate symmetric.[3] This implies that the first half of these frequency domain vectors contains all the necessary information and, hence, their second halves can be ignored. This reduces the computational complexity and memory requirement of the FBLMS algorithm by about 50%.

8.3.5 FBLMS Misadjustment Equations

Derivation of the misadjustment equations for the various implementations of the FBLMS algorithms is quite tedious and long. This is done in Appendix 8B. The derivations presented in Appendix 8B result in the following misadjustment equations:

$$\mathcal{M}^{\rm c}_{\rm FBLMS} \approx \mu N \phi_{xx}(0) \tag{8.67}$$

$$\mathcal{M}^{\rm u}_{\rm FBLMS} \approx \mu N' \phi_{xx}(0) \tag{8.68}$$

$$\mathcal{M}^{\rm cn}_{\rm FBLMS} \approx \mu_o N/N' \tag{8.69}$$

$$\mathcal{M}^{\rm un}_{\rm FBLMS} \approx \mu_o \tag{8.70}$$

In these equations, superscripts c and u refer to the constrained and unconstrained versions of the FBLMS algorithm, respectively, and the superscript "n" indicates that the step-normalization has been applied. We also note that, similar to Eqs. (6.64) and (8.13), (8.67)–(8.70) are valid only for misadjustment values of 10% or lower.

It can be immediately concluded from Eqs. (8.67)–(8.70) that the constrained FBLMS algorithm outperforms its unconstrained counterpart, in the sense that the former results in a lower misadjustment, for a given step-size parameter. Equivalently, for a given misadjustment, the constrained FBLMS algorithm converges faster than its unconstrained counterpart. The difference between the two algorithms is determined by the ratio $\frac{N}{N'}(=\frac{N}{N+L-1})$, which, in turn, is determined by the ratio $\frac{L}{N}$. Clearly, when $L \ll N$, $\frac{N}{N'} \approx 1$, then the difference between the constrained FBLMS algorithm and its unconstrained counterpart becomes insignificant. On the other hand, when L and N are comparable, the difference between the two algorithms will be significant.

8.3.6 Selection of the Block Length

Block processing of signals, in general, results in certain time delay at the system output. In many applications, this processing delay may be intolerable and hence it has to be minimized. It arises because a block of samples of input signal has to be collected before the processing of the data can begin. Consequently, the processing delay increases with block length. On the other hand, the per sample computational complexity of a block processing system varies with the block length, L. For values of L smaller than the filter length, N, per sample computational complexity of the FBLMS algorithm decreases as L increases. It reaches close to its minimum when $L \approx N$. Thus, in applications where

[3] A length M vector $\mathbf{u} = [u_0\, u_1 \cdots u_{M-1}]^{\rm T}$ is called *conjugate symmetric* when $u_i = u^*_{M-i}$, for $i = 0, 1, \cdots, \lfloor \frac{M}{2} \rfloor$, where $\lfloor \frac{M}{2} \rfloor$ denotes integer part of $\frac{M}{2}$.

the processing delay is not an issue, L is usually chosen close to N. The exact value of L depends on N. For a given N, one should choose L so that $N' = N + L - 1$ is an appropriate composite number and efficient FFT and IFFT algorithms can be used in the realization of the FBLMS algorithm. On the other hand, in applications where it is important to keep the processing delay small, one may need to strike a compromise between system complexity and processing delay. In such applications, an alternative implementation of the FBLMS algorithm, which is introduced in the next section, is found to be more efficient.

8.4 The Partitioned FBLMS Algorithm

When the filter length, N, is large and a block length, L, much smaller than N is used, an efficient implementation of the FBLMS algorithm can be derived by dividing (partitioning) the convolution sum of Eq. (8.17) into a number of smaller sums and proceeding as discussed below. The resulting implementation is called *partitioned FBLMS (PFBLMS)* algorithm.

The PFBLMS algorithm has apparently been discovered by a number of independent researchers and has been given different names: Asharif *et al.* (1986) and Asharif and Amano (1994) call it frequency bin adaptive filtering; Soo and Pang (1987, 1990) refer to it as multidelay FBLMS; and Sommen (1989) use the name PFBLMS.

Let us assume that $N = P \cdot M$, where P and M are integers, and note that the convolution sum of Eq. (8.17) may be written as

$$y(n) = \sum_{l=0}^{P-1} y_l(n) \tag{8.71}$$

where

$$y_l(n) = \sum_{i=0}^{M-1} w_{i+lM}\, x(n - lM - i) \tag{8.72}$$

To develop a frequency domain implementation of these convolutions, we choose a block length $L = M$ and divide the input data into blocks of length $2M$ samples such that the last M samples of, say, the kth block are same as the first M samples of the $(k + 1)$th block. Then, the convolution sum in Eq. (8.72) can be evaluated using the circular convolution of these data blocks with the appropriate weight vectors, having padded with M zeros. Using $x(kM + M - 1)$ to represent the newest sample in the input, we define the vectors

$$\mathbf{x}_{\mathcal{F},l}(k) = \text{FFT}([x((k-l)M - M) \quad x((k-l)M - M + 1) \quad \cdots \quad x((k-l)M + M - 1)]^{\text{T}}) \tag{8.73}$$

$$\mathbf{w}_{\mathcal{F},l}(k) = \text{FFT}([w_{lM}(k) \;\; w_{lM+1}(k) \;\; \cdots \;\; w_{lM+M-1}(k) \;\; \overbrace{0 \;\; 0 \;\; \cdots \;\; 0}^{M}]^{\text{T}}) \tag{8.74}$$

$$\mathbf{y}_l(k) = [y_l(kM) \quad y_l(kM+1) \quad \cdots \quad y_l(kM + M - 1)]^{\text{T}} \tag{8.75}$$

and note that

$$\mathbf{y}_l(k) = \text{the last } M \text{ elements of IFFT}(\mathbf{w}_{\mathcal{F},l}(k) \odot \mathbf{x}_{\mathcal{F},l}(k)) \tag{8.76}$$

where $\odot$ denotes multiplication on the element-wise basis, k, as before, is the block index, and l is the partition index. We also define

$$\mathbf{y}(k) = [y(kM)\ y(kM+1)\ \cdots\ y(kM+M-1)]^{\mathrm{T}} \tag{8.77}$$

and note that

$$\mathbf{y}(k) = \sum_{l=0}^{P-1} \mathbf{y}_l(k) \tag{8.78}$$

Furthermore, from Eq. (8.73), we note that

$$\mathbf{x}_{\mathcal{F},l}(k) = \mathbf{x}_{\mathcal{F},0}(k-l) \tag{8.79}$$

It may be noted that according to the derivations in the earlier sections, for a filter (here, partition) length of M and a block length of $L = M$, the frequency domain vectors of length $2M - 1$ are sufficient to perform the necessary convolutions in the frequency domain. Here, we are using vectors that are of length $2M$ as this greatly simplifies the implementation of the PFBLMS algorithm. In particular, we note that Eq. (8.79) holds only when $L = M$.

Substituting Eq. (8.76) in Eq. (8.78), interchanging the order of summation and IFFT, and using Eq. (8.79), we obtain

$$\mathbf{y}(k) = \text{the last } M \text{ elements of IFFT}\left(\sum_{l=0}^{P-1} \mathbf{w}_{\mathcal{F},l}(k) \odot \mathbf{x}_{\mathcal{F},0}(k-l)\right) \tag{8.80}$$

Using this result, the block diagram of the PFBLMS algorithm may be proposed, as depicted in Figure 8.4. Here, the delays, z^{-1}'s, are in the unit of block size and the thick lines represent frequency domain vectors. Also, for our later discussion, it may be remarked here that the implementation of the summation on the right-hand side of Eq. (8.80) can also be considered as a parallel bank of $2M$ transversal filters, each of length P, with the jth filter processing the frequency domain samples belonging to the jth frequency bin, for $j = 0, 1, \ldots, 2M - 1$.

The adaptation of the filter tap weights is done according to the recursions

$$\mathbf{w}_{\mathcal{F},l}(k+1) = \mathbf{w}_{\mathcal{F},l}(k) + \boldsymbol{\mu}(k) \odot \mathbf{x}_{\mathcal{F},0}(k-l) \odot \mathbf{e}_{\mathcal{F}}(k),$$
$$\text{for } l = 0, 1, \ldots, P-1 \tag{8.81}$$

where $\boldsymbol{\mu}(k)$ is the vector of the associated step-size parameters that may be normalized in a similar manner as Eq. (8.63),

$$\mathbf{e}_{\mathcal{F}}(k) = \text{FFT}\begin{bmatrix} \mathbf{d}(k) - \mathbf{y}(k) \\ \mathbf{0} \end{bmatrix} \tag{8.82}$$

$\mathbf{d}(k) = [d(kM)\ d(kM+1)\ \cdots\ d(kM+M-1)]^{\mathrm{T}}$, and $\mathbf{0}$ is the length M zero column vector.

Recursion (8.81) corresponds to the unconstrained PFBLMS algorithm. The constrained PFBLMS algorithm recursion is obtained by constraining the filter tap weights after every iteration of Eq. (8.81).

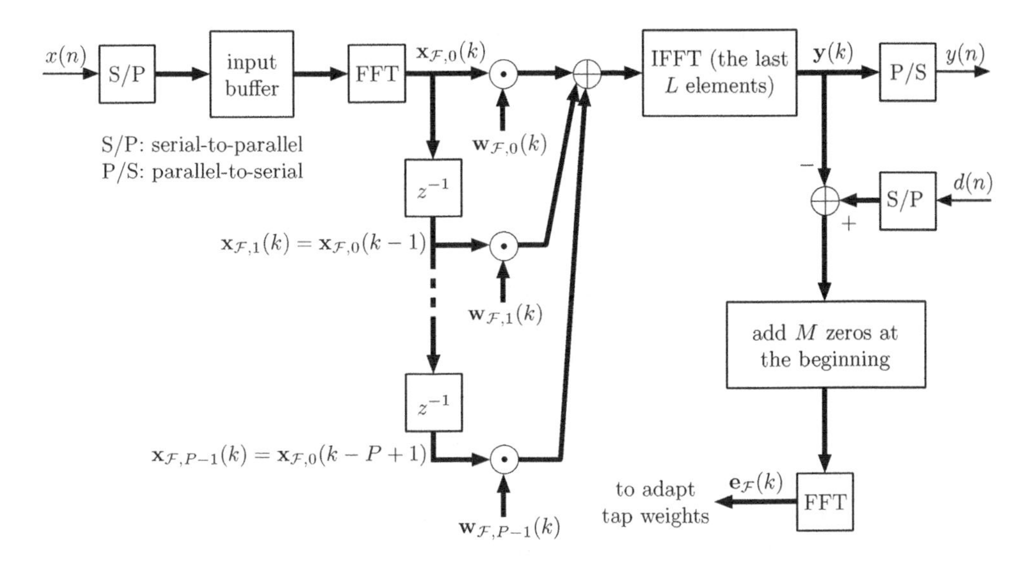

Figure 8.4 The partitioned FBLMS (PFBLMS) algorithm for $L = M$.

8.4.1 Analysis of the PFBLMS Algorithm

In this section, we analyze the convergence behavior of the PFBLMS algorithm. This analysis reveals that PFBLMS algorithm suffers from slow convergence and hence, we suggest some simple solutions to improve its convergence. The main emphasis of this section is convergence behavior of the unconstrained PFBLMS algorithm. However, we also make some comments on the behavior of the constrained PFBLMS algorithm.

From the analysis of the FBLMS algorithm, we recall that the frequency domain samples of input which belong to different frequency bins (i.e., the signal samples at the output of the first FFT in Figure 8.3) are approximately uncorrelated with one another and hence, the associated correlation matrix may be approximated by a diagonal matrix. The step-normalization is then used to equalize the time constants of the various modes of convergence of the algorithm.

Extending the above result to the PFBLMS structure, we find that in this case, there are $2M$ parallel transversal filters (one belonging to each frequency bin of the signal samples) whose associated input sequences are approximately uncorrelated with one another. Thus, a simple approach to analyze the PFBLMS algorithm is to assume that the transversal filters associated with each bin converge independent of one another so that we can concentrate on the convergence behavior of these as independent filters. We use the term *frequency bin filter* to refer to these independent filters. We note that the analysis of the PFBLMS algorithm based on the assumption of independent frequency bins is rather coarse. However, a more exact analysis of the PFBLMS algorithm will be quite involved and beyond the scope of this book.

From Eq. (8.73) and Figure 8.4, we note that the tap-input vector of the ith frequency bin filter is

$$\mathbf{x}_{\mathcal{F}}^{b_i}(k) = [x_{\mathcal{F},0,i}(k)\; x_{\mathcal{F},0,i}(k-1)\; \cdots\; x_{\mathcal{F},0,i}(k-P+1)]^{\mathrm{T}} \tag{8.83}$$

where $x_{\mathcal{F},0,i}(k)$ is the ith element of $\mathbf{x}_{\mathcal{F},0}(k)$. Convergence behavior of the ith frequency bin filter is then determined by the eigenvalue spread of the correlation matrix

$$\mathcal{R}_{xx}^{b_i} = E[\mathbf{x}_{\mathcal{F}}^{b_i}(k)\mathbf{x}_{\mathcal{F}}^{b_i\mathrm{H}}(k)] \tag{8.84}$$

or, equivalently, by its normalized version

$$\mathcal{R}_{xx}^{b_i,\mathrm{n}} = (\mathrm{diag}[\mathcal{R}_{xx}^{b_i}])^{-1}\mathcal{R}_{xx}^{b_i} \tag{8.85}$$

However, we note that when the input process, $x(n)$, is stationary, the diagonal elements of $\mathcal{R}_{xx}^{b_i}$ are all identical and thus, $\mathrm{diag}[\mathcal{R}_{xx}^{b_i}]$ is proportional to the identity matrix. Hence, $\mathcal{R}_{xx}^{b_i}$ and $\mathcal{R}_{xx}^{b_i,\mathrm{n}}$ have the same eigenvalue spread.

Now, observe the fact that the matrix $\mathcal{R}_{xx}^{b_i}$ is a subdiagonal part of the dual of matrix $\mathcal{R}_{xx}^{u}$, which was obtained in Section 8.3.2 while analyzing the unconstrained FBLMS algorithm (8.58). As a result, the following analysis is applicable only to the unconstrained PFBLMS algorithm as it is based on a study of the matrix $\mathcal{R}_{xx}^{b_i,\mathrm{n}}$. The constrained PFBLMS algorithm requires further attention and we will make some comments on its convergence behavior at the end of this subsection. A modified version of the constrained PFBLMS algorithm, with significantly less computational complexity, will be introduced in Section 8.4.5.

In order to keep the analysis simple, we consider the case where the input sequence, $x(n)$, is white. Even though this assumption simplifies the analysis greatly, the results obtained are still able to bring out the salient features of the algorithm. For example, the computer simulations given in the next section show that the conclusions drawn in this section remain valid even when $x(n)$ is highly colored.

The ith element of $\mathbf{x}_{\mathcal{F},0,i}(k)$ (i.e., the ith frequency bin sample of the filter input) is

$$x_{\mathcal{F},0,i}(k) = \sum_{m=0}^{2M-1} x(kM - M + m)\mathrm{e}^{-j2\pi im/2M} \tag{8.86}$$

When $x(n)$ is white, it is straightforward to show that

$$E[x_{\mathcal{F},0,i}(k-l)x_{\mathcal{F},0,i}^{*}(k-m)] = \begin{cases} 2M\sigma_x^2, & \text{for } \ l = m \\ (-1)^i \times M\sigma_x^2, & \text{for } \ l = m \pm 1 \\ 0, & \text{otherwise} \end{cases} \tag{8.87}$$

where σ_x^2 is the variance of $x(n)$. Using this result, we obtain

$$\mathcal{R}_{xx}^{b_i,\mathrm{n}} = \begin{bmatrix} 1 & \alpha_i & 0 & 0 & \cdots & 0 & 0 \\ \alpha_i & 1 & \alpha_i & 0 & \cdots & 0 & 0 \\ 0 & \alpha_i & 1 & \alpha_i & \cdots & 0 & 0 \\ \vdots & \vdots & \vdots & \vdots & \ddots & \vdots & \vdots \\ 0 & 0 & 0 & 0 & \cdots & \alpha_i & 1 \end{bmatrix} \tag{8.88}$$

where $\alpha_i = (-1)^i \times 0.5$.

The eigenvalues of $\mathcal{R}_{xx}^{b_i,\mathrm{n}}$ (which are independent of i) can be obtained numerically. These are presented in Table 8.2, for values of P in the range of 2 to 10. It is noted

Table 8.2 Eigenvalues of $\mathcal{R}_{xx}^{b_i,\mathrm{n}}$ for different number of partitions, P.

P	2	3	4	5	6	7	8	9	10
λ_0	1.500	1.707	1.809	1.866	1.901	1.924	1.940	1.951	1.959
λ_1	0.500	1.000	1.309	1.500	1.623	1.707	1.766	1.809	1.841
λ_2		0.293	0.691	1.000	1.222	1.383	1.500	1.588	1.655
λ_3			0.191	0.500	0.778	1.000	1.174	1.309	1.415
λ_4				0.134	0.376	0.617	0.826	1.000	1.142
λ_5					0.099	0.293	0.500	0.691	0.858
λ_6						0.076	0.234	0.412	0.585
λ_7							0.060	0.191	0.345
λ_8								0.049	0.159
λ_9									0.041
$\lambda_{\max}/\lambda_{\min}$	3	5.828	9.472	13.93	19.20	25.27	32.16	39.86	48.37

that these are widely spread and their dispersion increases significantly as P grows. This means that for large values of P, the PFBLMS algorithm may suffer from slow convergence and/or numerical instability, as in Eq. (8.88), $\mathcal{R}_{xx}^{b_i,\mathrm{n}}$ becomes badly ill-conditioned, for large P.

Observe from the PFBLMS structure shown in Figure 8.4 that the successive partitions of the input samples are 50% overlapped. The value of $|\alpha_i| = 0.5$ in Eq. (8.88), which in turn results in the large eigenvalue spread in $\mathcal{R}_{xx}^{b_i,\mathrm{n}}$ is a direct consequence of this 50% overlapping. Numerical studies show that this eigenvalue spread reduces as $|\alpha_i|$ decreases. Furthermore, $|\alpha_i|$ can be reduced by reducing the amount of overlap of the successive partitions of the input samples. This is easily achieved by choosing a block length, L, smaller than the partition length, M, as explained in the next section.

Before proceeding with this modification of the PFBLMS algorithm, we shall make some comments on the convergence behavior of the constrained PFBLMS algorithm. As was noted before, the correlation matrix $\mathcal{R}_{xx}^{b_i,\mathrm{n}}$ was the outcome of an analysis of the unconstrained PFBLMS algorithm. A detailed examination of the constrained PFBLMS algorithm is rather involved and beyond the scope of this book. As we will demonstrate through computer simulations later, the effect of overlapping of successive blocks is resolved when the tap weights of the filter are constrained. As a result, we find that the constrained PFBLMS algorithm does not have any convergence problem. It converges almost as fast as its nonpartitioned counterpart. These observations are in line with the theoretical findings that have presented in Chan (2000) and also in Chan and Farhang-Boroujeny (2001).

8.4.2 PFBLMS Algorithm with $M > L$

Assuming a block length L and a partition length M, define the vector

$$\tilde{\mathbf{x}}_0(k) = [x(kL-M)\, x(kL-M+1)\, \cdots\, x(kL+L-1)]^{\mathrm{T}} \tag{8.89}$$

Let us choose $M = pL$, where p is an integer. As we show later, this choice of L and M leads to an efficient implementation of the PFBLMS algorithm. We note that

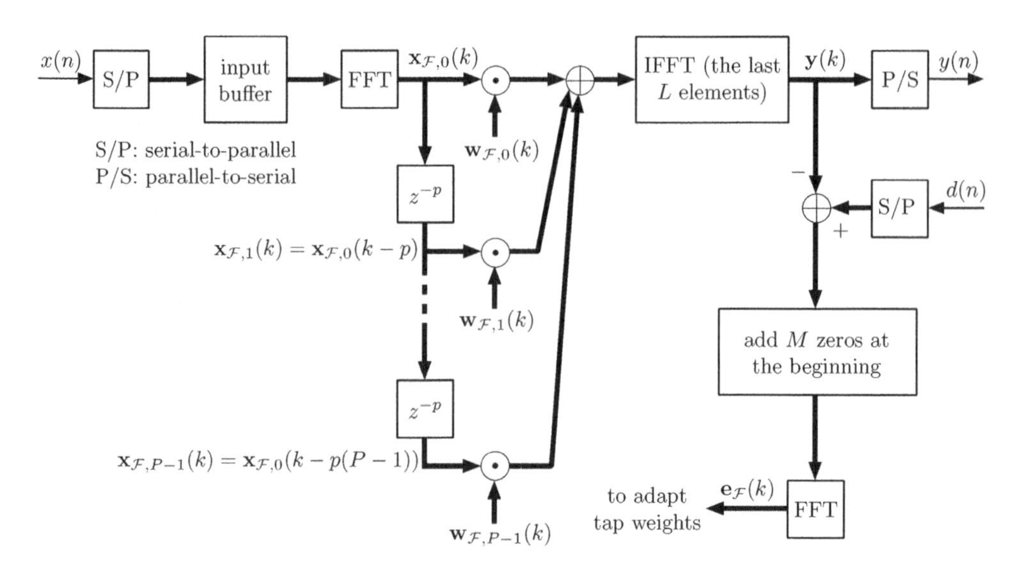

Figure 8.5 The PFBLMS algorithm for $M = pL$.

if we want to use the DFT to compute the partitions in Eq. (8.71), then $\tilde{\mathbf{x}}_0(k)$ corresponds to the vector of input samples associated with the first partition, that is, $y_0(n)$ in Eq. (8.71) with $n = kL + L - 1$. Observe that the first element of $\tilde{\mathbf{x}}_0(k)$ is $x(kL - M)$. Similarly, the vectors corresponding to the subsequent partitions start with samples $x(kL - 2M) = x((k-p)L - M)$, $x(kL - 3M) = x((k-2p)L - M)$, and so on. We thus find that $\tilde{\mathbf{x}}_l(k) = \tilde{x}_0(k - pl)$ or

$$\mathbf{x}_{\mathcal{F},l}(k) = \mathbf{x}_{\mathcal{F},0}(k - pl), \quad \text{for} \quad l = 1, 2, \ldots, P-1 \tag{8.90}$$

Using this result, Figure 8.5 depicts an implementation of the PFBLMS algorithm when $M = pL$. Comparing Figures 8.4 and 8.5, we find that the major difference between the two structures is that each delay unit in Figure 8.4 is replaced by p delay units in Figure 8.5. Table 8.3 gives a summary of the PFBLMS algorithm for the case where $M = pL$.

Following the notations used before, we note that, for the new arrangement in Figure 8.5,

$$\mathbf{x}_{\mathcal{F}}^{b_i}(k) = [x_{\mathcal{F},0,i}(k)\; x_{\mathcal{F},0,i}(k-p)\; \cdots\; x_{\mathcal{F},0,i}(k-(P-1)p)]^{\mathrm{T}} \tag{8.91}$$

Also,

$$x_{\mathcal{F},0,i}(k) = \sum_{m=0}^{M+L-1} x(kL - M + m)\mathrm{e}^{-j2\pi im/(M+L)} \tag{8.92}$$

Assuming that $x(n)$ is white, we obtain from Eq. (8.92)

$$E[x_{\mathcal{F},0,i}(k-pl)x^*_{\mathcal{F},0,i}(k-pm)] = \begin{cases} (M+L)\sigma_x^2, & \text{for } l = m \\ \mathrm{e}^{j\frac{2\pi p(m-l)i}{p+1}} L\sigma_x^2, & \text{for } l = m \pm 1 \\ 0, & \text{otherwise.} \end{cases} \tag{8.93}$$

Table 8.3 Summary of the PFBLMS algorithm.

Input: Tap-weight vectors, $\mathbf{w}_{\mathcal{F},l}(k)$, $l = 0, 1, \ldots, P-1$,
Extended input vector,
$\tilde{\mathbf{x}}_0(k) = [x(kL-M) \quad x(kL-M+1) \quad \cdots \quad x(kL+L-1)]^{\mathrm{T}}$,
The past frequency domain vectors of input, $\mathbf{x}_{\mathcal{F},0}(k-l)$, for $l = 1, 2, \ldots, (P-1)p$,
and desired output vector,
$\mathbf{d}(k) = [d(kL) \quad d(kL+1) \quad \cdots \quad d(kL+L-1)]^{\mathrm{T}}$.

Output: Filter output, $\mathbf{y}(k) = [y(kL)\ y(kL+1) \cdots y(kL+L-1)]^{\mathrm{T}}$,
Tap-weight vector update, $\mathbf{w}_{\mathcal{F},l}(k+1)$, $l = 0, 1, \ldots, P-1$.

---

1. Filtering:

$$\mathbf{x}_{\mathcal{F},0}(k) = \mathrm{FFT}(\tilde{x}_0(k))$$

$$\mathbf{y}(k) = \text{the last } L \text{ elements of IFFT}\left(\sum_{l=0}^{P-1} \mathbf{w}_{\mathcal{F},l}(k) \odot \mathbf{x}_{\mathcal{F},0}(k-pl)\right)$$

2. Error estimation:

$$\mathbf{e}(k) = \mathbf{d}(k) - \mathbf{y}(k)$$

3. Step-normalization:

for $i = 0$ to $M' - 1$

$$\hat{\sigma}^2_{x_{\mathcal{F},0,i}}(k) = \beta\hat{\sigma}^2_{x_{\mathcal{F},0,i}}(k-1) + (1-\beta)|x_{\mathcal{F},0,i}(k)|^2$$

$$\mu_i(k) = \mu_o/\hat{\sigma}^2_{x_{\mathcal{F},0,i}}(k)$$

$$\boldsymbol{\mu}(k) = [\mu_0(k) \quad \mu_1(k) \quad \cdots \quad \mu_{M'-1}(k)]^{\mathrm{T}}$$

5. Tap-weight adaptation:

$$\mathbf{e}_{\mathcal{F}}(k) = \mathrm{FFT}\left(\begin{bmatrix} \mathbf{0} \\ \mathbf{e}(k) \end{bmatrix}\right)$$

for $l = 0$ to $P - 1$

$$\mathbf{w}_{\mathcal{F},l}(k+1) = \mathbf{w}_{\mathcal{F},l}(k) + 2\boldsymbol{\mu}(k) \odot \mathbf{x}^*_{\mathcal{F},0}(k-pl) \odot \mathbf{e}_{\mathcal{F}}(k)$$

5. Tap-weight constraint:

for $l = 0$ to $P - 1$

$$\mathbf{w}_{\mathcal{F},l}(k+1) = \mathrm{FFT}\left(\begin{bmatrix} \text{first } M \text{ elements of IFFT}(\mathbf{w}_{\mathcal{F},l}(k+1)) \\ \mathbf{0} \end{bmatrix}\right)$$

Notes:

- M: partition length; L: block length; $M' = M + L$.
- $\mathbf{0}$ denotes column zero vectors with appropriate length to extend vectors to the length of M'.
- $\odot$ denotes element-wise multiplication of vectors.
- Step 5 is applicable only for the constrained PFBLMS algorithm.

Table 8.4 Eigenvalue spread, $\lambda_{\max}/\lambda_{\min}$, of $\mathcal{R}_{xx}^{b_i,\mathrm{n}}$ for $P = 10$ and different values of p.

p	1	2	3	4	5	6	7	8	9	10
$\lambda_{\max}/\lambda_{\min}$	48.37	4.55	2.84	2.25	1.94	1.75	1.63	1.54	1.47	1.42

Using this, we get

$$\mathcal{R}_{xx}^{b_i,\mathrm{n}} = \begin{bmatrix} 1 & \alpha_i & 0 & 0 & \cdots & 0 & 0 \\ \alpha_i^* & 1 & \alpha_i & 0 & \cdots & 0 & 0 \\ 0 & \alpha_i^* & 1 & \alpha_i & \cdots & 0 & 0 \\ \vdots & \vdots & \vdots & \vdots & \cdots & \vdots & \vdots \\ 0 & 0 & 0 & 0 & \cdots & \alpha_i^* & 1 \end{bmatrix} \tag{8.94}$$

where $\alpha_i = \frac{1}{p+1} e^{j\frac{2\pi pi}{p+1}}$.

The eigenvalue spread of $\mathcal{R}_{xx}^{b_i,\mathrm{n}}$ (which is independent of i) for values of p changing from 1 to 10 and a fixed value of $P = 10$ are given in Table 8.4. The results clearly show that reducing the overlap of successive partitions significantly improves the convergence behavior of the unconstrained PFBLMS algorithm.

8.4.3 PFBLMS Misadjustment Equations

The following results can be derived for the PFBLMS algorithm by following the same line of derivations as in Appendix 8B.

$$\mathcal{M}_{\mathrm{PFBLMS}}^{\mathrm{c}} \approx \mu P M \phi_{xx}(0) \tag{8.95}$$

$$\mathcal{M}_{\mathrm{PFBLMS}}^{\mathrm{u}} \approx \mu P (M + L) \phi_{xx}(0) \tag{8.96}$$

$$\mathcal{M}_{\mathrm{PFBLMS}}^{\mathrm{cn}} \approx \mu_o P \frac{M}{M + L} \tag{8.97}$$

$$\mathcal{M}_{\mathrm{PFBLMS}}^{\mathrm{un}} \approx \mu_o P \tag{8.98}$$

As in Section 8.5, here also we find that the constrained PFBLMS algorithm achieves a lower level of misadjustment compared to its unconstrained counterpart. The price paid for this is a higher computational complexity.

8.4.4 Computational Complexity and Memory Requirement

In this section, we give some figures indicating the computational complexity and memory requirement of the PFBLMS algorithm. Instead of specifying the exact number of multiplications and additions, we specify a macrofigure such as the number of butterflies for quantifying computational complexity as this may be more meaningful in the case of such algorithms. To estimate the memory requirement, we consider only its major blocks and ignore details such as the temporary memory locations required as these will depend on the DSP system used and also the efficiency of the code written. Furthermore, we only discuss the computational complexity of the unconstrained PFBLMS algorithm. The constrained PFBLMS algorithm has not been discussed here as its computational complexity will depend, to a great extent, on how the constraining step is implemented. We make some comments on this in the next subsection.

In the implementation of the unconstrained PFBLMS algorithm, processing of each data block requires two $(p+1)L$ $(=M+L)$ point FFTs and one IFFT of the same length. Assuming that the data signals are all real-valued, $(p+1)L$ is chosen a power of 2, and an efficient FFT algorithm such as the one used by Bergland (1968) is used, then $\frac{(p+1)L}{4}\log_2\frac{(p+1)L}{2}$ butterflies will have to be performed to complete each FFT. Computation of output samples in the frequency domain, that is, $\mathbf{x}_{\mathcal{F},0}(k-l)\odot\mathbf{w}_{\mathcal{F},l}(k)$, for $l=0,1,\ldots P-1$, and implementation of tap-weight adaptation recursion Eq. (8.81) require two-and-half complex multiplications and two complex additions per data point. Since step-size parameters are real-valued, multiplication of a gradient term with its step-size parameter is counted as half complex multiplication. Furthermore, step-normalization adds some more computations. To give a simple figure, we put all these computations (excluding the FFTs and IFFTs) together and roughly say that the complexity of processing of each data point in the frequency domain is equivalent to performing two butterflies. Noting that each partition of input samples, which consists of $(p+1)L$ real-valued samples in time domain, is converted to $(p+1)L/2$ complex-valued frequency domain samples and there are P such partitions, the total number of frequency domain samples is $(p+1)LP/2$. Adding these together and noting that L output samples are generated at the end of each block processing interval, we obtain the per-sample computational complexity of the unconstrained PFBLMS algorithm as

$$\begin{aligned} C &= \frac{(p+1)LP+\frac{3}{4}(p+1)L\log_2\frac{(p+1)L}{2}}{L} \\ &= (p+1)P+\frac{3}{4}(p+1)\log_2\frac{(p+1)L}{2} \end{aligned} \tag{8.99}$$

The memory requirements of the unconstrained and constrained PFBLMS algorithms are about the same. The number of frequency domain data samples (including the intermediate results in the z^{-p} delay units) is $(p(P-1)+1)(p+1)L$. We also need $(p+1)LP$ memory words to store the filter coefficients. Some additional storage for input, output, error samples, and step-size parameters is also required. Adding these together, the number of memory words required to implement the PFBLMS algorithm is approximately

$$\mathrm{S}=(p+1)^2LP \quad \text{words} \tag{8.100}$$

To get a feeling of the above numbers, we give the following example.

Example 8.2

Let us consider an acoustic echo canceler which has to cover an echo spread of at least 250 ms at the sampling rate of 8 kHz. It is recommended that the algorithm latency (delay) in delivering the echo-free samples shall not exceed 16 ms. To cover an echo spread of 250 ms, an adaptive filter with at least 2000 taps should be used as 250 ms is equivalent to 2000 samples at the sampling frequency of 8 kHz. To achieve a latency of less than 16 ms, $L=64$ is appropriate. Note that there will be a delay of L samples to collect a new block of input samples, and there will be an additional delay of up to one block period (i.e., L sample intervals) to calculate the corresponding block of output samples. This gives a total delay of up to $2L$ sample intervals, which for $L=64$ and the sampling

Table 8.5 Computational complexity and memory requirement of the unconstrained PFBLMS algorithm for the cases discussed in Example 8.2.

	Computational Complexity	Memory words
$p = 1$, $P = 32$	73	8192
$p = 3$, $P = 11$	65	11 264
$p = 7$, $P = 5$	88	20 480

rate of 8 kHz is equivalent to 16 ms. Table 8.5 gives a summary of the computational complexity and memory requirement of the unconstrained PFBLMS algorithm for $p = 1$, 3, and 7. These values of p result in $(p + 1)L$ being a power of 2 and, therefore, an efficient radix 2 FFT algorithm can be used. From these results, we note that $p = 3$ is a good compromise choice as it results in some reduction in computational complexity and, as demonstrated in Section 8.5, significant improvement in convergence behavior, at the cost of slight increase in memory.

8.4.5 Modified Constrained PFBLMS Algorithm

Our discussions on the PFBLMS algorithm, so far, suggest that the constrained PFBLMS algorithm is significantly more complicated than its unconstrained counterpart. This is because the tap weights of all partitions have to be constrained at the end of every iteration of the algorithm (see Step 5 in Table 8.3). McLaughlin (1996) has proposed a method that significantly reduces the computational complexity of the constrained PFBLMS algorithm, while its convergence behavior is almost unaffected. His method does not constrain the tap weights at the end of all iterations. In the context of a PFBLMS-based acoustic echo canceler, he has a special scheduling method for applying the constraint to the various partitions.

Chan (2000) (see also Chan and Farhang-Boroujeny (2001)) has analyzed the McLaughlin's constraining method and explained its excellent performance. In the context of a general constrained PFBLMS algorithm, the following tap-weight constraint scheduling scheme is suggested by Chan (2000). After every iteration of the PFBLMS algorithm, the tap weights of one or a few of the partitions are constrained on a rotational basis. For example, in the first iteration, the tap weights of the first partition is constrained. In the second iteration, the constraint operation is applied to the second partition. This process continues until all the partitions are constrained. The constraint operation then restarts with the first partition. Clearly, in cases where the number of partitions, P, is large, this simple approach can significantly reduce the computational complexity of the constrained PFBLMS algorithm. Here, to avoid excessive mathematical derivations, we limit ourselves to confirming the conclusions drawn by Chan (2000), numerically, through computer simulations.

8.5 Computer Simulations

In this section, we present some simulation results that confirm the theoretical results derived in the previous sections. These results also serve to enhance our understanding of the convergence behavior of the FBLMS and PFBLMS algorithms.

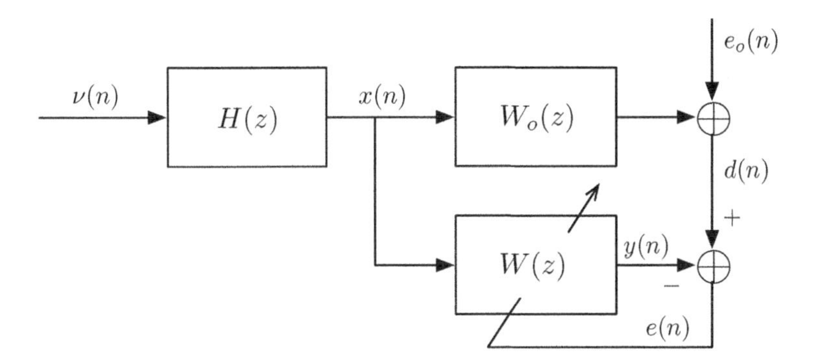

Figure 8.6 Adaptive modeling of an FIR plant.

We consider a modeling problem, as shown in Figure 8.6. The plant, $W_o(z)$, which is to be identified by the adaptive filter, $W(z)$, is assumed to be a FIR system with an impulse response stretching over 1985 samples. This choice of filter length allows us to use a FBLMS algorithm with $L = 64$ and $N' = N + L - 1 = 2^{11}$ for modeling $W_o(z)$ (note that $1985 = 2^{11} - 64 + 1$). $W(z)$ is assumed to have sufficient taps to model $W_o(z)$ perfectly. Two cases of the input, $x(n)$, are considered:

1. a white process.
2. a colored process that is generated by passing a white noise through a coloring filter with the transfer function

$$H(z) = 0.1 - 0.2z^{-1} - 0.3z^{-2} + 0.4z^{-3} + 0.4z^{-4} - 0.2z^{-5} - 0.1z^{-6}$$

Recall that this coloring filter is same as the filter $H_2(z)$ used in Chapter 7 (Section 7.6.4). The power spectral density of the process generated by this filter is shown in Figure 7.7. The samples of the plant impulse response, $w_{o,i}$'s, are chosen to be a set of identically independent random numbers, and they are normalized so that $\sum_i w_{o,i}^2 = 1$. The sequence $e_o(n)$ is an additive white Gaussian noise. It is independent of $x(n)$ and its variance is set equal to 0.001 for the simulations presented here. So the expected minimum MSE at the adaptive filter output is 0.001. Three cases of $p = 1$, 3, and 7 are considered. To completely cover the impulse response of the plant, P is chosen to be 32, 11, and 5, respectively, for these cases. For each case, the step-size parameter μ_o is chosen using the misadjustment equations given before so as to result in 10% misadjustment. The algorithms used are of the step-normalized type. Learning curves presented here are based on ensemble averages of 100 independent runs for each curve. The averaged curves are smoothed before being plotted.

Figure 8.7 shows the results of the simulations for white input. As expected, performance of the unconstrained PFBLMS algorithm is quite poor when the overlap is 50% among the successive partitions (i.e., the case $p = 1$) and improves as the overlap is reduced by increasing p. The case when no partitioning is applied, that is, corresponding to the FBLMS algorithm, is also shown for comparison.

Figure 8.8 repeats Figure 8.7 for the case when $x(n)$ is generated using the coloring filter $H(z)$. In this case, the eigenvalue spread of the correlation matrix of $x(n)$ can be as

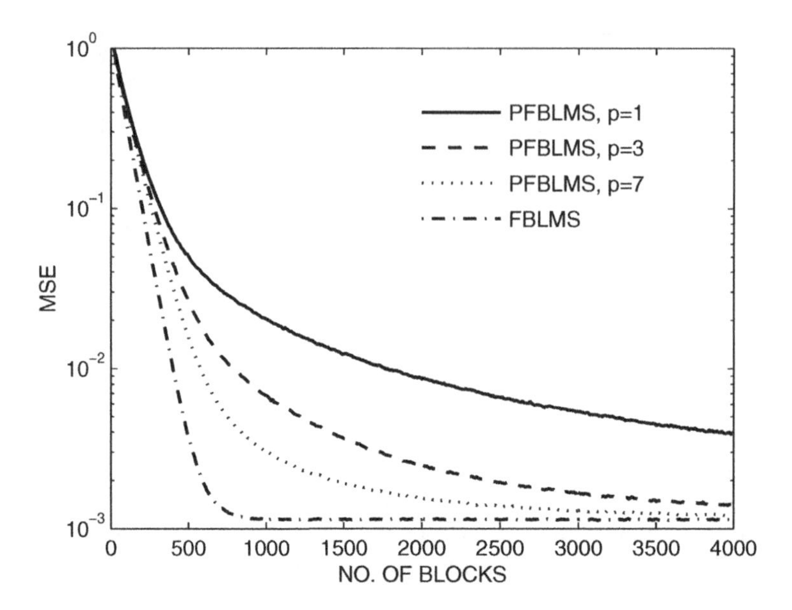

Figure 8.7 Learning curves of the FBLMS and PFBLMS algorithms with white input.

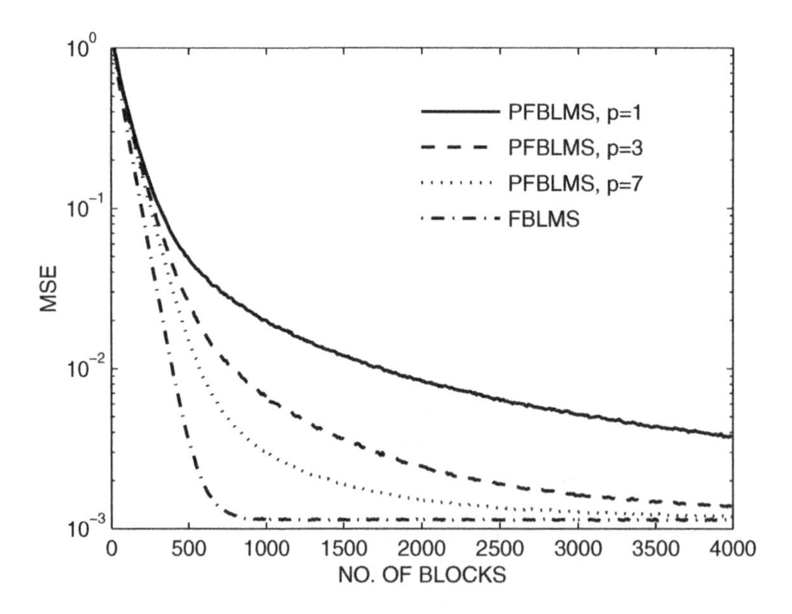

Figure 8.8 Learning curves of the FBLMS and PFBLMS algorithms for a colored input.

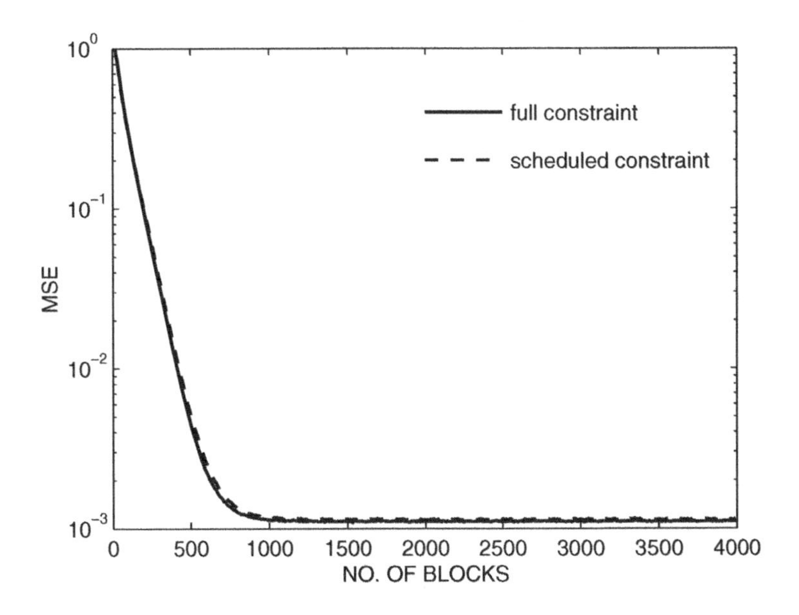

Figure 8.9 Learning curves of the constrained PFBLMS algorithm and its modified version.

high as 459. Here also we find that reducing the amount of overlap between successive partitions improves the performance of the PFBLMS algorithm. Furthermore, there is a very little difference between the results in Figures 8.7 and 8.8. This is in line with the theoretical results of the previous sections that predict that the step-normalized FBLMS and PFBLMS algorithms are insensitive to the power spectral density (eigenvalue spread) of the filter input.

Figure 8.9 compares the convergence performance of the constrained PFBLMS algorithm and one of its modified versions, with $p = 1$ and $P = 32$. The filter input is colored and is generated using the coloring filter $H(z)$. In the implementation of the modified PFBLMS algorithm, the tap-weight constraint operation is applied on rotational basis to only one of the partitions in each iteration. Observe from the results that even though each partition in the modified constrained PFBLMS algorithm is constrained only once in every 32 iterations, the resulting performance loss is negligible. Also, by direct inspection of the learning curves of Figure 8.9, we see that overlap of the partitions has no significant effect on the convergence behavior of the constrained PFBLMS algorithm. This is in view of the fact that there is only one dominant mode affecting the convergence behavior of the constrained PFBLMS algorithm, as can be seen from the learning curves.

Problems

P8.1 Consider the BLMS recursion Eq. (8.11). In Appendix 8A, it is shown that Eq. (8.11) can be rearranged as

$$\mathbf{v}(k+1) = \left(\mathbf{I} - 2\frac{\mu_{\mathrm{B}}}{L}\mathbf{X}^{\mathrm{T}}(k)\mathbf{X}(k)\right)\mathbf{v}(k) + 2\frac{\mu_{\mathrm{B}}}{L}\mathbf{X}^{\mathrm{T}}(k)\mathbf{e}_o(k)$$

where $\mathbf{v}(k) = \mathbf{w}(k) - \mathbf{w}_o$, $\mathbf{w}_o$ is the optimum tap-weight vector of the filter and $\mathbf{e}_o(k) = \mathbf{d}(k) - \mathbf{X}(k)\mathbf{w}_o$.

(i) Assuming $\mathbf{v}(k)$ and $\mathbf{X}(k)$ are independent of each other, show that

$$E[\mathbf{v}(k+1)] = (\mathbf{I} - 2\mu_B \mathbf{R})E[\mathbf{v}(k)]$$

where $\mathbf{R}$ is the correlation matrix of the filter tap inputs.

(ii) Use the result of (i) to obtain the time constants that control the convergence behavior of $E[\mathbf{v}(k)]$.

(iii) Based on the result obtained in (ii), justify the validity of Eq. (8.12).

P8.2 By direct application of Eqs. (8.21) and (8.26), confirm the identity (8.28).

P8.3 Define the time-reversed version of the vector $\mathbf{a} = [a_0\ a_1\ a_2\ \cdots\ a_{M-1}]^T$ as $\mathbf{a}^r = [a_0\ a_{M-1}\ a_{M-2}\ \cdots\ a_1]^T$.

(i) Show that if $\mathbf{a}_{\mathcal{F}}$ and $\mathbf{a}^r_{\mathcal{F}}$ are the DFTs of $\mathbf{a}$ and $\mathbf{a}^r$, respectively, then

$$\mathbf{a}^r_{\mathcal{F}} = \mathbf{a}^*_{\mathcal{F}}$$

where asterisk denotes complex conjugation.

(ii) Show that if $\mathbf{A}_c$ is a circular matrix as in Eq. (8.21), $\mathbf{A}_c^T$ is also a circular matrix. Compare the first columns of $\mathbf{A}_c$ and $\mathbf{A}_c^T$ and show that they are time-reversed versions of each other.

(iii) Use the above observation to give an alternative derivation of Eq. (8.32).

P8.4 By direct application of Eqs. (8.33), (8.34), (8.42), (8.43), and (8.45), show that Eq. (8.44) is just an alternative formulation of Eq. (8.11).

P8.5 In the derivation of the LMS algorithm, the instantaneous value of $e^2(n)$ was used as an estimate of the cost function $\xi = E[e^2(n)]$. Give a direct derivation of the BLMS algorithm by considering

$$\hat{\xi}_B(k) = \frac{1}{L}\sum_{i=0}^{L-1} e^2(kL+i)$$

as an estimate of the cost function ξ and running the steepest-descent recursion once after every L samples of the data.

P8.6 The estimate $\hat{\xi}_B(k)$ defined in Problem P8.5 may equivalently be written as

$$\hat{\xi}_B(k) = \frac{1}{L}\hat{\mathbf{e}}^T(k)\hat{\mathbf{e}}(k) \tag{P8.6.1}$$

where $\hat{\mathbf{e}}(k)$ the output error vector of the filter in the extended form, as defined by Eq. (8.43). Using the DFT properties, Eq. (P8.6.1) can be expressed in terms of $\mathbf{e}_{\mathcal{F}}(k) = \boldsymbol{\mathcal{F}}\hat{\mathbf{e}}(k)$ as

$$\hat{\xi}_B(k) = \frac{1}{LN'}\mathbf{e}^H_{\mathcal{F}}(k)\mathbf{e}_{\mathcal{F}}(k) \tag{P8.6.2}$$

where $N' = N + L - 1$ is the length of the vector $\hat{\mathbf{e}}(k)$.

To obtain the optimum frequency domain tap-weight vector $\mathbf{w}_{\mathcal{F}}$, the cost function $\xi(\mathbf{w}_{\mathcal{F}}) = E[\hat{\xi}_{\rm B}(k)]$ should be minimized. Accordingly, the optimum solution obtained by the constrained FBLMS algorithm is the one minimizing $\xi(\mathbf{w}_{\mathcal{F}})$, subject to the constraint $\mathcal{P}_{N,0}\mathbf{w}_{\mathcal{F}} = \mathbf{w}_{\mathcal{F}}$. On the other hand, the unconstrained FBLMS minimizes $\xi(\mathbf{w}_{\mathcal{F}})$ without imposing any constraint on $\mathbf{w}_{\mathcal{F}}$.

(i) Show that

$$\xi(\mathbf{w}_{\mathcal{F}}) = \frac{1}{LN'}(\mathbf{w}_{\mathcal{F}}^{\rm H}\mathcal{R}_{xx}^{u}\mathbf{w}_{\mathcal{F}} - \mathbf{p}_{\mathcal{F}}^{\rm H}\mathbf{w}_{\mathcal{F}} - \mathbf{w}_{\mathcal{F}}^{\rm H}\mathbf{p}_{\mathcal{F}} + E[\mathbf{d}_{\mathcal{F}}^{\rm H}\mathbf{d}_{\mathcal{F}}])$$

where $\mathcal{R}_{xx}^{u}$ is as defined in Eq. (8.57) and $\mathbf{p}_{\mathcal{F}} = E[\mathcal{X}_{\mathcal{F}}^{*}\mathcal{P}_{0,L}\mathbf{d}_{\mathcal{F}}]$.

(ii) Find the optimum value of $\mathbf{w}_{\mathcal{F}}$ that minimizes $\xi(\mathbf{w}_{\mathcal{F}})$. Show the nonsingularity of $\mathcal{R}_{xx}^{u}$ that exists is the necessary and sufficient condition for this solution to be unique.

(iii) It is understood that when a sufficiently small step-size parameter is used for both constrained and unconstrained FBLMS algorithms so that the misadjustments of the two algorithms can be ignored, the unconstrained FBLMS algorithm converges to a mean-squared error (MSE), which is less than or equal to what can be achieved by the constrained FBLMS algorithm. With the knowledge developed in this problem, how do you explain this?

P8.7 Starting with Eq. (P8.6.2) of the last problem, give a direct derivation of the unconstrained recursion (8.49).

P8.8 Consider a modeling problem with the desired output $d(n) = \mathbf{w}_{\rm o}^{\rm T}\mathbf{x}(n) + e_{\rm o}(n)$, where the length of $\mathbf{w}_{\rm o}$ is less than or equal to the length of the adaptive filter, N. Assume that the plant noise, $e_{\rm o}(n)$, and its input, $\mathbf{x}(n)$, are uncorrelated with each other. Under these conditions, it is understood that the constrained and unconstrained FBLMS algorithms converge to exactly the same solution. Using the result obtained in Problem P8.6 and assuming that the inverse of $\mathcal{R}_{xx}^{u}$ exists, give reasons that explain this. What is the common solution to which both the constrained and unconstrained FBLMS algorithms converge?

P8.9 Show that when the block length, L, is equal to the filter length, N, for a given misadjustment, the constrained FBLMS algorithm converges twice faster than its unconstrained counterpart. Support your answer by giving a careful consideration to the time constants associated with the two algorithms. Does your answer continue to hold if step-normalization is (i) used and (ii) not used?

P8.10 Consider the constrained FBLMS recursion (8.47). Show that when the block length, L, is one:

(i) $\mathcal{P}_{N,0} = \mathbf{P}_{N,0} = \mathbf{I}$, where $\mathbf{I}$ is the N-by-N identity matrix. Then, argue that the constrained and unconstrained FBLMS algorithms are the same.

(ii) the FBLMS recursion can be rearranged as

$$\mathbf{w}_{\mathcal{F}}(k+1) = \mathbf{w}_{\mathcal{F}}(k) + 2\mu\mathbf{\Gamma}\mathbf{x}_{\mathcal{F}}^{*}(k)e(k)$$

where $\mathbf{x}_{\mathcal{F}}(k)$ is the DFT of the first column of the circular matrix $\mathbf{X}_{\rm c}(k)$, as defined by Eq. (8.33), $e(k)$ is the scalar output error at time k, and $\mathbf{\Gamma}$

is the diagonal matrix consisting of the elements 1, $e^{j2\pi/N}$, $e^{j4\pi/N}$, ..., $e^{j2(N-1)\pi/N}$.

(iii)

$$y(k) = \text{the last term of} \quad \mathcal{F}^{-1}(\mathbf{w}_{\mathcal{F}}(k) \odot \mathbf{x}_{\mathcal{F}}(k))$$
$$= \mathbf{w}_{\mathcal{F}}^{\mathrm{T}}(k)\mathbf{x}_{\mathcal{F}}(k)$$

where $\odot$ denotes the element-wise multiplication of vectors.

(iv) Now, consider the TDLMS algorithm with $\mathcal{T} = \mathcal{F}$. Write down the equations corresponding to this case and compare them with the above results. Verify that the FBLMS algorithm with block length $L = 1$ is equivalent to the TDLMS algorithm with $\mathcal{T} = \mathcal{F}$.

P8.11 An alternative procedure for derivation of Eq. (8.61) is proposed in this problem.

(i) Show that

$$\mathbf{P}_{0,L} = \mathcal{F}\mathcal{P}_{0,L}^{*}\mathcal{F}^{-1}$$

and conclude that $\mathcal{P}_{0,L}$ is a circular matrix.

(ii) Show that

$$\text{First column of } \mathcal{P}_{0,L} = \frac{1}{N'}\mathcal{F}\mathbf{p}_{0,L}$$

where $\mathbf{p}_{0,L}$ is the column vector consisting of the diagonal elements of $\mathbf{P}_{0,L}$.

(iii) Considering the fact that $\mathcal{X}_{\mathcal{F}}(k)$ is a diagonal matrix, show that

$$\mathcal{X}_{\mathcal{F}}^{*}(k)\mathcal{P}_{0,L}\mathcal{X}_{\mathcal{F}} = \mathcal{P}_{0,L} \odot ([\mathbf{x}_{\mathcal{F}}(k)\mathbf{x}_{\mathcal{F}}^{\mathrm{H}}(k)]^{*})$$

where $\mathbf{x}_{\mathcal{F}}(k)$ is the column vector consisting of the diagonal elements of $\mathcal{X}_{\mathcal{F}}(k)$, and $\odot$ denotes element-wise multiplication of the matrices. Thus, show that

$$\mathcal{R}_{xx}^{u} = \mathcal{P}_{0,L} \odot \mathcal{R}_{xx}^{*}$$

where $\mathcal{R}_{xx} = E[\mathbf{x}_{\mathcal{F}}(k)\mathbf{x}_{\mathcal{F}}^{\mathrm{H}}(k)]$.

(iv) Assuming that the cross-correlation between different elements of the vector $\mathbf{x}_{\mathcal{F}}(k)$ are negligible, show that

$$\mathcal{R}_{xx} \approx N' \times \operatorname{diag}\left(\Phi_{xx}\left(e^{j\frac{2\pi\times 0}{N'}}\right), \Phi_{xx}\left(e^{j\frac{2\pi}{N'}}\right), \ldots, \Phi_{xx}\left(e^{j\frac{2\pi(N'-1)}{N'}}\right)\right).$$

(v) Using the results of (iii) and (iv), derive Eq. (8.61).

(vi) Do a thorough study of the elements of the matrix $\mathcal{P}_{0,L}$. In particular, verify that the largest (in magnitude) elements of $\mathcal{P}_{0,L}$ are its diagonal elements. Use your findings to conclude that $\mathcal{R}_{xx}^{u}$ is closer to diagonal than $\mathcal{R}_{xx}$, in the sense that the nondiagonal elements of the normalized matrix $(\operatorname{diag}[\mathcal{R}_{xx}^{u}])^{-1}\mathcal{R}_{xx}^{u}$ are smaller than the corresponding elements of $(\operatorname{diag}[\mathcal{R}_{xx}])^{-1}\mathcal{R}_{xx}$.

P8.12 Verify the results presented in Eq. (8.87).

P8.13 Verify the results presented in Eq. (8.93).

P8.14 For the case discussed in Example 8.2, evaluate the computational complexity and memory requirement of the FBLMS algorithm, for both the constrained and unconstrained cases, and compare your results with those given in Table 8.5.

P8.15 Discuss in detail why the selection of $M = pL$ results in less overlap among successive partitions as p increases.

P8.16 In the results presented in Table 8.5, we find that the unconstrained PFBLMS algorithm with $p = 3$ is less complex than the case where $p = 1$. Explore the contribution of various parts of the algorithm to find out why this is happening.

P8.17 Evaluate the computational complexities of the constrained PFBLMS implementation and its modified version, which were used to obtain the simulation results of Figure 8.9 and compare your results with those in Table 8.5.

Computer-Oriented Problems

P8.18 The MATLAB program "`blk_mdlg.m`" which was used to obtain the results of Example 8.1 is available on an accompanying website. Run this program and confirm the results of Figure 8.2. In addition, using this program, study the convergence behavior of the BLMS algorithm for the following choices of L and $\mathcal{M}_{BMLS}$ and discuss your findings.

L	$\mathcal{M}_{BLMS}$
$4N,\ 5N$	10%
$N,\ 2N,\ 3N,\ 4N,\ 5N$	5%
$N,\ 2N,\ 3N,\ 4N,\ 5N$	20%

P8.19 Consider a channel equalization problem similar to the one discussed in Section 6.4.2. Assume that the channel response is characterized by the transfer function

$$H(z) = 0.1 + 0.3z^{-1} + 0.6z^{-2} + z^{-3} + 0.5z^{-4} - 0.2z^{-5} + 0.1z^{-6}$$

the input data, $s(n)$, to the channel is binary and white, the channel noise, $\nu(n)$, is white and Gaussian, signal-to-noise ratio at the channel output is 30 dB, and equalizer length, N, and the delay, Δ, are set equal to 33 and 18, respectively. Develop a simulation program to study the performance of FBLMS algorithm in this application.

P8.20 Consider the channel equalization setup of problem P8.19. By running appropriate simulation programs, study the convergence behaviors of the conventional LMS algorithm and the TDLMS algorithm (with various transforms) and compare your results with those of the FBLMS algorithm.

P8.21 Consider a system modeling problem similar to the one in Figure 8.6. Assume that $W_o(z)$ is a FIR filter with $N = 1024$ taps, and the length $W(z)$ is chosen to be the same. The vector of coefficients of $W_o(z)$, $\mathbf{w}_o$, has entries that are complex-valued Gaussian random variables with the same variance and are normalized such that $\mathbf{w}_o^H \mathbf{w}_o = 1$. Also, assume that $\nu(n)$ is complex-valued Gaussian white process with variance of unity. Consider the following cases of the input coloring filter $H(z)$:

$$H(z) = H_1(z) = 1$$

$$H(z) = H_2(z) = 0.4 + z^{-1} + 0.4z^{-2}$$

$$H(z) = H_3(z) = \frac{1 + 1.2z^{-1}}{1 - 1.5z^{-1} + 0.56z^{-2}}$$

(i) For each case, evaluate the eigenvalue spread $\lambda_{max}/\lambda_{min}$ of the correlation matrix $\mathbf{R}$ of the input signal to $W(z)$. Note that as here N is very large, you may use the bounds maximum of $\Phi_{xx}(e^{j\omega})$ and minimum of $\Phi_{xx}(e^{j\omega})$ as good approximations to λ_{max} and λ_{min}, respectively.

(ii) Develop the necessary codes to compare the convergence behavior of LMS and FBLMS algorithms. In the case of FBLMS, consider both cases of the algorithm with and without step-normalization. For all cases, choose the step-size parameter μ for a misadjustment of 10%.

P8.22 Develop the necessary codes to implement the PFBLMS structure of Figure 8.4 for both unconstrained and constrained cases. Let $M = L = 64$ and present the learning curves of the algorithms when the step-size parameter μ is chosen to achieve 10% misadjustment. Compare your results with those of Problem P8.21.

P8.23 Develop the necessary codes to implement the PFBLMS structure of Figure 8.5 for both unconstrained and constrained cases. Let $L = 32$ and $M = 96$, and present the learning curves of the algorithms when the step-size parameter μ is chosen to achieve 10% misadjustment. Compare your results with those of Problems P8.21 and P8.22.

P8.24 Repeat Problems P8.22 and P8.23 when the constrained PFBLMS algorithm is replaced by its scheduled constrained version. Compare the results with those of the fully constrained version of the algorithm.

Appendix 8A: Derivation of a Misadjustment Equation for the BLMS Algorithm

In this appendix, we present a simple derivation of the misadjustment of the BLMS algorithm. This derivation is different from the one used for the conventional LMS algorithm in Chapter 6. Because of certain assumptions used here (such as the step-size parameter, μ_{B}, is small, adaptive filter models the plant almost exactly), this derivation is rather less accurate.

We start with recursion Eq. (8.11) and use the definition $\mathbf{v}(k) = \mathbf{w}(k) - \mathbf{w}_{\mathrm{o}}$, where $\mathbf{w}_{\mathrm{o}}$ is the optimum tap-weight vector of the filter, to obtain

$$\mathbf{v}(k+1) = \mathbf{v}(k) + 2\frac{\mu_{\mathrm{B}}}{L}\mathbf{X}^{\mathrm{T}}(k)\mathbf{e}(k) \tag{8A.1}$$

We also note that

$$\mathbf{e}(k) = \mathbf{d}(k) - \mathbf{X}(k)\mathbf{w}(k) = \mathbf{e}_{\mathrm{o}}(k) - \mathbf{X}(k)\mathbf{v}(k) \tag{8A.2}$$

where $\mathbf{e}_{\mathrm{o}}(k) = \mathbf{d}(k) - \mathbf{X}(k)\mathbf{w}_{\mathrm{o}}$ is the output error when the optimum tap-weight vector, $\mathbf{w}_{\mathrm{o}}$, is used.

Substituting Eq. (8A.2) in Eq. (8A.1), we get

$$\mathbf{v}(k+1) = \left(\mathbf{I} - 2\frac{\mu_{\mathrm{B}}}{L}\mathbf{X}^{\mathrm{T}}(k)\mathbf{X}(k)\right)\mathbf{v}(k) + 2\frac{\mu_{\mathrm{B}}}{L}\mathbf{X}^{\mathrm{T}}(k)\mathbf{e}_{\mathrm{o}}(k) \tag{8A.3}$$

Next, we multiply both sides of Eq. (8A.3) from the left by their respective transposes and expand to obtain

$$\begin{aligned}\mathbf{v}^{\mathrm{T}}(k+1)\mathbf{v}(k+1) = {} & \mathbf{v}^{\mathrm{T}}(k)\left(\mathbf{I} - 2\frac{\mu_{\mathrm{B}}}{L}\mathbf{X}^{\mathrm{T}}(k)\mathbf{X}(k)\right)^{2}\mathbf{v}(k) \\ & + 2\frac{\mu_{\mathrm{B}}}{L}\mathbf{e}_{\mathrm{o}}^{\mathrm{T}}(k)\mathbf{X}(k)\left(\mathbf{I} - 2\frac{\mu_{\mathrm{B}}}{L}\mathbf{X}^{\mathrm{T}}(k)\mathbf{X}(k)\right)\mathbf{v}(k) \\ & + 2\frac{\mu_{\mathrm{B}}}{L}\mathbf{v}^{\mathrm{T}}(k)\left(\mathbf{I} - 2\frac{\mu_{\mathrm{B}}}{L}\mathbf{X}^{\mathrm{T}}(k)\mathbf{X}(k)\right)\mathbf{X}^{\mathrm{T}}(k)\mathbf{e}_{\mathrm{o}}(k) \\ & + 4\frac{\mu_{\mathrm{B}}^{2}}{L^{2}}\mathbf{e}_{\mathrm{o}}^{\mathrm{T}}(k)\mathbf{X}(k)\mathbf{X}^{\mathrm{T}}(k)\mathbf{e}_{\mathrm{o}}(k)\end{aligned} \tag{8A.4}$$

Now, we follow the same line of derivation as in Chapter 6 (Section 6.3). We take expectation on both sides of Eq. (8A.4) and assume that $\mathbf{e}_{\mathrm{o}}(k)$ is zero-mean, $\mathbf{X}(k)$ and $\mathbf{e}_{\mathrm{o}}(k)$ are jointly Gaussian and uncorrelated with each others, and $\mathbf{v}(k)$ is independent of $\mathbf{X}(k)$ and $\mathbf{e}_{\mathrm{o}}(k)$. This results in

$$\begin{aligned}\|\mathbf{v}(k+1)\|^{2} = {} & E\left[\mathbf{v}^{\mathrm{T}}(k)\left(\mathbf{I} - 2\frac{\mu_{\mathrm{B}}}{L}\mathbf{X}^{\mathrm{T}}(k)\mathbf{X}(k)\right)^{2}\mathbf{v}(k)\right] \\ & + 4\frac{\mu_{\mathrm{B}}^{2}}{L^{2}}E[\mathbf{e}_{\mathrm{o}}^{\mathrm{T}}(k)\mathbf{X}(k)\mathbf{X}^{\mathrm{T}}(k)\mathbf{e}_{\mathrm{o}}(k)]\end{aligned} \tag{8A.5}$$

where $\|\mathbf{v}(k)\|^2 = E[\mathbf{v}^{\mathrm{T}}(k)\mathbf{v}(k)]$. The first term on the right-hand side of Eq. (8A.5) can be expanded as

$$E\left[\mathbf{v}^{\mathrm{T}}(k)\left(\mathbf{I} - 2\frac{\mu_{\mathrm{B}}}{L}\mathbf{X}^{\mathrm{T}}(k)\mathbf{X}(k)\right)^2\mathbf{v}(k)\right] = \|\mathbf{v}(k)\|^2 - 4\frac{\mu_{\mathrm{B}}}{L}E[\mathbf{v}^{\mathrm{T}}(k)\mathbf{X}^{\mathrm{T}}(k)\mathbf{X}(k)\mathbf{v}(k)]$$
$$+ 4\frac{\mu_{\mathrm{B}}^2}{L^2}E[\mathbf{v}^{\mathrm{T}}(k)\mathbf{X}^{\mathrm{T}}(k)\mathbf{X}(k)\mathbf{X}^{\mathrm{T}}(k)\mathbf{X}(k)\mathbf{v}(k)] \tag{8A.6}$$

To simplify this, we assume that μ_{B} is small so that the last term on the right-hand side of Eq. (8A.6) can be ignored. Furthermore, using the *independence assumption* between $\mathbf{v}(k)$ and $\mathbf{X}(k)$ and following the same line of argument as in Chapter 6, we obtain

$$E\left[\mathbf{v}^{\mathrm{T}}(k)\left(\mathbf{I} - 2\frac{\mu_{\mathrm{B}}}{L}\mathbf{X}^{\mathrm{T}}(k)\mathbf{X}(k)\right)^2\mathbf{v}(k)\right] \approx$$
$$\|\mathbf{v}(k)\|^2 - 4\frac{\mu_{\mathrm{B}}}{L}E[\mathbf{v}^{\mathrm{T}}(k)E[\mathbf{X}^{\mathrm{T}}(k)\mathbf{X}(k)]\mathbf{v}(k)] \tag{8A.7}$$

Now note from Eq. (8.4) that

$$E[\mathbf{X}^{\mathrm{T}}(k)\mathbf{X}(k)] = L\mathbf{R} \tag{8A.8}$$

where $\mathbf{R}$ is the N-by-N correlation matrix of the filter tap inputs. Substituting Eq. (8A.8) in Eq. (8A.7), we get

$$E\left[\mathbf{v}^{\mathrm{T}}(k)\left(\mathbf{I} - 2\frac{\mu_{\mathrm{B}}}{L}\mathbf{X}^{\mathrm{T}}(k)\mathbf{X}(k)\right)^2\mathbf{v}(k)\right] \approx$$
$$\|\mathbf{v}(k)\|^2 - 4\mu_{\mathrm{B}}E[\mathbf{v}^{\mathrm{T}}(k)\mathbf{R}\mathbf{v}(k)] \tag{8A.9}$$

To evaluate the second term on the right-hand side of Eq. (8A.5), we note that $\mathbf{e}_{\mathrm{o}}^{\mathrm{T}}(k)\mathbf{X}(k)\mathbf{X}^{\mathrm{T}}(k)\mathbf{e}_{\mathrm{o}}(k)$ is a scalar and use Eq. (6.23) to write

$$\mathbf{e}_{\mathrm{o}}^{\mathrm{T}}(k)\mathbf{X}(k)\mathbf{X}^{\mathrm{T}}(k)\mathbf{e}_{\mathrm{o}}(k) = \mathrm{tr}[\mathbf{e}_{\mathrm{o}}^{\mathrm{T}}(k)\mathbf{X}(k)\mathbf{X}^{\mathrm{T}}(k)\mathbf{e}_{\mathrm{o}}(k)]$$
$$= \mathrm{tr}[\mathbf{e}_{\mathrm{o}}(k)\mathbf{e}_{\mathrm{o}}^{\mathrm{T}}(k)\mathbf{X}(k)\mathbf{X}^{\mathrm{T}}(k)] \tag{8A.10}$$

Taking expectation on both sides of Eq. (8A.10) and noting that $\mathbf{e}_{\mathrm{o}}(k)$ and $\mathbf{X}(k)$ are independent of each other, we obtain

$$E[\mathbf{e}_{\mathrm{o}}^{\mathrm{T}}(k)\mathbf{X}(k)\mathbf{X}^{\mathrm{T}}(k)\mathbf{e}_{\mathrm{o}}(k)] = \mathrm{tr}[E[\mathbf{e}_{\mathrm{o}}(k)\mathbf{e}_{\mathrm{o}}^{\mathrm{T}}(k)]E[\mathbf{X}(k)\mathbf{X}^{\mathrm{T}}(k)]] \tag{8A.11}$$

Next, we assume that the elements of $\mathbf{e}_{\mathrm{o}}(k)$ are samples of a white noise process. This assumption is justified when the adaptive filter is long enough to model the plant almost exactly. This implies that

$$E[\mathbf{e}_{\mathrm{o}}(k)\mathbf{e}_{\mathrm{o}}^{\mathrm{T}}(k)] = \xi_{\min}\mathbf{I} \tag{8A.12}$$

where $\xi_{\min} = E[e_{\mathrm{o}}^2(n)]$ is the minimum MSE at the filter output, and the identity matrix $\mathbf{I}$ is L-by-L. Substituting Eq. (8A.12) in Eq. (8A.11), noting that $\mathrm{tr}[\mathbf{X}(k)\mathbf{X}^{\mathrm{T}}(k)] = \mathrm{tr}[\mathbf{X}^{\mathrm{T}}(k)\mathbf{X}(k)]$, and using Eq. (8A.8), we get

$$E[\mathbf{e}_{\mathrm{o}}^{\mathrm{T}}(k)\mathbf{X}(k)\mathbf{X}^{\mathrm{T}}(k)\mathbf{e}_{\mathrm{o}}(k)] = L\xi_{\min}\mathrm{tr}[\mathbf{R}] \tag{8A.13}$$

Substituting Eqs. (8A.13) and (8A.9) in Eq. (8A.5), we obtain

$$\|\mathbf{v}(k+1)\|^2 \approx \|\mathbf{v}(k)\|^2 - 4\mu_\mathrm{B} E[\mathbf{v}^\mathrm{T}(k)\mathbf{R}\mathbf{v}(k)] + 4\frac{\mu_\mathrm{B}^2}{L}\xi_{\min}\mathrm{tr}[\mathbf{R}] \tag{8A.14}$$

When the algorithm has converged and reached its steady state, $\|\mathbf{v}(k+1)\|^2 = \|\mathbf{v}(k)\|^2$. Using this in Eq. (8A.14), we obtain, in the steady state

$$E[\mathbf{v}^\mathrm{T}(k)\mathbf{R}\mathbf{v}(k)] \approx \frac{\mu_\mathrm{B}}{L}\xi_{\min}\mathrm{tr}[\mathbf{R}] \tag{8A.15}$$

We recall that the left-hand side of Eq. (8A.14) is equal to the excess MSE of the algorithm after its convergence (see Eq. (6.21) and the subsequent discussions in the same section). Thus, we obtain

$$\text{Excess MSE of the BLMS algorithm} \approx \frac{\mu_\mathrm{B}}{L}\xi_{\min}\mathrm{tr}[\mathbf{R}] \tag{8A.16}$$

Dividing this excess MSE by the minimum MSE, $\xi_{\min}$, we obtain Eq. (8.13).

Appendix 8B: Derivation of Misadjustment Equations for the FBLMS Algorithms

Let us start with the definition of misadjustment. We recall from Chapter 6 that for an adaptive algorithm, misadjustment is defined by the equation

$$\mathcal{M} = \frac{\xi_{\text{excess}}}{\xi_{\text{min}}} \tag{8B.1}$$

where ξ_{excess} is the excess MSE due to perturbation of the filter tap weights after the algorithm has reached its steady state, and ξ_{min} is the minimum MSE that could be achieved by the optimum tap weights. The excess MSE, as defined before (in Chapter 6), is given by the following equation and is evaluated after the convergence of the filter:

$$\xi_{\text{excess}} = E[(\mathbf{v}^{\text{T}}(n)\mathbf{x}(n))^2] \tag{8B.2}$$

Here, $\mathbf{v}(n) = \mathbf{w}(n) - \mathbf{w}_{\text{o}}$ is the tap-weight perturbation vector and thus, $\mathbf{v}^{\text{T}}(n)\mathbf{x}(n)$ is an associated error quantity.

In the case of FBLMS algorithm, where the perturbation vector $\mathbf{v}(k)$ varies only once every block, the excess MSE is defined as

$$\xi_{\text{excess}} = \frac{1}{L}E[(\mathbf{X}(k)\mathbf{v}(k))^{\text{T}}(\mathbf{X}(k)\mathbf{v}(k))] \tag{8B.3}$$

where $\mathbf{X}(k)\mathbf{v}(k)$ is the length L vector of error samples arising from the tap-weight perturbation $\mathbf{v}(k)$ during kth block.

If $\mathbf{w}_{\mathcal{F}}(k)$ in Eq. (8.41) is replaced by $\mathbf{v}_{\mathcal{F}}(k)$, where $\mathbf{v}_{\mathcal{F}}(k)$ is defined as in Eq. (8.55), the result would be the error due to the tap-weight error $\mathbf{v}_{\mathcal{F}}(k)$. Using this result, we obtain

$$\xi_{\text{excess}} = \frac{1}{L}E[(\mathbf{P}_{0,L}\mathcal{F}^{-1}\mathcal{X}_{\mathcal{F}}(k)\mathbf{v}_{\mathcal{F}}(k))^{\text{H}}(\mathbf{P}_{0,L}\mathcal{F}^{-1}\mathcal{X}_{\mathcal{F}}(k)\mathbf{v}_{\mathcal{F}}(k))] \tag{8B.4}$$

Note that the transpose operator "T" is replaced by the Hermitian transpose operator "H" in Eq. (8B.4) as the frequency domain variables are, in general, complex-valued. It should also be noted that the $\mathbf{v}_{\mathcal{F}}(k)$ in Eq. (8B.4) need not to be constrained, that is, the last $L-1$ samples of $\mathcal{F}^{-1}\mathbf{v}_{\mathcal{F}}(k)$ need not be zero. Accordingly, Eq. (8B.4) can be used for evaluating the excess MSE for both constrained and unconstrained FBLMS algorithms.

Rearranging the terms under the expectation in Eq. (8B.4), and noting that $\mathbf{P}_{0,L}^{\text{H}} = \mathbf{P}_{0,L}$, $\mathbf{P}_{0,L}^2 = \mathbf{P}_{0,L}$, and $(\mathcal{F}^{-1})^{\text{H}} = \frac{1}{N'}\mathcal{F}$, we obtain

$$\begin{aligned}\xi_{\text{excess}} &= \frac{1}{LN'}E[\mathbf{v}_{\mathcal{F}}^{\text{H}}(k)\mathcal{X}_{\mathcal{F}}^{*}(k)\mathcal{F}\mathbf{P}_{0,L}\mathcal{F}^{-1}\mathcal{X}_{\mathcal{F}}(k)\mathbf{v}_{\mathcal{F}}(k)] \\ &= \frac{1}{LN'}E[\mathbf{v}_{\mathcal{F}}^{\text{H}}(k)\mathcal{X}_{\mathcal{F}}^{*}(k)\mathcal{P}_{0,L}\mathcal{X}_{\mathcal{F}}(k)\mathbf{v}_{\mathcal{F}}(k)]\end{aligned} \tag{8B.5}$$

We recall that the length of $\mathbf{v}_{\mathcal{F}}(k)$ is $N' = N + L - 1$, and $\mathcal{X}_{\mathcal{F}}(k)$ is an N'-by-N' diagonal matrix. Assuming that $\mathbf{v}_{\mathcal{F}}(k)$ and $\mathcal{X}_{\mathcal{F}}(k)$ are independent of each other, we obtain from Eq. (8B.5)

$$\xi_{\text{excess}} = \frac{1}{LN'}E[\mathbf{v}_{\mathcal{F}}^{\text{H}}(k)\mathcal{R}_{xx}^{u}\mathbf{v}_{\mathcal{F}}(k)] \tag{8B.6}$$

where $\mathcal{R}_{xx}^{u} = E[\mathcal{X}_{\mathcal{F}}^{*}(k)\mathcal{P}_{0,L}\mathcal{X}_{\mathcal{F}}(k)]$, as defined in Eq. (8.57).

Proceeding in the same line of derivations as in Chapter 6 (Section 6.3), we obtain from Eq. (8B.6)

$$\xi_{\text{excess}} = \frac{1}{LN'}\text{tr}[\mathbf{K}_{\mathcal{F}}^{\text{H}}(k)\mathcal{R}_{xx}^{u}] \tag{8B.7}$$

where $\mathbf{K}_{\mathcal{F}}(k) = E[\mathbf{v}_{\mathcal{F}}(k)\mathbf{v}_{\mathcal{F}}^{\text{H}}(k)]$.

With the expressions Eqs. (8B.6) and (8B.7) for ξ_{excess}, we are now ready to proceed with the derivation of the excess MSE for the various implementations of the FBLMS algorithm.

Unconstrained FBLMS Algorithm Without Step-Normalization

We multiply both sides of Eq. (8.56) from the right by their respective Hermitian transposes, expand, take expectation on both sides, and use similar assumptions as those used in deriving Eq. (8A.6), to obtain

$$\begin{aligned}\|\mathbf{v}_{\mathcal{F}}(k+1)\|^2 &\approx \|\mathbf{v}_{\mathcal{F}}(k)\|^2 - 4\mu E[\mathbf{v}_{\mathcal{F}}^{\text{H}}(k)\mathcal{R}_{xx}^{u}\mathbf{v}_{\mathcal{F}}(k)] \\ &\quad + 4\mu^2 E[(\mathcal{X}_{\mathcal{F}}^{*}(k)\mathcal{P}_{0,L}\mathbf{e}_{\text{o},\mathcal{F}}(k))^{\text{H}}(\mathcal{X}_{\mathcal{F}}^{*}(k)\mathcal{P}_{0,L}\mathbf{e}_{\text{o},\mathcal{F}(k)})]\end{aligned} \tag{8B.8}$$

In the steady state, $\|\mathbf{v}_{\mathcal{F}}(k+1)\|^2 = \|\mathbf{v}_{\mathcal{F}}(k)\|^2$. Thus, when the algorithm has reached its steady state, we obtain from Eq. (8B.8)

$$\begin{aligned}E[\mathbf{v}_{\mathcal{F}}^{\text{H}}(k)\mathcal{R}_{xx}^{u}\mathbf{v}_{\mathcal{F}}(k)] &\approx \mu E[(\mathcal{X}_{\mathcal{F}}^{*}(k)\mathcal{P}_{0,L}\mathbf{e}_{\text{o},\mathcal{F}}(k))^{\text{H}}(\mathcal{X}_{\mathcal{F}}^{*}(k)\mathcal{P}_{0,L}\mathbf{e}_{\text{o},\mathcal{F}}(k))] \\ &\approx \mu E[(\mathcal{P}_{0,L}\mathbf{e}_{\text{o},\mathcal{F}(k)})^{\text{H}}\mathcal{X}_{\mathcal{F}}(k)\mathcal{X}_{\mathcal{F}}^{*}(k)(\mathcal{P}_{0,L}\mathbf{e}_{\text{o},\mathcal{F}}(k))]\end{aligned} \tag{8B.9}$$

We note that the last expectation in Eq. (8B.9) is a scalar and thus, using Eq. (6.23), it may be rearranged as

$$\begin{aligned}&E[(\mathcal{P}_{0,L}\mathbf{e}_{\text{o},\mathcal{F}}(k))^{\text{H}}\mathcal{X}_{\mathcal{F}}(k)\mathcal{X}_{\mathcal{F}}^{*}(k)(\mathcal{P}_{0,L}\mathbf{e}_{\text{o},\mathcal{F}}(k))] \\ &\quad = E[\text{tr}[(\mathcal{P}_{0,L}\mathbf{e}_{\text{o},\mathcal{F}}(k))^{\text{H}}\mathcal{X}_{\mathcal{F}}(k)\mathcal{X}_{\mathcal{F}}^{*}(k)(\mathcal{P}_{0,L}\mathbf{e}_{\text{o},\mathcal{F}}(k))]] \\ &\quad = \text{tr}[E[(\mathcal{P}_{0,L}\mathbf{e}_{\text{o},\mathcal{F}}(k))(\mathcal{P}_{0,L}\mathbf{e}_{\text{o},\mathcal{F}}(k))^{\text{H}}\mathcal{X}_{\mathcal{F}}(k)\mathcal{X}_{\mathcal{F}}^{*}(k)]] \\ &\quad = \text{tr}[E[(\mathcal{P}_{0,L}\mathbf{e}_{\text{o},\mathcal{F}}(k))(\mathcal{P}_{0,L}\mathbf{e}_{\text{o},\mathcal{F}}(k))^{\text{H}}]E[\mathcal{X}_{\mathcal{F}}(k)\mathcal{X}_{\mathcal{F}}^{*}(k)]]\end{aligned} \tag{8B.10}$$

where the last equality follows from the independence assumption. Furthermore, we note that

$$\begin{aligned}\mathcal{P}_{0,L}\mathbf{e}_{\text{o},\mathcal{F}}(k) &= \mathcal{F}\mathbf{P}_{0,L}\mathcal{F}^{-1}\mathbf{e}_{\text{o},\mathcal{F}}(k) \\ &= \mathcal{F}\tilde{\mathbf{e}}_{\text{o}}(k)\end{aligned} \tag{8B.11}$$

where $\tilde{\mathbf{e}}_{o}(k) = [0\ 0\ \cdots\ 0\ e_{\text{o}}(kL)\ e_{\text{o}}(kL+1)\ \cdots\ e_{\text{o}}(kL+L-1)]^{\text{T}}$ is the optimum output error vector in the extended form. Using Eq. (8B.11), we obtain

$$E[(\mathcal{P}_{0,L}\mathbf{e}_{\text{o},\mathcal{F}})(\mathcal{P}_{0,L}\mathbf{e}_{\text{o},\mathcal{F}})^{\text{H}}] = N'\mathcal{F}E[\tilde{\mathbf{e}}_{\text{o}}(k)\tilde{\mathbf{e}}_{\text{o}}^{\text{T}}(k)]\mathcal{F}^{-1} \tag{8B.12}$$

where we have noted that $\mathcal{F}^{\text{H}} = N'\mathcal{F}^{-1}$ and $\tilde{\mathbf{e}}_{\text{o}}^{\text{H}}(k)$ is replaced by $\tilde{\mathbf{e}}_{\text{o}}^{\text{T}}(k)$ as $\tilde{\mathbf{e}}_{o}(k)$ is assumed to be a real-valued vector. Assuming that the optimum error terms $e_{\text{o}}(kL)$, $e_{\text{o}}(kL+1), \ldots, e_{\text{o}}(kL+L-1)$ are samples of a white noise process with variance

$E[e_o^2(n)]$ and noting that $E[e_o^2(n)] = \xi_{\min}$, we get

$$E[\tilde{\mathbf{e}}_o(k)\tilde{\mathbf{e}}_o^T(k)] = \xi_{\min}\mathbf{P}_{0,L} \tag{8B.13}$$

where $\mathbf{P}_{0,L}$ is defined as in Eq. (8.36).

Substituting Eq. (8B.13) in Eq. (8B.12), we get

$$E[(\mathcal{P}_{0,L}\mathbf{e}_{o,\mathcal{F}}(k))(\mathcal{P}_{0,L}\mathbf{e}_{o,\mathcal{F}}(k))^H] = N'\xi_{\min}\mathcal{P}_{0,L} \tag{8B.14}$$

Using this result and the identity (6.23), we obtain

$$\begin{aligned}
\text{tr}[E[(\mathcal{P}_{0,L}\mathbf{e}_{o,\mathcal{F}}(k))(\mathcal{P}_{0,L}\mathbf{e}_{o,\mathcal{F}}(k))^H]E[\mathcal{X}_{\mathcal{F}}(k)\mathcal{X}_{\mathcal{F}}^*(k)]] \\
&= N'\xi_{\min}\text{tr}[E[\mathcal{P}_{0,L}\mathcal{X}_{\mathcal{F}}(k)\mathcal{X}_{\mathcal{F}}^*(k)]] \\
&= N'\xi_{\min}\text{tr}[E[\mathcal{X}_{\mathcal{F}}^*(k)\mathcal{P}_{0,L}\mathcal{X}_{\mathcal{F}}(k)]] \\
&= N'\xi_{\min}\text{tr}[\mathcal{R}_{xx}^u] \\
&= LN'^2\phi_{xx}(0)\xi_{\min}
\end{aligned} \tag{8B.15}$$

where the last equality follows from the identity

$$\begin{aligned}
\text{tr}[\mathcal{R}_{xx}^u] &= \text{tr}[\mathcal{F}\mathbf{R}_{xx}^u\mathcal{F}^{-1}] \\
&= \text{tr}[\mathcal{F}^{-1}\mathcal{F}\mathbf{R}_{xx}^u] \\
&= \text{tr}[\mathbf{R}_{xx}^u] = N'\phi_{xx}(0)
\end{aligned} \tag{8B.16}$$

which is obtained from Eqs. (8.57) and (8.60).

Substituting Eq. (8B.15) in Eq. (8B.10), and taking the result back to Eqs. (8B.6) through (8B.9), we obtain

$$\xi_{\text{excess}}^u = \mu N'\phi_{xx}(0)\xi_{\min} \tag{8B.17}$$

Substituting this result in Eq. (8B.1), we get the misadjustment for the unconstrained FBLMS algorithm without step-normalization as

$$\mathcal{M}_{\text{FBLMS}}^u = \mu N'\phi_{xx}(0) \tag{8B.18}$$

Unconstrained FBLMS algorithm with step-normalization

Following the same line of derivations as in Section 8.3.2, for the present case, we obtain

$$\begin{aligned}
\mathbf{v}_{\mathcal{F}}(k+1) = (\mathbf{I} - 2\mu_o\mathbf{\Lambda}^{-1}\mathcal{X}_{\mathcal{F}}^*(k)\mathcal{P}_{0,L}\mathcal{X}_{\mathcal{F}}(k))\mathbf{v}_{\mathcal{F}}(k) \\
+ 2\mu_o\mathbf{\Lambda}^{-1}\mathcal{X}_{\mathcal{F}}^*(k)\mathcal{P}_{0,L}\mathbf{e}_{o,\mathcal{F}}(k)
\end{aligned} \tag{8B.19}$$

where $\mathbf{\Lambda} = E[\mathcal{X}_{\mathcal{F}}(k)\mathcal{X}_{\mathcal{F}}^*(k)]$. Postmultiply both sides of Eq. (8B.19) by their respective Hermitian transposes, take expectation, assume that $\mathbf{e}_{0,\mathcal{F}}(k)$ is zero mean and independent of $\mathcal{X}_{\mathcal{F}}(k)$, and do some manipulations and approximation similar to what was done above, we obtain

$$\mathbf{K}_{\mathcal{F}}(k+1) \approx \mathbf{K}_{\mathcal{F}}(k) - 2\mu_{\text{o}}\mathbf{\Lambda}^{-1}\mathcal{R}_{xx}^{u}\mathbf{K}_{\mathcal{F}}(k) - 2\mu_{\text{o}}\mathbf{K}_{\mathcal{F}}(k)\mathcal{R}_{xx}^{u}\mathbf{\Lambda}^{-1}$$
$$+ 4\mu_{\text{o}}^{2}\mathbf{\Lambda}^{-1}E[\mathcal{X}_{\mathcal{F}}^{*}(k)(\mathcal{P}_{0,L}\mathbf{e}_{\text{o},\mathcal{F}}(k))(\mathcal{P}_{0,L}\mathbf{e}_{\text{o},\mathcal{F}}(k))^{\text{H}}\mathcal{X}_{\mathcal{F}}(k)]\mathbf{\Lambda}^{-1} \tag{8B.20}$$

where $\mathbf{K}_{\mathcal{F}}(k) = E[\mathbf{v}_{\mathcal{F}}(k)\mathbf{v}_{\mathcal{F}}^{\text{H}}(k)]$. Since $\mathbf{e}_{o,\mathcal{F}}(k)$ and $\mathcal{X}_{\mathcal{F}}(k)$ are independent, we get using Eq. (8B.14)

$$\begin{aligned}
&E[\mathcal{X}_{\mathcal{F}}^{*}(k)(\mathcal{P}_{0,L}\mathbf{e}_{\text{o},\mathcal{F}}(k))(\mathcal{P}_{0,L}\mathbf{e}_{\text{o},\mathcal{F}}(k))^{\text{H}}\mathcal{X}_{\mathcal{F}}(k)] \\
&\quad = E[\mathcal{X}_{\mathcal{F}}^{*}(k)E[(\mathcal{P}_{0,L}\mathbf{e}_{\text{o},\mathcal{F}}(k))(\mathcal{P}_{0,L}\mathbf{e}_{\text{o},\mathcal{F}}(k))^{\text{H}}]\mathcal{X}_{\mathcal{F}}(k)] \\
&\quad = N'\xi_{\min}E[\mathcal{X}_{\mathcal{F}}^{*}(k)\mathcal{P}_{0,L}\mathcal{X}_{\mathcal{F}}(k)] \\
&\quad = N'\xi_{\min}\mathcal{R}_{xx}^{u}
\end{aligned} \tag{8B.21}$$

We note that $\mathbf{\Lambda}$ is a diagonal matrix consisting of the estimates of the powers of the input signal samples in the frequency domain. Considering the spectral separation property of the DFT (see e.g., Oppenheim and Schafer (1975)), we obtain

$$\mathbf{\Lambda} \approx N' \times \text{diag}\left(\Phi_{xx}\left(\text{e}^{j\frac{2\pi\times 0}{N'}}\right),\ \Phi_{xx}\left(\text{e}^{j\frac{2\pi}{N'}}\right),\ \ldots,\ \Phi_{xx}\left(\text{e}^{j\frac{2\pi(N'-1)}{N'}}\right)\right) \tag{8B.22}$$

where $\Phi_{xx}(\text{e}^{j\omega})$ is the power spectral density of the underlying input process, $x(n)$. The factor N' in Eq. (8B.22) is the length of the DFT in the present case. Comparing Eq. (8B.22) with Eq. (8.61), we find that

$$\mathbf{\Lambda} \approx \frac{N'}{L}\mathcal{R}_{xx}^{u} \tag{8B.23}$$

Substituting Eqs. (8B.21) and (8B.23) in Eq. (8B.20), we get

$$\mathbf{K}_{\mathcal{F}}(k+1) \approx \mathbf{K}_{\mathcal{F}}(k) - 4\mu_{o}\frac{L}{N'}\mathbf{K}_{\mathcal{F}}(k) + 4\mu_{o}^{2}\xi_{\min}\frac{L^{2}}{N'}\mathcal{R}_{xx}^{u-1} \tag{8B.24}$$

In the steady state, when $\mathbf{K}_{\mathcal{F}}(k+1) = \mathbf{K}_{\mathcal{F}}(k)$, we obtain

$$\mathbf{K}_{\mathcal{F}}(k) \approx \mu_{o}\xi_{\min}L\mathcal{R}_{xx}^{u-1} \tag{8B.25}$$

Substituting Eq. (8B.25) in Eq. (8B.7), we get

$$\xi_{\text{excess}}^{\text{un}} \approx \mu_{o}\xi_{\min} \tag{8B.26}$$

Substituting this result in Eq. (8B.1), we obtain the misadjustment for the unconstrained FBLMS algorithm with step-normalization as

$$\mathcal{M}_{\text{FBLMS}}^{\text{un}} \approx \mu_{\text{o}} \tag{8B.27}$$

Constrained FBLMS algorithm without step-normalization

In this case, the FBLMS algorithm is an exact and fast implementation of the BLMS algorithm, that is, with a reduced computational complexity. Hence, the corresponding excess MSE is given by Eq. (8B.16). Noting that $\mathbf{R}$ is N-by-N and its diagonal elements are all equal to $\phi_{xx}(0)$, Eq. (8A.16) may also be written as

$$\xi_{\text{excess}}^{c} = \mu N \xi_{\min} \phi_{xx}(0) \tag{8B.28}$$

to be in line with the rest of the results in this appendix. Substituting this result in Eq. (8B.1), we obtain the corresponding misadjustment as

$$\mathcal{M}_{\text{FBLMS}}^{\text{cn}} \approx \mu N \phi_{xx}(0) \tag{8B.29}$$

Constrained FBLMS algorithm with step-normalization

Premultiplication of the gradient vector $\mathcal{X}_{\mathcal{F}}^{*}(k)\mathbf{e}_{\mathcal{F}}(k)$ by the matrix $\mathcal{P}_{N,0}$ implements the constraining step (8.47). Combining this step with step-normalization, we get the recursion

$$\begin{aligned}\mathbf{v}_{\mathcal{F}}(k+1) = (\mathbf{I} - 2\mu_{\text{o}}\boldsymbol{\Lambda}^{-1}\mathcal{P}_{N,0}\mathcal{X}_{\mathcal{F}}^{*}(k)\mathcal{P}_{0,L}\mathcal{X}_{\mathcal{F}}(k))\mathbf{v}_{\mathcal{F}}(k)\\ + 2\mu_{\text{o}}\boldsymbol{\Lambda}^{-1}\mathcal{P}_{N,0}\mathcal{X}_{\mathcal{F}}^{*}(k)\mathcal{P}_{0,L}\mathbf{e}_{\text{o},\mathcal{F}}(k)\end{aligned} \tag{8B.30}$$

analogous to Eq. (8B.19). Following the same line of derivations as in the case of Eq. (8B.19), we obtain

$$\begin{aligned}-2\mu_{\text{o}}\boldsymbol{\Lambda}^{-1}\mathcal{P}_{N,0}\mathcal{R}_{xx}^{u}\mathbf{K}_{\mathcal{F}}(k) - 2\mu_{\text{o}}\mathbf{K}_{\mathcal{F}}(k)\mathcal{R}_{xx}^{u}\mathcal{P}_{N,0}\boldsymbol{\Lambda}^{-1}\\ + 4\mu_{\text{o}}^{2}N'\xi_{\min}\boldsymbol{\Lambda}^{-1}\mathcal{P}_{N,0}\mathcal{R}_{xx}^{u}\mathcal{P}_{N,0}\boldsymbol{\Lambda}^{-1} = 0\end{aligned} \tag{8B.31}$$

We shall now solve this equation to find $\mathbf{K}_{\mathcal{F}}(k)$.

To proceed, let us define

$$\mathbf{G} = \boldsymbol{\Lambda}^{-1}\mathcal{P}_{N,0}\mathcal{R}_{xx}^{u} \tag{8B.32}$$

and note that $\mathbf{G}^{\text{H}} = \mathcal{R}_{xx}^{u}\mathcal{P}_{N,0}\boldsymbol{\Lambda}^{-1}$ as $\mathcal{P}_{N,0}$ is Hermitian and $\mathcal{R}_{xx}^{u}$ and $\boldsymbol{\Lambda}^{-1}$ are diagonal matrices. Using these, Eq. (8B.31) may be rearranged as

$$\mathbf{G}\mathbf{K}_{\mathcal{F}}(k) - \mu_{o}N'\xi_{\min}\mathbf{G}\mathcal{P}_{N,0}\boldsymbol{\Lambda}^{-1} + \mathbf{K}_{\mathcal{F}}(k)\mathbf{G}^{\text{H}} - \mu_{o}N'\xi_{\min}\boldsymbol{\Lambda}^{-1}\mathcal{P}_{N,0}\mathbf{G}^{\text{H}} = 0 \tag{8B.33}$$

or

$$\mathbf{G}(\mathbf{K}_{\mathcal{F}}(k) - \mu_{o}N'\xi_{\min}\mathcal{P}_{N,0}\boldsymbol{\Lambda}^{-1}) + (\mathbf{K}_{\mathcal{F}}(k) - \mu_{o}N'\xi_{\min}\boldsymbol{\Lambda}^{-1}\mathcal{P}_{N,0})\mathbf{G}^{\text{H}} = 0 \tag{8B.34}$$

General solution of Eq. (8B.37) turns out to be difficult. However, a trivial solution of that, which closely matches the simulation results, can be easily identified as

$$\mathbf{K}_{\mathcal{F}}(k) = \mu_{o}N'\xi_{\min}\mathcal{P}_{N,0}\boldsymbol{\Lambda}^{-1} \tag{8B.35}$$

Substituting Eq. (8.35) in Eq. (8B.7), we get

$$\xi_{\text{excess}}^{\text{cn}} = \mu_o \xi_{\min} \frac{1}{L} \text{tr}[\mathcal{P}_{N,0} \mathbf{\Lambda}^{-1} \mathcal{R}_{xx}^{u}] \tag{8B.36}$$

Using Eq. (8B.23) in Eq. (8B.36), we obtain

$$\xi_{\text{excess}}^{\text{cn}} = \mu_o \xi_{\min} \frac{1}{N'} \text{tr}[\mathcal{P}_{N,0}] \tag{8B.37}$$

Noting that

$$\text{tr}[\mathcal{P}_{N,0}] = \text{tr}[\mathcal{F}\mathbf{P}_{N,0}\mathcal{F}^{-1}] = \text{tr}[\mathcal{F}^{-1}\mathcal{F}\mathbf{P}_{N,0}] = \text{tr}[\mathbf{P}_{N,0}] = N$$

we get from Eq. (8B.37)

$$\xi_{\text{excess}}^{\text{cn}} = \mu_o \xi_{\min} \frac{N}{N'} \tag{8B.38}$$

Substituting Eq. (8B.38) in Eq. (8B.1), we obtain the misadjustment for constrained FBLMS algorithm with step-normalization as

$$\mathcal{M}_{\text{FBLMS}}^{\text{cn}} = \mu_o \frac{N}{N'} \tag{8B.39}$$

9

Subband Adaptive Filters

In the last two chapters, we have discussed two classes of LMS adaptive filtering algorithms that have improved convergence behavior compared to the conventional LMS algorithm. Convergence improvement in both classes was found to be a direct consequence of using orthogonal transforms for decomposing the filter input into a number of partially mutually exclusive bands. This was referred to as *band-partitioning*. Moreover, our study of transform domain adaptive filters in Chapter 7 clearly showed that the imperfect separation of the input signal into mutually exclusive bands is the main reason for the suboptimal convergence behavior of such filters.

In this chapter, we present another class of adaptive filters that also uses the concept of band-partitioning to improve the convergence behavior of LMS algorithm. This structure, which is called *subband adaptive filter*, is different from the transform domain adaptive filters in many ways. Firstly, the filters used for band-partitioning of the input signal are well-designed filters with *high* stop-band rejection, that is, very low side lobes. As a result, we find that the subband adaptive filters achieve higher degree of improvement in convergence compared to the transform domain adaptive filters of Chapter 7. Secondly, because of the high stop-band rejection, the subband signals can be *decimated* (down-sampled to a lower rate) before doing any filtering in subbands. Thirdly, implementation of subband filters at a decimated rate results in significant reduction in the computational complexity of the overall filter. However, this reduction is not as significant as what is usually achieved by the fast block LMS (FBLMS) algorithm of Chapter 8. We will make some comments on comparison of the subband adaptive structure and FBLMS algorithm in Section 9.11.

The subject of subband filtering is closely related to *multirate signal processing*. In a subband adaptive filter, the filter input is first partitioned into a set of subband signals through an analysis filter bank. These subband signals are then decimated to a lower rate and passed through a set of independent or partially independent adaptive filters that operate at the decimated rate. The outputs from these filters are subsequently combined together using a synthesis filter bank to reconstruct the fullband output of the overall filter. The DFT filter banks are commonly used for efficient realization of the analysis and synthesis filter banks. We thus start this chapter with a short review of the DFT filter banks and introduce the method of *weighted overlap–add* for efficient realization of these filter banks. We also discuss the conditions that should be imposed on the analysis and synthesis filters so that the reconstructed fullband signals have negligible distortion.

Adaptive Filters: Theory and Applications, Second Edition. Behrouz Farhang-Boroujeny.

For a deeper study on multirate signal processing, the reader may refer to Crochiere and Rabiner (1983) or Vaidyanathan (1993), for example.

Successful implementation of subband adaptive filters requires careful design of analysis and synthesis filters. Much of our effort in this chapter is thus devoted to the design of analysis and synthesis filters that are suitable for subband adaptive filtering.

9.1 DFT Filter Banks

Consider the case where a sequence, $x(n)$, has to be separated into a number of subbands. For this, we may start with a lowpass filter, $H(z)$, and proceed as follows. By passing $x(n)$ through $H(z)$, the low-frequency part of its spectrum is extracted. To extract any other part of the spectrum of $x(n)$, say, the part centered around the frequency $\omega = \omega_i$, we may shift the desired portion of the spectrum to the baseband (i.e., around $\omega = 0$) by multiplying $x(n)$ with the complex sinusoid $e^{-j\omega_i n}$, and then use the lowpass filter $H(z)$ to extract that. The filter $H(z)$, which is repeatedly used for extraction of different parts of the input spectrum, is called the *prototype* filter.

Using this method, a sequence, $x(n)$, can be partitioned into any set of arbitrary bands. As the separated subband signals are in baseband and have a smaller bandwidth than the original fullband signal, they have a lower Nyquist rate and thus may be decimated (down-sampled) to a lower rate before any further processing. Figure 9.1 depicts the steps required for partitioning a sequence $x(n)$ into M equally spaced subbands, centered at frequencies

$$\omega_i = \frac{2\pi i}{M}, \quad \text{for} \quad i = 0, 1, \ldots, M-1$$

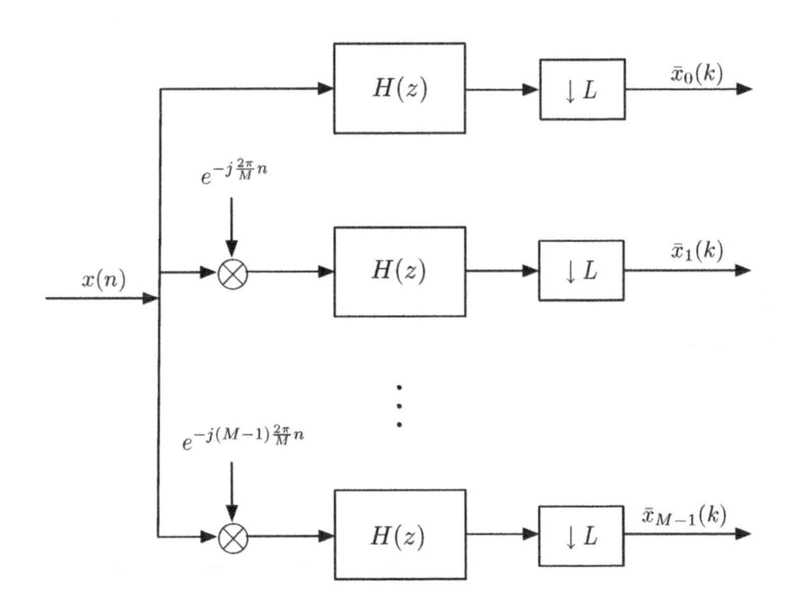

Figure 9.1 DFT analysis filter bank.

and decimating the subband signals using a decimation factor L. The structure of Figure 9.1 is known as *DFT analysis filter bank*, for reasons that will become clear shortly. In Figure 9.1, decimation is denoted by a downward arrow followed by the decimation factor, L. We may also note that, in Figure 9.1, the time index n is used for the fullband input sequence $x(n)$. In contrast, we use the time index k for subband sequences. These choices of time indices will be consistently followed throughout this chapter. Furthermore, the subband signals are represented by over-bar variables, such as $\bar{x}_i(k)$'s in Figure 9.1, so as to distinguish from fullband signals.

We may note that in the structure of Figure 9.1, there is no restriction on the bandwidth of the prototype filter, $H(z)$, the number of subbands, M, and the decimation factor L. Thus, there may be some overlap between different subbands. However, if L is chosen too large, the decimated subband signals may suffer from aliasing effects. Although aliasing is not desirable in most applications, we will later show that a small amount of aliasing may be beneficial in the implementation of subband adaptive filters.

A general procedure for efficient realization of the DFT filter banks, for any choice of L and M, is the *weighted overlap–add method*. When M is a multiple of L, a slightly different procedure that leads to the so-called *polyphase* filter bank structure may be more useful from the point of view of *computational complexity* (Crochiere and Rabiner, 1983; Vaidyanathan, 1993). Since M is not necessarily a multiple of L in most applications of subband adaptive filters, we discuss only the weighted overlap–add method in the rest of this section.

9.1.1 Weighted Overlap–Add Method for Realization of DFT Analysis Filter Banks

To begin with, let us define

$$W_M = e^{j(2\pi/M)}$$

where $j = \sqrt{-1}$. Then, the ith output of the DFT analysis filter bank may be expressed as[1]

$$\bar{x}_i(k) = \sum_{n=-\infty}^{\infty} h_n \tilde{x}_i(kL - n) \tag{9.1}$$

where

$$\tilde{x}_i(n) = x(n) W_M^{-in} \tag{9.2}$$

is the modulated version of the input, $x(n)$ (Figure 9.1). Replacing n by $-n$, Eq. (9.1) may be rearranged as

$$\bar{x}_i(k) = \sum_{n=-\infty}^{\infty} h_{-n} \tilde{x}_i(kL + n) \tag{9.3}$$

Substituting Eq. (9.2) in Eq. (9.3), we get

$$\bar{x}_i(k) = W_M^{-ikL} \sum_{n=-\infty}^{\infty} h_{-n} x(kL + n) W_M^{-in} \tag{9.4}$$

[1] We note that, in practice, the sequence h_n is always causal (i.e., $h_n = 0$, for $n < 0$) and has a finite duration. However, here, we let n to vary from $-\infty$ to $+\infty$, to keep the derivations simple.

Now, the method of *time aliasing* may be applied to the summation on the right-hand side of Eq. (9.4) for its evaluation in an efficient manner. To this end, we define the sequence

$$u_k(n) = h_{-n}x(kL+n) \tag{9.5}$$

and note that $u_k(n)$ is a windowed version of the input sequence, $x(n)$, the window being the time reverse of the prototype filter, h_n. Using Eq. (9.5) in Eq. (9.4), we get

$$\bar{x}_i(k) = W_M^{-ikL} \sum_{n=-\infty}^{\infty} u_k(n) W_M^{-in} \tag{9.6}$$

With change of variable $n = r + lM$ and noting that $W_M^{-ilM} = 1$, Eq. (9.6) may be rearranged as

$$\bar{x}_i(k) = W_M^{-ikL} \sum_{r=0}^{M-1} u_k^{\mathrm{a}}(r) W_M^{-ir} \tag{9.7}$$

where

$$u_k^{\mathrm{a}}(r) = \sum_{l=-\infty}^{\infty} u_k(r+lM), \quad \text{for} \quad r = 0, 1, \ldots, M-1 \tag{9.8}$$

We note that the M-point sequence $u_k^{\mathrm{a}}(r)$ is obtained by subdividing the sequence $u_k(n)$ into blocks of M samples and stacking and adding (i.e., time aliasing) these blocks.

From Eq. (9.7), we note that the subband signal samples, $\bar{x}_i(k)$, for $i = 0, 1, \ldots, M-1$, can be computed simultaneously, once the time-aliased sequence $u_k^{\mathrm{a}}(r)$ is obtained. This is done by applying an M-point DFT to the samples $u_k^{\mathrm{a}}(r)$, for $r = 0, 1, \ldots, M-1$, and multiplying the DFT outputs by the coefficients W_M^{-ikL}, as suggested in Eq. (9.7). Furthermore, computation of the DFT may be performed using an efficient FFT algorithm.

9.1.2 Weighted Overlap–Add Method for Realization of DFT Synthesis Filter Banks

Consider the case where the subband signals $\bar{y}_i(k)$, for $i = 0, 1, \ldots, M-1$, are to be synthesized to reconstruct the fullband signal $y(n)$. Also, assume that these subband signals are in baseband and at a decimated rate L times lower than the fullband rate. To generate $y(n)$, we may proceed as follows:

1. By appending $L-1$ zeros after every sample of subband signals, these signals are expanded to the fullband rate. This is referred to as *interpolation* and, accordingly, L is called the *interpolation factor*. Interpolation results in a set of fullband signals whose spectra consist of L repetitions of their associated baseband spectra (see e.g., Oppenheim and Schafer (1989),).
2. The repetitions of the baseband spectra are removed by the lowpass filter.
3. The lowpass filtered fullband signals are then shifted to their respective bands through appropriate modulators.

   The combination of Steps 1 to 3 can be mathematically expressed as

$$y_i(n) = W_M^{in} \sum_{k=-\infty}^{\infty} \bar{y}_i(k) g_{n-kL}, \quad \text{for} \quad i = 0, 1, \ldots, M-1 \tag{9.9}$$

where the sequence g_n is the impulse response of the lowpass filter, and the coefficients W_M^{in} are the modulating factors. The fact that the samples added in Step 1, to expand the subband signal sequences to fullband, are zero has been used to arrive at the special form of the summation on the right-hand side of Eq. (9.9). Verification of this is left to the reader as an exercise (Problem P9.1)

4. Finally, the fullband signals $y_i(n)$'s are added together to obtain the synthesized sequence

$$y(n) = \frac{1}{M}\sum_{i=0}^{M-1} y_i(n) \tag{9.10}$$

The factor $1/M$, in Eq. (9.10), is added for convenience.

Figure 9.2 presents the block diagram of a synthesis filter bank, where interpolation is denoted by an upward arrow followed by the interpolation factor, L.

To obtain an efficient realization of synthesis filter banks, we proceed as follows. Substituting Eq. (9.9) in Eq. (9.10) and rearranging, we obtain

$$y(n) = \sum_{k=-\infty}^{\infty} g_{n-kL}\frac{1}{M}\sum_{i=0}^{M-1} \bar{y}_i(k)W_M^{in} \tag{9.11}$$

Next, we define the following fullband sequence

$$\hat{y}_k(n) = g_n\frac{1}{M}\sum_{i=0}^{M-1} \bar{y}_i(k)W_M^{i(n+kL)} \tag{9.12}$$

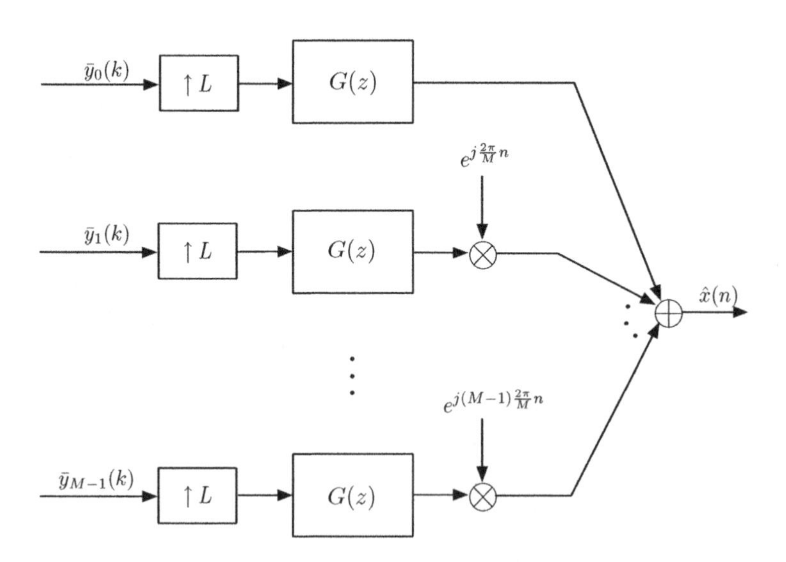

Figure 9.2 DFT synthesis filter bank.

Then, using Eq. (9.12), we can write Eq. (9.11) as

$$y(n) = \sum_{k=-\infty}^{\infty} \hat{y}_k(n - kL) \tag{9.13}$$

That is, the output sequence $y(n)$ is obtained by overlapping and adding the sequences $\hat{y}_k(n)$'s, thus the name overlap–add.

Equation (9.12) may also be written as

$$\hat{y}_k(n) = g_n \tilde{y}_k(n) \tag{9.14}$$

where

$$\tilde{y}_k(n) = \frac{1}{M} \sum_{i=0}^{M-1} [\bar{y}_i(k) W_M^{ikL}] W_M^{in} \tag{9.15}$$

Note that $\tilde{y}_k(n)$ is a periodic function of n with period M as W_M^{in} is periodic in n with period M, and the rest of the terms on the right-hand side of Eq. (9.15) are independent of n. Furthermore, it is straightforward to see that the values of $\tilde{y}_k(n)$, for $n = 0, 1, \ldots, M - 1$ (i.e., the first period of $\tilde{y}_k(n)$) are samples of the inverse DFT of the sequence $\bar{y}_i(k) W_M^{ikL}$, for $i = 0, 1, \ldots, M - 1$.

From the above observation, we may adopt the following procedure to generate the samples of the synthesized output sequence, $y(n)$:

1. Upon the receipt of the latest samples of the subband signals, say $\bar{y}_i(k)$, for $i = 0, 1, \ldots, M - 1$, we construct the vector

$$\bar{\mathbf{y}}(k) = [\bar{y}_0(k) \;\; \bar{y}_1(k) W_M^{kL} \;\; \bar{y}_2(k) W_M^{2kL} \cdots \bar{y}_{M-1}(k) W_M^{k(M-1)L}]$$

   and compute the inverse DFT of $\bar{\mathbf{y}}(k)$.
2. The result of this inverse DFT is repeated to generate a periodic sequence. This makes the sequence $\tilde{y}_k(n)$ of Eq. (9.15).
3. The sequence $\hat{y}_k(n)$ is obtained by multiplying the sequences $\tilde{y}_k(n)$ and g_n on an element-by-element basis, as in Eq. (9.14). Assuming that g_n is causal, $\hat{y}_k(n)$ will also be causal.
4. Finally, to generate the samples of $y(n)$, the sequence $\hat{y}_k(n)$ is added to a buffer holding the accumulated results of the previous iterations, that is, $\sum_{l=-\infty}^{k-1} \hat{y}_l(n - lL)$. The first L elements of the updated buffer are the samples $y(kL), y(kL + 1), \ldots, y(kL + L - 1)$ of the synthesized output. While these samples are being sent to the output, the content of the buffer is shifted and filled with zeros from its other end and becomes ready for stacking the next set of samples, that is, $\hat{y}_{k+1}(n)$, in the next iteration.

9.2 Complementary Filter Banks

In multirate signal processing, in general, analysis and synthesis filters need to satisfy certain conditions in order that the reconstructed fullband signals have no or, at least, insignificant distortion. For this to be true in subband adaptive filters, we find

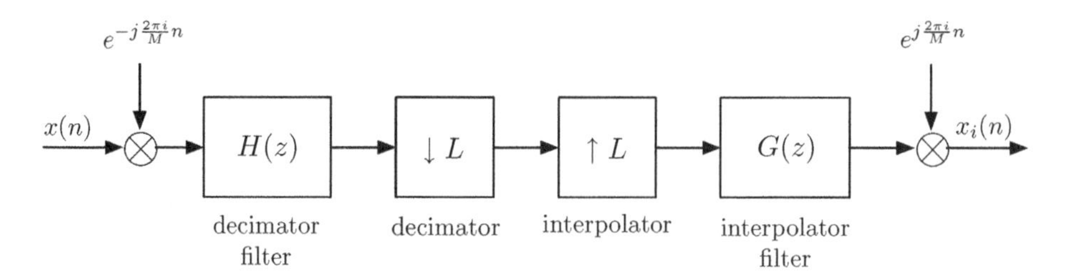

Figure 9.3 The ith channel of an M-band analysis–synthesis DFT filter bank.

that the combined responses of the analysis and synthesis filters should be those of a *complementary filter bank.*

To explain what do we mean by a complementary filter bank and also to derive the conditions required for a filter bank to be complementary, consider the ith channel (frequency band) of a pair of analysis–synthesis filter banks, as depicted in Figure 9.3. Figure 9.4 presents a set of plots showing the results of the various stages of Figure 9.3. Figure 9.4a shows a representative graph of the spectrum of the fullband input, $x(n)$. The portion of the spectrum of $x(n)$ that is centered around $\omega_i = 2\pi i/M$ is shifted to $\omega = 0$ and lowpass filtered through the decimator filter, $H(z)$. Let us choose $\omega_i = \pi/2$ and $L = 4$ for this example. Furthermore, let the lowpass filter $H(z)$ be an ideal filter with unit gain over the frequency range $-\pi/4 \le \omega \le \pi/4$ and zero elsewhere. Then, the spectrum of the output of $H(z)$ will be as shown in Figure 9.4b. The decimation, which compresses the output of $H(z)$ along the time axis, results in expansion of the spectrum along the frequency axis, as shown in Figure 9.4c. The interpolator, in contrast, expands the signal samples along the time axis and thus results in compression of the spectrum, as shown in Figure 9.4d. This leads to L repetitions of the spectrum of the decimated signal over the range $0 \le \omega \le 2\pi$. The interpolator filter, $G(z)$, selects the baseband part of the repeated spectrum and rejects its repetitions, thereby recovering back the lowpass spectrum of Figure 9.4b. Finally, the output of $G(z)$ is shifted to its respective band through a modulator. This results in a fullband signal $x_i(n)$, which is a bandpass-filtered portion of the input, $x(n)$, as shown in Figure 9.4e.

From the above example, we also note that the effect of the decimator–interpolator blocks in Figure 9.3 is to repeat the baseband spectrum of Figure 9.4b, as shown in Figure 9.4d. However, as these repetitions are in turn rejected by the synthesis filter, we may delete these blocks from Figure 9.3, without affecting its input–output relationship. Furthermore, one can easily show that the combination of the modulator stages (i.e., multiplication of input, $x(n)$, by $e^{-j2\pi in/M}$ and the interpolator filter output by $e^{j2\pi in/M}$) and the lowpass filters $H(z)$ and $G(z)$ is equivalent to the cascade of the bandpass filters $H(ze^{-j2\pi i/M})$ and $G(ze^{-j2\pi i/M})$ (Problem P9.2). In an M-band analysis–synthesis filter bank, there are M such pairs of filters in parallel, as shown in Figure 9.5. For a sequence $x(n)$ to pass through this bank of filters without distortion, the overall transfer function of the system should resemble that of a pure delay. That is

$$\sum_{i=0}^{M-1} F(ze^{-j2\pi i/M}) = z^{-\Delta} \tag{9.16}$$

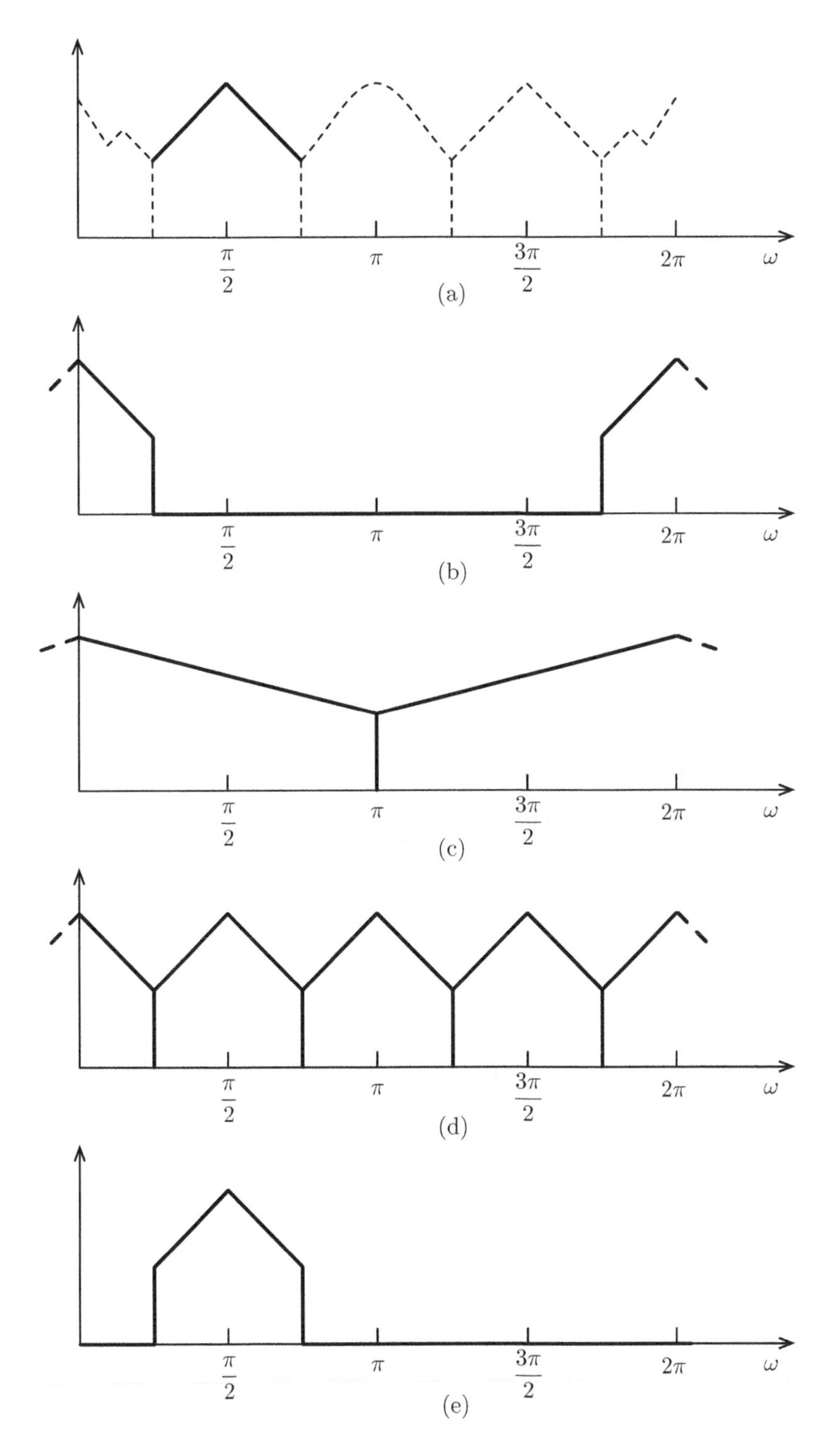

Figure 9.4 Spectra of the signal sequences at various stages of Figure 9.3: (a) input signal, $x(n)$, (b) decimator filter output, (c) decimator output, (d) interpolator output, and (e) final output, $x_i(n)$.

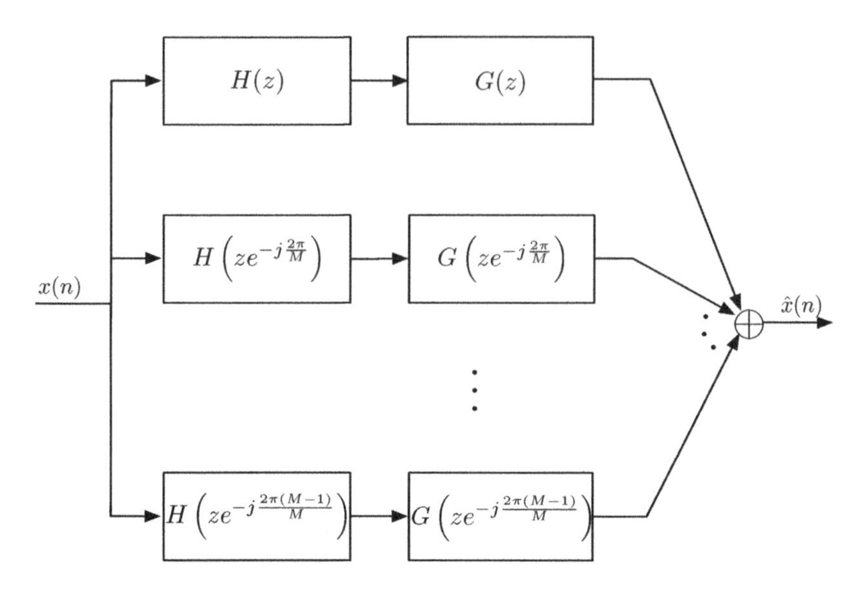

Figure 9.5 An equivalent block diagram of an M-band analysis–synthesis DFT filter bank.

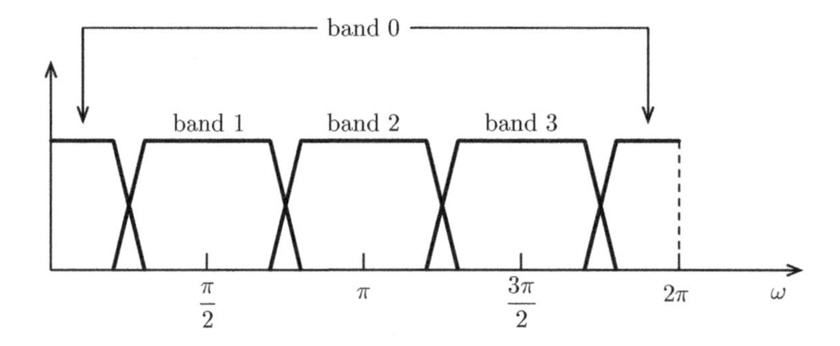

Figure 9.6 A pictorial representation of the concept of complementary filter banks.

where $F(z) = H(z)G(z)$, and Δ is the delay introduced by the cascade of the analysis and synthesis filters.

When Eq. (9.16) holds, we say that the filter bank is *complementary*. The complementary condition (9.16) implies that $\hat{x}(n) = x(n - \Delta)$. That is, the reconstructed signal, $\hat{x}(n)$, at the synthesis bank output is a delayed replica of the input, $x(n)$. Figure 9.6 gives a pictorial representation of the concept of complementary filters, where the magnitude responses of the filters $F(ze^{-j2\pi i/M}) = H(ze^{-j2\pi i/M})G(ze^{-j2\pi i/M})$ of a four-band filter bank are plotted, for $i = 0$, 1, 2, and 3. As shown, there is some overlap among neighboring filters. However, the filters are chosen so that the overall response adds up to unity across the fullband.

Figure 9.6 as well as Eq. (9.16) states the condition in frequency domain that should be satisfied for the filter bank to be complementary. In the design of complementary filter banks, however, we often find that it is more convenient to work with time domain constraints. So, we convert the constraint specified by Eq. (9.16) to its equivalent in time domain. For this, we define the sequence f_n as the inverse z-transform of $F(z)$. That is,

$$F(z) = \sum_{n=-\infty}^{\infty} f_n z^{-n} \tag{9.17}$$

Using Eq. (9.17) in Eq. (9.16) and rearranging, we obtain

$$\sum_{n=-\infty}^{\infty} \left(\sum_{i=0}^{M-1} \mathrm{e}^{-j\frac{2\pi in}{M}} \right) f_n z^{-n} = z^{-\Delta} \tag{9.18}$$

Furthermore, it is straightforward to show that

$$\sum_{i=0}^{M-1} \mathrm{e}^{-j\frac{2\pi in}{M}} = \begin{cases} M, & \text{when} \, n \, \text{is a multiple of M} \\ 0, & \text{otherwise} \end{cases} \tag{9.19}$$

Using Eq. (9.19) in Eq. (9.18), we find that Eq. (9.18) can only be satisfied when $\Delta = KM$, where K is a positive integer, and

$$f_n = \begin{cases} 1/M, & n = KM \\ 0, & n = \text{all multiples of} \, M \, \text{except} \, KM \\ \text{unspecified}, & \text{otherwise} \end{cases} \tag{9.20}$$

Thus, the value of K determines the *total delay* introduced by the filter bank.

9.3 Subband Adaptive Filter Structures

Figure 9.7 depicts the schematic of a commonly used structure of subband adaptive filters.[2] The adaptive filter is used to model a plant, $W_o(z)$. The input, $x(n)$, and the plant output, $d(n)$, are passed through a pair of identical analysis filter banks to be partitioned into M subbands and decimated to a rate that is $1/L$ of the fullband rate. The subband adaptive filters, $\bar{W}_i(z)$'s, are thus running at a rate that is only $1/L$ of the fullband rate. To generate the adaptive filter output in fullband, the outputs from the subband filters are combined together through a synthesis filter bank.

The subband adaptive filter structure presented in Figure 9.7 is referred to as *synthesis independent*, as the adaptation of the subband filters is independent of the synthesis filters. The assumption here is that the synthesis filters are ideal, in the sense that their stop-band attenuation is infinity and their cascade with the analysis filters results in a complementary filter bank. In practice, these ideal requirements can be satisfied only approximately. Hence, the synthesis-independent subband adaptive filters are bound to have some distortion. This distortion can be reduced using an alternative structure, which is known as *synthesis-dependent* subband adaptive filter. This is shown in Figure 9.8. The delay Δ is to account for the combined delay due to analysis and synthesis filters. In this structure, even though the filtering is still done in subbands, the computation of output

[2] The concept of subband adaptive filtering was first introduced by Furukawa (1984) and Kellermann (1984 and 1985).

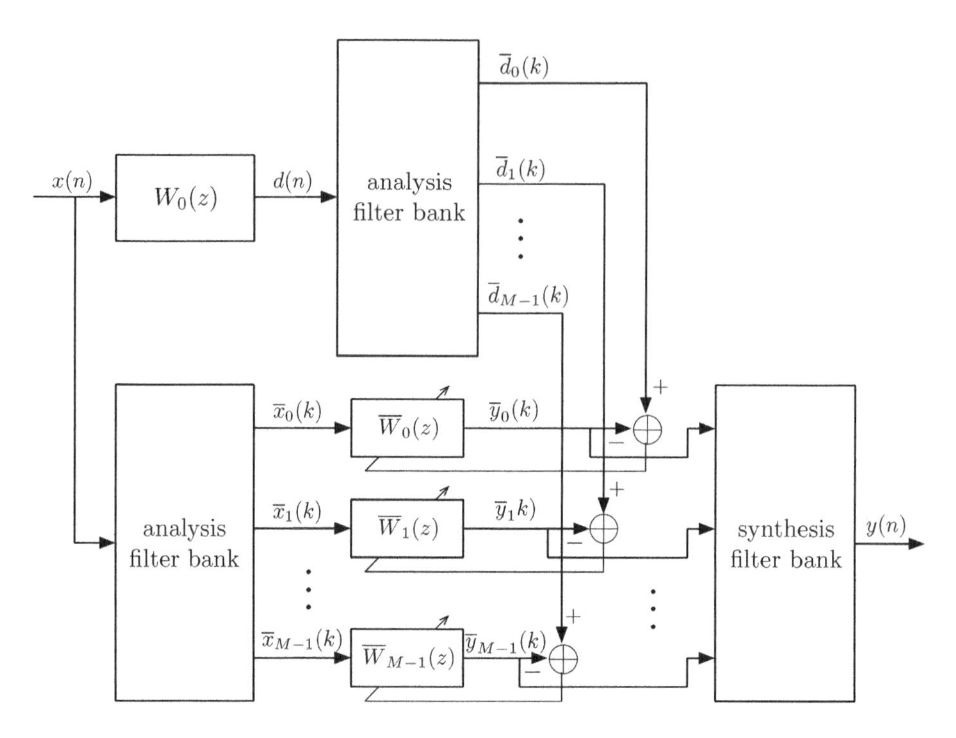

Figure 9.7 Subband adaptive filter (synthesis-independent structure).

error, $e(n)$, is done in fullband. The fullband error, $e(n)$, is subsequently partitioned into subbands using an analysis filter bank, and the subband errors, $\bar{e}_i(k)$'s, are used for the adaptation of the associated subband filters.

The synthesis-dependent structure, although resolves the distortion introduced by the synthesis filters, has some drawbacks, which hinder its application in practice (Sondhi and Kellermann, 1992). In particular, the cascade of synthesis and analysis filter banks in the adaptation loop introduces an undesirable delay that makes the filter more prone to instability. Furthermore, the presence of a delay in the adaptation loop increases the memory requirement of the filter (Problem P9.3). Because of these problems, the synthesis-dependent subband adaptive filter structure has been less popular than its synthesis-independent counterpart. Noting this, our emphasis in the rest of this chapter will be on the synthesis-independent structure. Nevertheless, most of the results we develop are applicable to the synthesis-dependent structure as well.

9.4 Selection of Analysis and Synthesis Filters

Design of analysis and synthesis filters with well-behaved responses is crucial to successful implementation of subband adaptive filters. In this section, we look into the basic requirements of the analysis and synthesis filters. We note that there are many requirements that should be taken into account while selecting these filters and hence a compromise has to be struck to achieve an acceptable design. As a result, it is very difficult to give any

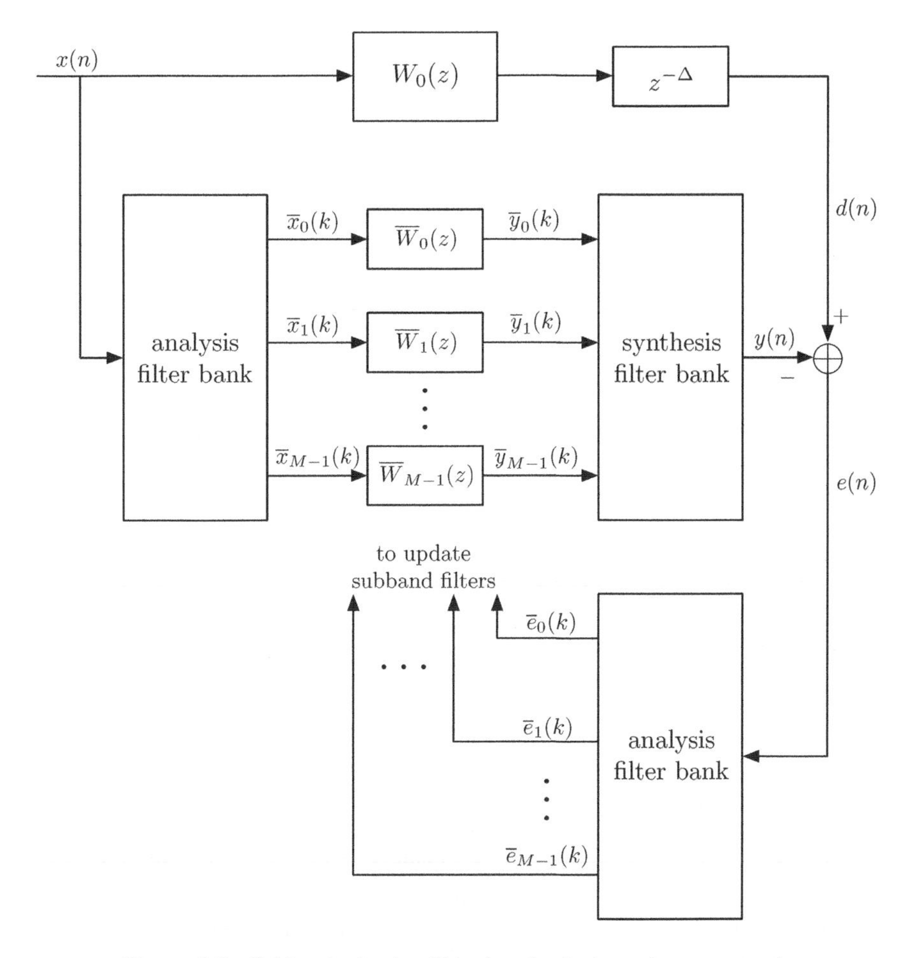

Figure 9.8 Subband adaptive filter (synthesis-dependent structure).

specific criterion whose optimization will lead to the optimum set of filters. Instead, we find it more appropriate to deal with this problem in a subjective manner, which would lead us to a number of specifications for a good compromise design.

As was noted earlier in Section 9.2, for the reconstructed output of a subband adaptive structure to have small distortion, the analysis and synthesis filters should form a complementary filter bank. There are many pairs of analysis and synthesis filter banks that satisfy the complementary condition. This provides some degrees of freedom, which may be used to facilitate the design and/or enhance the performance of the subband adaptive filters.

A first attempt may be to use the same prototype filter for both analysis and synthesis. Unfortunately, this leads to subband signals whose spectra vary and decay to some small values near the ends of their respective bands. This, in turn, will result in the inputs to the subband adaptive filters to be badly conditioned because of the low excitation levels near the band edges. Furthermore, from our discussions in the previous chapters, we know that such inputs will result in large eigenvalue spreads and thus poor convergence. This problem may be resolved as follows (Morgan (1995) and De Leon and Etter (1995)):

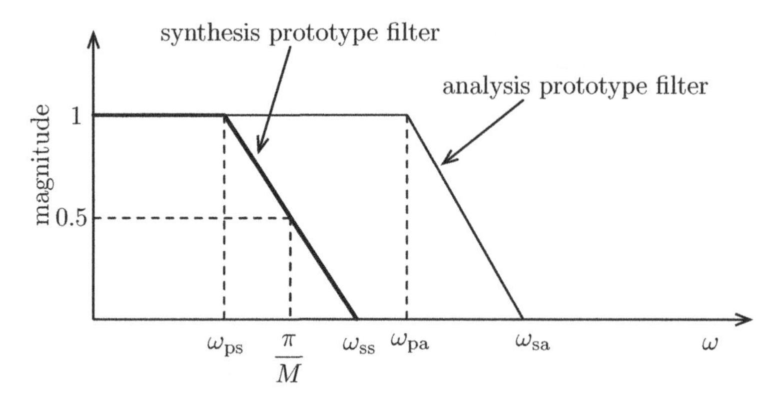

Figure 9.9 A possible choice of analysis and synthesis prototype filters that resolves the problem of slow convergence of subband adaptive filters.

Figure 9.9 presents a diagram showing a good choice of analysis and synthesis prototype filters that resolves the problem of slow convergence of subband adaptive filters. The analysis prototype filter is chosen such that it has a flat magnitude response and linear phase response (constant group-delay[3]) between zero and a frequency larger than or equal to ω_{ss}, where ω_{ss} is the beginning of the stop-band (i.e., the end of the transition band) of the synthesis prototype filter. Moreover, the synthesis filters are chosen to be complementary. The cascade of the analysis and synthesis filters will then be a complementary filter bank because in this case the multiplication (cascade) of the analysis and synthesis prototype filters is just the same as the synthesis prototype filter. The analysis filters introduce only a fixed delay in the overall response of the subband structure.

Next, we explain why the choice of the analysis and synthesis prototype filters, as shown in Figure 9.9, resolves the problem of poor convergence of subband adaptive filters. Assuming that the power spectral density of the fullband input, $x(n)$, does not vary significantly over each subband, using an analysis prototype filter similar to the one shown in Figure 9.9, would result in all decimated subband sequences to have approximately flat spectra over the range of frequencies $|\omega| \leq \omega_{pa}$. On the other hand, the band of interest over which matching between the frequency response of each subband adaptive filter and its associated desired response from the respective band of the plant should be achieved is $|\omega| < \omega_{ss}$, as frequencies beyond this are cut off by the synthesis filters (Figure 9.9). We may thus say that in a subband adaptive filter structure whose analysis and synthesis prototype filters are selected as shown in Figure 9.9, all the subband filters will be well excited over their respective bands of interest, and hence there will not be any slow mode, which may affect the convergence behavior of the overall filter.

Another consideration that should be noted in the implementation of subband adaptive filters, and hence in the design of analysis and synthesis filters, is the problem of delay (or latency) in the filter output, $y(n)$. This delay is caused by the analysis and synthesis filters. Minimization of this delay is exceedingly important as the maximum delay permitted in many applications is often very much limited. For instance, in the application of acoustic

[3] Group-delay of a system is defined as the derivative of its phase response with respect to the angular frequency, ω.

echo cancellation (AEC), around which most of the theory of subband adaptive filters have been developed, the maximum delay allowed is usually very minimal.[4] The factors that influence the delay introduced by the analysis and synthesis filters are the number of subbands, M, the decimation factor, L, the accuracy of analysis and synthesis filters (which may be defined in terms of their stop-band attenuation and pass-band ripple), and also the criterion used in designing analysis and synthesis filters. The last two issues are addressed in Section 9.7, where a method for designing analysis and synthesis filters with small delay is given.

The delay increases with number of subbands, M. On the other hand, we may recall from our previous discussion that the idea of subband adaptive filtering is to partition the input signal into a number of narrow bands such that the signal spectrum is approximately flat over each band, thus giving an implementation that does not suffer because of large eigenvalue spreads. Hence, from convergence point of view, larger values of M are preferred. The choice of the decimation factor, L, also affects the selection of the analysis and synthesis filters and hence the delay. In general, the delay increases with L, as well. On the other hand, the computational complexity of a subband adaptive structure decreases as L increases. Thus, a compromise has to be struck while choosing L and M.

From the above discussion, we find that the selection of the analysis and synthesis filters is not a straightforward or a clearly formulated problem. On the one hand, we should make sure that the delay introduced by the analysis and synthesis filters does not exceed a specified value. This is usually specified as one of the design requirements. On the other hand, we may choose the number of subbands, M, and the decimation factor, L, as large as possible, while designing analysis and synthesis filters with some (loosely defined) acceptable aspects. A procedure for the design of analysis and synthesis filters as well as for the selection of values of L and M and the other parameters of the subband adaptive filters is given in Sections 9.7 and 9.8.

9.5 Computational Complexity

Computational complexity of subband adaptive filters, in general, decreases as the decimation factor, L, increases. To explore the impact of L in reducing the computational complexity of a subband adaptive filter, let us consider the implementation of an adaptive filter whose fullband implementation requires N taps. We also assume that the filter input, $x(n)$, and the desired signal, $d(n)$, are real-valued. The number of taps required for each subband filter is then N/L, because in subbands, each sample interval is equivalent to L sample intervals in the fullband. We also note that although the input, $x(n)$, is real-valued, the subband signals are, in general, complex-valued. They also appear in complex-conjugate pairs, with the exceptions of bands 0 and $M/2$ (we assume that M is even), whose corresponding inputs are real-valued when the input, $x(n)$, is real-valued. Considering these two bands as one band with complex-valued input, the computational complexity of a subband adaptive filter with N real-valued fullband taps may be evaluated based on $(M/2)(N/L) = MN/2L$ complex-valued subband taps. Furthermore, we note that processing in subbands is done at a rate that is only $1/L$ of the fullband rate. Noting these

[4] In the ITU-T standard G.167, it is stated that "for end-to-end digital communications (e.g., wide-band teleconference systems), the delay shall be no more than 16 ms in each direction of speech transmission."

and counting each complex-valued tap as equivalent to four real-valued taps, we obtain

$$\frac{\text{Complexity of subband filter}}{\text{Complexity of fullband filter}} = \frac{2M}{L^2} \tag{9.21}$$

This result does not include the complexity of the analysis and synthesis filters. However, in applications where subband adaptive filters are found useful, the filter length, N, is usually very large, in the range of 1000 or above. For such values of N, the contribution from the complexity of analysis and synthesis filters is not that significant (usually in the range of 20% or less).

In typical designs, one of which is given in Section 9.9, we usually find that $L \approx M/2$. Thus, we obtain

$$\frac{\text{Complexity of subband filter}}{\text{Complexity of fullband filter}} \approx \frac{4}{L} \approx \frac{8}{M} \tag{9.22}$$

9.6 Decimation Factor and Aliasing

With the choice of the analysis and synthesis filters as shown in Figure 9.9, the largest value of L that may be used without causing aliasing of signal spectra over the bands of interest, that is, those selected by the synthesis filters, is given by

$$L_{\max} = \left\lfloor \frac{2\pi}{\omega_{ss} + \omega_{sa}} \right\rfloor \tag{9.23}$$

where $\lfloor x \rfloor$ denotes the largest integer smaller than or equal to x; ω_{sa} and ω_{ss}, as indicated in Figure 9.9, denote the ends of the transition bands of the analysis and synthesis prototype filters, respectively. This choice of L will result in some aliasing in the outputs of analysis filters. However, the aliased portions of the spectra are those that will be filtered out by the synthesis filters, and hence will not affect the fullband output of the filter. Figure 9.10 illustrates this, where we have plotted the magnitude responses of the analysis and synthesis filters after decimation. The selection of $L = L_{\max}$,

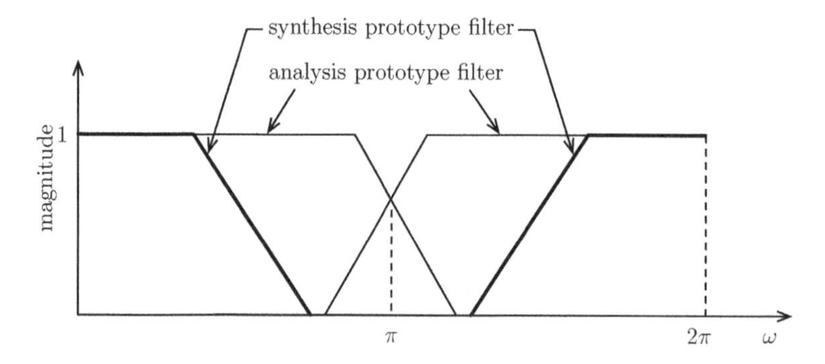

Figure 9.10 Magnitude responses of the analysis and synthesis filters after decimation, illustrating the fact that even though the decimated output samples of the analysis filters are aliased, the portion of the signal spectrum that is filtered by the synthesis filter is free of aliasing.

although may seem quite reasonable at the first glance, has some drawbacks when it comes to adaptation of the subband filters. It results in significant augmentation of the misadjustment, as explained next.

The Fourier transform of the desired signal, $\bar{d}_i(n)$, in the ith subband is given by

$$\bar{D}_i(e^{j\omega}) = \sum_{m=i-1}^{i+1} X(e^{j\frac{\omega-2\pi m}{L}})W_o(e^{j\frac{\omega-2\pi m}{L}})H(e^{j\frac{\omega-2\pi m}{L}}) \tag{9.24}$$

where $X(e^{j\omega})$ is the Fourier transform of the input, $x(n)$, and $W_o(e^{j\omega})$ and $H(e^{j\omega})$ are the frequency responses of the plant and the analysis prototype filter, respectively. The three terms contributing to the spectrum of $\bar{d}_i(n)$ are (i) the ith band spectrum, $m = i$, (ii) the aliased spectrum from the immediately following band, $m = i + 1$, and (iii) the aliased spectrum from the immediately preceding band, $m = i - 1$. The division of the frequency, ω, by L is due to the spectral expansion because of the L-fold decimation. Similarly, the Fourier transform of the output, $\bar{y}_i(k)$, of the ith subband filter is obtained as

$$\bar{Y}_i(e^{j\omega}) = \bar{W}_i(e^{j\omega}) \sum_{m=i-1}^{i+1} X(e^{j\frac{\omega-2\pi m}{L}})H(e^{j\frac{\omega-2\pi m}{L}}) \tag{9.25}$$

Using Eqs. (9.24) and (9.25), the Fourier transform of the subband error sequence $\bar{e}_i(k) = \bar{d}_i(k) - \bar{y}_i(k)$ is obtained as

$$\begin{aligned}\bar{E}_i(e^{j\omega}) &= \bar{D}_i(e^{j\omega}) - \bar{Y}_i(e^{j\omega})\\ &= \left[W_o(e^{j\frac{\omega-2\pi i}{L}}) - \bar{W}_i(e^{j\omega})\right] H(e^{j(\omega-2\pi i)/L})X(e^{j\frac{\omega-2\pi i}{L}})\\ &\quad+\left[W_o(e^{j\frac{\omega-2\pi(i+1)}{L}}) - \bar{W}_i(e^{j\omega})\right] H(e^{j\frac{\omega-2\pi(i+1)}{L}})X(e^{j\frac{\omega-2\pi(i+1)}{L}})\\ &\quad+\left[W_o(e^{j\frac{\omega-2\pi(i-1)}{L}}) - \bar{W}_i(e^{j\omega})\right] H(e^{j\frac{\omega-2\pi(i-1)}{L}})X(e^{j\frac{\omega-2\pi(i-1)}{L}})\end{aligned} \tag{9.26}$$

Inspection of Eq. (9.26) reveals that to minimize

$$E[|e_i(k)|^2] = \frac{1}{2\pi}\int_{-\pi}^{\pi} |\bar{E}_i(e^{j\omega})|^2 d\omega$$

$\bar{W}_i(e^{j\omega})$ has to be selected so that the three differences in the brackets on the right-hand side of Eq. (9.26) reduce to some small values. Moreover, we note that the frequency interval $-\pi \le \omega \le \pi$ may be divided into three distinct ranges. The first range, defined as $-\omega_1 \le \omega \le \omega_1$, is where there is no overlap between $H(e^{j\frac{\omega-2\pi i}{L}})$ and its shifted versions, $H(e^{j\omega-2\pi(i-1)/L})$ and $H(e^{j\omega-2\pi(i+1)/L})$. In this range, the last two terms on the right-hand side of Eq. (9.26) are zero as this range coincides with the stop-bands of $H(e^{j\omega-2\pi(i-1)/L})$ and $H(e^{j\frac{\omega-2\pi(i+1)}{L}})$. The first term can also be made small by choosing $\bar{W}_i(e^{j\omega})$ to be close to the plant response, $W_o(e^{j\frac{\omega-2\pi i}{L}})$, for $-\omega_1 \le \omega \le \omega_1$. It is important to note that a selection of $L \le L_{max}$ implies that $\omega_{ss} \le \omega_1$. This in turn means that the portions of the plant response that are picked up by the synthesis filters can be modeled well by the subband adaptive filters.

The second range, $\omega_1 < \omega \leq \pi$, is where the filters attached to the first two terms on the right-hand side of Eq. (9.26), that is, $H(e^{j\frac{\omega-2\pi i}{L}})$ and $H(e^{j\frac{\omega-2\pi(i+1)}{L}})$, overlap. In this range, for $|\bar{E}_i(e^{j\omega})|$ to reduce to a small value, $\bar{W}_i(e^{j\omega})$ has to match with two different parts of the plant response, namely $W_o(e^{j\frac{\omega-2\pi i}{L}})$ and $W_o(e^{j\frac{\omega-2\pi(i+1)}{L}})$, for $\omega_1 < \omega \leq \pi$. This, of course, is not possible as $W_o(e^{j\frac{\omega-2\pi i}{L}}) \neq W_o(e^{j\frac{\omega-2\pi(i+1)}{L}})$, in general. Similarly, in the third range also, where $-\pi \leq \omega < -\omega_1$, it may not be possible to reduce $|\bar{E}_i(e^{j\omega})|$ to a small value. As a result of these mismatches, $E[|e_i(k)|^2]$ may be very significant, even after the convergence of the subband adaptive filter. This will result in large perturbation of the tap weights because of the use of stochastic gradients, thereby increasing the misadjustment of these filters, unless a very small step-size is used to reduce the level of perturbations. But, reducing the step-size is undesirable, as it proportionately reduces the convergence rate of the adaptive filter. Another solution that has been proposed to solve this problem is to add cross-filters between the neighboring subbands (Gilloire and Vetterli, 1992).[5] However, this increases the system complexity, and hence not acceptable because the main goal of increasing L was to reduce the complexity. Yet another solution, which is found to be more appropriate than the others, is to select the decimation factor, L, so that the overlapping of the adjacent analysis filters is limited only to those portions of the analysis filter responses that are below a certain level (Farhang-Boroujeny and Wang, 1997). However, no fixed value may be specified for this "level." It is a loosely defined design parameter that can only be found experimentally. Thus, a compromise value of L could only be selected through a trial-and-error design process (Section 9.8).

9.7 Low-Delay Analysis and Synthesis Filter Banks

In this section, we present a method for designing analysis and synthesis filters with low group-delay.[6] As was noted before, design of low-delay filters is desirable in subband adaptive filters as it reduces the latency of the overall filter response.

9.7.1 Design Method

From our discussion in Section 9.4, we recall that the analysis and synthesis filters should have good attenuation in their stop-bands. The problem of designing an optimum FIR filter with maximum attenuation in the stop-band may be formulated as follows.

Consider an FIR filter with length N_a and tap weights given by the real-valued coefficient-vector

$$\mathbf{a} = [a_0 \ a_1 \cdots a_{N_a-1}]^{\mathrm{T}}$$

where the superscript T denotes the transposition. Then, the transfer function of the FIR filter is given by

$$A(e^{j\omega}) = \mathbf{a}^{\mathrm{T}}\boldsymbol{\Omega} \tag{9.27}$$

[5] It shall be noted that the purpose behind the use of cross-filters by Gilloire and Vetterli (1992) was to resolve the problem of perfect reconstruction, which is different from our aim here. Nevertheless, the concepts discussed there, with minor modifications, may also be applied to suit the implementation presented in this chapter.

[6] The design method presented here is from Farhang-Boroujeny and Wang (1997). It follows the idea of Mueller (1973) who used the same method for designing Nyquist filters for data transmission purpose. Vaidyanathan and Nguyen (1987) have also proposed a similar method (with some extensions) and called the *resulting designs eigenfilters*. However, neither Mueller nor Vaidyanathan and Nguyen emphasized on low-delay filters

where

$$\mathbf{\Omega} = [1 \ \mathrm{e}^{-j\omega} \ \mathrm{e}^{-j2\omega} \ \cdots \ \mathrm{e}^{-j(N_a-1)\omega}]^{\mathrm{T}}$$

Suppose we want this filter to have its stop-band to begin from ω_{s}. Then, the total energy in the stop-band is given by

$$E_{\mathrm{s}} = \frac{1}{2\pi}\int_{\omega_{\mathrm{s}}}^{2\pi-\omega_{\mathrm{s}}} |A(\mathrm{e}^{j\omega})|^2 \mathrm{d}\omega \tag{9.28}$$

Substituting Eq. (9.27) in Eq. (9.28), we obtain

$$E_{\mathrm{s}} = \mathbf{a}^{\mathrm{T}}\mathbf{\Phi}\mathbf{a} \tag{9.29}$$

where

$$\mathbf{\Phi} = \frac{1}{2\pi}\int_{\omega_{\mathrm{s}}}^{2\pi-\omega_{\mathrm{s}}} \mathbf{\Omega}\mathbf{\Omega}^{\mathrm{H}} \mathrm{d}\omega \tag{9.30}$$

and the superscript H denotes the Hermitian transposition. We note that $\mathbf{\Phi}$ is an N_a-by-N_a matrix whose klth element is

$$\phi_{kl} = \frac{1}{2\pi}\int_{\omega_{\mathrm{s}}}^{2\pi-\omega_{\mathrm{s}}} \mathrm{e}^{-j\omega(k-l)} \mathrm{d}\omega = \begin{cases} 1 - \frac{\omega_{\mathrm{s}}}{\pi} & k = l \\ -\frac{\sin[\omega_{\mathrm{s}}(k-l)]}{\pi(k-l)} & k \neq l \end{cases} \tag{9.31}$$

The optimum coefficients of the FIR filter are those that minimize the energy function E_{s} of Eq. (9.29). To prevent the trivial solution $a_i = 0$, for $i = 0, 1, \ldots, N_a - 1$, we impose the constraint $\mathbf{a}^{\mathrm{T}}\mathbf{a} = 1$. The problem of minimizing E_{s} with respect to the vector $\mathbf{a}$, subject to the constraint $\mathbf{a}^{\mathrm{T}}\mathbf{a} = 1$, is a standard eigenproblem whose optimum solution is the eigenvector of $\mathbf{\Phi}$, which corresponds to its minimum eigenvalue; see Property 7 of eigenvalues and eigenvectors in Chapter 4.

We recall that the synthesis filters have to be complementary. Moreover, as we shall see later, the complementary filters are also appropriate for use as analysis filters. To adopt the above procedure for designing the complementary filters, we recall the M-band complementary condition (9.20). This condition is repeated below in terms of the coefficients a_i, for $i = 0, 1, \ldots, N_a - 1$, of the filter $A(\mathrm{e}^{j\omega})$:

$$a_i = \begin{cases} 1/M, & i = KM \\ 0, & i = \text{all multiples of } M \text{ except } KM \\ \text{unspecified}, & \text{otherwise} \end{cases} \tag{9.32}$$

where K is a constant integer that determines the group-delay of the filter bank, as discussed in Section 9.2.

To satisfy the conditions stated in Eq. (9.32), we may simply drop those a_i's that have to be zero from Eq. (9.29). The new energy function to be minimized is then

$$\tilde{E}_{\mathrm{s}} = \tilde{\mathbf{a}}^{\mathrm{T}}\tilde{\mathbf{\Phi}}\tilde{\mathbf{a}}, \text{ subject to the constraint} \tilde{\mathbf{a}}^{\mathrm{T}}\tilde{\mathbf{a}} = 1 \tag{9.33}$$

where $\tilde{\mathbf{a}}$ is obtained from $\mathbf{a}$ by deleting those elements that should be constrained to zero, and $\tilde{\mathbf{\Phi}}$ is obtained from $\mathbf{\Phi}$ by deleting the corresponding rows and columns so to be made compatible with $\tilde{\mathbf{a}}$. The minimization of $\tilde{E}_{\mathrm{s}}$ also is an eigenproblem. Its solution is the eigenvector of $\tilde{\mathbf{\Phi}}$ that corresponds to its minimum eigenvalue. The desired vector $\mathbf{a}$ is obtained from $\tilde{\mathbf{\Phi}}$ by inserting the dropped out zeros in the appropriate locations. Finally, to satisfy the condition $a_{KM} = 1/M$ of Eq. (9.32), a simple scaling is applied to $\mathbf{a}$.

One may note that the above procedure does not specify any specific range of frequencies for the pass-band and transition band. Only the stop-band is specified. To be more accurate on this, we recall that in an M-band complementary filter bank, the frequency $\omega = \pi/M$ is located at the middle of the transition band of its prototype filter; see Figure 9.6 as an example. The stop-band of the prototype filter begins at $(1+\alpha)\pi/M$, where α, known as *roll-off factor*, determines the widths of the pass-band and transition band. The pass-band of the prototype filter is given as $0 \leq \omega \leq (1-\alpha)\pi/M$ and the transition band as $(1-\alpha)\pi/M < \omega < (1+\alpha)\pi/M$. The numerical examples given next and the supporting discussions show that, for the filters designed by the proposed method, the pass-, stop-, and transition bands will be clearly separated according to the above boundaries, once ω_{s} is set equal to $(1+\alpha)\pi/M$.

9.7.2 Filters Properties

In this subsection, we look at the main features of the filters designed by the method presented above. We recall that an M-band complementary filter bank with the prototype filter $A(\mathrm{e}^{j\omega})$ and the parameter K as specified in Eq. (9.32) satisfies the identity

$$\sum_{i=0}^{M-1} A(\mathrm{e}^{j(\omega-2\pi i/M)}) = \mathrm{e}^{-j\omega KM} \tag{9.34}$$

On the other hand, the design procedure given in the last subsection emphasizes only on the stop-band of the prototype filter $A(\mathrm{e}^{j\omega})$. But, there is no clear emphasis on how the pass-band and the transition band of $A(\mathrm{e}^{j\omega})$ are separated. Next, we show that the boundary between these two bands can be easily identified once the design parameters M and ω_{s} are known.

From our discussion in Section 9.2, we recall that the midpoint of the transition band of the prototype filter of an M-band complementary filter bank is $\omega = \pi/M$. Moreover, from the pictorial representation of Figure 9.6, it is straightforward to conclude that if ω_{p} and ω_{s} are, respectively, the end of pass-band and the beginning of stop-band of the prototype filter of an M-band filter bank, then

$$\frac{\pi}{M} = \frac{\omega_{\mathrm{p}} + \omega_{\mathrm{s}}}{2} \tag{9.35}$$

Hence, when M and ω_{s} are given, ω_{p} is obtained as

$$\omega_{\mathrm{p}} = \frac{2\pi}{M} - \omega_{\mathrm{s}} \tag{9.36}$$

In the discussion that follows, it is convenient to specify ω_{s} in terms of the midpoint frequency π/M as

$$\omega_{\mathrm{s}} = (1+\alpha)\frac{\pi}{M} \tag{9.37}$$

where α is a positive parameter that specifies the width of the transition band of the prototype filter of the filter bank as explained below. The parameter α, as noted above, is known as *roll-off factor*.

Substituting Eq. (9.37) in Eq. (9.36), we get

$$\omega_{\mathrm{p}} = (1-\alpha)\frac{\pi}{M} \tag{9.38}$$

Also, the width of the transition band of the prototype filter is obtained as

$$\omega_{\mathrm{s}} - \omega_{\mathrm{p}} = \frac{2\pi\alpha}{M} \tag{9.39}$$

Equation (9.34) explicitly states that the group-delay introduced by the filter bank is KM. In most subband adaptive filtering applications, we want to keep this delay as small as possible. On the other hand, the optimum K that results in maximum attenuation in the stop-band is obtained by choosing K so that KM is the nearest multiple of M to $N_a/2$. However, this delay is generally large and thus, we would instead strike a compromise between delay and stop-band attenuation. That is, we may accept a lower delay at the cost of lower stop-band attenuation.

For effective implementation of subband adaptive filters, it is important to understand the effect of reduced delay on the performance of the analysis and synthesis filters and its overall impact on the performance of the adaptive filter. This can be best understood through an example.

Figure 9.11 shows the magnitude and group-delay responses of three filters that have been designed by the above method. The filter length, N_a, the number of subbands, M, and the roll-off factor, α, are set equal to 97, 4, and 0.25, respectively, and the three designs are differentiated by the parameter K. The separation of the pass-band, transition band, and stop-band can be clearly seen in the responses. In particular, we note that the transition bands in the three designs are the same and match the band edges predicted by Eq. (9.37) and Eq. (9.38). We also note that the price to be paid for achieving reduced delay is lower stop-band attenuation, an undesirable boost in the magnitude response in the transition band, and group-delay distortion in the transition and stop bands. However, the magnitude and group-delay responses in the pass-band remain nearly undistorted. This is a desirable feature of this design method that makes it very appropriate for designing analysis as well as synthesis filters in the application of subband adaptive filtering.

9.8 A Design Procedure for Subband Adaptive Filters

Since there are many compromises to be made in the overall design of subband adaptive filters, it is very hard to suggest a simple design procedure for such filters. In this section, we present a procedure that the author has found useful in his research work. This procedure is iterative in nature and its application requires some experience. Hence, a novice needs to do some experiments with that before he/she can use it for actual design.

To choose all the parameters necessary for setting up a subband adaptive filter, one may take the following steps:

1. Choose a value for the number of subbands, M.
2. Choose an integer parameter, J, in the range of $\frac{1}{2}$ to $\frac{2}{3}$ of M, and select the pass-band, transition band, and stop-band of the analysis and synthesis prototype filters, as shown

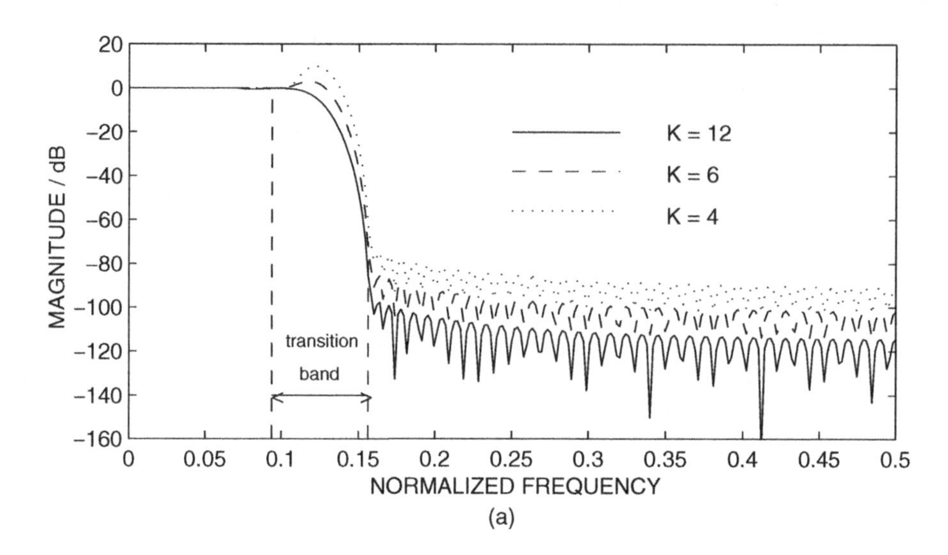

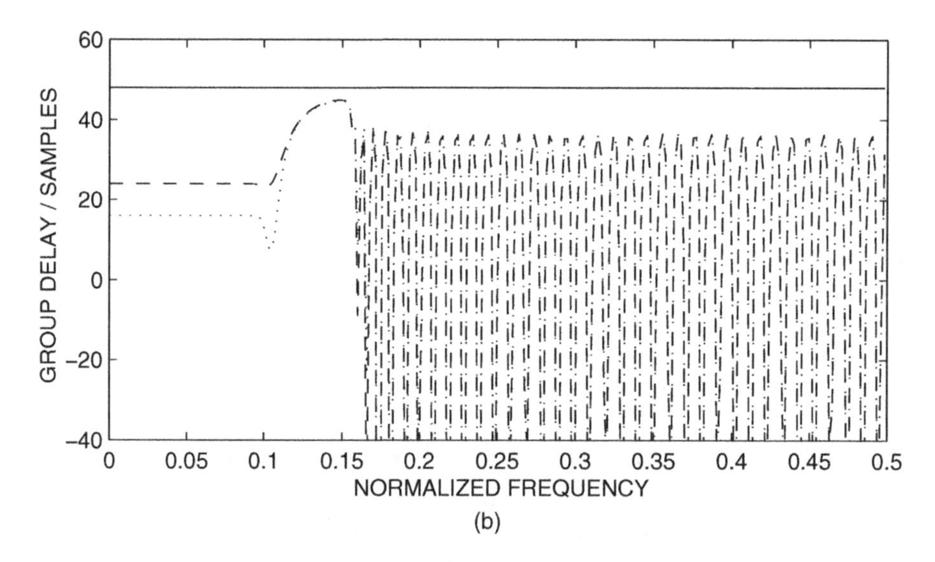

Figure 9.11 Magnitude and group-delay responses of three filters that have been designed by the method in Farhang-Boroujeny and Wang (1997).

in Figure 9.12. This determines the values of the roll-off factors α_a and α_s of the analysis and synthesis filters, respectively. Note that the midpoints of the transition bands of the analysis and synthesis prototype filters are π/J and π/M, respectively. We note that when the method of Section 9.7 is used to design the analysis and synthesis filters, these choices of the midpoints of the transition bands lead to analysis filters, which are J-band complementary, and synthesis filters, which are M-band complementary. Furthermore, the positions of these midpoints determine the range of the roll-off factors α_a and α_s of the analysis and synthesis filters, respectively. The range of possible values that α_a and α_s may take can easily be worked out by inspection from Figure 9.12, and noting that the pass-band and transition band of the synthesis filter should be covered by the pass-band of the analysis filter (see also Problem P9.3).

3. Choose values for the lengths of analysis and synthesis filters. Call these N_a and N_b, respectively. Also, select values for the parameters K of the analysis and synthesis filters. Call these K_a and K_s, respectively.
4. Using the parameters selected above, design the analysis and synthesis prototype filters by following the method presented in Section 9.7.1.
5. Evaluate the stop-band rejection of the prototype filters. If satisfactory, proceed with the next step. Otherwise, reselect one or a few of the parameters N_a, N_s, K_a, K_s, J, α_a, and α_s and redesign the prototype filters until the design is satisfactory.
6. Select a value of $L < J$ and evaluate the aliasing of the decimated subband signals. A limited amount of aliasing may be allowed. However, because of the reason discussed in Section 9.6, such aliasing should be relatively small.
7. Evaluate the design by putting the designed analysis and synthesis filters in the subband adaptive filter structure and running a typical simulation of some application. If the performance is not satisfactory, then the filters need to be redesigned for other choices of the parameters listed above.

While putting the designed analysis and synthesis filters in a subband structure, we should note that the analysis filters are J-band complementary with parameter $K = K_a$, while the synthesis filters are M-band complementary with parameter $K = K_s$. As a result, the group-delay introduced by the analysis filters is K_aJ and that from the synthesis filters is K_sM. Then, the net group-delay due to a direct cascade of the two filter banks would be

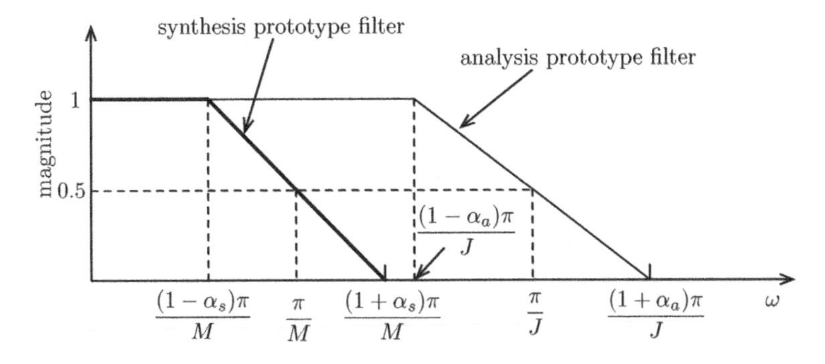

Figure 9.12 Definitions of the band edges in the analysis and synthesis prototype filters.

$K_aJ + K_sM$. On the other hand, according to our discussion in Section 9.2, the cascade of the analysis and synthesis filters must be M-band complementary so that the total group-delay is an integer multiple of M. This may be achieved either by selecting K_a and J so that $K_aJ + K_sM$ is an integer multiple of M, or by padding appropriate number of zero coefficients at the beginning of the analysis and/or synthesis filters so that the total group-delay is made an integer multiple of M. Accordingly, the following equation may be used to calculate the delay, Δ:

$$\Delta = \text{the first integer multiple of } M, \text{ which is greater than or equal to } K_aJ + K_sM \tag{9.40}$$

9.9 An Example

In this section, we discuss two design examples to demonstrate the effectiveness of the design technique that was introduced in the last two sections.[7] The aim is to see how far we can go in reducing the delay and the price that we pay for it.

The following common parameters are used in both the designs:

$$M = 32, \quad J = 19, \quad \alpha_a = \frac{3}{16} \quad \text{and} \quad \alpha_s = \frac{7}{19}$$

In the first design, we ignore the problem of delay and design the analysis and synthesis filters that result in maximum attenuation in their stop-bands. As was noted before, maximum stop-band attenuation is achieved when the delay introduced by each filter is about half of its respective length. We refer to this as the *conventional-delay design*. In the second design, a few attempts are made to obtain a pair of low-delay analysis and synthesis filters with stop-band attenuations comparable to those in the first design. This is called the *low-delay design*. To achieve similar stop-band attenuations with reduced delay, the lengths of the filters need to be chosen longer than their conventional-delay counterparts.

The two designs are summarized in Table 9.1. The value of the delay, Δ, for each design is calculated according to Eq. (9.40). Note that the low-delay design achieves a delay that is half of that of the conventional-delay design. This, as expected, is at the cost of increased filter lengths, N_a and N_s. In Table 9.1, E_a and E_s are the stop-band energies of the analysis and synthesis filters, respectively.

To illustrate the effect of decimation factor, L, on the overall performance of the filter, the designed low-delay analysis and synthesis filters are put into a subband structure that is used for modeling a 1600-tap plant. The plant response is that of an acoustical echo path of a normal size office room (Section 9.10). The plant input is assumed to be a white Gaussian noise. We also add some noise to the plant output. The normalized LMS (NLMS) algorithm of Section 6.6 is used for adaptation of the subband filters – also see Section 9.10. Figure 9.13 shows the learning curves of the subband adaptive filter for values of $L = 16$ to 19. $L = 16$ corresponds to the case where the decimated subband

[7] The design examples presented here and their application in the study of acoustic echo canceler is taken from Wang (1996).

Table 9.1 Summary of the two designs of analysis–synthesis prototype filters.

Parameters	Conventional delay	Low delay
K_a	5	3
K_s	3	1
N_a	191	289
N_s	193	353
Δ	192	96
E_a	1.4×10^{-6}	5.0×10^{-7}
E_s	4.1×10^{-7}	4.6×10^{-7}

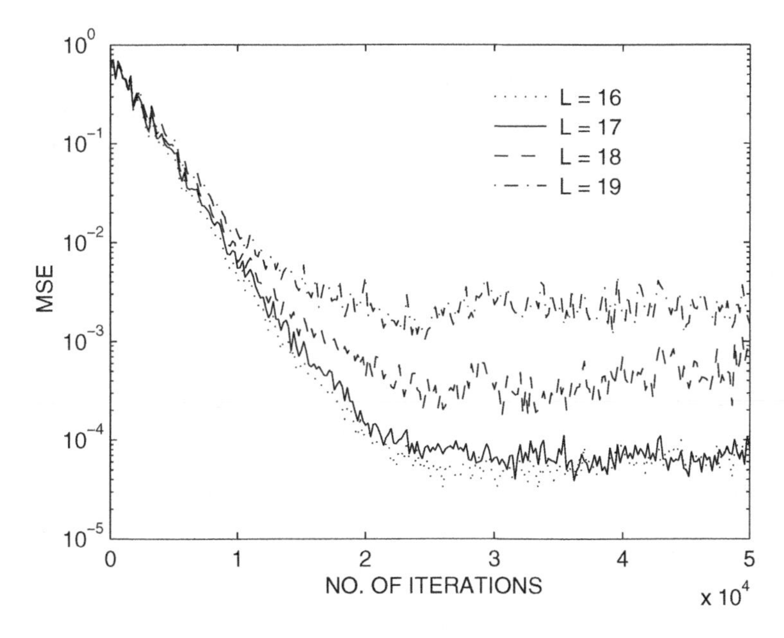

Figure 9.13 Learning curves of the subband adaptive filter for different values of the decimation factor, L.

signals do not suffer from any aliasing. On the contrary, $L = 19$ corresponds to the case where the decimated subband signals are fully aliased over the transition bands of their respective analysis filters. However, the signals in their pass-bands do not suffer from any serious aliasing, except that due to nonideal stop-band attenuations that are negligible. The case $L = 17$ does suffer from aliasing in transition bands, although relatively low. These results clearly confirm our earlier conjecture that in the selection of the decimation factor, L, a small amount of aliasing is acceptable.

9.10 Comparison with FBLMS Algorithm

Adaptive filtering in subbands has many similarities with the FBLMS algorithm[8] which was introduced in Chapter 8. Firstly, both these methods may be categorized under the class of block/parallel processing algorithms. As a result, both of these methods offer fast implementations of adaptive filters, that is, implementations with reduced complexity as compared to the non-block methods. Furthermore, they resolve the problem of slow convergence of the LMS algorithm. These advantages are obtained at the cost of certain processing delay at the filter output. Hence, at this point, it seems appropriate and essential to make some comments on the relative performance of the method of subband adaptive filtering and the FBLMS algorithm in terms of convergence behavior, computational complexity, and processing delay. However, a quantitative comparison of the two methods is not straightforward. Thus, in the rest of this section, we make an attempt to give some general comments on the above issues, leaving the discussion an open-ended one so that the reader can complete it by closely examining his/her specific application of interest.

We note that adaptation of each of the subband filters can be performed using any of the adaptive filtering algorithms that has been introduced so far or that will be introduced in the subsequent chapters. In the discussion that follows, for convenience, we assume that the NLMS algorithm is used for this. We thus use the term *subband NLMS algorithm* to refer to this implementation of subband adaptive structure.

Simulations and experiments show that both the subband NLMS and FBLMS algorithms are quite successful in decorrelating the samples of the filter input. By careful selection of their parameters, both these algorithms can be tuned to offer learning curves that are dominantly governed by a single mode of convergence.

Comparison of the two algorithms with respect to their computational complexity is also not straightforward. For a pair of designs with comparable convergence behavior and processing delay, one may use the number of operations per sample for comparing the computational complexities of the two algorithms. This, although often used in the literature for comparing different algorithms, does not seem to be fair in the present case because of the many structural differences between the two algorithms. For instance, the subband NLMS algorithm has a more regular structure than the FBLMS algorithm. On the other hand, in typical applications of interest, say adaptive filters with at least a few hundred fullband taps, we usually find that the FBLMS algorithm has lower operation count than the subband NLMS algorithm. Thus, to a great extent, the choice between the two algorithms depends on the available hardware/software platform. In software implementation on digital signal processors, the FBLMS algorithm is usually found to be more efficient than the subband NLMS algorithm. In contrast, the more regular structure of subband NLMS algorithm may make it a better choice in a custom chip design. In particular, we may note that the subband filter structure can easily be divided into a number of separate blocks. For instance, each of the analysis/synthesis filter banks and the subband filters of Figure 9.7 may be treated as a separate block in a multiprocessor chip.

[8] In this section, we use the term *FBLMS algorithm* in a general sense. It includes the FBLMS as well as partitioned FBLMS algorithms of the last chapter.

Delay is an adjustable parameter in the subband structure as well as FBLMS algorithm. In general, by allowing larger delay (up to certain limit), the complexities of both the methods can be reduced. The choice of the delay is usually limited by the system specification.

Problems

P9.1 Show that if a sequence $\bar{y}_i(k)$ is interpolated using an interpolation factor of L and passed through a filter with the impulse response g_n, the resulting output may be written as

$$y_i(n) = \sum_{k=-\infty}^{\infty} \bar{y}_i(k) g_{n-kL}$$

P9.2 Show that the following pairs of structures are equivalent:

(i) Case I:

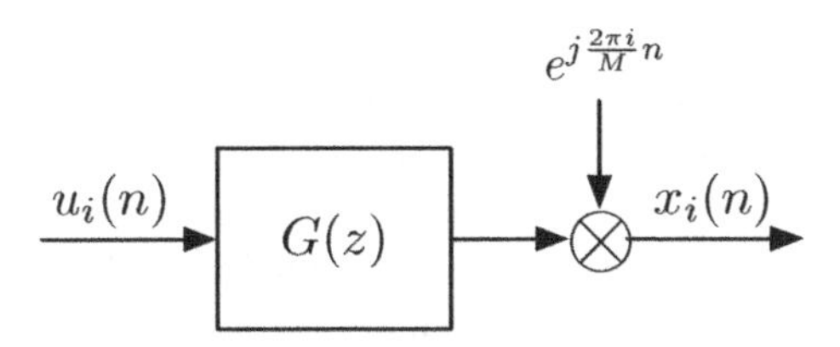

and

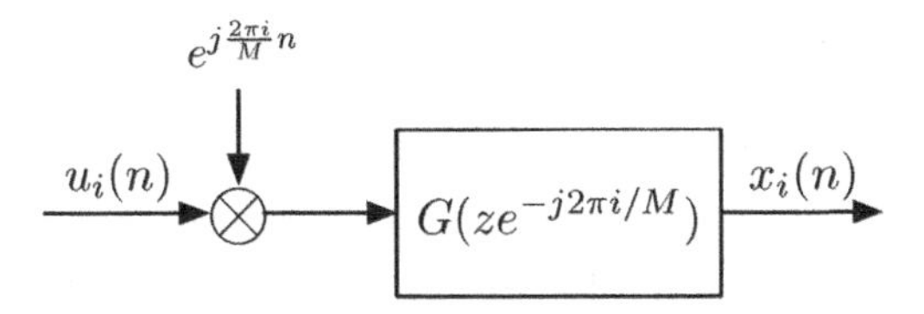

(ii) Case II:

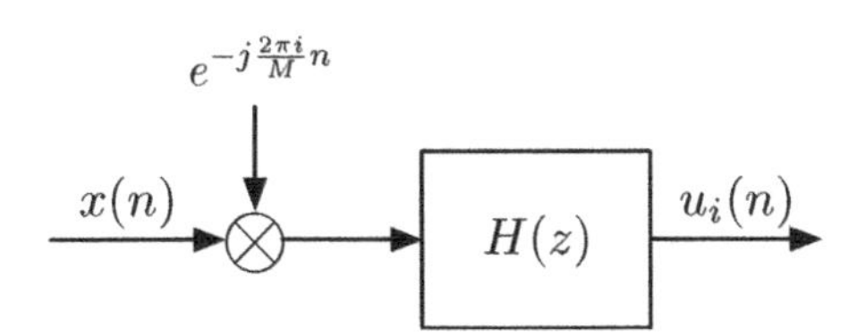

and

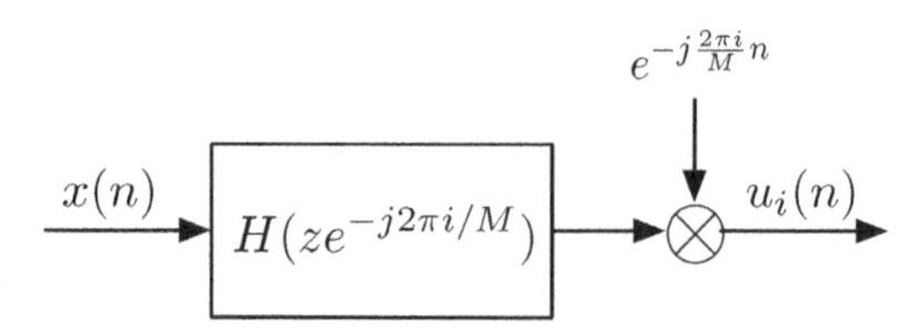

P9.3 Consider a transversal adaptive filter with tap-input and tap-weight vectors $\mathbf{x}(n)$ and $\mathbf{w}(n)$, respectively. To adapt $\mathbf{w}(n)$, we wish to use the delayed LMS recursion

$$\mathbf{w}(n+1) = \mathbf{w}(n) + 2\mu e(n-\Delta)\mathbf{x}(n-\Delta)$$

where Δ is a constant delay, $e(n) = d(n) - \mathbf{w}^{\mathrm{T}}(n)\mathbf{x}(n)$ is the output error, and $d(n)$ is the desired signal. Study the hardware/software implementation of this algorithm and discuss how the hardware/memory requirements of the filter vary with Δ.
Refer to the synthesis-dependent structure given in Figure 9.8. Note that there is some delay introduced by the synthesis–analysis filters in the path from subband filters outputs $\bar{y}_i(k)$'s to the subband errors $\bar{e}_i(k)$'s. Discuss how this delay leads to a set of delayed LMS recursions for adaptation of the subband filters $\bar{W}_i(z)$'s, and how this affects the hardware/memory requirements of the subband structure.

P9.4 Consider an M-band subband structure with parameter J as defined in Section 9.8. Show that the condition necessary for the pass-bands and transition bands of synthesis filters to be covered by the pass-bands of analysis filters is the following:

$$J\alpha_{\mathrm{s}} + M\alpha_{\mathrm{a}} \leq M - J$$

P9.5 Explore the validity of Eq. (9.23) in detail.

P9.6 Equations (9.21) and (9.22) are given for the subband adaptive filters with real-valued input. Derive similar equations for the case where the filter input and desired signal are complex-valued.

P9.7 In a pair of complementary analysis–synthesis filter banks, for each of the following set of parameters, determine the number of zeros needed to be added in front of either the analysis or synthesis filters such that their combination is a complementary M-band filter bank:

(i) $M = 4,\ J = 3,\ K_{\mathrm{a}} = 7,\ K_{\mathrm{s}} = 5$
(ii) $M = 64,\ J = 48,\ K_{\mathrm{a}} = 4,\ K_{\mathrm{s}} = 3$.

P9.8 Recall that in the realization of DFT analysis filters using weighted overlap–add method, at the last stage, we need to multiply the DFT outputs by the coefficients W_M^{-ikL}, for $k = 0, 1, \ldots, M-1$ (Eq. 9.7). On the other hand, in the realization of DFT synthesis filters using weighted overlap–add method, the subband signals should be multiplied by the coefficients W_M^{ikL}, for $k = 0, 1, \ldots, M - 1$, before the application of the DFT (Eq. 9.15). Carefully examine the structure of subband adaptive filter and show that these two operations may be deleted from the structure of a subband adaptive filter without affecting its performance.

Computer-Oriented Problems

P9.9 The MATLAB program "`eigenfir.m`" in the accompanying website can be used for designing complementary eigenfilters of the type discussed in Section 9.7.1. Use this program to design three filters with the following specifications:

(i) $N = 129,\ M = 4,\ \alpha = 0.25,\ K = 16$
(ii) $N = 129,\ M = 4,\ \alpha = 0.25,\ K = 8$
(iii) $N = 129,\ M = 4,\ \alpha = 0.25,\ K = 4$

For each design, confirm that the band edges are realized, as predicted in Section 9.7.2.

P9.10 Design a pair of analysis and synthesis prototype filters with the following parameters:

$$M = 16,\quad J = 10,\quad N_a = N_s = 257,\quad \alpha_a = \alpha_s = 0.15,\quad K_a = 3,\quad K_s = 4.$$

Put these into a subband structure and verify that the cascade of the analysis and synthesis filter banks is equivalent to a pure delay. For this, you may put a random sequence as input to the analysis filter bank and observe that the same sequence, with some delay, appears at the synthesis filter bank output. You may need to add an appropriate number of zeros at the beginning of the analysis or synthesis filters in order to get the right result from this experiment. Try your experiment for different values of the decimation factor $L = 7$, 8, 9, and 10. Do you observe any significant difference in the results? Explain your observation. Among these values of L, show that only $L = 7$ prevents aliasing of the subband signals.

P9.11 Use the analysis and synthesis filter banks of the last problem to realize an NLMS-based subband adaptive filter to model a plant with 500 fullband taps. Choose a set of independent random numbers with variance 0.01 as the samples of the plant impulse response. Also, add a Gaussian noise with variance 10^{-4} to the plant output as the plant noise. Run your program for different values of the decimation factor, L, and verify that the subband adaptive filter converges toward an MSE that is much larger than the minimum MSE when the aliasing of the subband signals is significant. To convince yourself that this is due to some excessive misadjustment, as discussed in Section 9.6, you can reduce the step-size parameter $\tilde{\mu}$ to some small value and let the NLMS algorithm to run over sufficient number of iterations and observe that it converges toward the expected minimum MSE. To confirm that subband adaptive filters are robust to the variation of the power spectral density of the input, try your experiment with white as well as colored inputs.

10

IIR Adaptive Filters

In our study of adaptive filters in the previous chapters, we always limited ourselves to filters with finite-impulse response (FIR). The main feature of FIR filters, which has made them the most attractive structure in the application of adaptive filters, is that they are nonrecursive. That is, the filter output is computed based on only a finite number of input samples. This, as we noted in the previous chapters, results in a quadratic mean-squared error (MSE) performance surface, allowing us to use any of the simple gradient-based algorithms for finding the optimum coefficients (tap weights) of the filter.

The use of *recursive* or infinite-impulse response (IIR) filters, on the other hand, have been less popular in the realization of adaptive filters for the following reasons:

1. IIR filters can easily become unstable because their poles may get shifted out of the unit circle (i.e., $|z| = 1$, in the z-plane) by the adaptation process.
2. The performance function (e.g., MSE as a function of filter coefficients) of an IIR filter, usually, has many local minima points.

The problem of instability is usually dealt with by checking the filter coefficients after each adaptation step and limiting them to the range that results in a stable transfer function. This, in general, is a difficult job and adds additional complexity, which in many cases becomes significant when the filter order is large. This additional complexity tends to nullify the computational advantage provided by the recursive nature of these filters.

Because of the multimodal nature of their performance surfaces, convergence of the IIR adaptive filters to their global minima is not guaranteed. The following approaches are usually used to deal with this problem:

(i) Local minima are usually observed when the criterion used to adjust the filter coefficients is MSE. A modification to this criterion leads to quadratic performance surfaces similar to those of FIR filters, thereby eliminating the problem of local minima. This modification results in a special implementation of IIR adaptive filters known as the *equation error method*. The details of this method are discussed in Section 10.2. In contrast, the conventional formulation of IIR adaptive filters based on the Wiener filter theory, which may suffer from the problem of local minima, is referred to as the *output error method*. This is discussed in Section 10.1.

(ii) For specific applications, we may limit ourselves to IIR transfer functions whose associated MSE performance surfaces are unimodal, that is, they have no local minima. In such cases, the use of output error method is the preferred choice as its convergence to the associated Wiener filter is guaranteed.

In this chapter, we discuss both the output error and equation error methods. As the performance of these methods are application dependent, we also present two case studies to highlight some of the implementation issues that one should consider while using IIR adaptive filters. The case studies that we have chosen are special applications, which demonstrate the efficiency of IIR adaptive filters when their structure and/or design criteria are wisely selected and, at the same time, some peculiar behaviors of such filters that are hard to predict, in general.

The first application that we consider is adaptive line enhancement. The problem of adaptive line enhancement was discussed earlier in Chapter 1, where we reviewed various applications of adaptive filters, and also in Chapter 6, as an example of application of the LMS algorithm. We used a transversal filter to implement the line enhancer. However, the problem of line enhancement may also be viewed as one of realizing/achieving narrow-band adaptive filters. But, to realize a narrow-band filter in transversal form, we would need very long filter lengths. In the example given in Chapter 6 (Section 6.4.3), we used a 30-tap transversal filter to achieve satisfactory enhancement of a single sinusoidal signal. On the contrary, as we will see later, a second-order IIR adaptive filter with four coefficients is sufficient for this problem. In Section 10.3, we introduce and study a special form of transfer function, which has been found very appropriate for realization of IIR line enhancers. This is a good representative example of the second approach cited earlier, showing how a wise choice of the transfer function in a specific application can lead to a unimodal performance function, thereby solving the problem of local minima of the output error method.

The second application of IIR adaptive filters that we discuss is equalization of magnetic recording channels. In the case of magnetic recording channels, realization of equalizers in digital form turns out to be very costly because of very high data rates (a few hundred megabits per second). To solve this problem, the general trend in the present industry is to use analog equalizers. We use the techniques presented in this chapter as tools to design analog equalizers for magnetic recording channels. This serves as a good representative example of the use of equation error method.

10.1 Output Error Method

Output error method results when the Wiener filter theory is made use of in a direct manner to develop algorithms for designing and/or adaptation of IIR filters. This can be best explained in the context of a system modeling problem as depicted in Figure 10.1. According to the Wiener theory, the coefficients of the recursive transfer function

$$W(z) = \frac{A(z)}{1 - B(z)} \tag{10.1}$$

where $A(z)$ and $B(z)$ are polynomials in z, are obtained by minimizing the *output error*, $e(n)$, in the mean-square sense. We thus need to find the global minimum of

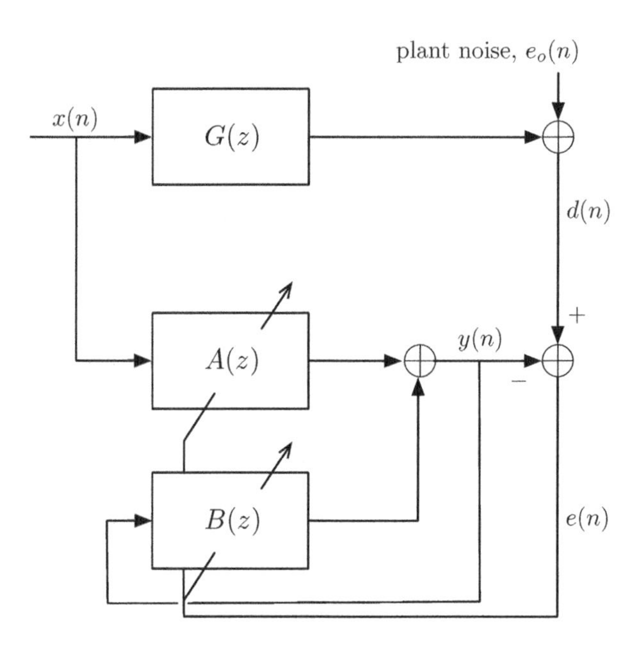

Figure 10.1 IIR adaptive filter with output error adaptation method.

the performance function $\xi = E[e^2(n)]$ in an adaptive manner. However, we note that the performance function ξ is, in general, a multimodal function of the coefficients of the filter $W(z)$, that is, ξ may have many local minima (see Chapter 3). This may lead to convergence of any gradient-based (such as LMS) algorithm to a suboptimal solution. In this section, we ignore this problem and simply develop an LMS algorithm for adaptation of the coefficients of $W(z)$. As the coefficients are obtained by minimizing the output error (in some sense), this approach is named "output error method." The use of this name, hence, is to emphasize on the special feature of the output error method as against the equation error method, which is based on a different criterion (see the following section).

Next, we develop an LMS algorithm for adaptation of the coefficients of IIR filters. To facilitate this, we define the time-varying transfer functions

$$A(z, n) = \sum_{i=0}^{N} a_i(n) z^{-i} \tag{10.2}$$

and

$$B(z, n) = \sum_{i=1}^{M} b_i(n) z^{-i} \tag{10.3}$$

and note that the output, $y(n)$, of the adaptive IIR filter is obtained according to the equation

$$y(n) = \sum_{i=0}^{N} a_i(n) x(n-i) + \sum_{i=1}^{M} b_i(n) y(n-i) \tag{10.4}$$

The LMS algorithm, for this case, can now be derived by following similar line of derivations as those given in the previous chapters in the case of FIR filters. In particular, we recall that the LMS algorithm makes use of the stochastic gradient vector given by

$$\hat{\nabla}(n) = \nabla_{\mathbf{w}} e^2(n) = 2e(n)\nabla_{\mathbf{w}} e(n) \tag{10.5}$$

where $\nabla_{\mathbf{w}}$ is the gradient operator with respect to the filter tap-weight vector, $\mathbf{w}(n)$, and

$$e(n) = d(n) - y(n) \tag{10.6}$$

is the output error. Here, the filter tap-weight vector $\mathbf{w}(n)$ is defined as

$$\mathbf{w}(n) = [a_0(n) \quad a_1(n) \ldots a_N(n) \quad b_1(n) \ldots b_M(n)]^{\mathrm{T}} \tag{10.7}$$

Substituting Eq. (10.6) in Eq. (10.5) and noting that $d(n)$ is independent of $\mathbf{w}(n)$, we obtain

$$\begin{aligned}\hat{\nabla}(n) &= -2e(n)\nabla_{\mathbf{w}} y(n) \\ &= -2e(n)\left[\frac{\partial y(n)}{\partial a_0(n)}\,\frac{\partial y(n)}{\partial a_1(n)} \cdots \frac{\partial y(n)}{\partial a_N(n)} \quad \frac{\partial y(n)}{\partial b_1(n)} \cdots \frac{\partial y(n)}{\partial b_M(n)}\right]^{\mathrm{T}}.\end{aligned} \tag{10.8}$$

The derivatives in Eq. (10.8) should be considered with special care, as $y(n)$ depends on its previous values, $y(n-1)$, $y(n-2)$, ...

From Eq. (10.4), we get

$$\frac{\partial y(n)}{\partial a_i(n)} = x(n-i) + \sum_{l=1}^{M} b_l(n)\frac{\partial y(n-l)}{\partial a_i(n)}, \quad \text{for } i = 0, 1, \ldots, N \tag{10.9}$$

and

$$\frac{\partial y(n)}{\partial b_i(n)} = y(n-i) + \sum_{l=1}^{M} b_l(n)\frac{\partial y(n-l)}{\partial b_i(n)}, \quad \text{for } i = 1, 2, \ldots, M \tag{10.10}$$

To proceed, it is convenient to define

$$\alpha_i(n) = \frac{\partial y(n)}{\partial a_i(n)}, \quad \text{for } i = 0, 1, \ldots, N \tag{10.11}$$

and

$$\beta_i(n) = \frac{\partial y(n)}{\partial b_i(n)}, \quad \text{for } i = 1, 2, \ldots, M \tag{10.12}$$

Assuming that the coefficients $a_i(n)$'s and $b_i(n)$'s vary slowly in time, we get

$$\frac{\partial y(n-l)}{\partial a_i(n)} \approx \frac{\partial y(n-l)}{\partial a_i(n-l)} = \alpha_i(n-l) \tag{10.13}$$

and

$$\frac{\partial y(n-l)}{\partial b_i(n)} \approx \frac{\partial y(n-l)}{\partial b_i(n-l)} = \beta_i(n-l) \tag{10.14}$$

for $l = 1, 2, \ldots, M$. Substituting Eqs. (10.13) and (10.14) in Eqs. (10.9) and (10.10), respectively, we get the following recursive equations for obtaining the successive samples of $\alpha_i(n)$s and $\beta_i(n)$s:

$$\alpha_i(n) = x(n-i) + \sum_{l=1}^{M} b_l(n)\alpha_i(n-l) \tag{10.15}$$

and

$$\beta_i(n) = y(n-i) + \sum_{l=1}^{M} b_l(n)\beta_i(n-l) \tag{10.16}$$

Using these results, the LMS recursion for adaptation of IIR filters may be summarized as

$$\mathbf{w}(n+1) = \mathbf{w}(n) + 2\mu e(n)\boldsymbol{\eta}(n) \tag{10.17}$$

where

$$\boldsymbol{\eta}(n) = [\alpha_0(n) \quad \alpha_1(n)\cdots\alpha_N(n) \quad \beta_1(n)\cdots\beta_M(n)]^{\mathrm{T}} \tag{10.18}$$

and $\alpha_i(n)$'s and $\beta_i(n)$'s are obtained recursively according to Eqs. (10.15) and (10.16) respectively.

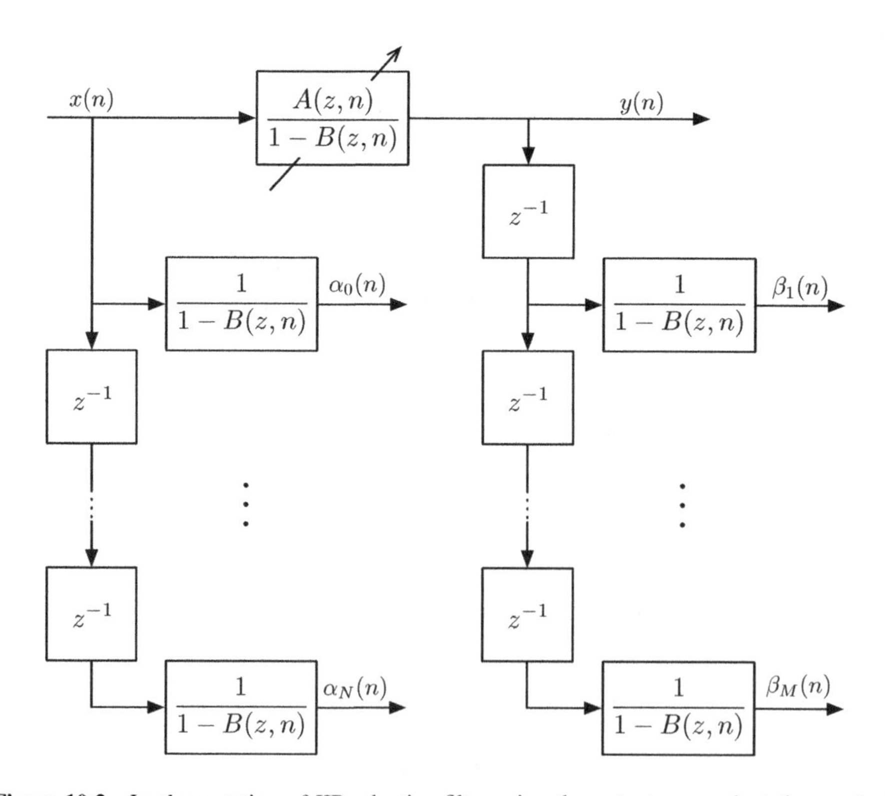

Figure 10.2 Implementation of IIR adaptive filter using the output error adaptation method.

Figure 10.2 depicts a block diagram showing the computations involved in calculating $\alpha_i(n)$'s and $\beta_i(n)$'s, according to Eqs. (10.15) and (10.16) respectively. From this diagram, we see that the computation of the elements of $\eta(n)$ requires parallel implementation of $M+N+1$ recursive filters with the same transfer function $1/(1-B(z,n))$, but different inputs, one for each element.

The diagram of Figure 10.2 can be greatly simplified, if we assume that the transfer function $1/(1-B(z,n))$ varies only slowly with time. Then, we may use the following approximations:

$$\frac{1}{1-B(z,n)} \approx \frac{1}{1-B(z,n-i)}, \quad \text{for } i = 1, 2, \ldots, \max(N, M-1) \tag{10.19}$$

where $\max(N, M-1)$ denotes maximum of N and $M-1$. This allows us to write from Eq. (10.15)

$$\alpha_i(n) \approx x(n-i) + \sum_{l=1}^{M} b_l(n-i)\alpha_i(n-l). \tag{10.20}$$

On the other hand, substituting i by 0 and n by $n-i$ in Eq. (10.15), we get

$$\alpha_0(n-i) = x(n-i) + \sum_{l=1}^{M} b_l(n-i)\alpha_0(n-i-l) \tag{10.21}$$

Now, comparing Eqs. (10.21) and (10.20), we note that $\alpha_i(n)$ and $\alpha_0(n-i)$ are generated based on the same input, $x(n-i)$, and approximately the same recursive equations. Thus, we get

$$\alpha_i(n) \approx \alpha_0(n-i), \quad \text{for } i = 1, 2, \ldots, N \tag{10.22}$$

Similarly, we obtain

$$\beta_i(n) \approx \beta_1(n-i+1), \quad \text{for } i = 2, 3, \ldots, M \tag{10.23}$$

Using these results, we obtain Figure 10.3 as an approximation to Figure 10.2. Note that in Figure 10.3, as opposed to Figure 10.2, we only need to use two filters with the transfer function $1/(1-B(z,n))$ to calculate $\alpha_0(n)$ and $\beta_1(n)$. The rest of values of $\alpha_i(n)$'s and $\beta_i(n)$'s are simply delayed versions of $\alpha_0(n)$ and $\beta_1(n)$, respectively. Table 10.1 gives a summary of the LMS algorithm, which follows Figure 10.3.

10.2 Equation Error Method

As was mentioned before, the main problem with the direct minimization of output error of an IIR filter is that the associated performance surface may have many local minima points, thereby resulting in convergence of the LMS algorithm to one of these local minima, which may not be the desired global minimum. This problem may be resolved by using the method of equation error as explained in the following section.

Figure 10.4 depicts a block diagram illustrating the principle behind the equation error method. Here, the error used to adapt the transfer functions $A(z)$ and $B(z)$ is

$$e'(n) = d(n) - y'(n) \tag{10.24}$$

Table 10.1 Summary of the output error LMS algorithm.

Input: Tap-weight vector,
$\mathbf{w}(n) = [a_0(n)\ a_1(n) \cdots a_N(n)\ b_1(n) \cdots b_M(n)]^{\mathrm{T}}$,
Input vector,
$\mathbf{u}(n) = [x(n)\ x(n-1) \cdots x(n-N)\ y(n-1) \cdots y(n-M)]^{\mathrm{T}}$,
the previous samples of $\alpha_0(n)$ and $\beta_1(n)$, and Desired output, $d(n)$.

Output: Filter output, $y(n)$,
Tap-weight vector update, $\mathbf{w}(n+1)$,
and the samples of $\alpha_0(n)$ and $\beta_1(n)$ for next iteration.

1. Filtering:
$y(n) = \mathbf{w}^{\mathrm{T}}(n)\mathbf{u}(n)$
2. Error estimation:
$e(n) = d(n) - y(n)$
3. $\boldsymbol{\eta}(n)$ update:
$\alpha_0(n) = x(n) + \sum_{l=1}^{M} b_l(n)\alpha_0(n-l)$
$\beta_1(n) = y(n) + \sum_{l=1}^{M} b_l(n)\beta_1(n-l)$
$\boldsymbol{\eta}(n) = [\alpha_0(n)\ \alpha_0(n-1) \cdots \alpha_0(n-N)\ \beta_1(n) \cdots \beta_1(n-M)]^{\mathrm{T}}$
4. Tap-weight vector adaptation:
$\mathbf{w}(n+1) = \mathbf{w}(n) + 2\mu e(n)\boldsymbol{\eta}(n)$

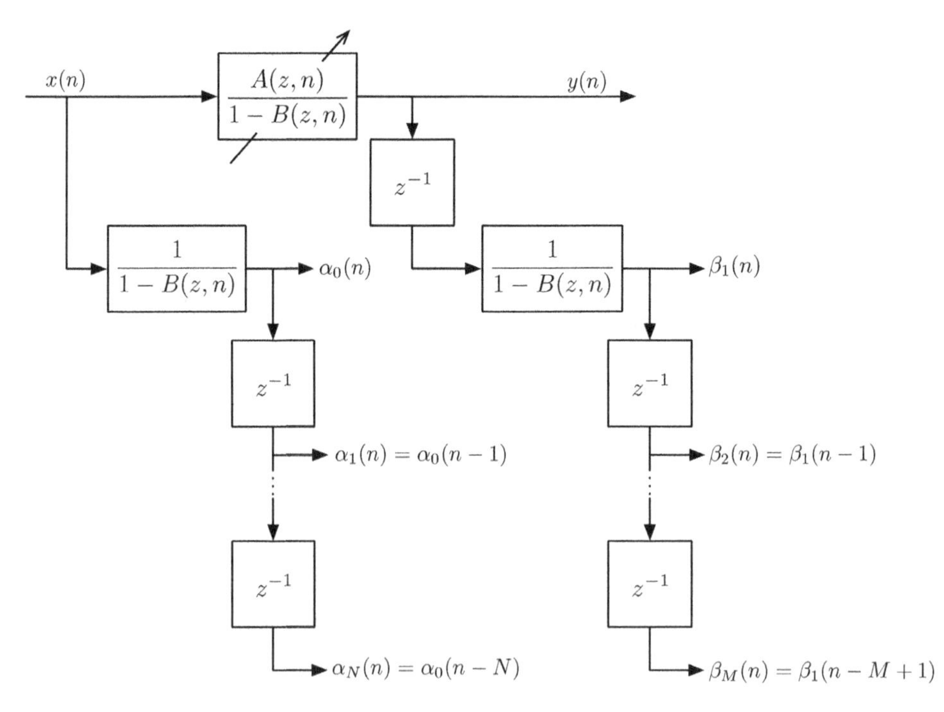

Figure 10.3 Simplified implementation of IIR adaptive filter using the output error adaptation method.

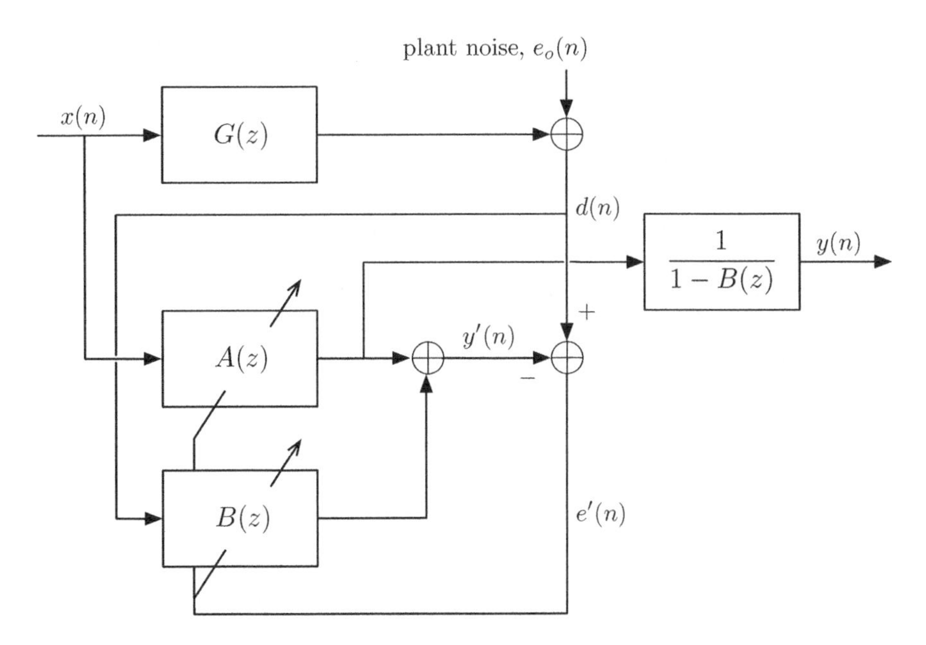

Figure 10.4 IIR adaptive filter using the equation error method.

where

$$y'(n) = \sum_{i=0}^{N} a_i(n)x(n-i) + \sum_{i=1}^{M} b_i(n)d(n-i) \tag{10.25}$$

This equation may be thought of as a modified version of Eq. (10.4). It is obtained by replacing the past samples of output, $y(n-1)$, $y(n-2)$, ..., in Eq. (10.4), by the past samples of the desired output, $d(n-1)$, $d(n-2)$, ... The name "equation error" refers to this difference in the equation used to calculate the error $e'(n)$, as against the exact value of the output error, $e(n)$.

The adoption of the equation error method is based on the following rationale. When the structure and order of an adaptive filter are correctly selected, one would expect $d(n) \approx y(n)$ upon adaptation of the filter. In that case, the difference between the error sequences $e(n)$ and $e'(n)$, when both have converged toward their optimum values, is expected to be small. Hence, we may expect the performance surfaces associated with the output error and equation error methods to have approximately the same global minimum points. The use of equation error is then preferred, as its associated performance surface will be unimodal, that is, does not have any local minimum. This unimodality results from the fact that the output of the filter $y'(n)$ in Eq. (10.25) is no more recursive in nature, that is, $y'(n)$ is effectively the output of a linear combiner with the tap-weight vector $\mathbf{w}(n)$, as defined by Eq. (10.7), and the tap-input vector

$$u'(n) = [x(n)\ x(n-1) \cdots x(n-N)\ d(n-1) \cdots d(n-M)]^{\mathrm{T}}. \tag{10.26}$$

Now, we present a study of the equation error method, which reveals its relationship as well as its difference with the output error method in a greater detail. For this study, we note that in the equation error method the criterion used for adaptation of the transfer functions $A(z)$ and $B(z)$ is

$$\xi' = E[e'^2(n)] \tag{10.27}$$

It is straightforward to show that this is a quadratic function of the coefficients of $A(z)$ and $B(z)$. This readily follows from the fact that the output $y'(n)$, as mentioned earlier, is the output of a linear combiner derived by the input sequences $x(n)$ and $d(n)$. Hence, convergence of the LMS or any other gradient-based algorithm, which may be used to find the optimum tap weights of $A(z)$ and $B(z)$, is guaranteed.

We obtain a better understanding of the equation error method by finding the relationship between the output and equation errors, $e(n)$ and $e'(n)$, respectively. This relationship can be easily arrived at using the z-transform approach. From Figure 10.1, we note that

$$D(z) = X(z)G(z) + E_o(z) \tag{10.28}$$

and

$$Y(z) = \frac{X(z)A(z)}{1 - B(z)} \tag{10.29}$$

where $D(z)$, $X(z)$, $E_o(z)$, and $Y(z)$ are the z transforms of the sequences $d(n)$, $x(n)$, $e_o(n)$, and $y(n)$, respectively. Then, because $e(n) = d(n) - y(n)$, we obtain from Eqs. (10.28) and (10.29)

$$\begin{aligned} E(z) &= D(z) - Y(z) \\ &= X(z)\left(G(z) - \frac{A(z)}{1 - B(z)}\right) + E_o(z) \end{aligned} \tag{10.30}$$

where $E(z)$ is the z transform of the sequence $e(n)$. On the other hand, from Figure 10.4, we note that $e'(n) = d(n) - y'(n)$, with $y'(n)$ as given in Eq. (10.25). Hence, we get

$$E'(z) = D(z) - D(z)B(z) - X(z)A(z) \tag{10.31}$$

where $E'(z)$ is the z transform of the sequence $e'(n)$. Substituting Eq. (10.28) in Eq. (10.31) and rearranging, we get

$$\begin{aligned} E'(z) &= X(z)[(1 - B(z))G(z) - A(z)] + [1 - B(z)]E_o(z) \\ &= \left[X(z)\left(G(z) - \frac{A(z)}{1 - B(z)}\right) + E_o(z)\right][1 - B(z)] \end{aligned} \tag{10.32}$$

Finally, comparing Eqs. (10.30) and (10.32), we obtain

$$E'(z) = E(z)(1 - B(z)) \tag{10.33}$$

This result shows that the equation error, $e'(n)$, is related to the output error, $e(n)$, through the transfer function $1 - B(z)$. In general, minimization of the mean-squared values of the output and equation errors, $e(n)$ and $e'(n)$, respectively, could lead to two different sets of tap weights for the IIR filter. However, the two solutions may be very close for certain cases. For instance, when $\xi' = E[e'^2(n)]$ converges to a very small value and $1 - B(z)$ is not very small for all values of z on the unit circle, $\xi = E[e^2(n)]$ would also be very small; thus, we expect both the output and equation error methods to converge to about the same solutions. On the other hand, when the minimum value of ξ' is large or $1 - B(z)$ is very small over a range of frequencies, the two solutions may be significantly different. The following example clarifies this concept further.

Example 10.1

Consider Figures 10.1 and 10.4. Let the plant $G(z)$ be given by

$$G(z) = \frac{1}{1 - 0.5z^{-1}}$$

and choose the modeling filter as

$$W(z) = \frac{a_0}{1 - b_1 z^{-1}}$$

Clearly, when the plant noise $e_o(n)$ is uncorrelated with the input, $x(n)$, the minimum MSE (Wiener) solution to this problem, that is, what we obtain by using the output error method (assuming that the global minimum of the corresponding mean-squared error function can be found), is $a_{0,\mathrm{o}} = 1$ and $b_{1,\mathrm{o}} = 0.5$. Here, the subscript "o" emphasizes that the coefficients are those of the optimum Wiener filter. The minimum MSE in this case is

$$\xi_{\min} = \sigma_{\mathrm{o}}^2$$

where $\sigma_{\mathrm{o}}^2 = E[e_o^2(n)]$.

To find the optimum values of a_0 and b_1 in the case of the equation error method, we note that the filter tap-input and tap-weight vectors are, respectively, $\mathbf{u}'(n) = [x(n)\ d(n-1)]^{\mathrm{T}}$ and $\mathbf{w} = [a_0\ b_1]^{\mathrm{T}}$ and the desired signal is $d(n)$. The optimum value of $\mathbf{w}$ is then obtained by solving the normal equation

$$\mathbf{R}\mathbf{w} = \mathbf{p} \tag{10.34}$$

where

$$\mathbf{R} = E[\mathbf{u}'(n)\mathbf{u}'^{\mathrm{T}}(n)] = \begin{bmatrix} E[x^2(n)] & E[x(n)d(n-1)] \\ E[d(n-1)x(n)] & E[d^2(n-1)] \end{bmatrix}$$

and

$$\mathbf{p} = \begin{bmatrix} E[d(n)x(n)] \\ E[d(n)d(n-1)] \end{bmatrix}$$

To facilitate evaluation of $\mathbf{R}$ and $\mathbf{p}$ and the subsequent calculations in this example, we assume that the input, $x(n)$, is white and has variance unity. This implies that the power spectral density of $x(n)$ is equal to 1 for all frequencies, that is, $\Phi_{xx}(z) = 1$. Also, $E[x^2(n)] = 1$. The rest of the elements of the correlation matrix $\mathbf{R}$ and the cross-correlation vector $\mathbf{p}$ are obtained by using the results of Chapter 2. For example,

$$\begin{aligned} E[d(n)x(n)] = \phi_{dx}(0) &= \frac{1}{2\pi j}\oint \Phi_{xx}(z)G(z)\frac{\mathrm{d}z}{z} \\ &= \frac{1}{2\pi j}\oint \frac{1}{1 - 0.5z^{-1}}\frac{\mathrm{d}z}{z} \end{aligned}$$

$$= \frac{1}{2\pi j} \oint \frac{dz}{z - 0.5}$$

$$= \text{residue of } \frac{1}{z - 0.5} \text{ at } z = 0.5$$

$$= 1$$

Similarly, we also obtain

$$E[d^2(n)] = \frac{4}{3} + \sigma_o^2, \quad E[x(n)d(n-1)] = E[d(n-1)x(n)] = 0$$

and

$$E[d(n)d(n-1)] = \frac{2}{3}$$

Substituting these results in Eq. (10.34) and solving for **w**, we obtain

$$a_{0,e} = 1 \quad \text{and} \quad b_{1,e} = \frac{1}{2} \cdot \frac{1}{1 + 3\sigma_o^2/4}$$

where the subscript "e" signifies that the solutions correspond to the equation error method. We note that, in this particular case, $a_{0,e}$ is unbiased, that is, it is equal to its optimum value. However, $b_{1,e}$ is different from its optimum value in the Wiener filter. The amount of bias in $b_{1,e}$ is

$$b_{1,e} - b_{1,o} = -\frac{1}{2} \cdot \frac{3\sigma_o^2/4}{1 + 3\sigma_o^2/4}$$

This bias is negligible when σ_o^2 is small. However, it becomes significant as σ_o^2 increases.

Further study of this example shows that when $x(n)$ is colored (nonwhite), both $a_{0,e}$ and $b_{1,e}$ are biased and the amount of bias, as we expect, increases with σ_o^2. This is left as an exercise for the reader (see Problem P10.2).

10.3 Case Study I: IIR Adaptive Line Enhancement

As was noted earlier in this chapter, adaptive line enhancement is a special problem that can be best solved by using IIR filters. In this section, we consider a special second-order IIR transfer function that was first proposed by David *et al.* (1983) and subsequently used and developed further by the same authors and others Ahmed *et al.* (1984); Cupo and Gitlin (1989); Hush *et al.* (1986); Regalia (1991); Cho and Lee (1993), and Farhang-Boroujeny and Wang (1997).

Figure 10.5 depicts the block diagram of the adaptive line enhancer (ALE) that we wish to study in this section. Here, $W(z)$ is an IIR filter with the transfer function

$$W(z) = \frac{(1-s)(w - z^{-1})}{1 - (1+s)wz^{-1} + sz^{-2}} \tag{10.35}$$

This is a narrow-band filter that may be used to extract a portion of the spectrum of the input, $x(n)$. When $x(n)$ is the sum of a narrow-band and a wide-band processes and $W(z)$ is centered around the narrow-band part of $x(n)$, the output of $W(z)$ will contain mainly

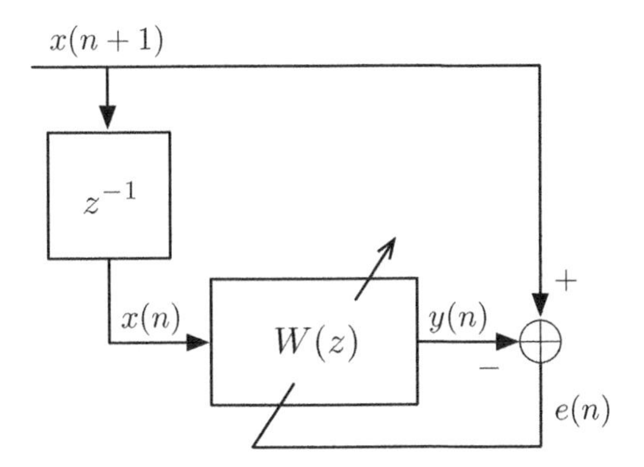

Figure 10.5 Adaptive line enhancer.

the narrow-band part of $x(n)$. The term line enhancer, thus, refers to the fact that the narrow-band part of $x(n)$, which may be considered as a spectral line, is enhanced in the sense that it is separated from the wide-band part of $x(n)$, which may be thought of as noise. In the following, we look into the details of the IIR ALE.

10.3.1 IIR ALE Filter, $W(z)$

As was noted earlier, the transfer function $W(z)$ of Eq. (10.35) is that of a narrow-band filter. Its bandwidth is controlled by the parameter s, which may select any value in the range from 0 to 1. Filters with really narrow bandwidth can be realized by choosing values of s very close to 1. The parameter w is related to the center frequency, θ, of the passband of $W(z)$ according to the following equation:

$$w = \cos\theta \tag{10.36}$$

Substituting Eq. (10.36) in Eq. (10.35) and evaluating $W(z)$ at $z = \mathrm{e}^{j\theta}$, that is, frequency response of $W(z)$ at the center of its passband, we obtain

$$W(\mathrm{e}^{j\theta}) = \mathrm{e}^{j\theta} \tag{10.37}$$

This shows that at $z = \mathrm{e}^{j\cos^{-1} w}$, $W(z) = z$ or, equivalently,

$$\text{at frequency } \omega = \cos^{-1} w, \quad z^{-1}W(z) = 1 \tag{10.38}$$

Noting that $z^{-1}W(z)$ is the transfer function between the input, $x(n)$, and the output, $y(n)$, of the line enhancer, the above result implies that the gain of the line enhancer to a sinusoid at frequency $\omega = \cos^{-1} w$ is exactly equal to 1. This interesting property of the IIR line enhancer of Eq. (10.35) becomes advantageous in applications of notch filtering and also when multiple stages of line enhancers are cascaded together to enhance multiple sinusoids (spectral lines). Application of the line enhancer structure of Figure 10.5 as a notch filter is obvious if we note that the transfer function between the input, $x(n)$, and the error, $e(n)$, is $1 - z^{-1}W(z)$ and according to Eq. (10.38), this has a null at $\omega = \cos^{-1} w$.

10.3.2 Performance Functions

To simplify our discussion, we assume that the input signal to the ALE is

$$x(n) = a\sin(\theta_o n) + \nu(n) \tag{10.39}$$

where a and θ_o are constants and $\nu(n)$ is a zero-mean white noise process with variance σ_ν^2. We refer to the first term in Eq. (10.39) as (desired) signal and $\nu(n)$ as noise. When $x(n)$ is given by Eq. (10.39), the performance function $\xi_w(s, w) = E[e^2(n)]$ of the IIR ALE, is given by the following equation (see Problem P10.3):

$$\xi_w(s, w) = \frac{a^2}{2}|1 - e^{-j\theta_o}W(e^{j\theta_o})|^2 + \frac{2\sigma_\nu^2}{1+s} \tag{10.40}$$

The subscript w in $\xi_w(s, w)$ signifies the fact that, as we see shortly, this is the performance function that is used to adjust w. In contrast, we define another performance function, $\xi_s(s, w)$, later, which is used for adapting the parameter s. Figure 10.6 shows a set of plots of $\xi_w(s, w)$ as a function of w when s is given different values. These plots correspond to the case where $\theta_o = \pi/3$, $a = \sqrt{2}$, and $\sigma_\nu^2 = 1$. Observe from these plots that the performance function $\xi_w(s, w)$ is a unimodal function of w. Its minimum corresponds to $w = \cos\theta_o$, irrespective of the value of s. This can be easily proved analytically and is left as an exercise for the reader. This observation suggests that if s is kept fixed, the optimum value of w can be obtained by using a gradient search method, such as the LMS

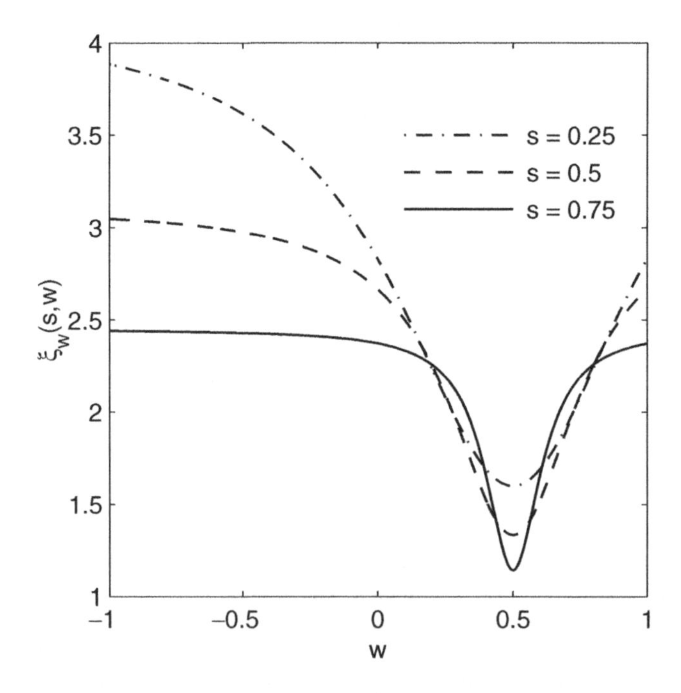

Figure 10.6 Plots of the performance function $\xi_w(s, w)$ for different values of s.

algorithm. Furthermore, note also from Figure 10.6 that if w is set to its optimum value, the minimum value of $\xi_w(s, w)$ reduces as s approaches 1. This clearly improves the performance of the ALE. On the other hand, when w is not close to its optimum value, increasing the value of s results in slowing down the convergence of w as the gradient of $\xi_w(s, w)$ is quite small when w is away from its optimum value and s is close to 1. To solve this problem, the parameter s may initially be given a smaller value and after or close to the convergence of w, it is changed to a larger value (Cho and Lee, 1993). To automate this, we need to find another performance function that allows us to quantify or detect the closeness of w to its optimum value. A possible performance function that may be used for this purpose is[1]

$$\xi_s(s, w) = E[y_s^2(n)] \tag{10.41}$$

where

$$y_s(n) = \sqrt{\frac{1+s}{1-s}}y(n) \tag{10.42}$$

The adaptation of the parameter s is done by maximizing $\xi_s(s, w)$ with respect to s. It is straightforward to show that

$$\xi_s(s, w) = \frac{a^2}{2} \cdot \frac{1+s}{1-s} \cdot |W(e^{j\theta_o})|^2 + \sigma_v^2. \tag{10.43}$$

Figure 10.7 shows the plots of $\xi_s(s, w)$, as a function of w, for $\theta_o = \pi/3$, $a = \sqrt{2}$, $\sigma_v^2 = 1$, and $s = 0.5$, 0.7, and 0.8. These plots clearly show that the performance function $\xi_s(s, w)$ is a proper choice for adjusting s. It perfectly satisfies the requirements stated earlier for changing s, namely, the maximization of $\xi_s(s, w)$ reduces s when w is far from its optimum value, and increases s as w approaches its optimum value. This can also be shown by observing the sign of $\partial\xi_s(s, w)/\partial s$ as w varies.

10.3.3 Simultaneous Adaptation of *s* and *w*

Following similar derivations to those given in Section 10.1, we obtain the algorithm presented in Table 10.2, for simultaneous adaptation of s and w. We refer to this as Algorithm 1, for our reference later. As in Table 10.2, henceforth we will use the notation $s(n)$ and $w(n)$ for s and w, respectively, as they vary with time because of adaptation. The derivations of the first four steps in Table 10.2 are straightforward. To derive the last two steps of the algorithm, we note that $s(n)$ is updated according to the recursive equation

$$s(n+1) = s(n) + \mu_s \cdot \frac{\partial y_s^2(n)}{\partial s(n)} \tag{10.44}$$

as our goal is to select $s(n)$ so that $\xi_s(s, w) = E[y_s^2(n)]$ is maximized. Note also from Eq. (10.42) that

$$y_s^2(n) = \frac{1+s(n)}{1-s(n)} \cdot y^2(n)$$

[1] The performance function $\xi_s(s, w)$ was first proposed by Farhang-Boroujeny and Wang (1997).

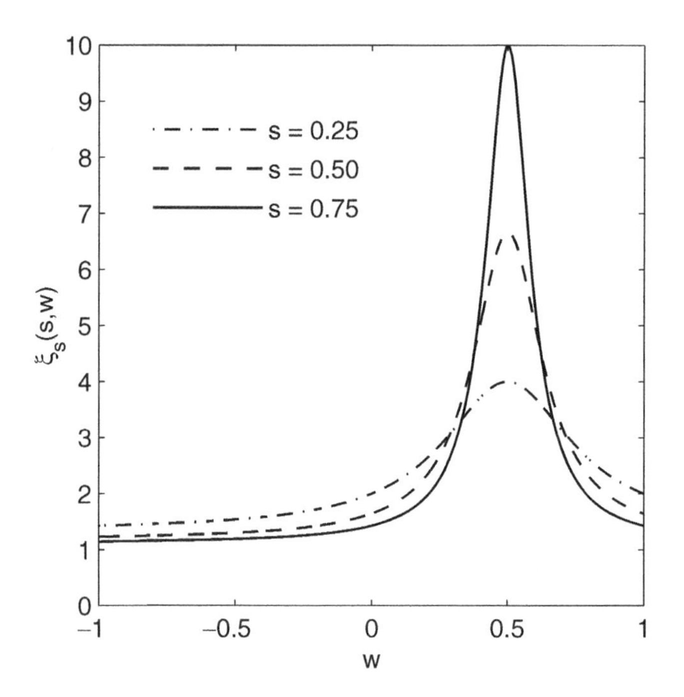

Figure 10.7 Plots of the performance function $\xi_s(s, w)$ for different values of s.

Table 10.2 Summary of adaptive IIR ALE (Algorithm 1).

$$y(n) = (1 + s(n))w(n)y(n-1) - s(n)y(n-2) + (1 - s(n))(w(n)x(n) - x(n-1))$$

$$e(n) = x(n+1) - y(n)$$

$$\alpha(n) = (1 + s(n))w(n)\alpha(n-1) - s(n)\alpha(n-2) + (1 + s(n))y(n-1) + (1 - s(n))x(n)$$

$$w(n+1) = w(n) + 2\mu_w e(n)\alpha(n)$$

$$\beta(n) = (1 + s(n))w(n)\beta(n-1) - s(n)\beta(n-2) - (w(n)e(n-1) - e(n-2))$$

$$s(n+1) = s(n) + 2\mu_s\left[\frac{1}{(1-s(n))^2}y^2(n) + \frac{1+s(n)}{1-s(n)}y(n)\beta(n)\right]$$

Definitions: $\alpha(n) = \dfrac{\partial y(n)}{\partial w(n)}$, $\beta(n) = \dfrac{\partial y(n)}{\partial s(n)}$

μ_w and μ_s are step-size parameters.

10.3.4 Robust Adaptation of w

Consider Figure 10.8a, where two plots of $\xi_w(s, w)$ are given corresponding to $\theta_o = \pi/3$ and $\pi/15$, with s fixed at 0.75. These plots show that the shape of the performance function $\xi_w(s, w)$ is sensitive to the value of θ_o. In particular, we note that when θ_o is close to 0, the function $\xi_w(s, w)$ is nearly flat over most of the values of w, except when w is very close to its optimum value. This would result in extremely slow convergence for any gradient-based algorithm, unless w is initialized close to its optimum value. The same sensitivity is observed when θ_o is close to π.

This problem may be solved if we let $w = \cos\theta$ in Eq. (10.35) and adapt θ instead of w. With this amendment, the plots of the performance function $\xi_w(s, \cos\theta)$, as a function of θ, are as shown in Figure 10.8b. We note that there is not much difference between the two plots in Figure 10.8b, as opposed to the pair in Figure 10.8a.

A robust implementation of the IIR ALE, which has reduced sensitivity to variations of θ_o, may thus be proposed by considering the change of variable $w = \cos\theta$ in Eq. (10.35) and adaptive adjustment of θ instead of w. Table 10.3 gives a summary of the resulting algorithm and is called Algorithm 2 for our reference later. This algorithm, although more complicated than Algorithm 1 (because of the involvement of the sine and cosine functions), has been found to be much more robust when θ_o is close to 0 or π (Farhang-Boroujeny and Wang, 1997).

10.3.5 Simulation Results

In this section, we study the performance of the algorithms given in Tables 10.2 and 10.3 using computer simulations. We also discuss a cascade implementation of the IIR ALE, which may be used for enhancement of multiple sinusoidal signals.

Figure 10.9 presents a set of plots that show convergence as well as tracking behavior of Algorithms 1 and 2, when the ALE input is a single sinusoid in additive white Gaussian noise, as in Eq. (10.39). The simulated scenario consists of $\sigma_\nu^2 = 0.5$ and a unit amplitude sinusoid with its angular frequency, $\theta_o(n)$, varying as shown in the figure. This corresponds to a signal-to-noise ratio (SNR) of 0 dB. The step-sizes are selected (empirically) according to the following equations:

$$\mu_s = 0.0005, \quad \mu_w(n) = 0.025(1 - s(n))^3, \quad \text{and } \mu_\theta(n) = 0.05(1 - s(n))^3$$

Note that the step-size parameters $\mu_w(n)$ and $\mu_\theta(n)$ are chosen to be time varying and are selected according to the present value of $s(n)$. This results in large step-sizes when $s(n)$ is small and small step-sizes as $s(n)$ approaches 1. The rationale behind this choice is the following. When $w(n)$ and $\theta(n)$ are far from their optimum values, $s(n)$ becomes small and hence it is better to use larger step-sizes to ensure faster convergence of the algorithm. On the other hand, when $w(n)$ and $\theta(n)$ are close to their optimum values, smaller step-sizes should be used to reduce the misadjustment of the algorithms. The above choices of $\mu_w(n)$ and $\mu_\theta(n)$ also compensate for the change of slope of the performance function $\xi_w(s, w)$ as $s(n)$ selects different values (see Figure 10.6). Thus, the equations proposed for adjusting $\mu_w(n)$ and $\mu_\theta(n)$ are based on these intuitions as well as a wide range of simulation tests. The parameter $s(n)$ is initialized to 0.25 in the beginning of each simulation and is allowed to vary in the range from 0.25 to 0.9. For Algorithm 1,

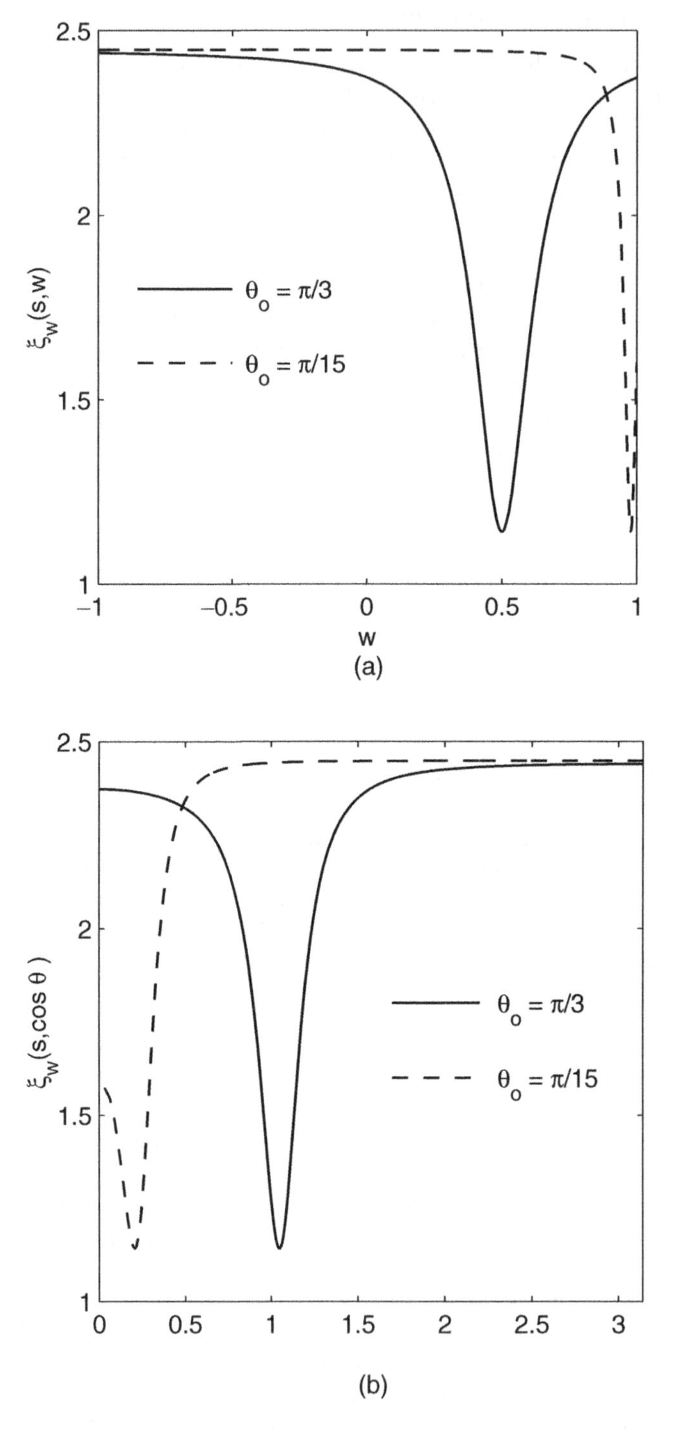

Figure 10.8 (a) Plots showing the variation of the performance function $\xi_w(s, w)$ as θ_o approaches 0 and (b) plots showing reduced sensitivity of the performance function $\xi_w(s, \cos\theta)$ to variations in θ_o.

Table 10.3 Summary of adaptive IIR ALE: Algorithm 2.

$$w(n) = \cos\theta(n)$$

$$w'(n) = \sin\theta(n)$$

$$y(n) = (1 + s(n))w(n)y(n-1) - s(n)y(n-2) + (1 - s(n))(w(n)x(n) - x(n-1))$$

$$e(n) = x(n+1) - y(n)$$

$$\alpha(n) = (1 + s(n))w(n)\alpha(n-1) - s(n)\alpha(n-2) - w'(n)[(1 + s(n))y(n-1) + (1 - s(n))x(n)]$$

$$\theta(n+1) = \theta(n) + 2\mu_\theta e(n)\alpha(n)$$

$$\beta(n) = (1 + s(n))w(n)\beta(n-1) - s(n)\beta(n-2) - (w(n)e(n-1) - e(n-2))$$

$$s(n+1) = s(n) + 2\mu_s\left[\frac{1}{(1 - s(n))^2}y^2(n) + \frac{1 + s(n)}{1 - s(n)}y(n)\beta(n)\right]$$

Definitions: $\alpha(n) = \dfrac{\partial y(n)}{\partial\theta(n)}$, $\beta(n) = \dfrac{\partial y(n)}{\partial s(n)}$

μ_θ and μ_s are step-size parameters.

the parameter $w(n)$ is initialized to 0 ($= \cos^{-1}(\pi/2)$) and is confined to the range from -0.999 to 0.999. Similarly, for Algorithm 2, $\theta(n)$ is initialized to $\pi/2$ and is confined to the range $\cos^{-1}(-0.999)$ to $\cos^{-1}(0.999)$. The results clearly show the superior performance of Algorithm 2. In particular, observe that as $\theta_o(n)$ approaches 0, its estimate, $\theta(n)$, becomes more noisy, when Algorithm 1 is used. On the contrary, Algorithm 2 is much more robust.

To enhance or extract multiple sinusoids, we may use a cascade of a few IIR ALEs as in Figure 10.10. This configuration corresponds to an L stage line enhancer, where each stage is responsible for enhancement of one single sinusoid. The output error from each stage is the input to the next stage. The enhanced narrow-band outputs, $y_i(n)$'s, from the successive stages are added together to obtain the final output, $y(n)$, of the line enhancer. The adaptation of multistage line enhancer begins with its first stage. The adaptation of the following stages begins once the previous stages have converged. Experiments have shown that this method works well (Cho and Lee, 1993). To decide on activating/deactivating the successive stages of the multistage IIR ALE we may use the parameters $s(n)$'s of the previous stages. We know that for each stage $s(n)$ increases and approaches 1 only when $w(n)$ (or $\theta(n)$) is near its optimum value. Thus, by comparing $s(n)$ of each stage with a threshold level, we may decide on activating or deactivating the adaptation of the following stage(s). This provides a very simple and effective mechanism for controlling the adaptation of the cascaded IIR ALE.

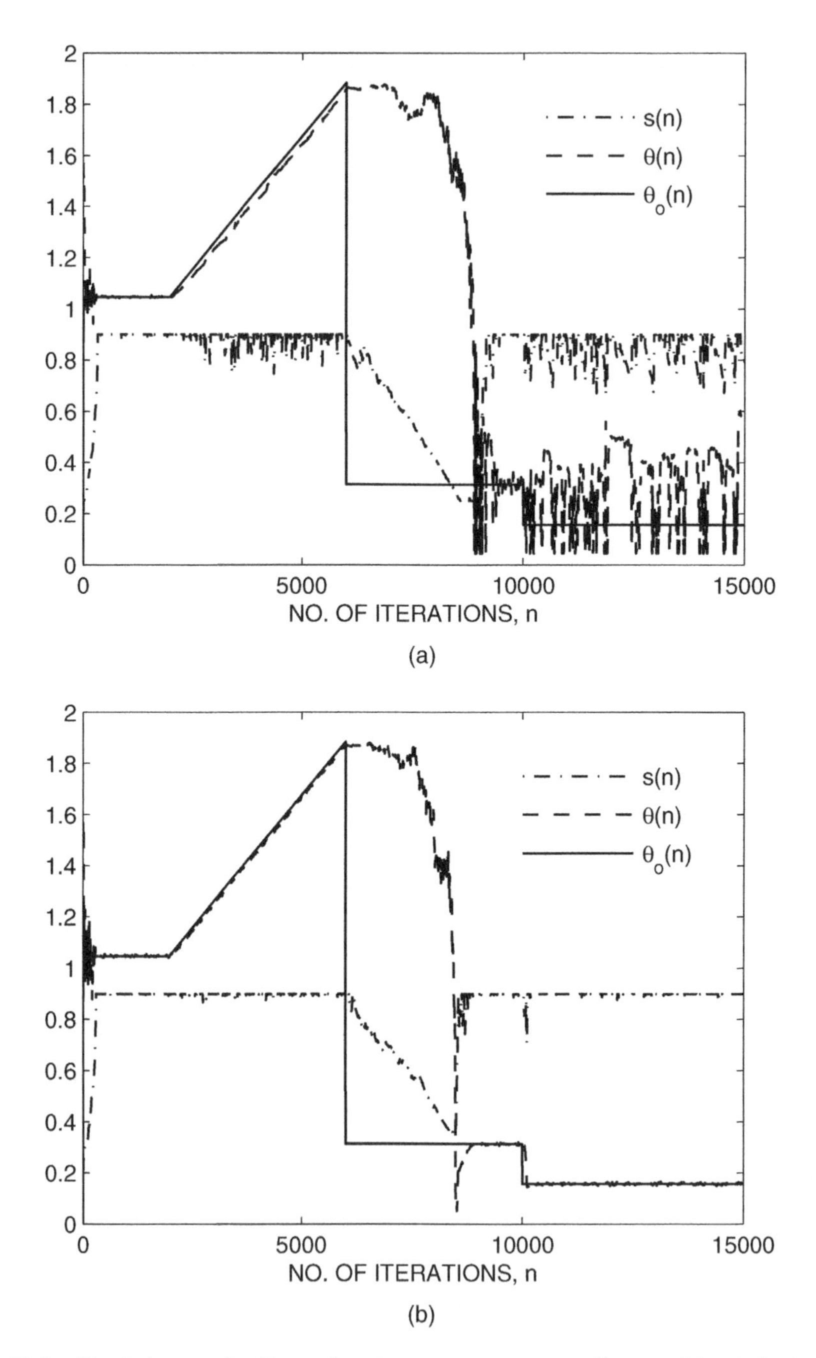

Figure 10.9 Simulation results illustrating the convergence as well as tracking behavior of the IIR ALE: (a) Algorithm 1 and (b) Algorithm 2.

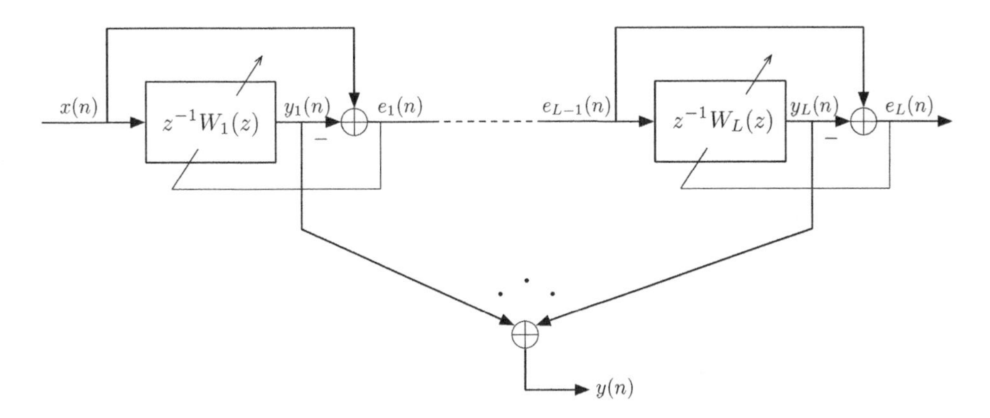

Figure 10.10 Cascaded IIR ALE for enhancement of multiple sinusoids buried in white noise.

Figure 10.11 illustrates the performance of the cascaded IIR ALE when Algorithm 2 is used. The input signal consists of the sum of four sinusoids in additive white Gaussian noise, and is given by

$$x(n) = \sin(\omega_1 n + \phi_1) + 2\sin(\omega_2 n + \phi_2) + 0.25\sin(\omega_3 n + \phi_3) + 0.5\sin(\omega_4 n + \phi_4) + \nu(n)$$

where ω_1, ω_2, ω_3, and ω_4 are equal to $\pi/1.8$, $\pi/3.5$, $\pi/6$, and $\pi/12$, respectively, ϕ_1 to ϕ_4 are random phases that are selected in the beginning of each simulation trial and remain fixed during that trial, and $\sigma_\nu^2 = 0.25$. This value corresponds to SNRs of 3, 9, -9, and -3 dB, respectively, for the individual sinusoids. As the energy of the input signals to successive stages of the ALE are different, the step-sizes of each stage are normalized to the energy of the input signal to that stage. The equations used for this purpose are $\mu_{s,i} = 0.0005/\hat{\sigma}_{x_i}^2$ and $\mu_{\theta,i}(n) = 0.01(1 - s(n))^3/\hat{\sigma}_{x_i}^2$. In these equations, i refers to the stage number, and $\hat{\sigma}_{x_i}^2$ is an estimate of the energy of the input signal, $x_i(n)$, to the ith stage. The following recursive equation is used for estimation of $\hat{\sigma}_{x_i}^2$

$$\hat{\sigma}_{x_i}^2(n) = 0.98\hat{\sigma}_{x_i}^2(n-1) + 0.02x_i^2(n).$$

The parameters $s_i(n)$'s are allowed to change between 0.25 and 0.9, and the threshold level used for activating or deactivating the adaptation of the following stages is set at 0.85. The results in Figure 10.11 show that this mechanism works very well. It may also be noted that the first stage is tuned to the strongest sinusoid (ω_2), and the last stage is tuned to the weakest one (ω_3). Such observation is intuitively sound. The MATLAB programs used to generate the results of this section are available on an accompanying website. The reader is encouraged to examine these programs and run further simulations to learn more about the line enhancer as well as the difficulties that one may encounter in using IIR adaptive filters. It would be also interesting to compare the behavior of FIR and IIR line enhancers. This is left as an exercise for interested readers.

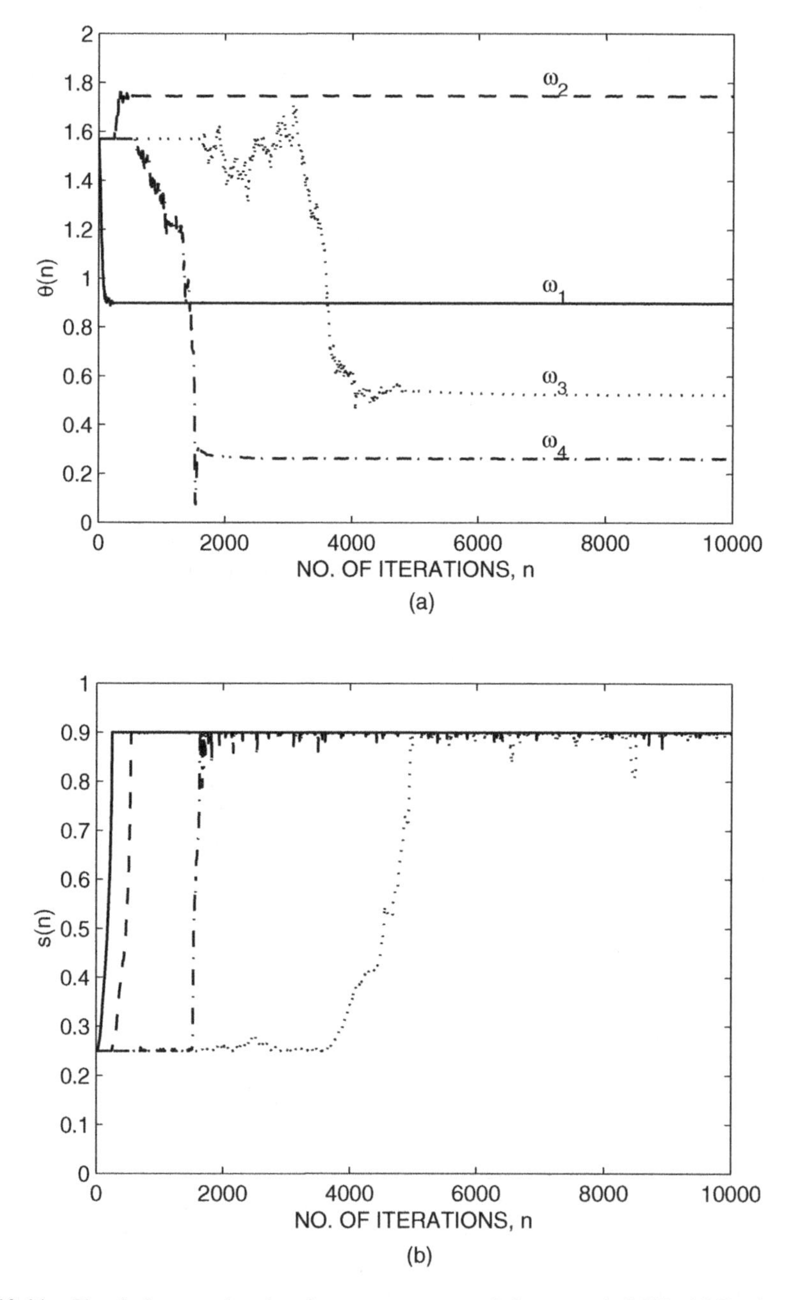

Figure 10.11 Simulation results showing convergence of the cascaded IIR ALE when used to detect/enhance multiple sinusoids: (a) angular frequencies and (b) $s(n)$ parameters. Algorithm 2 is used for the adaptation.

10.4 Case Study II: Equalizer Design for Magnetic Recording Channels

Figure 10.12 depicts the block diagram of the magnetic recording channel that we wish to address in this section. This channel is characterized by its continuous time impulse response, $h_{\mathrm{a}}(t)$. The subscript a is to emphasize that $h_{\mathrm{a}}(t)$ is an analog quantity, that is, it is a continuous function in amplitude as well as time, t. As was noted in Chapter 1 (Section 1.6.2), $h_{\mathrm{a}}(t)$ is also called *the dibit response* and is usually modeled as the superposition of positive and negative Lorentzian pulses, separated by one-bit interval, T. That is,

$$h_{\mathrm{a}}(t) = g_{\mathrm{a}}(t) - g_{\mathrm{a}}(t - T) \tag{10.45}$$

where $g_{\mathrm{a}}(t)$ is the Lorentzian pulse defined as

$$g_{\mathrm{a}}(t) = \frac{1}{1 + (2t/t_{50})^2} \tag{10.46}$$

and it is the response of the channel to a step input. The parameter t_{50}, which is the pulse-width of $g_{\mathrm{a}}(t)$ measured at 50% of its maximum amplitude, is an indicator of the *recording density*. The recording density, D, is specified by the ratio t_{50}/T. Clearly, higher density implies denser storage and vice versa.

The response of the channel to the data bits,[2] $s(n)$, is then

$$x_{\mathrm{a}}(t) = \sum_n s(n)h_{\mathrm{a}}(t - kT) + \nu_{\mathrm{a}}(t) \tag{10.47}$$

where $\nu_{\mathrm{a}}(t)$ is the channel noise. The detector assumes that its input is the convolution of the data bits, $s(n)$, with a known response, called target response, and it uses this information in doing the detection. Hence, our aim is to design an analog equalizer (filter) whose impulse response, $w_{\mathrm{a}}(t)$, when convolved with the dibit response, $h_{\mathrm{a}}(t)$,

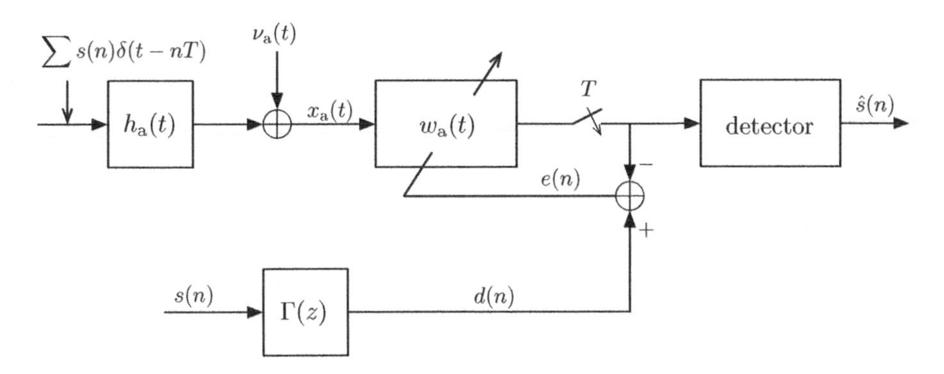

Figure 10.12 Model of a magnetic recording channel.

[2] The data bits $s(n)$ are assumed to take values $+1$ and -1.

matches the desired target response as close as possible. In particular, we are interested in matching the combined response of the channel and equalizer, that is,

$$\eta_{\mathrm{a}}(t) = \int_{-\infty}^{\infty} w_{\mathrm{a}}(\tau) h_{\mathrm{a}}(t-\tau)\mathrm{d}\tau \tag{10.48}$$

with the target response at sampling instants separated by the bit interval, T. As was noted in Chapter 1 (Section 1.6.2), the target response in magnetic channels is usually one of the class-IV partial responses characterized by the transfer functions

$$\Gamma(z) = z^{-\Delta}(1+z^{-1})^K(1-z^{-1}) \tag{10.49}$$

where z^{-1} represents one-bit delay, Δ is a parameter that takes care of the delays introduced by the channel and equalizer, and K is an integer greater than or equal to 1. The choice of K depends on the recording density, D. The value of K also determines the complexity of the detector. The commonly used values of K are 1, 2, and 3.

Next, we go through a sequence of discussions that lead us to a design methodology, using the results of this chapter as well as the previous chapters, for designing analog equalizers in the application of magnetic recording channels.

10.4.1 Channel Discretization

As all of the derivations in this book are based on sampled signals, we would like to replace the continuous time channel and equalizer impulse responses, $h_{\mathrm{a}}(t)$ and $w_{\mathrm{a}}(t)$, respectively, by their associated discrete-time counterparts. Define the sequences

$$h_i = h_{\mathrm{a}}(iT_{\mathrm{s}}) \quad \text{and} \quad w_i = w_{\mathrm{a}}(iT_{\mathrm{s}})$$

where T_{s} is the sampling period. When T_{s} is sufficiently small, we obtain, from Eq. (10.48)

$$\eta_i = \eta_{\mathrm{a}}(iT_{\mathrm{s}}) \approx T_{\mathrm{s}} \cdot (h_i \star w_i) \tag{10.50}$$

where $\star$ denotes convolution. The identity (10.50) follows from Eq. (10.48) by setting $t = iT_{\mathrm{s}}$ and approximating the integration on the right-hand side of Eq. (10.48) by a summation.

We note that the accuracy of the approximation used in Eq. (10.50) depends on the value of T_{s}. In practice, T_{s} has to be selected a few times smaller than the bit interval, T, for the results to be reasonably accurate. In the design procedure that we develop here we select T_{s} so that $T = LT_{\mathrm{s}}$, where L is an integer greater than 1. We call L the oversampling factor. Reasonable values of L are in the range of 4 to 10.

In the rest of our discussion, we assume that the time scale is normalized so that $T_{\mathrm{s}} = 1$. We also ignore the nonexactness of Eq. (10.50) and thus obtain

$$\eta_i = h_i \star w_i \tag{10.51}$$

as the discrete-time combined response of the channel and equalizer.

10.4.2 Design Steps

The following steps are taken in designing analog equalizers for magnetic recording channels:[3]

1. Using the sampled channel response, h_i, and the statistics of the channel noise (e.g., autocorrelation of the noise at the channel output), a fractionally tap-spaced FIR equalizer[4] is designed. The criterion that we use in this design is the mean-squared error between the signal samples at the equalizer output and the desired signal that is obtained by passing the data sequence, $s(n)$, through the target response $\Gamma(z)$, as in Figure 10.12.
2. A discrete-time IIR filter whose impulse response matches best with the designed FIR equalizer is found. The equation error method will be used to find this match.
3. The discrete-time IIR filter obtained in Step 2 is then converted into an equivalent analog filter, as the desired analog equalizer.

Next, we proceed with the details of the above steps.

10.4.3 FIR Equalizer Design

We define the equalizer tap-input and tap-weight vectors as

$$\mathbf{x}(n) = [x(n) \quad x(n-1) \cdots x(n-N+1)]^{\mathrm{T}} \tag{10.52}$$

and

$$\mathbf{w} = [w_0 \; w_1 \cdots w_{N-1}]^{\mathrm{T}} \tag{10.53}$$

respectively. The equalizer output is then

$$y(n) = \mathbf{w}^{\mathrm{T}}\mathbf{x}(n) \tag{10.54}$$

We note that the samples of the equalizer input, $x(n)$, and output, $y(n)$, are at T_{s} intervals. However, in the optimization of the equalizer tap weights, w_is, we are only interested in samples of $y(n)$ at $T = LT_{\mathrm{s}}$ intervals. Hence, we define the error as

$$e(n) = d(n) - y(nL) \tag{10.55}$$

and, accordingly, the performance function

$$\xi = E[e^2(n)] \tag{10.56}$$

where

$$d(n) = \sum_i \gamma_i s(n-i) \tag{10.57}$$

---

[3] The design procedure discussed here follows Mathew, Farhang-Boroujeny, and Wood (1997).

[4] The term fractionally tap-spaced equalizer refers to the fact that the spacing between the successive taps of the equalizer, w_i, is T_{s} which, as noted earlier, is a few times smaller than the bit interval, T.

and γ_is are the samples of the target response that are obtained by taking the inverse z-transform of $\Gamma(z)$ (see Figure 10.12). As an example, when $K = 2$,

$$\begin{aligned}\Gamma(z) &= z^{-\Delta}(1 + z^{-1})^2(1 - z^{-1}) \\ &= z^{-\Delta} + z^{-(\Delta+1)} - z^{-(\Delta+2)} - z^{-(\Delta+3)}\end{aligned}$$

and this gives

$$\gamma_i = \begin{cases} 1, & \text{for } i = \Delta \text{ and } \Delta + 1 \\ -1, & \text{for } i = \Delta + 2 \text{ for } \Delta + 3 \\ 0, & \text{otherwise} \end{cases}$$

We also note that the dibit response, $h_a(t)$, is noncausal. To come up with a realizable equalizer, we need to shift $h_a(t)$ to the right by a sufficient length, t_o, such that the remaining noncausal part of the shifted dibit could be ignored. This is done by replacing $h_a(t)$ with $h_a(t - t_o)$ in the earlier results and assuming that $h_a(t - t_o) = 0$, for $t < 0$. We also redefine the sampled dibit response, h_i, as

$$h_i = h_a(iT_s - t_o).$$

Furthermore, we note that the samples h_i for certain large values of i are small and thus these may also be ignored. Hence, for our further derivations, we define the dibit vector

$$\mathbf{h} = [h_0 \quad h_1 \cdots h_{M-1}]^{\mathrm{T}}$$

where M is a sufficiently large integer such that the values of h_i for $i \geq M$ are negligible. Also, for convenience of the derivations that follow, we assume that M is an integer multiple of the oversampling factor, L.

We now turn back to the performance function ξ, which was introduced earlier, in Eq. (10.56). Using Eq. (10.54) and Eq. (10.55) in Eq. (10.56), and solving the corresponding Wiener–Hopf equation, which follow from $\nabla_{\mathbf{w}}\xi = 0$, we get the optimum tap-weight vector of the desired FIR equalizer as

$$\mathbf{w}_{\text{opt}} = \mathbf{R}^{-1}\mathbf{p} \tag{10.58}$$

where $\mathbf{R} = E[\mathbf{x}(nL)\mathbf{x}^{\mathrm{T}}(nL)]$ and $\mathbf{p} = E[d(n)\mathbf{x}(nL)]$. Here, the definition of the column vector $\mathbf{x}(nL)$ follows Eq. (10.52).

To obtain an explicit expression for $\mathbf{w}_{\text{opt}}$, we note that the elements h_i are at T_s intervals, but the data bits, $s(n)$, and the target response, γ_i, are at $T = LT_s$ intervals. Noting this and assuming that N is an integer multiple of L, we obtain

$$\mathbf{x}(nL) = \mathbf{H}\mathbf{s}(n) + \boldsymbol{\nu}(nL) \tag{10.59}$$

where

$$\mathbf{s}(n) = [s(n) \quad s(n-1) \cdots s(n - N/L + 1)]^{\mathrm{T}} \tag{10.60}$$

$$\mathbf{H} = \begin{bmatrix} h_0 & h_L & h_{2L} & h_{3L} & \cdots & h_{M-L} & 0 & \cdots \\ 0 & h_{L-1} & h_{2L-1} & h_{3L-1} & \cdots & h_{M-L-1} & h_{M-1} & \cdots \\ 0 & h_{L-2} & h_{2L-2} & h_{3L-2} & \cdots & h_{M-L-2} & h_{M-2} & \cdots \\ \vdots & \cdots & \vdots & \ddots & \vdots & \vdots & \ddots & \\ 0 & h_0 & h_L & h_{2L} & \cdots & h_{M-2L} & h_{M-L} & \cdots \\ 0 & 0 & h_{L-1} & h_{2L-1} & \cdots & h_{M-2L-1} & h_{M-L-1} & \cdots \\ 0 & 0 & h_{L-2} & h_{2L-2} & \cdots & h_{M-2L-2} & h_{M-L-2} & \cdots \\ \vdots & \vdots & \cdots & \vdots & \ddots & \vdots & \vdots & \ddots \\ 0 & 0 & h_0 & h_L & \cdots & h_{M-3L} & h_{M-2L} & \cdots \\ \vdots & \vdots & \cdots & \vdots & \ddots & \vdots & \vdots & \ddots \end{bmatrix} \tag{10.61}$$

and $\boldsymbol{\nu}(nL)$ is the associated vector of samples of the channel noise, $\nu_a(t)$. We may also write Eq. (10.57) as

$$d(n) = \boldsymbol{\gamma}^{\mathrm{T}}\mathbf{s}(n) \tag{10.62}$$

where $\boldsymbol{\gamma}$ is the column vector consisting of the samples of the target response, $\boldsymbol{\gamma}_i$'s. The length of γ is appropriately selected by appending extra zeros at its end so that it would be compatible with $\mathbf{s}(n)$.

Using the above results and assuming that the binary process $s(n)$ and the noise process, $\nu(n)$, are white and independent of one another, we obtain

$$\mathbf{R} = \mathbf{H}\mathbf{H}^{\mathrm{T}} + \sigma_\nu^2\mathbf{I} \tag{10.63}$$

where σ_ν^2 is the variance of $\nu(n)$ and $\mathbf{I}$ is the identity matrix. In arriving at Eq. (10.63), we have also used the fact that $E[\mathbf{s}(n)\mathbf{s}^{\mathrm{T}}(n)] = \mathbf{I}$ because $s(n)$ is white with values ± 1. Similarly, we also obtain

$$\mathbf{p} = \mathbf{H}\boldsymbol{\gamma} \tag{10.64}$$

Substituting Eqs. (10.63) and (10.64) in Eq. (10.58), we obtain the following explicit equation for the desired optimum fractionally tap-spaced FIR equalizer:

$$\mathbf{w}_{\mathrm{opt}} = (\mathbf{H}\mathbf{H}^{\mathrm{T}} + \sigma_\nu^2\mathbf{I})^{-1}\mathbf{H}\boldsymbol{\gamma} \tag{10.65}$$

10.4.4 Conversion from FIR into IIR Equalizer

As the next step in designing analog equalizers, we need to find an IIR filter whose response closely matches the designed FIR equalizer. For this, we use the method of equation error that was discussed in Section 10.2. With reference to Figure 10.4, in the present context of magnetic recording, $x(n)$ is the channel output, $G(z)$ is the designed FIR equalizer, $e_o(n) = 0$, for all n, $A(z)$ and $B(z)$ are polynomials, which define the transfer function $W(z)$ of the desired IIR filter according to Eq. (10.1), and the unit delay z^{-1} is equivalent to one T_s interval. We can use the LMS or any other adaptive filtering algorithm to find the coefficients of $A(z)$ and $B(z)$. We may also adopt an analytical method and develop a closed-form solution for the coefficients of $A(z)$ and $B(z)$, or, use time averages to estimate the coefficients of the related Wiener–Hopf equation. We use the last method in the numerical examples discussed below, because of convenience.

10.4.5 Conversion from z Domain into s Domain

Among the different methods available for conversion between s-domain and z-domain transfer functions, we discuss the method of impulse invariance. To give a brief introduction to this method, we consider a causal continuous-time system with the transfer function

$$H_a(s) = \sum_i \frac{a_i}{s - s_i} \tag{10.66}$$

with s_i's being the poles of the system. The impulse response of this system is

$$h_a(t) = \begin{cases} \sum_i a_i e^{s_i t}, & \text{for } t \geq 0 \\ 0, & \text{otherwise} \end{cases} \tag{10.67}$$

Now, if we consider a discrete-time system whose unit-sample (impulse) response is given by the samples $h_a(0)$, $h_a(T_s)$, $h_a(2T_s)$, …, its transfer function will be

$$\begin{aligned} H(z) &= \sum_{k=-\infty}^{\infty} h_a(kT_s) z^{-k} \\ &= \sum_{k=0}^{\infty} \sum_i a_i e^{s_i k T_s} z^{-k} \\ &= \sum_i \sum_{k=0}^{\infty} a_i e^{s_i k T_s} z^{-k} \\ &= \sum_i \frac{a_i}{1 - e^{s_i T_s} z^{-1}} \end{aligned} \tag{10.68}$$

The reverse of this conversion is obvious. That is, if

$$H(z) = \sum_i \frac{a_i}{1 - z_i z^{-1}} \tag{10.69}$$

is the transfer function of a discrete-time system with unit-sample response h_n, the transfer function of the continuous-time system whose impulse response samples, at T_s intervals, is the sequence h_n, is

$$H_a(s) = \sum_i \frac{a_i}{s - (1/T_s) \ln\, z_i} \tag{10.70}$$

10.4.6 Numerical Results

To highlight some of the features of the design method that was developed earlier, we present some numerical results using the Lorentzian pulse (see Eqs. (10.45) and (10.46)) as the model for the magnetic recording channel. The measure used for evaluating the designed equalizer is the SNR at the detector input. It is defined as

$$\text{detection SNR} = \frac{\sum_i \gamma_i^2}{\sum_i (\eta_{iL} - \gamma_i)^2 + \sigma_\nu^2 \sum_i w_i^2} \tag{10.71}$$

The value of the variance of the channel noise, σ_ν^2, is selected based on another SNR that is defined at the equalizer input (or channel output) as

$$\text{SNR at the equalizer input} = \frac{\sum_i h_i^2}{\sigma_\nu^2} \tag{10.72}$$

It may be noted from Eq. (10.71) that the noise at the detector input is considered to be the sum of channel noise at the equalizer output and residual intersymbol interference (ISI). Further exploration of this definition is left as an exercise for interested readers.

We present results of many designs that are obtained for various choices of channel parameters. Evaluating these results, we find that, among different parameters, performance of the IIR equalizer is highly affected by the choice of the delays t_o and Δ. Furthermore, the choice of t_o is closely related to the value of Δ. In a good design, usually, $t_o = \Delta T + \tau$, where T (as defined before) is the bit interval and τ is a relatively small delay in the range from 0 to $3T$. The best value of τ, which results in maximum detection SNR, depends on the number of zeros and poles of the IIR equalizer, noise level in the channel, and recording density, D. The effect of these on the optimum value of τ is difficult to predict. It appears that the only way of finding the optimum τ is to design many IIR equalizers for different values of τ and choose the best among them. In Figure 10.13, we show how the detection SNR varies as a function of τ/T, for certain selected choices of recording density, D, SNR at the equalizer input, and number of poles and zeros of the IIR equalizer. The value of K is set at 2 in this set of results. These plots clearly indicate that the choice of τ is very critical in the final performance of the IIR equalizer.

Figure 10.14 shows an example of the equalized dibit response of the magnetic recording channel. This is obtained by passing the dibit response h_i through the designed IIR equalizer. The parameters used to obtain these results are: $D = 2.5$, $\tau/T = 0.3$, SNR at the equalizer input = 30 dB, IIR equalizer with 5 zeros and 6 poles, and $K = 2$. Observe that an almost perfect match between the equalizer output and the target response has been achieved here.

The MATLAB program that has been used to obtain these results is available on an accompanying website. It is called `iirdsgn.m`. The reader is encouraged to run this program for other designs for enhancing his understanding of the concepts that were discussed earlier.

10.5 Concluding Remarks

In this chapter, we discussed the problem of IIR adaptive filtering. We noted that unlike FIR adaptive filters, whose adaptation is a rather straightforward task, adaptive adjustment of IIR filters is, in general, a complicated problem. IIR adaptive filters can easily become unstable as their poles may get shifted out of the unit circle by the adaptation process. Or, they can get trapped in one of the local minima points because the performance surfaces of IIR filters are, in general, multimodal. We saw that these problems could be resolved by either limiting ourselves to applications where special transfer functions with unimodal performance surfaces could be used, or using the method of equation error, which leads to suboptimal solution.

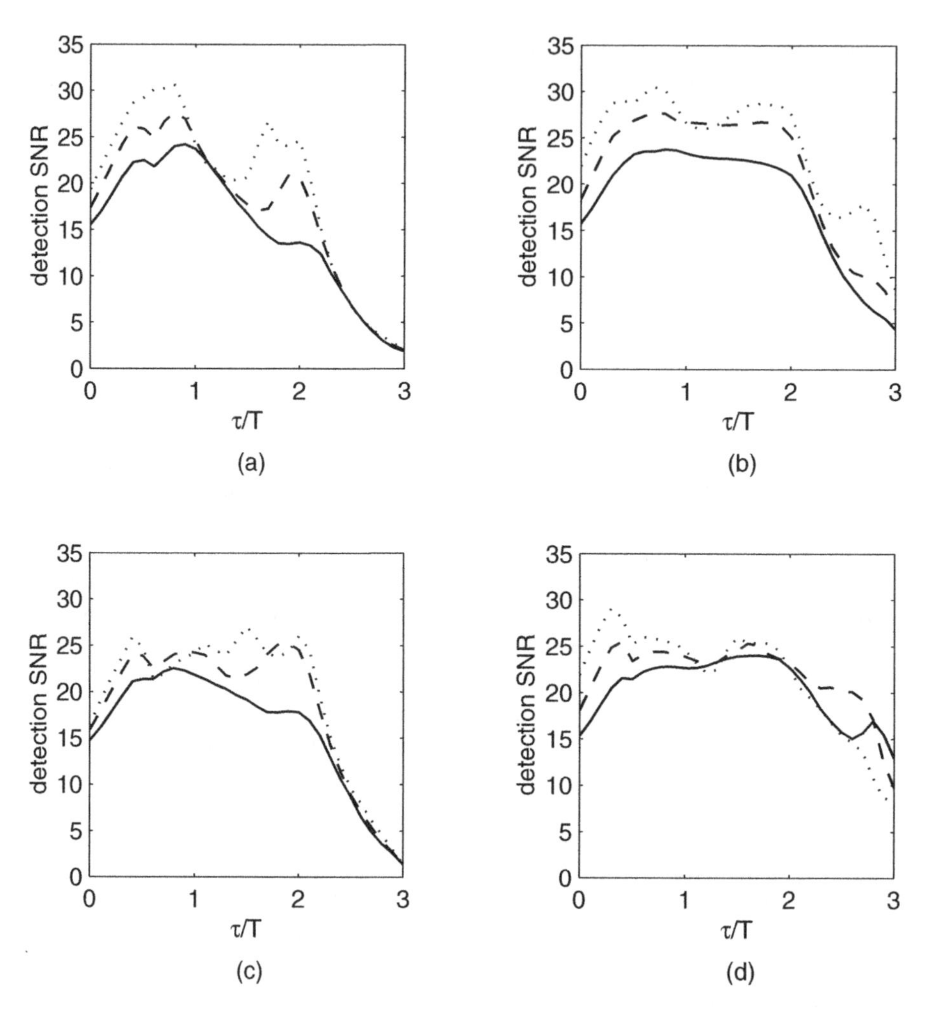

Figure 10.13 Performance of the IIR equalizers designed for different choices of the parameters: (a) 3 zeros and 4 poles, $D = 2$. (b) 5 zeros and 6 poles, $D = 2$. (c) 3 zeros and 4 poles, $D = 2.5$ and (d) 5 zeros and 6 poles, $D = 2.5$. The three plots in each case correspond to the following choices of channel SNR: 25 dB (—); 30 dB (– – –), 35 dB (· · · · ·).

We also presented two case studies, one for each of the above solutions. These studies showed some of the difficulties that one may encounter while dealing with IIR adaptive filters – problems that do not arise when FIR adaptive filters are used. In the first case study, we used a specific transfer function for realization of line enhancers. There were many considerations that we had to take note of before getting to our final solution. For instance, we saw that our initial transfer function gets into difficulties when the frequency of the sinusoid is close to 0 or π. For this, we found a specific solution, namely, replacement of the parameter w by $\cos\theta$ and adapting θ instead of w. We also had to take care of the parameter s of this structure in a very special way. In contrast to this, if we refer to Chapter 6 (Section 6.4.3) where we used an FIR filter to realize a

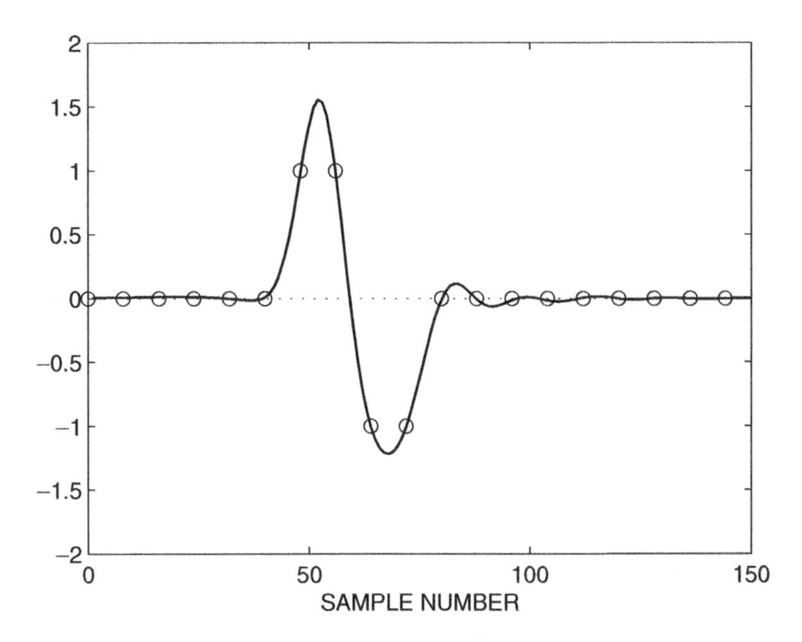

Figure 10.14 An example of the equalized dibit response of the magnetic recording channel for $K = 2$, $D = 2.5$, $\tau/T = 0.3$, channel SNR = 30 dB, and an IIR equalizer with 5 zeros and 6 poles. The circles correspond to the target response samples.

line enhancer, we find that none of the abovementioned kind of problems exist. The only point that we must consider while using an FIR filter is to include sufficient number of taps. As was noted in the beginning of this chapter, the main advantage of IIR adaptive filters, as compared with their FIR counterparts, is their lower order, which may lead to a lower computational complexity and hence reduction in the cost of implementation.

The second case study that we discussed was equalization of magnetic recording channels. In this application, we found that the optimum IIR equalizers are very sensitive to a delay parameter, τ. The results indicated that varying this delay even around its optimum value can significantly affect the performance of the resulting equalizer (Figure 10.13). Similar study for FIR equalizers shows that they do not exhibit such a level of sensitivity. In fact, FIR equalizers are very robust in this respect. A study of this problem is left as an exercise for the reader.

To conclude, our study in this chapter showed that although the IIR adaptive filters are attractive for some specific applications, it may not be possible to use them (directly) for any arbitrary application. This is unlike the FIR (transversal) adaptive filters, which are very versatile adaptive systems. While using IIR adaptive filters, special care has to be taken in the selection of transfer function and/or performance function depending upon the kind of application that we are dealing with.

At this point, we shall add that much of the research work in IIR adaptive filters has been carried out in the context of system modeling. Literature on this topic is much wider than what we could cover within the limits of a single chapter of this book. An excellent paper by John J. Shynk (1989) provides a good review of the fundamental work done in this subject. A good bibliography of the key references is also provided in this paper. Another interesting and classic reference is the book by Ljung and Söderström (1983).

Problems

P10.1 Start with Eq. (10.4) and use Eqs. (10.11) and (10.12) to give detailed derivations of Eqs. (10.15) and (10.16), respectively.

P10.2 For the modeling problem that was discussed in Example 10.1, obtain the values of $a_{0,\mathrm{o}}$, $b_{1,\mathrm{o}}$, $a_{0,\mathrm{e}}$, and $b_{1,\mathrm{e}}$ for the cases where the power spectral density of the input, $x(n)$, is given as:

(i)

$$\Phi_{xx}(\mathrm{e}^{j\omega}) = \frac{1}{|1 - 0.3\mathrm{e}^{-j\omega}|^2}$$

(ii)

$$\Phi_{xx}(\mathrm{e}^{j\omega}) = \frac{1}{|1 - 0.8\mathrm{e}^{-j\omega}|^2}$$

From this study, you should find that when $x(n)$ is colored, both $a_{0,\mathrm{e}}$ and $b_{1,\mathrm{e}}$ are biased with respect to the optimum Wiener coefficients $a_{0,\mathrm{o}}$ and $b_{1,\mathrm{o}}$. Explain how these biases are affected by the shape of $\Phi_{xx}(\mathrm{e}^{j\omega})$.

P10.3 Give a detailed derivation of Eq. (10.40).

P10.4 Consider the case where the input to the line enhancer of Figure 10.5 is the sum of a sinusoid and a white noise as in Eq. (10.39).

(i) Show that

$$E[y^2(n)] = \frac{a^2}{2}|W(\mathrm{e}^{j\theta_\mathrm{o}})|^2 + \sigma_\nu^2 \cdot \frac{1-s}{1+s}$$

(ii) For a given value of θ_o (say, $\theta_\mathrm{o} = \pi/3$), and a few values of s (say, $s = 0.25$, 0.5, and 0.75) plot $E[y^2(n)]$ as a function of w and observe that $E[y^2(n)]$ has only one maximum and this is achieved when $w = \cos\theta_\mathrm{o}$.

P10.5 Give a detailed derivation of Eq. (10.43).

P10.6 Give a detailed derivation of the LMS algorithm of Table 10.2.

P10.7 Give a detailed derivation of the LMS algorithm of Table 10.3.

P10.8 In the light of the result of Problem P10.4, adjustment of the parameter w of the line enhancer of Figure 10.5 may be done by maximizing the mean-squared value of the output $y(n)$. Develop an LMS algorithm, which works based on this principle. Also, develop another LMS algorithm that adapts $\theta = \cos^{-1}w$ instead of w, as in Algorithm 2 of Table 10.3.

P10.9 Give a formal proof of the fact that the performance function $\xi_w(s, w)$ of Eq. (10.40), for a given s, has only one minimum point and that corresponds to the value of $w = \cos\theta_\mathrm{o}$.

P10.10 Study the transfer function between the input, $x(n)$, and output, $y(n)$, of Figure 10.10 and show that this is that of a filter with L narrow bands.

P10.11 Study the transfer function between the input, $x(n)$, and output error, $e_L(n)$, of Figure 10.10 and show that this is that of a filter with L notches.

P10.12 In line enhancers, signal enhancement is defined as the ratio of the SNR at the enhancer output, to the SNR at its input. For the IIR ALE that was discussed in Section 10.3 show that when $x(n)$ is given by Eq. (10.39)

$$\text{Signal enhancement of IIR ALE} = \frac{1+s}{1-s}$$

P10.13 Work out a detailed derivation of Eq. (10.59).

P10.14 For the magnetic recording channel, which is discussed in Section 10.4, show that the mean-squared value of the sum of residual ISI and noise at the equalizer output is $\sum_i (\eta_{iL} - \gamma_i)^2 + \sigma_\nu^2 \sum_i w_i^2$. Thus, justify the use of definition Eq. (10.71).

P10.15 From our discussion in Section 10.2, we recall that the output error, $e(n)$, and equation error, $e'(n)$, are related through the transfer function $1 - B(z)$. Assuming that a relatively good estimate of $B(z)$ (say, $\hat{B}(z)$) is available, it is proposed that the setup of Figure P10.15 may be used to obtain better estimates of $A(z)$ and $B(z)$ compared with what could be achieved by the original equation error setup of Figure 10.4. Elaborate on this diagram and explain why this setup may give better estimates of $A(z)$ and $B(z)$, as compared with that in Figure 10.4. Also, develop an LMS algorithm for adaptation of $A(z)$ and $B(z)$ in this setup.

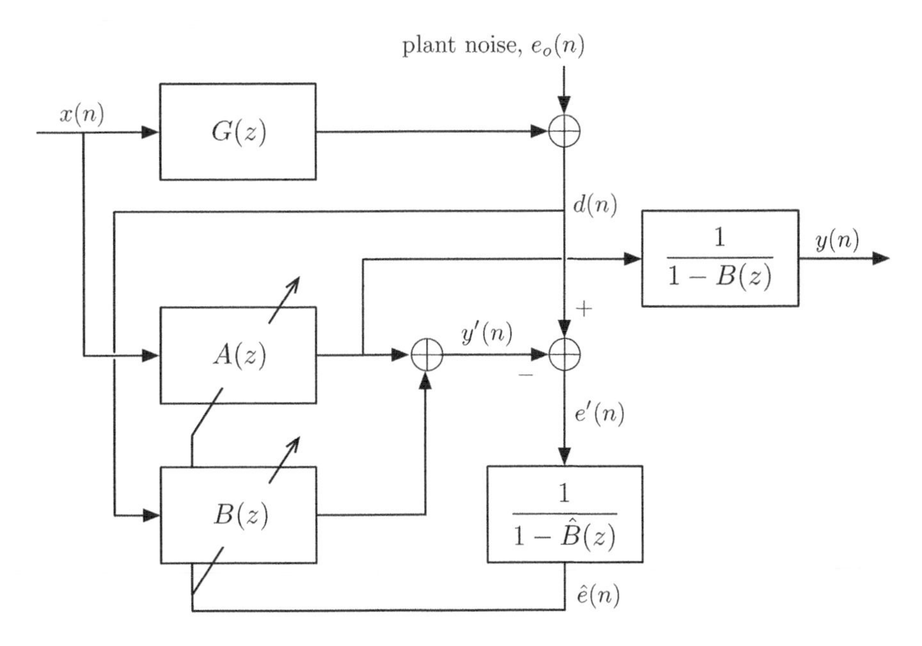

Figure P10.15

Computer-Oriented Problems

P10.16 Develop programs for implementation of the LMS algorithms of Problem P10.8. For an input signal consisting of a sinusoid in additive white noise, as in Eq. (10.39), compare the convergence behavior of these algorithms with Algorithms 1 and 2 of Tables 10.2 and 10.3, respectively. The MATLAB programs for Algorithms 1 and 2 are available on an accompanying website. As a bench mark for your comparisons, you may try to generate results similar to those in Figure 10.9.

P10.17 In magnetic recording, the best choice of the parameter K in the target response $\Gamma(z)$ depends on the recording density, D. In this exercise we study how the choice of K varies with D.

The program `iirdsgn.m` on an accompanying website allows you to design IIR equalizers in the application of magnetic recording. Different parameters of interest (such as recording density, channel SNR, and equalizer order) are inputs to the program. Use this program to design IIR equalizers for the densities $D = 1.5$ to 3, in steps of 0.25, and the choices of the parameter $K = 1, 2,$ and 3. Assume a channel SNR of 30 dB and an equalizer with 3 zeros and 4 poles, in all your designs. However, each design has to be optimized with respect to the delay, τ. The criterion for the optimum design is detection SNR. Tabulate your results and discuss how the choice of K varies with D.

11

Lattice Filters

In our discussions on FIR and IIR filters in the previous chapters, we always limited ourselves to implementation structures which were direct realization of their corresponding system functions. In this chapter, we introduce an alternative structure for realization of FIR and IIR filters. This new structure, which is called *lattice*, has a number of desirable properties that will become clear as we go along in this chapter. The lattice structure has most commonly been used for implementing linear predictors in the context of speech processing applications. Predictors may appear in two distinct forms: forward and backward. In a forward linear predictor, the aim is to estimate the present sample of a signal $x(n)$ in terms of a linear combination of its past samples $x(n-1), x(n-2), \ldots, x(n-m)$. This corresponds to one-step forward prediction of order m. In backward linear prediction, on the other hand, an estimate of $x(n-m)$ is obtained as a linear combination of the future samples $x(n), x(n-1), \ldots, x(n-m+1)$.

In this chapter, we start with a study of forward and backward linear predictors. We find that these two are closely related to each other. In particular, we introduce the so-called *order-update* equations which mean an $(m+1)$th-order linear prediction (forward or backward) of a signal sequence can be obtained as a linear combination of its mth-order forward and backward predictions. The order-update equations lead to a simple derivation of the lattice structure for forward and backward linear predictors. Other developments which follow this are the Levinson–Durbin algorithm (a computationally efficient procedure for solving Wiener–Hopf equations) and lattice structures for arbitrary FIR and IIR system functions. We also introduce the concept of autoregressive (AR) modeling of time series and use that for an efficient implementation of LMS–Newton algorithm.

Our discussion in this chapter is limited to the case where the filter tap weights, input and desired output are real-valued. Extension of this to the case of complex-valued signals is straightforward and is deferred to the problems at the end of this chapter.

11.1 Forward Linear Prediction

Figure 11.1 depicts the direct implementation of an mth-order forward linear predictor. A transversal filter with tap-input vector $\mathbf{x}_m(n-1) = [x(n-1)\ x(n-2) \cdots x(n-m)]^{\mathrm{T}}$ and tap-weight vector $\mathbf{a}_m = [a_{m,1}\ a_{m,2} \cdots a_{m,m}]^{\mathrm{T}}$ is used to obtain an estimate of the input sample $x(n)$. We use the subscript m in vectors $\mathbf{a}_m$ and $\mathbf{x}_m(n)$ and the elements of $\mathbf{a}_m$ to

Adaptive Filters: Theory and Applications, Second Edition. Behrouz Farhang-Boroujeny.

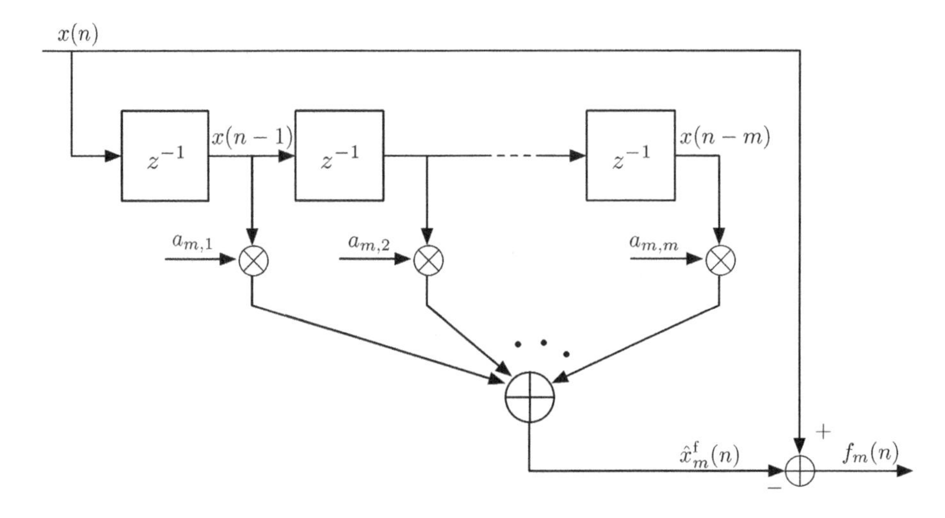

Figure 11.1 Forward linear predictor.

emphasize that the predictor order is m. The implementation structure of the type shown in Figure 11.1 is called the *transversal* or *tapped delay line predictor*, in contrast to the lattice structure which will be introduced later.

We assume that the input sequence, $x(n)$, is the realization of a stationary stochastic process. Furthermore, we assume that the predictor tap weights are optimized in the mean-square sense according to the Wiener filter theory. Thus, the optimum value of the predictor tap weights $a_{m,1}, a_{m,2}, \ldots, a_{m,m}$ are obtained by minimizing the function

$$P^{\mathrm{f}}_m = E[f^2_m(n)] \tag{11.1}$$

where

$$f_m(n) = x(n) - \hat{x}^{\mathrm{f}}_m(n) \tag{11.2}$$

is the forward prediction error and

$$\hat{x}^{\mathrm{f}}_m(n) = \sum_{i=1}^{m} a_{m,i}x(n-i) = \mathbf{a}^{\mathrm{T}}_m\mathbf{x}_m(n-1) \tag{11.3}$$

is the mth-order forward prediction of the input sample $x(n)$. This is a conventional Wiener filtering problem with the input vector $\mathbf{x}_m(n-1)$ and desired output $x(n)$. Hence, the corresponding Wiener–Hopf equation is obtained by direct substitution of $x(n)$ for $d(n)$ and $\mathbf{x}_m(n-1)$ for $\mathbf{x}(n)$ in Eqs. (3.10) and (3.11), and recalling Eq. (3.24). The result is

$$\mathbf{R}\mathbf{a}_{m,o} = \mathbf{r} \tag{11.4}$$

where $\mathbf{R} = E[\mathbf{x}_m(n-1)\mathbf{x}^{\mathrm{T}}_m(n-1)]$, $\mathbf{r} = E[x(n)\mathbf{x}_m(n-1)]$ and $\mathbf{a}_{m,o}$ denotes the optimum value of $\mathbf{a}_m$.

To simplify our notations in the discussion that follows, we assume that the predictor tap weights are always set to their optimum values and drop the extra subscript "o" from $\mathbf{a}_{m,o}$. Thus, Eq. (11.4) is simply written as

$$\mathbf{R}\mathbf{a}_m = \mathbf{r} \tag{11.5}$$

When the predictor tap weights are set according to Eq. (11.5), $P_m^{\rm f}$ is minimized and this can be obtained using Eq. (3.26) as

$$\begin{aligned} P_m^{\rm f} &= E[x^2(n)] - \mathbf{r}^{\rm T}\mathbf{a}_m \\ &= E[x^2(n)] - \mathbf{r}^{\rm T}\mathbf{R}^{-1}\mathbf{r} \end{aligned} \tag{11.6}$$

assuming that $\mathbf{R}$ is nonsingular.

For our later use, we define the autocorrelation function of the input process for lag k as

$$r(k) = E[x(n)x(n-k)] \tag{11.7}$$

Using this definition, we note that

$$\mathbf{R} = \begin{bmatrix} r(0) & r(1) & \cdots & r(m-1) \\ r(1) & r(0) & \cdots & r(m-2) \\ \vdots & \vdots & \ddots & \vdots \\ r(m-1) & r(m-2) & \cdots & r(0) \end{bmatrix} \tag{11.8}$$

and

$$\mathbf{r} = \begin{bmatrix} r(1) \\ r(2) \\ \vdots \\ r(m) \end{bmatrix} \tag{11.9}$$

We note that $\mathbf{R}$ and $\mathbf{r}$, with the exception of $r(0)$ and $r(m)$, share the same set of elements. This very close relationship between $\mathbf{R}$ and $\mathbf{r}$ is the key to many interesting properties of the linear predictors which will be focused in this chapter.

11.2 Backward Linear Prediction

Figure 11.2 depicts an mth-order backward linear predictor. A transversal filter with tap-input vector $\mathbf{x}_m(n) = [x(n)\ x(n-1) \cdots x(n-m+1)]^{\rm T}$ and tap-weight vector $\mathbf{g}_m = [g_{m,1}\ g_{m,2} \cdots g_{m,m}]^{\rm T}$ is used to obtain an estimate of the input sample $x(n-m)$. As in the forward prediction case, we assume that the backward predictor tap weights are optimized in the mean-square sense according to the Wiener filter theory. The optimum value of the predictor tap weights $g_{m,1}, g_{m,2}, \ldots, g_{m,m}$ are then obtained by minimizing the function

$$P_m^{\rm b} = E[b_m^2(n)] \tag{11.10}$$

where

$$b_m(n) = x(n-m) - \hat{x}_m^{\rm b}(n) \tag{11.11}$$

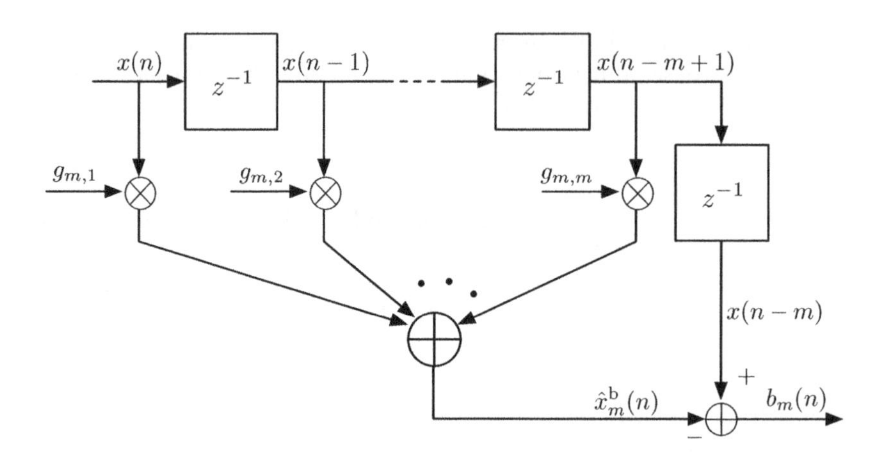

Figure 11.2 Backward linear predictor.

is the backward prediction error and

$$\hat{x}^{\rm b}_m(n) = \sum_{i=1}^{m} g_{m,i}x(n-i+1) = \mathbf{g}_m^{\rm T}\mathbf{x}_m(n) \tag{11.12}$$

is the mth-order backward prediction of the input sample $x(n-m)$. This is a conventional Wiener filtering problem with the input vector $\mathbf{x}_m(n)$ and desired output $x(n-m)$. Hence, the corresponding Wiener–Hopf equation is obtained by direct substitution of $x(n-m)$ for $d(n)$ and $\mathbf{x}_m(n)$ for $\mathbf{x}(n)$ in Eqs. (3.10) and (3.11), and recalling Eq. (3.24). The result is

$$\mathbf{R}\mathbf{g}_m = \mathbf{r}_{\rm b} \tag{11.13}$$

where $\mathbf{R} = E[\mathbf{x}_m(n)\mathbf{x}_m^{\rm T}(n)]$, $\mathbf{r}_{\rm b} = E[x(n-m)\mathbf{x}_m(n)]$.

Since $x(n)$ is stationary, the correlation matrix $\mathbf{R}$ in Eq. (11.13) is the same matrix as that in Eq. (11.5). However, the vector $\mathbf{r}_{\rm b}$ on the right-hand side of Eq. (11.13) is different from the vector $\mathbf{r}$ in Eq. (11.5). Using the definition (11.7), we obtain

$$\mathbf{r}_{\rm b} = \begin{bmatrix} r(m) \\ r(m-1) \\ \vdots \\ r(1) \end{bmatrix} \tag{11.14}$$

Comparing Eqs. (11.14) and (11.8), we note that $\mathbf{r}_{\rm b}$ is same as the vector $\mathbf{r}$ with its elements arranged in the reverse order.

When the tap weights of the backward predictor are optimized according to Eq. (11.13),

$$\begin{aligned} P_m^{\rm b} &= E[x^2(n-m)] - \mathbf{r}_{\rm b}^{\rm T}\mathbf{g}_m \\ &= E[x^2(n-m)] - \mathbf{r}_{\rm b}^{\rm T}\mathbf{R}^{-1}\mathbf{r}_{\rm b} \end{aligned} \tag{11.15}$$

11.3 Relationship Between Forward and Backward Predictors

We now show that there is a close relationship between the tap-weight vectors of the forward and backward linear predictors of a process $x(n)$. To see this, we substitute Eqs. (11.8) and (11.9) in Eq. (11.5) and write the result in scalar form as

$$\sum_{i=1}^{m} r(i-j)a_{m,i} = r(j), \quad \text{for } j = 1, 2, \ldots, m \tag{11.16}$$

where we have used the property $r(i-j) = r(j-i)$. Also, substitution of Eqs. (11.8) and (11.14) in to Eq. (11.13) gives

$$\sum_{i=1}^{m} r(i-j)g_{m,i} = r(m+1-j), \quad \text{for } j = 1, 2, \ldots, m \tag{11.17}$$

Next, we let $i = m+1-k$ and $j = m+1-l$ in Eq. (11.17) and use $r(k-l) = r(l-k)$, to obtain

$$\sum_{k=1}^{m} r(k-l)g_{m,m+1-k} = r(l), \quad \text{for } l = 1, 2, \ldots, m \tag{11.18}$$

Replacing k and l in Eq. (11.18) by i and j, respectively, and comparing the result with Eq. (11.16), we get

$$a_{m,i} = g_{m,m+1-i}, \quad \text{for } i = 1, 2, \ldots, m \tag{11.19}$$

or

$$g_{m,i} = a_{m,m+1-i}, \quad \text{for } i = 1, 2, \ldots, m \tag{11.20}$$

This result shows that *the optimum tap weights of the mth-order forward predictor of a wide sense stationary process $x(n)$ are the same as the optimum tap weights of the corresponding backward predictor, but in the reverse order*. Thus, we may write

$$f_m(n) = x(n) - \sum_{i=1}^{m} a_{m,i}x(n-i) \tag{11.21}$$

and

$$b_m(n) = x(n-m) - \sum_{i=1}^{m} a_{m,m+1-i}x(n-i+1) \tag{11.22}$$

11.4 Prediction-Error Filters

The forward predictor of Figure 11.1 uses an m-tap transversal filter to get an estimate of the present sample $x(n)$ of a sequence based on its past m samples $x(n-1), x(n-2), \ldots, x(n-m)$. The mth-order forward prediction-error filter for a sequence $x(n)$ is defined as the filter whose input is $x(n)$ and the forward prediction error $f_m(n)$ is its output. Figure 11.3 depicts a block schematic diagram showing how forward predictor and forward prediction-error filter are related.

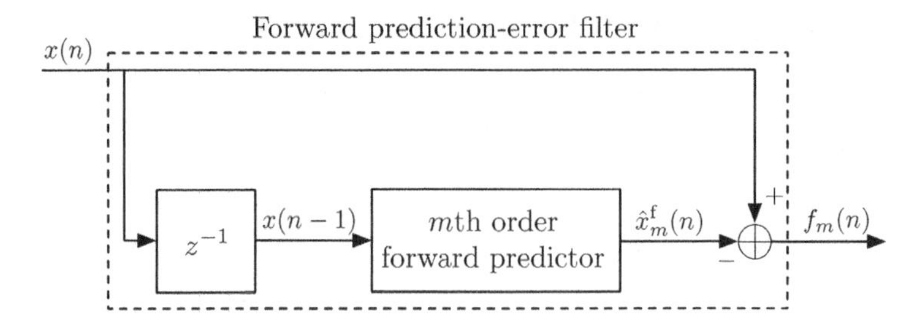

Figure 11.3 Block schematic diagram showing the relationship between forward predictor and forward prediction-error filter.

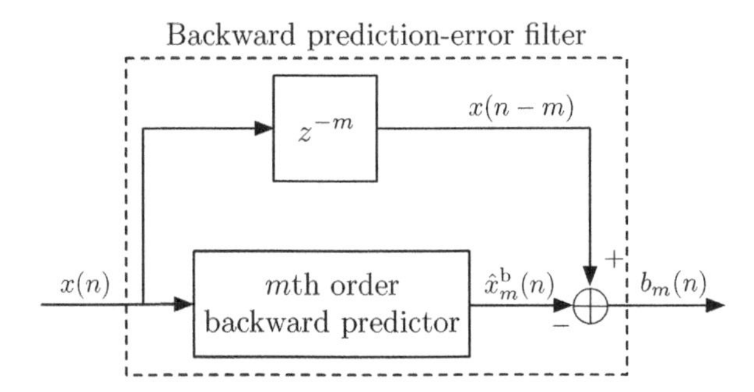

Figure 11.4 Block schematic diagram showing the relationship between backward predictor and backward prediction-error filter.

Similarly, the mth-order backward prediction-error filter of a sequence $x(n)$ is the one whose input is $x(n)$ and its output is the backward prediction error $b_m(n)$. Figure 11.4 depicts a block schematic diagram showing how backward predictor and backward prediction-error filter are related.

11.5 Properties of Prediction Errors

The forward and backward prediction errors possess certain properties that are fundamental to the development of lattice structures. These properties are reviewed in this section.

Property 1: *For any sequence $x(n)$, the forward and backward prediction errors of the same order have the same power. In other words,*

$$P_m^{\mathrm{b}} = P_m^{\mathrm{f}} \tag{11.23}$$

To show this, we note from Eq. (11.15),

$$P_m^{\mathrm{b}} = E[x^2(n-m)] - \sum_{i=1}^{m} g_{m,i} r(m+1-i) \tag{11.24}$$

Substituting Eq. (11.20) in Eq. (11.24) and noting that $E[x^2(n-m)] = E[x^2(n)]$ as $x(n)$ is stationary, we obtain

$$P_m^{\mathrm{b}} = E[x^2(n)] - \sum_{i=1}^{m} a_{m,m+1-i} r(m+1-i) \tag{11.25}$$

The proof of Eq. (11.23) is now complete since the right-hand side of Eq. (11.25) with $j = m+1-i$ is same as P_m^{f} given by Eq. (11.6).

This result shows that, *for a random process $x(n)$, the forward and backward predictors achieve the same level of minimum mean-squared error, when their tap weights are optimized.* Noting this, we drop the superscripts f and b from P_m^{f} and P_m^{b}, respectively, in the rest of this chapter.

Property 2: *For any sequence $x(n)$ and its mth-order forward prediction error $f_m(n)$*

$$E[f_m(n)x(n-k)] = 0, \quad \text{for } k = 1, 2, \ldots, m \tag{11.26}$$

This is easily proved by applying the principle of orthogonality to the forward predictor of Figure 11.1. Namely, the output error $f_m(n)$ is uncorrelated (orthogonal) with the samples $x(n-1), x(n-2), \ldots, x(n-m)$, at the filter (predictor) input.

Property 3: *For any sequence $x(n)$ and its mth-order backward prediction error $b_m(n)$*

$$E[b_m(n)x(n-k)] = 0, \quad \text{for } k = 0, 1, \ldots, m-1 \tag{11.27}$$

This is also proved by applying the principle of orthogonality to the backward predictor of Figure 11.1. Namely, the output error $b_m(n)$ is uncorrelated (orthogonal) with the samples $x(n), x(n-1), \ldots, x(n-m+1)$, at the filter (predictor) input.

Property 4: *The backward prediction errors $b_0(n), b_1(n), \ldots$ of a sequence $x(n)$ are always uncorrelated with one another. In other words, for any $k \neq l$,*

$$E[b_k(n)b_l(n)] = 0 \tag{11.28}$$

To show this, with no loss of generality, we assume that $k < l$ and substitute for $b_k(n)$ from Eq. (11.22). This gives

$$\begin{aligned} E[b_k(n)b_l(n)] &= E\left[\left(x(n-k) - \sum_{i=1}^{k} a_{k,k+1-i} x(n-i+1)\right) b_l(n)\right] \\ &= E[x(n-k)b_l(n)] - \sum_{i=0}^{k-1} a_{k,k-i} E[x(n-i))b_l(n)] \end{aligned} \tag{11.29}$$

Using Property 3 and noting that $k < l$, one finds that all the expectations on the right-hand side of Eq. (11.29) are zero. This completes the proof.

11.6 Derivation of Lattice Structure

In this section, we present a derivation of lattice structure for prediction-error filters. A distinct feature of lattice structure, as we will show in this section, is that it is a direct implementation of the *order-update equations* for computing mth-order forward and backward prediction errors from the forward and backward prediction errors of order $m-1$. This is not possible in the transversal structure case. To derive these order-update equations and thereby the structure of lattice filters, we start with the forward prediction error for an $(m+1)$th-order predictor:

$$f_{m+1}(n) = x(n) - \sum_{i=1}^{m+1} a_{m+1,i}x(n-i) \tag{11.30}$$

The summation on the right-hand side of Eq. (11.30) can be rearranged as

$$\sum_{i=1}^{m+1} a_{m+1,i}x(n-i) = \sum_{i=1}^{m} a_{m+1,i}x(n-i) + a_{m+1,m+1}x(n-m-1) \tag{11.31}$$

From Eq. (11.22), we get

$$x(n-m-1) = b_m(n-1) + \sum_{i=1}^{m} a_{m,m+1-i}x(n-i) \tag{11.32}$$

Substituting Eq. (11.32) in Eq. (11.31), we obtain

$$\begin{aligned}\sum_{i=1}^{m+1} a_{m+1,i}x(n-i) &= \sum_{i=1}^{m}(a_{m+1,i} + a_{m+1,m+1}a_{m,m+1-i})x(n-i) \\ &\quad + a_{m+1,m+1}b_m(n-1) \\ &= \sum_{i=1}^{m} a'_{m,i}x(n-i) + \kappa_{m+1}b_m(n-1)\end{aligned} \tag{11.33}$$

where

$$\kappa_{m+1} = a_{m+1,m+1} \tag{11.34}$$

and

$$a'_{m,i} = a_{m+1,i} + \kappa_{m+1}a_{m,m+1-i}, \quad \text{for } i = 1, 2, \ldots, m \tag{11.35}$$

The above development shows that any linear combination of the input samples $x(n-1), x(n-2), \ldots, x(n-m-1)$ can also be obtained as a linear combination of $x(n-1), x(n-2), \ldots, x(n-m)$ and $b_m(n-1)$. We also note that the summation on the left-hand side of Eq. (11.33) is the $(m+1)$th-order forward prediction of $x(n)$, that is, $\hat{x}^{\mathrm{f}}_{m+1}(n)$. We may thus argue that the estimate $\hat{x}^{\mathrm{f}}_{m+1}(n)$ can also be obtained as a linear combination of the past m samples of $x(n)$ and the backward prediction error $b_m(n-1)$, that is,

$$\hat{x}^{\mathrm{f}}_{m+1}(n) = \sum_{i=1}^{m} a'_{m,i}x(n-i) + \kappa_{m+1}b_m(n-1) \tag{11.36}$$

Then, the coefficients $a'_{m,i}$'s and κ_{m+1} can be obtained directly by minimizing the mean-squared of the estimation error

$$\begin{aligned} f_{m+1}(n) &= x(n) - \hat{x}^{\mathrm{f}}_{m+1}(n) \\ &= x(n) - \sum_{i=1}^{m} a'_{m,i}x(n-i) - \kappa_{m+1}b_m(n-1) \end{aligned} \tag{11.37}$$

To proceed, we define the vectors

$$\mathbf{z}(n) = [\mathbf{x}^{\mathrm{T}}_m(n-1)\ b_m(n-1)]^{\mathrm{T}} \tag{11.38}$$

and

$$\mathbf{w}_z = [\mathrm{a}'^{\mathrm{T}}_m \quad \kappa_{m+1}]^{\mathrm{T}} \tag{11.39}$$

where

$$\mathrm{a}'_m = [a'_{m,1}\ a'_{m,2} \cdots a'_{m,m}]^{\mathrm{T}} \tag{11.40}$$

and $\mathbf{x}_m(n-1)$ is as defined in Section 11.1. Using these definitions, Eq. (11.37) can be written as

$$f_{m+1}(n) = x(n) - \mathbf{w}^{\mathrm{T}}_z\mathbf{z}(n) \tag{11.41}$$

Then, the tap-weight vector $\mathbf{w}_z$ which minimizes $f_{m+1}(n)$ in the mean-square sense can be obtained from the corresponding Wiener–Hopf equation

$$\mathbf{R}_{zz}\mathbf{w}_z = \mathbf{p}_{xz} \tag{11.42}$$

where

$$\mathbf{R}_{zz} = E[\mathbf{z}(n)\mathbf{z}^{\mathrm{T}}(n)] \tag{11.43}$$

is the correlation matrix of the observation vector, $\mathbf{z}(n)$, and

$$\mathbf{p}_{xz} = E[x(n)\mathbf{z}(n)] \tag{11.44}$$

is the cross-correlation between $\mathbf{z}(n)$ and the desired output, $x(n)$.

Substituting Eq. (11.38) in Eq. (11.43) and using the definition (11.7) and Property 3 of prediction errors, that is, Eq. (11.27), we obtain

$$\mathbf{R}_{zz} = \begin{bmatrix} r(0) & r(1) & \cdots & r(m-1) & 0 \\ r(1) & r(0) & \cdots & r(m-2) & 0 \\ \vdots & \vdots & \ddots & \vdots & \\ r(m-1) & r(m-2) & \cdots & r(0) & 0 \\ 0 & 0 & \cdots & 0 & E[b^2_m(n-1)] \end{bmatrix} \tag{11.45}$$

We note that the m-by-m portion of the upper-left part of $\mathbf{R}_{zz}$ is nothing but the correlation matrix $\mathbf{R}$ of Eq. (11.8). Thus, we may write

$$\mathbf{R}_{zz} = \begin{bmatrix} \mathbf{R} & \mathbf{0}_m \\ \mathbf{0}^T_m & E[b^2_m(n-1)] \end{bmatrix} \tag{11.46}$$

where $\mathbf{0}_m$ denotes the length m zero column vector. Similarly, it is straightforward to show that

$$\mathbf{p}_{xz} = \begin{bmatrix} \mathbf{r} \\ E[x(n)b_m(n-1)] \end{bmatrix} \tag{11.47}$$

where $\mathbf{r}$ is the column vector defined in Eq. (11.9).

Substituting Eqs. (11.39), (11.46), and (11.47) in Eq. (11.42) and solving for κ_{m+1} and $\mathbf{a}'_m$, we obtain

$$\kappa_{m+1} = \frac{E[x(n)b_m(n-1)]}{E[b_m^2(n-1)]} \tag{11.48}$$

and

$$\mathbf{a}'_m = \mathbf{R}^{-1}\mathbf{r} \tag{11.49}$$

Comparing Eq. (11.49) with Eq. (11.5), one finds that

$$\mathbf{a}'_m = \mathbf{a}_m \tag{11.50}$$

Substituting this result in Eq. (11.37), we get

$$\begin{aligned} f_{m+1}(n) &= x(n) - \mathbf{a}_m^{\mathrm{T}}\mathbf{x}_m(n-1) - \kappa_{m+1}b_m(n-1) \\ &= f_m(n) - \kappa_{m+1}b_m(n-1) \end{aligned} \tag{11.51}$$

where use is made of Eq. (11.21). Thus, the $(m+1)$th-order forward prediction error can be obtained from the mth-order forward and backward prediction errors.

Following a similar procedure as above, a similar recursion for the backward prediction error can be derived. It is given by (Problem P11.1)

$$b_{m+1}(n) = b_m(n-1) - \kappa'_{m+1}f_m(n) \tag{11.52}$$

where

$$\kappa'_{m+1} = \frac{E[x(n-m-1)f_m(n)]}{E[f_m^2(n)]} \tag{11.53}$$

We now show that the two quantities given for κ_{m+1} in Eq. (11.48) and κ'_{m+1} in Eq. (11.53) are the same. Consider Eq. (11.48). Using Eq. (11.21), we can write

$$\begin{aligned} E[x(n)b_m(n-1)] &= E\left[\left(f_m(n) + \sum_{i=1}^{m} a_{m,i}x(n-i)\right)b_m(n-1)\right] \\ &= E[(f_m(n)b_m(n-1)] \\ &\quad + \sum_{i=1}^{m} a_{m,i}E[x(n-i)b_m(n-1)] \end{aligned} \tag{11.54}$$

We note from Eq. (11.27), with n replaced by $n-1$, that all the expectations under the summation on the right-hand side of Eq. (11.54) are 0. Thus, we obtain

$$E[x(n)b_m(n-1)] = E[f_m(n)b_m(n-1)] \tag{11.55}$$

Similarly, one can easily show that

$$E[x(n-m-1)f_m(n)] = E[f_m(n)b_m(n-1)] \tag{11.56}$$

The results in Eqs. (11.55) and (11.56) show that the numerators of the two expressions on the right-hand sides of Eqs. (11.48) and (11.53) are the same. The denominators of these expressions are also the same, as $E[f_m^2(n)] = P_m^{\rm f}$, $E[b_m^2(n-1)] = E[b_m^2(n)] = P_m^{\rm b}$, and according to Property 1 of prediction errors, $P_m^{\rm f} = P_m^{\rm b}$. Thus, we have established that the quantities κ_{m+1} in Eq. (11.48) and κ'_{m+1} in Eq. (11.53) are the same. This, of course, is true only when the predictors coefficients are optimum.

We may also write

$$\kappa_{m+1} = \frac{E[f_m(n)b_m(n-1)]}{\sqrt{E[f_m^2(n)]E[b_m^2(n-1)]}} \tag{11.57}$$

Thus, κ_{m+1} is the normalized correlation between the forward and backward errors $f_m(n)$ and $b_m(n-1)$. In fact, κ_{m+1} is known as *the partial correlation* (PARCOR) coefficient as it represents the correlation that remains between the forward and backward prediction errors. Using the Cauchy–Schwartz inequality[1] it is straightforward to show that the following inequality is always true:

$$|\kappa_{m+1}| \le 1 \tag{11.58}$$

We may also write

$$\kappa_{m+1} = \frac{E[f_m(n)b_m(n-1)]}{P_m} \tag{11.59}$$

since $E[f_m^2(n)] = E[b_m^2(n-1)] = P_m$, according to Property 1 of prediction errors.

Summarizing the above derived order-update equations for prediction errors, we have

$$f_{m+1}(n) = f_m(n) - \kappa_{m+1}b_m(n-1) \tag{11.60}$$

$$b_{m+1}(n) = b_m(n-1) - \kappa_{m+1}f_m(n) \tag{11.61}$$

where $m = 0, 1, 2, \ldots$, and κ_{m+1} is given by Eq. (11.57) or (11.59). Since $x(n)$ may be considered as the zeroth order forward or backward prediction errors, the initialization for the above recursions is given by

$$f_0(n) = b_0(n) = x(n) \tag{11.62}$$

The structure that implements the above recursions is called the lattice filter/predictor.

Figure 11.5a shows the lattice structure of an M-stage forward/backward predictor. Each stage has two inputs. These are the forward and backward prediction errors from the previous stage. The outputs of each stage are the forward and backward prediction errors of one order higher. These are calculated according to the order-update Eqs. (11.60)

[1] The Cauchy–Schwartz inequality states that for any set of numbers $\{a_i$ and b_i, for $i = 1, 2, \ldots, L\}$,

$$\left|\sum_{i=1}^{L} a_i b_i\right| \le \sqrt{\left(\sum_{i=1}^{L}|a_i|^2\right)\left(\sum_{i=1}^{L}|b_i|^2\right)}$$

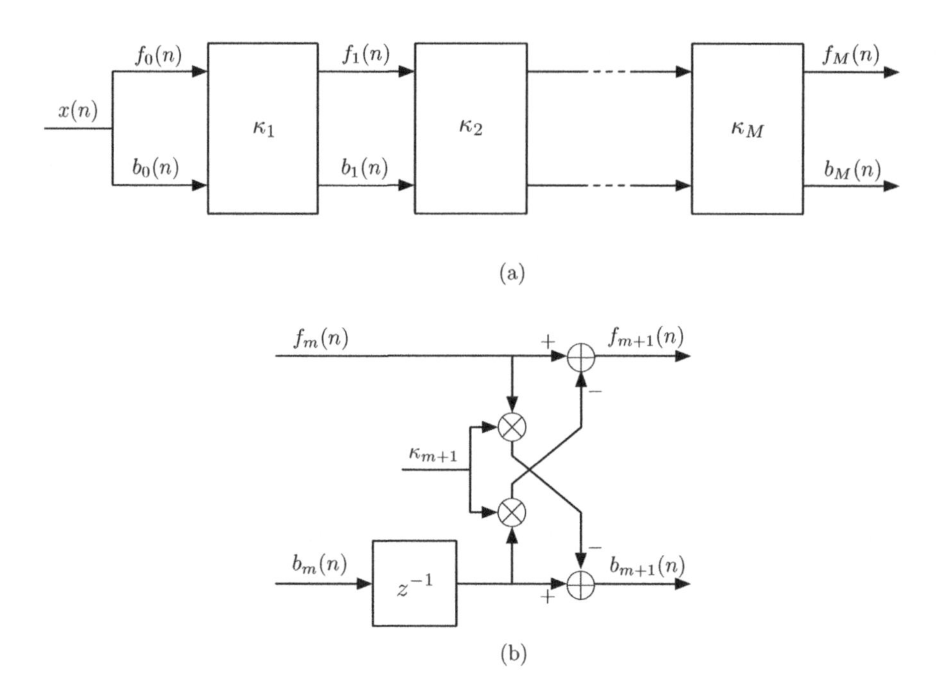

Figure 11.5 Lattice predictor: (a) overall structure and (b) details of Stage m.

and (11.61). The two inputs to Stage 1 are common and are equal to the predictor input $x(n)$. Figure 11.5b depicts the details of the mth stage of the lattice predictor. It follows from the order-update equations (11.60) and (11.61).

A special feature of the lattice predictor is that to obtain the Mth-order prediction errors, all prediction errors (forward and backward) of lower orders are also calculated. In other words, the Mth-order lattice predictor is a structure with a single input $x(n)$ and $2M + 2$ outputs $f_0(n)$, $b_0(n)$, $f_1(n)$, $b_1(n)$, …, $f_M(n)$ and $b_M(n)$. How many of these outputs will be used is application dependent. For example, in an Mth-order forward predictor where the final goal is $f_M(n)$, the rest of the prediction errors (with the exception of $b_M(n)$ which, in this case, can be dropped from the structure) are required only as intermediate signal sequences.

A useful relationship which we shall establish before ending this section is an order-update equation for the mean-squared value of the prediction errors. We note that

$$\begin{aligned}
P_{m+1} &= E[f_{m+1}^2(n)] \\
&= E\left[f_{m+1}(n)\left(x(n) - \sum_{i=1}^{m+1} a_{m+1,i}x(n-i)\right)\right] \\
&= E[f_{m+1}(n)x(n)] - \sum_{i=1}^{m+1} a_{m+1,i}E[f_{m+1}(n)x(n-i)]
\end{aligned} \tag{11.63}$$

But from Property 2 of prediction errors we know that all the expectations under the summation on the right-hand side of Eq. (11.63) are zero. Thus we obtain

$$P_{m+1} = E[f_{m+1}(n)x(n)] \tag{11.64}$$

Substituting for $f_{m+1}(n)$ and $x(n)$ in Eq. (11.64) from Eqs. (11.60) and (11.21), respectively, we get

$$\begin{aligned} P_{m+1} &= E\left[(f_m(n) - \kappa_{m+1} b_m(n-1))\left(f_m(n) + \sum_{i=1}^{m} a_{m,i} x(n-i)\right)\right] \\ &= E[f_m^2(n)] - \kappa_{m+1} E[f_m(n) b_m(n-1)] \end{aligned} \tag{11.65}$$

since $E[f_m(n)x(n-i)] = E[b_m(n-1)x(n-i)] = 0$, for $i = 1, 2, \ldots, m$ by Properties 2 and 3 of prediction errors. Substituting for $E[f_m(n)b_m(n-1)]$ from Eq. (11.59) and noting that $E[f_m^2(n)] = P_m$, we get

$$P_{m+1} = (1 - \kappa_{m+1}^2) P_m \tag{11.66}$$

This result shows that the mean-squared value of the prediction error decreases as the order of predictor increases. This, of course, is intuitively understandable. The contribution of each stage in reducing the prediction error is determined by its PARCOR coefficient, according to Eq. (11.66). A PARCOR coefficient with close to one magnitude reduces prediction error significantly. On the other hand, a PARCOR coefficient with small magnitude has little effect in improving (reducing) prediction error. Intuitively, we expect the prediction error to decrease rapidly for the first few stages and slowly for the later stages. This is equivalent to saying that the PARCOR coefficients are likely to be relatively larger (in magnitude) for the first few stages and drop to some values close to 0 at later stages.

11.7 Lattice as an Orthogonalization Transform

An important feature of the lattice predictor structure of Figure 11.5a, in the context of adaptive filters, is that it may be viewed as an orthogonalization transform. Furthermore, as we shall see later, the PARCOR coefficients which are central to the lattice structure can be obtained adaptively. So, the lattice predictor structure may be used for adaptive implementation of orthogonalization in the transform domain adaptive filters discussed in Chapter 7.

Before looking at the adaptive techniques for implementation of such orthogonalization, let us assume that the optimum PARCOR coefficients of the lattice structure are known and the corresponding prediction errors can be calculated. We also define the column vector

$$\mathbf{b}(n) = [b_0(n)\ b_1(n) \cdots b_{N-1}(n)]^{\mathrm{T}} \tag{11.67}$$

whose elements are the backward prediction errors of orders 0 to $N-1$. The vector $\mathbf{b}(n)$ may be obtained through the lattice predictor of Figure 11.5a, with $M = N-1$, or, equivalently, according to the equation

$$\mathbf{b}(n) = \mathbf{L}\mathbf{x}(n) \tag{11.68}$$

where

$$\mathbf{x}(n) = [x(n)\ x(n-1) \cdots x(n-N+1)]^{\mathrm{T}} \tag{11.69}$$

and

$$\mathbf{L} = \begin{bmatrix} 1 & 0 & 0 & \cdots & 0 & 0 \\ -a_{1,1} & 1 & 0 & \cdots & 0 & 0 \\ -a_{2,2} & -a_{2,1} & 1 & \cdots & 0 & 0 \\ \vdots & \vdots & \vdots & \ddots & \vdots & \vdots \\ -a_{N-1,N-1} & -a_{N-1,N-2} & -a_{N-1,N-3} & \cdots & -a_{N-1,1} & 1 \end{bmatrix} \tag{11.70}$$

with $a_{i,j}$ denoting the jth coefficient of the ith-order forward predictor. We note that the matrix $\mathbf{L}$ is invertible, since $\det(\mathbf{L}) \neq 0$ (Problem P11.3). Hence, Eq. (11.68) can also be written as

$$\mathbf{x}(n) = \mathbf{L}^{-1}\mathbf{b}(n) \tag{11.71}$$

Figure 11.6 depicts a block schematic obtained from Eq. (11.68). It consists of N prediction-error filters of orders 0 to $N-1$, in parallel. Compared to the lattice structure, this suggests a more direct way of converting (transforming) the input vector $\mathbf{x}(n)$ to the backward prediction errors $b_0(n), b_1(n), \ldots, b_{N-1}(n)$. It also resembles the idea of transformation in the context of the transform domain adaptive filters discussed in Chapter 7. However, it requires more computations compared to the lattice predictor. The lattice predictor requires only $2N$ multiplications and $2N$additions/subtractions for each updating all the backward prediction errors once, while the computational complexity of the direct implementation presented in Figure 11.6 is about $N^2/2$ multiplications and similar number of additions/subtractions for every input sample. However, in the following discussions we use Eq. (11.68) as an expression for the vector $\mathbf{b}(n)$, because it will help in developing certain theoretical results. In actual implementation, of course, one can use the corresponding lattice structure.

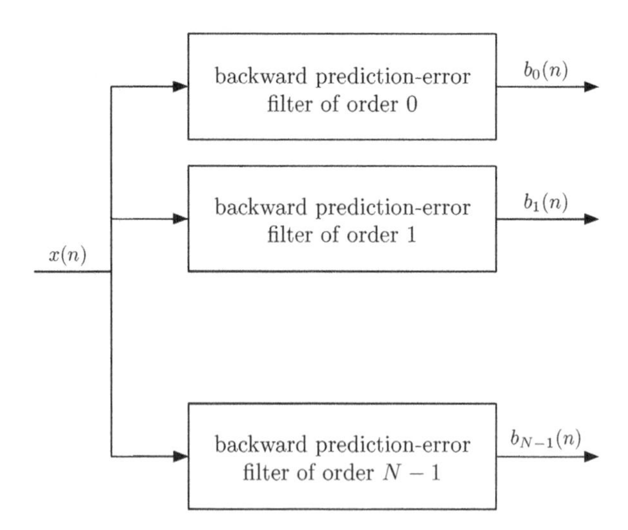

Figure 11.6 Block schematic diagram for a direct implementation of Eq. (11.68).

11.8 Lattice Joint Process Estimator

In the previous sections, our discussion on lattice structure was limited to its use as a prediction-error filter. In this section, we show how a general transversal filter, which is used to estimate a desired sequence $d(n)$ from another related sequence $x(n)$, can be implemented using lattice structure.

Consider a transversal filter with tap-input vector $\mathbf{x}(n)$, as in Eq. (11.69), tap-weight vector

$$\mathbf{w} = [w_0\ w_1 \cdots w_{N-1}]^{\mathrm{T}} \tag{11.72}$$

output

$$y(n) = \mathbf{w}^{\mathrm{T}}\mathbf{x}(n) \tag{11.73}$$

and desired output $d(n)$. Substituting Eq. (11.71) in Eq. (11.73), we obtain

$$y(n) = \mathbf{w}^{\mathrm{T}}\mathbf{L}^{-1}\mathbf{b}(n) \tag{11.74}$$

We define the column vector

$$\mathbf{c} = \mathbf{L}^{-\mathrm{T}}\mathbf{w} \tag{11.75}$$

where $\mathbf{L}^{-\mathrm{T}}$ is shorthand notation for $(\mathbf{L}^{\mathrm{T}})^{-1}$ or $(\mathbf{L}^{-1})^{\mathrm{T}}$ (note that $(\mathbf{L}^{\mathrm{T}})^{-1} = (\mathbf{L}^{-1})^{\mathrm{T}}$). Then, Eq. (11.74) simplifies as

$$y(n) = \mathbf{c}^{\mathrm{T}}\mathbf{b}(n) \tag{11.76}$$

This result show that the output $y(n)$ of the transversal filter can equivalently be obtained as a linear combination of the backward prediction errors. This also suggests an alternative structure for implementation of the system function of a transversal filter. Figure 11.7 depicts such an implementation. This is referred to as *lattice joint process estimator*.

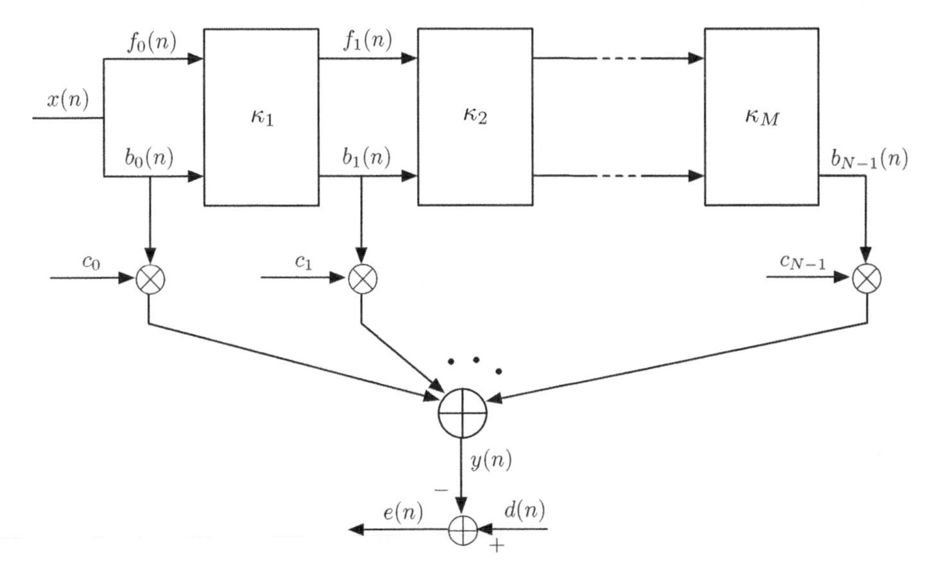

Figure 11.7 Lattice joint process estimator.

It consists of two distinct parts: the lattice predictor part and the linear combiner part. The lattice predictor part is used to convert the samples of input signal to backward prediction errors. The linear combiner part uses the backward prediction errors to obtain the filter output according to Eq. (11.76).

We note that once the backward predictor coefficients are known, the coefficients of the linear combiner part of the lattice joint process estimator are uniquely determined through Eq. (11.75), provided the tap-weight vector $\mathbf{w}$ of the corresponding transversal filter is known. Furthermore, the existence of $\mathbf{c}$ is guaranteed as $\mathbf{L}$ is invertible (Problem P11.3). The optimum values of the PARCOR coefficients in the lattice part of Figure 11.7 are determined from the statistics of the input sequence, $x(n)$.

11.9 System Functions

In this section, we present a system function view of lattice structure. This will be useful for our analyses in the following sections. We define $H_{f_m}(z)$ and $H_{b_m}(z)$ as the transfer functions relating the input sequence $x(n)$ and the mth-order forward and backward prediction errors $f_m(n)$ and $b_m(n)$, respectively. Then, it follows from Eqs. (11.60) and (11.61) that

$$H_{f_{m+1}}(z) = H_{f_m}(z) - \kappa_{m+1} z^{-1} H_{b_m}(z) \tag{11.77}$$

and

$$H_{b_{m+1}}(z) = z^{-1} H_{b_m}(z) - \kappa_{m+1} H_{f_m}(z) \tag{11.78}$$

These are order-update equations which may be used to obtain the system functions of forward and backward prediction-error filters of any order in terms of the system functions of one order lower prediction-error filters. The initial conditions to start these order-update equations are $H_{f_0}(z) = H_{b_0}(z) = 1$.

We also note that the system functions $H_{f_m}(z)$ and $H_{b_m}(z)$ may directly be realized using the expressions

$$H_{f_m}(z) = 1 - \sum_{i=1}^{m} a_{m,i} z^{-i} \tag{11.79}$$

and

$$H_{b_m}(z) = z^{-m} - \sum_{i=1}^{m} a_{m,m+1-i} z^{-i+1} \tag{11.80}$$

which follow directly from Eqs. (11.21) and (11.22), respectively. Also, for our later reference, we note that $H_{f_m}(z)$ and $H_{b_m}(z)$ are related according to the equation

$$H_{b_m}(z) = z^{-m} H_{f_m}(z^{-1}) \tag{11.81}$$

11.10 Conversions

From the results in the previous sections, and in particular the system functions presentation in the last section, one may conclude that there is a close relationship between the PARCOR coefficients, κ_m's, and the transversal predictor coefficients, $a_{m,i}$'s. In this section, we present procedures for conversion between these two sets of coefficients.

11.10.1 Conversion Between Lattice and Transversal Predictors

Given the PARCOR coefficients of a lattice predictor, the coefficients of the corresponding transversal structure can be calculated. This follows from the order-update equations (11.60) and (11.61) or (11.77) and (11.78). It is done by starting with the initial condition (system functions) $H_{f_0}(z) = H_{b_0}(z) = 1$ and iterating Eqs. (11.77) and (11.78) until the required order is reached. In particular, substituting Eqs. (11.79) and (11.80) in Eq. (11.77), we get

$$1 - \sum_{i=1}^{m+1} a_{m+1,i} z^{-i} = 1 - \sum_{i=1}^{m} a_{m,i} z^{-i} - \kappa_{m+1} z^{-1} \left(z^{-m} - \sum_{i=1}^{m} a_{m,m+1-i} z^{-i+1} \right) \quad (11.82)$$

Rearranging this, we obtain

$$\sum_{i=1}^{m+1} a_{m+1,i} z^{-i} = \sum_{i=1}^{m} (a_{m,i} - \kappa_{m+1} a_{m,m+1-i}) z^{-i} + \kappa_{m+1} z^{-m-1} \quad (11.83)$$

Equating the coefficients of similar powers of z on both sides of Eq. (11.83), we get

$$a_{m+1,i} = a_{m,i} - \kappa_{m+1} a_{m,m+1-i}, \quad \text{for } i = 1, 2, \ldots, m \quad (11.84)$$

and

$$a_{m+1,m+1} = \kappa_{m+1} \quad (11.85)$$

In order to obtain the coefficient of an Mth-order transversal predictor, we shall start with the initial condition $a_{0,0} = 1$ (equivalent to $H_{f_0}(z) = H_{b_0}(z) = 1$) and iterate Eqs. (11.84) and (11.85) M times. Table 11.1 summarizes this procedure.

Next, we derive a procedure for calculating the PARCOR coefficients $\kappa_1, \kappa_2, \ldots, \kappa_M$ from the coefficients $a_{M,1}, a_{M,2}, \ldots, a_{M,M}$ of an Mth-order transversal predictor. Consider Eq. (11.84) for a particular value of i and also when i is replaced by $m + 1 - i$. We get the following pair of simultaneous equations:

$$\begin{aligned} a_{m,i} - \kappa_{m+1} a_{m,m+1-i} &= a_{m+1,i} \\ a_{m,m+1-i} - \kappa_{m+1} a_{m,i} &= a_{m+1,m+1-i} \end{aligned} \quad (11.86)$$

Table 11.1 Conversion from lattice to transversal predictor.

Given: $\kappa_1, \kappa_2, \ldots, \kappa_M$
Required: $a_{M,1}, a_{M,2}, \ldots, a_{M,M}$

$a_{1,1} = \kappa_1$
for $m = 1$ to $M - 1$

$$a_{m+1,i} = a_{m,i} - \kappa_{m+1} a_{m,m+1-i}, \quad \text{for } i = 1, 2, \ldots, m$$
$$a_{m+1,m+1} = \kappa_{m+1}$$

end

Table 11.2 Conversion from transversal to lattice predictor (inverse Levinson–Durbin algorithm).

Given: $a_{M,1}, a_{M,2}, \ldots, a_{M,M}$
Required: $\kappa_1, \kappa_2, \ldots, \kappa_M$

$\kappa_M = a_{M,M}$
for $m = M-1,\ M-2, \ldots, 1$
$\quad a_{m,i} = \frac{a_{m+1,i}+\kappa_{m+1}a_{m+1,m+1-i}}{1-\kappa_{m+1}^2}, \quad \text{for } i = 1, 2, \ldots, m$
$\quad \kappa_m = a_{m,m}$
end

Solving these for $a_{m,i}$, we get

$$a_{m,i} = \frac{a_{m+1,i} + \kappa_{m+1}a_{m+1,m+1-i}}{1-\kappa_{m+1}^2}, \quad \text{for } i = 1, 2, \ldots, m \tag{11.87}$$

This with Eq. (11.85) suggest the procedure presented in Table 11.2 for calculating the PARCOR coefficients from the coefficients of the corresponding transversal predictor. This procedure is known as *inverse Levinson–Durbin algorithm*. It may be noted that although we are only interested in the PARCOR coefficients, the coefficients of the transversal filters of orders $m = 1, 2, \ldots, M-1$ are also obtained as intermediate results. Thus, given the coefficients of an Mth-order transversal predictor, the coefficients of the lower order transversal predictors are obtained by following the procedure provided in Table 11.2. In other words, given the last row of the matrix $\mathbf{L}$ of Eq. (11.70), for $M = N-1$, one can build the whole matrix $\mathbf{L}$ by following the inverse Levinson–Durbin algorithm.

11.10.2 Levinson–Durbin Algorithm

From Eq. (11.5), we note that the coefficients $a_{m,i}$'s of a transversal predictor are directly related to the autocorrelation function of its input. The well-known Levinson–Durbin algorithm is a computationally efficient procedure for solving the Wiener–Hopf equation (11.5) of the transversal predictor. It also provides the PARCOR coefficients of the corresponding lattice predictor. The efficiency is achieved by exploiting the fact that the input $x(n)$ is a stationary process. This will be clarified further at the end of this section. With the background that we have already developed, derivation of the Levinson–Durbin algorithm is straightforward. We note from Eq. (11.59)

$$\begin{aligned}\kappa_{m+1}P_m &= E[f_m(n)b_m(n-1)] \\ &= E\left[f_m(n)\left(x(n-m-1) - \sum_{i=1}^{m} a_{m,m+1-i}x(n-i)\right)\right] \\ &= E[f_m(n)x(n-m-1)]\end{aligned} \tag{11.88}$$

where the last equality follows from the identity $E[f_m(n)x(n-i)] = 0$, for $i = 1, 2, \ldots, m$. Substituting Eq. (11.21) in Eq. (11.88), we obtain

$$\kappa_{m+1} P_m = E\left[\left(x(n) - \sum_{i=1}^{m} a_{m,i} x(n-i)\right) x(n-m-1)\right]$$
$$= r(m+1) - \sum_{i=1}^{m} a_{m,i} r(m+1-i) \tag{11.89}$$

or

$$\kappa_{m+1} = \frac{r(m+1) - \sum_{i=1}^{m} a_{m,i} r(m+1-i)}{P_m} \tag{11.90}$$

Thus, given the autocorrelation coefficients $r(1), r(2), \ldots, r(m+1)$, the mth-order transversal predictor coefficients $a_{m,i}$'s, and P_m, one can calculate κ_{m+1} according to Eq. (11.90). The order-update equations (11.84) are then used to obtain the coefficients of the $(m+1)$th-order transversal predictor, $a_{m+1,i}$'s. Equation (11.66) is used to obtain P_{m+1} for the next iteration. Table 11.3 summarizes these results. This is called *Levinson–Durbin algorithm*. The most important feature of the Levinson–Durbin algorithm is its computational efficiency. Careful examination of Table 11.3 shows that the implementation of Levinson–Durbin algorithm requires about M^2 multiplications/divisions and the same number of additions/subtractions, where M is the order of predictor. This must be compared with M^3 which is the order of computations required for solving a system of M linear equations without exploiting the structure in the system. The special structure that is exploited here is the symmetric Toeplitz nature of the autocorrelation matrix $\mathbf{R}$. By definition, the autocorrelation matrix of any process (stationary or nonstationary) is symmetric. But, if the process $x(n)$ is stationary (at least wide sense), then the autocorrelation matrix becomes Toeplitz, in addition to being symmetric. That is, all the elements along any given diagonal are the same. For example, the kth subdiagonal and super-diagonal will be constituted by the autocorrelation at lag k.

Table 11.3 Levinson–Durbin algorithm.

Given: $r(0), r(1), \ldots, r(M)$
Required: $a_{M,1}, a_{M,2}, \ldots, a_{M,M}$
and $\kappa_1, \kappa_2, \ldots, \kappa_M$

$P_0 = r(0)$
$\kappa_1 = r(1)/P_0$
$a_{1,1} = \kappa_1$
$P_1 = (1 - \kappa_1^2) P_0$
for $m = 1$ to $M - 1$

$\kappa_{m+1} = \frac{r(m+1) - \sum_{i=1}^{m} a_{m,i} r(m+1-i)}{P_m}$

$a_{m+1,i} = a_{m,i} - \kappa_{m+1} a_{m,m+1-i}$, for $i = 1, 2, \ldots, m$

$a_{m+1,m+1} = \kappa_{m+1}$

$P_{m+1} = (1 - \kappa_{m+1}^2) P_m$

end

In fact, all the results that we have derived in this chapter are under this stationarity assumption on input $x(n)$.

In the above two sections, we have derived procedures for (i) conversion between the coefficients of lattice and transversal predictors and (ii) obtaining the coefficients of the lattice and transversal predictors from the autocorrelation values of the input process. Furthermore, given the power of the input sequence $x(n)$, that is, $P_0 = r(0) = E[x^2(n)]$, and the PARCOR coefficients $\kappa_1, \kappa_2, \ldots, \kappa_M$, one can develop a procedure to obtain the autocorrelation coefficients $r(1), r(2), \ldots, r(M)$ (Problem P11.6). The latter coefficients can also be obtained if P_0 and the coefficients $a_{M,1}, a_{M,2}, \ldots, a_{M,M}$ of an Mth-order transversal predictor are available (Problem P11.6). All these possible conversions show that the three sets of coefficients $(P_0, \kappa_1, \kappa_2, \ldots, \kappa_M)$, $(P_0, a_{M,1}, a_{M,2}, \ldots, a_{M,M})$ and $(r(0), r(1), \ldots, r(M))$ are three different representations of the same information. When M tends to infinity, these may be thought as an alternative representation of the power spectral density of the input process $x(n)$.

11.10.3 Extension of Levinson–Durbin Algorithm

The solution provided by Levinson–Durbin algorithm is only applicable to the case where the Wiener–Hopf equation to be solved corresponds to a predictor. In this section, the Levinson–Durbin algorithm is extended to the case of joint process estimator, as to handle the general case of estimating a signal from another related signal. Consider the Wiener–Hopf equation

$$\mathbf{R}\mathbf{w} = \mathbf{p} \tag{11.91}$$

where $\mathbf{R} = E[\mathbf{x}(n)\mathbf{x}^{\mathrm{T}}(n)]$, $\mathbf{x}(n)$ is the input vector as defined in Eq. (11.69), $\mathbf{p} = E[\mathbf{x}(n)d(n)]$, and $d(n)$ is the desired output of the estimator.

The solution to Eq. (11.91) consists of three steps:

Step 1: The conventional Levinson–Durbin algorithm of Table 11.3 is used to obtain the elements of the matrix $\mathbf{L}$ of Eq. (11.70). The PARCOR coefficients $\kappa_1, \kappa_2, \ldots, \kappa_{N-1}$ of the lattice predictor and the mean-squared values of the prediction errors, that is, $P_0, P_1, \ldots, P_{N-1}$, are also obtained in this process.

Step 2: The Wiener–Hopf equation corresponding to the linear combiner part of the lattice joint process estimator is built and solved. This gives the coefficient vector $\mathbf{c}$.

Step 3: The tap-weight vector $\mathbf{w}$ is obtained according to the equation

$$\mathbf{w} = \mathbf{L}^{\mathrm{T}}\mathbf{c} \tag{11.92}$$

This is obtained by premultiplying Eq. (11.75) on both side by $\mathbf{L}^{\mathrm{T}}$.

The Wiener–Hopf equation for $\mathbf{c}$ is (Figure 11.7)

$$\mathbf{R}_{bb}\mathbf{c} = \mathbf{p}_{db} \tag{11.93}$$

where $\mathbf{R}_{bb} = E[\mathbf{b}(n)\mathbf{b}^{\mathrm{T}}(n)]$ and $\mathbf{p}_{db} = E[d(n)\mathbf{b}(n)]$. Property 4 of prediction errors implies that the correlation matrix $\mathbf{R}_{bb}$ is diagonal. Furthermore, the diagonal elements of $\mathbf{R}_{bb}$ are the mean-squared values of the backward prediction errors $b_0(n), b_1(n), \ldots, b_{N-1}(n)$, that is, $P_0, P_1, \ldots, P_{N-1}$. Hence,

$$\mathbf{R}_{bb} = \mathrm{diag}(P_0, P_1, \ldots, P_{N-1}) \tag{11.94}$$

Also, the mth element of $\mathbf{p}_{db}$ is

$$\begin{aligned} p_{db}(m) &= E[d(n)b_m(n)] \\ &= E\left[d(n)\left(x(n-m) - \sum_{i=0}^{m-1} a_{m,m-i}x(n-i)\right)\right] \\ &= p(m) - \sum_{i=0}^{m-1} a_{m,m-i}\,p(i) \end{aligned} \tag{11.95}$$

where $p(i) = E[d(n)x(n-i)]$ is the ith element of the vector $\mathbf{p}$. Substituting Eqs. (11.94) and (11.95) in Eq. (11.93), we obtain

$$c_m = \begin{cases} \frac{p(0)}{P_0}, & m = 0 \\ \frac{p(m) - \sum_{i=0}^{m-1} a_{m,m-i}\,p(i)}{P_m}, & m = 1, 2, \ldots, N-1 \end{cases} \tag{11.96}$$

Table 11.4 Extended Levinson–Durbin algorithm.

Given: $\mathbf{R}$ and $\mathbf{p}$
Required: $\mathbf{w} = \mathbf{R}^{-1}\mathbf{p}$

$P_0 = r(0)$
$c_0 = \frac{p(0)}{P_0}$
$w_0 = c_0$
$\kappa_1 = r(1)/P_0$
$a_{1,1} = \kappa_1$
$P_1 = (1 - \kappa_1^2)P_0$
$c_1 = \frac{p(1) - a_{1,1}p(0)}{P_1}$
$w_0 = c_0 - a_{1,1}c_1$
$w_1 = c_1$
for $m = 1$ to $N-2$
  $\kappa_{m+1} = \frac{r(m+1) - \sum_{i=1}^{m} a_{m,i}r(m+1-i)}{P_m}$
  $a_{m+1,i} = a_{m,i} - \kappa_{m+1}a_{m,m+1-i}$, for $i = 1, 2, \ldots, m$
  $a_{m+1,m+1} = \kappa_{m+1}$
  $P_{m+1} = (1 - \kappa_{m+1}^2)P_m$
  $c_{m+1} = \frac{p(m+1) - \sum_{i=0}^{m} a_{m+1,m+1-i}\,p(i)}{P_{m+1}}$
  $w_i = w_i - a_{m+1,m+1-i}c_{m+1}$ for $i = 0, 1, \ldots, m$
  $w_{m+1} = c_{m+1}$
end

Table 11.4 summarizes the above results. The recursion for the transversal coefficients w_i follows from Eq. (11.92) (Problem P11.7). Here, the three steps listed above are combined together under a single "for loop." Careful examination of Table 11.4 shows that the computational complexity of the extended Levinson–Durbin algorithm is about $2N^2$ multiplications/divisions and similar number of additions/subtractions. If the joint process estimator is to be implemented in lattice form, then the computations reduce to $1.5N^2$, as Step 3 of the algorithm can then be ignored.

11.11 All-Pole Lattice Structure

The system functions that we have considered so far are all in the form of all-zero filters. In this section, we propose a lattice structure for implementation of an all-pole filter which is characterized by the system function

$$F(z) = \frac{1}{H_{f_M}(z)} = \frac{1}{1 - \sum_{i=1}^{M} a_{M,i} z^{-i}} \tag{11.97}$$

This choice of $F(z)$ is not restrictive except that it should be the system function of a stable system. This is the consequence of an important result of the theory of lattice filters which states a forward prediction-error filter is always minimum phase. This result, proof of which is beyond the scope of our discussion in this book, implies that the zeros of the system function $H_{f_M}(z)$ of any prediction-error filter are all less than one in magnitude. It can also be shown that for any arbitrary minimum phase system function $H_{f_M}(z)$, one can always find a process whose forward prediction-error filter is $H_{f_M}(z)$. Since $H_{f_M}(z)$ is a prediction-error filter, if we excite $F(z) = 1/H_{f_M}(z)$ with the Mth-order forward prediction error, $f_M(n)$, of $x(n)$, the output will be the original process $x(n)$. In other words, the system function which relates $f_M(n)$ and $x(n)$ is $F(z)$. With this in view, we recall the order-update equations (11.60) and (11.61) and rearrange (11.60) as

$$f_m(n) = f_{m+1}(n) + \kappa_{m+1} b_m(n-1) \tag{11.98}$$

Considering Eqs. (11.98) and (11.61), for values of $m = 0, 1, \ldots, M-1$, one can suggest the block diagram of Figure 11.8a. The detail of the mth stage of this block diagram is given in Figure 11.8b. The input sequence in Figure 11.8a is $f_M(n)$ and the generated output is $f_0(n) = x(n)$. We also recall that $b_0(n) = x(n)$. From our discussion above, this observation implies that the block diagram of Figure 11.8a is the lattice realization of $F(z)$.

We shall comment that although in the development of the lattice structure of the all-pole system function $F(z)$ we used $f_M(n)$ as the input, the choice of the input to the latter structure is not limited to $f_M(n)$. It can be any arbitrary input.

11.12 Pole-Zero Lattice Structure

In this section, we extend the all-pole lattice structure of the last section to an arbitrary system function $G(z)$ with M zeros and M poles. With no loss of generality, we let

$$G(z) = \frac{\sum_{i=0}^{M} w_i z^{-i}}{H_{f_M}(z)} \tag{11.99}$$

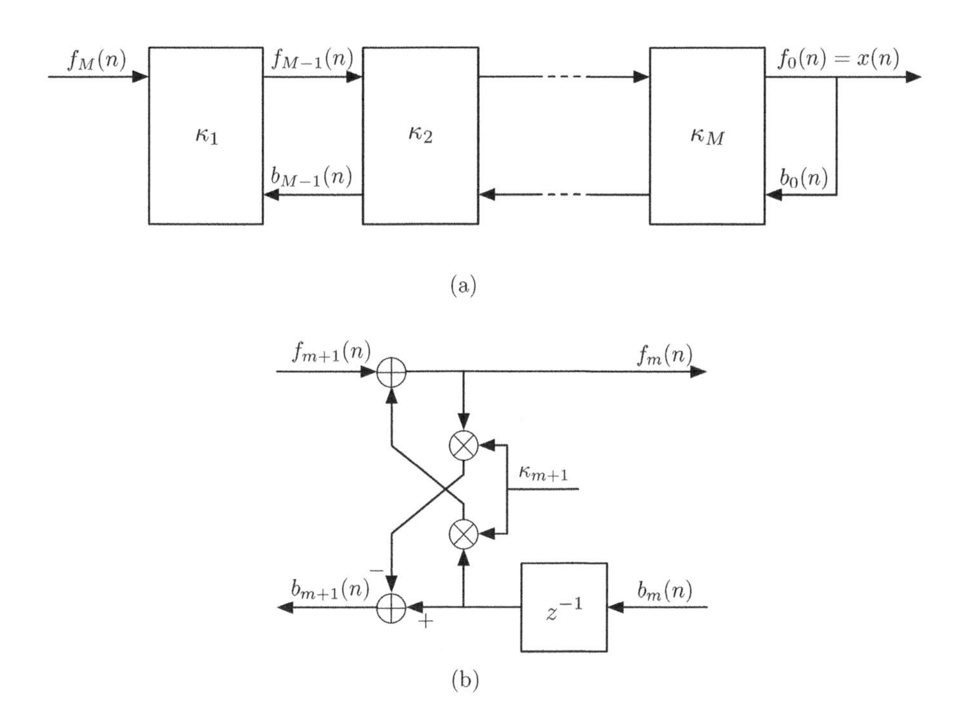

Figure 11.8 Lattice all-pole filter: (a) overall structure and (b) details of one stage.

We note that the denominator of $G(z)$ is assumed to be the system function of a prediction-error filter. This, as was noted in the last section, is not restrictive, as this condition only limits the poles of $G(z)$ to remain within the unit circle in the z-plane. In other words, the condition imposed on $G(z)$ is just to guarantee its stability.

To develop a lattice structure for $G(z)$, we first rearrange Eq. (11.75) as

$$\mathbf{w} = \mathbf{L}^{\mathrm{T}}\mathbf{c} \tag{11.100}$$

Next, we define $\mathbf{z} = [1 \quad z^{-1} \quad z^{-2} \cdots z^{-(N-1)}]^{\mathrm{T}}$, where z is the z-domain complex variable, multiply the transpose of both sides of Eq. (11.100) from right by $\mathbf{z}$ and replace $N-1$ by M, to obtain

$$W(z) = \sum_{i=0}^{M} c_i H_{b_i}(z) \tag{11.101}$$

where c_i's are the elements of vector $\mathbf{c}$,

$$W(z) = \sum_{i=0}^{M} w_i z^{-i} \tag{11.102}$$

and $H_{b_i}(z)$ is defined as in Eq. (11.80). Equation (11.101) shows that any arbitrary order M FIR system function $W(z)$ can equivalently be realized as a linear combination of the backward prediction-error filter system functions $H_{b_0}(z), H_{b_1}(z), \ldots, H_{b_M}(z)$.

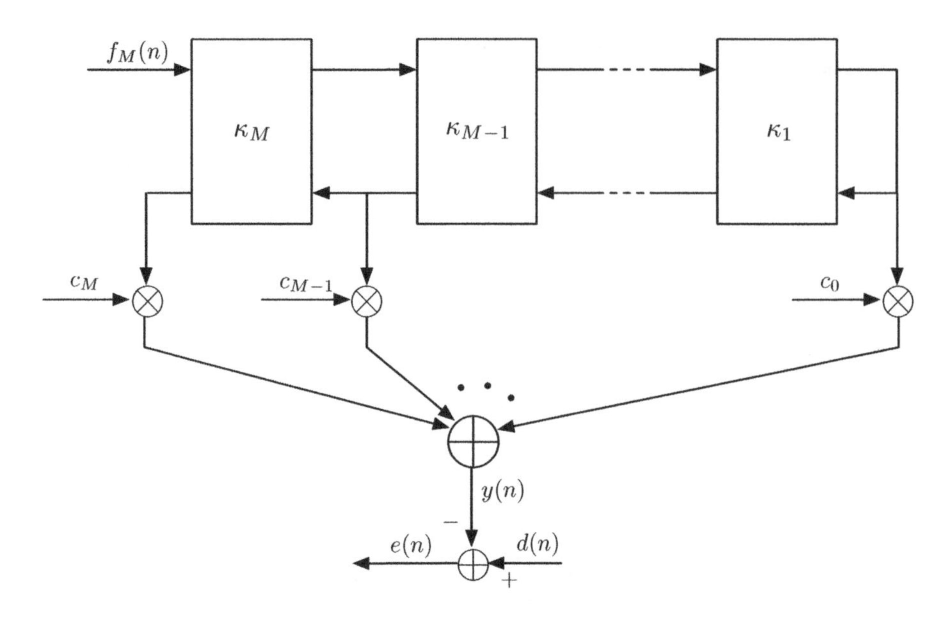

Figure 11.9 Pole-zero lattice for an arbitrary system function.

Using Eqs. (11.101) and (11.102), we obtain

$$G(z) = \sum_{i=0}^{M} c_i \frac{H_{b_i}(z)}{H_{f_M}(z)} \tag{11.103}$$

Furthermore, with reference to Figure 11.8a, we note that $H_{b_i}(z)/H_{f_M}(z)$ is the transfer function relating $f_M(n)$ and $b_i(n)$. It is obtained as the cascade of the transfer function between $f_M(n)$ and $x(n)$, that is, $1/H_{f_M}(z)$, and the transfer function between $x(n)$ and $b_i(n)$, that is, $H_{b_i}(z)$. Using these results, we obtain Figure 11.9 as lattice realization of the system function $G(z)$.

11.13 Adaptive Lattice Filter

In this section, an LMS algorithm for adaptive adjustment of the parameters of the lattice joint process estimator, given in Figure 11.7, is developed. Simulation results and some discussions on the performance of adaptive lattice filters are provided in the next section.

As the prediction error power becomes minimum when the predictor coefficients are chosen optimally, the optimum PARCOR coefficient κ_m of the mth stage of a lattice predictor is obtained by minimizing the cost function

$$\xi_{p,m} = E[f_m^2(n) + b_m^2(n)] \tag{11.104}$$

The cost function $\xi_{p,m}$ is equivalent to either of the cost functions $P_m^{\mathrm{f}} = E[f_m^2(n)]$ and $P_m^b = E[b_m^2(n)]$, since forward and backward predictors of the same order share the same set of coefficients and also the same level of minimum mean-squared error. By defining the cost function as in Eq. (11.104) help us to use both forward and backward prediction errors in the LMS algorithm, so that a lower misadjustment can be achieved.

The LMS algorithm for minimization of the cost function $\xi_{p,m}$ is implemented according to the recursive equation

$$\kappa_m(n+1) = \kappa_m(n) - \mu_{p,m}(n)\frac{\partial\hat{\xi}_{p,m}(n)}{\partial\kappa_m} \tag{11.105}$$

where $\mu_{p,m}(n)$ is the algorithm step-size and

$$\hat{\xi}_{p,m}(n) = f_m^2(n) + b_m^2(n) \tag{11.106}$$

is an estimate of the cost function $\xi_{p,m}$ based on the most recent samples of the forward and backward prediction errors. Substituting Eq. (11.106) in Eq. (11.105) and using Eqs. (11.60) and (11.61), we obtain

$$\kappa_m(n+1) = \kappa_m(n) + 2\mu_{p,m}(n)[f_m(n)b_{m-1}(n-1) + b_m(n)f_{m-1}(n)] \tag{11.107}$$

To assure fast convergence of the algorithm, the step-size $\mu_{p,m}(n)$ is normalized by the signal power at the input to the mth stage of the predictor. To estimate this power, we use the recursive equation

$$P_{m-1}(n) = \beta P_{m-1}(n-1) + 0.5(1-\beta)(f_{m-1}^2(n) + b_{m-1}^2(n-1)) \tag{11.108}$$

The normalized step-size parameter is then given by

$$\mu_{p,m}(n) = \frac{\mu_{p,o}}{P_{m-1}(n) + \epsilon} \tag{11.109}$$

where $\mu_{p,o}$ is an unnormalized step-size parameter common to all stages of the predictor, and ϵ is a small positive constant which is added to prevent instability of the algorithm when $P_{m-1}(n)$ assumes values close to 0.

The step-normalized LMS algorithm is also used for adaptation of c_i coefficients of the linear combiner part of the lattice joint process estimator. The derivation of this procedure is the same as the recursions developed in Chapter 7. The result is

$$\mathbf{c}(n+1) = \mathbf{c}(n) + 2\boldsymbol{\mu}_c e(n)\mathbf{b}(n) \tag{11.110}$$

where $e(n) = d(n) - y(n)$, $y(n)$ is obtained according to Eq. (11.76), and $\boldsymbol{\mu}_c$ is a diagonal matrix consisting the normalized step-size parameters

$$\mu_{c,m} = \frac{\mu_{c,o}}{P_m(n) + \epsilon}, \quad \text{for } m = 0, 1, \ldots, N-1 \tag{11.111}$$

where $\mu_{c,o}$ is an unnormalized step-size which, in general, may be different from $\mu_{p,o}$. Table 11.5 gives a summary of the above results.

Table 11.5 LMS algorithm for adaptive lattice joint process estimator.

Given: Estimator parameters:

$\kappa_1(n), \kappa_2(n) \cdots \kappa_{N-1}(n)$

and $\mathbf{c}(n) = [c_0(n)\ c_1(n) \cdots c_{N-1}(n)]^{\mathrm{T}}$,

the most recent input sample $x(n)$, desired output $d(n)$,

backward prediction error vector

$\mathbf{b}(n-1) = [b_0(n-1)\ b_1(n-1) \cdots b_{N-1}(n-1)]^{\mathrm{T}}$,

and power estimates $P_0(n-1),\ P_1(n-1), \ldots, P_{N-1}(n-1)$.

Required: Estimator parameter updates:

$\kappa_0(n+1),\ \kappa_1(n+1), \ldots, \kappa_{N-1}(n+1)$

and $\mathbf{c}(n+1) = [c_0(n+1)\ c_1(n+1) \ldots c_{N-1}(n+1)]^{\mathrm{T}}$,

backward prediction error vector

$\mathbf{b}(n) = [b_0(n)\ b_1(n) \cdots b_{N-1}(n)]^{\mathrm{T}}$,

and power estimates $P_0(n),\ P_1(n), \ldots, P_{N-1}(n)$.

*** Lattice Predictor Part ***

$f_0(n) = b_0(n) = x(n)$

$P_0(n) = \beta P_0(n-1) + 0.5(1-\beta)[f_0^2(n) + b_0^2(n-1)]$

for $m = 1$ to $N-1$

$f_m(n) = f_{m-1}(n) - \kappa_m(n) b_{m-1}(n-1)$

$b_m(n) = b_{m-1}(n-1) - \kappa_m(n) f_{m-1}(n)$

$\kappa_m(n+1) = \kappa_m(n) + \frac{2\mu_{p,o}}{P_{m-1}(n)+\epsilon}[f_{m-1}(n) b_m(n) + b_{m-1}(n-1) f_m(n)]$

$P_m(n) = \beta P_m(n-1) + 0.5(1-\beta)[f_m^2(n) + b_m^2(n-1)]$

end

*** Linear Combiner Part ***

$y(n) = \mathbf{c}^{\mathrm{T}}(n)\mathbf{b}(n)$

$e(n) = d(n) - y(n)$

$\boldsymbol{\mu}_c = \mu_{c,o}\mathrm{diag}((P_0(n)+\epsilon)^{-1}, (P_1(n)+\epsilon)^{-1}, \ldots, (P_{N-1}(n)+\epsilon)^{-1})$

$\mathbf{c}(n+1) = \mathbf{c}(n) + 2\boldsymbol{\mu}_c e(n)\mathbf{b}(n)$

11.13.1 Discussion and Simulations

Analysis of the convergence behavior of the LMS algorithm when applied to a lattice structure is rather difficult. In particular, in the case of lattice joint process estimator there are two sets of parameters which are being adapted simultaneously. The optimum values of the PARCOR coefficients, κ_m's, depend only on the statistics of the input signal. The optimum value of the coefficient vector $\mathbf{c}$ of the linear combiner part depends on the current values of PARCOR coefficients as well as the optimum value of the coefficient vector $\mathbf{w}$ in the original transversal filter, $\mathbf{w}_o$, viz., Eq. (11.75). An important point to be noted is that even if $\mathbf{w}_o$, that is, the optimum impulse response to be realized by the estimator, is fixed, any change in the PARCOR coefficients will require readjustment of the coefficient vector $\mathbf{c}$. This, as we will demonstrate by a simulation example, may lead to a significant increase in misadjustment.

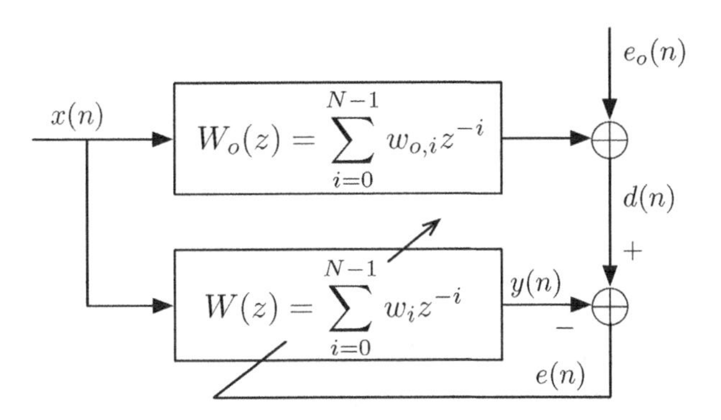

Figure 11.10 A modeling problem.

To demonstrate the above phenomenon, we consider the modeling problem shown in Figure 11.10. A plant $W_o(z)$ is to be modeled by an adaptive filter $W(z)$. The common input signal to the plant and adaptive filter, $x(n)$, is generated by passing a unit variance white Gaussian process, $\nu(n)$, through a coloring filter with the system function

$$H_1(z) = K_1 \frac{1 - 1.2z^{-1}}{1 - 1.2z^{-1} + 0.8z^{-2}} \tag{11.112}$$

The coefficient K_1 is set equal to 0.488. This results in a sequence with unit variance. For our later use, the colored process generated by $H_1(z)$ will be called $x_1(n)$. Figure 11.11 shows the power spectral density of $x_1(n)$ evaluated using

$$\Phi_{x_1 x_1}(e^{j\omega}) = \Phi_{\nu\nu}(e^{j\omega})|H_1(e^{j\omega})|^2 \tag{11.113}$$

with $\Phi_{\nu\nu}(e^{j\omega}) = 1$. Observe that $x_1(n)$ is highly colored and the eigenvalue spread of its corresponding correlation matrix can be as large as 338. This is obtained as the ratio of maximum to minimum value of the spectral density – see Chapter 4.

In this simulation example, we consider realizing the adaptive filter $W(z)$ in transversal as well as lattice (i.e., the lattice joint process estimator of Figure 11.7) forms. The plant and the adaptive filter are both selected to have a length of $N = 30$. This choice of N for the present input sequence results in an eigenvalue spread of 300. The variance of the additive white noise sequence $e_o(n)$ at the plant output is set equal to $\sigma_{e_o}^2 = 10^{-4}$. So, when the adaptive filter $W(z)$ is set to its optimum choice, $W_o(z)$, the resulting minimum mean-squared error will be $\sigma_{e_o}^2 = 10^{-4}$.

Figure 11.12 presents a pair of learning curves of the modeling problem. The curves correspond to transversal and lattice LMS, and each is an ensemble average of 50 independent runs. The final curves have been smoothed. In the case of transversal LMS, the step-size parameter, μ, is selected according to Eq. (6.64) to result in 10% misadjustment. For lattice LMS, the following parameters are used: $\epsilon = 0.02$, $\mu_{p,o} = 0.001$ (for the first 2500 iterations only) and $\mu_{c,o} = 0.1/N = 0.0033$. This choice of $\mu_{c,o}$ would result in a misadjustment of about 10% if the PARCOR coefficients perturbation, which arise due to

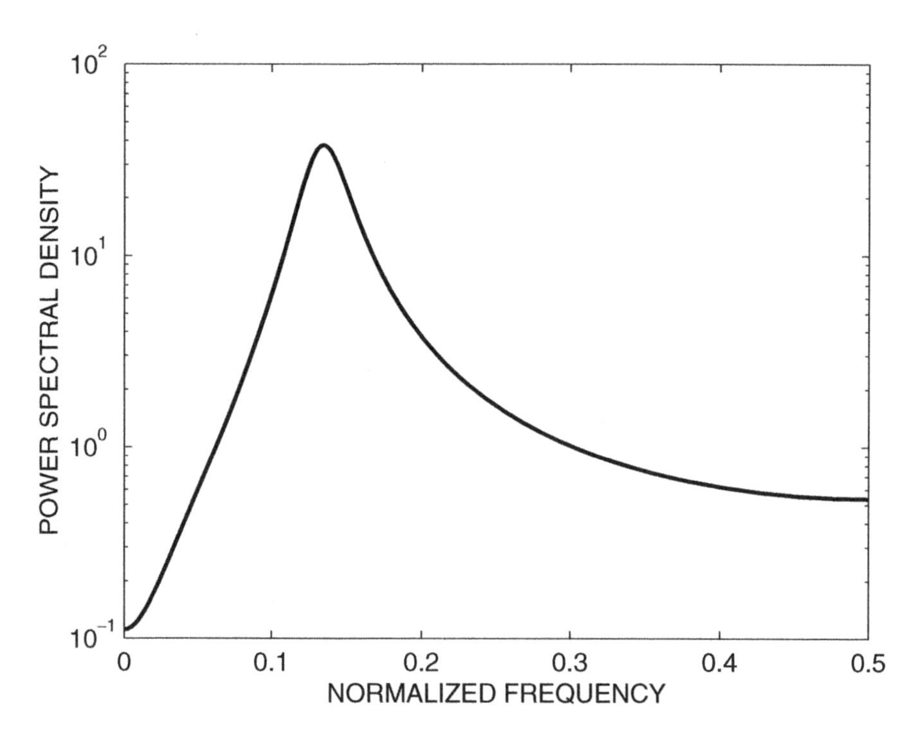

Figure 11.11 Power spectral density of the input process for the modeling problem simulations.

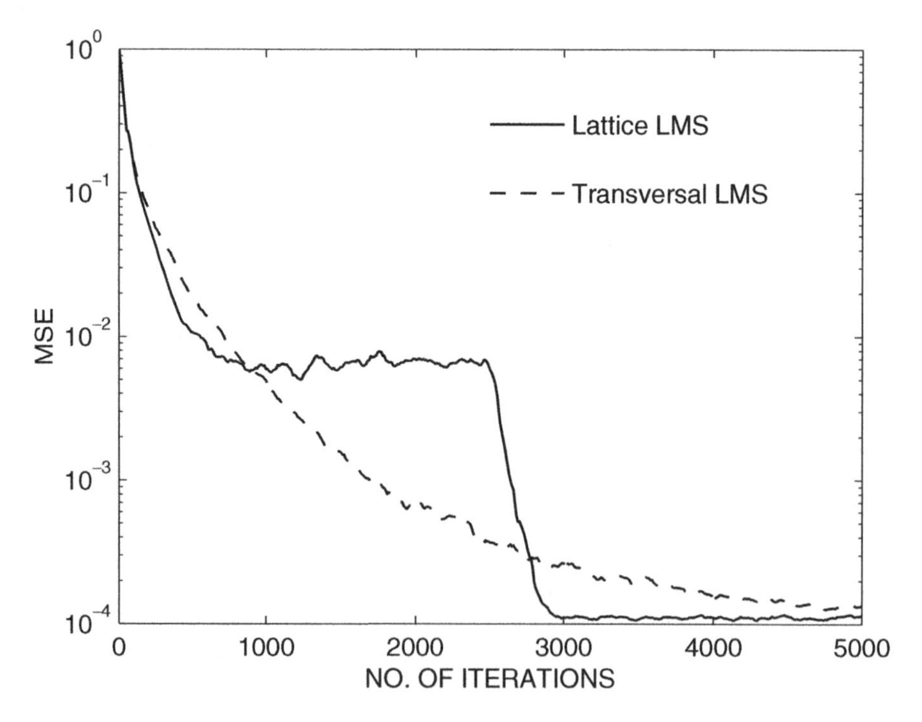

Figure 11.12 Learning curves showing convergence of the transversal and lattice LMS applied to modeling problem of Figure 11.10.

their adaptation, could be ignored. To demonstrate the effect of the perturbation of PARCOR coefficients, $\mu_{p,o}$ is forced to zero after the first 2500 iterations, so that the PARCOR coefficients remain fixed from iteration 2500. Observe that the perturbation of PARCOR coefficients has significant impact on the misadjustment of the lattice LMS algorithm. Once the PARCOR coefficients adjustment is stopped, we see a fast convergence of the algorithm with a misadjustment close to what we predicted before. At iteration 2500, the PARCOR coefficients are already near their optimal values and the backward prediction errors are almost uncorrelated with one another. This is why the lattice LMS converges faster than the transversal LMS, after iteration 2500.

The problem of PARCOR coefficients perturbation is a serious one which limits the application of adaptive lattice joint process estimator. As the above example demonstrated, unless the PARCOR coefficients adjustment is stopped after some initial convergence, the lattice LMS cannot be relied on as a good choice for improving convergence performance of adaptive filters. The large misadjustment arising from adaptation of PARCOR coefficients prohibit their applicability. The problem may be more serious when the input signal is nonstationary. In that case, the optimum PARCOR coefficients are time-varying, as they follow the time-varying statistics of the input. This, in turn, necessitates continuous adaptation of the PARCOR coefficients as well as the coefficient vector **c**. The inevitable lag in the adaptation of **c** will result in further increase of the mean-squared error. This in effect means higher misadjustment.

11.14 Autoregressive Modeling of Random Processes

A random process $x(n)$ is said to be AR of order M if it can be generated through a difference equation of the form

$$x(n) = \sum_{i=1}^{M} h_i x(n-i) + \nu(n) \tag{11.114}$$

where h_i's are AR coefficients and $\nu(n)$ is a zero-mean white noise process which is called *innovation* of $x(n)$. This implies that any new sample of the process, $x(n)$, is related to its previous M samples according to the summation on the right-hand side of Eq. (11.114). In addition, there is a new piece of information (namely, the innovation $\nu(n)$) in $x(n)$ which is uncorrelated to its previous samples. This, in turn, implies that the best linear prediction of $x(n)$ based on its past M samples is nothing but the summation on the right-hand side of Eq. (11.114). Moreover, the latter estimate cannot be improved by increasing the order of the predictor beyond M, as the portion of $x(n)$ which could not be estimated by the latter summation, that is, $\nu(n)$, has no correlation with the farther samples of $x(n)$, that is, $x(n-M-1), x(n-M-2), \ldots$ A procedure for analytical derivation of these results is discussed in Problem P11.26.

An AR process $x(n)$ can be characterized by its model which may be obtained by passing $x(n)$ through a forward (or backward) linear predictor and optimizing the predictor coefficients by minimizing the mean-squared error of its output. This results in a set of predictor coefficients which match the coefficients h_i of Eq. (11.114). In particular, if a predictor of order $M' \geq M$ is used, we obtain

$$a_{M',i} = \begin{cases} h_i, & 1 \leq i \leq M \\ 0, & M+1 \leq i \leq M' \end{cases} \tag{11.115}$$

and

$$P_{M'} = P_M = \sigma_\nu^2 \tag{11.116}$$

where σ_ν^2 is the variance of $\nu(n)$. An important point to be noted here is that the set of coefficients $a_{M,1}, a_{M,2}, \ldots, a_{M,M}$ and P_M provide sufficient information to obtain the autocorrelation function of $x(n)$ for any arbitrary lag. This directly follows from Eq. (11.114) with $h_i = a_{M,i}$. To see this, multiplying Eq. (11.114) on both sides by $x(n-k)$ and taking expectations, we obtain

$$r(k) = \sum_{i=1}^{M} a_{M,i} r(k-i), \quad k > 0 \tag{11.117}$$

Value of $r(0)$ can be obtained using Eq. (11.66) as

$$r(0) = \left(\prod_{i=1}^{M} (1-\kappa_i^2) \right)^{-1} P_M \tag{11.118}$$

where the PARCOR coefficients can be obtained using the inverse Levinson–Durbin algorithm discussed in Section 11.10.

It may also be noted that the estimated AR coefficients may be used to obtain the power spectral density of $x(n)$ according to the following equation

$$\Phi_{xx}(e^{j\omega}) = P_M |H_{AR}(e^{j\omega})|^2 \tag{11.119}$$

where it has been noted that $\Phi_{\nu\nu}(e^{j\omega}) = \sigma_\nu^2 = P_M$, and

$$H_{AR}(z) = \frac{1}{1 - \sum_{i=1}^{M} a_{M,i} z^{-i}} \tag{11.120}$$

It is also instructive to note that the process $x(n)$ can be reconstructed by passing its innovation $\nu(n)$ through $H_{AR}(z)$.

Although many of the practically arising processes may not be truly AR, AR modeling of arbitrary processes for the purpose of spectral estimation has been found to be quite effective, provided that a sufficiently large order is considered. Usually, one finds that a model order in the range 5 to 10 is more than sufficient to get an acceptable estimate of the power spectral density for most of the processes encountered in practice.

In the context of adaptive filters, the above results have the following implication. *The correlation matrix of the input process to an adaptive transversal filter may be characterized by an AR model whose order may be much less than the order of the adaptive filter.* This, as we shall see in the next section, may effectively be used to improve the performance of adaptive filters, at very little computational cost.

11.15 Adaptive Algorithms Based on Autoregressive Modeling

The LMS–Newton algorithm was introduced in Chapter 7, as a method to solve the eigenvalue spread problem of adaptive filters whose inputs were colored. In this section, the results of the previous sections are used to propose two efficient implementations of the LMS–Newton algorithm. We assume that the input sequence to the adaptive filter can be modeled as an AR process whose order may be kept much lower than the adaptive filter length. The two implementations (referred to as *Algorithm 1* and *Algorithm 2*) differ in their structural complexity. The first algorithm, which will be an exact implementation of the LMS–Newton algorithm, if the AR modeling assumption is accurate, is structurally complicated and fits best into a DSP-based implementation. On the other hand, the second algorithm is structurally simple and is tailored more toward VLSI custom chip design.

We recall that the LMS–Newton algorithm recursion for an adaptive filter with real-valued input is

$$\mathbf{w}(n+1) = \mathbf{w}(n) + 2\mu e(n)\hat{\mathbf{R}}_{xx}^{-1}\mathbf{x}(n) \tag{11.121}$$

where $\mathbf{w}(n) = [w_0(n) \quad w_1(n) \cdots w_{N-1}]^{\mathrm{T}}$ is the filter tap-weight vector, $\mathbf{x}(n) = [x(n)\; x(n-1) \cdots x(n-N+1)]^{\mathrm{T}}$ is the filter input vector, $\hat{\mathbf{R}}_{xx}$ is an estimate of the input correlation matrix $\mathbf{R}_{xx} = E[\mathbf{x}(n)\mathbf{x}^{\mathrm{T}}(n)]$, μ is the algorithm step-size parameter, $e(n) = d(n) - y(n)$ is the measured error at the filter output, $d(n)$ is the desired output and $y(n) = \mathbf{w}^{\mathrm{T}}(n)\mathbf{x}(n)$ is the filter output.

It may be noted that, here, we have added the subscript "xx" to $\mathbf{R}_{xx}$ to emphasize that it corresponds to the input vector $\mathbf{x}(n)$. We follow this notation in the rest of this chapter, as we need to refer to a number of different correlation matrices.

To implement the LMS–Newton algorithm, one needs to calculate $\hat{\mathbf{R}}_{xx}^{-1}\mathbf{x}(n)$ for each update of recursion (11.121). A trivial way would be to obtain an estimate of $\mathbf{R}_{xx}^{-1}$ first, and then perform the matrix by vector multiplication $\hat{\mathbf{R}}_{xx}^{-1}\mathbf{x}(n)$. This, of course, is inefficient and, therefore, an alternative solution has to be found. Here, we propose an efficient method for direct updating of the vector $\hat{\mathbf{R}}_{xx}^{-1}\mathbf{x}(n)$, without estimating $\mathbf{R}_{xx}^{-1}$. For this, we note that the vector $\mathbf{x}(n)$ may be converted to the vector $\mathbf{b}(n) = [b_0(n) \quad b_1(n) \cdots b_{N-1}(n)]^{\mathrm{T}}$, made up of the backward prediction errors of $x(n)$ for the predictors of orders 0 to $N-1$. The vectors $\mathbf{x}(n)$ and $\mathbf{b}(n)$ are related according to Eq. (11.68). We also recall that the elements of $\mathbf{b}(n)$, that is, the backward prediction errors $b_0(n), b_1(n), \ldots, b_{N-1}(n)$, are uncorrelated with one another. This means that the correlation matrix $\mathbf{R}_{bb} = E[\mathbf{b}(n)\mathbf{b}^{\mathrm{T}}(n)]$ is diagonal, and, therefore, evaluation of its inverse is trivial. Furthermore, using Eq. (11.68), we obtain

$$\mathbf{R}_{bb} = E[\mathbf{L}\mathbf{x}(n)(\mathbf{L}\mathbf{x}(n))^{\mathrm{T}}] = \mathbf{L}\mathbf{R}_{xx}\mathbf{L}^{\mathrm{T}} \tag{11.122}$$

Inverting both sides of Eq. (11.122) and pre- and postmultiplying the result by $\mathbf{L}^{\mathrm{T}}$ and $\mathbf{L}$, respectively, we obtain

$$\mathbf{R}_{xx}^{-1} = \mathbf{L}^{\mathrm{T}}\mathbf{R}_{bb}^{-1}\mathbf{L} \tag{11.123}$$

Next, we define $\mathbf{u}(n) = \mathbf{R}_{xx}^{-1}\mathbf{x}(n)$ and substitute for $\mathbf{R}_{xx}^{-1}$ from Eq. (11.123), to obtain

$$\begin{aligned}\mathbf{u}(n) &= \mathbf{L}^{\mathrm{T}}\mathbf{R}_{bb}^{-1}\mathbf{L}\mathbf{x}(n) \\ &= \mathbf{L}^{\mathrm{T}}\mathbf{R}_{bb}^{-1}\mathbf{b}(n)\end{aligned} \tag{11.124}$$

This result is fundamental to the derivation of the algorithms that follow.

In the rest of this section, for the sake of convenience, we shall use the notation $\mathbf{u}(n)$ even when $\mathbf{R}_{xx}$ is replaced by its estimate, $\hat{\mathbf{R}}_{xx}$.

11.15.1 Algorithms

Algorithm 1: Implementation of Eq. (11.123) requires a mechanism for converting the vector of input samples, $\mathbf{x}(n)$, to the vector of backward prediction error samples, $\mathbf{b}(n)$. A lattice predictor may be used for efficient implementation of this mechanism. Moreover, if one assumes that the input sequence, $x(n)$, can be modeled as an AR process of order $M < N$, then, a lattice predictor with order M will suffice, and the matrix $\mathbf{L}$ and vector $\mathbf{b}(n)$ take the following forms:

$$\mathbf{L} = \begin{bmatrix} 1 & 0 & \cdots & 0 & 0 & \cdots & 0 & 0 & \cdots & 0 \\ -a_{1,1} & 1 & \cdots & 0 & 0 & \cdots & 0 & 0 & \cdots & 0 \\ \vdots & \vdots & \ddots & \vdots & \vdots & \ddots & \vdots & \vdots & \ddots & \vdots \\ -a_{M,M} & -a_{M,M-1} & \cdots & 1 & 0 & \cdots & 0 & 0 & \cdots & 0 \\ 0 & -a_{M,M} & \cdots & -a_{M,1} & 1 & \cdots & 0 & 0 & \cdots & 0 \\ \vdots & \vdots & \ddots & \vdots & \vdots & \ddots & \vdots & \vdots & \ddots & \vdots \\ 0 & 0 & \cdots & 0 & 0 & \cdots & -a_{M,M} & -a_{M,M-1} & \cdots & 1 \end{bmatrix} \quad (11.125)$$

and

$$\mathbf{b}(n) = [b_0(n) \quad b_1(n) \cdots b_M(n) \quad b_M(n-1) \cdots b_M(n-N+M+1)]^{\mathrm{T}} \quad (11.126)$$

The special structure in rows $M+1$ to N of $\mathbf{L}$ and elements $M+1$ to L of $\mathbf{b}(n)$ follows from Eq. (11.115).

In certain applications, such as acoustic echo cancellation, a value of M much smaller than N may be used. In such cases, the computational burden of updating $\mathbf{b}(n)$ would be negligible when compared with the total computational complexity of the whole system, as only the first $M+1$ samples of $\mathbf{b}(n)$ require updating. The rest of the elements of $\mathbf{b}(n)$ are delayed versions of $b_M(n)$. Multiplication of $\mathbf{R}_{bb}^{-1}$ by $\mathbf{b}(n)$ (according to Eq. (11.124)) also requires only a small amount of computation. It involves estimation of the powers of $b_0(n)$ through $b_M(n)$ and normalization of these samples by their power estimates.

Multiplication of $\mathbf{L}^{\mathrm{T}}$ by $\mathbf{R}_{bb}^{-1}\mathbf{b}(n)$, to complete computation of $\mathbf{u}(n)$ (according to Eq. (11.124)), however, is more involved, since a structure such as lattice is not applicable. It requires estimation of the elements of $\mathbf{L}$ and direct multiplication of $\mathbf{L}^{\mathrm{T}}$ by $\mathbf{R}_{bb}^{-1}\mathbf{b}(n)$. Considering the forms of $\mathbf{L}$ and $\mathbf{b}(n)$, one finds that only the first $M+1$ and the last M elements of $\mathbf{L}^{\mathrm{T}}\mathbf{R}_{bb}^{-1}\mathbf{b}(n)$ need to be computed. The remaining elements of $\mathbf{L}^{\mathrm{T}}\mathbf{R}_{bb}^{-1}\mathbf{b}(n)$ are delayed versions of its $(M+1)$th element.

The following procedure may be used for estimating the elements of $\mathbf{L}$, that is, the coefficients of the predictors of orders 1 to M. An LMS-based adaptive lattice predictor is used to obtain the PARCOR coefficients $\kappa_1, \kappa_2, \ldots, \kappa_M$ of $x(n)$. The conversion algorithm of Table 11.1 is then used to obtain the predictor coefficients of orders 1 to M. Table 11.6 summarizes this procedure.

Table 11.6 An LMS-based procedure for implementation of Algorithm 1.

Given: Parameter vectors $\boldsymbol{\kappa}(n) = [\kappa_1(n)\ \kappa_2(n) \cdots \kappa_{M-1}(n)]^{\mathrm{T}}$
and $\mathbf{w}(n) = [w_0(n)\ w_1(n) \cdots w_{N-1}(n)]^{\mathrm{T}}$,
data vectors $\mathbf{x}(n)$, $\mathbf{b}(n-1)$ and $\mathbf{u}(n-1)$, desired output $d(n)$,
and power estimates $P_0(n-1),\ P_1(n-1), \ldots, P_M(n-1)$.

Required: Vector updates $\boldsymbol{\kappa}(n)$, $\mathbf{w}(n+1)$, $\mathbf{b}(n)$ and $\mathbf{u}(n)$,
and power estimate updates $P_0(n),\ P_1(n), \ldots, P_M(n)$.

⋆⋆⋆ Lattice Predictor ⋆⋆⋆

$f_0(n) = b_0(n) = x(n)$
$P_0(n) = \beta P_0(n-1) + 0.5(1-\beta)[f_0^2(n) + b_0^2(n-1)]$
for $m = 1$ to M
$\quad f_m(n) = f_{m-1}(n) - \kappa_m(n) b_{m-1}(n-1)$
$\quad b_m(n) = b_{m-1}(n-1) - \kappa_m(n) f_{m-1}(n)$
$\quad \kappa_m(n+1) = \kappa_m(n) + \frac{2\mu_{p,o}}{P_{m-1}(n)+\epsilon}[f_{m-1}(n) b_m(n) + b_{m-1}(n-1) f_m(n)]$
$\quad P_m(n) = \beta P_m(n-1) + 0.5(1-\beta)[f_m^2(n) + b_m^2(n-1)]$
$\quad$ if $|\kappa_m(n)| > \gamma$, $\kappa_m(n) = \kappa_m(n-1)$
end

⋆⋆⋆ Conversion from Lattice to Transversal ⋆⋆⋆

$\bar{P}_0(n) = P_0(n)$
$a_{1,1} = \kappa(1)$
$\bar{P}_1(n) = (1 - \kappa_1^2(n)) \bar{P}_0(n)$
for $m = 1$ to $M-1$
$\quad a_{m+1,j}(n) = a_{m,j}(n) - \kappa_{m+1}(n) a_{m,m+1-j}(n)$, for $j = 1,\ 2, \ldots, m$
$\quad a_{m+1,m+1}(n) = \kappa_{m+1}(n)$
$\quad \bar{P}_{m+1}(n) = (1 - \kappa_{m+1}^2(n)) \bar{P}_m(n)$
end

⋆⋆⋆ $\mathbf{u}(n)$ update ⋆⋆⋆

$u_j(n) = u_{j-1}(n-1)$, for $j = M+1,\ M+2, \ldots, N-M-1$,
where $u_j(n)$ is the jth element of $\mathbf{u}(n)$.

$$\mathbf{b}_{\mathrm{h}}(n) = \left[\frac{b_0(n)}{\bar{P}_0(n)}\ \frac{b_1(n)}{\bar{P}_1(n)} \cdots \frac{b_M(n)}{\bar{P}_M(n)}\ \frac{b_M(n-1)}{\bar{P}_M(n)} \cdots \frac{b_M(n-M)}{\bar{P}_M(n)}\right]^{\mathrm{T}}$$

$[u_0(n)\ u_1(n) \cdots u_M(n)]^{\mathrm{T}} = \mathbf{L}_{\mathrm{tl}}^{\mathrm{T}} \mathbf{b}_{\mathrm{h}}(n)$,
where $\mathbf{L}_{\mathrm{tl}}$ is the $(2M+1)$-by-$(M+1)$ top-left part of $\mathbf{L}$.

$\mathbf{b}_{\mathrm{t}}(n) = \bar{P}_M^{-1}(n-N+M+1)[b_M(n-N+2M)\ b_M(n-N+2M-1)$
$\cdots b_M(n-N+M+1)]^{\mathrm{T}}$
$[u_{N-M}(n)\ u_{N-M+1}(n) \cdots u_{N-1}(n)]^{\mathrm{T}} = \mathbf{L}_{\mathrm{br}}^{\mathrm{T}} \mathbf{b}_{\mathrm{t}}(n)$,
where $\mathbf{L}_{\mathrm{br}}$ is the M-by-M bottom-right part of $\mathbf{L}$.

⋆⋆⋆ Filtering and tap-weight vector adaptation ⋆⋆⋆

$y(n) = \mathbf{w}^{\mathrm{T}}(n)\mathbf{x}(n)$
$e(n) = d(n) - y(n)$
$\mathbf{w}(n+1) = \mathbf{w}(n) + 2\mu e(n)\mathbf{u}(n)$

We have the following comments regarding the algorithm presented in Table 11.6. The role of the constant ϵ in the PARCOR updating equation is to ensure stability of the algorithm when $P_m(n)$ droops to very small values. Also, at every iteration, the PARCOR coefficients are constrained to lie within a maximum magnitude γ. The predictor coefficients, $a_{m,j}$'s, and the backward errors obtained through the lattice predictor are used to update the first $M+1$ and the last M elements of the vector $\mathbf{u}(n)$. The iteration suggested by Eq. (11.66) is used to obtain the estimates of the backward error powers. These are denoted as $\bar{P}_m(n)$'s in Table 11.6. The power estimates obtained in the lattice predictor part of the algorithm, that is, $P_m(n)$'s, could also be used. However, experiments have shown that the use of $\bar{P}_m(n)$'s results in a more reliable algorithm. The vectors $\mathbf{b}_\mathrm{h}(n)$ and $\mathbf{b}_\mathrm{t}(n)$ denote the backward error vectors which correspond to the input samples at the head and the end tail parts, respectively, of the tap-delay-line filter. When the input signal to the filter is stationary, the elements of $\mathbf{b}_\mathrm{t}(n)$ can be obtained by delaying the output $b_M(n)$ of the lattice predictor at the head of $\mathbf{x}(n)$. This has been our assumption in Table 11.6. When the filter input is nonstationary and the filter length, N, is large, one may have to use a separate predictor for the samples at the tail of $\mathbf{x}(n)$.

Algorithm 2: The Algorithm 1, although low in computational complexity, is structurally complicated, as the implementation of the Levinson–Durbin algorithm and ordering of the manipulated data is not straightforward. This would not be much of a problem if a DSP processor is used. Therefore, Algorithm 1 is suitable for software implementation. However, if one is interested in a custom chip implementation, he may use Algorithm 2 proposed below which has a much simpler structure compared to Algorithm 1.

The reason why Algorithm 1 is not simple is because it needs to update the parameters of the lattice and transversal predictors of orders 1 to M at each time instant. Furthermore, only the middle samples in $\mathbf{u}(n)$ could be obtained as delayed versions of earlier samples. In Algorithm 2, we overcome these problems by extending the input and tap-weight vectors, $\mathbf{x}(n)$ and $\mathbf{w}(n)$, to the vectors

$$\mathbf{x}_\mathrm{E}(n) = [x(n+M) \cdots x(n+1)\; x(n) \cdots x(n-N+1) \cdots x(n-N-M+1)]^\mathrm{T}$$

and

$$\mathbf{w}_\mathrm{E}(n) = [w_{-M}(n) \cdots w_{-1}(n) \quad w_0(n) \cdots w_{N-1}(n) \cdots w_{N+M-1}(n)]^\mathrm{T}$$

respectively, and apply an LMS–Newton algorithm similar to Eq. (11.121) for updating $\mathbf{w}_\mathrm{E}(n)$. Since the tap weights of the original filter correspond to $w_0(n)$ through $w_{N-1}(n)$, the first M and last M elements of $\mathbf{w}_\mathrm{E}(n)$ may be frozen at 0. This can easily be done by initializing these weights to zero and assigning a zero step-size parameter to all of them. If this is done, the computation of the first M and last M elements of $\mathbf{L}^\mathrm{T}\mathbf{R}_{bb}^{-1}\mathbf{L}\mathbf{x}_\mathrm{E}(n)$ (with appropriate dimensions for $\mathbf{L}$ and $\mathbf{R}_{bb}$) is immaterial and may be ignored. This results in the following recursive equation for updating the adaptive filter tap weights:

$$\mathbf{w}(n+1) = \mathbf{w}(n) + 2\mu e(n)\mathbf{u}_\mathrm{a}(n) \tag{11.127}$$

where $\mathbf{w}(n)$ is the filter tap-weight vector as defined above, and

$$\mathbf{u}_\mathrm{a}(n) = \mathbf{L}_2\mathbf{R}_{bb}^{-1}\mathbf{L}_1\mathbf{x}_\mathrm{E}(n) \tag{11.128}$$

Here, $\mathbf{R}_{bb}$ is a diagonal matrix compatible to the column vector $\mathbf{L}_1\mathbf{x}_{\mathrm{E}}(n)$ and the diagonal elements of $\mathbf{R}_{bb}$ are estimates of the powers of the elements of the latter vector. The matrices $\mathbf{L}_1$ and $\mathbf{L}_2$ are given by

$$\mathbf{L}_1 = \begin{bmatrix} -a_{M,M} & -a_{M,M-1} & \cdots & 1 & 0 & \cdots & 0 & \cdots & 0 & 0 \\ 0 & -a_{M,M} & \cdots & -a_{M,1} & 1 & \cdots & 0 & \cdots & 0 & 0 \\ \vdots & \vdots & \ddots & \vdots & \vdots & \ddots & \vdots & \ddots & \vdots & \vdots \\ 0 & 0 & \cdots & 0 & 0 & \cdots & -a_{M,M} & \cdots & -a_{M,1} & 1 \end{bmatrix} \tag{11.129}$$

and

$$\mathbf{L}_2 = \begin{bmatrix} 1 & -a_{M,1} & \cdots & -a_{M,M} & 0 & \cdots & 0 & \cdots & 0 & 0 \\ 0 & 1 & \cdots & -a_{M,M-1} & -a_{M,M} & \cdots & 0 & \cdots & 0 & 0 \\ \vdots & \vdots & \ddots & \vdots & \vdots & \ddots & \vdots & \ddots & \vdots & \vdots \\ 0 & 0 & \cdots & 0 & 0 & \cdots & 1 & \cdots & -a_{M,M-1} & -a_{M,M} \end{bmatrix} \tag{11.130}$$

The dimension of $\mathbf{L}_1$ and $\mathbf{L}_2$ are $(N+M)$-by-$(N+2M)$ and N-by-$(N+M)$, respectively. The number of rows in $\mathbf{L}_1$ is only N as we do not want to compute the first M and last M elements of $\mathbf{L}^{\mathrm{T}}\mathbf{R}_{bb}^{-1}\mathbf{L}\mathbf{x}_{\mathrm{E}}(n)$.

Inspection of Eq. (11.128) reveals that each updating of $\mathbf{u}_{\mathrm{a}}(n)$ requires only the updating of the first element of the vector $\mathbf{R}_{bb}^{-1}\mathbf{L}_1\mathbf{x}_{\mathrm{E}}(n)$, and, then, the first element of the final result, $\mathbf{u}_{\mathrm{a}}(n)$. The rest of the elements of the two vectors are delayed versions of their first elements. Putting these together, Figure 11.13 depicts a complete structure of Algorithm 2. It consists of a backward prediction-error filter $H_{b_M}(z)$ whose coefficients, $a_{M,i}(n)$'s, are updated using an adaptive algorithm. The time index "n" is added to these coefficients to emphasize their variability in time and their adaptation as input statistics may change. Any adaptive algorithm may be used for adjustment of these coefficients. The successive output samples from the backward prediction-error filter, that is, $b_M(n+M)$, make the elements of the column vector $\mathbf{L}_1\mathbf{x}_{\mathrm{E}}(n)$. Multiplication of $b_M(n+M)$ by the inverse of an estimate of its energy, denoted as $P_M^{-1}(n+M)$ in Figure 11.13, gives an update of $\mathbf{R}_{bb}^{-1}\mathbf{L}_1\mathbf{x}_{\mathrm{E}}(n)$. Finally, filtering of the latter result by the next filter, whose coefficients are duplicates of those of the backward prediction-error filter in reverse order, provides the samples of the sequence $u_{\mathrm{a}}(n)$, that is, the elements of the vector $\mathbf{u}_{\mathrm{a}}(n)$. It is also instructive to note that according to Eq. (11.81), the latter is noting but the forward equivalent of the backward prediction-error filter $H_{b_M}(z)$.

One may note that the filter output, $y(n)$, is obtained at the time when $x(n+M)$ is available at the input of Figure 11.13. This is equivalent to saying that there is a delay of M samples at the filter output as compared with the reference input. Although this delay could easily be prevented by shifting the delay box, z^{-M}, from the filter input to its output, we avoid this here to keep the analysis given in the next section as simple as possible. Shifting the delay box to the filter output introduces a delay into the adjustment loop of the filter. The result would then be a delayed LMS algorithm which is known to be inferior to its nondelayed version (Chapter 6). However, in the cases of interest, when $M \ll N$, the difference between the two algorithms is negligible.

Table 11.7 presents an implementation of Algorithm 2 that follows Figure 11.13, closely, except that we assume the input to the backward prediction-error filter to be $x(n)$ instead of $x(n+M)$. In this implementation, the backward prediction-error filter

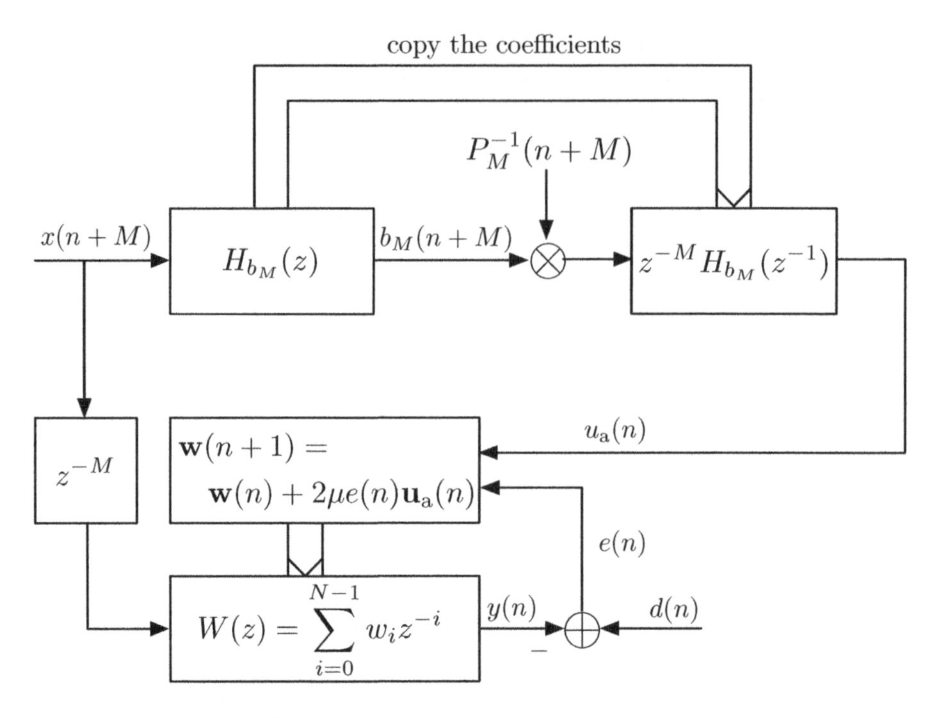

Figure 11.13 Block schematic diagram depicting Algorithm 2.

$H_{b_M}(z)$ is implemented in lattice form. Also note that the power-normalization factor $P_M^{-1}(n+M)$ is shifted to the output of the filter $z^{-M}H_{b_M}(z^{-1})$. Experiments have shown that this amendment results in a more reliable algorithm.

11.15.2 Performance Analysis

An analysis that reveals the differences between Algorithms 1 and 2 is presented. We assume that the input process, $x(n)$, is AR of order less than or equal to M. The predictors coefficients, $\{a_{i,j}$, for $i = 1, 2, \ldots, M$ and $j = 1, 2, \ldots, i\}$, and the corresponding mean-squared prediction error for different orders (i.e., the diagonal elements of $\mathbf{R}_{bb}$) are assumed to be known. In practice, when $M \ll N$, these assumptions are acceptable with a good approximation, as in that case the predictors coefficients will converge much faster than the adaptive filter tap weights and they will be jittering near their optimum setting after an initial transient. With these assumptions, one finds that $\mathbf{u}(n)$ is an exact estimate of $\mathbf{R}_{xx}^{-1}\mathbf{x}(n)$ and, therefore, Algorithm 1 will be an exact implementation of the ideal LMS–Newton algorithm, for which some theoretical results were presented in Chapter 7. We consider these results here as a base which determines the best performance that one may expect from Algorithm 1. Moreover, comparison of these results with what would be achieved by Algorithm 2, under the same ideal conditions, gives a good measure of the performance loss of Algorithm 2 as a result of simplification made in its structure.

Table 11.7 An LMS-based procedure for the implementation of Algorithm 2.

Given: Parameter vectors $\boldsymbol{\kappa}(n) = [\kappa_1(n)\ \kappa_2(n) \cdots \kappa_{M-1}(n)]^{\mathrm{T}}$
and $\mathbf{w}(n) = [w_0(n)\ w_1(n) \cdots w_{N-1}(n)]^{\mathrm{T}}$,
data vectors $\mathbf{x}(n)$, $\mathbf{b}(n-1)$ and $\mathbf{u}_{\mathrm{a}}(n-1)$, desired output $d(n)$,
and power estimates $P_0(n-1),\ P_1(n-1), \ldots, P_M(n-1)$.

Required: Vector updates $\boldsymbol{\kappa}(n)$, $\mathbf{w}(n+1)$, $\mathbf{b}(n)$ and $\mathbf{u}_{\mathrm{a}}(n)$,
and power estimate updates $P_0(n),\ P_1(n), \ldots, P_M(n)$.

⋆⋆⋆ Lattice Predictor Part ⋆⋆⋆

$f_0(n) = b_0(n) = x(n)$
$P_0(n) = \beta P_0(n-1) + 0.5(1-\beta)[f_0^2(n) + b_0^2(n-1)]$
for $m = 1$ to M
  $f_m(n) = f_{m-1}(n) - \kappa_m(n) b_{m-1}(n-1)$
  $b_m(n) = b_{m-1}(n-1) - \kappa_m(n) f_{m-1}(n)$
  $\kappa_m(n+1) = \kappa_m(n) + \frac{2\mu_{p,o}}{P_{m-1}(n)+\epsilon}[f_{m-1}(n) b_m(n) + b_{m-1}(n-1) f_m(n)]$
  $P_m(n) = \beta P_m(n-1) + 0.5(1-\beta)[f_m^2(n) + b_m^2(n-1)]$
  if $|\kappa_m(n)| > \gamma$, $\kappa_m(n) = \kappa_m(n-1)$
end

⋆⋆⋆ $\mathbf{u}_{\mathrm{a}}(n)$ update ⋆⋆⋆

$u_{\mathrm{a}}(n-j) = u_{\mathrm{a}}(n-j+1)$, for $j = N-1,\ N-2, \ldots, 2$
$f'_0(n) = b'_0(n) = b_M(n)$
for $m = 1$ to $M-1$
  $f'_m(n) = f'_{m-1}(n) - \kappa_m(n) b'_{m-1}(n-1)$
  $b'_m(n) = b'_{m-1}(n-1) - \kappa_m(n) f'_{m-1}(n)$
end
$u_{\mathrm{a}}(n) = (P_M(n) + \epsilon)^{-1}(f'_{M-1}(n) - \kappa_M(n) b'_{M-1}(n-1))$

⋆⋆⋆ Filtering and tap-weight vector adaptation ⋆⋆⋆

$y(n) = \mathbf{w}^{\mathrm{T}}(n)\mathbf{x}(n-M)$
$e(n) = d(n-M) - y(n)$
$\mathbf{w}(n+1) = \mathbf{w}(n) + 2\mu e(n)\mathbf{u}_{\mathrm{a}}(n)$

Under the ideal conditions stated above, the following results of the ideal LMS–Newton algorithm (presented in Chapter 7) are applicable to Algorithm 1.

- The algorithm does not suffer from any eigenvalue spread problem. It has only one mode of convergence which is characterized by the time constant

$$\tau = \frac{1}{4\mu} \tag{11.131}$$

- For small values of the step-size parameter, μ, its misadjustment is given by the equation

$$\mathcal{M}_1 = \frac{\mu N}{1 - \mu(N+2)} \tag{11.132}$$

- To guarantee the stability of the algorithm, its step-size parameter should remain within the limits

$$0 < \mu < \frac{1}{N+2} \tag{11.133}$$

The derivation of the above results has been based on a number of assumptions which we shall also assume here before proceeding to the analysis of Algorithm 2. A modeling problem such as Figure 11.10 is considered and the following assumptions are made:

1. The input samples, $x(n)$, and the desired output samples, $d(n)$, consist of jointly Gaussian-distributed random variables for all n.
2. At time n, $\mathbf{w}(n)$ is independent of the input vector $\mathbf{x}(n)$ and the desired output sample $d(n)$.
3. Noise samples $e_o(n)$, for all n, are zero-mean and uncorrelated with the input samples, $x(n)$.

The validity of the second assumption is justified for small values of μ, as discussed in Chapter 6. For the analysis of Algorithm 2, we extend these assumptions by replacing $\mathbf{x}(n)$ with $\mathbf{x}_{\mathrm{E}}(n)$, so that it extends to include the independence of $\mathbf{u}_{\mathrm{a}}(n)$ with $\mathbf{w}(n)$.

Now, we proceed with an analysis of Algorithm 2. First, we present an analysis of the convergence of $\mathbf{w}(n)$ in the mean which gives a result similar to Eq. (11.131). Next we analyze the convergence of $\mathbf{w}(n)$ in the variance to obtain the misadjustment of the algorithm. This analysis also reveals the effect of replacing $\mathbf{u}(n)$ by $\mathbf{u}_{\mathrm{a}}(n)$.

Convergence of Tap-Weight Vector in the Mean: We look at the convergence of $E[\mathbf{w}(n)]$ as n increases. To this end, we note that

$$e(n) = d(n) - \mathbf{w}^{\mathrm{T}}(n)\mathbf{x}(n) = e_o(n) - \mathbf{v}^{\mathrm{T}}(n)\mathbf{x}(n) \tag{11.134}$$

where $\mathbf{v}(n) = \mathbf{w}(n) - \mathbf{w}_o$ is the weight-error vector and from Figure 11.10, we have noted that $d(n) = \mathbf{w}_o^{\mathrm{T}}\mathbf{x}(n) + e_o(n)$. Substituting Eq. (11.133) in Eq. (11.127), we get

$$\mathbf{v}(n+1) = (\mathbf{I} - 2\mu\mathbf{u}_{\mathrm{a}}(n)\mathbf{x}^{\mathrm{T}}(n))\mathbf{v}(n) + 2\mu e_o(n)\mathbf{u}_{\mathrm{a}}(n) \tag{11.135}$$

where $\mathbf{I}$ denotes the identity matrix with appropriate dimension. Taking expectation and using the Assumptions 2 and 3 listed above, we obtain

$$E[\mathbf{v}(n+1)] = (\mathbf{I} - 2\mu E[\mathbf{u}_{\mathrm{a}}(n)\mathbf{x}^{\mathrm{T}}(n)])E[\mathbf{v}(n)] \tag{11.136}$$

To evaluate $E[\mathbf{u}_{\mathrm{a}}(n)\mathbf{x}(n)]$, we first define

$$\mathbf{u}_{\mathrm{E}}(n) = \mathbf{R}_{x_{\mathrm{E}}x_{\mathrm{E}}}^{-1}\mathbf{x}_{\mathrm{E}}(n) \tag{11.137}$$

where $\mathbf{R}_{x_{\mathrm{E}}x_{\mathrm{E}}} = E[\mathbf{x}_{\mathrm{E}}(n)\mathbf{x}_{\mathrm{E}}^{\mathrm{T}}(n)]$, and note that postmultiplying Eq. (11.137) by $\mathbf{x}_{\mathrm{E}}^{\mathrm{T}}(n)$ and taking expectation on both sides gives

$$E[\mathbf{u}_{\mathrm{E}}(n)\mathbf{x}_{\mathrm{E}}^{\mathrm{T}}(n)] = \mathbf{I} \tag{11.138}$$

This shows that the cross-correlation between the elements of $\mathbf{u}_{\mathrm{E}}(n)$ and $\mathbf{x}_{\mathrm{E}}(n)$ that are at the same position are unity and equal to zero for the other elements of the two vectors.

Clearly, this is also applicable to the elements of $\mathbf{u}_\mathrm{a}(n)$ and $\mathbf{x}(n)$, as they are truncated versions of $\mathbf{u}_\mathrm{E}(n)$ and $\mathbf{x}_\mathrm{E}(n)$, respectively. Thus,

$$E[\mathbf{u}_\mathrm{a}(n)\mathbf{x}^\mathrm{T}(n)] = \mathbf{I} \tag{11.139}$$

and therefore,

$$E[\mathbf{v}(n+1)] = (1-2\mu)E[\mathbf{v}(n)]. \tag{11.140}$$

This shows that, similar to Algorithm 1, Algorithm 2 is also governed by a single mode of convergence. Furthermore, the time constant Eq. (11.131) is also applicable to Algorithm 2.

Convergence of Tap-Weight Vector in the Mean Square: We first develop a recursive equation for the time evolution of the correlation matrix of the weight-error vector $\mathbf{v}(n)$, which is defined as $\mathbf{K}(n) = E[\mathbf{v}(n)\mathbf{v}^\mathrm{T}(n)]$. For this, we find the outer products of the left- and right-hand sides of Eq. (11.135) and take expectation on both sides of the resulting equation. Then, using Assumptions 2 and 3 listed above, we obtain

$$\begin{aligned}
\mathbf{K}(n+1) &= E[(\mathbf{I}-2\mu\mathbf{u}_\mathrm{a}(n)\mathbf{x}^\mathrm{T}(n))\mathbf{K}(n)(\mathbf{I}-2\mu\mathbf{x}(n)\mathbf{u}_\mathrm{a}^\mathrm{T}(n))] \\
&\qquad + 4\mu^2\xi_{\min}\mathbf{R}_{u_\mathrm{a}u_\mathrm{a}} \\
&= \mathbf{K}(n) - 2\mu E[\mathbf{u}_\mathrm{a}(n)\mathbf{x}^\mathrm{T}(n)]\mathbf{K}(n) - 2\mu\mathbf{K}(n)E[\mathbf{x}(n)\mathbf{u}_\mathrm{a}^\mathrm{T}(n)] \\
&\qquad + 4\mu^2 E[\mathbf{u}_\mathrm{a}(n)\mathbf{x}^\mathrm{T}(n)\mathbf{K}(n)\mathbf{x}(n)\mathbf{u}_\mathrm{a}^\mathrm{T}(n)] + 4\mu^2\xi_{\min}\mathbf{R}_{u_\mathrm{a}u_\mathrm{a}} \\
&= (1-4\mu)\mathbf{K}(n) + 4\mu^2 E[\mathbf{u}_\mathrm{a}(n)\mathbf{x}^\mathrm{T}(n)\mathbf{K}(n)\mathbf{x}(n)\mathbf{u}_\mathrm{a}^\mathrm{T}(n)] \\
&\qquad + 4\mu^2\xi_{\min}\mathbf{R}_{u_\mathrm{a}u_\mathrm{a}}
\end{aligned} \tag{11.141}$$

where $\xi_{\min} = E[e_\mathrm{o}^2(n)]$ is the minimum mean-squared error at the adaptive filter output and $\mathbf{R}_{u_\mathrm{a}u_\mathrm{a}} = E[\mathbf{u}_\mathrm{a}(n)\mathbf{u}_\mathrm{a}^\mathrm{T}(n)]$.

The second term on the right-hand side of Eq. (11.141) can be evaluated by following a procedure similar to the one given in Chapter 6, Appendix 6A, for the case of the conventional LMS algorithm. This results in (Appendix 11A for the derivation)

$$E[\mathbf{u}_\mathrm{a}(n)\mathbf{x}^\mathrm{T}(n)\mathbf{K}(n)\mathbf{x}(n)\mathbf{u}_\mathrm{a}^\mathrm{T}(n)] = \mathbf{R}_{u_\mathrm{a}u_\mathrm{a}}\mathrm{tr}[\mathbf{K}(n)\mathbf{R}_{xx}] + 2\mathbf{K}(n) \tag{11.142}$$

Using this result in Eq. (11.141), we obtain

$$\begin{aligned}
\mathbf{K}(n+1) &= (1-4\mu+8\mu^2)\mathbf{K}(n) + 4\mu^2\mathbf{R}_{u_\mathrm{a}u_\mathrm{a}}\mathrm{tr}[\mathbf{K}(n)\mathbf{R}_{xx}] \\
&\qquad + 4\mu^2\xi_{\min}\mathbf{R}_{u_\mathrm{a}u_\mathrm{a}}
\end{aligned} \tag{11.143}$$

Next, we recall from Chapter 6 that the excess mean-squared error of an adaptive filter with input and weight-error correlation matrices $\mathbf{R}_{xx}$ and $\mathbf{K}(n)$, respectively, is given by

$$\xi_{ex}(n) = \mathrm{tr}[\mathbf{K}(n)\mathbf{R}_{xx}] \tag{11.144}$$

Postmultiplying Eq. (11.143) on both sides by $\mathbf{R}_{xx}$ and equating the traces of the two sides of the resulting equation, we obtain

$$\begin{aligned}
\xi_{ex}(n+1) &= (1-4\mu+4\mu^2(\mathrm{tr}[\mathbf{R}_{u_\mathrm{a}u_\mathrm{a}}\mathbf{R}_{xx}]+2))\xi_{ex}(n) \\
&\qquad + 4\mu^2\xi_{\min}\mathrm{tr}[\mathbf{R}_{u_\mathrm{a}u_\mathrm{a}}\mathbf{R}_{xx}]
\end{aligned} \tag{11.145}$$

From Eq. (11.145), we note that the convergence of Algorithm 2 is guaranteed if

$$|1 - 4\mu + 4\mu^2(\mathrm{tr}[\mathbf{R}_{u_a u_a}\mathbf{R}_{xx}] + 2)| < 1 \tag{11.146}$$

This gives

$$0 < \mu < \frac{1}{\mathrm{tr}[\mathbf{R}_{u_a u_a}\mathbf{R}_{xx}] + 2} \tag{11.147}$$

Also, when $n \to \infty$, $\xi(n+1) = \xi(n)$. Using this in Eq. (11.145), we obtain

$$\mathcal{M}_2 = \frac{\xi_{ex}(\infty)}{\xi_{\min}} = \frac{\mu\mathrm{tr}[\mathbf{R}_{u_a u_a}\mathbf{R}_{xx}]}{1 - \mu(\mathrm{tr}[\mathbf{R}_{u_a u_a}\mathbf{R}_{xx}] + 2)} \tag{11.148}$$

This is the misadjustment equation for Algorithm 2. The above results reduce to those of Algorithm 1 if $\mathbf{R}_{u_a u_a}$ is replaced by $\mathbf{R}_{uu} = E[\mathbf{u}(n)\mathbf{u}^{\mathrm{T}}(n)]$ and one notes that $\mathbf{R}_{uu} = \mathbf{R}_{xx}^{-1}$.

In view of Eqs. (11.132) and (11.148), a good measure for comparing Algorithms 1 and 2 is the ratio

$$\gamma = \frac{\mathrm{tr}[\mathbf{R}_{u_a u_a}\mathbf{R}_{xx}]}{N} \tag{11.149}$$

A value of $\gamma > 1$ indicates that Algorithm 1 performs better than Algorithm 2. Furthermore, the larger the value of γ, the greater would be the loss in replacing Algorithm 1 by Algorithm 2. However, if $\gamma \approx 1$, then the two algorithms perform about the same.

An evaluation of the parameter γ is provided in Appendix 11B. It is shown that γ is always greater than unity. This means that there is always a penalty to be paid for the simplification made in replacing the vector $\mathbf{u}(n)$ of Algorithm 1, by the vector $\mathbf{u}_{\mathrm{a}}(n)$ of Algorithm 2. The amount of loss depends on the statistics of the input process, $x(n)$, and the filter length N. Fortunately, the evaluation provided in Appendix 11B shows that γ approaches one as N increases. This means that the difference between the two algorithms may be insignificant for long filters. Numerical examples which verify this are given next.

11.15.3 Simulation Results and Discussion

We present some simulation results using the input process $x_1(n)$, which was introduced in Section 11.14, and also two other processes, $x_2(n)$ and $x_3(n)$, which are generated using the coloring filters

$$H_2(z) = \frac{K_2}{1 - 0.650z^{-1} + 0.693z^{-2} - 0.220z^{-3} + 0.309z^{-4} - 0.177z^{-5}} \tag{11.150}$$

and

$$H_3(z) = \frac{K_3}{1 - 2.059z^{-1} + 2.312z^{-2} - 1.893z^{-3} + 1.148z^{-4} - 0.293z^{-5}} \tag{11.151}$$

respectively. The coefficients K_2 and K_3 are selected equal to 0.7208 and 0.2668, respectively, to normalize the resulting processes to unit power. We note that $x_2(n)$ and $x_3(n)$ are AR processes, but $x_1(n)$ is not.

To verify the theoretical results presented above, we start with some experiments using the AR processes $x_2(n)$ and $x_3(n)$. Figure 11.14 shows the power spectral densities of

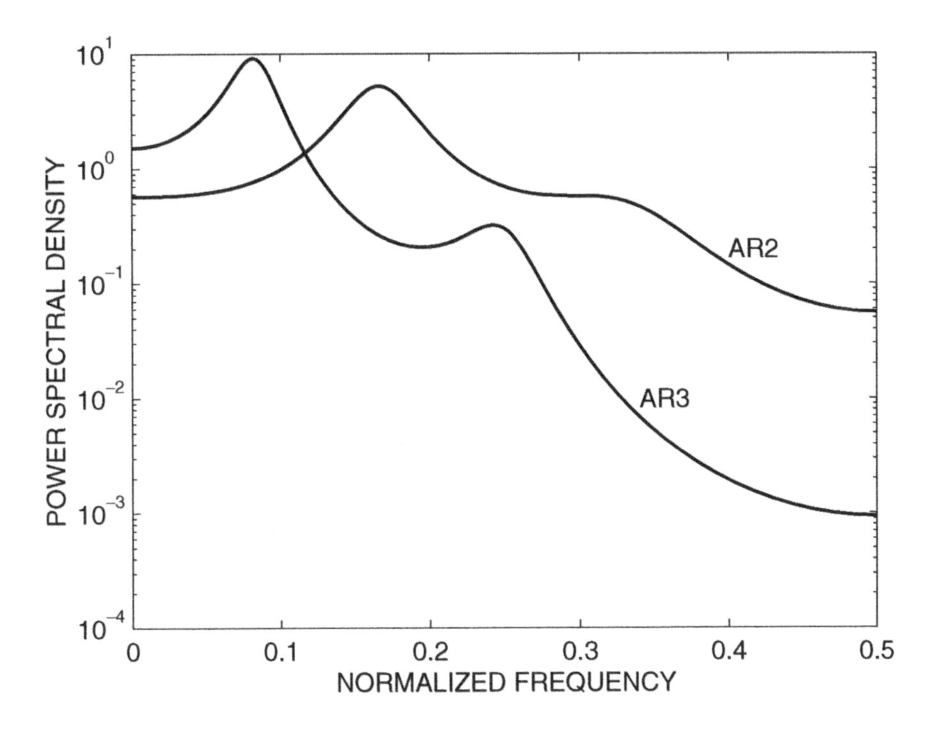

Figure 11.14 Power spectral densities of $x_2(n)$ (AR2) and $x_3(n)$ (AR3).

$x_2(n)$ and $x_3(n)$. From Chapter 4, we recall that the eigenvalue spread of the correlation matrix of a process is asymptotically determined by the maximum and minimum of its power spectral density. Noting this, we find that the eigenvalue spread of $x_2(n)$ is in the range of 100 and that of $x_3(n)$ can be as large as 10,000. This shows that $x_3(n)$ is a very badly conditioned process and one should expect difficulties in estimating the inverse of its correlation matrix.

To shed light on the differences between Algorithm 1 and Algorithm 2, we first present some simulation results for the case when the exact models of the AR inputs are known *a priori*. In this case, Algorithm 1 will be an exact implementation of the LMS–Newton algorithm and gives a good base for further comparisons. Figure 11.15 shows the variation of the parameter γ as a function of the filter length, N, for $x_2(n)$ and $x_3(n)$. As one may expect, the process $x_3(n)$, which is suffering from a serious eigenvalue spread problem, shows higher sensitivity toward replacing Algorithm 1 by Algorithm 2. However, as N increases, γ approaches 1 and, therefore, the two algorithms are expected to perform about the same.

Figures 11.16 and 11.17 show the simulation results for the inputs $x_2(n)$ and $x_3(n)$ and a filter length $N = 30$. These results as well as those presented in the rest of this section are averaged over 50 independent runs. The results are then smoothed so that the various curves could be distinguished. The step-size parameter, μ, is selected equal to $0.1/N$, for all the results. This, according to Eq. (11.132), results in about 10% misadjustment for Algorithm 1. According to the results of Figure 11.15 and Eqs. (11.132) and (11.148),

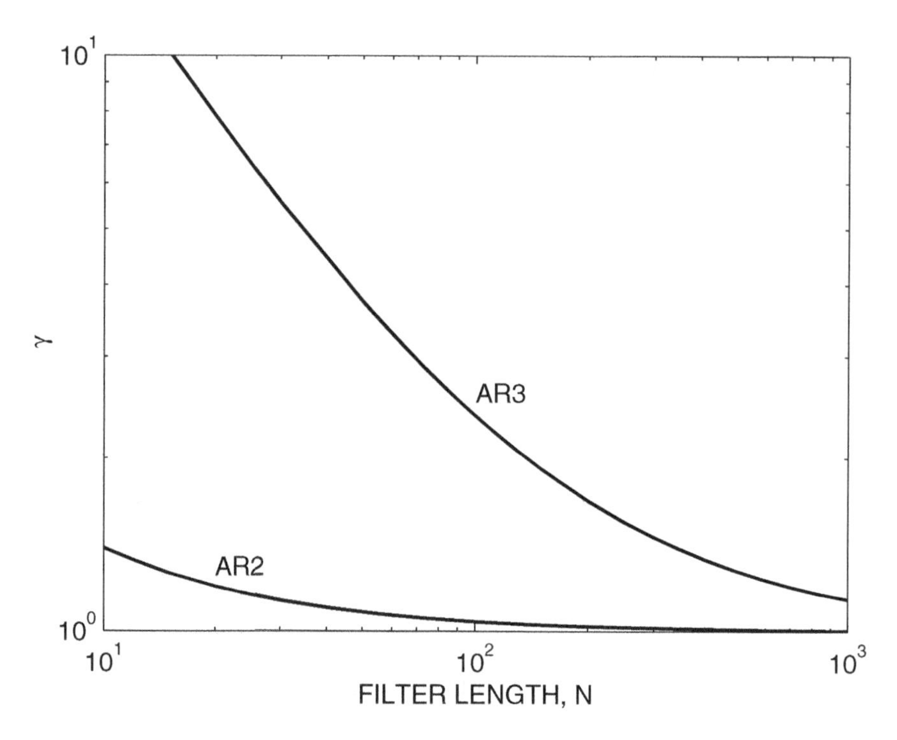

Figure 11.15 Variation of the parameter γ as a function of filter length for $x_2(n)$ (AR2) and $x_3(n)$ (AR3).

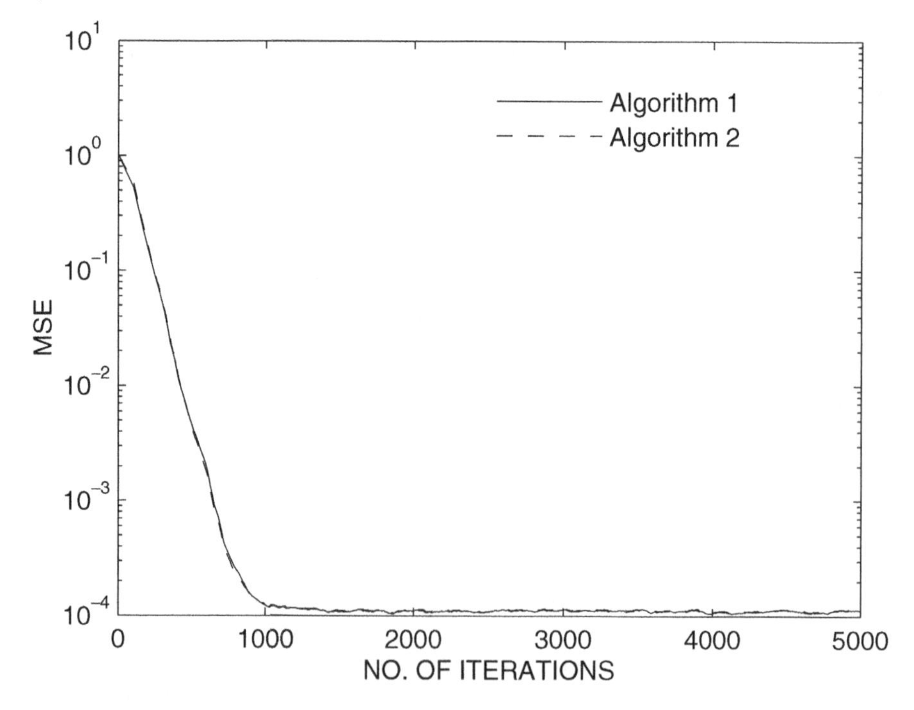

Figure 11.16 MSE versus iteration number for $x_2(n)$, $N = 30$ and with the AR model of input assumed known.

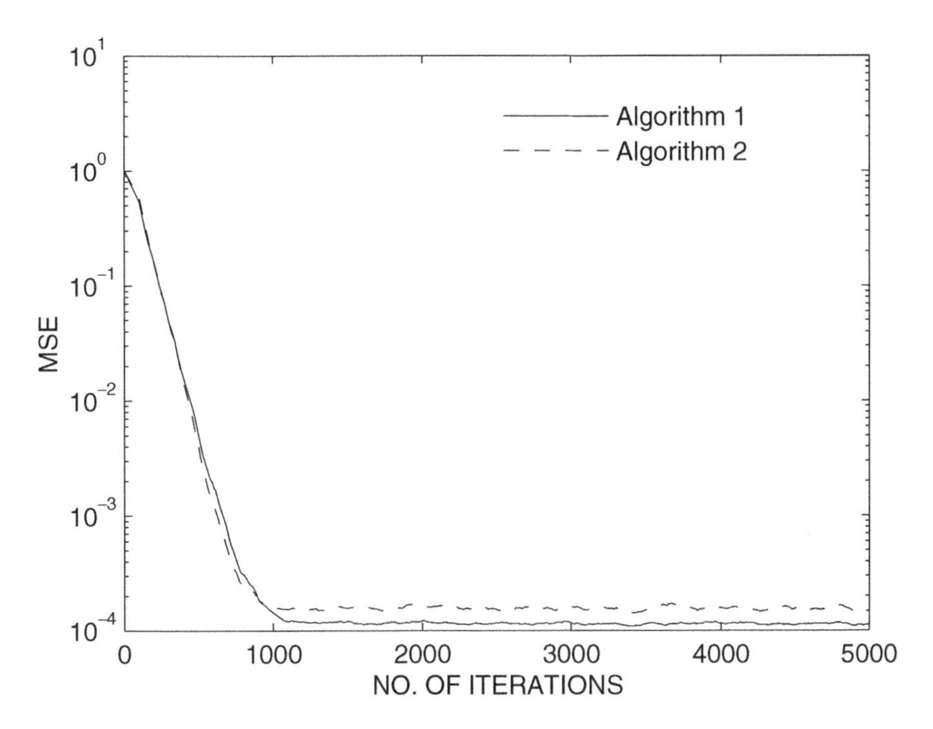

Figure 11.17 MSE versus iteration number for $x_3(n)$, $N = 30$ and with the AR model of input assumed known.

both algorithms should approach to about the same misadjustment, in the case of $x_2(n)$. However, their performance may be significantly different, in the case of $x_3(n)$. To be more exact, from the data used for generating Figure 11.15, we have $\gamma = 1.13$, for $x_2(n)$, and $\gamma = 5.57$, for $x_3(n)$, for $N = 30$. Using these and Eqs. (11.132) and (11.148), we obtain the following:

For $x_2(n)$: $\frac{\mathcal{M}_2}{\mathcal{M}_1} = 1.147.$

For $x_3(n)$: $\frac{\mathcal{M}_2}{\mathcal{M}_1} = 11.32.$

Careful examination of the numerical values which have been obtained by simulations show that for the $x_2(n)$ process $\frac{\mathcal{M}_2}{\mathcal{M}_1} = 1.152$. This matches well with the above ratio. However, for the $x_3(n)$ process the simulation results give $\frac{\mathcal{M}_2}{\mathcal{M}_1} = 3.85$. This which does not match the above theoretical ratio, may be explained as follows. Careful examination of the numerical results in simulations reveals that there are only a few terms in $\mathbf{u}_a(n)$ that have a major effect on the degradation of Algorithm 2, when compared with Algorithm 1. These terms, which greatly disturb the first and last few elements of the tap-weight vector $\mathbf{w}(n)$, are so large that their contribution violates the independence assumption 2 of the previous section. As a result, the theoretical derivation that led to Eq. (11.148) may not be valid unless the step-size parameter μ is set to a very small value so that the latter assumption could be justified. Nevertheless, the developed theory is able to predict

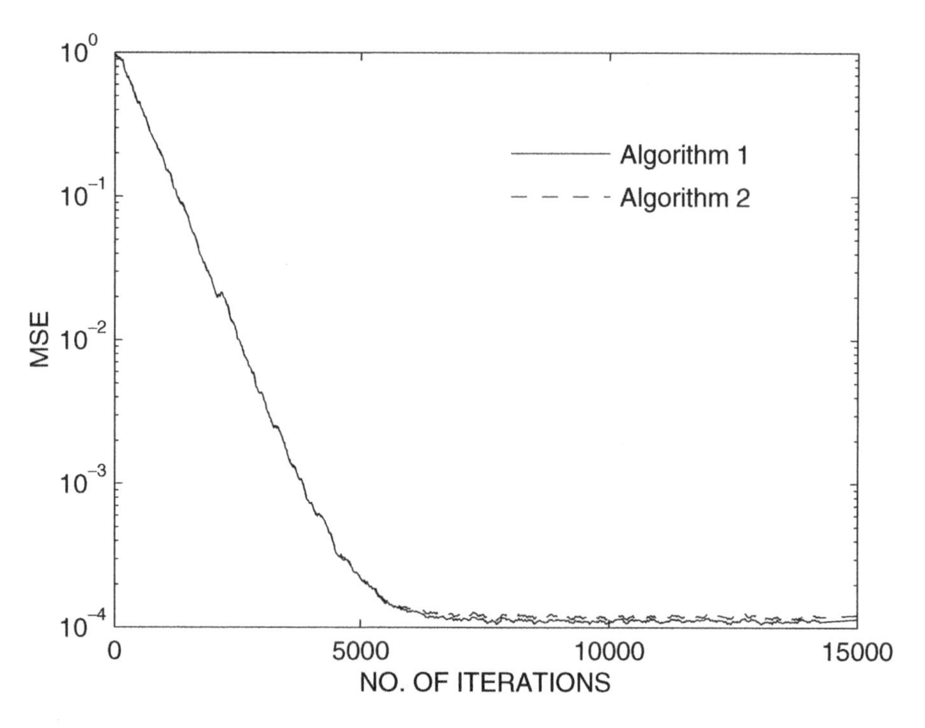

Figure 11.18 MSE versus iteration number for $x_3(n)$, $N = 200$ and with the AR model of input assumed known.

conditions under which Algorithm 2 is more likely to go unstable; namely, when the adaptive filter input is highly colored.

In order to support, the prediction made by the theory that the two algorithms perform about the same for long filters, we present another simulation example with the process $x_3(n)$ as the filter input. This time we increase the length of the filter, N, to 200. Figure 11.18 shows the results of this test. For this scenario, the theory gives $\frac{\mathcal{M}_2}{\mathcal{M}_1} = 1.69$ and simulation gives $\frac{\mathcal{M}_2}{\mathcal{M}_1} = 1.64$, a good match, as was predicted.

Next, the simulation results of more realistic cases when the input process is unknown and its model has to be estimated along with the adaptive filter tap weights, are presented. We present some results for Algorithm 2. The simulation program that we use follows Table 11.7. The following parameters are used: $\beta = 0.95$, $\gamma = 0.9$, $\epsilon = 0.02$, $\mu_{\text{p},o} = 0.01$, $\mu = 0.1/N$, and $N = 30$. Figure 11.19a, b, and c shows the simulation results for the processes $x_1(n)$, $x_2(n)$, and $x_3(n)$, respectively. The results are given for the conventional LMS algorithm and Algorithm 2, for the cases where the order of AR model, M, is set equal to 1, 2, 3, and 5. The results clearly show the improvement achieved by the AR modeling. We note that for $x_2(n)$ and $x_3(n)$, even a first-order modeling of the input processes results in significant improvement in convergence, compared to the conventional LMS algorithm. However, for $x_1(n)$ a modeling order of 2 or above is required to achieve some improvement.

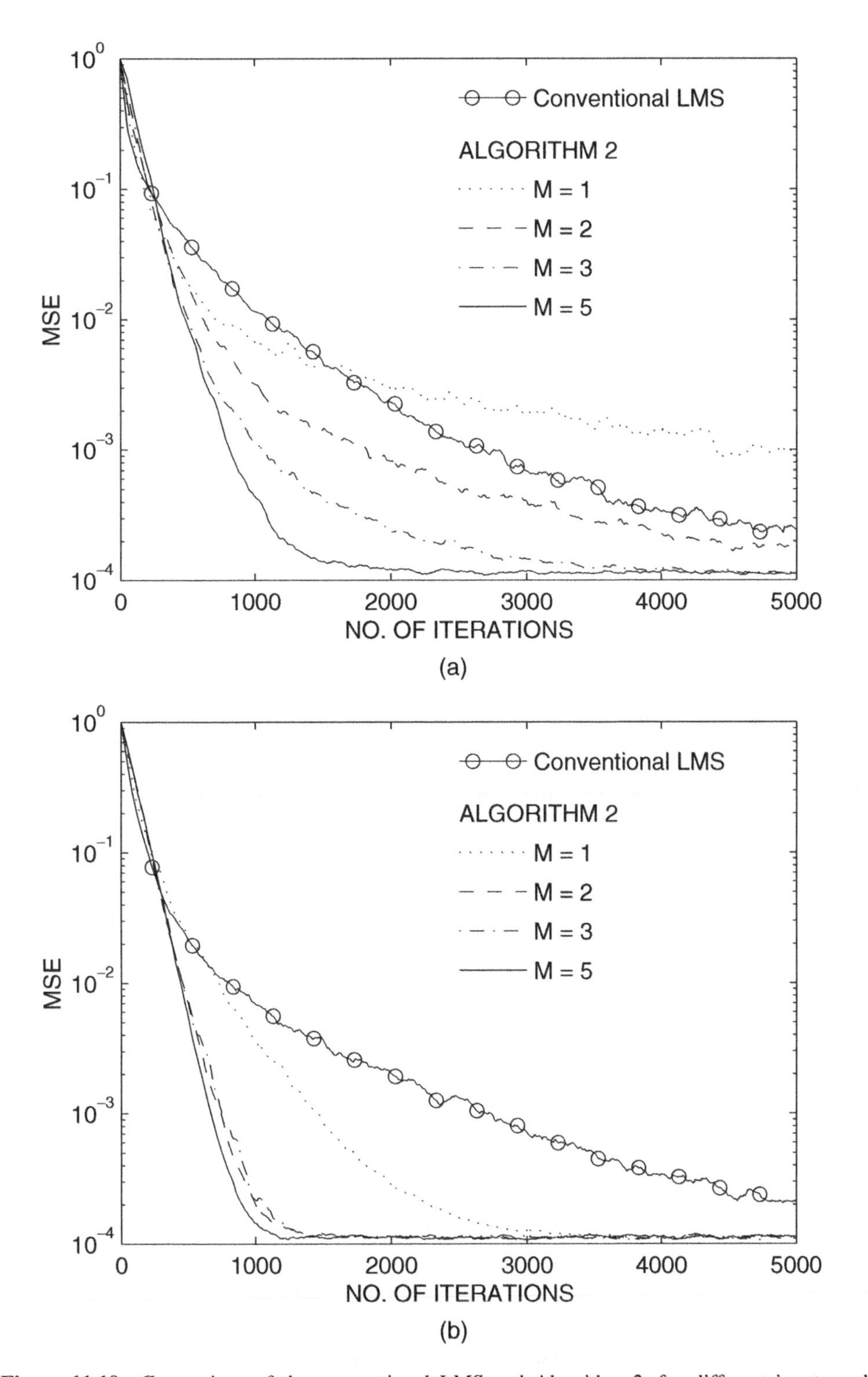

Figure 11.19 Comparison of the conventional LMS and Algorithm 2, for different inputs and various orders of AR model: (a) input process $x_1(n)$, (b) input process $x_2(n)$, and (c) input process $x_3(n)$.

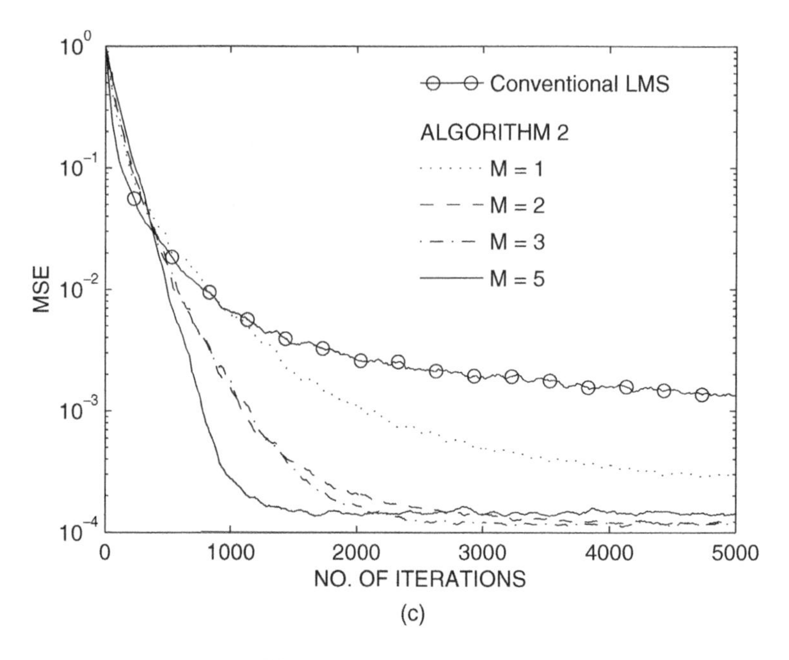

Figure 11.19 *(continued)*

Problems

P11.1 Give a detailed proof of Eq. (11.52) and find that such proof leads to Eq. (11.53).

P11.2 Define the m-by-m matrix $\mathbf{J}$ whose ijth element is 1 for $j = m - i + 1$ and 0 for all other $i, j \in \{1, 2, \ldots, m\}$. This is called an *exchange matrix*. Show that if $\mathbf{R}$ is the correlation matrix of a stationary stochastic process, then $\mathbf{JRJ} = \mathbf{R}$. Use this result to derive an alternate proof for Eq. (11.19) or Eq. (11.20).

P11.3 Using the procedure mentioned in Problem P4.8, show that for the matrix $\mathbf{L}$ as defined in Eq. (11.70)

$$\det(\mathbf{L}) = 1$$

Comment on the invertibility of $\mathbf{L}$.

P11.4 Give a detailed derivation of Eq. (11.81).

P11.5 Consider the order-update equations (11.60) and (11.61). Show that the optimization of κ_{m+1} in either of the two equations for minimization of the corresponding higher order errors in the mean-square sense gives Eq. (11.59).

P11.6 Equation (11.90) may be rearranged as

$$r(m+1) = P_m \kappa_{m+1} + \sum_{i=1}^{m} a_{m,i} r(m+1-i)$$

Use this result to develop procedures for the following:

1. Conversion of the set of coefficients $(P_0, \kappa_1, \kappa_2, \ldots, \kappa_M)$ to $(r(0), r(1), \ldots, r(M))$.
2. Conversion of the set of coefficients $(P_0, a_{M,1}, a_{M,2}, \ldots, a_{M,M})$ to $(r(0), r(1), \ldots, r(M))$.

P11.7 Derive the recursion used for obtaining the transversal predictor coefficients w_i in Table 11.4.

P11.8 Give the lattice equivalent of the forward prediction-error filter which is characterized by the system function

$$H_{f_4}(z) = 1 - 0.5z^{-2} + 0.5z^{-3} + 0.25z^{-4}$$

P11.9 Give the transversal equivalent of the third-order forward and backward prediction-error filters of a process which is characterized by the PARCOR coefficients

$$\kappa_1 = 0.8$$
$$\kappa_2 = 0.5$$
$$\kappa_3 = -0.2$$

P11.10 Find the lattice realization of the system function

$$H(z) = \frac{1}{1 - az^{-1}}$$

P11.11 Find the lattice realization of the system function

$$H(z) = \frac{1}{1 - 1.5z^{-1} + 0.56z^{-2}}$$

P11.12 Find the lattice realization of the system function

$$H(z) = \frac{1 + z^{-1} + 2z^{-2}}{1 - 1.2z^{-1} + 0.5z^{-2}}$$

P11.13 Use the Levinson–Durbin algorithm to find the fourth-order transversal and lattice predictors of a process $x(n)$, which is characterized by the correlation coefficients

$$r(0) = 5, \qquad r(1) = 3, \qquad r(2) = -1, \qquad r(3) = 2, \qquad r(4) = -0.5$$

P11.14 Use the Levinson–Durbin algorithm to solve the system of equations

$$\begin{bmatrix} 1.0 & 0.8 & -0.5 & 0.2 \\ 0.8 & 1.0 & 0.8 & -0.5 \\ -0.5 & 0.8 & 1.0 & 0.8 \\ 0.2 & -0.5 & 0.8 & 1.0 \end{bmatrix} \begin{bmatrix} w_0 \\ w_1 \\ w_2 \\ w_3 \end{bmatrix} = \begin{bmatrix} 0.8 \\ -0.5 \\ 0.2 \\ 0 \end{bmatrix}$$

P11.15 Use the extended Levinson–Durbin algorithm to solve the system of equations

$$\begin{bmatrix} 1.0 & 0.8 & -0.5 & 0.2 \\ 0.8 & 1.0 & 0.8 & -0.5 \\ -0.5 & 0.8 & 1.0 & 0.8 \\ 0.2 & -0.5 & 0.8 & 1.0 \end{bmatrix} \begin{bmatrix} w_0 \\ w_1 \\ w_2 \\ w_3 \end{bmatrix} = \begin{bmatrix} 0.5 \\ 1 \\ 0.5 \\ 0 \end{bmatrix}$$

P11.16 Consider the case where the input $x(n)$ to a lattice predictor is an AR process obtained by passing a unit variance white noise through a system with the transfer function

$$H(z) = \frac{1}{1 - 1.1z^{-1} + 0.3z^{-2}}$$

(i) Find the correlation coefficients $r(0), r(1), r(2), \ldots$ of $x(n)$.
(ii) Using the results of (i), find the PARCOR coefficients κ_i of the lattice predictor and show that $\kappa_i = 0$, for $i > 2$.
(iii) Could you predict the fact that $\kappa_i = 0$, for $i > 2$, without performing the calculations in (ii)? Explain.
(iv) The result in (ii) implies that the backward prediction error $b_2(n)$ is a white process. Prove this. What is the variance of $b_2(n)$?

P11.17 Consider the case where the input $x(n)$ to a lattice predictor is a moving average (MA) process obtained by passing a unit variance white noise through a system with the transfer function

$$H(z) = 1 - 1.1z^{-1} + 0.3z^{-2}$$

(i) Find the correlation coefficients $r(0), r(1), r(2), \ldots$ of $x(n)$.
(ii) Using the results of (i), find the PARCOR coefficients κ_i of the lattice predictor, for $i = 1$ through 6.
(iii) You should have found in (ii) that $|\kappa_i|$ decreases as i increases. Using this observation, can you argue that as i increase, $b_i(n)$ approaches a white noise with unit variance? Explain.

P11.18 Repeat problem P11.17 for the case where

$$H(z) = \frac{1 - 1.1z^{-1} + 0.3z^{-2}}{1 - 0.3z^{-1}}$$

P11.19 Consider an AR process which is described by the difference equation

$$x(n) = 0.6x(n-1) + 0.67x(n-2) + 0.36x(n-3) + \nu(n)$$

where $\nu(n)$ is a zero-mean white noise process with variance of unity.

(i) Find the system function $H(z)$ which relates $\nu(n)$ and $x(n)$.
(ii) Show that the poles of $H(z)$ are 0.9, -0.8, and 0.6.
(iii) Find the power of $x(n)$.
(iv) Find the PARCOR coefficients κ_1, κ_2, and κ_3 of $x(n)$.
(v) Find the prediction-error powers P_1, P_2, and P_3 of $x(n)$.
(vi) Comment on the values of κ_m and P_m, for values of $m \geq 4$.
(vii) Using the procedure developed in the Problem P11.6, find the autocorrelation coefficients $r(0), r(1), \ldots, r(5)$ of $x(n)$.

P11.20 For a real-valued process $x(n)$ with mth-order forward and backward prediction errors $f_m(n)$ and $b_m(n)$, respectively, prove the following results.

(i) $E[f_5(n)f_3(n-2)] = 0$
(ii) $E[b_5(n)b_2(n-2)] = 0$
(iii) $E[f_m(n)x(n)] = E[f_m^2(n)]$
(iv) $E[b_m(n)x(n-m)] = E[b_m^2(n)]$
(v) For $0 < k < m$, $E[f_m(n)f_{m-k}(n-k)] = 0$
(vi) For $0 < k < m$, $E[b_m(n)b_{m-k}(n-k)] = E[b_m^2(n)]$

P11.21 For a real-valued process $x(n)$ with mth-order forward and backward prediction errors $f_m(n)$ and $b_m(n)$, respectively, find the range of i for which the following results hold.

(i) $E[f_m(n)f_{m-k}(n-i)] = 0$
(ii) $E[b_m(n)b_{m-k}(n-i)] = 0$
(iii) $E[f_m(n)b_{m-k}(n-i)] = 0$
(iv) $E[b_m(n)f_{m-k}(n-i)] = 0$

P11.22 Consider a complex-valued process $x(n)$.

(i) Using the principle of orthogonality, derive the Wiener–Hopf equation that gives the coefficients $a_{m,i}$ of order m forward linear predictor of $x(n)$.
(ii) Repeat (i) to derive the coefficients $g_{m,i}$ of order m backward linear predictor of $x(n)$.
(iii) Show that

$$g_{m,i} = a^*_{m,m+1-i} \quad \text{for } i = 1, 2, \ldots, m$$

P11.23 In the case of complex-valued signals, the order-update equations (11.60) and (11.61) take the following forms:

$$f_{m+1}(n) = f_m(n) - \kappa_{m+1} b_m(n-1)$$

and

$$b_{m+1}(n) = b_m(n-1) - \kappa^*_{m+1} f_m(n)$$

where the asterisk denotes complex conjugation. Give a detailed proof of these equations and show that

$$\kappa_{m+1} = \frac{E[f_m(n)b^*_m(n-1)]}{P_m}$$

where $P_m = E[|f_m(n)|^2] = E[|b_m(n-1)|^2]$.

P11.24 For the case of complex-valued signals, propose a lattice joint process estimator similar to Figure 11.7 and develop an LMS algorithm for its adaptation.

P11.25 For a complex-valued process $x(n)$ prove the following properties:

(i) $E[f_m(n)x^*(n-k)] = 0, \quad \text{for } 1 \le k \le m$
(ii) $E[b_m(n)x^*(n-k)] = 0, \quad \text{for } 0 \le k \le m-1$
(iii) $E[b_k(n)b^*_l(n)] = 0, \quad \text{for } k \ne l$

P11.26 Consider the difference equation (11.114) and note that it may be written as

$$x(n) = \mathbf{x}_M^{\mathrm{T}}(n-1)\mathbf{h} + \nu(n) \tag{P11.26.1}$$

where $\mathbf{x}_M(n) = [x(n-1)\ x(n-2) \cdots x(n-M)]^{\mathrm{T}}$ and $\mathbf{h} = [h_1\ h_2 \cdots h_M]^{\mathrm{T}}$.

(i) Starting with Eq. (P11.26.1) show that the vector $\mathbf{h}$ is related to the autocorrelation coefficients $r(0), r(1), \ldots, r(M)$ of $x(n)$ according to the equation

$$\mathbf{R}_M \mathbf{h} = \mathbf{r}_M$$

where

$$\mathbf{R}_M = \begin{bmatrix} r(0) & r(1) & \cdots & r(M-1) \\ r(1) & r(0) & \cdots & r(M-2) \\ \vdots & \vdots & \ddots & \vdots \\ r(M-1) & r(M-2) & \cdots & r(0) \end{bmatrix}$$

and

$$\mathbf{r}_M = [r(1)\ r(2) \cdots r(M)]^{\mathrm{T}}$$

(ii) Show that for any m

$$r(m) = \sum_{i=1}^{M} h_i r(m-i)$$

(iii) By combining the results of (i) and (ii) show that for any $M' > M$,

$$\mathbf{R}_{M'} \mathbf{h}' = \mathbf{r}_{M'}$$

where $\mathrm{h}' = \begin{bmatrix} \mathbf{h} \\ \mathbf{0} \end{bmatrix}$ and $\mathbf{0}$, here, is the length $M' - M$ zero column vector.

(iv) Use the above results to justify the validity of the results presented in Eqs. (11.115) and (11.116).

P11.27 Suggest a lattice structure for realization of the transfer function $W(z)$ of the IIR line enhancer of Section 10.3 and obtain its coefficients in terms of the parameters s and w.

P11.28 Give a detailed derivation of Eq. (11.123).

P11.29 Give a derivation of Eq. (11.147) from Eq. (11.146).

P11.30 Recall that the unconstrained PFBLMS algorithm of Chapter 8 converges very slowly when the partition length, M, and the block length, L, are equal. In Section 8.4, it was noted the slow convergence of PFBLMS algorithm can be improved by choosing M a few times larger than L. We also noticed that the frequency domain processing involved in the implementation of the PFBLMS algorithm may be viewed as a parallel bank of a number of transversal filters, each belonging to one of the frequency bins. Furthermore, when the number of frequency bins is large, these filters operate (converge) almost independent of one another. Noting these, the following alternative solution may be proposed to improve the convergence behavior of PFBLMS algorithm. We may keep $L = M$ and use a lattice structure for decorrelating the samples of the input signal at each frequency bin. Explore this solution. In particular, note that when $L = M$, the autocorrelation coefficients of the signal samples at various frequency bins are known *a priori*. Explain how these known information can be exploited in the proposed implementation.

Computer-Oriented Problems

P11.31 Write a simulation program to confirm the results presented in Figure 11.12. If you are looking for a short-cut and you have access to the MATLAB software package, you may study and use the program `ltc_mdlg.m` on the accompanying website. Also, run `ltc_mdlg.m` or your program for the following values of the step-size parameters $\mu_{p,o}$ and $\mu_{c,o}$ and observe the impact of those on the performance of the algorithm. Comment on your observations.

$\mu_{p,o}$	$\mu_{c,o}$
0.01	0.003
0.001	0.010
0.001	0.001
0.0001	0.003

P11.32 Consider a process $x(n)$ which is characterized by the difference equation

$$x(n) = 1.2x(n-1) - 0.8x(n-2) + \nu(n) + \alpha\nu(n-1)$$

where α is a parameter and $\nu(n)$ is a zero-mean unit variance white process.

(i) Derive an equation for the power spectral density $\Phi_{xx}(e^{j\omega})$, of $x(n)$, and plot that for values of $\alpha = 0$, 0.5, 0.8, and 0.95.

(ii) The autocorrelation coefficients $r(0), r(1), \ldots$ of $x(n)$ can be numerically obtained by evaluating the inverse discrete Fourier transform (DFT) of samples of $\Phi_{xx}(e^{j\omega})$ taken at equally spaced intervals. The number of samples of $\Phi_{xx}(e^{j\omega})$ used for this purpose should be large enough to give an accurate result. Use this method to obtain the autocorrelation coefficients of $x(n)$.

(iii) Use the results of (ii) to obtain the system functions of the backward prediction-error filters of $x(n)$ for predictor orders of 2, 5, 10, and 20, and values of $\alpha = 0$, 0.5, 0.8, and 0.95.

(iv) Plot the power spectral density $\Phi_{b_m b_m}(e^{j\omega})$ of the order m backward prediction error $b_m(n)$ for values of $m = 2$, 5, 10, and 20 and comment on your observations. In particular, you should find that for all values of α, $b_m(n)$ becomes closer to white as m increases. However, for values of α closer to unity, a larger m is required to obtain a white $b_m(n)$. Explain this observation.

P11.33

(i) On the basis of algorithms provided in Tables 11.6 and 11.7, develop and run simulation programs to confirm the results presented in Figures 11.16–11.19. If you are looking for a short-cut and you have access to the MATLAB software package, you may start with the programs `ar_m_al1.m` and `ar_m_al2.m` on the accompanying website. Note that these programs give only the core of your implementations. You need to study and amend them accordingly to get the results that you are looking for.

(ii) Use the process $x(n)$ of P11.32 as the adaptive filter input and try that for values of $\alpha = 0$, 0.5, 0.8, and 0.95, and various values of the AR modeling order, that is, the parameter M. Comment on your observations. To get further insight, you may need to evaluate the parameter γ of Eq. (11.149) for each case. Write a program to generate γ. For this you need to calculate the autocorrelation functions of $x(n)$ and $u_{\rm a}(n)$ and use them to build the matrices $\mathbf{R}_{xx}$ and $\mathbf{R}_{u_{\rm a}u_{\rm a}}$ required in Eq. (11.149). These can conveniently be obtained by calculating the inverse DFT of the samples of the power spectral densities of the corresponding processes as discussed in Part (ii) of Problem P11.32.

Appendix 11A: Evaluation of $E[\mathbf{u}_a(n)\mathbf{x}^T(n)\mathbf{K}(n)\mathbf{x}(n)\mathbf{u}_a^T(n)]$

First, we note that

$$\mathbf{x}^{\mathrm{T}}(n)\mathbf{K}(n)\mathbf{x}(n) = \sum_{i=0}^{N-1}\sum_{j=0}^{N-1} x(n-i)x(n-j)k_{ij}(n) \tag{11A.1}$$

is a scalar, with $k_{ij}(n)$ denoting the ijth element of $\mathbf{K}(n)$. Also, $\mathbf{C}(n) \triangleq \mathbf{u}_a(n)\mathbf{x}^{\mathrm{T}}(n)\mathbf{K}(n)\mathbf{x}(n)\mathbf{u}_a^{\mathrm{T}}(n)$ is an N-by-N matrix whose lmth element is

$$c_{lm}(n) = u_a(n-l)u_a(n-m)\sum_{i=0}^{N-1}\sum_{j=0}^{N-1} x(n-i)x(n-j)k_{ij}(n) \tag{11A.2}$$

Taking statistical expectation of $c_{lm}(n)$, we obtain

$$E[c_{lm}(n)] = \sum_{i=0}^{N-1}\sum_{j=0}^{N-1} E[u_a(n-l)u_a(n-m)x(n-i)x(n-j)]k_{ij}(n) \tag{11A.3}$$

Now, recall the assumption that the input samples $x(n)$ are a set of jointly Gaussian random variables. Since for any set of real-valued jointly Gaussian random variables x_1, x_2, x_3, and x_4

$$E[x_1x_2x_3x_4] = E[x_1x_2]E[x_3x_4] + E[x_1x_3]E[x_2x_4] + E[x_1x_4]E[x_2x_3] \tag{11A.4}$$

we obtain

$$E[u_a(n-l)u_a(n-m)x(n-i)x(n-j)] = r_{u_a u_a}^{lm} r^{ij} + \delta(l-i)\delta(m-j) + \delta(l-j)\delta(m-i) \tag{11A.5}$$

where $\delta(\cdot)$ is the Kronecker delta function, and $r_{u_a u_a}^{lm}$ and r^{ij} are the lmth and ijth elements of the correlation matrices $\mathbf{R}_{u_a u_a}$ and $\mathbf{R}_{xx}$, respectively. In deriving this result, we have noted that

$$E[u_a(n-l)x(n-i)] = \begin{cases} 1 & l = i \\ 0 & l \neq l \end{cases} \tag{11A.6}$$

Substituting Eq. (11A.5) in Eq. (11A.3) and noting that $k_{lm}(n) = k_{ml}(n)$, we obtain

$$\begin{aligned} E[c_{lm}(n)] &= r_{u_a u_a}^{lm} \sum_{i=0}^{N-1}\sum_{j=0}^{N-1} r_x^{ij} k_{ij}(n) + 2k_{lm}(n) \\ &= r_{u_a u_a}^{lm} \mathrm{tr}[\mathbf{K}(n)\mathbf{R}_{xx}] + 2k_{lm}(n) \end{aligned} \tag{11A.7}$$

for $l = 0, 1, \ldots, N-1$ and $m = 0, 1, \ldots, N-1$.

Combining these elements to construct the matrix $E[\mathbf{C}(n)] = E[\mathbf{u}_a(n)\mathbf{x}^{\mathrm{T}}(n)\mathbf{K}(n)\mathbf{x}(n) \times \mathbf{u}_a^{\mathrm{T}}(n)]$, we get Eq. (11.142).

Appendix 11B: Evaluation of the parameter γ

To evaluate γ, we proceed as follow:

$$\begin{aligned}\text{tr}[\mathbf{R}_{u_a u_a}\mathbf{R}_{xx}] &= \text{tr}[E[\mathbf{u}_a(n)\mathbf{u}_a^T(n)]\mathbf{R}_{xx}] \\ &= E[\text{tr}[\mathbf{u}_a(n)\mathbf{u}_a^T(n)\mathbf{R}_{xx}]]. \end{aligned} \tag{11B.1}$$

We note that for any pair of matrices **A** and **B** with dimensions N_1-by-N_2 and N_2-by-N_1, respectively, $\text{tr}[\mathbf{AB}] = \text{tr}[\mathbf{BA}]$. Using this in Eq. (11B.1), we may write

$$\text{tr}[\mathbf{R}_{u_a u_a}\mathbf{R}_{xx}] = E[\mathbf{u}_a^T(n)\mathbf{R}_{xx}\mathbf{u}_a(n)] \tag{11B.2}$$

Note that the trace function has been dropped from the right-hand side of Eq. (11B.2), as $\mathbf{u}_a^T(n)\mathbf{R}_{xx}\mathbf{u}_a(n)$ is a scalar.

Next, we recall that the correlation matrix $\mathbf{R}_{xx}$ may be decomposed as (see Eq. (4.23))

$$\mathbf{R}_{xx} = \sum_{i=0}^{N-1} \lambda_i \mathbf{q}_i \mathbf{q}_i^T \tag{11B.3}$$

where λ_i's and $\mathbf{q}_i$'s are the eigenvalues and eigenvectors, respectively, of $\mathbf{R}_{xx}$. Using Eq. (11B.3) in Eq. (11B.2), we obtain

$$\text{tr}[\mathbf{R}_{u_a u_a}\mathbf{R}_{xx}] = \sum_{i=0}^{N-1} \lambda_i \eta_i \tag{11B.4}$$

where $\eta_i = E[(\mathbf{q}_i^T\mathbf{u}_a(n))^2]$

Now, we shall analyze the terms $\lambda_i\eta_i$. For this, we refer to Eq. 11B.1 which depicts a procedure for measuring $\lambda_i\eta_i$ through a sequence of filtering and averaging procedures. The AR process $x(n)$ is generated by passing its innovation, $\nu(n)$, through its model transfer function

$$H_{AR}(e^{j\omega}) = \frac{1}{1 - \sum_{m=1}^{M} a_{M,i}e^{-jm\omega}} \tag{11B.5}$$

The innovation $\nu(n)$ is a white noise process with variance σ_ν^2. Passing $x(n)$ through the eigenfilter $Q_i(e^{j\omega})$ (the FIR filter whose coefficients are the elements of the eigenvector $\mathbf{q}_i$) generates a signal whose mean-squared value is equal to λ_i. On the other hand, according to Figure 11.13, the sequence $u_a(n)$ is generated from $b_M(n+M)$ by first multiplying that by σ_b^{-2} (the inverse of the variance of $b_M(n+M)$) and then passing the result through an FIR filter with the transfer function $z^{-M}H_{b_M}(z^{-1})$ which is nothing but $1/H_{AR}(e^{j\omega})$. Passing $u_a(n)$ through the eigenfilter $Q_i(e^{j\omega})$ generates the samples of the sequence $\mathbf{q}_i^T\mathbf{u}_a(n)$ whose mean-squared value is then measured. Accordingly, following Figure 11B.1, one will find that

$$\lambda_i = \frac{1}{2\pi}\int_0^{2\pi} \sigma_\nu^2 |H_{AR}(e^{j\omega})|^2 |Q_i(e^{j\omega})|^2 d\omega \tag{11B.6}$$

and

$$\eta_i = \frac{1}{2\pi}\int_0^{2\pi} \sigma_b^2 \frac{\sigma_b^{-4}}{|H_{AR}(e^{j\omega})|^2} |Q_i(e^{j\omega})|^2 d\omega \tag{11B.7}$$

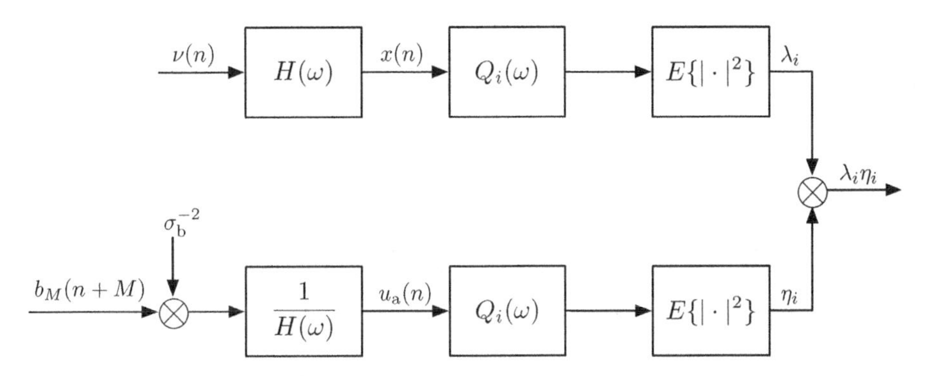

Figure 11B.1 Procedure for the evaluation of $\lambda_i \eta_i$.

We also note that the innovation process $\nu(n)$ and the backward prediction error, $b_M(n)$ (or equivalently $b_M(n+M)$), are statistically the same. This implies, $\sigma_b^2 = \sigma_\nu^2$. Noting this, Eqs. (11B.6) and (11B.7) give

$$\lambda_i \eta_i = \left(\frac{1}{2\pi}\right)^2 \left(\int_0^{2\pi} |H_{\mathrm{AR}}(\mathrm{e}^{j\omega})|^2 |Q_i(\mathrm{e}^{j\omega})|^2 \mathrm{d}\omega\right) \left(\int_0^{2\pi} \frac{1}{|H_{\mathrm{AR}}(\mathrm{e}^{j\omega})|^2} |Q_i(\mathrm{e}^{j\omega})|^2 \mathrm{d}\omega\right) \tag{11B.8}$$

Equation (11B.8) is in an appropriate form that may be used to give some argument with regard to the value of $\lambda_i \eta_i$ and the overall summation in Eq. (11B.4).

Now, if $f(x)$ and $g(x)$ are two arbitrary functions with finite energy in the interval (a, b), then the Cauchy–Schwartz inequality states that

$$\left|\int_a^b f(x) g(x) \mathrm{d}x\right|^2 \leq \left(\int_a^b |f(x)|^2 \mathrm{d}x\right) \left(\int_a^b |g(x)|^2 \mathrm{d}x\right) \tag{11B.9}$$

with the equality valid when $f(x) = \alpha g(x)$, α being a scalar. Using this, Eq. (11B.8) gives

$$\lambda_i \eta_i \geq \left(\frac{1}{2\pi} \int_0^{2\pi} |Q_i(\mathrm{e}^{j\omega})|^2 \mathrm{d}\omega\right)^2 \tag{11B.10}$$

Noting that $Q_i(\mathrm{e}^{j\omega})$ is a normalized eigenfilter in the sense that $\mathbf{q}_i^{\mathrm{T}} \mathbf{q}_i = 1$, the right-hand side of Eq. (11B.10) is always equal to unity (Chapter 4). Using this result in Eq. (11B.4) and recalling the definition of the parameter γ, we obtain

$$\gamma \geq 1 \tag{11B.11}$$

A particular case of interest for which the inequality (11B.10) (and thus Eq. (11B.11)) will be converted to equality is when $|Q_i(\mathrm{e}^{j\omega})|^2$ is an impulse function in the form $2\pi\delta(\omega - \omega_i)$. In fact, this happens to be nearly the case as the filter length, N, increases to a large value. With this argument one can say that the above inequalities will all be close to equalities when the filter length, N, is large.

12

Method of Least-Squares

The problem of filter design for estimating a desired signal based on another signal can be formulated from either a statistical point of view or deterministic point of view, as was mentioned in Chapter 1. The Wiener filter and its adaptive version (LMS algorithm and its derivatives) belong to the statistical framework as their design is based on minimizing a statistical quantity, *the mean-squared error (MSE)*. So far, all our discussions have been limited to this statistical class of algorithms. In the next two chapters, we are going to consider the second class of algorithms that are derived based on the *method of least-squares*, which belongs to the deterministic framework. We have noted that the class of LMS-based algorithms is very wide and covers a large variety of algorithms, each having some merits over the others. The class of least-squares-based algorithms is also equally wide. Current literature contains a large number of scientific papers that report a diverse range of least-squares-based adaptive filtering algorithms.

We recall that, in the derivation of the LMS algorithm, the goal was to minimize the *mean-squared* of the estimation error. In the *method of least-squares*, on the other hand, at any time instant $n > 0$, the adaptive filter parameters (tap weights) are calculated so that the quantity

$$\zeta(n) = \sum_{k=1}^{n} \rho_n(k) e_n^2(k) \tag{12.1}$$

is minimized, and hence the name *least-squares*. In Eq. (12.1), $k = 1$ is the time at which the algorithm starts, $e_n(k)$, for $k = 1, 2, \ldots, n$, are the samples of error estimates that would be obtained if the filter is run from time $k = 1$ to n using the set of filter parameters that are computed at time "n," and $\rho_n(k)$ is a weighting function whose role will be discussed later. Thus, in the method of least-squares, the filter parameters are optimized using all the observations from the time the filter begins until the present time and minimizing the sum of squared values of the error samples of the filter output. Clearly, this is a deterministic optimization of the filter parameters based on the observed data.

An insightful interpretation of the method of least-squares is its *curve fitting* property. Consider a curve whose samples are the desired output samples of the adaptive filter. In the same manner, samples of the filter output (given some input sequence) can be considered to constitute another curve. Then, the problem of choosing the filter parameters to find the best fit between these two curves boils down to the method of least-squares, if we define

Adaptive Filters: Theory and Applications, Second Edition. Behrouz Farhang-Boroujeny.

the *best fit* as one which minimizes a weighted sum of squared values of the differences between the samples of the two curves.

In this book, our discussion on the method of least-squares is rather limited. In this chapter, we first present a formulation of the problem of least-squares for a linear combiner and discuss some of its properties. We also introduce *the standard recursive least-squares* (RLS) *algorithm* as an example of the class of least-squares-based adaptive filtering algorithms. Some results that compare the LMS and RLS algorithms are also given in this chapter. In the next chapter, we present the development of *fast RLS algorithms* that are computationally more efficient than the standard RLS algorithm, for recursive implementation of the method of least-squares.

12.1 Formulation of Least-Squares Estimation for a Linear Combiner

Consider a linear adaptive filter with the observed real-valued input vector $\mathbf{x}(n) = [x_0(n)\, x_1(n) \cdots x_{N-1}(n)]^{\mathrm{T}}$, tap-weight vector $\mathbf{w}(n) = [w_0(n)\, w_1(n) \cdots\ w_{N-1}(n)]^{\mathrm{T}}$, and desired output $d(n)$. The filter output is obtained as the inner product of $\mathbf{w}(n)$ and $\mathbf{x}(n)$, that is, $\mathbf{w}^{\mathrm{T}}(n)\mathbf{x}(n)$. Note that, here, we have not specified any particular structure for the elements of the input vector $\mathbf{x}(n)$. The elements of $\mathbf{x}(n)$ may be successive samples of a particular input process, as it happens in the case of transversal filters, or may be samples of a parallel set of input sources, as in the case of antenna arrays.

In *the method of least-squares*, at time instant "n," we choose $\mathbf{w}(n)$ so that the summation (12.1) is minimized. We define

$$y_n(k) = \mathbf{w}^{\mathrm{T}}(n)\mathbf{x}(k), \quad \text{for} \quad k = 1, 2, \ldots, n \tag{12.2}$$

as the filter output generated using the tap-weight vector $\mathbf{w}(n)$. The corresponding estimation error would then be

$$e_n(k) = d(k) - y_n(k) \tag{12.3}$$

Thus, we note from Eqs. (12.1) and (12.2) that the addition of the subscript "n" to the samples of filter output, $y_n(k)$, and the error estimates, $e_n(k)$, is to emphasize that these quantities are computed using the solution, $\mathbf{w}(n)$, at instant n that is obtained by minimizing the weighted sum of error squares over *all* the instants up to "n."

To keep our derivations simple, we assume that the weighting function $\rho_n(k)$ is equal to 1, for all values of k in the first three sections of this chapter. We also adopt a matrix/vector formulation of the problem.

We define the following vectors:

$$\mathbf{d}(n) = [d(1)\, d(2) \cdots d(n)]^{\mathrm{T}} \tag{12.4}$$

$$\mathbf{y}(n) = [y_n(1)\, y_n(2) \cdots y_n(n)]^{\mathrm{T}} \tag{12.5}$$

and

$$\mathbf{e}(n) = [e_n(1)\, e_n(2) \cdots e_n(n)]^{\mathrm{T}} \tag{12.6}$$

We also define the matrix of observed input samples as

$$\mathbf{X}(n) = [\mathbf{x}(1)\, \mathbf{x}(2) \cdots \mathbf{x}(n)] \tag{12.7}$$

Then, using Eqs. (12.2) and (12.3) in Eqs. (12.4)–(12.7), we get

$$\mathbf{y}(n) = \mathbf{X}^{\mathrm{T}}(n)\mathbf{w}(n) \tag{12.8}$$

and

$$\mathbf{e}(n) = \mathbf{d}(n) - \mathbf{y}(n) \tag{12.9}$$

Furthermore, with $\rho_n(k) = 1$, for all k, Eq. (12.1) can be written as

$$\zeta(n) = \mathbf{e}^{\mathrm{T}}(n)\mathbf{e}(n) \tag{12.10}$$

Substituting Eqs. (12.8) and (12.9) in Eq. (12.10), we obtain

$$\zeta(n) = \mathbf{d}^{\mathrm{T}}(n)\mathbf{d}(n) - 2\boldsymbol{\theta}^{\mathrm{T}}(n)\mathbf{w}(n) + \mathbf{w}^{\mathrm{T}}(n)\boldsymbol{\Psi}(n)\mathbf{w}(n) \tag{12.11}$$

where

$$\boldsymbol{\Psi}(n) = \mathbf{X}(n)\mathbf{X}^{\mathrm{T}}(n) \tag{12.12}$$

and

$$\boldsymbol{\theta}(n) = \mathbf{X}(n)\mathbf{d}(n) \tag{12.13}$$

Setting the gradient of $\zeta(n)$ with respect to the tap-weight vector $\mathbf{w}(n)$ equal to zero and following the same line of derivations as in the case of Wiener filters (Chapter 3), we obtain

$$\boldsymbol{\Psi}(n)\hat{\mathbf{w}}(n) = \boldsymbol{\theta}(n) \tag{12.14}$$

where $\hat{\mathbf{w}}(n)$ is the estimate of filter tap-weight vector in the *least-squares sense*. Equation (12.14) is known as the *normal equation for a linear least-squares filter*. It results in the following least-squares solution

$$\hat{\mathbf{w}}(n) = \boldsymbol{\Psi}^{-1}(n)\boldsymbol{\theta}(n) \tag{12.15}$$

Substituting Eq. (12.15) in Eq. (12.11), the minimum value of $\zeta(n)$ is obtained as

$$\begin{aligned}\zeta_{\min}(n) &= \mathbf{d}^{\mathrm{T}}(n)\mathbf{d}(n) - \boldsymbol{\theta}^{\mathrm{T}}(n)\boldsymbol{\Psi}^{-1}(n)\boldsymbol{\theta}(n) \\ &= \mathbf{d}^{\mathrm{T}}(n)\mathbf{d}(n) - \boldsymbol{\theta}^{\mathrm{T}}(n)\hat{\mathbf{w}}(n)\end{aligned} \tag{12.16}$$

12.2 Principle of Orthogonality

We recall that in the case of Wiener filters, the optimized output error, $e_{\mathrm{o}}(n)$, is orthogonal to the filter tap inputs, in the sense that the following identities hold:

$$E[e_{\mathrm{o}}(n)x_i(n)] = 0, \quad \text{for} \quad i = 0, 1, \ldots, N-1 \tag{12.17}$$

where $x_i(n)$ is the ith element of the tap-input vector $\mathbf{x}(n)$, and $E[\cdot]$ denotes the statistical expectation. This was called *the principle of orthogonality* for Wiener filters. Similar result can also be derived in the case of linear least-squares estimation by following the same line of derivations as those given in Chapter 3 (Section 3.3).

Using Eqs. (12.6) and (12.10), we obtain

$$\frac{\partial \zeta(n)}{\partial w_i(n)} = 2\sum_{k=1}^{n} e_n(k)\frac{\partial e_n(k)}{\partial w_i(n)} \tag{12.18}$$

Using the identity

$$e_n(k) = d(k) - \sum_{i=0}^{N-1} w_i(n)x_i(k) \tag{12.19}$$

to evaluate the second factor on the right-hand side of Eq. (12.18), we get

$$\frac{\partial \zeta(n)}{\partial w_i(n)} = -2\sum_{k=1}^{n} e_n(k)x_i(k) \tag{12.20}$$

Furthermore, we note that when $\mathbf{w}(n) = \hat{\mathbf{w}}(n)$,

$$\frac{\partial \zeta(n)}{\partial w_i(n)} = 0, \quad \text{for} \quad i = 0, 1, \ldots, N-1 \tag{12.21}$$

Using Eqs. (12.20) and (12.21), we find that when $\mathbf{w}(n) = \hat{\mathbf{w}}(n)$, the following identities hold:

$$\sum_{k=1}^{n} \hat{e}_n(k)x_i(k) = 0, \quad \text{for} \quad i = 0, 1, \ldots, N-1 \tag{12.22}$$

where $\hat{e}_n(k)$ is the optimized estimation error in the least-squares sense. This result, which is equivalent to Eq. (12.17), is known as the *principle of orthogonality* in the least-squares formulation. We define the vectors

$$\hat{\mathbf{e}}(n) = [\hat{e}_n(1)\,\hat{e}_n(2)\,\cdots\,\hat{e}_n(n)]^{\mathrm{T}} \tag{12.23}$$

and

$$\mathbf{x}_i(n) = [x_i(1)\,x_i(2)\,\cdots\,x_i(n)]^{\mathrm{T}} \tag{12.24}$$

and note that Eq. (12.22) may also be expressed in terms of these vectors as

$$\hat{\mathbf{e}}^{\mathrm{T}}(n)\mathbf{x}_i(n) = 0, \quad \text{for} \quad i = 0, 1, \ldots, N-1 \tag{12.25}$$

We note that the left-hand side of Eq. (12.25) is the *inner product* of $\hat{\mathbf{e}}(n)$ and $\mathbf{x}_i(n)$, thus the name *principle of orthogonality*.

Comparison of Eqs. (12.17) and (12.25) reveals that the definition of orthogonality is in terms of statistical averages in Wiener filtering, whereas it is in terms of inner products of data vectors in the case of least-squares estimation. By dividing both sides of Eq. (12.22) by n, we see that we can also use time averages to define orthogonality in the least-squares case:

$$\frac{1}{n}\sum_{k=1}^{n} \hat{e}_n(k)x_i(k) = 0, \quad \text{for} \quad i = 0, 1, \ldots, N-1 \tag{12.26}$$

An immediate corollary to the principle of orthogonality is that *when the tap weights of a filter are optimized in the least-squares sense, the filter output and its optimized estimation error are orthogonal.* That is,

$$\hat{\mathbf{e}}^{\mathrm{T}}(n)\hat{\mathbf{y}}(n) = 0 \tag{12.27}$$

where $\hat{\mathbf{y}}(n)$ is the vector of the output samples of the filter when $\mathbf{w}(n) = \hat{\mathbf{w}}(n)$. This follows immediately if we note that Eq. (12.8) may also be written as

$$\mathbf{y}(n) = \sum_{i=0}^{N-1} w_i(n)\mathbf{x}_i(n) \tag{12.28}$$

and use the identity (12.25) to obtain Eq. (12.27).

Example 12.1

Consider the case where $n = 3$,

$$\mathbf{X}(3) = \begin{bmatrix} 2 & 1 & 0 \\ 1 & 2 & 0.1 \end{bmatrix}$$

and

$$\mathbf{d}(3) = \begin{bmatrix} 1 \\ -1 \\ 0 \end{bmatrix}$$

We wish to find $\hat{\mathbf{w}}(3)$, $\hat{\mathbf{y}}(3)$, and $\hat{\mathbf{e}}(3)$ and to confirm the principle of orthogonality. We have

$$\boldsymbol{\Psi}(3) = \mathbf{X}(3)\mathbf{X}^{\mathrm{T}}(3) = \begin{bmatrix} 5 & 4 \\ 4 & 5.01 \end{bmatrix}$$

and

$$\boldsymbol{\theta}(3) = \mathbf{X}(3)\mathbf{d}(3) = \begin{bmatrix} 1 \\ -1 \end{bmatrix}$$

Thus,

$$\hat{\mathbf{w}}(3) = \begin{bmatrix} 5 & 4 \\ 4 & 5.01 \end{bmatrix}^{-1} \begin{bmatrix} 1 \\ -1 \end{bmatrix} = \frac{1}{9.05}\begin{bmatrix} 5.01 & -4 \\ -4 & 5 \end{bmatrix}\begin{bmatrix} 1 \\ -1 \end{bmatrix} = \frac{1}{9.05}\begin{bmatrix} 9.01 \\ -9 \end{bmatrix}$$

$$\begin{aligned}\hat{\mathbf{y}}(3) &= \hat{w}_0(3)\mathbf{x}_0(3) + \hat{w}_1(3)\mathbf{x}_1(3) \\ &= \frac{9.01}{9.05}\begin{bmatrix} 2 \\ 1 \\ 0 \end{bmatrix} - \frac{9}{9.05}\begin{bmatrix} 1 \\ 2 \\ 0.1 \end{bmatrix} = \frac{1}{9.05}\begin{bmatrix} 9.02 \\ -8.99 \\ -0.9 \end{bmatrix}\end{aligned}$$

and

$$\hat{\mathbf{e}}(3) = \mathbf{d}(3) - \hat{\mathbf{y}}(3) = \begin{bmatrix} 1 \\ -1 \\ 0 \end{bmatrix} - \frac{1}{9.05}\begin{bmatrix} 9.02 \\ -8.99 \\ -0.9 \end{bmatrix} = \frac{1}{9.05}\begin{bmatrix} 0.03 \\ -0.06 \\ 0.90 \end{bmatrix}$$

We can now confirm the principle of orthogonality by noting that

$$\hat{\mathbf{e}}^{\mathrm{T}}(3)\mathbf{x}_0(3) = 0$$

and, also,

$$\hat{\mathbf{e}}^{\mathrm{T}}(3)\mathbf{x}_1(3) = 0$$

12.3 Projection Operator

An alternative interpretation to the solution of least-squares problem can be given using the concept of *projection operator*. Projection of a 1-by-n vector $\mathbf{d}(n)$ into the subspace spanned by a set of vectors $\mathbf{x}_0(n)$, $\mathbf{x}_1(n), \ldots, \mathbf{x}_{N-1}(n)$ is a vector $\hat{\mathbf{d}}(n)$ with the following properties:

1. The vector $\hat{\mathbf{d}}(n)$ is obtained as a linear combination of the vectors $\mathbf{x}_0(n)$, $\mathbf{x}_1(n), \ldots, \mathbf{x}_{N-1}(n)$.
2. Among all the vectors in the subspace spanned by $\mathbf{x}_0(n)$, $\mathbf{x}_1(n), \ldots, \mathbf{x}_{N-1}(n)$, the vector $\hat{\mathbf{d}}(n)$ has the minimum Euclidean distance from $\mathbf{d}(n)$.
3. The difference $\mathbf{d}(n) - \hat{\mathbf{d}}(n)$ is a vector that is orthogonal to the subspace spanned by $\mathbf{x}_0(n)$, $\mathbf{x}_1(n), \ldots, \mathbf{x}_{N-1}(n)$

We may note that the least-squares estimate $\hat{\mathbf{y}}(n)$ satisfies the three properties listed above. Namely, we note from Eq. (12.28) that $\hat{\mathbf{y}}(n)$ is also obtained as a linear combination of the vectors $\mathbf{x}_0(n)$, $\mathbf{x}_1(n), \ldots, \mathbf{x}_{N-1}(n)$. Furthermore, obtaining $\hat{\mathbf{y}}(n)$ by minimizing $\hat{\mathbf{e}}^{\mathrm{T}}(n)\hat{\mathbf{e}}(n)$, where $\hat{\mathbf{e}}(n) = \mathbf{d}(n) - \hat{\mathbf{y}}(n)$, is equivalent to minimizing the Euclidean distance between $\mathbf{d}(n)$ and $\hat{\mathbf{y}}(n)$. Also, from the principle of orthogonality, the error vector $\hat{\mathbf{e}}(n) = \mathbf{d}(n) - \hat{\mathbf{y}}(n)$ is orthogonal to the vectors $\mathbf{x}_0(n)$, $\mathbf{x}_1(n), \ldots, \mathbf{x}_{N-1}(n)$. We thus conclude that $\hat{\mathbf{y}}(n)$ *is nothing but the projection of* $\mathbf{d}(n)$ *into the subspace spanned by the vectors* $\mathbf{x}_0(n)$, $\mathbf{x}_1(n), \ldots, \mathbf{x}_{N-1}(n)$.

We also note that from Eq. (12.8)

$$\hat{\mathbf{y}}(n) = \mathbf{X}^{\mathrm{T}}(n)\hat{\mathbf{w}}(n) \tag{12.29}$$

Substituting Eqs. (12.12) and (12.13) in Eq. (12.15) and the result in Eq. (12.29), we obtain

$$\hat{\mathbf{y}}(n) = \mathbf{P}(n)\mathbf{d}(n) \tag{12.30}$$

where

$$\mathbf{P}(n) = \mathbf{X}^{\mathrm{T}}(n)(\mathbf{X}(n)\mathbf{X}^{\mathrm{T}}(n))^{-1}\mathbf{X}(n) \tag{12.31}$$

Consequently, the matrix $\mathbf{P}(n)$ is known as *the projection operator*.

Using Eq. (12.30), we find that the optimized error vector $\hat{\mathbf{e}}(n) = \mathbf{d}(n) - \hat{\mathbf{y}}(n)$ can be expressed as

$$\hat{\mathbf{e}}(n) = [\mathbf{I} - \mathbf{P}(n)]\mathbf{d}(n) \tag{12.32}$$

where $\mathbf{I}$ is the identity matrix of the same dimension as $\mathbf{P}(n)$. As a result, the matrix $\mathbf{I} - \mathbf{P}(n)$ is referred to as *the orthogonal complement projection operator*.

12.4 Standard Recursive Least-Squares Algorithm

The least-squares solution provided by Eq. (12.15) is of very little interest in actual implementation of adaptive filters as it requires that all the past samples of the input as well as the desired output be available at every iteration. Furthermore, the number of operations needed to calculate $\hat{\mathbf{w}}(n)$ grows proportional to n as the number of columns of $\mathbf{X}(n)$ and the length of $\mathbf{d}(n)$ grow with n. These problems are solved by employing recursive methods. In this section, as an example of recursive methods, we present the *standard RLS algorithm*.

12.4.1 RLS Recursions

In the standard RLS algorithm (or just "RLS" algorithm, for short), the weighting factor $\rho_n(k)$ is chosen as

$$\rho_n(k) = \lambda^{n-k}, \quad k = 1, 2, \ldots, n \tag{12.33}$$

where λ is a positive constant close to, but smaller than, 1. The ordinary method of least-squares, discussed in the previous sections, corresponds to the case of $\lambda = 1$. The parameter λ is known as the *forgetting factor*. Clearly, when $\lambda < 1$, the weighting factors defined by Eq. (12.33) give more weightage to the recent samples of the error estimates (and thus to the recent samples of the observed data) compared to the old ones. In other words, *the choice of $\lambda < 1$ results in a scheme that puts more emphasis on the recent samples of the observed data and tends to forget the past.* This is exactly what one may wish when he/she develops an adaptive algorithm with some tracking capability. Roughly speaking, $1/(1-\lambda)$ is a measure of the *memory* of the algorithm. The case of $\lambda = 1$ corresponds to *infinite memory*.

Substituting Eq. (12.33) in Eq. (12.1) and using the vector/matrix notations of Section 12.1, we obtain

$$\zeta(n) = \mathbf{e}^{\mathrm{T}}(n)\boldsymbol{\Lambda}(n)\mathbf{e}(n) \tag{12.34}$$

where $\boldsymbol{\Lambda}(n)$ is the diagonal matrix consisting of the weighting factors $1, \lambda, \lambda^2, \ldots$, that is,

$$\boldsymbol{\Lambda}(n) = \begin{bmatrix} \lambda^{n-1} & 0 & 0 & \cdots & 0 \\ 0 & \lambda^{n-2} & 0 & \cdots & 0 \\ 0 & 0 & \lambda^{n-3} & \cdots & 0 \\ \vdots & \vdots & \vdots & \ddots & \vdots \\ 0 & 0 & 0 & \cdots & 1 \end{bmatrix} \tag{12.35}$$

Following the same line of derivations that led to Eq. (12.15), we obtain the minimizer of $\zeta(n)$ in Eq. (12.34) as

$$\hat{\mathbf{w}}(n) = \boldsymbol{\Psi}_\lambda^{-1}(n)\boldsymbol{\theta}_\lambda(n) \tag{12.36}$$

where

$$\boldsymbol{\Psi}_\lambda(n) = \mathbf{X}(n)\boldsymbol{\Lambda}(n)\mathbf{X}^{\mathrm{T}}(n) \tag{12.37}$$

and

$$\boldsymbol{\theta}_\lambda(n) = \mathbf{X}(n)\boldsymbol{\Lambda}(n)\mathbf{d}(n) \tag{12.38}$$

On substituting Eqs. (12.4) and (12.7) in Eqs. (12.37) and (12.38), and expanding the summations, we get

$$\begin{aligned}\mathbf{\Psi}_\lambda(n) = \mathbf{x}(n)\mathbf{x}^{\mathrm{T}}(n) + \lambda\mathbf{x}(n-1)\mathbf{x}^{\mathrm{T}}(n-1)\\ +\lambda^2\mathbf{x}(n-2)\mathbf{x}^{\mathrm{T}}(n-2) + \cdots\end{aligned} \tag{12.39}$$

and

$$\begin{aligned}\boldsymbol{\theta}_\lambda(n) = \mathbf{x}(n)d(n) + \lambda\mathbf{x}(n-1)d(n-1)\\ +\lambda^2\mathbf{x}(n-2)d(n-2) + \cdots\end{aligned} \tag{12.40}$$

respectively. Using Eqs. (12.39) and (12.40), it is straightforward to see that $\mathbf{\Psi}_\lambda(n)$ and $\boldsymbol{\theta}_\lambda(n)$ can be obtained recursively as

$$\mathbf{\Psi}_\lambda(n) = \lambda\mathbf{\Psi}_\lambda(n-1) + \mathbf{x}(n)\mathbf{x}^{\mathrm{T}}(n) \tag{12.41}$$

and

$$\boldsymbol{\theta}_\lambda(n) = \lambda\boldsymbol{\theta}_\lambda(n-1) + \mathbf{x}(n)d(n) \tag{12.42}$$

respectively. These two recursions and the following result of matrix algebra form the *basis* for the derivation of the RLS algorithm.

For an arbitrary nonsingular N-by-N matrix $\mathbf{A}$, any N-by-1 vector $\mathbf{a}$ and a scalar α,

$$(\mathbf{A} + \alpha\mathbf{a}\mathbf{a}^{\mathrm{T}})^{-1} = \mathbf{A}^{-1} - \frac{\alpha\mathbf{A}^{-1}\mathbf{a}\mathbf{a}^{\mathrm{T}}\mathbf{A}^{-1}}{1 + \alpha\mathbf{a}^{\mathrm{T}}\mathbf{A}^{-1}\mathbf{a}} \tag{12.43}$$

This identity, which was also used for some other derivations in Chapter 6, is a special form of the *matrix inversion lemma.*

We let $\mathbf{A} = \lambda\mathbf{\Psi}_\lambda(n-1)$, $\mathbf{a} = \mathbf{x}(n)$ and $\alpha = 1$ to evaluate the inverse of $\mathbf{\Psi}_\lambda(n) = \lambda\mathbf{\Psi}_\lambda(n-1) + \mathbf{x}(n)\mathbf{x}^{\mathrm{T}}(n)$. This results in the following recursive equation for updating the inverse of $\mathbf{\Psi}_\lambda(n)$:

$$\mathbf{\Psi}_\lambda^{-1}(n) = \lambda^{-1}\mathbf{\Psi}_\lambda^{-1}(n-1) - \frac{\lambda^{-2}\mathbf{\Psi}_\lambda^{-1}(n-1)\mathbf{x}(n)\mathbf{x}^{\mathrm{T}}(n)\mathbf{\Psi}_\lambda^{-1}(n-1)}{1 + \lambda^{-1}\mathbf{x}^{\mathrm{T}}(n)\mathbf{\Psi}_\lambda^{-1}(n-1)\mathbf{x}(n)} \tag{12.44}$$

To simplify the subsequent steps, we define the column vector

$$\mathbf{k}(n) = \frac{\lambda^{-1}\mathbf{\Psi}_\lambda^{-1}(n-1)\mathbf{x}(n)}{1 + \lambda^{-1}\mathbf{x}^{\mathrm{T}}(n)\mathbf{\Psi}_\lambda^{-1}(n-1)\mathbf{x}(n)} \tag{12.45}$$

The vector $\mathbf{k}(n)$ is referred to as the *gain vector* for reasons that will become apparent later in this section.

Substituting Eq. (12.45) in Eq. (12.44), we obtain

$$\mathbf{\Psi}_\lambda^{-1}(n) = \lambda^{-1}(\mathbf{\Psi}_\lambda^{-1}(n-1) - \mathbf{k}(n)\mathbf{x}^{\mathrm{T}}(n)\mathbf{\Psi}_\lambda^{-1}(n-1)) \tag{12.46}$$

By rearranging Eq. (12.45), we get

$$\mathbf{k}(n) = \lambda^{-1}(\mathbf{\Psi}_\lambda^{-1}(n-1) - \mathbf{k}(n)\mathbf{x}^{\mathrm{T}}(n)\mathbf{\Psi}_\lambda^{-1}(n-1))\mathbf{x}(n) \tag{12.47}$$

Using Eq. (12.46) in Eq. (12.47), we obtain

$$\mathbf{k}(n) = \boldsymbol{\Psi}_\lambda^{-1}(n)\mathbf{x}(n) \tag{12.48}$$

Next, we substitute Eq. (12.42) in Eq. (12.36) and expand to obtain

$$\begin{aligned}\hat{\mathbf{w}}(n) &= \lambda\boldsymbol{\Psi}_\lambda^{-1}(n)\boldsymbol{\theta}_\lambda(n-1) + \boldsymbol{\Psi}_\lambda^{-1}(n)\mathbf{x}(n)d(n)\\ &= \lambda\boldsymbol{\Psi}_\lambda^{-1}(n)\boldsymbol{\theta}_\lambda(n-1) + \mathbf{k}(n)d(n)\end{aligned} \tag{12.49}$$

where the last equality is obtained using Eq. (12.48). Substituting Eq. (12.46) in Eq. (12.49), we get

$$\begin{aligned}\hat{\mathbf{w}}(n) &= \boldsymbol{\Psi}_\lambda^{-1}(n-1)\boldsymbol{\theta}_\lambda(n-1) - \mathbf{k}(n)\mathbf{x}^{\mathrm{T}}(n)\boldsymbol{\Psi}_\lambda^{-1}(n-1)\boldsymbol{\theta}_\lambda(n-1) + \mathbf{k}(n)d(n)\\ &= \hat{\mathbf{w}}(n-1) - \mathbf{k}(n)\mathbf{x}^{\mathrm{T}}(n)\hat{\mathbf{w}}(n-1) + \mathbf{k}(n)d(n)\\ &= \hat{\mathbf{w}}(n-1) + \mathbf{k}(n)(d(n) - \hat{\mathbf{w}}^{\mathrm{T}}(n-1)\mathbf{x}(n))\end{aligned} \tag{12.50}$$

Finally, we define

$$\hat{e}_{n-1}(n) = d(n) - \hat{\mathbf{w}}^{\mathrm{T}}(n-1)\mathbf{x}(n) \tag{12.51}$$

and use this in Eq. (12.50) to obtain

$$\hat{\mathbf{w}}(n) = \hat{\mathbf{w}}(n-1) + \mathbf{k}(n)\hat{e}_{n-1}(n) \tag{12.52}$$

This is the recursion used by the RLS algorithm to update $\hat{\mathbf{w}}(n)$. The amount of change to be made in the tap weights at the nth iteration is determined by the product of the *estimation error* $\hat{e}_{n-1}(n)$ and the *gain vector* $\mathbf{k}(n)$.

From Eq. (12.51), we note that $\hat{e}_{n-1}(n)$ is the estimation error at time "n" based on the tap-weight vector estimated at time "$n-1$", $\hat{\mathbf{w}}(n-1)$. Hence, $\hat{e}_{n-1}(n)$ is referred to as the *a priori* estimation error. On the other hand, the *a posteriori* estimation error is given by

$$\hat{e}_n(n) = d(n) - \hat{\mathbf{w}}^{\mathrm{T}}(n)\mathbf{x}(n) \tag{12.53}$$

which would be obtained if the *current* least-squares estimate of the filter tap weights, that is, $\hat{\mathbf{w}}(n)$, was used to calculate the filter output.

Equations (12.45), (12.51), (12.52), and (12.46), in this order, describe one iteration of the standard RLS algorithm.

12.4.2 Initialization of the RLS Algorithm

Actual implementation of the RLS algorithm requires proper initialization of $\boldsymbol{\Psi}_\lambda(0)$ and $\hat{\mathbf{w}}(0)$ before the start of the algorithm. In particular, we note that the matrix

$$\boldsymbol{\Psi}_\lambda(n) = \sum_{k=1}^{n} \lambda^{n-k}\mathbf{x}(k)\mathbf{x}^{\mathrm{T}}(k) \tag{12.54}$$

for values of n smaller than the filter length, N, has a rank which is less than its dimension, N. This implies that the inverse of $\boldsymbol{\Psi}_\lambda(n)$ does not exist for $n < N$. A simple and commonly used solution to this problem is to start the RLS algorithm with an initial setting of

$$\mathbf{\Psi}_\lambda(0) = \delta \mathbf{I} \tag{12.55}$$

where δ is a small positive constant. Then, iterating the recursive equation (12.41), we obtain

$$\mathbf{\Psi}_\lambda(n) = \sum_{k=1}^{n} \lambda^{n-k} \mathbf{x}(k)\mathbf{x}^{\mathrm{T}}(k) + \delta\lambda^n \mathbf{I} \tag{12.56}$$

We observe that, for $\lambda < 1$, the effect of $\mathbf{\Psi}_\lambda(0)$ reduces exponentially as n increases. Thus, this initialization of $\mathbf{\Psi}_\lambda(0)$ has very little effect on the steady-state performance of the RLS algorithm. Furthermore, the effect of $\mathbf{\Psi}_\lambda(0)$ on the convergence behavior of the RLS algorithm can be minimized by choosing a very small value for δ.

As for the initialization of the filter tap weights, it is common practice to set

$$\hat{\mathbf{w}}(0) = \mathbf{0} \tag{12.57}$$

where $\mathbf{0}$ is the N-by-1 zero vector. However, setting $\hat{\mathbf{w}}(0)$ equal to an arbitrary nonzero vector, also, does not result in any significant effect on the convergence and steady-state behavior of the RLS algorithm, provided that the elements of $\hat{\mathbf{w}}(0)$ are not very large. A study of the effect of a nonzero selection of $\hat{\mathbf{w}}(0)$ is discussed in Problem P12.6.

12.4.3 Summary of the Standard RLS Algorithm

Table 12.1 presents a summary of the standard RLS algorithm. This is one of the few possible implementations of the RLS algorithm. It exploits the special form of the gain vector, $\mathbf{k}(n)$, to simplify its computation using an intermediate vector $\mathbf{u}(n) = \mathbf{\Psi}_\lambda^{-1}(n-1)\mathbf{x}(n)$. We also note that by multiplying the numerator and denominator of the right-hand side of Eq. (12.45) by λ and using this definition of $\mathbf{u}(n)$, we obtain

$$\mathbf{k}(n) = \frac{1}{\lambda + \mathbf{x}^{\mathrm{T}}(n)\mathbf{u}(n)} \mathbf{u}(n)$$

Careful examination of Table 12.1 reveals that the computational complexity of this implementation is mainly determined by:

1. Computation of the vector $\mathbf{u}(n)$
2. Computation of $\mathbf{x}^{\mathrm{T}}(n)\mathbf{\Psi}_\lambda^{-1}(n-1)$ in $\mathbf{\Psi}_\lambda^{-1}(n)$ update equation
3. Computation of the outer product of $\mathbf{k}(n)$ and $\mathbf{x}^{\mathrm{T}}(n)\mathbf{\Psi}_\lambda^{-1}(n-1)$ in $\mathbf{\Psi}_\lambda^{-1}(n)$ update equation.
4. Subtraction of the two terms within the brackets in $\mathbf{\Psi}_\lambda^{-1}(n)$ update equation, and scaling of the result by λ^{-1}

Each of these steps require N^2 multiplications. In addition, steps 1, 2, and 4 require N^2 additions/subtractions, each. This brings up the total computational complexity of the RLS algorithm of Table 12.1 to about $4N^2$ multiplications and $3N^2$ additions/subtractions.

Table 12.1 Summary of the standard RLS algorithm (version I).

Input:	Tap-weight vector estimate, $\hat{\mathbf{w}}(n-1)$, Input vector, $\mathbf{x}(n)$, desired output, $d(n)$, and the matrix $\boldsymbol{\Psi}_\lambda^{-1}(n-1)$
Output:	Filter output, $\hat{y}_{n-1}(n)$, tap-weight vector update, $\hat{\mathbf{w}}(n)$, and the updated matrix $\boldsymbol{\Psi}_\lambda^{-1}(n)$

1. Computation of the gain vector:

$$\mathbf{u}(n) = \boldsymbol{\Psi}_\lambda^{-1}(n-1)\mathbf{x}(n)$$

$$\mathbf{k}(n) = \frac{1}{\lambda + \mathbf{x}^{\mathrm{T}}(n)\mathbf{u}(n)}\mathbf{u}(n)$$

2. Filtering:

$$\hat{y}_{n-1}(n) = \hat{\mathbf{w}}^{\mathrm{T}}(n-1)\mathbf{x}(n)$$

3. Error estimation:

$$\hat{e}_{n-1}(n) = d(n) - \hat{y}_{n-1}(n)$$

4. Tap-weight vector adaptation:

$$\hat{\mathbf{w}}(n) = \hat{\mathbf{w}}(n-1) + \mathbf{k}(n)\hat{e}_{n-1}(n)$$

5. $\boldsymbol{\Psi}_\lambda^{-1}(n)$ update:

$$\boldsymbol{\Psi}_\lambda^{-1}(n) = \lambda^{-1}(\boldsymbol{\Psi}_\lambda^{-1}(n-1) - \mathbf{k}(n)[\mathbf{x}^{\mathrm{T}}(n)\boldsymbol{\Psi}_\lambda^{-1}(n-1)])$$

The fact that $\boldsymbol{\Psi}_\lambda(n)$ is a symmetric matrix can be used to reduce the computational complexity of the RLS algorithm. Using this, we find that $\mathbf{x}^{\mathrm{T}}(n)\boldsymbol{\Psi}_\lambda^{-1}(n-1) = (\boldsymbol{\Psi}_\lambda^{-1}(n-1)\mathbf{x}(n))^{\mathrm{T}} = \mathbf{u}^{\mathrm{T}}(n)$. The last step of Table 12.1 may then be simplified as

$$\boldsymbol{\Psi}_\lambda^{-1}(n) = \lambda^{-1}(\boldsymbol{\Psi}_\lambda^{-1}(n-1) - \mathbf{k}(n)\mathbf{u}^{\mathrm{T}}(n)) \tag{12.58}$$

This amendment, although logical and precise, results in a useless implementation when applied in practice. Computer simulations and theoretical analysis (Verhaegen, 1989) show that this amended version of the RLS algorithm is numerically unstable. This behavior of the RLS algorithm is due to the roundoff error accumulation, which makes $\boldsymbol{\Psi}_\lambda^{-1}(n-1)$ nonsymmetric. This in turn invalidates the assumption $\mathbf{x}^{\mathrm{T}}(n)\boldsymbol{\Psi}_\lambda^{-1}(n-1) = \mathbf{u}^{\mathrm{T}}(n)$, which was used to introduce the above amendment.

To resolve this problem and come up with an efficient and stable implementation of the RLS algorithm, one may compute only the upper or lower triangular part of $\boldsymbol{\Psi}_\lambda^{-1}(n)$ according to Eq. (12.58) and copy the result to obtain the rest of elements of $\boldsymbol{\Psi}_\lambda^{-1}(n)$ to preserve its symmetric structure (Verhaegen, 1989; Yang, 1994). Table 12.2 presents a summary of this implementation of the RLS algorithm. Here, the operator Tri{·} signifies that the computation of $\boldsymbol{\Psi}_\lambda^{-1}(n)$ is based on either the upper or lower triangular parts. Clearly, this results in significant saving in computations as a large portion of the

Table 12.2 Summary of the standard RLS algorithm (version II).

Input:	Tap-weight vector estimate, $\hat{\mathbf{w}}(n-1)$, Input vector, $\mathbf{x}(n)$, desired output, $d(n)$, and the matrix $\mathbf{\Psi}_\lambda^{-1}(n-1)$
Output:	Filter output, $\hat{y}_{n-1}(n)$, Tap-weight vector update, $\hat{\mathbf{w}}(n)$, and the updated matrix $\mathbf{\Psi}_\lambda^{-1}(n)$

1. Computation of the gain vector:

$$\mathbf{u}(n) = \mathbf{\Psi}_\lambda^{-1}(n-1)\mathbf{x}(n)$$

$$\mathbf{k}(n) = \frac{1}{\lambda + \mathbf{x}^{\mathrm{T}}(n)\mathbf{u}(n)}\mathbf{u}(n)$$

2. Filtering:

$$\hat{y}_{n-1}(n) = \hat{\mathbf{w}}^{\mathrm{T}}(n-1)\mathbf{x}(n)$$

3. Error estimation:

$$\hat{e}_{n-1}(n) = d(n) - \hat{y}_{n-1}(n)$$

4. Tap-weight vector adaptation:

$$\hat{\mathbf{w}}(n) = \hat{\mathbf{w}}(n-1) + \mathbf{k}(n)\hat{e}_{n-1}(n)$$

5. $\mathbf{\Psi}_\lambda^{-1}(n)$ update:

$$\mathbf{\Psi}_\lambda^{-1}(n) = \mathrm{Tri}\,\{\lambda^{-1}(\mathbf{\Psi}_\lambda^{-1}(n-1) - \mathbf{k}(n)\mathbf{u}^{\mathrm{T}}(n))\}$$

complexity of the RLS algorithm arises from computation of $\mathbf{\Psi}_\lambda^{-1}(n)$. Basically, the computational complexity of steps 3 and 4 above (corresponding to $\mathbf{\Psi}_\lambda^{-1}(n)$ update) is halved and step 2 is eliminated. This brings down the computational complexity of the RLS algorithm in Table 12.2 to about $2N^2$ multiplications and $1.5N^2$ additions/subtractions, which is about half of that of the algorithm of Table 12.1.

From, the above discussion, one may perceive the potential problem of the RLS algorithm. It is indeed true that *roundoff errors may accumulate and result in undesirable behavior for any algorithm that works based on some recursive update equations.* This statement is general and applicable to all LMS and least-squares-based algorithms. However, the problem turns out to be more serious in the case of least-squares-based algorithms. See Cioffi (1987) for an excellent qualitative discussion on the roundoff error in various adaptive filtering algorithms. *Engineers who use adaptive filtering algorithms should be aware of this potential problem and must evaluate the algorithms on this issue before going for their practical implementations.*

12.5 Convergence Behavior of the RLS Algorithm

In this section, we study the convergence behavior of the RLS algorithm in the context of a system modeling problem. As the plant, we consider a linear multiple regressor characterized by the equation

$$d(n) = \mathbf{w}_{\mathrm{o}}^{\mathrm{T}}\mathbf{x}(n) + e_{\mathrm{o}}(n) \tag{12.59}$$

where $\mathbf{w}_o$ is the regressor tap-weight vector, $\mathbf{x}(n)$ is the tap-input vector, $e_o(n)$ is the plant noise, and $d(n)$ is the plant output. The noise samples, $e_o(n)$, are assumed to be zero-mean and white, and independent of the input samples, $\mathbf{x}(n)$. The tap-input vector $\mathbf{x}(n)$ is also applied to an adaptive filter whose tap-weight vector, $\mathbf{w}(n)$, is adapted so that the difference between its output, $y(n) = \mathbf{w}^{\mathrm{T}}(n)\mathbf{x}(n)$, and the plant output, $d(n)$, is minimized in the least-squares sense.

The derivations that follow use the vector/matrix formulation adopted in the previous sections. In particular, we note that with the definitions (12.4) and (12.7),

$$\mathbf{d}(n) = \mathbf{X}^{\mathrm{T}}(n)\mathbf{w}_o + \mathbf{e}_o(n) \tag{12.60}$$

where $\mathbf{e}_o(n) = [e_o(1)\, e_o(2) \cdots e_o(n)]^{\mathrm{T}}$.

12.5.1 Average Tap-Weight Behavior of the RLS Algorithm

We show that the least-squares estimate $\hat{\mathbf{w}}(n)$ is an unbiased estimate of the tap-weight vector $\mathbf{w}_o$.

From Eqs. (12.36)–(12.38), we obtain

$$\hat{\mathbf{w}}(n) = (\mathbf{X}(n)\mathbf{\Lambda}(n)\mathbf{X}^{\mathrm{T}}(n))^{-1}\mathbf{X}(n)\mathbf{\Lambda}(n)\mathbf{d}(n) \tag{12.61}$$

Substituting Eq. (12.60) in Eq. (12.61), and using Eq. (12.37), we get

$$\hat{\mathbf{w}}(n) = \mathbf{w}_o + \mathbf{\Psi}_\lambda^{-1}(n)\mathbf{X}(n)\mathbf{\Lambda}(n)\mathbf{e}_o(n) \tag{12.62}$$

Taking expectation on both sides of Eq. (12.62) and recalling that $\mathbf{X}(n)$ and $\mathbf{e}_o(n)$ are independent of each other, we obtain

$$\begin{aligned} E[\hat{\mathbf{w}}(n)] &= \mathbf{w}_o + E[\mathbf{\Psi}_\lambda^{-1}(n)\mathbf{X}(n)]\mathbf{\Lambda}(n)E[\mathbf{e}_o(n)] \\ &= \mathbf{w}_o \end{aligned} \tag{12.63}$$

where the last equality follows from the fact that $e_o(n)$ is a zero-mean process, that is, $E[e_o(n)] = 0$, for all values of n. This result shows that $\hat{\mathbf{w}}(n)$ is an unbiased estimate of $\mathbf{w}_o$.

The above derivation does not include the effect of initialization, that is, $\mathbf{\Psi}^{-1}(0) = \delta^{-1}\mathbf{I}$, which is required for proper operation of the RLS algorithm. This initialization introduces some bias in $\hat{\mathbf{w}}(n)$ that is proportional to δ and decreases as n increases (Problem P12.3).

12.5.2 Weight-Error Correlation Matrix

Let us define the weight-error vector

$$\hat{\mathbf{v}}(n) = \hat{\mathbf{w}}(n) - \mathbf{w}_o \tag{12.64}$$

From Eq. (12.62), we obtain

$$\hat{\mathbf{v}}(n) = \mathbf{\Psi}_\lambda^{-1}(n)\mathbf{X}(n)\mathbf{\Lambda}(n)\mathbf{e}_o(n) \tag{12.65}$$

We also define the weight-error correlation matrix

$$\hat{\mathbf{K}}(n) = E[\hat{\mathbf{v}}(n)\hat{\mathbf{v}}^{\mathrm{T}}(n)] \tag{12.66}$$

Substituting Eq. (12.65) in Eq. (12.66) and noting that $(\boldsymbol{\Psi}_\lambda^{-1}(n))^{\mathrm{T}} = \boldsymbol{\Psi}_\lambda^{-1}(n)$ and $\boldsymbol{\Lambda}^{\mathrm{T}}(n) = \boldsymbol{\Lambda}(n)$, we obtain

$$\hat{\mathbf{K}}(n) = E[\boldsymbol{\Psi}_\lambda^{-1}(n)\mathbf{X}(n)\boldsymbol{\Lambda}(n)\mathbf{e}_{\mathrm{o}}(n)\mathbf{e}_{\mathrm{o}}^{\mathrm{T}}(n)\boldsymbol{\Lambda}(n)\mathbf{X}^{\mathrm{T}}(n)\boldsymbol{\Psi}_\lambda^{-1}(n)] \tag{12.67}$$

Recalling the independence of $e_{\mathrm{o}}(n)$ and $\mathbf{x}(n)$, from Eq. (12.67), we obtain

$$\hat{\mathbf{K}}(n) = E[\boldsymbol{\Psi}_\lambda^{-1}(n)\mathbf{X}(n)\boldsymbol{\Lambda}(n)E[\mathbf{e}_{\mathrm{o}}(n)\mathbf{e}_{\mathrm{o}}^{\mathrm{T}}(n)]\boldsymbol{\Lambda}(n)\mathbf{X}^{\mathrm{T}}(n)\boldsymbol{\Psi}_\lambda^{-1}(n)] \tag{12.68}$$

Since $e_{\mathrm{o}}(n)$ is a white noise process,

$$E[\mathbf{e}_{\mathrm{o}}(n)\mathbf{e}_{\mathrm{o}}^{\mathrm{T}}(n)] = \sigma_{\mathrm{o}}^2\mathbf{I} \tag{12.69}$$

where σ_{o}^2 is the variance of $e_{\mathrm{o}}(n)$, and $\mathbf{I}$ is the identity matrix with appropriate dimension. Finally, substituting Eq. (12.69) in Eq. (12.48), we get

$$\hat{\mathbf{K}}(n) = \sigma_{\mathrm{o}}^2 E[\boldsymbol{\Psi}_\lambda^{-1}(n)\boldsymbol{\Psi}_{\lambda^2}(n)\boldsymbol{\Psi}_\lambda^{-1}(n)] \tag{12.70}$$

where $\boldsymbol{\Psi}_{\lambda^2}(n) = \mathbf{X}(n)\boldsymbol{\Lambda}^2(n)\mathbf{X}^{\mathrm{T}}(n)$.

Rigorous evaluation of Eq. (12.70) is a difficult task. Hence, we make the following assumptions to facilitate an approximate evaluation of $\hat{\mathbf{K}}(n)$:

1. The observed input vectors $\mathbf{x}(1), \mathbf{x}(2), \ldots, \mathbf{x}(n)$ constitute the samples of an ergodic process. Thus, the time averages may be used instead of the ensemble averages.
2. The forgetting factor λ is very close to 1.
3. The time "n" at which $\hat{\mathbf{K}}(n)$ is evaluated is large.

We note from Eq. (12.39) that $\boldsymbol{\Psi}_\lambda(n)$ is a weighted sum of the outer products $\mathbf{x}(n)\mathbf{x}^{\mathrm{T}}(n)$, $\mathbf{x}(n-1)\mathbf{x}^{\mathrm{T}}(n-1)$, $\mathbf{x}(n-2)\mathbf{x}^{\mathrm{T}}(n-2), \ldots$. Thus, considering the above assumptions, we find that

$$\boldsymbol{\Psi}_\lambda(n) \approx \frac{1-\lambda^n}{1-\lambda}\mathbf{R} \tag{12.71}$$

where $\mathbf{R} = E[\mathbf{x}(n)\mathbf{x}^{\mathrm{T}}(n)]$ is the correlation matrix of the input.

Substituting Eq. (12.71) in Eq. (12.70), we obtain

$$\begin{aligned}\hat{\mathbf{K}}(n) &= \sigma_{\mathrm{o}}^2\left(\frac{1-\lambda}{1-\lambda^n}\right)^2 \times \frac{1-\lambda^{2n}}{1-\lambda^2}\mathbf{R}^{-1} \\ &= \sigma_{\mathrm{o}}^2\frac{1-\lambda}{1+\lambda}\cdot\frac{1+\lambda^n}{1-\lambda^n}\mathbf{R}^{-1}\end{aligned} \tag{12.72}$$

In the steady state, that is, when $n \to \infty$, we obtain from Eq. (12.72)

$$\hat{\mathbf{K}}(\infty) = \sigma_{\mathrm{o}}^2\frac{1-\lambda}{1+\lambda}\mathbf{R}^{-1} \tag{12.73}$$

12.5.3 Learning Curve

From the summary of the RLS algorithm in Table 12.1 (or Table 12.2), we find that the filter output at time "n" is obtained according to the equation

$$\hat{y}_{n-1}(n) = \hat{\mathbf{w}}^{\mathrm{T}}(n-1)\mathbf{x}(n) \tag{12.74}$$

Accordingly, the *learning curve* of the RLS algorithm is defined in terms of the *a priori* error $\hat{e}_{n-1}(n)$ as

$$\hat{\xi}_{n-1}(n) = E[\hat{e}_{n-1}^2(n)] \tag{12.75}$$

To evaluate $\hat{\xi}_{n-1}(n)$, we proceed as follows.

Using Eqs. (12.64) and (12.59), we obtain

$$\begin{aligned}\hat{e}_{n-1}(n) &= d(n) - \hat{\mathbf{w}}^{\mathrm{T}}(n-1)\mathbf{x}(n)\\ &= d(n) - \mathbf{w}_{\mathrm{o}}^{\mathrm{T}}\mathbf{x}(n) - \hat{\mathbf{v}}^{\mathrm{T}}(n-1)\mathbf{x}(n)\\ &= e_{\mathrm{o}}(n) - \hat{\mathbf{v}}^{\mathrm{T}}(n-1)\mathbf{x}(n)\end{aligned} \tag{12.76}$$

Substituting Eq. (12.76) in Eq. (12.75), we obtain

$$\begin{aligned}\hat{\xi}_{n-1}(n) &= E[(e_{\mathrm{o}}(n) - \hat{\mathbf{v}}^{\mathrm{T}}(n-1)\mathbf{x}(n))^2]\\ &= E[e_{\mathrm{o}}^2(n)] - 2E[\hat{\mathbf{v}}^{\mathrm{T}}(n-1)\mathbf{x}(n)e_{\mathrm{o}}(n)]\\ &\quad + E[\hat{\mathbf{v}}^{\mathrm{T}}(n-1)\mathbf{x}(n)\mathbf{x}^{\mathrm{T}}(n)\hat{\mathbf{v}}(n-1)]\end{aligned} \tag{12.77}$$

where the last term is obtained by noting that $\hat{\mathbf{v}}^{\mathrm{T}}(n-1)\mathbf{x}(n) = \mathbf{x}^{\mathrm{T}}(n)\hat{\mathbf{v}}(n-1)$.

To simplify the evaluation of the last two terms on the right-hand side of Eq. (12.77), we make use of the assumptions we have made on $e_{\mathrm{o}}(n)$ and $\mathbf{x}(n)$. First, $e_{\mathrm{o}}(n)$ is a zero-mean white process and second, $e_{\mathrm{o}}(n)$ and $\mathbf{x}(n)$ are independent. Consequently, $e_{\mathrm{o}}(n)$ is also independent of $\hat{\mathbf{v}}(n-1)\mathbf{x}(n)$ as $\hat{\mathbf{v}}(n-1)$ depends only on the past observations, which includes only the past samples of $e_{\mathrm{o}}(n)$. Noting these, we obtain

$$E[\hat{\mathbf{v}}^{\mathrm{T}}(n-1)\mathbf{x}(n)e_{\mathrm{o}}(n)] = E[\hat{\mathbf{v}}^{\mathrm{T}}(n-1)\mathbf{x}(n)]E[e_{\mathrm{o}}(n)] = 0 \tag{12.78}$$

as $E[e_{\mathrm{o}}(n)] = 0$.

To simplify the third term on the right-hand side of Eq. (12.77), we also assume that $\hat{\mathbf{v}}(n-1)$ and $\mathbf{x}(n)$ are independent. This is similar to the independence assumption in the case of LMS algorithm (Chapter 6). Strictly speaking, the latter assumption is hard to justify as $\hat{\mathbf{v}}(n-1)$ depends on the past samples of $\mathbf{x}(n)$, and the current $\mathbf{x}(n)$ may not be independent of its past samples. However, when n is large, $\hat{\mathbf{v}}(n-1)$, which is determined by a large number of past observations of $\mathbf{x}(n)$, only weakly depends on the present sample of $\mathbf{x}(n)$. This is because the older samples of $\mathbf{x}(n)$ are only loosely dependent on the present $\mathbf{x}(n)$ unless $\mathbf{x}(n)$ contains one or more significant narrow-band components. We exclude such special cases in our study here and assume that $\hat{\mathbf{v}}(n-1)$ and $\mathbf{x}(n)$ are independent of each other. This is referred to as the *independence assumption* in analogy with the independence assumption used in the case of LMS algorithm. However, we note that unlike the LMS algorithm, for which the independence assumption could be used for all (small and large) values of the time index n, in the case of RLS algorithm, the independence assumption is valid only for large values of n.

Using the independence assumption, we get

$$\begin{aligned}E[\hat{\mathbf{v}}^{\mathrm{T}}(n-1)\mathbf{x}(n)\mathbf{x}^{\mathrm{T}}(n)\hat{\mathbf{v}}(n-1)] &= E[\hat{\mathbf{v}}^{\mathrm{T}}(n-1)E[\mathbf{x}(n)\mathbf{x}^{\mathrm{T}}(n)]\hat{\mathbf{v}}(n-1)]\\ &= E[\hat{\mathbf{v}}^{\mathrm{T}}(n-1)\mathbf{R}\hat{\mathbf{v}}(n-1)]\end{aligned} \tag{12.79}$$

Next, we note that $E[\hat{\mathbf{v}}^{\mathrm{T}}(n-1)\mathbf{R}\hat{\mathbf{v}}(n-1)]$ is a scalar and proceed as follows:

$$\begin{aligned} E[\hat{\mathbf{v}}^{\mathrm{T}}(n-1)\mathbf{R}\hat{\mathbf{v}}(n-1)] &= \mathrm{tr}[E[\hat{\mathbf{v}}^{\mathrm{T}}(n-1)\mathbf{R}\hat{\mathbf{v}}(n-1)]] \\ &= E[\mathrm{tr}[\hat{\mathbf{v}}^{\mathrm{T}}(n-1)\mathbf{R}\hat{\mathbf{v}}(n-1)]] \\ &= E[\mathrm{tr}[\hat{\mathbf{v}}(n-1)\hat{\mathbf{v}}^{\mathrm{T}}(n-1)\mathbf{R}]] \\ &= \mathrm{tr}[E[\hat{\mathbf{v}}(n-1)\hat{\mathbf{v}}^{\mathrm{T}}(n-1)]\mathbf{R}] \\ &= \mathrm{tr}[\hat{\mathbf{K}}(n-1)\mathbf{R}] \end{aligned} \tag{12.80}$$

where tr[·] denotes the trace of the indicated matrix. In the derivation of Eq. (12.80), we have used the linearity property of the expectation and trace operators, the definition (12.66), and the identity

$$\mathrm{tr}[\mathbf{AB}] = \mathrm{tr}[\mathbf{BA}]$$

which is valid for any pair of N-by-M and M-by-N, $\mathbf{A}$ and $\mathbf{B}$ matrices, respectively.

Substituting Eqs. (12.80) and (12.78) in Eq. (12.77), we get

$$\hat{\xi}_{n-1}(n) = \xi_{\min} + \mathrm{tr}[\hat{\mathbf{K}}(n-1)\mathbf{R}] \tag{12.81}$$

where $\xi_{\min} = E[e_{\mathrm{o}}^2(n)]$ is the minimum MSE of the filter which is achieved when a perfect estimate of $\mathbf{w}_{\mathrm{o}}$ is available. Substituting Eq. (12.72) in Eq. (12.81), we obtain

$$\hat{\xi}_{n-1}(n) = \xi_{\min} + \frac{1-\lambda}{1+\lambda} \bullet \frac{1+\lambda^{n-1}}{1-\lambda^{n-1}} \bullet N\xi_{\min} \tag{12.82}$$

This describes the learning curve of the RLS algorithm. Note that we made the assumption of n being large in the derivation of Eq. (12.82). Thus, Eq. (12.82) can predict the behavior of the RLS algorithm only after certain initial transient period. Some comments on the behavior of the RLS algorithm during its initial transient period will be given later in this section.

At this point, it is instructive that we elaborate on the behavior of the RLS algorithm, as predicted by Eq. (12.82). We note that the second term on the right-hand side of Eq. (12.82) is a positive value, indicating the deviation of $\hat{\xi}_{n-1}(n)$ from $\xi_{\min}$. This term converges toward its final value as n grows. The speed at which this term converges is determined by the exponential term λ^{n-1}, or equivalently λ^n. Accordingly, we define the time constant τ_{RLS} associated with the RLS algorithm using the following equation:

$$\lambda^n = \mathrm{e}^{-n/\tau_{\mathrm{RLS}}} \tag{12.83}$$

Solving this for τ_{RLS}, we obtain

$$\tau_{\mathrm{RLS}} = -\frac{1}{\ln \lambda} \tag{12.84}$$

To simplify this, we use the following approximation:

$$\ln(1+x) \approx x, \quad \text{for} \quad |x| \ll 1 \tag{12.85}$$

We note that $0 < 1-\lambda \ll 1$ as λ is smaller than, but close to, 1. Using this in Eq. (12.85), we get

$$\ln \lambda = \ln(1-(1-\lambda)) \approx -(1-\lambda) \tag{12.86}$$

Substituting Eq. (12.86) in Eq. (12.84), we obtain

$$\tau_{\mathrm{RLS}} \approx \frac{1}{1-\lambda} \tag{12.87}$$

We thus note that the convergence behavior of the RLS algorithm is controlled by only a single mode of convergence. Unlike the LMS algorithm whose convergence behavior is affected by the eigenvalues of the correlation matrix, $\mathbf{R}$, of the filter input, the above results show that the convergence behavior of the RLS algorithm is independent of the eigenvalues of $\mathbf{R}$. This may be explained by substituting Eq. (12.48) in the RLS recursion (12.52) to obtain

$$\hat{\mathbf{w}}(n) = \hat{\mathbf{w}}(n-1) + \mathbf{\Psi}_{\lambda}^{-1}(n)\hat{e}_{n-1}(n)\mathbf{x}(n) \tag{12.88}$$

When n is large, we may use the approximation (12.71) in Eq. (12.88) to get

$$\hat{\mathbf{w}}(n) = \hat{\mathbf{w}}(n-1) + \frac{1-\lambda}{1-\lambda^n}\mathbf{R}^{-1}\hat{e}_{n-1}(n)\mathbf{x}(n) \tag{12.89}$$

When n is large such that $\lambda^n \ll 1$, the above recursion simplifies to

$$\hat{\mathbf{w}}(n) = \hat{\mathbf{w}}(n-1) + (1-\lambda)\mathbf{R}^{-1}\hat{e}_{n-1}(n)\mathbf{x}(n) \tag{12.90}$$

Letting $\lambda = 1 - 2\mu$, we find that this is nothing but the LMS–Newton algorithm, which was introduced in Chapter 7 (Section 7.5), and its convergence behavior was found to be independent of the eigenvalues of $\mathbf{R}$.

At this point, once again, we shall remind the reader that most of the results developed above are valid only when n is large. In particular, the similarity between RLS and LMS–Newton algorithms, which was noted above is valid in the sense that for large values of n, the RLS recursion approaches an update equation which is similar to the LMS–Newton recursion. However, this does not mean that the RLS and LMS–Newton algorithms have the same convergence behavior as the convergence behaviors of the two algorithms are completely different when n is small. This is further illustrated through the simulation results, which are presented in Section 12.5.5.

12.5.4 Excess MSE and Misadjustment

As in Chapter 6, we define the excess MSE of the RLS algorithm as the difference between its steady-state MSE and the minimum achievable MSE. In other words, we define

$$\text{Excess MSE} = \lim_{n\to\infty} \hat{\xi}_{n-1}(n) - \xi_{\min} \tag{12.91}$$

Using Eq. (12.82) in Eq. (12.91), we obtain

$$\text{Excess MSE} = \frac{1-\lambda}{1+\lambda}N\xi_{\min} \tag{12.92}$$

As in the case of LMS algorithm, the misadjustment of the RLS algorithm is given by

$$\mathcal{M}_{\mathrm{RLS}} = \frac{\text{Excess MSE}}{\xi_{\min}} \tag{12.93}$$

Substituting Eq. (12.92) in Eq. (12.93), we obtain

$$\mathcal{M}_{\text{RLS}} = \frac{1-\lambda}{1+\lambda}N \tag{12.94}$$

12.5.5 *Initial Transient Behavior of the RLS Algorithm*

Much of the benefit of the RLS algorithm is attributed to the fact that it shows very fast convergence when it is started from a rest condition with the initial values of $\mathbf{w}(0) = \mathbf{0}$ and $\mathbf{\Psi}^{-1}(0) = \delta^{-1}\mathbf{I}$. This fast convergence is observed only after the first N samples of the input and desired output sequences are processed. In typical implementations of the RLS algorithm, we always find that the MSE of adaptive filter converges to a level close to its minimum value within a small number of iterations (usually two to three times the filter length) and then it proceeds with a fine-tuning process that may last much longer before the MSE reaches its steady-state value. The initial transient behavior of the RLS algorithm can be best explained through a numerical example (computer simulation).

As a numerical example, here, we apply the RLS algorithm to the system modeling setup of Section 6.4.1. Thus, a comparison of the RLS and LMS algorithms can be made. Figure 12.1 presents the schematic diagram of the modeling setup. The common input, $x(n)$, to the plant, $W_{\text{o}}(z)$, and adaptive filter, $W(z)$, is obtained by passing a unit variance white Gaussian sequence, $\nu(n)$, through a filter with the system function $H(z)$. The plant noise, as before, is denoted by $e_{\text{o}}(n)$. It is assumed to be a white noise process independent of $x(n)$.

For our experiments, as in Section 6.4.1, we select $\sigma_{\text{o}}^2 = 0.001$ and

$$W_{\text{o}}(z) = \sum_{i=0}^{7} z^{-i} - \sum_{i=8}^{14} z^{-i} \tag{12.95}$$

The length of the adaptive filter, N, is chosen equal to the length of $W_{\text{o}}(z)$, that is, $N = 15$. Also, we present results of simulations for two choices of input that are characterized by

$$H(z) = H_1(z) = 0.35 + z^{-1} - 0.35z^{-2} \tag{12.96}$$

and

$$H(z) = H_2(z) = 0.35 + z^{-1} + 0.35z^{-2} \tag{12.97}$$

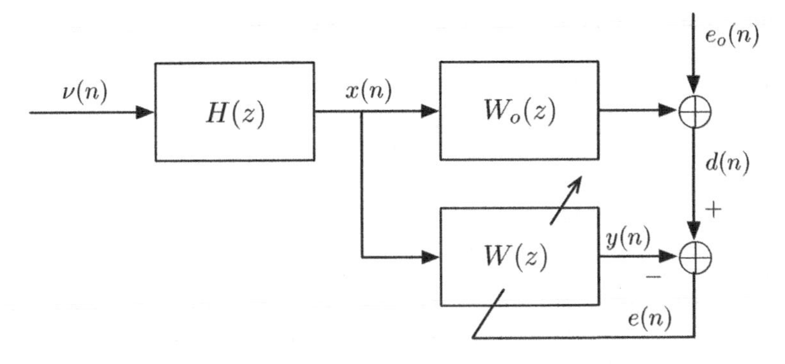

Figure 12.1 Adaptive modeling of an FIR plant.

The first choice results in an input, $x(n)$, whose corresponding correlation matrix has an eigenvalue spread of 1.45. This is close to white input. On the contrary, the second choice of $H(z)$ results is a highly colored input with an associated eigenvalue spread of 28.7 (Section 6.4.1).

Figure 12.2a and b shows the learning curves of the RLS algorithm for the two choices of the input. Each plot is obtained by averaging over 100 independent simulation runs. In all the runs, the RLS algorithm was started with zero initial tap weights, and the parameter δ (used to initialize $\boldsymbol{\Psi}^{-1}(0) = \delta^{-1}\mathbf{I}$) was set to 0.0001. The forgetting factor λ was chosen according to Eq. (12.94) to achieve a misadjustment of 10%. From Figure 12.2, we see that the convergence behavior of the RLS algorithm is independent of the eigenvalue spread of the correlation matrix, $\mathbf{R}$, of the filter input. This is in line with the theoretical predictions of the last section. Furthermore, we find that the learning curves of the RLS algorithm are quite different from the learning curves of the LMS algorithm – compare the learning curves of Figure 12.2a and b with their LMS counterparts in Figure 6.7a and b, respectively.

Each learning curve of the RLS algorithm may be divided into three distinct parts:

1. During the first N iterations (N is the filter length), the MSE remains almost unchanged at a high level.
2. The MSE converges at a very fast rate once the iteration number, n, exceeds N.
3. After this period of fast convergence, the RLS algorithm converges toward its steady state at a much slower rate.

The three separate parts of the learning curves of the RLS algorithm may be explained as follows.

During the first $N - 1$ iterations, that is, when $n < N$, there are infinite number of possible choices of the tap-weight vector $\hat{\mathbf{w}}(n)$, which satisfy the set of equations

$$\hat{\mathbf{w}}^{\mathrm{T}}(n)\mathbf{x}(k) = d(k), \quad \text{for} \quad k = 1, 2, \ldots, n \tag{12.98}$$

as there are less number of equations than the number of unknown tap weights. This ambiguity in the solution of the least-squares problem is manifested in the coefficient matrix $\boldsymbol{\Psi}(n)$ whose rank remains less than its dimension, N. This rank deficiency problem, as suggested before, is solved by initializing $\boldsymbol{\Psi}(0)$ to a positive definite matrix, $\delta\mathbf{I}$, which will result in a full-rank $\boldsymbol{\Psi}(n)$, and thus, a solution for $\hat{\mathbf{w}}(n)$ with no ambiguity. However, the resulting solution, although satisfies Eq. (12.98) within a very good approximation, may not give an accurate estimate of the true tap weights of the plant, $\mathbf{w}_{\mathrm{o}}$. As a result, one finds that during the first $N - 1$ iterations while the *a posteriori* error, $\hat{e}_n(n)$, is very small, the *a priori* error, $\hat{e}_{n-1}(n)$, may still be large. This initial behavior of the RLS algorithm also explains why in the definition of its learning curve we use $\hat{e}_{n-1}(n)$ and not $\hat{e}_n(n)$. Clearly, the latter does not reflect the fact that the least-squares estimate $\hat{\mathbf{w}}(n)$ may be far from its true value, $\mathbf{w}_{\mathrm{o}}$.

On the contrary, when $n > N$, the number of equations in Eq. (12.98) is more than the number of unknown tap weights, which we wish to estimate. In that case, Eq. (12.98) cannot be satisfied exactly. However, a least-squares solution can be found without any ambiguity. The accuracy of the estimate $\hat{\mathbf{w}}(n)$ of $\mathbf{w}_{\mathrm{o}}$ depends on the level of the plant noise and also the number of observed points, that is, the iteration number n. In particular, in the

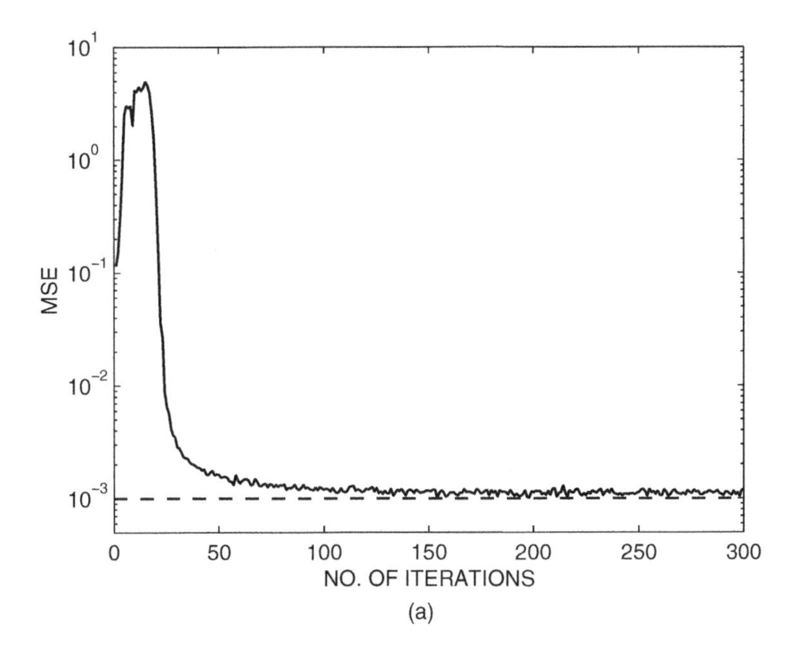

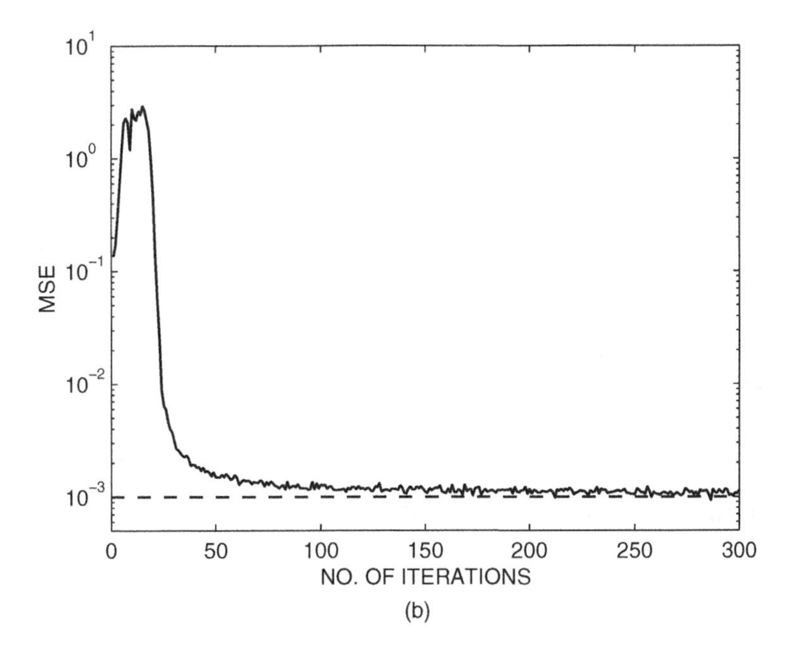

Figure 12.2 Learning curves of the RLS algorithm for the modeling problem of Figure 12.1: (a) $H(z) = H_1(z)$ and (b) $H(z) = H_2(z)$. In both cases, the forgetting factor λ is chosen according to Eq. (12.94) for 10% misadjustment.

case where the plant noise, $e_o(n)$, is zero, Eq. (12.98) can be satisfied exactly, for all values of k, by choosing $\hat{\mathbf{w}}(n) = \mathbf{w}_o$. This clearly is the least-squares solution to the problem, as it results in $\zeta_\lambda(n) = 0$, which is the minimum achievable value of the cost function $\zeta_\lambda(n)$. The RLS algorithm, which is designed to minimize the cost function $\zeta_\lambda(n)$, will find this optimum estimate of $\hat{\mathbf{w}}(n)$, once there are enough samples of the observed input and desired output such that the filter tap weights could be found without any ambiguity. This explains the sharp drop of the learning curve of the RLS algorithm when n exceeds N.

The last part of the learning curve of the RLS algorithm, which decays exponentially, matches the results of Section 12.5.3.

Problems

P12.1 The observed samples of the input to a three-tap filter are

$$\mathbf{x}(1) = \begin{bmatrix} 1 \\ -1 \\ 0 \end{bmatrix}, \quad \mathbf{x}(2) = \begin{bmatrix} -2 \\ 1 \\ -1 \end{bmatrix}, \quad \mathbf{x}(3) = \begin{bmatrix} 1 \\ 1 \\ 1 \end{bmatrix}, \quad \mathbf{x}(4) = \begin{bmatrix} 0 \\ -1 \\ -1 \end{bmatrix}$$

(i) Find the projection and the orthogonal complement projection operators of the set of observed input vectors.

(ii) Using the results of (i), find the least-squares estimate $\hat{\mathbf{y}}(4)$ of the desired vector

$$\mathbf{d}(4) = [1 \ 2 \ -1 \ -1]^{\mathrm{T}}$$

Also, obtain the associated estimation error vector $\mathbf{e}_4(4)$. To check on the accuracy of your results, evaluate $\hat{\mathbf{y}}^{\mathrm{T}}(4)\mathbf{e}_4(4)$ and show that it is equal to zero.

(iii) Repeat (ii) for

$$\mathbf{d}(4) = [0 \ -1 \ 1 \ -1]^{\mathrm{T}}$$

P12.2 Repeat Problem P12.1 when the observed samples of input are $\mathbf{x}(1)$, $\mathbf{x}(2)$, and $\mathbf{x}(3)$, as given there, and we wish to obtain the least-squares estimates, $\hat{\mathbf{y}}(3)$, of

(i) $\mathbf{d}(3) = [1 \ 1 \ 1]^{\mathrm{T}}$

(ii) $\mathbf{d}(3) = [1 \ -1 \ 2]^{\mathrm{l}}\mathrm{T}$

P12.3 Show that the initialization $\mathbf{\Psi}_\lambda^{-1}(0) = \delta^{-1}\mathbf{I}$ will introduce a bias in $\hat{\mathbf{w}}(n)$ which is given by

$$\Delta\hat{\mathbf{w}}(n) = E[\hat{\mathbf{w}}(n)] - \mathbf{w}_o = -\delta\lambda^n E[\mathbf{\Psi}_\lambda^{-1}(n)]\mathbf{w}_o$$

Simplify this result for the case when $\lambda = 1$ and n is large.

P12.4 Show that when $(\mathbf{X}^{\mathrm{T}}(N))^{-1}$ exists,

$$(\mathbf{X}^{\mathrm{T}}(N))^{-1} = (\mathbf{X}(N)\mathbf{X}^{\mathrm{T}}(N))^{-1}\mathbf{X}(N),$$

and thus conclude that the solution provided by the equation $\mathbf{X}^{\mathrm{T}}(N)\mathbf{w}(N) = \mathbf{d}(N)$ and the least-squares solution $\hat{\mathbf{w}}(N) = (\mathbf{X}(N)\mathbf{X}^{\mathrm{T}}(N))^{-1}\mathbf{X}(N)\mathbf{d}(N)$ are the same. What would be the *a posteriori* estimation error $e_N(N)$?

P12.5 Show that in the RLS algorithm, when $\lambda = 1$, and the iteration number, n, is large,

$$\mathcal{M}_{RLS} = \frac{N}{n}$$

where N is the filter length.

P12.6 Consider the case where the RLS algorithm begins with a nonzero $\hat{\mathbf{w}}(0)$ and $\boldsymbol{\Psi}_\lambda(0) = \delta\mathbf{I}$.

(i) Show that this initial condition is equivalent to solving the problem of least-squares according to the following procedure:
- Let $\boldsymbol{\Psi}_\lambda(0) = \delta\mathbf{I}$ and $\boldsymbol{\theta}_\lambda(0) = \delta\hat{\mathbf{w}}(0)$.
- Update $\boldsymbol{\Psi}_\lambda(n)$ and $\boldsymbol{\theta}_\lambda(n)$ using the recursions (12.41) and (12.42).
- Calculate $\hat{\mathbf{w}}(n)$ using the equation $\hat{\mathbf{w}}(n) = \boldsymbol{\Psi}_\lambda^{-1}(n)\boldsymbol{\theta}_\lambda(n)$.

(ii) Use the result of (i) to show that when δ is small, a nonzero choice of $\hat{\mathbf{w}}(0)$ has no significant effect on the convergence behavior of the RLS algorithm.

P12.7 Give a detailed derivation of Eq. (12.71).

P12.8 In some modeling applications, the tap-weight misalignment defined as

$$\eta(n) = E[(\mathbf{w}(n) - \mathbf{w}_o)^{\mathrm{T}}(\mathbf{w}(n) - \mathbf{w}_o)]$$

may be of interest.

(i) Using Eq. (12.73), find the steady-state misalignment of the RLS algorithm, $\eta_{\mathrm{RLS}}(\infty)$, and show that it is a function of the eigenvalues of the correlation matrix $\mathbf{R}$. How does the eigenvalue spread of $\mathbf{R}$ affect $\eta_{\mathrm{RLS}}(\infty)$?

(ii) Refer to Chapter 6, Section 6.3.3, and show that for the LMS algorithm,

$$\eta_{\mathrm{LMS}}(n) = \text{sum of elements of the vector } \mathbf{k}'(n)$$

Then, starting with Eq. (6.55), show that

$$\eta_{\mathrm{LMS}}(\infty) = \mu\xi_{\min}\frac{\sum_{i=0}^{N-1}\frac{1}{1-2\mu\lambda_i}}{1-\sum_{i=0}^{N-1}\frac{\mu\lambda_i}{1-2\mu\lambda_i}}$$

For the case where $\mu\lambda_i \ll 1$, for all values of i, simplify this result and show that $\eta_{\mathrm{LMS}}(\infty)$ depends only on $\sum_{i=0}^{N-1}\lambda_i = \mathrm{tr}[\mathbf{R}]$, and thus is independent of the eigenvalue spread of $\mathbf{R}$.

(iii) Discuss on the significance of your observations in (i) and (ii).

P12.9 Consider the modified least-squares cost function

$$\zeta_a(n) = \sum_{k=1}^{n}\lambda^{n-k}e_n^2(k) + \lambda^n K(\hat{\mathbf{w}}(n) - \hat{\mathbf{w}}(0))^{\mathrm{T}}(\hat{\mathbf{w}}(n) - \hat{\mathbf{w}}(0))$$

where $\hat{\mathbf{w}}(0) \neq \mathbf{0}$ is the initial tap-weight vector and K is a constant.

(i) Use this cost function to derive a modified RLS algorithm.

(ii) Obtain an expression for the learning curve of the filter and discuss the convergence behavior of the proposed RLS algorithm, for small and large values of K.

P12.10 Formulate the problem of modeling a memoryless nonlinear system with input $x(n)$ and output

$$y(n) = ax^3(n) + bx^2(n) + cx(n)$$

where a, b, and c are the unknown coefficients of the system that should be found in the least-squares sense.

P12.11 Repeat Problem P12.10 for the case where

$$y(n) = ax^3(n) + bx^2(n) + cx(n) + d$$

and a, b, c, and d are the unknown coefficients of the system.

Computer-Oriented Problems

P12.12 The MATLAB program used to obtain the simulation results of Figure 12.2 is available on an accompanying website. This is written based on version I of the RLS algorithm, as in Table 12.2, and is called `rlsI.m`. Run this program (or develop and run your own program) to verify the results of Figure 12.2. Also, to gain a better insight into the behavior of the RLS algorithm, try the following runs:

(i) Run `rlsI.m` for $\hat{\mathbf{w}}(0) = 0$ and values of $\delta = 0.001, 0.01, 0.1$, and 1, and compare your results with those in Figure 12.2.
(ii) Run `rlsI.m` for $\delta = 0.0001$ and a few (randomly selected) nonzero values of $\hat{\mathbf{w}}(0)$. Comment on your observation.
(iii) Repeat (ii) for values of $\delta = 0.001, 0.01, 0.1$, and 1.

P12.13 Develop a simulation program to study the variation of the *a posteriori* MSE, $\xi_n(n) = E[\hat{e}_n^2(n)]$, of the RLS algorithm in the case of the modeling problem of Figure 12.1. Comment on your observation.

P12.14 Develop a simulation program to study the convergence behavior of the RLS algorithm when applied to the channel equalization problem of Section 6.4.2. Compare your results with those of the LMS algorithm in Figure 6.9a and b.

P12.15 Consider the case where a sequence $x(n)$ is generated by passing a unit variance white noise through a filter with the transfer function

$$H(z) = \frac{1}{1 - 1.6z^{-1} + 0.65z^{-2}}$$

By taking $x(n)$ as input to a forward linear predictor of order N whose coefficients are adapted using the RLS algorithm, study the convergence of the predictor coefficients for $N = 2, 3$, and 4 and the choices of the forgetting factor $\lambda = 0.9, 0.95$, and 0.99 as n varies from 1 to 50. Discuss your observation.

13

Fast RLS Algorithms

The standard recursive least-squares (RLS) algorithm, which was introduced in Chapter 12, has a computational complexity that grows proportional to the square of the length of the filter. For long filters, this may be unacceptable. In the past, many researchers have attempted to solve this drawback of the least-squares method and have come-up with a variety of elegant solutions. These solutions, whose computational complexity grows proportional to the length of the filter, are commonly referred to as *fast RLS algorithms*.

In this chapter, we review the underlying principles that are fundamental to the development of fast RLS algorithms. Our intention by no means is to cover the whole spectrum of the fast RLS algorithms. A thorough treatment of these algorithms requires many more pages than what is allocated to this topic in this book. Moreover, such a treatment is beyond the scope of this book, whose primary aim is to serve as an introductory text book on adaptive filters. Our aim is to put together the basic relations on which most fast RLS algorithms are built. Once these basic concepts are grabbed by the reader, he/she would feel comfortable to proceed with reading the more advanced topics on this subject (see Haykin (1991, 1996) and Kalouptsidis and Theodoridis (1993) for more extensive treatment of the RLS algorithms).

All the fast RLS algorithms benefit from the order-update and time-update equations similar to those introduced in Chapter 11. In other words, the fast RLS algorithms combine the concepts of prediction and filtering in an elegant way to come-up with computationally efficient implementations. Among these implementations, RLS lattice (RLSL) algorithm appears to be numerically the most robust implementation. The fast transversal RLS (FTRLS) algorithm (also known as *fast transversal filter* – FTF), on the other hand, is an alternative solution that has minimum number of operation count among all the present RLS algorithms.

In this chapter, our emphasis is on the RLSL algorithm. The derivation of the RLSL algorithm leads to a number of order- and time-update equations, which are fundamental to the derivation of the whole class of fast RLS algorithms. We also present the FTRLS algorithm as a by-product of these equations. Since lattice structure is closely related to forward and backward linear predictors, we begin with some preliminary discussion on these predictors.

Adaptive Filters: Theory and Applications, Second Edition. Behrouz Farhang-Boroujeny.

13.1 Least-Squares Forward Prediction

Consider the mth-order forward transversal predictor shown in Figure 13.1. The tap-weight vector $\mathbf{a}_m(n) = [a_{m,1}(n) \ a_{m,2}(n) \ \cdots \ a_{m,m}(n)]^{\mathrm{T}}$ is optimized in the least-squares sense over the entire observation interval $k = 1, 2, \ldots, n$. Accordingly, in the forward transversal predictor, the observed tap-input vectors are $\mathbf{x}_m(0)$, $\mathbf{x}_m(1)$, ..., $\mathbf{x}_m(n-1)$, where $\mathbf{x}_m(k) = [x(k) \ x(k-1) \ \cdots \ x(k-m+1)]^{\mathrm{T}}$, and the desired output samples are $x(1)$, $x(2)$, ..., $x(n)$. The *normal equations* of the forward transversal predictor are then obtained as (Chapter 12)

$$\mathbf{\Psi}_m(n-1)\mathbf{a}_m(n) = \boldsymbol{\psi}_m^{\mathrm{f}}(n) \tag{13.1}$$

where

$$\mathbf{\Psi}_m(n) = \sum_{k=1}^{n} \lambda^{n-k}\mathbf{x}_m(k)\mathbf{x}_m^{\mathrm{T}}(k) \tag{13.2}$$

$$\psi_m^{\mathrm{f}}(n) = \sum_{k=1}^{n} \lambda^{n-k}x(k)\mathbf{x}_m(k-1) \tag{13.3}$$

and λ is the forgetting factor. The least-squares sum of the estimation errors is then given by

$$\zeta_m^{ff}(n) = \sum_{k=1}^{n} \lambda^{n-k} f_{m,n}^2(k) \tag{13.4}$$

where

$$f_{m,n}(k) = x(k) - \mathbf{a}_m^{\mathrm{T}}(n)\mathbf{x}_m(k-1) \tag{13.5}$$

The sequence $f_{m,n}(k)$ is known as the *a posteriori prediction error* as its computation is based on the latest value of the predictor tap-weight vector $\mathbf{a}_m(n)$. In contrast, the *a priori*

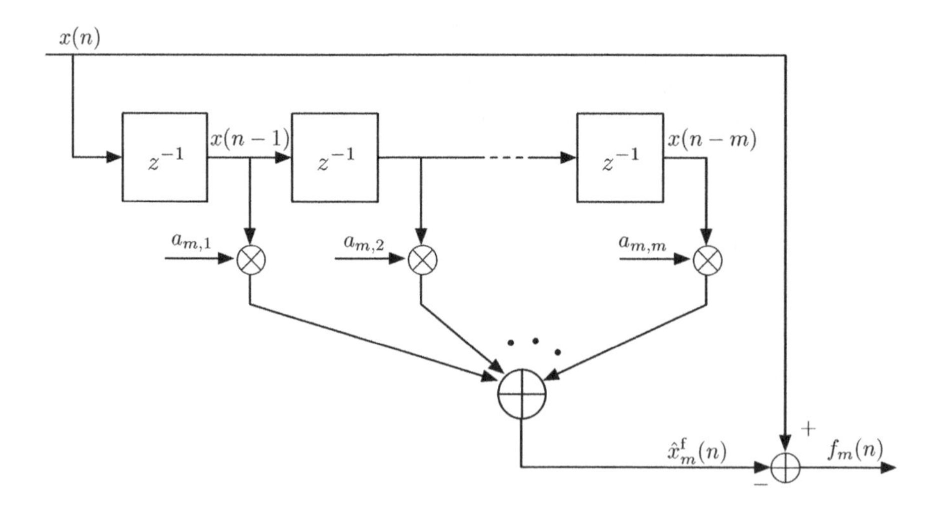

Figure 13.1 Transversal forward predictor.

prediction error of the forward predictor is defined as

$$f_{m,n-1}(k) = x(k) - \mathbf{a}_m^{\mathrm{T}}(n-1)\mathbf{x}_m(k-1) \tag{13.6}$$

where $\mathbf{a}_m(n-1)$ is the previous value of the predictor tap-weight vector.

We recall that according to the *principle of orthogonality,* the *a posteriori* estimation error, $f_{m,n}(k)$, and the predictor tap-input vector, $\mathbf{x}_m(k-1)$, are orthogonal, in the sense that

$$\sum_{k=1}^{n} \lambda^{n-k} f_{m,n}(k)\mathbf{x}_m(k-1) = \mathbf{0} \tag{13.7}$$

Substituting Eqs. (13.5) and (13.7) in Eq. (13.4), one can show that (Problem P13.1)

$$\zeta_m^{ff}(n) = \sum_{k=1}^{n} \lambda^{n-k} x^2(k) - \mathbf{a}_m^{\mathrm{T}}(n)\boldsymbol{\psi}_m^{\mathrm{f}}(n) \tag{13.8}$$

This result could also be obtained by inserting the relevant variables in Eq. (12.16).

Application of the *standard RLS algorithm* for adapting the forward transversal predictor results in the following recursion:

$$\mathbf{a}_m(n) = \mathbf{a}_m(n-1) + \mathbf{k}_m(n-1) f_{m,n-1}(n) \tag{13.9}$$

where $f_{m,n-1}(n)$ is the latest sample of the *a priori* estimation error of the forward predictor and $\mathbf{k}_m(n-1)$ is the present *gain vector* of the algorithm. The time index "$n-1$" in the gain vector here follows the predictor tap-input whose latest value is $\mathbf{x}_m(n-1)$. The use of this notation also keeps our notations consistent, as we proceed with similar formulations for the backward transversal predictor. Furthermore, following the results of Chapter 12, Eq. (12.48), the gain vector of the forward transversal predictor is obtained as

$$\mathbf{k}_m(n-1) = \boldsymbol{\Psi}_m^{-1}(n-1)\mathbf{x}_m(n-1) \tag{13.10}$$

At this point, we may note that there are a few differences between some of our notations here and those in Chapter 12. Since we will be making frequent use of the order-update equations in the derivations of this chapter, the order of the predictor, m, is explicitly reflected on all the variables. On the other hand, we have dropped the subscript λ (the forgetting factor) from the correlation matrix $\boldsymbol{\Psi}_m(n)$ and the vector $\boldsymbol{\psi}_m(n)$ to simplify the notations. However, the subscript m has been added to them to indicate their dimensions. The *hat*-sign that was used to refer to optimized tap-weight vectors and prediction errors is also dropped here in order to simplify the notations.

13.2 Least-Squares Backward Prediction

Figure 13.2 depicts an mth-order backward transversal predictor. Here, the predictor tap-weight and tap-input vectors are $\mathbf{g}_m(n) = [g_{m,1}(n)\ g_{m,2}(n)\ \cdots\ g_{m,m}(n)]^{\mathrm{T}}$ and $\mathbf{x}_m(k) = [x(k)\ x(k-1)\ \cdots\ x(k-m+1)]^{\mathrm{T}}$, respectively. The tap-weight vector, $\mathbf{g}_m(n)$, of the predictor is optimized in the least-squares sense over the entire observation interval $k = 1, 2, \ldots, n$. The samples of the desired output are $x(1-m)$, $x(2-m)$, …, $x(n-m)$.

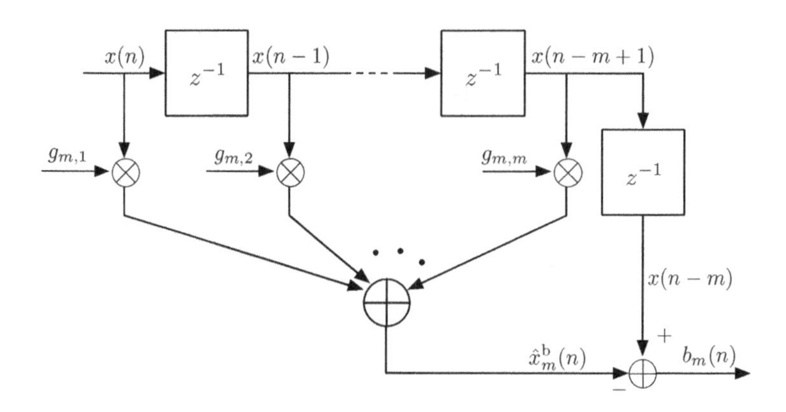

Figure 13.2 Transversal backward predictor.

Accordingly, the *normal equations* associated with the backward transversal predictor are obtained as (Chapter 12)

$$\boldsymbol{\Psi}_m(n)\mathbf{g}_m(n) = \boldsymbol{\psi}^{\rm b}_m(n) \tag{13.11}$$

where $\boldsymbol{\Psi}_m(n)$ is given in Eq. (13.2), and

$$\boldsymbol{\psi}^{\rm b}_m(n) = \sum_{k=1}^{n} \lambda^{n-k} x(k-m)\mathbf{x}_m(k) \tag{13.12}$$

The least-squares sum of the estimation errors is then given by

$$\zeta^{bb}_m(n) = \sum_{k=1}^{n} \lambda^{n-k} b^2_{m,n}(k) \tag{13.13}$$

where

$$b_{m,n}(k) = x(k-m) - \mathbf{g}^{\rm T}_m(n)\mathbf{x}_m(k) \tag{13.14}$$

The sequence $b_{m,n}(k)$ is the *a posteriori estimation error* of the backward predictor. In contrast, the *a priori prediction error* is defined as

$$b_{m,n-1}(k) = x(k-m) - \mathbf{g}^{\rm T}_m(n-1)\mathbf{x}_m(k) \tag{13.15}$$

where $\mathbf{g}_m(n-1)$ is the previous value of the predictor tap-weight vector.

We recall that according to the *principle of orthogonality,* the *a posteriori* estimation error, $b_{m,n}(k)$, and the predictor tap-input vector, $\mathbf{x}_m(k)$, are orthogonal, in the sense that

$$\sum_{k=1}^{n} \lambda^{n-k} b_{m,n}(k)\mathbf{x}_m(k) = \mathbf{0} \tag{13.16}$$

Substituting Eqs. (13.14) and (13.16) in Eq. (13.13), one can show that (Problem P13.2)

$$\zeta^{bb}_m(n) = \sum_{k=1}^{n} \lambda^{n-k} x^2(k-m) - \mathbf{g}^{\rm T}_m(n)\boldsymbol{\psi}^{\rm b}_m(n) \tag{13.17}$$

This result could also be obtained by inserting relevant variables in Eq. (12.16).

Application of the *standard RLS algorithm* for adapting the backward transversal predictor results in the following recursion:

$$\mathbf{g}_m(n) = \mathbf{g}_m(n-1) + \mathbf{k}_m(n) b_{m,n-1}(n) \tag{13.18}$$

where $b_{m,n-1}(n)$ is the latest sample of the *a priori* estimation error of the backward predictor and $\mathbf{k}_m(n)$ is the present *gain vector* of the algorithm, given by

$$\mathbf{k}_m(n) = \mathbf{\Psi}_m^{-1}(n)\mathbf{x}_m(n) \tag{13.19}$$

It is instructive to note that the gain vector of an mth-order forward predictor is equal to the previous value of the gain vector of the associated backward predictor. This, clearly, follows from the fact that the tap-input vectors of forward predictor, $\mathbf{x}_m(k-1)$, and backward predictor, $\mathbf{x}_m(k)$, are one sample apart, for a given time instant k.

13.3 Least-Squares Lattice

Figure 13.3 depicts the schematic of a lattice joint process estimator. This is similar to the lattice joint process estimator presented in Chapter 11 (Figure 11.7). However, there are some changes made in the notations here so that they can serve our discussion in this chapter better. From the last chapter, we recall that in the least-squares optimization, at any instant of time, say n, the filter parameters are optimized based on the observed data samples from time 1 to n, so that a weighted sum of the error squares is minimized. With this view of the problem, in Figure 13.3, the time index of the signal sequences is chosen to be k, and at time n, k is varied from 1 to n. Furthermore, following the notations in the last two sections, the estimation errors are labeled with two subscripts. The first subscript denotes the filter/predictor length/order. The second subscript indicates the length, n, of the observed data. The lattice PARCOR coefficients, $\kappa_m^{\mathrm{f}}(n)$ and $\kappa_m^{\mathrm{b}}(n)$, and the regressor coefficients, $c_m(n)$, are also labeled with time index n to emphasize that they are optimized in the least-squares sense on the basis of the data samples up to time n. In addition, to facilitate the derivation of the RLS lattice in the next section, the summer at the output of the joint process estimator is divided into a set of distributed adders so that the estimation errors of order 1 to N (denoted as $e_{1,n}(k)$ through $e_{N,n}(k)$) can be obtained in a sequential manner, as will be explained later.

From Chapter 11, we recall that when the input sequence, $x(n)$, is stationary, and the lattice coefficients are optimized to minimize the mean-squared errors of the forward and backward predictors, the PARCOR coefficients κ_m^{f} and κ_m^{b} are found to be equal. However, we note that this is not the case when the optimization of the lattice coefficients is based on least-squares criteria. Noting this, throughout this chapter, we keep the superscripts f and b on κ_m^{f} and κ_m^{b}, respectively, to differentiate between the two.

To optimize the coefficients of the lattice joint process estimator, the following sums are minimized, simultaneously:

$$\zeta_m^{ff}(n) = \sum_{k=1}^{n} \lambda^{n-k} f_{m,n}^2(k), \quad \text{for} \quad m = 1, 2, \ldots, N-1 \tag{13.20}$$

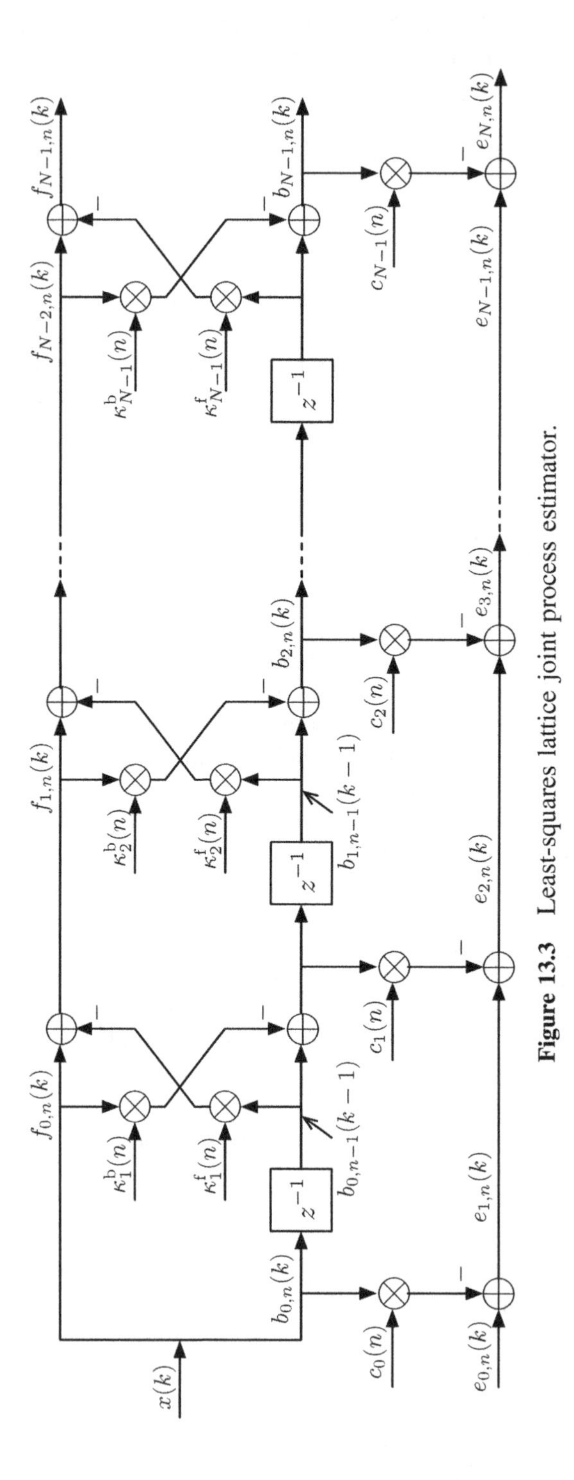

Figure 13.3 Least-squares lattice joint process estimator.

$$\zeta_m^{bb}(n) = \sum_{k=1}^{n} \lambda^{n-k} b_{m,n}^2(k), \quad \text{for} \quad m = 1, 2, \ldots, N-1 \tag{13.21}$$

$$\zeta_m^{ee}(n) = \sum_{k=1}^{n} \lambda^{n-k} e_{m,n}^2(k), \quad \text{for} \quad m = 1, 2, \ldots, N \tag{13.22}$$

where $f_{m,n}(k)$ and $b_{m,n}(k)$ are the *a posteriori* estimation errors as defined before, and similarly $e_{m,n}(k)$ is defined as the *a posteriori estimation error of the length m joint process estimator.* Note from Figure 13.3 that there are effectively N forward and N backward predictors of order 1 to N, as well as N joint process estimators of length 1 to N, optimized simultaneously. Furthermore, it is important to note that the least-squares sums Eqs. (13.20)–(13.22) are independent of whether the predictors and joint process estimators are implemented in transversal or lattice forms. We will exploit this fact to simplify the derivations that follow by switching between the equations derived for the transversal and lattice forms to arrive at the desired results.

In Chapter 11, we discussed a number of properties of the lattice structure. In particular, we noted that the backward prediction errors of different orders are orthogonal (uncorrelated) with one another. This and many other properties of the lattice structure, which were discussed in Chapter 11 based on *stochastic averages,* are equally applicable to the lattice structure of Figure 13.3, where optimization of the filter coefficients is done based on *time averages* (because of the least-squares optimization). The most important properties of the least-squares lattice that are relevant to the derivation of the RLSL algorithm in the next section are the following:

1. At time n, the PARCOR coefficients $\kappa_m^{\mathrm{f}}(n)$ and $\kappa_m^{\mathrm{b}}(n)$ of the successive stages of the lattice structure can be optimized sequentially as follows. We first note the sequences at the tap inputs of the first stage (i.e., the signals multiplied by the PARCOR coefficients $\kappa_1^{\mathrm{f}}(n)$ and $\kappa_1^{\mathrm{b}}(n)$) are $\mathrm{f}_{0,n}(k)$ and $b_{0,n-1}(k-1)$, for $k = 1, 2, \ldots, n$ (Figure 13.3). Using these sequences, the coefficients $\kappa_1^{\mathrm{f}}(n)$ and $\kappa_1^{\mathrm{b}}(n)$ are optimized so that the output sequences $f_{1,n}(k)$ and $b_{1,n}(k)$ of the first stage are minimized in least-squares sense. Next, we note that the tap inputs in the second stage are $f_{1,n}(k)$ and $b_{1,n-1}(k-1)$. Accordingly, considering the sequences $f_{1,n}(k)$ and $b_{1,n-1}(k-1)$, for $k = 1, 2, \ldots, n$, as tap inputs to the second stage, the coefficients $\kappa_2^{\mathrm{f}}(n)$ and $\kappa_2^{\mathrm{b}}(n)$ are optimized so that the output sequences $f_{2,n}(k)$ and $b_{2,n}(k)$ of the second stage are minimized in least-squares sense. This process continues for the rest of the stages as well. The above process leads to the following equations, which are the bases for derivation of the RLS algorithm, as explained in the next section:

$$\kappa_m^{\mathrm{f}}(n) = \frac{\sum_{k=1}^{n} \lambda^{n-k} f_{m-1,n}(k) b_{m-1,n-1}(k-1)}{\sum_{k=1}^{n} \lambda^{n-k} b_{m-1,n-1}^2(k-1)} \tag{13.23}$$

and

$$\kappa_m^{\mathrm{b}}(n) = \frac{\sum_{k=1}^{n} \lambda^{n-k} f_{m-1,n}(k) b_{m-1,n-1}(k-1)}{\sum_{k=1}^{n} \lambda^{n-k} f_{m-1,n}^2(k)} \tag{13.24}$$

for $m = 1, 2, \ldots, N-1$.

2. Once the PARCOR coefficients are optimized, the backward prediction errors $b_{0,n}(k)$, $b_{1,n}(k)$, ..., $b_{N,n}(k)$ are orthogonal with one another, in the sense that

$$\sum_{k=1}^{n} \lambda^{n-k} b_{i,n}(k) b_{j,n}(k) = 0 \tag{13.25}$$

for any pair of unequal i and j in the range of 0 to $N-1$.

3. The regressor coefficients $c_0(n)$, $c_1(n)$, ..., $c_{N-1}(n)$ may also be optimized in a sequential manner. That is, first $c_0(n)$ is optimized by minimizing $\zeta_1^{ee}(n)$. We then hold $c_0(n)$, run the sequence $\{x(1), x(2), \ldots, x(n)\}$ through the first stage of the lattice, and optimize $c_1(n)$ so that $\zeta_2^{ee}(n)$ is minimized. This process continues for the rest of the joint process estimator coefficients as well. To summarize, the regressor coefficients $c_0(n)$, $c_1(n)$, ..., $c_{N-1}(n)$ are obtained according to the following equations:

$$c_m(n) = \frac{\sum_{k=1}^{n} \lambda^{n-k} e_{m,n}(k) b_{m,n}(k)}{\sum_{k=1}^{n} \lambda^{n-k} b_{m,n}^2(k)} \tag{13.26}$$

for $m = 0, 1, \ldots, N-1$.

Equations (13.23), (13.24), and (13.26), although fundamental in providing a clear understanding of the underlying principles in the development of the least-squares lattice algorithm, cannot be used for the computation of the lattice coefficients in an adaptive application, as their computational complexity grows with the number of data samples, n. As in the case of the standard RLS algorithm, the problem is solved by finding a set of equations that update the filter coefficients in a recursive manner. This is the subject of the next section.

13.4 RLSL Algorithm

In this section, we go through a systematic step-by-step procedure to develop the recursions necessary for the derivation of the RLSL algorithm. The development of the RLSL algorithm involves a large number of variables, compared to any of the algorithms that we have derived/discussed thus far in this book. Because of this, it is often difficult for a novice to the topic to follow these equations. Thus, choosing the right set of notations that reduce this burden is crucial to the development of a readable material on this topic. Bearing this in mind, our discussion on the RLSL algorithm begins with an introduction to the notations and some preliminaries. The derivations will be followed thereafter.

13.4.1 Notations and Preliminaries

In the past few sections, we introduced a number of notations for formulating the least-squares solutions in the cases of forward and backward transversal predictors as well as lattice joint process estimator. Here, we introduce some more notations and also some new definitions that are necessary for the derivations that follow.

Prewindowing of Input Data: Throughout the discussion in the remainder of this chapter, we assume that the samples of input signal, $x(k)$, are all-zero for values of $k \leq 0$. This assumption on input signal is known as *prewindowing*. In this book, we do not consider other variations of the fast RLS algorithms that are based on other types of windowing methods (Honig and Messerschmitt, 1984; Alexander, 1986a,b).

A Priori and A Posteriori Estimation Errors: We noted that the subscript n in the sequences $f_{m,n}(k)$, $b_{m,n}(k)$, and $e_{m,n}(k)$ denotes that they are *a posteriori* estimation errors. The term *a posteriori* signifies that the errors are obtained using the filter (predictor) coefficients, which have been optimized using the past as well as the present samples of the input and desired output, that is, $x(k)$ and $d(k)$, for $k = 1, 2, \ldots, n$. In other words, the *a posteriori* estimation errors are obtained when the lattice coefficients $\kappa_m^{\mathrm{f}}(n)$'s, $\kappa_m^{\mathrm{b}}(n)$'s, and $c_m(n)$'s, as given by Eqs. (13.23), (13.24), and (13.26), are chosen. In contrast, if we compute these estimation errors using the last values of the joint process estimator coefficients, that is, $\kappa_m^{\mathrm{f}}(n-1)$'s, $\kappa_m^{\mathrm{b}}(n-1)$'s, and $c_m(n-1)$'s, then the resulting errors are known as the *a priori*. We use the notations $f_{m,n-1}(k)$, $b_{m,n-1}(k)$, and $e_{m,n-1}(k)$ (with the subscripts $n-1$ signifying the use of the last values of the lattice coefficients, $\kappa_m^{\mathrm{f}}(n-1)$'s, $\kappa_m^{\mathrm{b}}(n-1)$'s, and $c_m(n-1)$'s) to refer to the *a priori* estimation errors.

Conversion Factor, $\gamma_m(n)$: A key result to the development of the fast RLS algorithms is the following relationship:

$$\frac{e_{m,n-1}(n)}{e_{m,n}(n)} = \frac{b_{m,n-1}(n)}{b_{m,n}(n)} = \frac{f_{m,n}(n+1)}{f_{m,n+1}(n+1)} \tag{13.27}$$

Note that in Eq. (13.27) the numerators are the *a priori* estimation errors and the denominators are the *a posteriori* estimation errors. To appreciate this relationship, we note that the tap-input vector to the length m joint process estimator at time n is $\mathbf{x}_m(n) = [x(n)\ x(n-1)\ \cdots\ x(n-m+1)]^{\mathrm{T}}$. This is also the tap-input vector to the mth-order backward predictor at time n and that of the forward predictor at time $n+1$. Noting this, it appears that, in general, the ratio of the *a priori* and *a posteriori* estimation errors depends only on the tap-input vector of the filter (predictor or joint process estimator). This ratio, which will be discussed in detail later, is called *conversion factor*, denoted as $\gamma_m(n)$.

Least-Squares Error Sums, $\zeta_m^{ee}(n)$, $\zeta_m^{ff}(n)$, and $\zeta_m^{bb}(n)$: We recall that $\zeta_m^{ee}(n)$, $\zeta_m^{ff}(n)$, and $\zeta_m^{bb}(n)$, as defined in Eqs. (13.22), (13.20), and (13.21), respectively, refer to the least-squares error sums of the joint process estimator of length m, and forward and backward predictors of order m. Note also that all these are based on the *a posteriori estimation errors*.

Cross-correlations, $\zeta_m^{fb}(n)$ and $\zeta_m^{be}(n)$: The summation in the numerator of Eq. (13.23) (and also Eq. (13.24)) may be defined as the (deterministic) *cross-correlation* between the forward and backward prediction errors, $f_{m-1,n}(k)$ and $b_{m-1,n-1}(k-1)$. Similarly, the summation in the numerator of Eq. (13.26) may be called the *cross-correlation* between the backward prediction error, $b_{m,n}(k)$, and the joint process estimation error, $e_{m,n}(k)$.

Accordingly, we define

$$\zeta_m^{fb}(n) = \sum_{k=1}^{n} \lambda^{n-k} b_{m,n-1}(k-1) f_{m,n}(k) \tag{13.28}$$

and

$$\zeta_m^{be}(n) = \sum_{k=1}^{n} \lambda^{n-k} e_{m,n}(k) b_{m,n}(k) \tag{13.29}$$

To follow the same terminology, we may refer to the least-squares sums $\zeta_m^{ff}(n)$, $\zeta_m^{bb}(n)$, and $\zeta_m^{ee}(n)$ as *autocorrelations.*

Using Eqs. (13.20), (13.21), (13.22), (13.28), and (13.29), the set of Eqs. (13.23), (13.24), and (13.26) are written as

$$\kappa_m^{\mathrm{f}}(n) = \frac{\zeta_{m-1}^{fb}(n)}{\zeta_{m-1}^{bb}(n-1)} \tag{13.30}$$

$$\kappa_m^{\mathrm{b}}(n) = \frac{\zeta_{m-1}^{fb}(n)}{\zeta_{m-1}^{ff}(n)} \tag{13.31}$$

and

$$c_m(n) = \frac{\zeta_m^{be}(n)}{\zeta_m^{bb}(n)} \tag{13.32}$$

We later develop a set of equations for recursive updating of the auto- and cross-correlations that were just defined. *The updated auto- and cross-correlation will then be substituted into Eqs.* (13.30)–(13.32) *for the computation of the lattice coefficients at every iteration.*

Augmented Normal Equations for Forward and Backward Prediction: Using the definition (13.2), $\mathbf{\Psi}_{m+1}(n)$ may be extended as

$$\mathbf{\Psi}_{m+1}(n) = \begin{bmatrix} \psi^{00}(n) & \boldsymbol{\psi}_m^{f\mathrm{T}}(n) \\ \boldsymbol{\psi}_m^{\mathrm{f}}(n) & \mathbf{\Psi}_m(n-1) \end{bmatrix} \tag{13.33}$$

where $\boldsymbol{\psi}_m^{\mathrm{f}}(n)$ is defined by Eq. (13.3) and

$$\psi^{00}(n) = \sum_{k=1}^{n} \lambda^{n-k} x^2(k) \tag{13.34}$$

Using Eqs. (13.33) and (13.34), Eqs. (13.1) and (13.8) may be combined together to obtain

$$\mathbf{\Psi}_{m+1}(n)\tilde{\mathbf{a}}_m(n) = \begin{bmatrix} \zeta_m^{ff}(n) \\ \mathbf{0}_m \end{bmatrix} \tag{13.35}$$

where

$$\tilde{\mathbf{a}}_m(n) = \begin{bmatrix} 1 \\ -\mathbf{a}_m(n) \end{bmatrix} \tag{13.36}$$

$\mathbf{0}_m$ denotes the m-by-1 zero vector, and $\mathbf{a}_m(n)$ is the tap-weight vector of the transversal forward predictor, optimized in the least-squares sense. Equation (13.35) is known as the *augmented normal equation for forward predictor of order m.*

The matrix $\boldsymbol{\Psi}_{m+1}(n)$ may also be extended as

$$\boldsymbol{\Psi}_{m+1}(n) = \begin{bmatrix} \boldsymbol{\Psi}_m(n) & \boldsymbol{\psi}_m^{\mathrm{b}}(n) \\ \boldsymbol{\psi}_m^{bT}(n) & \psi^{mm}(n) \end{bmatrix} \tag{13.37}$$

where $\boldsymbol{\psi}_m^{\mathrm{b}}(n)$ is defined by Eq. (13.12) and

$$\psi^{mm}(n) = \sum_{k=1}^{n} \lambda^{n-k} x^2(k-m) \tag{13.38}$$

Using Eqs. (13.37) and (13.38), Eqs. (13.11) and (13.17) may be combined together to obtain

$$\boldsymbol{\Psi}_{m+1}(n)\tilde{\mathbf{g}}_m(n) = \begin{bmatrix} \mathbf{0}_m \\ \zeta_m^{bb}(n) \end{bmatrix} \tag{13.39}$$

where

$$\tilde{\mathbf{g}}_m(n) = \begin{bmatrix} -\mathbf{g}_m(n) \\ 1 \end{bmatrix} \tag{13.40}$$

and $\mathbf{g}_m(n)$ is the tap-weight vector of the transversal backward predictor, optimized in the least-squares sense. Equation (13.39) is known as the *augmented normal equation for backward predictor of order m.*

13.4.2 Update Recursion for the Least-Squares Error Sums

Consider an N-tap transversal filter with the tap-input vector $\mathbf{x}(k) = [x(k)\ x(k-1)\ \cdots\ x(k-N+1)]^{\mathrm{T}}$ and desired output $d(k)$. From Chapter 12, we recall that the least-squares error sum of the filter at time n is [1]

$$\zeta_{\min}(n) = \mathbf{d}^{\mathrm{T}}(n)\boldsymbol{\Lambda}(n)\mathbf{d}(n) - \boldsymbol{\theta}^{\mathrm{T}}(n)\hat{\mathbf{w}}(n) \tag{13.41}$$

where $\hat{\mathbf{w}}(n)$ is the optimized tap-weight vector of the filter, $\mathbf{d}(n) = [d(1)\ d(2)\ \cdots\ d(n)]^{\mathrm{T}}$,

$$\boldsymbol{\theta}(n) = \sum_{k=1}^{n} \lambda^{n-k} d(k)\mathbf{x}(k)$$

and $\boldsymbol{\Lambda}(n)$ is the diagonal matrix consisting of the powers of the forgetting factor, λ, as defined in Eq. (12.35). Substituting Eqs. (12.42), (12.52), and

$$\mathbf{d}^{\mathrm{T}}(n)\boldsymbol{\Lambda}(n)\mathbf{d}(n) = d^2(n) + \lambda\mathbf{d}^{\mathrm{T}}(n-1)\boldsymbol{\Lambda}(n-1)\mathbf{d}(n-1)$$

[1] It may be noted that Eq. (13.41) is similar to Eq. (12.16) with the forgetting factor, λ, included in the results. In addition, to be consistent with the rest of our notations in this chapter, the subscript λ has been dropped from vectors and matrices.

in Eq. (13.41) and rearranging, we obtain

$$\begin{aligned}\zeta_{\min}(n) &= \lambda(\mathbf{d}^{\mathrm{T}}(n-1)\boldsymbol{\Lambda}(n-1)\mathbf{d}(n-1) - \boldsymbol{\theta}^{\mathrm{T}}(n-1)\hat{\mathbf{w}}(n-1)) \\ &\quad + d(n)(d(n) - \hat{\mathbf{w}}^{\mathrm{T}}(n-1)\mathbf{x}^{\mathrm{T}}(n)) - \mathbf{x}^{\mathrm{T}}(n)\mathbf{k}(n)e_{n-1}(n)d(n) \\ &\quad - \lambda\boldsymbol{\theta}^{\mathrm{T}}(n-1)\mathbf{k}(n)e_{n-1}(n) \\ &= \lambda\zeta_{\min}(n-1) + d(n)e_{n-1}(n) \\ &\quad - (\mathbf{x}(n)d(n) + \lambda\boldsymbol{\theta}(n-1))^{\mathrm{T}}\mathbf{k}(n)e_{n-1}(n) \end{aligned} \tag{13.42}$$

where we have noted that $d(n) - \hat{\mathbf{w}}^{\mathrm{T}}(n-1)\mathbf{x}(n)$ is the *a priori* estimation error $e_{n-1}(n)$. Furthermore, from Eq. (12.42), we note that $\mathbf{x}(n)d(n) + \lambda\boldsymbol{\theta}(n-1) = \boldsymbol{\theta}(n)$. Using this result and Eq. (12.48), we obtain

$$\begin{aligned}(\mathbf{x}(n)d(n) + \lambda\boldsymbol{\theta}(n-1))^{\mathrm{T}}\mathbf{k}(n) &= \boldsymbol{\theta}^{\mathrm{T}}(n)\boldsymbol{\Psi}^{-1}(n)\mathbf{x}(n) \\ &= \hat{\mathbf{w}}^{\mathrm{T}}(n)\mathbf{x}(n) \end{aligned} \tag{13.43}$$

where we have noted that $\boldsymbol{\theta}^{\mathrm{T}}(n)\boldsymbol{\Psi}^{-1}(n) = (\boldsymbol{\Psi}^{-1}(n)\boldsymbol{\theta}(n))^{\mathrm{T}} = \hat{\mathbf{w}}^{\mathrm{T}}(n)$, as $\boldsymbol{\Psi}^{-1}(n)$ is symmetrical. Substituting Eq. (13.43) in Eq. (13.42) and rearranging, we obtain

$$\zeta_{\min}(n) = \lambda\zeta_{\min}(n-1) + e_n(n)e_{n-1}(n) \tag{13.44}$$

where $e_n(n) = d(n) - \hat{\mathbf{w}}^{\mathrm{T}}(n)\mathbf{x}(n)$ is the *a posteriori* estimation error. Thus, to update $\zeta_{\min}(n-1)$, we only need to know the *a priori* and *a posteriori* estimation errors at instant n, that is, $e_{n-1}(n)$ and $e_n(n)$, respectively.

Recursion Eq. (13.44) can readily be applied to update the least-squares error sums of the forward and backward predictors as well as the joint process estimator. The results are

$$\zeta_m^{ff}(n) = \lambda\zeta_m^{ff}(n-1) + f_{m,n}(n)f_{m,n-1}(n) \tag{13.45}$$

$$\zeta_m^{bb}(n) = \lambda\zeta_m^{bb}(n-1) + b_{m,n}(n)b_{m,n-1}(n) \tag{13.46}$$

$$\zeta_m^{ee}(n) = \lambda\zeta_m^{ee}(n-1) + e_{m,n}(n)e_{m,n-1}(n) \tag{13.47}$$

where $f_{m,n-1}(n)$, $f_{m,n}(n)$, $b_{m,n-1}(n)$, $b_{m,n}(n)$, $e_{m,n-1}(n)$, and $e_{m,n}(n)$, as defined before, are the associated *a priori* and *a posteriori* estimation errors.

We note that the above update equations involve the use of both the *a priori* and *a posteriori* estimation errors. We next see that with the aid of the conversion factor, $\gamma_m(n)$, the above recursions may only be written in terms of either the *a priori* or *a posteriori* estimation errors.

13.4.3 Conversion Factor

Using recursion Eq. (12.52), the *a posteriori* estimation error $e_n(n)$ of a transversal filter with the least-squares optimized tap-weight vector $\hat{\mathbf{w}}(n)$, tap-input vector $\mathbf{x}(n)$, and desired output $d(n)$ may be expanded as

$$\begin{aligned}
e_n(n) &= d(n) - \hat{\mathbf{w}}^{\mathrm{T}}(n)\mathbf{x}(n) \\
&= d(n) - (\hat{\mathbf{w}}(n-1) + \mathbf{k}(n)e_{n-1}(n))^{\mathrm{T}}\mathbf{x}(n) \\
&= d(n) - \hat{\mathbf{w}}^{\mathrm{T}}(n-1)\mathbf{x}(n) - \mathbf{k}^{\mathrm{T}}(n)\mathbf{x}(n)e_{n-1}(n) \\
&= e_{n-1}(n) - \mathbf{k}^{\mathrm{T}}(n)\mathbf{x}(n)e_{n-1}(n) \\
&= (1 - \mathbf{k}^{\mathrm{T}}(n)\mathbf{x}(n))e_{n-1}(n)
\end{aligned} \tag{13.48}$$

where $e_{n-1}(n) = d(n) - \hat{\mathbf{w}}^{\mathrm{T}}(n-1)\mathbf{x}(n)$ is the *a priori* estimation error. We note that the *a priori* and *a posteriori* estimation errors, $e_{n-1}(n)$ and $e_n(n)$, respectively, are related by the factor $1 - \mathbf{k}^{\mathrm{T}}(n)\mathbf{x}(n)$. This is called *conversion factor*, as mentioned in Section 13.4.1, and is denoted by $\gamma(n)$. Thus,

$$\gamma(n) = 1 - \mathbf{k}^{\mathrm{T}}(n)\mathbf{x}(n) \tag{13.49}$$

Substituting Eq. (12.48) in Eq. (13.49), we also obtain

$$\gamma(n) = 1 - \mathbf{x}^{\mathrm{T}}(n)\boldsymbol{\Psi}^{-1}(n)\mathbf{x}(n) \tag{13.50}$$

An interesting interpretation of $\gamma(n)$ whose study is left for the reader in Problem P13.8 reveals that $\gamma(n)$ is a positive quantity less than or equal to 1. Another interesting property of $\gamma(n)$ is seen by noting that for a given forgetting factor, λ, $\boldsymbol{\Psi}(n)$ depends only on the input samples to the filter. Accordingly, $\gamma(n)$ can be found once the observed tap-input vectors to the filter are known. The following cases are then identified:

1. In an mth-order forward predictor, with the observed tap-input vectors $\mathbf{x}_m(0)$, $\mathbf{x}_m(1)$, ..., $\mathbf{x}_m(n-1)$ (with $\mathbf{x}_m(0) = 0$, because of prewindowing the input data), the conversion factor is recognized as

$$\begin{aligned}
\gamma_m(n-1) &= 1 - \mathbf{x}_m^{\mathrm{T}}(n-1)\boldsymbol{\Psi}^{-1}(n-1)\mathbf{x}_m(n-1) \\
&= 1 - \mathbf{k}_m^{\mathrm{T}}(n-1)\mathbf{x}_m(n-1)
\end{aligned} \tag{13.51}$$

   where $\mathbf{k}_m(n-1)$ is the gain vector of the forward predictor, as was identified before (Section 13.1). Accordingly, the *a priori* and *a posteriori* estimation errors $f_{m,n-1}(n)$ and $f_{m,n}(n)$ of the forward predictor are related according to the following equation:

$$f_{m,n}(n) = \gamma_m(n-1)f_{m,n-1}(n) \tag{13.52}$$

2. Similarly, in an mth-order backward predictor, with the observed tap-input vectors $\mathbf{x}_m(1)$, $\mathbf{x}_m(2)$, ..., $\mathbf{x}_m(n)$, the conversion factor is recognized as

$$\begin{aligned}
\gamma_m(n) &= 1 - \mathbf{x}_m^{\mathrm{T}}(n)\boldsymbol{\Psi}_m^{-1}(n)\mathbf{x}_m(n) \\
&= 1 - \mathbf{k}_m^{\mathrm{T}}(n)\mathbf{x}_m(n)
\end{aligned} \tag{13.53}$$

   Accordingly, the *a priori* and *a posteriori* estimation errors $b_{m,n-1}(n)$ and $b_{m,n}(n)$ of the backward predictor are related according to the following equation:

$$b_{m,n}(n) = \gamma_m(n)b_{m,n-1}(n) \tag{13.54}$$

3. The observed tap-input vectors to an m-tap joint process estimator are $\mathbf{x}_m(1), \mathbf{x}_m(2), \ldots, \mathbf{x}_m(n)$. Since these are similar to the tap-input vectors to the mth-order backward predictor, the conversion factor of the m-tap joint process estimator is also $\gamma_m(n)$, given by Eq. (13.53). Accordingly, the *a priori* and *a posteriori* estimation errors $e_{m,n-1}(n)$ and $e_{m,n}(n)$ of the joint process estimator are related according to the following equation:

$$e_{m,n}(n) = \gamma_m(n) e_{m,n-1}(n) \tag{13.55}$$

13.4.4 Update Equation for Conversion Factor

First, we show that

$$\boldsymbol{\Psi}_{m+1}^{-1}(n) = \begin{bmatrix} \boldsymbol{\Psi}_m^{-1}(n) & \mathbf{0}_m \\ \mathbf{0}_m^{\mathrm{T}} & \mathbf{0} \end{bmatrix} + \frac{1}{\zeta_m^{bb}(n)} \tilde{\mathbf{g}}_m(n) \tilde{\mathbf{g}}_m^{\mathrm{T}}(n) \tag{13.56}$$

To this end, we multiply the right-hand side of Eq. (13.56) by $\boldsymbol{\Psi}_{m+1}(n)$ and show that the result is the $(m+1)$-by-$(m+1)$ identity matrix. The two separate expressions arising from this multiplication are

$$\mathbf{A} = \boldsymbol{\Psi}_{m+1}(n) \begin{bmatrix} \boldsymbol{\Psi}_m^{-1}(n) & \mathbf{0}_m \\ \mathbf{0}_m^{\mathrm{T}} & 0 \end{bmatrix} \tag{13.57}$$

and

$$\mathbf{B} = \frac{1}{\zeta_m^{bb}(n)} \boldsymbol{\Psi}_{m+1}(n) \tilde{\mathbf{g}}_m(n) \tilde{\mathbf{g}}_m^{\mathrm{T}}(n) \tag{13.58}$$

Substituting Eq. (13.37) in Eq. (13.57), we obtain

$$\begin{aligned} \mathbf{A} &= \begin{bmatrix} \boldsymbol{\Psi}_m(n) & \boldsymbol{\psi}_m^{\mathrm{b}}(n) \\ \boldsymbol{\psi}_m^{b\mathrm{T}}(n) & \psi^{mm}(n) \end{bmatrix} \begin{bmatrix} \boldsymbol{\Psi}_m^{-1}(n) & \mathbf{0}_m \\ \mathbf{0}_m^{\mathrm{T}} & 0 \end{bmatrix} \\ &= \begin{bmatrix} \mathbf{I}_m & \mathbf{0}_m \\ \boldsymbol{\psi}_m^{b\mathrm{T}}(n) \boldsymbol{\Psi}_m^{-1}(n) & 0 \end{bmatrix} \end{aligned} \tag{13.59}$$

where $\mathbf{I}_m$ is the m-by-m identity matrix. Furthermore, recalling that $\boldsymbol{\Psi}_m^{-1}(n)$ is a symmetric matrix and using Eq. (13.11), we obtain

$$\boldsymbol{\psi}_m^{b\mathrm{T}}(n) \boldsymbol{\Psi}_m^{-1}(n) = (\boldsymbol{\Psi}_m^{-1}(n) \boldsymbol{\psi}_m^{\mathrm{b}}(n))^{\mathrm{T}} = \mathbf{g}_m^{\mathrm{T}}(n) \tag{13.60}$$

Substituting Eq. (13.60) in Eq. (13.59), we obtain

$$\mathbf{A} = \begin{bmatrix} \mathbf{I}_m & \mathbf{0}_m \\ \mathbf{g}_m^{\mathrm{T}}(n) & 0 \end{bmatrix} \tag{13.61}$$

In addition, substituting Eq. (13.39) in Eq. (13.58), we obtain

$$\mathbf{B} = \begin{bmatrix} \mathbf{0}_m \\ 1 \end{bmatrix} [-\mathbf{g}_m^{\mathrm{T}}(n) \;\; 1] = \begin{bmatrix} 0 & \cdots & 0 & 0 \\ \vdots & \ddots & \vdots & \vdots \\ 0 & \cdots & 0 & 0 \\ \multicolumn{3}{c}{-\mathbf{g}_m^{\mathrm{T}}(n)} & 1 \end{bmatrix} \tag{13.62}$$

Adding the results of Eqs. (13.61) and (13.62), we obtain

$$\mathbf{A} + \mathbf{B} = \mathbf{I}_{m+1}.$$

This completes the proof of Eq. (13.56).

Premultiplying and postmultiplying Eq. (13.56) by $\mathbf{x}_{m+1}^{\mathrm{T}}(n)$ and $\mathbf{x}_{m+1}(n)$, respectively, and noting that

$$\mathbf{x}_{m+1}(n) = \begin{bmatrix} \mathbf{x}_m(n) \\ x(n-m) \end{bmatrix} \quad \text{and} \quad \mathbf{k}_m(n) = \mathbf{\Psi}_m^{-1}(n)\mathbf{x}_m(n)$$

we obtain

$$\mathbf{x}_{m+1}^{\mathrm{T}}(n)\mathbf{k}_{m+1}(n) = \mathbf{x}_m^{\mathrm{T}}(n)\mathbf{k}_m(n) + \frac{(\tilde{\mathbf{g}}_m^{\mathrm{T}}(n)\mathbf{x}_{m+1}(n))^2}{\zeta_m^{bb}(n)} \tag{13.63}$$

Using Eq. (13.40), we obtain

$$\begin{aligned} \tilde{\mathbf{g}}_m^{\mathrm{T}}(n)\mathbf{x}_{m+1}(n) &= x(n-m) - \mathbf{g}_m^{\mathrm{T}}(n)\mathbf{x}_m(n) \\ &= b_{m,n}(n) \end{aligned} \tag{13.64}$$

Finally, substituting Eq. (13.64) in Eq. (13.63), subtracting both sides of the result from 1, and recalling Eq. (13.53), we obtain

$$\gamma_{m+1}(n) = \gamma_m(n) - \frac{b_{m,n}^2(n)}{\zeta_m^{bb}(n)} \tag{13.65}$$

13.4.5 Update Equation for Cross-Correlations

The recursions that are remaining to complete the derivation of the RLSL algorithm are the update equations for cross-correlations $\zeta_m^{fb}(n)$ and $\zeta_m^{be}(n)$.

Recall the RLS recursion for the mth-order forward transversal predictor

$$\mathbf{a}_m(n) = \mathbf{a}_m(n-1) + \mathbf{k}_m(n-1)f_{m,n-1}(n) \tag{13.66}$$

where $\mathbf{k}_m(n-1)$ and $f_{m,n-1}(n)$, as defined earlier, are the gain vector and *a priori* estimation error of the forward predictor, respectively. The samples of the *a posteriori* estimation error of the forward predictor, for $k = 1, 2, \ldots, n$, are given by

$$f_{m,n}(k) = x(k) - \mathbf{a}_m^{\mathrm{T}}(n)\mathbf{x}_m(k-1) \tag{13.67}$$

Substituting Eq. (13.66) in Eq. (13.67) and rearranging, we obtain

$$f_{m,n}(k) = f_{m,n-1}(k) - \mathbf{k}_m^{\mathrm{T}}(n-1)\mathbf{x}_m(k-1)f_{m,n-1}(n) \tag{13.68}$$

where

$$f_{m,n-1}(k) = x(k) - \mathbf{a}_m^{\mathrm{T}}(n-1)\mathbf{x}_m(k-1) \tag{13.69}$$

for $k = 1, 2, \ldots, n$ are samples of the *a priori* forward prediction error.

In addition, recall the RLS recursion for the mth-order backward predictor

$$\mathbf{g}_m(n) = \mathbf{g}_m(n-1) + \mathbf{k}_m(n)b_{m,n-1}(n) \tag{13.70}$$

where $\mathbf{k}_m(n)$ and $b_{m,n-1}(n)$, as defined earlier, are the gain vector and *a priori* estimation error of the backward predictor, respectively. The samples of the *a posteriori* estimation error of the backward predictor, for $k = 0, 1, \ldots, n$, are given by

$$b_{m,n}(k) = x(k-m) - \mathbf{g}_m^{\mathrm{T}}(n)\mathbf{x}_m(k) \tag{13.71}$$

Substituting Eq. (13.70) in Eq. (13.71) and rearranging, we obtain

$$b_{m,n}(k) = b_{m,n-1}(k) - \mathbf{k}_m^{\mathrm{T}}(n)\mathbf{x}_m(k)b_{m,n-1}(n) \tag{13.72}$$

where

$$b_{m,n-1}(k) = x(k-m) - \mathbf{g}_m^{\mathrm{T}}(n-1)\mathbf{x}_m(k) \tag{13.73}$$

for $k = 1, 2, \ldots, n$, are samples of the *a priori* backward prediction error.

Next, substituting Eqs. (13.68) and (13.72) in Eq. (13.28) and expanding, we obtain

$$\begin{aligned}\zeta_m^{fb}(n) = {} & \sum_{k=1}^{n} \lambda^{n-k} f_{m,n-1}(k)b_{m,n-2}(k-1) \\ & -\mathbf{k}_m^{\mathrm{T}}(n-1)b_{m,n-2}(n-1)\sum_{k=1}^{n}\lambda^{n-k} f_{m,n-1}(k)\mathbf{x}_m(k-1) \\ & -\mathbf{k}_m^{\mathrm{T}}(n-1)f_{m,n-1}(n)\sum_{k=1}^{n}\lambda^{n-k} b_{m,n-2}(k-1)\mathbf{x}_m(k-1) \\ & +\mathbf{k}_m^{\mathrm{T}}(n-1)\mathbf{\Psi}_m(n-1)\mathbf{k}_m(n-1)f_{m,n-1}(n)b_{m,n-2}(n-1)\end{aligned} \tag{13.74}$$

where, to obtain the last term, we have noted that $\mathbf{k}_m^{\mathrm{T}}(n-1)\mathbf{x}_m(n-1) = \mathbf{x}_m^{\mathrm{T}}(n-1)\mathbf{k}_m(n-1)$ and also

$$\sum_{k=1}^{n}\lambda^{n-k}\mathbf{x}_m(k-1)\mathbf{x}_m^{\mathrm{T}}(k-1) = \sum_{k=1}^{n-1}\lambda^{n-1-k}\mathbf{x}_m(k)\mathbf{x}_m^{\mathrm{T}}(k) = \mathbf{\Psi}_m(n-1)$$

as $\mathbf{x}_m(0) = \mathbf{0}$ because of prewindowing.

We treat the four terms on the right-hand side of Eq. (13.74) separately:

- *First term:* We note that

$$\begin{aligned}&\sum_{k=1}^{n}\lambda^{n-k} f_{m,n-1}(k)b_{m,n-2}(k-1) \\ &\quad = \lambda\sum_{k=1}^{n-1}\lambda^{n-1-k} f_{m,n-1}(k)b_{m,n-2}(k-1) + f_{m,n-1}(n)b_{m,n-2}(n-1) \\ &\quad = \lambda\zeta_m^{fb}(n-1) + f_{m,n-1}(n)b_{m,n-2}(n-1)\end{aligned} \tag{13.75}$$

- *Second term:* We first note that

$$\sum_{k=1}^{n} \lambda^{n-k} f_{m,n-1}(k)\mathbf{x}_m(k-1)$$
$$= \lambda \sum_{k=1}^{n-1} \lambda^{n-1-k} f_{m,n-1}(k)\mathbf{x}_m(k-1) + f_{m,n-1}(n)\mathbf{x}_m(n-1)$$
$$= f_{m,n-1}(n)\mathbf{x}_m(n-1)$$

where the last equality follows from Eq. (13.7), with n replaced by $n-1$. Using this result, we obtain

$$\mathbf{k}_m^{\mathrm{T}}(n-1)b_{m,n-2}(n-1)\sum_{k=1}^{n} \lambda^{n-k} f_{m,n-1}(k)\mathbf{x}_m(k-1)$$
$$= \mathbf{k}_m^{\mathrm{T}}(n-1)\mathbf{x}_m(n-1) f_{m,n-1}(n) b_{m,n-2}(n-1) \tag{13.76}$$

- *Third term:* Using the change of variable $l = k-1$, we obtain

$$\sum_{k=1}^{n} \lambda^{n-k} b_{m,n-2}(k-1)\mathbf{x}_m(k-1) = \sum_{l=0}^{n-1} \lambda^{n-1-l} b_{m,n-2}(l)\mathbf{x}_m(l)$$
$$= \sum_{l=1}^{n-1} \lambda^{n-1-l} b_{m,n-2}(l)\mathbf{x}_m(l)$$
$$= \lambda \sum_{l=1}^{n-2} \lambda^{n-2-l} b_{m,n-2}(l)\mathbf{x}_m(l) + b_{m,n-2}(n-1)\mathbf{x}_m(n-1)$$
$$= b_{m,n-2}(n-1)\mathbf{x}_m(n-1)$$

where we have used $\mathbf{x}_m(0) = \mathbf{0}$ (because of prewindowing) in the second step and Eq. (13.16) with n replaced by $n-2$ for the last step. Using this result, we obtain

$$\mathbf{k}_m^{\mathrm{T}}(n-1) f_{m,n-1}(n) \sum_{k=1}^{n} \lambda^{n-k} b_{m,n-1}(k-1)\mathbf{x}_m(k-1)$$
$$= \mathbf{k}_m^{\mathrm{T}}(n-1)\mathbf{x}_m(n-1) f_{m,n-1}(n) b_{m,n-2}(n-1) \tag{13.77}$$

- *Fourth Term:* Using Eq. (13.10), we obtain

$$\mathbf{\Psi}_m(n-1)\mathbf{k}_m(n-1) = \mathbf{x}_m(n-1)$$

Thus,

$$\mathbf{k}_m^{\mathrm{T}}(n-1)\mathbf{\Psi}_m(n-1)\mathbf{k}_m(n-1) f_{m,n-1}(n) b_{m,n-2}(n-1)$$
$$= \mathbf{k}_m^{\mathrm{T}}(n-1)\mathbf{x}_m(n-1) f_{m,n-1}(n) b_{m,n-2}(n-1) \tag{13.78}$$

Substituting Eqs. (13.75), (13.76), (13.77), and (13.78) in Eq. (13.74) and rearranging, we obtain

$$\zeta_m^{fb}(n) = \lambda\zeta_m^{fb}(n-1) + (1 - \mathbf{k}_m^{\mathrm{T}}(n-1)\mathbf{x}_m(n-1))f_{m,n-1}(n)b_{m,n-2}(n-1) \quad (13.79)$$

Next, noting that $1 - \mathbf{k}_m^{\mathrm{T}}(n-1)\mathbf{x}_m(n-1) = \gamma_m(n-1)$ and $\gamma_m(n-1)b_{m,n-2}(n-1) = b_{m,n-1}(n-1)$, according to Eqs. (13.53) and (13.54), respectively, Eq. (13.79) can be simplified as

$$\zeta_m^{fb}(n) = \lambda\zeta_m^{fb}(n-1) + f_{m,n-1}(n)b_{m,n-1}(n-1) \quad (13.80)$$

Following a similar line of derivations, we also obtain

$$\zeta_m^{be}(n) = \lambda\zeta_m^{be}(n-1) + e_{m,n-1}(n)b_{m,n-1}(n) \quad (13.81)$$

We have now developed all the basic equations/recursions necessary for implementation of the RLSL algorithms.

13.4.6 RLSL Algorithm Using A Posteriori Errors

Table 13.1 lists a possible implementation of the RLSL algorithm that uses the *a posteriori* estimation errors. For every iteration, the algorithm begins with the initial values of $f_{0,n}(n)$, $b_{0,n}(n)$, $e_{0,n}(n)$, and $\gamma_0(n)$, as inputs to the first stage and proceeds with updating the successive stages of the lattice in a *for loop*. The operations in this loop may be divided into those related to forward and backward predictions and the operations related to the filtering. In the prediction part, the recursive Eqs. (13.45), (13.46), and (13.80) are used to update $\zeta_m^{ff}(n)$, $\zeta_m^{bb}(n)$, and $\zeta_m^{fb}(n)$, respectively. Here, we have also used Eqs. (13.52) and (13.54) to write the recursions in terms of only the *a posteriori* estimation errors, $f_{m,n}(n)$ and $b_{m,n}(n)$. The results of these recursions are then used to calculate the PARCOR coefficients $\kappa_{m+1}^{\mathrm{f}}(n)$ and $\kappa_{m+1}^{\mathrm{b}}(n)$ according to Eqs. (13.30) and (13.31), respectively. This follows with the order-update equations for the computation of the *a posteriori* estimation errors of the forward and backward predictors. These follow from Figure 13.3 – see also Chapter 11. The filtering is done in a similar way using the recursion Eq. (13.47) and Eqs. (13.32) and (13.55). Finally, the conversion factor $\gamma_m(n)$ is updated according to recursion Eq. (13.65).

Theoretically, the auto- and cross-correlations $\zeta_m^{ff}(n)$, $\zeta_m^{bb}(n)$ should be initialized to zero. However, as such initialization results in division by zeros during the first few iterations of the algorithm, $\zeta_m^{ff}(0)$ and $\zeta_m^{bb}(0)$, for $m = 0, 1, \ldots, N-1$, are initialized to a small positive number, δ, to prevent this numerical difficulties. The cross-correlations $\zeta_m^{fb}(0)$ and $\zeta_m^{be}(0)$ are initialized to the value of zero.

13.4.7 RLSL Algorithm with Error Feedback

The RLSL algorithm given in Table 13.1 uses the auto- and cross-correlations of the input signals to the successive stages of lattice to calculate the coefficients $\kappa_m^{\mathrm{f}}(n)$, $\kappa_m^{\mathrm{b}}(n)$, and $c_m(n)$, according to Eqs. (13.30), (13.31), and (13.32), respectively. Alternatively, we can develop a set of recursive equations for updating the coefficients $\kappa_m^{\mathrm{f}}(n)$, $\kappa_m^{\mathrm{b}}(n)$, and $c_m(n)$. This leads to an alternative implementation of the RLSL algorithm that has been found to

Table 13.1 RLSL algorithm using the *a posteriori* estimation errors.

Input: Latest sample of input, $x(n)$,
Past values of
the backward *a posteriori* estimation errors, $b_{m,n-1}(n-1)$,
the auto- and cross-correlations, $\zeta_m^{ff}(n-1)$, $\zeta_m^{bb}(n-1)$, $\zeta_m^{fb}(n-1)$, and $\zeta_m^{be}(n-1)$,
the conversion factors, $\gamma_m(n-1)$,
for $m = 0, 1, \ldots, N-1$.

Output: The updated values of
the backward *a posteriori* estimation errors, $b_{m,n}(n)$
the auto- and cross-correlations, $\zeta_m^{ff}(n)$, $\zeta_m^{bb}(n)$, $\zeta_m^{fb}(n)$ and $\zeta_m^{be}(n)$,
the conversion factors, $\gamma_m(n)$,
for $m = 0, 1, \ldots, N-1$.
The lattice coefficients are also available at the end of each iteration.

$f_{0,n}(n) = b_{0,n}(n) = x(n)$
$e_{0,n}(n) = d(n)$
$\gamma_0(n) = 1$
for $m = 0$ to $N-1$

$$\zeta_m^{ff}(n) = \lambda\zeta_m^{ff}(n-1) + \frac{f_{m,n}^2(n)}{\gamma_m(n-1)}$$

$$\zeta_m^{bb}(n) = \lambda\zeta_m^{bb}(n-1) + \frac{b_{m,n}^2(n)}{\gamma_m(n)}$$

$$\zeta_m^{fb}(n) = \lambda\zeta_m^{fb}(n-1) + \frac{f_{m,n}(n)b_{m,n-1}(n-1)}{\gamma_m(n-1)}$$

$$\kappa_{m+1}^{\mathrm{f}}(n) = \frac{\zeta_m^{fb}(n)}{\zeta_m^{bb}(n-1)}$$

$$\kappa_{m+1}^{\mathrm{b}}(n) = \frac{\zeta_m^{fb}(n)}{\zeta_m^{ff}(n)}$$

$$f_{m+1,n}(n) = f_{m,n}(n) - \kappa_{m+1}^{\mathrm{f}}(n)b_{m,n-1}(n-1)$$

$$b_{m+1,n}(n) = b_{m,n-1}(n-1) - \kappa_{m+1}^{\mathrm{b}}(n)f_{m,n}(n)$$

$$\zeta_m^{be}(n) = \lambda\zeta_m^{be}(n-1) + \frac{e_{m,n}(n)b_{m,n}(n)}{\gamma_m(n)}$$

$$c_m(n) = \frac{\zeta_m^{be}(n)}{\zeta_m^{bb}(n)}$$

$$e_{m+1,n}(n) = e_{m,n}(n) - c_m(n)b_{m,n}(n)$$

$$\gamma_{m+1}(n) = \gamma_m(n) - \frac{b_{m,n}^2(n)}{\zeta_m^{bb}(n)}$$

end

be *less sensitive* to numerical errors as compared with the algorithm of Table 13.1 (Ling, 1993).

Table 13.2 gives a summary of this alternative implementation of the RLSL algorithm. We note that here all the errors are the *a priori* ones, while in Table 13.1, all the equations are in terms of the *a posteriori* errors. In addition, the update equations of the cross-correlations $\zeta_m^{fb}(n)$ and $\zeta_m^{be}(n)$ have been deleted in Table 13.2, as they are no longer required. Instead, there are three recursions for time-updating the coefficients of

Table 13.2 RLSL algorithm using the *a priori* estimation errors with error feedback.

Input: Latest sample of input, $x(n)$,
Past values of
the backward *a priori* estimation errors, $b_{m,n-2}(n-1)$,
the autocorrelations, $\zeta_m^{ff}(n-1)$, $\zeta_m^{bb}(n-1)$,
the lattice coefficients $\kappa_{m+1}^{\mathrm{f}}(n-1)$, $\kappa_{m+1}^{\mathrm{b}}(n-1)$ and $c_m(n-1)$,
the conversion factors, $\gamma_m(n-1)$,
for $m = 0, 1, \ldots, N-1$.

Output: The updated values of
the backward *a priori* estimation errors, $b_{m,n-1}(n)$
the autocorrelations, $\zeta_m^{ff}(n)$, $\zeta_m^{bb}(n)$,
the lattice coefficients $\kappa_{m+1}^{\mathrm{f}}(n)$, $\kappa_{m+1}^{\mathrm{b}}(n)$, and $c_m(n)$,
the conversion factors, $\gamma_m(n)$,
for $m = 0, 1, \ldots, N-1$.

$f_{0,n-1}(n) = b_{0,n-1}(n) = x(n)$
$e_{0,n-1}(n) = d(n)$
$\gamma_0(n) = 1$
for $m = 0$ to $N-1$

$$\zeta_m^{ff}(n) = \lambda\zeta_m^{ff}(n-1) + \gamma_m(n-1)f_{m,n-1}^2(n)$$

$$\zeta_m^{bb}(n) = \lambda\zeta_m^{bb}(n-1) + \gamma_m(n)b_{m,n-1}^2(n)$$

$$f_{m+1,n-1}(n) = f_{m,n-1}(n) - \kappa_{m+1}^{\mathrm{f}}(n-1)b_{m,n-2}(n-1)$$

$$b_{m+1,n-1}(n) = b_{m,n-2}(n-1) - \kappa_{m+1}^{\mathrm{b}}(n-1)f_{m,n-1}(n)$$

$$\kappa_{m+1}^{\mathrm{f}}(n) = \kappa_{m+1}^{\mathrm{f}}(n-1) + \frac{\gamma_m(n-1)b_{m,n-2}(n-1)}{\zeta_m^{bb}(n-1)}f_{m+1,n-1}(n)$$

$$\kappa_{m+1}^{\mathrm{b}}(n) = \kappa_{m+1}^{\mathrm{b}}(n-1) + \frac{\gamma_m(n-1)f_{m,n-1}(n)}{\zeta_m^{ff}(n)}b_{m+1,n-1}(n)$$

$$e_{m+1,n-1}(n) = e_{m,n-1}(n) - c_m(n-1)b_{m,n-1}(n)$$

$$c_m(n) = c_m(n-1) - \frac{\gamma_m(n)b_{m,n-1}(n)}{\zeta_m^{bb}(n)}e_{m+1,n-1}(n)$$

$$\gamma_{m+1}(n) = \gamma_m(n) - \frac{\gamma_m^2(n)b_{m,n-1}^2(n)}{\zeta_m^{bb}(n)}$$

end

the lattice. Next, we explain the derivation of one of these recursions as an example. The other two can be derived by following the same line of derivation.

Recall that

$$\kappa_{m+1}^{\mathrm{f}}(n) = \frac{\zeta_m^{fb}(n)}{\zeta_m^{bb}(n-1)} \tag{13.82}$$

Substituting Eq. (13.80) in Eq. (13.82), we get

$$\kappa_{m+1}^{\mathrm{f}}(n) = \frac{\lambda\zeta_m^{fb}(n-1)}{\zeta_m^{bb}(n-1)} + \frac{f_{m,n-1}(n)b_{m,n-1}(n-1)}{\zeta_m^{bb}(n-1)} \tag{13.83}$$

However,

$$\begin{aligned}\frac{\lambda\zeta_m^{fb}(n-1)}{\zeta_m^{bb}(n-1)} &= \frac{\zeta_m^{fb}(n-1)}{\zeta_m^{bb}(n-2)}\cdot\frac{\lambda\zeta_m^{bb}(n-2)}{\zeta_m^{bb}(n-1)}\\ &= \kappa_{m+1}^{\mathrm{f}}(n-1)\frac{\lambda\zeta_m^{bb}(n-2)}{\zeta_m^{bb}(n-1)}\\ &= \kappa_{m+1}^{\mathrm{f}}(n-1)\frac{\zeta_m^{bb}(n-1)-b_{m,n-1}(n-1)b_{m,n-2}(n-1)}{\zeta_m^{bb}(n-1)}\\ &= \kappa_{m+1}^{\mathrm{f}}(n-1)-\frac{\kappa_{m+1}^{\mathrm{f}}(n-1)b_{m,n-1}(n-1)b_{m,n-2}(n-1)}{\zeta_m^{bb}(n-1)} \qquad (13.84)\end{aligned}$$

where we have used Eq. (13.46) to replace $\lambda\zeta_m^{bb}(n-2)$ by $\zeta_m^{bb}(n-1)-b_{m,n-1}(n-1)b_{m,n-2}(n-1)$. Substituting Eq. (13.84) in Eq. (13.83) and rearranging, we obtain

$$\begin{aligned}\kappa_{m+1}^{\mathrm{f}}(n) &= \kappa_{m+1}(n-1)+\frac{b_{m,n-1}(n-1)}{\zeta_m^{bb}(n-1)}(f_{m,n-1}(n)-\kappa_{m+1}^{\mathrm{f}}(n-1)b_{m,n-2}(n-1))\\ &= \kappa_{m+1}^{\mathrm{f}}(n-1)+\frac{b_{m,n-1}(n-1)f_{m+1,n-1}(n)}{\zeta_m^{bb}(n-1)} \qquad (13.85)\end{aligned}$$

Finally, using Eq. (13.54) to convert the *a posteriori* estimation error $b_{m,n-1}(n-1)$ to its equivalent *a priori* estimation error $b_{m,n-2}(n-1)$, in Eq. (13.85), we get

$$\kappa_{m+1}^{\mathrm{f}}(n)=\kappa_{m+1}^{\mathrm{f}}(n-1)+\frac{\gamma_m(n-1)b_{m,n-2}(n-1)}{\zeta_m^{bb}(n-1)}f_{m+1,n-1}(n) \qquad (13.86)$$

which is the recursion used in Table 13.2, for adaptation of $\kappa_{m+1}^{\mathrm{f}}(n)$. Following the same line of derivation, we can also obtain the recursions associated with the adaptation of $\kappa_{m+1}^{\mathrm{b}}(n)$ and $c_m(n)$. This is left to the reader as exercises.

13.5 FTRLS Algorithm

The FTF or FTRLS algorithm is another alternative numerical technique for solving the least-squares problem. The main advantage of the FTRLS algorithm is its reduced computational complexity as compared with other available solutions, such as the standard RLS and RLSL algorithms. Table 13.3 summarizes the number of operations (additions, multiplications, and divisions) required in each iteration of the standard RLS algorithm (Table 12.2), the two versions of RLSL algorithm presented in Tables 13.1 and 13.2, and also the two versions of FTRLS algorithm that will be discussed in this section, as an indication of their computational complexity.[2] We note that as the filter length, N, increases, the standard RLS becomes a rather expensive algorithm, as its computational complexity grows proportional to the square of the filter length. On the other hand, the

[2] We note that the number of operations, in general, may not be a fair measure in comparing various algorithms. A fair comparison would only be possible if the platform over which the algorithms are implemented is known *a priori*. For example, in hardware implementation, the modular structure of the RLSL may be very beneficial when a pipe-line structure is considered (Ling, 1993).

Table 13.3 Computational complexity of various RLS algorithm.

Algorithm	No. of $+$, $\times$, and $\div$ (added)
RLS (Table 12.2)	$3.5N^2$
RLSL (Table 13.1)	$28N$
RLSL (Table 13.2)	$31N$
FTRLS (Table 13.4)	$14N$
FTRLS (stabilized)	$18N$

computational complexities of RLSL and FTRLS algorithms grow only linearly with filter length. In addition, we find that the FTRLS algorithm has only about half the complexity of the RLSL algorithm. However, unfortunately, such a significant reduction in the complexity of the FTRLS algorithm does not come for free. Computer simulations and also theoretical studies have shown that the FTRLS algorithms are, in general, highly sensitive to roundoff error accumulation. Precautions have to be taken to deal with this problem to prevent the algorithm from becoming unstable. It is generally suggested that the algorithms should be reinitialized once a sign of instability is observed (Eleftheriou and Falconer, 1987; Cioffi and Kailath, 1984). To reduce the chance of instability in the FTRLS algorithm, a new version that is more robust against roundoff error accumulation has been proposed by Slock and Kailath (1988, 1991). This is called *stabilized fast transversal recursive least-squares* (SFTRLS) algorithm. However, studies show that even the SFTRLS algorithm has some limitations in the sense that it becomes unstable when the forgetting factor, λ, is not close enough to 1. This definitely limits the applicability of the FTRLS algorithm in cases where smaller values of λ should be used to achieve fast tracking (see Chapter 14).

13.5.1 Derivation of the FTRLS Algorithm

The FTRLS algorithm, basically, takes advantage of the interrelationships that exist between the forward and backward predictors as well as the joint process estimator when they share the same set of input samples. In particular, in the development of the RLSL algorithm in Section 13.4, we found that the forward and backward predictors and also the joint process estimator share the same conversion factor and gain vector. These properties led to a number of order- and time-update equations that were eventually put together to obtain the RLSL algorithm. In the RLSL algorithm, the problem of prediction and filtering (joint process estimation) is solved for orders of 1 to N, simultaneously. In cases where the goal is to solve the problem only for a filter of length N, this solution clearly has many redundant elements, which may unnecessarily complicate the solution. Accordingly, a set of equations that are limited to order N predictors and also to a length N filter (joint process estimator) may give a more efficient solution. This is the main essence of the FTRLS algorithm, when it is viewed as an improvement to the RLSL algorithm.

To have a clear treatment of the FTRLS algorithm, we proceed with the derivations of the necessary recursions separated into three sections, namely forward prediction, backward prediction, and filtering.

Forward Prediction

Consider an Nth-order forward transversal predictor with tap-weight vector $\mathbf{a}_N(n)$ and tap-input vector $\mathbf{x}_N(k-1) = [x(k-1)\ x(k-2)\ \cdots\ x(k-N)]^{\mathrm{T}}$, for $k = 1, 2, \ldots, n$. The RLS recursion for adaptive adjustment of $\mathbf{a}_N(n)$ is

$$\mathbf{a}_N(n) = \mathbf{a}_N(n-1) + \mathbf{k}_N(n-1) f_{N,n-1}(n) \tag{13.87}$$

where $\mathbf{k}_N(n-1)$ is the gain vector of the adaptation as defined in Eq. (13.10), and $f_{N,n-1}(n)$ is the *a priori* estimation error of the forward predictor.

Let us define the normalized gain vector

$$\bar{\mathbf{k}}_N(n) = \frac{\mathbf{k}_N(n)}{\gamma_N(n)} \tag{13.88}$$

where $\gamma_N(n)$ is the conversion factor as defined before. Substituting Eqs. (13.88) and (13.52) in Eq. (13.87), we get

$$\begin{aligned}\mathbf{a}_N(n) &= \mathbf{a}_N(n-1) + \bar{\mathbf{k}}_N(n-1)\gamma_N(n-1) f_{N,n-1}(n) \\ &= \mathbf{a}_N(n-1) + \bar{\mathbf{k}}_N(n-1) f_{N,n}(n)\end{aligned} \tag{13.89}$$

where $f_{N,n}(n)$ is the *a posteriori* estimation error of the forward predictor. Furthermore, using the definition (13.36), we may rewrite Eq. (13.89) as

$$\tilde{\mathbf{a}}_N(n) = \tilde{\mathbf{a}}_N(n-1) - \begin{bmatrix} 0 \\ \bar{\mathbf{k}}_N(n-1) \end{bmatrix} f_{N,n}(n) \tag{13.90}$$

Next, we note that

$$\mathbf{\Psi}_{N+1}^{-1}(n) = \begin{bmatrix} 0 & \mathbf{0}_N^{\mathrm{T}} \\ \mathbf{0}_N & \mathbf{\Psi}_N^{-1}(n-1) \end{bmatrix} + \frac{1}{\zeta_N^{ff}(n)} \tilde{\mathbf{a}}_N(n)\tilde{\mathbf{a}}_N^{\mathrm{T}}(n) \tag{13.91}$$

This identity, which appears similar to Eq. (13.56), can also be proved in the same way as Eq. (13.56). This is left to the reader as an exercise. Postmultiplying Eq. (13.91) by $\mathbf{x}_{N+1}(n)$, recalling Eqs. (13.36), (13.5), (13.19), and (13.88), and noting that

$$\mathbf{x}_{N+1}(n) = \begin{bmatrix} x(n) \\ \mathbf{x}_N(n-1) \end{bmatrix},$$

we obtain

$$\gamma_{N+1}(n)\bar{\mathbf{k}}_{N+1}(n) = \gamma_N(n-1) \begin{bmatrix} 0 \\ \bar{\mathbf{k}}_N(n-1) \end{bmatrix} + \frac{f_{N,n}(n)}{\zeta_N^{ff}(n)} \tilde{\mathbf{a}}_N(n) \tag{13.92}$$

Substituting Eq. (13.90) in Eq. (13.92) and rearranging, we obtain

$$\begin{aligned}\gamma_{N+1}(n)\bar{\mathbf{k}}_{N+1}(n) = &\left(\gamma_N(n-1) - \frac{f_{N,n}^2(n)}{\zeta_N^{ff}(n)}\right) \begin{bmatrix} 0 \\ \bar{\mathbf{k}}_N(n-1) \end{bmatrix} \\ &+ \frac{f_{N,n}(n)}{\zeta_N^{ff}(n)} \tilde{\mathbf{a}}_N(n-1)\end{aligned} \tag{13.93}$$

On the other hand, post- and premultiplying Eq. (13.91) by $\mathbf{x}_{N+1}(n)$ and $\mathbf{x}_{N+1}^{\mathrm{T}}(n)$, respectively, subtracting both sides of the result from unity, and recalling Eq. (13.53), we obtain

$$\gamma_{N+1}(n) = \gamma_N(n-1) - \frac{f_{N,n}^2(n)}{\zeta_N^{ff}(n)} \tag{13.94}$$

Substituting Eq. (13.94) in Eq. (13.93) and dividing both sides of the result by $\gamma_{N+1}(n)$, we get

$$\bar{\mathbf{k}}_{N+1}(n) = \begin{bmatrix} 0 \\ \bar{\mathbf{k}}_N(n-1) \end{bmatrix} + \frac{f_{N,n}(n)}{\gamma_{N+1}(n)\zeta_N^{ff}(n)} \tilde{\mathbf{a}}_N(n-1) \tag{13.95}$$

Moreover, combining Eqs. (13.94) and (13.45), it is straightforward to show that (Problem P13.17)

$$\gamma_{N+1}(n)\zeta_N^{ff}(n) = \lambda\gamma_N(n-1)\zeta_N^{ff}(n-1) \tag{13.96}$$

Finally, substituting Eq. (13.96) in Eq. (13.95) and using Eq. (13.52), we obtain

$$\bar{\mathbf{k}}_{N+1}(n) = \begin{bmatrix} 0 \\ \bar{\mathbf{k}}_N(n-1) \end{bmatrix} + \lambda^{-1} \frac{f_{N,n-1}(n)}{\zeta_N^{ff}(n-1)} \tilde{\mathbf{a}}_N(n-1) \tag{13.97}$$

This recursion gives a time as well as order-update of the normalized gain vector. Next, we develop another recursion that keeps the time index of the normalized gain vector fixed at n, but reduces its length from $N+1$ to N. This also leads to a time update of the tap-weight vector of the backward predictor.

Backward Prediction

Consider Eq. (13.56) with $m = N$. Then, postmultiplying it by $\mathbf{x}_{N+1}(n)$ and recalling Eqs. (13.40) and (13.88), we obtain

$$\gamma_{N+1}(n)\bar{\mathbf{k}}_{N+1}(n) = \gamma_N(n) \begin{bmatrix} \bar{\mathbf{k}}_N(n) \\ 0 \end{bmatrix} + \frac{b_{N,n}(n)}{\zeta_N^{bb}(n)} \tilde{\mathbf{g}}_N(n) \tag{13.98}$$

Equating the last elements of the vectors on both sides of Eq. (13.98) and rearranging, we obtain

$$\bar{\mathbf{k}}_{N+1,N+1}(n) = \frac{b_{N,n}(n)}{\gamma_{N+1}(n)\zeta_N^{bb}(n)} \tag{13.99}$$

where $\bar{\mathbf{k}}_{N+1,N+1}(n)$ denotes the last element of $\bar{\mathbf{k}}_{N+1}(n)$. On the other hand, combining Eqs. (13.46) and (13.65), and replacing m by N, it is straightforward to show that (Problem P13.18)

$$\gamma_{N+1}(n)\zeta_N^{bb}(n) = \lambda\gamma_N(n)\zeta_N^{bb}(n-1) \tag{13.100}$$

Substituting Eq. (13.100) in Eq. (13.99), recalling Eq. (13.54), and rearranging the result, we get

$$b_{N,n-1}(n) = \lambda\zeta_N^{bb}(n-1)\bar{\mathbf{k}}_{N+1,N+1}(n) \tag{13.101}$$

Furthermore, solving Eq. (13.100) for $\gamma_N(n)$ and using Eq. (13.46), we obtain

$$\begin{aligned}\gamma_N(n) &= \frac{\zeta_N^{bb}(n)}{\lambda\zeta_N^{bb}(n-1)}\gamma_{N+1}(n)\\ &= \left(\frac{\lambda\zeta_N^{bb}(n-1)}{\zeta_N^{bb}(n)}\right)^{-1}\gamma_{N+1}(n)\\ &= \left(1-\frac{b_{N,n}(n)b_{N,n-1}(n)}{\zeta_N^{bb}(n)}\right)^{-1}\gamma_{N+1}(n)\end{aligned} \tag{13.102}$$

Substituting Eq. (13.99) in Eq. (13.102), we get

$$\gamma_N(n) = (1 - b_{N,n-1}(n)\gamma_{N+1}(n)\bar{k}_{N+1,N+1}(n))^{-1}\gamma_{N+1}(n) \tag{13.103}$$

Moreover, we note that the update recursion Eq. (13.18) (with $m = N$) may be rearranged as

$$\tilde{\mathbf{g}}_N(n) = \tilde{\mathbf{g}}_N(n-1) - \begin{bmatrix}\bar{\mathbf{k}}_N(n)\\ 0\end{bmatrix} b_{N,n}(n) \tag{13.104}$$

Substituting Eq. (13.104) in Eq. (13.98) and rearranging, we obtain

$$\begin{aligned}\gamma_{N+1}(n)\bar{\mathbf{k}}_{N+1}(n) &= \left(\gamma_N(n) - \frac{b_{N,n}^2(n)}{\zeta_N^{bb}(n)}\right)\begin{bmatrix}\bar{\mathbf{k}}_N(n)\\ 0\end{bmatrix}\\ &\quad + \frac{b_{N,n}(n)}{\zeta_N^{bb}(n)}\tilde{\mathbf{g}}_N(n-1)\end{aligned} \tag{13.105}$$

Finally, substituting Eq. (13.65) with $m = N$ in Eq. (13.105), dividing both sides of the result by $\gamma_{N+1}(n)$, and using Eq. (13.99), we obtain

$$\begin{bmatrix}\bar{\mathbf{k}}_N(n)\\ 0\end{bmatrix} = \bar{\mathbf{k}}_{N+1}(n) - \bar{k}_{N+1,N+1}(n)\tilde{\mathbf{g}}_N(n-1) \tag{13.106}$$

With this recursion, we recover back the updated value of the gain vector in the right order, N. We thus can proceed with the next iteration of predictions and also use $\bar{\mathbf{k}}_N(n)$ for adaptation of the tap-weight vector, $\hat{\mathbf{w}}_N(n)$, of an adaptive filter with tap-input vector $\mathbf{x}_N(n)$, as explained below.

Filtering

Having obtained the normalized gain vector $\bar{\mathbf{k}}_N(n)$, the following equations may be used for adaptation of the tap-weight vector, $\hat{\mathbf{w}}_N(n)$, of an adaptive filter with tap-input vector $\mathbf{x}_N(n)$.

We first obtain the *a priori* estimation error

$$e_{N,n-1}(n) = d(n) - \hat{\mathbf{w}}_N^{\mathrm{T}}(n-1)\mathbf{x}_N(n) \tag{13.107}$$

Then, we calculate the corresponding *a posteriori* estimation error

$$e_{N,n}(n) = \gamma_N(n)e_{N,n-1}(n) \tag{13.108}$$

Finally, the update of the tap-weight vector of the adaptive filter is done according to the recursion

$$\hat{\mathbf{w}}_N(n) = \hat{\mathbf{w}}_N(n-1) + \bar{\mathbf{k}}_N(n)e_{N,n}(n) \tag{13.109}$$

We note that Eq. (13.109) is the same as the recursion Eq. (12.52). The only difference is that Eq. (13.109) is written in terms of the normalized gain vector $\bar{\mathbf{k}}_N(n)$ and, as a result, the *a priori* estimation error $e_{N,n-1}(n)$ is replaced by the *a posteriori* estimation error $e_{N,n}(n)$.

13.5.2 Summary of the FTRLS Algorithm

In Table 13.4, we present a summary of the FTRLS algorithm by collecting together the relevant equations from Section 13.4 and some of the new results that were developed in this section.

As mentioned before, the FTRLS algorithm may experience numerical instability. To deal with this problem, it has been noted that the sign of the expression

$$\beta(n) = 1 - b_{N,n-1}(n)\gamma_{N+1}(n)\bar{\mathbf{k}}_{N+1,N+1}(n) \tag{13.110}$$

is a good indication of the state of the algorithm with regard to its numerical instability. From Eq. (13.103), we note that $\beta(n) = \gamma_{N+1}(n)/\gamma_N(n)$ and this has to be always positive, as the conversion factors, $\gamma_N(n)$ and $\gamma_{N+1}(n)$, are nonnegative quantities (Problem P13.8). However, studies have shown that the FTRLS algorithm has some unstable modes that are not excited when infinite precision is assumed for arithmetic. Under finite precision arithmetics, these unstable modes receive some excitation, which will lead to some misbehavior of the algorithm and eventually resulting in its divergence. In particular, it has been noted that the quantity $\beta(n)$ becomes negative just before divergence of the algorithm occurs (Cioffi and Kailath, 1984). For this reason, $\beta(n)$ is called *rescue variable*, and it is suggested that once a negative value of $\beta(n)$ is observed, the normal execution of the FTRLS algorithm must be stopped and it should be restarted. In that case, the latest values of the filter coefficients may be used for a soft-reinitialization of the algorithm (see Cioffi and Kailath (1984) for the reinitialization procedure).

13.5.3 Stabilized FTRLS Algorithm

Further developments in the FTRLS algorithm has shown that the use of a special *error feedback* mechanism can greatly stabilize the FTRLS algorithm. It has been noted that by introducing *computational redundancy* by computing certain quantities in different ways, one can make specific measurements of the numerical errors present. These measurements can then be *fed back* to modify the dynamics of error propagation such that the unstable modes of the FTRLS algorithm are stabilized (Slock and Kailath, 1988, 1991). The quantities that have been identified to be appropriate for this purpose are the backward prediction error $b_{N,n-1}(n)$, the conversion factor $\gamma_{N+1}(n)$, and the last element of

Table 13.4 The FTRLS algorithm.

Input: Tap-input vector $\mathbf{x}_{N+1}(n-1)$, Desired output $d(n)$,
Tap-weight vectors $\tilde{\mathbf{a}}_N(n-1)$, $\tilde{\mathbf{g}}_N(n-1)$ and $\hat{\mathbf{w}}_N(n-1)$,
Normalized gain vector $\bar{\mathbf{k}}_N(n-1)$, and Least-squares sums $\zeta_N^{ff}(n-1)$ and $\zeta_N^{bb}(n-1)$.
Output: The updated values of
$\tilde{\mathbf{a}}_N(n)$, $\tilde{\mathbf{g}}_N(n)$, $\hat{\mathbf{w}}_N(n)$, $\bar{\mathbf{k}}_N(n)$, $\zeta_N^{ff}(n)$ and $\zeta_N^{bb}(n)$.

Prediction:

$$f_{N,n-1}(n) = \tilde{\mathbf{a}}_N^{\mathrm{T}}(n-1)\mathbf{x}_{N+1}(n)$$

$$f_{N,n}(n) = \gamma_N(n-1)f_{N,n-1}(n)$$

$$\zeta_N^{ff}(n) = \lambda\zeta_N^{ff}(n-1) + f_{N,n}(n)f_{N,n-1}(n)$$

$$\gamma_{N+1}(n) = \lambda\frac{\zeta_N^{ff}(n-1)}{\zeta_N^{ff}(n)}\gamma_N(n-1)$$

$$\bar{\mathbf{k}}_{N+1}(n) = \begin{bmatrix} 0 \\ \bar{\mathbf{k}}_N(n-1) \end{bmatrix} + \lambda^{-1}\frac{f_{N,n-1}(n)}{\zeta_N^{ff}(n-1)}\tilde{\mathbf{a}}_N(n-1)$$

$$\tilde{\mathbf{a}}_N(n) = \tilde{\mathbf{a}}_N(n-1) - \begin{bmatrix} 0 \\ \bar{\mathbf{k}}_N(n-1) \end{bmatrix} f_{N,n}(n)$$

$$b_{N,n-1}(n) = \lambda\zeta_N^{bb}(n-1)\bar{k}_{N+1,N+1}(n)$$

$$\beta(n) = 1 - b_{N,n-1}(n)\gamma_{N+1}(n)\bar{k}_{N+1,N+1}(n) \qquad \text{(rescue variable)}$$

$$\gamma_N(n) = \beta^{-1}(n)\gamma_{N+1}(n)$$

$$b_{N,n}(n) = \gamma_N(n)b_{N,n-1}(n)$$

$$\zeta_N^{bb}(n) = \lambda\zeta_N^{bb}(n-1) + b_{N,n}(n)b_{N,n-1}(n)$$

$$\begin{bmatrix} \bar{\mathbf{k}}_N(n) \\ 0 \end{bmatrix} = \bar{\mathbf{k}}_{N+1}(n) - \bar{k}_{N+1,N+1}(n)\tilde{\mathbf{g}}_N(n-1)$$

$$\tilde{\mathbf{g}}_N(n) = \tilde{\mathbf{g}}_N(n-1) - \begin{bmatrix} \bar{\mathbf{k}}_N(n) \\ 0 \end{bmatrix} b_{N,n}(n)$$

Filtering:

$$e_{N,n-1}(n) = d(n) - \hat{\mathbf{w}}_N^{\mathrm{T}}(n-1)\mathbf{x}_N(n)$$

$$e_{N,n}(n) = \gamma_N(n)e_{N,n-1}(n)$$

$$\hat{\mathbf{w}}_N(n) = \hat{\mathbf{w}}_N(n-1) + \bar{\mathbf{k}}_N(n)e_{N,n}(n)$$

the normalized gain vector $\bar{\mathbf{k}}_{N+1}(n)$, that is, the three quantities used in computation of $\beta(n)$ in Eq. (13.110). Slock and Kailath (1991) have proposed an elegant procedure for exploiting these redundancies in the FTRLS algorithm and have come up with a stabilized version of the FTRLS algorithm. However, as was noted before, even the stabilized FTRLS algorithm has to be treated with some special care, which makes it rather restrictive in applications. In particular, it has been found that the stability of the SFTRLS can only be guaranteed when the forgetting factor, λ, is chosen very close to 1. As a rule of thumb, it is suggested that λ should be kept within the range

$$1 - \frac{1}{2N} < \lambda < 1 \tag{13.111}$$

where N is the length of the filter.

Problems

P13.1 Starting with Eq. (13.4) and using the principle of orthogonality derive Eq. (13.8). In addition, by inserting the relevant variables in Eq. (12.16), suggest an alternative derivation of Eq. (13.8).

P13.2 Following similar line of derivations as those in Problem P13.1, suggest two methods for derivation of Eq. (13.17).

P13.3 Work out the detail of derivations of the augmented normal Eqs. (13.35) and (13.39).

P13.4 Give a detailed derivation of Eqs. (13.23) and (13.24).

P13.5 Using the principle of orthogonality, prove Eq. (13.25).

P13.6 Consider the *a posteriori* forward and backward prediction errors $f_{i,n}(k)$ and $b_{j,n}(k)$, respectively, of a real-valued and prewindowed signal sequence $x(k)$. Prove the following results:

(i)

$$\sum_{k=1}^{n} \lambda^{n-k} f_{6,n}(k) f_{5,n}(k-1) = 0$$

(ii)

$$\sum_{k=1}^{n} \lambda^{n-k} f_{m,n}(k) x(k) = \sum_{k=1}^{n} \lambda^{n-k} f_{m,n}^2(k)$$

(iii)

$$\sum_{k=1}^{n} \lambda^{n-k} b_{m,n}(k) x(k-m) = \sum_{k=1}^{n} \lambda^{n-k} b_{m,n}^2(k)$$

(iv) For $0 < l < m$,

$$\sum_{k=1}^{n} \lambda^{n-k} b_{m-l,n-l}(k-l) f_{m,n}(k) = 0$$

P13.7 Consider the *a priori* forward and backward prediction errors $f_{i,n-1}(k)$ and $b_{j,n-1}(k)$ and also the associated *a posteriori* errors $f_{i,n}(k)$ and $b_{j,n}(k)$, respectively, of a real-valued and prewindowed signal sequence $x(k)$. Prove the following results:

(i)

$$\sum_{k=1}^{n} \lambda^{n-k} b_{m-1,n-1}(k-1) f_{m,n}(k) = 0$$

(ii)

$$\sum_{k=1}^{n} \lambda^{n-k} f_{m-1,n}(k) b_{m,n}(k) = 0$$

P13.8 Consider a linear adaptive filter with tap-input vectors $\mathbf{x}(1), \mathbf{x}(2), \ldots, \mathbf{x}(n)$, and desired output sequence

$$d(k) = \begin{cases} 1, & k = n \\ 0, & k = 1, 2, \ldots, n-1 \end{cases}$$

Find the least-squares error sum of this filter and show that it is equal to the conversion factor $\gamma(n)$ as given by Eq. (13.49) or (13.50). Then, prove that

$$0 \leq \gamma(n) \leq 1$$

P13.9 Show that in a lattice structure, at any instant of time, n,

$$\gamma_{m+1}(n) \leq \gamma_m(n)$$

P13.10 Prove that

$$\frac{\zeta_m^{ff}(n)}{\zeta_{m-1}^{ff}(n)} = \frac{\zeta_m^{bb}(n)}{\zeta_{m-1}^{bb}(n-1)} = 1 - \kappa_m^{\mathrm{f}}(n)\kappa_m^{\mathrm{b}}(n)$$

P13.11 Use the result of Problem P13.10 to derive the following update equations for the least-squares sums $\zeta_m^{ff}(n)$ and $\zeta_m^{bb}(n)$:

$$\zeta_m^{ff}(n) = \zeta_{m-1}^{ff}(n) - \frac{(\zeta_{m-1}^{fb}(n))^2}{\zeta_{m-1}^{bb}(n-1)}$$

and

$$\zeta_m^{bb}(n) = \zeta_{m-1}^{bb}(n-1) - \frac{(\zeta_{m-1}^{fb}(n))^2}{\zeta_{m-1}^{ff}(n)}$$

P13.12 Obtain the normal equation that results from the least-square optimization of the *a posteriori* estimation error $e_{N,n}(k)$ of the joint process estimator of Figure 13.3 and show that this leads to the following set of independent equations:

$$c_m(n) = \frac{\sum_{k=1}^{n} \lambda^{n-k} b_{m,n}(k) d(k)}{\sum_{k=1}^{n} \lambda^{n-k} b_{m,n}^2(k)}$$

for $m = 0, 1, \ldots, N-1$. Then, use the orthogonality of the backward errors $b_{m,n}(k)$, for $m = 0, 1, \ldots, N$, to convert these equations to those given in Eq. (13.26).

P13.13 Give a detailed proof of Eq. (13.81).

P13.14 Derive the update equations of $\kappa^{\mathrm{b}}_{m+1}(n)$ and $c_m(n)$ that have appeared in Table 13.2.

P13.15 Prove the following identity

$$\mathbf{\Psi}^{-1}_{m+1}(n) = \begin{bmatrix} 0 & \mathbf{0}^{\mathrm{T}}_m \\ \mathbf{0}_m & \mathbf{\Psi}^{-1}_m(n-1) \end{bmatrix} + \frac{1}{\zeta^{ff}_m(n)} \tilde{\mathbf{a}}_m(n)\tilde{\mathbf{a}}^{\mathrm{T}}_m(n)$$

P13.16 Show that

$$\gamma_m(n) = \gamma_m(n-1) + \frac{b^2_{m,n}(n)}{\zeta^{bb}_m(n)} - \frac{f^2_{m,n}(n)}{\zeta^{ff}_m(n)}$$

P13.17 Give a detailed derivations of Eq. (13.96).

P13.18 Give a detailed derivations of Eq. (13.100).

P13.19 Use the result of Problems P13.17 and P13.18 to obtain a time-update equation relating $\gamma_m(n)$ and $\gamma_m(n-1)$.

P13.20 Explore the possibility of rearranging the recursions/equations in Table 13.1 in terms of *a priori* estimation errors. Thus, suggest an alternative implementation RLSL algorithm using the *a priori* estimation errors.

14

Tracking

Our study of adaptive filters so far has been based on the assumption that the filter input and its desired output are jointly stationary processes. Under this condition, the correlation matrix, **R**, of the filter input and the cross-correlation vector, **p**, between the filter input and its desired output are fixed quantities. Consequently, the performance surface of the filter is also fixed, with its minimum point given by the Wiener–Hopf solution $\mathbf{w}_o = \mathbf{R}^{-1}\mathbf{p}$. Comparison between different algorithms, thus, would be based on their convergence behavior. In this context, superior algorithms are those with shorter convergence time.

In this chapter, we study another important aspect of adaptive filters. In many applications, the underlying processes are nonstationary. As a result, the Wiener–Hopf solution, $\mathbf{w}_o = \mathbf{R}^{-1}\mathbf{p}$, varies with time, since **R** and **p** are time varying. In such a situation, adaptive algorithm is expected to not only adapt the filter tap weights to a neighborhood of their optimum values, but also follow the variations of the optimum tap weights. The latter, which is the subject of this chapter, is known as *tracking*.

Before we start our study on tracking, we would like to remark that there is a clear distinction between convergence and tracking. *Convergence is a transient phenomenon.* It refers to the behavior of a system (here, an adaptive filter) when it starts from an arbitrary initial condition and undergoes a transient period before it reaches its steady state. *Tracking,* on the other hand, *is a steady-state phenomenon.* It refers to the behavior of a system in following variations of its surrounding environment, after it has reached its steady state. An algorithm with good convergence properties does not necessarily possess a fast tracking capability and vice versa. Part of our effort in this chapter is to clarify this *seemingly* unusual behavior of adaptive algorithms.

14.1 Formulation of the Tracking Problem

Much of the works related to the tracking behavior of adaptive filters is done in the context of the modeling problem depicted in Figure 14.1. The plant is a linear multiple regressor characterized by the equation

$$d(n) = \mathbf{w}_o^{\mathrm{T}}(n)\mathbf{x}(n) + e_o(n) \tag{14.1}$$

where $\mathbf{x}(n) = [x_0(n) \; x_1(n) \; \cdots \; x_{N-1}(n)]^{\mathrm{T}}$ is the tap-input vector, $\mathbf{w}_o(n) = [w_{o,0}(n) \; w_{o,1}(n) \; \cdots \; w_{o,N-1}(n)]^{\mathrm{T}}$ is the plant tap-weight vector, $e_o(n)$ is the plant noise, and

Adaptive Filters: Theory and Applications, Second Edition. Behrouz Farhang-Boroujeny.

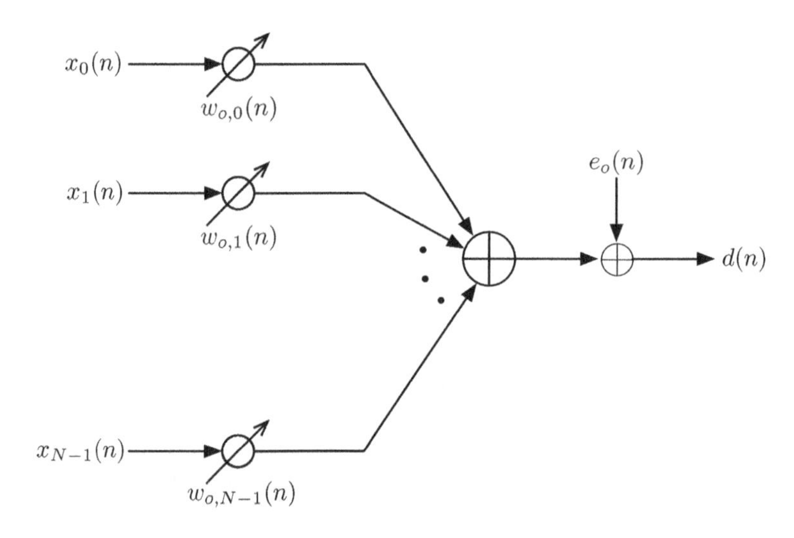

Figure 14.1 Linear multiple regressor.

$d(n)$ is the plant output. The presence of the time index n in $\mathbf{w}_o(n)$ is to emphasize that the plant tap-weight vector is time variant. This is unlike the notation $\mathbf{w}_o$ that was used in the previous chapters to represent fixed plant weights. The role of the adaptive algorithm is to follow the variations in $\mathbf{w}_o(n)$.

The time-varying tap-weight vector $\mathbf{w}_o(n)$ is chosen to be a *multivariate random-walk process* characterized by the difference equation

$$\mathbf{w}_o(n+1) = \mathbf{w}_o(n) + \boldsymbol{\epsilon}_o(n) \tag{14.2}$$

where $\boldsymbol{\epsilon}_o(n)$ is the *process noise vector.*

The following assumptions are made throughout this chapter:

1. The sequences $e_o(n)$, $\boldsymbol{\epsilon}_o(n)$, and $\mathbf{x}(n)$ are zero-mean and stationary random processes.
2. The sequences $e_o(n)$, $\boldsymbol{\epsilon}_o(n)$, and $\mathbf{x}(n)$ are statistically independent of one another.
3. The successive increments, $\boldsymbol{\epsilon}_o(n)$, of the plant tap weights, are independent. However, the elements of $\boldsymbol{\epsilon}_o(n)$, for a given n, may be statistically dependent.
4. At time n, the tap-weight vector $\mathbf{w}(n)$ of the adaptive filter is statistically independent of $e_o(n)$ and $\mathbf{x}(n)$.

The validity of the last assumption (which is known as *independence assumption*) is justified only for small values of the step-size parameter(s) of the adaptation algorithm (Chapter 6, Section 6.2). This is assumed to be true throughout our discussions in this chapter.

14.2 Generalized Formulation of LMS Algorithm

In this section, we present a generalized formulation of the LMS (least-mean-square) Algorithm, which can be used for a unified study of the tracking behavior of various adaptive algorithms. The LMS recursion that we consider is

$$\mathbf{w}(n+1) = \mathbf{w}(n) + 2\boldsymbol{\mu} e(n)\mathbf{x}(n) \tag{14.3}$$

where $e(n) = d(n) - y(n)$ is the output error, $y(n) = \mathbf{w}^{\mathrm{T}}(n)\mathbf{x}(n)$ is the filter output, $\mathbf{w}(n)$ and $\mathbf{x}(n)$ are the tap-weight and tap-input vectors, respectively, and $\boldsymbol{\mu}$ is a diagonal matrix consisting of the step-size parameters corresponding to various taps of the filter. These parameters, which are called μ_i, $i = 0, 1, \ldots, N-1$, are assumed fixed in our analysis. Furthermore, to keep Eq. (14.3) in its most general form, we follow the modeling problem of Figure 14.1 and choose $\mathbf{x}(n) = [x_0(n)\ x_1(n)\ \cdots\ x_{N-1}(n)]^{\mathrm{T}}$. This allows for the possibility that the tap inputs may not correspond to those from a tapped-delay-line.

The algorithms that are covered by Eq. (14.3) are:

Conventional LMS Algorithm: By choosing $\boldsymbol{\mu} = \mu\mathbf{I}$, where μ is a scalar step-size parameter and $\mathbf{I}$ is the N-by-N identity matrix, Eq. (14.3) reduces to the conventional LMS recursion.

TDLMS algorithm: The recursion (14.3) will be that of the TDLMS (transform domain least-mean-square) algorithm, if $\mathbf{x}(n)$ is replaced by $\mathbf{x}_{\mathcal{T}}(n) = \mathcal{T}\mathbf{x}(n)$, where $\mathcal{T}$ is a transformation matrix and $\mathbf{x}(n)$ is the filter tap-input vector before transformation. Moreover, we choose the normalized step-size parameters as (Chapter 7)

$$\mu_i = \frac{\mu'}{E[x_{\mathcal{T},i}^2(n)]}, \quad \text{for} \quad i = 0, 1, \ldots, N-1 \tag{14.4}$$

where μ' is a common scalar, $x_{\mathcal{T},i}(n)$ is the ith element of $\mathbf{x}_{\mathcal{T}}(n)$, and $E[\cdot]$ denotes statistical expectation.

In actual implementation of the TDLMS algorithm, the values of $E[x_{\mathcal{T},i}^2(n)]$ are estimated through time averaging. However, to simplify our discussion, we assume that such averages are known *a priori*; thus, μ_i's are fixed in our study here.

Ideal LMS-Newton Algorithm: From Chapter 7, we recall that the ideal LMS-Newton algorithm is equivalent to the TDLMS with $\mathcal{T}$ replaced by the Karhunen Loéve transform (KLT) of the input process. Thus, the analysis that we do for the TDLMS algorithm can be immediately applied to evaluate the tracking behavior of the ideal LMS-Newton algorithm.

RLS Algorithm: In Chapter 12 we found that when the RLS algorithm has undergone a large number of iterations so that it has reached its steady state, it can be approximated by the LMS-Newton recursion (12.90). This implies that the tracking behavior of the RLS and LMS-Newton algorithms are about the same, as tracking refers to the steady-state phase of the algorithms.

14.3 MSE Analysis of the Generalized LMS Algorithm

In this section, we consider the performance of the generalized LMS recursion Eq. (14.3) and derive an expression for its steady-state mean-squared error (MSE). Our discussion is in the context of the modeling problem introduced in Section 14.1. Our derivations here are similar to those in Chapter 6, where the convergence behavior of the LMS algorithm was analyzed (Section 6.3). However, to overcome the analytical difficulties arising from the use of different step-size parameters at various taps, we make some further approximations.

We note that

$$\begin{aligned} e(n) &= d(n) - \mathbf{w}^{\mathrm{T}}(n)\mathbf{x}(n) \\ &= d(n) - \mathbf{x}^{\mathrm{T}}(n)\mathbf{w}(n) \\ &= d(n) - \mathbf{x}^{\mathrm{T}}(n)\mathbf{w}_{\mathrm{o}}(n) - \mathbf{x}^{\mathrm{T}}(n)[\mathbf{w}(n) - \mathbf{w}_{\mathrm{o}}(n)] \\ &= e_{\mathrm{o}}(n) - \mathbf{x}^{\mathrm{T}}(n)\mathbf{v}(n) \end{aligned} \tag{14.5}$$

where $\mathbf{v}(n) = \mathbf{w}(n) - \mathbf{w}_{\mathrm{o}}(n)$ is the weight-error vector, and from Eq. (14.1), $e_{\mathrm{o}}(n) = d(n) - \mathbf{x}^{\mathrm{T}}(n)\mathbf{w}_{\mathrm{o}}(n)$. Substituting Eq. (14.5) in Eq. (14.3) and using Eq. (14.2), we obtain

$$\mathbf{v}(n+1) = (\mathbf{I} - 2\boldsymbol{\mu}\mathbf{x}(n)\mathbf{x}^{\mathrm{T}}(n))\mathbf{v}(n) + 2\boldsymbol{\mu}e_{\mathrm{o}}(n)\mathbf{x}(n) - \boldsymbol{\epsilon}_{\mathrm{o}}(n) \tag{14.6}$$

where $\mathbf{I}$ is the identity matrix. Next, we multiply both sides of Eq. (14.6) from right by their respective transposes, take statistical expectation of the results, and expand to obtain

$$\begin{aligned} \mathbf{K}(n+1) = \mathbf{K}(n) &- 2\boldsymbol{\mu}E[\mathbf{x}(n)\mathbf{x}^{\mathrm{T}}(n)\mathbf{v}(n)\mathbf{v}^{\mathrm{T}}(n)] \\ &-2E[\mathbf{v}(n)\mathbf{v}^{\mathrm{T}}(n)\mathbf{x}(n)\mathbf{x}^{\mathrm{T}}(n)]\boldsymbol{\mu} \\ &+4\boldsymbol{\mu}E[\mathbf{x}(n)\mathbf{x}^{\mathrm{T}}(n)\mathbf{v}(n)\mathbf{v}^{\mathrm{T}}(n)\mathbf{x}(n)\mathbf{x}^{\mathrm{T}}(n)]\boldsymbol{\mu} \\ &+2E[(\mathbf{I} - 2\boldsymbol{\mu}\mathbf{x}(n)\mathbf{x}^{\mathrm{T}}(n))\mathbf{v}(n)\mathbf{x}^{\mathrm{T}}(n)e_{\mathrm{o}}(n)]\boldsymbol{\mu} \\ &+2\boldsymbol{\mu}E[e_{\mathrm{o}}(n)\mathbf{x}(n)\mathbf{v}^{\mathrm{T}}(n)(\mathbf{I} - 2\mathbf{x}^{\mathrm{T}}(n)\mathbf{x}(n)\boldsymbol{\mu})] \\ &+4\boldsymbol{\mu}E[|e_{\mathrm{o}}(n)|^2\mathbf{x}(n)\mathbf{x}^{\mathrm{T}}(n)]\boldsymbol{\mu} \\ &-E[(\mathbf{I} - 2\boldsymbol{\mu}\mathbf{x}(n)\mathbf{x}^{\mathrm{T}}(n))\mathbf{v}(n)\boldsymbol{\epsilon}_{\mathrm{o}}^{\mathrm{T}}(n)] \\ &-E[\boldsymbol{\epsilon}_{\mathrm{o}}(n)\mathbf{v}^{\mathrm{T}}(n)(\mathbf{I} - 2\mathbf{x}^{\mathrm{T}}(n)\mathbf{x}(n)\boldsymbol{\mu})] \\ &-2\boldsymbol{\mu}E[e_{\mathrm{o}}(n)\mathbf{x}(n)\boldsymbol{\epsilon}_{\mathrm{o}}^{\mathrm{T}}(n)] \\ &-2E[e_{\mathrm{o}}(n)\boldsymbol{\epsilon}_{\mathrm{o}}^{\mathrm{T}}(n)\mathbf{x}^{\mathrm{T}}(n)]\boldsymbol{\mu} \\ &+E[\boldsymbol{\epsilon}_{\mathrm{o}}(n)\boldsymbol{\epsilon}_{\mathrm{o}}^{\mathrm{T}}(n)] \end{aligned} \tag{14.7}$$

where $\mathbf{K}(n) = E[\mathbf{v}(n)\mathbf{v}^{\mathrm{T}}(n)]$. According to assumptions 1–4 of Section 14.1, $e_{\mathrm{o}}(n)$ is zero-mean and independent of $\mathbf{x}(n)$, $\mathbf{v}(n)$ and $\boldsymbol{\epsilon}_{\mathrm{o}}(n)$. The independence of $e_{\mathrm{o}}(n)$ and $\mathbf{v}(n) = \mathbf{w}(n) - \mathbf{w}_{\mathrm{o}}(n)$ follows from the fact that $e_{\mathrm{o}}(n)$ is independent of $\mathbf{w}(n)$ (Assumption 4) and $\boldsymbol{\epsilon}_{\mathrm{o}}(n)$ (Assumption 2). Consequently, the 5th, 6th, 10th, and 11th terms on the right-hand side of Eq. (14.7) become zero. Similarly, the eighth and ninth terms on the right-hand side of Eq. (14.7) are also zero since $\boldsymbol{\epsilon}_{\mathrm{o}}(n)$ is zero-mean and independent of $\mathbf{x}(n)$ and $\mathbf{v}(n)$. The independence of $\boldsymbol{\epsilon}_{\mathrm{o}}(n)$ and $\mathbf{v}(n)$ follows from the fact that $\mathbf{v}(n)$ is affected only by the past values of $\boldsymbol{\epsilon}_{\mathrm{o}}(n)$ and according to Assumption 3, $\boldsymbol{\epsilon}_{\mathrm{o}}(n)$ is independent of its past observations. Furthermore, the independence of $\mathbf{x}(n)$ and $\mathbf{v}(n)$ implies that

$$\begin{aligned} E[\mathbf{x}(n)\mathbf{x}^{\mathrm{T}}(n)\mathbf{v}(n)\mathbf{v}^{\mathrm{T}}(n)] &= E[\mathbf{x}(n)\mathbf{x}^{\mathrm{T}}(n)]E[\mathbf{v}(n)\mathbf{v}^{\mathrm{T}}(n)] \\ &= \mathbf{R}\mathbf{K}(n) \end{aligned} \tag{14.8}$$

and

$$E[\mathbf{v}(n)\mathbf{v}^{\mathrm{T}}(n)\mathbf{x}(n)\mathbf{x}^{\mathrm{T}}(n)] = E[\mathbf{v}(n)\mathbf{v}^{\mathrm{T}}(n)]E[\mathbf{x}(n)\mathbf{x}^{\mathrm{T}}(n)] = \mathbf{K}(n)\mathbf{R} \quad (14.9)$$

where $\mathbf{R} = E[\mathbf{x}(n)\mathbf{x}^{\mathrm{T}}(n)]$. Assumption 2 implies that

$$E[|e_{\mathrm{o}}(n)|^2\mathbf{x}(n)\mathbf{x}^{\mathrm{T}}(n)] = \sigma_{e_{\mathrm{o}}}^2\mathbf{R} \quad (14.10)$$

where $\sigma_{e_{\mathrm{o}}}^2 = E[|e_{\mathrm{o}}(n)|^2]$ is the variance of the zero-mean random variable $e_{\mathrm{o}}(n)$. Finally, considering the independence of $\mathbf{v}(n)$ and $\mathbf{x}(n)$ and assuming that the elements of $\mathbf{x}(n)$ are Gaussian-distributed and following a similar line of derivations as that which led to Eq. (6.39) (Appendix 6A), we obtain

$$E[\mathbf{x}(n)\mathbf{x}^{\mathrm{T}}(n)\mathbf{v}(n)\mathbf{v}^{\mathrm{T}}(n)\mathbf{x}(n)\mathbf{x}^{\mathrm{T}}(n)] = \mathbf{R}\mathrm{tr}[\mathbf{R}\mathbf{K}(n)] + 2\mathbf{R}\mathbf{K}(n)\mathbf{R} \quad (14.11)$$

Using these results in Eq. (14.7), we obtain

$$\mathbf{K}(n+1) = \mathbf{K}(n) - 2\boldsymbol{\mu}\mathbf{R}\mathbf{K}(n) - 2\mathbf{K}(n)\mathbf{R}\boldsymbol{\mu} + 4\boldsymbol{\mu}\mathbf{R}\boldsymbol{\mu}\mathrm{tr}[\mathbf{R}\mathbf{K}(n)] + 8\boldsymbol{\mu}\mathbf{R}\mathbf{K}(n)\mathbf{R}\boldsymbol{\mu} + 4\sigma_{e_{\mathrm{o}}}^2\boldsymbol{\mu}\mathbf{R}\boldsymbol{\mu} + \mathbf{G} \quad (14.12)$$

where

$$\mathbf{G} = E[\boldsymbol{\epsilon}_{\mathrm{o}}(n)\boldsymbol{\epsilon}_{\mathrm{o}}^{\mathrm{T}}(n)] \quad (14.13)$$

is the correlation matrix of the plant tap-weight increments.

Next, we recall that

$$\xi_{\mathrm{ex}}(n) = E[(\mathbf{v}^{\mathrm{T}}(n)\mathbf{x}(n))^2] \quad (14.14)$$

where $\xi_{\mathrm{ex}}(n)$ is the excess MSE at time n. Using the independence of $\mathbf{v}(n)$ and $\mathbf{x}(n)$ and following the same line of derivations that led to Eq. (6.26), we obtain

$$E[(\mathbf{v}^{\mathrm{T}}(n)\mathbf{x}(n))^2] = \mathrm{tr}[\mathbf{R}\mathbf{K}(n)] \quad (14.15)$$

Substituting this result in Eq. (14.14), we obtain

$$\xi_{\mathrm{ex}}(n) = \mathrm{tr}[\mathbf{R}\mathbf{K}(n)] \quad (14.16)$$

Since all the underlying processes are assumed to be stationary (Assumption 1 in Section 14.1), $\mathbf{K}(n)$ and $\xi_{\mathrm{ex}}(n)$ will be independent of n in the steady state. Hence, the time index n is dropped from $\mathbf{K}(n)$ and $\xi_{\mathrm{ex}}(n)$ henceforth.

Premultiplying Eq. (14.12) on both sides by $\frac{1}{2}\boldsymbol{\mu}^{-1}$, taking the trace, and assuming that the algorithm has reached its steady state so that $\mathbf{K}(n+1) = \mathbf{K}(n) = \mathbf{K}$, we obtain

$$\mathrm{tr}[\mathbf{R}\mathbf{K}] + \mathrm{tr}[\boldsymbol{\mu}^{-1}\mathbf{K}\mathbf{R}\boldsymbol{\mu}] = 2\mathrm{tr}[\mathbf{R}\boldsymbol{\mu}]\mathrm{tr}[\mathbf{R}\mathbf{K}] + 4\mathrm{tr}[\mathbf{R}\mathbf{K}\mathbf{R}\boldsymbol{\mu}] + 2\sigma_{e_{\mathrm{o}}}^2\mathrm{tr}[\mathbf{R}\boldsymbol{\mu}] + \frac{1}{2}\mathrm{tr}[\boldsymbol{\mu}^{-1}\mathbf{G}] \quad (14.17)$$

Next, using the identity $\mathrm{tr}[\mathbf{AB}] = \mathrm{tr}[\mathbf{BA}]$, which is true for any pair of M-by-N and N-by-M matrices $\mathbf{A}$ and $\mathbf{B}$, we get $\mathrm{tr}[\mathbf{R}\boldsymbol{\mu}] = \mathrm{tr}[\boldsymbol{\mu}\mathbf{R}]$, $\mathrm{tr}[\mathbf{RKR}\boldsymbol{\mu}] = \mathrm{tr}[\boldsymbol{\mu}\mathbf{RKR}]$ and

$$\mathrm{tr}[\boldsymbol{\mu}^{-1}\mathbf{K}\mathbf{R}\boldsymbol{\mu}] = \mathrm{tr}[\mathbf{R}\boldsymbol{\mu}\boldsymbol{\mu}^{-1}\mathbf{K}] = \mathrm{tr}[\mathbf{R}\mathbf{K}]$$

Using these and Eq. (14.16) in Eq. (14.17), we obtain

$$2\xi_{\rm ex} = 2{\rm tr}[\boldsymbol{\mu}\mathbf{R}]\xi_{\rm ex} + 4{\rm tr}[\boldsymbol{\mu}\mathbf{RKR}] + 2\sigma_{e_o}^2{\rm tr}[\boldsymbol{\mu}\mathbf{R}] + \frac{1}{2}{\rm tr}[\boldsymbol{\mu}^{-1}\mathbf{G}] \tag{14.18}$$

To arrive at a mathematically tractable result, we assume that the term tr$[\boldsymbol{\mu}\mathbf{RKR}]$ in Eq. (14.18) can be ignored. Numerical examples and computer simulations show that when N (the filter length) is large tr$[\boldsymbol{\mu}\mathbf{RKR}]$ is usually at least an order of magnitude smaller than tr$[\boldsymbol{\mu}\mathbf{R}]\xi_{\rm ex}$. See Problem P14.1 for more exposure over this approximation. This leads to the following result:

$$\xi_{\rm ex} = \frac{1}{1-{\rm tr}[\boldsymbol{\mu}\mathbf{R}]}\left(\sigma_{e_o}^2{\rm tr}[\boldsymbol{\mu}\mathbf{R}] + \frac{1}{4}{\rm tr}[\boldsymbol{\mu}^{-1}\mathbf{G}]\right) \tag{14.19}$$

Using Eq. (14.19) to evaluate the misadjustment of the generalized LMS algorithm, we obtain

$$\mathcal{M} = \frac{\xi_{\rm ex}}{\xi_{\rm min}} = \frac{1}{1-{\rm tr}[\boldsymbol{\mu}\mathbf{R}]}\left({\rm tr}[\boldsymbol{\mu}\mathbf{R}] + \frac{1}{4}\sigma_{e_o}^{-2}{\rm tr}[\boldsymbol{\mu}^{-1}\mathbf{G}]\right) \tag{14.20}$$

where $\xi_{\rm min} = \sigma_{e_o}^2$ is the minimum MSE of the filter that is obtained when $\mathbf{w}(n) = \mathbf{w}_o(n)$.

To relate this result to the results of the previous chapters, let us consider the case of conventional LMS algorithm. In this case, $\boldsymbol{\mu} = \mu\mathbf{I}$, where μ is a scalar step-size parameter. Substituting $\boldsymbol{\mu}$ by $\mu\mathbf{I}$ in Eq. (14.20), we obtain

$$\mathcal{M}_{\rm LMS} = \frac{1}{1-\mu{\rm tr}[\mathbf{R}]}\left(\mu{\rm tr}[\mathbf{R}] + \frac{1}{4}\sigma_{e_o}^{-2}\mu^{-1}{\rm tr}[\mathbf{G}]\right) \tag{14.21}$$

It is instructive to note that when the plant tap-weight, $\mathbf{w}_o$, is time invariant, the correlation matrix $\mathbf{G}$ is zero, as $\boldsymbol{\epsilon}_o(n)$ is zero for all values of n. Thus, we obtain

$$\mathcal{M}_{\rm LMS} = \frac{\mu{\rm tr}[\mathbf{R}]}{1-\mu{\rm tr}[\mathbf{R}]} \tag{14.22}$$

This is exactly the result that we obtained in Chapter 6 – see Eq. (6.63). This observation shows that Eq. (14.20) is in fact a generalization of similar results that were obtained in the previous chapters. This includes the effect of plant variation and also the use of different step-size parameters at various taps. Moreover, we see that when the plant is time varying, there are two distinct terms contributing to the misadjustment of the LMS algorithm. Accordingly, we may write

$$\mathcal{M} = \mathcal{M}_1 + \mathcal{M}_2 \tag{14.23}$$

where

$$\mathcal{M}_1 = \frac{{\rm tr}[\boldsymbol{\mu}\mathbf{R}]}{1-{\rm tr}[\boldsymbol{\mu}\mathbf{R}]} \tag{14.24}$$

and

$$\mathcal{M}_2 = \frac{\sigma_{e_o}^{-2}}{4}\cdot\frac{{\rm tr}[\boldsymbol{\mu}^{-1}\mathbf{G}]}{1-{\rm tr}[\boldsymbol{\mu}\mathbf{R}]} \tag{14.25}$$

With reference to recursion (14.6) and the subsequent derivations, we find that $\mathcal{M}_1$ originates from the term $2\boldsymbol{\mu}e_o(n)\mathbf{x}(n)$ on the right-hand side of Eq. (14.6). This, clearly, is contributed by the plant noise, $e_o(n)$. Similarly, one finds that $\mathcal{M}_2$ is a direct contribution of the plant tap-weight increments $\boldsymbol{\epsilon}_o(n)$. Accordingly, $\mathcal{M}_1$ is called *the noise misadjustment*, and $\mathcal{M}_2$ is referred to as *the lag misadjustment*. We note that the noise misadjustment decreases as the step-size parameters, μ_i's, decrease. On the other hand, a smaller lag misadjutment is achieved by increasing the step-size parameters. Thus, it becomes necessary to find a compromise choice of the step-size parameters which will result in the right balance between the noise and lag misadjustments. This is the subject of the next section.

14.4 Optimum Step-Size Parameters

To derive a set of equations for the optimum step-size parameters that minimize the excess MSE and thus the misadjustment of the LMS algorithm, we first expand Eq. (14.19) to obtain

$$\xi_{\text{ex}} = \frac{1}{1 - \sum_i \mu_i \sigma_{x_i}^2} \sum_{i=0}^{N-1} \left(\mu_i \sigma_{e_o}^2 \sigma_{x_i}^2 + \frac{1}{4\mu_i} \sigma_{\epsilon_{o,i}}^2 \right) \tag{14.26}$$

where $\sigma_{x_i}^2$ and $\sigma_{\epsilon_{o,i}}^2$ are, respectively, the variances of $x_i(n)$ and the ith element of $\boldsymbol{\epsilon}_o(n)$, that is, the diagonal elements of the respective correlation matrices, $\mathbf{R}$ and $\mathbf{G}$.

The optimum values of the step-size parameters are obtained by setting the derivatives of ξ_{ex} with respect to μ_i's equal to zero. Solving the set of simultaneous equations

$$\frac{\partial \xi_{\text{ex}}}{\partial \mu_i} = 0, \quad \text{for} \quad i = 0, 1, \ldots, N-1 \tag{14.27}$$

we obtain (Problem P14.4)

$$\mu_{o,i} = \frac{\sigma_{\epsilon_{o,i}}}{2\sigma_{x_i}\sqrt{\xi_{\text{ex,o}} + \sigma_{e_o}^2}} \qquad i = 0, 1, \ldots, N-1 \tag{14.28}$$

where the subscript "o" is added to $\mu_{o,i}$'s to emphasize that they are the optimum values of the step-size parameters. Moreover, $\xi_{\text{ex,o}}$ refers to the excess MSE when the optimum step-size parameters, $\mu_{o,i}$'s, are used. This solution, of course, is not complete, as $\xi_{\text{ex,o}}$ depends on $\mu_{o,i}$'s. To complete the solution, we define $\eta = \sqrt{\xi_{\text{ex,o}} + \sigma_{e_o}^2}$ and replace Eq. (14.28) in Eq. (14.26). This results in a second-order equation in η whose solutions are

$$\eta = \frac{\sum_i \sigma_{\epsilon_{o,i}} \sigma_{x_i} \pm \sqrt{\left(\sum_i \sigma_{\epsilon_{o,i}} \sigma_{x_i}\right)^2 + 4\sigma_{e_o}^2}}{2}$$

Noting that η cannot be negative, we find that

$$\eta = \frac{\sum_i \sigma_{\epsilon_{o,i}} \sigma_{x_i} + \sqrt{\left(\sum_i \sigma_{\epsilon_{o,i}} \sigma_{x_i}\right)^2 + 4\sigma_{e_o}^2}}{2} \tag{14.29}$$

is the only acceptable solution of η. With this, we obtain

$$\mu_{o,i} = \frac{\sigma_{\epsilon_{o,i}}}{2\eta\sigma_{x_i}}, \quad \text{for} \quad i = 0, 1, \ldots, N-1 \tag{14.30}$$

It is instructive to note that Eq. (14.30) is intuitively sound. It suggests that those taps that have larger tap perturbation should be given larger step-size parameters. It also suggests normalization of the step-size parameters proportional to the inverse of the signal level at various taps. However, this normalization is different from the one commonly used in the step-normalized algorithms, where μ_i is selected proportional to the inverse of signal power at the respective tap, that is, proportional to $1/\sigma_{x_i}^2$. Moreover, Eq. (14.30) suggests that the step-size parameters should be reduced as the error level at the filter output increases – note that η^2 is equal to the MSE of the filter after it has converged.

The validity of Eq. (14.30) is subject to the condition that the optimum step-size parameters remain in a range that does not result in instability of the algorithm. For the case of the conventional LMS algorithm, where a single step-size parameter, μ, is employed, a useful and practically applicable upper bound for μ is the one derived in Chapter 6 and repeated below for convenience (6.74):

$$\mu < \frac{1}{3\mathrm{tr}[\mathbf{R}]} \tag{14.31}$$

or, equivalently,

$$\mu\mathrm{tr}[\mathbf{R}] < \frac{1}{3} \tag{14.32}$$

This result can be extended to the generalized LMS recursion Eq. (14.3) as follows.

Consider the recursion Eq. (14.3) and define $\tilde{\mathbf{x}}(n) = \boldsymbol{\mu}^{1/2}\mathbf{x}(n)$, where $\boldsymbol{\mu}^{1/2}$ is the diagonal matrix consisting of the square roots of the diagonal elements of $\boldsymbol{\mu}$. Then, multiplying both sides of Eq. (14.6) from left by $\boldsymbol{\mu}^{-1/2}$ and, also, defining $\tilde{\mathbf{v}}(n) = \boldsymbol{\mu}^{-1/2}\mathbf{v}(n)$ and $\tilde{\boldsymbol{\epsilon}}_{\mathrm{o}}(n) = \boldsymbol{\mu}^{-1/2}\boldsymbol{\epsilon}_{\mathrm{o}}(n)$, we obtain

$$\tilde{\mathbf{v}}(n+1) = (\mathbf{I} - 2\tilde{\mathbf{x}}(n)\tilde{\mathbf{x}}^{\mathrm{T}}(n))\tilde{\mathbf{v}}(n) + 2e_{\mathrm{o}}(n)\tilde{\mathbf{x}}(n) - \tilde{\boldsymbol{\epsilon}}_{\mathrm{o}}(n) \tag{14.33}$$

The recursion Eq. (14.33) is similar to the conventional LMS recursion with $\mu = 1$. Accordingly, Eq. (14.32) can be applied. Hence, we find that the stability of Eq. (14.33), and thus Eq. (14.6) or, equivalently, the recursion Eq. (14.3), is guaranteed if

$$\mathrm{tr}[\tilde{\mathbf{R}}] < 1/3 \tag{14.34}$$

where

$$\begin{aligned}\tilde{\mathbf{R}} &= E[\tilde{\mathbf{x}}(n)\tilde{\mathbf{x}}^{\mathrm{T}}(n)] \\ &= E[\boldsymbol{\mu}^{1/2}\mathbf{x}(n)\mathbf{x}^{\mathrm{T}}(n)\boldsymbol{\mu}^{1/2}] \\ &= \boldsymbol{\mu}^{1/2}E[\mathbf{x}(n)\mathbf{x}^{\mathrm{T}}(n)]\boldsymbol{\mu}^{1/2} \\ &= \boldsymbol{\mu}^{1/2}\mathbf{R}\boldsymbol{\mu}^{1/2}\end{aligned} \tag{14.35}$$

Substituting this result in Eq. (14.34) and noting that $\mathrm{tr}[\boldsymbol{\mu}^{1/2}\mathbf{R}\boldsymbol{\mu}^{1/2}] = \mathrm{tr}[\boldsymbol{\mu}\mathbf{R}]$ (according to identity $\mathrm{tr}[\mathbf{AB}] = \mathrm{tr}[\mathbf{BA}]$), we get

$$\mathrm{tr}[\boldsymbol{\mu}\mathbf{R}] < \frac{1}{3} \tag{14.36}$$

This is a sufficient condition that may be imposed on the algorithm step-size parameters, μ_i's, to guarantee the stability of the generalized LMS recursion Eq. (14.3).

When Eq. (14.36) holds, the minimum excess MSE of the filter, $\xi_{\rm ex,o}$, is obtained by substituting Eq. (14.30) in Eq. (14.26). This gives

$$\xi_{\rm ex,o} = \frac{1}{2 - \frac{1}{\eta}\sum_i \sigma_{\epsilon_{\rm o},i}\sigma_{x_i}} \left(\frac{\sigma_{e_{\rm o}}^2}{\eta} + \eta\right) \sum_i \sigma_{\epsilon_{\rm o},i}\sigma_{x_i} \tag{14.37}$$

Substituting for η from Eq. (14.29) in Eq. (14.37), we get

$$\xi_{\rm ex,o} = \frac{\sum_i \sigma_{\epsilon_{\rm o},i}\sigma_{x_i} + \sqrt{\left(\sum_i \sigma_{\epsilon_{\rm o},i}\sigma_{x_i}\right)^2 + 4\sigma_{e_{\rm o}}^2}}{2} \sum_i \sigma_{\epsilon_{\rm o},i}\sigma_{x_i} \tag{14.38}$$

14.5 Comparisons of Conventional Algorithms

In this section, we compare tracking behavior of various versions of the LMS algorithm in the context of the modeling problem discussed in the last few sections. Noting that the tracking behaviors of the RLS and LMS-Newton algorithms are about the same, the comparisons also cover the RLS algorithm. *The indicator of better tracking behavior (performance) is lower steady-state excess MSE.*

In order to prevent diverting into many possible cases, we concentrate on the comparison of the direct implementation of a transversal filter, using the LMS algorithm, and its implementation in transform domain. We note that for a transversal filter $\mathbf{x}(n) = [x(n)\, x(n-1)\, \cdots\, x(n-N+1)]^{\rm T}$, and for its transform-domain implementation $\mathbf{x}(n)$ is replaced by $\mathbf{x}_{\mathcal{T}}(n) = \mathcal{T}\mathbf{x}(n)$, where $\mathcal{T}$ is an orthonormal transformation matrix satisfying the condition[1]

$$\mathcal{T}\mathcal{T}^{\rm T} = \mathbf{I} \tag{14.39}$$

where $\mathbf{I}$ is the identity matrix.

In addition, if $\boldsymbol{\epsilon}_{\rm o}(n)$ represents the plant tap-weight increments in its transversal form, the corresponding increments in the transform domain are given by

$$\boldsymbol{\epsilon}_{\mathcal{T},{\rm o}}(n) = \mathcal{T}\boldsymbol{\epsilon}_{\rm o}(n) \tag{14.40}$$

We also define $\mathbf{R}_{\mathcal{T}} = E[\mathbf{x}_{\mathcal{T}}(n)\mathbf{x}_{\mathcal{T}}^{\rm T}(n)]$ and $\mathbf{G}_{\mathcal{T}} = E[\boldsymbol{\epsilon}_{\mathcal{T},{\rm o}}(n)\boldsymbol{\epsilon}_{\mathcal{T},{\rm o}}^{\rm T}(n)]$ and note that

$$\mathbf{R}_{\mathcal{T}} = \mathcal{T}\mathbf{R}\mathcal{T}^{\rm T} \tag{14.41}$$

and

$$\mathbf{G}_{\mathcal{T}} = \mathcal{T}\mathbf{G}\mathcal{T}^{\rm T} \tag{14.42}$$

The ith diagonal elements of $\mathbf{R}_{\mathcal{T}}$ and $\mathbf{G}_{\mathcal{T}}$ are denoted as $\sigma_{x_{\mathcal{T},i}}^2$ and $\sigma_{\epsilon_{\mathcal{T},{\rm o},i}}^2$, respectively. Moreover, to simplify our discussion, yet with no loss of generality, we assume that the input sequence to the transversal filter is normalized to unit power, that is, $\sigma_{x_i}^2 = E\{|x(n-i)|^2\} = 1$, for $i = 0, 1, \ldots, N-1$. Then, the orthonormality of $\mathcal{T}$, that is, the

[1] To avoid complex-valued coefficients/variables in our formulations in this chapter, we only consider transformations with real-valued coefficients.

condition (14.39), implies that

$$\sum_{i=0}^{N-1} \sigma_{x_{\mathcal{T},i}}^2 = \sum_{i=0}^{N-1} \sigma_{x_i}^2 = N \tag{14.43}$$

We note that in the case of the conventional LMS algorithm, a single step-size parameter, μ, is used for all taps. On the other hand, in the case of TDLMS algorithm, different step-size parameters are used for various taps and they are selected according to Eq. (14.4). Furthermore, for a fixed misadjustment, say $\mathcal{M}$, we have ((6.64) and (7.31))

$$\mu = \frac{\mathcal{M}}{\mathrm{tr}[\mathbf{R}]} = \frac{\mathcal{M}}{\sum_i \sigma_{x_i}^2} \tag{14.44}$$

and

$$\mu' = \frac{\mathcal{M}}{N} \tag{14.45}$$

Thus, in the light of Eq. (14.43), we find that $\mu' = \mu$ in this case.

Using the above results in Eq. (14.26), the excess MSE of the conventional LMS and TDLMS algorithms are obtained as

$$\xi_{\mathrm{ex}}(\mathrm{LMS}) = \frac{1}{1-\mu N}\left(\mu N \sigma_{e_o}^2 + \frac{1}{4\mu}\sum_{i=0}^{N-1} \sigma_{\epsilon_{o,i}}^2\right) \tag{14.46}$$

and

$$\xi_{\mathrm{ex}}(\mathrm{TDLMS}) = \frac{1}{1-\mu N}\left(\mu N \sigma_{e_o}^2 + \frac{1}{4\mu}\sum_{i=0}^{N-1} \sigma_{\epsilon_{\mathcal{T},o,i}}^2 \sigma_{x_{\mathcal{T},i}}^2\right) \tag{14.47}$$

respectively. In arriving at Eqs. (14.46) and (14.47), we made use of the assumption $\sigma_{x_i}^2 = 1$, for $i = 0, 1, \ldots, N-1$, along with Eqs. (14.43) and (14.4).

Now, let us consider a few specific cases:

Case 1: $\mathbf{G} = \sigma_{\epsilon_o}^2 \mathbf{I}$ *and* $\mathbf{R}$ *is arbitrary.*

Substituting $\mathbf{G} = \sigma_{\epsilon_o}^2 \mathbf{I}$ in Eq. (14.42) and recalling Eq. (14.39), we obtain $\mathbf{G}_{\mathcal{T}} = \sigma_{\epsilon_o}^2 \mathbf{I}$, which implies that

$$\sigma_{\epsilon_{\mathcal{T},o,i}}^2 = \sigma_{\epsilon_{o,i}}^2 = \sigma_{\epsilon_o}^2, \quad \text{for} \quad i = 0, 1, \ldots, N-1 \tag{14.48}$$

Substituting Eqs. (14.48) and (14.43) in Eq. (14.47), we obtain

$$\begin{aligned}\xi_{\mathrm{ex}}(\mathrm{TDLMS}) &= \frac{1}{1-\mu N}\left(\mu N \sigma_{e_o}^2 + \frac{1}{4\mu}\sigma_{\epsilon_o}^2 \sum_{i=0}^{N-1} \sigma_{x_{\mathcal{T},i}}^2\right) \\ &= \frac{N(\mu\sigma_{e_o}^2 + \frac{1}{4\mu}\sigma_{\epsilon_o}^2)}{1-\mu N}\end{aligned} \tag{14.49}$$

From this result, we see that when $\mathbf{G} = \sigma_{\epsilon_o}^2 \mathbf{I}$, $\xi_{\mathrm{ex}}(\mathrm{TDLMS})$ is independent of $\mathcal{T}$. Furthermore, with $\mathbf{G} = \sigma_{\epsilon_o}^2 \mathbf{I}$, Eq. (14.46) also simplifies to Eq. (14.49).

This, in turn, means that independent of the transformation used, the tracking performance of the TDLMS algorithm remains similar to that of the conventional LMS algorithm. Furthermore, noting that the LMS-Newton algorithm is equivalent to the TDLMS algorithm when KLT is used as its transformation, this conclusion also applies to the comparison of the conventional LMS and LMS-Newton algorithms. Moreover, noting that the RLS and LMS-Newton algorithms have similar tracking behavior (Section 14.2), we may also add that in this case the conventional LMS and RLS algorithms have similar tracking behavior.

Case 2: $\mathbf{R} = \mathbf{I}$ *and* $\mathbf{G}$ *is arbitrary.*

Using $\mathbf{R} = \mathbf{I}$ in Eq. (14.41), we find that $\mathbf{R}_{\mathcal{T}}$ is also equal to the identity matrix. Thus,

$$\sigma_{x_{\mathcal{T},i}}^2 = 1, \quad \text{for} \quad i = 0, 1, \ldots, N-1$$

Using this in Eq. (14.47), we obtain

$$\xi_{\text{ex}}(\text{TDLMS}) = \frac{1}{1-\mu N}\left(\mu N \sigma_{e_\text{o}}^2 + \frac{1}{4\mu}\sum_{i=0}^{N-1}\sigma_{\epsilon_{\mathcal{T},\text{o},i}}^2\right) \tag{14.50}$$

Now,

$$\sum_{i=0}^{N-1}\sigma_{\epsilon_{\mathcal{T},\text{o},i}}^2 = \text{tr}[\mathbf{G}_{\mathcal{T}}] = \text{tr}[\mathcal{T}\mathbf{G}\mathcal{T}^\text{T}] = \text{tr}[\mathcal{T}^\text{T}\mathcal{T}\mathbf{G}]$$

$$= \text{tr}[\mathbf{G}] = \sum_{i=0}^{N-1}\sigma_{\epsilon_{\text{o},i}}^2 \tag{14.51}$$

where we have used Eqs. (14.39) and (14.42), and the identity $\text{tr}[\mathbf{AB}] = \text{tr}[\mathbf{BA}]$. Substituting Eq. (14.51) in Eq. (14.50), we obtain

$$\xi_{\text{ex}}(\text{TDLMS}) = \frac{1}{1-\mu N}\left(\mu N \sigma_{e_\text{o}}^2 + \frac{1}{4\mu}\sum_{i=0}^{N-1}\sigma_{\epsilon_{\text{o},i}}^2\right) \tag{14.52}$$

Comparing this with Eq. (14.46), we find that in this case also, irrespective of the transformation $\mathcal{T}$, there is no difference between the tracking behaviors of the conventional LMS and TDLMS algorithms. Thus, all the conclusions drawn for Case 1 continue to hold for Case 2 also, that is, *the conventional LMS, TDLMS, LMS-Newton, and RLS algorithms all have similar tracking behavior.*

Case 3: $\mathbf{R}$ *and* $\mathbf{G}$ *are arbitrary.*

From Eq. (14.47), we note that to study the variation of the excess MSE of the TDLMS algorithm for different choices of $\mathcal{T}$, we need to study the summation

$$\sum_{i=0}^{N-1}\sigma_{\epsilon_{\mathcal{T},\text{o},i}}^2\sigma_{x_{\mathcal{T},i}}^2 \tag{14.53}$$

Moreover, we note that the orthonormality of $\mathcal{T}$, that is, the identity $\mathcal{T}\mathcal{T}^\text{T} = \mathbf{I}$, implies that the summations $\sum_i \sigma_{x_{\mathcal{T},i}}^2$ and $\sum_i \sigma_{\epsilon_{\mathcal{T},\text{o},i}}^2$ are independent of $\mathcal{T}$. However, the individual terms under the summations, that is, $\sigma^2_{x_{\mathcal{T},i}}$'s and $\sigma^2_{\epsilon_{\mathcal{T},o,i}}$'s, vary with $\mathcal{T}$. Thus, while the summations $\sum_i \sigma^2_{x_{\mathcal{T},i}}$ and $\sum_i \sigma^2_{\epsilon_{\mathcal{T},o,i}}$ are fixed for different choices of $\mathcal{T}$, the distributions of the terms $\sigma^2_{x_{\mathcal{T},i}}$'s and $\sigma^2_{\epsilon_{\mathcal{T},o,i}}$'s vary with $\mathcal{T}$. These distributions also depend on the correlation matrices $\mathbf{R}$ and $\mathbf{G}$. When $\mathbf{R}$ and $\mathbf{G}$ are arbitrary, these distribution are also arbitrary. As a result, we find that *when no prior information about* $\mathbf{R}$ *and* $\mathbf{G}$ *is available, nothing can be said about the summation* Eq. (14.53), *and thus no specific comment can be made about the tracking behavior of various algorithms.* The following numerical example clarifies this further.

Let

$$\mathbf{R} = \begin{bmatrix} 1.0 & 0.5 \\ 0.5 & 1.0 \end{bmatrix} \quad \text{and} \quad \mathbf{G} = \begin{bmatrix} 0.0010 & 0.0008 \\ 0.0008 & 0.0100 \end{bmatrix}$$

In addition, define

$$\mathcal{T} = \begin{bmatrix} \cos\theta & \sin\theta \\ -\sin\theta & \cos\theta \end{bmatrix}$$

This is an arbitrary 2-by-2 orthonormal transformation matrix that varies with θ. Table 14.1 presents the summary of the results that we have obtained for the LMS and TDLMS algorithms for two choices of $\theta = \pi/8$ and $\pi/4$. It is noted that in the case of $\theta = \pi/8$, the TDLMS shows a better tracking behavior than the LMS algorithm – compare the summations in the last line of Table 14.1. However, the LMS algorithm behaves better when $\theta = \pi/4$ is chosen. Incidentally, here, $\theta = \pi/4$ makes $\mathcal{T}$ correspond to the KLT of the filter input, for which the TDLMS algorithm is also equivalent to the LMS-Newton algorithm.

The comparisons given above assume that we have no information about the correlation matrix $\mathbf{G}$ of the plant tap-weight increments. Thus, the optimum step-size parameters derived in the last section Eq. (14.30) could not be used. In Section 14.7, we show that the optimum step-size parameters can, in fact, be obtained adaptively using the variable step-size least mean square (VSLMS) algorithm introduced in Chapter 6 (Section 6.7). Noting this, we consider using the optimum step-size parameters given by Eq. (14.30) and present some more comparisons of the various algorithms in the next section.

Table 14.1 Comparison of the conventional LMS and TDLMS for a numerical example.

	TDLMS	
LMS	$\theta = \pi/8$	$\theta = \pi/4$
$\sigma^2_{x_0} = 1.0000$	$\sigma^2_{x_{\mathcal{T},0}} = 1.3536$	1.5000
$\sigma^2_{x_1} = 1.0000$	$\sigma^2_{x_{\mathcal{T},1}} = 0.6464$	0.5000
$\sigma^2_{\epsilon_0} = 0.0010$	$\sigma^2_{\epsilon_{\mathcal{T},o,0}} = 0.0029$	0.0063
$\sigma^2_{\epsilon_1} = 0.0100$	$\sigma^2_{\epsilon_{\mathcal{T},o,1}} = 0.0081$	0.0047
$\sum_i \sigma^2_{x_i}\sigma^2_{\epsilon_i} = 0.0110$	$\sum_i \sigma^2_{x_{\mathcal{T},i}}\sigma^2_{\epsilon_{\mathcal{T},o,i}} = 0.0091$	0.0118

14.6 Comparisons Based on Optimum Step-Size Parameters

From the theoretical results of Section 14.4 and the definitions of $\mathbf{R}_{\mathcal{T}}$ and $\mathbf{G}_{\mathcal{T}}$, in the last section, we recall that when the optimum step-size parameters given by Eq. (14.30) are used, the excess MSE of the TDLMS algorithm is given by Eq. (14.38)

$$\xi_{\mathrm{ex,o}}(\mathrm{TDLMS}) = \frac{\Gamma_{\mathcal{T}} + \sqrt{\Gamma_{\mathcal{T}}^2 + 4\sigma_{e_o}^2}}{2}\Gamma_{\mathcal{T}} \tag{14.54}$$

where

$$\Gamma_{\mathcal{T}} = \sum_{i=0}^{N-1} \sigma_{\epsilon_{\mathcal{T},\mathrm{o},i}} \sigma_{x_{\mathcal{T},i}} \tag{14.55}$$

We note that $\Gamma_{\mathcal{T}}$ is a function of $\mathbf{R}$, $\mathbf{G}$, and $\mathcal{T}$.

We also note that when no transformation has been applied, but the optimum step-size parameters are used for different taps, the excess MSE of the LMS algorithm is given by

$$\xi_{\mathrm{ex,o}}(\mathrm{LMS}) = \frac{\Gamma_{\mathrm{I}} + \sqrt{\Gamma_{\mathrm{I}}^2 + 4\sigma_{e_o}^2}}{2}\Gamma_{\mathrm{I}} \tag{14.56}$$

where

$$\Gamma_{\mathrm{I}} = \sum_{i=0}^{N-1} \sigma_{\epsilon_{\mathrm{o},i}} \sigma_{x_i} \tag{14.57}$$

Clearly, to achieve the best tracking performance of the TDLMS algorithm, one should find the matrix $\mathcal{T}$, which minimizes $\Gamma_{\mathcal{T}}$. A general solution to this problem appears to be difficult. We thus limit ourselves to a few particular cases whose study is found to be instructive. The following lemma is widely used in the study of the cases that follows.

Lemma 14.1 *Consider the diagonal matrix* $\mathbf{\Lambda} = \mathrm{diag}\,(\lambda_0, \lambda_1, \ldots, \lambda_{N-1})$, *where* λ_i*'s are all real and nonnegative. If* $\mathcal{T}$ *is an orthonormal matrix, that is,* $\mathcal{T}\mathcal{T}^T = \mathbf{I}$, *and* $\mathbf{S} = \mathcal{T}\mathbf{\Lambda}\mathcal{T}^{\mathrm{T}}$, *then the following inequality always holds:*

$$\sum_{i=0}^{N-1} \sqrt{\lambda_i} \leq \sum_{i=0}^{N-1} \sqrt{s_{ii}} \tag{14.58}$$

where s_{ii} *is the* i*th diagonal element of* $\mathbf{S}$.

Proof. We first note that for $x \geq 0$, $f(x) = \sqrt{x}$ is a concave function. In addition, according to the theory of the convex functions, Rockafellar (1970), if $f(x)$ is a concave function and $\zeta_0, \zeta_1, \ldots, \zeta_{N-1}$ are a set of nonnegative numbers that satisfy $\sum_{i=0}^{N-1} \zeta_i = 1$, then for any set of numbers $x_0, x_1, \ldots, x_{N-1}$ in the domain of $f(x)$, the following inequality holds

$$\sum_{i=0}^{N-1} \zeta_i f(x_i) \leq f\left(\sum_{i=0}^{N-1} \zeta_i x_i\right) \tag{14.59}$$

Next, we note that

$$s_{ii} = \sum_{l=0}^{N-1} \lambda_l \tau_{il}^2 \tag{14.60}$$

where τ_{il} is the ilth element of $\mathcal{T}$. In addition, the orthonormality of $\mathcal{T}$ implies that

$$\sum_{i=0}^{N-1} \tau_{il}^2 = 1 \tag{14.61}$$

Choosing $\zeta_i = \tau_{il}^2$, $x_i = \lambda_i$, and $f(x) = \sqrt{x}$ in Eq. (14.59), and using Eq. (14.60), we obtain

$$\sum_{l=0}^{N-1} |\tau_{il}|^2 \sqrt{\lambda_l} \leq \sqrt{s_{ii}} \tag{14.62}$$

Summing up both sides of Eq. (14.62) over $i = 0, 1, \ldots, N-1$ and using Eq. (14.61) completes the proof.

We are now ready to consider a few specific cases:

Case 1: $\mathbf{R} = \mathbf{I}$ *and* $\mathbf{G}$ *is an arbitrary diagonal matrix.*
The assumption $\mathbf{R} = \mathbf{I}$ and the orthonormality of $\mathcal{T}$ implies that $\mathbf{R}_{\mathcal{T}} = \mathbf{I}$. Thus,

$$\sigma_{x_i}^2 = \sigma_{x_{\mathcal{T},i}}^2 = 1, \quad \text{for} \quad i = 0, 1, \ldots, N-1 \tag{14.63}$$

Using Eq. (14.63) in Eqs. (14.57) and (14.55), we get

$$\Gamma_{\mathrm{I}} = \sum_{i=0}^{N-1} \sigma_{\epsilon_{\mathrm{o},i}} \tag{14.64}$$

and

$$\Gamma_{\mathcal{T}} = \sum_{i=0}^{N-1} \sigma_{\epsilon_{\mathcal{T},\mathrm{o},i}} \tag{14.65}$$

respectively. On the other hand, noting that $\mathbf{G}$ is a diagonal matrix consisting of the elements $\sigma_{\epsilon_{\mathrm{o},0}}^2, \sigma_{\epsilon_{\mathrm{o},1}}^2, \ldots, \sigma_{\epsilon_{\mathrm{o},N-1}}^2$, the diagonal elements of $\mathbf{G}_{\mathcal{T}} = \mathcal{T}\mathbf{G}\mathcal{T}^{\mathrm{T}}$ are $\sigma_{\epsilon_{\mathcal{T},\mathrm{o},0}}^2, \sigma_{\epsilon_{\mathcal{T},\mathrm{o},1}}^2, \ldots, \sigma_{\epsilon_{\mathcal{T},\mathrm{o},N-1}}^2$, and using the above Lemma, we find that

$$\Gamma_{\mathrm{I}} \leq \Gamma_{\mathcal{T}} \tag{14.66}$$

Using Eq. (14.66) in Eqs. (14.54) and (14.56), we find that in this case

$$\xi_{\mathrm{ex,o}}(\mathrm{LMS}) \leq \xi_{\mathrm{ex,o}}(\mathrm{TDLMS}) \tag{14.67}$$

That is, *when* $\mathbf{R} = \mathbf{I}$ *and* $\mathbf{G}$ *is diagonal, and the optimum step-size parameters,* $\mu_{\mathrm{o},i}$*'s, are used; there is no transformation that can improve the tracking behavior of the LMS algorithm.*

Case 2: $\mathbf{R} = \mathbf{I}$ *and* $\mathbf{G}$ *is arbitrary.*
Let $\mathcal{T} = \mathcal{T}_{\mathrm{o}}$ be the orthonormal transform which results in a diagonal matrix $\mathbf{G}_{\mathcal{T}_{\mathrm{o}}} = \mathcal{T}_{\mathrm{o}}\mathbf{G}\mathcal{T}_{\mathrm{o}}^{\mathrm{T}}$. Using $\mathcal{T} = \mathcal{T}_{\mathrm{o}}$ leads to a TDLMS algorithm in which $\mathbf{R}_{\mathcal{T}_{\mathrm{o}}} = \mathcal{T}_{\mathrm{o}}\mathbf{R}\mathcal{T}_{\mathrm{o}}^{\mathrm{T}} = \mathbf{I}$ (since $\mathbf{R} = \mathbf{I}$), and $\mathbf{G}_{\mathcal{T}_{\mathrm{o}}}$ is diagonal. This is similar to Case 1, above.

Hence, for the same reason as in Case 1, we can argue that *the choice of* $\mathcal{T} = \mathcal{T}_o$ *results in a TDLMS algorithm with optimum tracking behavior.*

Case 3: $\mathbf{G} = \sigma_{\epsilon_o}^2 \mathbf{I}$ *and* $\mathbf{R}$ *is arbitrary.*
Following the same line of reasoning as in Case 2, we find that here the optimum transform, $\mathcal{T}_o$, which result in a TDLMS algorithm with the best tracking behavior is the one that result in a diagonal $\mathbf{R}_{\mathcal{T}_o} = \mathcal{T}_o \mathbf{R} \mathcal{T}_o^{\mathrm{T}}$. That is, here, *the optimum transform, for achieving best tracking, is the KLT.*

14.7 VSLMS: An Algorithm with Optimum Tracking Behavior

The variable step-size least mean square (VSLMS) algorithm was introduced in Chapter 6, on the basis of an intuitive understanding of the behavior of the LMS algorithm. In this section, we present a formal derivation[2] of the VSLMS algorithm as an adaptive filtering scheme with optimal tracking behavior.

14.7.1 Derivation of VSLMS Algorithm

From the results presented in Section 14.3, we observe that in a time-varying environment, the steady-state MSE of an adaptive filter vary with the step-size parameters. Moreover, a study of the excess MSE, ξ_{ex}, shows that it is a convex function of the step-size parameters, μ_i's, when these vary over a range that does not result in instability (14.26). This implies that the MSE

$$\xi = E[e^2(n)] = \xi_{\min} + \xi_{\mathrm{ex}}$$

is also a convex function of the step-size parameters. With this concept in mind, one may suggest the following gradient search method for finding the optimum step-size parameters, $\mu_{o,i}$'s, of the LMS algorithm:

$$\mu_i(n) = \mu_i(n-1) - \rho \frac{\partial \xi}{\partial \mu_i(n-1)} \tag{14.68}$$

where ρ is a small positive adaptation parameter.

In analogy with the LMS algorithm, the stochastic version of the gradient recursion (14.68) is

$$\mu_i(n) = \mu_i(n-1) - \rho \frac{\partial e^2(n)}{\partial \mu_i(n-1)} \tag{14.69}$$

Recalling that

$$\begin{aligned} e(n) &= d(n) - \mathbf{w}^{\mathrm{T}}(n)\mathbf{x}(n) \\ &= d(n) - \sum_{l=0}^{N-1} w_l(n) x_l(n) \end{aligned} \tag{14.70}$$

[2] The derivation presented here first appeared in Mathews and Xie (1990).

we obtain

$$\frac{\partial e^2(n)}{\partial \mu_i(n-1)} = 2e(n)\frac{\partial e(n)}{\partial \mu_i(n-1)}$$
$$= -2e(n)x_i(n)\frac{\partial w_i(n)}{\partial \mu_i(n-1)} \quad (14.71)$$

where we have noted that in the summation $\sum_l w_l(n)x_l(n)$ only $w_i(n)$ varies with $\mu_i(n-1)$. The tap weight $w_i(n)$ is related to $\mu_i(n-1)$ according to the LMS recursion

$$w_i(n) = w_i(n-1) + 2\mu_i(n-1)e(n-1)x_i(n-1) \quad (14.72)$$

Substituting Eq. (14.72) in Eq. (14.71), and defining

$$g_i(n) = -2e(n)x_i(n) \quad (14.73)$$

we get

$$\frac{\partial e^2(n)}{\partial \mu_i(n-1)} = -g_i(n)g_i(n-1) \quad (14.74)$$

Finally, substituting Eq. (14.74) in Eq. (14.69), we obtain

$$\mu_i(n) = \mu_i(n-1) + \rho g_i(n)g_i(n-1) \quad (14.75)$$

for $i = 0, 1, \ldots, N-1$. This is the recursion Eq. (6.138), which was introduced in Chapter 6.

From Eq. (14.75), we note that the step-size parameters, $\mu_i(n)$'s, settle near their steady-state values when

$$E[g_i(n)g_i(n-1)] = 0, \quad \text{for} \quad i = 0, 1, \ldots, N-1 \quad (14.76)$$

Accordingly, a rigorous proof of the optimality of the VSLMS algorithm may be given by solving Eq. (14.76) for a set of unknown step-size parameters and showing that its solution matches the optimum parameters as given in Eq. (14.30). In fact, some researchers have proved this to be true for some specific cases (Mathews and Xie, 1993; Farhang-Boroujeny and Gazor, 1994). Here, we ignore such derivations as they are lengthy and a general solution turns out to be difficult to derive. Instead, we rely on simulations to verify the optimality of the VSLMS algorithm (Section 14.7.4).

14.7.2 Variations and Extensions

Sign Update Equation

Recall from Chapter 6 that the stochastic gradient terms, $g_i(n)$ and $g_i(n-1)$, in Eq. (14.75) may be replaced by their respective signs, to obtain

$$\mu_i(n) = \mu_i(n-1) + \rho\text{sign}[g_i(n)] \cdot \text{sign}[g_i(n-1)]$$
$$= \mu_i(n-1) + \rho\text{sign}[g_i(n)g_i(n-1)] \quad (14.77)$$

This may be referred to as *sign* update equation, in analogy with the LMS sign algorithm (Section 6.5).

Multiplicative versus Linear Increments

Other variations of the step-size update equations that have been proposed in the literature are (Farhang-Boroujeny and Gazor, 1994)

$$\mu_i(n) = (1 + \rho g_i(n) g_i(n-1))\mu_i(n-1) \tag{14.78}$$

and its sign update version

$$\mu_i(n) = (1 + \rho \text{sign}[g_i(n) g_i(n-1)])\mu_i(n-1) \tag{14.79}$$

For easy reference, we refer to Eqs. (14.75) and (14.77) as *step-size update equations with linear increments* and Eqs. (14.78) and (14.79) as *step-size update equations with multiplicative increments*. We make some comments on the performance of the linear and multiplicative increments later in Section 14.7.4.

Clearly, for a small ρ, Eq. (14.78) reaches it steady state when Eq. (14.76) is satisfied. This shows that both Eqs. (14.75) and (14.78) converge to the same set of step-size parameters. Similarly, Eqs. (14.77) and (14.79) converge to the same set of step-size parameters. Furthermore, *when* $g_i(n)g_i(n-1)$ *has a symmetrical distribution around its mean* (a case likely to happen, at least approximately, in most of applications) *and* ρ *is small, all of these step-size update equations converge to the same set of parameters.*

VSLMS Algorithm with a Common Step-Size Parameter

In many applications, to keep the complexity of the filter low, we are often interested in using a common step-size parameter, $\mu(n)$, for all the filter taps. Following a similar line of derivations as in Eqs. (14.68)–(14.75), the following recursion can be easily derived (Mathews and Xie, 1993):

$$\mu(n) = \mu(n-1) + \rho e(n) e(n-1) \mathbf{x}^{\mathrm{T}}(n) \mathbf{x}(n-1) \tag{14.80}$$

The sign version of this recursion may be given as

$$\mu(n) = \mu(n-1) + \rho \text{sign}[e(n) e(n-1) \mathbf{x}^{\mathrm{T}}(n) \mathbf{x}(n-1)] \tag{14.81}$$

These are recursions with linear increments. Extension of these recursions to those with multiplicative increments is straightforward.

VSLMS Algorithm for Complex-Valued Case

For filters with complex-valued input, following a similar line of derivation as that led to Eq. (14.75) results in the following recursion (Problem P14.7):

$$\mu_i(n) = \mu_i(n-1) + \rho(g_{i,\mathrm{R}}(n) g_{i,\mathrm{R}}(n-1) + g_{i,\mathrm{I}}(n) g_{i,\mathrm{I}}(n-1)) \tag{14.82}$$

Here, the subscripts R and I refer to the real and imaginary parts of $g_i(n)$ and $g_i(n-1)$, and

$$g_i(n) = -2e^*(n) x_i(n) \tag{14.83}$$

The sign version of Eq. (14.82) is

$$\mu_i(n) = \mu_i(n-1) + \rho(\text{sign}[g_{i,\text{R}}(n)g_{i,\text{R}}(n-1)] + \text{sign}[g_{i,\text{I}}(n)g_{i,\text{I}}(n-1)]) \tag{14.84}$$

Extensions of these results to update equations with multiplicative increments and also to the case where a common step-size parameter is used for all the filter taps are straightforward.

The last point to be noted here is that *the step-size parameters should always be limited to a range that satisfies the stability requirement of the LMS algorithm.* Equation (14.36) specifies the condition that must be satisfied by the step-size parameters to guarantee stability. However, how this is implemented in actual practice to limit the step-size parameters can vary. In Section 14.7.4, we discuss a possible way of limiting the step-size parameters.

14.7.3 Normalization of the Parameter ρ

In the update equations (14.75) and (14.78), and similar equations for complex-valued case, the step-size increments are proportional to the size of $g_i(n)g_i(n-1)$. This might be inappropriate when signal levels vary significantly with time. This results in fast and slow variations of the step-size parameters, depending on the level of input signal, $x_i(n)$, and also the output error, $e(n)$. To keep a more uniform control (variation) of the step-size parameters that do not depend on signal levels, we adopt a step-normalization technique similar to the one used in the LMS algorithm. Here, the adaptation parameter ρ is replaced by $\rho_i(n)$, which is obtained according to the following equation:

$$\rho_i(n) = \frac{\rho_\text{o}}{\hat{\sigma}_g^2(n) + \psi} \tag{14.85}$$

where ρ_o is an un-normalized parameter common to all taps, $\hat{\sigma}_g^2(n)$ is an estimate of $E[|g_i(n)|^2]$ that may be obtained through the recursion

$$\hat{\sigma}_g^2(n) = \beta\hat{\sigma}_g^2(n-1) + (1-\beta)|g_i(n)|^2 \tag{14.86}$$

where β is a forgetting factor close to but less than 1, and ψ is a positive constant that prevents possible instability of the algorithm when $\hat{\sigma}_g^2(n)$ is small.

In the case of Eq. (14.80), the normalization is done with respect to $E[e^2(n)\mathbf{x}^\text{T}(n)\mathbf{x}(n)]$. This can also be estimated through a time-averaging recursion similar to Eq. (14.86).

14.7.4 Computer Simulations

In this section, we present some simulation results to illustrate the optimal tracking behavior of the VSLMS algorithm. As an example, we consider the application of VSLMS algorithm in the identification of a multipath communication channel.[3] The channel is

[3] The simulation results presented here are simplified versions of those presented by the author in Farhang-Boroujeny and Gazor (1994). Here, we consider the case where all variables are real-valued. In Farhang-Boroujeny and Gazor (1994), all the variables are assumed to be complex-valued. However, the conclusions derived from the results of this section as well as those in Farhang-Boroujeny and Gazor (1994) are similar.

assumed to have two distinct paths with a continuous-time impulse response

$$h_t(t_o) = a_1(t_o)p(t - \tau_1(t_o)) + a_2(t_o)p(t - \tau_2(t_o)) \tag{14.87}$$

where t_o is the time at which the channel response is given (measured), $\tau_1(t_o)$ and $\tau_2(t_o)$ are the path delays, $a_1(t_o)$ and $a_2(t_o)$ are the path gains, t is the continuous time variable, $p(t)$ is the raised-cosine pulse with 50% roll-off factor given by

$$p(t) = \frac{\sin(\pi t/T_s)}{\pi t/T_s} \cdot \frac{\cos(\pi t/2T_s)}{1 - (t/T_s)^2} \tag{14.88}$$

and T_s is the symbol interval. As explicitly indicated in Eq. (14.87), the path delays and gains may vary with time, for example, they depend on time t_o at which the channel response is given. An adaptive filter is used to track these variations.

Figure 14.2 depicts the simulation setup that is used here. The discrete-time channel model

$$W_o(z) = \sum_{i=0}^{N-1} w_{o,i}(n)z^{-i} \tag{14.89}$$

is related to the continuous-time channel response $h_t(t_o)$ as below:

$$w_{o,i}(n) = h_{iT_s}(nT_s), \quad \text{for} \quad i = 0, 1, \ldots, N-1 \tag{14.90}$$

where nT_s is the time at which the discrete response of the channel is measured.

For all simulations, we keep $\tau_1(nT_s)$ fixed at the value of $2T_s$, but let $\tau_2(nT_s)$ vary at a constant rate from $4T_s$ to $14T_s$ over every simulation run that takes 100 000 iterations (equivalent to 100 000 T_s s). The discrete-time samples of path gains $a_1(nT_s)$ and $a_2(nT_s)$ are generated independently by passing two independent unit-variance white Gaussian processes through single pole low-pass filters with the system function

$$H(z) = \frac{\sqrt{1-\alpha^2}}{1-\alpha z^{-1}} \tag{14.91}$$

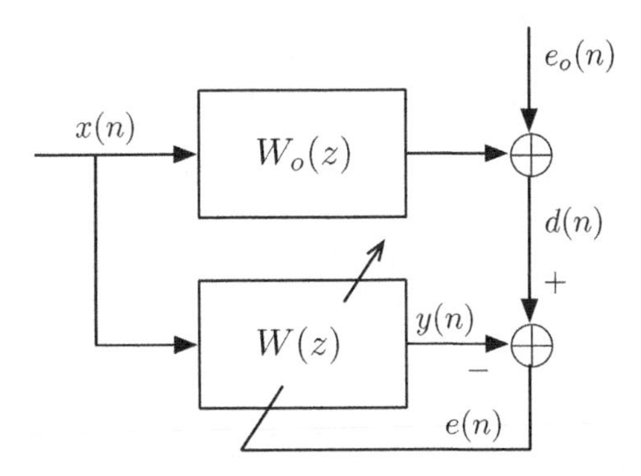

Figure 14.2 Modeling of a multipath communication channel.

where the parameter α is related to the channel fade rate, f_d, and the symbol-rate, $f_s = 1/T_s$, according to the following equation:

$$\alpha = 1 - \frac{\pi f_d}{f_s} \tag{14.92}$$

For typical values of f_d that allow the VSLMS algorithm to follow variations of the channel, the variations of $a_1(nT_s)$ and $a_2(nT_s)$ very closely approximate a random walk model similar to the one used to develop the analytical results of this chapter (Eweda, 1994); see also Problem P14.8. This approximate realization of random walk prevents indefinite increase in the path gains that would otherwise happen if we had used the random walk model of Section 14.1.

The following parameters are used in all the simulations that follow. The channel length, N, is set equal to 16. The same length is also assumed for the channel model (adaptive filter). The tails of the raised-cosine pulses associated with the two paths that lie outside the range set by channel length are truncated. A fade rate of $f_d = f_s/2400$ is assumed. In the implementation of the sign update equations with multiplicative increments, we choose $\rho = 0.002$. For the conventional update Eqs. (14.75) and (14.78), the parameter ρ is normalized as discussed in Section 14.7.3. The following parameters are used:

- for Eq. (14.75), $\beta = 0.95$, $\psi = 0.001$, and $\rho_o = 0.0002$
- for Eq. (14.78), $\beta = 0.95$, $\psi = 0.001$, and $\rho_o = 0.002$

The data sequence, $s(n)$, at the channel and adaptive filter input is a binary zero-mean white random process, taking values ± 1. The channel noise, $e_o(n)$, is a zero-mean white Gaussian process with variance $\sigma_{e_o}^2 = 0.02$. This choice of $\sigma_{e_o}^2$ results in an average signal-to-noise ratio of 20 dB at the channel output.

The step-size parameters are checked at the end of every iteration of the algorithm and limited to stay within a range that satisfies Eq. (14.36). When Eq. (14.36) is not satisfied, all the step-size parameters are scaled down by the same factor such that $\mathrm{tr}[\boldsymbol{\mu}\mathbf{R}]$ reduces to its upper bound 1/3. The step-size parameters are also hard limited to the minimum value of 0.001.

Next, we present a number of results comparing the relative performance of various implementations of the step-size adaptation in the present application. These results also serve to show optimal tracking behavior of the VSLMS algorithm.

Figure 14.3 presents a typical result comparing the performance of the update Eqs. (14.75) and (14.78), that is, the recursions with linear and multiplicative increments, respectively. These and the subsequent results of this section are based on single simulation runs; that is, no ensemble averaging is used. However, time averaging with a moving rectangular window is used to smoothen the plots. We note that adaptation based on multiplicative increments results in lower steady-state MSE, that is, a superior tracking behavior. This may be explained by noting that the variation of step-size parameters follows a geometrical progression with multiplicative increments and hence it can react much faster to changes than its linear counter part. Because of this observation, the rest of the simulation results are given only for step-size updates with multiplicative increments.

Figure 14.4 shows a set of curves comparing the tracking behaviors of the VSLMS algorithm and the LMS algorithm with optimum step-size parameters. The optimum step-size parameters of the LMS algorithm are obtained according to Eq. (14.30), as explained later. Results of both the conventional step-size update Eq. (14.78) and its sign version,

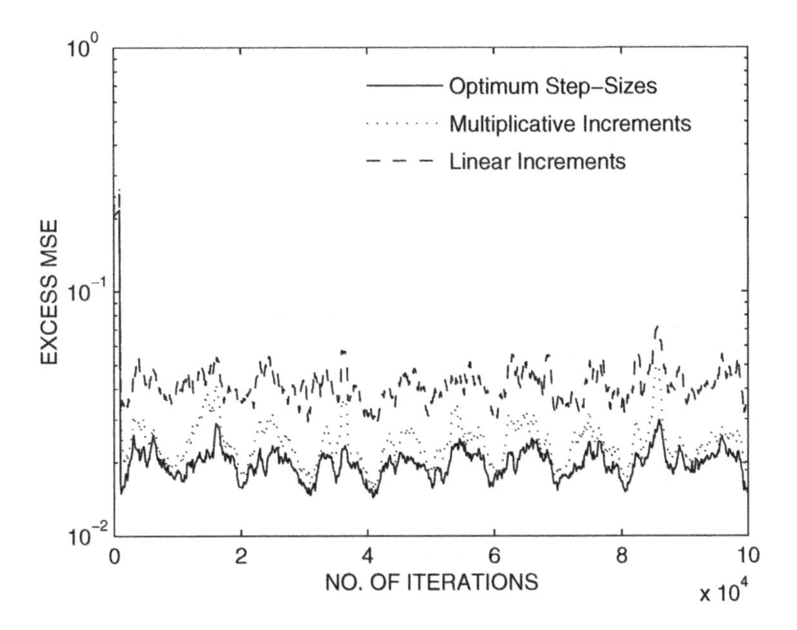

Figure 14.3 A typical simulation result comparing the VSLMS algorithm with linear and multiplicative increments.

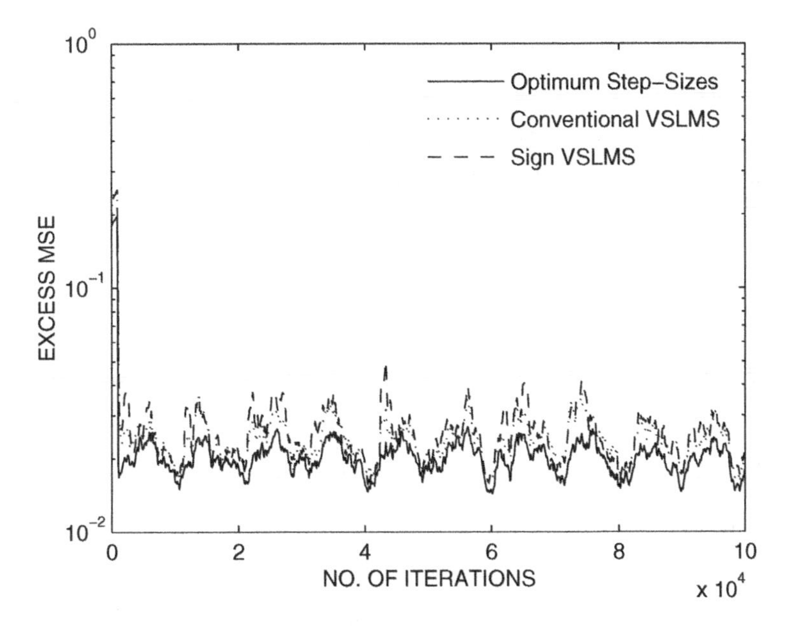

Figure 14.4 A typical simulation result comparing the LMS algorithm with optimum step-size parameters and the VSLMS algorithm.

Eq. (14.79), are presented. We note that both implementations of the VSLMS algorithm converge to about the same excess MSE as the case where optimal step-size parameters are used. This clearly illustrates the optimal tracking behavior of the VSLMS algorithm, as was predicted earlier in this section. We also note that there is very little difference between the behavior of the recursion Eq. (14.78) and its sign counterpart, Eq. (14.79).

Figure 14.5 presents the results showing how the VSLMS algorithm track variation of $\mu_{o,10}(n)$, that is, the optimum step-size parameter of the 10th tap of the adaptive filter. The results are given for the recursion Eq. (14.78) and also its sign counterpart, Eq. (14.79). The results show that the VSLMS algorithm converges to the optimum step-size parameters, thus achieving a close to optimum tracking behavior. Further experiments have confirmed this optimum performance even when the adaptive filter input is colored (Mathews and Xie, 1993; Farhang-Boroujeny and Gazor, 1994).

Computation of the optimum step-size parameters, which are used to obtain the results of Figures 14.4 and 14.5, is carried out by finding σ_{x_i} and $\sigma_{\epsilon_{o,i}}$ first, and then substituting them in Eq. (14.30). Noting that the filter input is a binary sequence, we get $\sigma_{x_i} = 1$. To evaluate $\sigma_{\epsilon_{o,i}}$, we recall from Eq. (14.91) that the path gains, $a_1(nT_s)$ and $a_2(nT_s)$, are generated using the recursions

$$a_k((n+1)T_s) = \alpha a_k(nT_s) + \sqrt{1-\alpha^2}\nu_k(n+1), \quad \text{for} \quad k = 1,\ 2 \tag{14.93}$$

where $\nu_1(n+1)$ and $\nu_2(n+1)$ are two independent unit-variance zero-mean Gaussian white sequences. Assuming that α is smaller but close to 1, we find that

$$1 - \alpha \ll \sqrt{1-\alpha^2}$$

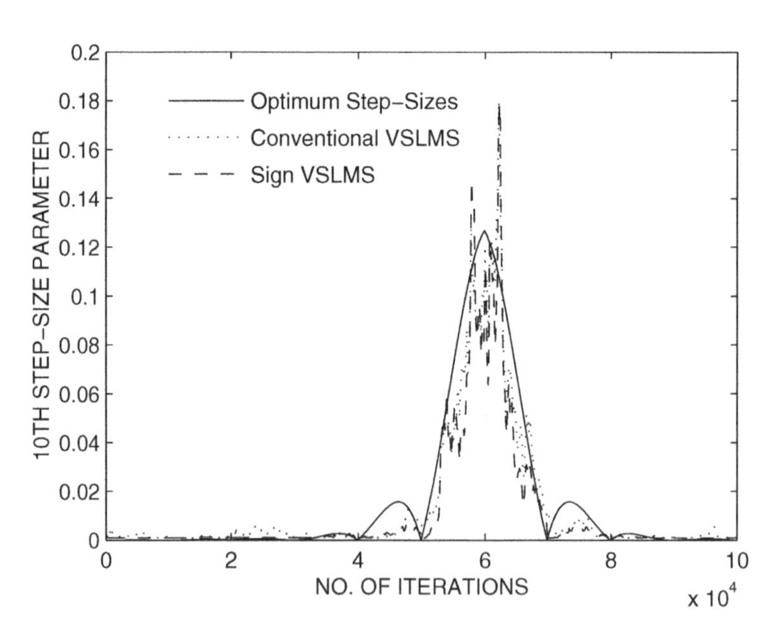

Figure 14.5 A typical simulation result showing that the VSLMS algorithm closely track the optimum step-size parameters given by Eq. (14.30).

Thus, for $k = 1$ and 2, we obtain

$$\begin{aligned} a_k((n+1)T_s) - a_k(nT_s) &= -(1-\alpha)a_k(nT_s) + \sqrt{1-\alpha^2}\nu_k(n+1) \\ &\approx \sqrt{1-\alpha^2}\nu_k(n+1) \end{aligned} \tag{14.94}$$

where the latter approximation is (statistically) justified by noting that $a_k(nT_s)$ and $\nu_k(n+1)$ are two random variables with the same variance. On the other hand, from Eq. (14.2) we get $\boldsymbol{\epsilon}_o(n) = \mathbf{w}_o(n+1) - \mathbf{w}_o(n)$, or

$$\epsilon_{o,i}(n) = w_{o,i}(n+1) - w_{o,i}(n) \tag{14.95}$$

Next, substituting Eq. (14.90) in Eq. (14.95), and assuming that the path delays $\tau_1(nT_s)$ and $\tau_2(nT_s)$ vary very slowly in time so that their variations over the span of the channel length (NT_s seconds) can be ignored, we obtain using Eq. (14.94)

$$\begin{aligned} \epsilon_{o,i}(n) = \sqrt{1-\alpha^2}[&\nu_1(n+1)p(iT_s - \tau_1(nT_s)) \\ &+\nu_2(n+1)p(iT_s - \tau_2(nT_s))] \end{aligned} \tag{14.96}$$

for $i = 0, 1, \ldots, N-1$. Using Eq. (14.96), and recalling that $\nu_1(n+1)$ and $\nu_2(n+1)$ are unit-variance independent processes, we obtain

$$\sigma^2_{\epsilon_{o,i}}(n) = (1-\alpha^2)(p^2(iT_s - \tau_1(nT_s)) + p^2(iT_s - \tau_2(nT_s))) \tag{14.97}$$

or

$$\sigma_{\epsilon_{o,i}}(n) = \sqrt{(1-\alpha^2)(p^2(iT_s - \tau_1(nT_s)) + p^2(iT_s - \tau_2(nT_s)))} \tag{14.98}$$

14.8 RLS Algorithm with Variable Forgetting Factor

In this section, we extend the idea of VSLMS algorithm to propose an RLS algorithm with variable forgetting factor. To this end, we recall from the last chapter that in the steady state, the RLS algorithm is approximately equivalent to the LMS-Newton algorithm. In particular, when the filter input is stationary, in the steady state, the RLS recursion is approximately equivalent to the recursion

$$\hat{\mathbf{w}}(n) = \hat{\mathbf{w}}(n-1) + (1-\lambda(n))\hat{e}_{n-1}(n)\mathbf{R}^{-1}\mathbf{x}(n) \tag{14.99}$$

(see Eq. (12.90)). Note that here we have added the time index n to the forgetting factor $\lambda(n)$ to emphasize that it may vary with time.

Starting with Eq. (14.99) and following the same line of derivations as those used to derive the VSLMS algorithm, the following update equation is obtained for adaptive adjustment of $\lambda(n)$:

$$\lambda(n) = \lambda(n-1) - \rho z(n) \tag{14.100}$$

where

$$z(n) = (2\hat{e}_{n-1}(n)\mathbf{x}(n))^{\mathrm{T}}(2\hat{e}_{n-2}(n-1)\mathbf{R}^{-1}\mathbf{x}(n-1)) \tag{14.101}$$

A more robust update equation for adaptation of $\lambda(n)$ is obtained by defining

$$\beta(n) = 1 - \lambda(n) \tag{14.102}$$

and noting that Eq. (14.100) can equivalently be written as

$$\beta(n) = \beta(n-1) + \rho z(n) \tag{14.103}$$

Moreover, from our experience with the VSLMS algorithm, we may suggest using multiplicative increments instead of linear increments, and also replacing $z(n)$ by its sign, to obtain the following recursion:

$$\beta(n) = (1 + \rho\text{sign}[z(n)])\beta(n-1) \tag{14.104}$$

To get a more easily usable expression for $z(n)$, from Chapter 12, we recall that in the steady state

$$\mathbf{\Psi}_\lambda(n) \approx \frac{1}{1-\lambda(n)}\mathbf{R} \tag{14.105}$$

Rearranging Eq. (14.105) and replacing n by $n-1$, we obtain

$$\mathbf{R}^{-1} \approx \frac{1}{1-\lambda(n-1)}\mathbf{\Psi}_\lambda^{-1}(n-1) \tag{14.106}$$

Substituting Eq. (14.106) in Eq. (14.101) and using the definition (12.48), we obtain

$$\begin{aligned} z(n) &= \hat{e}_{n-1}(n)\hat{e}_{n-2}(n-1)\frac{\mathbf{x}^{\mathrm{T}}(n)\mathbf{\Psi}_\lambda^{-1}(n-1)\mathbf{x}(n-1)}{1-\lambda(n-1)} \\ &= \hat{e}_{n-1}(n)\hat{e}_{n-2}(n-1)\frac{\mathbf{x}^{\mathrm{T}}(n)\mathbf{k}(n-1)}{1-\lambda(n-1)} \end{aligned} \tag{14.107}$$

where $\mathbf{k}(n)$ is the gain vector of the RLS algorithm. Taking sign of $z(n)$ and recalling that $1-\lambda(n)$ is positive, as $\lambda(n) < 1$, we get

$$\text{sign}[z(n)] = \text{sign}[\hat{e}_{n-1}(n)\hat{e}_{n-2}(n-1)] \times \text{sign}[\mathbf{x}^{\mathrm{T}}(n)\mathbf{k}(n-1)] \tag{14.108}$$

Using the above results, Table 14.2 presents a summary of implementation of the RLS algorithm with variable forgetting factor – compare this with Table 12.2. Note that after every iteration, the forgetting factor, $\lambda(n)$, is checked and limited to some preselected values, λ^+ and λ^-.

14.9 Summary

In this chapter, we studied tracking behavior of various adaptive filtering algorithms in the context of a system modeling problem. We introduced and analyzed a generalized formulation of the LMS algorithm that could cover most of the algorithms that have been discussed in the previous chapters. The general conclusion derived from the analysis is that convergence and tracking are two different phenomena, and hence should be treated separately. We found that the algorithms that were previously introduced to speed up the convergence of adaptive filters do not necessarily have superior tracking behavior. We presented cases where the conventional LMS algorithm, which has the slowest convergence behavior, has better tracking behavior than those with more complicated structures, such as TDLMS or even RLS algorithm.

We also considered the VSLMS algorithm (of Chapter 6) as an adaptive filtering scheme with optimum tracking. The optimal tracking behavior of the VSLMS algorithm was

Table 14.2 Summary of the RLS algorithm with variable forgetting factor.

Input:	Tap-weight vector estimate, $\hat{\mathbf{w}}(n-1)$, Input vector, $\mathbf{x}(n)$, Desired output, $d(n)$, Forgetting factor, $\lambda(n-1)$, Gain vector $\mathbf{k}(n-1)$, and the matrix $\boldsymbol{\Psi}_{\lambda}^{-1}(n-1)$.
Output:	Filter output, $\hat{y}_{n-1}(n)$, Tap-weight vector update, $\hat{\mathbf{w}}(n)$, Forgetting factor $\lambda(n)$, and the updated matrix $\boldsymbol{\Psi}_{\lambda}^{-1}(n)$.

1. Computation of the gain vector:

$$\mathbf{u}(n) = \boldsymbol{\Psi}_{\lambda}^{-1}(n-1)\mathbf{x}(n)$$

$$\mathbf{k}(n) = \frac{1}{\lambda(n-1) + \mathbf{x}^{\mathrm{T}}(n)\mathbf{u}(n)}\mathbf{u}(n)$$

2. Filtering:

$$\hat{y}_{n-1}(n) = \hat{\mathbf{w}}^{\mathrm{T}}(n-1)\mathbf{x}(n)$$

3. Error estimation:

$$\hat{e}_{n-1}(n) = d(n) - \hat{y}_{n-1}(n)$$

4. Tap-weight vector adaptation:

$$\hat{\mathbf{w}}(n) = \hat{\mathbf{w}}(n-1) + \mathbf{k}(n)\hat{e}_{n-1}(n)$$

5. $\boldsymbol{\Psi}_{\lambda}^{-1}(n)$ update:

$$\boldsymbol{\Psi}_{\lambda}^{-1}(n) = \mathrm{Tri}\ \{\lambda^{-1}(n-1)(\boldsymbol{\Psi}_{\lambda}^{-1}(n-1) - \mathbf{k}(n)\mathbf{u}^{\mathrm{T}}(n))\}$$

6. $\lambda(n)$ update:

$$\beta(n) = \{1 + \rho \mathrm{sign}\ [\hat{e}_{n-1}(n)\hat{e}_{n-2}(n-1)] \times \mathrm{sign}\ [\mathbf{x}^{\mathrm{T}}(n)\mathbf{k}(n-1)]\}\beta(n-1)$$

$$\lambda(n) = 1 - \beta(n)$$

$$\text{if}\quad \lambda(n) > \lambda^{+},\quad \lambda(n) = \lambda^{+} \text{ and } \beta(n) = 1 - \lambda^{+}$$

$$\text{if}\quad \lambda(n) < \lambda^{-},\quad \lambda(n) = \lambda^{-} \text{ and } \beta(n) = 1 - \lambda^{-}$$

confirmed through computer simulations. The idea of the VSLMS algorithm was also extended to the RLS algorithm by introducing a similar adaptive mechanism for controlling its forgetting factor (memory length).

Most of the present literature on tracking behavior of adaptive filters and also our discussion in this chapter are limited to the case where the adaptive filter input is a stationary process.[4] Only the desired output of the filter has been assumed to be nonstationary. We may thus say that the treatment of the problem of tracking in the present

[4] An exception is the work of Macchi and Bershad (1991) where they have compared the tracking performance of the LMS and RLS algorithms in recovering a chirped sinusoid. This is an example where the adaptive filter input is nonstationary.

literature is rather immature, and much more work need to be done on this very important topic.

Problems

P14.1 In the derivation of Eq. (14.19), we assumed that $\text{tr}[\boldsymbol{\mu}\mathbf{RKR}] \ll \text{tr}[\boldsymbol{\mu}\mathbf{R}]\xi_{\text{ex}}$. There, we remarked that on average this assumption becomes more accurate (valid) as the filter length, N, increases. In this problem, we examine the validity of the assumption for a few specific cases.

(i) Show that in the case where $\mathbf{R} = \mathbf{I}$ ($\mathbf{I}$ is the identity matrix) and $\boldsymbol{\mu} = \mu\mathbf{I}$, that is, a single scalar step-size parameter is used for all the taps,

$$\frac{\text{tr}[\boldsymbol{\mu}\mathbf{RKR}]}{\text{tr}[\boldsymbol{\mu}\mathbf{R}]\xi_{\text{ex}}} = \frac{1}{N} \tag{P14.1.1}$$

(ii) Consider the case where $\mathbf{R}$ is diagonal and the elements of $\boldsymbol{\mu}$ are chosen according to Eq. (14.4). Show that in this case also Eq. (P14.1.1) holds.

(iii) Consider the case where $\boldsymbol{\mu} = \mu\mathbf{I}$, but $\mathbf{R}$ and $\mathbf{G}$ are arbitrary correlation matrices. Use the decomposition $\mathbf{R} = \mathbf{Q\Lambda Q}^{\text{T}}$ (Eq. (4.19) of Chapter 4), to show that

$$\text{tr}[\boldsymbol{\mu}\mathbf{RKR}] = \mu \sum_{i=0}^{N-1} \lambda_i^2 k'_{ii}$$

and

$$\text{tr}[\boldsymbol{\mu}\mathbf{R}]\xi_{\text{ex}} = \mu \left(\sum_{i=0}^{N-1} \lambda_i \right) \left(\sum_{i=0}^{N-1} \lambda_i k'_{ii} \right)$$

where λ_i and k'_{ii} are the ith diagonal elements of the matrices $\boldsymbol{\Lambda}$ and $\mathbf{K}' = \mathbf{Q}^{\text{T}}\mathbf{KQ}$, respectively. Then, study the ratio

$$\frac{\text{tr}[\boldsymbol{\mu}\mathbf{RKR}]}{\text{tr}[\boldsymbol{\mu}\mathbf{R}]\xi_{\text{ex}}}$$

and find how the distribution of λ_i's and k'_{ii}'s affect this ratio.

P14.2 In the case of conventional LMS algorithm, that is, where $\boldsymbol{\mu} = \mu\mathbf{I}$ with μ being a scalar, show that

(i)

$$\xi_{\text{ex}} = \frac{1}{1 - \mu\text{tr}[\mathbf{R}]} \left(\mu\sigma_{e_o}^2 \text{tr}[\mathbf{R}] + \frac{1}{4}\mu^{-1}\text{tr}[\mathbf{G}] \right)$$

(ii) Assuming that $\mu\text{tr}[\mathbf{R}] \ll 1$ for the range of interest of μ, show that the optimum value of μ that minimizes ξ_{ex} is

$$\mu_o \approx \frac{1}{2\sigma_{e_o}} \sqrt{\frac{\text{tr}[\mathbf{G}]}{\text{tr}[\mathbf{R}]}}$$

(iii) Show that when $\mu = \mu_o$, the noise and lag misadjustments of the LMS algorithm are equal.

(iv) Show that the minimum value of ξ_{ex} is given by

$$\xi_{\text{ex,o}} \approx \sigma_{e_o} \sqrt{\text{tr}[\mathbf{R}]\text{tr}[\mathbf{G}]}$$

P14.3 A shortcoming of the result of the last problem is that when σ_{e_o} is very small (i.e., the plant noise is very small), the calculated optimum step-size parameter, μ_o, may become excessively large, resulting in an unstable LMS algorithm. Recalculate μ_o and $\xi_{\text{ex,o}}$ without imposing the condition $\mu\text{tr}[\mathbf{R}] \ll 1$ and show that the results are as follows:

$$\mu_o = \frac{\sqrt{\text{tr}[\mathbf{G}]/\text{tr}[\mathbf{R}]}}{\sqrt{\text{tr}[\mathbf{G}]\text{tr}[\mathbf{R}]} + \sqrt{\text{tr}[\mathbf{G}]\text{tr}[\mathbf{R}] + 4\sigma_{e_o}^2}}$$

and

$$\xi_{\text{ex,o}} = \frac{\left(\sqrt{\text{tr}[\mathbf{G}]\text{tr}[\mathbf{R}]} + \sqrt{\text{tr}[\mathbf{G}]\text{tr}[\mathbf{R}] + 4\sigma_{e_o}^2}\right)\sqrt{\text{tr}[\mathbf{G}]\text{tr}[\mathbf{R}]}}{2}$$

Simplify these results when the plant noise is zero.

P14.4 Show that the solution of Eq. (14.27) is Eq. (14.28).

P14.5 Give a detailed derivations of Eq. (14.29).

P14.6 Give a detailed derivation of Eq. (14.80).

P14.7 This problem aims at giving a derivation of the VSLMS algorithm for complex-valued case.

(i) Show that Eq. (14.82) can also be written as

$$\mu_i(n) = \mu_i(n-1) + \rho\Re[g_i(n)g_i^*(n-1)] \tag{P14.7.1}$$

where $\Re[x]$ denotes real part of x.

(ii) Following a similar line of derivation to those in Section 14.7.1, give a detailed derivation of Eq. (P14.7.1).

P14.8 Consider the Markovian process, $w(n)$, generated through the recursive equation

$$w(n) = \alpha w(n-1) + \nu(n)$$

where α is a parameter in the range of -1 to $+1$, and $\nu(n)$ is a stationary white noise.

(i) Define the sequence $\epsilon(n) = w(n) - w(n-1)$ and show that

$$\frac{\phi_{\epsilon\epsilon}(k)}{\phi_{\epsilon\epsilon}(0)} = \frac{\alpha - 1}{\alpha + 1}\alpha^{|k|-1}\sigma_\nu^2$$

where $\phi_{\epsilon\epsilon}(k)$ is the autocorrelation function of $\epsilon(n)$.

(ii) Use the result of (i) to argue that the results presented in this chapter are also valid (within a good approximation) when the tap weights of the plant in the modeling problem of Figure 14.1 vary according to a Markovian model with a parameter α smaller, but close to 1.

P14.9 Consider the LMS-Newton recursion

$$\mathbf{w}(n+1) = \mathbf{w}(n) + 2\mu\mathbf{R}^{-1}e(n)\mathbf{x}(n)$$

when applied for tracking in the modeling problem of Section 14.1.

(i) Show that

$$\xi_{\text{ex}} = \frac{1}{1-\mu N}\left(\mu\sigma_{e_o}^2 N + \frac{\mu^{-1}}{4}\text{tr}[\mathbf{RG}]\right)$$

(ii) Let μ_o denote the optimum value of μ, which results in minimum ξ_{ex}. Assuming that the plant changes very slowly so that $\mu_o N \ll 1$, show that

$$\mu_o = \frac{1}{2\sigma_{e_o}}\sqrt{\text{tr}[\mathbf{RG}]}$$

(iii) Obtain an approximate expression for $\xi_{\text{ex,o}}$, that is, minimum value of ξ_{ex}, and compare that with your results in Part (iv) of Problem P14.2.

(iv) Repeat Parts (ii) and (iii) for the case where the condition $\mu_o N \ll 1$ does not hold.

P14.10 Suggest a variable step-size implementation of the LMS-Newton algorithm of Table 11.7.

P14.11 Give a detailed derivation of Eq. (14.101).

Computer-Oriented Problems

P14.12 Consider a two-tap modeling problem with the parameters as specified in Case 3 of Section 14.5. Develop a simulation program for implementation of the LMS and TDLMS algorithms in this case. By running your program confirm that the predictions made through the numerical results of Table 14.1 are consistent with simulations. That is, the choice of $\theta = \pi/8$ results in the least steady-state MSE, and $\theta = \pi/4$ results in the maximum steady-state MSE.

Tip: To generate a random vector $\mathbf{x}(n)$ with a correlation matrix $\mathbf{R}$, you may proceed as follows. First, find a square matrix $\mathbf{L}$ that satisfies $\mathbf{R} = \mathbf{L}^{\text{T}}\mathbf{L}$. You may then generate $\mathbf{x}(n)$ according to the equation $\mathbf{x}(n) = \mathbf{L}^{\text{T}}\mathbf{u}(n)$, where $\mathbf{u}(n)$ is a random vector with the correlation matrix equal to the identity matrix. If you are using MATLAB, a convenient way of obtaining a square matrix $\mathbf{L}$ that satisfies $\mathbf{R} = \mathbf{L}^{\text{T}}\mathbf{L}$ is by using the function "**chol**" that finds Cholesky factorization of $\mathbf{R}$. The same method can be used to generate a random vector $\boldsymbol{\epsilon}_o(n)$ with a correlation matrix $\mathbf{G}$.

P14.13 The MATLAB simulation program used to generate the results of Figures 14.3–14.5 is available on an accompanying website. It is called "`vs_mdlg.m`." Experiment with this program and confirm the results of Figures 14.3–14.5. Try also other variations of the simulation parameters to gain a better understanding of the behavior of various implementations of the VSLMS algorithm.

P14.14 Develop a simulation program to study the convergence behavior of the VSLMS algorithm in the channel modeling application that was discussed in Section 14.7.4, when a common step-size parameter is used for all taps. Compare your results with those of Figures 14.3–14.5, and discuss your observations.

P14.15 Develop a simulation program to study the convergence behavior of the RLS algorithm with variable forgetting factor (Table 14.2) in the channel modeling application that was discussed in Section 14.7.4. Compare your results with those of Figures 14.3–14.5 and also your results in Problem P14.14. Discuss your observations.

15

Echo Cancellation

The presence of echoes in telephone line networks is an inevitable phenomenon that, unless removed, can be very annoying to the calling parties. Two types of echoes are encountered: (i) echoes arising from imbalanced hybrid bridges in the circuits, referred to as *hybrid echoes* (*HEs*), and (ii) echoes arising from acoustical echoes between loudspeakers/earphones and microphones, referred to as *acoustic echoes* (AEs). In Chapter 1, Section 1.6.4, a few introductory comments related to these two types of echo cancellers were presented. The purpose of this chapter is to dig into the details and discuss the various adaptive filtering algorithms that have been proved useful in the implementation of echo cancellers. Moreover, the chapter addresses the problems of double-talk and positive feedback in echo loops, both of which can have detrimental effects on the operation of echo cancellers. In addition, a large part of this chapter is devoted to the issue of stereophonic AE cancellation, which has its own theoretical and practical peculiarities that need special treatment.

15.1 The Problem Statement

Figure 15.1 presents a schematic diagram in which the various sources of echo in a telephone line network are identified. Hybrid bridges/circuits that act as an interface between the bidirectional two-wire lines and the single direction four-wire lines are spread in central switching offices as well as at the subscriber premises. HEs will exist (to some degree) at all the hybrid bridges. Long duration echoes can occur when an echo happens at a hybrid bridge and later reflects back from another hybrid bridge at some distance. This can lead to echoes with durations of as much as a few hundreds of milliseconds. However, the more common echoes have durations of less than 30 ms; reflections from very far hybrids are likely to be significantly attenuated and thus are usually negligible.

The acoustic link between the loudspeaker(s) and the microphone(s) in a teleconferencing setup results in echoes whose duration (depending on the room size and the furniture type in the room as well as the walls material) can stretch over a few hundreds of milliseconds, that is, an order of magnitude larger than the HEs.

We recall from Chapter 1 that echo cancellation may be cast as a modeling problem. However, there are a number of practical issues that make echo cancellation a nontrivial modeling problem. The long duration of the echo response translates into an adaptive

Adaptive Filters: Theory and Applications, Second Edition. Behrouz Farhang-Boroujeny.

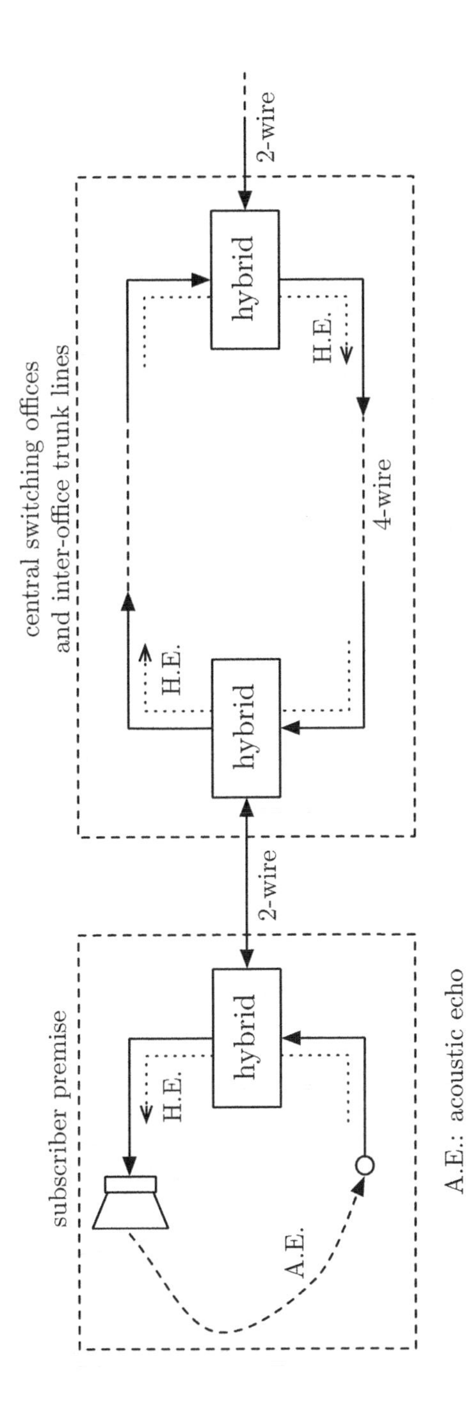

Figure 15.1 Schematic diagram presenting various sources of echo in a telephone line network.

filter with a relatively large number of tap weights. For example, if we assume that the signals have been sampled at a (minimum) rate of 8 kilo samples per second, a HE with a time span of 30 ms requires an adaptive filter with $30 \times 8 = 240$ taps to be modeled. This may increase by an order of magnitude for modeling an AE in a typical conference room. Clearly, the implementation of an adaptive filter with a large number of taps incurs a significant cost in computational complexity. Hence, adaptation algorithms such as variations of the least-squares methods are too demanding and may be unacceptable for the implementation of echo cancellers, particularly for the implementation of AE cancellers.

Another common problem of echo cancellers is that the underlying processes are speech signals. Speech signals are hard to deal with when applied as inputs to adaptive filters. Firstly, they are nonstationary processes with a very wide dynamic power range and a time-varying power spectral density. Secondly, their power spectral density over the band of interest varies significantly, and this clearly (Chapter 4) leads to a very wide eigenvalues spread of the correlation matrix of the input to the echo canceller (a transversal adaptive filter), and hence, can significantly affect its performance.

On the other hand, the residual power of the output error in an echo canceller can vary significantly, owing to the presence of the so-called *double-talk* intervals. The term double-talk, as its name applies, refers to the cases where both parties on the two sides of the telephone line simultaneously speak. In that case, the residual error, ideally, is equal to the speech signals from the near-end speaker(s) only. This concept, in the context of an AE canceller, is depicted in Figure 15.2. The incoming signal, $x(n)$, is the speech

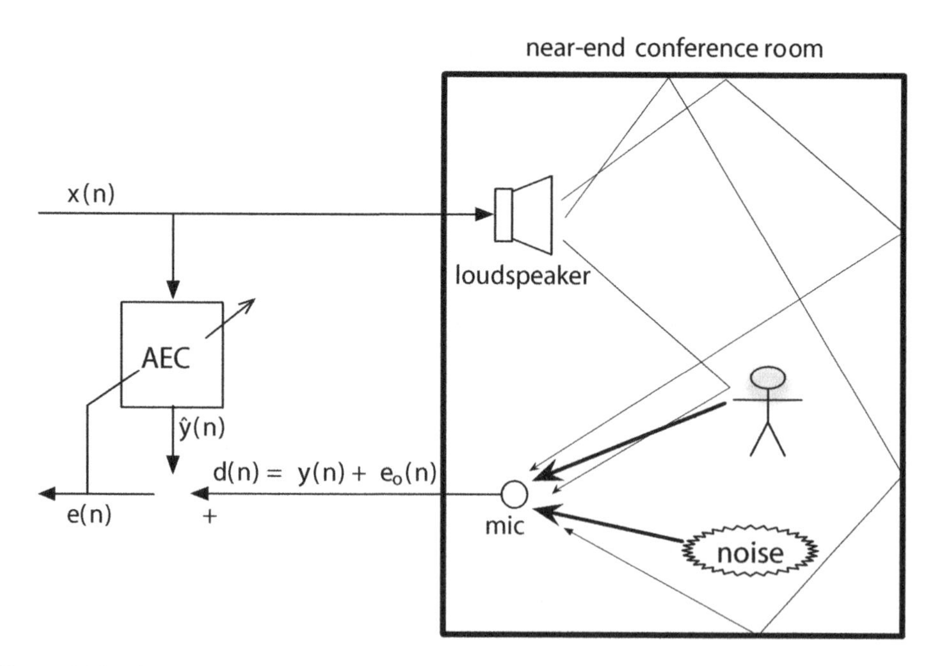

Figure 15.2 Schematic diagram presenting an acoustic echo cancellation setup and the problem of double-talk.

signal from a far-end room. The desired signal, $d(n)$, is the summation of the echo of the incoming signal, $y(n)$, and the other sounds activities within the near-end room, $e_o(n)$. The goal is to remove the echoed copies of $x(n)$, that is, $y(n)$, from $d(n)$ and transmit the residual signal $e(n) = e_o(n)$. Note that this reflects the remaining sound activities in the near-end room. Note that the notation $e_o(n)$ follows the notation of plant noise in a modeling problem, as appeared in many occasions throughout this book (Figure 6.5).

A transversal adaptive filter is used to model the echo paths and generate a replica of the echoed incoming signal in the near-end room. Obviously, when the coefficients of the adaptive filter are set perfectly, the output error $e(n) = e_o(n)$. However, we note that perfect setting of the coefficients of the adaptive filter is not possible, because any adaptive filter is bound to suffer from some misadjustment (or, equivalently, misalignment of its coefficients). Such misadjustment/misalignment increases with the size (say, the mean-squared value) of the residual error $e(n)$. In the absence of the double-talk, assuming that the background noise in the near-end room is small, $e(n)$ will be small and, hence, the adaptive filter (echo canceller) will suffer from a small misadjustment/misalignment. On the other hand, as soon as a double-talk period starts, $e(n)$ increases significantly and therefore, if the echo canceller adaptation continues (as before), its coefficients may be badly disturbed and hence $e(n)$ will be contaminated with a significant level of residual echo. In other words, the presence of double-talk can badly disturb the operation of the echo canceller. To avoid this problem, a practical echo canceller should be equipped with a mechanism that identifies the presence of double-talk and takes a necessary action to guarantee a correct operation of the system.

We note that although, here, we presented the problem of double-talk in the context of AE cancellers, the same problem exists in HE cancellers. Moreover, the solutions that will be discussed in Section 15.3 are applicable to both cases.

In Figure 15.1, the loop consisting of the AE and the HE can be the source of another problem in a teleconferencing setup. When the loop gain at a particular frequency, say, f_0, is greater than 1 and the total phase shift within the loop is some factor of 2π, the combination of AE and HE will make an unstable loop that begins oscillating with an increasing amplitude at the frequency f_0. This phenomenon, which also commonly occurs in any microphone-amplifier setup, is known as *howling*. Thus, AE cancellers should also be equipped with a howling suppression mechanism. The common methods that may be used for howling suppression are discussed in Section 15.4.

15.2 Structures and Adaptive Algorithms

The LMS (least-mean-square) algorithms, such as the normalized LMS and variable step-size LMS, unfortunately, may be too slow for adaptation of echo cancellers. This is because the input to an echo canceller is usually a speech signal that, as noted above, is nonstationary and highly colored. Throughout this book, we introduced a number of algorithms that were aimed for resolving the problem of slow convergence of LMS algorithm when it is subject to a colored/highly correlated input. In this section, we revisit these algorithms and provide some new comments that emphasize on their applications to echo cancellation. We also present numerical results that compare these algorithms in an AE cancellation setup. We exclude the recursive least-squares (RLS) and RLS lattice algorithms in our study, because we believe their relatively high complexity makes them a weak choice in the application of interest in this chapter.

15.2.1 Normalized LMS (NLMS) Algorithm

Recall from Chapter 6, Section 6.6, that in the NLMS (normalized least-mean-square) algorithm the filter tap weights are adjusted using the recursive equation

$$\mathbf{w}(n+1) = \mathbf{w}(n) + \frac{\tilde{\mu}}{\mathbf{x}^{\mathrm{T}}(n)\mathbf{x}(n) + \psi} e(n)\mathbf{x}(n) \tag{15.1}$$

where $\tilde{\mu}$ is the algorithm step-size parameter, and ψ is a small constant that is added to avoid possible numerical instability of algorithm when $\mathbf{x}^{\mathrm{T}}(n)\mathbf{x}(n)$ approaches 0. Also, here,

$$e(n) = d(n) - \hat{y}(n) \tag{15.2}$$

and

$$\hat{y}(n) = \mathbf{w}(n)\mathbf{x}(n) \tag{15.3}$$

where

$$\mathbf{w}(n) = [w_0(n) \;\; w_1(n) \;\; \cdots \;\; w_{N-1}(n)]^{\mathrm{T}} \tag{15.4}$$

and

$$\mathbf{x}(n) = [x(n) \;\; x(n-1) \cdots x(n-N+1)]^{\mathrm{T}} \tag{15.5}$$

Although for most of the studies in the previous chapters we considered mean-squared error (MSE) as a good measure of convergence of the various adaptive algorithms, one may find that an alternative measure of convergence that may prove useful, particularly in the case of AE cancellers, is the *misalignment* of the adaptive tap weights from their optimum values, defined as

$$\zeta(n) = E[\|\mathbf{v}(n)\|^2] \tag{15.6}$$

where $\|\mathbf{v}(n)\|^2 = \mathbf{v}^{\mathrm{T}}(n)\mathbf{v}(n)$, $\mathbf{v}(n) = \mathbf{w}(n) - \mathbf{w}_{\mathrm{o}}$, and $\mathbf{w}_{\mathrm{o}}$ is the optimum tap-weight vector of the filter that we strive to find. It turns out that finding an expression for the optimum value of $\tilde{\mu}$ that results in the fastest convergence of $\zeta(n)$ is rather straightforward.

Subtracting $\mathbf{w}_{\mathrm{o}}$ from both sides of Eq. (15.1) and letting $\psi = 0$, we obtain

$$\mathbf{v}(n+1) = \mathbf{v}(n) + \frac{\tilde{\mu}}{\mathbf{x}^{\mathrm{T}}(n)\mathbf{x}(n)} e(n)\mathbf{x}(n) \tag{15.7}$$

Assuming that $\mathbf{x}(n)$ and $\mathbf{v}(n)$ are known, premultiplying each side of Eq. (15.7) by its transpose and taking expectation, we obtain

$$E[\|\mathbf{v}(n+1)\|^2] = \|\mathbf{v}(n)\|^2 + \tilde{\mu}^2 \frac{E[e^2(n)]}{\mathbf{x}^{\mathrm{T}}(n)\mathbf{x}(n)} + 2\tilde{\mu} \frac{E[e(n)\mathbf{v}^{\mathrm{T}}(n)\mathbf{x}(n)]}{\mathbf{x}^{\mathrm{T}}(n)\mathbf{x}(n)} \tag{15.8}$$

Note that because we assume $\mathbf{x}(n)$ and $\mathbf{v}(n)$ are known, the expectation on the right-hand side is applied only to the terms containing the output error $e(n)$. We also note that in an echo canceller, $e(n)$ is the signal sent to the far-end room. Moreover, it may be written as

$$\begin{aligned} e(n) &= d(n) - \hat{y}(n) \\ &= e_o(n) + \mathbf{w}_o^T\mathbf{x}(n) - \mathbf{w}^T(n)\mathbf{x}(n) \\ &= e_o(n) - \mathbf{v}^T(n)\mathbf{x}(n) \end{aligned} \tag{15.9}$$

where $e_o(n)$ is the signal corresponding to the sound activities in the near-end room, excluding the far-end signal that is broadcast in the room through the loudspeaker. We also note that $\mathbf{v}^T(n)\mathbf{x}(n)$ is an undesirable residual echo. Assuming that $e_o(n)$ has a zero mean, and defining

$$r(n) = \mathbf{v}^T(n)\mathbf{x}(n) \tag{15.10}$$

Equation (15.8) reduces to

$$E[\|\mathbf{v}(n+1)\|^2] = \|\mathbf{v}(n)\|^2 + \tilde{\mu}^2\frac{E[e^2(n)]}{\mathbf{x}^T(n)\mathbf{x}(n)} - 2\tilde{\mu}\frac{r^2(n)}{\mathbf{x}^T(n)\mathbf{x}(n)} \tag{15.11}$$

The goal is to find the value of $\tilde{\mu}$ that minimizes $E[\|\mathbf{v}(n+1)\|^2]$. This, obviously, is obtained by solving the equation

$$\frac{\partial E[\|\mathbf{v}(n+1)\|^2]}{\partial \tilde{\mu}} = 0 \tag{15.12}$$

This leads to

$$\begin{aligned} \tilde{\mu}_{\rm opt}(n) &= \frac{r^2(n)}{E[e^2(n)]} \\ &= \frac{r^2(n)}{E[e_o^2(n)] + r^2(n)} \end{aligned} \tag{15.13}$$

where the second identity is obtained using Eq. (15.80) and recalling that $e_o(n)$ and $r(n) = \mathbf{v}^T(n)\mathbf{x}(n)$ are uncorrelated.

It is important to make some comments on the significance of the optimum step-size parameter (15.13).

- When $E[e_o^2(n)] = 0$, that is, when $e_o(n) = 0$ one finds that $\tilde{\mu}_{\rm opt}(n) = 1$. In other words, if one thinks of the echo canceller as a modeling problem just as the one in Figure 6.5, when there is no noise at the output of the plant, $\tilde{\mu}_{\rm opt}(n) = 1$ leads to an optimum performance of the NLMS algorithm.
- The optimum step-size parameter decreases as $e_o(n)$ increases in power.
- At the beginning of adaptation, when $\mathbf{w}(n) = 0$, $r^2(n)$ is generally large and thus, unless $E[e_o^2(n)]$ is also large, $\tilde{\mu}_{\rm opt}(n)$ will be approximately equal to 1.
- As the algorithm iterates, $r^2(n)$ reduces, on average, and thus, in the presence of some nonzero $E[e_o^2(n)]$, $\tilde{\mu}_{\rm opt}(n)$ decreases. This, clearly, is a sound strategy for reducing the filter misadjustment and/or misalignment.
- In case $\mathbf{w}_o$ changes, $r^2(n)$ increases and, hence, $\tilde{\mu}_{\rm opt}(n)$ will increase to track the change of $\mathbf{w}_o$.

Despite its appealing form, unfortunately, the implementation of Eq. (15.13) is not straightforward. This is because the quantities $r^2(n)$ and $E[e_o^2(n)]$ are not available. In order to resolve this problem and still come up with a solution that follows the same concept, we replace $r^2(n)$ in Eq. (15.13) by its expected value, $E[r^2(n)]$. Moreover, assuming that $e_o(n)$ and $r(n)$ are uncorrelated processes, one finds that

$$E[e^2(n]) = E[(e_o(n) - r(n))^2] = E[e_o^2(n)] + E[r^2(n)] \tag{15.14}$$

Considering the above point, we introduce

$$\tilde{\mu}_{\text{opt}}(n) \approx \frac{E[r^2(n)]}{E[e^2(n)]} \tag{15.15}$$

This result may be further rearranged as

$$\tilde{\mu}_{\text{opt}}(n) \approx 1 - \frac{E[e_o^2(n)]}{E[e^2(n)]} \tag{15.16}$$

Next, we recall that the signal at the microphone output is $d(n) = e_o(n) + y(n)$, where $y(n)$, as defined before, is the echo signal from the loudspeaker. Assuming that $e_o(n)$ and $y(n)$ are a pair of uncorrelated processes, and the echo canceller has converged, thus, the estimate $\hat{y}(n) \approx y(n)$, we obtain

$$E[e_o^2(n)] \approx E[d^2(n)] - E[\hat{y}^2(n)] \tag{15.17}$$

Substituting Eq. (15.17) in Eq. (15.16), we get

$$\tilde{\mu}_{\text{opt}}(n) \approx 1 - \frac{E[d^2(n)] - E[\hat{y}^2(n)]}{E[e^2(n)]} \tag{15.18}$$

Unfortunately, the approximation (15.18) works only when the echo canceller has converged and, thus, $y(n) \approx \hat{y}(n)$. It does not work, or may become an adverse algorithm when it has not converged and $y(n)$ and $\hat{y}(n)$ are significantly different. In particular, when the echo canceller starts with the initial tap weights 0, $\hat{y}(n) = 0$, $e(n) = d(n)$, and consequently $E[d^2(n)] = E[e^2(n)]$. This in turn implies $\tilde{\mu}_{\text{opt}}(n)$ given by Eq. (15.18) is 0; hence, the adaptation will install and the echo canceller never converges. To resolve this problem, we suggest using the following variable step-size parameter

$$\tilde{\mu}(n) = 1 - \eta \frac{E[\hat{y}^2(n)]}{E[d^2(n)]} \tag{15.19}$$

where $0 < \eta < 1$ is a design parameter. This, although may not be considered as an optimum choice, has the following appealing properties.

- At the start of the algorithm, when $\hat{y}(n) = 0$ (or approximately equal to 0), $\tilde{\mu}(n) = 1$ will result in the fastest convergence of the NLMS algorithm. This is in line with the comments made above for the optimum step-size parameter given by Eq. (15.13).

- As the algorithm converges and $\hat{y}(n)$ approaches $y(n)$, assuming that $e_o(n)$ is small, $d(n) \approx \hat{y}(n)$, hence, $E[\hat{y}^2(n)] \approx E[d^2(n)]$ and therefore $\tilde{\mu}(n)$ approaches $1-\eta$. This will allow the algorithm to converge to a misadjustment proportional to $1-\eta$. Hence, the design parameter η can be chosen to control the misadjustment of the algorithm as desired.

The estimates of $E[d^2(n)]$ and $E[\hat{y}^2(n)]$ can be obtained by taking the short-term averages of the most recent samples of $d(n)$ and $\hat{y}(n)$, respectively. For instance, the short-term averages may be calculated using the following recursive equations.

$$\sigma_{\mathrm{d}}^2(n) = \alpha\sigma_{\mathrm{d}}^2(n-1) + (1-\alpha)d^2(n) \tag{15.20}$$

$$\sigma_{\hat{y}}^2(n) = \alpha\sigma_{\hat{y}}^2(n-1) + (1-\alpha)\hat{y}^2(n) \tag{15.21}$$

Accordingly, we obtain

$$\tilde{\mu}(n) \approx 1 - \eta\frac{\sigma_{\hat{y}}^2(n)}{\sigma_{\mathrm{d}}^2(n)} \tag{15.22}$$

Owing to the errors in the estimates $\sigma_{\mathrm{d}}^2(n)$ and $\sigma_{\hat{y}}^2(n)$, $\tilde{\mu}_{\mathrm{opt}}(n)$ may become negative (when $\sigma_{\mathrm{d}}^2(n) < \eta\sigma_{\hat{y}}^2(n)$) and thus result in a divergence of the algorithm. To resolve this problem, one may choose to set $\tilde{\mu}(n)$ equal to 0, if Eq. (15.22) leads to a negative $\tilde{\mu}(n)$.

Table 15.1 summarizes the NLMS algorithm that was introduced in this chapter. We refer to this as variable step-size normalized least-mean-square (VSNLMS) algorithm, to differentiate it from its original version that was presented in Chapter 6, Section 6.6.

15.2.2 *Affine Projection LMS (APLMS) Algorithm*

In Chapter 6, we introduced the APLMS (affine projection least-mean-square) algorithm as a generalization to the NLMS algorithm. We also noted that the APLMS algorithm may be thought as an approximation to the LMS-Newton algorithm, and hence, has a superior convergence behavior over the NLMS algorithm. However, an APLMS algorithm may suffer from a larger misadjustment/misalignment than its NLMS counterpart. Irrespective of this potential problem, a number of researchers have considered the APLMS algorithm a better choice than the NLMS algorithm in the implementation of echo cancellers. Numerical examples that substantiate this claim are presented in Section 15.2.6.

To bring the APLMS algorithm on par with its VSNLMS counterpart, we also develop a variable step-size version of it. To this end, we recall the APLMS recursion (6.133), let $\psi = 0$, and note that using the definition $\mathbf{v}(n) = \mathbf{w}(n) - \mathbf{w}_o$, we obtain

$$\mathbf{v}(n+1) = \mathbf{v}(n) + \tilde{\mu}\mathbf{X}(n)(\mathbf{X}^{\mathrm{T}}(n)\mathbf{X}(n))^{-1}\mathbf{e}(n) \tag{15.23}$$

where, as defined in Chapter 6,

$$\mathbf{X}(n) = [\mathbf{x}(n)\ \ \mathbf{x}(n-1)\cdots\mathbf{x}(n-M+1)] \tag{15.24}$$

$$\mathbf{e}(n) = \mathbf{d}(n) - \mathbf{X}^{\mathrm{T}}(n)\mathbf{w}(n) \tag{15.25}$$

and

$$\mathbf{d}(n) = [d(n)\ \ d(n-1)\cdots d(n-M+1)]^{\mathrm{T}} \tag{15.26}$$

Recall that M is an integer parameter and usually $M \ll N$.

Table 15.1 Summary of the variable step-size normalized LMS (VSNLMS) algorithm for echo cancellation.

Input:	Tap-weight vector, $\mathbf{w}(n)$, input vector, $\mathbf{x}(n)$, desired output, $d(n)$, and power estimates $\sigma_{\mathrm{d}}^2(n-1)$, $\sigma_{\hat{y}}^2(n-1)$, and $\sigma_e^2(n-1)$
Output:	Tap-weight vector update, $\mathbf{w}(n+1)$, power estimates $\sigma_{\mathrm{d}}^2(n)$, $\sigma_{\hat{y}}^2(n)$, and $\sigma_e^2(n)$ The echo canceller output is $e(n)$.

1. Filtering:

$$\hat{y}(n) = \mathbf{w}^{\mathrm{T}}(n)\mathbf{x}(n)$$

2. Error estimation:

$$e(n) = d(n) - \hat{y}(n)$$

3. Step-size calculation:

$$\sigma_{\mathrm{d}}^2(n) = \alpha\sigma_{\mathrm{d}}^2(n-1) + (1-\alpha)d^2(n)$$

$$\sigma_{\hat{y}}^2(n) = \alpha\sigma_{\hat{y}}^2(n-1) + (1-\alpha)\hat{y}^2(n)$$

$$\sigma_e^2(n) = \alpha\sigma_e^2(n-1) + (1-\alpha)e^2(n)$$

$$\tilde{\mu}(n) = 1 - \frac{\sigma_{\mathrm{d}}^2(n) - \sigma_{\hat{y}}^2(n)}{\sigma_e^2(n)}$$

if $\tilde{\mu}(n) > 1$, let $\tilde{\mu}(n) = 1$

if $\tilde{\mu}(n) < 0$, let $\tilde{\mu}(n) = 0$

4. Tap-weight vector adaptation:

$$\mathbf{w}(n+1) = \mathbf{w}(n) + \frac{\tilde{\mu}(n)}{\mathbf{x}^{\mathrm{T}}(n)\mathbf{x}(n) + \psi}e(n)\mathbf{x}(n)$$

Following the same lines of argument that led to the derivation of Eq. (15.11) from Eq. (15.8), we obtain from Eq. (15.23)

$$E[\|\mathbf{v}(n+1)\|^2] = \|\mathbf{v}(n)\|^2 + \mu^2 E[\mathbf{e}^{\mathrm{T}}(n)(\mathbf{X}^{\mathrm{T}}(n)\mathbf{X}(n))^{-1}\mathbf{e}(n)] - 2\mu\mathbf{r}^{\mathrm{T}}(n)(\mathbf{X}^{\mathrm{T}}(n)\mathbf{X}(n))^{-1}\mathbf{r}(n) \tag{15.27}$$

where $\mathbf{r}(n) = \mathbf{X}^{\mathrm{T}}(n)\mathbf{v}(n)$ is the vector of residual error, an extension of Eq. (15.10).

Now solving Eq. (15.12), for the present case, we obtain

$$\mu_{\mathrm{opt}}(n) = \frac{\mathbf{r}^{\mathrm{T}}(n)(\mathbf{X}^{\mathrm{T}}(n)\mathbf{X}(n))^{-1}\mathbf{r}(n)}{E[\mathbf{e}^{\mathrm{T}}(n)(\mathbf{X}^{\mathrm{T}}(n)\mathbf{X}(n))^{-1}\mathbf{e}(n)]} \tag{15.28}$$

This is somewhat similar to the first line in Eq. (15.13); however, it has a more complex form because of the presence of the matrix $(\mathbf{X}^{\mathrm{T}}(n)\mathbf{X}(n))^{-1}$. Nevertheless, the comments made following Eq. (15.13) are also applicable here. Noting this and the fact that a direct implementation of Eq. (15.28) is not possible in practice, we propose to use Eq. (15.22) for

Table 15.2 Summary of the variable step-size APLMS (VSAPLMS) algorithm for echo cancellation.

Input: Tap-weight vector, $\mathbf{w}(n)$,
input vector, $\mathbf{x}(n)$, Desired output, $d(n)$,
and power estimates $\sigma_{\mathrm{d}}^2(n-1)$, $\sigma_{\hat{y}}^2(n-1)$, and $\sigma_e^2(n-1)$

Output: Tap-weight vector update, $\mathbf{w}(n+1)$,
power estimates $\sigma_{\mathrm{d}}^2(n)$, $\sigma_{\hat{y}}^2(n)$, and $\sigma_e^2(n)$
The echo canceller output is $e(n)$.

1. Filtering:

$$\hat{\mathbf{y}}(n) = \mathbf{X}^{\mathrm{T}}(n)\mathbf{v}(n),\ \hat{y}(n) = \text{the first element of } \hat{\mathbf{y}}(n)$$

2. Error estimation:

$$\mathbf{e}(n) = \mathbf{d}(n) - \hat{\mathbf{y}}(n),\ e(n) = \text{the first element of } \mathbf{e}(n)$$

3. Step-size calculation:

$$\sigma_{\mathrm{d}}^2(n) = \alpha\sigma_{\mathrm{d}}^2(n-1) + (1-\alpha)d^2(n)$$

$$\sigma_{\hat{y}}^2(n) = \alpha\sigma_{\hat{y}}^2(n-1) + (1-\alpha)\hat{y}^2(n)$$

$$\sigma_e^2(n) = \alpha\sigma_e^2(n-1) + (1-\alpha)e^2(n)$$

$$\tilde{\mu}(n) = 1 - \frac{\sigma_{\mathrm{d}}^2(n) - \sigma_{\hat{y}}^2(n)}{\sigma_e^2(n)}$$

if $\tilde{\mu}(n) > 1$, let $\tilde{\mu}(n) = 1$

if $\tilde{\mu}(n) < 0$, let $\tilde{\mu}(n) = 0$

4. Tap-weight vector adaptation:

$$\mathbf{w}(n+1) = \mathbf{w}(n) + \tilde{\mu}(n)\mathbf{X}(n)(\mathbf{X}^{\mathrm{T}}(n)\mathbf{X}(n) + \psi\mathbf{I})^{-1}\mathbf{e}(n)$$

the APLMS algorithm as well and thus summarize the variable step-size affine projection least mean square (VSAPLMS) algorithm as in Table 15.2.

15.2.3 Frequency Domain Block LMS Algorithm

The frequency domain/fast block least mean square (FBLMS) algorithm was introduced in Chapter 8. It was noted that the step-normalization at different frequency bins will resolve the problem of eigenvalue spread and thus results in a fast converging algorithm with a single mode of convergence. We also noted that for very long adaptive filters, such as those of interest in this chapter, the FBLMS algorithm may introduce an unacceptably long latency at the output; that is, the output samples are generated with a significant delay. The partitioned frequency domain/fast block least mean square (PFBLMS) was thus proposed as a method of reducing the latency, while no noticeable degradation in performance could be observed.

Here, to cope with the nonstationary nature of the speech signals, we consider a version of the PFBLMS algorithm that follows the NLMS algorithm formulation for the step-normalization at each frequency bin. We also adopt the method that was developed for the NLMS to control the step-size parameter of the PFBLMS algorithm. A summary of this version of PFBLMS algorithm, which we call variable step-size normalized partitioned frequency domain/fast block least mean square (VSNPFBLMS) algorithm, is presented in Table 15.3.

15.2.4 Subband LMS Algorithm

The subband LMS algorithm is another form of the LMS algorithm that has been proposed as an effective method of implementing high performance AE cancellers. As discussed in Chapter 9, to resolve and equalize the various modes of convergence of the LMS algorithm, the input signal, $x(n)$, and the desired output signal, $d(n)$, are partitioned into a number of narrowband signals. The narrowband partitions from $x(n)$ and $d(n)$ are matched using a set of subband adaptive filters. The output of these filters are combined through a synthesis filter bank to construct a full-band (filtered) signal that matches $d(n)$.

We also recall from Chapter 9 that there are two possible structures for the subband adaptive filters: the *synthesis-independent* structure, Figure 9.7, and the *synthesis-dependent* structure, Figure 9.8. The latter may provide a more accurate output; however, the delay in the adaption loop results in a less stable and also a slower adaptation algorithm (see the discussion in Section 9.3). Because of these problems, the synthesis-dependent subband adaptive filters are less popular. For echo cancellers also, we limit our attention to the synthesis-independent structure.

We also note that in the case of echo cancellers, the output signal of interest is the error signal $e(n)$. Noting this, the synthesis-independent structure of Figure 9.7 is modified as in Figure 15.3. The subband AE cancellation results that are presented in Section 15.2.6 are those of this structure. The tap weights of the subband adaptive filters $\overline{W}_0(z)$ through $\overline{W}_{M-1}(z)$ are adjusted using the NLMS algorithm using a common, and possibly time-varying, step-size parameter $\tilde{\mu}(n)$.

15.2.5 LMS-Newton Algorithm

The last modification to the LMS algorithm that was presented in the previous chapters was an approximation to the LMS-Newton algorithm based on autoregressive modeling of the input signal $x(n)$. In Section 11.15, two version of this approximation were presented: Algorithm 1 and Algorithm 2. We also noted that although Algorithm 1 was a better approximation to the LMS-Newton algorithm, Algorithm 2 was more appealing because of its simple structure.

We note that the use of the approximate LMS-Newton algorithms of Section 11.15 in the application of echo cancellers is very appealing. Firstly, the vast studies of speech signals in the past have proved the fact that autoregressive modeling is an excellent fit to speech signals. Moreover, a model order of 8 to 12 (typically, 10) is commonly used for speech signal in the celebrated linear predictive coding systems. Secondly, the number of taps in typical AE cancellers is in excess of 1000. These imply that linear modeling part constitute a negligible portion of the complexity of the approximate LMS-Newton algorithms. Hence, the proposed LMS-Newton algorithms have a complexity that is only marginally higher than that of the conventional LMS/NLMS algorithm.

Table 15.3 Summary of the variable step-size normalized PFBLMS (VSNPFBLMS) algorithm for echo cancellation.

Input: Tap-weight vectors, $\mathbf{w}_{\mathcal{F},l}(k)$, $l = 0, 1, \ldots, P-1$,
extended input vector,
$\tilde{\mathbf{X}}_0(k) = [x(kL-M)\ x(kL-M+1) \cdots x(kL+L-1)]^{\mathrm{T}}$,
the past frequency domain vectors of input, $\mathbf{x}_{\mathcal{F},0}(k-l)$, for $l = 1, 2, \ldots, (P-1)p$,
and the desired output vector,
$\mathbf{d}(k) = [d(kL)\ d(kL+1) \cdots d(kL+L-1)]^{\mathrm{T}}$.

Output: Tap-weight vector update, $\mathbf{w}_{\mathcal{F},l}(k+1)$, $l = 0, 1, \ldots, P-1$,
power estimates $\sigma_{\mathrm{d}}^2(k)$, $\sigma_{\hat{y}}^2(k)$, and $\sigma_e^2(k)$
The echo canceller output is $e(n)$.

1. Filtering:

$$\mathbf{x}_{\mathcal{F},0}(k) = \mathrm{FFT}(\tilde{\mathbf{X}}_0(k))$$

$$\hat{\mathbf{y}}(k) = \text{the last } L \text{ elements of IFFT}\left(\sum_{l=0}^{P-1} \mathbf{w}_{\mathcal{F},l}(k) \odot \mathbf{x}_{\mathcal{F},0}(k-pl)\right)$$

2. Error estimation:

$$\mathbf{e}(k) = \mathbf{d}(k) - \hat{\mathbf{y}}(k)$$

3. Step-size calculation:

$$\sigma_{\mathrm{d}}^2(k) = \alpha\sigma_{\mathrm{d}}^2(k-1) + (1-\alpha)(\mathbf{d}^{\mathrm{H}}(k)\mathbf{d}(k))$$

$$\sigma_{\hat{y}}^2(k) = \alpha\sigma_{\hat{y}}^2(k-1) + (1-\alpha)(\hat{\mathbf{y}}^{\mathrm{H}}(k)\hat{\mathbf{y}}(k))$$

$$\sigma_e^2(k) = \alpha\sigma_e^2(k-1) + (1-\alpha)(\mathbf{e}^{\mathrm{H}}(n)\mathbf{e}(k))$$

$$\tilde{\mu}(k) = 1 - \frac{\sigma_{\mathrm{d}}^2(k) - \sigma_{\hat{y}}^2(k)}{\sigma_e^2(k)}$$

if $\tilde{\mu}(k) > 1$, let $\tilde{\mu}(k) = 1$
if $\tilde{\mu}(k) < 0$, let $\tilde{\mu}(k) = 0$
for $i = 0$ to $M' - 1$

$$\mu_i(k) = \tilde{\mu}(k) / \sum_{l=0}^{P-1} x_{\mathcal{F},0,i}^2(k-pl)$$

$$\boldsymbol{\mu}(k) = [\mu_0(k)\ \mu_1(k) \cdots \mu_{M'-1}(k)]^{\mathrm{T}}$$

4. Tap-weight adaptation:

$$\mathbf{e}_{\mathcal{F}}(k) = \mathrm{FFT}\left(\begin{bmatrix} \mathbf{0} \\ \mathbf{e}(k) \end{bmatrix}\right)$$

for $l = 0$ to $P-1$

$$\mathbf{w}_{\mathcal{F},l}(k+1) = \mathbf{w}_{\mathcal{F},l}(k) + 2\boldsymbol{\mu}(k) \odot \mathbf{x}_{\mathcal{F},0}^*(k-pl) \odot \mathbf{e}_{\mathcal{F}}(k)$$

5. Tap-weight constraint:

for $l = 0$ to $P-1$

$$\mathbf{w}_{\mathcal{F},l}(k+1) = \mathrm{FFT}\left(\begin{bmatrix} \text{first } M \text{ elements of IFFT}(\mathbf{w}_{\mathcal{F},l}(k+1)) \\ \mathbf{0} \end{bmatrix}\right)$$

Notes:

- M: partition length; L: block length; $M' = M + L$.
- $\mathbf{0}$ denotes column zero vectors with appropriate length to extend vectors to the length of M'.
- $\odot$ denotes elementwise multiplication of vectors.
- Step 5 is applicable only for the constrained PFBLMS algorithm.

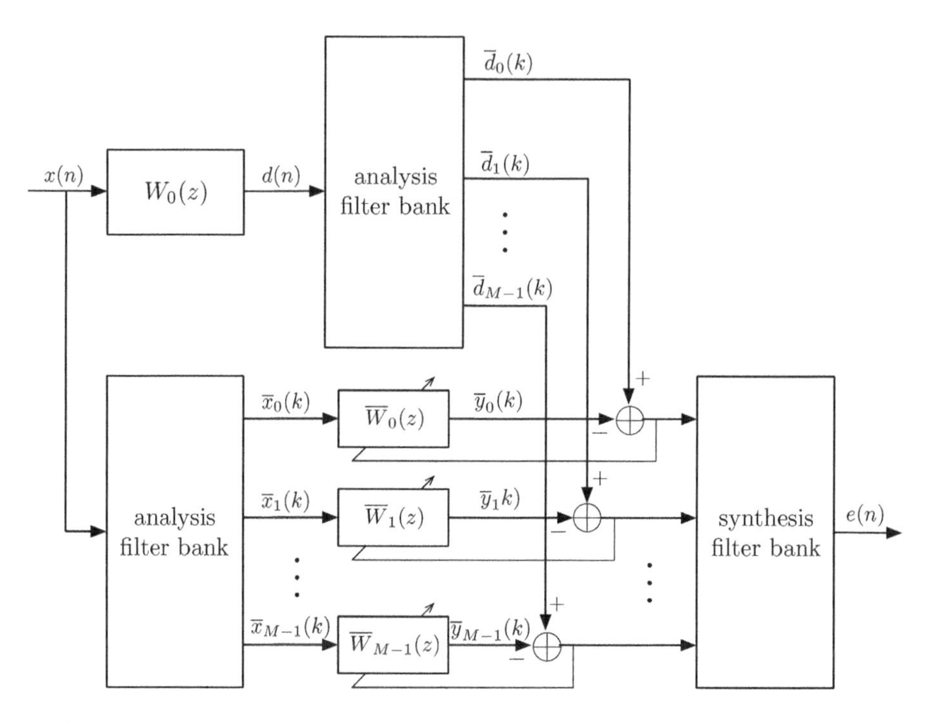

Figure 15.3 Schematic diagram of a subband synthesis-independent echo canceller.

Here, because of its simple structure, we limit our discussion to Algorithm 2 of Section 11.15. Also, to take care of the nonstationary nature of the speech signals, we develop a normalized version of the algorithm. Referring back to Chapter 11, Table 11.7, we recall that the tap-weight update equation

$$\mathbf{w}(n+1) = \mathbf{w}(n) + 2\mu e(n)\mathbf{u}_{\mathrm{a}}(n) \tag{15.29}$$

where as discussed in Section 11.15, $\mathbf{u}_{\mathrm{a}}(n)$ is an approximation to $\mathbf{R}^{-1}\mathbf{x}(n-M)$. Following the same line of derivation as those that led to Eq. (6.119), we begin with defining the *a posteriori* error

$$e^{+}(n) = d(n) - \mathbf{w}^{\mathrm{T}}(n+1)\mathbf{x}(n-M) \tag{15.30}$$

Substituting Eq. (15.29) in Eq. (15.30), replacing μ by $\mu(n)$, and rearranging, we obtain

$$e^{+}(n) = (1 - 2\mu(n)\mathbf{u}_{\mathrm{a}}^{\mathrm{T}}(n)\mathbf{x}(n-M))e(n) \tag{15.31}$$

Next, as in the case of the NLMS algorithm, minimizing $(e^{+}(n))^2$ with respect to $\mu(n)$ results in

$$\mu(n) = \frac{1}{2\mathbf{u}_{\mathrm{a}}^{\mathrm{T}}(n)\mathbf{x}(n-M)} \tag{15.32}$$

This forces $e^{+}(n)$ to 0. Substituting Eq. (15.32) in Eq. (15.29), we obtain

$$\mathbf{w}(n+1) = \mathbf{w}(n) + \frac{1}{\mathbf{u}_{\mathrm{a}}^{\mathrm{T}}(n)\mathbf{x}(n-M)} e(n)\mathbf{u}_{\mathrm{a}}(n) \tag{15.33}$$

This is the dual of the NLMS algorithm of Eq. (6.107). Following the same argument as in Section 6.6, one may replace Eq. (15.33) by its regularized form

$$\mathbf{w}(n+1) = \mathbf{w}(n) + \frac{\tilde{\mu}}{\mathbf{u}_\mathrm{a}^\mathrm{T}(n)\mathbf{x}(n-M) + \psi} e(n)\mathbf{u}_\mathrm{a}(n) \tag{15.34}$$

where $\tilde{\mu}$ is a step-size parameter, taking values in the range of greater than 0 and less than or equal to 1, and ψ is a small positive constant that avoids the algorithm numerical instability when $\mathbf{u}_\mathrm{a}^\mathrm{T}(n)\mathbf{x}(n-M)$ approaches 0.

We note that considering the identity $\mathbf{u}_\mathrm{a}^\mathrm{T}(n)\mathbf{x}(n-M) = \mathbf{x}^\mathrm{T}(n-M)\mathbf{u}_\mathrm{a}(n)$ and using the approximation $\mathbf{u}_\mathrm{a}(n) \approx \mathbf{R}^{-1}\mathbf{x}(n-M)$, we obtain

$$\mathbf{u}_\mathrm{a}^\mathrm{T}(n)\mathbf{x}(n-M) \approx \mathbf{x}^\mathrm{T}(n-M)\mathbf{R}^{-1}\mathbf{x}(n-M) \tag{15.35}$$

Since $\mathbf{R}$ and, hence, $\mathbf{R}^{-1}$ is a positive definite matrix, $\mathbf{u}_\mathrm{a}^\mathrm{T}(n)\mathbf{x}(n-M)$ is most likely a positive number. However, given that Eq. (15.35) is an approximation, still there is a low chance that $\mathbf{u}_\mathrm{a}^\mathrm{T}(n)\mathbf{x}(n-M)$ be a negative number and thus the algorithm takes a step in a wrong direction. To avoid this problem, one may choose not to update $\mathbf{w}(n)$ when $\mathbf{u}_\mathrm{a}^\mathrm{T}(n)\mathbf{x}(n-M)$ is negative. Also, one may choose to adjust $\mu(n)$ as in the NLMS/APLMS algorithm.

Considering the above changes/modifications, the variable step-size NLMS-Newton algorithm that we suggest for echo cancellation is listed in Table 15.4.

15.2.6 Numerical Results

In this section, we present some numerical results to evaluate and compare the relative performance of the various echo cancellation algorithms that were presented in the previous sections. The experiments are for the AE cancellation setup of Figure 15.2. Although our study will concentrate on an AE canceller setup, the conclusions derived are equally applicable to HE cancellers as well.

As the far-end signal $x(n)$, we have selected eight pieces of speech signals from the National Public Radio (NPR) in United States, and from the British Broadcasting Corporation (BBC) Radio in United Kingdom. Male and female speakers are selected to make sure that there are some diversity in the speech samples. These sampled signals are available on the accompanying website, as MATLAB .mat files. They are named "`speech1.mat`" through "`speech8.mat`." The sampling rate is set at 8820 Hz. For the results presented in this section, we have used "`speech1.mat`." However, the reader may repeat the experiments with other speech signals. The MATLAB codes for all the echo cancellers algorithms are available on the accompanying website.

The near-end room echo response is generated randomly by first choosing the random column vector $\mathbf{g}$ consisting of the elements

$$g_n = (n + N_0)^{-2}\eta(n), \quad n = 0, 1, \ldots, N-1 \tag{15.36}$$

where $\eta(n)$ is a Gaussian random sequence with unit variance and independent samples. The vector $\mathbf{g}$ is then normalized to obtain the desired echo response vector

$$\mathbf{h} = \frac{\mathbf{g}}{\sqrt{\mathbf{g}^\mathrm{T}\mathbf{g}}} \tag{15.37}$$

Table 15.4 A variable step-size NLMS-Newton algorithm for echo cancellers.

Given: Parameter vectors $\boldsymbol{\kappa}(n) = [\kappa_1(n)\ \kappa_2(n) \cdots \kappa_{M-1}(n)]^{\mathrm{T}}$
and $\mathbf{w}(n) = [w_0(n) w_1(n) \cdots w_{N-1}(n)]^{\mathrm{T}}$,
data vectors $\mathbf{x}(n)$, $\mathbf{b}(n-1)$, and $\mathbf{u}_{\mathrm{a}}(n-1)$, desired output $d(n)$,
and power estimates $P_0(n-1), P_1(n-1), \ldots, P_M(n-1)$

Required: Vector updates $\boldsymbol{\kappa}(n)$, $\mathbf{w}(n+1)$, $\mathbf{b}(n)$, and $\mathbf{u}_{\mathrm{a}}(n)$,
and power estimate updates $P_0(n), P_1(n), \ldots, P_M(n)$
The echo canceller output is $e(n)$.

⋆⋆⋆ Lattice Predictor Part ⋆⋆⋆

$f_0(n) = b_0(n) = x(n)$
$P_0(n) = \beta P_0(n-1) + 0.5(1-\beta)[f_0^2(n) + b_0^2(n-1)]$
for $m = 1$ to M
$f_m(n) = f_{m-1}(n) - \kappa_m(n) b_{m-1}(n-1)$
$b_m(n) = b_{m-1}(n-1) - \kappa_m(n) f_{m-1}(n)$

$$\kappa_m(n+1) = \kappa_m(n) + \frac{2\mu_{p,o}}{P_{m-1}(n) + \epsilon}[f_{m-1}(n) b_m(n) + b_{m-1}(n-1) f_m(n)]$$

$P_m(n) = \beta P_m(n-1) + 0.5(1-\beta)[f_m^2(n) + b_m^2(n-1)]$
if $|\kappa_m(n)| > \gamma$, $\kappa_m(n) = \kappa_m(n-1)$
end

⋆⋆⋆ $\mathbf{u}_{\mathrm{a}}(n)$ update ⋆⋆⋆

$u_{\mathrm{a}}(n-j) = u_{\mathrm{a}}(n-j+1)$, for $j = N-1, N-2, \ldots, 2$
$f_0'(n) = b_0'(n) = b_M(n)$
for $m = 1$ to $M-1$
$f_m'(n) = f_{m-1}'(n) - \kappa_m(n) b_{m-1}'(n-1)$
$b_m'(n) = b_{m-1}'(n-1) - \kappa_m(n) f_{m-1}'(n)$
end
$u_{\mathrm{a}}(n) = (P_M(n) + \epsilon)^{-1}(f_{M-1}'(n) - \kappa_M(n) b_{M-1}'(n-1))$

⋆⋆⋆ Filtering and tap-weight vector adaptation ⋆⋆⋆

1. Filtering:
$$\hat{y}(n) = \mathbf{w}^{\mathrm{T}}(n)\mathbf{x}(n-M)$$
2. Error estimation:
$$e(n) = d(n-M) - \hat{y}(n)$$
3. Step-size calculation: $\sigma_{\mathrm{d}}^2(n) = \alpha\sigma_{\mathrm{d}}^2(n-1) + (1-\alpha)d^2(n)$
$$\sigma_{\hat{y}}^2(n) = \alpha\sigma_{\hat{y}}^2(n-1) + (1-\alpha)\hat{y}^2(n)$$
$$\sigma_e^2(n) = \alpha\sigma_e^2(n-1) + (1-\alpha)e^2(n)$$
$$\tilde{\mu}(n) = 1 - \frac{\sigma_{\mathrm{d}}^2(n) - \sigma_{\hat{y}}^2(n)}{\sigma_e^2(n)}$$
if $\tilde{\mu}(n) > 1$, let $\tilde{\mu}(n) = 1$
if $\tilde{\mu}(n) < 0$, let $\tilde{\mu}(n) = 0$
4. Tap-weight vector adaptation:
$$\mathbf{w}(n+1) = \mathbf{w}(n) + \frac{\tilde{\mu}(n)}{\mathbf{u}_{\mathrm{a}}^{\mathrm{T}}(n)\mathbf{x}(n-M) + \psi} e(n)\mathbf{u}_{\mathrm{a}}(n)$$
(ignore this step if $\mathbf{u}_{\mathrm{a}}^{\mathrm{T}}(n)\mathbf{x}(n-M) < 0$)

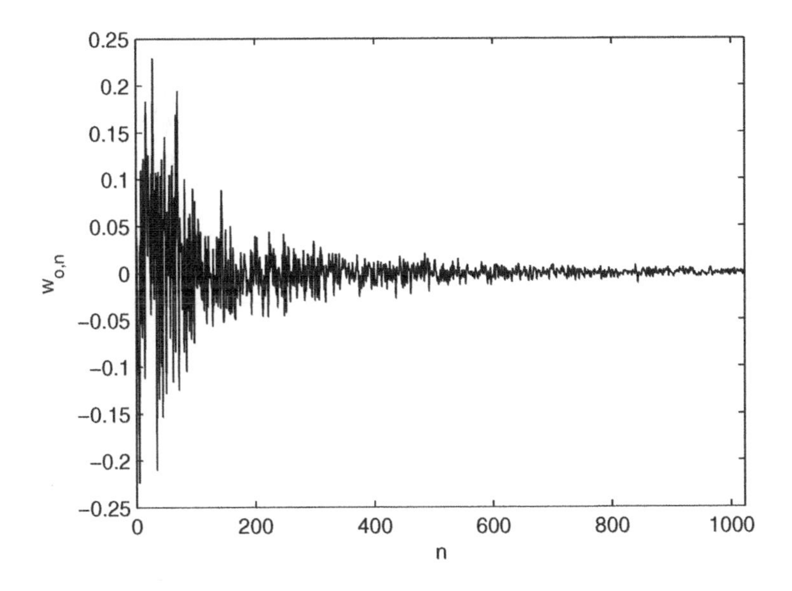

Figure 15.4 A sample example of the echo response h_n.

For the results presented in this section, we have chosen $N_0 = 150$ and $N = 1024$. The echoed signal is thus given by

$$y(n) = h_n \star x(n) \tag{15.38}$$

where h_n's are the elements of $\mathbf{h}$ and $\star$ denotes convolution. A sample example of the echo response h_n is presented in Figure 15.4. We have used this response for generation of all the results in this section.

We use the echo return loss enhancement (ERLE), defined as

$$\text{ERLE} = 10\ \log_{10}\left\{\frac{E[y^2(n)]}{E[r^2(n)]}\right\} \tag{15.39}$$

to evaluate the performance of the various adaptive algorithms. Recall that $r(n) = y(n) - \hat{y}(n)$ is the residual echo within the error/transmit signal $e(n)$. Ideally, the ERLE should approach infinity as the estimated echo $\hat{y}(n)$ approaches the true echo $y(n)$. However, the nonzero value of $e(n)$, in the steady state, leads to a perturbation of the echo canceller tap weights around their steady-state value, and hence, does not allow the ERLE to grow indefinitely. Recall that in the context of adaptive filters theory, the perturbation of the tap weights in the steady state is measured by misadjustment, $\mathcal{M}$, which is defined as the ratio of excess MSE over the minimum mean-squared error (MMSE). Furthermore, we note that the presence of a double-talk may increase the MMSE significantly; thus, in the presence of a double-talk, the perturbation of the echo canceller tap weights can be very large. This, in turn, may reduce the ERLE significantly.

Figure 15.5 presents a set of ERLE plots that compare the various adaptive algorithms. These results have been collected using the MATLAB codes available on the

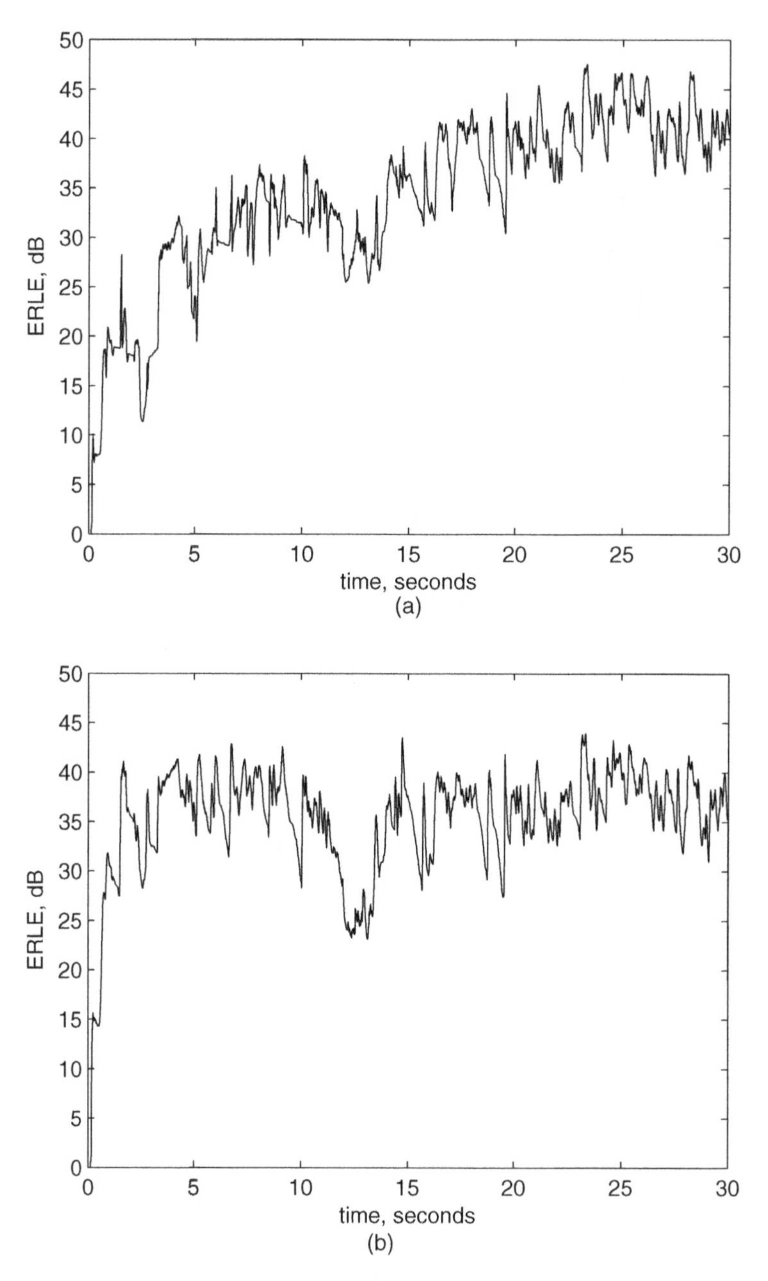

Figure 15.5 Echo return loss enhancement (ERLE) of the various algorithms: (a) NLMS, (b) APLMS, (c) FBLMS, (d) SBLMS, and (e) LMS-Newton.

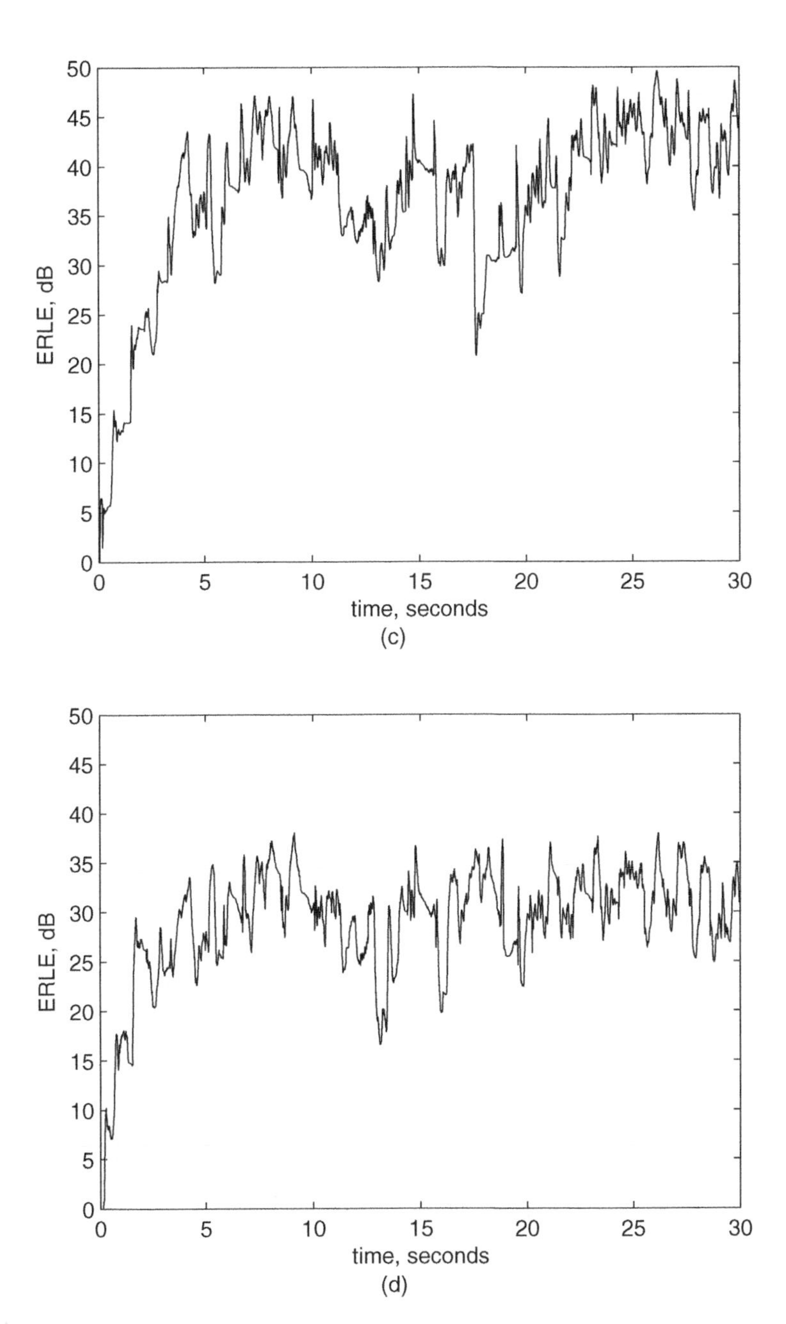

Figure 15.5 *(continued)*

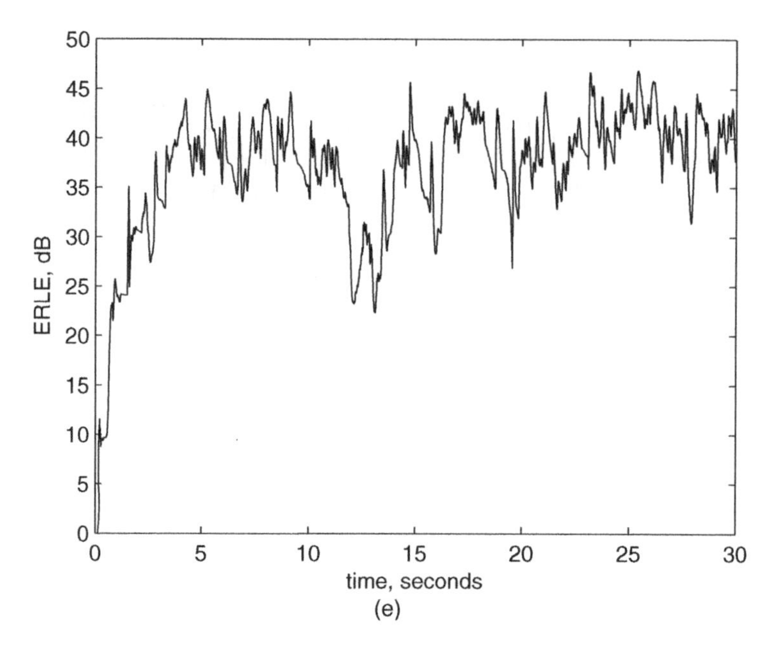

Figure 15.5 (*continued*)

Table 15.5 The parameters of the various algorithms that have been used for the results presented in Figure 15.5.

Algorithm	$\tilde{\mu}$	ψ	N	M	J	L	α_a	α_s	K_a	K_s	N_a	N_s	β	γ	ϵ	$\mu_{p,o}$
NLMS	0.5	0.1	1024	–	–	–	–	–	–	–	–	–	–	–	–	–
APLMS	0.5	0.1	1024	5	–	–	–	–	–	–	–	–	–	–	–	–
FBLMS	0.5	0.1	1024	64	–	64	–	–	–	–	–	–	–	–	–	–
SBLMS	0.5	0.1	1024	32	19	–	3/16	7/19	5	3	191	193	–	–	–	–
LMS-Newton	0.5	0.1	1024	8	–	–	–	–	–	–	–	–	0.98	0.9	0.05	0.001

accompanying website (www.wiley.com/go/adaptive_filters). For the results presented in Figure 15.5, the far-end signal is "`speech1.mat`." Here, we have kept the variable step-size part of the algorithms inactive; that is, we have used a fixed step-size parameter $\tilde{\mu}$. The parameters of the various algorithms that have been used for the results presented in Figure 15.5 are listed in Table 15.5. There is no near-end (double-talk) signal. However, a background Gaussian noise with a standard deviation of $\sigma_o = 0.001$ has been added at the microphone. This is about 45 dB below the near-end echoed signal that reaches the microphone. Also, as an alternative way of observing the performance of the various algorithms, the error (transmit) signal $e(n)$ produced by the examined algorithms, along with the microphone signal (i.e., before the echo cancellation), are presented in Figure 15.6.

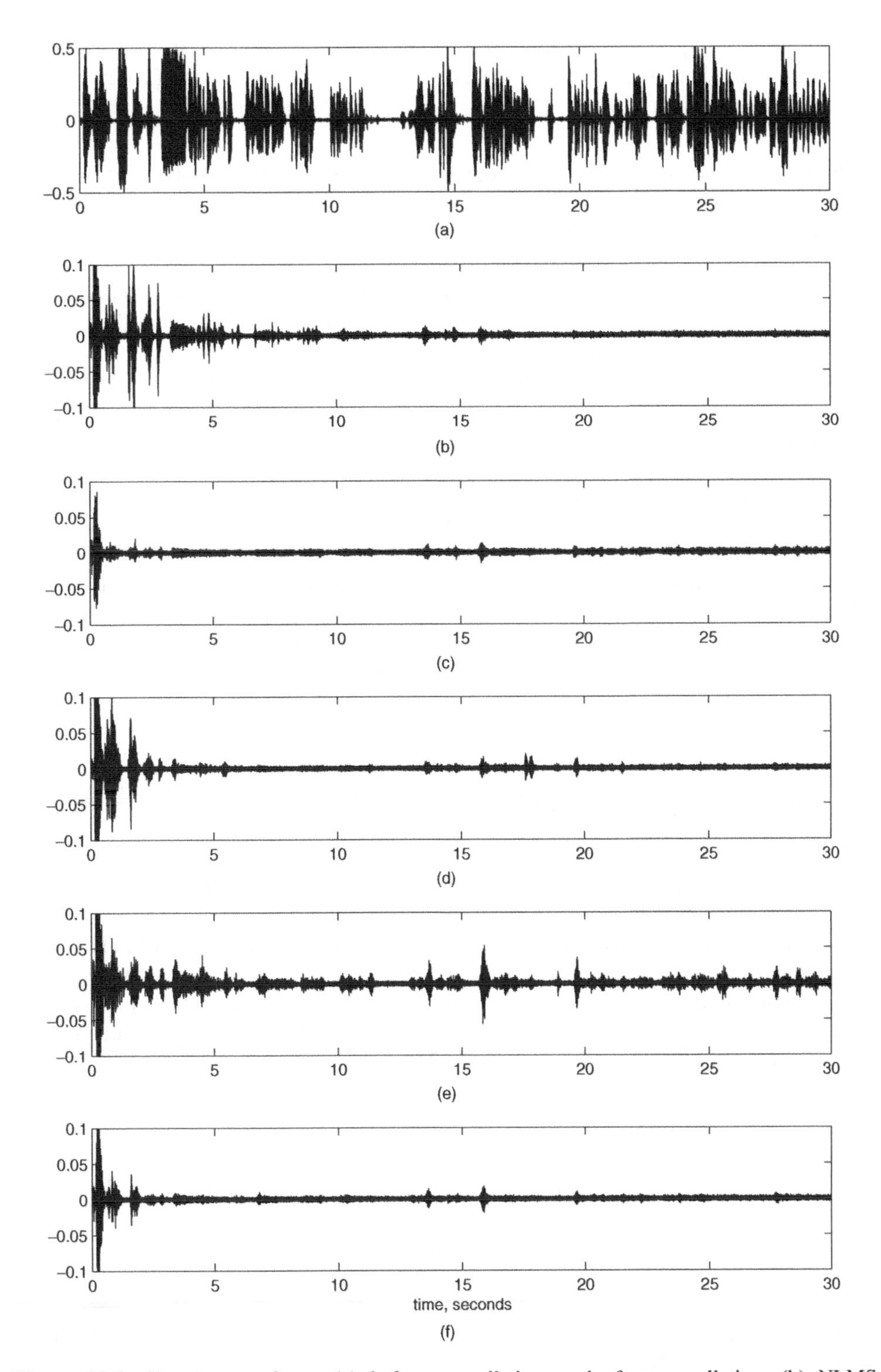

Figure 15.6 Signal wave forms (a) before cancellation, and after cancellation: (b) NLMS, (c) APLMS, (d) FBLMS, (e) SBLMS, and (f) LMS-Newton. Note that the vertical scale (amplitude) in (a) is different from the rest.

From the results presented in Figures 15.5 and 15.6, and further tests that have been performed with other samples of speech signals as well as other choices of the parameters of the various algorithms, the following observations have been made.

- The APLMS has the fastest rate of convergence. This may be understood, if we recall that the APLMS algorithm lies between the conventional LMS algorithm and the RLS algorithm. Its performance approaches that of the RLS algorithm, as the parameter M approaches the filter length N. However, because of the reasons discussed in Chapter 6, the APLMS algorithm suffers from a relatively high misadjustment. This drawback of the APLMS algorithm is reflected in its ERLE, which is lower than those of the NLMS, LMS-Newton, and FBLMS algorithm.
- The LMS-Newton algorithm, on the other hand, although has similar or better convergence compared to the rest of the algorithms, achieves the highest level of ERLE. Moreover, a subjective test that may be performed by listening to the error signal $e(n)$ confirms the fact that the LMS-Newton has a superior performance over all the algorithms that are tested here.
- The subband-LMS algorithm appears to be the most inferior adaptation algorithm in terms of the achievable ERLE. It has given an ERLE that is about 5 to 10 dB below the rest of the algorithms. This excessive loss in performance is due to imperfect filters that have been used for subband analysis and synthesis. The performance of the algorithm could be improved by using better filters; however, such improvement will be at a cost of an excessive latency in the echo canceller output that may be unacceptable.
- The presence of some slow modes of convergence in the NLMS algorithm is very evident.

15.3 Double-Talk Detection

Double-talk detection algorithms fall under the general category of *correlation/coherence-based* schemes. In this section, we present the general idea behind this class of algorithms and provide the details of an implementation of them. This presentation follows the results that were first reported by in Gänsler *et al.* (1996). The work presented in Benesty *et al.* (2000) gives a different point of view of the same concept. The additional study performed here reveals that a practically appealing implementation of the correlation/coherence-based schemes lead to a double-talk detection strategy that is closely related to the variable step-size NLMS that was developed in Section 15.2.1.

15.3.1 Coherence Function

The *coherence function* between a pair of random processes $u(n)$ and $v(n)$ is defined as

$$c_{vu}(e^{j\omega}) = \frac{|\Phi_{vu}(e^{j\omega})|^2}{\Phi_{uu}(e^{j\omega})\Phi_{vv}(e^{j\omega})} \tag{15.40}$$

where $\Phi_{uu}(e^{j\omega})$, $\Phi_{vv}(e^{j\omega})$, and $\Phi_{vu}(e^{j\omega})$ are the power and cross-power spectral density functions defined in Chapter 2.

When $u(n)$ and $v(n)$ are related through a linear time invariant system (e.g., $v(n)$ is the output of a system with input $u(n)$ and transfer function $H(e^{j\omega})$), using the results presented in Chapter 2, one will find that

$$\Phi_{vu}(e^{j\omega}) = \Phi_{uu}(e^{j\omega})H(e^{j\omega}) \tag{15.41}$$

and

$$\Phi_{vv}(e^{j\omega}) = \Phi_{uu}(e^{j\omega})|H(e^{j\omega})|^2 \tag{15.42}$$

Substituting Eqs. (15.41) and (15.42) in Eq. (15.40) and simplifying the result, we will find that

$$c_{vu}(e^{j\omega}) = 1 \tag{15.43}$$

This identity may be thought as a way of reflecting the fact that $u(n)$ an $v(n)$ are coherently related through a linear time invariant transfer function. In that case, we say there is a *full coherence* between $u(n)$ and $v(n)$.

15.3.2 Double-Talk Detection Using the Coherence Function

Consider the echo cancellation setup of Figure 15.2. The microphone signal $d(n)$ consists of two components: (i) the echo signal $y(n)$, obtained from passing the far-end signal $x(n)$ through the near-end room acoustic response (a linear system), and (ii) the summation of noise and double-talk, lumped together as $e_o(n)$. In the absence of double-talk, $e_o(n)$ may be negligible and thus one may argue that there will be an almost full coherence between $d(n)$ and $x(n)$. Hence,

$$c_{dx}(e^{j\omega}) \approx 1, \quad \text{in the absence of double-talk} \tag{15.44}$$

In the presence of double-talk, $\Phi_{dd}(e^{j\omega})$ increases and, thus, $c_{dx}(e^{j\omega})$ may drop significantly below 1.

In Gänsler *et al.* (1996), the authors have noted that a high-performance spectrum analysis technique should be used to evaluate $c_{dx}(e^{j\omega})$ at a number of discrete frequencies and accordingly form a decision variable

$$\mathcal{D}_1 = \frac{1}{L}\sum_{i=0}^{L-1} \hat{c}_{dx}(\mathrm{e}^{j\omega_i}) \tag{15.45}$$

where $\hat{c}_{dx}(\mathrm{e}^{j\omega_i})$ is the estimate of $c_{dx}(\mathrm{e}^{j\omega})$ at $\omega = \omega_i$, and ω_i, for $0 \leq i < L$, are the set of discrete frequencies. If $\mathcal{D}_1$ is greater than a certain threshold, it is assumed that there is no double-talk; otherwise, it is assumed that a double-talk exists.

15.3.3 Numerical Evaluation of the Coherence Function

In Section 3.6.2, as a side discussion, we presented a numerical method for calculating the power spectral density of a process and its cross-power spectral density with another process. This was done through a filtering step that extracts narrowband portions of the underlying processes in the vicinity of $\omega = \omega_i$. The extracted signal portions are then auto and/or cross-correlated to compute the desired power and cross-power spectral densities (Figure 3.7). These steps, when combined with an additional processing step to obtain a sample value of $\hat{c}_{dx}(\mathrm{e}^{j\omega})$ at $\omega = \omega_i$, are presented in Figure 15.7. The first blocks on the left-hand side of the figure are narrowband filters centered around $\omega = \omega_i$. The blocks $E[\cdot]$ denoted the expected values. However, in practice, it is assumed that the involved processes are ergodic and thus the statistical averages are replaced by time averages. The combiner box takes $\Phi_{dd}(\mathrm{e}^{j\omega_i})$, $\Phi_{dx}(\mathrm{e}^{j\omega_i})$, and $\Phi_{xx}(\mathrm{e}^{j\omega_i})$ as inputs and calculates

$$c_{dx}(\mathrm{e}^{j\omega_i}) = \frac{|\Phi_{dx}(\mathrm{e}^{j\omega_i})|^2}{\Phi_{xx}(\mathrm{e}^{j\omega_i})\Phi_{dd}(\mathrm{e}^{j\omega_i})} \tag{15.46}$$

An effective implementation of Eq. (15.46) requires averaging over sufficient samples of $|d_i(n)|^2$, $d_i(n)x_i^*(n)$, and $|x_i(n)|^2$, through the $E[\cdot]$ boxes. Also, for the averages to be effective, the samples from each of the processes $|d_i(n)|^2$, $d_i(n)x_i^*(n)$, and $|x_i(n)|^2$ should be independent of one another. Moreover, for an echo canceller to react fast enough to the appearance of a double talk, the signal samples $d_i(n)$ and $x_i(n)$ should be obtained from a minimum number of samples of $d(n)$ and $x(n)$, respectively. This is a classical spectrum analysis problem for which the best solution is the so-called *multitaper spectral estimator* of Thomson (1982). Here, in order to avoid diverting too much into the theory of the multitaper method, we only briefly discuss the basic ideas behind it and to provide some guidelines present a few numerical results. The derivation and details of the multitaper method are deferred to Appendix 15A at the end of this chapter.

The multitaper method is effectively a filter-bank-based signal analysis tool. Figure 15.8 presents the block diagram of a filter-bank-based spectral estimator for a signal $x(n)$. The filters $H_0(e^{j\omega})$, $H_1(e^{j\omega})$, ..., $H_{L-1}(e^{j\omega})$ make a bank of filters that decomposes the input signal $x(n)$ into L narrowband signals $x_0(n)$, $x_1(n)$, ..., $x_{L-1}(n)$; see Figure 15.9. For $i = 0, 1, \ldots, L-1$, the respective power estimator takes a time-average of $|x_i(n)|^2$, as an estimate of the signal power over the ith band of the filter bank. Normalizing the power estimates with respect to the width of the bands results in estimates of the power spectral density of $x(n)$, $\Phi_{xx}(\mathrm{e}^{j\omega_i})$, for $i = 0, 1, \ldots, L-1$.

The filter bank structure of Figure 15.8 consists of L parallel filters that may be centered at frequencies $\omega_i = \frac{2\pi i}{L}$, $i = 0, 1, \ldots, L-1$. This choice of the frequencies ω_i will allow an efficient implementation of the L filters in a polyphase structure. The polyphase structure, which is also introduced in Appendix 15A, has a complexity which is that of one of the L filters and one fast Fourier transform (FFT) of size L. The polyphase structure is developed based on the fact that the set of filters $H_0(e^{j\omega})$, $H_1(e^{j\omega})$, ..., $H_{L-1}(e^{j\omega})$ all originate from the same filter, call *prototype filter*. Here, the prototype filter is $H_0(e^{j\omega})$, and the rest of the filters are obtained from this filter through a modulation (i.e., a shift of the filter frequency response across the frequency axis).

Another feature of the multitaper method, from which the name multitaper has originated from, is that for each instant of time, n, a set of samples of $x_i(n)$, for each i, are generated based on a set of prototype filters. To differentiate these prototype filters and their respective subband signals, we refer to them as $H_0^{(k)}(e^{j\omega})$ and $x_i^{(k)}(n)$, for $k = 1, 2, \ldots, K$, where K is the number of prototype filters. Hence, the power averages are made across time, at the time instants $n_1, n_2, \ldots, n_P$, the samples of the power spectral

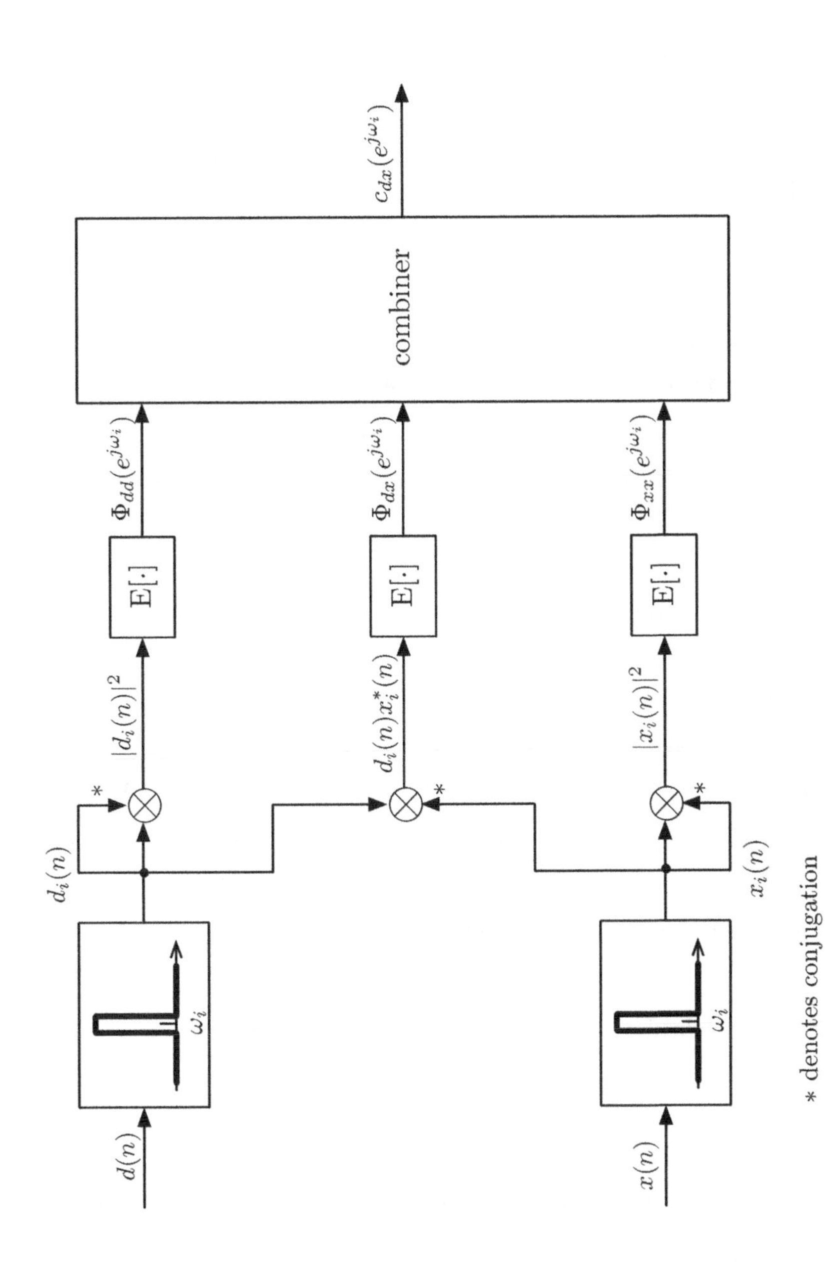

* denotes conjugation

Figure 15.7 Procedure for calculation the coherence function $\hat{c}_{dx}(e^{j\omega})$ at $\omega = \omega_i$.

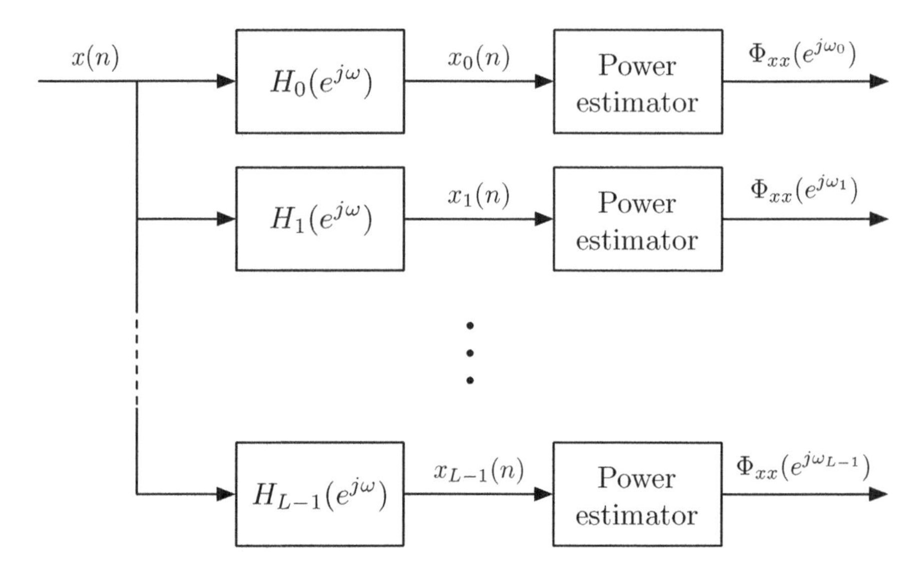

Figure 15.8 Filter-bank-based spectrum analyzer.

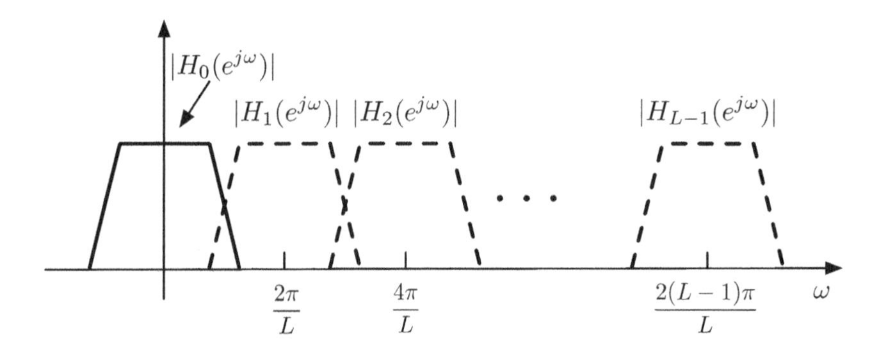

Figure 15.9 Presentation of the filters in a filter bank spectrum analyzer.

density of $x(n)$ is obtained as

$$\Phi_{xx}(e^{j\omega_i}) = \frac{q}{KP}\sum_{k=1}^{K}\sum_{p=1}^{P}|x_i^{(k)}(n_p)|^2 \tag{15.47}$$

where q is a normalization constant related to the bandwidth of prototype filter in the filter bank.

Similarly, we obtain

$$\Phi_{dd}(e^{j\omega_i}) = \frac{q}{KP}\sum_{k=1}^{K}\sum_{p=1}^{P}|d_i^{(k)}(n_p)|^2 \tag{15.48}$$

and

$$\Phi_{dx}(\mathrm{e}^{j\omega_i}) = \frac{q}{KP}\sum_{k=1}^{K}\sum_{p=1}^{P} d_i^{(k)}(n_p)x_i^{(k)*}(n_p) \tag{15.49}$$

Finally, Eqs. (15.47), (15.48), and (15.49) are substituted in Eq. (15.46) to calculate $c_{dx}(\mathrm{e}^{j\omega_i})$. We also note that after substituting Eqs. (15.47), (15.48), and (15.49) in Eq. (15.46), the normalizing factor q will cancel and hence its value become irrelevant.

15.3.4 Power-Based Double-Talk Detectors

Here, we derive an alternative decision number that proves instructive in relating the coherence function, defined in this section, with the variable step-size algorithms that were introduced in the various subsections of Section 15.2. Substituting $u(n)$ by $x(n)$, $v(n)$ by $d(n) = y(n) + e_\mathrm{o}(n)$, and $H(e^{j\omega})$ by $W_\mathrm{o}(e^{j\omega})$ in Eqs. (15.40), (15.41), and (15.42), and rearranging the result, we obtain

$$\begin{aligned} c_{dx}(e^{j\omega}) &= \frac{\Phi_{xx}(e^{j\omega})|W_\mathrm{o}(e^{j\omega})|^2}{\Phi_{dd}(e^{j\omega})} \\ &= \frac{\Phi_{yy}(e^{j\omega})}{\Phi_{dd}(e^{j\omega})} \end{aligned} \tag{15.50}$$

In the absence of double-talk, $\Phi_{dd}(e^{j\omega}) = \Phi_{yy}(e^{j\omega})$ and thus $c_{dx}(e^{j\omega}) = 1$. On the other hand, in the presence of double-talk, $\Phi_{dd}(e^{j\omega}) = \Phi_{yy}(e^{j\omega}) + \Phi_{e_\mathrm{o}e_\mathrm{o}}(e^{j\omega}) > \Phi_{yy}(e^{j\omega})$ and thus $c_{dx}(e^{j\omega}) < 1$. Hence, as $\Phi_{yy}(e^{j\omega})$ and $\Phi_{dd}(e^{j\omega})$ are real and positive functions of frequency ω, we may define

$$\mathcal{D}_2 = \frac{\frac{1}{2\pi}\int_{-\pi}^{\pi}\Phi_{yy}(e^{j\omega})\mathrm{d}\omega}{\frac{1}{2\pi}\int_{-\pi}^{\pi}\Phi_{dd}(e^{j\omega})\mathrm{d}\omega} \tag{15.51}$$

as an alternative decision number, with similar properties as $\mathcal{D}_1$. Namely, $\mathcal{D}_2 \approx 1$ in the absence of double-talk, and $\mathcal{D}_2$ is significantly smaller than 1 in the presence of double-talk. Moreover, using the Parseval's relation (2.60), Eq. (15.51) may be written as

$$\mathcal{D}_2 = \frac{E[y^2(n)]}{E[d^2(n)]} \tag{15.52}$$

Since $y(n)$ is not available, it may appear that $\mathcal{D}_2$, as given by Eq. (15.53), is a useless decision number. However, noting that upon the convergence of the echo canceller $\hat{y}(n) \approx y(n)$, one may adopt an estimate of $\mathcal{D}_2$ given by

$$\hat{\mathcal{D}}_2 = \frac{E[\hat{y}^2(n)]}{E[d^2(n)]} \tag{15.53}$$

Moreover, $E[d^2(n)]$ and $E[\hat{y}^2(n)]$ can be replaced by their estimates $\sigma_{\hat{\mathrm{y}}}^2(n)$ and $\sigma_{\mathrm{d}}^2(n)$ that may be obtained recursively using Eqs. (15.20) and (15.21), respectively.

The decision number (15.53) may be used as follows. If $\hat{\mathcal{D}}_2$ is greater than a threshold γ, set the step-size parameter μ (or $\tilde{\mu}$) to a large value, for example, set $\tilde{\mu} = 1$. If $\hat{\mathcal{D}}_2 \leq \gamma$, set the step-size parameter μ (or $\tilde{\mu}$) to a small value (possibly, 0).

On the other hand, referring to Eq. (15.19), we may find that the variable step-size LMS algorithms that were introduced in the various subsections of Section 15.2 also use $\hat{\mathcal{D}}_2$, however, in an opposite way. When $E[\hat{y}^2(n)]/E[d^2(n)]$ is small, $\tilde{\mu}(n)$ is set to a value close to 1 (a large step-size). Moreover, when $E[\hat{y}^2(n)]/E[d^2(n)]$ is close to 1, $\tilde{\mu}(n)$ converges to $1 - \eta$ (a smaller step-size). This clearly is not what one wishes to do after the convergence of the echo canceller and in the presence of the double-talk.

Considering the above discussion, one may propose the following procedure for dealing with double-talks. The step-size parameter $\tilde{\mu}(n)$ is set according to Eq. (15.19) for the first few seconds, after the activation of the echo canceller, and switches to the following setting afterwards:

$$\tilde{\mu}(n) = \begin{cases} \tilde{\mu}_1 & \text{if } \hat{\mathcal{D}}_2 > \gamma \\ \tilde{\mu}_2 & \text{if } \hat{\mathcal{D}}_2 \leq \gamma \end{cases} \tag{15.54}$$

where $\tilde{\mu}_1$ is for the case when there is no double talk, and $\tilde{\mu}_2$ is a small (possibly 0) step-size.

We note that the computation of the decision number $\hat{\mathcal{D}}_2$ is much simpler than $\mathcal{D}_1$. However, the use of $\hat{\mathcal{D}}_2$ can be problematic, if the echo canceller, because of any reason, diverges away from its optimum setting. A divergence of the echo canceller from its optimum setting can result in a value of $\hat{\mathcal{D}}_2$ smaller than γ, which, if Eq. (15.54) is used, results in a selection $\tilde{\mu}(n) = \tilde{\mu}_2$. This, in turn, can lead to a stalling or a very slow convergence of the algorithm. The decision number $\mathcal{D}_1$, on the other hand, is independent of the state of the echo canceller and, thus, never faces the latter problem.

15.3.5 Numerical Results

To develop a more in-depth understanding of the double-talk detector mechanisms that were discussed above, we present the results of an experiment. In this experiment, we start with a speech signal with a duration of 80 s. This is treated as the signal $x(n)$ is the AE cancellation setup of Figure 15.2. This is passed through a room acoustic with the same response as the one in Figure 15.4. A double-talk speech signal is then added in the intervals of (15 to 30), (45 to 50), and (60 to 70) s to form $d(n)$. It is assumed that there is no noise in the room.

To evaluate the decision number $\mathcal{D}_2$, we assume that the echo canceller is perfectly matched to the room acoustic, and hence $\hat{y}(n) = y(n)$. The mean-squared values of $d(n)$ and $\hat{y}(n)$ are then calculated based on Eqs. (15.20) and (15.21), respectively, with $\alpha = 0.999$.

The decision number $\mathcal{D}_1$ is evaluated by implementing multiple copies of the block diagram shown in Figure 15.13, using K multitaper prototype filters. The quantities $|d_i(n)|^2$, $d_i(n)x_i^*(n)$, and $|x_i(n)|^2$ are averaged across different prototype filters and across the time. To implement these efficiently, polyphase filter banks are adopted (see Appendix 15A, for the details). The number of subbands in the filter banks is set equal to $L = 256$. The length of each prototype filter is KL. With $K = 8$, this results in a length $KL = 2048$. The

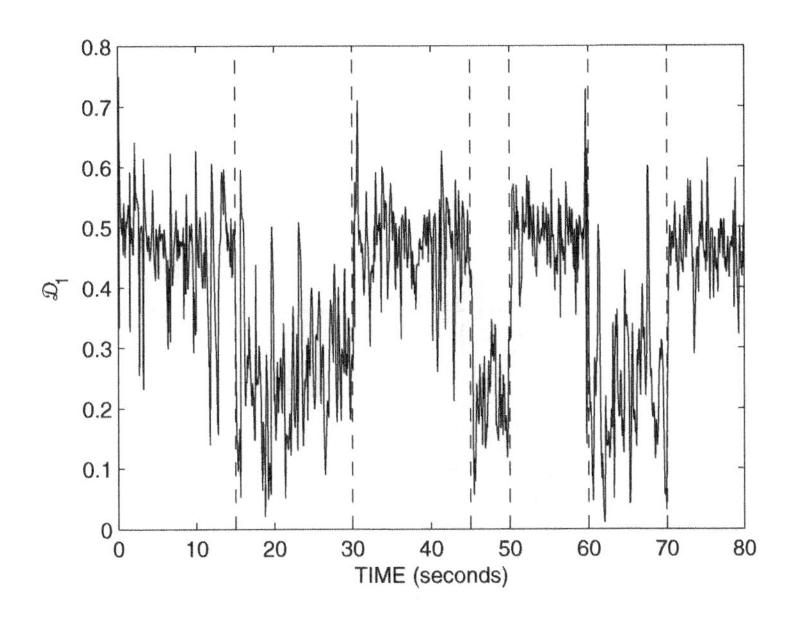

Figure 15.10 Variation of the decision number $\mathcal{D}_1$ in a double-talk detector. $L = 256$.

MATLAB program used to generate the results that are presented in Figures 15.10 and 15.11 is available on the accompanying website. It is called "`DTDexperiment.m`."

Both results in Figures 15.10 and 15.11 supply some information that may be considered as sufficient in conveying the presence of the double-talk intervals – the intervals between each pair of the vertical dashed lines. The decision number $\mathcal{D}_2$ has more distinct edges showing the transition into and out of each double-talk periods. The decision number $\mathcal{D}_1$, on the other hand, more clearly switched to a lower level during each double-talk period.

From the discussion above (and also by going through the MATLAB program "`DTDexperiment.m`"), it is obvious that the generation of $\mathcal{D}_1$ requires a lot more computational complexity than the generation of $\mathcal{D}_2$. However, to compute $\mathcal{D}_2$, the assumption is made that the echo canceller has already converged. Unfortunately, this may not be always true. Hence, in practice, the decision number $\mathcal{D}_2$ is less reliable than $\mathcal{D}_1$. In other words, we pay a very high price (in terms of computational complexity) to implement a more reliable double-talk detector.

A reader may recall from Eq. (15.46) that in the absence of double-talk, the coherence function $c_{dx}(e^{j\omega})$ should be equal to 1, for any ω; thus $\mathcal{D}_1$ should be also equal to 1. This is not the case in Figure 15.10. This difference from the theory is related to the fact that the width of the narrow-band filters used to generate $d_i(n)$ and $x_i(n)$ is not sufficiently narrow. An interested reader may run the MATLAB program "`DTDexperiment.m`," available on the accompanying website, with a larger value of the parameter L, to see how $\mathcal{D}_1$ improves. However, we should also note that increasing L results in further delay in the detection of a double-talk. This is because as L increases, the length of the filters in the filter banks increases and this in turn increases the filter bank (group) delay. Therefore, the choice of L is a compromise that should be made by the system designer. Figure 15.12 presents the result of repeating the experiment that lead to Figure 15.10, with

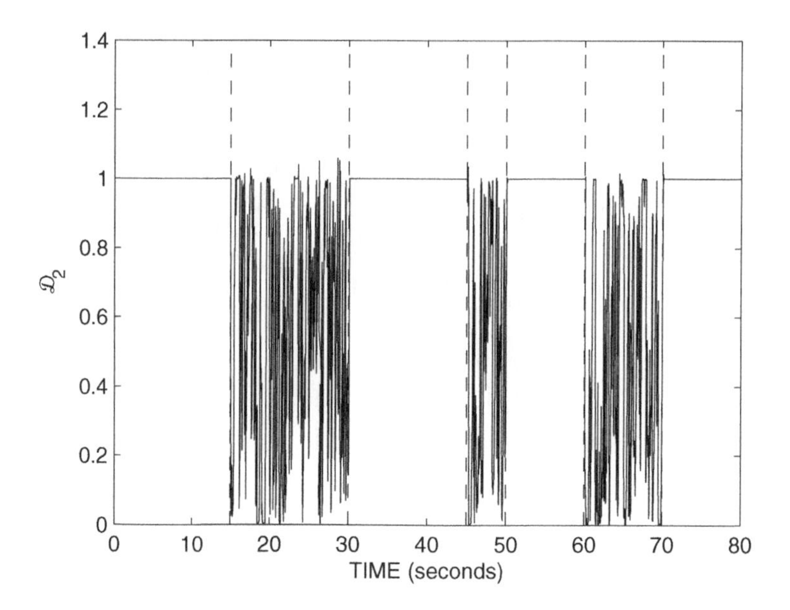

Figure 15.11 Variation of the decision number $\mathcal{D}_2$ is a double-talk detector.

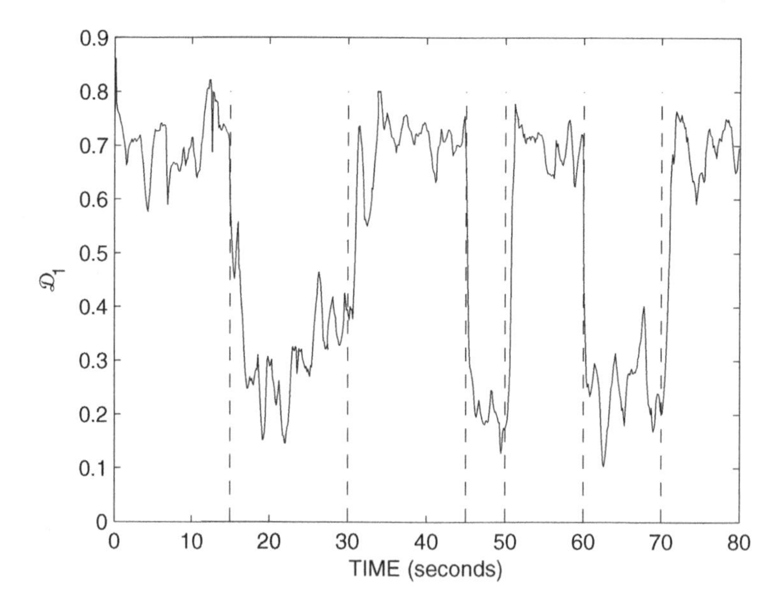

Figure 15.12 Variation of the decision number $\mathcal{D}_1$ is a double-talk detector. $L = 1024$.

L increased from 256 to 1024. The improvement in the result is clear. At the same time, a delay in the detection of transitions into a double-talk period and out of a double-talk period is more clearly seen.

15.4 Howling Suppression

As discussed earlier, howling can occur as a result of a possible positive feedback between the AE and the HE (Figure 15.1). Two methods of howling suppression have been developed and discussed in the literature. These methods are reviewed in the following sections.

15.4.1 Howling Suppression Through Notch Filtering

A howling phenomena is characterized by a sine-wave (line spectra) in the signal at any point of the AE–HE loop. Hence, a method of suppressing howling would be to deploy a signal analysis technique that evaluates the spectral content of the signal at an appropriate point in the AE–HE loop and adds a notch filter at the point in frequency that a howling is perceived. In a system that is equipped with an echo canceller, the most appropriate point in the loop is at the AE canceller output, $e(n)$.

Variety of schemes can be suggested to establish this task. A trivial scheme that we have examined and found effective is to perform signal analysis and notch filtering jointly by taking the following steps.

1. Take a block of length L of the signal samples and apply an FFT to them to evaluate the spectral content of the signal.
2. Over successive blocks evaluate the measured spectra and search for any spectral line that may be growing in amplitude.
3. Introduce and apply a real gain $0 < a < 1$ at the frequency bin(s) where howling is perceived.
4. Reduce the gain a if the howling persists.
5. Convert back the frequency domain samples to the time domain.

One may find this method particularly appealing in cases where the AE canceller is implemented using the FBLMS algorithm; as in that case, the tasks of converting the samples to the frequency domain and back to the time domain are already part of the system.

15.4.2 Howling Suppression by Spectral Shift

The ITU-T Recommendation G.167, the standard for AE controllers, has recommended the following method to avoid howling. Slightly shift the spectrum of the transmit signal $e(n)$. The maximum frequency shift allowed is 5 Hz.

To understand how this mechanism works, consider the feedback system presented in Figure 15.13. It consists of a microphone, a loudspeaker, the room acoustics, and any electronic circuits in the AE–HE loop. We model the loudspeaker and the microphone as blocks with ideal transfer functions of unity. The room acoustics plus any electronic

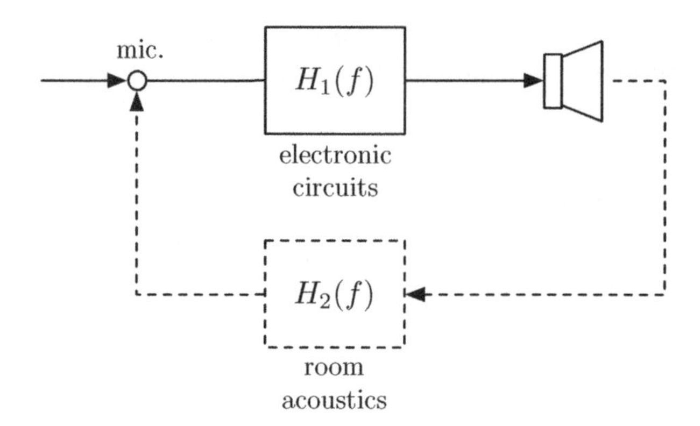

Figure 15.13 An acoustic-electronic feedback system that may result in howling.

circuits in the loop are collectively modeled by a transfer function $H(f) = H_1(f)H_2(f)$. Then, howling will occur at a frequency $f = f_0$, if $|H(f_0)| > 1$ and $\angle(H(f_0)) = 2k\pi$, for some integer k.

If the ITU-T spectral shift recommendation is included in Figure 15.13, the resulting system may be modeled as in Figure 15.14. Now, assuming that the spectrum shifter introduces a frequency shift of Δf, the loop gain of the system in Figure 15.14 will be $H(f + \Delta f)$. The following *plausible* statement have been made in the literature to explain why spectral shift suppresses howling. Any frequency component that propagates through the loop, starting from a point, returns back to the same point with a different frequency. Hence, no single frequency component will get a chance of growing in amplitude as it circles in the loop and, therefore, howling is avoided.

The frequency shifter in Figure 15.14 is implemented as follows. We first note that a direct modulation of $u(t)$ according to the equation $u_{\mathrm{m}}(t) = 2u(t)\cos(2\pi\Delta ft)$ does not work, because $U_{\mathrm{m}}(f) = U(f + \Delta f) + U(f - \Delta f)$; that is, $u_{\mathrm{m}}(t)$ will contain two replicas of $U(f)$. In other words, a positive and a negative shift of the spectrum is produced. The ITU-T recommendation is to convert $U(f)$ to $U(f + \Delta f)$ or $U(f - \Delta f)$, only. This can be established by following the structure presented in Figure 15.15. The block labeled as $H_{\mathrm{VSB}}(f)$ is a vestigial side-band (VSB) filter that removes the part of the spectrum of $u(t)$ at the negative side of the frequency axis. The resulting signal, $u_{\mathrm{VSB}}(t)$, is mixed with the complex sine-wave $e^{j2\pi\Delta ft}$; hence, its spectrum is shifted to the right by Δf. Taking the real part of the result and applying a gain of 2 results in a signal $v(t)$,

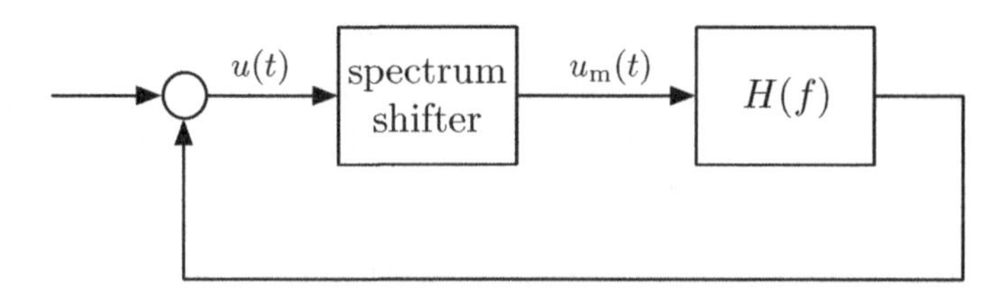

Figure 15.14 Block diagram of Figure 15.13 with an added spectrum shifter to avoid howling.

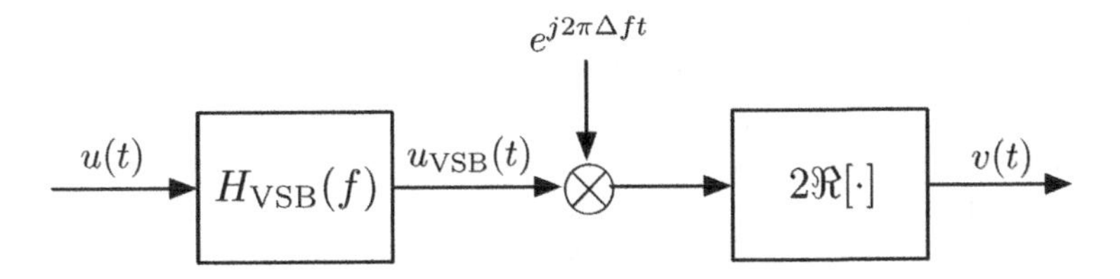

Figure 15.15 Block diagram of spectrum shifter.

whose Fourier transform, $V(f)$, is related to $U(f)$ according to the following equation

$$V(f) = \begin{cases} U(f + \Delta f), & f > 0 \\ U(f - \Delta f), & f < 0 \end{cases} \tag{15.55}$$

When $u(t)$ is replaced by its sampled version, $u(n)$, $H_{\text{VSB}}(f)$ should be also implemented in discrete time. We use $h_{\text{VSB},n}$ and $H_{\text{VSB}}(e^{j\omega})$ to denote the impulse response and the frequency response of this discrete time filter. Figure 15.16 presents a possible magnitude response of $h_{\text{VSB},n}$ and the procedure that we suggest for its design. We start with the design of a *half-band filter* $H_{\text{HB}}(e^{j\omega})$. This is called *half-band filter* because its passband occupies half of the full-band $0 \leq \omega \leq \pi$ (equivalently, $-\pi/2 \leq \omega \leq \pi/2$). A

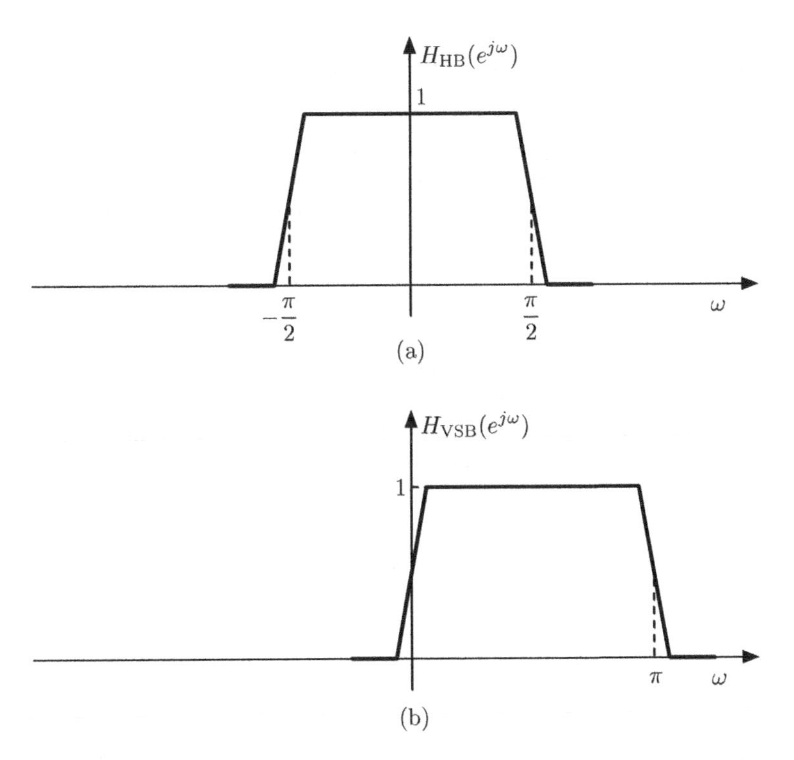

Figure 15.16 Construction of $H_{\text{VSB}}(e^{j\omega})$ from a half-band filter $H_{\text{HB}}(e^{j\omega})$.

nice and relevant to implementation property that half-band filters have is that almost half of their coefficients are 0. More specifically, $h_{\mathrm{HB},n} = 0$, for $n = \pm 2, \pm 4, \ldots$ This reduces the complexity of their implement by one-half. Once $H_{\mathrm{HB}}(e^{j\omega})$ is designed, it can be converted to the desired filter $H_{\mathrm{VSB}}(e^{j\omega})$ by modulating its coefficients with the sequence $\{j^n\} = \{\ldots, j, -1, -j, 1, j, -1, -j, \ldots\}$. Noting these points, given a half-band design $h_{\mathrm{HB},n}$, the coefficients of the desired filter $h_{\mathrm{VSB},n}$ are complex-valued, with the real and imaginary parts

$$h^{\mathrm{R}}_{\mathrm{VSB},n} = \begin{cases} h_{\mathrm{HB},n} & n = 0 \\ 0, & n \neq 0 \end{cases} \tag{15.56}$$

and

$$h^{\mathrm{I}}_{\mathrm{VSB},n} = \begin{cases} 0, & n = 0, \pm 2, \pm 4, \ldots \\ (-1)^{\frac{n-1}{2}} h_{\mathrm{HB},n}, & n = 1, 3, \ldots \end{cases} \tag{15.57}$$

respectively.

To help the reader to develop a better understanding of the impact of a spectral shift on the quality of speech signals, the MATLAB code "`specshift.m`," on the accompanying website, takes a speech signal, play it, then, introduce a shift in its spectrum and replay. The amount of spectral shift is a parameter. This will allow a curious reader to better understand what is the range of an acceptable spectral shift and why a maximum frequency shift 5 Hz has been imposed in the ITU-T Recommendation G.167.

15.5 Stereophonic Acoustic Echo Cancellation

The AE canceller setup that was presented in Figure 15.2 and was discussed in detail in the previous section is a monophonic system. Stereophonic AE cancellers have also been proposed and developed. Figure 15.17 presents the details of a stereophonic AE canceller setup. Compared to its monophonic counterpart, stereophonic teleconferencing provides special information that help a listener to distinguish the relative position of the speaker in the far-end teleconferencing room.

There are two microphones and two loudspeakers in each of the near-end and far-end teleconferencing rooms. Hence, considering the echo cancellation by the near-end room, there should be a pair of echo cancellers for removing echo signals from each of the microphones. A similar pair of echo cancellers should also be deployed by the far-end room. In Figure 15.17, for brevity and clarity of presentation, only one of the echo cancellers by the near-end room is shown. The relevant signals and signal paths/echoes are also shown. The speech signal, $s(n)$, from a person in the far-end room passes through the acoustic impulse responses $g_{1,n}$ and $g_{2,n}$ before reaching the pair of microphones in the far-end room. Assuming ideal transmission lines, one will find that

$$x_1(n) = g_{1,n} \star s(n) \tag{15.58}$$

and

$$x_2(n) = g_{2,n} \star s(n) \tag{15.59}$$

The signals $x_1(n)$ and $x_2(n)$, after being broadcast in the near-end room, will be picked up by the pair of microphones in the room. Here, the discussion is limited to the signal

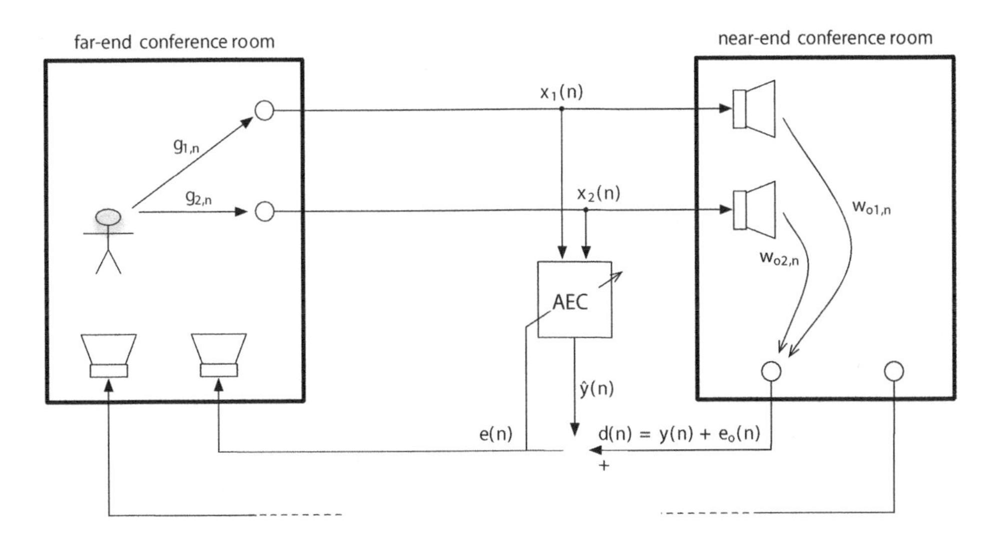

Figure 15.17 A stereophonic acoustic echo cancellation setup.

picked up by the microphone on the left-hand side. The room echo response between the loudspeakers in the near-end room and this microphone are represented by the impulse responses $w_{o1,n}$ and $w_{o2,n}$. The AEC inputs are $x_1(n)$ and $x_2(n)$, and its output is given by

$$\hat{y}(n) = \mathbf{w}_1^{\mathrm{T}}(n)\mathbf{x}_1(n) + \mathbf{w}_2^{\mathrm{T}}(n)\mathbf{x}_2(n) \tag{15.60}$$

where

$$\mathbf{w}_1(n) = [w_{1,0}(n)\ \ w_{1,1}(n) \cdots w_{1,N-1}(n)]^{\mathrm{T}} \tag{15.61}$$

$$\mathbf{w}_2(n) = [w_{2,0}(n)\ \ w_{2,1}(n) \cdots w_{2,N-1}(n)]^{\mathrm{T}} \tag{15.62}$$

$$\mathbf{x}_1(n) = [x_1(n)\ \ x_1(n-1) \cdots x_1(n-N+1)]^{\mathrm{T}} \tag{15.63}$$

$$\mathbf{x}_2(n) = [x_2(n)\ \ x_2(n-1) \cdots x_2(n-N+1)]^{\mathrm{T}} \tag{15.64}$$

Defining

$$\mathbf{w}(n) = [w_{1,0}(n)\ \ w_{2,0}(n)\ \ w_{1,1}(n)\ \ w_{2,1}(n) \cdots w_{1,N-1}(n)\ \ w_{2,N-1}]^{\mathrm{T}} \tag{15.65}$$

and

$$\mathbf{x}(n) = [x_1(n)\ x_2(n)\ x_1(n-1)\ x_2(n-1)\ \ \cdots\ \ x_1(n-N+1)\ x_2(n-N+1)]^{\mathrm{T}} \tag{15.66}$$

Equation (15.60) may be rearranged as

$$\hat{y}(n) = \mathbf{w}^{\mathrm{T}}(n)\mathbf{x}(n) \tag{15.67}$$

This, clearly, has the same form as Eq. (15.3), with apparently a minor difference that the length of the vectors $\mathbf{w}(n)$ and $\mathbf{x}(n)$ has been doubled. Noting this, one may argue that all the algorithms that were discussed in Section 15.2 for the monophonic AE cancellers may be readily extended to their stereophonic counterparts. However, as discussed below, this is not really the case. There are certain peculiar behaviors in stereophonic AE cancellers that need to be taken care of through special methods that will be introduced in the subsequent parts of this section.

15.5.1 The Fundamental Problem

There is a fundamental problem with the stereophonic AE cancellers that distinguishes it from its monophonic counterpart. In the latter, under a relatively weak constraints on the spectrum of the input signal to the echo canceller, the tap-weight vector $\mathbf{w}(n)$, on mean, converges toward the room echo response $\mathbf{w}_o$, if the length of $\mathbf{w}(n)$ is greater than or equal to the length of $\mathbf{w}_o$. Surprisingly, this is not the case in a stereophonic echo canceller. In the sequel, we explain this peculiar behavior of the stereophonic echo cancellers from the frequency domain point of view and from the time domain point of view.

To present the frequency domain point of view of the problem, we recall that the definition (15.40) and note that the coherence function between $x_1(n)$ and $x_2(n)$ is defined as

$$c_{x_1x_2}(e^{j\omega}) = \frac{|\Phi_{x_1x_2}(e^{j\omega})|^2}{\Phi_{x_1x_1}(e^{j\omega})\Phi_{x_2x_2}(e^{j\omega})} \tag{15.68}$$

Moreover, recalling Eqs. (15.58) and (15.59) and using the relevant identities from Chapter 2 (also, see Problem P2.14), one may find that

$$\Phi_{x_1x_1}(e^{j\omega}) = \Phi_{ss}(e^{j\omega})|G_1(e^{j\omega})|^2 \tag{15.69}$$

$$\Phi_{x_2x_2}(e^{j\omega}) = \Phi_{ss}(e^{j\omega})|G_2(e^{j\omega})|^2 \tag{15.70}$$

$$\Phi_{x_1x_2}(e^{j\omega}) = \Phi_{ss}(e^{j\omega})G_1(e^{j\omega})G_2^*(e^{j\omega}) \tag{15.71}$$

Substituting these results in Eq. (15.68), one will find that

$$c_{x_1x_2}(e^{j\omega}) = 1 \tag{15.72}$$

which may be interpreted as $x_1(n)$ and $x_2(n)$ are coherently fully correlated. This, as demonstrated below, has a consequence that the tap-weight vector $\mathbf{w}(n)$ of an AE canceller may not converge to the desired unique solution $\mathbf{w} = \mathbf{w}_o$.

Clearly, a trivial as well as desirable choice for the pair of tap-weight vectors of the echo canceller, after convergence, are $\hat{\mathbf{w}}_{o1} = \mathbf{w}_{o1}$ and $\hat{\mathbf{w}}_{o2} = \mathbf{w}_{o2}$. In the frequency domain, these are written as

$$\hat{W}_{o1}(e^{j\omega}) = W_{o1}(e^{j\omega}) \tag{15.73}$$

and

$$\hat{W}_{o2}(e^{j\omega}) = W_{o2}(e^{j\omega}) \tag{15.74}$$

respectively. This choice, obviously, leads to the estimate $\hat{y}(n) = y(n)$. Hence, it also leads to the desired output error

$$e(n) = e_o(n) \tag{15.75}$$

A fundamental question to ask here is "are Eqs. (15.73) and (15.74) the only choices of $W_{o1}(e^{j\omega})$ and $W_{o2}(e^{j\omega})$, respectively, that result in the desirable result (15.75)?".

We answer this question by providing a few choices that show that the above solution is not unique. We first note that the choice (15.73) and (15.74) implies that

$$\begin{aligned}\hat{Y}(e^{j\omega}) &= S(e^{j\omega})(G_1(e^{j\omega})\hat{W}_{o1}(e^{j\omega}) + G_2(e^{j\omega})\hat{W}_{o2}(e^{j\omega})) \\ &= S(e^{j\omega})(G_1(e^{j\omega})W_{o1}(e^{j\omega}) + G_2(e^{j\omega})W_{o2}(e^{j\omega})) = Y(e^{j\omega})\end{aligned} \quad (15.76)$$

On the other hand, one can easily show that the alternative choice

$$\hat{W}_{o1}(e^{j\omega}) = W_{o1}(e^{j\omega}) + \alpha G_2(e^{j\omega}) \quad (15.77)$$

and

$$\hat{W}_{o2}(e^{j\omega}) = W_{o2}(e^{j\omega}) - \alpha G_1(e^{j\omega}) \quad (15.78)$$

for any value of α, also lead to the identity $\hat{Y}(e^{j\omega}) = Y(e^{j\omega})$, hence the desirable result (15.75). Surprisingly, there are more choices that also lead to Eq. (15.69). Couple of such choices are

$$\hat{W}_{o1}(e^{j\omega}) = \alpha\left(W_{o1}(e^{j\omega}) + \frac{G_2(e^{j\omega})}{G_1(e^{j\omega})}W_{o2}(e^{j\omega})\right) \quad (15.79)$$

and

$$\hat{W}_{o2}(e^{j\omega}) = (1-\alpha)\left(W_{o2}(e^{j\omega}) + \frac{G_1(e^{j\omega})}{G_2(e^{j\omega})}W_{o1}(e^{j\omega})\right) \quad (15.80)$$

or

$$\hat{W}_{o1}(e^{j\omega}) = \alpha W_{o1}(e^{j\omega}) + (1-\alpha)\frac{G_2(e^{j\omega})}{G_1(e^{j\omega})}W_{o2}(e^{j\omega}) \quad (15.81)$$

and

$$\hat{W}_{o2}(e^{j\omega}) = \alpha W_{o2}(e^{j\omega}) + (1-\alpha)\frac{G_1(e^{j\omega})}{G_2(e^{j\omega})}W_{o1}(e^{j\omega}) \quad (15.82)$$

for any value of α.

This observation shows that there are an infinite set of choices for $\hat{W}_{o1}(e^{j\omega})$ and $\hat{W}_{o2}(e^{j\omega})$ that lead to the desirable echo canceller output error (15.75). Moreover, examining of any of the pair of choices ((15.77), (15.78)), ((15.79), (15.80)) and ((15.81), (15.82)), one may note that the presence of these extended solutions is a consequence of the fact that $x_1(n)$ and $x_2(n)$ both originate from the same source, $s(n)$. This, in turn, relates to the full coherence relation between $x_1(n)$ and $x_2(n)$, that is, the identity (15.72).

The nonunique solutions presented by the pairs ((15.77), (15.78)), ((15.79), (15.80)) and ((15.81), (15.82)) suffer from the following problematic property. They depend on the acoustic signal propagations, $G_1(e^{j\omega})$ and $G_2(e^{j\omega})$, in the far-end room. This is problematic, particularly here, because in a stereophonic teleconferencing setup $G_1(e^{j\omega})$ and $G_2(e^{j\omega})$ vary frequently as different people (at different locations in the far-end room) talk during different time periods. Hence, special measures have to be taken to avoid this potential problem of stereophonic AE cancellers. Such measures are introduced in Section 15.2.2.

The pairs ((15.77), (15.78)), ((15.79), (15.80)), and ((15.81), (15.82)) may be viewed as unconstrained solutions to the problem of stereophonic AE cancellation. In particular, the last two pairs may not be feasible in practice because divisions by $G_1(e^{j\omega})$ and $G_2(e^{j\omega})$

can lead to noncausal and/or unstable systems. Nevertheless, when $g_{1,n}$ and $g_{2,n}$ have finite durations that are shorter than or equal to the lengths of the echo canceller tapped-delay lines, namely the length of $w_{1,n}$ and $w_{2,n}$, Eqs. (15.77) and (15.78) are realizable and they have the time-domain equivalent of

$$\hat{\mathbf{w}}_{o1} = \mathbf{w}_{o1} + \alpha \mathbf{g}_2 \tag{15.83}$$

and

$$\hat{\mathbf{w}}_{o2} = \mathbf{w}_{o2} - \alpha \mathbf{g}_1 \tag{15.84}$$

respectively. Obviously, $\mathbf{g}_1$ and $\mathbf{g}_2$, the vectors with the elements of $g_{1,n}$ and $g_{2,n}$, respectively.

Finally, we note that, in practice, the acoustic impulse responses $g_{1,n}$ and $g_{2,n}$, as well as $w_{o1,n}$ and $w_{o2,n}$ are usually longer than the echo canceller length, albeit, with very low energy in their tails. Noting this, strictly speaking, one may find that the alternative solutions such as those in Eqs. (15.83) and (15.84) do not really minimize the output error. They may be viewed only as approximation to the underlying Wiener–Hopf equation. From what we learnt in the previous chapters, the discussion in this section leads to a conclusion that may be cast into the following form.

In a stereophonic AE cancellation setup, the correlation matrix $\mathbf{R} = E[\mathbf{x}(n)\mathbf{x}^{\mathrm{T}}(n)]$, where $\mathbf{x}(n)$ is defined as in Eq. (15.66), is usually is a highly ill-conditioned. That is, it has a set of eigenvalues that may be very close to 0. As a result, the echo canceller suffers from some very slow modes of convergence. This, in turn, results in convergence of the echo canceller to a set of coefficients that may not match the samples of the conference room acoustic response. This leads to a misalignment of the echo canceller coefficients, where the misalignment, as presented before, is defined as Eq. (15.6). Moreover, the misalignment depends on the acoustic responses, $g_{1,n}$ and $g_{2,n}$, in the far-end room. As noted earlier, this is problematic because a change of speaker in the far-end room results in a dramatic change in the responses $g_{1,n}$ and $g_{2,n}$ and hence, the echoes may be heard momentarily in the far-end room. This of course is unacceptable and solutions that minimize misalignment are necessary for satisfactory operation of any stereophonic AE canceller. Such solutions are discussed next.

To remind the reader on how a close to zero eigenvalue, arising from a highly ill-conditioned correlation matrix $\mathbf{R}$ in an adaptive filter, leads to a significant misalignment is the converged coefficients, Figure 15.18 presents an example of the performance surface of a 2-tap adaptive filter when one of the eigenvalues of $\mathbf{R}$ is 0. As seen, starting from any point on the performance surface, a gradient algorithm converges to the nearest point on the line marked with $\xi = \xi_{\min}$, the point $(\hat{w}_{o0}, \hat{w}_{o1})$, and remains there because there is no gradient to push the tap weights $\hat{w}_{o0}$ and $\hat{w}_{o1}$ toward the desired optimum point (w_{o0}, w_{o1}).

15.5.2 Reducing Coherence Between $x_1(n)$ and $x_2(n)$

As discussed earlier, the coherence between $x_1(n)$ and $x_2(n)$ results in a correlation matrix $\mathbf{R} = E[\mathbf{x}(n)\mathbf{x}^{\mathrm{T}}(n)]$ with a number of close to zero eigenvalues. As a result, the echo canceller may converge to a point with a possibly significant misalignment between the converged tap weights and the true tap weights of the room responses. To solve this

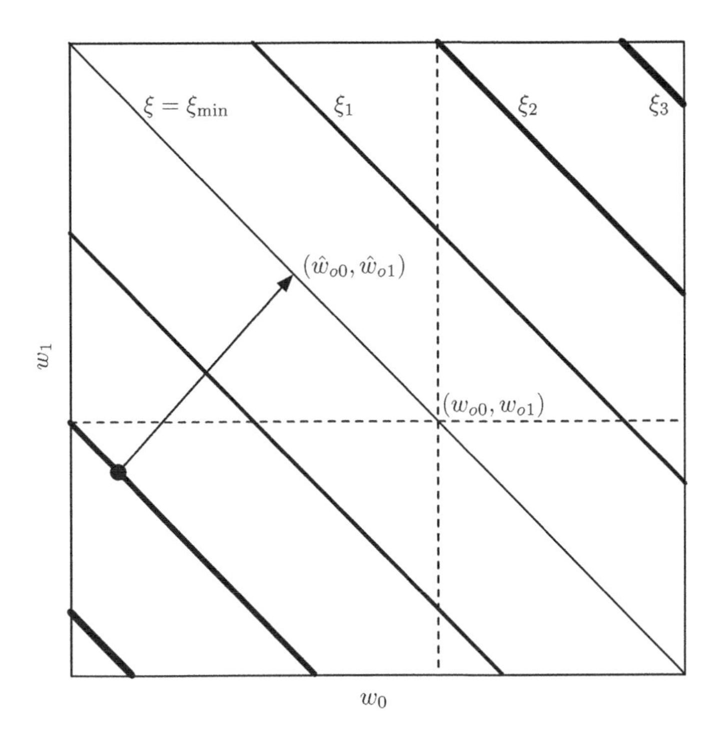

Figure 15.18 Demonstration of a performance surface when the underlying correlation matrix has a zero eigenvalue.

problem, one should introduce some changes in $x_1(n)$ and $x_2(n)$ so that the coherence between them reduces. Here, we introduce three methods for achieving this reduction.

Coherence reduction through additive noise

A trivial method of reducing coherence between $x_1(n)$ and $x_2(n)$ that was proposed in the early days of development of AE cancellers was to add a pair of independent white noise processes to $x_1(n)$ and $x_2(n)$. This addition is done at a point before $x_1(n)$ and $x_2(n)$ are being passed to the echo canceller and the loudspeakers in the near-end room. This method replaces the correlation matrix $\mathbf{R}$ by $\mathbf{R} + \sigma_n^2\mathbf{I}$, where σ_n^2 is the variance of the added noise samples. According to the results presented in Chapter 4, this boosts up all the eigenvalues of the underlying correlation matrix by σ_n^2, and thus resolves the problem of the close to zero eigenvalues of $\mathbf{R}$ – the main source of misalignment. However, unfortunately, this method was found to be unsuccessful because successful operation of it requires addition of an unacceptably high noise power that would be audible and thus annoying (Sondhi, Morgan, and Hall, 1995).

Coherence Reduction Through Leaky LMS Algorithm

The *leaky LMS algorithm* is an intelligent technique that, without adding any noise to the input to an LMS-based adaptive filter, increases the eigenvalues of the underlying correlation matrix by controllable constant. The leaky LMS algorithm was presented in Chapter 6 under an exercise problem at the end of the chapter. Here, for the completeness of our discussion, we present the leaky LMS algorithm in the context of an AE canceller. The use of the leaky LMS algorithm to reduce the misalignment in AE cancellers was first suggested in Hoya *et al.* (1998).

The leaky LMS algorithm works based on the recursion

$$\mathbf{w}(n+1) = \beta\mathbf{w}(n) + 2\mu e(n)\mathbf{x}(n) \tag{15.85}$$

where β is a constant slightly smaller than 1. The properties of this algorithm can be best understood by looking at the convergence of the mean of $\mathbf{w}(n)$ as Eq. (15.85) iterates.

In the case of the stereophonic AE canceller of Figure 15.17,

$$\begin{aligned} e(n) &= d(n) - y(n) \\ &= e_o(n) + y(n) - \hat{y}(n) \\ &= e_o(n) + (\mathbf{w}_o - \mathbf{w}(n))^{\mathrm{T}}\mathbf{x}(n) \end{aligned} \tag{15.86}$$

Substituting Eq. (15.86) in Eq. (15.85), taking expectations on both sides of the result, and recalling the *independence assumption* of Section 6.3, and defining $\overline{\mathbf{w}}(n) = E[\mathbf{w}(n)]$, we obtain

$$\begin{aligned} \overline{\mathbf{w}}(n+1) &= \beta\overline{\mathbf{w}}(n) - 2\mu\mathbf{R}\overline{\mathbf{w}}(n) + 2\mu\mathbf{p} \\ &= (\beta\mathbf{I} - 2\mu\mathbf{R})\overline{\mathbf{w}}(n) + 2\mu\mathbf{p} \\ &= (\mathbf{I} - 2\mu\mathbf{R}')\overline{\mathbf{w}}(n) + 2\mu\mathbf{p} \end{aligned} \tag{15.87}$$

where

$$\mathbf{R}' = \mathbf{R} + \frac{1-\beta}{2\mu}\mathbf{I} \tag{15.88}$$

and

$$\mathbf{p} = \mathbf{R}\mathbf{w}_o \tag{15.89}$$

Recalling the derivations in Chapters 5 and 6, one may deduce from Eq. (15.87) that

- The modes of convergence of the leaky LMS algorithm are determined by the set of eigenvalues of $\mathbf{R}'$, viz.,

  $$\lambda_i' = \lambda_i + \frac{1-\beta}{2\mu} \tag{15.90}$$

  where λ_i's are the eigenvalues of $\mathbf{R}$. Hence, assuming that $\frac{1-\beta}{2\mu}$ is sufficiently large, this avoids the presence of any very slow mode of convergence in the algorithm.
- As $n \to \infty$, $\overline{\mathbf{w}}(n)$ converges to

  $$\begin{aligned} \hat{\mathbf{w}}_o &= \mathbf{R}'^{-1}\mathbf{p} \\ &= \mathbf{R}'^{-1}\mathbf{R}\mathbf{w}_o \end{aligned} \tag{15.91}$$

  This shows that $\hat{\mathbf{w}}_o \neq \mathbf{w}_o$; hence, the leaky LMS bounds to suffer from some misalignment.

Using Eq. (15.91), the misalignment of the leaky LMS algorithm when $\mathbf{w}(n)$ has approached $\hat{\mathbf{w}}_\text{o}$ is obtained as

$$\begin{aligned}\zeta &= (\mathbf{w}_\text{o} - \mathbf{R}'^{-1}\mathbf{R}\mathbf{w}_\text{o})^\text{T}(\mathbf{w}_\text{o} - \mathbf{R}'^{-1}\mathbf{R}\mathbf{w}_\text{o}) \\ &= \mathbf{w}_\text{o}^\text{T}(\mathbf{I} - \mathbf{R}'^{-1}\mathbf{R})^2\mathbf{w}_\text{o}\end{aligned} \tag{15.92}$$

Recalling that according to the unitary singularity transformation, $\mathbf{R}$ and $\mathbf{R}'$ may be expanded as

$$\mathbf{R} = \mathbf{Q}\mathbf{\Lambda}\mathbf{Q}^\text{T} \tag{15.93}$$

and

$$\mathbf{R}' = \mathbf{Q}\mathbf{\Lambda}'\mathbf{Q}^\text{T} \tag{15.94}$$

where $\mathbf{\Lambda}$ and $\mathbf{\Lambda}'$ are the diagonal matrices of the eigenvalues of $\mathbf{R}$ and $\mathbf{R}'$, respectively, and $\mathbf{Q}$ is the associated eigenvectors matrix. Also, note that $\mathbf{R}$ and $\mathbf{R}'$ share the same set of eigenvectors. Substituting Eqs. (15.93) and (15.94) in Eq. (15.92), recalling the identity $\mathbf{Q}^\text{T}\mathbf{Q} = \mathbf{I}$ and the definition

$$\mathbf{Q} = [\mathbf{q}_0\ \mathbf{q}_1 \cdots \mathbf{q}_{2N-1}] \tag{15.95}$$

and rearranging the result, we obtain

$$\zeta = \left(\frac{1-\beta}{2\mu}\right)^2 \times \sum_{i=0}^{2N-1} \frac{(\mathbf{q}_i^\text{T}\mathbf{w}_\text{o})^2}{\left(\lambda_i + \frac{1-\beta}{2\mu}\right)^2} \tag{15.96}$$

This result shows that, on one hand, we should choose β sufficiently smaller than 1 ($1 - \beta$ large) to avoid slow modes of convergence. On the other hand, we should choose β sufficiently close to 1 ($1 - \beta$ small) to reduce the misalignment.

Coherence Reduction Through Exclusive Maximum Tap-Selection

The exclusive maximum (XM) tap-selection method is an adaptation strategy that is applicable to the various adaptive algorithms, including NLMS, affine projection LMS, and RLS algorithms. The basic idea is to apply each adaptation step only to those taps with the larger amplitude tap inputs. Khong and Naylor (2005, 2007) have studied the application of the XM tap-selection method to stereophonic AE cancellers and noted that this method has some success in reducing the misalignment. The rationale here is that at each iteration of the adaptive algorithm, the set of tap inputs selected from the two channels are randomly unaligned, and this results in a lower coherency among the signals from the two channels that are used for adaptation.

Coherence Reduction Through Nonlinearity

It has been noted that introduction of some nonlinearity into a speech signal to a great degree is perceptually undetectable. On the other hand, addition of such nonlinearity reduces the coherence between a pair of signals in a stereophonic AE canceller, and thus resolves the problem of a large misalignment. Morgan, Hall, and Benesty (2001) investigated several types of nonlinearities for stereophonic AE cancellers and noted that they all have similar effect in reducing the coherence between the processed signals. Among the types of nonlinearities evaluated, half-wave rectifier, defined as

$$\tilde{x}(n) = x(n) + \alpha \frac{x(n) + |x(n)|}{2} \tag{15.97}$$

where α is a constant that determines the amount of the added nonlinearity, was recognized as the simplest to implement and the one with minimal effects on the speech quality. Moreover, various authors have studied the combinations of nonlinearities and various adaptive filtering algorithms, for example, the leaky LMS and XM tap-section methods, and have demonstrated that such combinations can lead to systems with very good/acceptable misalignment performance.

15.5.3 The LMS-Newton Algorithm for Stereophonic Systems

In Section 15.2, a number of adaptive filter structures and algorithms were introduced for monophonic echo cancellers. Namely, the normalized LMS, affine projection LMS, frequency domain block LMS, subband LMS, and efficient implementation of the LMS-Newton algorithm. Among these, we found that affine projection LMS algorithm have the fastest rate of convergence; however, it settles a some lower ERLE compared to the rest of the algorithms (except the subband LMS algorithm, whose lower ERLE was attributed to the imperfect responses of the analysis and synthesis filters). The algorithms presented in Tables 15.1 and 15.2 are directly applicable to the stereophonic case, simply by redefining the tap-weight vector $\mathbf{w}(n)$ and the tap-input vector $\mathbf{x}(n)$ according to Eqs. (15.66) and (15.65), respectively. Development of the frequency domain block LMS algorithm for stereophonic AE cancellers also is not difficult. However, the extension of the LMS-Newton algorithm is not that straightforward.

Noting that the LMS-Newton algorithm has superior performance over the rest of the algorithms, in the rest of this chapter, we present a full development of the LMS-Newton algorithm for stereophonic AE cancellation. This requires a thorough understanding of the two-channel lattice predictor and how this can be used to obtain an estimate of the inverse of the correlation matrix $\mathbf{R}$ based on autoregressive modeling. This development is parallel to the single-channel lattice structures and algorithms that were presented in Chapter 11. Also, we will find that the application of the half-wave rectifier, Eq. (15.97), to improve on the misalignment of the LMS-Newton stereophonic echo canceller requires some special treatment. The details of this treatment will also be presented. The results presented in this section have been developed in Rao and Farhang-Boroujeny (2009) and (2010).

Two-Channel Lattice Predictor

Recall the one-channel lattice predictor that was developed in Chapter 11. In particular, recall the order-update equations (11.60) and (11.61) of the mth stage of the predictor. In the case of the two-channel predictor, the duals of Eqs. (11.60) and (11.61) are

$$\mathbf{f}_m(n) = \mathbf{f}_{m-1}(n) - \boldsymbol{\kappa}_m^{\mathrm{T}} \mathbf{b}_{m-1}(n-1) \tag{15.98}$$

and

$$\mathbf{b}_m(n) = \mathbf{b}_{m-1}(n-1) - \boldsymbol{\kappa}_m \mathbf{f}_{m-1}(n) \tag{15.99}$$

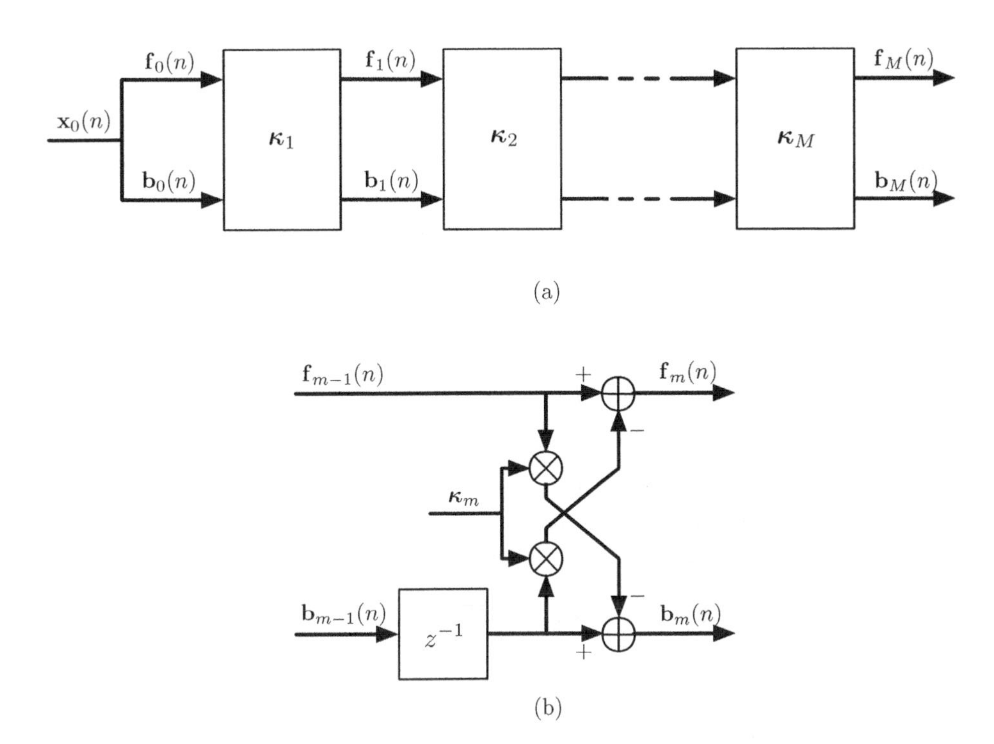

Figure 15.19 Details of Stage m of a two-channel lattice predictor. All the signals are 2-by-1 vectors and the partial correlation (PARCOR) coefficients, κ_m, are 2-by-2 matrices.

where $\mathbf{f}_m(n) = [f_{1,m}(n)\ f_{2,m}(n)]^{\mathrm{T}}$ and $\mathbf{b}_m(n) = [b_{1,m}(n)\ b_{2,m}(n)]^{\mathrm{T}}$ are the 2×1 forward and backward prediction-error vectors, respectively, and the 2×2 reflection coefficient matrix κ_m is defined as

$$\kappa_m = \begin{bmatrix} \kappa_{11,m} & \kappa_{12,m} \\ \kappa_{21,m} & \kappa_{22,m} \end{bmatrix} \tag{15.100}$$

Figure 15.19 depicts a two-channel lattice predictor structure, and the details of its mth stage. The lattice predictor is initialized with $\mathbf{f}_0(n) = \mathbf{b}_0(n) = \mathbf{x}_0(n)$, where

$$\mathbf{x}_0(n) = \begin{bmatrix} x_1(n) \\ x_2(n) \end{bmatrix} \tag{15.101}$$

As in the case of single-channel, here also a simple gradient adaptive algorithm can be used to compute $\kappa_m(n)$ in a recursive manner. The reflection coefficients of the mth cell can be chosen so as to minimize the sum of the instantaneous forward and backward prediction-errors of the corresponding cell, that is, $f_m^2(n) + b_m^2(n)$. This leads to the following LMS recursions

$$\kappa_{ij,m}(n+1) = \kappa_{ij,m}(n) - \frac{2\mu_{p,o}}{P_{ij,m-1}(n) + \varepsilon}$$

$$\times[f_{j,m}(n)b_{i,m-1}(n-1)+f_{j,m-1}(n)b_{i,m}(n)] \tag{15.102}$$
$$\text{for } (i,j)=\{(1,1),(1,2),(2,1),(2,2)\}$$

where $\mu_{\text{p,o}}$ is the adaptation step-size parameter, and ε is a constant added to prevent gradient noise amplification, when any of the power terms $P_{ij,m-1}(n)$ are small. The power terms are estimated using the following recursive equations

$$P_{ij,m}(n)=\beta P_{ij,m}(n-1)+0.5(1-\beta)[b_{i,m-1}^2(n-1)+f_{j,m-1}^2(n)] \tag{15.103}$$
$$\text{for } (i,j)=\{(1,1),(1,2),(2,1),(2,2)\}$$

where β is a constant close to but smaller than 1.

Upon the convergence of the PARCOR coefficient matrices $\boldsymbol{\kappa}_m$'s, the backward prediction errors $\mathbf{b}_m(n)$ will constitute a set of 2-by-1 uncorrelated random vectors. Following a similar line of thought that led to the lattice joint process estimator of Figure 11.7, an AE canceller may be implemented by passing the backward prediction errors $\mathbf{b}_m(n)$ to a linear combiner. Hence, the input to the linear combiner will be the vector

$$\begin{aligned}\mathbf{b}(n)&=\left[\mathbf{b}_0^{\mathrm{T}}(n)\ \mathbf{b}_1^{\mathrm{T}}(n)\ \cdots\ \mathbf{b}_{N-1}^{\mathrm{T}}(n)\right]^{\mathrm{T}}\\&=\left[b_{1,0}(n)\ b_{2,0}(n)\ b_{1,1}(n)\ b_{2,2}(n)\ \cdots\ b_{1,N-1}(n)\ b_{2,N-1}(n)\right]^{\mathrm{T}}\end{aligned} \tag{15.104}$$

This is a vector of length $2N$.

If we use $\mathbf{c}(n)$ to denote the tap-weight vector of the $2N$-tap linear combiner at the time instant n, the output of the lattice joint process estimator (here, the estimated echo signal) is given by

$$\hat{y}(n)=\mathbf{c}^{\mathrm{T}}(n)\mathbf{x}(n) \tag{15.105}$$

The output error is then obtained as

$$e(n)=d(n)-\hat{y}(n) \tag{15.106}$$

and $\mathbf{c}(n)$ may be updated using the recursive equation

$$\mathbf{c}(n+1)=\mathbf{c}(n)+2\mu\mathbf{R}_{bb}^{-1}\mathbf{b}(n)e(n) \tag{15.107}$$

where $\mathbf{R}_{bb}=E[\mathbf{b}(n)\mathbf{b}^{\mathrm{T}}(n)]$. In the case of single-channel lattice joint process estimator, $\mathbf{R}_{bb}$ is a diagonal matrix; hence, $\mathbf{R}_{bb}^{-1}$ is trivially obtained by inverting the diagonal elements of $\mathbf{R}_{bb}$. This is not the case in a two-channel joint process estimator as $E[\mathbf{b}_m(n)\mathbf{b}_j^{\mathrm{T}}(n)]=\mathbf{0}$, for $m\neq j$, but $\mathbf{R}_{b_mb_m}=E[\mathbf{b}_m(n)\mathbf{b}_m^{\mathrm{T}}(n)]$ is not a diagonal matrix. In other words, although $\mathbf{b}_m(n)$ and $\mathbf{b}_j(n)$, $m\neq j$, are uncorrelated, thus $\mathbf{R}_{b_mb_j}=\mathbf{0}$, the elements of $\mathbf{b}_m(n)$, for any m, are not necessarily uncorrelated. Accordingly, one should note that

$$\mathbf{R}_{bb}=\begin{bmatrix}\mathbf{R}_{b_0b_0} & \mathbf{0} & \cdots & \mathbf{0}\\ \mathbf{0} & \mathbf{R}_{b_1b_1} & \cdots & \mathbf{0}\\ \vdots & \vdots & \vdots & \vdots\\ \mathbf{0} & \mathbf{0} & \cdots & \mathbf{R}_{b_{N-1}b_{N-1}}\end{bmatrix} \tag{15.108}$$

and, hence,

$$\mathbf{R}_{bb}^{-1} = \begin{bmatrix} \mathbf{R}_{b_0b_0}^{-1} & \mathbf{0} & \cdots & \mathbf{0} \\ \mathbf{0} & \mathbf{R}_{b_1b_1}^{-1} & \cdots & \mathbf{0} \\ \vdots & \vdots & \vdots & \vdots \\ \mathbf{0} & \mathbf{0} & \cdots & \mathbf{R}_{b_{N-1}b_{N-1}}^{-1} \end{bmatrix} \tag{15.109}$$

Assuming that the PARCOR coefficients have converged to their optimum values and have been kept fixed, the update Eq. (15.107) is effectively an LMS-Newton algorithm and thus its behavior is governed by a single mode of convergence. However, in the particular case of echo cancellers, where the PARCORs should be continuously adopted to follow variation of the statistics of the input process (a speech signal), the lattice joint process estimator suffers from a significant misadjustment, and thus renders its application less attractive. The reader is referred to Section 11.13.1 for a more detailed discussion on this topic. In the sequel, we follow similar derivations to those in Section 11.15 to develop the dual of the algorithms 1 and 2 for the case of stereophonic AE cancellers, that is, the two-channel case.

Algorithms

Two versions of the LMS-Newton algorithm based on autoregressive modeling were proposed for the single-channel case in Section 11.15. These were based on the fact that the input sequence (a speech signal) to the adaptive filter can be modeled as an autoregressive process of order M, where M can be much smaller than the filter length N. This results in an efficient way of updating the term $\mathbf{R}^{-1}\mathbf{x}(n)$ of the LMS-Newton update Eq. (7.32) without having to estimate $\mathbf{R}^{-1}$, where $\mathbf{R} = E[\mathbf{x}(n)\mathbf{x}^{\mathrm{T}}(n)]$. Here, we derive the LMS-Newton algorithms for the stereophonic setup/the two-channel case. The reader may note that some of the derivations here are very similar to those presented in Section 11.15. However, differences also exist.

As in Section 11.15, in the rest of this chapter, we also add the subscript xx to $\mathbf{R}$ when a reference is made to the correlation matrix of $\mathbf{x}(n)$. This is to differentiate $\mathbf{R}_{xx}$ from the other correlation matrices that we also encounter in the sequel.

For a stereophonic system, we note that $\mathbf{b}(n)$ can be expressed as

$$\mathbf{b}(n) = \mathbf{L}\mathbf{x}(n) \tag{15.110}$$

where $\mathbf{x}(n)$ is the $2N \times 1$ filter input vector, defined by Eq. (15.66), $\mathbf{b}(n)$ is the $2N \times 1$ vector of backward prediction errors, defined by Eq. (15.98), and $\mathbf{L}$ is the $2N \times 2N$ transformation matrix and has the form

$$\mathbf{L} = \begin{bmatrix} \mathbf{I} & \mathbf{0} & \mathbf{0} & \cdots & \mathbf{0} & \mathbf{0} \\ -\mathbf{G}_{1,1} & \mathbf{I} & \mathbf{0} & \cdots & \mathbf{0} & \mathbf{0} \\ -\mathbf{G}_{2,1} & -\mathbf{G}_{2,2} & \mathbf{I} & \cdots & \mathbf{0} & \mathbf{0} \\ \vdots & \vdots & \vdots & \ddots & \vdots & \vdots \\ -\mathbf{G}_{N-1,1} & -\mathbf{G}_{N-1,2} & -\mathbf{G}_{N-1,3} & \cdots & -\mathbf{G}_{N-1,N-1} & \mathbf{I} \end{bmatrix} \tag{15.111}$$

where $\mathbf{I}$ is a 2×2 identity matrix, $\mathbf{0}$ is a 2×2 zero matrix, and each $\mathbf{G}_{m,j}$ is a 2×2 backward predictor coefficient matrix. The transformation (15.110) provides a one-to-one correspondence between the input vector $\mathbf{x}(n)$ and the backward prediction-error vector $\mathbf{b}(n)$. From Eq. (15.110), it follows that

$$\mathbf{R}_{xx}^{-1} = \mathbf{L}^{\mathrm{T}}\mathbf{R}_{bb}^{-1}\mathbf{L} \tag{15.112}$$

Next, we recall the ideal LMS-Newton update equation

$$\mathbf{w}(n+1) = \mathbf{w}(n) + 2\mu\mathbf{R}_{xx}^{-1}\mathbf{x}(n)e(n) \tag{15.113}$$

and note that using Eqs. (15.110) and (15.112) in Eq. (15.113), one will obtain

$$\mathbf{w}(n+1) = \mathbf{w}(n) + 2\mu\mathbf{u}(n)e(n) \tag{15.114}$$

where

$$\mathbf{u}(n) = \mathbf{L}^{\mathrm{T}}\mathbf{R}_{bb}^{-1}\mathbf{b}(n) \tag{15.115}$$

The significance of Eq. (15.114) is that the computation of $\mathbf{u}(n)$ according to Eq. (15.115) can be performed at a low complexity. In the sequel, following a similar line of derivations to those in Chapter 11, we present two implementations of the LMS-Newton algorithms based on Eqs. (15.114) and (15.115). These are referred to as Algorithm 1 and Algorithm 2.

Algorithm 1: This algorithm will involve the direct implementation of Eq. (15.115) through the use of a lattice predictor. If we assume that the input sequence $\mathbf{x}(n)$ is an autoregressive process of order $M \ll N$, a lattice predictor of order M is sufficient. The matrix $\mathbf{L}$ then takes the form of

$$\mathbf{L} = \begin{bmatrix} \mathbf{I} & \mathbf{0} & \cdots & \mathbf{0} & \mathbf{0} \cdots & \mathbf{0} & \mathbf{0} & \cdots & \mathbf{0} \\ -\mathbf{G}_{1,1} & \mathbf{I} & \cdots & \mathbf{0} & \mathbf{0} \cdots & \mathbf{0} & \mathbf{0} & \cdots & \mathbf{0} \\ \vdots & \vdots & \ddots & \vdots & \vdots \ddots & \vdots & \vdots & \ddots & \vdots \\ -\mathbf{G}_{M,1} & -\mathbf{G}_{M,2} & \cdots & \mathbf{I} & \mathbf{0} \cdots & \mathbf{0} & \mathbf{0} & \cdots & \mathbf{0} \\ \mathbf{0} & -\mathbf{G}_{M,1} & \cdots & -\mathbf{G}_{M,M} & \mathbf{I} \cdots & \mathbf{0} & \mathbf{0} & \cdots & \mathbf{0} \\ \vdots & \vdots & \ddots & \vdots & \vdots \ddots & \vdots & \vdots & \ddots & \vdots \\ \mathbf{0} & \mathbf{0} & \cdots & \mathbf{0} & \mathbf{0} \cdots & -\mathbf{G}_{M,1} & -\mathbf{G}_{M,2} & \cdots & \mathbf{I} \end{bmatrix} \tag{15.116}$$

and the vector $\mathbf{b}(n)$ takes the form of

$$\begin{aligned} \mathbf{b}(n) = [&b_{1,0}(n), b_{2,0}(n), b_{1,1}(n), b_{2,1}(n), \ldots, b_{1,M}(n), b_{2,M}(n), b_{1,M}(n-1), \\ &b_{2,M}(n-1), \ldots, b_{1,M}(n-L+M+1), b_{2,M}(n-L+M+1)]^{\mathrm{T}} \end{aligned} \tag{15.117}$$

The special structure of $\mathbf{b}(n)$ in Eq. (15.117) requires us to update only the first $2(M+1)$ elements of $\mathbf{b}(n)$. The remaining elements are just the delayed versions of $\mathbf{b}_M(n) = [b_{1,M}(n)\ b_{2,M}(n)]^{\mathrm{T}}$.

We first consider the multiplication of $\mathbf{b}(n)$ by $\mathbf{R}_{bb}^{-1}$. It involves the estimation of the correlation matrices $\mathbf{R}_{b_0b_0}$ through $\mathbf{R}_{b_Mb_M}$. These may be obtained using the recursive equations

$$\mathbf{R}_{b_mb_m}(n+1) = \beta\mathbf{R}_{b_mb_m}(n) + (1-\beta)\mathbf{b}_m(n)\mathbf{b}_m^{\mathrm{T}}(n) \tag{15.118}$$

These estimates are used to construct

$$\mathbf{R}_{bb}^{-1} = \begin{bmatrix} \mathbf{R}_{b_0b_0}^{-1} & \mathbf{0} & \cdots & \mathbf{0} & \mathbf{0} & \cdots \\ \mathbf{0} & \mathbf{R}_{b_1b_1}^{-1} & \cdots & \mathbf{0} & \mathbf{0} & \cdots \\ \vdots & \vdots & \ddots & \vdots & \vdots & \ddots \\ \mathbf{0} & \mathbf{0} & \cdots & \mathbf{R}_{b_Mb_M}^{-1} & \mathbf{0} & \vdots \\ \mathbf{0} & \mathbf{0} & \cdots & \mathbf{0} & \mathbf{R}_{b_Mb_M}^{-1} & \vdots \\ \vdots & \vdots & \ddots & \vdots & \vdots & \ddots \end{bmatrix} \tag{15.119}$$

We note that the $M+1$th diagonal element of $\mathbf{R}_{bb}^{-1}$ repeats for its subsequent diagonal elements. We also note that in a typical AE canceller, M can be chosen to be as small as 8 and inverting $M+1$ matrices of size 2×2 constitutes only a small percentage of AE canceller complexity.

To complete the computation of $\mathbf{u}(n)$, we now have to multiply $\mathbf{L}^{\mathrm{T}}$ by $\mathbf{R}_{bb}^{-1}\mathbf{b}(n)$. A close examination of the structures of the $\mathbf{L}$ matrix and $\mathbf{b}(n)$ vector described by Eqs. (15.116) and (15.117), respectively, will reveal that in order to compute $\mathbf{u}(n)$, only the first $2(M+1)$ and the last $2M$ elements of $\mathbf{u}(n)$ need to be computed. The remaining elements of $\mathbf{u}(n)$ are the delayed versions of its $(2M+1)$th and $(2M+2)$nd elements. The elements of $\mathbf{L}$ can be estimated using the two-channel Levinson–Durbin algorithm – an extension of the single-channel Levinson–Durbin algorithm. This extension is presented in Appendix I of this chapter. Note that only the coefficients of prediction filters of order 1 to M need to be computed. Accordingly, we formulate the Algorithm 1 as

1. Run the lattice predictor of order M using the Eqs. (15.98), (15.99), (15.102), and (15.103) to obtain the reflection coefficients and the backward prediction errors.
2. Run the two-channel Levinson–Durbin algorithm of Appendix 15B to convert the reflection coefficients to the backward predictor coefficients $\mathbf{G}_{m,j}$'s. The equations for this particular step are shown below for completeness.

$$\begin{aligned}
&\mathbf{A}_{1,1}(n) = \quad \boldsymbol{\kappa}_1^{\mathrm{T}}(n) \\
&\mathbf{G}_{1,1}(n) = \quad \boldsymbol{\kappa}_1(n) \\
&\text{for} \quad m = 1 : M-1 \\
&\quad \mathbf{A}_{m+1,j}(n) = \mathbf{A}_{m,j}(n) - \boldsymbol{\kappa}_{m+1}^{\mathrm{T}}(n)\mathbf{G}_{m,j}(n)\ ; \ j = 1, 2, \ldots, m \\
&\quad \mathbf{A}_{m+1,m+1}(n) = \quad \boldsymbol{\kappa}_{m+1}^{\mathrm{T}}(n) \\
&\quad \mathbf{G}_{m+1,j+1}(n) = \mathbf{G}_{m,j}(n) - \boldsymbol{\kappa}_{m+1}(n)\mathbf{A}_{m,j}(n)\ ; \ j = 1, 2, \ldots, m \\
&\quad \mathbf{G}_{m+1,1}(n) = \quad \boldsymbol{\kappa}_{m+1}(n)
\end{aligned}$$

where each $\mathbf{A}_{m,j}$ is a 2×2 forward predictor coefficient matrix.[1] The above results may be substituted in Eq. (15.116) to form the transformation matrix L.

[1] It is important to note that in the single-channel lattice, the forward and backward predictor coefficients are related according to the equation $a_{m,j} = g_{m,m+1-j}$; see Chapter 11. However, such a relationship does not hold

3. Compute the elements that are the delayed versions of the $(2M+1)$th and $(2M+2)$nd elements of $\mathbf{u}(n)$ from the previous iteration as

$$u_{1,j}(n) = u_{1,j-1}(n-1)\;;\; j = M+1, M+2, \ldots, L-M-1$$
$$u_{2,j}(n) = u_{2,j-1}(n-1)\;;\; j = M+1, M+2, \ldots, L-M-1$$

   Note $\mathbf{u}_j(n) = [u_{1,j}(n)\; u_{2,j}(n)]^{\mathrm{T}}$ is the jth 2×1 vector component of $\mathbf{u}(n)$.
4. Compute the first $2(M+1)$ elements of $\mathbf{u}(n)$ as

$$[u_{1,0}(n), u_{2,0}(n), u_{1,1}(n), u_{2,1}(n), \ldots, u_{1,M}(n), u_{2,M}(n)]^{\mathrm{T}} = \mathbf{L}_{\mathrm{tl}}^{\mathrm{T}}\mathbf{b}_{\mathrm{h}}(n)$$

   where

$$\mathbf{b}_{\mathrm{h}}(n) = \begin{bmatrix} \mathbf{R}_{b_0b_0}^{-1}(n)\mathbf{b}_0(n) \\ \mathbf{R}_{b_1b_1}^{-1}(n)\mathbf{b}_1(n) \\ \vdots \\ \mathbf{R}_{b_Mb_M}^{-1}(n)\mathbf{b}_M(n) \\ \mathbf{R}_{b_Mb_M}^{-1}(n)\mathbf{b}_M(n-1) \\ \vdots \\ \mathbf{R}_{b_Mb_M}^{-1}(n)\mathbf{b}_M(n-M) \end{bmatrix}$$

   and $\mathbf{L}_{\mathrm{tl}}$ denotes the top-left part of $\mathbf{L}$ having dimension $2(2M+1) \times 2(M+1)$.
5. Compute the last $2M$ elements of $\mathbf{u}(n)$ as

$$[u_{1,(L-M)}(n), u_{2,(L-M)}(n), \ldots, u_{1,L-1}(n), u_{2,L-1}(n)]^{\mathrm{T}} = \mathbf{L}_{\mathrm{br}}^{\mathrm{T}}\mathbf{b}_{\mathrm{t}}(n)$$

   where

$$\mathbf{b}_{\mathrm{t}}(n) = \begin{bmatrix} \mathbf{R}_{b_Mb_M}^{-1}(n-L+M+1)\mathbf{b}_M(n-L+2M) \\ \mathbf{R}_{b_Mb_M}^{-1}(n-L+M+1)\mathbf{b}_M(n-L+2M-1) \\ \vdots \\ \mathbf{R}_{b_Mb_M}^{-1}(n-L+M+1)\mathbf{b}_M(n-L+M+1) \end{bmatrix}$$

   and $\mathbf{L}_{\mathrm{br}}$ is the bottom-right part of $\mathbf{L}$ having dimension $2M \times 2M$, then the last $2M$ elements of $\mathbf{u}(n)$ are
6. Finally, compute the adaptive filter output $\hat{y}(n) = \mathbf{w}^{\mathrm{T}}(n)\mathbf{x}(n)$, the error signal $e(n) = d(n) - \hat{y}(n)$ and update the filter taps using Eq. (15.114).

To implement the lattice predictor using equations, we require $25M + 5$ multiplications. The two-channel Levinson-Durbin algorithm requires $8M(M-1)$ multiplications. We further need $6M^2 + 26M + 8$ multiplications to update $\mathbf{u}(n)$. Finally, $4N$ multiplications are required to adaptively update the transversal filter coefficients. Hence, in order to implement the fast LMS-Newton Algorithm 1, we require a total of $14M^2 + 43M + 13 + 4N$ multiplications. The number of the required additions is about the same.

in a two-channel lattice. Thus, some simplifications that are applicable to single-channel lattice equations are inapplicable to the two-channel case. Consequently, direct mimicking of the results of Chapter 11 is not possible here. This is why we present a fresh derivation of Algorithms 1 and 2.

Typically, M can take a value of 8 and the adaptive filter length N may be 1500, for a medium size office room. With these numbers, each update of $\mathbf{u}(n)$ would make up only 17% of the total computational complexity of the AE canceller.

Algorithm 2: The two-channel LMS-Newton Algorithm 1 is structurally complicated despite having reasonably low-computational complexity. Manipulating the data is not all that straightforward and hence it is more suitable for implementing using software. We now propose an alternate algorithm that is computationally less complex and can be easily implemented in hardware.

If we look at the $\mathbf{L}$ matrix given in Eq. (15.116), we observe that only the first $2(M+1)$ rows of this matrix are uniquely represented. The remaining rows are just the delayed versions of the $(2M+1)$th and $(2M+2)$nd rows given by

$$[-\mathbf{G}_{M,1} \quad -\mathbf{G}_{M,2} \quad \cdots \quad -\mathbf{G}_{M,M} \;\; \mathbf{I} \;\; 0 \;\; \cdots \quad 0]$$

As discussed in Chapter 11, for the single-channel case, Algorithm 1 may be greatly simplified by extending the input and tap-weight vectors $\mathbf{x}(n)$ and $\mathbf{w}(n)$, to the followings vectors

$$\begin{aligned}\mathbf{x}_{\mathrm{E}}(n) = [&x_1(n+M), x_2(n+M), \ldots, x_1(n+1), x_2(n+1), x_1(n), x_2(n), \ldots,\\ &x_1(n-L+1), x_2(n-L+1), \ldots, x_1(n-L-M+1), x_2(n-L-M+1)]^{\mathrm{T}}\end{aligned} \quad (15.120)$$

and

$$\begin{aligned}\mathbf{w}_{\mathrm{E}}(n) = [&w_{1,-M}(n), w_{2,-M}(n), \ldots, w_{1,-1}(n), w_{2,-1}(n), w_{1,0}(n), w_{2,0}(n), \ldots,\\ &w_{1,L-1}(n), w_{2,L-1}(n), \ldots, w_{1,L+M-1}(n), w_{2,L+M-1}(n)]^{\mathrm{T}}\end{aligned} \quad (15.121)$$

respectively, and applying Eq. (15.114) to update the extended tap-weight vector $\mathbf{w}_{\mathrm{E}}(n)$. We also need to appropriately take care of the dimensions of $\mathbf{L}$ and $\mathbf{R}_{bb}$. Moreover, because we are interested only in the tap-weights corresponding to $w_{1,0}(n)$, $w_{2,0}(n)$, $w_{1,1}(n)$, $w_{2,1}(n)$, $\ldots$, $w_{1,L-1}(n)$, $w_{2,L-1}(n)$, the first $2M$ and the last $2M$ elements of the extended tap-weight vector can be permanently set to zero. This will also remove the computation of the first $2M$ and the last $2M$ elements of $\mathbf{L}^{\mathrm{T}}\mathbf{R}_{bb}^{-1}\mathbf{L}\mathbf{x}_{\mathrm{E}}(n)$. This leads to the following update equation:

$$\mathbf{w}(n+1) = \mathbf{w}(n) + \mu\mathbf{u}_{\mathrm{a}}(n)e(n) \quad (15.122)$$

where

$$\mathbf{u}_{\mathrm{a}}(n) = \mathbf{L}_2\mathbf{R}_{bb}^{-1}(n)\mathbf{L}_1\mathbf{x}_{\mathrm{E}}(n) \quad (15.123)$$

Also, $\mathbf{L}_1$ is a $2(L+M) \times 2(L+2M)$ matrix defined as

$$\mathbf{L}_1 = \begin{bmatrix} -\mathbf{G}_{M,1} & -\mathbf{G}_{M,2} & \cdots & \mathbf{I} & \mathbf{0} \cdots & \mathbf{0} & \cdots & \mathbf{0} & \mathbf{0} \\ \mathbf{0} & -\mathbf{G}_{M,1} & \cdots & -\mathbf{G}_{M,M} & \mathbf{I} \cdots & \mathbf{0} & \cdots & \mathbf{0} & \mathbf{0} \\ \vdots & \vdots & \ddots & \vdots & \vdots \;\ddots & \vdots & \ddots & \vdots & \vdots \\ \mathbf{0} & \mathbf{0} & \cdots & \mathbf{0} & \mathbf{0} \cdots & -\mathbf{G}_{M,1} & \cdots & -\mathbf{G}_{M,M} & \mathbf{I} \end{bmatrix} \quad (15.124)$$

and $\mathbf{L}_2$ is a $2L \times 2(L+M)$ matrix defined as

$$\mathbf{L}_2 = \begin{bmatrix} \mathbf{I} & -\mathbf{G}_{M,M}^{\mathrm{T}} & \cdots & -\mathbf{G}_{M,1}^{\mathrm{T}} & \mathbf{0} & \cdots & \mathbf{0} & \cdots & \mathbf{0} & \mathbf{0} \\ \mathbf{0} & \mathbf{I} & \cdots & -\mathbf{G}_{M,2}^{\mathrm{T}} & -\mathbf{G}_{M,1}^{\mathrm{T}} & \cdots & \mathbf{0} & \cdots & \mathbf{0} & \mathbf{0} \\ \vdots & \vdots & \ddots & \vdots & \vdots & \ddots & \vdots & \ddots & \vdots & \vdots \\ \mathbf{0} & \mathbf{0} & \cdots & \mathbf{0} & \mathbf{0} & \cdots & \mathbf{I} & \cdots & -\mathbf{G}_{M,2}^{\mathrm{T}} & -\mathbf{G}_{M,1}^{\mathrm{T}} \end{bmatrix}. \tag{15.125}$$

On examining Eq. (15.123), we can see that it is only necessary to update the first 2×1 element vector of $\mathbf{R}_{bb}^{-1}\mathbf{L}_1\mathbf{x}_{\mathrm{E}}(n)$ and then the first 2×1 element vector of the final result, $\mathbf{u}_{\mathrm{a}}(n)$. The remaining elements will be the delayed versions of these first two elements.

On the other hand, recall that the forward and backward prediction errors are given as

$$\mathbf{f}_m(n) = \mathbf{x}(n) - \sum_{j=1}^{m} \mathbf{A}_{m,j}\mathbf{x}(n-j) \tag{15.126}$$

and

$$\mathbf{b}_m(n) = \mathbf{x}(n-m) - \sum_{j=1}^{m} \mathbf{G}_{m,j}\mathbf{x}(n-j+1) \tag{15.127}$$

respectively. It is well known that in a single-channel lattice $g_{m,j} = a_{m,m+1-j}$, $j = 1, 2, \ldots, m$, and this relationship between the forward and backward predictor coefficients was used to derive the single-channel LMS-Newton algorithm 1 in Chapter 11. In a two-channel lattice, $\mathbf{G}_{m,j} = \mathbf{A}_{m,m+1-j}^{\mathrm{T}}$ for only $j = 1$ (refer to Eqs. (15B-6) and (15B-9) in Appendix 15B). But on the basis of the perspective gained from extensive experimentation, in Rao and Farhang-Boroujeny (2009), it has been found that this relationship also approximately holds true for $j = 2, 3, \ldots$ Hence, we may write

$$\mathbf{G}_{m,j} \approx \mathbf{A}_{m,m+1-j}^{\mathrm{T}},\ j = 1, 2, \ldots \tag{15.128}$$

The main motivation behind introducing this approximation is to use the transposed backward predictor coefficients in reverse order to estimate the forward prediction errors, as was done in the case of single-channel in Chapter 11. Consequently, we rewrite Eq. (15.126) as

$$\mathbf{f}'_m(n) \approx \mathbf{f}_m(n) = \mathbf{x}(n) - \sum_{j=1}^{m} \mathbf{G}_{m,m+1-j}^{\mathrm{T}}\mathbf{x}(n-j) \tag{15.129}$$

From Eqs. (15.124) and (15.127), we recognize that the filtering of the input vector $\mathbf{x}(n+M)$ through a backward prediction-error filter to obtain $\mathbf{b}_M(n+M)$ is equivalent to evaluating $\mathbf{L}_1\mathbf{x}_{\mathrm{E}}(n)$. The backward prediction-error vector $\mathbf{b}_M(n+M)$ is normalized with $\mathbf{R}_{b_M b_M}^{-1}(n+M)$. This will give us an update of $\mathbf{R}_{bb}^{-1}(n+M)\mathbf{L}_1\mathbf{x}_{\mathrm{E}}(n)$. We then use the normalized backward prediction-error vector $\mathbf{R}_{bb}^{-1}(n+M)\mathbf{b}_M(n+M)$ as an input to a filter whose coefficients are the transposed duplicates of the backward prediction-error filter in reverse order. We recognize from Eqs. (15.125) and (15.129) that this filter turns out to be the forward prediction-error filter, assuming that Eq. (15.128) holds. As a result, the output of the forward prediction-error filter will provide us with the samples of the vector $\mathbf{u}_{\mathrm{a}}(n) = \mathbf{L}_2\mathbf{R}_{bb}^{-1}(n)\mathbf{L}_1\mathbf{x}_{\mathrm{E}}(n)$. Thus, we can see that the approximation introduced in Eq. (15.128) facilitates the development of an algorithm that can be efficiently implemented on hardware. At the same time, over a wide range of experiments, this algorithm has been found to satisfactorily exhibit the fast converging characteristics of the LMS-Newton algorithm.

In a practical implementation, we choose the original input to the predictor filter to be $\mathbf{x}(n)$ and not $\mathbf{x}(n+M)$. To account for this delay, the desired signal $d(n)$ is delayed by M samples to be time aligned with $\mathbf{u}_{\mathrm{a}}(n)$. This will result in a delayed LMS algorithm whose performance is very close to its nondelayed version when $M \ll L$. Moreover, as the power terms are assumed to be time invariant over the length of the prediction filters, the normalization block is moved to the output of the forward prediction-error filter. Accordingly, we formulate Algorithm 2 as:

1. Run the lattice predictor of order M using Eqs. (15.98), (15.99), (15.102), and (15.103) to obtain the reflection coefficients and the backward prediction-errors.
2. Compute the elements of $\mathbf{u}_{\mathrm{a}}(n)$ that are the delayed versions of its first two elements.

$$u_{\mathrm{a}_{1,j}}(n) = u_{\mathrm{a}_{1,j-1}}(n-1)\ ;\ j = L-1, L-2, \ldots, 1$$

$$u_{\mathrm{a}_{2,j}}(n) = u_{\mathrm{a}_{2,j-1}}(n-1)\ ;\ j = L-1, L-2, \ldots, 1$$

3. Run the lattice predictor of order M (reflection coefficients have already been computed in Step 1) with $\mathbf{b}_M(n)$ as the input to obtain the forward prediction-error vector $\mathbf{f}'_M(n)$.
4. Compute the first two elements of $\mathbf{u}_{\mathrm{a}}(n)$ to be the first two elements of $\mathbf{f}'_M(n)$ premultiplied with the 2×2 normalization matrix $\mathbf{R}_{b_M b_M}^{-1}(n)$.
5. Finally, compute the adaptive filter output $\hat{y}(n) = \mathbf{w}^{\mathrm{T}}(n)\mathbf{x}(n)$, the error signal $e(n) = d(n) - \hat{y}(n)$ and update the filter taps using Eq. (15.122).

This particular version of the LMS-Newton algorithm is computationally less intensive when compared to Algorithm 1. To implement the lattice predictor, we require $25M + 5$ multiplications. Updating $\mathbf{u}_{\mathrm{a}}(n)$ using the forward prediction-error filter requires a further $8M + 8$ multiplications. If we include the adaptive transversal filter update, Algorithm 2 requires a total of $33M + 13 + 4N$ multiplications. Thus, for $M = 8$ and $N = 1500$, updating $\mathbf{u}_{\mathrm{a}}(n)$ constitutes only 4% of the total complexity of the AE canceller.

Appendix 15A: Multitaper method

Multitaper method is a nonparametric spectrum analysis technique. It may be thought as a generalization of the celebrated periodogram spectral estimation. In this appendix, we start with a brief review of the periodogram spectral estimation method. The shortcomings of the periodogram spectral estimation method are then highlighted, and the multitaper method is introduced as a way of overcoming these shortcomings. Efficient implementation of the multitaper method, using polyphase structures, is also reviewed for completeness of the discussion.

Periodogram Spectral Estimation

The periodogram spectral estimator is the most basic and simplest member of the class of nonparametric spectral estimators. It obtains an estimate of the spectrum $\Phi_{xx}(e^{j\omega})$ of a random process $x(n)$, based on M samples of one realization of it, as

$$\hat{\Phi}_{xx}(\mathrm{e}^{j\omega_i}) = \left|\sum_{k=0}^{M-1} h_{i,k}x(n-k)\right|^2 \tag{15A.1}$$

where $\{x(n-k), k = 0, 1, \ldots, M-1\}$ is the sample set, $h_{i,n} = w_n \mathrm{e}^{j\omega_i n}$, and w_n is a window function. Clearly, if w_n's are chosen such that they are the coefficients of a finite-impulse response low-pass filter, $h_{i,n}$ will be a band-pass filter with the center frequency ω_i. Also, if we choose $\omega_i = 2\pi i/L$, $i = 0, 1, \ldots, L-1$, the set of filters $h_{i,n}$ define a L-band filter bank with the prototype filter $h_n = w_n$.

Substituting $h_{i,n} = w_n \mathrm{e}^{j\omega_i n}$ and $\omega_i = 2\pi i/M$ (i.e., assuming that $M = L$) in Eq. (15A.1) and defining $w_k x(n-k) = u_k$, we obtain

$$\hat{\Phi}_{xx}(\mathrm{e}^{j\omega_i}) = \left|\sum_{k=0}^{M-1} u_k \mathrm{e}^{j\frac{2\pi}{M}ik}\right|^2 \tag{15A.2}$$

One may note that the summation on the right-hand side of Eq. (15A.2) is the DFT of the sequence u_k. Moreover, assuming that M is properly chosen, this summation can be efficiently implemented using an FFT algorithm. This, in turn, implies that the filter bank associated with the periodogram spectral estimator can be efficiently implemented by weighting the input samples, using the weighting function w_n, and applying an FFT to the windowed samples.

In its simplest form $w_n = 1$, for $n = 0, 1, \ldots, M-1$. This is a rectangular window that is characterized with a sinc magnitude response. The sinc pulse is not desirable for spectral estimators. This is because its relatively large side lobes result in a significant spectral leakage among different frequency bands. By replacing the rectangular window with a window function that smoothly decays on both sides (a *taper*), a prototype filter with much smaller side lobes is obtained. There exists a wide range of window functions from which one may choose. Among them Hamming, Hanning and Blackman are the most popular and widely used window functions. They are, respectively, defined as

$$\text{Hamming:}\quad w_n = 0.54 - 0.46\cos\frac{2\pi n}{M-1} \tag{15A.3}$$

$$\text{Hanning:}\quad w_n = 0.5 - 0.5\cos\frac{2\pi n}{M-1} \tag{15A.4}$$

$$\text{Blackman:}\quad w_n = 0.42 - 0.5\cos\frac{2\pi n}{M-1} + 0.08\cos\frac{4\pi n}{M-1} \tag{15A.5}$$

The magnitude of frequency responses of these window functions, for $M = 32$, along with that of rectangular window, are presented in Figure 15A.1. Comparing the responses, we find that (i) the rectangular window has the narrowest main lobe (equal to $2/M$) while its side lobes are largest in magnitude; (ii) Hamming and Hanning windows achieve much lower side lobes at the cost of a wider main lobe (equal to $4/M$); (iii) Blackman window further improves the side lobes at the cost of further expansion of the width of the main lobe (equal to $6/M$).

The above window functions are very limited in controlling the width of the main lobe and the size of the side lobes of the frequency response. The size of the side lobes and the width of the main lobe are determined by the window type, and once a window type is selected, one can control only the width of the main lobe by changing the window length M. Moreover, the spectral sample estimates $\hat{\Phi}_{xx}(e^{j\omega_i})$, for $i = 0, 1, \ldots, M-1$, obtained according to Eq. (15A.1) are very coarse, because no averaging is applied. Clearly, the estimates can be smoothen, by averaging samples from successive blocks of $x(n)$. However, this results in some latency in the estimates, which, in the case of double-talk detectors, is undesirable because, to avoid possible divergence of the AE canceller,

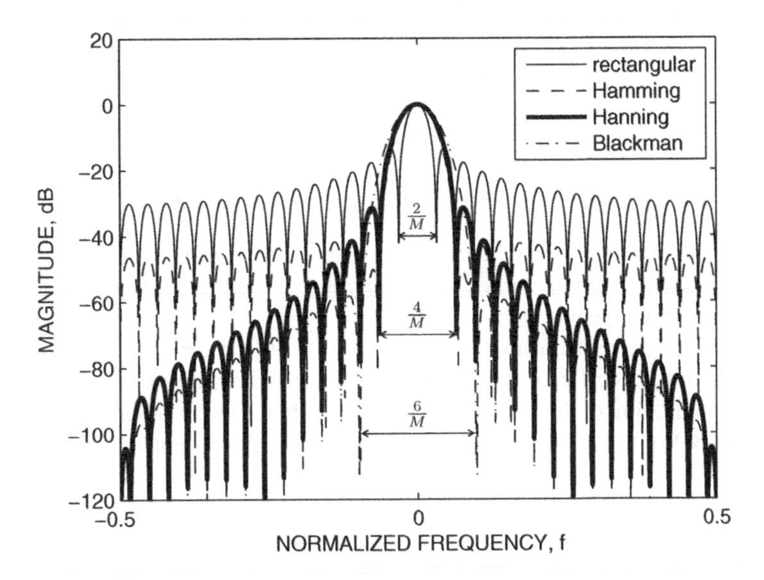

Figure 15.A.1 Frequency responses of various window functions for $M = 32$. The responses are normalized to a peak of 0 dB.

one wishes to detect any double-talk as soon as it starts. The multitaper uses multiple windows over the same block of the input signal, say, hence, the name *multitaper*, and thus allows smoothening the power spectral (cross-power spectral) estimates with a minimum latency. Furthermore, the multitaper method follows a systematic method in selecting a set of optimum window functions, called *prolate sequences*.

Prolate Sequences: The Optimal Window Functions for Multitaper Method

The process of choosing a window w_n with a target main lobe width Δf and minimizing the side lobes may be formulated as the following optimization problem:

Given a bandwidth Δf, design a low-pass FIR filter of length M whose main lobe is within the range $(-\Delta f/2, \Delta f/2)$ and has the minimum stopband energy.

The coefficients of the designed filter, w_n, constitute a sequence that is called prolate. It may be also viewed as a window function and thus referred to as *prolate window*.

Moreover, one may extend the above optimization and set the goal as finding a set of $K > 1$ *orthogonal* window functions (hence, the name *multitaper*) whose main lobes are within the range $(-\Delta f/2, \Delta f/2)$ and whose stopband energies are minimized. One may realize that this optimization problem can be solved using the minimax theorem of Chapter 4, by taking the following steps:

- Let $x(n)$ be a random process with power spectral density

$$\Phi(f) = \begin{cases} 1 & -\Delta f/2 \le f \le \Delta f/2 \\ 0 & \text{otherwise} \end{cases} \tag{15A.6}$$

- Construct the M-by-M correlation matrix $\mathbf{R}$ of $x(n)$. This is the symmetric Toeplitz matrix whose first row consists of the correlation coefficients of $x(n)$. These are obtained as

$$\phi(k) = \Delta f \operatorname{sinc}(\Delta f\, k), \quad \text{for } k = 0, 1, \ldots, M-1 \tag{15A.7}$$

- The first K eigenvectors of $\mathbf{R}$, that is, $\mathbf{q}_0, \mathbf{q}_1, \ldots \mathbf{q}_{K-1}$, are the desired prolate sequences.

The window functions obtained according to this procedure are often called *prolate sequences*. The term *Slepian sequences* is a synonym to the *prolate sequences*; hence, both are used interchangeably.

Let us elaborate. We recall from Chapter 4 that $\lambda_0 = E[|\mathbf{q}_0^{\mathrm{T}}\mathbf{x}(n)|^2]$ is the output power of an FIR filter with the coefficient vector $\mathbf{q}_0$. Moreover, according to the minimax theorem, $\mathbf{q}_0$ is selected to maximize the output power of this filter. On the other hand, when $x(n)$ is chosen to satisfy Eq. (15A.6), using the Rayleigh's relation (Chapter 2), one will find that

$$\begin{aligned}\lambda_0 &= \max_{Q_0(f)} \int_{-0.5}^{0.5} |Q_0(f)|^2 \Phi(f)\mathrm{d}f \\ &= \max_{Q_0(f)} \int_{-\Delta f/2}^{\Delta f/2} |Q_0(f)|^2 \mathrm{d}f\end{aligned}$$

$$= 1 - \min_{Q_0(f)} \int_{\Delta f/2}^{1-\Delta f/2} |Q_0(f)|^2 \mathrm{d}f \tag{15A.8}$$

where the last identity follows from the Paseval's identity $\mathbf{q}_0^{\mathrm{T}}\mathbf{q}_0 = \int_0^1 |Q_0(f)|^2 \mathrm{d}f = 1$. Accordingly, one may find that $\mathbf{q}_0$ is the coefficient vector of a filter whose stopband energy $\int_{\Delta f/2}^{1-\Delta f/2} |Q_0(f)|^2 \mathrm{d}f$ is minimized; that is, it attains the maximum attenuation of the side lobes.

Following the same argument, $\mathbf{q}_1$ will also be the coefficient vector of a low-pass filter whose stopband begins at $\Delta f/2$ and achieves the minimum stopband energy, subject to the constraint $\mathbf{q}_0^{\mathrm{T}}\mathbf{q}_1 = 0$. Clearly, this will result in a filter whose stopband attenuation will not be as good as that of $\mathbf{q}_0$. Proceeding further, one finds that subsequent filters, $\mathbf{q}_2, \mathbf{q}_3, \ldots$, will experience more loss in their stopband attenuation because of more constraints.

From the above discussions, the prolate sequences define the coefficients of a set of prototype filters with certain optimal properties. In particular, the good stopband behavior of these filters makes them a desirable candidate in the application of nonparametric spectral estimation, particularly, in applications where a wide spectral dynamic range is required. It is also worth noting that the orthogonality condition imposed on the prolate filters is to assure that under the condition where $\Phi_{xx}(e^{j\omega})$ variation over each subband is negligible, the set of outputs from various filter banks that correspond to the same subband will be uncorrelated. Hence, averaging the energy/cross-correlation of the signals from the filter banks results in power spectral/cross-power spectral estimates with a minimum variance.

An Example of Prolate Sequences

The numerical example given here is to provide a more in-depth understanding of the properties of the prolate sequences as a set of multitaper window functions. We choose $M = 64$ and set $\Delta f = 1/L$, where $L = 8$ is the number of points (approximated by frequency bands) at which power/cross-power spectra of the underlying signals has to be estimated. We also choose the first $K = M/L = 8$ prolate sequences obtained according to the procedure mentioned above.

Figure 15A.2 presents the magnitude responses of the filters associated with these sequences. This figure reveals the following facts.

- Only the first few filters have good stopband attenuation.
- Thomson (1982) has identified the number of useful prolate filters as $K - 1$.
- As is evident from the results of Figure 15A.2, the stopband responses of the prolate filters deteriorate very fast in the higher numbered filters. Here, $\mathbf{q}_6$ attains a stopband response of −20 to −25 dB. The Kth prolate filter $\mathbf{q}_7$ has a stopband response of less than −20 dB.

The number of $K - 1$ prolate filters, predicted by Thomson, is a soft limit. For some cases, the use of the first K prolate filters may be also acceptable. Experiments that we have carried out for double-talk detection have convinced us that the prediction made by Thomson is also applicable for a fair estimate of $c_{dx}(e^{j\omega})$. The results that are presented in Figures 15.10 and 15.12 are based on K prolate filters.

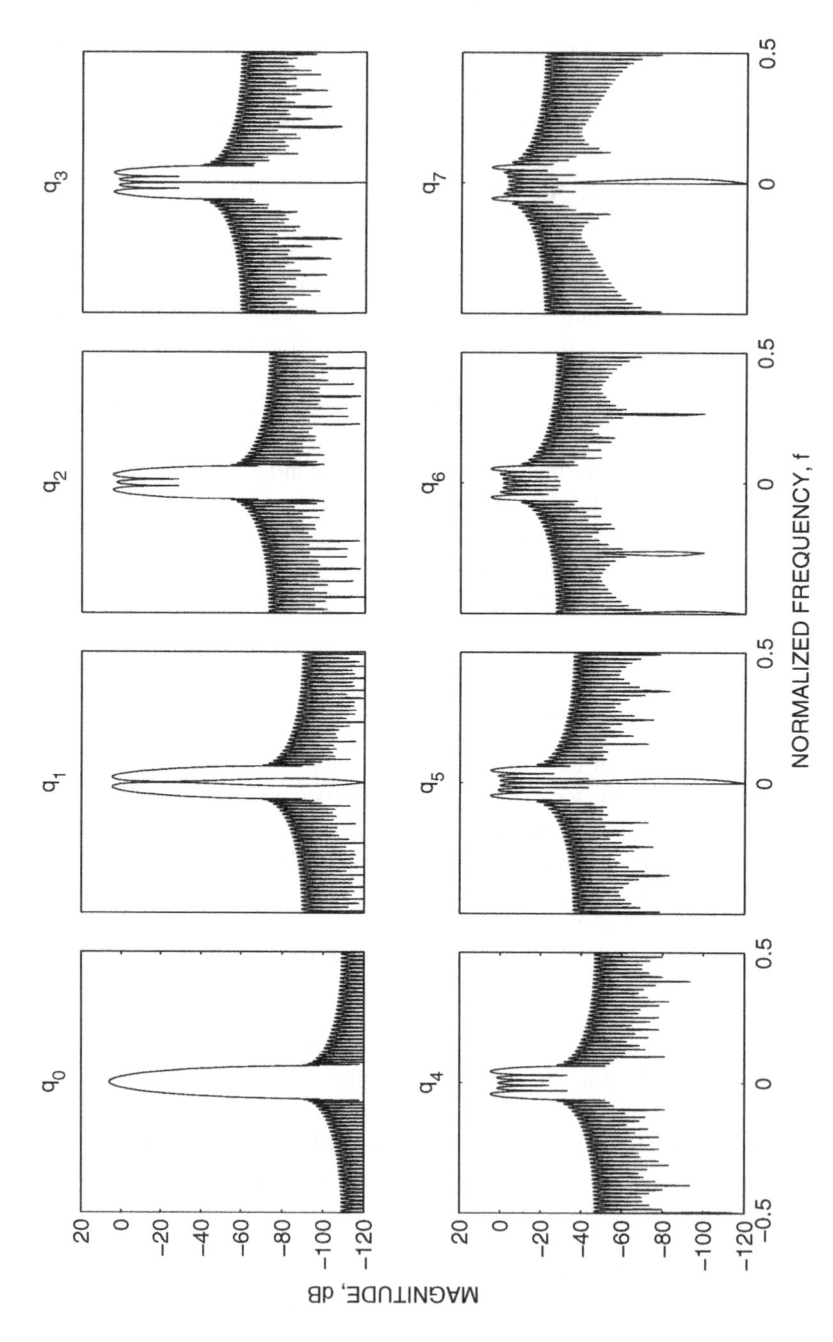

Figure 15A.2 Magnitude response of a set of prolate sequences/filters.

The results presented in Figure 15A.2 have been obtained using the MATLAB program "`prolates.m`," available on the accompanying website. Interested readers are encouraged to run this program for different choices of the parameters to see their effects.

Polyphase Filter Banks

According to the discussion in Section 15.3.3, an implementation of the multitaper method requires realization of a multiple set of filter banks. An example of such filter banks was presented in Figure 15.14. We also recall that a filter bank is naturally implemented based on a prototype filter. Here, we consider a filter bank with a prototype filter

$$H_0(z) = H(z) = \sum_{n=0}^{M-1} h_n z^{-n} \tag{15A.9}$$

Here, we are interested in implementing the set of filters

$$H_m(z) = \sum_{n=0}^{M-1} h_n e^{j\frac{2\pi mn}{L}} z^{-n} \tag{15A.10}$$

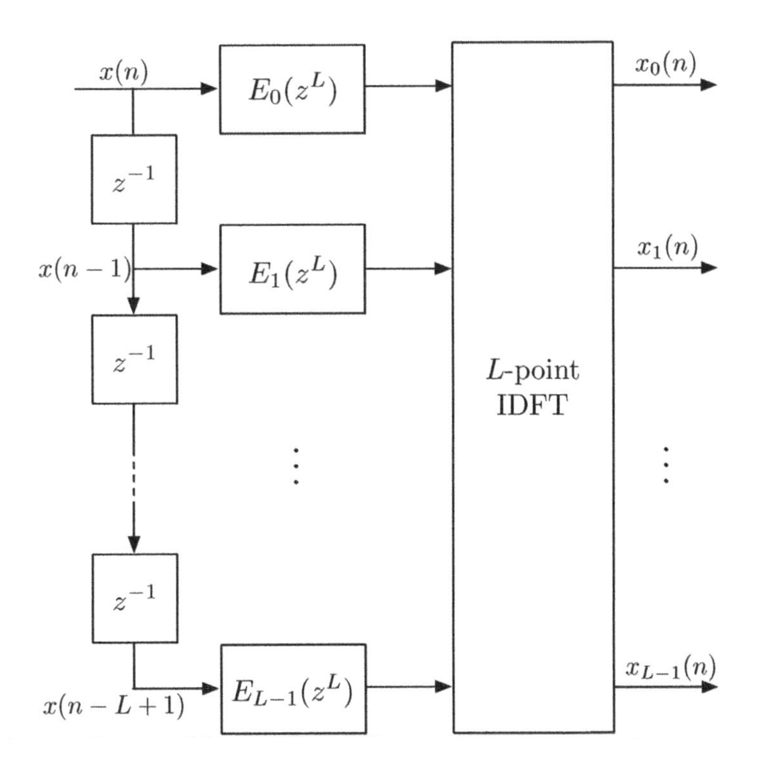

Figure 15A.3 Polyphase filter bank.

for $m = 0, 1, \ldots, L-1$. The set of filters (15A.10) can be implemented efficiently in a common structure, called *polyphase*. To develop such a polyphase structure, we assume that $K = M/L$ is an integer, let

$$n = kL + l \tag{15A.11}$$

and note that the range $0 \leq n \leq M-1$ will be covered by letting $k = 0, 1, \ldots, K-1$ and $l = 0, 1, \ldots, L-1$. Substituting Eq. (15A.11) in Eq. (15A.10), we obtain

$$H_m(z) = \sum_{l=0}^{L-1} z^{-l} E_l(z^L) e^{j\frac{2\pi ml}{L}} \tag{15A.12}$$

where

$$E_l(z) = \sum_{k=0}^{K-1} h_{kL+l} z^{-k} \tag{15A.13}$$

is the lth *polyphase component* of $H(z)$.

Using Eq. (15A.12), one will find that the set of the transfer functions $H_m(z)$, for $m = 0, 1, \ldots, L-1$, can be jointly implemented using the polyphase structure presented in Figure 15A.3. One may note that the computation of each set of output samples in this structure involves M multiplications and (approximately) M additions and one L-point IDFT operation. Clearly, the latter may be implemented efficiently using an FFT. Finally, we note that in the implementation of a multitaper, the prolate sequences $\mathbf{q}_0$, $\mathbf{q}_1$, ... are used as the prototype filters, and a number of independent filter banks are implemented.

Appendix 15B: Derivation of the Two-Channel Levinson–Durbin Algorithm

If $\mathbf{b}_0(n)$, $\mathbf{b}_1(n)$, ..., $\mathbf{b}_{L-1}(n)$ forms an orthogonal basis set for $\mathbf{x}(n)$, $\mathbf{x}(n-1)$, ..., $\mathbf{x}(n-L+1)$, then the backward prediction-error vectors are given by

$$\mathbf{b}_m(n) = \mathbf{x}(n-m) - \sum_{j=1}^{m} \mathbf{G}_{m,j}\mathbf{x}(n-j+1) \tag{15B.1}$$

Similarly, the forward prediction-error vectors are given by

$$\mathbf{f}_m(n) = \mathbf{x}(n) - \sum_{j=1}^{m} \mathbf{A}_{m,j}\mathbf{x}(n-j) \tag{15B.2}$$

We also recall the update equations

$$\mathbf{f}_{m+1}(n) = \mathbf{f}_m(n) - \boldsymbol{\kappa}_{m+1}^{\mathrm{T}}\mathbf{b}_m(n-1) \tag{15B.3}$$

and

$$\mathbf{b}_{m+1}(n) = \mathbf{b}_m(n-1) - \boldsymbol{\kappa}_{m+1}\mathbf{f}_m(n) \tag{15B.4}$$

respectively.

Using Eqs. (15B.1) and (15B.2), we can expand Eq. (15B.4) as

$$\mathbf{x}(n-m-1) - \sum_{j=0}^{m} \mathbf{G}_{m+1,j+1}\mathbf{x}(n-j) = \mathbf{x}(n-m-1) - \sum_{j=1}^{m} \mathbf{G}_{m,j}\mathbf{x}(n-j) - \boldsymbol{\kappa}_{m+1}\left(\mathbf{x}(n) - \sum_{j=1}^{m} \mathbf{A}_{m,j}\mathbf{x}(n-j)\right) \tag{15B.5}$$

On equating the coefficients of $\mathbf{x}(n-j)$, we obtain

$$\mathbf{G}_{m+1,1} = \boldsymbol{\kappa}_{m+1} \tag{15B.6}$$

$$\mathbf{G}_{m+1,j+1} = \mathbf{G}_{m,j} - \boldsymbol{\kappa}_{m+1}\mathbf{A}_{m,j}; \quad j = 1, 2, \ldots, m \tag{15B.7}$$

Similarly, we can work with the forward prediction-errors and use Eqs. (15B.1) and (15B.2) to expand Eq. (15B.3) as

$$\mathbf{x}(n) - \sum_{j=1}^{m+1} \mathbf{A}_{m+1,j}\mathbf{x}(n-j) = \mathbf{x}(n) - \sum_{j=1}^{m} \mathbf{A}_{m,j}\mathbf{x}(n-j) - \boldsymbol{\kappa}_{m+1}^{\mathrm{T}}\left(\mathbf{x}(n-m-1) - \sum_{j=1}^{m} \mathbf{G}_{m,j}\mathbf{x}(n-j)\right) \tag{15B.8}$$

Once again, equating coefficients of $\mathbf{x}(n-j)$ will lead to the following result

$$\mathbf{A}_{m+1,m+1} = \boldsymbol{\kappa}^{\mathrm{T}}_{m+1} \tag{15B.9}$$

$$\mathbf{A}_{m+1,j} = \mathbf{A}_{m,j} - \boldsymbol{\kappa}^{\mathrm{T}}_{m+1}\mathbf{G}_{m,j}; \quad j = 1, 2, \ldots, m \tag{15B.10}$$

Equations (15B.6), (15B.7), (15B.9), and (15B.10) constitute the two-channel Levinson–Durbin algorithm that may be used to convert the reflection coefficients to the predictor coefficients.

16

Active Noise Control

Acoustic noise occurs in different forms in variety of environments. Clearly, such noises are undesirable and methods of reducing them are always sought. Conventionally, the use of passive materials to absorb an acoustic noise or contain it within an enclosure has been considered. However, these methods generally found unsuccessful is suppressing the low frequency portions of acoustic signals. This is because for an absorber to be effective its thickness should be comparable or larger than the wavelength of the acoustic signal that it meant to be suppressed. Moreover, at low frequencies, the acoustic wavelength grows very fast. For instance, while, in air, the wavelength of a 5 kHz acoustic signal is 6.8 cm, it increases to 3.4 m at 100 Hz.

The term *active noise control* (ANC) refers to the methods where acoustic (and, also, hydroacoustic) noise signals are canceled using a combination of electromechanical systems. ANC methods are built based on the fact that an acoustic signal/pressure can be nullified by introducing an acoustic pressure of opposite direction. To put this in a right prospect and also to be able to discuss the advantages as well as the limitations of ANC methods, in this chapter, we present two (important) examples of ANC systems.

The first case of interest is the problem of noise cancellation in air ducts, for example, air conditioning ducts and exhaust pipes in cars. This case is often exemplified by the noise cancellation setup that was presented in Figure 1.20 and for convenience of reference is repeated here in Figure 16.1.

The second case that we consider is when a primary acoustical source has generated a signal (noise) that reaches a point P in space with a pressure of $p(t)$. A microphone measures this pressure and through an ANC filter instructs a secondary source (a canceling loudspeaker) at a point near P to broadcast the same signal, but with an opposite sign. This concept is presented in Figure 16.2. We note that when the primary source is at some location far away from P, the cancellation will be limited to the points near P, that is, a silent zone around the error microphone is generated. This is because the acoustic pressure resulting from the secondary source vanishes in space relatively fast, while that of the primary source remains nearly intact at the points surrounding P. This may be understood, if one notes that the sound pressure resulting from a source in free space at a distance r is proportional to r^{-1}.

The setups presented in Figures 16.1 and 16.2 are different in a number of ways. While the setup in Figure 16.2 is that of an acoustic noise cancellation in a three-dimensional

Adaptive Filters: Theory and Applications, Second Edition. Behrouz Farhang-Boroujeny.

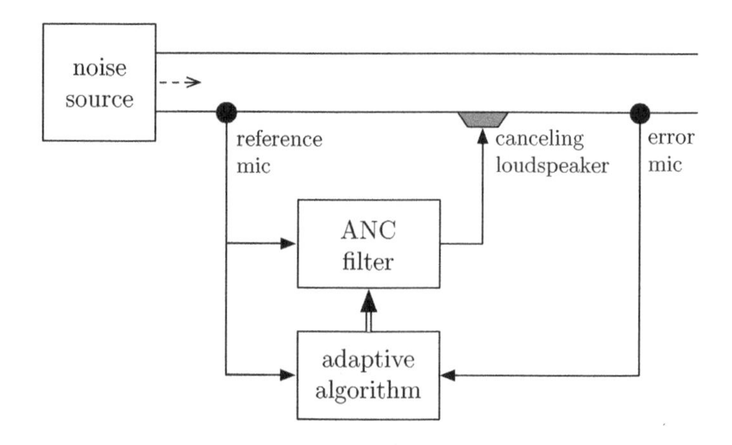

Figure 16.1 ANC in an air duct.

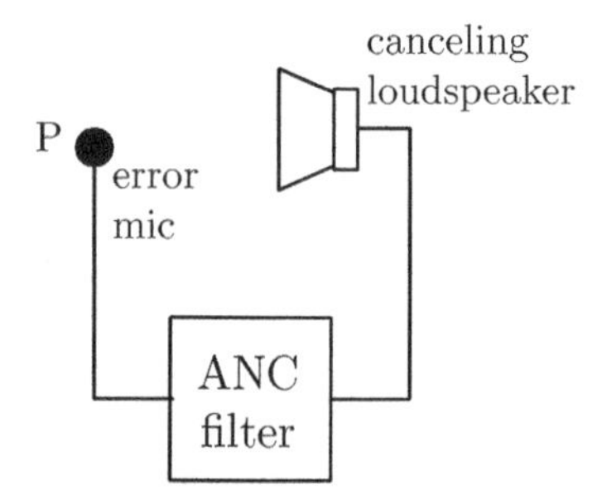

Figure 16.2 ANC surrounding a point P.

space, the one in Figure 16.1 may be thought of as a noise cancellation in a one-dimensional sound field along a pipe/duct, that is, a one-dimensional space. As a result, in Figure 16.1, once the acoustic signal/pressure is suppressed at a point along the duct, it will be also suppressed at all points to the right-hand side of that point. The ANC setup in Figure 16.2, on the other hand, is only able to generate a silent zone surrounding the point P. The setup in Figure 16.1 consists of a canceling loudspeaker, a reference microphone, and an error microphone. The setup in Figure 16.2, on the other hand, consists of only a canceling loudspeaker and an error microphone. There is no reference microphone in Figure 16.2. Accordingly, the setup in Figure 16.1 may be categorized as an open loop system in which the signal to the canceling loudspeaker is generated by filtering the signal picked up by the reference microphone. The feedback from the error microphone is used to evaluate the residual error that will be used for the adaptation of the ANC filter. In Figure 16.2, on the other hand, the signal picked up by the error microphone, which partly comes from the canceling loudspeaker, is the input to the ANC filter that derives the canceling loudspeaker. This is clearly a closed loop system.

In the rest of this chapter, we elaborate on the details of the implementation of the open loop ANC setup of Figure 16.1 as well as the closed loop ANC setup of Figure 16.2. We also introduce the multichannel ANC systems where multiple reference/error microphones and canceling loudspeakers are used to improve the performance of ANC systems. The multichannel extension of Figure 16.2, in particular, is important as it allows expansion of the size of the silent zone. Many acoustic noises originate from rotary electromechanical systems, such as motors, engines, compressors, fans, and propellers. Such systems lead to an acoustic noise that is periodic, hence, consists of a fundamental sine wave and its harmonics. ANC systems that are designed to cancel sinusoidal/periodic noises are referred to as *narrowband*, because of obvious reasons. Other types of ANC systems are referred to as *broadband*.

16.1 Broadband Feedforward Single-Channel ANC

Broadband feedforward single-channel ANC systems are often exemplified by the ANC in an air duct, that is, the case presented in Figure 16.1. Figure 16.3 has repeated the same figure and has added to it the pertinent transfer functions related to the system design. The acoustic path between the points near the reference microphone and the error microphone is denoted by $P(z)$. This we refer to as the *primary path*. There are also two *secondary* paths between the canceling loudspeaker and the reference and error microphones, denoted as $S_1(z)$ and $S_2(z)$, respectively. The ANC filter has to be adapted so that the sum of the acoustic signals at the error microphone vanishes to zero. We also note that although in practice the underlying acoustic paths are analog in nature, here, we have chosen to present them with the discrete transfer functions $P(z)$, $S_1(z)$, and $S_2(z)$, for consistency of our presentation with the rest of the chapters in this book as well as the notational convenience.

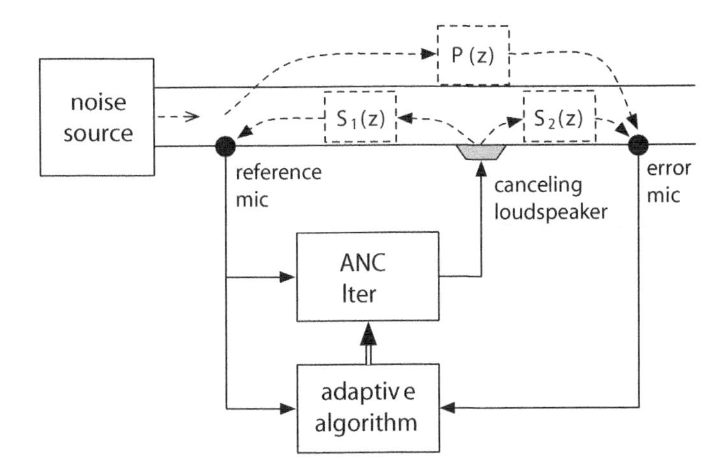

Figure 16.3 Details of ANC in an air duct.

16.1.1 System Block Diagram in the Absence of the Secondary Path $S_1(z)$

At this stage of the system development, to simplify the discussion, the secondary path from the canceling loudspeaker to the reference microphone, $S_1(z)$, is assumed to be absent. We add this and accordingly modify the proposed ANC system later, in Section 16.1.4.

Figure 16.4 presents a block diagram of the broadband feedforward single-channel ANC system of Figure 16.3, in the absence of $S_1(z)$. The adaptive filter $W(z)$ denotes the transfer function of the ANC filter. The error microphone is denoted by a summing node whose inputs are $d(n)$ and $y(n)$. This arrangement is slightly different from the conventional adaptive filter structures that have been presented so far in this book. We note that this has no significant practical implication since with this new arrangement, the tap weights of the adaptive filter $W(z)$ converge to the negative of their respective values if a minus sign was added to the left-side input of the summing node.

One may realize that if the secondary path $S_2(z)$ is removed from Figure 16.4, it reduces to the system modeling problems that were introduced in the various parts of the previous chapter, for example, Figure 6.5. In that case, the error signal $e(n)$ vanishes to zero, if $W(z)$ is chosen such that $W(z) + P(z) = 0$, or if we let

$$W(z) = -P(z) \tag{16.1}$$

The presence of $S_2(z)$ in Figure 16.4 leads to a different modeling problem whose solution, for reducing $e(n)$ to zero, should satisfy the following equation

$$W(z)S_2(z) + P(z) = 0 \tag{16.2}$$

Solving Eq. (16.2) for $W(z)$, we obtain

$$W(z) = -\frac{P(z)}{S_2(z)} \tag{16.3}$$

In practice, Eq. (16.3) may not be a feasible solution. First, because $1/S_2(z)$ may not be a stable system. Second, even when $1/S_2(z)$ is a stable system, because of the reasons discussed in Chapter 3, the choice of a recursive filter for $W(z)$ is usually undesirable.

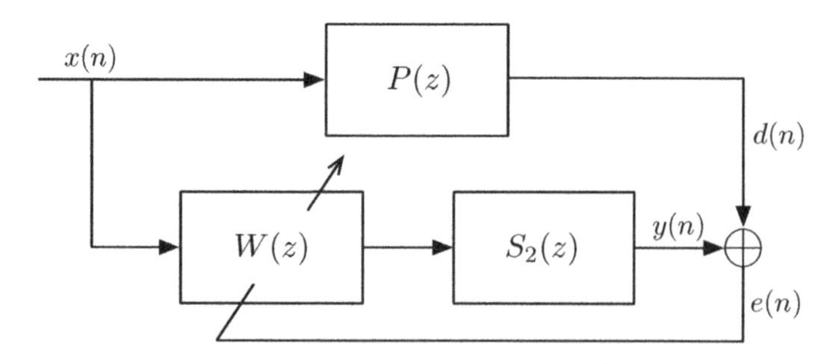

Figure 16.4 A simplified block diagram of the air duct ANC when the secondary path $S_1(z)$ is removed. The input $x(n)$ is the noise signal at the reference microphone.

Hence, in practice, the solution (16.3) is approximated by an (FIR) filter that may be adapted using variety of the algorithms that were discussed in the previous chapters. However, we note that the presence of the transfer function $S_2(z)$ between the adaptive filter $W(z)$ and the error signal $e(n)$ requires a careful revisit and modification of any adaptive algorithm that one selects to use. Such development for the case of the LMS algorithm is presented next.

16.1.2 Filtered-X LMS Algorithm

Filtered-X LMS algorithm refers to a version of the LMS algorithm that is obtained by applying the latter to the modeling problem of Figure 16.4. The choice of the prefix *filtered-X* will become evident as we go through the development of the algorithm.

As in the case of the conventional LMS algorithm, we begin with the instantaneous cost function $\hat{\xi}(n) = e^2(n)$ and use this in the implementation of the steepest-descent update

$$\mathbf{w}(n+1) = \mathbf{w}(n) - \mu \nabla e^2(n) \tag{16.4}$$

We note that, here,

$$\begin{aligned} e(n) &= d(n) + y(n) \\ &= d(n) + \mathbf{w}^{\mathrm{T}}(n)\mathbf{x}'(n) \end{aligned} \tag{16.5}$$

where $\mathbf{w}(n)$ is the tap-weight vector of the adaptive filter $W(z)$ at the time instant n, and the second line is obtained by noting that $y(n)$ is obtained by passing $x(n)$ through the cascade of $W(z)$ and $S_2(z)$, swapping the order of $W(z)$ and $S_2(z)$, and defining $x'(n)$ as a *filtered* version of $x(n)$ obtained from passing $x(n)$ through $S_2(z)$. Clearly, the prefix *filtered-X* conveys this rearrangement of the underlying signal sequences. Moreover, assuming that the ANC filter has a length of N, we follow the convention of the previous chapters and define the filtered-input and tap-weight vectors

$$\mathbf{x}'(n) = [x'(n)\ x'(n-1)\ \cdots\ x'(n-N+1)]^{\mathrm{T}} \tag{16.6}$$

and

$$\mathbf{w}(n) = [w_0(n)\ w_1(n)\ \cdots\ w_{N-1}(n)]^{\mathrm{T}} \tag{16.7}$$

respectively.

Substituting Eq. (16.5) in Eq. (16.4), the filtered-X LMS update is obtained as

$$\mathbf{w}(n+1) = \mathbf{w}(n) - 2\mu e(n)\mathbf{x}'(n) \tag{16.8}$$

Moreover, the block diagram presented in Figure 16.5 depicts the above developments in a clear-to-understand form. In this diagram, $\hat{S}_2(z)$ is an estimate of $S_2(z)$ that may also be obtained through an LMS algorithm, before the activation of the ANC filter, $W(z)$.

16.1.3 Convergence Analysis

Convergence behavior of the filtered-X LMS algorithm is very similar to that of the LMS algorithm. To provide some insight, we note that under a slow adaptation, that

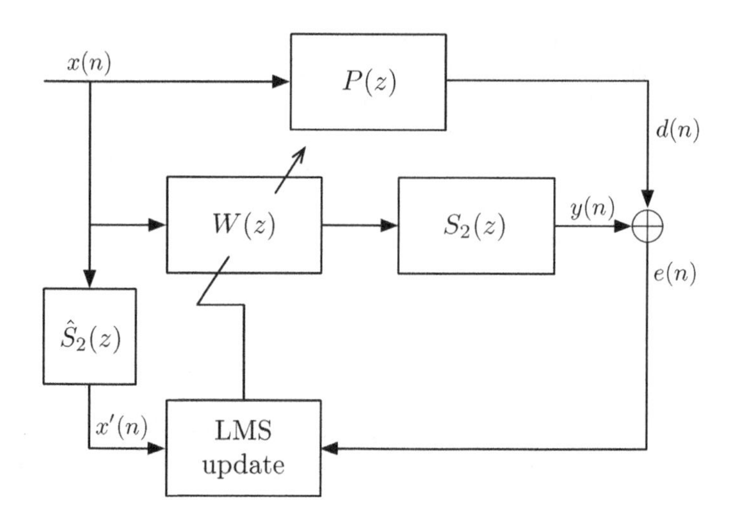

Figure 16.5 Detail of implementation of the filtered-X LMS algorithm.

is, when the step-size parameter μ is sufficiently small (e.g., when μ is selected to achieve a misadjustment of 10% or less), over any short period of time $W(z)$ may be assumed to be a linear and time-invariant system. Thus, if the duration of the impulse response of $S_2(z)$ can be considered as short, one may swap the blocks $W(z)$ and $S_2(z)$ in Figure 16.5. Moreover, if we assume that $\hat{S}_2(z) = S_2(z)$, Figure 16.5 may be rearranged as in Figure 16.6.

Considering Figure 16.6, we have an adaptive filter with the input signal $x'(n)$ and a desired output $d(n)$. From the discussion in Chapter 6, one may recall that the convergence behavior of the LMS algorithm is mostly determined by the statistics of its input signal.

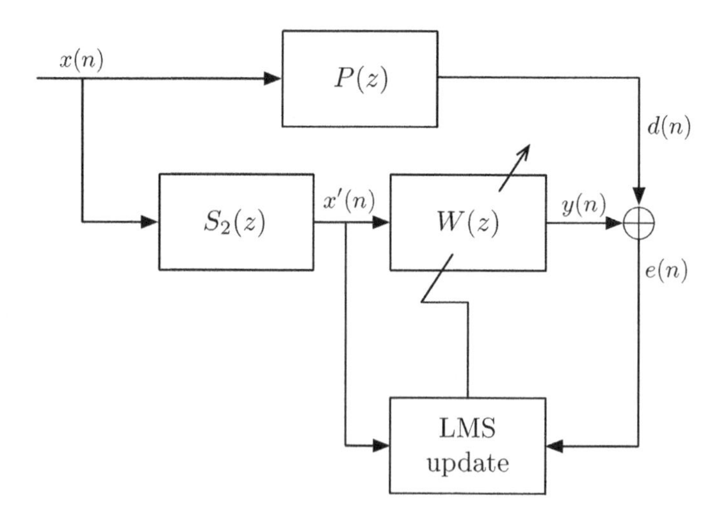

Figure 16.6 An equivalent to the block diagram of Figure 16.5.

More specifically, the convergence behavior of the LMS algorithm is controlled by a number of time constants whose values are given by

$$\tau_i = \frac{1}{4\mu\lambda_i}, \quad \text{for } i = 0, 1, \ldots, N-1 \tag{16.9}$$

where λ_i's are the eigenvalues of the correlation matrix $\mathbf{R}$ of the tap inputs of the adaptive filter. Here, the input to the adaptive filter is $x'(n)$ and, thus,

$$\mathbf{R} = E[\mathbf{x}'(n)\mathbf{x}'^{\mathrm{T}}(n)] \tag{16.10}$$

According to the analysis presented in Chapter 6, one may say that a sure convergence of the LMS algorithm is guaranteed if

$$0 < \mu < \frac{1}{3\mathrm{tr}[\mathbf{R}]} \tag{16.11}$$

Although this range of μ works pretty well in most cases of ANC systems, for some cases the upper limit of the range, $1/3\mathrm{tr}[\mathbf{R}]$, may not be tight enough. Going back to Figure 16.4, one may note that the presence of the secondary path between the ANC filter and the error microphone may introduce a significant delay in the adaptation loop and such delay, like in any feedback loop, can be a source of instability. In the context of ANC adaptation, this may mean the instability (i.e., divergence) of the filtered-X LMS algorithm. Hence, when the delay introduced by $S_2(z)$ is significantly large, one may reduce the upper bond of the stability range (16.11). Alternatively, by positioning the error microphone closer to canceling loudspeaker, one may reduce the delay in the adaptation loop and thus reduce the chance of instability of the filtered-X LMS algorithm.

Yet, the ANC systems, in general, suffer from a number of other problems that should be carefully considered in their design and implementation. The noise signal $x(n)$ picked up by the reference microphone may be highly colored. More coloring will be added to it after passing through the secondary path estimate $\hat{S}_2(z)$. This in turn implies that the correlation matrix $\mathbf{R} = E[\mathbf{x}'(n)\mathbf{x}'^{\mathrm{T}}(n)]$ whose eigenvalues determine the modes of convergence of the filtered-X LMS algorithm, may have a number of eigenvalues that are close to zero. Hence, filtered-X LMS algorithm may converge very slowly or diverge away from the desired optimum solution within the subspace that corresponds to the near zero eigenvalues of $\mathbf{R}$. This problem may be resolved by replacing the update equation (16.8) by its leaky version

$$\mathbf{w}(n+1) = \beta\mathbf{w}(n) - 2\mu e(n)\mathbf{x}'(n) \tag{16.12}$$

where β is constant smaller than, but close to, 1. An interested reader may refer to Section 15.5.2 for a detailed discussion of the leaky LMS algorithm.

16.1.4 Adding the Secondary Path $S_1(z)$

Figure 16.7 presents a complete block diagram equivalent to Figure 16.3. This is an improved diagram over the one presented in Figure 16.4, where the secondary path $S_1(z)$

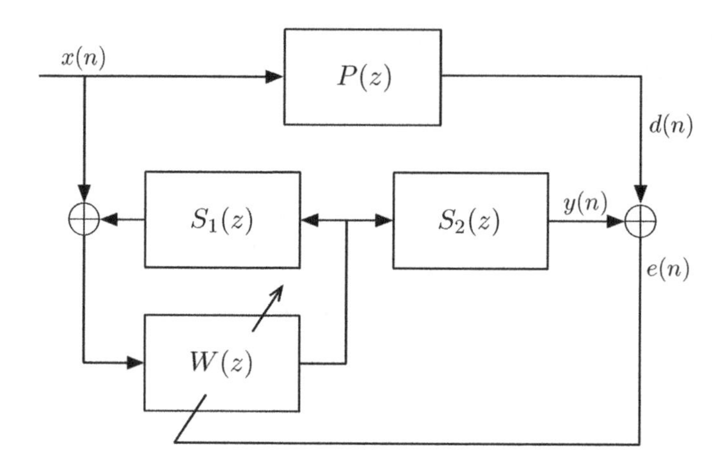

Figure 16.7 An equivalent to the block diagram of Figure 16.5.

was ignored. The transfer function between the input $x(n)$ and the output error $e(n)$ may be evaluated and shown to be

$$\frac{E(z)}{X(z)} = P(z) + \frac{W(z)}{1 - S_1(z)W(z)}S_2(z) \tag{16.13}$$

Letting $e(n) = 0$, thus, $E(z) = 0$, and solving Eq. (16.13) for $W(z)$, we obtain

$$W(z) = \frac{P(z)}{P(z)S_1(z) - S_2(z)} \tag{16.14}$$

As discussed in the case of Eq. (16.3), the realization of the ANC filter $W(z)$ of Figure 16.7 in an IIR form as predicted by Eq. (16.14) is also problematic. Hence, an FIR solution should be sought. However, unfortunately, unlike Eq. (16.3), the approximation of Eq. (16.14) by an FIR is not possible because of the following reason. In practice, the canceling loudspeaker and the error microphone are placed at some distance away from the reference microphone. The error microphone, on the other hand, is placed at a point near the canceling loudspeaker. This arrangement results in high order transfer functions for the primary path $P(z)$ and the secondary path $S_1(z)$, and a significantly lower order transfer function for the secondary path $S_2(z)$. Therefore, one may note that while the denominator of the right-hand side of Eq. (16.3) contains a low order polynomial, the denominator of the right-hand side of Eq. (16.14) contains a significantly higher order polynomial. This, in turn, implies that while the transfer function (16.3) can be reasonably approximated by an FIR filter, this may not be possible for the transfer function (16.14), unless a very high order FIR filter is used. To resolve this problem, the following solution has been adopted in most of the ANC systems. An estimate of the secondary path $S_1(z)$, say, $\hat{S}_1(z)$, is obtained and used to generate a model of $S_1(z)$ and remove its effect digitally as presented in Figure 16.8. One may note that this is very similar to the acoustic echo cancellation setup that was presented in Chapter 15.

One may note that when $\hat{S}_1(z) = S_1(z)$, the block diagram presented in Figure 16.8 reduces to the one in Figure 16.4, hence, all the development subsequent to the presentation of the latter figure are also applicable to the former. In particular, the addition of

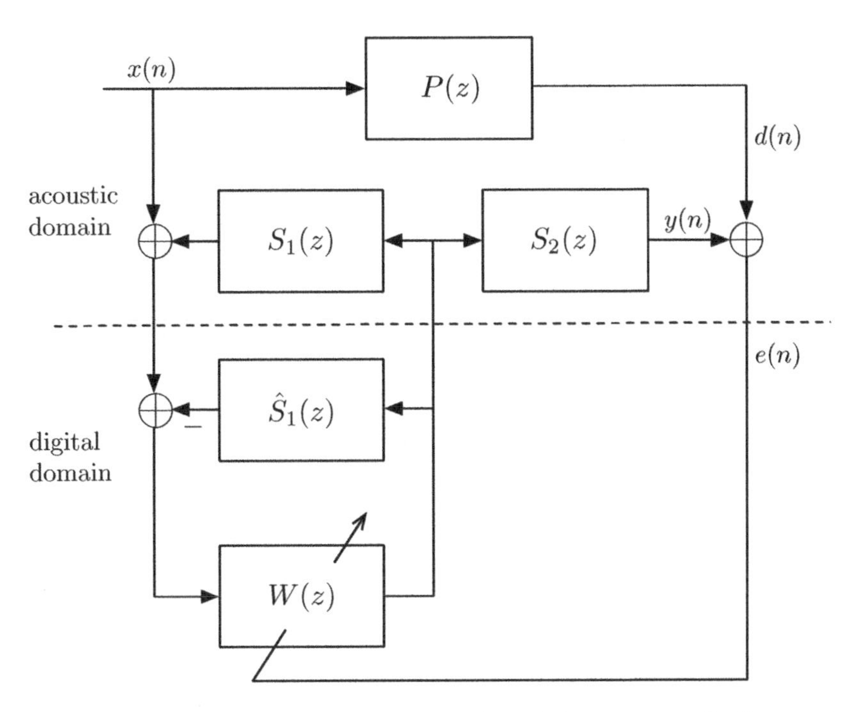

Figure 16.8 Block diagram showing how the effect of the secondary/feedback path $S_1(z)$ is removed.

the filtered-X LMS to Figure 16.8 is a straightforward task. Doing so, and replacing the acoustic blocks in Figure 16.8, by the actual components (the duct, the loudspeaker, and the microphones), a complete ANC system for air ducts is obtained. This is presented in Figure 16.9.

The ANC system presented in Figure 16.9 has two modes of operation: (i) offline processing and (ii) online processing. In the offline mode, the input to the canceling loudspeaker comes from a broadband noise source and the secondary paths $S_1(z)$ and $S_2(z)$ are estimated using an adaptive algorithm, for example, an LMS algorithm. Once the convergence of $\hat{S}_1(z)$ and $\hat{S}_2(z)$ are complete, the ANC is switched to the online mode where the adaptation of the ANC filter $W(z)$ begins.

16.2 Narrowband Feedforward Single-Channel ANC

In many applications of ANC, the underlying noise sources and, thus, the generated noise signals are periodic. Such periodic noise signals are generated by rotary machines, such as engines, motors, fans, and compressors. Recalling from the Fourier series theory that any periodic signal can be expanded as a summation of a number of sine waves, consisting of a fundamental and its harmonic components, one may adopt the adaptive filtering structure presented in Figure 16.10 which, without any loss of generality, is given for ANC in an air duct. A tachometer/sensor detects the fundamental period/frequency of the noise source. This information is passed to a signal generator to generate a periodic signal with the same fundamental frequency as the source. The generated signal is passed to an adaptive

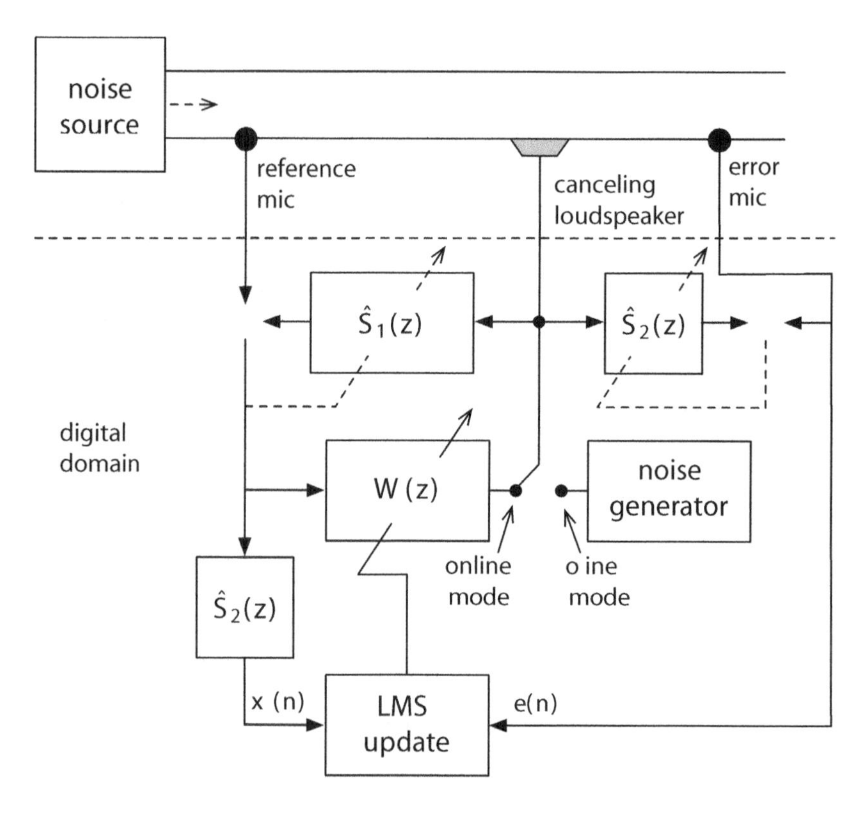

Figure 16.9 Complete ANC system for an air duct.

filter (the ANC filter) which shapes its spectrum to match that of the noise amplitude and phase (with an opposite sign) at the canceling loudspeaker. The error microphone picks the residual noise and instructs the adaptation of the ANC filter.

Comparing Figure 16.10 with its broadband counterpart in Figure 16.1, one may observe the following fundamental differences. In Figure 16.1, the broadband noise is picked up by the reference microphone. In Figure 16.10, on the other hand, the period/frequency of the periodic noise is obtained through a nonacoustic device. A reference signal is then generated accordingly. In both cases (Figures 16.1 and 16.10), the ANC filter shapes the spectrum of the reference signal such that a correct anti-noise is generated at the canceling loudspeaker. However, the feedback from the canceling loudspeaker to the reference microphone, which can be a major source of instability, has no counterpart in the case of narrowband ANC systems. Hence, the narrowband ANC systems are less prone to instability problems and thus are easier to design and implement.

16.2.1 Waveform Synthesis Method

The waveform synthesis method constructs a sign-inverted version of the periodic noise, viz., the anti-noise, at the canceling loudspeaker through the following mechanism. The ANC filter is chosen to be an FIR filter with a length equal to one period of the

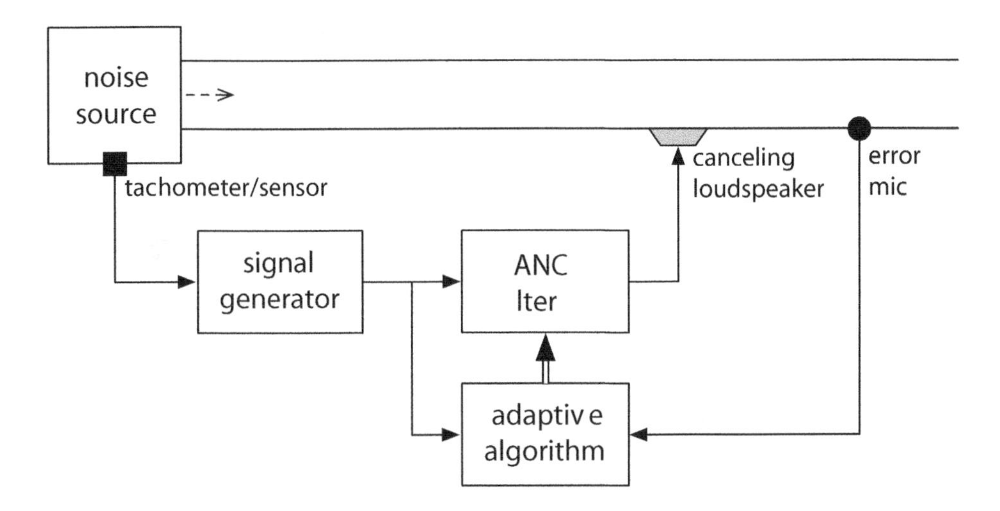

Figure 16.10 Adaptive filtering structure for narrowband ANC in an air duct.

noise/anti-noise. It is exited by a sequence of impulses at a spacing equal to one period of the anti-noise, and its coefficients are adjusted to converge toward the samples of the anti-noise. This concept is presented in Figure 16.11.

Analysis with the Secondary Path Excluded

To gain some insight to the operation of the waveform synthesis method, we ignore the secondary path between the canceling loudspeaker output and the error microphone and accordingly present the equivalent block diagram of Figure 16.12. We assume that the periodic noise has a period of N samples. Hence, at the time instant n, the ANC filter transfer function may be written as

$$W(z) = \sum_{i=0}^{N-1} w_i(n) z^{-i} \tag{16.15}$$

The input to $W(z)$ is

$$x(n) = \begin{cases} 1, & n = \text{any integer multiple of } N \\ 0, & \text{otherwise} \end{cases} \tag{16.16}$$

The desired signal $d(n)$ in the adaptive filtering setup of Figure 16.12 is the noise signal that reaches the error microphone. The adder presents the error microphone.

The setup presented in Figure 16.12 has the following interesting property. When the LMS algorithm is used to adapt the coefficients of $W(z)$ and the block diagram presented in Figure 16.12 is treated as a systems with input $d(n)$ and output $e(n)$, it behaves like a linear time-invariant system. We prove this by showing that $d(n)$ and $e(n)$ are related through a difference equation with a set of time-invariant coefficients.

We first note that the LMS recursion in Figure 16.12 is obtained as

$$\mathbf{w}(n+1) = \mathbf{w}(n) - 2\mu e(n)\mathbf{x}(n) \tag{16.17}$$

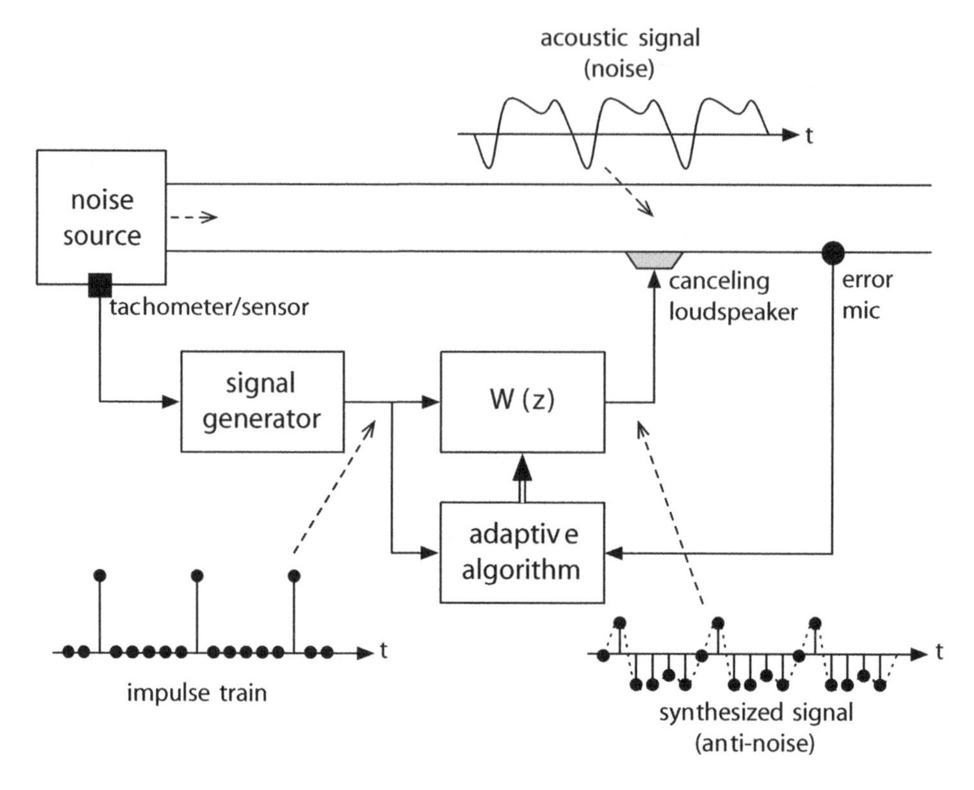

Figure 16.11 Waveform synthesis method applied to a narrowband ANC in an air duct.

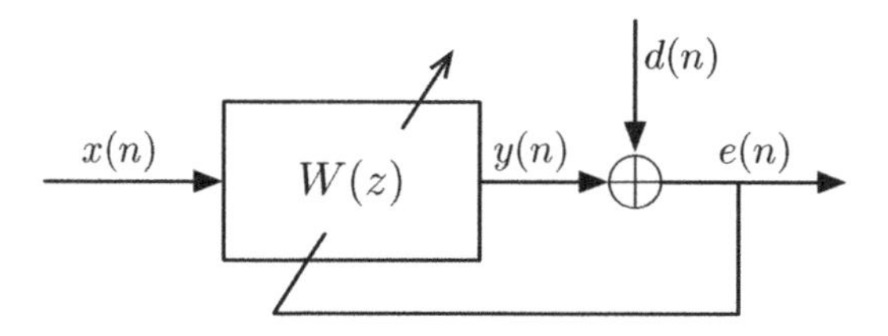

Figure 16.12 An equivalent diagram of Figure 16.11 when the secondary path between the canceling loudspeaker output and the error microphone is ignored.

Next, we note that for a given n,

$$\mathbf{x}(n) = \begin{bmatrix} 0 \\ \vdots \\ 0 \\ 1 \\ 0 \\ \vdots \\ 0 \end{bmatrix} \tag{16.18}$$

where the term "1" is at the ith position and $i \triangleq i(n) = (n \text{ modulo } N)$. This implies that at the time instant n only the $i(n)$th element of $\mathbf{w}(n)$ will be updated, hence, Eq. (16.17) may be written as

$$w_{i(n)}(n+1) = w_{i(n)}(n) - 2\mu e(n) \tag{16.19}$$

On the other hand, recalling

$$y(n) = \sum_{i=0}^{N-1} w_i(n)x(n) \tag{16.20}$$

and using Eq. (16.18), one will find that

$$y(n) = w_{i(n)}(n) \tag{16.21}$$

Moreover, we note that the updated tap weight $w_{i(n)}(n+1)$ of Eq. (16.19) remains unchanged until the time instant $n+N$ and, hence,

$$y(n+N) = w_{i(n+N)}(n+N) = w_{i(n)}(n+1) \tag{16.22}$$

Using Eqs. (16.21) and (16.22) in Eq. (16.17), we obtain

$$y(n+N) = y(n) - 2\mu e(n) \tag{16.23}$$

Moreover, considering the fact that $d(n) + y(n) = e(n)$ and $d(n+N) + y(n+N) = e(n+N)$, Eq. (16.23) may be rearranged as

$$d(n+N) - d(n) = e(n+N) - (1-2\mu)e(n) \tag{16.24}$$

Finally, replacing n by $n-N$ in Eq. (16.24), we obtain

$$d(n) - d(n-N) = e(n) - (1-2\mu)e(n-N) \tag{16.25}$$

This is the linear difference equation that relates $d(n)$ and $e(n)$.

Taking the z-transforms of both sides of Eq. (16.25) and rearranging the result, we obtain

$$H(z) = \frac{E(z)}{D(z)} = \frac{1-z^{-N}}{1-(1-2\mu)z^{-N}} \tag{16.26}$$

This is the transfer function that relates the original noise, $d(n)$, and its suppressed version, $e(n)$. $H(z)$ has N zeros at the equally spaced angle positions $0, 2\pi/N, 4\pi/N, \ldots, 2(N-1)\pi/N$ on the unit circle, $|z| = 1$, and N poles at the same angles, but on the circle $|z| = \sqrt[N]{1-2\mu}$. These poles and zeros are presented in Figure 16.13, for $N = 8$. The half-power/3 dB bandwidth of each null is given by

$$\text{BW} \approx 2\left(1 - \sqrt[N]{1-2\mu}\right) \tag{16.27}$$

The reader may refer to the discussion regarding Eq. (16.52), in the following, for derivation of a similar equation.

Figure 16.14 presents a plot of the magnitude response of $H(z)$, when $N = 8$ and $\mu = 0.4$. As seen, $H(z)$ is a multinotch filter with N notch frequencies in the interval

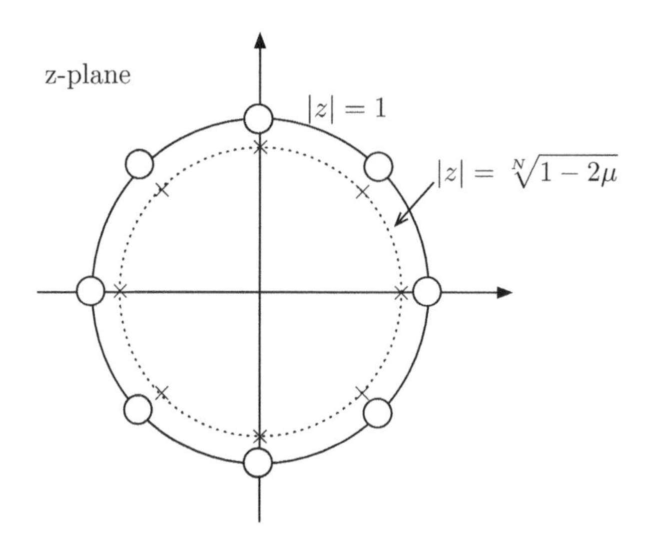

Figure 16.13 Zeros and poles arrangement of $H(z)$, for $N = 8$.

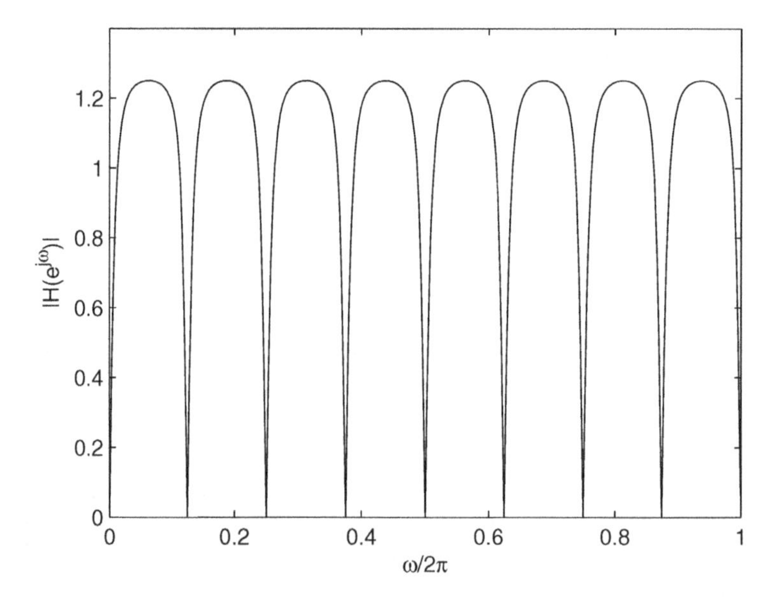

Figure 16.14 An example of magnitude response of the multinotch filter $H(z)$, for $N = 8$ and $\mu = 0.2$.

$0 \leq \omega < 2\pi$. Also, the step-size parameter μ determines the bandwidth of each notch, according to Eq. (16.27). A smaller μ results in a narrower band notch.

In some applications, it is desirable to keep some residual of the noise; that is, noise cancellation should be purposefully kept imperfect. This can be easily achieved by replacing the LMS recursion (16.17) by its leaky counterpart, viz.,

$$\mathbf{w}(n+1) = \beta\mathbf{w}(n) - 2\mu e(n)\mathbf{x}(n) \tag{16.28}$$

where β is a constant close to, but smaller than 1. Starting with Eq. (16.28), and following a similar line of derivation to those that led to Eq. (16.26), we obtain

$$H(z) = \frac{E(z)}{D(z)} = \frac{1 - \beta z^{-N}}{1 - (1 - 2\mu)z^{-N}} \tag{16.29}$$

To understand the impact of the leakage parameter β on the magnitude response of $H(z)$, we note that for $z = \mathrm{e}^{j\omega}$ and values of $\omega_k = 2k\pi/N$, for $k = 0, 1, \ldots, N-1$,

$$|H(\mathrm{e}^{j\omega_k})| = \frac{1 - \beta}{2\mu} \tag{16.30}$$

When $\beta = 1$, that is, when the conventional LMS recursion (16.17) is used, $|H(\mathrm{e}^{j\omega_k})| = 0$. That is, the nulls are perfect. On the other hand, when $\beta < 1$, the gain of the nonideal nulls is given by Eq. (16.30). Clearly, to have meaningful nulls, the condition $|H(\mathrm{e}^{j\omega_k})| \ll 1$ or, equivalently,

$$1 - \beta \ll 2\mu \tag{16.31}$$

should hold.

Analysis with the Secondary Path $S_2(z)$ Included

Consider the case where the secondary path between the canceling loudspeaker and the error microphone, $S_2(z)$, is taken into account and the filtered-X LMS algorithm is used for the adaptation of $W(z)$. Following the same line of argument to the one that led to Figure 16.6, here, we obtain the equivalent block diagram of Figure 16.15. Also shown in Figure 16.15 are the power spectral density $\Phi_{xx}(\mathrm{e}^{j\omega})$ of the signal generator output (a

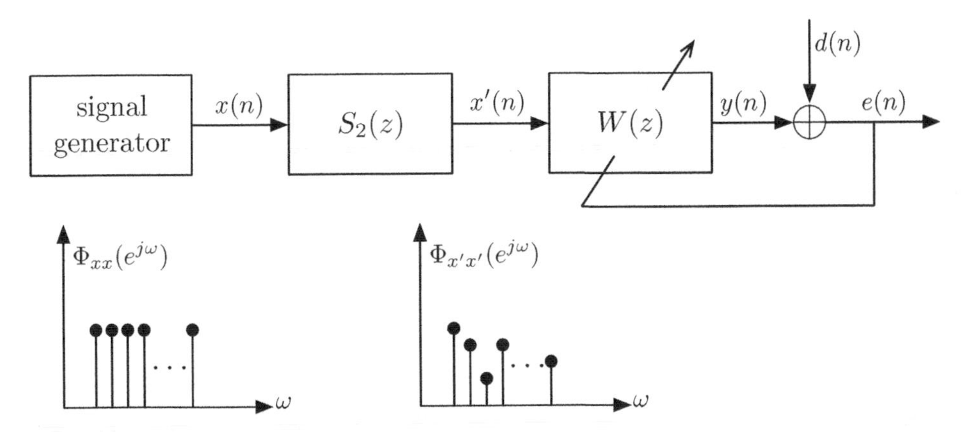

Figure 16.15 An equivalent block diagram of the narrowband ANC in an air duct when the secondary path $S_2(z)$ is considered.

periodic set of impulses; see Figure 16.11) as well as the power spectral density $\Phi_{x'x'}(e^{j\omega})$ of the secondary path output when it is subject to the input $x(n)$. In time domain, the latter will be a periodic sequence with the period of N samples.

Now comparing Figure 16.15 with Figure 16.12, one will find that while the input to $W(z)$ in the latter has a flat spectra at all of its frequency components (the fundamental and harmonics), in Figure 16.15 the input to $W(z)$ is spectrally shaped by the secondary path, $S_2(z)$. This variation of the spectral components of the input to $W(z)$ has an impact on the transfer function $H(z)$. Recalling from Chapter 6 that the variation of the signal power at the input of an LMS-based adaptive filter is equivalent to scaling its step-size parameter with the input signal power, one may intuitively argue that the transfer function $H(z)$ in the case of Figure 16.15 is obtained from Eq. (16.26) by replacing μ with $\mu S_2(z)S_2(z^{-1})$. This leads to

$$H(z) = \frac{E(z)}{D(z)} = \frac{1 - z^{-N}}{1 - (1 - 2\mu S_2(z)S_2(z^{-1}))z^{-N}} \tag{16.32}$$

Note that when z varies on the unit circle

$$H(e^{j\omega}) = \frac{1 - e^{-jN\omega}}{1 - (1 - 2\mu|S_2(e^{j\omega})|^2)e^{-jN\omega}} \tag{16.33}$$

The transfer function (16.32) can also be obtained through a mathematical derivation. One such method starts with the ANC block diagram presented in Figure 16.16a. Assuming that $\hat{S}_2(z) = S_2(z)$, the block diagram shown in Figure 16.16a can be rearranged as in Figure 16.16b. Next, we note that as in the case of Eq. (16.21), here,

$$y'(n) = w_{i(n)}(n) \tag{16.34}$$

Moreover,

$$\begin{aligned} w_{i(n)}(n+N) &= w_{i(n+N)}(n+N) \\ &= w_{i(n)}(n) - 2\mu(e'(n) \star s(n)) \end{aligned} \tag{16.35}$$

where

$$s(n) = s_2(n) \star s_2(-n) \tag{16.36}$$

and $s_2(n)$ is the inverse z-transform of $S_2(z)$. Substituting Eq. (16.34) in Eq. (16.35), we obtain

$$y'(n+N) = y'(n) - 2\mu(e'(n) \star s(n)) \tag{16.37}$$

Moreover, noting that $y'(n) = e'(n) - d'(n)$, Eq. (16.37) may be written as

$$e'(n+N) - d'(n+N) = e'(n) - d'(n) - 2\mu(e'(n) \star s(n)) \tag{16.38}$$

Taking z-transform on both sides of Eq. (16.38) and recalling Eq. (16.36), we obtain

$$z^N E'(z) - z^N D'(z) = E'(z) - D'(z) - 2\mu E'(z)S_2(z)S_2(z^{-1}) \tag{16.39}$$

Moreover, from Figure 16.16b, one may observe that

$$E'(z) = \frac{E(z)}{S_2(z)} \tag{16.40}$$

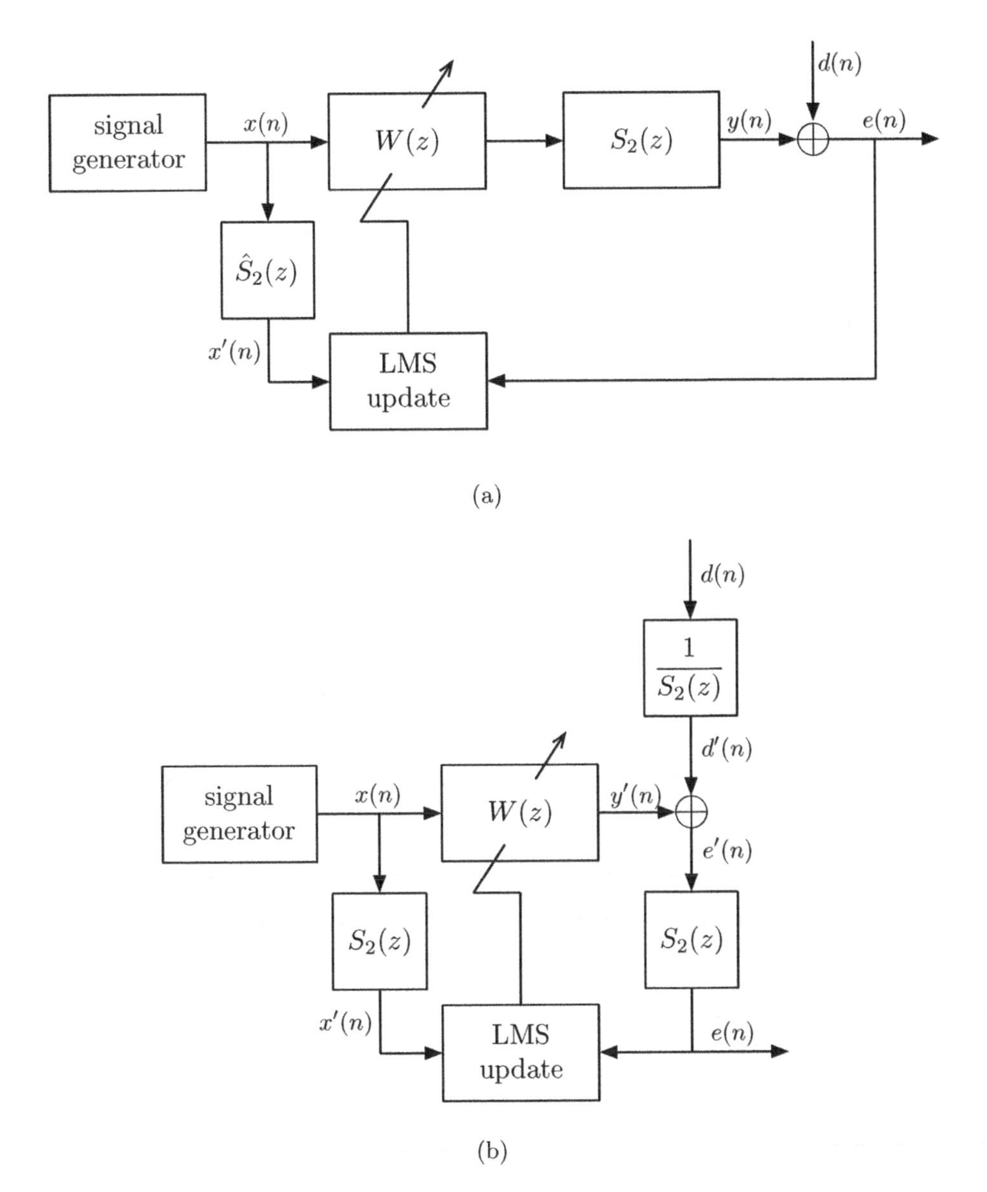

Figure 16.16 (a) Filtered-X LMS algorithm applied to a narrowband ANC system and (b) a rearrangement of the system when $\hat{S}_2(z) = S_2(z)$.

and

$$D'(z) = \frac{D(z)}{S_2(z)} \tag{16.41}$$

Substituting Eqs. (16.40) and (16.41) in Eq. (16.39), one obtains Eq. (16.32).

To develop a more in-depth understanding of the effect of the secondary path $S_2(z)$ on the performance of the multinotch filter $H(z)$ of Eq. (16.32), Figure 16.17 presents a

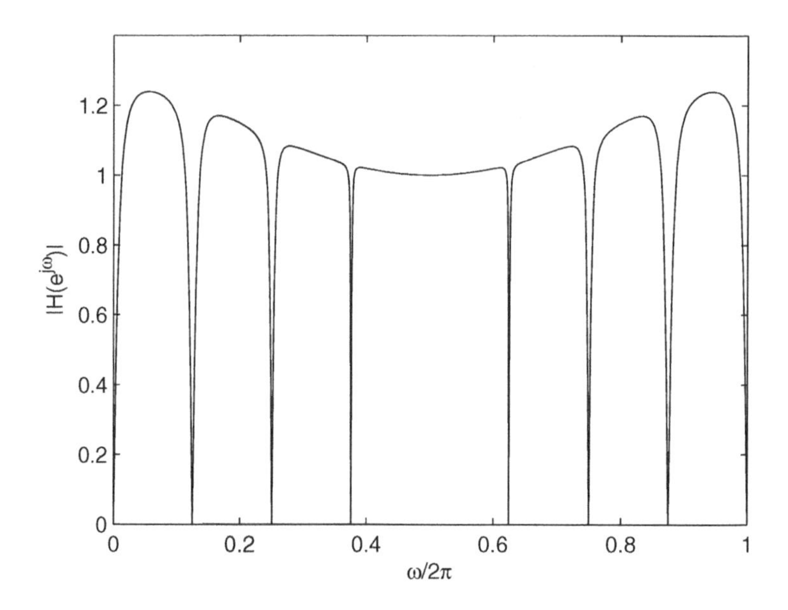

Figure 16.17 An example of magnitude response of the multinotch filter $H(z)$, for $N = 8$ and $\mu = 0.2$, and in the presence of a secondary path $S_2(z) = 0.5 + 0.5z^{-1}$.

counterpart of Figure 16.14, when a secondary path $S_2(z) = 0.5 + 0.5z^{-1}$ has been added. To explain the magnitude response plot of Figure 16.17, we note that, here,

$$|S_2(e^{j\omega})|^2 = \cos^2\omega \tag{16.42}$$

Moreover, we note that $\cos^2\omega$ is equal to 1 at $\omega = 0$ and decreases as ω varies between 0 and π. It reaches a minimum of zero at $\omega = \pi$, and increases toward 1, as ω varies between π and 2π. The variation of $S_2(e^{j\omega})$ has the following impact on the magnitude response of $H(z)$. Near zero, where $|S_2(e^{j\omega})|^2 \approx 1$, there is very little difference between the plot in Figure 16.17 and its counterpart in Figure 16.14. On the other hand, as ω approaches π, thus, $|S_2(e^{j\omega})|^2$ decreases, the bandwidths of the notches decrease. At $\omega = \pi$, where $|S_2(e^{j\omega})|^2 = 0$, the notch disappears from the response. This observation shows that when the secondary path $S_2(z)$ has a null at any of the harmonics of the noise, the ANC system discussed in this section is incapable of removing that harmonic.

Synchronization

The underlying assumption made in the above derivations was that the rate of the input samples to the ANC filter $W(z)$ was synchronized with the fundamental frequency of the noise such that each period of the noise was exactly equal to N samples. Such synchronization can be achieved in practice by a proper design of the tachometer. The tachometer should be designed so that it generates N pulses per each full rotation of the noise generating rotor, for example, by placing N equally spaced sensors (magnets) around the shaft of the rotor.

16.2.2 Adaptive Notch Filters

In the cases where each period of the noise is long (i.e., it has a low fundamental frequency), hence, the number of the taps in the ANC filter $W(z)$ is large, waveform synthesis method may be too complex to implement. In such cases, an alternative solution that directly synthesizes the fundamental and harmonic components of the noise may be the preferred choice. Figure 16.18 presents this solution for the case where the noise is a single tone at the frequency ω_0. A two-tap linear combiner with the tap inputs $x_0(n) = \cos(\omega_0 n)$ and $x_1(n) = \sin(\omega_0 n)$ and the coefficients $w_0(n)$ and $w_1(n)$ is used to synthesize a sign-reversed replica of the tonal noise of the acoustic signal

$$d(n) = a\cos(\omega_0 + \phi) + \nu(n) \tag{16.43}$$

In Eq. (16.43), a and ϕ are, respectively, the amplitude and phase of the tonal noise and $\nu(n)$ is a broadband process/noise. Clearly, the setup presented in Figure 16.18 is only capable of canceling the tonal component of $d(n)$, leaving its broadband component, $\nu(n)$, intact.

The adder on the acoustic side of Figure 16.18 refers to the error microphone. The error signal $e(n)$ from the microphone is fed back to the left side of the figure for adaptation of the tap weights w_0 and w_1. Upon convergence of the tap weight w_0 and w_1, they approach their optimal values

$$w_{0,o} = a\cos\phi \tag{16.44}$$

and

$$w_{1,o} = -a\sin\phi \tag{16.45}$$

This setting of w_0 and w_1 results in the residual error $e(n) = \nu(n)$. Thus, as in the case of waveform synthesis method, the setup in Figure 16.18 acts like a notch with a perfect notch at $\omega = \omega_0$. In fact, a careful analysis of the signal flows in Figure 16.18, presented in

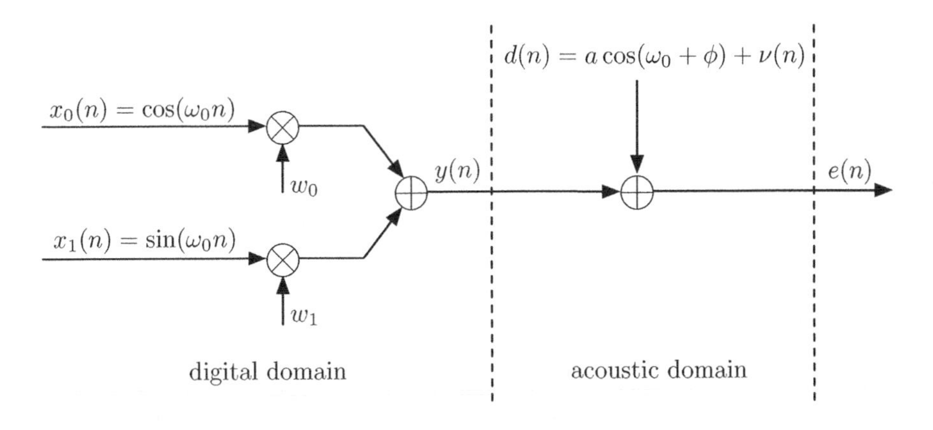

Figure 16.18 A notch filtering setup with the notch frequency of ω_0.

Widrow *et al.* (1975), reveals that $d(n)$ and $e(n)$ are related through the transfer function

$$H(z) = \frac{1 - 2z^{-1}\cos\omega_0 + z^{-2}}{1 - 2(1-\mu)z^{-1}\cos\omega_0 + (1-2\mu)z^{-2}} \tag{16.46}$$

An alternative derivation of the same result can be found in Glover (1977); see also Elliott and Nelson (1993). In Appendix 16A, a proof of Eq. (16.46) that follows the work of Glover (1977) is presented.

The transfer function (16.46) has a pair of zeros at

$$z = \mathrm{e}^{\pm j\omega_0} \tag{16.47}$$

which implies that $H(z)$ has a perfect notch at the angular frequency $\omega = \omega_0$. The poles of $H(z)$ are located at

$$z = (1-\mu)\cos\omega_0 \pm j\sqrt{(1-2\mu) - (1-\mu)^2\cos^2\omega_0} \tag{16.48}$$

These poles are inside the unit circle at a radial distance

$$r = \sqrt{1-2\mu} \tag{16.49}$$

from the center of the unit circle. They are also located at the angular position

$$\theta = \pm\cos^{-1}\left(\frac{1-\mu}{\sqrt{1-2\mu}}\cos\omega_0\right) \tag{16.50}$$

Since typically, in practice, $\mu \ll 1$, and for such cases the approximation $\sqrt{1-2\mu} \approx 1-\mu$ holds, one will find that $r \approx 1-\mu$ and $\theta \approx \pm\omega_0$, hence, Eq. (16.48) reduces to

$$z \approx (1-\mu)\mathrm{e}^{\pm j\omega_0} \tag{16.51}$$

The zeros and poles given in Eqs. (16.47) and (16.51), respectively, are presented in Figure 16.19a. Also shown in this figure is the half-power bandwidth of the notch bands. This concept is further clarified in Figure 16.19b, where an example of the magnitude response of $H(z)$ is presented over the range $0 \le \omega \le \pi$. The half-power/3 dB bandwidth of the notch filter (16.46) is seen from the presented diagrams to be

$$\mathrm{BW} \approx 2\mu \tag{16.52}$$

Adding the Secondary Path $S_2(z)$

When the secondary path $S_2(z)$ is included in Figure 16.18, the filtered-X LMS algorithm should be used for adaptation of the tap weights w_0 and w_1. Moreover, recalling the filtered-X LMS algorithm, $x_0(n)$ and $x_1(n)$ should be passed through the estimate $\hat{S}_2(z)$ of the secondary path $S_2(z)$ to generate the respective filtered signals $x_0'(n)$ and $x_1'(n)$ which subsequently will be used for adaptation of the tap weights w_0 and w_1. On the other hand, as $x_0(n)$ and $x_1(n)$ are sinusoidal signals, $x_0'(n)$ and $x_1'(n)$ can be obtained from $x_0(n)$ and $x_1(n)$, respectively, by introducing a change in amplitude and a phase

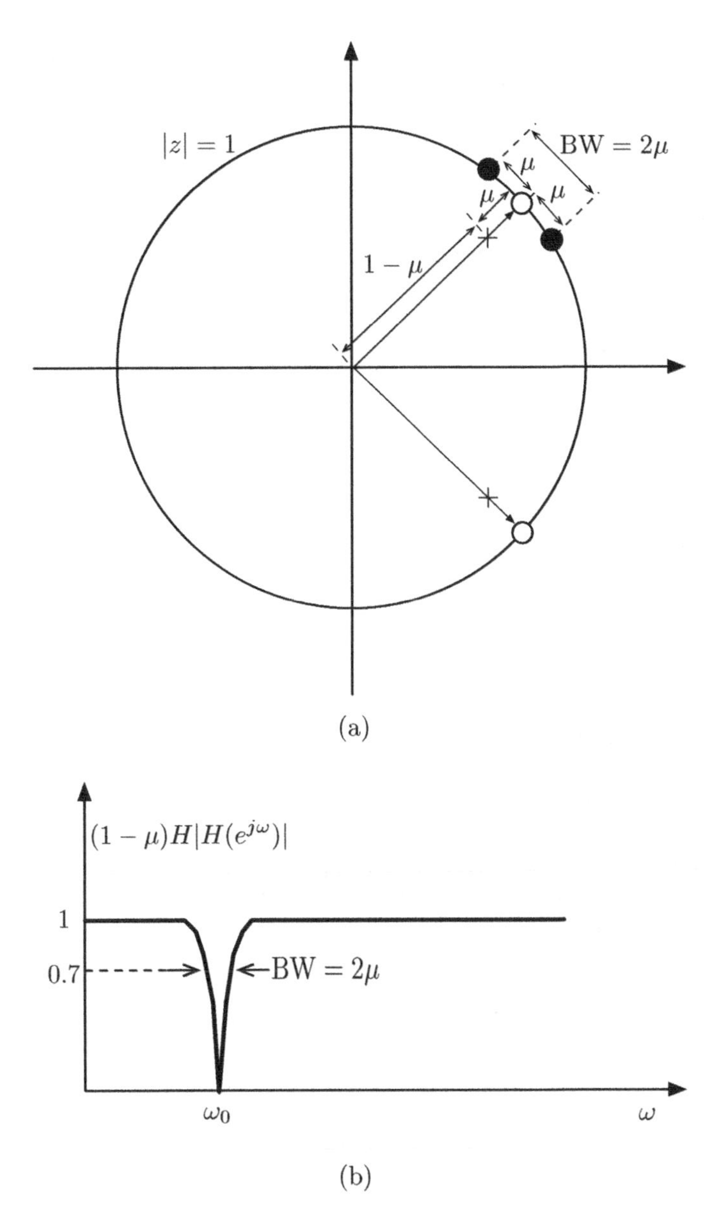

Figure 16.19 (a) Location of zeros and poles of the transfer function (16.46) and (b) its magnitude response.

shift. The phase shift can be realized by delaying the signals $x_0(n)$ and $x_1(n)$, say, by Δ samples, where Δ is equal to the ratio of the desired delay in seconds divide by T_s (the sampling period), rounded to the nearest integer. The amplitude change can be compensated for by adjusting the step-size parameter of the LMS algorithm; see Chapter 6

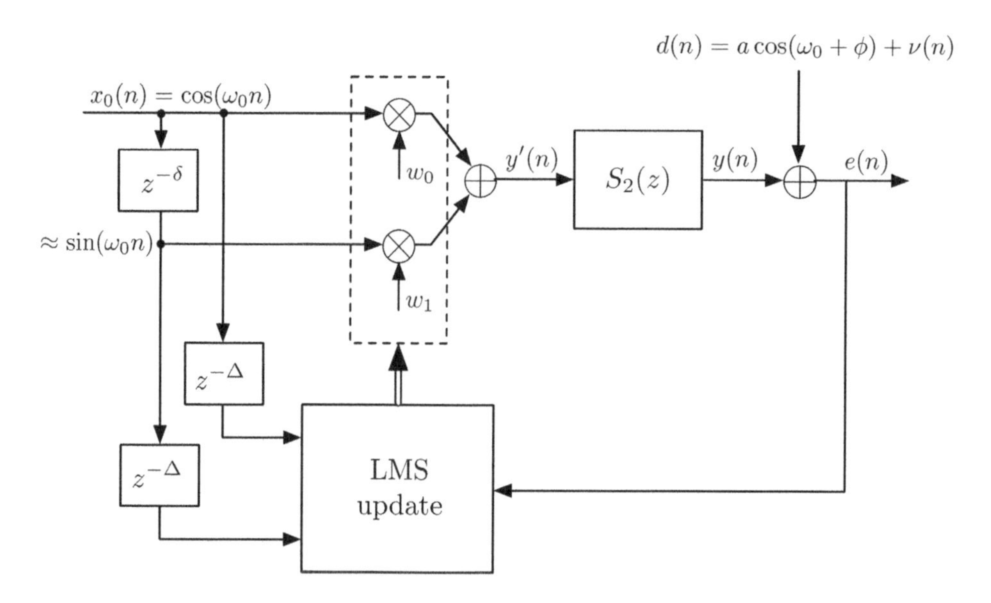

Figure 16.20 Block diagram of a notch filter with the secondary path $S_2(z)$ included.

for the relationship between the step-size parameter in the LMS algorithm and the signal level at the adaptive filter input.

Figure 16.20 presents a block diagram of a notch filter setup with the secondary path included, and the filtered-X LMS algorithm used for adaptation of the filter coefficients. In this figure, we have also used a delay of δ samples to generate $\sin(\omega_0 n)$ from $\cos(\omega_0 n)$. One may note that the assumption $x_0(n-\delta) = x_1(n)$ may not be accurate, unless the equation $\omega_0\delta = \frac{\pi}{2}$ is satisfied for some integer δ. However, we note that a slight mismatch of $x_0(n-\delta)$ and the optimum choice of $x_1(n) = \sin(\omega_0 n)$ has no significant impact on the performance of the ANC system.

In Appendix 16B, the relationship between $d(n)$ and $e(n)$ of Figure 16.20 has been studied, and it has been shown when $\omega_0\delta = \frac{\pi}{2}$ holds, $d(n)$ and $e(n)$ are related through the transfer function

$$H(z) = \frac{1 - 2z^{-1}\cos\omega_0 + z^{-2}}{1 - 2z^{-1}\cos\omega_0 + z^{-2} + 2\mu z^{-1}S_2(z)(z^{-1}\cos(\omega_0 - \theta) - \cos\theta)} \tag{16.53}$$

where $\theta = \omega_0\Delta$.

Extension to Multiple Notches

The block diagram presented in Figure 16.20 can be extended to introduce multiple notches in the response between $d(n)$ and $e(n)$. Such an extension for the case where the notches are introduced at the frequencies ω_0 and ω_1 is presented in Figure 16.21. Extension to the cases with more than two notch frequencies follows the same concept and thus is obvious.

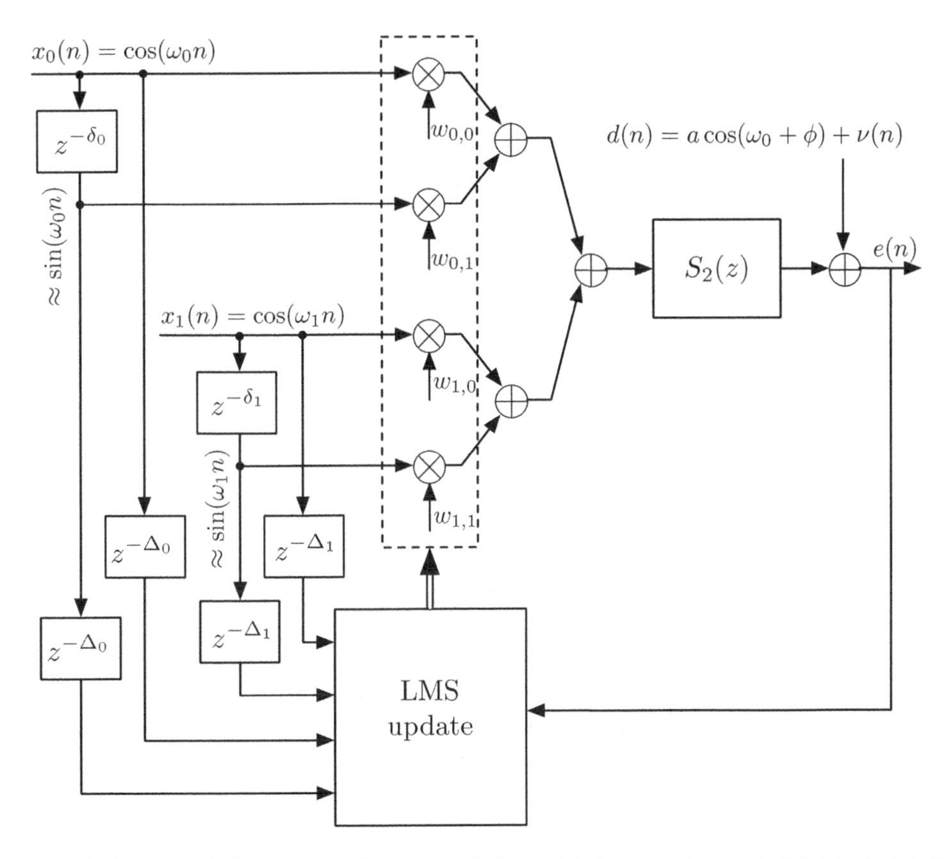

Figure 16.21 Block diagram of a double notch filter with the secondary path $S_2(z)$ included. The notch frequencies are at ω_0 and ω_1.

16.3 Feedback Single-Channel ANC

Consider the feedback ANC system of Figure 16.2, and let the acoustical link between the canceling loudspeaker and the error microphone be denoted as $S(z)$. This leads to the block diagram of Figure 16.22. This is a feedback system with the open loop transfer function

$$\begin{aligned} G(z) &= -\frac{Y(z)}{E(z)} \\ &= -W(z)S(z) \end{aligned} \tag{16.54}$$

and the closed loop transfer function

$$\begin{aligned} H(z) &= \frac{E(z)}{D(z)} \\ &= \frac{1}{1+G(z)} \\ &= \frac{1}{1-W(z)S(z)} \end{aligned} \tag{16.55}$$

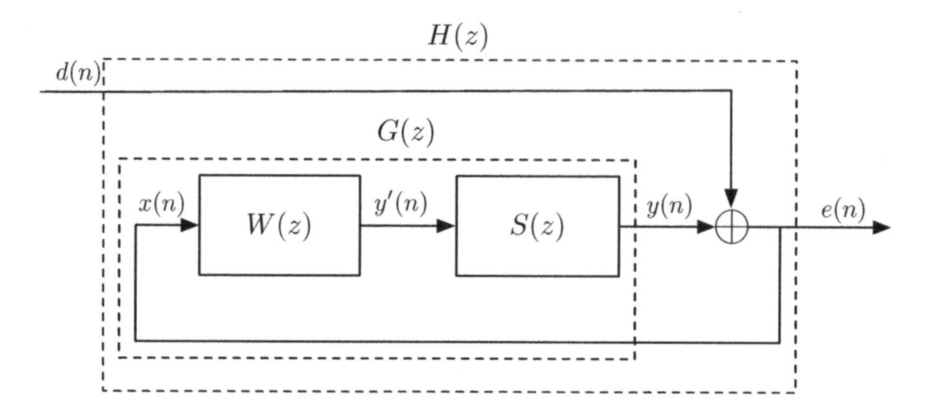

Figure 16.22 Block diagram of a feedback single-channel ANC system.

From the theory of control systems, we recall that to avoid possible instability of the system presented in Figure 16.22 one should make sure that if at any frequency the phase of $G(e^{j\omega})$ reaches 180°, $|G(e^{j\omega})|$ remains strictly smaller than 1. Ideally, if one can choose $W(z)$ such that the phase of $G(e^{j\omega})$ remains around zero for all values of ω, it will be assured that the ANC will be a strictly stable system. In that case, to reduce $e(n)$ to a value close to zero, all one needs to do is to add a real and positive high gain to $W(z)$. This should be obvious since, according to Eq. (16.55),

$$|E(e^{j\omega})| = \frac{|D(e^{j\omega})|}{|1 + G(e^{j\omega})|} \tag{16.56}$$

However, the task of reaching the condition $\angle(G(e^{j\omega})) \approx 0^\circ$, for all values ω, may not be achievable in practice.

The most common method of handling the feedback ANC problem is based on the following philosophy/intuition. The noise signal $d(n)$ that we wish to reconstruct a sign-reversed version of it at the point $y(n)$ in Figure 16.22 is not available. On the other hand, we may recall from the theory of Wiener filters (particularly, the principle of correlation discussed in Chapter 3) that to predict a good estimate of $d(n)$ from a process $x(n)$, $x(n)$ should be highly correlated with $d(n)$. But, we just said $d(n)$ is not available. This puzzle can be solved by adopting the setup presented in Figure 16.23. This setup assumes that an estimate $\hat{S}(z)$ of $S(z)$ is available and passes the output of $W(z)$, $y'(n)$, through this estimate to obtain an estimate $\hat{y}(n)$ of $y(n)$. Moreover, noting that

$$d(n) = e(n) - y(n) \tag{16.57}$$

an estimate $\hat{d}(n)$ of $d(n)$ is obtained by subtracting $\hat{y}(n)$ from $e(n)$. So,

$$\begin{aligned} x(n) &= \hat{d}(n) \\ &= e(n) - \hat{y}(n) \end{aligned} \tag{16.58}$$

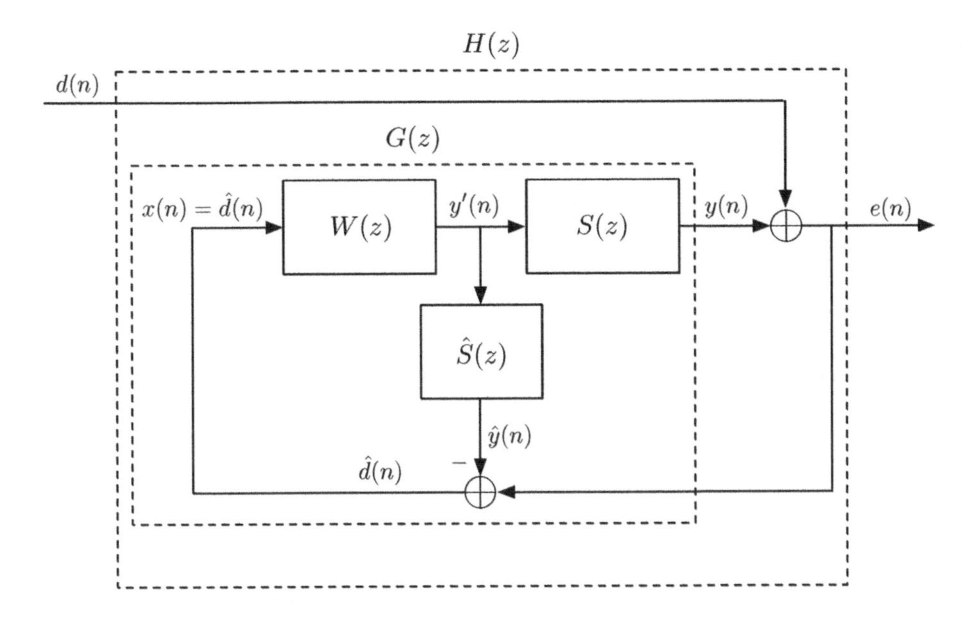

Figure 16.23 A setup for implementation of the feedback ANC system of Figure 16.22.

Analysis

We may note that when $\hat{d}(n)$ is a good estimate of $d(n)$, to reduce the residual error $e(n)$, $W(z)$ has to be set such that

$$W(z)S(z) = -1 \tag{16.59}$$

This will lead to $y(n) = -\hat{d}(n)$ and, thus, $e(n) = d(n) - \hat{d}(n)$ will be a small error signal. One may also note that Eq. (16.59) implies that $W(z) = -\frac{1}{S(z)}$. This setting of $W(z)$ may not be possible in practice as $\frac{1}{S(z)}$ may be a noncausal transfer function. Hence, in practice, one has to resort to an approximation $W(z) \approx -\frac{1}{S(z)}$. Thus, noise suppression in feedback ANC systems will usually be a partial one.

More insight can be developed by deriving the transfer function $H(z)$ between $d(n)$ and $e(n)$. To this end, we first note that

$$\frac{Y'(z)}{E(z)} = \frac{W(z)}{1 + W(z)\hat{S}(z)} \tag{16.60}$$

Next,

$$\begin{aligned} G(z) &= \frac{Y(z)}{E(z)} \\ &= \frac{Y'(z)}{E(z)}S(z) \\ &= \frac{W(z)S(z)}{1 + W(z)\hat{S}(z)} \end{aligned} \tag{16.61}$$

Finally, we obtain

$$
\begin{aligned}
H(z) &= \frac{1}{1 - G(z)} \\
&= \frac{1 + W(z)\hat{S}(z)}{1 + W(z)(\hat{S}(z) - S(z))}
\end{aligned}
\tag{16.62}
$$

When $\hat{S}(z) = S(z)$, Eq. (16.62) reduces to

$$
H(z) = 1 + W(z)S(z) \tag{16.63}
$$

Moreover, if $1/S(z)$ is stable and causal, and $W(z)$ is set equal to $-1/S(z)$, perfect suppression of $d(n)$ will occur. As the stability and causality of $1/S(z)$ usually is not satisfied in practice, a compromise choice that minimizes the residual error $e(n)$ should be found. The filtered-X LMS algorithm (discussed below) searches for such a solution.

The Inverse Modeling Points of View

In the ANC system presented in Figure 16.23, in the ideal case where $\hat{S}(z) = S(z)$, one finds that $\hat{d}(n) = d(n)$ and accordingly Figure 16.23 simplifies to Figure 16.24. Note that in this figure, we have also changed the order of the cascade of $W(z)$ and $S(z)$ to $S(z)$ and $W(z)$.

One may compare Figure 16.24 with Figure 3.9 and note that this is an *inverse modeling* (or, channel equalization) problem. Moreover, as the inverse of $S(z)$ may not be a causal system, the desired inverse modeling solution $W(z) = -1/S(z)$ may only be attainable to a certain degree. Typical noise reductions that have been reported in the literature are usually in the order of 10–15 dB.

Filtered-X LMS Algorithm

Application of the filtered-X LMS algorithm to adaptation of $W(z)$ in Figure 16.23 is a rather straightforward task. Figure 16.25 presents a block diagram of the filtered-X LMS algorithm when applied to the feedback ANC system of Figure 16.23. As in the case of the feedforward ANC systems, here also it is assumed that the estimate $\hat{S}(z)$ of the secondary path $S(z)$ is obtained offline.

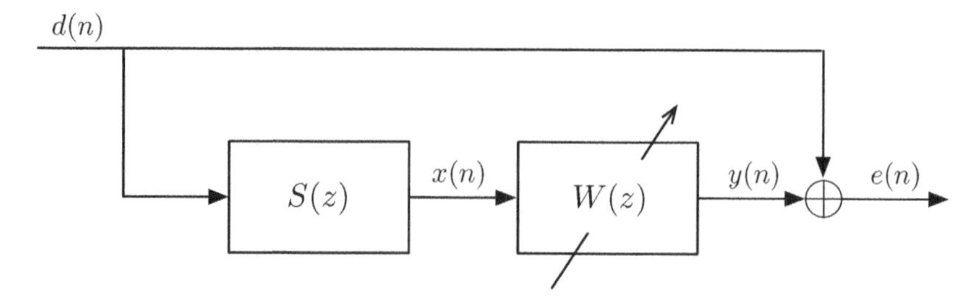

Figure 16.24 The equivalent bock diagram to Figure 16.23 when $\hat{S}(z) = S(z)$.

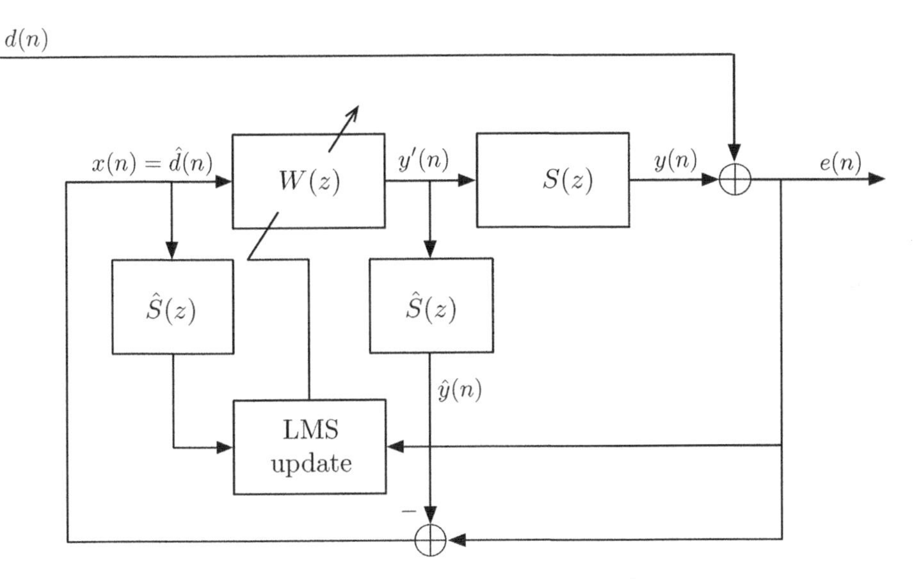

Figure 16.25 Filtered-X LMS algorithm implementation of the feedback ANC system of Figure 16.23.

16.4 Multichannel ANC Systems

The single-channel ANC systems that were introduced in the previous sections of this chapter only work when the noise cancellation is limited to a narrow air duct (like an engine exhaust) or in a small zone in the vicinity of the error microphone (e.g., in the case of feedback ANC systems of Figure 16.2). In many applications, where the volume in which the noise has to be suppressed is large (including wide air ducts), it may be necessary to adopt ANC systems that use multiple reference microphones/sensors, multiple canceling loudspeakers, and/or multiple error microphones. An air duct exemplifying this scenario is presented in Figure 16.26.

Figure 16.26 may be viewed as a dual of Figure 16.9. The difference is that here all the signals are vector signals, thus, the lines connecting different blocks/transfer functions are drawn thicker. The transfer functions connecting different signal vectors have coefficients that are matrices. For example, assuming an N coefficient ANC filter $\mathbf{W}(z)$, it will have the following form:

$$\mathbf{W}(z) = \sum_{i=0}^{N-1} \mathbf{W}_i z^{-i} \tag{16.64}$$

where $\mathbf{W}_i$'s are the coefficient matrices of size $M_\mathrm{r} \times M_\mathrm{c}$, and M_r and M_c are the number of reference microphones and the number of canceling loudspeakers, respectively. The error signal $\mathbf{e}(n)$ is an $M_\mathrm{e} \times 1$ vector and the filtered-X LMS algorithm minimizes the cost function

$$\xi = E[\mathbf{e}^\mathrm{T}(n)\mathbf{e}(n)] \tag{16.65}$$

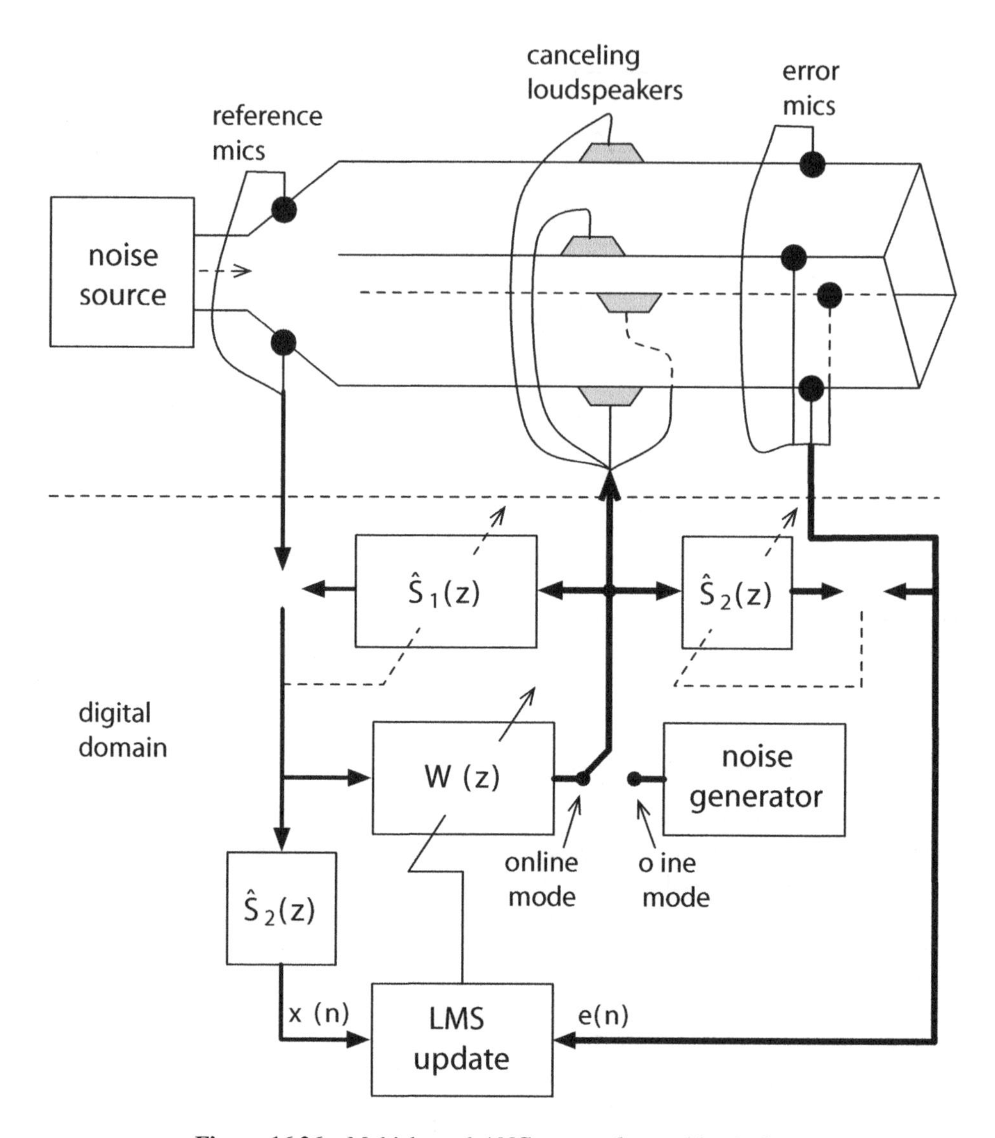

Figure 16.26 Multichannel ANC system for a wide air duct.

16.4.1 MIMO Blocks/Transfer Functions

The ANC filter $\mathbf{W}(z)$ and other blocks in Figure 16.26 are referred to as *multi-input multi-output* (MIMO) systems. To gain a more clear understanding of a MIMO system, we present the two-input two-output system presented in Figure 16.27. There are four links that connect the two inputs $x_0(n)$ and $x_1(n)$ and the two outputs $y_0(n)$ and $y_1(n)$. Let these links be expressed by the transfer functions

$$H_{00}(z) = \frac{Y_0(z)}{X_0(z)} = a_{00} + b_{00}z^{-1} \tag{16.66}$$

$$H_{01}(z) = \frac{Y_1(z)}{X_0(z)} = a_{01} + b_{01}z^{-1} \tag{16.67}$$

$$H_{10}(z) = \frac{Y_0(z)}{X_1(z)} = a_{10} + b_{10}z^{-1} \tag{16.68}$$

$$H_{11}(z) = \frac{Y_1(z)}{X_1(z)} = a_{11} + b_{11}z^{-1} \tag{16.69}$$

From these, one will find that

$$\mathbf{y}(z) = \mathbf{H}(z)\mathbf{x}(z) \tag{16.70}$$

where $\mathbf{x}(z) = \begin{bmatrix} X_0(z) \\ X_1(z) \end{bmatrix}$ $\mathbf{y}(z) = \begin{bmatrix} Y_0(z) \\ Y_1(z) \end{bmatrix}$ and

$$\mathbf{H}(z) = \begin{bmatrix} H_{00}(z) & H_{01}(z) \\ H_{10}(z) & H_{11}(z) \end{bmatrix} = \begin{bmatrix} a_{00} & a_{01} \\ a_{10} & a_{11} \end{bmatrix} + \begin{bmatrix} b_{00} & b_{01} \\ b_{10} & b_{11} \end{bmatrix} z^{-1} \tag{16.71}$$

16.4.2 Derivation of the LMS Algorithm for MIMO Adaptive Filters

To develop an insight to the understanding of how an LMS algorithm may be developed for minimization of the cost function (16.65), here, we present the development of an LMS algorithm for the adaptive filtering setup that is presented in Figure 16.28. Once this is understood, it should be a straightforward task for an interested reader to develop similar algorithms in different contexts, including LMS algorithms for offline adaptation of $\hat{\mathbf{S}}_1(z)$ and $\hat{\mathbf{S}}_2(z)$ as well as a filtered-X LMS algorithm for adaptation of the ANC filter $\mathbf{W}(z)$ in the system setup of Figure 16.26.

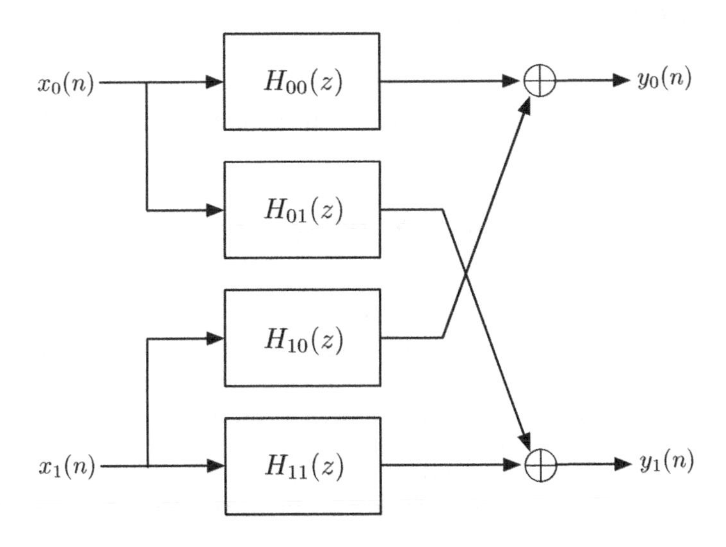

Figure 16.27 Block diagram of a 2 × 2 MIMO system.

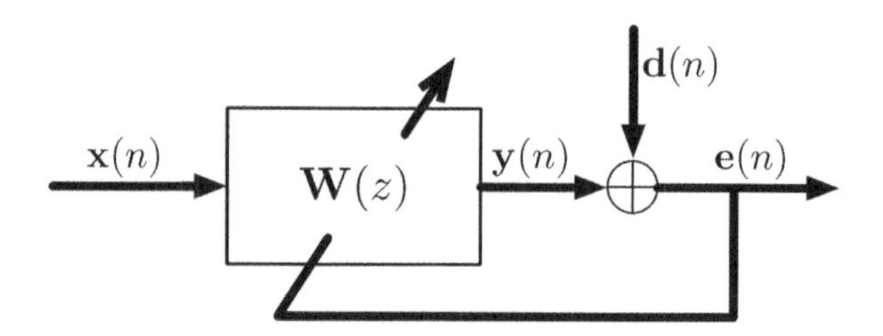

Figure 16.28 A multi-input multi-output adaptive filter.

The input $\mathbf{x}(n)$ to the adaptive filter $\mathbf{W}(z)$ is a vector of size $M_{\mathrm{i}} \times 1$. The output $\mathbf{y}(n)$ of $\mathbf{W}(z)$ and the desired output $\mathbf{d}(n)$ are vectors of size $M_{\mathrm{o}} \times 1$. Accordingly, the coefficients $\mathbf{W}_i$ of $\mathbf{W}(z)$ are matrices of size $M_{\mathrm{o}} \times M_{\mathrm{i}}$. Moreover,

$$\mathbf{y}(n) = \sum_{i=0}^{N-1} \mathbf{W}_i \mathbf{x}(n-i) \tag{16.72}$$

Also,

$$\mathbf{e}(n) = \mathbf{d}(n) - \mathbf{y}(n) \tag{16.73}$$

To proceed, we note that the cost function (16.65) may be expanded as

$$\xi = \sum_{k=0}^{M_{\mathrm{O}}-1} \xi_k \tag{16.74}$$

where

$$\xi_k = E[e_k^2(n)] \tag{16.75}$$

and $e_k(n)$ is the kth element of $\mathbf{e}(n)$. Moreover, we note that

$$e_k(n) = d_k(n) - y_k(n) \tag{16.76}$$

where $d_k(n)$ and $y_k(n)$ are the kth elements of $\mathbf{d}(n)$ and $\mathbf{y}(n)$, respectively. In addition, from Eq. (16.72), one may find that

$$y_k(n) = \sum_{i=0}^{N-1} \mathbf{W}_i^{(k)} \mathbf{x}(n-i) \tag{16.77}$$

where $\mathbf{W}_i^{(k)}$ denotes the kth row of $\mathbf{W}_i$. Substituting Eq. (16.77) in Eq. (16.76), and the result in Eq. (16.75), one will find that minimization of ξ_k involves optimization of the k rows of the coefficient matrices $\mathbf{W}_i$'s. This, in turn, implies that the subcost functions ξ_k in Eq. (16.74), for each k, can be optimized independently, because each depends on a separate set of the elements of $\mathbf{W}_i$'s.

Before proceeding with the derivation of an LMS algorithm for minimization of ξ_k, it is also instructive to derive an expression for ξ_k. From the above discussions, one may note that

$$e_k(n) = d_k(n) - \bar{\mathbf{w}}_k^{\mathrm{T}} \bar{\mathbf{x}}(n) \tag{16.78}$$

where

$$\bar{\mathbf{w}}_k = [\mathbf{W}_0^{(k)}\ \mathbf{W}_1^{(k)} \cdots \mathbf{W}_{N-1}^{(k)}]^{\mathrm{T}} \tag{16.79}$$

and

$$\bar{\mathbf{x}}(n) = \begin{bmatrix} \mathbf{x}(n) \\ \mathbf{x}(n-1) \\ \vdots \\ \mathbf{x}(n-N+1) \end{bmatrix} \tag{16.80}$$

Note that the vectors $\bar{\mathbf{w}}_k$ and $\bar{\mathbf{x}}(n)$ are column vectors of length NM_{i}.

One may now note that Eq. (16.78) has the same form as Eq. (3.7). Hence, following a line of derivations similar to those in Chapter 3, one will find the optimum value of $\bar{\mathbf{w}}_k$ that minimizes ξ_k as

$$\bar{\mathbf{w}}_{k,\mathrm{o}} = \bar{\mathbf{R}}^{-1}\bar{\mathbf{p}} \tag{16.81}$$

where $\bar{\mathbf{R}} = E[\bar{\mathbf{x}}(n)\bar{\mathbf{x}}^{\mathrm{T}}(n)]$ and $\bar{\mathbf{p}} = E[\bar{\mathbf{x}}(n)d_k(n)]$. Also, the LMS algorithm for the adaptation of $\bar{\mathbf{w}}_k$ is trivially obtained as

$$\bar{\mathbf{w}}_k(n+1) = \bar{\mathbf{w}}_k(n) - 2\mu e_k(n)\bar{\mathbf{x}}(n) \tag{16.82}$$

Repeating Eq. (16.82) for $k = 0, 1, \ldots, M_{\mathrm{o}} - 1$, we will have the update equations for all of the coefficients of $\mathbf{W}(z)$.

Appendix 16A: Derivation of Eq. (16.46)

Using the LMS algorithm to adapt the coefficients w_0 and w_1 in Figure 16.18, we have the following update equation for w_0:

$$w_0(n+1) = w_0(n) - 2\mu e(n)x_0(n) \tag{16A.1}$$

Substituting

$$x_0(n) = \cos(\omega_0 n) = \frac{e^{j\omega_0 n} + e^{-j\omega_0 n}}{2} \tag{16A.2}$$

in Eq. (16A.1), we obtain

$$w_0(n+1) = w_0(n) - \mu e(n)(e^{j\omega_0 n} + e^{-j\omega_0 n}) \tag{16A.3}$$

Taking the z-transform on both sides of Eq. (16A.3), we get

$$zW_0(z) = W_0(z) - \mu(E(ze^{-j\omega_0}) + E(ze^{j\omega_0})) \tag{16A.4}$$

where we have noted that $w_0(n+1) \Leftrightarrow zW_0(z)$, $e(n)e^{j\omega_0 n} \Leftrightarrow E(ze^{-j\omega_0 n})$ and $e(n)e^{-j\omega_0 n} \Leftrightarrow E(ze^{j\omega_0 n})$. Equation (16A.4) may be rearranged as

$$W_0(z) = -\frac{\mu}{z-1}(E(ze^{-j\omega_0}) + E(ze^{j\omega_0})) \tag{16A.5}$$

Similarly, using the LMS update equation of w_1 and recalling the identity

$$x_1(n) = \sin(\omega_0 n) = \frac{e^{j\omega_0 n} - e^{-j\omega_0 n}}{2j} \tag{16A.6}$$

we obtain

$$W_1(z) = -\frac{\mu}{z-1}\,\frac{E(ze^{-j\omega_0}) - E(ze^{j\omega_0})}{j} \tag{16A.7}$$

On the other hand,

$$y(n) = w_0(n)\cos(\omega_0 n) + w_1(n)\sin(\omega_0 n) \tag{16A.8}$$

Taking z-transform on both sides of Eq. (16A.8), and recalling Eqs. (16A.2) and (16A.6), we obtain

$$Y(z) = \frac{W_0(ze^{-j\omega_0}) + W_0(ze^{j\omega_0})}{2} + \frac{W_1(ze^{-j\omega_0}) - W_1(ze^{j\omega_0})}{2j} \tag{16A.9}$$

Using Eqs. (16A.5) and (16A.7), after some straightforward manipulations, one will find that Eq. (16A.9) converts to

$$Y(z) = \frac{-2\mu z\cos\omega_0 + 2\mu}{z^2 - 2\cos\omega_0 + 1}E(z) \tag{16A.10}$$

Finally, noting that $y(n) = e(n) - d(n)$, hence, $Y(z) = E(z) - D(z)$, Eq. (16.46) is obtained from Eq. (16A.10).

Appendix 16B: Derivation of Eq. (16.53)

Consider the ANC system presented in Figure 16.20. Let $\omega_0\delta = \frac{\pi}{2}$, thus the lower input to the linear combiner in Figure 16.20 will be exactly equal to $\sin(\omega_0 n)$. Also, we let $\omega_0\Delta = \theta$. These lead to the following update equations for updating w_0 and w_1:

$$w_0(n+1) = w_0(n) - \mu e(n)(e^{j\omega_0 n}e^{j\theta} + e^{-j\omega_0 n}e^{-j\theta}) \tag{16B.1}$$

and

$$w_1(n+1) = w_1(n) + j\mu e(n)(e^{j\omega_0 n}e^{j\theta} - e^{-j\omega_0 n}e^{-j\theta}) \tag{16B.2}$$

respectively. Starting with Eqs. (16B.1) and (16B.2) and following the same line of derivations that led to Eq. (16A.10), here, we obtain

$$Y'(z) = \frac{-2\mu z\cos(\omega_0 - \theta) + 2\mu\cos\theta}{z^2 - 2\cos\omega_0 + 1}E(z) \tag{16B.3}$$

We also note from Figure 16.20 that

$$Y(z) = Y'(z)S_2(z) \tag{16B.4}$$

Substituting Eq. (16B.3) in Eq. (16B.4) and recalling that $Y(z) = E(z) - D(z)$, one will arrive at Eq. (16.53).

17

Synchronization and Equalization in Data Transmission Systems

Channel equalization is one of the most widely explored topics in the field of adaptive filters. The earliest form of adaptive filters was suggested in the 1960s as the developments on digital modems began. A significant portion of the theory of adaptive filters in those early days was developed in the context of adaptive equalizers and then applied to other applications. In addition, the adaptive equalizers had their own specific problems that had to be tackled separately. For instance, synchronization and time alignment of the receiver with the transmit signal was not a trivial task at the beginning. It was almost a decade after the invention of the adaptive equalizers that the more practical methods of synchronization (such as cyclic equalizers discussed in Section 17.6) were developed. It also took a number of years before those working on the channel equalizers could appreciate the advantages of fractionally spaced equalizers over their symbol-spaced counterparts (Section 17.4). Moreover, further advancement in the theory of adaptive equalizers led to blind adaptation methods that would allow adaptation of equalizer coefficients without transmitting any pilot/training symbols. The capacity achieving maximum-likelihood detectors and soft equalizers have also been introduced. Another important development is an elegant signaling format that allows equalization in the frequency domain. These varieties of equalization methods and the synchronization steps that are necessary to allow their successful operation are presented in some detail in this chapter.

In this chapter, we need to bounce forth and back between the continuous and discrete time signals. We differentiate a continuous time signal and its discrete time counterpart using the argument "t" (or τ) to denote continuous time and the argument "n" (or m) when reference is made to a discrete time index. For example, whereas we use $x(t)$ to represent a continuous time signal, the samples of $x(t)$ at the time instants $t = nT_s$, where T_s is the sampling interval, are represented by the sequence $x(n)$. Of course, this is a misnomer of notations; however, as long as we remain consistent in the use of notations, there should be no confusion. As for the frequency domain signals, the Fourier transform of a continuous time signal $x(t)$ is denoted by $X(\Omega)$. For a discrete time sequence, we keep the notation that was used in the previous chapters, that is, $X(e^{j\omega})$ denotes the discrete time Fourier transform of the sequence $x(n)$. In other words, whereas Ω is used to denote the frequency of a continuous time sine-wave $x(t) = \sin(\Omega t)$, ω is used to denote the normalized frequency (with respect to the sampling frequency $f_s = 1/T_s$) of

Adaptive Filters: Theory and Applications, Second Edition. Behrouz Farhang-Boroujeny.

its sampled version $x(n) = \sin(\Omega n T_s) = \sin(\omega n)$. One may also note that the absolute Ω and the normalized ω frequencies are related according to the equation

$$\omega = \frac{\Omega}{f_s} \tag{17.1}$$

17.1 Continuous Time Channel Model

In the study of adaptive equalizers, the channel is often modeled in its equivalent baseband and discrete time form, even though the actual transmission happens over a passband radio frequency (RF) and in continuous time form. In this section, we present a development of such a channel model.

Figure 17.1 presents the block diagram of a digital quadrature-amplitude modulation (QAM) communication system. The transmit data symbols, $s(n)$, are streamed to a sequence of impulses at the interval of T. We refer to T as *symbol-space*. This is presented by the continuous time signal

$$s(t) = \sum_n s(n)\delta(t - nT) \tag{17.2}$$

where $s(n)$ are data symbols from a QAM constellation.

In Figure 17.1, the sequence of the symbol modulated impulses, represented by the continuous time signal $s(t)$, is passed through a transmit pulse-shaping filter with the impulse response $p_{\mathrm{T}}(t)$. This results in a band-limited baseband signal, which subsequently modulates a carrier with the frequency Ω_c. Multiplication by $e^{j\Omega_c t}$ shifts the baseband signal spectrum to the carrier band, and the block $\Re[\cdot]$ takes the real part of the result, thus produces a duplicate image spectrum around $-\Omega_c$ as well. This leads to

$$X_{\mathrm{QAM}}(\Omega) = S(\Omega - \Omega_c)P_{\mathrm{T}}(\Omega - \Omega_c) + S^*(-\Omega - \Omega_c)P_{\mathrm{T}}^*(-\Omega - \Omega_c) \tag{17.3}$$

where $X_{\mathrm{QAM}}(\Omega)$ is the Fourier transform of $x_{\mathrm{QAM}}(t)$. Using Eq. (17.3), we obtain

$$X'_{\mathrm{QAM}}(\Omega) = \left(S(\Omega - \Omega_c)P_{\mathrm{T}}(\Omega - \Omega_c) + S^*(-\Omega - \Omega_c)P_{\mathrm{T}}^*(-\Omega - \Omega_c)\right)C(\Omega) \tag{17.4}$$

where $X'_{\mathrm{QAM}}(\Omega)$ and $C(\Omega)$ are the Fourier transforms of $x'_{\mathrm{QAM}}(t)$ and $c(t)$, respectively. At the receiver side, the multiplication of $x'_{\mathrm{QAM}}(t)$ by $e^{-j\Omega_c t}$ shifts the spectrum to the

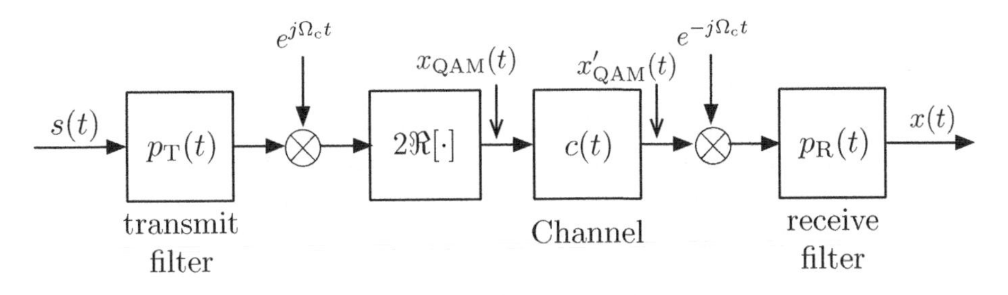

Figure 17.1 Block diagram of the channel model in a digital QAM communication system.

left, by Ω_c, and the receive filter removes the other portion of the signal spectrum that is now located around $-2\Omega_c$. These steps lead to

$$X(\Omega) = S(\Omega)H(\Omega) \tag{17.5}$$

where

$$H(\Omega) = P_T(\Omega)P_R(\Omega)C(\Omega + \Omega_c) \tag{17.6}$$

This shows that the combination of all the blocks in the transmission path from the input $s(t)$ to the output $x(t)$ is characterized by the transfer function $H(\Omega)$ given by Eq. (17.6).

We also note that the common choice for both the pulse-shaping filters $p_T(t)$ and $p_R(t)$ is the square-root raised-cosine pulse shape

$$p_{srrc}(t) = \frac{1}{\sqrt{T}} \frac{\sin\left((1-\alpha)\frac{\pi t}{T}\right) + \frac{4\alpha t}{T}\cos\left((1+\alpha)\frac{\pi t}{T}\right)}{\frac{\pi t}{T}\left(1 - \left(\frac{4\alpha t}{T}\right)^2\right)} \tag{17.7}$$

where the parameter $0 < \alpha \leq 1$ is referred to as roll-off factor. The Fourier transform of $p_{srrc}(t)$ is

$$P_{srrc}(\Omega) = \begin{cases} \sqrt{T}, & |\Omega| \leq \frac{(1-\alpha)\pi}{T} \\ \sqrt{T}\cos\left[\frac{T}{4\alpha}\left(|\Omega| - \frac{(1-\alpha)\pi}{T}\right)\right], & \frac{(1-\alpha)\pi}{T} \leq |\Omega| \leq \frac{(1+\alpha)\pi}{T} \\ 0, & \text{otherwise} \end{cases} \tag{17.8}$$

Note that $P_{srrc}(\Omega)$ spans over the frequency band $-\frac{(1+\alpha)\pi}{T} < \Omega < \frac{(1+\alpha)\pi}{T}$.

In the absence of channel (or when the channel is ideal), $H(\Omega) = P_T(\Omega)P_R(\Omega) = |P_{srrc}(\Omega)|^2 \triangleq P_{rc}(\Omega)$, where $P_{rc}(\Omega)$ is referred to as *raised-cosine pulse shape*. It is straightforward to obtain, from Eq. (17.8),

$$P_{rc}(\Omega) = \begin{cases} T, & |\Omega| \leq \frac{(1-\alpha)\pi}{T} \\ \frac{T}{2}\left\{1 + \cos\left[\frac{T}{2\alpha}\left(|\Omega| - \frac{(1-\alpha)\pi}{T}\right)\right]\right\}, & \frac{(1-\alpha)\pi}{T} \leq |\Omega| \leq \frac{(1+\alpha)\pi}{T} \\ 0, & \text{otherwise} \end{cases} \tag{17.9}$$

Moreover, in the time domain,

$$p_{rc}(t) = \text{sinc}(t/T)\frac{\cos(\pi\alpha t/T)}{1 - 4\alpha^2 t^2/T^2} \tag{17.10}$$

where $\text{sinc}(t/T) = \frac{\sin(\pi t/T)}{\pi t/T}$. Note that when the channel is ideal and the received signal is sampled at the correct timing phase, $h(t) = p_{rc}(t) = 1$ at $t = 0$ and is zero for any other integer multiple of T; hence, it is a Nyquist pulse and thus results in an intersymbol interference (ISI) free transmission. The presence of channel usually results in an $h(t)$, which is not Nyquist and thus equalization is needed to remove ISI.

Using $h(t)$ to denote the inverse Fourier transform of $H(\Omega)$, Figure 17.1 simplifies to Figure 17.2. Now considering Eq. (17.2), one will find that

$$x(t) = \sum_n s(n)h(t - nT) \tag{17.11}$$

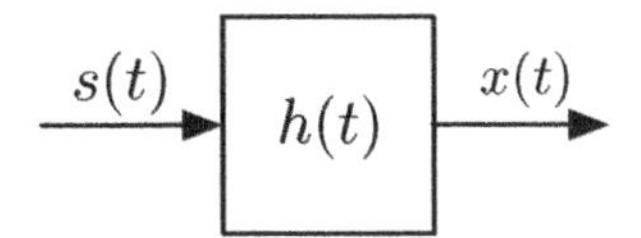

Figure 17.2 The simplified diagram of the channel model of a digital QAM communication system.

In the next section, we use Eq. (17.11) to develop a discrete time channel model that will be used in the rest of this chapter. However, before proceeding with the next section, it will be instructive to present an example of a multipath wireless channel.

Example 17.1

A multipath passband communication channel that operates at the carrier frequency 1 GHz has two paths with the gains of 1 and 0.5 and the delays of 0 and 0.75 μs, respectively. The transmit and receive filters are square-root raised-cosine pulse shapes designed for $T = 1\ \mu s$ and $\alpha = 0.5$.

(i) Derive an equation for the equivalent baseband channel transfer function $H(\Omega)$.
(ii) Find $H(\Omega)$ if the delay of the second path changes to 0.75025 μs.
(iii) In both cases present the plots of the magnitude and phase response of $H(\Omega)$.
(iv) In both cases obtain and plot the impulse response of the baseband equivalent of the passband channel.

Solution:

(i) The impulse response of the channel is $c(t) = \delta(t) + 0.5\delta(t - 0.75)$. Hence, $C(\Omega) = 1 + 0.5\mathrm{e}^{-j0.75\Omega}$. Substituting this in Eq. (17.6), we obtain

$$H(\Omega) = (1 + 0.5\mathrm{e}^{-j0.75(\Omega+\Omega_c)})P_{\mathrm{rc}}(\Omega) = (1 + 0.5\mathrm{e}^{-j0.75\Omega})P_{\mathrm{rc}}(\Omega)$$

where $P_{\mathrm{rc}}(\Omega)$ is the raised-cosine pulse shape (17.9) with $T = 1\ \mu s$ and $\alpha = 0.5$ and we have noted that, for $\Omega_c = 2\pi \times (10^3\ \mathrm{MHz})$, $\mathrm{e}^{j0.75\Omega_c} = 1$.

(ii) Following the same line of derivation as in (i), in this case, we obtain

$$\begin{aligned} H(\Omega) &= (1 + 0.5\mathrm{e}^{-j0.5\pi}\mathrm{e}^{-j0.75025\Omega})P_{\mathrm{rc}}(\Omega) \\ &= (1 - 0.5j\mathrm{e}^{-j0.75025\Omega})P_{\mathrm{rc}}(\Omega) \end{aligned}$$

(iii) The plots are shown in Figure 17.3. As seen, a small change in one of the path delays can significantly affect the baseband equivalent of the channel response.

(iv) Taking the inverse Fourier transforms of the results in (i) and (ii), we obtain the following:

$$\text{for case (i),} \quad h(t) = p_{\mathrm{rc}}(t) + 0.5p_{\mathrm{rc}}(t - 0.75)$$

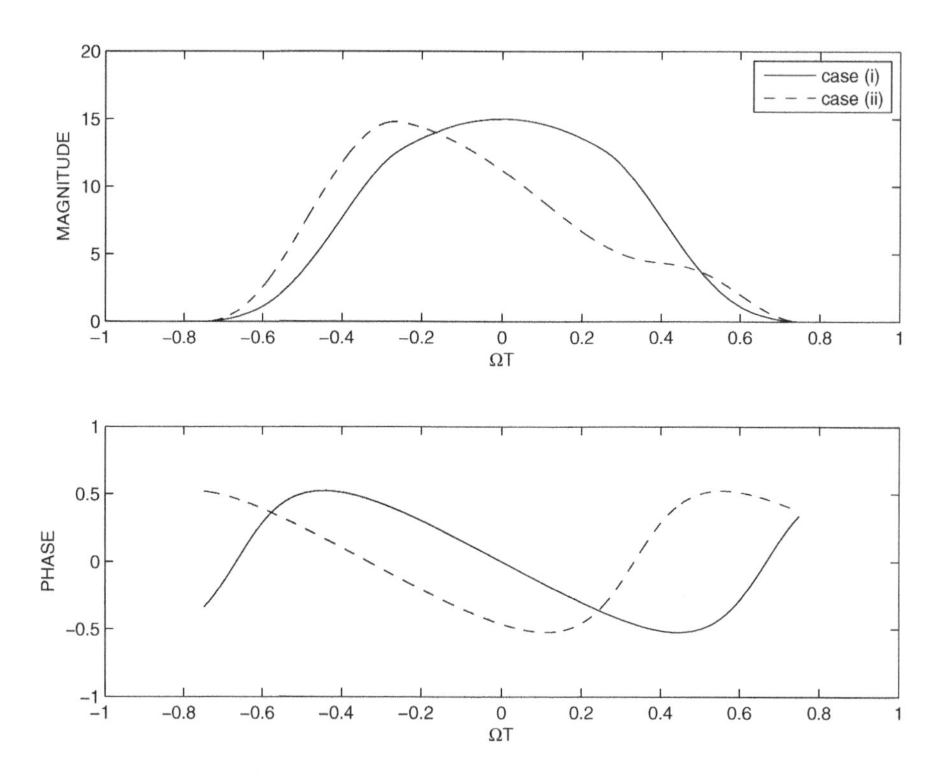

Figure 17.3 The equivalent baseband channel response of two examples of a two-path wireless channel, in the frequency domain.

and

$$\text{for case (ii),} \quad h(t) = p_{\text{rc}}(t) - 0.5j\, p_{\text{rc}}(t - 0.75025)$$

These are plotted in Figure 17.4. Interesting to note here is that whereas case (i) leads to a real-valued impulse response, case (ii) has a complex-valued impulse response. We note that, in general, $h(t)$ is expected to be a complex-valued function of time. Another important observation here is that a relatively small change in the delay of one of the paths has a profound impact on the response of equivalent baseband channel both in the time and frequency domain.

Generalizing the above example, if the channel is a multipath channel that is characterized by the impulse response

$$c(t) = \sum_{i=0}^{M-1} a_i \delta(t - \tau_i) \tag{17.12}$$

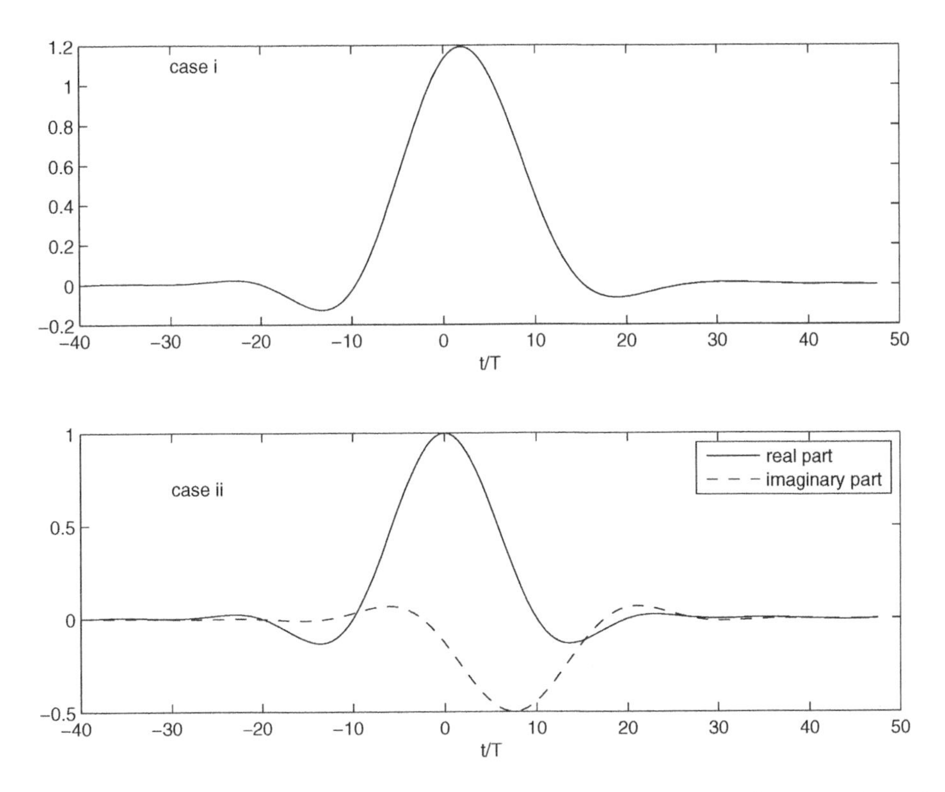

Figure 17.4 The equivalent baseband channel response of two examples of a two-path wireless channel, in the time domain.

where M is the number of paths, and a_i and τ_i are the ith path gain and delay, respectively, the baseband equivalent of the channel is obtained as

$$h(t) = \sum_{i=0}^{M-1} a_i e^{-j\Omega_c \tau_i} p_{rc}(t - \tau_i) \tag{17.13}$$

17.2 Discrete Time Channel Model and Equalizer Structures

Consider the case where the continuous time signal $x(t)$ is sampled at a rate L times faster than the symbol rate $1/T$, and use $x(n)$ to denote the sample value of $x(t)$ at $t = nT/L$. Also, using $h(n)$ to denote the sample value of $h(t)$ at $t = nT/L$, one will obtain from Eq. (17.11)

$$x(n) = \sum_{m} s(m)h(n - mL) \tag{17.14}$$

Moreover, if we assume that the channel contains some additive noise (which we ignored in the presentations of Figure 17.1, for simplicity of the derivations), one may modify Eq. (17.14) as

$$x(n) = \sum_{m} s(m)h(n - mL) + \nu(n) \tag{17.15}$$

where $\nu(n)$ is the channel noise.

In a digital communication system, the samples $x(n)$ are passed through an adaptive equalizer whose coefficients are adjusted to undo the distortion caused by the channel and thus deliver (almost) ISI free estimates of $s(n)$ at its output. Considering the channel equation (17.15), and adding the equalizer after the channel model, the discrete-time system model of a digital communication system is obtained as the one presented in Figure 17.5. The interpolator block adds $L - 1$ zeros after each symbol $s(n)$, changing the sample rate from $1/T$ to L/T. Passing the upsampled sequence through $H(z)$ and adding the channel noise, $\nu(n)$, is to implement Eq. (17.15). The equalizer $W(z)$ is an FIR filter with the tap weights $w_0, w_1, \ldots, w_{N-1}$, and the estimates $\hat{s}(n - \Delta)$ of the transmitted data symbols, where Δ is the delay caused by the combination of the channel and equalizer, are obtained by taking the decimated samples at the output of $W(z)$.

17.2.1 Symbol-Spaced Equalizer

If one chooses to use an equalizer whose tap spacing is equal to the spacing T of the transmitted symbols, Figure 17.5 reduces to Figure 17.6. This corresponds to the case where the interpolation and decimation factor L is equal to 1. It may be noted that the system setup in Figure 17.6 is similar to the equalizer setup that was presented in Chapter 6 (Section 6.4.2). Because of obvious reasons, in this setup, $W(z)$ is referred to as a *symbol-spaced equalizer*.

It is also worth noting that because the transmit signal has a transmission bandwidth that is larger than $1/T$ (e.g., when $p_T(t)$ is square-root raised-cosine pulse shape with a

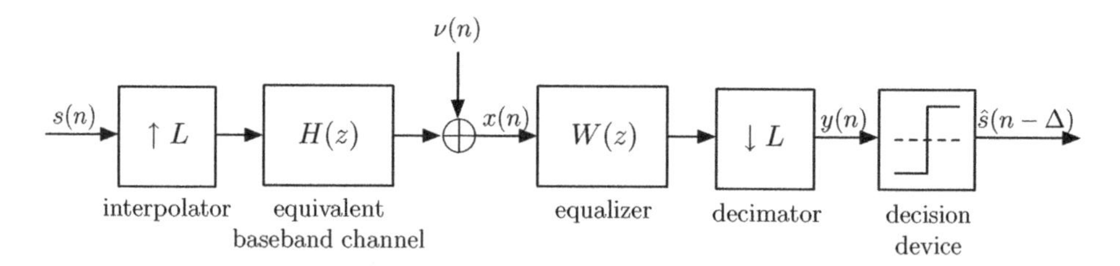

Figure 17.5 The discrete-time system model of a digital communication system.

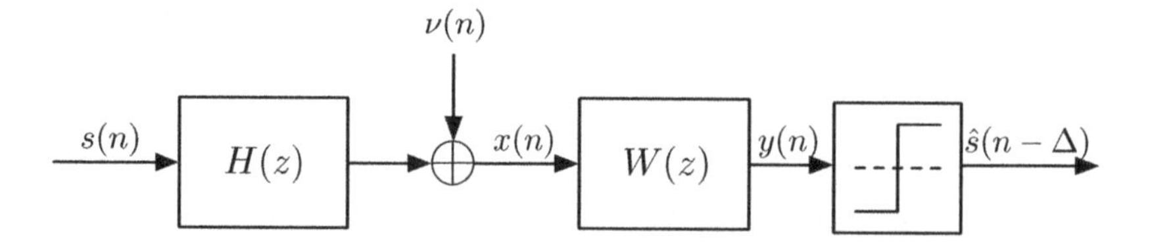

Figure 17.6 A digital communication system with a symbol-spaced equalizer.

roll-off factor α, the transmission bandwidth is $(1+\alpha)/T$), the sampled signal $x(n)$ (the equalizer input) is subject to aliasing. This, as discussed later (Section 17.4), may result in a poor performance if the input signal to the equalizer is sampled at a timing phase that results in a significant spectral attenuation within the aliased band.

17.2.2 Fractionally Spaced Equalizer

As it was noted earlier, the symbol-spaced equalizers are subject to aliasing and an improper choice of the timing phase of the samples to the equalizer may lead to significant performance degradation. This problem is avoided by adopting an equalizer whose input is sampled at a rate above the Nyquist. Such equalizers are referred to as *fractionally spaced.* As, in practice, $0 < \alpha \leq 1$, an equalizer tap-spacing $T/2$ is always sufficient to avoid aliasing. However, when $\alpha < 1$, one may set the tap-spacing equal to some value between $T/2$ and T. For example, when $\alpha = 0.5$, an equalizer tap-spacing $2T/3$ is sufficient to avoid aliasing. Also, in this case, one may prefer the spacing $2T/3$ over $T/2$, because for a fixed time span of the equalizer, the choice of $2T/3$ results in 4/3 $(= (2/3)/(1/2))$ times less number of tap weights.

Implementation of a fractionally spaced equalizer with the tap-spacing $T/2$ is straightforward and follows the block diagram of Figure 17.5, with $L = 2$. Figure 17.7 presents the detail of implementation of this equalizer. The equalizer taps are at the spacing $T/2$. Moreover, as the equalizer output is decimated twofold, the equalizer output needs to be calculated only for even values of n.

When the tap-spacing is a less trivial fraction of T, for example, it is equal to KT/L, for a pair of integers K and L, the equalizer detail is only slightly more involved than that in Figure 17.7. The detail of a fractionally spaced equalizer with the tap-spacing $2T/3$ is presented in Figure 17.8. This should serve as an example that can be easily extended to other cases as well. Here, the samples $x(n)$ are spaced at $T/3$, and the tap-spacing $2T/3$ is achieved by replacing each single delay block z^{-1} by a doubly delay block z^{-2}. Also, samples of output are calculated for values of n that are integer multiple of 3.

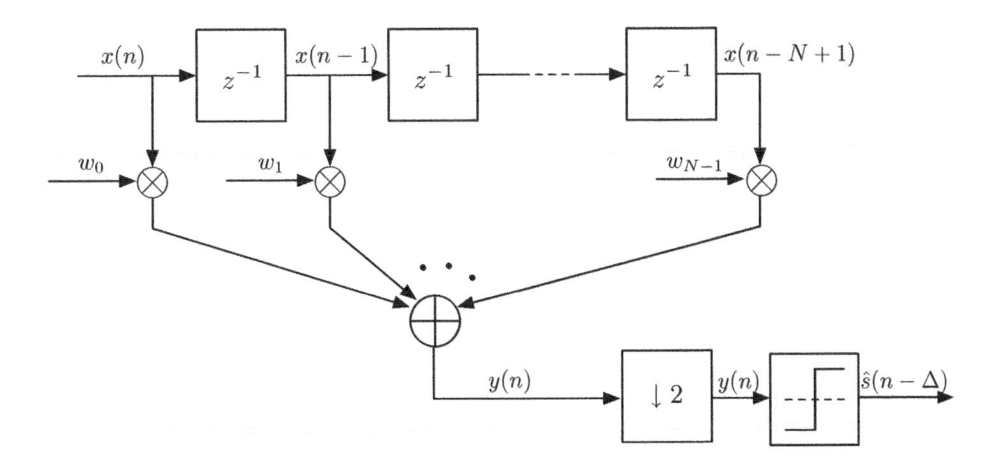

Figure 17.7 Details of a fractionally spaced equalizer with the tap-spacing $T/2$.

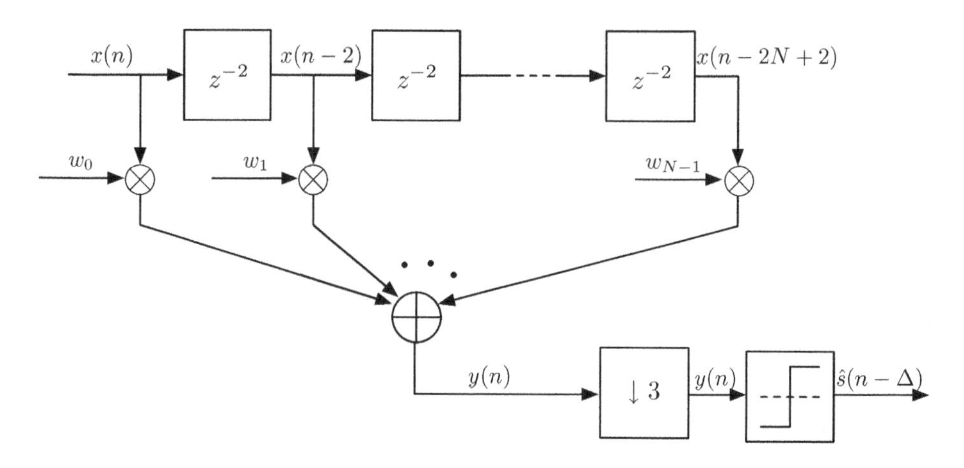

Figure 17.8 Details of a fractionally spaced equalizer with the tap-spacing $2T/3$.

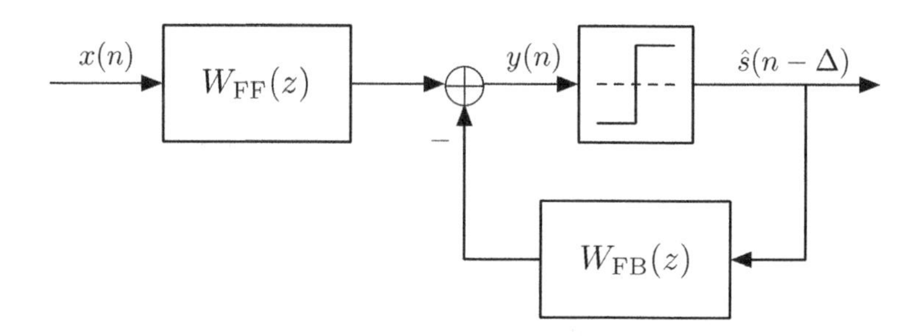

Figure 17.9 Block diagram of a decision feedback equalizer.

17.2.3 Decision Feedback Equalizer

The symbol-spaced and the fractionally spaced equalizer structures that were presented earlier belong to the class of linear equalizers/filters. The decision feedback (DF) equalizers, on the other hand, belong to the class of nonlinear filters. Figure 17.9 presents a block diagram of a DF equalizer. It consists of a feedforward filter $W_{\rm FF}(z)$ and a feedback filter $W_{\rm FB}(z)$. The feedforward filter $W_{\rm FF}(z)$ is a linear (transversal) filter and can be a symbol-spaced or fractionally spaced one. It equalizes the channel so that its response up to the decision device has no precursor ISI, but may contain some postcursor ISI. The postcursor ISI is canceled by passing the output of the decision device through the feedback filter with the transfer function

$$W_{\rm FB}(z) = \sum_{i=1}^{M} w_{{\rm FB},i} z^{-i} \tag{17.16}$$

where $w_{{\rm FB},i}$ are the feedback tap weights and M is the number of feedback taps.

When the channel contains some deep fade(s) in its amplitude response, a linear equalizer has to compensate for the fade(s). This, along with amplifying the desired (data) signal, will amplify the channel noise as well. In other words, equalization also leads to a *noise enhancement*, thus, may result in a significant degradation of the receiver performance. The DF equalizers take advantage of the degree of freedom provided by the feedback filter $W_{FB}(z)$ and avoid such noise enhancement. Hence, a DF equalizer, expectedly, should perform better than its linear equalizer counterpart. However, the DF equalizers suffer from another potential problem that sometimes leads to inferior performance when they are compared with their linear counterparts. Any decision error at the output of the decision device feeds back through the feedback filter and can result in more errors. This phenomenon, which is known as *error propagation*, is a serious problem that in many cases keeps the engineers lukewarm of adopting the DF equalizers for their applications.

17.3 Timing Recovery

Before the implementation of any equalizer, the received signal should be sampled at a rate that is an integer multiple of symbol rate $1/T$, (for example, at an interval $T_s = T/L$ or, equivalently, at a rate $f_s = L/T$), and at a correct timing phase. This process is often referred to as *timing recovery*. As in any adaptive algorithm, a timing recovery algorithm is developed based on a cost function whose optimization leads to the desired timing information. There are two classes of timing recovery methods: (i) nondata aided methods; and (ii) data-aided methods. The discussion in this section is limited to the nondata-aided methods. Interested readers should refer to the more detailed texts for a broader discussion on the subject of timing recovery; for example, see Farhang-Boroujeny (2010) and Meyr *et al.* (1998).

The nondata-aided timing recovery algorithms are commonly developed based on a cost function that is built based on the following property of the digital data signals. The power of the signal samples taken at the symbol interval T is a periodic function of the timing phase, τ, with the period of T. This power value is defined as a function, say, $\rho(\tau)$, and τ is adapted to maximize $\rho(\tau)$. As discussed later, variety of algorithms can be devised for this maximization, and such maximization, although not always, in many cases leads to a near optimum timing phase for the symbol-spaced equalizers. The fractionally spaced equalizers as we shall see later are insensitive to timing phase.

17.3.1 Cost Function

Ignoring the channel noise and recalling the equivalent baseband model of Figure 17.2, we obtain

$$x(t) = \sum_{n=-\infty}^{\infty} s(n)h(t - nT) \tag{17.17}$$

where $s(n)$'s are data symbols and T is the baud/symbol interval. The data symbols are in general from a complex-valued constellation.

The key idea in development of nondata-aided timing recovery stems from the following observation. If $s(n)$'s are a set of independent and identically distributed symbols with mean of zero and variance of σ_s^2, we obtain the ensemble mean-squared/power of $x(t)$ as

$$\rho(t) \triangleq E[|x(t)|^2] = \sigma_s^2 \sum_{n=-\infty}^{\infty} |h(t-nT)|^2 \tag{17.18}$$

By direct inspection, we note that $\rho(t)$ is a periodic signal with period of T. It thus can be expanded using Fourier series as

$$\rho(t) = \sum_{n=-\infty}^{\infty} \rho_n e^{j2\pi nt/T} \tag{17.19}$$

where ρ_n's are the Fourier series coefficients given by

$$\rho_n = \frac{1}{T}\int_0^T \rho(t) e^{-j2\pi nt/T} dt \tag{17.20}$$

Substituting Eq. (17.18) in Eq. (17.20), we obtain

$$\begin{aligned}\rho_n &= \frac{1}{T}\int_0^T \left(\sigma_s^2 \sum_{n=-\infty}^{\infty} |h(t-nT)|^2\right) e^{-j2\pi nt/T} dt \\ &= \frac{\sigma_s^2}{T} \sum_{n=-\infty}^{\infty} \int_0^T |h(t-nT)|^2 e^{-j2\pi nt/T} dt \\ &= \frac{\sigma_s^2}{T} \int_{-\infty}^{\infty} |h(t)|^2 e^{-j2\pi nt/T} dt\end{aligned} \tag{17.21}$$

where the second line follows by changing the order of the integral and summation, and the third line follows by introducing the change of variable $t - nT$ to t and noting that the resulting integrals add up to a single integral over the range $-\infty < t < \infty$.

Next, from the theory of Fourier transform, we recall that for any pair of functions $x(t)$ and $y(t)$,

$$\mathcal{F}[x(t)y(t)] = \frac{1}{2\pi} X(\Omega) \star Y(\Omega) \tag{17.22}$$

where $\mathcal{F}[\cdot]$ denotes Fourier transform, $\star$ denotes convolution, and $X(\Omega)$ and $Y(\Omega)$ are the Fourier transforms of $x(t)$ and $y(t)$, respectively. Also, we note that in Eq. (17.21), $\int_{-\infty}^{\infty} |h(t)|^2 e^{-j2\pi nt/T} dt$ is the Fourier transform of $|h(t)|^2 = h(t)h^*(t)$ at $\Omega = 2\pi n/T$. Hence, substituting $x(t) = h(t)$ and $y(t) = h^*(t)$ in Eq. (17.22), and noting that $\mathcal{F}[h^*(t)] = H^*(-\Omega)$, we get

$$\begin{aligned}\rho_n &= \frac{\sigma_s^2}{2\pi T} H(\Omega) \star H^*(-\Omega)|_{\Omega=\frac{2\pi n}{T}} \\ &= \frac{\sigma_s^2}{2\pi T} \int_{-\infty}^{\infty} H(\Omega) H^*\left(\Omega - \frac{2\pi n}{T}\right) d\Omega\end{aligned} \tag{17.23}$$

From Eq. (17.23), the following observations are made:

- $$\rho_0 = \frac{\sigma_s^2}{2\pi T}\int_{-\infty}^{\infty}|H(\Omega)|^2 \mathrm{d}\Omega \tag{17.24}$$

  is a real and positive number.

- $$\rho_1 = \frac{\sigma_s^2}{2\pi T}\int_{-\infty}^{\infty} H(\Omega)H^*\left(\Omega - \frac{2\pi}{T}\right)\mathrm{d}\Omega \tag{17.25}$$

  and

$$\begin{aligned}\rho_{-1} &= \frac{\sigma_s^2}{2\pi T}\int_{-\infty}^{\infty} H(\Omega)H^*\left(\Omega + \frac{1}{T}\right)\mathrm{d}\Omega \\ &= \frac{\sigma_s^2}{2\pi T}\int_{-\infty}^{\infty} H\left(\Omega - \frac{1}{T}\right)H^*(\Omega)\mathrm{d}\Omega \\ &= \rho_1^*\end{aligned} \tag{17.26}$$

- In almost all practical channels, the excess bandwidth of the transmit pulse-shaping filter is less than 100%; e.g., when $p_T(t)$ is a square-root raised-cosine pulse shape with a role of factor $\alpha \leq 1$. In such cases, $H(\Omega) = 0$ for $|\Omega| > 2\pi/T$. Using this, one finds that $\rho_n = 0$, for $|n| > 1$.

These observations imply that

$$\begin{aligned}\rho(t) &= \rho_0 + \rho_1 \mathrm{e}^{j2\pi t/T} + \rho_1^* \mathrm{e}^{-j2\pi t/T} \\ &= \rho_0 + 2|\rho_1|\cos\left(\frac{2\pi}{T}t + \angle\rho_1\right)\end{aligned} \tag{17.27}$$

where $|\rho_1|$ and $\angle\rho_1$ are the amplitude and phase of ρ_1, respectively.

Next, we note that, for a given timing phase τ, $E[|x(\tau + nT)|^2] = \rho(\tau)$, for $\forall n$. Hence, we may define

$$\rho(\tau) = E[|x(\tau + nT)|^2] \tag{17.28}$$

as the *timing recovery cost function.* From Eq. (17.27), it is obvious that $\rho(\tau)$ is a periodic function of τ with a period of T, the maximum value of $\rho_0 + 2|\rho_1|$, and the minimum value of $\rho_0 - 2|\rho_1|$. Figure 17.10 presents a typical plot of one period of $\rho(\tau)$ over the interval $-\frac{T}{2} \leq \tau \leq \frac{T}{2}$. Either the minima or maxima points of $\rho(\tau)$ can provide a reference point to which one may choose to lock the symbol clock rate. This, as presented later, allows one to develop algorithms that keep the receiver synchronized with the incoming data symbols.

17.3.2 The Optimum Timing Phase

When the equalizer is a fractionally spaced one, its performance is independent of the timing phase. Hence, there will be no optimum timing phase. All one may choose to do is to lock to a point of the cost function $\rho(\tau)$; the maximum and the minimum points of

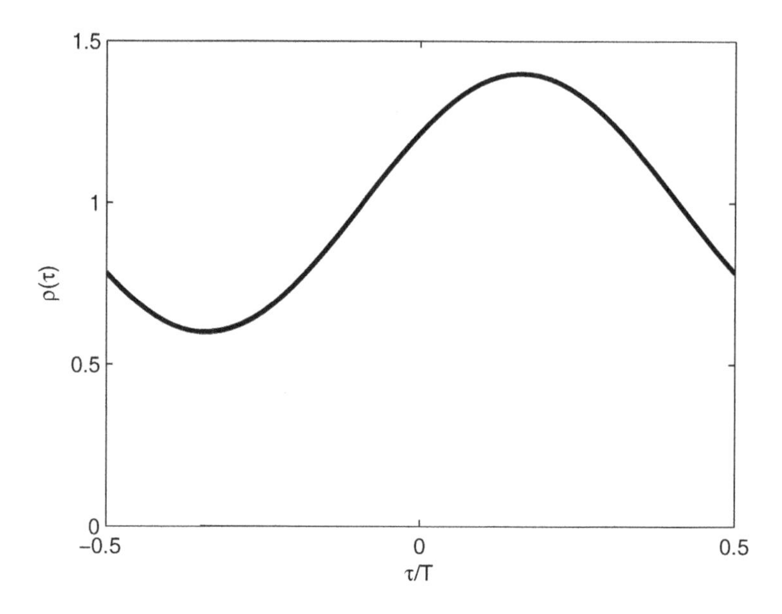

Figure 17.10 A typical plot of one period of the timing recovery cost function $\rho(\tau)$.

the cost function are usually easier points to lock to. However, when the equalizer is a symbol-spaced one, its performance varies significantly with the timing phase τ. It turns out that often (but not necessarily always) a good estimate of the optimum timing phase is obtained by maximizing the cost function $\rho(\tau)$.

When the equalizer is a symbol-spaced one and the signal samples at the equalizer input are sampled at a timing phase τ, the equivalent discrete time baseband impulse response of the channel is given by the sequence

$$h(n,\tau) = h(nT + \tau) \tag{17.29}$$

Let $H(\mathrm{e}^{j\omega},\tau)$ denote the Fourier transform of the sample sequence $h(n,\tau)$, and note that $H(\mathrm{e}^{j\omega},\tau)$ is the frequency response of the channel.

Next, we let $\tau = 0$ and recall from the theory of sampling that the frequency response $H(\mathrm{e}^{j\omega},0)$ is related to its continuous time counterpart $H(\Omega)$, that is, the Fourier transform of $h(t)$, as

$$H(\mathrm{e}^{j\omega},0) = \frac{1}{T}\sum_{k=-\infty}^{\infty} H\left(\frac{\omega - 2k\pi}{T}\right) \tag{17.30}$$

We also take note of the following points. $H(\mathrm{e}^{j\omega},0)$ is a periodic function ω with the period of 2π. Considering the common case where the pulse-shaping filter $p_{\mathrm{T}}(t)$ has an excess bandwidth of 100% or smaller, only the adjacent terms on the right-hand side of Eq. (17.30) overlap. Hence,

$$H(\mathrm{e}^{j\omega},0) = \frac{1}{T}\left(H\left(\frac{\omega}{T}\right) + H\left(\frac{\omega - 2\pi}{T}\right)\right), \quad \text{for } 0 \le \omega \le 2\pi. \tag{17.31}$$

Moreover, one may note that changing the timing phase from 0 to τ, in the frequency domain, has the impact of replacing $H(\Omega)$ by $e^{j\Omega\tau}H(\Omega)$. This, in turn, implies that

$$H(e^{j\omega},\tau)=\frac{1}{T}e^{j\omega\tau/T}\left(H\left(\frac{\omega}{T}\right)+e^{-j2\pi\tau/T}H\left(\frac{\omega-2\pi}{T}\right)\right),\quad \text{for } 0\le\omega\le 2\pi \tag{17.32}$$

Now, for the purpose of a simple demonstration, let us consider the case where the channel is ideal, that is, $c(t)=\delta(t)$, and note that in this case Eq. (17.6) implies that $H(\Omega)=P_{\mathrm{T}}(\Omega)P_{\mathrm{R}}(\Omega)$ is a Nyquist, for example, a raised-cosine, pulse. Figure 17.11 presents a set of plots of the magnitude response $|H(e^{j\omega},\tau)|$ as τ varies in the range of 0 to $T/2$. When $\tau=0$, $|H(e^{j\omega},\tau)|=\frac{1}{T}$, for $0\le\omega\le 2\pi$. For $\tau=\frac{T}{2}$, $|H(e^{j\omega},\tau)|$ experiences a null at $\omega=\pi$. For other values of τ in the range of 0 to $T/2$, $|H(e^{j\omega},\tau)|$ lies between these two extreme cases.

Figure 17.12 presents a block diagram of the discrete time baseband equivalent of the channel pertinent to our discussion in this section. The input to the channel is a random process consisting the transmitted data symbols $s(n)$. Assuming that the data symbols have a zero mean, are independent of each other, and have a variance of unity, one will

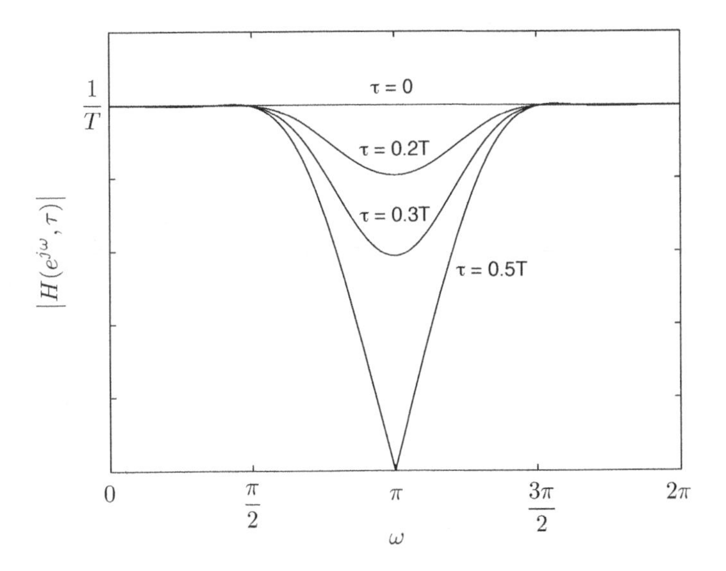

Figure 17.11 Visualization of Eq. (17.32).

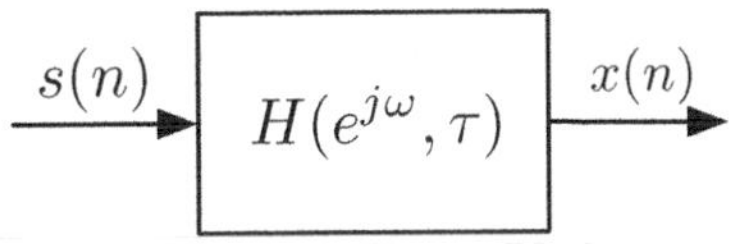

$s(n)$ → $H(e^{j\omega},\tau)$ → $x(n)$

Figure 17.12 Channel model for generating the signal $x(n)$ that feeds a symbol-spaced equalizer. $H(e^{j\omega},\tau)$ is given by Eq. (17.32).

find the power spectral density of the channel output as

$$\Phi_{xx}(e^{j\omega}, \tau) = |H(e^{j\omega}, \tau)|^2 \tag{17.33}$$

Also, recalling Eq. (2.60),

$$\begin{aligned}\rho(\tau) &= \frac{1}{2\pi}\int_{-\pi}^{\pi} \Phi_{xx}(e^{j\omega}, \tau)d\omega \\ &= \frac{1}{2\pi T^2}\int_{-\pi}^{\pi} |H(e^{j\omega}, \tau)|^2 d\omega \end{aligned} \tag{17.34}$$

where $\rho(\tau) = E[|x(n)|^2]$ denotes the signal power at the equalizer input for a timing phase τ. Recall that this is the timing recovery cost function that was defined earlier.

Next, considering Eq. (17.34), one may find that for the present case, where the channel is ideal, $\rho(\tau)$ finds its maximum value at $\tau = 0$ and it reduces as τ varies from 0 to $T/2$, reaching its minimum at $\tau = T/2$. This can be easily understood, if one notices from Figure 17.11 that the area under $|H(e^{j\omega}, \tau)|^2$ is at its maximum when $\tau = 0$ and reduces as τ varies from 0 to $T/2$. Moreover, it may be noted that when $\tau = 0$, $h(n, \tau)$ reduces to an impulse in the discrete time, n. Hence, the input to the equalizer is free of ISI, and there is no need for equalization. When $\tau \neq 0$, the deep in the magnitude of $|H(e^{j\omega}, \tau)|$ has to be compensated for by an equalizer with a gain larger than one around $\omega = \pi$. This, clearly, results in some noise enhancement, and thus, results in a poorer performance of the receiver. We may hence conclude that when the channel is ideal, the optimum timing phase is the one that avoids any attenuation in the aliased response $H(e^{j\omega}, \tau)$ of the channel. One may imagine that this conclusion is true even in the cases where the channel is nonideal, viz., a choice of timing phase that results in a significant attenuation of $H(e^{j\omega}, \tau)$, over the aliased portion of the spectrum, may lead to a significant noise enhancement, thus a poor performance of the receiver. However, unfortunately, as the numerical results presented in Section 17.4.2 show, the maximization of the cost function $\rho(\tau)$ does not necessarily avoid nulls in the aliased spectra and hence cannot always avoid possible poor performance of the symbol-spaced equalizers.

17.3.3 Improving the Cost Function

Figure 17.13 presents plots of the cost function $\rho(\tau)$ for an ideal channel where $H(\Omega) = P(\Omega)$ and $P(\Omega)$ is the Fourier transform of a raised-cosine pulse shape. The plots are given for three values of the roll-off factor $\alpha = 0.25$, 0.5, and 1. An important point to note here is that the variation of $\rho(\tau)$ with τ reduces as α decreases. Also, in an adaptive setting, a stochastic gradient (similar to the one in LMS algorithm) is used to search for the timing phase that maximizes $\rho(\tau)$. In addition, we note that the variance of a stochastic gradient is approximately proportional to the magnitude of the underlying signal power. Relating these points, one may argue that the stochastic gradients used for timing recovery become less reliable as α decreases.

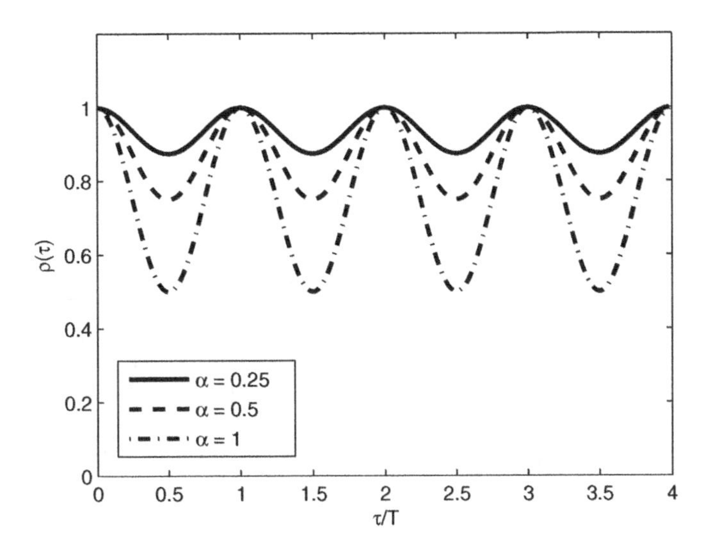

Figure 17.13 Plots of the timing recovery cost function $\rho(\tau)$ for an ideal channel and a raised-cosine pulse shape with three values of roll-off factor α.

To overcome the above problem, we proceed with an intuitive reasoning of why the gradient of $\rho(\tau)$ reduces with α and from there suggest a method of modifying the cost function $\rho(\tau)$ such that it will be less dependent on α. Referring to Figure 17.11, one finds that the variation of the received signal power as a function of timing phase, τ, is a direct consequence of augmentation or cancelation of the aliased signal components as τ varies. Moreover, if we note that the amount of aliased components reduces with α, it becomes obvious that the variation of $\rho(\tau)$ with τ reduces with α. Extending this argument, we may suggest, to obtain a cost function that will be less dependent on α, one should only concentrate on the signal power over the band of the aliased components. This can be achieved easily by passing the received signal through a bandpass filter that is centered around $1/2T$ and choosing the output power of this filter as the timing recovery cost function.

Figure 17.14 presents a set of plots of a modified cost function that is obtained by taking the T-spaced samples of the received signal and passing them through a single pole high-pass filter with the transfer function

$$B(z) = \frac{\sqrt{1-\beta^2}}{1+\beta z^{-1}} \tag{17.35}$$

where $0 < \beta < 1$ determines the bandwidth of the filter and the factor $\sqrt{1-\beta^2}$ is to normalize the power gain of the filter, for a white input, to unity. We refer to this cost function as $\rho_\beta(\tau)$. We also note that $\rho(\tau)$ can be thought as a special case of $\rho_\beta(\tau)$, which is obtained by choosing $\beta = 0$. For the results presented in Figure 17.14, β is set equal to 0.95. As expected, unlike $\rho(\tau)$, which varies significantly with α, $\rho_\beta(\tau)$, for β close to 1, remains nearly the same for all values of α.

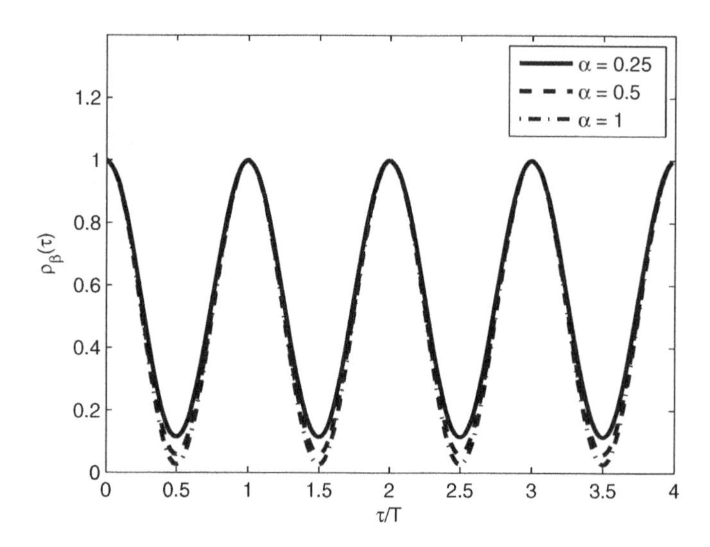

Figure 17.14 Plots of the improved timing recovery cost function for an ideal channel and a raised-cosine pulse shape with three values of the roll-off factor α. The parameter β is set equal to 0.95.

17.3.4 Algorithms

Application of the cost function $\rho(\tau)$ (or, equivalently, $\rho_\beta(\tau)$) leads to a variety of timing recovery/tracking algorithms. Here, we present two timing recovery algorithms that are built based on this cost function.

17.3.5 Early-Late Gate Timing Recovery

Early-late gate is one of the most common methods of timing recovery and can be applied to a variety of cost functions. To develop an early-late gate timing recovery algorithm based on the cost function $\rho(\tau)$, we proceed as follows.

We set the goal of timing recovery to choose a timing phase $\tau = \tau_{\text{opt}}$, which maximizes $\rho(\tau)$. Figure 17.15 presents the pertinent components relevant to the early-late gate timing recovery. We note that when $\tau = \tau_{\text{opt}}$ and $\delta\tau$ is a timing phase deviation, $\rho(\tau + \delta\tau) - \rho(\tau - \delta\tau) = 0$. On the other hand, for a nonoptimum timing phase τ and a small $\delta\tau$, we note that $\rho(\tau + \delta\tau) - \rho(\tau - \delta\tau) > 0$, when $\tau < \tau_{\text{opt}}$, and $\rho(\tau + \delta\tau) - \rho(\tau - \delta\tau) < 0$, when $\tau > \tau_{\text{opt}}$. The former case is demonstrated in Figure 17.15.

In the light of the above observation, one may propose the following update equation for adaptive adjustment of the timing phase:

$$\tau(n+1) = \tau(n) + \mu(\rho(\tau(n) + \delta\tau) - \rho(\tau(n) - \delta\tau)) \tag{17.36}$$

where μ is a step-size parameter. Moreover, we note that, in practice, the cost function $\rho(\tau)$ is not available and only could be estimated based on the observed signal samples, say, by taking the average of the squares of a few recent samples of $x(t)$. Or, we may

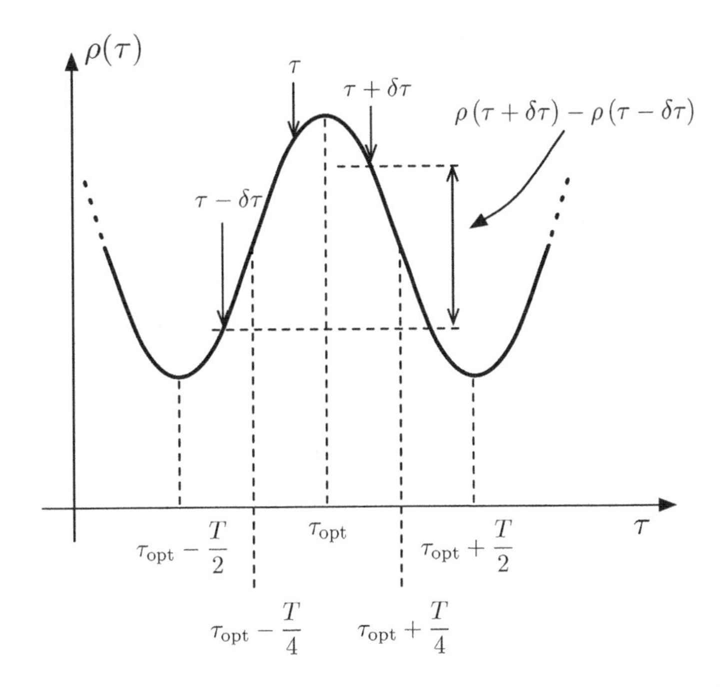

Figure 17.15 A demonstration of the early-late timing recovery Information.

follow the philosophy of the LMS algorithm and simply use $|x(t)|^2$ at $t = \tau + nT$ as an estimate of $\rho(\tau)$. Applying such coarse estimates, we obtain the update equation

$$\tau(n+1) = \tau(n) + \mu(|x(\tau(n) + \delta\tau + nT)|^2 - |x(\tau(n) - \delta\tau + nT)|^2). \tag{17.37}$$

The equations used for the realization of the early-late gate timing recovery with the modified cost function are summarized as follows:

$$x_1(n) = \sqrt{1-\beta^2}x(\tau(n) + \delta\tau + nT) - \beta x_1(n-1) \tag{17.38}$$

$$x_{-1}(n) = \sqrt{1-\beta^2}x(\tau(n) - \delta\tau + nT) - \beta x_{-1}(n-1) \tag{17.39}$$

$$\tau(n+1) = \tau(n) + \mu(|x_1(n)|^2 - |x_{-1}(n)|^2). \tag{17.40}$$

Note that $x_1(n)$ and $x_{-1}(n)$ are, respectively, sequences obtained by passing the signal samples $x(\tau(n) + \delta\tau + nT)$ and $x(\tau(n) - \delta\tau + nT)$ through the high-pass filter $B(z)$.

To explore the performance of the timing recovery recursion (17.37) (and Eq. (17.38)), we use the MATLAB script that is listed below. This is also available on the accompanying website under the name "`TxRxQAM.m`." This script that will be extended and used throughout the rest of this chapter for a number experiments contains the following features/components:

- As the continuous time cannot be simulated on a computer, a dense set of samples of a signal is used as a representation of its continuous time version. Here, L sample per each symbol interval is considered as dense. In the below script, we have set L equal to 100.

- The data symbols $s(n)$ are randomly selected from a four-point QAM constellation, with the alphabet $\{+1+j, -1+j, -1-j, +1-j\}$.
- Samples of the square-root raised-cosine pulse shape Eq. (17.7) are taken as the impulse response of the transmit and receive filters, $p_{\rm T}(t)$, and $p_{\rm R}(t)$, respectively. This is generated by calling the function "`sr_cos_p.m`," which is available on the accompanying website.
- The sequence of the data symbols $s(n)$ are up-sampled by a factor of L, before being passed through $p_{\rm T}(t)$, by adding $L-1$ zeros after each symbol. This task is taken care of by the function "`expander.m`," also available on the accompanying website.
- The modulator, channel, and demodulator are implemented according to the sequence of the blocks in Figure 17.1.
- The samples of the equivalent baseband impulse response of the transmission system from the input $s(t)$ to the output $x(t)$ can be obtained by letting $s(t) = \delta(t)$ (equivalently, letting $s(n) = 1$, for $n = 0$, and $s(n) = 0$, otherwise) and decimating the samples of $x(t)$ to the desired rate.
- The last segment of the MATLAB script (an attachment to "`TxRxQAM.m`," here) takes the continuous time signal $x(t)$ (having its samples at the dense grid of time points) and apply the early-late gate timing recovery algorithm to it. On the accompanying website, the MATLAB script "`TxRxQAM.m`" with this attachment is named "`TxRxQAMELG.m`."

```
%
% This MATLAB script provides an skeleton for simulation of a quadratue
% amplitude modulated (QAM) transmission system.
%
%%% PARAMETERS %%%
T=0.0001;          % Symbol/baud period
L=100;             % Number of samples per symbol period
Ts=T/L;            % A dense sampling interval (approx to continuous time)
fc=100000;         % Carrier frequency at the transmitter
Dfc=0;             % Carrier frequency offset
                   % (carrier at transmitter minus carrier at receiver)
phic=0;            % Carrier phase offset
alpha=0.5;         % Roll-off factor for the square-root raised cosine filter
sigma_v=0;         % Standard deviation of channel noise
c=1;               % Channel impulse response
%%% 4QAM symbols %%%
N=1000; s=sign(randn(N,1))+1i*sign(randn(N,1));
%%% Transmitter filterring %%%
pT=sr_cos_p(6*L,L,alpha);        % Transmit filter
xbbT=conv(expander(s,L),pT);     % Band-limited baseband transmit signal
%%% MODULATION %%%
t=[0:length(xbbT)-1]'*Ts;        % dense set of time points
xT=2*real(exp(i*2*pi*fc*t).*xbbT);
%%% CHANNEL (including aditive noise) %%%
xR=conv(c,xT); xR=xR+sigma_v*randn(size(xR)); % Received signal
%%% DEMODULATION %%%
t=[0:length(xR)-1]'*Ts;          % dense set of time points
xbbR=exp(-i*(2*pi*(fc+Dfc)*t-phic)).*xR;
```

```
%%% Receiver filtering  %%%
pR=pT; x=conv(xbbR,pR);
%%%%%%%%%%%%%%%%%%%%%%%%%%%%%%%%%%%%%%%%%%%%%%
%%% Early-late gate timing recovery %%%
%%%%%%%%%%%%%%%%%%%%%%%%%%%%%%%%%%%%%%%%%%%%%%
beta=0;              % Should be changed for the modified algorithm
mu0=0.005;           % Step-size parameter
dtau=12;             % (Δ tau) times L
mu=mu0*(L/4)/dtau;   % Adjusted step-size
kk=1;xp=0;xm=0;
start=5*L+1;         % To drop the transient of x(t) at the beginning
tau=0.3*ones(1,floor((length(x)-start)/L)); % Initialize the timing offset
            % The timing offset tau is adjusted as the algorithm proceeds.
for k=start:L:length(tau)*L
    tauT=round(tau(kk)*L);
    xp=sqrt(1-beta^2)*x(k+tauT+dtau)-beta*xp;
    xm=sqrt(1-beta^2)*x(k+tauT-dtau)-beta*xm;
    tau(kk+1)=tau(kk)+mu*(abs(xp)^2-abs(xm)^2);
    kk=kk+1;
end
figure, axes('position',[0.1 0.25 0.7 0.5]), plot(tau(1:kk),'k')
xlabel('Iteration Number, n'), ylabel('\tau(n)')
```

Figure 17.16 presents a set of plots of variation of τ as the early-late gate timing recovery operates according to the above MATLAB script. The results are for choices of $\delta\tau = T/L$, $10T/L$, $20T/L$, and $25T/L$ (respectively, dtau = 1, 10, 20, and 25, in the MATLAB script). From these results, and further experiments that may be performed,

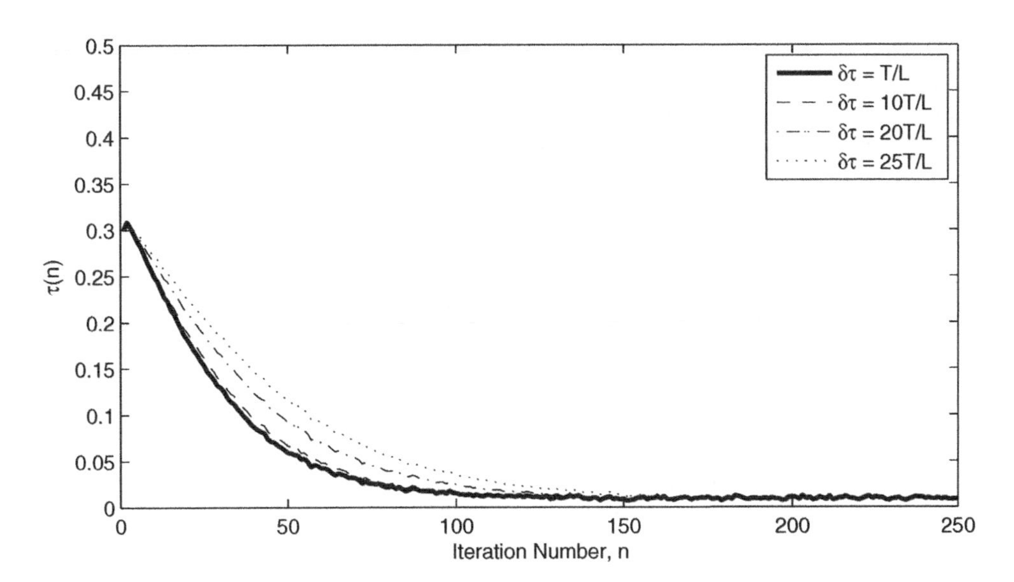

Figure 17.16 Plots of the timing phase update of the early-late gate timing recovery algorithm. The parameters used are $\beta = 0$, $\mu_0 = 0.01$, and four choices of $\delta\tau$. Each plot is an ensemble average of 100 independent runs.

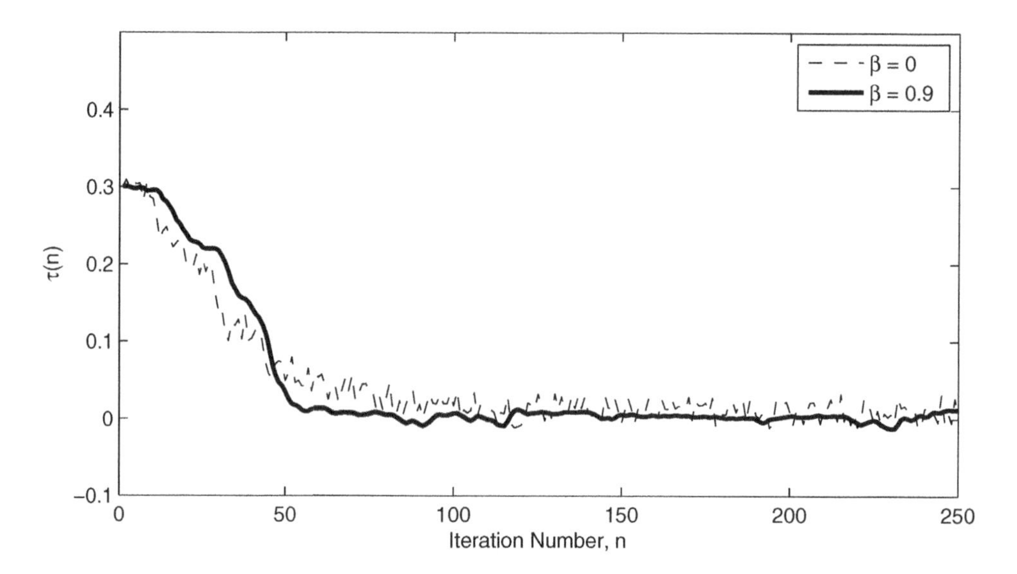

Figure 17.17 Comparison of the original and the improved early-late gate timing recovery algorithm for a single run. The parameters for the original algorithm are $\beta = 0$, $\mu_0 = 0.01$, and $\delta\tau = T/L$. The parameters for the improved algorithm are $\beta = 0.9$, $\mu_0 = 0.0025$, and $\delta\tau = T/L$.

one should observe that the convergence behavior of the early-late gate timing recovery algorithm remains independent of the parameter $\delta\tau$ over a relatively wide range of values within the interval T/L (a relatively small value) to $10T/L$ $(= T/10)$. However, as $\delta\tau$ increases and approaches $25T/L$ $(= T/4)$, some degradation in the convergence rate is observed. This observation can be explained, if we note that the ratio $(\rho(\tau(n) + \delta\tau) - \rho(\tau(n) - \delta\tau))/\delta\tau$ reduces as $\delta\tau$ increases (Figure 17.15).

Figure 17.17 presents a pair of the learning curves of $\tau(n)$ for the early-late gate timing recovery with $\beta = 0$ and its modified version with $\beta = 0.9$. The rest of the parameters have been chosen so that the cases have similar convergence rate. As seen, the choice $\beta = 0.9$ leads to an algorithm with significantly less jitters after it has converged. This result, of course, is in line with our discussion in Section 17.3.3.

17.3.6 Gradient-Based Algorithm

Using the gradient algorithm, the following recursion may be used to find the timing phase τ that maximizes the timing recovery cost function $\rho(\tau)$,

$$\tau(n + 1) = \tau(n) + \mu \frac{\partial \rho(\tau)}{\partial \tau} \tag{17.41}$$

where μ is a step-size parameter.

The early-late gate timing recovery recursion (17.36) (and, thus, Eq. (17.37)) may also be thought as a gradient-based algorithm, where the approximation

$$\frac{\partial \rho(\tau)}{\partial \tau} \approx \frac{|x(\tau(n)+\delta\tau+nT)|^2 - |x(\tau(n)-\delta\tau+nT)|^2}{2\delta\tau} \tag{17.42}$$

is used in Eq. (17.41) and the ratio $\mu/(2\delta\tau)$ is redefined as the step-size parameter. We also note that in the early-late gate timing recovery algorithm, each iteration requires three samples of $x(t)$; the desired sample $x(nT+\tau)$ and the lagged samples $x(\tau(n)+\delta\tau+nT)$ and $x(\tau(n)-\delta\tau+nT)$.

Here, we present a lower complexity timing recovery algorithm that operates based on only two samples $x(\tau(n)+nT)$ and $x(\tau(n)+nT+T/2)$ for each update of $\tau(n)$. We begin with using Eq. (17.28) to obtain

$$\frac{\partial \rho(\tau)}{\partial \tau} = -\frac{4\pi}{T}|\rho_1|\sin\left(\frac{2\pi}{T}\tau + \angle\rho_1\right) \tag{17.43}$$

Also, it has been shown in Farhang-Boroujeny (1994b) that when β is close to, but smaller than, 1,

$$E[\Re\{x_0(\tau(n)+nT)x_1^*(\tau(n)+nT+T/2)\}] = -k\sin\left(\frac{2\pi}{T}\tau + \angle\rho_1\right) \tag{17.44}$$

where $x_0(\tau(n)+nT)$ and $x_1(\tau(n)+nT+T/2)$, respectively, are the signal sequences obtained by passing $x(\tau(n)+nT)$ and $x(\tau(n)+nT+T/2)$ through the transfer function $B(z)$ of Eq. (17.35), and k is a positive constant. Following the same approach as the one used in the LMS algorithm and, also, in Eq. (17.37), we use $\Re\{x_0(\tau(n)+nT)x_1^*(\tau(n)+nT+T/2)\}$ as a stochastic estimate proportional to the gradient $\partial\rho(\tau)/\partial\tau$ in Eq. (17.41). This leads to the update equation

$$\tau(n+1) = \tau(n) + \mu\Re\{x_0(\tau(n)+nT)x_1^*(\tau(n)+nT+T/2)\} \tag{17.45}$$

The above algorithm can be implemented according to the following MATLAB script:

```
%%%%%%%%%%%%%%%%%%%%%%%%%%%%%%%%%%%%%%
%%% Gradient-based timing recovery %%%
%%%%%%%%%%%%%%%%%%%%%%%%%%%%%%%%%%%%%%
beta=0.9;             % Should be changed for the modified algorithm
mu=0.0025;            % Step-size parameter
kk=1;xp=0;xm=0;
start=5*L+1;          % To drop the transient of x(t) at the beginning
tau=0.3*ones(1,floor((length(x)-start)/L)); % Initialize the timing offset
            % The timing offset tau is adjusted as the algorithm proceeds.
for k=start:L:length(tau)*L
    tauT=round(tau(kk)*L);
    xp=sqrt(1-beta^2)*x(k+tauT)-beta*xp;
    xm=sqrt(1-beta^2)*x(k+tauT+L/2)-beta*xm;
    tau(kk+1)=tau(kk)+mu*real(xp*xm');
    kk=kk+1;
end
```

On the accompanying website, the MATLAB script "TxRxQAM.m" with this attachment is called "TxRxQAMGB.m."

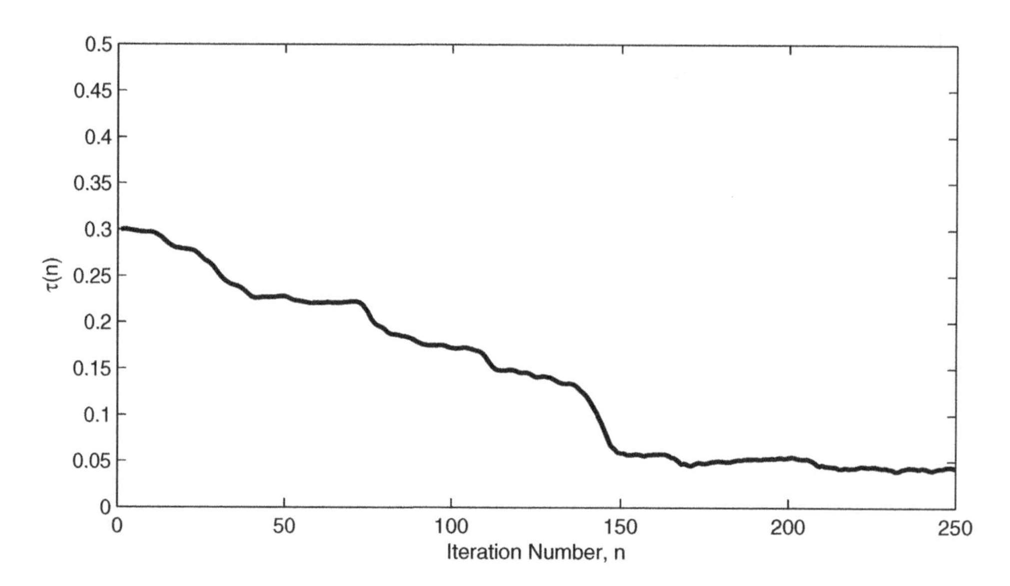

Figure 17.18 A plot of the timing phase update of the gradient-based timing recovery algorithm. The parameters used are $\beta = 0.9$ and $\mu = 0.0025$.

Figure 17.18 presents a typical plot of $\tau(n)$ when the gradient-based algorithm is used. The simulation setup is similar to the one used to generate Figures 17.16 and 17.17 and the parameters used here are $\beta = 0.9$ and $\mu = 0.0025$. From this result, we may observe that the gradient-based timing recovery algorithm is somewhat slower than the early-late gate algorithm proposed earlier. This clearly is the price paid for a less complex algorithm.

17.4 Equalizers Design and Performance Analysis

When one has access to the statistics of the transmitted data symbols, the channel response, and the statistical characteristics of the channel noise, it is possible to evaluate and study the equalizer performance based on the Wiener filters theory that was developed in Chapter 3. In this section, we take this approach to develop some insight into the performance of the symbol-spaced, fractionally spaced, and DF equalizers. Adaptation algorithms for equalizer design as well as design strategies that jointly perform the tasks of carrier frequency and timing phase recovery are presented in the following sections.

17.4.1 Wiener–Hopf Equation for Symbol-Spaced Equalizers

To develop the Wiener–Hopf equations for symbol-spaced equalizers, we begin with the system model of Figure 17.19. This follows the block diagram shown in Figure 17.6. In addition, here, we have shown the details of how the noise sequence $\nu(n)$ is generated. The sequence $\nu(n)$ is obtained by passing the channel noise $\nu_c(n)$ through the receiver front-end filter, $p_R(n)$. The latter is usually a square-root raised-cosine filter matched to its counterpart at the transmitter and is usually run at a rate that is a multiple of the

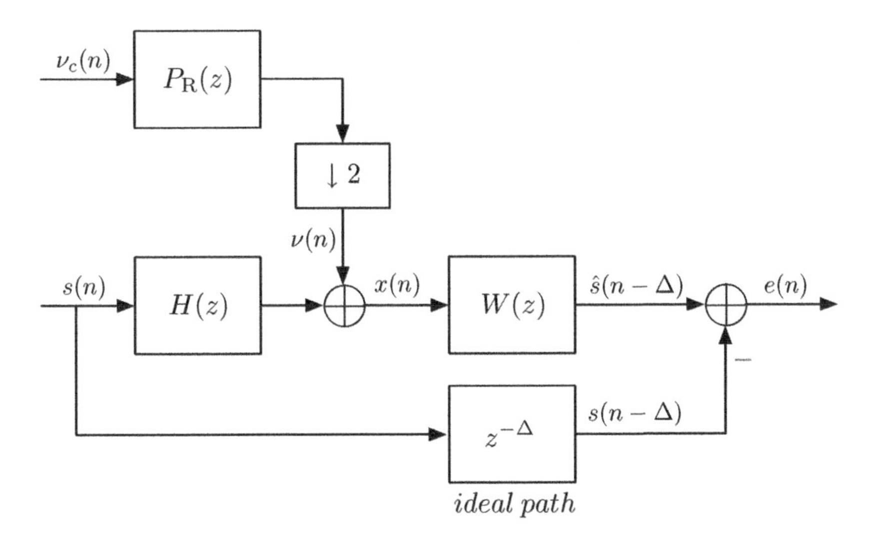

Figure 17.19 System setup for a symbol-spaced equalizer.

symbol rate. In Figure 17.19, we have assumed that $p_R(n)$ is run at twice the symbol rate. We model $\nu_c(n)$ as a white sequence with variance $\sigma_{\nu_c}^2$. The transmit data stream $s(n)$ is also modeled as a white sequence with variance σ_s^2.

The equalizer tap weights w_n should be selected to minimize output error $e(n)$. The error $e(n)$ originates from two independent sources: the data symbols $s(n)$, and the channel noise $\nu_c(n)$. We may thus write $e(n) = e^{(s)}(n) + e^{(\nu_c)}(n)$, where $e^{(s)}(n)$ and $e^{(\nu_c)}(n)$ are the errors that originate from $s(n)$ and $\nu_c(n)$, respectively. The power spectral density $\Phi_{ee}(z)$ of $e(n)$, thus, can be written as

$$\Phi_{ee}(z) = \Phi_{ee}^{(s)}(z) + \Phi_{ee}^{(\nu_c)}(z) \tag{17.46}$$

where $\Phi_{ee}^{(s)}(z)$ and $\Phi_{ee}^{(\nu_c)}(z)$ are the power spectral densities of $e^{(s)}(n)$ and $e^{(\nu_c)}(n)$, respectively. $\Phi_{ee}^{(s)}(z)$ is obtained straightforwardly, if we note that the transfer function between $s(n)$ and $e(n)$ is $H(z)W(z) - z^{-\Delta}$ and apply Eq. (2.80). This leads to

$$\Phi_{ee}^{(s)}(z) = \Phi_{ss}(z)|H(z)W(z) - z^{-\Delta}|^2 \tag{17.47}$$

To obtain $\Phi_{ee}^{(\nu_c)}(z)$, we note that the path between $\nu_c(n)$ and $e^{(\nu_c)}(n)$ can be redrawn as in Figure 17.20. This is obtained by making use of the theory of multirate signal processing (Vaidyanathan, 1993) and noting that the cascade of $P_R(z)$ and the twofold decimator can be replaced by the corresponding polyphase structure, as presented in Figure 17.20. Here, $P_R^0(z)$ and $P_R^1(z)$ are the polyphase components of $P_R(z)$. They are obtained by separating the odd and even ordered powers of z in $P_R(z)$ and expanding it as

$$P_R(z) = P_R^0(z^2) + z^{-1}P_R^1(z^2). \tag{17.48}$$

Also, the sequences $\nu_c^0(n)$ and $\nu_c^1(n)$ are the polyphase components of $\nu_c(n)$, defined as

$$\{\nu_c^0(n)\} = \{\ldots, \nu_c(-2), \nu_c(0), \nu_c(2), \ldots\} \tag{17.49}$$

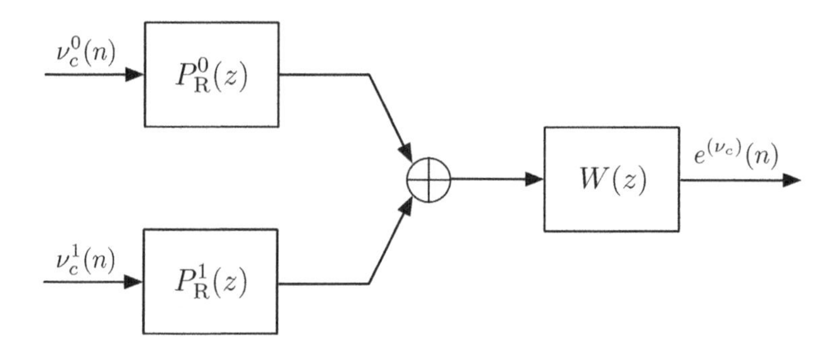

Figure 17.20 The path between $\nu_c(n)$ and the output error $e(n)$.

and

$$\{\nu_c^1(n)\} = \{\ldots, \nu_c(-1), \nu_c(1), \nu_c(3), \ldots\}. \tag{17.50}$$

Next, we note that as $\nu_c(n)$ is a white process, $\nu_c^0(n)$ and $\nu_c^1(n)$ are two uncorrelated sequences with the same power spectral density to that of $\nu_c(n)$. Hence, applying Eq. (2.80), one obtains

$$\Phi_{ee}^{(\nu_c)}(z) = \Phi_{\nu_c\nu_c}(z)(|P_R^0(z)W(z)|^2 + |P_R^1(z)W(z)|^2) \tag{17.51}$$

Combining Eqs. (17.47) and (17.51), and noting that $\Phi_{ss}(z) = \sigma_s^2$ and $\Phi_{\nu_c\nu_c}(z) = \sigma_{\nu_c}^2$ (as $s(n)$ and $\nu_c(n)$ are white processes), we get

$$\Phi_{ee}(z) = \sigma_s^2|H(z)W(z) - z^{-\Delta}|^2 + \sigma_{\nu_c}^2(|P_R^0(z)W(z)|^2 + |P_R^1(z)W(z)|^2) \tag{17.52}$$

On the other hand, one may note that

$$\begin{aligned}\xi &= E[|e(n)|^2] \\ &= \frac{1}{2\pi j}\oint \Phi_{ee}(z)\frac{dz}{z} \\ &= \sigma_s^2\frac{1}{2\pi j}\oint |H(z)W(z) - z^{-\Delta}|^2\frac{dz}{z} + \sigma_{\nu_c}^2\frac{1}{2\pi j}\oint |P_R^0(z)W(z)|^2\frac{dz}{z} \\ &\quad + \sigma_{\nu_c}^2\frac{1}{2\pi j}\oint |P_R^1(z)W(z)|^2\frac{dz}{z}\end{aligned} \tag{17.53}$$

Following the principle of Wiener filtering, to design $W(z)$, we set the goal of minimizing the mean-squared error ξ. To this end, we first convert Eq. (17.53) to its time-domain equivalent. For this purpose, we define the error sequences $e_s(n)$, $e_{\nu_c}^0(n)$, and $e_{\nu_c}^1(n)$ whose z transforms are, respectively,

$$E_s(z) = H(z)W(z) - z^{-\Delta} \tag{17.54}$$

$$E_{\nu_c}^0(z) = P_R^0(z)W(z) \tag{17.55}$$

and

$$E_{\nu_c}^1(z) = P_R^1(z)W(z) \tag{17.56}$$

and note that

$$e_s(n) = h(n) \star w_n - d(n) \tag{17.57}$$

$$e^0_{\nu_c}(n) = p^0_R(n) \star w_n \tag{17.58}$$

and

$$e^1_{\nu_c}(n) = p^1_R(n) \star w_n \tag{17.59}$$

where $\star$ denotes linear convolution and $d(n)$ is the inverse z transform of $z^{-\Delta}$. We note that unlike $e^{(s)}(n)$ and $e^{(\nu_c)}(n)$, which have indefinite length and are power signals, the sequences $e_s(n)$, $e^0_{\nu_c}(n)$, and $e^1_{\nu_c}(n)$, defined here, may have finite duration and are energy signals. These, respectively, are the impulse responses between the input sequences $s(n)$, $\nu^0_c(n)$, and $\nu^1_c(n)$ and the output error sequence $e(n)$. Moreover, using the Parseval's relation, the time-domain equivalent of Eq. (17.53) is obtained as

$$\xi = \sigma_s^2 \sum_n |e_s(n)|^2 + \sigma_{\nu_c}^2 \sum_n |e^0_{\nu_c}(n)|^2 + \sigma_{\nu_c}^2 \sum_n |e^1_{\nu_c}(n)|^2 \tag{17.60}$$

Next, we define the column vectors

$$\mathbf{e}_s = \begin{bmatrix} e_s(0) \\ e_s(1) \\ e_s(2) \\ \vdots \end{bmatrix}, \quad \mathbf{e}^0_{\nu_c} = \begin{bmatrix} e^0_{\nu_c}(0) \\ e^0_{\nu_c}(1) \\ e^0_{\nu_c}(2) \\ \vdots \end{bmatrix}, \quad \mathbf{e}^1_{\nu_c} = \begin{bmatrix} e^1_{\nu_c}(0) \\ e^1_{\nu_c}(1) \\ e^1_{\nu_c}(2) \\ \vdots \end{bmatrix}$$

$$\mathbf{w} = \begin{bmatrix} w_0^* \\ w_1^* \\ w_2^* \\ \vdots \end{bmatrix} \quad \text{and} \quad \mathbf{d} = \begin{bmatrix} d(0) \\ d(1) \\ d(2) \\ \vdots \end{bmatrix} \tag{17.61}$$

Note that, here, we have followed the convention set in Chapter 3 and have defined the elements of the tap-weight vector $\mathbf{w}$ as the conjugates of the tap weights of the equalizer. Also, whereas the index for the tap weights appears as a subscript (e.g., w_i), for the time sequences such as $e_s(n)$ the time index n is enclosed in parentheses.

Using the definitions (17.61), one finds that Eqs. (17.57), (17.58), and (17.59) can be expanded, respectively, as

$$\mathbf{e}_s = \mathbf{H}\mathbf{w}^* - \mathbf{d} \tag{17.62}$$

$$\mathbf{e}^0_{\nu_c} = \frac{\sigma_{\nu_c}}{\sigma_s}\mathbf{P}^0\mathbf{w}^* \tag{17.63}$$

and

$$\mathbf{e}^1_{\nu_c} = \frac{\sigma_{\nu_c}}{\sigma_s}\mathbf{P}^1\mathbf{w}^* \tag{17.64}$$

where

$$\mathbf{H} = \begin{bmatrix} h(0) & 0 & 0 & \cdots \\ h(1) & h(0) & 0 & \cdots \\ h(2) & h(1) & h(0) & \cdots \\ \vdots & \vdots & \vdots & \ddots \end{bmatrix} \tag{17.65}$$

$$\mathbf{P}^0 = \begin{bmatrix} p_R^0(0) & 0 & 0 & \cdots \\ p_R^0(1) & p_R^0(0) & 0 & \cdots \\ p_R^0(2) & p_R^0(1) & p_R^0(0) & \cdots \\ \vdots & \vdots & \vdots & \ddots \end{bmatrix} \tag{17.66}$$

and

$$\mathbf{P}^1 = \begin{bmatrix} p_R^1(0) & 0 & 0 & \cdots \\ p_R^1(1) & p_R^1(0) & 0 & \cdots \\ p_R^1(2) & p_R^1(1) & p_R^1(0) & \cdots \\ \vdots & \vdots & \vdots & \ddots \end{bmatrix} \tag{17.67}$$

Also, if we define

$$\mathbf{e} = \begin{bmatrix} \mathbf{e}_s \\ \mathbf{e}_{\nu_c}^0 \\ \mathbf{e}_{\nu_c}^1 \end{bmatrix} \quad \text{and} \quad \mathbf{Q} = \begin{bmatrix} \mathbf{H} \\ \frac{\sigma_{\nu_c}}{\sigma_s}\mathbf{P}^0 \\ \frac{\sigma_{\nu_c}}{\sigma_s}\mathbf{P}^1 \end{bmatrix}$$

Eqs. (17.62), (17.63), and (17.64) can be combined as

$$\mathbf{e} = \mathbf{Q}\mathbf{w}^* - \mathbf{d} \tag{17.68}$$

where, here, $\mathbf{d}$ is the vector $\mathbf{d}$ defined before with a number of zeros appended at the end of it to have the same length as $\mathbf{Q}\mathbf{w}^*$. Using Eq. (17.68), Eq. (17.60) reduces to

$$\xi = \sigma_s^2 \mathbf{e}^{\mathrm{H}} \mathbf{e} \tag{17.69}$$

where the superscript H denotes Hermitian.

Substituting Eq. (17.68) in Eq. (17.69) and expanding, we obtain

$$\xi = \sigma_s^2(\mathbf{w}^{\mathrm{T}}(\mathbf{Q}^{\mathrm{H}}\mathbf{Q})\mathbf{w}^* - \mathbf{w}^{\mathrm{T}}(\mathbf{Q}^{\mathrm{H}}\mathbf{d}) - (\mathbf{Q}^{\mathrm{H}}\mathbf{d})^{\mathrm{H}}\mathbf{w}^* + \mathbf{d}^{\mathrm{H}}\mathbf{d}) \tag{17.70}$$

As ξ is real-valued, applying a conjugate sign to both sides of Eq. (17.70) leads to

$$\xi = \sigma_s^2(\mathbf{w}^{\mathrm{H}}(\mathbf{Q}^{\mathrm{T}}\mathbf{Q}^*)\mathbf{w} - \mathbf{w}^{\mathrm{H}}(\mathbf{Q}^{\mathrm{T}}\mathbf{d}^*) - (\mathbf{Q}^{\mathrm{T}}\mathbf{d}^*)^{\mathrm{H}}\mathbf{w} + \mathbf{d}^{\mathrm{H}}\mathbf{d}). \tag{17.71}$$

This is similar to the cost function expressions derived in Chapter 3 and elsewhere in this book. It is thus straightforward to show that the optimum tap weight vector of the equalizer (i.e., the minimizer of ξ) is given by

$$\mathbf{w}_o = \mathbf{R}^{-1}\mathbf{p} \tag{17.72}$$

where $\mathbf{R} = \mathbf{Q}^{\mathrm{T}}\mathbf{Q}^*$ and $\mathbf{p} = \mathbf{Q}^{\mathrm{T}}\mathbf{d}^*$. Also, the minimum mean-squared error, that is, the value of ξ when $\mathbf{w} = \mathbf{w}_o$, is obtained as

$$\xi_{\min} = \sigma_s^2(1 - \mathbf{w}_o^{\mathrm{H}}\mathbf{p}) \tag{17.73}$$

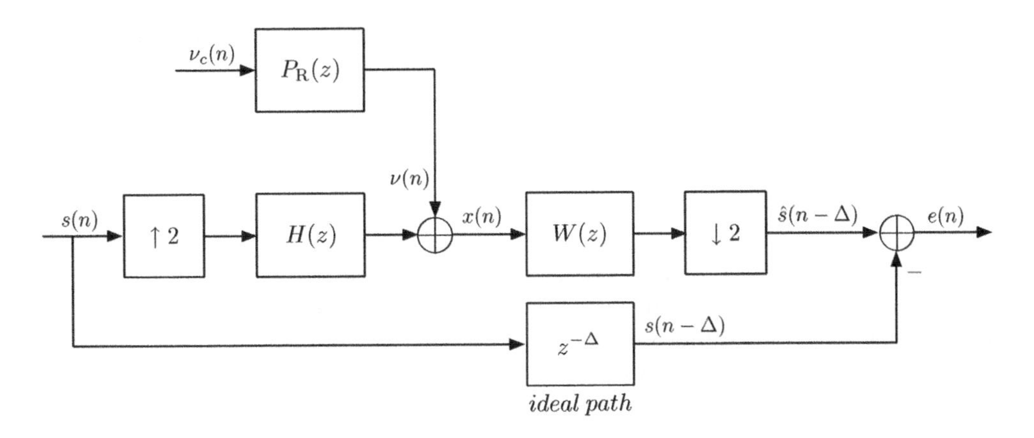

Figure 17.21 System setup for a half symbol-spaced equalizer.

Fractionally Spaced Equalizer

For brevity and clarity of presentation, here, we limit our discussion to a particular case of fractionally spaced equalizer where the equalizer taps are at one-half of symbol spacing, that is, when $L = 2$ and $K = 1$. Figure 17.21 presents a system setup that may be used for the study of a fractionally spaced equalizer with symbol spacing of half a symbol interval.

Following Figure 17.21 and a similar vector formulation as in the case of symbol-spaced equalizer, here, we obtain

$$\mathbf{e}_\mathrm{s} = \mathbf{H}\mathbf{w} - \mathbf{d} \tag{17.74}$$

and

$$\mathbf{e}_{\nu_\mathrm{c}} = \frac{\sigma_{\nu_\mathrm{c}}}{\sigma_\mathrm{s}}\mathbf{P}\mathbf{w} \tag{17.75}$$

with

$$\mathbf{H} = \begin{bmatrix} h(0) & 0 & 0 & 0 & 0 & \cdots \\ h(2) & h(1) & h(0) & 0 & 0 & \cdots \\ h(4) & h(3) & h(2) & h(1) & h(0) & \cdots \\ \vdots & \vdots & \vdots & \vdots & \vdots & \ddots \end{bmatrix} \tag{17.76}$$

and

$$\mathbf{P} = \begin{bmatrix} p_\mathrm{R}(0) & 0 & 0 & \cdots \\ p_\mathrm{R}(1) & p_\mathrm{R}(0) & 0 & \cdots \\ p_\mathrm{R}(2) & p_\mathrm{R}(1) & p_\mathrm{R}(0) & \cdots \\ \vdots & \vdots & \vdots & \ddots \end{bmatrix} \tag{17.77}$$

Moreover, if we define

$$\mathbf{e} = \begin{bmatrix} \mathbf{e}_\mathrm{s} \\ \mathbf{e}_{\nu_\mathrm{c}} \end{bmatrix} \quad \text{and} \quad \mathbf{Q} = \begin{bmatrix} \mathbf{H} \\ \frac{\sigma_{\nu_\mathrm{c}}}{\sigma_\mathrm{s}}\mathbf{P} \end{bmatrix}$$

Eqs. (17.74) and (17.75) can be combined to obtain an equation similar to Eq. (17.68), which will then lead to similar results to those in Eqs. (17.69), (17.71), (17.72), and (17.73).

Decision Feedback Equalizer

Figure 17.22 presents a system setup that may be used for the study of a DF equalizer. For simplicity of presentation, we have chosen to limit our discussion here to the case where the feedforward filter is a symbol-spaced one. The extension of the results to fractionally spaced equalizers is not difficult and is left as an exercise for the interested readers.

Following Figure 17.22, one will find that, here, Eqs. (17.63) and (17.64) should be modified as

$$\mathbf{e}_{\nu_c}^0 = \frac{\sigma_{\nu_c}}{\sigma_s}\mathbf{P}^0\mathbf{w}_{\mathrm{FF}}^* \tag{17.78}$$

and

$$\mathbf{e}_{\nu_c}^1 = \frac{\sigma_{\nu_c}}{\sigma_s}\mathbf{P}^1\mathbf{w}_{\mathrm{FF}}^* \tag{17.79}$$

respectively, and Eq. (17.57) as

$$e_s(n) = h(n) \star w_{\mathrm{FF},n} - \delta(n - \Delta - 1) \star w_{\mathrm{FB},n} - d(n). \tag{17.80}$$

Moreover, the vector formulation of Eq. (17.80) is obtained as

$$\mathbf{e}_s = \begin{bmatrix}\mathbf{H} & -\mathbf{G}\end{bmatrix}\begin{bmatrix}\mathbf{w}_{\mathrm{FF}}^* \\ \mathbf{w}_{\mathrm{FB}}^*\end{bmatrix} - \mathbf{d} \tag{17.81}$$

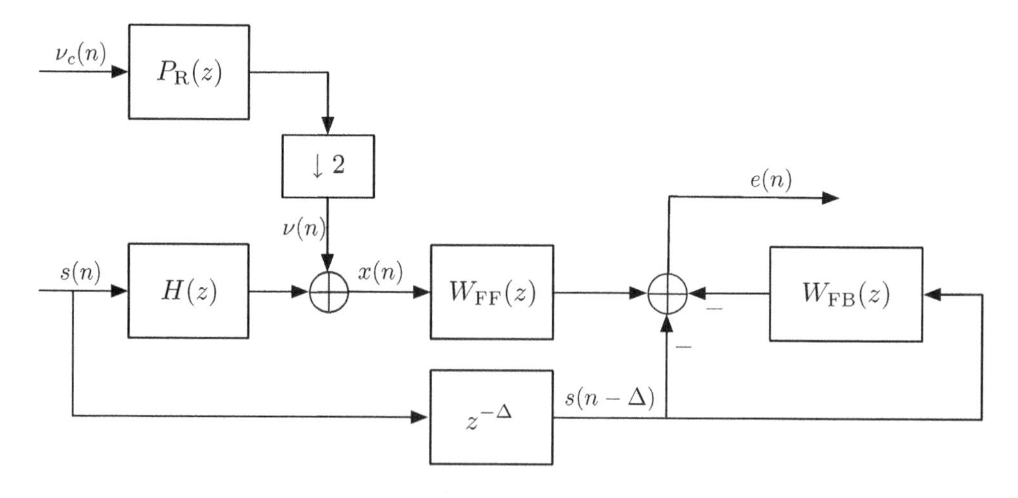

Figure 17.22 System setup for a decision feedback equalizer.

where $\mathbf{H}$ and $\mathbf{d}$ have the same definitions as before, $\mathbf{w}_{\mathrm{FF}}$ and $\mathbf{w}_{\mathrm{FB}}$ are the tap-weight vectors of the feedforward and feedback filters, respectively, and

$$\mathbf{G} = \begin{bmatrix} 0 & \cdots & 0 \\ \vdots & \ddots & \vdots \\ 0 & \cdots & 0 \\ 1 & \cdots & 0 \\ \vdots & \ddots & \vdots \\ 0 & \cdots & 1 \end{bmatrix} \tag{17.82}$$

The top part of $\mathbf{G}$ is a zero matrix with M columns (where M, as defined earlier, is the length of the feedback filter, $\mathbf{w}_{\mathrm{FB}}$) and $\Delta + 1$ rows, and the bottom part of $\mathbf{G}$ is an identity matrix of size $M \times M$.

Next, if we define

$$\mathbf{e} = \begin{bmatrix} \mathbf{e}_{\mathrm{s}} \\ \mathbf{e}_{\nu_{\mathrm{c}}}^{0} \\ \mathbf{e}_{\nu_{\mathrm{c}}}^{1} \end{bmatrix} \quad \text{and} \quad \mathbf{Q} = \begin{bmatrix} \mathbf{H} & -\mathbf{G} \\ \frac{\sigma_{\nu_{\mathrm{c}}}}{\sigma_{\mathrm{s}}}\mathbf{P}^{0} & 0 \\ \frac{\sigma_{\nu_{\mathrm{c}}}}{\sigma_{\mathrm{s}}}\mathbf{P}^{1} & 0 \end{bmatrix}$$

Eqs. (17.78), (17.79), and (17.81) can be combined to obtain an equation similar to Eq. (17.68), which will then lead to similar results to those in Eqs. (17.69), (17.71), (17.72), and (17.73).

17.4.2 Numerical Examples

The MATLAB script "`equalizer_eval.m`," below, also available on the accompanying website, puts together the above results and evaluates the performance of a symbol-spaced equalizer, a half symbol-spaced equalizer, and a DF equalizer with a symbol-spaced feedforward filter. The baseband equivalent channel $h(t)$ is evaluated using Eq. (17.13). For both cases, the timing phases of the samples to equalizer are varied over one symbol interval. As explained earlier, the fractionally spaced equalizer should be insensitive to the timing phase. The symbol-spaced equalizer, on the other hand, is expected to show some sensitivity to the timing phase. The sensitivity of the DF equalizer to timing phase is examined and commented on through numerical examples.

We note that the performance of an equalizer depends on the delay Δ. However, when the equalizer length is sufficiently long, the following choice of Δ provides a near optimal performance:

$$\Delta = \frac{1}{2}(\text{length of channel} + \text{length of equalizer}) \tag{17.83}$$

whereas here, the lengths are in the unit of symbol interval. Hence, whereas a symbol-space equalizer with N tap weights has a length of N symbol intervals, a half symbol-space equalizer with N tap weights has a length of $N/2$ symbol intervals. The number of feedback taps in the DF filter $W_{\mathrm{FB}}(z)$, M, is chosen to remove all the postcurser samples of the channel response at the output of the feedforward filter, $W_{\mathrm{FF}}(z)$. The reader is encouraged to read carefully through the MATLAB script "`equalizer_eval.m`" to see how the parameters Δ and M are calculated. The reader is also encouraged to explore the script to understand how the various equations have been programmed.

```
%
% This script evaluates the minimum mean-squared error (MMSE) of the
% symbol-spaced and half symbol-spaced equalizers. The equivalent baseband
% channel h(t) is obtained according to (17.13). The MMSE values are then
% calculated, following the formulations in Section 17.4.
%
clear all,close all
%% parameters %%
T=0.0001; L=100; Ts=T/L; fs=1/Ts; fc=100000;
alpha=0.5; sigmas=1; sigmanuc=0.01;
c=[0.5 zeros(1,60) 1 zeros(1,137) 0.3];      % More common channel
c=[1 zeros(1,67) 0.75 zeros(1,145) 0.4];  % Less common channel
c=c/sqrt(c*c');
pT=sr_cos_p(16*L,L,alpha)';
pR=pT;
%% Construction of the equivalent baseband channel %%
p=conv(pT,pR);
c=c.*exp(-j*2*pi*[0:length(c)-1]*Ts*fc);
h=conv(c,p);
pR=sqrt(L/2)*pR(1:L/2:end);
%%
%%%%%%%%%%%%%%%%%%%%%%%%%%
%  T spaced equalizer   %
%%%%%%%%%%%%%%%%%%%%%%%%%%
N=21;        % equalizer length = N = 21
for k=1:L
    h0=h(k:L:end);
    Delta=round((length(h(1:L:end))+N)/2);
    C=toeplitz([h0 zeros(1,N-1)],[h0(1) zeros(1,N-1)]);
    P0=toeplitz([pR(1:2:end) zeros(1,N-1)],[pR(1) zeros(1,N-1)]);
    P1=toeplitz([pR(2:2:end) zeros(1,N-1)],[pR(2) zeros(1,N-1)]);
    Q=[C; (sigmanuc/sigmas)*P0; (sigmanuc/sigmas)*P1];
    d=[zeros(Delta,1); 1;zeros(length(Q(:,1))-(Delta+1),1)];
    Ryy=(Q'*Q+1e-14*eye(N));
    pyd=(Q'*d);
    w=Ryy\pyd;
    mmse0(k)=sigmas^2*real(1-w'*pyd);
    spower(k)=sum(abs(h0).^2);
end
%%
%%%%%%%%%%%%%%%%%%%%%%%%%%
%  T/2 spaced equalizer  %
%%%%%%%%%%%%%%%%%%%%%%%%%%
N=21;
hf=h(1:L/2:end);Delta=round((length(hf)+N)/4);
for k=1:L
    hf=h(k:L/2:end);
    C=toeplitz([hf zeros(1,N-1)],[hf(1) zeros(1,N-1)]); C=C(1:2:end,:);
    P=toeplitz([pR zeros(1,N-1)],[pR(1) zeros(1,N-1)]);
    Q=[C; (sigmanuc/sigmas)*P];
    d=[zeros(Delta,1); 1;zeros(length(Q(:,1))-(Delta+1),1)];
    Ryy=(Q'*Q+1e-6*eye(N));
    pyd=(Q'*d);
```

```
    w=Ryy\pyd;
    mmsef1(k)=sigmas^2*real(1-w'*pyd);
end
%%
%%%%%%%%%%%%%%%%%%%%%%%%%%%%%%%%
%  decision fedback equalizer   %
%%%%%%%%%%%%%%%%%%%%%%%%%%%%%%%%
N=21;        % FF filter length
for k=1:L
    h0=h(k:L:end);
    Delta=round((length(h(1:L:end))+N)/2);
    C=toeplitz([h0 zeros(1,N-1)],[h0(1) zeros(1,N-1)]);
    M=length(h0)+N-1-Delta-1;  % FB filter length
    P0=toeplitz([pR(1:2:end) zeros(1,N-1)],[pR(1) zeros(1,N-1)]);
    P1=toeplitz([pR(2:2:end) zeros(1,N-1)],[pR(2) zeros(1,N-1)]);
    Q=[[C [zeros(Delta+1,M); eye(M)]]; ...
        [(sigmanuc/sigmas)*P0 zeros(length(P0(:,1)),M)]; ...
        [(sigmanuc/sigmas)*P1 zeros(length(P1(:,1)),M)]];
    d=[zeros(Delta,1); 1;zeros(length(Q(:,1))-(Delta+1),1)];
    Ryy=(Q'*Q+1e-14*eye(N+M));
    pyd=(Q'*d);
    w=Ryy\pyd;
    mmsedf(k)=sigmas^2*real(1-w'*pyd);
end
```

Figures 17.23 and 17.24 present the results of execution of "`equalizer_eval.m`" for two examples of the channel $c(t)$, as listed in the above script. The first example, labeled as "more common channel," is a case where the timing phase that maximizes $\rho(\tau)$ leads to an equivalent baseband channel for which the symbol-spaced equalizer performs well. Most of the channels in practice have this behavior. The second example, labeled as "less common channel," is a case where the timing phase that maximizes $\rho(\tau)$ leads to an equivalent baseband channel for which the symbol-spaced equalizer performs poorly. The results presented in Figures 17.23 and 17.24 include (i) Panel (a): the signal power at the input of a symbol-spaced (T-spaced) equalizer as a function of the timing phase τ; (ii) Panel (b): MMSE (minimum mean-squared error) results of a 21-tap T-spaced equalizer, a 21-tap $T/2$-spaced equalizer, a 31-tap $T/2$-spaced equalizer, and a DF equalizer with a 21-tap T-spaced feedforward filter; (iii) Panel (c): the magnitude response of the baseband equivalent channel seen by the T-spaced equalizer for the case when τ is optimally selected to minimize MMSE, and the worse choice of τ that results in the maximum MMSE.

From the results presented in Figures 17.23 and 17.24 and many more that one can try (by changing the channel, $c(t)$, and the other system parameters in "`equalizer_eval.m`"), the following observations are made:

- As one would expect, the performance of the symbol-spaced equalizer can degrade significantly for some choices of the timing phase. The fractionally spaced equalizer as well as the DF equalizer, on the other hand, are both almost insensitive to the timing phase.
- The performance of the symbol-spaced equalizer usually degrades near the timing phase where the signal power is minimum. It also has a near optimum performance when the timing phase is chosen to maximize the signal power, that is, the cost function $\rho(\tau)$.

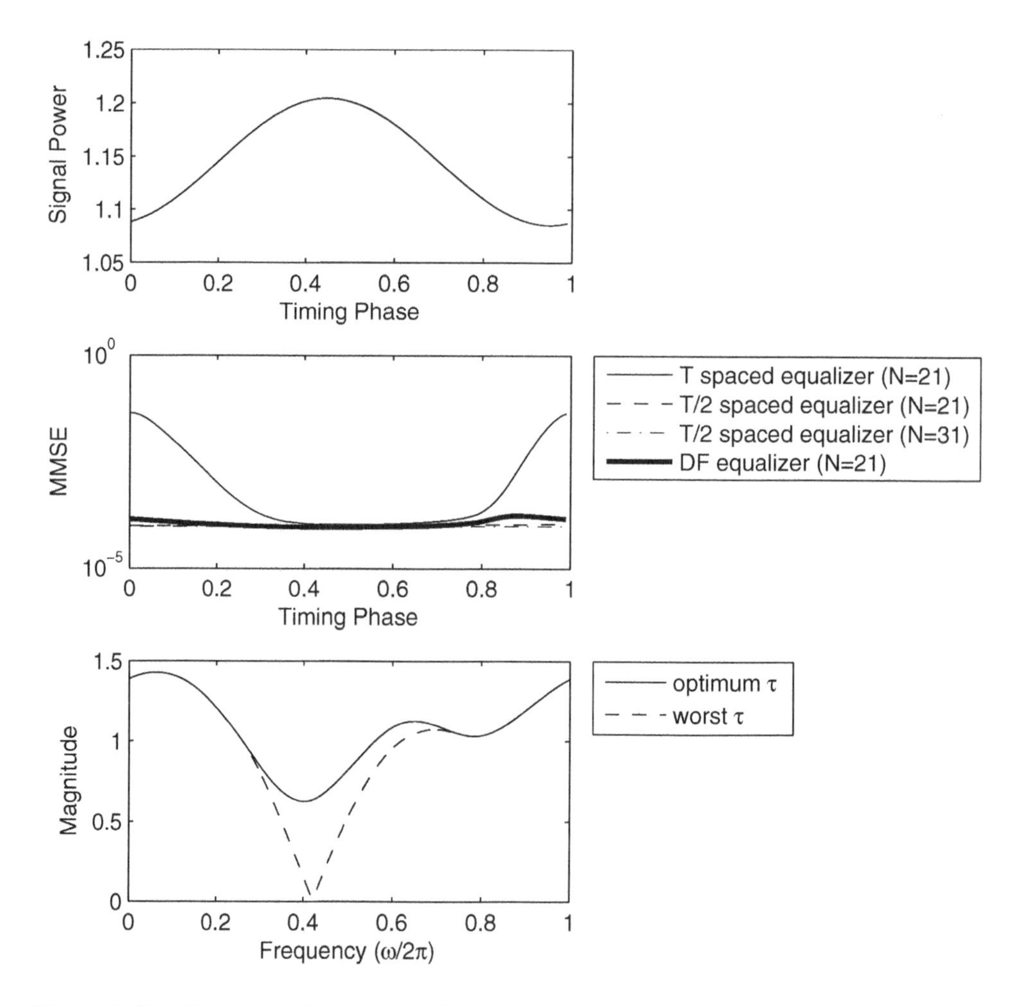

Figure 17.23 The results of execution of "`equalizer_eval.m`" for the more common channel.

However, this rule is not general. For some cases that one may encounter occasionally, it is found that the symbol-spaced equalizer performs very poorly when the timing phase is selected for the power maximization.

- The fractionally spaced equalizer performs equal to or better than the symbol-spaced equalizer, when both are allowed to have the same time span. For the results presented in Figures 17.23 and 17.24, two cases of the fractionally spaced equalizer are considered. These have time spans that are 50% and 75% of that of the symbol-spaced equalizer, respectively.
- As was predicted before, assuming no error propagation, the DF equalizer always performs better than its linear counterpart. Also, as observed, there is no significant difference between a fractionally spaced equalizer and its DF counterpart. Hence, one may conclude that because of the potential problem of error propagation in DF equalizers, the choice of a fractionally spaced equalizer should always be preferred.

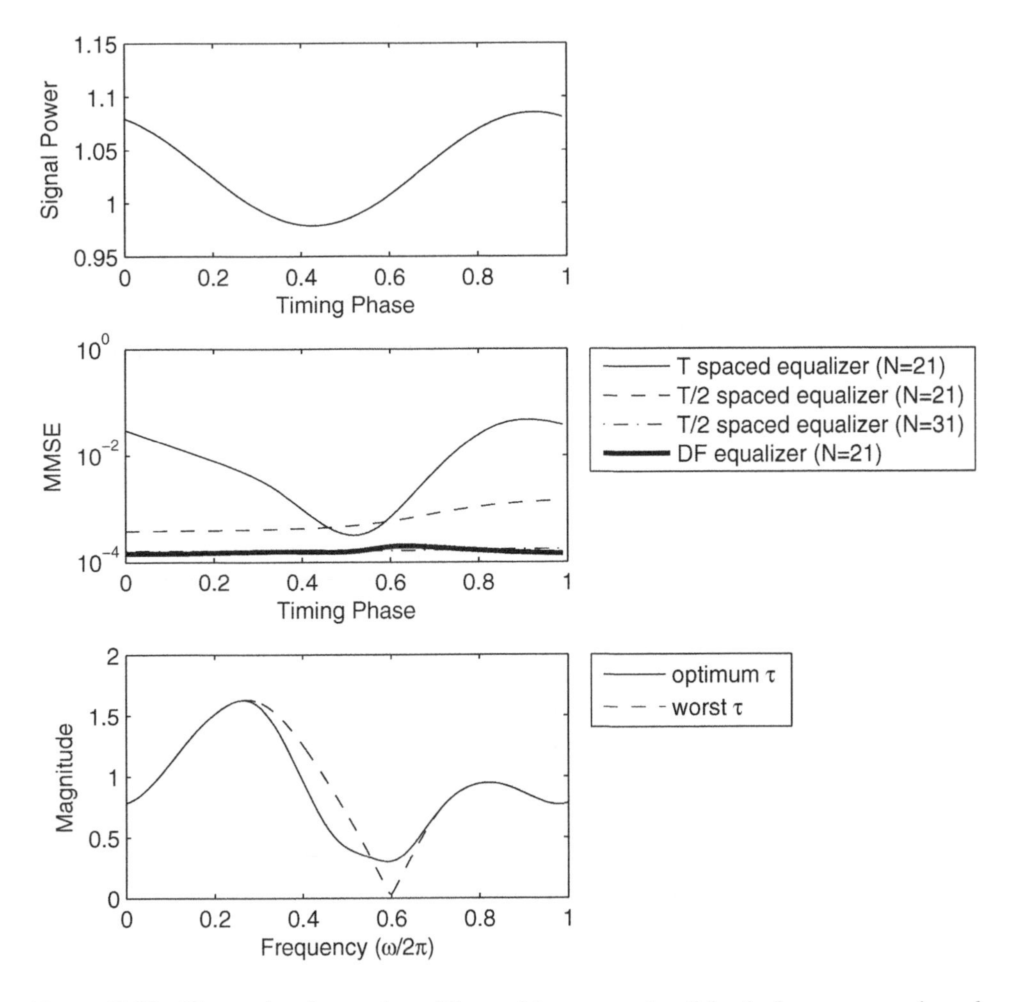

Figure 17.24 The results of execution of "`equalizer_eval.m`" for the less common channel.

17.5 Adaptation Algorithms

Obviously, any of the various adaptive algorithms that were developed in the previous chapters can be directly applied to channel equalizers. However, in choosing an adaptive algorithm, one should also take note of the following points:

- The power spectral density of the input to an equalizer may vastly vary, depending on the channel impulse response; see the examples presented in Section 6.4.2. Hence, the performance of the LMS algorithm in channel equalization is highly dependent on the channel and may vary significantly from channel to channel.
- The RLS algorithm, on the other hand, performs independently of the channel. However, besides its much higher computational complexity, it may suffer from the numerical

stability problem, especially for channels that are poorly conditioned, that is, the cases where the eigenvalue spread of the underlying correlation matrix is large.

- The intermediate algorithms, such as the affine projection algorithm (of Section 6.7) and the LMS-Newton algorithm (of Section 11.15) may thus be good compromised choices for adaptation of channel equalizers.
- For the fractionally spaced equalizers, one should take note of the following point. The input signal to the equalizer is oversampled above its Nyquist rate. This means that the power spectral density of the input signal to the equalizer over higher portion of its frequency band reaches zero. This, in turn, translates to the fact that the underlying correlation matrix is poorly conditioned. Hence, the problems mentioned in the first two items above may be more pertinent in the case of the fractionally spaced equalizers.

17.6 Cyclic Equalization

Training symbols that are known to the receiver are often used at the beginning of a communication session to synchronize the receiver carrier and symbol clock to the incoming signal and to adjust the equalizer coefficients to a point near their optimal values. It turns out that if the training sequence is selected to be periodic and have a period equal to the length of the equalizer, the equalizer tap weights can be obtained almost instantly. Moreover, when such training sequences are used, simple mechanism can be developed for fast acquisition of the carrier frequency and symbol clock of the received signal. Furthermore, any phase offset in the carrier will be taken care of by the equalizer. In addition, the use of the periodic training sequences allows adoption of a simple mechanism for selection of the time delay Δ and, thus, alignment of the data symbol sequences between transmitter and receiver.

17.6.1 Symbol-Spaced Cyclic Equalizer

Let the periodic sequence $\ldots, s(N-1), s(0), s(1), s(2), \ldots, s(N-1), s(0), \ldots$ be transmitted through a channel with the symbol-spaced complex baseband equivalent response $h(n)$. Ignoring the channel noise, this periodic input to the channel results in a periodic output $x(n)$ with the same period. Now consider an equalizer setup with the input $x(n)$ and desired output $s(n)$. As, here, $s(n)$ and $x(n)$ are periodic, one may pick a period of samples of $x(n)$, with an arbitrary starting point, and put them in a tapped-delay line/shift register with its output connected back to its input. A similar shift register is also used to keep one cycle of $s(n)$. An equalizer whose tap weights are adjusted to match its output $y(n)$ with $s(n)$ is then constructed as shown in Figure 17.25.

The adaptation algorithm in Figure 17.25 can be any of the known adaptive algorithms, including the LMS, NLMS, APLMS or RLS. Also, once a cycle of $x(n)$ is received, the adaptation process can begin and run for sufficient number of iterations for the equalizer to converge. To put this in a mathematical framework, we define the column vectors

$$\mathbf{x}_0 = [x(n)\; x(n-1) \cdots x(n-N+1)]^{\mathrm{T}}$$

and

$$\mathbf{w} = [w_0\; w_1 \cdots w_{N-1}]^{\mathrm{H}}$$

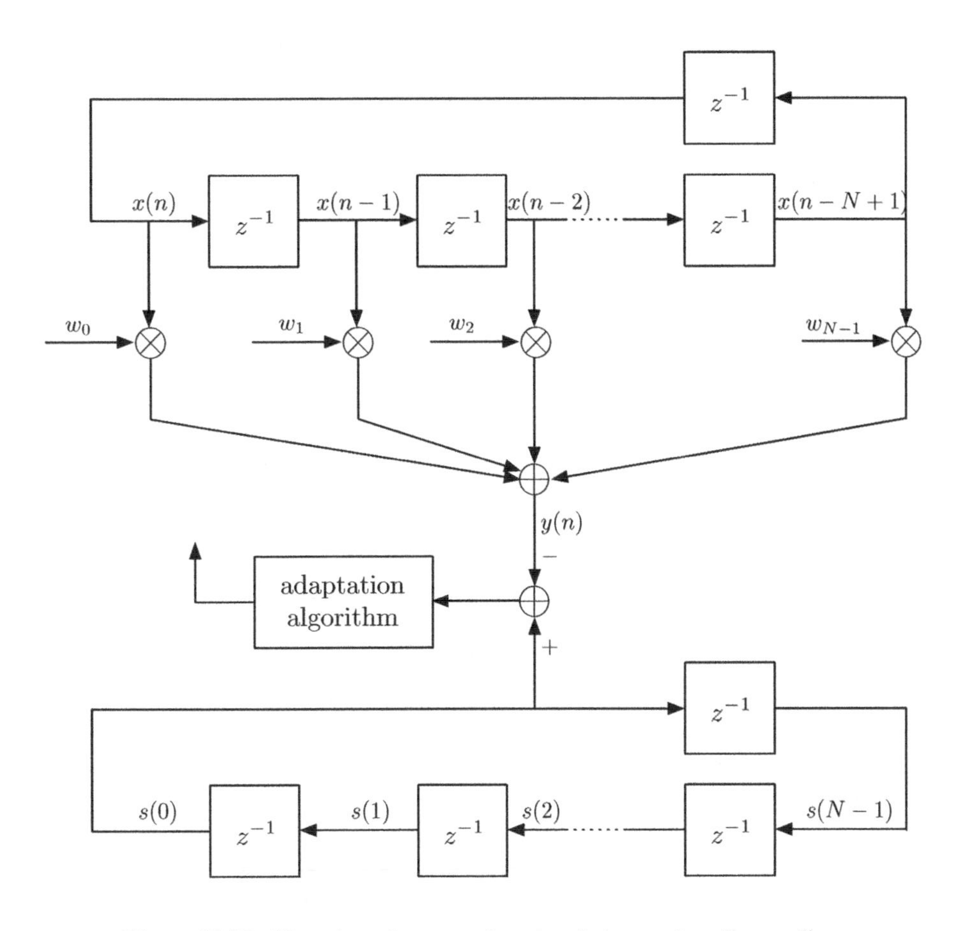

Figure 17.25 The adaptation setup for a symbol-spaced cyclic equalizer.

where "T" and "H" denote transpose and Hermitian (i.e., conjugate transpose), respectively. Also, if we define $\mathbf{x}_i$ as the circularly shifted version of $\mathbf{x}_0$ after i shifts and use the LMS algorithm for tap-weight adaptation, the repetition of the following loop will converge to a tap-weight vector that closely approximates the optimum vector $\mathbf{w}_o$. Here, the tap-weight vector $\mathbf{w}$ is initialized to an arbitrary value $\mathbf{w}(0)$. In practice, the common choice for $\mathbf{w}(0)$ is the zero vector.

for $i = 0, 1, 2, \ldots$
$$e(i) = s(i \mod N) - \mathbf{w}^{\mathrm{H}}(i)\mathbf{x}_i$$
$$\mathbf{w}(i+1) = \mathbf{w}(i) + 2\mu e^*(i)\mathbf{x}_i$$
end

In the above loop, "$i \mod N$" reads i modulo N, which means the integer remainder of i divided by N. It thus generates the ordered indices of the periodic sequence $s(0)$, $s(1)$, $\ldots$, $s(N-1)$, $s(0)$, $s(1)$, $\ldots$

Alternatively, one may set up the problem of finding the optimum choice of the tap-weight vector $\mathbf{w}$ as a least-squares problem with the cost function

$$\zeta = \|\mathbf{e}\|^2 = \mathbf{e}^{\mathrm{H}}\mathbf{e} \tag{17.84}$$

where $\mathbf{e} = [e(0)\ e(1)\ e(2) \cdots e(N-1)]^{\mathrm{T}}$. To proceed with a solution to this least-squares problem, we note that by conjugating both sides of the first line in the above "for loop," we get

$$s^*(i) - \mathbf{x}_i^{\mathrm{H}}\mathbf{w} = e^*(i), \quad i = 0, 1, 2, \ldots, N-1 \tag{17.85}$$

Combining these equations, we obtain

$$\mathbf{s}^* - \mathbf{X}^{\mathrm{H}}\mathbf{w} = \mathbf{e}^* \tag{17.86}$$

where $\mathbf{X}$ is an $N \times N$ matrix whose columns are the equalizer tap-input vectors $\mathbf{x}_0$, $\mathbf{x}_1$, $\mathbf{x}_2$, ..., $\mathbf{x}_{N-1}$. It is also interesting to note that, in the expanded form,

$$\mathbf{X} = \begin{bmatrix} x(n) & x(n-N+1) & x(n-N+2) & \cdots & x(n-1) \\ x(n-1) & x(n) & x(n-N+1) & \cdots & x(n-2) \\ x(n-2) & x(n-1) & x(n) & \cdots & x(n-3) \\ \vdots & \vdots & \vdots & \ddots & \vdots \\ x(n-N+1) & x(n-N+2) & x(n-N+3) & \cdots & x(n) \end{bmatrix} \tag{17.87}$$

Now recall that the optimum value of $\mathbf{w}$ (in the cyclic equalizer) is the one that minimizes the Euclidean norm of the vector $\mathbf{e}$. On the other hand, as the choice of $\mathbf{w} = (\mathbf{X}^{\mathrm{H}})^{-1}\mathbf{s}^*$ results in $\mathbf{e} = 0$, which has the trivial minimum Euclidean norm of zero, one may argue that the optimum value of $\mathbf{w}$ is obtained simply by solving the equation

$$\mathbf{X}^{\mathrm{H}}\mathbf{w} = \mathbf{s}^*. \tag{17.88}$$

We also note that the solution to this problem is unique when $\mathbf{X}^{\mathrm{H}}$ (or, equivalently, $\mathbf{X}$) is full rank, or, in other words, the inverse of $\mathbf{X}^{\mathrm{H}}$ exists.

When $\mathbf{X}^{\mathrm{H}}$ has a lower rank than its size, the Eq. (17.88) is underdetermined and thus does not have a unique solution. One method of dealing with this problem is to proceed as follows. By multiplying Eq. (17.88) from left by $\mathbf{X}$, we obtain

$$(\mathbf{X}\mathbf{X}^{\mathrm{H}})\mathbf{w} = \mathbf{X}\mathbf{s}^*. \tag{17.89}$$

This is still an underdetermined equation. However, as the coefficient matrix $(\mathbf{X}\mathbf{X}^{\mathrm{H}})$ is Hermitian, it is possible to modify Eq. (17.89) to a determined equation with a unique solution. This is done by replacing the coefficient matrix $(\mathbf{X}\mathbf{X}^{\mathrm{H}})$ by $(\mathbf{X}\mathbf{X}^{\mathrm{H}} + \epsilon\mathbf{I})$, where ϵ is a small positive constant. Hence, the cyclic equalizer tap weights may be obtained by solving the equation

$$(\mathbf{X}\mathbf{X}^{\mathrm{H}} + \epsilon\mathbf{I})\mathbf{w} = \mathbf{X}\mathbf{s}^*. \tag{17.90}$$

It is also important to note that even in the cases where $\mathbf{X}$ is full rank and, thus, Eq. (17.88) has a unique solution, the use of Eq. (17.90) is recommended. This is because the additive/channel noise in $x(n)$ will introduce a bias on the equalizer tap weights that may be very destructive when the equalizer is applied to the rest of the received signal samples. The addition of $\epsilon\mathbf{I}$ may be thought as a regularization step that moderates such a bias.

A Low Complexity Method for Solving the Cyclic Equalizer Equation

Because of the special form of the matrix **X**, there is a low-complexity method for solving Eq. (17.90). We note that **X** is a circular matrix, and recall from the theory of circular matrices presented in Chapter 8 that when **X** is a circular matrix, it can be expanded as

$$\mathbf{X} = \mathcal{F}^{-1}\mathcal{X}\mathcal{F} \tag{17.91}$$

where $\mathcal{F}$ is the DFT transformation matrix and $\mathcal{X}$ is the diagonal matrix whose diagonal elements are obtained by taking the DFT of the first column of **X**. Substituting Eq. (17.91) in Eq. (17.90) and noting that $\mathcal{F}\mathcal{F}^{-1} = \mathbf{I}$ and

$$\mathbf{X}^{\mathrm{H}} = \mathcal{F}^{-1}\mathcal{X}^{*}\mathcal{F} \tag{17.92}$$

we get

$$\mathcal{F}^{-1}(\mathcal{X}\mathcal{X}^{*} + \epsilon\mathbf{I})\mathcal{F}\mathbf{w} = \mathcal{F}^{-1}\mathcal{X}\mathcal{F}\mathbf{s}^{*} \tag{17.93}$$

Multiplying this equation from left by $\mathcal{F}$ and rearranging the result, we obtain

$$\mathbf{w} = \mathcal{F}^{-1}(\mathcal{X}\mathcal{X}^{*} + \epsilon\mathbf{I})^{-1}\mathcal{X}\mathcal{F}\mathbf{s}^{*} \tag{17.94}$$

Following Eq. (17.94), the computation of the cyclic equalizer tap weights can be carried out by taking the following steps:

1. Compute the DFTs of the vectors $\mathbf{s}^{*}$ and $\mathbf{x}_0$, that is, compute $\mathcal{F}\mathbf{s}^{*}$ and $\mathcal{F}\mathbf{x}_0$. Point-wise multiply the elements of the two DFT results. This gives the vector $\mathcal{X}\mathcal{F}\mathbf{s}^{*}$ in Eq. (17.94).
2. Point-wise divide the elements of the result of Step 1 by the elements of the vector $|\mathcal{F}\mathbf{x}_0|^2 + \epsilon$. The result will be the vector $(\mathcal{X}\mathcal{X}^{*} + \epsilon\mathbf{I})^{-1}\mathcal{X}\mathcal{F}\mathbf{s}^{*}$
3. Taking the inverse DFT of the result of Step 2 gives the desired tap-weight vector **w**.

To gain a better understanding of the behavior of the cyclic equalizer, we present some numerical examples. The MATLAB script "`CyclicEqT_eval.m`," available on the accompanying website, allows one to evaluate the performance of the symbol-spaced cyclic equalizers. It evaluates the equalizer tap weights according to Eq. (17.90) and substitutes the result in Eq. (17.71) to obtain the MSE that will be achieved when an arbitrary data sequence is passed through the cascade of channel and equalizer. The MATLAB script "`CyclicEqT_eval.m`" also evaluates the MMSE according to Eq. (17.73), for comparison. The results of such comparisons for channels c_1 and c_2 (introduced in Section 17.4.2) and two choices of the timing phase for each are presented in Table 17.1. Here, we have set $\sigma_{\nu_c} = 0.01$, $\alpha = 0.5$, the equalizer length $N = 32$, and the results are based on one randomly selected cycle of the received signal.

The results presented in Table 17.1 reveal that the cyclic equalizer achieves some level of equalization. However, the resulting MSE, for some cases (particularly, when the timing phase is wrong and MMSE is relatively high) can be over an order of magnitude higher than the minimum achievable MSE. In the sequel, we discuss a number of fixes to this problem.

Table 17.1 Performance comparison of the symbol-spaced cyclic equalizer with the achievable MMSE (the optimum equalizer).

Channel	MMSE	MSE of Cyclic Equalizer
c_1, $\tau = T/2$	0.000105	0.000196
c_1, $\tau = 0$	0.0290	0.3217
c_2, $\tau = T/2$	0.000253	0.000654
c_2, $\tau = 0.9T$	0.0299	0.6634

Equalizer Design via Channel Estimation

As noted earlier, direct computation of the equalizer tap weights, through the use of an adaptive algorithm (e.g., the LMS algorithm) or using Eq. (17.90) may result in an inaccurate design. An alternative method that results in better designs is to first identify the channel and also obtain an estimate of the variance of the channel noise and then use the design equations of Section 17.4.1 to obtain an estimate of the equalizer tap weights.

The channel identification setup here will follow the same structure as the equalizer structure in Figure 17.25, with the sequences $x(n)$ and $s(n)$ switched. Figure 17.26 presents a block diagram of such a channel estimator. The samples of impulse response of the equivalent baseband channel are the tap weights $h(0)$ through $h(N-1)$, which should be found using an adaptive approach or through the solution of a system of equations, similar to the procedures suggested earlier for finding the equalizer tap weights, w_n.

There are two advantages to the approach of equalizer design via channel identification:

1. In most of the practical channels, the duration of the impulse response $h(n)$ is limited and can be well approximated within the specified cyclic length N samples. The optimum equalizer whose role is to realize the inverse of the channel transfer function, usually, has much longer length. Strictly speaking, the optimum equalizer has an infinite length; often called an *unconstrained equalizer* (Chapter 3). In practice, by limiting the length of the equalizer to N, one should apply a constraint to the optimization while designing the equalizer. The cyclic equalizer does not perform such optimization. As it only considers the signal samples over a grid of frequencies with N samples over the range of $0 \leq \omega \leq 2\pi$, it simply obtains a set of equalizer tap weights, which, in the absence of channel noise, are time aliased samples of the unconstrained equalizer, following the sampling theorem. By adopting the method of equalizer design via channel identification, the constrained optimization is readily applied.
2. When the LMS (or NLMS) algorithm is used to adjust the tap weights of the equalizer, the cyclic equalizer may experience some very slow modes of convergence and, thus, may need many thousands of iterations before it converges. The slow convergence happens in cases where there exists a deep fade over a portion of the equivalent baseband channel; a situation that is caused by the channel and thus is out of the designer's control. In the case of the channel estimation approach of Figure 17.26, on the other hand, the designer has full control over the choice of the sequence $s(n)$, whose

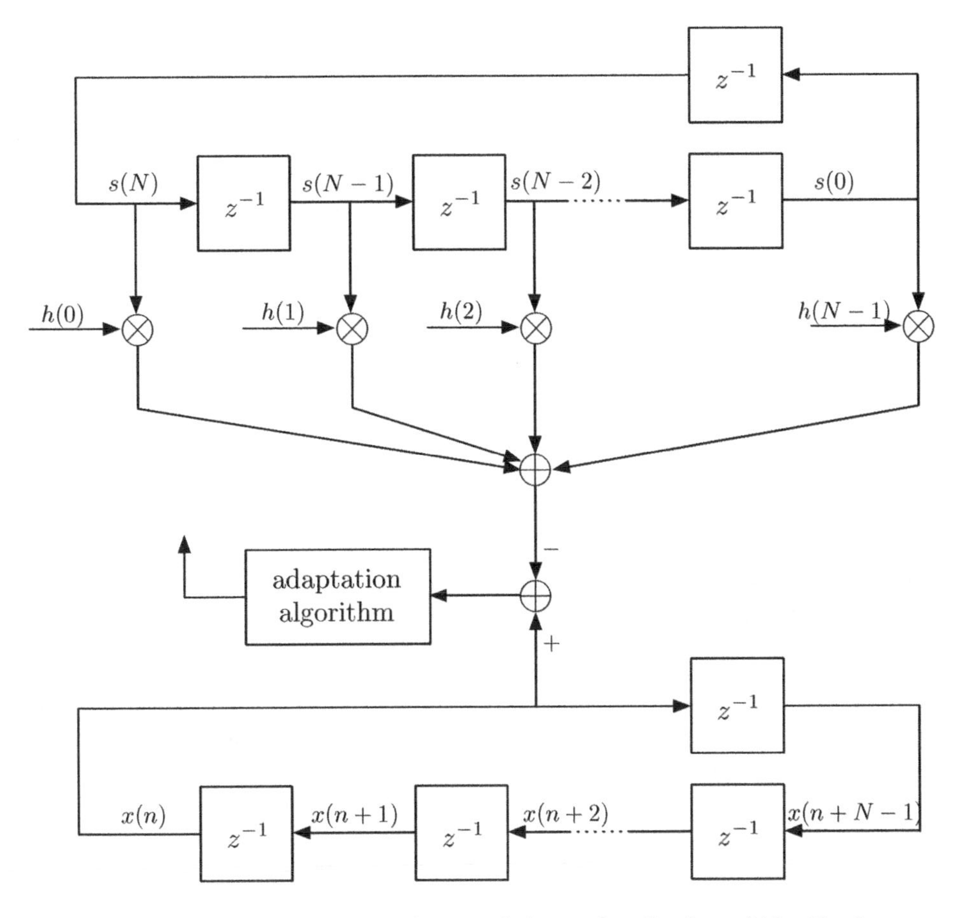

Figure 17.26 The system setup for a symbol-spaced cyclic channel identification.

correlation properties determine the convergence behavior of the LMS algorithm; see further elaborations later.

Selection of Pilot Sequence

Next, we discuss the desirable properties of the pilot sequence $s(n)$, and present a class of pilot sequences that hold such properties. To this end, we note that the dual of Eq. (17.88), here, is

$$\mathbf{S}^{\mathrm{H}}\mathbf{h} = \mathbf{x}^{*} \tag{17.95}$$

where

$$\mathbf{x} = [x(n)\ x(n+1) \cdots x(n+N-1)]^{\mathrm{T}} \tag{17.96}$$

$$\mathbf{h} = [h(0)\ h(1) \cdots h(N-1)]^{\mathrm{H}} \tag{17.97}$$

and

$$\mathbf{S} = \begin{bmatrix} s(N-1) & s(0) & s(1) & \cdots & s(N-2) \\ s(N-2) & s(N-1) & s(0) & \cdots & s(N-3) \\ s(N-3) & s(N-2) & s(N-1) & \cdots & s(N-4) \\ \vdots & \vdots & \vdots & \ddots & \vdots \\ s(0) & s(1) & s(2) & \cdots & s(N-1) \end{bmatrix} \tag{17.98}$$

Multiplying Eq. (17.95) from the left by $\mathbf{S}$ and solving for $\mathbf{h}$, we get

$$\mathbf{h} = (\mathbf{S}\mathbf{S}^{\mathrm{H}})^{-1}(\mathbf{S}\mathbf{x}^*) \tag{17.99}$$

This solution would become trivial, if we had chosen the pilot sequence $s(n)$ such that $\mathbf{S}$ was an orthogonal matrix, that is, if $\mathbf{S}\mathbf{S}^{\mathrm{H}} = K\mathbf{I}$, where K is a constant and $\mathbf{I}$ is the identity matrix. In that case, Eq. (17.99) reduces to

$$\mathbf{h} = \frac{1}{K}\mathbf{S}\mathbf{x}^*, \tag{17.100}$$

which shows that the channel impulse response vector $\mathbf{h}$ can be obtained by premultiplying $\mathbf{x}^*$ with the matrix $\frac{1}{K}\mathbf{S}$.

In addition to the orthogonality property of $\mathbf{S}$, in practice, it is desirable to choose a set of $s(n)$s with the same amplitudes, so that the transmit power is uniformly spread across time. It turns out that such sequences exist. They are called *polyphase codes* (Chu, 1972). They exist for any length, N. A particular construction of polyphase codes that we use here follows the formula

$$s(n) = \begin{cases} e^{j\pi n^2/N}, & \text{for } N \text{ even} \\ e^{j\pi n(n+1)/N}, & \text{for } N \text{ odd} \end{cases} \tag{17.101}$$

for $n = 0, 1, \ldots, N-1$. The MATLAB function "`CycPilot.m`" on the accompanying website can be used to generate pilot sequences based on the formula (17.101).

Impact of the Channel Noise

In the equations given earlier, for simplicity of derivations, the channel noise was ignored. If the channel noise is included, Eq. (17.95) becomes

$$\mathbf{S}^{\mathrm{H}}\mathbf{h} + \mathbf{v}^* = \mathbf{x}^* \tag{17.102}$$

where $\mathbf{v}$ is the noise vector associated with $\mathbf{x}$. Multiplying Eq. (17.102) from the left by $\frac{1}{K}\mathbf{S}$, we get

$$\hat{\mathbf{h}} = \mathbf{h} + \frac{1}{K}\mathbf{S}\mathbf{v}^* = \frac{1}{K}\mathbf{S}\mathbf{x}^* \tag{17.103}$$

where $\hat{\mathbf{h}}$ is a noisy estimate of $\mathbf{h}$.

One method of improving the estimate of $\mathbf{h}$ is to transmit multiple periods of pilot symbols and replace the vector $\mathbf{x}$ by its average obtained by averaging over multiple periods.

Estimation of the Variance of the Channel Noise

To obtain an estimate of the variance of the channel noise, we take the following approach. As discussed earlier, many operations at the receiver can be greatly simplified by sending a few cycles of the periodic pilot sequence $s(n)$. Assuming that we have been able to identify a portion of the received signal sequence $x(n)$ that is associated with the periodic sequence $s(n)$ and note that

$$x(n) = h(n) \star s(n) + \nu(n) \tag{17.104}$$

one finds that the first term on the right-hand side of Eq. (17.104) is periodic. Hence,

$$z(n) = x(n) - x(n+N) = \nu(n) - \nu(n+N). \tag{17.105}$$

Now, if we assume that $\nu(n)$ and $\nu(n+N)$ are uncorrelated, a time average of $|z(n)|^2$, obviously, gives an estimate of $2\sigma_\nu^2$.

Comparisons

To give of an idea of the performance difference of a direct cyclic equalizer design and its indirect counterpart where an estimate of the channel is used to design the equalizer, we present some numerical results. The results are presented in Table 17.2. This is an extension of Table 17.1, by adding the last column. As was predicted earlier, the results clearly show a superior performance of the indirect method. In the indirect case, the two periods of $x(n)$ that are used in Eq. (17.105) are averaged to obtain a less noisy version of $\mathbf{x}$ before application of Eq. (17.100). Further comparisons are left to the reader.

It is also worth noting that whereas in the direct method the equalizer length must be equal to the length of the pilot symbols, when the indirect method is adopted one has the freedom of choosing any arbitrary length for the equalizer.

17.6.2 Fractionally Spaced Cyclic Equalizer

Fractionally spaced cyclic equalizer structures are obtained by simple modifications to their symbol-spaced counterparts. Figure 17.27 presents a system setup for a half symbol-spaced direct cyclic equalizer. Note that as here the rate of the samples of $x(n)$ is twice

Table 17.2 Performance comparison of the cyclic equalizers for direct and indirect setting.

		MSE of Cyclic Equalizer	
Channel	MMSE	Direct	Indirect
c_1, $\tau = T/2$	0.000105	0.000196	0.000164
c_1, $\tau = 0$	0.0290	0.3217	0.0292
c_2, $\tau = T/2$	0.000253	0.000654	0.0003743
c_2, $\tau = 0.9T$	0.0299	0.6634	0.0351

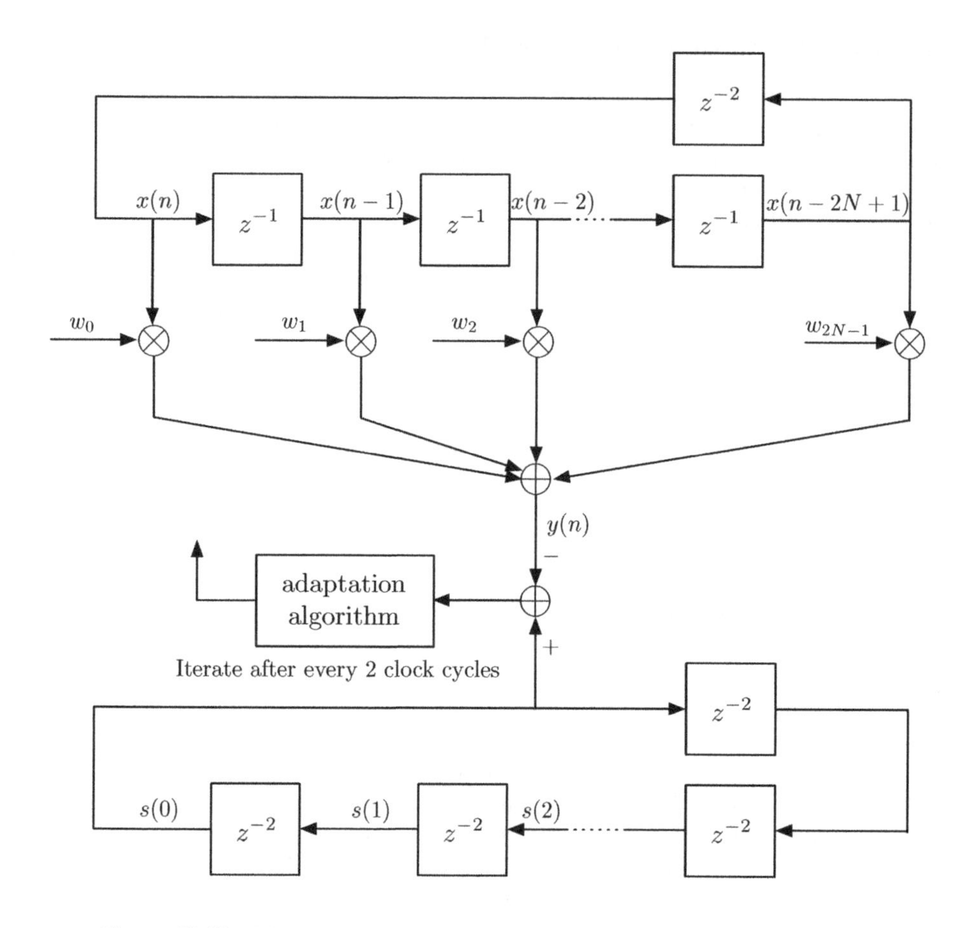

Figure 17.27 The adaptation setup for a half symbol-spaced cyclic equalizer.

that of $s(n)$, the delay between the pilot symbols $s(n)$ is z^{-2}. Also, because it is only after every two clock cycles that one new sample of $s(n)$ will be available for comparison with the equalizer output, the adaptation algorithm should be iterated after every two clock cycles. Moreover, as the cyclic equalizer should have the same time span as pilot symbols, the equalizer length here is $2N$.

In the case of indirect cyclic equalizer, multiple samples of the channel impulse response should be estimated per symbol interval. This can be easily done by setting up a set of parallel channel estimators similar to the one in Figure 17.26. If the samples of the channel impulse response at a spacing T/L are required, L parallel channel estimators will be used. Each channel estimator operates on a set of symbol-spaced samples of $x(n)$ that correspond to one of the L time phases. The estimated samples of the channel responses will be interleaved to construct the desired response.

The general conclusions that were derived in the case of the symbol-spaced equalizer also apply here. In particular, direct computation of the equalizer tap weights using an equation such as Eq. (17.88) or Eq. (17.89) leads to results that are relatively far from the optimum tap weights. On the other hand, an indirect method that uses an estimate of the channel response and noise variance to calculate the equalizer tap weights results in much improved results.

17.6.3 *Alignment of $s(n)$ and $x(n)$*

So far, we have assumed that the receiver is able to identify the beginning of each cycle of the received signal that matches with the transmitted pilots. Actually, this assumption may never be true in practice. In a practical receiver, one can only obtain a coarse estimate of the boundaries of the beginning and end of the cyclic preamble.

When $s(n)$ and $x(n)$ are not time-aligned, the estimated equalizer tap weights or the channel impulse response will be replaced by their time-shifted replicas. Also, as the signals are periodic with period of N samples, these shifts will be cyclic, that is, the samples rotate within a vector of length N ($2N$ in the case of the half symbol-spaced equalizer). To remove the ambiguity caused by such rotation, the samples associated with the equalizer tap weights or the channel impulse response are rotated so that the larger tap weights/samples be positioned around the middle of the estimated vector. This method, in most cases, works pretty well, because both equalizer tap weights and the samples of the channel impulse response are sequences that grow from zero to a maximum (roughly at the middle) and decay to zero afterward.

17.6.4 *Carrier and Timing Phase Acquisition and Tracking*

Our discussion in this section so far was based on the assumption that the received cyclic signal was free of any frequency offset and in the case of symbol-spaced equalizer a proper timing phase was selected *a priori*. However, in practice, it is the task of the receiver to find a proper timing phase and to compensate for any residual carrier frequency offset in the demodulated received baseband signal.

Carrier Acquisition and Tracking

As in the case of the noise variance estimation, here also we assume that the cyclic preamble consists of a few cycles of the pilot sequence $s(n)$. In that case, in the presence of a carrier frequency offset $\Delta\omega_c$, the demodulated signal samples $x(n)$ are replaced by

$$x'(n) = e^{j\Delta\omega_c n} x(n) + \nu(n). \tag{17.106}$$

Note that we have ignored the rotation of the noise samples $\nu(n)$ that may be caused by the carrier frequency offset, as such rotation does not affect the statistics of the result. Recalling that during the cyclic preamble (assuming that the samples are taken at symbol rate, $1/T$) $x(n) = x(n+N)$ and assuming that the noise samples $\nu(n)$ are small enough, one may argue that

$$\sum_{n=n_0}^{K+n_0-1} x'(n+N)x'^{*}(n) \approx e^{j\Delta\omega_c N} \sum_{n=n_0}^{K+n_0-1} |x(n)|^2 \tag{17.107}$$

where n_0 is the starting point and K is the number of sample pairs that are used. Solving Eq. (17.107) for $\Delta\omega_c$, we obtain

$$\Delta\omega_c \approx \frac{1}{N}\angle\left(\frac{\sum_{n=n_0}^{K-n_0-1} x'(n+N)x'^*(n)}{\sum_{n=n_0}^{K-n_0-1} |x(n)|^2}\right) \tag{17.108}$$

where $\angle(\cdot)$ denotes the angle of.

Timing Phase Acquisition and Tracking

For the purpose of timing phase tuning, we should have access to the samples of the received baseband signal at a rate higher than its Nyquist rate. Let $x(n)$ denote a sequence of signal samples at a rate L/T. To select a good timing phase, one may take one of the following steps (other alternatives are also possible):

1. If the equalizer is a fractionally spaced one, any choice of the timing phase, as long as it is maintained throughout the payload, will work well. A convenient method thus may be to examine all (quantized) timing phases of 0, T/L, $2T/L$, ..., $(L-1)T/L$, evaluate the signal power over one or more cycles of the cyclic preamble, and choose the one that results in the maximum signal power. Subsequently, either the early-late gate or the gradient-based timing recovery that was presented earlier in this chapter may be used to maintain the timing phase.
2. A similar method may be used in the case of a symbol-spaced equalizer. However, as demonstrated earlier, the choice of timing phase based on the power maximization does not always lead to an acceptable equalizer performance.
3. For the symbol-spaced equalizers, an alternative to Step 2 is to first identify the channel impulse response at the rate L/T samples per second. Then, evaluate the channel magnitude response when sampled at the symbol rate for different timing phases. From these choices, pick the one with the least variation in the magnitude response and set the timing phase accordingly. To make sure that this timing phase will be maintained during the payload, identify the offset of this timing phase with respect to the one that results in the power maximization. Moreover, during the payload, one may choose to adopt either the early-late gate or the gradient-based timing recovery to make sure that no timing drift occurs.

17.7 Joint Timing Recovery, Carrier Recovery, and Channel Equalization

In a receiver where the three functions of timing recovery, carrier recovery, and channel equalization are performed concurrently, these will interact. In particular, the channel equalizer can compensate for any phase offset and a residual (but, small) carrier frequency offset. The phase offset θ is compensated simply by multiplying all the equalizer tap weights with the same factor $e^{-j\theta}$. Also, a carrier frequency offset Δf_c will be automatically compensated for when the equalizer adaptation is fast enough to track the time varying carrier phase $2\pi\Delta f_c n T_s$.

The equalizer may also adapt to compensate for any gradual timing phase drift. In fact, when a fractionally spaced equalizer is used in a receiver, it is possible to completely remove the timing recovery block and let the equalizer track the timing phase drift. In such an implementation, as the timing phase drifts, the equalizer tap weights also drift to the right or left (depending on the direction of the timing phase drift). If the receiver runs for a long time, the timing phase drift may result in an excessive shift of the equalizer coefficients to the right or left and result in a significant degradation in performance. Fortunately, this problem can be easily fixed. While running the adaptation algorithm, one may keep track of the largest tap weight of the equalizer. If it drifts too far from the center, the equalizer tap weights are shifted to the right or left to move the tap weight with the largest magnitude at the middle. See also the discussion in Section 17.6.3.

17.8 Maximum Likelihood Detection

The equalization methods that were discussed in the previous sections may be classified as members of suboptimum detectors. An optimum detector can be constructed when one has perfect knowledge of the channel impulse response and the statistical properties of the channel noise. Such a detector searches over all possible combinations of the transmitted data symbols sequence and finds the one that matches best with the received signal. However, the detector complexity grows exponentially with the number of the transmitted data symbols and, thus, its implementation becomes prohibitive in practice. Fortunately, with a negligible loss in performance, the optimal detector can be implemented by using an elegant method called *Viterbi algorithm*. The Viterbi algorithm also has a complexity that grows exponentially with the length of the channel impulse response. Hence, for a channel with long duration, even the use of Viterbi algorithm may be insufficient to provide a feasible implementation. To solve this problem, and yet achieve a near optimum performance, it is often proposed to design and use an equalizer that limits the channel duration to a finite length. Later, we present a formulation for the design of such equalizers, which we refer to as *channel memory truncation*.

As for the details of the Viterbi algorithm and other soft equalization (some of which are discussed in the next section) the interested readers should refer to the more relevant texts, such as Lin and Costello (2004). The idea here is only to bring to the attention of readers these more advanced techniques that in recent years have become common practice in the industry.

Channel Memory Truncation

Following a similar line of thoughts to those in Section 17.4, we consider the system setup presented in Figure 17.28 for the study of the channel memory truncation in the case of a symbol-spaced equalizer. Here, $D(z) = \sum_{i=0}^{L-1} d_i z^{-i}$ is a desired transfer function of a limited length L. Both the equalizer $W(z)$ and the desired response $D(z)$ are unknown and should be optimized jointly to minimize the mean square of the output error, $e(n)$.

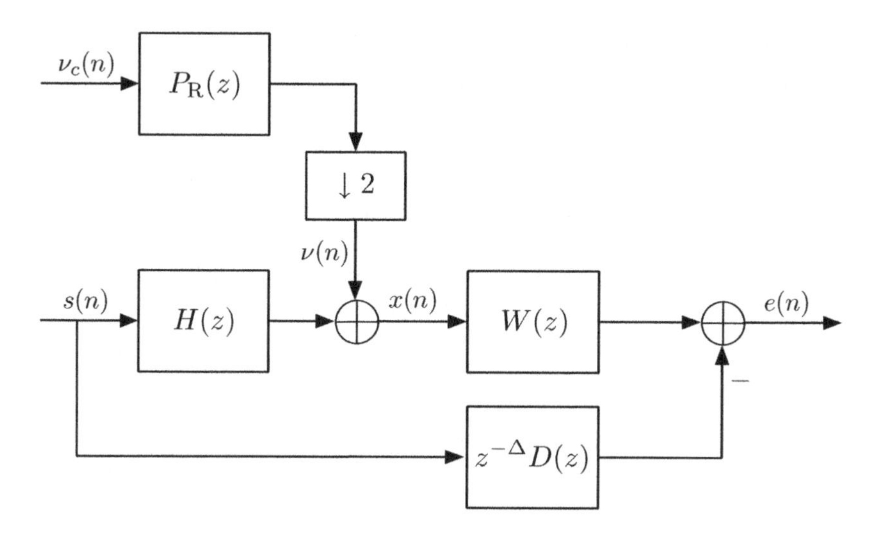

Figure 17.28 System setup for a symbol-spaced equalizer for channel memory truncation.

Following the same line of derivations to those presented in Section 17.4.1, it is not difficult to show that Eq. (17.69) is also applicable here, with

$$\mathbf{d} = \left[\overbrace{0\,0\cdots0}^{\Delta\text{zeros}}\ d_0\ d_1\cdots d_{L-1}\ 0\cdots\right]^{\mathrm{T}} \tag{17.109}$$

Moreover, Eqs. (17.71) and (17.72) also follow. Assuming $\mathbf{d}$ is given, substituting Eq. (17.72) in Eq. (17.71), and recalling that $\mathbf{R} = \mathbf{Q}^{\mathrm{T}}\mathbf{Q}^*$ and $\mathbf{p} = \mathbf{Q}^{\mathrm{T}}\mathbf{d}^*$, after some straightforward manipulations, we obtain

$$\xi(\mathbf{w}_\mathrm{o}) = \sigma_\mathrm{s}^2\mathbf{d}^{\mathrm{H}}[\mathbf{I} - \mathbf{Q}(\mathbf{Q}^{\mathrm{H}}\mathbf{Q})^{-1}\mathbf{Q}^{\mathrm{H}}]\mathbf{d} \tag{17.110}$$

Next, we should find $\mathbf{d}$ that minimizes $\xi(\mathbf{w}_\mathrm{o})$. To this end, we first note that Eq. (17.110) may be reduced to

$$\xi(\mathbf{w}_\mathrm{o}) = \sigma_\mathrm{s}^2\mathbf{d}_\mathrm{r}^{\mathrm{H}}\mathbf{A}\mathbf{d}_\mathrm{r} \tag{17.111}$$

where $\mathbf{d}_\mathrm{r}$ is the vector $\mathbf{d}$ after removing all zeros from its beginning and end, that is,

$$\mathbf{d}_\mathrm{r} = [d_0\ d_1\cdots d_{L-1}]^{\mathrm{T}} \tag{17.112}$$

and $\mathbf{A}$ is a submatrix of $\mathbf{I} - \mathbf{Q}(\mathbf{Q}^{\mathrm{H}}\mathbf{Q})^{-1}\mathbf{Q}^{\mathrm{H}}$, obtained by keeping the rows and columns that are compatible with $\mathbf{d}_\mathrm{r}$. Moreover, as for any $\mathbf{d}_\mathrm{r}$, $\xi(\mathbf{w}_\mathrm{o})$ is always greater than or equal to zero, the matrix $\mathbf{A}$ is a positive definite matrix. Also, to avoid the trivial choice of $\mathbf{d}_\mathrm{r} = 0$ that sets $\xi(\mathbf{w}_\mathrm{o})$ equal to its trivial minimum of zero, we impose the unity length constraint $\|\mathbf{d}_\mathrm{r}\|^2 = \mathbf{d}_\mathrm{r}^{\mathrm{H}}\mathbf{d}_\mathrm{r} = 1$ to $\mathbf{d}_\mathrm{r}$. With this constraint, recalling the minimax theorem that was introduced in Chapter 4, one will find that the optimum choice of $\mathbf{d}_\mathrm{r}$ is the eigenvector of $\mathbf{A}$ that is associated with its minimum eigenvalue. Hence, we may write

$$\mathbf{d}_{\mathrm{r,o}} = \text{the eigenvector of } \mathbf{A} \quad \text{associated with } \lambda_{\min}(\mathbf{A}) \tag{17.113}$$

where $\lambda_{\min}(\mathbf{A})$ means the minimum eigenvalue of $\mathbf{A}$. Moreover, substituting this solution in Eq. (17.110), one will find that

$$\xi_{\min} = \xi(\mathbf{w}_{\rm o}, \mathbf{d}_{\rm o}) \tag{17.114}$$

$$= \sigma_{\rm s}^2 \lambda_{\min}(\mathbf{A}) \tag{17.115}$$

LMS Algorithm for Channel Memory Truncation

The channel memory truncation concept that was developed earlier was first introduced in Falconer and Magee (1973). In this work, the authors also proposed an LMS algorithm for joint adaptation of the equalizer coefficients w_i and the desired (truncated) response d_i. They viewed the combination of $W(z)$ and $z^{-\Delta}D(z)$ of Figure 17.28 as a linear combiner with inputs from the samples of $x(n)$ and $s(n)$. The LMS algorithm is applied to minimize the mean squared value of the output error $e(n)$. Clearly, the direct application of the LMS algorithm results in the trivial and unacceptable solution $\mathbf{w} = \mathbf{0}$ and $\mathbf{d}_{\rm r} = \mathbf{0}$. To avoid this solution, Falconer and Magee have suggested that $\mathbf{d}_{\rm r}$ should be renormalized to the length of unity after each iteration. That is, the updated vector $\mathbf{d}_{\rm r}(n)$, after each iteration of the LMS algorithm, is normalized as

$$\mathbf{d}_{\rm r}(n) \leftarrow \frac{\mathbf{d}_{\rm r}(n)}{\sqrt{\mathbf{d}_{\rm r}^{\rm H}(n)\mathbf{d}_{\rm r}(n)}} \tag{17.116}$$

17.9 Soft Equalization

The sequence estimator that was discussed in the previous section searches for the best sequence of data symbols that if convolved with the channel impulse response results in a signal that best matches the received signal. Hence, the name maximum-likelihood detector has been adopted. The more modern communication systems take a different approach, which may be thought as a modification to the former.

An information sequence $a(n)$ of $N_{\rm a}$ bits, before transmission, is coded and converted to $N_{\rm b} > N_{\rm a}$ coded bits, $b(n)$. The coded bits, $b(n)$, are related to the uncoded bits $a(n)$ and are correlated with one another through the coding scheme. At the front-end of the receiver, a detector, known as *soft equalizer*, evaluates the probability of each coded bit for being 0 or 1 and delivers a number called *log-likelihood ratio* (LLR). For a given coded bit $b(n)$, its LLR value is defined as

$$\lambda(b(n)) = \ln \frac{P(b(n) = 0|\mathbf{x}, \mathbf{h}, \sigma_\nu^2)}{P(b(n) = 1|\mathbf{x}, \mathbf{h}, \sigma_\nu^2)} \tag{17.117}$$

where $P(\cdot|\cdot)$ is the conditional probability function, $\mathbf{x}$ is the vector of the received signal, $\mathbf{h}$ is the vector of channel impulse response, and σ_ν^2 is the channel noise variance.

The information collected by the soft equalizer is passed to a soft decoder that takes into account the correlation between the coded bits and improves on the LLR values. The difference between the new LLR values and their counterparts from the soft equalizer makes a set of new information that was unknown to the soft equalizer. This information, that is called *extrinsic*, is passed to the soft equalizer, to obtain more accurate values of LLR whose extrinsic part will be passed to the soft decoder for further processing. This

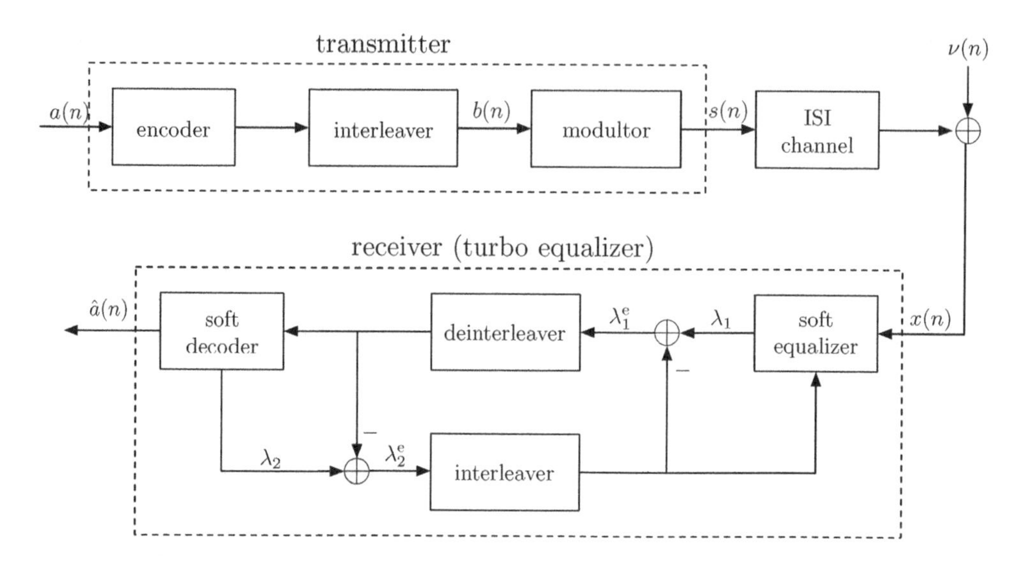

Figure 17.29 A transceiver communication system equipped with a turbo receiver.

exchange of information between the soft equalizer and decoder continues for a number of iterations to converge. This process is often referred to as *turbo iteration*, and the corresponding receiver is called *turbo equalizer*. After convergence of turbo iterations, the decoder decides on its best guess of the uncoded (information) bits, $a(n)$.

Figure 17.29 presents a block diagram of a transceiver communication system that is equipped with a turbo equalizer. The LLR values out of the soft equalizer and the soft decoder are called λ_1 and λ_2, respectively. The extrinsic values are identified by the superscript "e," λ_1^e, and λ_2^e, respectively. There is also a block called *interleaver* between the encoder and modulator at the transmitter. This block shuffles the coded bits before being passed to the modulator. This is to make sure that the neighboring bits in the transmitted sequence are uncorrelated with one another. This process will be undone in transferring the LLR values from the soft equalizer to the soft decoder, and will be reverted again in transferring the LLR values from the soft decoder to the soft equalizer.

Any discussion on the coding and soft decoding is beyond the scope of this book. Here, we only concentrate on the methods of soft equalization. A soft equalizer, as presented in Figure 17.29, has two inputs; the received signal $x(n)$ and the *a priori* information (the extrinsic LLR values) from the soft decoder. There are a variety of the methods in the literature that discuss different implementations of the soft equalizers. Two of these methods are presented in the sequel. The first method is a modified form of the conventional MMSE linear equalizer, which will take into account the *a priori* information λ_2^e to improve on the estimates (the LLR values) of the coded bits $b(n)$. The second method takes a statistical approach for combining the *a priori* information λ_2^e and the received signal samples for updating the LLR values at each iteration of the turbo equalizer. The references cited later are for an interested reader to begin with and explore the related literature further.

17.9.1 Soft MMSE Equalizer

The material presented in this section is a simplified version of the soft MMSE equalizer of Tüchler *et al.* (2002). The presentation in this work considers the case where the modulator in Figure 17.29 converts the coded bits $b(n)$ to a sequence of QAM symbols, $s(n)$. Here, we present the derivations for the case where the modulator converts the coded bits $b(n)$ to the binary phase-shift keying (BPSK) symbols that take values of -1, for $b(n) = 0$, and $+1$, for $b(n) = 1$. We also limit our discussion to the case where the channel impulse response and the equalizer coefficients are real-valued.

When there exists some *a priori* information about the transmitted symbols $s(n)$, the Wiener filter estimator will be biased and thus the equalizer output (assuming a linear, symbol-spaced, or fractionally spaced equalizer) should be written as

$$y(n) = \mathbf{w}^{\mathrm{T}}\mathbf{x}(n) + b \tag{17.118}$$

where b is the bias. Hence, here, we are interested in minimizing the cost function

$$\xi = E[(s(n-\Delta) - y(n))^2] \tag{17.119}$$

where Δ, as defined before, is the delay caused by the combination of the channel and equalizer. To proceed, we define

$$\tilde{\mathbf{w}} = \begin{bmatrix} \mathbf{w} \\ b \end{bmatrix} \tag{17.120}$$

and

$$\tilde{\mathbf{x}}(n) = \begin{bmatrix} \mathbf{x}(n) \\ 1 \end{bmatrix} \tag{17.121}$$

and note that Eq. (17.118) can be written as

$$y(n) = \tilde{\mathbf{w}}^{\mathrm{T}}\tilde{\mathbf{x}}(n) \tag{17.122}$$

Using Eq. (17.122) in Eq. (17.119) and expanding the results, we obtain

$$\xi = E[|s(n-\Delta)|^2] - \tilde{\mathbf{w}}^{\mathrm{T}}\tilde{\mathbf{p}} - \tilde{\mathbf{p}}^{\mathrm{T}}\tilde{\mathbf{w}} + \tilde{\mathbf{w}}^{\mathrm{T}}\tilde{\mathbf{R}}\tilde{\mathbf{w}} \tag{17.123}$$

where $\tilde{\mathbf{R}} = E[\tilde{\mathbf{x}}(n)\tilde{\mathbf{x}}^{\mathrm{T}}(n)]$ and $\tilde{\mathbf{p}} = E[s(n-\Delta)\tilde{\mathbf{x}}(n)]$. Using Eq. (17.121), one will find that

$$\tilde{\mathbf{R}} = \begin{bmatrix} \mathbf{R} & E[\mathbf{x}(n)] \\ E[\mathbf{x}^{\mathrm{T}}(n)] & 1 \end{bmatrix} \tag{17.124}$$

and

$$\tilde{\mathbf{p}} = \begin{bmatrix} \mathbf{p} \\ E[s(n-\Delta)] \end{bmatrix} \tag{17.125}$$

Following similar derivations to those in Chapter 3, minimization of the cost function Eq. (17.123) leads to the Wiener–Hopf equation $\tilde{\mathbf{R}}\tilde{\mathbf{w}}_{\mathrm{o}} = \tilde{\mathbf{p}}$, which, using Eqs. (17.124) and (17.125), can be written as

$$\begin{bmatrix} \mathbf{R} & E[\mathbf{x}(n)] \\ E[\mathbf{x}^{\mathrm{T}}(n)] & 1 \end{bmatrix} \begin{bmatrix} \mathbf{w}_{\mathrm{o}} \\ b_{\mathrm{o}} \end{bmatrix} = \begin{bmatrix} \mathbf{p} \\ E[s(n-\Delta)] \end{bmatrix} \tag{17.126}$$

Note that, here, the subscript "o" has been added to emphasize that the parameters are set at their optimum values.

Solving Eq. (17.126) for $\mathbf{w}_o$ and b_o, we obtain

$$\mathbf{w}_o = (\mathbf{R} - E[\mathbf{x}(n)]E[\mathbf{x}^T(n)])^{-1}(\mathbf{p} - E[s(n-\Delta)]E[\mathbf{x}(n)]) \tag{17.127}$$

and

$$b_o = E[s(n-\Delta)] - \mathbf{w}_o^T E[\mathbf{x}(n)] \tag{17.128}$$

Note that the nonzero mean-value terms $E[\mathbf{x}(n)]$ and $E[s(n-\Delta)]$ arise as a result of the *a priori* extrinsic information from the soft decoder. When there is no such *a priori* information, $E[\mathbf{x}(n)]$ and $E[s(n-\Delta)]$ are both zero. Hence, $b_o = 0$, thus, the estimate Eq. (17.118) will be unbiased and Eq. (17.127) reduces to the familiar solution $\mathbf{w}_o = \mathbf{R}^{-1}\mathbf{p}$.

Using Eqs. (17.127) and (17.128), respectively, for $\mathbf{w}$ and b in Eq. (17.118), the optimum estimate of $s(n-\Delta)$ that takes into account the *a priori* information from the decoder is obtained as

$$\begin{aligned} y_o(n) = {} & (\mathbf{p} - E[s(n-\Delta)]E[\mathbf{x}(n)])^T(\mathbf{R} - E[\mathbf{x}(n)]E[\mathbf{x}^T(n)])^{-1}\mathbf{x}(n) \\ & + E[s(n-\Delta)] - \mathbf{w}_o^T E[\mathbf{x}(n)] \end{aligned} \tag{17.129}$$

We also note that given the LLR values λ_2^e from the soft decoder, $E[s(n-\Delta)]$ is calculated as

$$E[s(n-\Delta)] = +1 \times P(s(n-\Delta) = +1) - 1 \times P(s(n-\Delta) = -1) \tag{17.130}$$

where $P(s(n-\Delta) = +1)$ and $P(s(n-\Delta) = -1)$ are related according to

$$\ln \frac{P(s(n-\Delta) = -1)}{P(s(n-\Delta) = +1)} = \lambda_2^e(s(n-\Delta)). \tag{17.131}$$

Also, noting that $P(s(n-\Delta) = +1) + P(s(n-\Delta) = -1) = 1$, one will find that

$$P(s(n-\Delta) = -1) = \frac{e^{\lambda_2^e(s(n-\Delta))}}{1 + e^{\lambda_2^e(s(n-\Delta))}} \tag{17.132}$$

and

$$P(s(n-\Delta) = +1) = \frac{1}{1 + e^{\lambda_2^e(s(n-\Delta))}} \tag{17.133}$$

Substituting Eqs. (17.132) and (17.133) into Eq. (17.130), we obtain

$$E[s(n-\Delta)] = \frac{1 - e^{\lambda_2^e(s(n-\Delta))}}{1 + e^{\lambda_2^e(s(n-\Delta))}}. \tag{17.134}$$

To obtain $E[\mathbf{x}(n)]$, we note that

$$\mathbf{x}(n) = \mathbf{H}\mathbf{s}(n) + \mathbf{v}(n) \tag{17.135}$$

where

$$\mathbf{H} = \begin{bmatrix} h(L-1) & h(L-2) & \cdots & h(0) & 0 & \cdots & 0 \\ 0 & h(L-1) & h(L-2) & \cdots & h(0) & \cdots & 0 \\ & & & \ddots & & & \\ 0 & \cdots & 0 & h(L-1) & h(L-2) & \cdots & h(0) \end{bmatrix} \tag{17.136}$$

$\mathbf{s}(n)$ is the vector of the transmitted symbols compatible with $\mathbf{H}$, and the vector $\mathbf{v}(n)$ contains the channel noise samples. Assuming that the channel noise has a zero mean, from Eq. (17.135), one finds that

$$E[\mathbf{x}(n)] = \mathbf{H}E[\mathbf{s}(n)]. \tag{17.137}$$

We also note that the elements of $E[\mathbf{s}(n)]$ can be calculated using similar equations to Eq. (17.134).

Next, we use $y_o(n)$ to calculate a new value of the LLR according to

$$\lambda_1(s(n-\Delta)) = \ln \frac{P(s(n-\Delta)=-1)}{P(s(n-\Delta)=+1)} \tag{17.138}$$

where $P(s(n-\Delta)=+1)$ and $P(s(n-\Delta)=-1)$ are related according to the equation

$$P(s(n-\Delta)=+1) - P(s(n-\Delta)=-1) = y_o(n) \tag{17.139}$$

and also we note that

$$P(s(n-\Delta)=+1) + P(s(n-\Delta)=-1) = 1. \tag{17.140}$$

Solving Eqs. (17.139) and (17.140) for $P(s(n-\Delta)=+1)$ and $P(s(n-\Delta)=-1)$ and substituting the results in Eq. (17.138), we obtain

$$\lambda_1(s(n-\Delta)) = \ln \frac{1-y_o(n)}{1+y_o(n)}. \tag{17.141}$$

Finally,

$$\lambda_1^e(s(n-\Delta)) = \lambda_1(s(n-\Delta)) - \lambda_2^e(s(n-\Delta)) \tag{17.142}$$

is calculated and passed to the soft decoder for the next iteration.

17.9.2 Statistical Soft Equalizer

The soft MMSE equalizer obtains a linear estimate (an average) of data symbols according to Eq. (17.129) and use that to calculate the corresponding LLR values according to Eqs. (17.141) and (17.142).

The accurate estimate of the LLR value λ_1 of the data bit $b(n)$ is given by

$$\lambda_1(b(n)) = \ln \frac{P(b(n)=0|\mathbf{x}, \mathbf{h}, \sigma_\nu^2, \boldsymbol{\lambda}_2^e)}{P(b(n)=1|\mathbf{x}, \mathbf{h}, \sigma_\nu^2, \boldsymbol{\lambda}_2^e)} \tag{17.143}$$

where $\mathbf{x}$ is the vector of the received signal, $\mathbf{h}$ is the vector of channel impulse response, σ_ν^2 is the channel noise variance, and $\boldsymbol{\lambda}_2^{\mathrm{e}}$ is the vector of extrinsic LLR values of all the transmitted data bits from the soft decoder.

Here, also, we assume that the modulator converts the coded bits $b(n)$ to the BPSK symbols that take values of -1, for $b(n) = 0$, and $+1$, for $b(n) = 1$. Considering this direct mapping between $b(n)$ and $s(n)$, in the rest of this section, we use $b(n)$ and $s(n)$ interchangeably, as appropriate. For instance, we use $b(n)$ when reference is to be made to LLR values, and use $s(n)$ when the received signal samples are to be calculated according to the channel equation

$$x(n) = \sum_{i=0}^{L-1} h(i)s(n-i) + \nu(n) \tag{17.144}$$

In the sequel, for brevity, we drop $\mathbf{h}$ and σ_ν^2 from the equations and thus, to begin, we rewrite Eq. (17.143) as

$$\lambda_1(b(n)) = \ln \frac{P(b(n) = 0|\mathbf{x}, \boldsymbol{\lambda}_2^{\mathrm{e}})}{P(b(n) = 1|\mathbf{x}, \boldsymbol{\lambda}_2^{\mathrm{e}})}. \tag{17.145}$$

This may be further written as

$$\lambda_1(s(n)) = \ln \frac{\sum\limits_{\mathbf{b}:b(n)=0} P(\mathbf{b}|\mathbf{x}, \boldsymbol{\lambda}_2^{\mathrm{e}})}{\sum\limits_{\mathbf{b}:b(n)=1} P(\mathbf{b}|\mathbf{x}, \boldsymbol{\lambda}_2^{\mathrm{e}})} \tag{17.146}$$

where $\mathbf{b}$ is the vector of the transmitted symbols and "$\mathbf{b} : b(n) = 0$" and "$\mathbf{b} : b(n) = 1$" means the summations are over all possible combinations of the information bits, subject to $b(n) = 0$ and $b(n) = 1$, respectively.

To proceed, we recall the Bayes' theorem, which for a pair of events A and B is formulated as

$$P(A|B) = \frac{P(B|A)P(A)}{P(B)}. \tag{17.147}$$

Using the Bayes' theorem, Eq. (17.146) can be rearranged as

$$\begin{aligned}
\lambda_1(s(n)) &= \ln \frac{\sum\limits_{\mathbf{b}:b(n)=0} \frac{p(\mathbf{x}|\mathbf{b},\boldsymbol{\lambda}_2^{\mathrm{e}})P(\mathbf{b}|\boldsymbol{\lambda}_2^{\mathrm{e}})}{P(\mathbf{x}|\boldsymbol{\lambda}_2^{\mathrm{e}})}}{\sum\limits_{\mathbf{b}:b(n)=1} \frac{p(\mathbf{x}|\mathbf{b},\boldsymbol{\lambda}_2^{\mathrm{e}})P(\mathbf{b}|\boldsymbol{\lambda}_2^{\mathrm{e}})}{P(\mathbf{x}|\boldsymbol{\lambda}_2^{\mathrm{e}})}} \\
&= \ln \frac{\sum\limits_{\mathbf{b}:b(n)=0} p(\mathbf{x}|\mathbf{b})P(\mathbf{b}|\boldsymbol{\lambda}_2^{\mathrm{e}})}{\sum\limits_{\mathbf{b}:b(n)=1} p(\mathbf{x}|\mathbf{b})P(\mathbf{b}|\boldsymbol{\lambda}_2^{\mathrm{e}})}
\end{aligned} \tag{17.148}$$

where, in the second line, the condition "$\mathbf{b}, \boldsymbol{\lambda}_2^{\mathrm{e}}$" has been reduced to "$\mathbf{b}$," because when $\mathbf{b}$ is specified, the *a priori* information $\boldsymbol{\lambda}_2^{\mathrm{e}}$ becomes irrelevant. In Eq. (17.148), we have used the common notations where $P(\cdot)$ refers to the probability mass of a discrete random variable and $p(\cdot)$ denotes a probability density of a continuous random variable. We also note that

$$p(\mathbf{x}|\mathbf{b}) = \prod_k p(x(k)|\mathbf{s}(k-L+1:k))$$
$$\propto \prod_k \exp\left(-\frac{1}{\sigma_v^2}\left|x(k) - \sum_{i=0}^{L-1} h(i)s(k-i)\right|^2\right)$$
$$= \exp\left(-\frac{1}{\sigma_v^2}\sum_k\left|x(k) - \sum_{i=0}^{L-1} h(i)s(k-i)\right|^2\right). \tag{17.149}$$

On the other hand,

$$P(\mathbf{b}|\boldsymbol{\lambda}_2^{\mathrm{e}}) = \prod_k P(b(k)|\lambda_2^{\mathrm{e}}(b(k)). \tag{17.150}$$

Moreover, as in Eq. (17.132), here,

$$P(b(k)=0|\lambda_2^{\mathrm{e}}(b(k))) = \frac{\mathrm{e}^{\lambda_2^{\mathrm{e}}(b(k))}}{1+\mathrm{e}^{\lambda_2^{\mathrm{e}}(b(k))}} = \frac{\mathrm{e}^{\lambda_2^{\mathrm{e}}(b(k))/2}}{\mathrm{e}^{-\lambda_2^{\mathrm{e}}(b(k))/2}+\mathrm{e}^{\lambda_2^{\mathrm{e}}(b(k))/2}} \tag{17.151}$$

and

$$P(b(k)=1|\lambda_2^{\mathrm{e}}(b(k))) = \frac{1}{1+\mathrm{e}^{\lambda_2^{\mathrm{e}}(b(k))}} = \frac{\mathrm{e}^{-\lambda_2^{\mathrm{e}}(b(k))/2}}{\mathrm{e}^{-\lambda_2^{\mathrm{e}}(b(k))/2}+\mathrm{e}^{\lambda_2^{\mathrm{e}}(b(k))/2}}. \tag{17.152}$$

Combining these results, we obtain

$$P(b(k)|\lambda_2^{\mathrm{e}}(b(k))) = \frac{\mathrm{e}^{(-1)^{b(k)}\lambda_2^{\mathrm{e}}(b(k))/2}}{\mathrm{e}^{-\lambda_2^{\mathrm{e}}(b(k))/2}+\mathrm{e}^{\lambda_2^{\mathrm{e}}(b(k))/2}}. \tag{17.153}$$

Next, substituting Eqs. (17.149) and (17.150) in Eq. (17.148) and using Eq. (17.153), we obtain

$$\lambda_1(b(n)) = \ln\frac{\sum_{\mathbf{b}:b(n)=0}\exp\left(-\frac{1}{\sigma_v^2}\sum_k\left|x(k)-\sum_{i=0}^{L-1}h(i)s(k-i)\right|^2\right)}{\sum_{\mathbf{b}:b(n)=1}\exp\left(-\frac{1}{\sigma_v^2}\sum_k\left|x(k)-\sum_{i=0}^{L-1}h(i)s(k-i)\right|^2\right)}$$
$$+\ln\frac{\prod_k\frac{\mathrm{e}^{(-1)^{b(k)}\lambda_2^{\mathrm{e}}(b(k))/2}}{\mathrm{e}^{-\lambda_2^{\mathrm{e}}(b(k))/2}+\mathrm{e}^{\lambda_2^{\mathrm{e}}(b(k))/2}}}{\prod_k\frac{\mathrm{e}^{(-1)^{b(k)}\lambda_2^{\mathrm{e}}(b(k))/2}}{\mathrm{e}^{-\lambda_2^{\mathrm{e}}(b(k))/2}+\mathrm{e}^{\lambda_2^{\mathrm{e}}(b(k))/2}}}$$
$$= \ln\frac{\sum_{\mathbf{b}:b(n)=0}\exp\left(-\frac{1}{\sigma_v^2}\sum_k\left|x(k)-\sum_{i=0}^{L-1}h(i)s(k-i)\right|^2+\sum_k(-1)^{b(k)}\frac{\lambda_2^{\mathrm{e}}(b(k))}{2}\right)}{\sum_{\mathbf{b}:b(n)=1}\exp\left(-\frac{1}{\sigma_v^2}\sum_k\left|x(k)-\sum_{i=0}^{L-1}h(i)s(k-i)\right|^2+\sum_k(-1)^{b(k)}\frac{\lambda_2^{\mathrm{e}}(b(k))}{2}\right)}. \tag{17.154}$$

Once $\lambda_1(b(n))$ is calculated, it may be used to obtain $\lambda_1^{\rm e}(b(n) = \lambda_1(b(n)) - \lambda_2^{\rm e}(b(n))$, which is then passed to the soft decoder for the next iteration.

The main difficulty with the computation of $\lambda_1(b(n))$ according to Eq. (17.154) is that the number of the terms under the summations over "$\mathbf{b} : b(n) = 0$" and "$\mathbf{b} : b(n) = 1$" grow exponentially with the length of $\mathbf{b}$. With the typical length of $\mathbf{b}$ in the order of at least a few hundreds or, maybe, a few thousands, this clearly results in a prohibitive complexity. The key idea in development of statistical soft equalizers is that in practice only a small subset of the choices of $\mathbf{b}$ have significant contributions to the terms under the summations over "$\mathbf{b} : b(n) = 0$" and "$\mathbf{b} : b(n) = 1$." Hence, a good estimate of $\lambda_1(b(n))$ may be obtained, if one can find and limit the summations to these important choices of $\mathbf{b}$. An approach that has been successfully adopted in the literature uses a search method called *Gibbs sampler*. The Gibbs sampler is a particular implementation of a Marov chain Monte Carlo (MCMC) simulation that randomly walks through the subsets of "$\mathbf{b} : b(n) = 0$" and "$\mathbf{b} : b(n) = 1$" that have significant contributions to the computation of $\lambda_1(b(n))$. Once these subsets are obtained, they will be expanded (as explained later) and used to calculate the LLR values $\lambda_1(b(n))$. We refer to the combinations of these two steps as *MCMC equalizer*. We also call the set of the choices of $\mathbf{b}$ obtained through the Gibbs sampler the important sample set $\mathcal{I}$.

Step 1: Gibbs Sampler

Let us assume that a total of $N_{\rm b}$ bits are transmitted, hence, $\mathbf{b} = \{b(0), b(1), \ldots, b(N_{\rm b} - 1)\}$. The Gibbs sampler starts with a random (or a deterministic) initialization of $\mathbf{b} = \mathbf{b}^{(0)}$ and iterates through the elements of $\mathbf{b}$ according to the following rule.

During the ith iteration, starting with $\mathbf{b}^{(i-1)} = \{b^{(i-1)}(0), b^{(i-1)}(1), \ldots, b^{(i-1)}(N_{\rm b} - 1)\}$, $\mathbf{b}^{(i)}$ is updated by taking the following steps for each bit. Assuming that $b(0)$ through $b(n-1)$ have been updated, we define

$$\overline{\mathbf{b}}_n = \{b^{(i)}(0), \ldots, b^{(i)}(n-1), b^{(i-1)}(n+1), \ldots, b^{(i-1)}(N_{\rm b} - 1)\}$$

and, for $a = 0, 1$, we let

$$\mathbf{b}_n^a = \{b^{(i)}(0), \ldots, b^{(i)}(n-1), a, b^{(i-1)}(n+1), \ldots, b^{(i-1)}(N_{\rm b} - 1)\}$$

The selection of the bit $b^{(i)}(n)$ is based on the conditional probability distribution $\gamma_{\rm a}$, for $a = 0, 1$, where

$$\gamma_{\rm a} = P(b(n) = a|\overline{\mathbf{b}}_n, \mathbf{y}, \lambda_2^e(b(n))).$$

That is, we choose $b(n) = 0$ with the probability γ_0 and $b(n) = 1$ with the probability γ_1.

Noting that $b(n)$ maps to $s(n)$, we have

$$\begin{aligned}
\gamma_{\mathrm{a}} &= P(b(n) = a|\overline{\mathbf{b}}_n, \mathbf{x}, \lambda_2^e(b(n))) \\
&\propto p(\mathbf{x}|\mathbf{b}_n^a)P(b(n) = a) \\
&= p(\mathbf{x}|\mathbf{s}_n^a)P(b(n) = a) \\
&= \prod_{i=0}^{N_{\mathrm{b}}+L-2} p(x(i)|\mathbf{s}_m^a(i-L+1:i))P(b(n) = a) \\
&= \left\{ \prod_{i=0}^{n-1} p(x(i)|\mathbf{s}_n^a(i-L+1:i)) \prod_{i=n+L}^{N_{\mathrm{b}}+L-2} p(x(i)|\mathbf{s}_n^a(i-L+1:i)) \right\} \\
&\quad \times \prod_{i=n}^{n+L-1} p(x(i)|\mathbf{s}_n^a(i-L+1:i))P(b(n) = a)
\end{aligned} \tag{17.155}$$

where $\mathbf{s}_n^a$ is obtained from a direct mapping of $\mathbf{b}_n^a$. Moreover, because for $i \le n-1$ and $i \ge n+L$, the vector $\mathbf{s}_n^a(i-L+1:i)$ is independent of a, Eq. (17.155) reduces to

$$\begin{aligned}
\gamma_{\mathrm{a}} &\propto \prod_{i=n}^{n+L-1} p(x(i)|\mathbf{s}_n^a(i-L+1:i))P(b(n) = a) \\
&= C \cdot \exp\left\{ \sum_{i=n}^{n+L-1} -\frac{1}{2\sigma_v^2}\left| x(i) - \sum_{l=0}^{L-1} h(l)s^a(i-l) \right|^2 + (-1)^a \frac{\lambda_2^{\mathrm{e}}(b(n))}{2} \right\}
\end{aligned} \tag{17.156}$$

where C is a scaling constant to ensure that $\gamma_0 + \gamma_1 = 1$.

Step 2: Computation of LLR Values

Assume that the Gibbs sampler has produced the important sample set $\mathcal{I}$. Each element in $\mathcal{I}$ is a bit vector of length N_{b}. As bit $b(n)$ is mapped to symbol $s(n)$, the received signal samples that are affected by $b(n)$ are $\mathbf{x}(n:n+L-1)$. As $\mathbf{x}(n:n+L-1)$ depends only on bits $\{b(l),\ n_1 = n-L+1 \le l \le n+L-1 = n_2\}$, we find that when computing the output LLR for $b(n)$, it is sufficient to truncate each sequence in $\mathcal{I}$ to take into account only bits $\{b(l),\ n_1 \le l \le n_2\}$. We denote the set that contains the truncated sequences by $\mathcal{I}_{n_1:n_2}$. For each $0 \le n \le N_{\mathrm{b}} - 1$, we construct a larger set $\mathcal{I}_{n_1:n_2}^n$, which includes all sequences in $\mathcal{I}_{n_1:n_2}$, together with new sequences that are obtained by flipping the nth bit of each sequence in $\mathcal{I}_{n_1:n_2}$. Repetitious sequences are removed from $\mathcal{I}_{n_1:n_2}^n$. Furthermore, we let $\mathcal{I}_{n_1:n_2}^{n,0}$ and $\mathcal{I}_{n_1:n_2}^{n,1}$ denote subsequences in $\mathcal{I}_{n_1:n_2}^n$ whose nth bit equals 0 and 1, respectively. The LLR value for bit $b(n)$ is then computed as

$$\lambda_1(b(n)) = \ln \frac{\sum_{\mathbf{b}_{n_1:n_2} \in \mathcal{I}_{n_1:n_2}^{n,0}} p(\mathbf{x}(n:n+L-1)|\mathbf{b}_{n_1:n_2}) \prod_{l=n_1}^{n_2} P(b(l))}{\sum_{\mathbf{b}_{n_1:n_2} \in \mathcal{I}_{n_1:n_2}^{n,1}} p(\mathbf{x}(n:n+L-1)|\mathbf{b}_{n_1:n_2}) \prod_{l=n_1}^{n_2} P(b(l))} \tag{17.157}$$

The performance of the MCMC equalizer is dependent on the quality of samples in the important set $\mathcal{I}$. In practice, a Gibbs sampler may require many iterations to converge to its stationary distribution. This is called *burn-in period*. As a result, including the burn-in period in the implementation of the Gibbs sampler may increase the complexity significantly. In Farhang-Boroujeny, Zhu and Shi (2006), it has been shown empirically that the formulas such as Eq. (17.157) still work well if the stationary distribution of the underlying Markov chain is replaced by a uniform distribution over the significant samples. To obtain samples with this uniform distribution, it has also been noted in Farhang-Boroujeny, Zhu and Shi (2006) that a set of parallel Gibbs samplers with no burn-in period and a small number of iterations are more effective than using a single Gibbs sampler with many iterations. Uniform distribution is obtained by removing the repeated samples in $\mathcal{I}$.

In order to further guide the reader in understanding the above procedure, the MATLAB script "`MCMCTest.m`" and the accompanying function "`MCMCEq.m`" on the accompanying website are provided.

Parallel Implementation of MCMC Equalizer

The Gibbs sampler that was presented earlier runs over a complete packet of received signal $x(0)$ through $x(L+N_\mathrm{b}-2)$, equivalently, the information bits $b(0)$ through $b(N_\mathrm{b}-1)$, sequentially, one bit at a time. This sequential processing is repeated over a number of iterations for each Gibbs sampler on sufficient number of samples of $\mathbf{b}$. Also, as noted earlier, a few Gibbs samplers may be run in parallel to obtain a more diverse (a balanced) set of samples of $\mathbf{b}$ with uniform distribution.

From an implementation point of view, sequential processing requires a large number of clock cycles that is proportional to the sequence length N_b. This (with typical values of N_b in an order of a few thousands), in turn, means a long processing time or, equivalently, a long delay in the detection path. This, of course, is undesirable and should be avoided if possible.

This problem can be solved by introducing a parallel implementation of the Gibbs sampler that begins with partitioning the transmitted symbol vector $\mathbf{s}$ in to a number of smaller vectors (Peng, Chen, and Farhang-Boroujeny, 2010), say, $\mathbf{s}_1, \mathbf{s}_2, \ldots$, that we call subsequences. Then, instead of running a single Gibbs sampler over the entire sequence of $\mathbf{s}$, we run parallel Gibbs samplers, one for each subsequence. Within each subsequence, the Gibbs sampler updates the symbols sequentially from left to right. Owing to the channel memory, the LLR value of a given symbol in a subsequence may depend on the values of some symbols belonging to other subsequences. To speed up the convergence rate of the Gibbs sampler, it has been empirically found in Peng, Chen, and Farhang-Boroujeny (2010) that it will be helpful to synchronize the parallel Gibbs samplers and have the updated symbol values available to the neighboring subsequences. To show how the parallel implementation works, we examine the example shown in Figure 17.30 where $N_\mathrm{b}=8$ and $\mathbf{s}$ is partitioned into two subsequences. Hence, two Gibbs samplers are run in parallel. The symbol updating process of these two Gibbs samplers is shown in Table 17.3. At the time 0, the samples are initialized randomly to $\mathbf{s}^{(0)}$. Then at time i, two new samples (one for each subsequence) are updated in parallel, conditioned upon

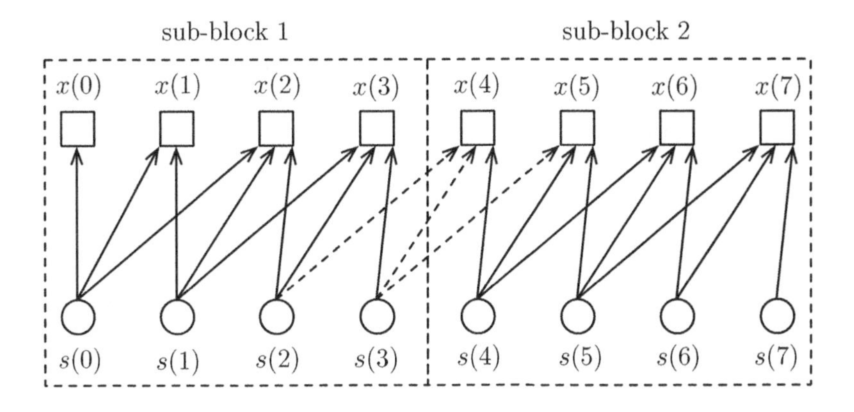

Figure 17.30 Partitioning of symbol sequence of length $N = 8$ to two subsequences, each of length 4. The channel has a length of $L = 3$. The arrows show the dependency of the received signal samples $x(n)$ to the transmitted symbols $s(n)$. The dashed arrows show the inter subsequence dependency.

Table 17.3 Samples updating in parallel Gibbs sampler.

	Gibbs Sampler 1		Gibbs Sampler 1	
t	Update	Conditioned upon	Update	Conditioned upon
0	Initialize $\mathbf{s}^{(0)} = \{s^{(0)}(0), s^{(0)}(1), s^{(0)}(2), s^{(0)}(3), s^{(0)}(4), s^{(0)}(5), s^{(0)}(6), s^{(0)}(7)\}$			
1	$s^{(1)}(0)$	$s^{(0)}(1)$, $s^{(0)}(2)$	$s^{(1)}(4)$	$s^{(0)}(2)$, $s^{(0)}(3)$, $s^{(0)}(5)$, $s^{(0)}(6)$
2	$s^{(1)}(1)$	$s^{(1)}(0)$, $s^{(0)}(2)$, $s^{(0)}(3)$	$s^{(1)}(5)$	$s^{(0)}(3)$, $s^{(1)}(4)$, $s^{(0)}(6)$, $s^{(0)}(7)$
3	$s^{(1)}(2)$	$s^{(1)}(0)$, $s^{(1)}(1)$, $s^{(0)}(3)$, $s^{(1)}(4)$	$s^{(1)}(6)$	$s^{(1)}(4)$, $s^{(1)}(5)$, $s^{(0)}(7)$
4	$s^{(1)}(3)$	$s^{(1)}(1)$, $s^{(1)}(2)$, $s^{(1)}(4)$, $s^{(1)}(5)$	$s^{(1)}(7)$	$s^{(1)}(5)$, $s^{(1)}(6)$

the already updated symbols. The reader should carefully examine Table 17.3 to understand how parallel processing and the connection between the adjacent subsequences are being handled.

17.9.3 Iterative Channel Estimation and Data Detection

In the derivations made earlier, it was assumed that the channel response $\mathbf{h}$ was known perfectly. In practice, this is not the case. An initial estimate of the channel may be obtained through use of a training sequence that is transmitted at the beginning of each packet, as discussed in Section 17.6, or may be repeated periodically when the channel varies significantly over the duration of each packet. Such initial estimate(s) of the channel is (are) used to run the soft equalizer to produce the first set of LLR values, which will be passed to the soft decoder. The LLR values generated by the soft decoder are then used to refine the channel estimate before the next iteration of the soft equalizer. This process is repeated for both turbo equalizer and the channel estimator to converge and hence the best possible estimate of the transmitted data is obtained.

The following procedure explains how the LLR values (the soft information) from the soft decoder are used to refine the channel estimate. To keep the discussion simple, we assume, as before, that the modulator converts the coded bits $b(n)$ to the BPSK symbols $s(n)$ that take values of -1, for $b(n) = 0$, and $+1$, for $b(n) = 1$. We also note that following the same line of thought that led to Eqs. (17.151) and (17.152), one obtains

$$P(s(n) = -1|\lambda_2(b(n))) = \frac{e^{\lambda_2(b(n))/2}}{e^{-\lambda_2(b(n))/2} + e^{\lambda_2(b(n))/2}} \tag{17.158}$$

and

$$P(s(n) = +1|\lambda_2(b(n))) = \frac{e^{-\lambda_2(b(n))/2}}{e^{-\lambda_2(b(n))/2} + e^{\lambda_2(b(n))/2}}. \tag{17.159}$$

Note that here we use $\lambda_2(\cdot)$ (not the extrinsic information $\lambda_2^e(\cdot)$).

On the other hand, we recall the least-squares cost function ζ, of Chapter 12. When the sequence $s(n)$ is perfectly known, the channel estimate $\hat{\mathbf{h}}$ is obtained by minimizing

$$\zeta = \sum_n |x(n) - \mathbf{h}^{\mathrm{T}}\mathbf{s}(n)|^2 \tag{17.160}$$

where $\mathbf{h} = [h(0)\ h(1) \cdots h(L-1)]^{\mathrm{T}}$ and $\mathbf{s}(n) = [s(n)\ s(n-1) \cdots s(n-L+1)]^{\mathrm{T}}$. When the sequence $s(n)$ is stochastic, as is the case here, Eq. (17.160) should be replaced by

$$\zeta = \sum_n E[|x(n) - \mathbf{h}^{\mathrm{T}}\mathbf{s}(n)|^2] \tag{17.161}$$

where the expectation is over the elements of the sequence $s(n)$. Expanding Eq. (17.161), we obtain

$$\zeta = \sum_n |x(n)|^2 - 2\mathbf{h}^{\mathrm{T}} \sum_n E[\mathbf{s}(n)]x(n) - \mathbf{h}^{\mathrm{T}} \left(\sum_n E\left[\mathbf{s}(n)\mathbf{s}^{\mathrm{T}}(n)\right] \right) \mathbf{h} \tag{17.162}$$

Hence,

$$\hat{\mathbf{h}} = \left(\sum_n E\left[\mathbf{s}(n)\mathbf{s}^{\mathrm{T}}(n)\right] \right)^{-1} \left(\sum_n E[\mathbf{s}(n)]x(n) \right). \tag{17.163}$$

In Eq. (17.163), the elements of $\mathbf{s}(n)$ are independent random variables with the statistics given as in Eqs. (17.158) and (17.159). Accordingly, the ith element of $E[\mathbf{s}(n)]$ is obtained as

$$\begin{aligned}(E[\mathbf{s}(n)])_i &= E[s(n-i)] \\ &= (-1) \times P(s(n) = -1|\lambda_2(b(n))) + (+1) \times P(s(n) = +1|\lambda_2(b(n)))\end{aligned} \tag{17.164}$$

Also, the ijth element of $E[\mathbf{s}(n)\mathbf{s}^{\mathrm{T}}(n)]$ is obtained as (since $s(n-i)$ and $s(n-j)$ are independent)

$$(E[\mathbf{s}(n)\mathbf{s}^{\mathrm{T}}(n)])_{ij} = \begin{cases} 1, & i = j \\ E[s(n-i)]E[s(n-j)], & i \neq j \end{cases} \tag{17.165}$$

where $E[s(n-i)]$ and $E[s(n-j)]$ are computed according to Eq. (17.164).

17.10 Single-Input Multiple-Output Equalization

In certain wireless applications where the channel suffers from a severe ISI and/or a high level of noise and interference, the received signal is picked up by a number of antennas that are spatially spread apart sufficiently to observe a set of independent copies of the transmitted signal. The channels formed in this way are often referred to as *single-input multiple-output* (SIMO). Moreover, the receiver performance gain achieved as a result of processing more than one copies of the received signal is called *diversity gain*.

In the electromagnetic wireless channels, the use of multiple receive antennas often falls under the category of smart antennas. As discussed in Chapter 6 (and is developed further in the next chapter), by combining the signals from various antennas, one can form an antenna directivity that sees the desired signal and introduces nulls in the directions of interfering signals. Moreover, SIMO channels have been found very useful and are widely used in the underwater acoustic (UWA) communication channels where electromagnetic waves are replaced by acoustic waves. This is because electromagnetic waves cannot propagate far enough in water. The UWA channels are characterized by long impulse responses with many nulls/fades within the band of interest. The presence of nulls, obviously, degrades the equalizer performance. The rationale for the use SIMO setup in this application is that it is very unlikely that all the transmission paths between a transmit antenna and a few receive antennas undergo a fade at the same frequency. Hence, the use of a SIMO setup provides enough diversity in the transmission paths that assures adequate energy will be received at all the frequencies over the transmission band.

To develop an insight to the concept of diversity gain, consider the SIMO channel presented in Figure 17.31. Also, let us consider the simplified case where the channel gains between the transmit antenna and all the M receive antennas are the same and equal to one. However, the channel noise at the receive antennas are a set of independent identically distributed (iid) random processes. Hence, the received signals $x_0(n)$ through

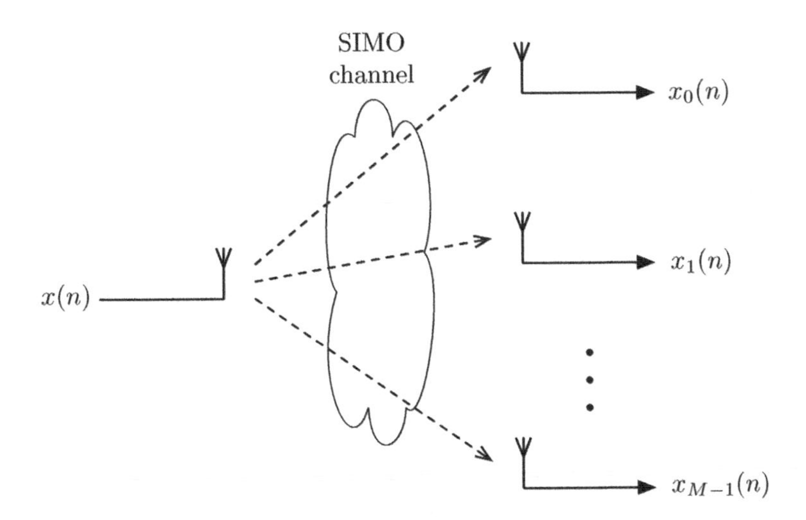

Figure 17.31 Single-input multiple-output channel.

$x_{M-1}(n)$ are given as

$$x_i(n) = x(n) + \nu_{\rm i}(n) \tag{17.166}$$

where $\nu_j(n)$, for $i = 0, 1, \ldots, M-1$, are a set of iid processes. Assuming that the processes $x(n)$ and $\nu_j(n)$ have zero mean, and using σ_x^2 and σ_ν^2 to denote the respective variances/powers, the signal-to-noise ratio (SNR) at each receive antenna is obtained as

$$\rho_{\rm r} = \frac{\sigma_x^2}{\sigma_\nu^2} \tag{17.167}$$

As the noise components $\nu_{\rm i}(n)$ are uncorrelated with one another, one may argue that an optimum receiver takes the average of the received signals $x_0(n)$ through $x_{M-1}(n)$ as the best estimate of the transmitted signal $x(n)$, viz.,

$$\begin{aligned}\hat{x}(n) &= \frac{1}{M}\sum_{i=0}^{M-1} x_i(n) \\ &= x(n) + \frac{1}{M}\sum_{i=0}^{M-1} \nu_i(n)\end{aligned} \tag{17.168}$$

Now, one may note that the variance of the noise term on the right-hand side of Eq. (17.168) is equal to σ_ν^2/M and, hence, the SNR at the receiver output (i.e., after averaging) is obtained as

$$\rho_{\rm o} = \frac{\sigma_x^2}{\sigma_\nu^2/M} = M\rho_{\rm r} \tag{17.169}$$

In other words, the SNR after receiver processing improves by a gain of M, the number of the receive antennas.

Next, let us consider the case where the channel has a gain of g_i between the transmit antenna and the ith receive antenna. Hence,

$$x_i(n) = g_i x(n) + \nu_i(n) \tag{17.170}$$

In this more general case, we still assume that the noise components $\nu_i(n)$ have the same variance, equal to σ_ν^2. We wish to find a combining gain vector $\mathbf{w}$ for obtaining the best *unbiased* estimate and of $x(n)$ as

$$\hat{x}(n) = \mathbf{w}^{\rm H}\mathbf{x}(n) \tag{17.171}$$

where $\mathbf{x}(n) = [x_0(n)\ x_1(n) \cdots x_{M-1}(n)]^{\rm T}$. The unbiased estimate means that the mean of $\hat{x}(n)$ must be equal to $x(n)$. This, in turn, implies that the difference $x(n) - \hat{x}(n)$ should be minimized subject to the constraint $\mathbf{w}^{\rm H}\mathbf{g} = 1$, where $\mathbf{g} = [g_0\ g_1 \cdots g_{M-1}]^{\rm T}$.

Now, defining the cost function

$$\begin{aligned}\xi &= E[|x(n) - \hat{x}(n)|^2] \\ &= E[|x(n) - \mathbf{w}^{\rm H}\mathbf{x}(n)|^2]\end{aligned} \tag{17.172}$$

and attempting to minimize it subject to the constraint $\mathbf{w}^{\rm H}\mathbf{g} = 1$, one will find that this is the same constraint minimization problem as the one that we solved in Section 6.11, using the method of Lagrange multipliers. Following the same line of derivations, here, one will find the following solution

$$\mathbf{w}_{\mathrm{o}}^{\mathrm{c}} = \frac{1}{\mathbf{g}^{\mathrm{H}}\mathbf{g}}\mathbf{g} \tag{17.173}$$

where the superscript "c" is to emphasize that the solution is obtained under the constraint condition.

The solution (17.173) leads to the following estimate of the desired signal, $x(n)$:

$$\hat{x}(n) = \frac{\mathbf{g}^{\mathrm{H}}\mathbf{x}(n)}{\mathbf{g}^{\mathrm{H}}\mathbf{g}}. \tag{17.174}$$

This has the following interpretation. The estimate $\hat{x}(n)$ is obtained as a weighted average of the elements of the received signal vector $\mathbf{x}(n)$ with the weighting factors that are equal to the channel path gains between the transmit antenna and each of the receive antennas. In other words, the paths with higher gains are given higher weights, obviously, because they lead to less noisy signal components at the receiver. Consequently, the estimator (17.174) is called the *maximum ratio combiner*.

Further generalization of the maximum ratio combiner may be suggested for the case where the noise (plus interference) components $\nu_i(n)$ have different powers/variances. The result will be that the combiner weight at each tap is proportional to the SNR (the square root of SNR, to be more exact) at that tap, as one would expect; see Problem P17 at the end of this chapter.

When the channel is a broadband one, and is hence characterized by a frequency dependent gain across the band of interest, the received signal at each antenna should be passed through a transversal filter with several taps. Adding the outputs of these transversal filters to form a single output form a multiple-input single-output (MISO) equalizer. Clearly, any of the adaptive algorithms that have been presented in the previous chapters of this book can be applied for adaptation of the MISO equalizer as well. The adapted equalizer, after adaptation, forms a frequency-dependent maximum ratio combiner that for each frequency considers the SNR values at the various receive antennas and performs the combining accordingly. We note that for a finite length equalizer, the maximum ratio combining is only achieved to the extent allowed by the limited equalizer length. In other words, only an approximate maximum ratio combiner can be realized. Problem P17, at the end of this chapter, guides the reader to develop further insight to this property of MISO equalizers through a numerical example.

17.11 Frequency Domain Equalization

Frequency domain equalization method that is presented in this section has been proposed as a method of reducing the complexity and, for some scenarios, fast adaptation of the equalizer to channel variations. As discussed later, the frequency domain equalizer makes use of certain properties of the discrete Fourier transform (DFT) to design a data packet structure that when transmitted over an ISI channel, the ISI effects could be removed through a simple processing in the frequency domain.

17.11.1 Packet Structure

Because of reasons that will become clear later, to facilitate the frequency domain equalization, the stream of transmit data symbols is partitioned into blocks of size N symbols. A guard interval of a length equal to the duration of the channel impulse response or greater is inserted between every two consecutive blocks. Moreover, each guard interval is filled up by the symbols from the end of the succeeding data block. This concept is presented in Figure 17.32.

To be more explicit, we note that, in Figure 17.32a, the Nth block contains the vector

$$\mathbf{s}(n) = \begin{bmatrix} s(nN) \\ s(nN+1) \\ \vdots \\ s(nN+N-1) \end{bmatrix} \tag{17.175}$$

After the insertion of the guard interval and copying the symbols from the end of the succeeding block, the result will be an extended block whose content is the vector

$$\bar{\mathbf{s}}(n) = \begin{bmatrix} s(nN+N-N_{\mathrm{g}}) \\ \vdots \\ s(nN+N-1) \\ s(nN) \\ s(nN+1) \\ \vdots \\ s(nN+N-1) \end{bmatrix} \tag{17.176}$$

where N_{g} is the length of the guard interval.

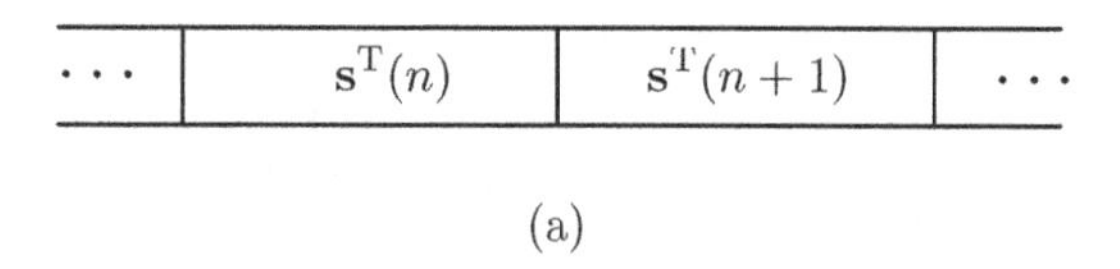

(a)

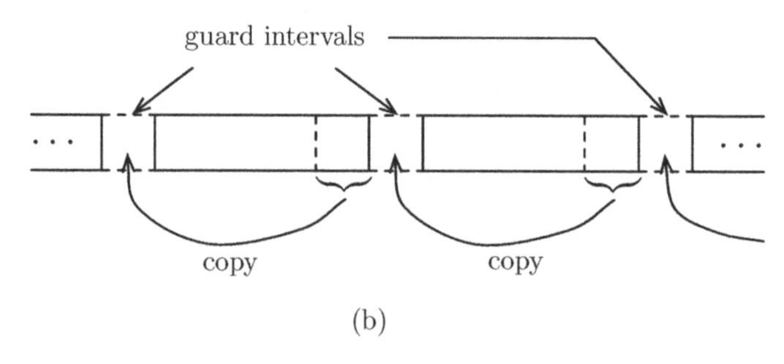

(b)

Figure 17.32 Packet structure for frequency domain equalization: (a) data stream before insertion of guard intervals and (b) data stream after insertion of guard intervals.

One may note that $\bar{\mathbf{s}}_n$ may be thought of as a length $N + N_g$ extraction from a periodic sequence whose one period is $\mathbf{s}_n$. We will take advantage of this structure of $\bar{\mathbf{s}}_n$, in the next section, to develop a low-complexity frequency domain equalizer. Alternatively, one may say, the vector $\bar{\mathbf{s}}_n$ is obtained by copying the last N_g elements of $\mathbf{s}_n$ to its beginning as a cyclic prefix (CP). The name *CP* is thus used to refer to the data symbols that are used to fill in the guard interval between each pair of successive data blocks.

17.11.2 Frequency Domain Equalizer

To develop the structure of the desired frequency domain equalizer, consider the case that an isolated (and extended) block $\bar{\mathbf{s}}_n$ is passed through a channel with the impulse response

$$\mathbf{h} = [h(0)\ h(1) \cdots h(L-1)]^{\mathrm{T}} \tag{17.177}$$

The received signal, excluding the channel noise, will be the vector $\bar{\mathbf{x}}_n$ obtained by convolving $\bar{\mathbf{s}}_n$ and the vector of the channel impulse response, $\mathbf{h}$. The vector $\bar{\mathbf{x}}_n$ has a length of $N + N_g + L - 1$. Also, because of the presence of the CP in $\bar{\mathbf{s}}_n$, it is not difficult to show that the elements $(N_g + 1)$st through $(N_g + N)$th of $\bar{\mathbf{x}}_n$ (stored in a vector $\mathbf{x}_n$) are obtained by circularly convolving the vector $\mathbf{s}_n$ and $\mathbf{h}$. Mathematically, this is written as

$$\mathbf{H}\mathbf{s}_n = \mathbf{x}_n \tag{17.178}$$

where $\mathbf{H}$ is a circular matrix whose first column is obtained by appending a number of zeros to the end of $\mathbf{h}$ to extend its length to N.

Next, we argue that Eq. (17.178) is still valid when a sequence of cyclic prefixed blocks $\bar{\mathbf{s}}_n$ are transmitted. This is because the CP of each block absorbs the transient of the channel and the effect of the previous block will be faded out before we reach the CP stripped part of each block, namely, $\mathbf{x}_n$.

Recalling the theory of circular matrices, which was presented in Chapter 8, one may note that the circular matrix $\mathbf{H}$ can be expanded as

$$\mathbf{H} = \mathcal{F}^{-1}\mathcal{H}\mathcal{F} \tag{17.179}$$

where $\mathcal{F}$ is the DFT transformation matrix and $\mathcal{H}$ is the diagonal matrix whose diagonal elements are obtained by taking the DFT of the first column of $\mathbf{H}$. Substituting Eq. (17.179) in Eq. (17.178) and rearranging the result, we obtain

$$\mathbf{s}_n = \mathcal{F}^{-1}\mathcal{H}^{-1}\mathcal{F}\mathbf{x}_n. \tag{17.180}$$

This result has the following interpretation. Given a cyclic stripped block of the received signal, $\mathbf{x}_n$, the associated data symbol vector, $\mathbf{s}_n$, can be obtained by taking the following steps:

1. Take the DFT of $\mathbf{x}_n$.
2. Point multiply the elements of the result from Step 1 by the inverse of the elements of the DFT of the first column of $\mathbf{H}$.
3. Take the inverse DFT of the result in Step 2.

The following comments regarding the earlier steps are in order. Step 1 gives the vector $\mathcal{F}\mathbf{x}_n$. Step 2 calculates the vector $\mathcal{H}^{-1}\mathcal{F}\mathbf{x}_n$. This is equivalent to saying a single-tap frequency domain equalizer is applied to each of the frequency domain elements of $\mathcal{F}\mathbf{x}_n$. Step 3 converts the equalized signal samples from the frequency domain to the time domain.

We also take note of the following points.

- The addition of the guard interval avoids interference among the data symbol blocks $\mathbf{s}_n$.
- The use of the CP allows application of a trivial single-tap equalizer per frequency domain sample.
- The presence of the guard interval/CP takes up a fraction of time, which otherwise could be used for data transmission. This constitutes an overhead, which results in a loss in the use of the available spectrum. This is quantified by the efficiency loss $N_{\rm g}/(N+N_{\rm g})$ or, equivalently the efficiency $N/(N+N_{\rm g})$.
- The single-tap equalizer coefficients are the inverse of the samples of the N-point DFT of the channel impulse response. Hence, to initialize/track the equalizer coefficients, the channel impulse response should be estimated/tracked. The method of cyclic equalization/channel estimation that was introduced in Section 17.6 can also be applied here. For tracking, one may adopt a decision-directed method, where the decisions from each block are used to update the channel estimate after processing of each block $\mathbf{x}_n$. For the fast varying channels, where the decision-directed method may fail, an alternative method (an extension of cyclic channel estimation) is presented in the next section.
- The single-tap equalizer coefficients that are suggested in Eq. (17.180) (also Step 2 of the earlier procedure) ignore the channel noise and hence its possible amplification at the frequency points where the channel gain is low. To avoid such noise enhancement, the common solution is to adopt an MMSE equalizer. In that case, Eq. (17.180) is replaced by

$$\mathbf{s}_n = \mathcal{F}^{-1}(\mathcal{H}\mathcal{H}^* + \epsilon\mathbf{I})^{-1}\mathcal{H}\mathcal{F}\mathbf{x}_n \tag{17.181}$$

where ϵ is the noise power over signal power at the equalizer input.

17.11.3 Packet Structure for Fast Tracking

As discussed earlier, the cyclic equalization/channel estimation method that was introduced in Section 17.6 may be used to initialize the coefficients of the equalizer, at the beginning of a communication session/a packet, and subsequently a decision-directed method may be used to track variations of the channel for the rest of the session/packet. However, in applications where channel variation is too fast and/or because of channel noise, decision errors may lead to a failure of the decision-directed method, regular training symbols may be inserted in the transmit signal. An interesting method of inserting such trainings has been discussed in Falconer *et al.* (2002). Here, we elaborate on this method.

The packet structure presented in Figure 17.32 may be modified as the one presented in Figure 17.33. Here, each data block $\mathbf{s}_n$ is preceded with a training block (TB) and succeeded with the same TB. The TB is a fixed block of known symbols to both transmitter

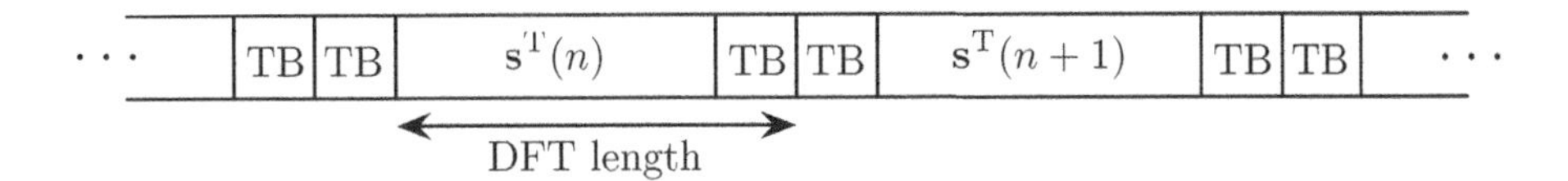

Figure 17.33 Packet structure for fast tracking in frequency domain equalizers.

and receiver. The length of TB should be equal to or greater than the duration of the channel impulse response. Each data block $\mathbf{s}_n$ combined with the succeeding TB has the length of the DFT used for the frequency domain equalization. Accordingly, the concatenation of $\mathbf{s}_n$ and its succeeding TB plays the same role as the data block $\mathbf{s}_n$ in Figure 17.32. Moreover, the TB preceding $\mathbf{s}_n$ plays the role of the CP for the block consisting of $\mathbf{s}_n$ and its succeeding TB. In addition and interestingly, in a pair of two adjacent TBs, the first TB acts as a CP to the second TB. Therefore, the received version of the second TB will be the circular convolution of the transmitted TB and the channel impulse response. Consequently, a cyclic channel estimator, similar to the one in Figure 17.26, can be used to estimate the channel impulse response $\mathbf{h}$. Finally, we note that this clever choice of the packet structure comes at a price. As part of each data block should be given up in favor of a TB, its efficiency reduces from $N/(N+N_g)$ to $(N-N_g)/(N+N_g)$.

17.11.4 Summary

Figure 17.34 presents a summary of our findings in this section. To facilitate the implementation of a computationally efficient frequency domain equalizer, the transmit data stream is partitioned into blocks of N symbols, each. Each block is cyclic prefixed from samples from the end of the block. This extended stream is then passed through the channel. The CP samples act as a guard interval that keep the data block free of interblock interference. They also play a key role in facilitating the task of frequency domain equalization. Because of the choice of the CP samples, the received signal vectors $\mathbf{x}_n$ is the circular convolution of the symbol vector $\mathbf{s}_n$ and the channel impulse response vector $\mathbf{h}$. Consequently, frequency domain equalization can be performed by taking the DFT of each block $\mathbf{x}_n$, applying a single-tap equalizer per DFT output, and converting the results back to the time domain through an IDFT.

17.12 Blind Equalization

In certain applications, for example, when a continuous stream of data is being broadcasted, there may be no training/preamble for tuning of the receiver to the incoming signal. Equalizer adaptation in such applications that should be performed without any access to a desired signal is referred to as *blind equalization*. In the literature, three classes of blind equalization have been proposed. They are as follows:

1. *Higher-Order Statistics-Based Methods*: These methods obtain an estimate of the impulse response of the equivalent baseband channel from the estimates of the third or higher-order statistics of the received signal. Subsequently, using equations similar

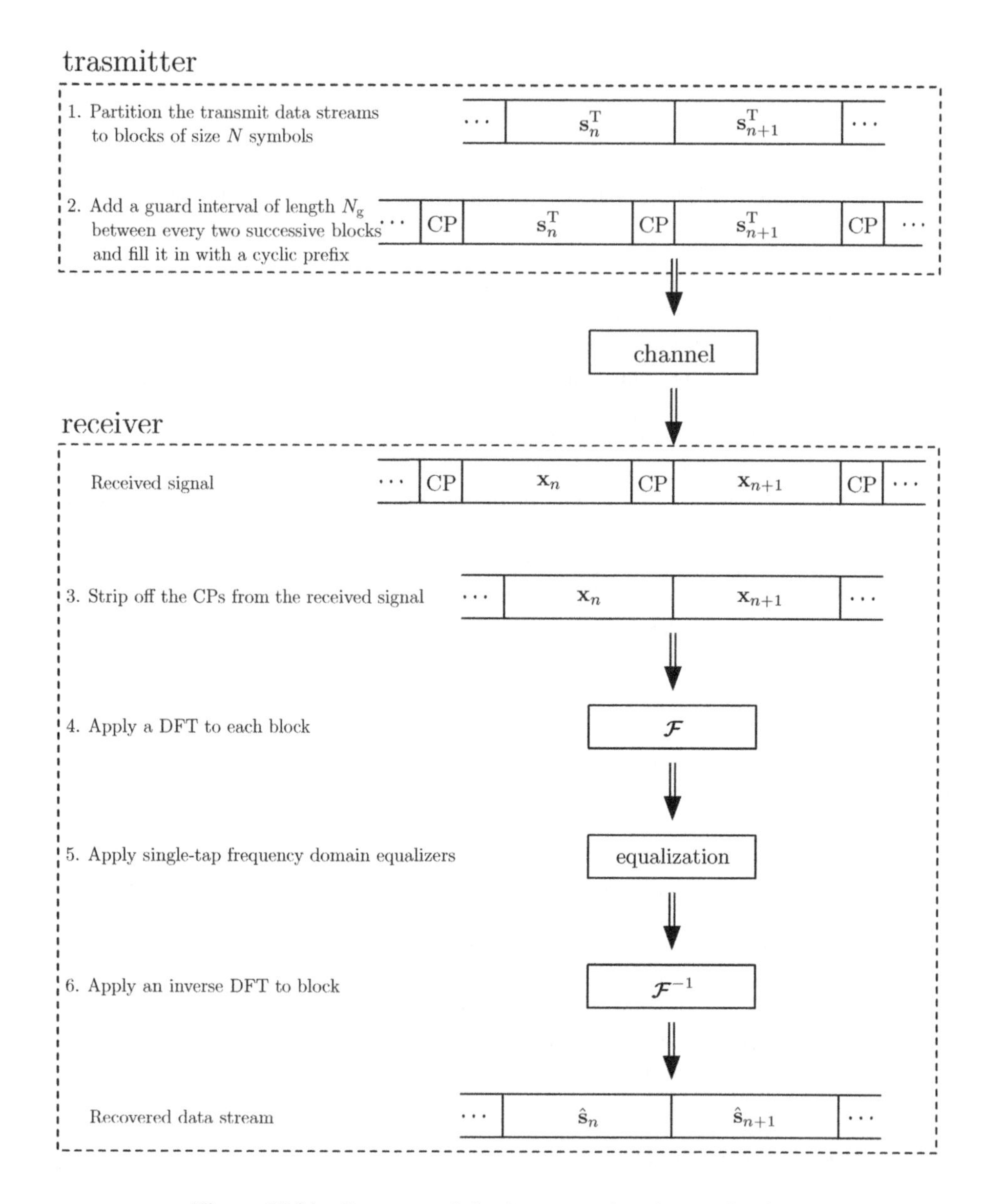

Figure 17.34 Summary of the frequency domain equalization.

to those developed in Section 17.4, the equalizer coefficients are calculated based on the estimated channel.

2. *Cyclostationary (Second-Order) Statistics-Based Methods*: These methods take note of the fact that in a digital communication system, the received signal is cyclostationary and its statistics repeat after every symbol interval. This property allows one to obtain an estimate of the impulse response of the equivalent baseband channel from the

estimates of the second-order statistics of the received signal. The equalizer coefficients are then calculated based on the estimated channel.

3. *Kurtosis-Based Methods*: These methods take note that the quality of the output of the equalizer can be measured by its kurtosis, a proper measure of its probability distribution. The kurtosis-based methods have also been categorized under the class of Bussgang algorithms (Haykin, 1996).

From these, the third method is more widely used in practice. In this section, we limit our discussion to the kurtosis based blind equalizers with an attempt to provide some intuitive understanding of how they work.

17.12.1 Examples of Kurtosis

The common kurtosis that has been used for channel equalization is defined as follows. For a random variable x with the probability distribution $p_x(x)$, the lth kurtosis of x is defined as

$$\kappa_l = \frac{E[|x|^{2l}]}{(E[|x|^l])^2} \tag{17.182}$$

where $E[\cdot]$ denotes the expected value. We also recall that

$$E[|x|^l] = \int_{-\infty}^{\infty} |x|^l p_x(x)\mathrm{d}x. \tag{17.183}$$

In the case where x is complex-valued, $p_x(x)$ spans over the real and imaginary parts of x and accordingly the integration on the right-hand side of Eq. (17.183) will be a double integral.

The common choices of l for blind equalizers are $l = 1$ and 2. Let us concentrate on the choice of $l = 2$. Figure 17.35 presents a few choices of the distribution function $p_x(x)$ and their associated kurtosis κ_2. When x is normally distributed, $\kappa_2 = 3$. One may also observe that κ_2 exceeds this value when $p_x(x)$ has a sharper peak and/or longer tails. On the other hand, for distributions with a more flat peak and/or shorter tails, $\kappa_2 < 3$. These observations may lead one to conclude that kurtosis provides a measure of the *peakedness* of the probability distribution of a random variable.

To develop further insight, which may prove useful for the rest of our discussions in this section, consider a few choices of pulse amplitude modulated (PAM) symbols. When the symbols $s(n)$ belong to a two-level PAM with equiprobable alphabet $\{+1, -1\}$, it is straightforward to show that (when $x = s(n)$) $\kappa_2 = 1$. When the symbols $s(n)$ belong to a four-level PAM with equiprobable alphabet $\{+3, +1, -1, -3\}$, one can show that $\kappa_2 = 1.64$. When the symbols $s(n)$ belong to an eight-level PAM with equiprobable alphabet $\{+7, +5, +3, +1, -1, -3, -5, -7\}$, one can show that $\kappa_2 = 1.7619$. Here, we make the following observations.

- As the alphabet size of the PAM symbols increases, its kurtosis approaches the value of 1.8, which is that of the uniform distribution; see Figure 17.35.
- Kurtosis κ_2 is a decreasing function as the number of symbols/level in a PAM decreases.

Similar observations are made when x is a complex-valued random variable.

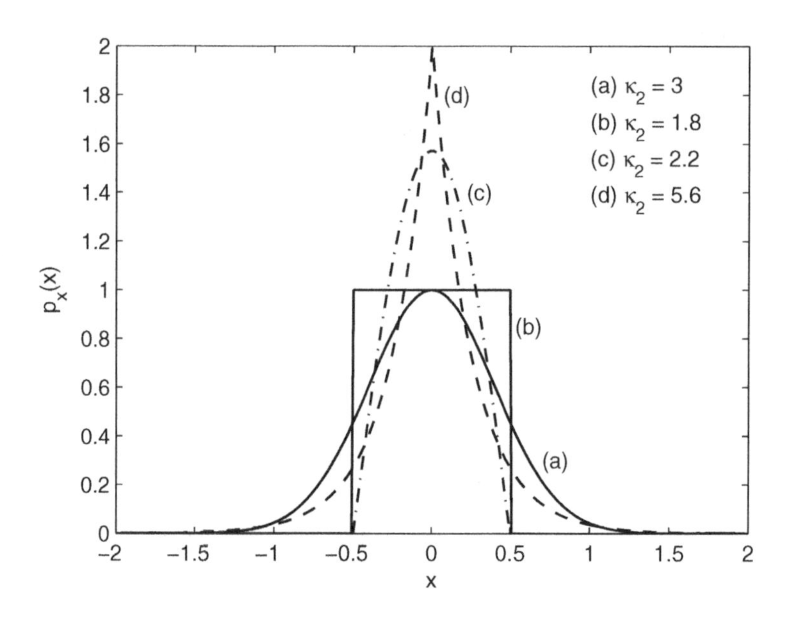

Figure 17.35 Probability distribution functions of a few random variables and their associated kurtosis values. (a) Gaussian distribution, $p_x(x) = \mathrm{e}^{-\pi x^2}$; (b) Uniform distribution, $p_x(x) = 1$ in the interval $-0.5 \leq x \leq 0.5$; (c) Half-cosine distribution, $p_x(x) = \frac{\pi}{2}\cos(\pi x)$ in the interval $-0.5 \leq x \leq 0.5$; (d) Exponential distribution, $p_x(x) = 2\mathrm{e}^{-4|x|}$.

17.12.2 Cost Function

Following the earlier observations, we may argue that to develop an adaptive blind equalization algorithm one can evaluate the kurtosis of the equalizer output and adjust the equalizer coefficients in the direction that reduces the kurtosis. The rationale here is that an unequalized output that suffers from a significant level of ISI has a Gaussian-like distribution, hence, a large kurtosis. As the adaptation proceeds and the equalizer coefficients are adjusted to reduce kurtosis of the resulting output, it will expectedly converge in the direction where the final outputs will be those of the transmitted PAM/QAM symbols, as the PAM/QAM symbols offer the smallest kurtosis.

Unfortunately, because of the rather complex form of the kurtosis expression (being the ratio of two expectations), the derivation of a kurtosis-based adaptation algorithm may not lead to a fruitful product. Instead, the following cost function has been suggested for development of the practical blind equalizers.

$$\xi = E[(|y(n)|^l - \gamma)^2] \tag{17.184}$$

where $y(n)$ is the equalizer output and γ is a constant whose value will be derived later.

In the case where the data symbols $s(n)$ are chosen from an alphabet where $|s(n)|$ is a constant, say, equal to r, the choice of the cost function

$$\xi = E[(|y(n)|^l - r^l)^2] \tag{17.185}$$

and its minimization leads to an equalizer output that satisfies $|y(n)| = r$, because this reduces ξ to zero; an absolute minimum for ξ. Of course, this ideal scenario can be only possible when the channel noise is zero and the equalizer length approaches infinity. Nevertheless, this argument provides some convincing reasoning on why the cost function (17.184) can be effective.

In the more general case where the data symbols $s(n)$ are from a general QAM constellation, where $|s(n)|$ varies with n, a dual to Eq. (17.185) is the cost function

$$\xi' = E[(|y(n)|^l - |s(n)|^l)^2]. \tag{17.186}$$

This of course is not a practical cost function, as it requires the knowledge of $|s(n)|$, a quantity that is unknown to the receiver by the virtue of the blind equalization. Instead, in Eq. (17.184), one may choose the parameter γ such that when ξ is minimized, $y(n)$ converges toward a process with the same statistical characteristics as the transmit symbols $s(n)$.

Recall that

$$y(n) = \mathbf{w}^{\mathrm{H}}(n)\mathbf{x}(n) \tag{17.187}$$

where $\mathbf{w}(n)$ and $\mathbf{x}(n)$ are the equalizer tap-weight vector and tap-input vector, respectively. Also, using Eq. (17.187), we obtain

$$|y(n)|^2 = \mathbf{w}^{\mathrm{H}}(n)\mathbf{x}(n)\mathbf{x}^{\mathrm{H}}(n)\mathbf{w}(n). \tag{17.188}$$

Substituting Eq. (17.188) in Eq. (17.184), we get

$$\xi = E[((\mathbf{w}^{\mathrm{H}}(n)\mathbf{x}(n)\mathbf{x}^{\mathrm{H}}(n)\mathbf{w}(n))^{l/2} - \gamma)^2]. \tag{17.189}$$

Upon convergence of the blind equalizer, when ξ is minimized, the equality

$$\nabla_{\mathbf{w}}^{\mathrm{C}}\xi = \mathbf{0} \tag{17.190}$$

should be satisfied. In Eq. (17.190), following the notation introduced in Chapter 3, $\nabla_{\mathbf{w}}^{\mathrm{C}}\xi$ denotes the gradient of ξ with respect to the complex-valued vector $\mathbf{w}$. Moreover, recalling from Chapter 3 the identity $\nabla_{\mathbf{w}}^{\mathrm{C}}\mathbf{w}^{\mathrm{H}}\mathbf{R}\mathbf{w} = 2\mathbf{R}\mathbf{w}$, which is valid for any arbitrary matrix $\mathbf{R}$, substituting Eq. (17.189) in Eq. (17.190) and letting $\mathbf{w}(n) = \mathbf{w}_{\mathrm{o}}$, we obtain

$$E[2(l/2)\mathbf{x}(n)\mathbf{x}^{\mathrm{H}}(n)\mathbf{w}_{\mathrm{o}}(\mathbf{w}_{\mathrm{o}}^{\mathrm{H}}\mathbf{x}(n)\mathbf{x}^{\mathrm{H}}(n)\mathbf{w}_{\mathrm{o}})^{(l-2)/2}((\mathbf{w}_{\mathrm{o}}^{\mathrm{H}}\mathbf{x}(n)\mathbf{x}^{\mathrm{H}}(n)\mathbf{w}_{\mathrm{o}})^{l/2} - \gamma)] = \mathbf{0}. \tag{17.191}$$

Next, multiplying Eq. (17.191) from left by $\mathbf{w}_{\mathrm{o}}^{\mathrm{H}}$, substituting $\mathbf{w}_{\mathrm{o}}^{\mathrm{H}}\mathbf{x}(n)$ by $y_{\mathrm{o}}(n)$ (the output $y(n)$ when $\mathbf{w}(n)$ is replaced by its optimum value $\mathbf{w}_{\mathrm{o}}$), and removing the constant factor $2(l/2)$, we get

$$E[|y_{\mathrm{o}}(n)|^l(|y_{\mathrm{o}}(n)|^l - \gamma)] = 0. \tag{17.192}$$

Solving this equation for γ, we obtain

$$\gamma = \frac{E[|y_{\mathrm{o}}(n)|^{2l}]}{E[|y_{\mathrm{o}}(n)|^l]}. \tag{17.193}$$

Upon convergence of the equalizer tap weights, assuming that they have converged to the desired values, $y_o(n) \approx s(n)$, hence, one may argue that the choice of

$$\gamma = \frac{E[|s(n)|^{2l}]}{E[|s(n)|^{l}]} \tag{17.194}$$

is reasonable.

17.12.3 Blind Adaptation Algorithm

Starting with the cost function ξ and following the LMS algorithm philosophy to replace ξ by its coarse estimate $\hat{\xi} = (|y(n)|^l - r^l)^2$ in a steepest descent recursive equation, we obtain

$$\mathbf{w}(n+1) = \mathbf{w}(n) - l\mu y^*(n)|y(n)|^{l-2}(|y(n)|^l - \gamma)\mathbf{x}(n) \tag{17.195}$$

where μ is a step-size parameter. For the choice of $l = 1$, Eq. (17.195) reduces to

$$\mathbf{w}(n+1) = \mathbf{w}(n) - \mu\frac{y^*(n)}{|y(n)|}(|y(n)| - \gamma)\mathbf{x}(n) \tag{17.196}$$

with $\gamma = E[|s(n)|^2]/E[|s(n)|]$. For the choice of $l = 2$, on the other hand, we obtain

$$\mathbf{w}(n+1) = \mathbf{w}(n) - 2\mu y^*(n)(|y(n)|^2 - \gamma)\mathbf{x}(n) \tag{17.197}$$

with $\gamma = E[|s(n)|^4]/E[|s(n)|^2]$.

One may note that the choice of $l = 2$ has a simpler form than the choice of $l = 1$. This is because the computation of $|y(n)| = \sqrt{|y(n)|^2} = \sqrt{y_R^2(n) + y_I^2(n)}$ is more involved than the computation of $|y(n)|^2 = y_R^2(n) + y_I^2(n)$ and besides, there is a divide by $|y(n)|$ in Eq. (17.196), which is absent in Eq. (17.197). Moreover, in practice $l = 2$ works better than the case of $l = 1$. The programs "`Blind_equalizer_l1.m`" and "`Blind_equalizer_l2.m`" for experimenting with the blind equalizers with choices of $l = 1$ and $l = 2$, respectively, are available on the accompanying website of this book.

Problems

P17.1 A naive pulse shape that satisfies the Nyquist condition is $p(t) = p_T(t) \star p_R(t) = \Lambda(t/T)$, where

$$\Lambda\left(\frac{t}{T}\right) = \begin{cases} 1 + t/T, & -T < t \leq 0 \\ 1 - t/T, & 0 < t < T \\ 0, & \text{otherwise.} \end{cases}$$

For this choice of $p(t)$ and an ideal channel $c(t) = \delta(t)$, consider the received signal

$$x(t) = \sum_{n=-\infty}^{\infty} s(n)p(t - nT).$$

(i) Assuming that the data symbols $s(n)$ are independent of one another and $E[|s(n)|^2] = \sigma_s^2$, evaluate and obtain an expression for

$$\rho(\tau) = E[|x(nT + \tau)|^2]$$

for $0 \leq \tau \leq T$.

(ii) Present a plot of $\rho(\tau)$. You should find that in contrast to the fundamental results presented in Section 17.3, where $\rho(\tau)$ was a biased sine wave, here, $\rho(\tau)$ has a different form. Explain what the source of this discrepancy is.

(iii) What is the value of τ that maximizes $\rho(\tau)$. For this choice of τ, find samples of $x(t)$ at the sampling times $nT + \tau$ and show such choice results in zero ISI. It is, thus, the optimum timing phase.

P17.2 Run the MATLAB script "`TxRxQAMELG.m`," available on the accompanying website, for a number of choices of the parameters μ, $\delta\tau$, and β, to gain a more in-depth understanding of the early-late gate timing recovery algorithm, and to confirm the results presented in Figures 17.16 and 17.17.

P17.3 Run the MATLAB script "`TxRxQAMGB.m`," available on the accompanying website, for a number of choices of the parameters μ and β, to gain a more in-depth understanding of the gradient-based gate timing recovery algorithm, and to confirm the result presented in Figure 17.18.

P17.4 (i) Following the same line of derivations to those in Section 17.4, derive the relevant equations for designing a DF equalizer with a half symbol-spaced feedfoward filter.

(ii) Add a new section to the end of the MATLAB script "`equalizer_eval.m`" to numerically examine the results of the derivations of (i).

(iii) By examining the modified script, confirm the results presented in Figures 17.23 and 17.24 and compare them with those of the fractionally spaced DF equalizer that you developed.

P17.5 Run the MATLAB script "`equalizer_eval.m`" on the accompanying website to confirm the results presented in Section 17.4.2. Run the script for a few additional choices of channel (that you choose) to develop further insight to the performance of the various equalizers.

P17.6 Develop a MATLAB program to study the convergence behavior of the LMS algorithm when applied to the channel equalization problem. Examine and present the learning curves of the algorithm for the more common and the less common channels that are listed in the MATLAB script "`equalizer_eval.m`." In particular, examine the equalizer lengths mentioned in Figures 17.23 and 17.24 and confirm that for the various choices of the timing phase, the mean-squared error of the equalizer converges toward the values presented in these figures.

P17.7 Repeat Problem P17.6 when the RLS algorithm is used for the adaptation of the equalizer tap weights.

P17.8 Repeat Problem P17.6 when the LMS-Newton algorithm is used for the adaptation of the equalizer tap weights.

P17.9 Repeat Problem P17.6 when the affine projection LMS algorithm is used for the adaptation of the equalizer tap weights.

P17.10 Present a detailed derivation of Eq. (17.111).

P17.11 Consider the system setup of Figure 17.28. Let the channel noise be absent and the impulse response of the channel be

$$\mathbf{h} = [0.1\ \ -0.2\ 0.3\ 0.7\ 1\ 0.8\ \ -0.5\ \ -0.2\ 0.1]^{\mathrm{T}}$$

(i) Find the equalizer vector $\mathbf{w}$ that truncates the channel response to a length of $L = 3$. Let $\Delta = 5$. To confirm your solution compare the sequences $h(n) \star w_n$ and d_n.

(ii) Develop an LMS algorithm for joint adaptation of $\mathbf{w}$ and $\mathbf{d}_{\mathrm{r}}$ and confirm that it converges to the solution you obtained in (i).

P17.12 Present a derivation of the solution (17.173).

P17.13 Figure 17.26 presents a structure for identifying the channel impulse response at the spacing T. Suggest a method of modifying this structure for identifying the channel impulse response at the spacing T/L, for an arbitrary integer L.

P17.14 Starting with the MATLAB script "`CyclicEq_eval.m`" on the accompanying website, extend it to confirm the results presented in Table 17.2.

P17.15 Develop a MATLAB code to obtain an estimate of the equivalent baseband impulse response of a channel by sending a single symbol $s(n) = \delta(n)$ and taking the received signal samples at the required rate, say, at the rate of $1/T$. Compare the result with what you obtain through the use of the MATLAB script "`CyclicEq_eval.m`" and confirm that both give the same channel estimate when channel noise is absent.

P17.16 In Figure 17.31, let the received signals $x_k(n)$ be related to the transmit signal $x(n)$ according to the equations

$$x_k(n) = g_k x(n) + \nu_k(n), \quad \text{for} \quad k = 0, 1, \ldots, M-1$$

where $\nu_k(n)$ is a white noise with variance σ_k^2. We wish to design an unbiased estimator for $x(n)$ using the received signal samples $x_k(n)$, for $k = 0, 1, \ldots, M-1$. To this end, one may first multiply both sides of the above equation by $1/\sigma_k$ to obtain

$$x'_k(n) = g'_k x(n) + \nu'_k(n), \quad \text{for} \quad k = 0, 1, \ldots, M-1$$

where $x'(n) = x(n)/\sigma_k$, $g'_k = g_k/\sigma_k$, and the noise terms $\nu'_k(n)$, for all k, have unit variance.

(i) Design the maximum ratio combiner for the unbiased minimum variance estimate of $x(n)$ from the signal samples $x'_k(n)$.
(ii) Using the result of (i) find the maximum ratio combiner for the unbiased minimum variance estimate of $x(n)$ from the signal samples $x_k(n)$.

P17.17 Consider a SIMO channel with two receive antennas. Figure P17.13 presents a block diagram of the complete communication system, including a multiple-input single-output (MISO) equalizer at the receiver. Let the channel noise processes, $\nu_0(n)$ and $\nu_1(n)$, be identically independent white noise with the variance σ_ν^2. Also, assume that the transmit data sequence $s(n)$ is a white random sequence with variance σ_s^2.

(i) Assuming that the equalizers $W_0(z)$ and $W_1(z)$ are unconstrained in duration show that the optimum choices of them that minimize the mean-squared error $E[|s(n) - \hat{s}(n)|^2]$ are given by

$$W_{0,o}(z) = \frac{\sigma_s^2 H_0^*(z)}{\sigma_s^2(|H_0(z)|^2 + |H_1(z)|^2) + \sigma_\nu^2}$$

and

$$W_{1,o}(z) = \frac{\sigma_s^2 H_1^*(z)}{\sigma_s^2(|H_0(z)|^2 + |H_1(z)|^2) + \sigma_\nu^2}$$

Hint: Refer to Chapter 3 to recall how to derive unconstrained Wiener–Hopf equations.

(ii) Simplify $W_{0,o}(z)$ and $W_{1,o}(z)$ for the case where $H_0(z) = 1 + z^{-1}$, $H_1(z) = 1 - z^{-1}$, $\sigma_s^2 = 1$, and $\sigma_\nu^2 = 0.1$.

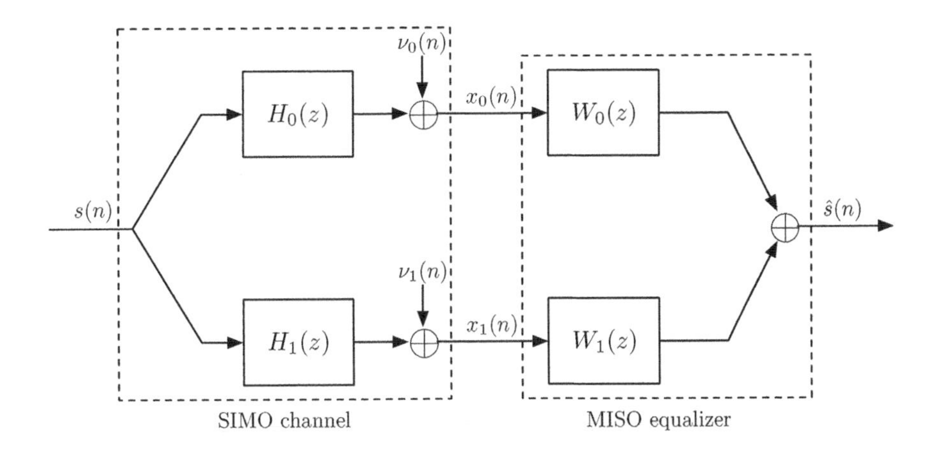

Figure P17.13

(iii) By presenting the plots of the magnitude responses of $W_{0,o}(z)$, and $W_{1,o}(z)$ and comparing them with those of $H_0(z)$ and $H_1(z)$, respectively, explain how maximum ratio combining is established in the numerical case of (ii).

P17.18 Following the steps summarized in Figure 17.34, develop a MATLAB script to implement a communication system that is equipped with a frequency domain equalizer and test its correct operation.

P17.19 The MATLAB script "`Blind_equalizer_l1.m`," on the accompanying website, allows you to test the performance of the kurtosis based equalizer for the case where the parameter $l = 1$. To explore the convergence behavior of the stochastic adaptation algorithm (17.195), examine this program for various constellation sizes and the step-size parameter μ. Note that for larger constellations, convergence may occur only after many hundreds of thousands or even millions of iterations. Hence, work patiently.

P17.20 Repeat Problem P17.19, when "`Blind_equalizer_l1.m`" is replaced by "`Blind_equalizer_l2.m`," also available on the accompanying website.

18

Sensor Array Processing

Sensor array processing has broad applications in diverse fields such as wireless communications, microphone arrays, radar, seismic and underwater explorations, radio astronomy, and radio surviellence. Direction of arrival (DOA) estimation and beam forming are two basic areas of sensor array processing that are common to its diverse applications. In this chapter, we review a number of signal processing techniques that over the past few decades have been developed for DOA estimation and/or adaptive beam forming.

Sensor array processing techniques have been developed and applied to both narrowband and broadband signals. However, the underlying concepts (which are common to both narrowband and broadband cases) are more easily understood when presented in the context of narrowband processing methods and algorithms. Noting this, most of the developments in this chapter are first presented in the context of narrowband signals. Extensions, which will be presented in the later parts of the chapter, will then be straightforward.

Implementation of a sensor array involves design and fabrication of sensor elements and the attached electronics, placement of the sensors according to a desired geometric topology, and development of a proper signal processing algorithm. The subject of this chapter is related to the latter, that is, development of signal processing algorithms for sensor arrays. In this development, often idealistic assumptions are made about the fabricated sensors and their position accordingly to the assumed geometric topology. This may not be the case in practice. Hence, some tolerance should be admitted for each fabricated sensor element. Also, the placement of the sensor elements according to the desired geometric topology may not be realized exactly. Clearly, these variations have to be considered in the development of the signal processing algorithms. In other words, robustness has to be considered when an algorithm is developed for a sensor array. In this chapter, we first develop a number of signal processing algorithms/techniques, assuming that all the sensor elements match their assumed, idealized model and also are positioned exactly according to a desired geometric topology. The issue of robustness is discussed in a later section of the chapter, where some modifications to the basic algorithms/techniques that result in robust sensor arrays are also presented.

Adaptive Filters: Theory and Applications, Second Edition. Behrouz Farhang-Boroujeny.

18.1 Narrowband Sensor Arrays

18.1.1 Array Topology and Parameters

The simplest and the most considered array geometric topology is the so-called *linear array*. In a linear array, the array elements are positioned on a straight line and often the elements are equally spaced. Figure 18.1 presents one such topology with M elements. The elements are spaced at a distance of l m. Following the presentations in Chapters 3 and 6, here too we have chosen to present the array elements as omni-directional antennas. However, we note that this is only for convenience and the results presented in this chapter are applicable to the variety of applications mentioned at the beginning of this chapter. Clearly, other array geometric topologies are also possible. Two examples are presented in Figures 18.2 and 18.3. These topologies may be found useful in cases where the space for placement of the sensor elements is limited.

In Figure 18.1, it is assumed that a plane-wave signal is impinging the sensor elements from an angle of θ. In this section, we assume that $x(n) = \alpha(n)\cos(\omega_0 n + \phi)$ is a narrowband signal centered around frequency ω_0. The fact that $x(n)$ is narrowband implies that $\alpha(n)$ varies slowly with the time index n. Moreover, in Figure 18.1, $\underline{x}_0(n)$ through $\underline{x}_{M-1}(n)$ are phasor representations of the received signal at the array element outputs. As discussed in Section 6.10, such phasor signals are recovered at each sensor element output and will be available to the sensor array processor, that is, the algorithms that are discussed in this chapter. The parameters w_0 through w_{M-1} are a set of complex-valued coefficients that we refer to as the array gains. Accordingly, we define the array *gain vector*

$$\mathbf{w} = [w_0^* \ w_1^* \ \cdots \ w_{M-1}^*]^{\mathrm{H}} \tag{18.1}$$

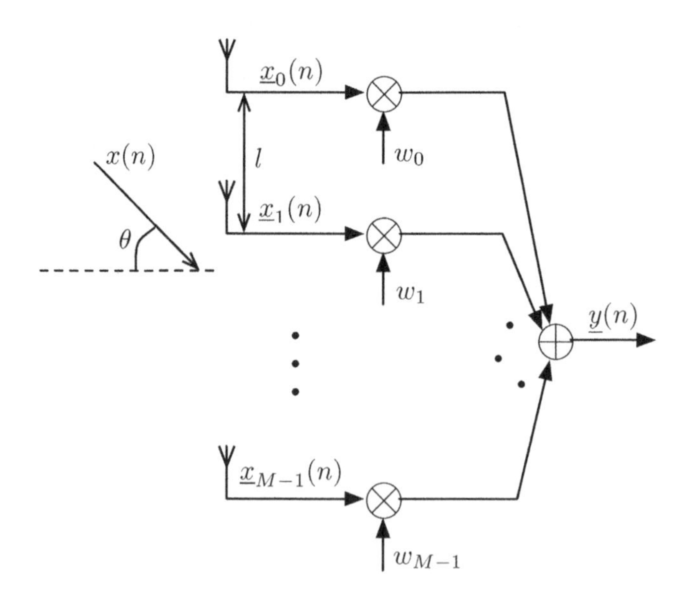

Figure 18.1 Linear geometric topology for sensor arrays.

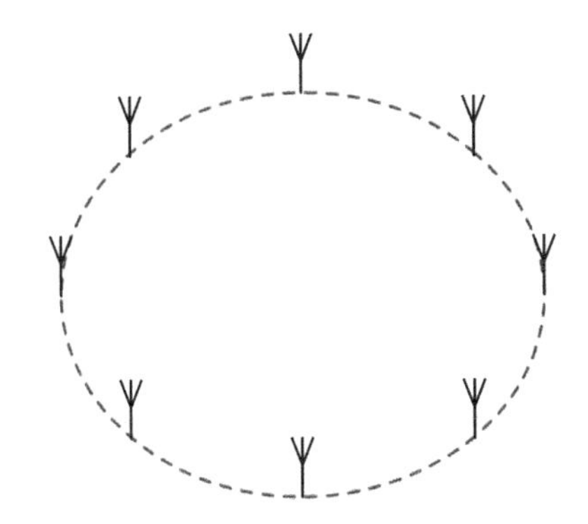

Figure 18.2 Circular geometric topology for sensor arrays.

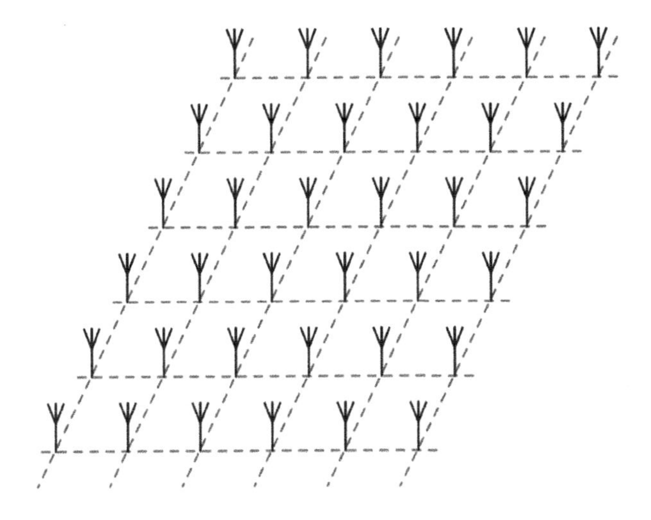

Figure 18.3 Regular grid geometric topology for sensor arrays.

where the superscript H denotes complex-conjugate transpose or Hermitian. Also, we define the array *phasor vector* (baseband signal vector)

$$\mathbf{x}(n) = [\underline{x}_0(n)\ \underline{x}_1(n)\ \cdots\ \underline{x}_{M-1}(n)]^{\mathrm{T}} \tag{18.2}$$

where the superscript T denotes transpose.

Let the phasor signal at the sensor element k be written as

$$\underline{x}_k(n) = \alpha(n)\mathrm{e}^{j\phi_k} \tag{18.3}$$

Considering the fact that for the case presented in Figure 18.1, the plane-wave signal $x(n)$ reaches the element 0 first and $(l \sin\theta)/c$ samples later reach the element 1, where c is the propagation speed. Similarly, for the rest of the elements, one will find that

$$\phi_k - \phi_{k-1} = \frac{l \sin\theta}{c} \times \frac{\omega_0}{T} \tag{18.4}$$

where T is the sampling period. The reader should refer to Section 6.4.4 for details of this derivation. Moreover, if we assume that the spacing between the sensor elements, l, is equal to one half of the wavelength of the plane-wave, Eq. (18.4) reduces to

$$\phi_k - \phi_{k-1} = \pi \sin\theta \tag{18.5}$$

Hence, $\underline{x}_0(n) = \alpha(n)\mathrm{e}^{j\phi_0}$, $\underline{x}_1(n) = \alpha(n)\mathrm{e}^{j(\phi_0+\pi\sin\theta)}, \cdots,$ and $\underline{x}_{M-1}(n) = \alpha(n)$ $\mathrm{e}^{j(\phi_0+(M-1)\pi\sin\theta)}$. Using these results, one will find that

$$\mathbf{x}(n) = \underline{x}_0(n)\mathbf{s}(\theta) \tag{18.6}$$

where

$$\mathbf{s}(\theta) = \begin{bmatrix} 1 \\ \mathrm{e}^{j\pi\sin\theta} \\ \vdots \\ \mathrm{e}^{j(M-1)\pi\sin\theta} \end{bmatrix} \tag{18.7}$$

The vector $\mathbf{s}(\theta)$ is called the *steering vector*.

The array output, after applying the gains w_i's, is given by

$$y(n) = \mathbf{w}^{\mathrm{H}}\mathbf{x}(n) \tag{18.8}$$

Substituting Eq. (18.6) in Eq. (18.8), we obtain,

$$y(n) = \mathcal{G}(\theta)\underline{x}_0(n) \tag{18.9}$$

where

$$\begin{aligned} \mathcal{G}(\theta) &= \mathbf{w}^{\mathrm{H}}\mathbf{s}(\theta) \\ &= \sum_{i=0}^{M-1} w_i \mathrm{e}^{ji\pi\sin\theta} \end{aligned} \tag{18.10}$$

is called *array gain* for a narrowband signal arriving at the angle θ. Note that, as $\underline{x}_0(n) = \alpha(n)\mathrm{e}^{j\phi_0}$, within a phase error, $\phi_0 - \phi$, $\underline{x}_0(n)$ is the same as $\alpha(n)$. Hence, $\mathcal{G}(n)$ accurately reflects the gain imposed on the incoming signal after passing through the array and summed up through the array gains w_i's. In particular, if we let $\phi_0 = \phi$, Eq. (18.6) simplifies to

$$\mathbf{x}(n) = \alpha(n)\mathbf{s}(\theta) \tag{18.11}$$

18.1.2 Signal subspace, noise subspace, and spectral factorization

Let a set of L plane-wave signals $x_0(n), x_1(n), \cdots, x_{L-1}(n)$ with the direction angles θ_0, $\theta_1, \cdots, \theta_{L-1}$, respectively, are received by a linear array as in Figure 18.4. This clearly is an extension to Figure 18.1. Recalling Eq. (18.11), the received signal vector at the array outputs is given by

$$\mathbf{x}(n) = \sum_{k=0}^{L-1} \alpha_k(n)\mathbf{s}(\theta_k) + \mathbf{v}(n) \tag{18.12}$$

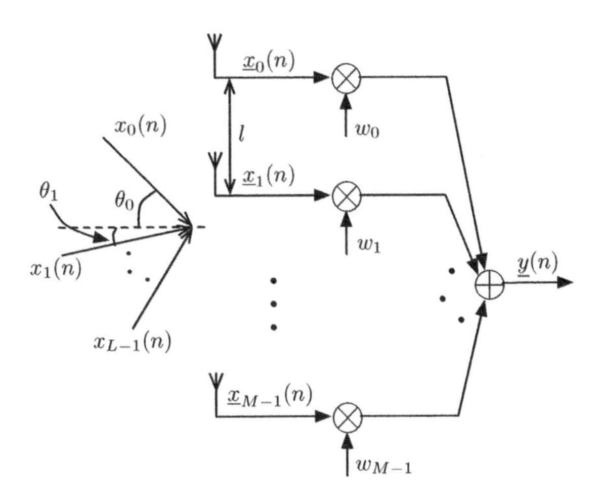

Figure 18.4 A linear array receiving L plane-wave signals.

where $\mathbf{v}(n)$ is an additive noise vector, and $\mathbf{s}(\theta_0)$, $\mathbf{s}(\theta_1)$, $\cdots$, $\mathbf{s}(\theta_{L-1})$ are the steering vectors of the respective signals, viz.,

$$\mathbf{s}(\theta_k) = \begin{bmatrix} 1 \\ e^{j\pi \sin\theta_k} \\ \vdots \\ e^{j(M-1)\pi \sin\theta_k} \end{bmatrix}, \qquad k = 0, 1, \cdots, L-1 \tag{18.13}$$

We assume that the elements of $\mathbf{v}(n)$ are a set of zero-mean identically distributed Gaussian random variables. Hence,

$$E[\mathbf{v}(n)\mathbf{v}^{\mathrm{H}}(n)] = \sigma_\nu^2 \mathbf{I} \tag{18.14}$$

where σ_ν^2 is the variance of each element of $\mathbf{v}(n)$ and $\mathbf{I}$ is the identity matrix.

Using Eqs. (18.12) and (18.14) and assuming that the baseband signals $\alpha_k(n)$, for $k = 0, 1, \cdots, L-1$, are independent of one another, the correlation matrix of the array signal vector, $\mathbf{R} = E[\mathbf{x}(n)\mathbf{x}^{\mathrm{H}}(n)]$, is obtained as

$$\mathbf{R} = \mathbf{R}_s + \mathbf{R}_\nu \tag{18.15}$$

where

$$\mathbf{R}_s = \sum_{k=0}^{L-1} P_k \mathbf{s}(\theta_k)\mathbf{s}^{\mathrm{H}}(\theta_k) \tag{18.16}$$

is the portion of $\mathbf{R}$ arising from the signal components,

$$\mathbf{R}_\nu = \sigma_\nu^2 \mathbf{I} \tag{18.17}$$

is the portion of $\mathbf{R}$ arising from the noise components, and $P_k = E[|\alpha_k(n)|^2]$.

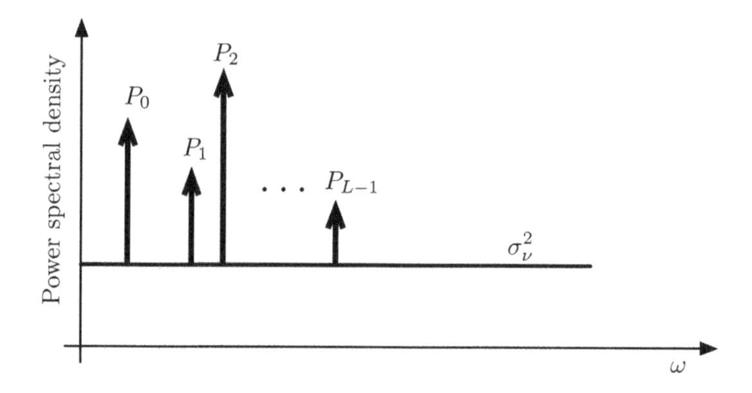

Figure 18.5 Power spectral density of the process $\{x(n)\}$ consisting of a white noise plus a multi-tone signal.

A thorough understanding of an eigenanalysis of the correlation matrix $\mathbf{R}$ is essential and very helpful to our discussion in the rest of this chapter. To this end, consider the case where $L = 1$ and note that, for this case, $\mathbf{R}$ has the same form as the correlation matrix $\mathbf{R}$ in Example 4.2 (of Chapter 4). Moreover, recall that for the latter, $\mathbf{R}$ is the correlation matrix associated with a random process $\{x(n)\}$ whose power spectral density is presented in Figure 4.3. For the more general case of Eq. (18.15), one may think of $\mathbf{R}$ as the correlation matrix associated with a random process $\{x(n)\}$, whose power spectral density is presented in Figure 18.5, where the tones are located at the frequencies $\omega_k = \pi \sin\theta_k$, for $k = 0, 1, \cdots, L-1$ and has the respective power levels $P_0, P_1, \cdots, P_{L-1}$.

Next, recalling the minimax theorem that was introduced in Chapter 4 and the related discussions in the same chapter, one will find that if $L < M$ and the eigenvalues of $\mathbf{R}$ are written in the descending order $\lambda_0, \lambda_1, \cdots, \lambda_{M-1}$, the last $M - L$ eigenvalues of $\mathbf{R}$ are all the same and equal to σ_ν^2. Alternatively, one may note that $\mathbf{R}_s$ is a rank L matrix of size $M \times M$; hence $M - L$ of its eigenvalues are equal to zero. If we call the nonzero eigenvalues of $\mathbf{R}_s$, $\lambda_{s,0}$ through $\lambda_{s,L-1}$, we will find that the eigenvalues of $\mathbf{R}$ are

$$\lambda_i = \begin{cases} \lambda_{s,i} + \sigma_\nu^2, & 0 \le i \le L-1 \\ \sigma_\nu^2, & L \le i \le M-1 \end{cases} \tag{18.18}$$

Moreover, we note that the set of steering vectors $\mathbf{s}(\theta_k)$ span a subspace of the M dimensional Euclidean space. We call this subspace the *signal subspace*, because of obvious reasons. Accordingly, the subspace orthogonal to the signal subspace is called the *noise subspace*.

Another important point that is derived from the above observations is that the first L eigenvectors of $\mathbf{R}$ are in the signal subspace and, hence, those that are associated with the eigenvalues σ_ν^2 reside in the noise subspace. Finally, recalling the unitary similarity transformation that was introduced in Chapter 4, $\mathbf{R}$ may be expanded as

$$\mathbf{R} = \mathbf{Q}\mathbf{\Lambda}\mathbf{Q}^{\mathrm{H}} \tag{18.19}$$

where $\boldsymbol{\Lambda}$ is the diagonal matrix of the eigenvalues of $\mathbf{R}$, listed in Eq. (18.18), and the matrix $\mathbf{Q}$ is made-up of a set of unit-length orthogonal eigenvectors of $\mathbf{R}$. Moreover, the first L columns of $\mathbf{Q}$ expand the signal subspace of $\mathbf{R}$ and the rest of the columns of $\mathbf{Q}$ expand the noise subspace of $\mathbf{R}$. Also, because of reasons that will become clear shortly, Eq. (18.19) is often referred to as *spectral factorization.*

18.1.3 Direction of Arrival Estimation

Given the observed samples of the phasor signals in Figure 18.4, that is, the sample vectors $\mathbf{x}(n)$ for a range of n, we wish to find the DOA of plane-wave signals $x_0(n)$, $x_1(n)$, $\cdots$, $x_{L-1}(n)$. When stated in the context of the presentation in Figure 18.5, the goal is to find the location of the spectral components. This, clearly, can be thought of as a spectral estimation problem. In the context of this chapter, this may be referred to as a *spatial* spectral estimation problem. The challenge here is that each observed sample of $\mathbf{x}(n)$ has a limited length equal to the array size. This limited, and possibly very short, length of $\mathbf{x}(n)$ reduces the resolution of a conventional spectral estimator, if such estimators are to be used. Hence, special signal analysis schemes that are tuned to the structure of the observed vector $\mathbf{x}(n)$, hence, expectedly, will improve the accuracy of the results, should be adopted. In the sequel, we introduce a number of such signal analysis schemes/algorithms. These algorithms all start with an estimate of $\mathbf{R}$ or its inverse. Such estimates can be obtained by averaging over a number of samples $\mathbf{x}(n)$ in an interval $n_1 \leq n \leq n_2$, viz.,

$$\hat{\mathbf{R}} = \frac{\sum_{n_1}^{n_2} \mathbf{x}(n)\mathbf{x}^{\mathrm{H}}(n)}{n_2 - n_1 + 1} \tag{18.20}$$

When the DOAs are changing over time and one wishes to track such changes, the estimate of $\mathbf{R}$ can be updates using the recursive update equation

$$\hat{\mathbf{R}}(n+1) = \lambda\mathbf{R}(n) + (1-\lambda)\mathbf{x}(n)\mathbf{x}^{\mathrm{H}}(n) \tag{18.21}$$

where λ is a forgetting factor, a number close to but smaller than one. Also, when needed, the estimate of $\mathbf{R}^{-1}$ can be updated using the procedure that was developed in Chapter 12, using the matrix inversion lemma.

With this background, we are now ready to present the details of a number of DOA algorithms. A reader that may be interested in a wider variety of the DOA algorithms may refer to, for example, Godara (1997).

Bartlett Method

The most straightforward implementation of this class of algorithms is to calculate the spatial spectrum of $\mathbf{x}(n)$, simply, by finding the inner product of the samples $\mathbf{x}(n)$ with the steering vector $\mathbf{s}(\theta)$, squaring the result and averaging over the range $n_1 \leq n \leq n_2$, viz.,

$$S(\theta) = \frac{\sum_{n_1}^{n_2} |\mathbf{s}^{\mathrm{H}}(\theta)\mathbf{x}(n)|^2}{n_2 - n_1 + 1} \tag{18.22}$$

This procedure is called *Bartlett* method, a name that has been borrowed from the spectral estimation literature. Note that the spatial spectrum $S(\theta)$ is a function of DOA, the angle θ. Recalling Eq. (18.20), Eq. (18.22) can be rearranged as

$$S(\theta) = \mathbf{s}^{\mathrm{H}}(\theta)\hat{\mathbf{R}}\mathbf{s}(\theta) \tag{18.23}$$

The DOAs are estimated by presenting the spatial spectrum $S(\theta)$ as a function of θ and finding the peaks of it.

Looking at this algorithm in the context of signal analysis and classical spectral estimation, it is equivalent of passing the underlying signal through a bank of filters that are built based on a prototype filter with a rectangular impulse response. The signal powers at the outputs of the filter bank are taken as an estimate of the power spectral density of the input signal to the filter bank. A plot of the magnitude response of the prototype filter (which makes the zeroth subband of the filter bank) and a modulated version of it (which makes another subband of the filter bank) are presented in Figure 18.6. Considering these plots, one may note that this direct signal analysis suffers from two major problems:

- The relatively large side lobes of the prototype filter magnitude response results in a leakage of the signal power of each tone to other portions of the frequency band. In the context of DOA estimation, this results in some interference among the signals arriving from different directions, hence, some error/bias in the DOA estimations.
- For a linear array, the width of the main lobe of the prototype filter response is proportional to the inverse of the length of the array. Hence, unless the array size is large, this method suffers from a low resolution in distinguishing between a pair of sources with close DOAs.

To clarify these points, we consider a ten-element array that impinges with three plane-wave signals from the angles $\theta_0 = 25^\circ$, $\theta_1 = 28^\circ$ and $\theta_2 = -30^\circ$. We assume that $P_0 = P_1 = P_2 = 1$, and $\sigma_\nu^2 = 0.1$. Taking 100 randomly and independently generated samples of $\mathbf{x}(n)$, $\hat{\mathbf{R}}$ is calculated according to Eq. (18.20) and the result is substituted in Eq. (18.23)

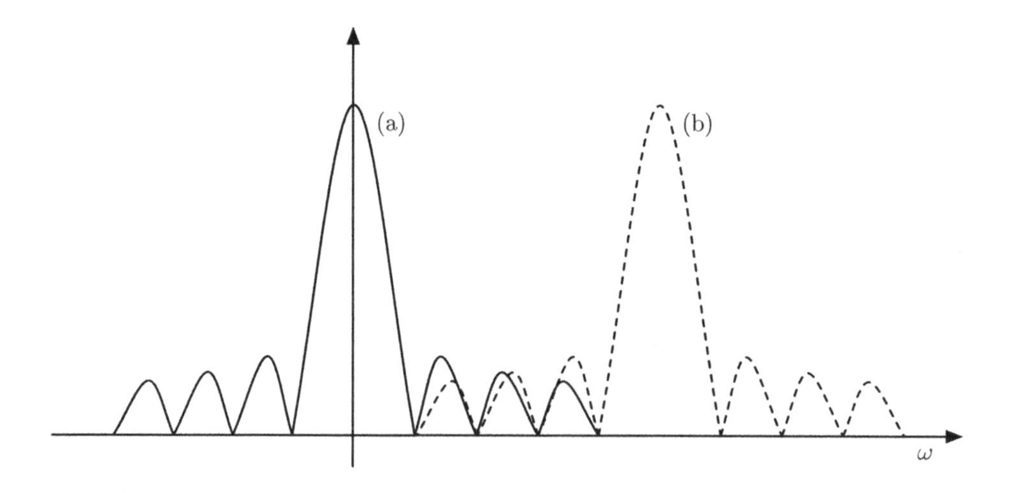

Figure 18.6 Magnitude response of subband filters in a filter bank system with Bartlett prototype filter. (a) Prototype filter (the 0th subband). (b) An arbitrary subband.

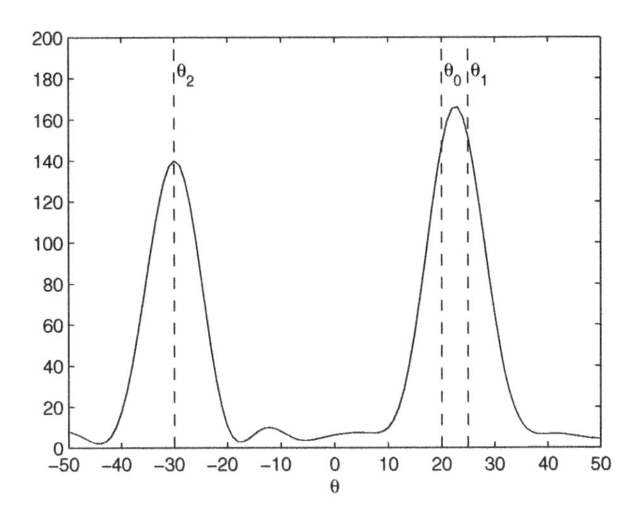

Figure 18.7 Spatial spectrum $S(\theta)$ according to the Bartlett method, for a linear array of size 10 elements, when impinges with three equal power plane-wave signals from the angles $\theta_0 = 25^\circ$, $\theta_1 = 28^\circ$ and $\theta_2 = -30^\circ$. The signals are at 20 dB above the noise level.

to evaluate the spatial spectrum $S(\theta)$. The result is presented in Figure 18.7. As observed, $S(\theta)$ has a clear maximum at $\theta = \theta_2$; however, the responses arising from the DOAs θ_0 and θ_1 have overlapped with each other and thus may be confused with a single DOA at the mid-point between θ_0 and θ_1.

MVDR Algorithm

MVDR, or Minimum Variance Distortionless Response, is an elegant DOA estimation method that resolves the low resolution problem of the Bartlett method by taking the following approach. For each angle of arrival, θ, an optimum tap-weight vector $\mathbf{w}_o(\theta)$ that has a gain of unity in the look direction (angle θ) and minimizes the signal power from all other directions is found and the spatial spectrum

$$S(\theta) = \mathbf{w}_o^{\mathrm{H}}(\theta)\hat{\mathbf{R}}\mathbf{w}_o^{\mathrm{H}}(\theta) \tag{18.24}$$

is calculated accordingly. Subsequently, the DOAs are estimated by locating the peaks of $S(\theta)$.

To find $S(\theta)$ here, we may take the following steps. We wish to solve the following constraint optimization problem.

Find $\mathbf{w}$ that minimizes the quadratic function $\mathbf{w}^{\mathrm{H}}\hat{\mathbf{R}}\mathbf{w}$, subject to the constraint $\mathbf{s}^{\mathrm{H}}(\theta)\mathbf{w} = 1$.

This problem can be easily solved by using the method of Lagrange multipliers and taking the steps presented in Section 6.11. It leads to the solution

$$\mathbf{w}_o(\theta) = \frac{\hat{\mathbf{R}}^{-1}\mathbf{s}(\theta)}{\mathbf{s}^{\mathrm{H}}(\theta)\hat{\mathbf{R}}^{-1}\mathbf{s}(\theta)} \tag{18.25}$$

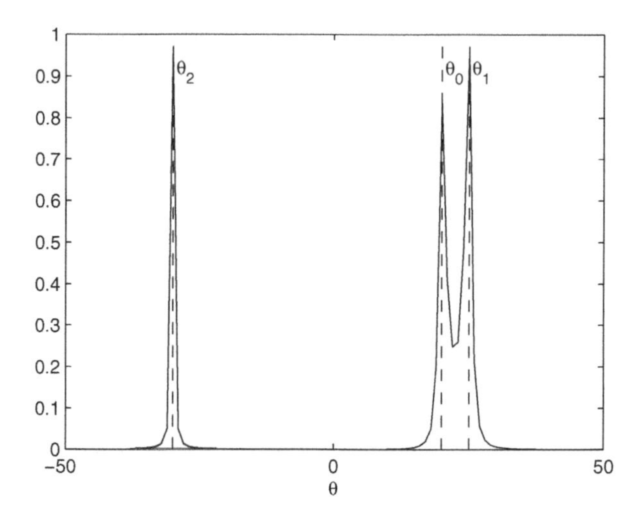

Figure 18.8 Spatial spectrum $S(\theta)$ according to the MVDR algorithm, for a linear array of size 10 elements, when impinged with three equal power plane-wave signals from the angles $\theta_0 = 25^\circ$, $\theta_1 = 28^\circ$, and $\theta_2 = -30^\circ$. The signals are at 20 dB above the noise level.

Substituting Eq. (18.25) into Eq. (18.24), we obtain

$$S(\theta) = \frac{1}{\mathbf{s}^{\mathrm{H}}(\theta)\hat{\mathbf{R}}^{-1}\mathbf{s}(\theta)} \tag{18.26}$$

Figure 18.8 presents the result of applying the MVDR algorithm to finding the DOA of sources for the case that was introduced above. Comparing this result with the one presented in Figure 18.7, one can see that the MVDR algorithm performs significantly better than the Bartlett method. It has a surprisingly high resolution and identifies all DOAs accurately. Nevertheless, we should note that MVDR algorithm may not perform as well when the impinging signals are comparable with the noise level. The MUSIC algorithm that is introduced in the following section solves this problem to a great degree.

MUSIC Algorithm

MUSIC, or MUltiple SIgnal Classification, is another elegant algorithm that can also achieve high resolution DOA estimation. Moreover, it outperforms MVDR, particularly in more noisy environments. As we will find, excellent performance of MUSIC algorithm is attributed to the fact that it makes almost maximum use of the structure of the sampled vector $\mathbf{x}(n)$ in estimating the DOAs.

Recalling the signal structure $\mathbf{x}(n)$ in Eq. (18.12) and the spectral factorization (18.19), MUSIC algorithm is implemented by taking the following steps.

1. Given the samples of $\mathbf{x}(n)$, calculate the $\hat{\mathbf{R}}$ according to Eq. (18.20).
2. Find the eigenvalues $\lambda_0, \lambda_1, \cdots, \lambda_{M-1}$ of $\hat{\mathbf{R}}$ and the respective eigenvectors $\mathbf{q}_0, \mathbf{q}_1, \cdots, \mathbf{q}_{M-1}$. Also assume that these are sorted such that $\lambda_0 \geq \lambda_1 \geq \cdots \geq \lambda_{M-1}$.

3. If the number of plane-wave signals, L, impinging the array are known, form the $M \times (M - L)$ noise subspace matrix $\mathbf{Q}_\nu$ that consists of $\mathbf{q}_{M-L}$ through $\mathbf{q}_{M-1}$ at its columns.

   If L is unknown, find an estimate of it by identifying the last eigenvalues of $\hat{\mathbf{R}}$ that are approximately equal. The number of these eigenvalues is equal to $M - L$. Once this is found, form the noise subspace matrix $\mathbf{Q}_\nu$.
4. Evaluate the spatial spectrum

$$S(\theta) = \frac{1}{\|\mathbf{Q}_\nu^{\mathrm{H}}\mathbf{s}(\theta)\|^2} \tag{18.27}$$

5. The L largest peaks $S(\theta)$ are the estimates of DOAs.

To explain how MUSIC algorithm makes maximum use of the sampled vectors $\mathbf{x}(n)$, we first note that any steering vector $\mathbf{s}(\theta)$ that matches one of the wave-plane signals will be orthogonal to the noise subspace. Hence, it will be orthogonal to the columns of $\mathbf{Q}_\nu$ and, thus, $\|\mathbf{Q}_\nu^{\mathrm{H}}\mathbf{s}(\theta)\|^2$ will be a small quantity. This in turn implies that the spatial spectrum (18.27) will have a large amplitude for any steering vector that falls within the signal subspace.

Figure 18.9 presents the result of applying the MUSIC algorithm to finding the DOA of sources for the case that was introduced above and tested with Bartlett and MVDR algorithm. We previously found that MVDR algorithm significantly performs better than Bartlett algorithm. Comparing the results of Figure 18.8 and Figure 18.9, it may appear that MUSIC algorithm outperforms MVDR algorithm only by a small margin. However, if the noise variance is increased, the difference between the two algorithms will become more obvious. The result of one such experiment in which $\sigma_\nu^2 = 0.01$ is replaced by the

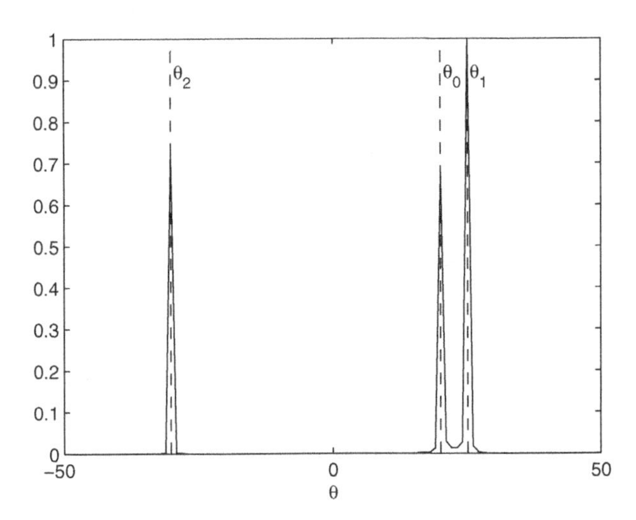

Figure 18.9 Spatial spectrum $S(\theta)$ according to the MUSIC algorithm, for a linear array of size 10 elements, when impinged with three equal power plane-wave signals from the angles $\theta_0 = 25^\circ$, $\theta_1 = 28^\circ$ and $\theta_2 = -30^\circ$. The signals are at 20 dB above the noise level.

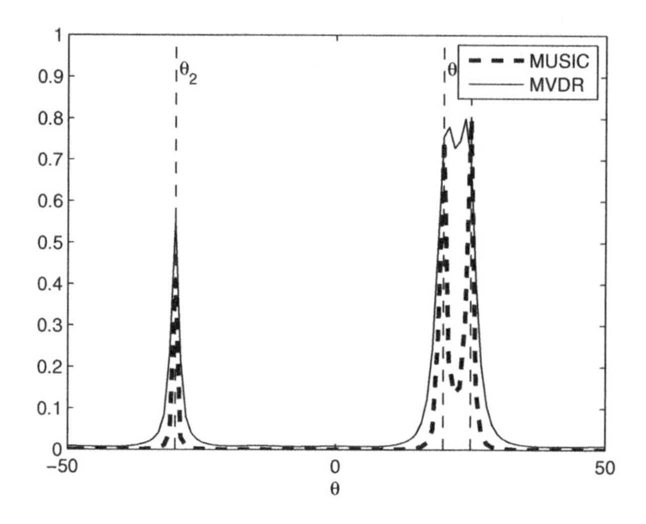

Figure 18.10 Comparison of the MVDR and MUSIC algorithms in a high noise environment.

increased value $\sigma_\nu^2 = 0.1$ is presented in Figure 18.10. The difference is obvious and significantly in favor of the MUSIC algorithm. In fact, in this case, the MVDR algorithm fails to clearly distinguish θ_0 and θ_1.

18.1.4 Beamforming Methods

Consider the linear array of Figure 18.1. As discussed above, when this array is impinged by a plane-wave from the direction θ, it will have a gain $\mathcal{G}(\theta)$ that is given by Eq. (18.10). A plot of $|\mathcal{G}(\theta)|$ as a function of the angle of arrival θ is called *array* or *beam pattern.* The problem of beamforming is that of selecting the tap-weight vector $\mathbf{w}$ that results in a beam pattern with some desired characteristics.

Conventional Beamforming

In the conventional beamforming, tuned to a *look* direction θ_0, we simply set

$$\mathbf{w} = \frac{1}{M}\mathbf{s}(\theta_0) \tag{18.28}$$

where $\mathbf{s}(\theta)$ is the steering vector, defined in Eq. (18.7). Substituting Eq. (18.28) in Eq. (18.10), we obtain

$$\mathcal{G}(\theta) = \frac{1}{M}\sum e^{ji\pi(\sin\theta - \sin\theta_0)} \tag{18.29}$$

Example 18.1

Figure 18.11 presents a plot of the beam pattern of Eq. (18.29), when $M = 10$ and $\theta_0 = 45^\circ$. As expected, the beam pattern is directed towards the look direction $\theta = \theta_0$.

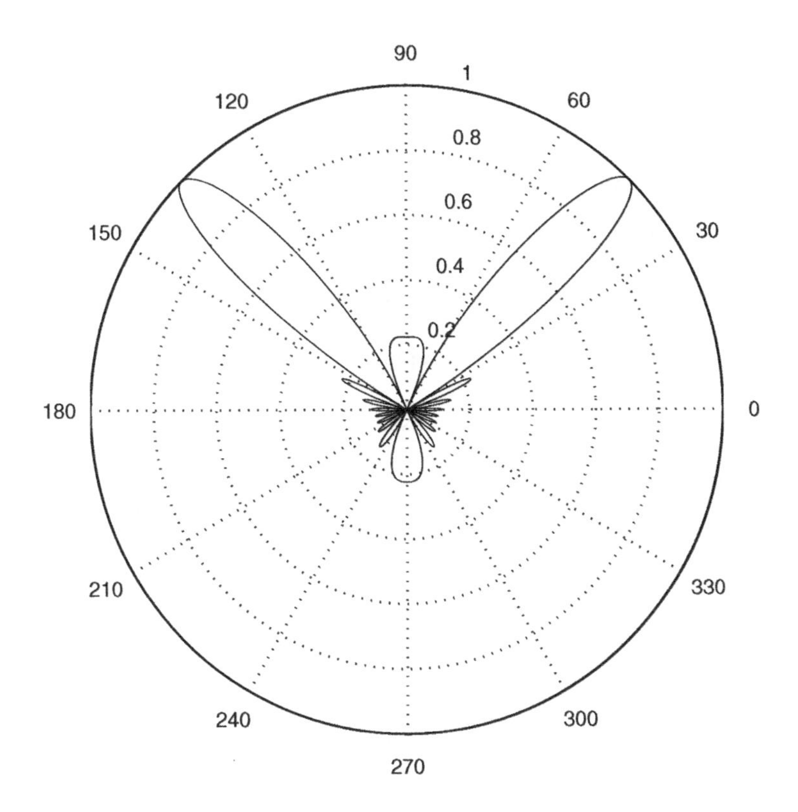

Figure 18.11 Beam pattern of a conventional beamformer, for a ten element linear array.

It has a gain of unity in this direction, and its gain is smaller than unity in other directions. Moreover, as Eq. (18.29) implies, here, the beam pattern is symmetric with respect to the angle 90° and, hence, it also sees a plane-wave that may be arriving from the angle $90^\circ + 45^\circ = 135^\circ$.

The conventional beamforming satisfies certain optimality, and thus its derivations with that respect may prove instructive. Let us assume that the array is impinged with a plane-wave with the angle of arrival θ_0 and there is no other impinging signal. However, there is an additive noise with variance σ_ν^2 at each element of the array. For this setup, the beamformer tap-weight vector $\mathbf{w}$ can be optimally chosen as follows. Find $\mathbf{w}$ that minimizes the cost function

$$\xi = \sigma_\nu^2 \mathbf{w}^{\mathrm{H}} \mathbf{w} \tag{18.30}$$

subject to the constraint

$$\mathbf{w}^{\mathrm{H}} \mathbf{s}(\theta_0) = 1 \tag{18.31}$$

This constraint minimization will make perfect sense, if we note that the cost function ξ is the noise variance at the beamformer output, and the constraint (18.31) allows the plane-wave signal to go through the beamformer with a gain of unity. This clearly maximizes the signal-to-noise ratio at the beamformer output and, thus, is optimum in that sense.

Solution to the above constraint minimization can be easily found by using the method of the Lagrange multipliers, following the same procedure as the one in Section 6.11. Such procedure leads to Eq. (18.28).

Null-Steering Beamforming

The conventional beamformer selects a beam pattern with the gain of unity in the desired look direction, without any concern of other possible impinging signals to the array. Although, as demonstrated above, the conventional beamformer is optimum when the desired signal is the only impinging plane-wave to the array, it may behave poorly in the presence of other impinging signals. Hence, more elegant choices of $\mathbf{w}$ should be searched for.

A trivial solution is to select $\mathbf{w}$ so that nulls are introduced in the direction of interfering signals, while the array gain is constrained to unity in the desired look direction. This is mathematically formulated as follows. If θ_0 is the desired look direction, and θ_1, θ_2, $\cdots$, θ_L are the directions of the interference signals, $\mathbf{w}$ has to be selected to satisfy the following equations simultaneously:

$$\mathbf{w}^{\mathrm{H}}\mathbf{s}(\theta_0) = 1 \tag{18.32}$$

and

$$\mathbf{w}^{\mathrm{H}}\mathbf{s}(\theta_k) = 0, \quad \text{for } k = 1, 2, \cdots, L \tag{18.33}$$

Rearranging the left-hand sides of these equations in the form of $\mathbf{s}^{\mathrm{H}}(\theta_k)\mathbf{w}$, for $k = 0, 1, \cdots, L$, and combining the results, we obtain

$$\mathbf{S}^{\mathrm{H}}\mathbf{w} = \mathbf{e}_0 \tag{18.34}$$

where

$$\mathbf{S} = \begin{bmatrix} | & | & & | \\ \mathbf{s}(\theta_0) & \mathbf{s}(\theta_1) & \cdots & \mathbf{s}(\theta_L) \\ | & | & & | \end{bmatrix} \tag{18.35}$$

is a matrix of size $M \times (L+1)$, and $\mathbf{e}_0 = [1\ 0\ 0\ \cdots\ 0]^{\mathrm{T}}$ is a column vector of size $L+1$.

When $L+1 > M$, and the columns of $\mathbf{S}$ are linearly independent, Eq. (18.34) is overdetermined and, hence, it has no solution. When $L+1 = M$, and the columns of $\mathbf{S}$ are linearly independent, the solution to Eq. (18.34) is

$$\mathbf{w} = (\mathbf{S}^{\mathrm{H}})^{-1}\mathbf{e}_0 \tag{18.36}$$

Recalling the definition of $\mathbf{e}_0$, here, $\mathbf{w}$ is equal to the first column of the inverse of $\mathbf{S}^{\mathrm{H}}$.

The more common case is when $L+1 < M$. In this case, Eq. (18.34) is underdetermined and thus has no unique solution. To limit the solution to a unique one, it is common to reformulate it as follows. We find $\mathbf{w}$ that minimizes the cost function (18.30), subject to the constraints specified by Eqs. (18.32) and (18.33). This clearly minimizes the noise variance at the beamformer output but produces nulls in the direction of the undesired plane-waves impinging the array and lets the desired signal to pass through the array with the gain of unity.

This constraint minimization can be easily solved by using the method of Lagrange multipliers. Because of multiple constraints, here we need an extension to the procedure that was presented in Section 6.11. We define

$$\xi^{c} = \sigma_{\nu}^{2}\mathbf{w}^{H}\mathbf{w} + (\mathbf{S}^{H}\mathbf{w} - \mathbf{e}_{0})^{H}\boldsymbol{\lambda} \tag{18.37}$$

where $\boldsymbol{\lambda} = [\lambda_0\ \lambda_1\ \cdots\ \lambda_L]^{T}$ is a vector of Lagrange multipliers. The optimum value of $\mathbf{w}$, which we call $\mathbf{w}_{ns}$ (where 'ns' stands for null-steering), is obtained by simultaneous solution of

$$\nabla_{\mathbf{w}}^{C}\xi^{c} = 2\sigma_{\nu}^{2}\mathbf{w}_{ns} + 2\mathbf{S}\boldsymbol{\lambda} = \mathbf{0} \tag{18.38}$$

and

$$\nabla_{\boldsymbol{\lambda}}^{C}\xi^{c} = (\mathbf{S}^{H}\mathbf{w}_{ns} - \mathbf{e}_{0})^{*} = \mathbf{0} \tag{18.39}$$

From Eq. (18.38), we obtain

$$\mathbf{w}_{ns} = -\frac{1}{\sigma_{\nu}^{2}}\mathbf{S}\boldsymbol{\lambda} \tag{18.40}$$

Substituting Eq. (18.40) in Eq. (18.39), we get

$$\boldsymbol{\lambda} = -\sigma_{\nu}^{2}(\mathbf{S}^{H}\mathbf{S})^{-1}\mathbf{e}_{0} \tag{18.41}$$

Finally, substituting Eq. (18.41) in Eq. (18.40), we obtain

$$\mathbf{w}_{ns} = \mathbf{S}(\mathbf{S}^{H}\mathbf{S})^{-1}\mathbf{e}_{0} \tag{18.42}$$

It is interesting to note that the optimum constraint tap-weight vector $\mathbf{w}_{ns}$ is independent of the noise variance, σ_{ν}^{2}.

Example 18.2

For the example that was studied to evaluate the DOA estimations above, we have generated the beam pattern of the array when it is set to receive the plane-wave arriving from the angle 20°, while rejecting those arriving from angles 25° and -30°. The beam pattern of the designed beamformer is presented in Figure 18.12. Careful examination of this figure reveals that, as one would expect, the array gain in the look direction 20° is unity, and it is zero in the directions 25° and -30° (note that $-30^{\circ} \equiv 330^{\circ}$).

Optimal Beamforming

In a null-steering beamformer, the design is performed to minimize the noise variance at the array output, while perfect nulls are introduced in the directions of undesired plane-waves. One may argue that this may not be a good design. The optimum design is the one that balances between the minimization of noise and the residual interference from the plane-wave signals that are impinging the array from the directions different from the look direction. This optimum design, which is often referred to as Capon beamforming

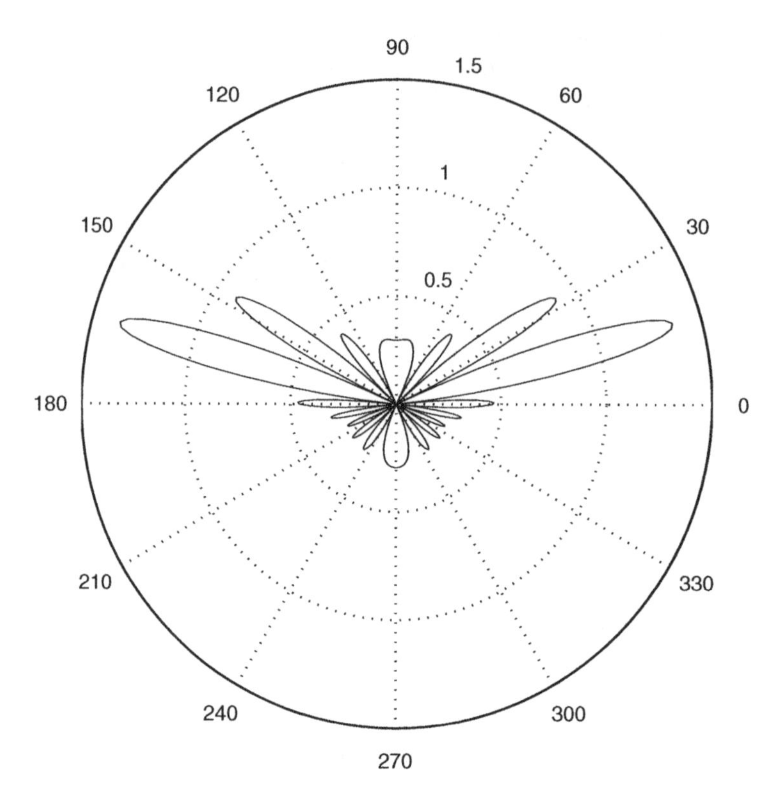

Figure 18.12 Beam pattern of the null-steering beamformer, for a ten element linear array. It is designed for the look direction 20°. There are interferers in the directions 25° and −30°.

(Capon, 1969), is carried out by minimizing the cost function

$$\begin{aligned}\xi &= E[|\underline{y}(n)|^2] \\ &= \mathbf{w}^{\mathrm{H}}\mathbf{R}\mathbf{w} \quad \text{subject to the constraint} \quad \mathbf{w}^{\mathrm{H}}\mathbf{s}(\theta_0) = 1.\end{aligned} \tag{18.43}$$

Recall that $\underline{y}(n)$ is the array output, and $\mathbf{R}$ is the correlation matrix of the signal samples at the array inputs. Also, recall the structure of $\mathbf{R}$ as presented in Eq. (18.15).

The constraint minimization (18.43), also, can be solved using the method of Lagrange multipliers. The result is

$$\mathbf{w}_{\mathrm{o}}^{\mathrm{c}} = \frac{\mathbf{R}^{-1}\mathbf{s}(\theta_0)}{\mathbf{s}^{\mathrm{H}}(\theta_0)\mathbf{R}^{-1}\mathbf{s}(\theta_0)} \tag{18.44}$$

The optimum/Capon beamformer is also called minimum variance distortionless response (MVDR), as it delivers a distortionless copy of the signal coming from the look direction while minimizing the variance of the summation of noise and any residual from interfering signals coming from other directions. Since the level of the desired signal is kept while variance of noise plus interference is minimized, one may say that the optimum beamformer maximizes the signal-to-interference-plus-noise ratio (SINR) at its output.

Example 18.3

A beam pattern of the optimum beamformer for the three plane-wave case that has been considered in our previous studies in this chapter, is presented in Figure 18.13, for the case where $\sigma_\nu^2 = 1$. The result is very similar to the one in Figure 18.12. The difference between the two beam patterns is minor and can be seen only by presenting the two plots in a cartesian coordinate. These plots are presented in Figure 18.14. As noted, the optimum beamformer reduces the gain of the array for most of the values of θ. This reduces the noise variance at the array output. However, the reduced noise at the array output is at the cost of a slight shift of the null at $\theta = 25^\circ$. The choice of $\mathbf{w}$, by design is to balance between joint suppression of noise and interference, caused by the undesirable plane-waves.

Further experiments reveal the obvious fact that as σ_ν^2 decreases, the optimum beamformer approaches the null-steering beamformer. On the other hand, when the undesirable

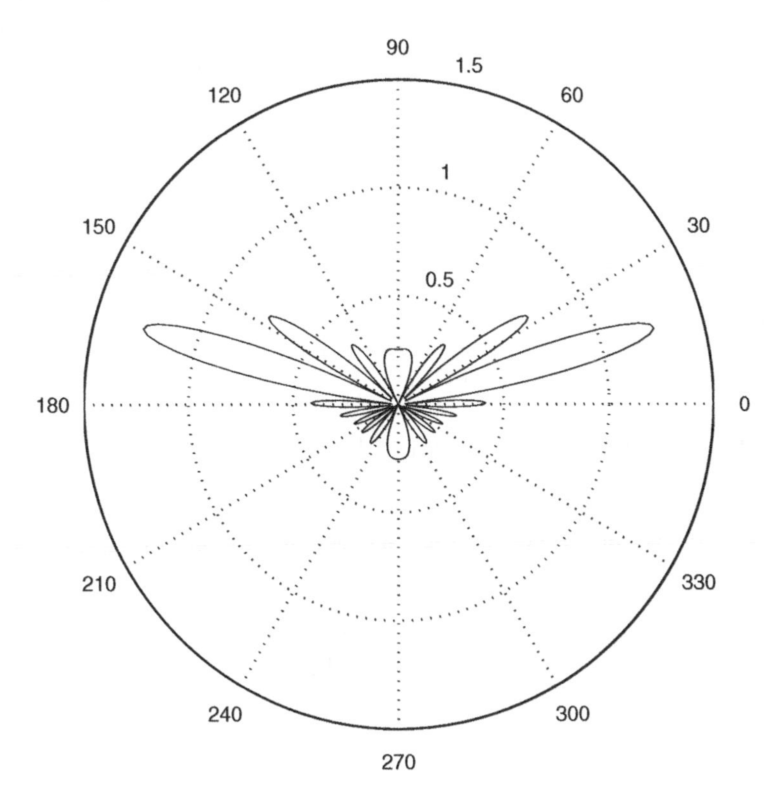

Figure 18.13 Beam pattern of the optimum beamformer for a ten element linear array. It is designed for the look direction 20°. There are interferers in the directions 25° and -30° and there is an additive noise at each element with variance $\sigma_\nu^2 = 1$.

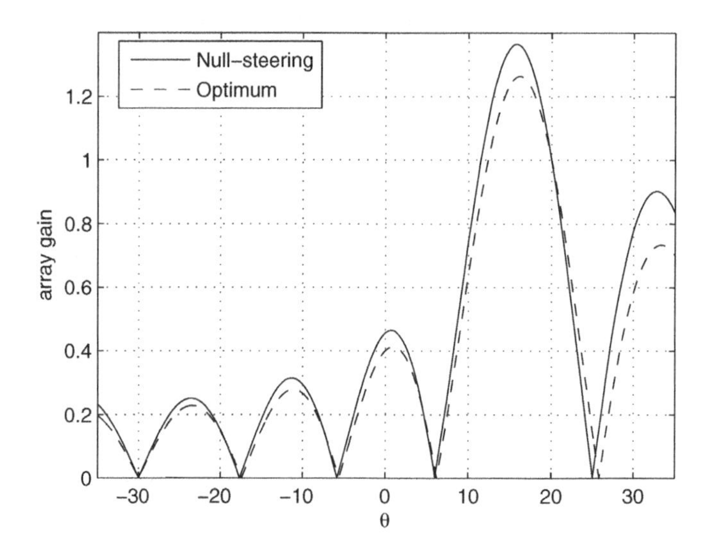

Figure 18.14 Beam patterns of the null-steering and optimum beamformer in cartesian coordinates for a ten element linear array. It is designed for the look direction 20°. There are interferers in the directions 25° and -30° and there is an additive noise at each element with variance $\sigma_\upsilon^2 = 1$.

plane-waves have significantly lower amplitudes compared to the noise variance, the optimum beamformer approaches the conventional beamformer.

In comparing the optimum beamformer with the null-steering beamformer, we note that while the latter requires us to know the directions of the desired and interfering signals, the former needs us to know only the direction of the desired signal. This makes the design and implementation of an adaptive filter for adjustment of the optimum beamformer coefficients a straightforward task. One can simply use the constrained LMS algorithm that was developed in Section 6.11, of Chapter 6.

The above observation seems to imply that the optimum beamformer is superior to the null-steering beamformer both with respect to performance and implementation. This statement, unfortunately, remains true only if the look direction is known perfectly or, at least, within a very good precision. If the assumed look direction is not very close to the angle of arrival of the desired user, the optimum beamformer will interpret the desired user as another interference and will attempt to remove it from the output of the array. The null-steering method, on the other hand, generates a main lobe around the look direction and thus the generated beam pattern can tolerate some variation of the angle of arrival of the desired plane-wave. In other words, the null-steering method is more robust than the optimum beamforming; however, it requires more information and may be more complex to implement.

Beam-Space Beamforming

The beamforming techniques that have been proposed so far are collectively referred to as element-space beamforming/processing, as they generate an array output by combining the signals directly taken from the array elements. Beam-space beamforming, in contrast, as its name implies, combines signals from a number of already-formed beams (beamed signals) to obtain the array output.

Figure 18.15 presents a block diagram of the beam-space beamformer. The signals from the array elements are combined together according to the conventional beamforming method first to enhance the signal impinging the array from the look direction (angle of arrival of θ_0). We call this the primary beamformer. It has the fixed coefficient vector $\mathbf{p} = \mathbf{s}(\theta_0)$. The auxiliary beam generator block, also called blocking beamformer, contains $L-1$ beamformers that all have a null to remove the desired signal while enhancing the signals impinging the array from the other directions. The coefficient vectors of these beamformers are the columns of a matrix $\mathbf{B}$ that we refer to as blocking matrix. The beamed signals $\underline{y}_1(n)$ through $\underline{y}_{L-1}(n)$ are thus generated as

$$\mathbf{y}(n) = \mathbf{B}\mathbf{x}(n) \tag{18.45}$$

where $\mathbf{x}(n) = [\underline{x}_0(n)\ \underline{x}_1(n)\ \cdots\ \underline{x}_{L-1}(n)]^{\mathrm{T}}$ and $\mathbf{y}(n) = [\underline{y}_1(n)\ \cdots\ \underline{y}_{L-1}(n)]^{\mathrm{T}}$. The beamed signals are then passed through an adaptive linear combiner, with coefficients

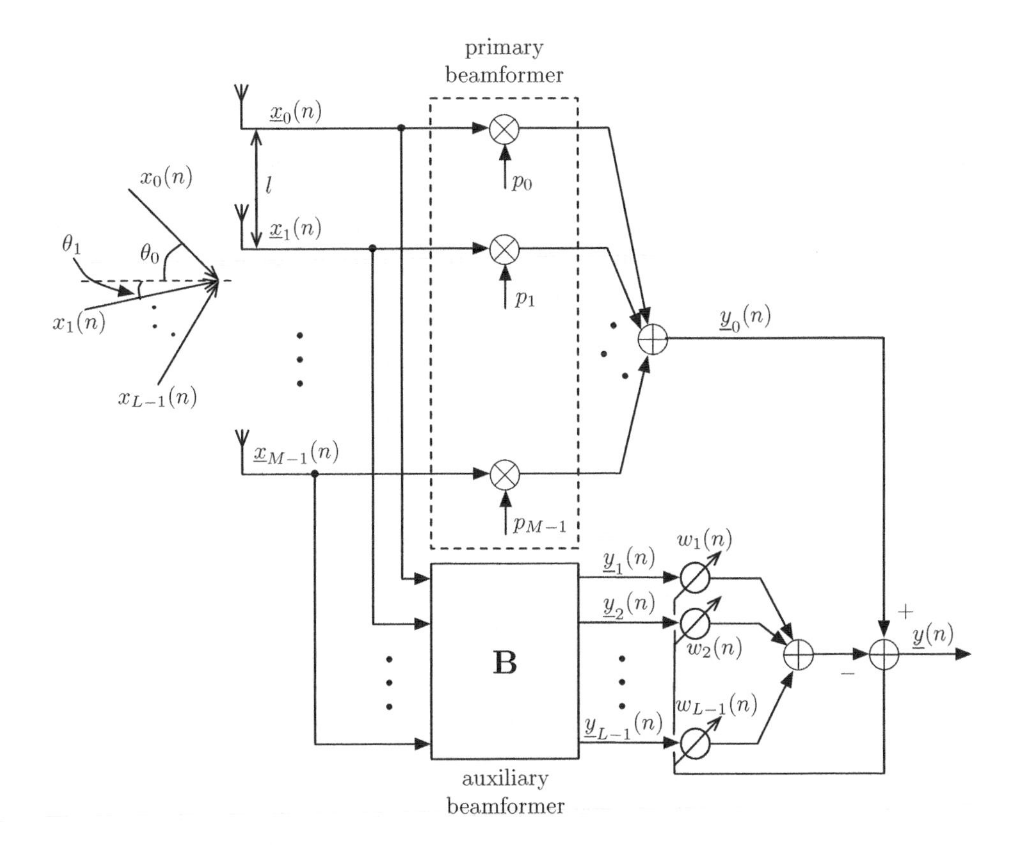

Figure 18.15 Beam-space beamformer.

$w_1(n)$ through $w_{L-1}(n)$, whose output is compared with $\underline{y}_0(n)$ and its tap weights are adapted to minimize the mean-squared value of the sum of undesirable signals and noise at the array output $\underline{y}(n)$. LMS or any other adaptive algorithm may be used for adaptation of the coefficients $w_1(n)$ through $w_{L-1}(n)$.

The following criteria are considered to set the columns of the blocking matrix **B**:

- Columns of **B** must be orthogonal to the primary beamformer coefficient vector **p**. This is to avoid the presence of any signal from the desired look direction in the beamed signals $\underline{y}_1(n)$ through $\underline{y}_{L-1}(n)$.
- Columns of **B** should be orthogonal with one another. This reduces the correlation between the beamed signals $\underline{y}_1(n)$ through $\underline{y}_{L-1}(n)$, and hence, improves the convergence behavior of the underlying adaptive linear combiner.

Various methods can be adopted to satisfy these criteria. Here, we propose one method that benefits from the eigenanalysis results that were presented in Chapter 4.

Assuming that the directions of the desired signal and other plane-wave signals that impinge the array are known and given by the steering vectors $\mathbf{s}(\theta_0)$, $\mathbf{s}(\theta_1)$, $\cdots$, $\mathbf{s}(\theta_{L-1})$, we construct the correlation matrix

$$\mathbf{R} = K\mathbf{s}(\theta_0)\mathbf{s}^{\mathrm{H}}(\theta_0) + \sum_{l=1}^{L-1} \mathbf{s}(\theta_l)\mathbf{s}^{\mathrm{H}}(\theta_l) + \sigma_\nu^2\mathbf{I} \tag{18.46}$$

By choosing the coefficient K to a large value, the first eigenvector of **R** that corresponds to its largest eigenvalue will approach $\mathbf{s}(\theta_0)$ as K increases. This follows from the minimax theorem that was introduced in Chapter 4. We let the primary beamformer coefficient vector **p** be equal to this first eigenvector of **R**. Any subset of size $L-1$ of the rest of the eigenvectors of **R**, as columns of **B**, satisfies the two criteria that were mentioned above. Furthermore, to make sure that the plane-wave signals from the directions θ_1 through θ_{L-1} will be present in the beamed signals $\underline{y}_1(n)$ through $\underline{y}_{L-1}(n)$, one should choose the second to Lth eigenvectors of **R** as the columns of the blocking matrix **B**. This is also a consequence of the properties that are derived from the minimax theorem. Further exploration related to this procedure of setting the columns of **B** is deferred, and is left as an exercise at the end of this chapter.

18.2 Broadband Sensor Arrays

Broadband processing of sensor array signals are built based on the same principles as the narrowband processing techniques that have been presented so far in this chapter. To find the DOAs of multiple broadband signals, the signal at each element is decomposed into a set of narrowband signals, through a bank of filters. The narrowband signals from the same subband of all the array elements are then taken as a set, and any of the DOA methods that were presented in Section 18.1.3 may be used to find the DOA of the various impinging signals to the array. Moreover, the results from different subbands may be averaged to improve on the accuracy of the estimated DOAs. For the rest of this section, we assume that the necessary DOAs have already been obtained and thus concentrate on a number of beamforming methods.

18.2.1 Steering

Recall that in narrowband beamformers, the demodulated signals at the array elements could be represented by phasors. The associated phasors to a plane-wave signal in an array are differentiated by their phase angles that are determined by the array topology and the DOA of the signal. For instance, for the linear array presented in Figure 18.1, these phase angles are quantified by the steering vector $\mathbf{s}(\theta)$ of Eq. (18.7). We also recall that the phase angles that are reflected in $\mathbf{s}(\theta)$ are to realize the delays between arrival time of the plane-wave to the various elements of the array.

For broadband signals, unfortunately, the phasor representation of the signals is not applicable; hence, the delays cannot be realized through a simple steering vector. Steering of broadband signals is performed by introducing time delays at the array elements and selecting these delays such that the desired signal components impinging the array elements are time-aligned after at the delay blocks outputs. This concept is presented in Figure 18.16. For a signal $x(n)$ that impinges the array from an angle θ, its replicas at the output of the delay blocks will be time-aligned, when

$$\tau_k = \frac{kl \sin\theta}{c}, \quad \text{for} \quad k = 0, 1, \cdots, M-1 \tag{18.47}$$

where c is the propagation speed. Signals impinging the array from the other angles will not be time-aligned. Hence, when τ_0 through τ_{M-1} are set according to Eq. (18.47), $x(n)$ will be enhanced at the array output, $y(n)$, while signals impinging the array from other directions will not. This will result in a beam pattern that has a main lobe for DOA of θ.

Since the processing of array signals is usually performed digitally, demodulator followed by analog-to-digital converter (ADC) circuitries are inserted at the antennas' output. The sampled signals are passed through a bank of shift-registers from which the delayed

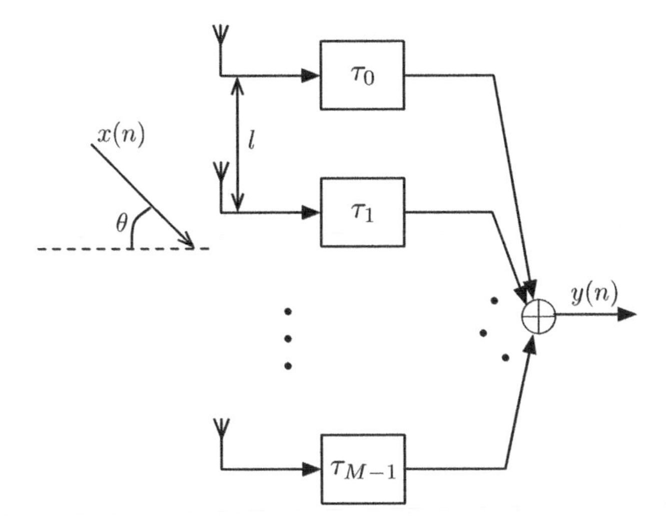

Figure 18.16 Steering in broadband beamformer.

signals are tapped out. Interpolation may also be performed over the samples from the shift-registers to realize the desired delays with a sufficient accuracy. Such an implementation also allows one to obtain multiple signal samples from each shift-register, and hence, obtain signal samples for more than one steering direction.

18.2.2 Beamforming Methods

Using the digital delay implementation that was introduced above, signal samples for different steering directions can be easily obtained. In the rest of this section, we present the extensions of the conventional and optimum narrowband beamformers that were introduced in Section 18.1.4, to broadband signals. Direct extensions of null-steering and beam-space beamformers to broadband signals are not possible. However, they may be implemented if a broadband signal is partitioned to a sufficient number of narrowband signals, and each narrowband signal is treated as a phasor.

Conventional Beamforming

For narrowband signals, the conventional beamforming was established by multiplying the phasor signals $\underline{x}_0(n)$ through $\underline{x}_{M-1}(n)$ by the conjugate of the element of the steering vector, $\mathbf{s}(\theta)$. This is equivalent to time-alining the signals received by the array elements for the desired look direction. Applying the same concept to broadband signals, one will find that a conventional beamformer is implemented, simply by adding a scaling factor $\frac{1}{M}$ at the output of Figure 18.16. This is presented in Figure 18.17.

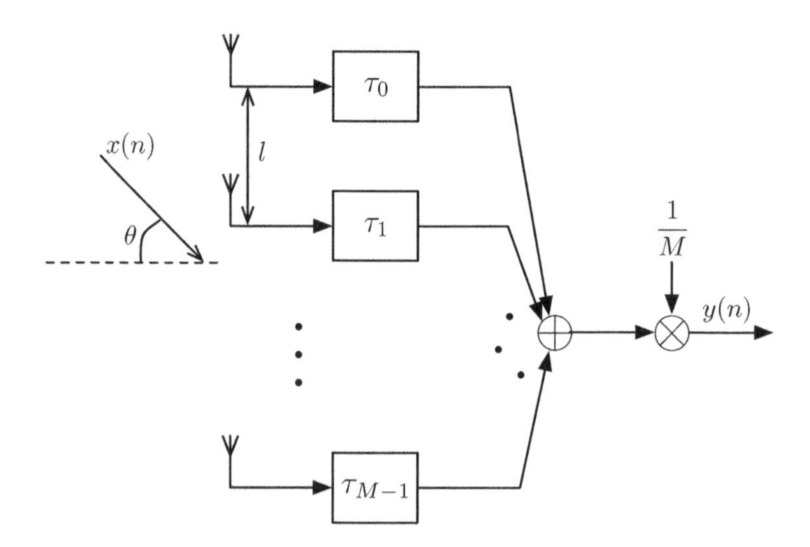

Figure 18.17 Conventional beamformer for broadband signals.

Optimal Beamforming

For broadband signals, an optimal beamformer can be implemented according to the structure presented in Figure 18.18. This structure, which was originally proposed by Frost (1972), has been explored and extended by many other researchers as well, for example, Er and Cantoni (1983) and Buckley and Griffiths (1986).

In Figure 18.18, $W_0(z)$ through $W_{M-1}(z)$ are a set of adaptive FIR filters, each of length N. The signal of interest, $x_0(n)$, impinging the array from the look direction θ_0, is time-aligned at the output of the delay blocks τ_0 through τ_{M-1}. This, in turn, implies that for the desired signal, the array is characterized by the transfer function

$$C(z) = \sum_{i=0}^{N-1} W_i(z) \tag{18.48}$$

A trivial choice for $C(z)$ may be $C(z) = z^{-\Delta}$, where Δ is a fixed delay. In this case, if $W_0(z)$ through $W_{M-1}(z)$ are constrained to satisfy Eq. (18.48) and $C(z) = z^{-\Delta}$, a delay replica of $x_0(n)$ will appear at the array output. Alternatively, $C(z)$ may be selected to apply some filtering to $x(n)$, for example, to reduce any out-of-band additive noise that the channel has added to the received signal. Here, we denote such a choice of $C(z)$ by $C_0(z)$ and adapt $W_0(z)$ through $W_{M-1}(z)$ to minimize the cost function

$$\xi = E[|y(n)|^2] \tag{18.49}$$

subject to the constraint

$$\sum_{i=0}^{N-1} W_i(z) = C_0(z) \tag{18.50}$$

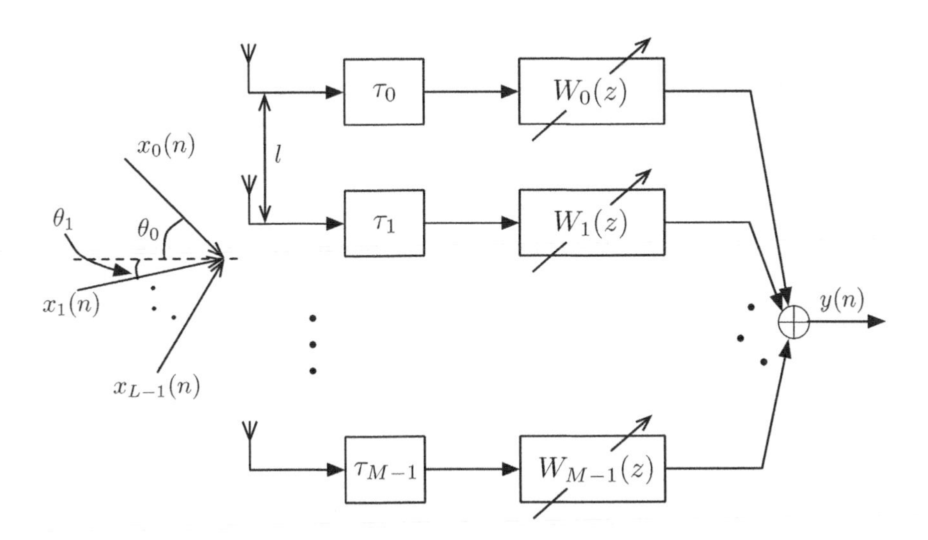

Figure 18.18 A structure for an optimal broadband beamformer.

The solution to this constrained minimization can be derived as follows. Let

$$\mathbf{w} = [\mathbf{w}_0^{\mathrm{H}} \quad \mathbf{w}_1^{\mathrm{H}} \quad \cdots \quad \mathbf{w}_{M-1}^{\mathrm{H}}]^{\mathrm{H}} \tag{18.51}$$

where $\mathbf{w}_i = [w_{i,0} \ w_{i,1} \ \cdots \ w_{i,N-1}]^{\mathrm{H}}$, for $i = 0, 1, \cdots, M-1$, is the tap-weight vector of $W_i(z)$. Also, we use the vector $\mathbf{c}_0 = [c_{0,0} \ c_{0,1} \ \cdots \ c_{0,N-1}]^{\mathrm{H}}$ to denote the tap-weight vector associated with $C_0(z)$. Moreover, we define $\mathbf{x}(n)$ as the vector of tap values associated with $\mathbf{w}$. Accordingly,

$$y(n) = \mathbf{w}^{\mathrm{H}}\mathbf{x}(n) \tag{18.52}$$

and the constraint (18.50) may be written as

$$\mathbf{C}\mathbf{w} = \mathbf{c}_0 \tag{18.53}$$

where

$$\mathbf{C} = \begin{bmatrix}\mathbf{I} & \mathbf{I} & \cdots & \mathbf{I}\end{bmatrix} \tag{18.54}$$

and $\mathbf{I}$ is the identity matrix of size N, and there are M identity matrix in $\mathbf{C}$. Substituting Eq. (18.52) in Eq. (18.49), we obtain

$$\xi = \mathbf{w}^{\mathrm{H}}\mathbf{R}\mathbf{w} \tag{18.55}$$

where $\mathbf{R} = E[\mathbf{x}(n)\mathbf{x}^{\mathrm{H}}(n)]$.

To minimize the cost function ξ subject to the constraint (18.53), we use the method of Lagrange multipliers and accordingly define

$$\xi^{\mathrm{c}} = \mathbf{w}^{\mathrm{H}}\mathbf{R}\mathbf{w} + (\mathbf{C}\mathbf{w} - \mathbf{c}_0)^{\mathrm{H}}\boldsymbol{\lambda} \tag{18.56}$$

where $\boldsymbol{\lambda}$ is a vector of Lagrange multipliers. Following the same derivations to those that led to Eq. (18.42), here we obtain

$$\mathbf{w}_{\mathrm{o}}^{\mathrm{c}} = \mathbf{R}^{-1}\mathbf{C}^{\mathrm{T}}(\mathbf{C}\mathbf{R}^{-1}\mathbf{C}^{\mathrm{T}})^{-1}\mathbf{c}_0 \tag{18.57}$$

As in the case of constrained LMS algorithm that was developed in Chapter 6 and was presented in Table 6.6, here also a similar LMS algorithm can be developed. This is left as an exercise problem at the end of this chapter. The resulting recursions will be

$$\mathbf{w}^{+}(n) = \mathbf{w}(n) - 2\mu y^{*}(n)\mathbf{x}(n) \tag{18.58}$$

and

$$\mathbf{w}(n+1) = \mathbf{w}^{+}(n) + \frac{1}{M}\mathbf{C}^{\mathrm{T}}(\mathbf{c}_0 - \mathbf{C}\mathbf{w}^{+}(n)) \tag{18.59}$$

We refer to these update equations as LMS-Frost algorithm.

Subband Beamforming

A broadband signal may be partitioned into a number of narrowband signals through an analysis filter bank (AFB). The subband signals may then be processed to match a certain target. The processed subband signals are subsequently combined together to reconstruct the processed broadband desired signal. This concept was studied thoroughly and the respective adaptive filters were developed in Chapter 9.

In the context of beamformers, to benefit from the narrowband beamforming techniques that were developed in Section 18.1, the number of subbands should be selected sufficiently large so that each subcarrier signal can be modeled by a flat spectrum over the band that is nonzero, and hence the phasor representation will be applicable. Once this is established, any of the narrowband beamforming techniques can be applied to each set of subcarrier signals separately. Figure 18.19 presents the structure of subband beamformer. The signal from each antenna is passed through an AFB, and the same indexed subband signals from the AFBs are passed to a narrowband beamformer. The outputs from the narrowband beamformers are combined together using a synthesis filter bank (SFB).

18.3 Robust Beamforming

So far, in this chapter, we have made the idealized assumption that the antenna elements are placed at their expected positions precisely, for example, in a linear array, and the antenna elements are placed along a straight line and are equally spaced at a distance of one half of the wavelength of the plane-wave impinging the array. As noted at the

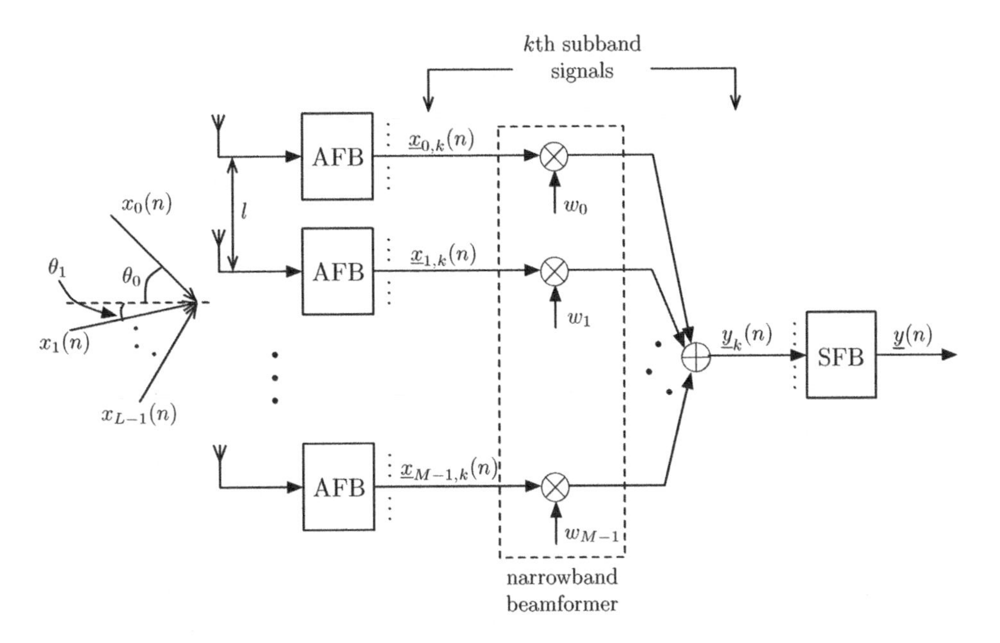

Figure 18.19 Subband beamforming.

beginning of this chapter, these idealized assumptions are usually not true. In the context of narrowband beamformers, the assumption that the steering vector of a plane-wave impinging the array is given by Eq. (18.7) will become inaccurate. The following example provides some insight into the impact of such wrong assumption.

Example 18.4

Consider a 10-element linear array that has been designed for element spacing of one half of the wavelength of impinging signals. However, owing to an implementation error, the spacing between the antennas turned out to be 5% shorter than the desired length. Study the generated optimum array pattern for the look direction of 45°, assuming the steering vector is calculated based on the assumption that the antennas spacing is that of the original design. Also, assume that in addition to the desired plane-wave signal, there are two interfering signals with powers of $P_1 = P_2 = 1$ impinging the array from angles 75° and -30°. The desired signal has the power of $P_0 = 4$. There is also a background white noise at each array element with variance $\sigma_\nu^2 = 0.01$.

Solution: The steering vector for the look direction, calculated based on the assumption that the element spacing is one half of the wavelength, is calculated as

$$\mathbf{s}_\mathrm{o}(\theta_0) = \begin{bmatrix} 1 \\ \mathrm{e}^{j\pi \sin 45^\circ} \\ \vdots \\ \mathrm{e}^{j9\pi \sin 45^\circ} \end{bmatrix} \tag{18.60}$$

The steering vectors for the interfering signals as well as that for the desired signal, for the element spacing of 95% of one half of the wavelength, on the other, are

$$\mathbf{s}(\theta) = \begin{bmatrix} 1 \\ \mathrm{e}^{j0.95\pi \sin \theta} \\ \vdots \\ \mathrm{e}^{j0.95\times 9\pi \sin \theta} \end{bmatrix}, \quad \text{for} \quad \theta = 45^\circ, 75^\circ \text{and} -30^\circ \tag{18.61}$$

Using these, we calculate the correlation matrix

$$\mathbf{R} = \sum_{k=0}^{2} P_k \mathbf{s}(\theta_k)\mathbf{s}^\mathrm{H}(\theta_k) + \sigma_\nu^2 \mathbf{I}$$

and use the result to find the optimum coefficients of the optimum beamformer according to Eq. (18.44) with $\mathbf{s}(\theta_0)$ substituted with $\mathbf{s}_\mathrm{o}(\theta_0)$. Subsequently, the array gain is obtained using Eq. (18.10) with $\mathbf{w}$ replaced by $\mathbf{w}_\mathrm{o}^\mathrm{c}$ and for $\mathbf{s}(\theta)$ given by Eq. (18.61).

For the numerical values given in this example, Figure 18.20 presents the beam pattern of the implemented array. As seen, the design generates an undesirable null in the look direction $\theta_0 = 45^\circ$, meaning that the desired signal will be blocked by the array.

To explain this behavior of the array and develop further insight into the problem, we note that the steering vector $\mathbf{s}_\mathrm{o}(\theta_0)$ given by Eq. (18.60) for the implemented array corresponds to the look direction θ_0', which is obtained by solving the equation

$$0.95\pi \sin \theta_0' = \pi \sin 45^\circ$$

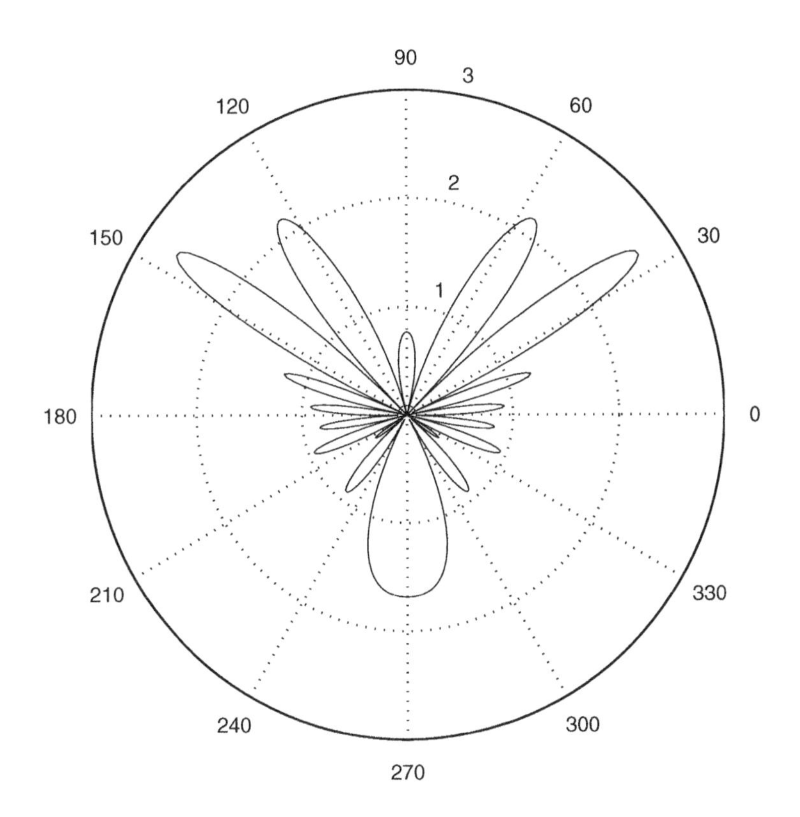

Figure 18.20 Beam pattern of the implemented array of Example 18.4 with inaccurate knowledge of the elements spacing.

This gives $\theta_0' = 48.1^\circ$ and this, in turn, means, by design, the designed array should have a gain of one at the angle 48.1°. Careful examination of Figure 18.20 reveals that this indeed is the case.

One more observation that is developed by exploring Figure 18.20 is that to block the signal arriving from the angle 45°, while assuring a gain of unity in the direction 48.1°, the generated beam pattern has developed relatively large lobes. These large side lobes results in some noise amplification at the array output. Hence, the array output suffers from both cancellation of the desired signal as well as an increase in the background noise. These are of course undesirable and measures to avoid them should be taken.

To further see the deviation of the faulty design of Figure 18.20 from an optimum design, Figure 18.21 presents the generated beam pattern when the exact spacing of the array elements are taken into account in forming the steering vector for the look direction. Comparing this beam pattern with the one in Figure 18.20, one will observe that a perfect main lobe is generated in the look direction, and there is no significant lobes in other directions. That is, only the desired signal is allowed to pass through the array. The plane wave signals from other directions and noise are greatly suppressed.

The above example clearly shows that a small deviation of the installed array from its original specifications can result in a serious failure of the optimum beamformer.

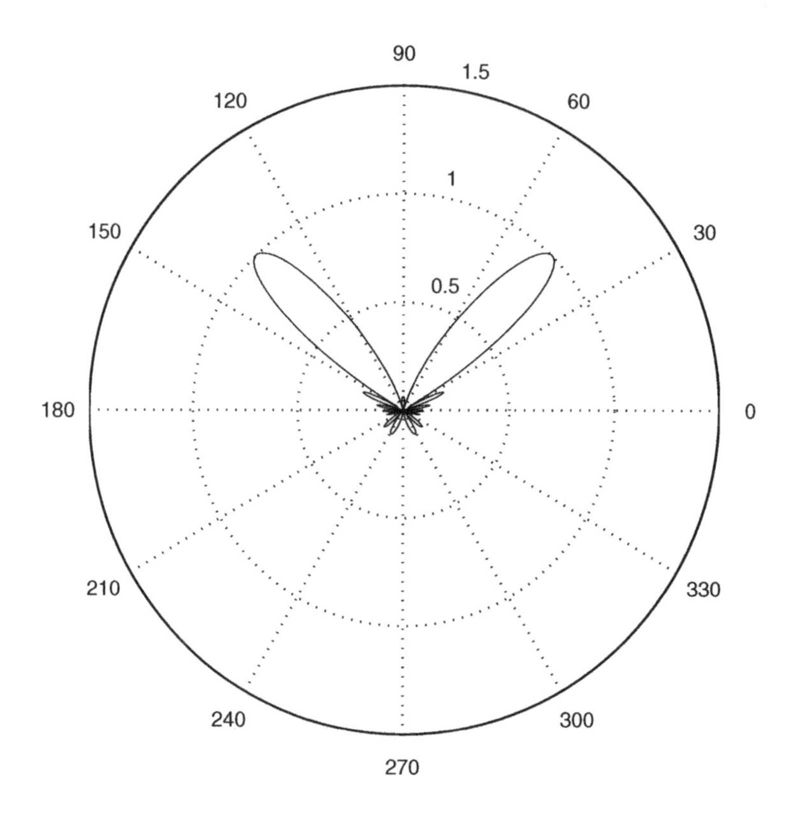

Figure 18.21 Beam pattern of the implemented array of Example 18.4 with accurate knowledge of the elements spacing.

One immediate solution to avoid such failure is to impose additional restrictions when designing array tap weights w_i. For instance, some researchers have proposed to add additional constraints to flatten the array gain around the look direction. This, for instance, is achieved by restricting the derivatives of the array gain to remain close to zero around the look direction, for example, (Er and Cantoni, 1983; Buckley and Griffiths, 1986), and (Stoica, Wang, and Li, 2003). Others have introduced methods that take into account some uncertainty in the steering vector that defines the look direction and accordingly optimize the array coefficient vector $\mathbf{w}$. Here, to develop some insight into the operation of the class of robust beamformers, we first introduce the concept of soft constraint and study the resulting solutions. This paves the way for an in-depth understanding of a few robust beamforming methods from the literature that will be presented subsequently.

18.3.1 Soft-Constraint Minimization

Recall that in the narrowband optimum beamformer, the beamformer output power is minimized subject to the constraint of array unity gain in the look direction. Mathematically, this is formulated as in Eq. (18.43) and leads to the solution (18.44).

One can also arrive at the same solution by minimizing the amended cost function

$$\xi = \mathbf{w}^{\mathrm{H}}\mathbf{R}\mathbf{w} + K|\mathbf{w}^{\mathrm{H}}\mathbf{s}(\theta_0) - 1|^2 \tag{18.62}$$

and finding the result as $K \to \infty$. This result can be easily proved as follows.

We first note that Eq. (18.62) can be expanded as

$$\xi = \mathbf{w}^{\mathrm{H}}(\mathbf{R} + K\mathbf{s}(\theta_0)\mathbf{s}^{\mathrm{H}}(\theta_0))\mathbf{w} - K\mathbf{w}^{\mathrm{H}}\mathbf{s}(\theta_0) - K\mathbf{s}^{\mathrm{H}}(\theta_0)\mathbf{w} + K \tag{18.63}$$

Next, letting the gradient of ξ with respect to $\mathbf{w}$ equal to zero and solving for $\mathbf{w}$, we obtain

$$\mathbf{w} = (\mathbf{R} + K\mathbf{s}(\theta_0)\mathbf{s}^{\mathrm{H}}(\theta_0))^{-1}K\mathbf{s}(\theta_0) \tag{18.64}$$

Using the matrix inversion lemma (that was introduced in Chapter 6), one finds that

$$(\mathbf{R} + K\mathbf{s}(\theta_0)\mathbf{s}^{\mathrm{H}}(\theta_0))^{-1} = \mathbf{R}^{-1} - \frac{K\mathbf{R}^{-1}\mathbf{s}(\theta_0)\mathbf{s}^{\mathrm{H}}(\theta_0)\mathbf{R}^{-1}}{1 + K\mathbf{s}^{\mathrm{H}}(\theta_0)\mathbf{R}^{-1}\mathbf{s}(\theta_0)} \tag{18.65}$$

Substituting Eq. (18.65) in Eq. (18.64) and simplifying the result, we get

$$\mathbf{w} = \frac{K\mathbf{R}^{-1}\mathbf{s}(\theta_0)}{1 + K\mathbf{s}^{\mathrm{H}}(\theta_0)\mathbf{R}^{-1}\mathbf{s}(\theta_0)} \tag{18.66}$$

Finally, as $K \to \infty$, Eq. (18.66) simplifies to

$$\mathbf{w} = \frac{\mathbf{R}^{-1}\mathbf{s}(\theta_0)}{\mathbf{s}^{\mathrm{H}}(\theta_0)\mathbf{R}^{-1}\mathbf{s}(\theta_0)} \tag{18.67}$$

This, clearly, has the same form as Eq. (18.44).

Alternatively, one can find the coefficient vector $\mathbf{w}$ that minimizes ξ for some positive large value of K. This solution may be thought as the one that minimizes the cost function $\xi = \mathbf{w}^{\mathrm{H}}\mathbf{R}\mathbf{w}$ subject to $\mathbf{w}^{\mathrm{H}}\mathbf{s}(\theta_0) - 1 \approx 0$. Hence, this approach may be referred to *soft-constraint minimization*.

An advantage of soft-constraint minimization is that it allows one to easily extend it to multiple constraints, and still solves the problem straightforwardly. For instance, to broaden the main lobe of beam pattern in the look direct, one may choose to minimize the cost function

$$\xi = \mathbf{w}^{\mathrm{H}}\mathbf{R}\mathbf{w} + K\sum_k |\mathbf{w}^{\mathrm{H}}\mathbf{s}(\theta_0 + \delta_k) - 1|^2 \tag{18.68}$$

where δ_k are a set of perturbation angles around zero that are added to make sure that the main lobe of the beam pattern remains close to one around the look direction θ_0.

Example 18.5

Consider the beamforming scenario that was discussed in Example 18.4. To avoid introduction of a null in the beam pattern at the angle $\theta_0 = 45^\circ$, we let the perturbation δ_k in Eq. (18.68) be a Gaussian random variable with standard deviation σ_δ. Also, we let $K = 1$ and run the summation on the right-hand side of Eq. (18.68) over 10,000 random choices of δ_k. Also, let each array element be subject to an independent white noise with variance $\sigma_\nu^2 = 0.01$. Study the generated beam pattern for $\sigma_\delta = 2^\circ$.

Solution: The results can be generated using the following MATLAB code. The variable names used here are similar to those defined above.

```
%
% Example 18.5
%
M=10;
P0=4;
P1=1;
P2=1;
sigma_d=2;
sigmanu2=0.01;
theta0=45*pi/180;
theta1=75*pi/180;
theta2=-30*pi/180;
so=exp(1i*pi*sin(theta0)*[0:M-1]');
s0=exp(1i*0.95*pi*sin(theta0)*[0:M-1]');
s1=exp(1i*0.95*pi*sin(theta1)*[0:M-1]');
s2=exp(1i*0.95*pi*sin(theta2)*[0:M-1]');
R=P0*s0*s0'+P1*s1*s1'+P2*s2*s2'+sigmanu2*eye(M);
p=0;
for k= 1:10000
    theta=(45+sigma_d*randn)*pi/180;
    so=exp(1i*pi*sin(theta)*[0:M-1]');
    R=R+so*so';
    p=p+so;
end
woc=R\p;
for k=0:360
    theta=k*pi/180;
    G(k+1)=woc'*exp(1j*0.95*pi*[0:M-1]'*sin(theta));
end
theta=[0:360]*pi/180;
figure(1),polar(theta,abs(G))
figure(2),plot(theta*180/pi,abs(G))
```

The results are presented in Figure 18.22. Note that an array gain of around one is generated over the angles of 35° to 60°, allowing a robust performance for some variation of the position of the array elements. There are also nulls in the direction of the interference signals impinging the array from the directions $\theta_1 = 75^{\circ}$ and $\theta_2 = -30^{\circ}$. Nevertheless, the array gain has increased significantly at other directions. This, as discussed before, may result in a significant noise enhancement at the array output.

18.3.2 Diagonal Loading Method

In Example 18.5, we observed that to keep the array gain flat around the desired look direction, it may increase significantly in other directions. We may also recall that, in Example 18.4 (Figure 18.20), to keep the array gain in the wrongly set look direction equal to one, while to generate a null in the correct look direction (arising from the erroneous set-up of the array), large gains appear in the beam pattern. These problems are resolved and a robust/more robust array is generated if a *regularization* term $\gamma \mathbf{w}^{\mathrm{H}}\mathbf{w}$

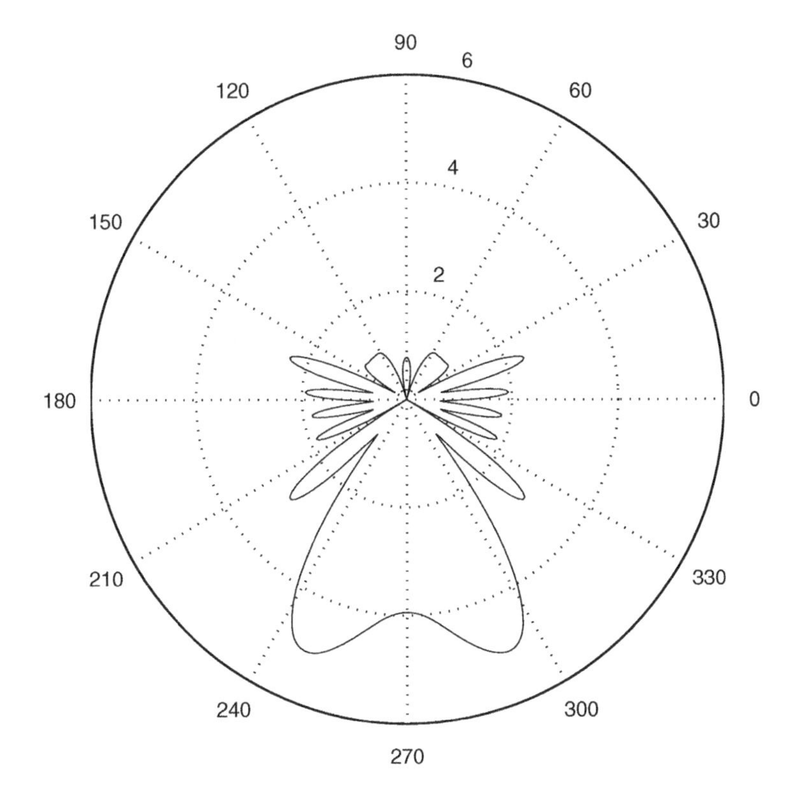

Figure 18.22 Beam pattern of implemented array of Example 18.5.

is added to the cost functions ξ in Eqs. (18.43), (18.68), or other similar equations. This is equivalent to artificially increasing the background noise in the array. This results in a significantly improved beam pattern. The following examples clarify this point.

Example 18.6

For the case presented in Example 18.5, when diagonal loading is added, $\mathbf{w}$ is found by minimizing the cost function

$$\xi = \mathbf{w}^{\mathrm{H}}\mathbf{R}\mathbf{w} + K\sum_{k} |\mathbf{w}^{\mathrm{H}}\mathbf{s}(\theta_0 + \delta_k) - 1|^2 + \gamma \mathbf{w}^{\mathrm{H}}\mathbf{w} \qquad (18.69)$$

Keeping the parameters listed in the MATLAB code in Example 18.5 and finding $\mathbf{w}$ that minimizes ξ in Eq. (18.69), we obtain an array whose beam pattern is presented in Figure 18.23. Here, the parameter γ is set equal to one. As seen, the large beam pattern gains have significantly reduced; take note of the amplitude scales.

Diagonal loading also helps in improving the robustness of other designs. A few problems at the end of this chapter provide additional insights.

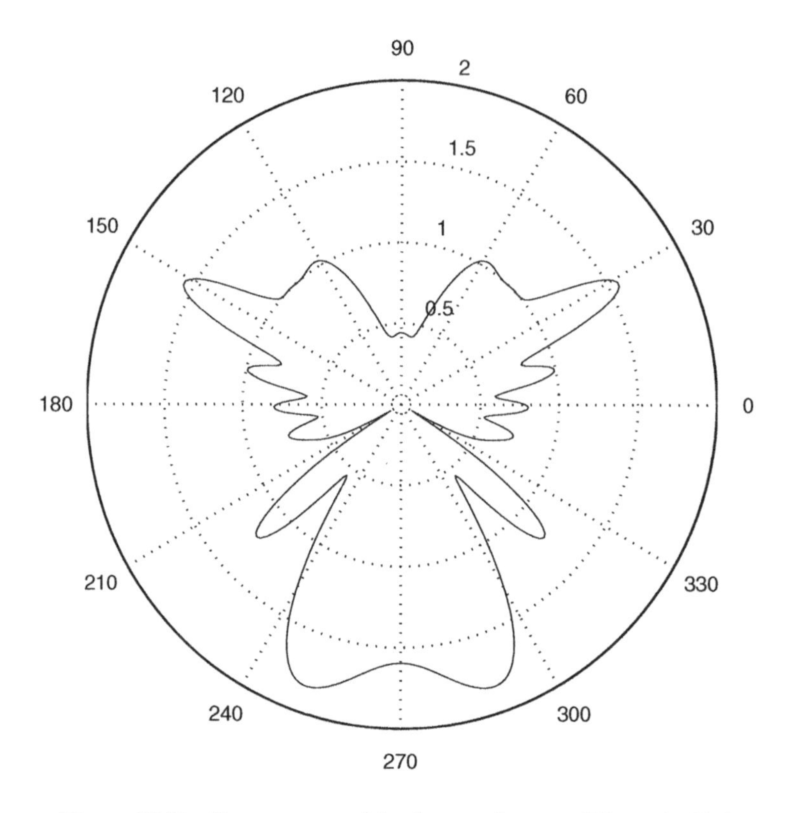

Figure 18.23 Beam pattern of implemented array of Example 18.6.

18.3.3 Methods Based on Sample Matrix Inversion

Recall the optimum beamformer design formulation (18.43) and note that the constraint $\mathbf{w}^{\mathrm{H}}\mathbf{s}(\theta_0) = 1$ may be also written as

$$\mathbf{w}^{\mathrm{H}}\mathbf{R}_{ss}\mathbf{w} = 1 \tag{18.70}$$

where $\mathbf{R}_{ss} = \mathbf{s}(\theta_0)\mathbf{s}^{\mathrm{H}}(\theta_0)$. Accordingly, an alternative formulation of the optimum beamformer may be presented as

$$\min_{\mathbf{w}} \mathbf{w}^{\mathrm{H}}\mathbf{R}\mathbf{w}, \quad \text{subject to the constraint} \quad \mathbf{w}^{\mathrm{H}}\mathbf{R}_{ss}\mathbf{w} = 1 \tag{18.71}$$

Furthermore, it is instructive to note that the constraint $\mathbf{w}^{\mathrm{H}}\mathbf{R}_{ss}\mathbf{w} = 1$ is the equivalent of choosing $\mathbf{w}$ for a unity array gain in the look direction. This constrains the (desired) signal power to a fixed value. On the other hand, $\mathbf{w}^{\mathrm{H}}\mathbf{R}\mathbf{w}$ is equal to the signal plus interference and noise power at the array output. Moreover, assuming that the desired signal is uncorrelated with interference signals and noise, one may decompose $\mathbf{R}$ as $\mathbf{R}_{\mathrm{s}} + \mathbf{R}_{\mathrm{i}}$, where $\mathbf{R}_{\mathrm{s}}$ is the part that arises from the signal of interest and $\mathbf{R}_{\mathrm{i}}$ is the part that arises from interference signals and noise. In addition, $\mathbf{R}_{\mathrm{s}} = \eta\mathbf{R}_{ss}$, where η is some constant related to the level of the signal of interest. Finally, one may define the SINR at the array output as

$$\mathrm{SINR} = \frac{\mathbf{w}^{\mathrm{H}}\mathbf{R}_{\mathrm{s}}\mathbf{w}}{\mathbf{w}^{\mathrm{H}}\mathbf{R}_{\mathrm{i}}\mathbf{w}} = \eta\frac{\mathbf{w}^{\mathrm{H}}\mathbf{R}_{ss}\mathbf{w}}{\mathbf{w}^{\mathrm{H}}\mathbf{R}_{\mathrm{i}}\mathbf{w}} \tag{18.72}$$

It is straightforward to show that SINR can be maximized, equivalently, by maximizing the ratio

$$\frac{\mathbf{w}^{\mathrm{H}}\mathbf{R}_{ss}\mathbf{w}}{\mathbf{w}^{\mathrm{H}}\mathbf{R}\mathbf{w}}.$$

This clearly is equivalent to the constrained minimization problem (18.71).

We use the method of Lagrange multipliers to solve Eq. (18.71). To this end, we define

$$\xi^{\mathrm{c}} = \mathbf{w}^{\mathrm{H}}\mathbf{R}\mathbf{w} + \lambda(1 - \mathbf{w}^{\mathrm{H}}\mathbf{R}_{ss}\mathbf{w}) \tag{18.73}$$

where λ is a Lagrange multiplier. Letting the gradient of ξ^{c} with respect to $\mathbf{w}$ equal to zero, one finds that the minimizer of ξ^{c} should satisfy the following equation

$$\mathbf{R}\mathbf{w} = \lambda\mathbf{R}_{ss}\mathbf{w}. \tag{18.74}$$

This may be viewed as a generalized eigen-problem. The solution of the constrained minimizer of $\mathbf{w}^{\mathrm{H}}\mathbf{R}\mathbf{w}$ is one of the M eigenvalue–eigenvector pairs $(\lambda, \mathbf{w})$ that satisfies Eq. (18.74). Multiplying Eq. (18.74) from left by $\mathbf{w}^{\mathrm{H}}$, we obtain

$$\mathbf{w}^{\mathrm{H}}\mathbf{R}\mathbf{w} = \lambda\mathbf{w}^{\mathrm{H}}\mathbf{R}_{ss}\mathbf{w} = \lambda \tag{18.75}$$

Noting that $\mathbf{R}$ is a correlation matrix and, thus, is positive definite, all the eigenvalues that satisfy Eq. (18.74) are positive. Also, as the goal is to minimize $\mathbf{w}^{\mathrm{H}}\mathbf{R}\mathbf{w}$, we are interested in the eigenvector of Eq. (18.74) that is associated with its minimum eigenvalue.

Next, rearranging Eq. (18.74) as

$$(\mathbf{R}^{-1}\mathbf{R}_{ss})\mathbf{w} = \frac{1}{\lambda}\mathbf{w} \tag{18.76}$$

we conclude that the optimum $\mathbf{w}$ that minimizes $\mathbf{w}^{\mathrm{H}}\mathbf{R}\mathbf{w}$, subject to the constraint $\mathbf{w}^{\mathrm{H}}\mathbf{R}_{ss}\mathbf{w} = 1$ (or, equivalently, $\mathbf{w}^{\mathrm{H}}\mathbf{s}(\theta_0) = 1$), is given by

$$\mathbf{w}_{\mathrm{o}}^{\mathrm{c}} = \mathcal{P}\ \{\mathbf{R}^{-1}\mathbf{R}_{ss}\} \tag{18.77}$$

where $\mathcal{P}\{\cdot\}$ refers to the *principal eigenvector* of a matrix, that is, the one that is associated with its largest eigenvalue. The solution, (18.77) when normalized to satisfy the constraint $\mathbf{w}^{\mathrm{H}}\mathbf{R}_{ss}\mathbf{w} = 1$, will be the same as Eq. (18.44). Next, we discuss practical applications of Eq. (18.77) in designing a few robust beamformers.

Given the steering vector $\mathbf{s}(\theta_0)$ and the measured correlation matrix $\hat{\mathbf{R}}$ according to Eq. (18.20) or Eq. (18.21), one can replace the estimated matrix in Eq. (18.77) to obtain an estimate of $\mathbf{w}_{\mathrm{o}}^{\mathrm{c}}$ as

$$\hat{\mathbf{w}}_{\mathrm{o}}^{\mathrm{c}} = \mathcal{P}\ \{\hat{\mathbf{R}}^{-1}\mathbf{R}_{ss}\} \tag{18.78}$$

This method is often referred to as the sample matrix inversion (SMI).

The SMI method, if implemented according to Eq. (18.78), suffers from the same sensitivity problems as the original optimum beamformer. One common method of reducing this sensitivity, and thus to arrive at a robust design, is to replace $\mathbf{R}_{ss}$ by

$$\tilde{\mathbf{R}}_{ss} = \mathbf{T} \odot \mathbf{R}_{ss} \tag{18.79}$$

where $\mathbf{T}$ is a properly chosen Toeplitz matrix and $\odot$ denotes element-wise multiplication. Common choices of $\mathbf{T}$ are

$$[\mathbf{T}]_{m,n} = \mathrm{e}^{-\alpha|m-n|} \tag{18.80}$$

and

$$[\mathbf{T}]_{m,n} = \mathrm{e}^{-\alpha(m-n)^2} \tag{18.81}$$

where α is a positive parameter. $\mathbf{T}$ is called tapered matrix because of obvious reasons.

Application of tapered matrix to both $\mathbf{R}$ and $\mathbf{R}_{ss}$ have also been introduced by some researchers. This method, which is referred to as covariance matrix taper (CMT), replaces $\hat{\mathbf{R}}$ and $\mathbf{R}_{ss}$ in Eq. (18.78) by

$$\tilde{\mathbf{R}} = \mathbf{T} \odot \hat{\mathbf{R}} \quad \text{and} \quad \tilde{\mathbf{R}}_{ss} = \mathbf{T} \odot \mathbf{R}_{ss} \tag{18.82}$$

respectively. A detailed review of these methods and their extensions can be found in (Shahbazpanahi *et al.*, 2003).

Problems

P18.1 The MATLAB code used to generate the result of Figure 18.7 is available on the accompanying website. It is called `'DOABartlett.m'`.

(i) Examine this code and explain how it relates to the mathematical equations in the text.
(ii) By running `'DOABartlett.m'`, confirm the result presented in Figure 18.7.
(iii) By making necessary changes to `'DOABartlett.m'`, examine the results for the following parameters and discuss your observations.

(a) $\theta_0 = 10^\circ, \theta_1 = 30^\circ, \theta_2 = -30^\circ, P_0 = P_1 = P_2 = 1$, and $\sigma_\nu^2 = 0.01$.
(b) $\theta_0 = 10^\circ, \theta_1 = 30^\circ, \theta_2 = -30^\circ, P_0 = P_2 = 1, P_1 = 0.01$, and $\sigma_\nu^2 = 0.01$.

P18.2 The MATLAB code used to generate the result of Figure 18.8 is available on the accompanying website. It is called `'DOAMVDR.m'`.

(i) Examine this code and explain how it relates to the mathematical equations in the text.
(ii) By running `'DOAMVDR.m'`, confirm the result presented in Figure 18.8.
(iii) By making necessary changes to `'DOAMVDR.m'`, examine the results for the following parameters and discuss your observations.

(a) $\theta_0 = 20^\circ, \theta_1 = 25^\circ, \theta_2 = -30^\circ, P_0 = P_1 = P_2 = 1$, and $\sigma_\nu^2 = 1$.

(b) $\theta_0 = 20^\circ$, $\theta_1 = 25^\circ$, $\theta_2 = -30^\circ$, $P_0 = P_2 = 1$, $P_1 = 0.01$, and $\sigma_\nu^2 = 0.01$.
(c) $\theta_0 = 20^\circ, \theta_1 = 25^\circ, \theta_2 = -30^\circ$, $P_0 = P_2 = 1$, $P_1 = 0.01$, and $\sigma_\nu^2 = 0.1$.

P18.3 The MATLAB code used to generate the result of Figure 18.9 is available on the accompanying website. It is called `'DOAMUSIC.m'`.

(i) Examine this code and explain how it relates to the mathematical equations in the text.
(ii) By running `'DOAMUSIC.m'`, confirm the result presented in Figure 18.9.
(iii) By making necessary changes to `'DOAMUSIC.m'`, examine the results for the following parameters and discuss your observations.

(a) $\theta_0 = 20^\circ$, $\theta_1 = 25^\circ$, $\theta_2 = -30^\circ$, $P_0 = P_1 = P_2 = 1$, and $\sigma_\nu^2 = 0.1$.
(b) $\theta_0 = 20^\circ$, $\theta_1 = 25^\circ$, $\theta_2 = -30^\circ$, $P_0 = P_1 = P_2 = 1$, and $\sigma_\nu^2 = 1$.
(c) $\theta_0 = 20^\circ, \theta_1 = 25^\circ, \theta_2 = -30^\circ$, $P_0 = P_2 = 1$, $P_1 = 0.1$, and $\sigma_\nu^2 = 0.01$.
(d) $\theta_0 = 20^\circ$, $\theta_1 = 25^\circ$, $\theta_2 = -30^\circ$, $P_0 = P_2 = 1$, $P_1 = 0.1$, and $\sigma_\nu^2 = 0.1$.
(e) $\theta_0 = 20^\circ$, $\theta_1 = 25^\circ$, $\theta_2 = -30^\circ$, $P_0 = P_2 = 1$, $P_1 = 0.5$, and $\sigma_\nu^2 = 0.1$.

P18.4 Show that the solution to the constraint minimization defined by Eqs. (18.30) and (18.31) is $\mathbf{w} = \frac{1}{M}\mathbf{s}(\theta_0)$.

P18.5 The MATLAB code that has been used to generate Figure 18.12 is available on the accompanying website. It is called `'NullSteeringBF.m'`.

(i) Examine this code and explain how it relates to the mathematical equations in the text.
(ii) By running `'NullSteeringBF.m'`, confirm the result presented in Figure 18.12.
(iii) Modify `'NullSteeringBF.m'` to steer the beam in the direction of θ_1. Examine the modified code to confirm that the desired result is obtained.
(iv) Repeat Part (iii) for the steered beam in the direction of θ_2.

P18.6 Using the method of Lagrange multiplier, prove Eq. (18.44).

P18.7 Consider a 10-element linear array with three narrowband signals impinging it with the angles of arrivals of $\theta_0 = 20^\circ$, $\theta_1 = 30^\circ$, and $\theta_2 = -10^\circ$. Considering the discussions around Eqs. (18.45) and (18.46), find the blocking matrix for the following cases.

(i) With $K = 1$ and $\sigma_\nu^2 = 0.1$.
(ii) With $K = 10$ and $\sigma_\nu^2 = 0.1$.
(iii) With $K = 100$ and $\sigma_\nu^2 = 0.1$.
(iv) With $K = 1000$ and $\sigma_\nu^2 = 0.1$.
(v) With $K = 1000$ and $\sigma_\nu^2 = 0.001$.
(vi) With $K = 1000$ and $\sigma_\nu^2 = 10$.
(vii) Compare the above results and discuss your observations. In particular, study the array gain for the primary and the secondary beamformers

P18.8 Present a derivation of the LMS-Frost algorithm, given by Eqs. (18.58) and (18.59).

P18.9 The LMS-Frost algorithm in the original formulation of Frost (1972) appears as

$$\mathbf{w}(n+1) = \left(\mathbf{I} - \frac{1}{M}\mathbf{C}^{\mathrm{T}}\mathbf{C}\right)(\mathbf{w}(n)\text{-}2\mu y^*(n)\mathbf{x}(n)) + \frac{1}{M}\mathbf{C}^{\mathrm{T}}\mathbf{c}_0.$$

Show that this is the same as Eqs. (18.58) and (18.59).

P18.10 The MATLAB code presented in Example 18.5 is available on the accompanying website. It is called `'Ex18_5.m'`. Run this code to confirm the result presented in Figure 18.23. Also, modify the code to study the following cases and discuss the observed results.

(i) The case presented in Example 18.6.
(ii) The case where $\sigma_\nu^2 = 0$ and there is no regularization (parameter γ).
(iii) The case where $\sigma_\nu^2 = 1$ and there is no regularization.
(iv) The case where $\sigma_\nu^2 = 10$ and there is no regularization.

P18.11 Repeat Problem P18.10 when σ_δ is reduced to 1°.

P18.12 Show that when the solution (18.77) is normalized to satisfy the constraint $\mathbf{w}^{\mathrm{H}}\mathbf{R}_{ss}\mathbf{w} = 1$, it will lead to Eq. (18.44).

P18.13 Show that the choice of $\mathbf{w}$ that maximizes the SINR (18.72) also maximizes the ratio

$$\frac{\mathbf{w}^{\mathrm{H}}\mathbf{R}_{ss}\mathbf{w}}{\mathbf{w}^{\mathrm{H}}\mathbf{R}\mathbf{w}}.$$

19

Code Division Multiple Access Systems

Code division multiple access (CDMA), one of the common technologies in cellular communications, has it roots in spread spectrum technique, a signaling method that was initially developed for secure communications in military applications. In the context of cellular communications, using CDMA, a number of users communicate with a base station over the same frequency band. However, to allow separation of the users' information, each user is assigned a different *signature*, called *spreading code*. By careful selection of users' spreading codes and/or effective signal processing, it is possible to extract information of each user while the interference from other users (called, *multiple-access interferenceor* or MAI) and noise are minimized. In this chapter, a number of adaptive signal processing techniques that may be used for reduction of MAI in CDMA systems are introduced and some of their aspects are analyzed.

19.1 CDMA Signal Model

Figure 19.1 presents the process of generating a CDMA signal. Each information symbol, say, $s_0(n)$, from a user (here, we have chosen the 0th user to reference to) is first multiplied by a spreading code sequence, $\{c_{0,i},\ i = 0, 1, \ldots, L-1\}$. Each element of the spreading code sequence, $c_{0,i}$, is referred to as a chip. Also, because of reasons that will become clear later, the length of the spreading code sequence, L, is referred to as *spreading gain*. The information symbols $s_0(n)$ are coming at a spacing T, hence, have a rate of $1/T$. Assuming that $s_0(n)$ and $c_{0,i}$ are binary, after spreading, the generated sequence is also a binary, however, has a rate L times faster than $s_0(n)$, that is, L/T. The block $p_{\mathrm{T}}(t)$ is the transmit filter. Following the principles developed in Chapter 17, here, $p_{\mathrm{T}}(t)$ is a square-root Nyquist filter set to match the rate of its input sequence, L/T. At the channel output, noise, and signals from other users (similarly generated CDMA signals) are added. At the receiver input, the received signal is passed through a filter $p_{\mathrm{R}}(t)$ that is matched to the transmit filter, that is, $p_{\mathrm{R}}(t) = p_{\mathrm{T}}(-t)$. The output of this matched filter is then sampled at a rate $1/T_s \geq L/T$. The result is the signal sequence $x(n)$, which has to be processed to extract the transmitted information symbols. The emphasis of this chapter

Adaptive Filters: Theory and Applications, Second Edition. Behrouz Farhang-Boroujeny.

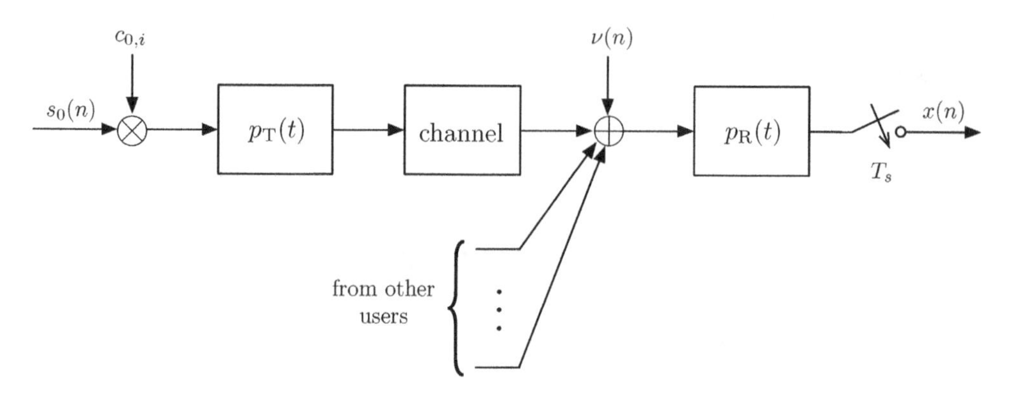

Figure 19.1 CDMA signal model.

is on various processing techniques that are applied to the signal sequence $x(n)$, for this purpose.

19.1.1 Chip-Spaced Users-Synchronous Model

Let us consider the simple case where the channel is absent and signals from all users are perfectly synchronized. Also, assume that the output of the receive filter $p_R(t)$ is perfectly sampled at the middle of each chip. Under this perfect condition, the vector of the signal samples over the interval of the nth information symbol $s_0(n)$ is obtained as

$$\mathbf{x}(n) = s_0(n)\mathbf{c}_0 + \sum_{k=1}^{K-1} \sqrt{P_k} s_k(n)\mathbf{c}_k + \mathbf{v}(n) \tag{19.1}$$

where P_k is signal power of kth user, it is assumed that $P_0 = 1$, K is the number of users,

$$\mathbf{c}_k = [c_{k,0}\ c_{k,1}\ \cdots\ c_{k,L-1}]^{\mathrm{T}} \tag{19.2}$$

$$\mathbf{x}(n) = [x_0(n)\ x_1(n)\ \cdots\ x_{L-1}(n)]^{\mathrm{T}} \tag{19.3}$$

and $\mathbf{v}(n)$ is the noise vector. We assume that the elements of $\mathbf{v}(n)$ are a set of identically independent zero-mean Gaussian random variables with variance σ_ν^2. Hence, $\mathbf{v}(n)$ has the covariance matrix

$$E[\mathbf{v}(n)\mathbf{v}^{\mathrm{T}}(n)] = \sigma_\nu^2\mathbf{I} \tag{19.4}$$

where $\mathbf{I}$ is the identity matrix.

Under the condition that the spreading codes $\mathbf{c}_0, \mathbf{c}_1, \ldots,$ and $\mathbf{c}_{K-1}$ are a set of orthogonal vectors, that is, $\mathbf{c}_k^{\mathrm{T}}\mathbf{c}_l = 0$, for $k \neq l$, an estimate of $s_0(n)$ can be trivially obtained as

$$\begin{aligned}\hat{s}_0(n) &= \frac{\mathbf{c}_0^{\mathrm{T}}\mathbf{x}(n)}{\|\mathbf{c}_0\|^2} \\ &= s_0(n) + \nu'(n)\end{aligned} \tag{19.5}$$

where $\|\mathbf{c}_0\|^2 = \mathbf{c}_0^{\mathrm{T}}\mathbf{c}_0$ and $\nu'(n) = \frac{\mathbf{c}^{\mathrm{T}}\mathbf{v}(n)}{\|\mathbf{c}_0\|^2}$. Note that under this condition, the MAI (i.e., interference from other users) is completely removed.

To develop more insight to the process of data recovery equation (19.5), we note that in Eq. (19.1), the signal-to-noise ratio (SNR) can be calculated as

$$\begin{aligned}\mathrm{SNR}_{\mathrm{in}} &= \frac{E[\|s_0(n)\mathbf{c}_0\|^2]}{E[\|\mathbf{v}(n)\|^2]} \\ &= \frac{\|\mathbf{c}_0\|^2 E[s_0^2(n)]}{L\sigma_\nu^2} = \frac{\|\mathbf{c}_0\|^2}{L\sigma_\nu^2}\end{aligned} \tag{19.6}$$

On the other hand, considering Eq. (19.5), at the estimator output,

$$\begin{aligned}\mathrm{SNR}_{\mathrm{out}} &= \frac{E[s_0^2(n)]}{E[\nu'^2(n)]} \\ &= \frac{1}{\sigma_\nu^2/\|\mathbf{c}_0\|^2} = \frac{\|\mathbf{c}_0\|^2}{\sigma_\nu^2}\end{aligned} \tag{19.7}$$

Next, we define the *processing gain* of a CDMA system as

$$\mathcal{G} = \frac{\mathrm{SNR}_{\mathrm{out}}}{\mathrm{SNR}_{\mathrm{in}}} \tag{19.8}$$

Substituting Eqs. (19.6) and (19.7) in Eq. (19.8), we obtain

$$\begin{aligned}\mathcal{G} &= \frac{\|\mathbf{c}_0\|^2/\sigma_\nu^2}{\|\mathbf{c}_0\|^2/L\sigma_\nu^2} \\ &= L.\end{aligned} \tag{19.9}$$

That is, under the ideal synchronized condition assumed here, the processing gain of CDMA is equal to the number of chips per information symbol.

The presence of a channel adds the channel impulse response $h_{0,i}$ to the path between $s_0(n)$ and $x(n)$. As a result, if a signal vector $\mathbf{x}(n)$ of length L of $x(n)$ is extracted, the counterpart equation to Eq. (19.1) may be written as

$$\mathbf{x}(n) = s_0(n)\mathbf{c}_0' + \sum_{k=1}^{K-1} \sqrt{P_k} s_k(n)\mathbf{c}_k' + \mathbf{v}(n) \tag{19.10}$$

where $\mathbf{c}_k'$, for $k = 0, 1, \ldots, K-1$, are obtained by first convolving $c_{k,i}$ and $h_{k,i}$, and then truncating the result to the length L. One may also note that the presence of the channel results in some interference among adjacent symbols. In this chapter, we assume that the length of each channel is much shorter than the length of the spreading code, hence, ignore such interference in Eq. (19.10) and similar equations that will be developed later. Moreover, one should note that although the vectors $\mathbf{c}_k$ are real-valued and $p_{\mathrm{T}}(t)$ and $p_{\mathrm{R}}(t)$ are real functions of time, the vectors $\mathbf{c}_k'$ are not, because the channel impulse responses $h_{k,i}$ are, in general, complex-valued. In the discussion that follows, for brevity of notations, we remove the prime signs from the vectors $\mathbf{c}_k'$ in Eq. (19.10); however, it is understood that unlike the vectors $\mathbf{c}_k$ in Eq. (19.1), these are complex-valued and most likely nonorthogonal. Also, for the future developments, it is convenient to present the following vector-matrix formulation of Eq. (19.10):

$$\mathbf{x}(n) = \mathbf{C}\mathbf{s}(n) + \mathbf{v}(n) \tag{19.11}$$

where

$$\mathbf{s}(n) = [s_0(n)\ s_1(n)\ \cdots\ s_{K-1}(n)]^{\mathrm{T}} \tag{19.12}$$

is a column vector of length K, and

$$\mathbf{C} = \begin{bmatrix} | & | & & | \\ \mathbf{c}_0 & \sqrt{P_1}\mathbf{c}_1 & \cdots & \sqrt{P_{K-1}}\mathbf{c}_{K-1} \\ | & | & & | \end{bmatrix} \tag{19.13}$$

is an $L \times K$ matrix.

As in presence of channel the columns of $\mathbf{C}$ may be nonorthogonal, a trivial detector like Eq. (19.5) will not be able to remove MAI. Hence, the main problem that is addressed in this chapter is the following. What is a good linear estimator in the form of

$$\hat{s}_0(n) = \mathbf{w}^{\mathrm{H}}\mathbf{x}(n) \tag{19.14}$$

where $\mathbf{w}$ is a tap-weight vector, which results in a reasonable estimate of $s_0(n)$ that balances between the residual MAI and noise. Note that in Eq. (19.14), $\mathbf{w}$ and $\mathbf{x}(n)$ are complex-valued vectors. Moreover, following the notations that were first introduced in Chapter 3, we define

$$\mathbf{x}(n) = [x_0(n)\ x_1(n)\ \cdots\ x_{L-1}(n)]^{\mathrm{T}} \tag{19.15}$$

and

$$\mathbf{w} = [w_0\ w_1\ \cdots\ w_{L-1}]^{\mathrm{H}} \tag{19.16}$$

where the superscript H denotes complex-conjugate transpose, or Hermitian.

19.1.2 Chip-Spaced Users-Asynchronous Model

Consider the case where the users are not symbol synchronized. We assume that the received signal samples are symbol synchronized with the desired user (i.e., user 0), and the users 1 through $K-1$ have integer time offsets L_1 through L_k, respectively, with respect to the user 0. We note that although L_k's can be either positive or negative, here, we limit L_k's to positive values only, as this greatly simplifies the equations, without causing any loss in the generality of our conclusions.

When $0 \le L_k \le L-1$, for $k = 1, 2, \ldots, K-1$, Eq. (19.10), with the vectors $\mathbf{c}'_k$ replaced by $\mathbf{c}_k$, converts to

$$\mathbf{x}(n) = s_0(n)\mathbf{c}_0 + \sum_{k=1}^{K-1} \sqrt{P_k}(s_k(n)\mathbf{c}_k^{\mathrm{u}} + s_k(n-1)\mathbf{c}_k^{\mathrm{l}}) + \mathbf{v}(n) \tag{19.17}$$

where $\mathbf{c}_k^{\mathrm{u}}$ is a length L vector with its first $L - L_k$ elements equal to those of $\mathbf{c}_k$ and the rest of its elements equal to 0. On the other hand, $\mathbf{c}_k^{\mathrm{l}}$ is a length L vector with its first $L - L_k$ elements equal to 0 and the last L_k elements equal to those of $\mathbf{c}_k$.

The vector-matrix formulation (19.11) is also applicable to Eq. (19.17), with the following modifications to the definitions of the vector $\mathbf{s}(n)$ and the matrix $\mathbf{C}$:

$$\mathbf{s}(n) = [s_0(n)\ s_1(n)\ s_1(n-1)\ \cdots\ s_{K-1}(n)\ s_{K-1}(n-1)]^{\mathrm{T}} \tag{19.18}$$

and

$$\mathbf{C} = \begin{bmatrix} | & | & | & & | & | \\ \mathbf{c}_0 & \sqrt{P_1}\mathbf{c}_1^{\mathrm{u}} & \sqrt{P_1}\mathbf{c}_1^{\mathrm{l}} & \cdots & \sqrt{P_{K-1}}\mathbf{c}_{K-1}^{\mathrm{u}} & \sqrt{P_{K-1}}\mathbf{c}_{K-1}^{\mathrm{l}} \\ | & | & | & & | & | \end{bmatrix} \tag{19.19}$$

19.1.3 Fractionally Spaced Model

The CDMA signal model, when the received signal is sampled at a rate faster than the chip rate (hence, fractionally spaced samples are obtained), leads to an equation similar to Eq. (19.11). The difference only lies in the fact that the samples of $\mathbf{x}(n)$ and $\mathbf{s}(n)$ are fractionally spaced. The columns of $\mathbf{C}$ are also samples of the spreading codes that are fractionally spaced. Hence, the detectors that are introduced below are applicable to both symbol-spaced and fractionally spaced as well as users-synchronous and users-asynchronous cases. However, there is an advantage in using the fractionally spaced samples. The receiver performance remains insensitive to small variations of the timing phase, say, within a chip period. Hence, fractionally spaced CDMA receivers may perform more robustly when compared with their chip-spaced counterparts. This concept follows the same principles to those that were discussed in Chapter 17 with regard to the symbol-spaced versus fractionally spaced equalizer.

19.2 Linear Detectors

A linear detector finds an estimate of the desired user symbol $s_0(n)$ according to the equation

$$\hat{s}_0(n) = \mathbf{w}^{\mathrm{H}}\mathbf{x}(n) \tag{19.20}$$

where $\mathbf{w}$ is a tap-weight vector that is chosen based on a selected criterion. In this section, we review three criteria for the selection of the tap-weight vector $\mathbf{w}$ and develop the relevant detectors.

The detectors that are presented later require one or more of the following information to obtain the desired estimate of the tap-weight vector $\mathbf{w}$.

1. The spreading code of the desired user, the vector $\mathbf{c}_0$.
2. The spreading code of the interfering users, the vectors $\mathbf{c}_1$ through $\mathbf{c}_{K-1}$.
3. The symbol boundaries of the desired user.
4. The symbol boundaries of the interfering users.
5. The received amplitudes of the interfering users, $\sqrt{P_k}$, for $k = 1, 2, \ldots, K-1$. We assume $P_0 = 1$.
6. The channel noise statistics.
7. The training sequences that may be used for initial adaptation of the weight vector $\mathbf{w}$.

In the sequel, we refer to the above items as requirements (1) through (7).

19.2.1 Conventional Detector: The Matched Filter Detector

The most trivial linear detector sets

$$\mathbf{w} = \mathbf{c}_0/\|\mathbf{c}_0\|^2 \tag{19.21}$$

This may be thought as a filtering operation matched to the desired user signal. Its implementation only needs the requirements (1) and (3) of the above list.

We note that because the spreading codes $\mathbf{c}_0$, $\mathbf{c}_1$, ..., and $\mathbf{c}_{K-1}$ in Eq. (19.11) are the original spreading codes after passing through the respective channels, they are most likely nonorthogonal, even if the original spreading codes were orthogonal. Hence, the matched filter output is obtained as

$$\hat{s}_0(n) = s_0(n) + \sum_{k=1}^{K-1} \frac{\mathbf{c}_0^{\mathrm{H}}\mathbf{c}_k}{\|\mathbf{c}_0\|^2} s_k(n) + \nu_{\mathrm{out}}(n) \tag{19.22}$$

where

$$\nu_{\mathrm{out}}(n) = \frac{\mathbf{c}_0^{\mathrm{H}}\mathbf{v}(n)}{\|\mathbf{c}_0\|^2} \tag{19.23}$$

is the noise sample at the detector output. Also, note that the second term on the right-hand side of Eq. (19.22) is the interference due to the other users. Thus, it is referred to as MAI (multiple-access interference).

We note that the knowledge of the exact value of $\mathbf{c}_0$ requires one to know the spreading code of the desired user at both the transmitter as well as the respective channel impulse response. However, the latter is usually unavailable. Hence, in practice, $\mathbf{c}_0$ in Eq. (19.21) is often replaced by the desired user spreading code at the respective transmitter. This clearly incurs some loss in performance, as the receiver is not exactly matched to the received waveform. This problem may be solved by using the so-called *rake receiver.* A rake receiver consists of a set of filters matched to the spreading code of the desired user, but with different delays to extract the signal energy from the different paths of the channel impulse response. The outputs of these filters are combined together after applying a set of weight factors to obtain an output that optimally combines the received signals from different paths. In any case, the conventional/matched filter detector suffers from some level of MAI, which makes it inferior to the rest of the detectors that are presented in the remaining parts of this section.

19.2.2 Decorrelator Detector

The decorrelator detector ignores the noise term in Eq. (19.11) and solves the equation

$$\mathbf{C}\mathbf{s}(n) = \mathbf{x}(n) \tag{19.24}$$

for the vector $\mathbf{s}(n)$ and extracts the desired user symbol from the calculated solution. This solution is obtained by first multiplying Eq. (19.24) from left by $\mathbf{C}^{\mathrm{H}}$ and the result, from left, by $\left(\mathbf{C}^{\mathrm{H}}\mathbf{C}\right)^{-1}$. These steps lead to the solution

$$\hat{\mathbf{s}}(n) = \left(\mathbf{C}^{\mathrm{H}}\mathbf{C}\right)^{-1}\mathbf{C}^{\mathrm{H}}\mathbf{x}(n) \tag{19.25}$$

We note that in the absence of channel noise, $\mathbf{v}(n)$, $\hat{\mathbf{s}}(n) = \mathbf{s}(n)$. Hence, the decorrelator detector separates the users' symbols perfectly and, thus, its output will be free of MAI. If $\mathbf{v}(n)$ is nonzero and is included in the above equations, one will find that

$$\hat{\mathbf{s}}(n) = \mathbf{s}(n) + \mathbf{v}_{\text{out}}(n) \tag{19.26}$$

where

$$\mathbf{v}_{\text{out}}(n) = \left(\mathbf{C}^{\text{H}}\mathbf{C}\right)^{-1}\mathbf{C}^{\text{H}}\mathbf{v}(n) \tag{19.27}$$

For the user of interest, we have

$$\hat{s}_0(n) = s_0(n) + \nu_{0,\text{out}}(n) \tag{19.28}$$

One may also note from the above results that the tap-weight vector $\mathbf{w}$ of the decorrelator detector, for the desired user, is equal to the first column of $\left[\left(\mathbf{C}^{\text{H}}\mathbf{C}\right)^{-1}\mathbf{C}^{\text{H}}\right]^{\text{H}}$. This, in turn, implies that

$$\mathbf{w} = \mathbf{C} \times \left(\text{the first column of } \left(\mathbf{C}^{\text{H}}\mathbf{C}\right)^{-1}\right) \tag{19.29}$$

Although the decorrelator detector removes MAI completely, it suffers from possible noise enhancement problem. To explain this, using Eq. (19.27) and recalling Eq. (19.4), one will find that

$$E[\mathbf{v}_{\text{out}}(n)\mathbf{v}_{\text{out}}^{\text{H}}(n)] = \sigma_\nu^2\left(\mathbf{C}^{\text{H}}\mathbf{C}\right)^{-1} \tag{19.30}$$

The diagonal element of this matrix is equal to the noise variance at the decorrelator detector outputs, with the first one of them being that of the desired user. The size of these variances depends on the condition number (the spread of eigenvalues) of the matrix $\mathbf{C}^{\text{H}}\mathbf{C}$. If this matrix has a large condition number, the diagonal elements of $\left(\mathbf{C}^{\text{H}}\mathbf{C}\right)^{-1}$ will be a set of large numbers, and the decorrelator detector will suffer from a noise enhancement problem.

Another problem that the decorrelator detector may face in practice is that to find $\mathbf{w}$, one needs an accurate estimate of $\mathbf{C}$, and this may not be readily available. Moreover, the estimation of $\mathbf{C}$ over a channel may not be a straightforward task. Hence, an exact implementation of a decorrelator detector can be a difficult task to achieve in practice. See Section 19.3 for more exposure.

Both problems of noise enhancement and exact implementation of a decorrelator detector make this method unattractive in practice. The MMSE detector, that is introduced next, and the blind detector, that is discussed subsequently, are significantly less complex to implement and yet perform better. Noting these, one may find that the decorrelator detectors are only attractive from a historical point of view.

19.2.3 Minimum Mean-Squared Error (Optimal) Detector

The minimum mean-squared error (MMSE) detector, as its name implies, is constructed based on the cost function

$$\xi = E[|s_0(n) - \hat{s}_0(n)|^2] \tag{19.31}$$

The tap-weight vector $\mathbf{w}$ is set/adapted to minimize ξ. Recalling Eq. (19.14), Eq. (19.31) can be written as

$$\xi = E[|s_0(n) - \mathbf{w}^{\mathrm{H}}\mathbf{x}(n)|^2] \tag{19.32}$$

This, clearly, is a typical Wiener filtering problem, which was thoroughly studied in Chapter 3. It, thus, has the solution

$$\mathbf{w}_{\mathrm{o}} = \mathbf{R}^{-1}\mathbf{p} \tag{19.33}$$

where $\mathbf{R} = E[\mathbf{x}(n)\mathbf{x}^{\mathrm{H}}(n)]$ and $\mathbf{p} = E[\mathbf{x}(n)s_0^*(n)]$, because, here, the desired output of the Wiener filter is $s_0(n)$. Moreover, the MMSE of the detector is obtained as

$$\xi_{\min} = 1 - \mathbf{p}^{\mathrm{H}}\mathbf{R}^{-1}\mathbf{p} \tag{19.34}$$

Using Eq. (19.11) and assuming that the information symbols $s_0(n)$ through $s_{K-1}(n)$ are independent of one another and they are also uncorrelated with the noise vector $\mathbf{v}(n)$, one will find that

$$\mathbf{p} = \mathbf{c}_0 \tag{19.35}$$

and

$$\mathbf{R} = \mathbf{C}\mathbf{C}^{\mathrm{H}} + \sigma_\nu^2\mathbf{I} \tag{19.36}$$

where $\mathbf{C}$ is given according to Eq. (19.13) or Eq. (19.19) depending on the system being users synchronous or users asynchronous.

Substituting Eqs. (19.36) and (19.35) into Eqs. (19.33) and (19.34), we obtain, respectively,

$$\mathbf{w}_{\mathrm{o}} = \left(\mathbf{C}\mathbf{C}^{\mathrm{H}} + \sigma_\nu^2\mathbf{I}\right)^{-1}\mathbf{c}_0 \tag{19.37}$$

and

$$\xi_{\min} = 1 - \mathbf{c}_0^{\mathrm{H}}\left(\mathbf{C}\mathbf{C}^{\mathrm{H}} + \sigma_\nu^2\mathbf{I}\right)^{-1}\mathbf{c}_0 \tag{19.38}$$

It is also instructive to note that the MMSE detector output is given by

$$\begin{aligned}\hat{s}_0(n) &= \mathbf{w}_{\mathrm{o}}^{\mathrm{H}}\mathbf{x}(n) \\ &= \left(\mathbf{c}_0^{\mathrm{H}}\left(\mathbf{C}\mathbf{C}^{\mathrm{H}} + \sigma_\nu^2\mathbf{I}\right)^{-1}\mathbf{c}_0\right)s_0(n) + \sum_{k=1}^{K-1}\left(\mathbf{c}_0^{\mathrm{H}}\left(\mathbf{C}\mathbf{C}^{\mathrm{H}} + \sigma_\nu^2\mathbf{I}\right)^{-1}\mathbf{c}_k\right)s_k(n) \\ &\quad + \mathbf{c}_0^{\mathrm{H}}\left(\mathbf{C}\mathbf{C}^{\mathrm{H}} + \sigma_\nu^2\mathbf{I}\right)^{-1}\mathbf{v}(n)\end{aligned} \tag{19.39}$$

Noting that the first term on the right-hand side of Eq. (19.39) is the desired signal, and the rest of the terms are interference and noise, the signal-to-interference-plus-noise ratio (SINR) at the detector output is obtained as

$$\mathrm{SINR} = \frac{\left|\mathbf{c}_0^{\mathrm{H}}\left(\mathbf{C}\mathbf{C}^{\mathrm{H}} + \sigma_\nu^2\mathbf{I}\right)^{-1}\mathbf{c}_0\right|^2}{\sum_{k=1}^{K-1}\left|\mathbf{c}_0^{\mathrm{H}}\left(\mathbf{C}\mathbf{C}^{\mathrm{H}} + \sigma_\nu^2\mathbf{I}\right)^{-1}\mathbf{c}_k\right|^2 + \sigma_\nu^2\mathbf{c}_0^{\mathrm{H}}\left(\mathbf{C}\mathbf{C}^{\mathrm{H}} + \sigma_\nu^2\mathbf{I}\right)^{-2}\mathbf{c}_0} \tag{19.40}$$

One may infer from Eq. (19.37) that the computation of $\mathbf{w}_{\mathrm{o}}$, hence, the implementation of the MMSE detector needs requirements (1) through (6) of the above list. In practice, where $\mathbf{w}$ is adjusted through an adaptive algorithm, one only needs to be aware of the symbol boundaries of the desired user, that is, the requirement (3) only. In addition, at the beginning of each communication session a training sequence that is known to the receiver should be transmitted by the transmitter of the desired user; the requirement (7). The receiver uses these training symbols to find an estimate of $\mathbf{w}_o$, through an adaptive algorithm. Once a good estimate of $\mathbf{w}_o$ is obtained, the system is switched to a decision-directed mode for further tuning of the detector and/or tracking of the variations in the channel condition. Either of the LMS algorithm or RLS algorithm or any of their variations may be used. More details are given in Section 19.3.

Another point that is worth noting here is that under the condition where the channel noise is absent, hence, $\sigma_\nu^2 = 0$, the MMSE detector reduces to the decorrelator detector. This can be explained simply by noting that when the channel noise is absent, the cost function ξ of Eq. (19.32) can be reduced to 0 by choosing $\mathbf{w}$ such that MAI is completely removed. This, of course, is nothing but the decorrelator detector. A mathematical proof of this concept is left as a problem at the end of this chapter.

The final note here is that the MMSE detector is also referred to as *optimal detector*. Obviously, it is optimal in the sense that it minimizes the sum of the residual MAI and noise power at its output. Moreover, as discussed below, the MMSE detector may be seen as an optimal detector in the sense that it maximizes the SINR at its output.

19.2.4 Minimum Output Energy (Blind) Detector

Consider a detector that is set to minimize the cost function

$$\xi = E[|\mathbf{w}^{\mathrm{H}}\mathbf{x}(n)|^2] \tag{19.41}$$

subject to the constraint

$$\mathbf{w}^{\mathrm{H}}\mathbf{c}_0 = 1 \tag{19.42}$$

Such a detector is called mean output energy (MOE) detector, because of the use of the cost function (19.41).

Rearranging Eq. (19.11) as

$$\mathbf{x}(n) = \mathbf{c}_0 s_0(n) + \mathbf{C}'\mathbf{s}'(n) + \mathbf{v}(n), \tag{19.43}$$

where $\mathbf{C}'$ is obtained from $\mathbf{C}$ after removing its first column, and $\mathbf{s}'(n)$ is obtained from $\mathbf{s}(n)$ after removing its first element, and substituting Eq. (19.43) in Eq. (19.41), after applying the constraint $\mathbf{w}^{\mathrm{H}}\mathbf{c}_0 = 1$, we obtain

$$\xi = E[|s_0(n) + \mathbf{w}^{\mathrm{H}}\mathbf{x}'(n)|^2] \tag{19.44}$$

where $\mathbf{x}'(n) = \mathbf{C}'\mathbf{s}'(n) + \mathbf{v}(n)$. Assuming that $s_0(n)$ is uncorrelated with the other users' data and channel noise, Eq. (19.44) can be rearranged as

$$\xi = E[|s_0(n)|^2] + E[|\mathbf{w}^{\mathrm{H}}\mathbf{x}'(n)|^2] \tag{19.45}$$

As the first term on the right-hand side of Eq. (19.45) is a constant (as mentioned before, here, we assume that $E[|s_0(n)|^2] = 1$), independent of $\mathbf{w}$, the vector $\mathbf{w}$ that minimizes

$E[|\mathbf{w}^{\mathrm{H}}\mathbf{x}'(n)|^2]$ is also the minimizer of the cost function ξ (of course, subject to the constraint $\mathbf{w}^{\mathrm{H}}\mathbf{c}_0 = 1$). On the other hand, noting that $E[|\mathbf{w}^{\mathrm{H}}\mathbf{x}'(n)|^2]$ arises from MAI and noise, one may argue that minimum output energy (MOE) detector maximizes the SINR at the output $y(n) = \mathbf{w}^{\mathrm{H}}\mathbf{x}(n)$.

Next, we note that Eq. (19.45) can be rearranged as

$$\xi = 1 + \mathbf{w}^{\mathrm{H}}\mathbf{R}'\mathbf{w} \tag{19.46}$$

where $\mathbf{R}' = E[\mathbf{x}'(n)\mathbf{x}'^{\mathrm{H}}(n) = \mathbf{C}'\mathbf{C}'^{\mathrm{H}} + \sigma_v^2\mathbf{I}$. Using the method of Lagrange multipliers, to minimize $\mathbf{w}^{\mathrm{H}}\mathbf{R}'\mathbf{w}$ subject to the constraint $\mathbf{w}^{\mathrm{H}}\mathbf{c}_0 = 1$, one will find that, the optimum tap-weight vector of MOE is given by

$$\mathbf{w}_{\mathrm{o}} = \frac{\mathbf{R}'^{-1}\mathbf{c}_0}{\mathbf{c}_0^{\mathrm{H}}\mathbf{R}'^{-1}\mathbf{c}_0} \tag{19.47}$$

Also,

$$\mathbf{w}_{\mathrm{o}}^{\mathrm{H}}\mathbf{R}'\mathbf{w}_{\mathrm{o}} = \frac{1}{\mathbf{c}_0^{\mathrm{H}}\mathbf{R}'^{-1}\mathbf{c}_0} \tag{19.48}$$

Moreover, the MOE SINR is obtained as

$$\mathrm{SINR} = \mathbf{c}_0^{\mathrm{H}}\mathbf{R}'^{-1}\mathbf{c}_0 \tag{19.49}$$

Another instructive observation that is discussed in Problem P19.3, at the end of this chapter, is that the right-hand sides of Eqs. (19.40) and (19.49) are equal. This in turn implies that the MMSE and MOE detectors are equivalent in the sense that after the optimum setting of their tap weights both result in the same SINR value. However, the adaptation algorithms that they use, as discussed in the following section, are significantly different.

The MOE detector is often referred to as *blind detector* because of the following reason. In order to estimate the vector $\mathbf{w}$ that minimizes the cost function (19.41) subject to the constraint (19.42), one does not require any training sequence. It only requires the desired user spreading vector $\mathbf{c}_0$ and the symbol boundaries of the desired user, that is, requirements (1) and (3) of the list that was presented at the beginning of this section.

To develop further insight to the properties of MOE detector, we proceed with a study of its properties under a few specific cases.

Noise Free Case

Assuming that the channel noise $\mathbf{v}(n)$ is absent, Eq. (19.11) may be expanded as

$$\mathbf{x}(n) = \mathbf{c}_0 s_0(n) + \sum_{k=1}^{K-1} \mathbf{c}_k s_k(n) \tag{19.50}$$

where the first term on the right-hand side is the desired signal and the second term is MAI. Assuming that $K \leq N$, the MAI belongs to a subspace of dimension $K - 1$ (or smaller), expanded by the spreading code vectors $\mathbf{c}_1$ through $\mathbf{c}_{K-1}$. We refer to this as *MAI subspace.* The MOE detector can completely remove MAI, by choosing a tap-weight vector $\mathbf{w}$ that is orthogonal to the MAI subspace. Moreover, one may write $\mathbf{c}_0$ as $\mathbf{c}_0 = \mathbf{c}_{0,1} + \mathbf{c}_{0,2}$, where $\mathbf{c}_{0,2}$ is the projection of $\mathbf{c}_0$ into the MAI subspace.

We note that as $\mathbf{c}_{0,2}$ belongs to the MAI subspace and $\mathbf{w}$ is orthogonal to this subspace, the constraint (19.42) reduces to

$$\mathbf{w}^{\mathrm{H}}\mathbf{c}_{0,1} = 1 \tag{19.51}$$

Accordingly, one will find that, in this case, the tap-weight vector

$$\mathbf{w}_{\mathrm{o}} = \frac{\mathbf{c}_{0,1}}{\|\mathbf{c}_{0,1}\|^2} \tag{19.52}$$

satisfies the constraint (19.42) and removes MAI completely. In addition, one may realize that when $K < N$, this solution is not unique. Any addition to $\mathbf{w}_{\mathrm{o}}$ of Eq. (19.52) that is orthogonal to the spreading vectors $\mathbf{c}_0$ through $\mathbf{c}_{K-1}$ may also be seen as a solution that removes MAI completely and satisfies the constraint (19.42). The solution (19.52) would be unique in the sense that among all the solutions, it has the minimum norm. Hence, we refer to it as the *minimum norm solution*.

Next, we present an alternative method of calculating the minimum norm tap-weight vector $\mathbf{w}$ that removes MAI and satisfies the constraint (19.42). From the earlier discussion, we infer that the desired tap-weight vector belongs to the subspace spanned by the columns of the spreading code matrix $\mathbf{C}$. Hence, it may be written in the canonical form

$$\mathbf{w} = \mathbf{C}\mathbf{w}' \tag{19.53}$$

where $\mathbf{w}'$ is a length K vector. Considering Eq. (19.53), Eq. (19.42) may be written as

$$\mathbf{w}'^{\mathrm{H}}\mathbf{c}'_0 = 1 \tag{19.54}$$

where $\mathbf{c}'_0 = \mathbf{C}^{\mathrm{H}}\mathbf{c}_0$. With these, one may use the following procedure to design an MOE detector.

Minimize the cost function

$$\xi = E[|\mathbf{w}^{\mathrm{H}}\mathbf{x}(n)|^2] = E[|\mathbf{w}'^{\mathrm{H}}\mathbf{C}^{\mathrm{H}}\mathbf{x}(n)|^2]$$

subject to the constraint (19.54).

Recalling Eq. (19.24), here, we will find that

$$\xi = \mathbf{w}'^{\mathrm{H}}\mathbf{S}\mathbf{w}' \tag{19.55}$$

where $\mathbf{S} = (\mathbf{C}^{\mathrm{H}}\mathbf{C})^2$. Note that $\mathbf{w}'$ has the length of K and $\mathbf{S}$ is a $K \times K$ matrix. Using the method of Lagrange multipliers, one will find that minimization of Eq. (19.55), subject to the constraint (19.54), has the following solution

$$\mathbf{w}'_{\mathrm{o}} = \frac{\mathbf{S}^{-1}\mathbf{c}'_0}{\mathbf{c}'^{\mathrm{H}}_0\mathbf{S}^{-1}\mathbf{c}'_0} \tag{19.56}$$

Also,

$$\xi_{\min} = \frac{1}{\mathbf{c}'^{\mathrm{H}}_0\mathbf{S}^{-1}\mathbf{c}'_0} \tag{19.57}$$

As MAI is completely removed, what remains at the MOE detector is $s_0(n)$. Hence, $\xi_{\min} = E[|s_0(n)|^2] = 1$. Using this result and Eq. (19.57), one will find that $\mathbf{c}_0'^{\mathrm{H}}\mathbf{S}^{-1}\mathbf{c}_0' = 1$. This, in turn, implies

$$\mathbf{w}_o' = \mathbf{S}^{-1}\mathbf{c}_0' \tag{19.58}$$

Finally, using this result in Eq. (19.53), we obtain

$$\mathbf{w}_o = \mathbf{C}\mathbf{w}_o' \tag{19.59}$$

Spreading Code Mismatch

In practice, because of the presence of the channel, the spreading codes of the desired and other users undergo some distortion before reaching the receiver. Hence, the MOE tap-weight vector $\mathbf{w}$ can only be obtained based on an estimate of the spreading code of the desired user, $\mathbf{c}_0$ (after considering the channel effect). Here, we do not consider the receiver uncertainty of other users' spreading codes, as the adaptation algorithms that are introduced later, need not know such estimates.

Let the estimate of $\mathbf{c}_0$ be $\hat{\mathbf{c}}_0$. The MOE detector attempts to minimize the cost function ξ of Eq. (19.41) subject to the constraint

$$\mathbf{w}^{\mathrm{H}}\hat{\mathbf{c}}_0 = 1 \tag{19.60}$$

Here, also, when the channel noise is absent, the minimum norm solution $\mathbf{w}$ to this problem belongs to the subspace spanned by the columns of the spreading code matrix $\mathbf{C}$. Hence, Eq. (19.53) through Eq. (19.56), with $\mathbf{c}_0$ replaced $\hat{\mathbf{c}}_0$, are applicable and, thus,

$$\mathbf{w}' = \frac{\mathbf{S}^{-1}\hat{\mathbf{c}}_0'}{\hat{\mathbf{c}}_0'^{\mathrm{H}}\mathbf{S}^{-1}\hat{\mathbf{c}}_0'} \tag{19.61}$$

where $\hat{\mathbf{c}}_0' = \mathbf{C}^{\mathrm{H}}\hat{\mathbf{c}}_0$. Note that here the identity $\xi_{\min} = 1$ may not hold, hence, the simplification (19.58) is not applicable.

In Honig *et al.* (1995), the authors have discussed the impact of the spreading code mismatch on the performance of a CDMA system. The general conclusion there is that under code mismatch, the MOE detector may also remove part of the energy of the desired user, hence, some degradation in the system performance should be expected. To remedy this problem, it is proposed in Honig *et al.* (1995) that one may start the receiver by first running an MOE detector adaptive algorithm for a coarse adjustment of the tap-weight vector $\mathbf{w}$. Once $\mathbf{w}$ is adjusted to some degree and the output from the MOE detector can be considered as reliable, one can switch to a decision-directed mode adaptive MMSE detector. The adaptive MMSE detector converges to near the optimum solution (19.37) without any need to know $\mathbf{c}_0$.

Including the Channel Noise

In the presence of channel noise, $\mathbf{w}$ is no longer limited to the subspace spanned by the columns of $\mathbf{C}$. In that case, the MOE detector has the same solution as the optimum detector. Hence, the MOE leads to the solution (19.37) and (19.38). Note that inserting

$\sigma_\nu^2 = 0$ in Eqs. (19.37) and (19.38) does not lead to any fruitful result, as the rank N matrix $\mathbf{CC}^{\mathrm{H}} + \sigma_\nu^2 \mathbf{I}$ converts to a rank $K < N$ matrix $\mathbf{CC}^{\mathrm{H}}$. It was in the light of this mathematical problem that the minimum norm solution discussed earlier was derived.

19.2.5 Soft Detectors

In Chapter 17, Section 17.9, it was noted that the optimum receivers could be implemented using the methods of soft equalization. We also introduced the methods of soft MMSE equalizer and statistical soft equalizer. These methods are extendable to CDMA systems as well and somewhat are similar to their equalizer counterparts. Interested readers are encouraged to refer to Wang and Poor (1999) and Farhang-Boroujeny, Zhu, and Shi (2006) for details.

19.3 Adaptation Methods

Recall that in the previous section we listed seven requirements and noted that each of the detectors discussed there needs a few of these requirements. In this section, we review further features of the discussed detector with emphasis on the adaptive methods related to their implementation.

19.3.1 Conventional Detector

The conventional/matched filter detector needs the requirements (1) and (3), only. However, compared to other detectors, it performs poorly. Strictly speaking, the vector $\mathbf{c}_0$ in requirement (1) refers to the spreading code of the desired user at the receiver side (obtained by convolving the corresponding spreading code at the transmitter and the channel impulse response). However, in practice, as channel is unknown (and is time varying), $\mathbf{c}_0$, at the receiver, is replaced by its counterpart at the transmitter, and this incurs further loss in performance.

19.3.2 Decorrelator Detector

The decorrelator detector, on the other hand, needs to have access to the requirements (1)–(4). Moreover, the spreading code vectors $\mathbf{c}_0$ through $\mathbf{c}_{K-1}$ must be those at the receiver side. Hence, for the purpose of implementation, $\mathbf{c}_0$ through $\mathbf{c}_{K-1}$ should be estimated, through use of some training/pilot symbols and by adopting an adaptive algorithm. Either of LMS or RLS algorithms and their alternative forms that we presented in the previous chapters may be used for this purpose. Moreover, as wireless channels are time varying, training symbols should be inserted at regular intervals between the data symbols. Overall, the process of estimating the spreading codes $\mathbf{c}_0$ through $\mathbf{c}_{K-1}$ is a rather expensive process, both in terms of computational complexity and the use of channel resources. The latter refers to the fact that the time used for transmission of training symbols could otherwise be used for transmission of data. In light of these difficulties and the fact that both MMSE and MOE detector are simpler to implement and are superior in performance, the decorrelator detector is of interest only because of historical and theoretical reasons.

19.3.3 MMSE Detector

The adaptation of MMSE detector requires transmission of a training sequence by the desired user (requirement 7) and the knowledge of the symbol boundaries of the desired user (requirement 3). Once this information is available, any adaptation algorithm may be used to adjust the tap-weight vector $\mathbf{w}$.

In an adaptive setting, the convergence of the MMSE detector is determined and/or directly linked to the eigenvalues distribution of the correlation matrix $\mathbf{R} = \mathbf{C}\mathbf{C}^{\mathrm{H}} + \sigma_\nu^2\mathbf{I}$. In particular, when the number of users, K, is smaller than the processing gain, N, (which is usually the case), and also the noise power, σ_ν^2, is significantly smaller than the users' signal powers, $\mathbf{R}$ will have a large eigenvalue spread. In this case, any LMS algorithm will suffer from slow convergence, and any RLS algorithm may experience difficulties because of potential round-off noise problems. The problem in LMS algorithm may be solved by adopting a leaky LMS algorithm. For the RLS type algorithms, a number of solutions that are beyond the scope of this book have been proposed; see Sayed (2003) for a broad discussion on these types of algorithms.

Another point that deserves some attention is the choice of the initial value of $\mathbf{w}$, say $\mathbf{w}(0)$, before starting an adaptive LMS algorithm. We note that the N-by-N matrix $\mathbf{C}\mathbf{C}^{\mathrm{H}}$ is a rank K matrix, thus, its $N - K$ smaller eigenvalues are 0. Moreover, the eigenvectors associated with the nonzero eigenvalues of this matrix belong to the subspace expanded by the columns of $\mathbf{C}$. In addition, assuming that the nonzero eigenvalues of $\mathbf{C}\mathbf{C}^{\mathrm{H}}$ are λ_0, $\lambda_1, \ldots,$ and λ_{K-1}, the eigenvalues of the correlation matrix $\mathbf{R} = E[\mathbf{x}(n)\mathbf{x}^{\mathrm{H}}(n)] = \mathbf{C}\mathbf{C}^{\mathrm{H}} + \sigma_\nu^2\mathbf{I}$ are $\lambda_0 + \sigma_\nu^2$, $\lambda_1 + \sigma_\nu^2$, $\ldots$, $\lambda_{K-1} + \sigma_\nu^2$, σ_ν^2, $\ldots$, and σ_ν^2. Clearly, these eigenvalues determine the convergence rates of the various modes of convergence of the MMSE detector, and the last $N - K$ eigenvalues are those associated with the subspace orthogonal to the signal subspace, that is, the subspace expanded by the columns of the spreading code matrix $\mathbf{C}$. Accordingly, to avoid the slow modes of convergence in an LMS algorithm that may be used for adaptation of $\mathbf{w}$, $\mathbf{w}(0)$ should be a vector within the signal subspace. Clearly, the most trivial solution that satisfies this condition is $\mathbf{w}(0) = 0$. Also, as discussed in Lim *et al.* (2000), it may be also helpful to reset $\mathbf{w}$ to $\mathbf{0}$, every time a user leaves/joins the network, to avoid slow modes of convergence of the MMSE detector.

19.3.4 MOE Detector

The adaptation of MOE detector requires the spreading code of the desired user (requirement 1) and the knowledge of the symbol boundaries of the desired user (requirement 3). Once this information is available, any constrained adaptation algorithm may be used to adjust the tap-weight vector $\mathbf{w}$. The most common algorithm used in practice is the constrained LMS algorithm (or its normalized version) that was introduced in Section 6.11. Also, as noted earlier, to take care of the receiver inaccuracy in its knowledge of the spreading code of the desired user (because the presence of channel changes it), it is common to run the constrained LMS algorithm for an initial training of the MOE detector. Once the tap-weight vector $\mathbf{w}$ reaches a point where the transmitted symbols $s_0(n)$ are detected correctly, with a good probability, the receiver is switched to an MMSE detector and the symbol decisions made by the detector are used to fine tune $\mathbf{w}$ and track possible variations of the channel.

19.3.5 Soft Detectors

The implementation of a soft detector needs the requirements (1) through (6). The requirements (1), (2), (5), and (6) may be replaced by the requirement (7), where the latter is used to obtain the estimates of the formers. It is also possible to ignore requirement (7) and adapt some blind channel identification method to find the spreading codes of the users, requirements (1) and (2). One such method is presented in Liu and Xu (1996), where the authors have developed a MUSIC-like algorithm for blind detection of the spreading codes $\mathbf{c}_0$ through $\mathbf{c}_{K-1}$.

It is also worth noting that a soft detector when combined with a channel decoder in a turbo receiver, in the same manner as the concept of the soft equalization that was introduced in Chapter 17 (Figure 17.35), leads to a receiver with optimum performance. Such receivers can approach the channel capacity bound and, thus, it has become exceedingly popular in recent years (after introduction of the third and fourth generations of the wireless systems). We will get to more details of such receivers in the next chapter, in the context of MIMO systems (i.e., transceivers that are equipped with multiple antennas at both transmitter and receiver sides of a communication channel).

Problems

P19.1 Recall that the tap-weight vectors of the decorrelator and MMSE detectors are, respectively, given by Eqs. (19.29) and (19.37). Show that under the condition where $\sigma_\nu^2 \to 0$, Eq. (19.37) reduces to Eq. (19.29).
Hint: Consider using the matrix inversion lemma formula

$$(\mathbf{B}^{-1} + \mathbf{C}\mathbf{D}^{-1}\mathbf{C}^{\mathrm{H}})^{-1} = \mathbf{B} - \mathbf{B}\mathbf{C}(\mathbf{D} + \mathbf{C}^{\mathrm{H}}\mathbf{B}\mathbf{C})^{-1}\mathbf{C}^{\mathrm{H}}\mathbf{B}$$

and also the identity

$$(\mathbf{I} + \mathbf{A})^{-1} = \mathbf{I} - (\mathbf{I} + \mathbf{A})^{-1}\mathbf{A}$$

P19.2 Derive the results presented in Eqs. (19.47) and (19.49).

P19.3 This problem attempts to provide further insight to the similarities and differences of the MMSE and MOE detectors, through numerical examples. Develop a MATLAB code that generates a random spreading code matrix $\mathbf{C}$, according to "C=randn(N,K)" and performs the following studies.

(i) Assuming $N = 10$ and $K = 3$ and $\sigma_\nu^2 = 0.01$, evaluate the SINR of the MMSE detector according to Eq. (19.40) and the SINR of the MOE detector according to Eq. (19.49). By running your code, confirm that both SINR values are the same all the times.

(ii) For both detectors, find the optimum tap-weight vector $\mathbf{w}_{\mathrm{o}}$ according to Eqs. (19.37) and (19.47), respectively. The answers should not be exactly the same. However, they are closely related. Explain your observation. In particular, if we denote the ith element of $\mathbf{w}_{\mathrm{o}}$ of the MMSE detector by $w_{\mathrm{o},i}^{\mathrm{mmse}}$ and its counterpart from the MOE detector by $w_{\mathrm{o},i}^{\mathrm{moe}}$, the following holds for all values of i:

$$\frac{w_{\mathrm{o},i}^{\mathrm{mmse}}}{w_{\mathrm{o},i}^{\mathrm{moe}}} = \frac{1}{1 + 1/\mathrm{SINR}}$$

(iii) The above numerical results can be also proven analytically, though the derivations are hefty. An interested reader may attempt such proofs.

P19.4 Given the spreading code matrix **C**, present a numerical procedure to decompose $\mathbf{c}_0$ to $\mathbf{c}_0 = \mathbf{c}_{0,1} + \mathbf{c}_{0,2}$ and find the minimum norm solution (19.52).
Hint: Consider using the projection operator concept introduced in Section 12.3.

P19.5 Starting with a random generation of the spreading code matrix **C** according to "C=randn(N,K)" and recalling the solution to Problem P19.4, develop a MATLAB code to find $\mathbf{w}_o$ numerically according to Eqs. (19.52) and (19.59), and confirm that both lead to the same result.

20

OFDM and MIMO Communications

Orthogonal frequency division multiplexing (OFDM) and communication systems that use multiple antennas at transmitter and receiver sides of the link (known as MIMO, for *multiple-input multiple-output*) heavily benefit from the various signal processing techniques, in general, and adaptive filters, in particular. In this chapter, we present a review of these methods and discuss the role of adaptive filtering techniques in implementing these systems and improving their performance.

20.1 OFDM Communication Systems

The OFDM belongs to the broader class of multicarrier communication systems. The general concept is to divide a fast stream of data into a number of much slower substreams to reduce/avoid intersymbol interference (ISI); see Farhang-Boroujeny (2011) for a tutorial introduction to the broad class of multicarrier systems. In OFDM, the ISI is completely removed by inserting a guard interval equal to or longer than the duration of the channel impulse response before each block of data symbols. It is common to refer to each block of data symbols as an OFDM symbol. Hence, each OFDM symbol consists of a set of data symbols, each of which may be from a QAM constellation.

20.1.1 The Principle of OFDM

In OFDM, the transmitter begins with generating the complex baseband signal vector $\mathbf{x}(n)$ by taking the inverse DFT (IDFT) of the symbol vector $\mathbf{s}(n) = [s_0(n)s_1(n)\cdots s_{N-1}(n)]^{\mathrm{T}}$, viz.,

$$\mathbf{x}(n) = \mathcal{F}^{-1}\mathbf{s}(n) \tag{20.1}$$

Adaptive Filters: Theory and Applications, Second Edition. Behrouz Farhang-Boroujeny.

where $\mathcal{F}$ is the DFT matrix and thus $\mathcal{F}^{-1}$ is the IDFT matrix. Noting that

$$\mathcal{F}^{-1} = \frac{1}{N}\begin{bmatrix} 1 & 1 & 1 & \cdots & 1 \\ 1 & e^{j\frac{2\pi}{N}\times 1} & e^{j\frac{4\pi}{N}\times 1} & \cdots & e^{j\frac{2(N-1)\pi}{N}\times 1} \\ 1 & e^{j\frac{2\pi}{N}\times 2} & e^{j\frac{4\pi}{N}\times 2} & \cdots & e^{j\frac{2(N-1)\pi}{N}\times 2} \\ \vdots & \vdots & \vdots & \ddots & \vdots \\ 1 & e^{j\frac{2\pi}{N}\times (N-1)} & e^{j\frac{4\pi}{N}\times (N-1)} & \cdots & e^{j\frac{2(N-1)\pi}{N}\times (N-1)} \end{bmatrix} \tag{20.2}$$

one finds that Eq. (20.1) can be expanded as

$$\mathbf{x}(n) = \sum_{k=0}^{N-1} \mathbf{x}_k(n) \tag{20.3}$$

where

$$\mathbf{x}_k(n) = \frac{1}{N} s_k(n)\mathbf{f}_k, \quad \text{for } k = 0, 1, 2, \ldots, N-1 \tag{20.4}$$

and

$$\mathbf{f}_k = \begin{bmatrix} 1 \\ e^{j\frac{2\pi k}{N}\times 1} \\ e^{j\frac{2\pi k}{N}\times 2} \\ \vdots \\ e^{j\frac{2\pi k}{N}\times (N-1)} \end{bmatrix} \tag{20.5}$$

This observation shows that the kth symbol $s_k(n)$ modulates a complex carrier at frequency $f_k = \frac{2\pi k}{N}$. The scale factor $\frac{1}{N}$ comes from the definition of the IDFT and may be removed without any major impact on the results. Also, we refer to the $\mathbf{f}_k$'s as *modulator vectors*.

For the purpose of transmission, the complex-valued sequence $x(n)$, obtained by concatenating the OFDM symbol vectors $\mathbf{x}(n)$, is modulated to an RF band. At the receiver, through a demodulation process, $x(n)$ or a distorted version of it is recovered; distortion, here, is caused by the channel and, thus, is linear. Assuming an ideal channel, an undistorted replica of $x(n)$ is recovered and, thus, the application of a DFT to each frame (i.e., each symbol vector) of $x(n)$ recovers the data symbols $s_0(n), s_1(n), \ldots, s_{N-1}(n)$. However, the sequence $x(n)$ often has a high rate, and the presence of a multipath channel, in general, results in a significant level of interference among the samples of $x(n)$, similar to ISI in single carrier communications (Chapter 17). Thus, some sort of equalization is needed to combat the channel distortion. In the case of single carrier communications, as discussed in Chapter 17, it is common to use an adaptive transversal equalizer with several taps. This has the disadvantages of high computational complexity and slow convergence. One of the key features that has made OFDM popular in broadband communication systems is that the very special structure of the OFDM symbols allows a very low complexity equalization method. Also, for reasons that are explained below, the OFDM equalizer does not suffer from any slow convergence problem.

The approach that OFDM designers have taken to combat channel effect is based on the following observation. Each subcarrier in OFDM is a constant amplitude sinusoidal tone

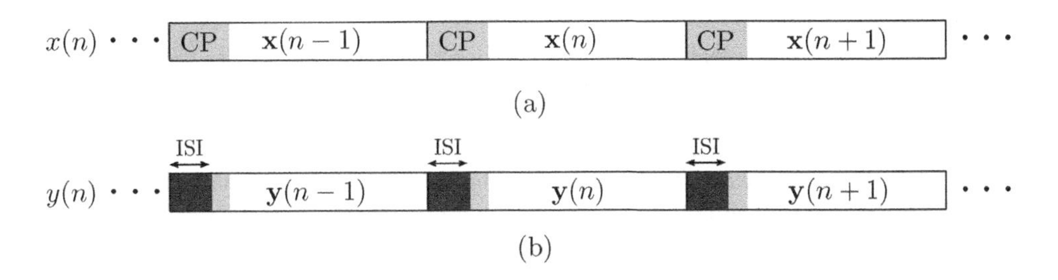

Figure 20.1 An OFDM signal with the added cyclic prefix samples: (a) transmitted signal $x(n)$ and (b) received signal $y(n)$. The presence of cyclic prefix samples avoid ISI among adjacent OFDM symbols.

that is modulated by the respective data symbol $s_k(n)$. Moreover, the frequencies of these tones are selected such that their summation at the transmitter can be generated through an IDFT and at the receiver they can be separated through a DFT. To perform the latter step, the DFT is applied to a length N sequence of the samples of the received signal. On the other hand, we note that each tone in the transmit signal undergoes a transient, which is equal to the duration of the channel impulse response, before reaching its steady state (i.e., becoming a modulated tone with the respective frequency) at the receiver. Hence, to allow separation of the data symbols at the receiver through a DFT, the tones at the transmitter should be extended from the length N to a length $N + N_{\mathrm{cp}}$, where N_{cp} is a duration equal to or longer than the duration of the channel impulse response. Given the form of the tones (i.e., Eq. 20.5), this extension can be made by taking the last N_{cp} samples of $\mathbf{x}(n)$ and appending to its beginning. The added samples are thus called *cyclic prefix* or CP, in short. The subscript "CP" on N_{cp} is clearly selected to reflect this method of extending $\mathbf{x}(n)$.

Figure 20.1 presents an OFDM signal sequence at the transmitter and its counterpart at the receiver. At the transmitter, a CP is added to each OFDM symbol, $\mathbf{x}(n)$. The CP acts as a guard interval that absorbs the channel impulse response. That is, channel transient is absorbed within the CP, leaving the last part of each OFDM symbol a summation of the subcarrier tones with a similar form to Eq. (20.3). More specifically, if we call the last N samples of the nth OFDM symbol at the receiver $\mathbf{y}(n)$, it can be expanded as

$$\mathbf{y}(n) = \sum_{k=0}^{N-1} \mathbf{y}_k(n) \tag{20.6}$$

where

$$\mathbf{y}_k(n) = H(\omega_k) \times \frac{1}{N} s_k(n)\mathbf{f}_k, \quad \text{for } k = 0, 1, 2, \ldots, N-1 \tag{20.7}$$

$H(\omega)$ is the channel frequency response, and $\omega_k = 2\pi k/N$ is the frequency of the kth subcarrier. Note that this result follows as each tone after passing through channel and reaching its steady state is affected by a gain equal to the channel frequency response at the respective frequency.

Considering Eq. (20.6), the data symbols $s_k(n)$, for $k = 0, 1, \ldots, N-1$, can be recovered from $\mathbf{y}(n)$ by taking the DFT of $\mathbf{y}(n)$ and applying a set of single-tap equalizers with the gains of $1/H(\omega_k)$, for $k = 0, 1, \ldots, N-1$, to the respective outputs. It is this trivial method of equalization that has made OFDM so popular. This method of equalization is often referred to as *frequency domain equalization*.

Summarizing the above result, an OFDM communication system is constructed as follows. At the transmitter each OFDM symbol is constructed by taking the IDFT of a vector consisting of N data symbols, and appending a CP to its beginning. The generated OFDM symbols are sequentially transmitted through the channel. At the receiver, symbol boundaries of each OFDM symbol are identified and the CP part of it is removed. The CP-stripped OFDM symbols are then passed through a DFT and the DFT outputs are equalized using the single-tap values $w_k = 1/H(\omega_k)$, for $k = 0, 1, \ldots, N-1$. These give a set of estimates of the transmitted data symbols. Clearly, in practice, the IDFT and DFT operations are implemented using the computationally efficient IFFT and FFT algorithms.

In the above derivation, we ignored the channel noise whose presence, inevitably, results in some inaccuracy in the transmitted symbol estimates. So, at the receiver, only noisy estimates $\tilde{s}_k(n)$ of $s_k(n)$ are obtained. We also assumed that the receiver was aware of the exact carrier frequency in the received signal, thus, the received OFDM symbols $\mathbf{y}(n)$ were free of any carrier offset. Moreover, we assumed that the channel had a finite duration shorter than or equal to the length of CP, and the channel $h(n)$ was known at the receiver, thus, perfect equalization was possible. Furthermore, it was assumed that the CP position could be identified by the receiver. Clearly, none of these tasks can be perfectly established at any receiver. In the rest of this section, we discuss a few basic concepts that are commonly used to obtain some estimates of the above information from the received signal $y(n)$, hence, perform the tasks of, carrier recovery, timing synchronization, and channel equalization.

To facilitate the tasks of carrier recovery, timing synchronization, and channel equalization, it has become a standard practice to partition the transmitted information into segments of several hundred or several thousand bytes and transmit one segment at a time. These segments are referred to as *packets*. Each packet is preambled with some training symbols to synchronize the receiver with the incoming packet and also to set up the frequency domain equalizer coefficients. For longer packets, additional pilot symbols are added in between the data symbols for tracking possible variations of channel, as well as carrier and timing drifts.

20.1.2 Packet Structure

Here, as a typical example, we present the packet structure proposed in the IEEE 802.11a standard and discuss how it is used to take care of carrier recovery, timing synchronization, and channel equalization. Figure 20.2 presents this packet structure. It begins with a preamble consisting of 10 short training symbols and 2 long training symbols. A signal field symbol and OFDM data symbols (the payload) then follow.

In 802.11a, the IFFT/FFT length is equal to 64. Therefore, the maximum number of subcarriers is also 64. From these, only 52 subcarriers carry data/training symbols. They are subcarrier numbers from -26 to -1 and from 1 to 26. The rest of the subcarriers, including the 0th subcarrier, are set equal to 0.

The short training, effectively, is a collection of 12 tones that are located at the center of the subcarriers $\mathcal{S} = \{-24, -20, -16, -12, -8, -4, 4, 8, 12, 16, 20, 24\}$. The symbol

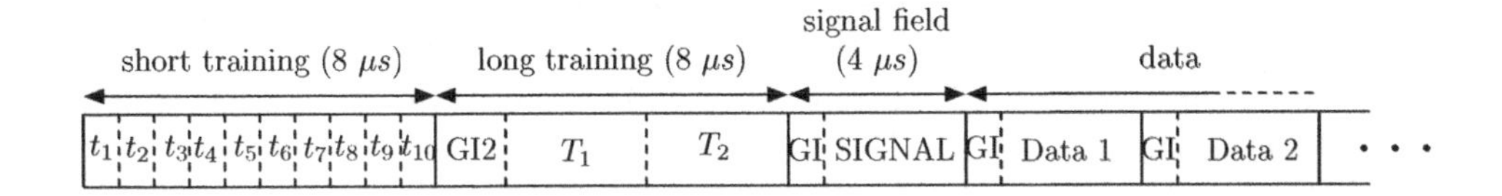

Figure 20.2 Packet structure of IEEE 802.11a.

Table 20.1 The symbol values for short training symbols in IEEE 802.11a.

k	$X_{s,k}$
−24	$\sqrt{\frac{13}{6}}(1+j)$
−20	$-\sqrt{\frac{13}{6}}(1+j)$
−16	$\sqrt{\frac{13}{6}}(1+j)$
−12	$-\sqrt{\frac{13}{6}}(1+j)$
−8	$-\sqrt{\frac{13}{6}}(1+j)$
−4	$\sqrt{\frac{13}{6}}(1+j)$
4	$-\sqrt{\frac{13}{6}}(1+j)$
8	$-\sqrt{\frac{13}{6}}(1+j)$
12	$\sqrt{\frac{13}{6}}(1+j)$
16	$\sqrt{\frac{13}{6}}(1+j)$
20	$\sqrt{\frac{13}{6}}(1+j)$
24	$\sqrt{\frac{13}{6}}(1+j)$

values, $X_s(k)$, corresponding to these tones are listed in Table 20.1. The short symbols are thus generated according to the equation

$$x_{\text{short}}(t) = g_{\text{short}}(t) \sum_{k \in \mathcal{S}} X_s(k) e^{j2\pi k \Delta_c t} \tag{20.8}$$

where

$$g_{\text{short}}(t) = \begin{cases} 1, & 0 \leq t \leq 8\mu s \\ 0, & \text{otherwise} \end{cases} \tag{20.9}$$

and Δ_c is the spacing between the adjacent subcarriers. In 802.11a, if all 64 subcarriers were used, the total bandwidth would be 20 MHz. Hence, the subcarrier spacing is $\frac{20}{64} =$ 0.3125 MHz. Substituting this value in Eq. (20.8), one finds that $x_{\text{short}}(t)$ over the interval $0 \leq t \leq 8\mu s$ is periodic and repeats 10 times. It is also worth noting that the factor $\sqrt{\frac{13}{6}}$ in the symbol values listed in Table 20.1 is to equalize the power of short training symbols to a level equal to that of the rest of the symbols in the packet.

The long training consists of two full-length OFDM symbols (3.2 μs each) with an extended CP of length of 1.6 μs. This is twice the length of the CP in the succeeding OFDM symbols. The long training symbols are generated using the following equation

$$x_{\text{long}}(t) = g_{\text{long}}(t) \sum_{k=-26}^{26} X_l(k) e^{j2\pi k \Delta_c t} \tag{20.10}$$

where

$$g_{\text{long}}(t) = \begin{cases} 1, & 8 < t \leq 16\mu s \\ 0, & \text{otherwise} \end{cases} \tag{20.11}$$

and the frequency domain symbol values $X_l(k)$ for $-26 \leq k \leq 26$ are given by

$$\begin{aligned} \{X_l(k)\} = \{&1, 1, -1 - 1, 1, 1, -1, 1, -1, 1, 1, 1, 1, 1, -1, -1, 1, 1, -1, \\ &1, -1, 1, 1, 1, 1, 0, 1, -1, -1, 1, 1, -1, \\ &1, -1, 1, -1, -1, -1, -1, -1, 1, 1, -1, -1, 1, -1, 1, -1, 1, 1, 1, 1\} \end{aligned}$$

Note that at DC, that is, for $k = 0$, $X_l(k) = 0$. The preamble is formed by appending $x_{\text{short}}(t)$ and $x_{\text{long}}(t)$.

The signal field is an OFDM symbol that carries information about the rest of the packet. The information content of the signal field includes the number of information bits in the packet and the data rate.

20.1.3 Carrier Acquisition

Recall the carrier acquisition method that was introduced in Section 17.6.4 in relation to cyclic equalizer. Because the preamble (both short training and long training) is periodic, in IEEE 802.11a, similar equations to Eq. (17.106) through Eq. (17.108) are applicable. The receiver should first identify the short training by looking for a periodic signal with period of 0.8 ms. It should then remove the transient part of the short training, simply by ignoring the first cycle of it, and apply Eq. (17.108) to the rest of its cycles to obtain an estimate of the carrier offset. The identified carrier offset is applied to correct the rest of the packet. This process is called *coarse carrier acquisition*. It brings the carrier offset to a value smaller than one-half of the carrier spacing in the OFDM data symbols.

Further tuning of the carrier is performed using the long training. Here, first, the timing acquisition method that is discussed below identifies the boundaries of the long training symbols T_1 and T_2 (Figure 20.2). Then, as T_1 and T_2 form two periods of the same signal, Eq. (17.108) is applied to obtain an estimate of the residual carrier offset. This is referred to as *fine carrier acquisition*, because of obvious reasons. This method of using periodic sequences to identify the carrier frequency offset was first introduced by Moose (1994) for OFDM systems. It thus often referred to as *Moose's carrier acquisition* method.

20.1.4 Timing Acquisition

Given a packet structure similar to that of IEEE 802.11a, the following algorithm may be used to find the boundaries of the training symbols T_1 and T_2. Once these are found, the rest of the symbols in the packet follow simply by keeping track of the signal sample indices.

The timing acquisition method discussed here was first introduced in Schmidl and Cox (1997). Note that the long training consist of two similar length 64 sequences that are preappended by a length 32 cyclic prefix. Accordingly, the autocorrelation function of the received signal $y(n)$, viz.,

$$r_{yy}(n) = \frac{1}{N}\sum_{k=0}^{N-1} y(n+k)y^*(n+N+k) \tag{20.12}$$

will find its peak when n is over the interval where $y(n+k) = y(n+N+k)$. This is when the above summation is run over the terms that belong to the long training part of the preamble. The magnitude of $r_{yy}(n)$ begins to drop as soon as the last term of $y(n+N+k)$, that is, $y(n+2N-1)$, drops out of the range of the long training symbol T_2. Accordingly, the following algorithm may be used to obtain the samples at the boundaries of OFDM symbols within a received packet:

- Explore $|r_{yy}(n)|$ for successive values of n.
- When n is at the beginning of the long training, $|r_{yy}(n)|$ should reach a peak and stay flat for 32 consecutive values. It will then begin to drop.
- Identify the value of n for which $|r_{yy}(n)|$ begins to drop. Add $2N = 128$ to this value of n to obtain the first sample of the signal field OFDM symbol.
- The succeeding data symbols begin at a period of $N + N_{\text{cp}} = 80$ samples.

To help the reader to examine the above concept, as well as the equalization method that is introduced below, a MATLAB script, named "`OFDM.m`," is provided on the accompanying website. This code provides a skeleton for generation of a packet of OFDM data sequence and the corresponding preamble. Adding additional lines to this code to obtain $r_{yy}(n)$ and plotting a sample result, we obtain Figure 20.3. It may come as a surprise to the reader that there are two similar top flats in Figure 20.3. This will not be a surprise, if one notes that the short training part of the preamble that has a period of 16 samples may be also thought as a periodic sequence with any integer multiple of 16, which of course will include the period of 64. The reader may also note that even though the above algorithm suggested the use of the fourth corner of the plot of $|r_{yy}(n)|$ as the reference point for locating the OFDM symbol boundaries, any of the corner points of $|r_{yy}(n)|$ may be used. Or one may choose to exploit all four corner points of the plot to obtain a more accurate reference for locating the symbol boundaries.

20.1.5 Channel Estimation and Frequency Domain Equalization

Once the OFDM symbols boundary points are identified and the received signal samples are frequency compensated, one may take either of the training symbols T_1 and T_2 or their

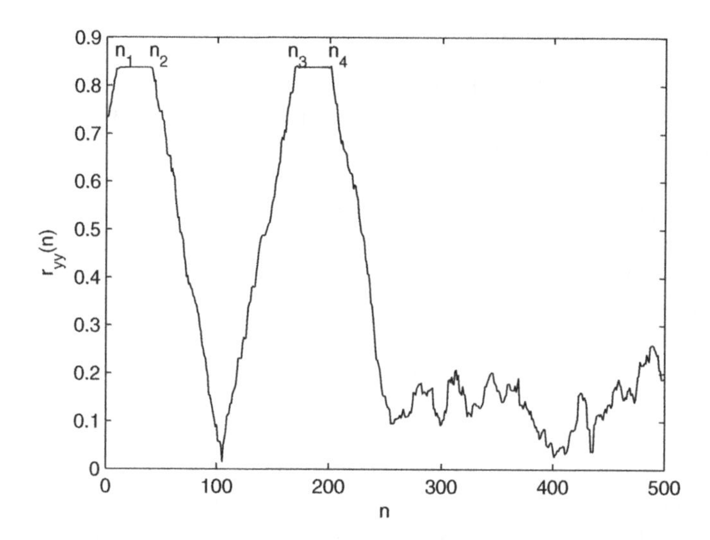

Figure 20.3 A sample plot of $|r_{yy}(n)|$. Points that may be used for timing acquisition are the top corners of the trapezoidal shapes on the plot. Their positions are marked as the time indices n_1, n_2, n_3, and n_4.

average, and find a channel response $\mathbf{h} = [h_0 h_1 \cdots h_{L-1}]^{\mathrm{T}}$ of a finite length L that when convolved with the long training best matches the received signal. The result is referred to as $\hat{\mathbf{h}}$. Note that in common practice L is set equal to N_{cp} and $N_{\mathrm{cp}} < N$. In the sequel, we discuss two approaches of finding $\hat{\mathbf{h}}$. Once $\hat{\mathbf{h}}$ is obtained, the frequency domain equalizer coefficients are calculated by taking the N point DFT of $\hat{\mathbf{h}}$ (after appending zeros to its end to expand it to the length of N) to obtain the channel frequency response $\hat{H}(\omega_k)$, for $k = 0, 1, \ldots, N-1$, and set $w_k = 1/\hat{H}(\omega_k)$.

Maximum Likelihood (ML) Channel Estimator

The ML channel estimator begins with the following channel model, which is inferred from the OFDM principle presented in Section 20.1.1:

$$\mathbf{z} = \mathbf{S}\mathbf{B}\mathbf{h} + \mathbf{v} \tag{20.13}$$

Here, $\mathbf{S}$ is the diagonal matrix containing the training symbols at its diagonal. Hence, assuming that there are M training symbols, $\mathbf{S}$ is a diagonal matrix of size $M \times M$. When Eq. (20.13) is used in connection to the long training symbols in IEEE 802.11a, $M = 52$ and diagonal elements of $\mathbf{S}$ are those listed under $\{X_l(k)\}$ above. The matrix $\mathbf{B}$ in Eq. (20.13) is to transform $\mathbf{h}$ to the channel frequency domain gains $H(\omega_k)$. It, thus, is of size $M \times L$ and its mlth element is given by

$$b_{ml} = \mathrm{e}^{-j2\pi k_m l/N} \tag{20.14}$$

where k_m is the index corresponding the mth training symbol. For the IEEE 802.11a, k_m takes the values of -26 to 26, excluding 0. The vectors $\mathbf{v}$ and $\mathbf{z}$, both of size $M \times 1$, contain the elements of the DFT of channel noise samples and the received signal samples that correspond to the training symbols. Also, the reader should note that we have dropped the time index n for convenience.

Given the vector $\mathbf{z}$, the ML channel estimator selects the estimate of $\mathbf{h}$ by minimizing the cost function

$$\xi_{\mathrm{ML}} = \|\mathbf{z} - \mathbf{SBh}\|^2 \tag{20.15}$$

where $\|\cdot\|$ denotes the norm of a vector, that is, for a vector $\mathbf{a}$, $\|\mathbf{a}\| = \sqrt{\mathbf{a}^{\mathrm{H}}\mathbf{a}}$. Under the condition that the elements of $\mathbf{v}$ are a set of independent identically distributed Gaussian random variables, minimization of ξ_{ML} results in the most likely value of $\mathbf{h}$, that is, the choice that maximizes the *a posteriori* probability $p(\mathbf{h}|\mathbf{z})$. Hence, the name ML channel estimator is to reflect this fact.

Minimization of the cost function ξ_{ML} is that of a least-squares problem. When $L \leq M$, it is an underdetermined problem, that is, there are more parameters (the number elements of $\mathbf{h}$) than the number of sample points (the number of elements of $\mathbf{z}$). In this case, one or more solutions to $\mathbf{h}$ can be found to match the vectors $\mathbf{z}$ and $\mathbf{SBh}$ perfectly.

A case of more interest in practice is when $M > L$. In this case, the problem is that of a overdetermined problem, and following the procedure presented in Chapter 12, one finds the following estimate minimizes ξ_{ML}:

$$\hat{\mathbf{h}} = \left(\mathbf{B}^{\mathrm{H}}\mathbf{S}^{\mathrm{H}}\mathbf{SB}\right)^{-1}\mathbf{B}^{\mathrm{H}}\mathbf{S}^{\mathrm{H}}\mathbf{z} \tag{20.16}$$

Assuming that the training symbols are from a PSK constellation with the magnitude of unity, $\mathbf{S}^{\mathrm{H}}\mathbf{S} = \mathbf{I}$ and Eq. (20.16) reduces to

$$\hat{\mathbf{h}} = \left(\mathbf{B}^{\mathrm{H}}\mathbf{B}\right)^{-1}\mathbf{B}^{\mathrm{H}}\mathbf{S}^{\mathrm{H}}\mathbf{z} \tag{20.17}$$

Linear Minimum Mean-Squared Error (LMMSE) Channel Estimator

The LMMSE channel estimator finds an estimate of $\mathbf{h}$ according to the equation

$$\hat{\mathbf{h}} = \mathbf{W}^{\mathrm{H}}\mathbf{z} \tag{20.18}$$

where

$$\mathbf{W} = [\mathbf{w}_0\mathbf{w}_1 \cdots \mathbf{w}_{L-1}] \tag{20.19}$$

is the estimator coefficient matrix, and $\mathbf{w}_0$ through $\mathbf{w}_{L-1}$ are column vectors of size M. The coefficient matrix is found by minimizing the cost function

$$\xi_{\mathrm{LMMSE}} = E[\|\hat{\mathbf{h}} - \mathbf{h}\|^2] \tag{20.20}$$

Substituting Eqs. (20.18) and (20.19) in Eq. (20.20), and rearranging the result, we obtain

$$\xi_{\mathrm{LMMSE}} = \sum_{l=0}^{L-1} E[|\mathbf{w}_l^{\mathrm{H}}\mathbf{z} - h_l|^2] \tag{20.21}$$

Furthermore, we note that the problem of minimization ξ_{LMMSE} can be broken into L simpler problems of minimization of the individual terms under the summation on the right-hand side of Eq. (20.21). Moreover, each of the latter problems has the familiar form that was first presented in Chapter 3 and has the solution

$$\mathbf{w}_{l,o} = \mathbf{R}_{zz}^{-1}\mathbf{p}_l, \quad \text{for } l = 0, 1, \ldots, L-1 \tag{20.22}$$

where, here, $\mathbf{R}_{zz} = E[\mathbf{z}\mathbf{z}^{\mathrm{H}}]$ and $\mathbf{p}_l = E[\mathbf{z}h_l^*]$. Combining the results in Eq. (20.22), we get

$$\mathbf{W}_o = \mathbf{R}_{zz}^{-1}\mathbf{P} \tag{20.23}$$

where

$$\begin{aligned}\mathbf{P} &= [\mathbf{p}_0\ \mathbf{p}_1 \cdots \mathbf{p}_{L-1}] \\ &= E[\mathbf{z}\mathbf{h}^{\mathrm{H}}]\end{aligned} \tag{20.24}$$

To develop some insight to the properties of the LMMSE channel estimator, we recall that in Eq. (20.13) the matrices $\mathbf{S}$ and $\mathbf{B}$ are deterministic, and the vectors $\mathbf{h}$ and $\mathbf{v}$ are unknown random variables. Hence,

$$\mathbf{R}_{zz} = \mathbf{S}\mathbf{B}\mathbf{R}_{hh}\mathbf{B}^{\mathrm{H}}\mathbf{S}^{\mathrm{H}} + \mathbf{R}_{\nu\nu} \tag{20.25}$$

where $\mathbf{R}_{hh} = E[\mathbf{h}\mathbf{h}^{\mathrm{H}}]$ and we have defined $\mathbf{R}_{\nu\nu} = E[\mathbf{v}\mathbf{v}^{\mathrm{H}}]$. Also, assuming that $\mathbf{h}$ and $\mathbf{v}$ are independent random vectors,

$$\mathbf{P} = \mathbf{S}\mathbf{B}\mathbf{R}_{hh} \tag{20.26}$$

From the results presented in Eqs. (20.23), (20.25), (20.26), and (20.16) (or Eq. (20.17)), we make the following observations:

- While the ML channel estimator does not take into account the statistics of $\mathbf{h}$, the LMMSE channel estimator optimization makes use of such statistics.
- The ML channel estimator makes the assumption that the elements of $\mathbf{v}$ are Gaussian and independent of one another but does not care about their variance. The LMMSE channel estimator, on the other hand, does not require that the elements of $\mathbf{v}$ be Gaussian and/or independent of one another, but makes use of the correlation among the elements of $\mathbf{v}$ to improve on the accuracy of the estimated channel.

Clearly, when the channel and noise statistics, $\mathbf{R}_{hh}$ and $\mathbf{R}_{\nu\nu}$, respectively, are known, as the LMMSE estimator makes use of them, it naturally performs superior to the ML estimator that ignores these information.

20.1.6 *Estimation of* $\mathbf{R}_{hh}$ and $\mathbf{R}_{\nu\nu}$

In practice, on the basis of empirical measurements, channel models are built and accordingly $\mathbf{R}_{hh}$ is precalculated. Additional adjustment to $\mathbf{R}_{hh}$ may also be made if SNR at the receiver could be measured.

In the case of IEEE 802.11a, its packet format allows the following trivial method of estimating $\mathbf{R}_{\nu\nu}$. Consider the channel model (20.13) and let

$$\mathbf{z}(n) = \mathbf{S}\mathbf{B}\mathbf{h} + \mathbf{v}(n), \quad \text{for } n = 1, 2 \tag{20.27}$$

denote the two long training symbols of a received data packet. Note that $\mathbf{S}$ is not time indexed, because the same symbols are transmitted for both $n = 1$ and $n = 2$. Next, if we define the measurable vector

$$\mathbf{u} = \mathbf{z}(1) - \mathbf{z}(2) \tag{20.28}$$

Equation (20.27) implies that

$$\mathbf{u} = \mathbf{v}(1) - \mathbf{v}(2) \tag{20.29}$$

Assuming that $\mathbf{v}(1)$ and $\mathbf{v}(2)$ are independent vectors, computation of the correlation coefficients of the elements of $\mathbf{u}$ leads to an estimated correlation matrix $\hat{\mathbf{R}}_{uu}$ which will be twice the respective estimate $\hat{\mathbf{R}}_{\nu\nu}$ of $\mathbf{R}_{\nu\nu}$. Hence, to obtain an estimate of $\mathbf{R}_{\nu\nu}$, one may use Eq. (20.28) to calculate $\mathbf{u}$. This is then used to obtain an estimate $\hat{\mathbf{R}}_{uu}$ of $\mathbf{R}_{uu}$ and then

$$\hat{\mathbf{R}}_{\nu\nu} = \frac{1}{2}\hat{\mathbf{R}}_{uu} \tag{20.30}$$

is calculated.

20.1.7 Carrier-Tracking Methods

Most standards have adopted the use of pilot symbols for carrier tracking. For instance, in IEEE 802.11a, the subcarrier numbers 7, 27, 43, and 57 are continuously filled in with pilot symbols that are known to the receiver. It has also been noted that the unused 0th subcarrier and the unused subcarriers at the two sides of the OFDM band may be treated as subcarriers with known null transmit symbols. These are commonly referred to as *virtual subcarriers*. In the section "pilot-aided carrier-tracking algorithms," below, we treat pilot and virtual subcarriers similarly and present a related carrier-tracking algorithm.

There are also several other carrier-tracking algorithms that make use of the redundancy introduced by CP. These fall under the general class of blind carrier-tracking methods. Examples of algorithms related to these methods are presented in the section "blind carrier-tracking algorithms."

Pilot-Aided Carrier-Tracking Algorithms

In the absence of carrier phase and frequency offset, the nth cyclic prefix striped received OFDM symbol is given by Eq. (20.6). When there is a phase offset ϕ_0 and a carrier frequency offset $\Delta\omega_c$, the nth cyclic prefix striped received OFDM symbol is obtained as

$$\mathbf{y}(n) = e^{j\phi_0}\boldsymbol{\Phi}(\Delta\omega_c)\sum_{k=0}^{N-1}\frac{H(\omega_k)}{N}s_k(n)\mathbf{f}_k + \mathbf{v}(n) \tag{20.31}$$

where $\mathbf{v}(n)$ is the channel noise vector and

$$\boldsymbol{\Phi}(\Delta\omega_c) = \begin{bmatrix} 1 & 0 & 0 & \cdots & 0 \\ 0 & e^{j\Delta\omega_c} & 0 & \cdots & 0 \\ 0 & 0 & e^{j2\Delta\omega_c} & \cdots & 0 \\ \vdots & \vdots & \vdots & \ddots & \vdots \\ 0 & 0 & 0 & \cdots & e^{j(N-1)\Delta\omega_c} \end{bmatrix} \tag{20.32}$$

To proceed, we note that the phase offset factor $e^{j\phi_0}$ is a constant that can be absorbed in the single-tap equalizers tap weights. Hence, we remove the factor from the right-hand side of Eq. (20.31) for the rest of our discussions. With this modification, Eq. (20.31) reduces to

$$\mathbf{y}(n) = \mathbf{\Phi}(\Delta\omega_c) \sum_{k=0}^{N-1} \frac{H(\omega_k)}{N} s_k(n)\mathbf{f}_k + \mathbf{v}(n) \tag{20.33}$$

We note that $\mathbf{\Phi}(\Delta\omega_c)$ is an undesired factor that has been introduced as a result of the frequency offset. Moreover, premultiplying $\mathbf{y}(n)$ by $\mathbf{\Phi}^*(\Delta\omega_c) = \mathbf{\Phi}(-\Delta\omega_c)$ will result in the carrier offset free vector

$$\begin{aligned}\tilde{\mathbf{y}}(n) &= \mathbf{\Phi}(-\Delta\omega_c)\mathbf{y}(n) \\ &= \sum_{k=0}^{N-1} \frac{H(\omega_k)}{N} s_k(n)\mathbf{f}_k + \mathbf{\Phi}(-\Delta\omega_c)\mathbf{v}(n)\end{aligned} \tag{20.34}$$

To develop an adaptive algorithm for correcting the carrier frequency offset, we define the vector

$$\check{\mathbf{y}}(n) = \mathbf{\Phi}(\varphi)\mathbf{y}(n) \tag{20.35}$$

and search for the value of φ that minimizes the cost function

$$J(\varphi) = \sum_{k\in\mathcal{P}} E\left[\left|\mathbf{f}_l^{\mathrm{H}}\check{\mathbf{y}}(n) - H(\omega_k)s_l(n)\right|^2\right] \tag{20.36}$$

where $\mathcal{P}$ is the set of subcarrier numbers of the pilot symbols. An LMS-type algorithm can now be implemented by replacing $J(\varphi)$ with its coarse estimate

$$\hat{J}(\varphi) = \sum_{k\in\mathcal{P}} \left|\mathbf{f}_l^{\mathrm{H}}\check{\mathbf{y}}(n) - H(\omega_k)s_l(n)\right|^2 \tag{20.37}$$

and running the gradient update equation

$$\varphi(n+1) = \varphi(n) - \mu\frac{\partial \hat{J}(\varphi)}{\partial\varphi} \tag{20.38}$$

Replacing Eq. (20.37) in Eq. (20.38) and noting that

$$\frac{\partial\mathbf{\Phi}(\varphi)}{\partial\varphi} = j\mathbf{\Lambda}\mathbf{\Phi}(\varphi) \tag{20.39}$$

where

$$\mathbf{\Lambda} = \begin{bmatrix} 0 & 0 & 0 & \cdots & 0 \\ 0 & 1 & 0 & \cdots & 0 \\ 0 & 0 & 2 & \cdots & 0 \\ \vdots & \vdots & \vdots & \ddots & \vdots \\ 0 & 0 & 0 & \cdots & (N-1) \end{bmatrix}$$

we obtain

$$\varphi(n+1) = \varphi(n) - 2\mu\sum_{k\in\mathcal{P}} \Re\left\{j\mathbf{f}_l^{\mathrm{H}}\mathbf{\Lambda}\check{\mathbf{y}}(n)\left(\mathbf{f}_l^{\mathrm{H}}\check{\mathbf{y}}(n) - H(\omega_k)s_l(n)\right)^*\right\} \tag{20.40}$$

The update equation (20.40) can be improved further by adding the unused subcarriers to the list of pilots. Here, the pilot symbols are equal to zero. As mentioned above, the unused subcarriers are called virtual subcarriers and, accordingly, the related carrier recovery algorithms are referred to as virtual subcarriers-based carrier recovery methods. Note that even if there are no pilot symbols, the virtual/null subcarriers alone may be used for carrier tracking. In that case, Eq. (20.40) converts to

$$\varphi(n+1) = \varphi(n) - 2\mu \sum_{k\in\mathcal{V}} \Re\left\{ j\mathbf{f}_l^{\mathrm{H}} \mathbf{\Lambda} \check{\mathbf{y}}(n) \left(\mathbf{f}_l^{\mathrm{H}} \check{\mathbf{y}}(n)\right)^* \right\} \tag{20.41}$$

where $\mathcal{V}$ is the set of subcarrier numbers of the virtual subcarriers. Clearly, the update equations (20.40) and (20.41) may also be combined.

We note that an implementation of Eq. (20.40) requires the knowledge of the channel coefficients $H(\omega_k)$ that should be replaced by their estimates. On the other hand, if we limit ourselves to the virtual subcarriers, where Eq. (20.41) is applicable, the channel knowledge is not required. Yet, there are other methods that can be used to track the carrier frequency without even using the knowledge of the position of pilot or virtual subcarriers. These methods are referred to as *blind carrier-tracking algorithms*.

Blind Carrier-Tracking Algorithms

There are several blind carrier-tracking algorithms in the literature. In this section, we present one of them. The presented algorithm takes note of the fact that the addition of cyclic prefix to OFDM symbols add a regular variation (ripple) to its spectrum that may be deployed to develop a simple carrier-tracking algorithm. The blind carrier-tracking method that is presented here is due to Talbot (2008); also see Talbot and Farhang-Boroujeny (2008).

OFDM spectrum

As discussed earlier, an OFDM signal is constructed by concatenating a set of cyclic prefixed symbols. Let us use $\mathbf{x}^{\mathrm{cpd}}(n)$ to denote the nth cyclic prefixed symbol of the OFDM signal $x(n)$. Following similar presentations to those of Eq. (20.3) through Eq. (20.5), here, we have

$$\mathbf{x}^{\mathrm{cpd}}(n) = \sum_{k=0}^{N-1} \mathbf{x}_k^{\mathrm{cpd}}(n) \tag{20.42}$$

where

$$\mathbf{x}_k^{\mathrm{cpd}}(n) = \frac{1}{N} s_k(n) \mathbf{f}_k^{\mathrm{cpd}}, \quad \text{for } k = 0, 1, 2, \ldots, N-1 \tag{20.43}$$

and

$$\mathbf{f}_k^{\mathrm{cpd}} = \begin{bmatrix} e^{j\frac{2\pi k}{N}\times(N-N_{\mathrm{cp}})} \\ \vdots \\ e^{j\frac{2\pi k}{N}\times(N-1)} \\ 1 \\ e^{j\frac{2\pi k}{N}\times 1} \\ e^{j\frac{2\pi k}{N}\times 2} \\ \vdots \\ e^{j\frac{2\pi k}{N}\times(N-1)} \end{bmatrix} \tag{20.44}$$

We are interested in evaluating the power spectral density of $x(n)$. To this end, we first assume that $s_k(n)$'s are independent across the frequency axis (the index k), and note that this implies

$$\Phi_{xx}(\omega) = \sum_{k \in \mathcal{K}} \Phi_{x_k x_k}(\omega) \tag{20.45}$$

where $\mathcal{K}$ is the set of subcarrier indices of the active (data and pilot) subcarriers, $\Phi_{xx}(\omega)$, following the notations in Chapter 2, denotes the PSD of $x(n)$, and, similarly, $\Phi_{x_k x_k}(\omega)$ denotes the PSD of the sequence $x_k(n)$ obtained by concatenating the $\mathbf{x}_k^{\text{cpd}}(n)$'s.

Next, to evaluate $\Phi_{x_k x_k}(\omega)$, we note that

$$x_k(n) = \frac{1}{N} \sum_{l=-\infty}^{\infty} s_k(l) g(n - l(N + N_{\text{cp}})) \mathrm{e}^{j2\pi(n - l(N+N_{\text{cp}}) - N_{\text{cp}})k/N} \tag{20.46}$$

where $g(n)$ is referred to as *symbol shaping window function*. It is a rectangular window that confines the duration of each OFDM symbol to $N + N_{\text{cp}}$ samples. More specifically,

$$g(n) = \begin{cases} 1, & -N_{\text{cp}} \leq n \leq N - 1 \\ 0, & \text{otherwise} \end{cases} \tag{20.47}$$

Equation (20.46) implies that $x_k(n)$ is obtained by passing the symbol sequence $s_k(n)$ through the system shown in Figure 20.4.

Following Figure 20.4, the PSD of $x_k(n)$ is obtained as

$$\Phi_{x_k x_k}(\omega) = \frac{\sigma_s^2}{(N + N_{\text{cp}})N^2} \left| G(\mathrm{e}^{j(\omega - \frac{2\pi k}{N})}) \right|^2 \tag{20.48}$$

where σ_s^2 is the variance of $s_k(n)$ and $G(\omega)$ is the Fourier transform of $g(n)$. In Eq. (20.48), the frequency shift of $\frac{k}{N}$ in $G(\mathrm{e}^{j(\omega - \frac{2\pi k}{N})})$ is a consequence of the modulating factor $\mathrm{e}^{j\frac{2\pi k}{N}(n - N_{\text{cp}})}$ behind $g(n)$ in Figure 20.4, and the factor $1/N^2$ arises because of the factor $1/N$ on the front of $g(n)$. The additional factor $1/(N + N_{\text{cp}})$ is included because there is only one $s_k(n)$ for every $N + N_{\text{cp}}$ samples of $x_k(n)$, hence, its energy averages out over $N + N_{\text{cp}}$ samples.

Substituting Eq. (20.48) in Eq. (20.45), we obtain

$$\Phi_{xx}(\omega) = \frac{\sigma_s^2}{(N + N_{\text{cp}})N^2} \sum_{k \in \mathcal{K}} \left| G(\mathrm{e}^{j(\omega - \frac{2\pi k}{N})}) \right|^2 \tag{20.49}$$

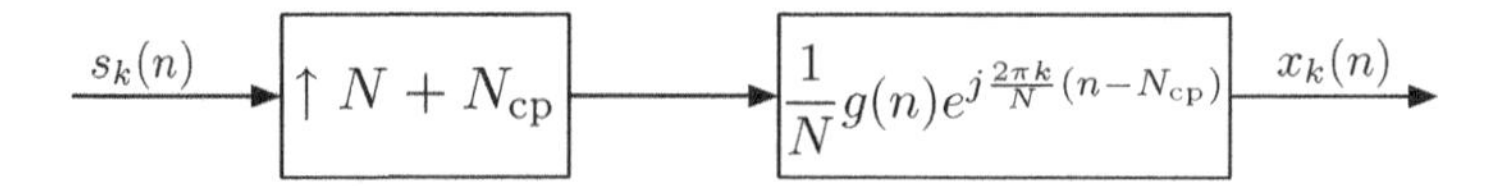

Figure 20.4 The process of generating $x_k(n)$ from $s_k(n)$.

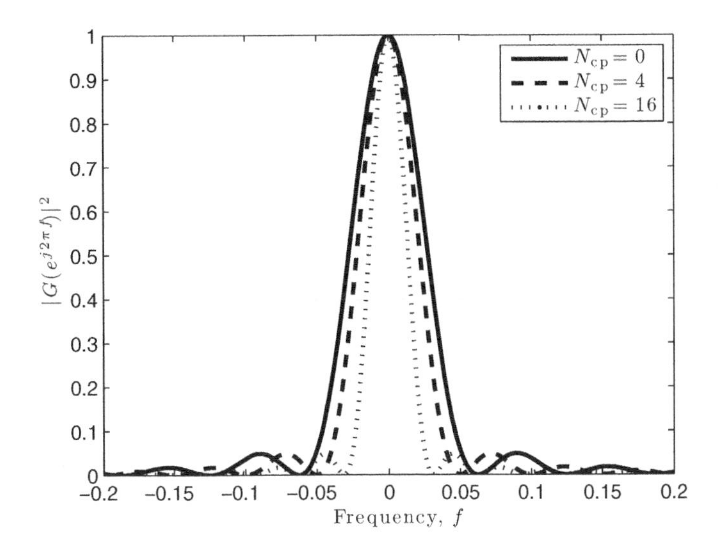

Figure 20.5 The impact of the cyclic prefix length on the variation of the width of main lobe of $G(\omega)$.

Also, it is straightforward to show that for the rectangular window (20.47) (see Problem P20.5, at the end of this chapter)

$$|G(\omega)|^2 = \sum_{n=-(N+N_{cp}-1)}^{N+N_{cp}-1} (N + N_{cp} - |n|) \cos(n\omega) \tag{20.50}$$

Figure 20.5 presents a set of plots of $|G(\omega)|^2$, normalized to the maximum amplitude of unity. The parameter N is set equal to 16 and N_{cp} (the length of CP) is given the values of 0, 4, and 16. We note that for a fixed N, as N_{cp} increases, the width of the main lobe of $|G(\omega)|^2$ decreases. On the other hand, the subcarrier spacing remains constant at $1/N$. Hence, Eq. (20.49) implies that, for a fixed N, increasing the length of CP induces ripples with larger amplitudes in the in-band region of OFDM PSD. This is shown in Figure 20.6 for the case where $N = 16$, there are 10 active subcarriers at frequencies $f = \pm\frac{1}{N}, \pm\frac{2}{N}, \ldots, \pm\frac{5}{N}$, and N_{cp} is given the values of 0, 4, and 16. The magnitude of different spectra are also normalized to the maximum of unity to allow a quantitative comparison.

Impact of Carrier Offset

The spectral effect of a carrier offset is, obviously, a shift of the signal spectrum. For example, if the OFDM signal corresponding to the PSD for the case of $N_{cp} = 4$ in Figure 20.6 experiences a carrier offset of $\Delta f_c = \Delta\omega_c/2\pi = 0.375/N$, then the normalized OFDM PSD will be the one shown in Figure 20.7. As the spectral shift is directly related to the size of carrier offset, Δf_c, a measurement of the spectral shift is a measurement of the carrier offset.

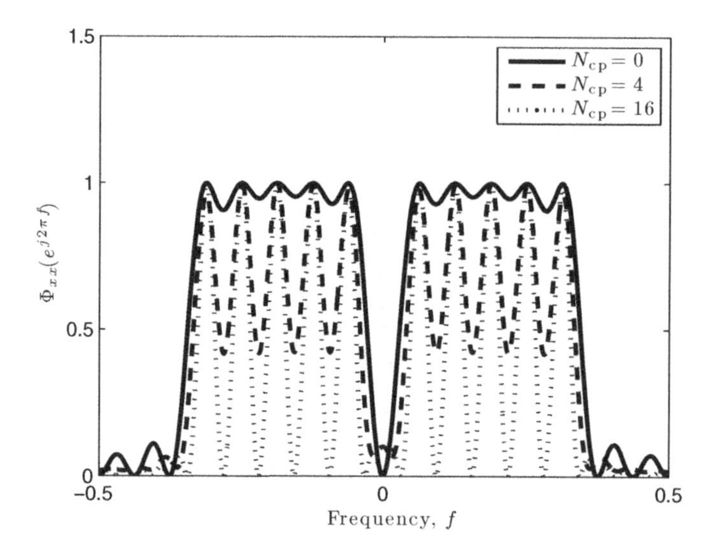

Figure 20.6 Impact of the cyclic prefix length on the variation of the PSD of an OFDM signal.

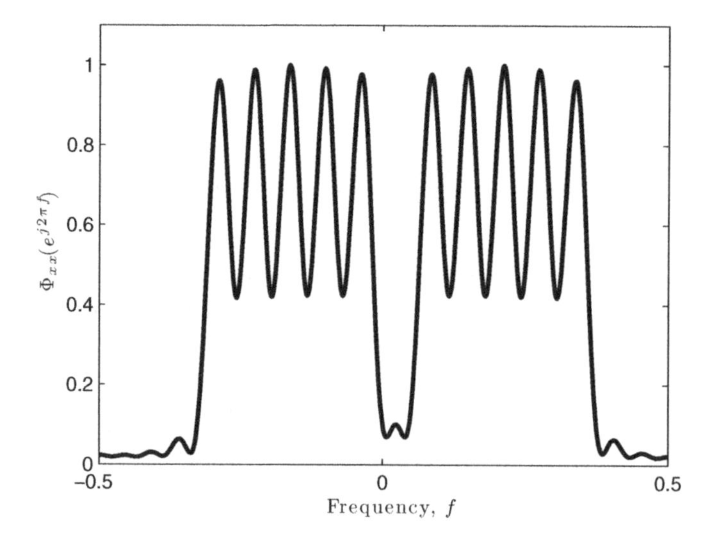

Figure 20.7 OFDM PSD for a frequency offset $\Delta f_c = 0.375/N$; $N = 16$, there are 10 active subcarriers, and $N_{\mathrm{cp}} = 4$.

Locating the Spectral Peaks

To locate the spectral peaks in the OFDM PSD, we adopt the use of a comb filter $F(z)$ in which comb-spacing is equal to the subcarrier spacing and filter nulls are placed halfway between the subcarrier locations. The normalized frequency response of one such filter, corresponding to a subcarrier spacing 1/16, that is, when $N = 16$, is shown in Figure 20.8. Suppose we shift the nulls of this filter by an amount ϵ, and at the same time preserve

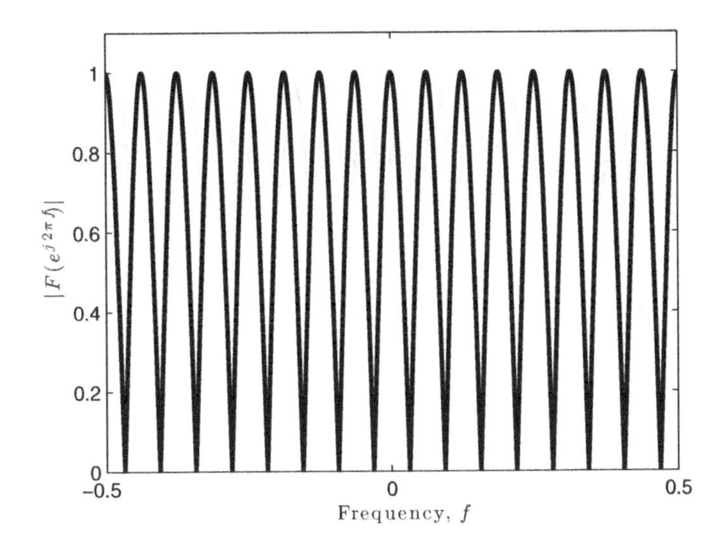

Figure 20.8 An example of the frequency response of the comb filter $F(z)$.

the spacing between the nulls. This is equivalent to circularly shifting the filter response over the normalized frequency range 0 to 1 (or, equivalently, the range −0.5 to 0.5).

We pass the demodulated received signal $y(n)$ through the comb filter $F(z)$ and call its output $y_{\mathrm{f}}(n)$. If we measure the power of $y_{\mathrm{f}}(n)$ as a function of ϵ, we will obtain a plot similar to the one shown in Figure 20.9. We can see that the power of the output consists of many peaks and troughs as we vary ϵ across the signal bandwidth. However, if we choose to look at the portion of the plot where ϵ is in the range ± 0.5 of the subcarrier spacing around the true carrier offset, that is, $-\frac{1}{2N} < \epsilon < \frac{1}{2N}$, we will find that the power of $y_{\mathrm{f}}(n)$ will have only one maximum.

With this in mind, we choose to adapt $F(z)$ to maximize the cost function

$$\xi(n, \epsilon) = E\left[|y_{\mathrm{f}}(n)|^2\right] \tag{20.51}$$

Obviously, this maximization will be achieved when the passbands of the comb filter $F(z)$ align with the OFDM PSD, and the total shift ϵ of the filter is an estimate of the carrier offset. It should be also noted that the proposed tracking algorithm relies on the ability of an acquisition algorithm to provide a carrier offset estimate within one-half of the subcarrier spacing.

The Comb Filter

One possibility for the comb filter is to choose an FIR filter whose zeros are on the unit-circle exactly halfway between all subcarriers as shown in Figure 20.10a and b. The placement of the data and null subcarriers are also shown for convenience.

Figure 20.10a corresponds to the case where there is no carrier offset. In this case, the zeros that are aligned with the subcarriers are located at

$$z_k = \mathrm{e}^{j(2\pi k/N + \pi/N)}, \quad \text{for } k = 0, 1, \ldots, N-1 \tag{20.52}$$

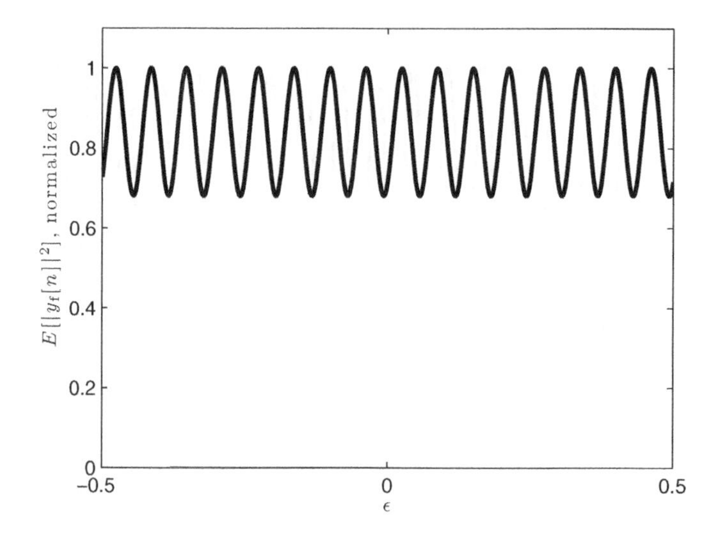

Figure 20.9 Variation of the power of $y_f(n)$ as ϵ varies.

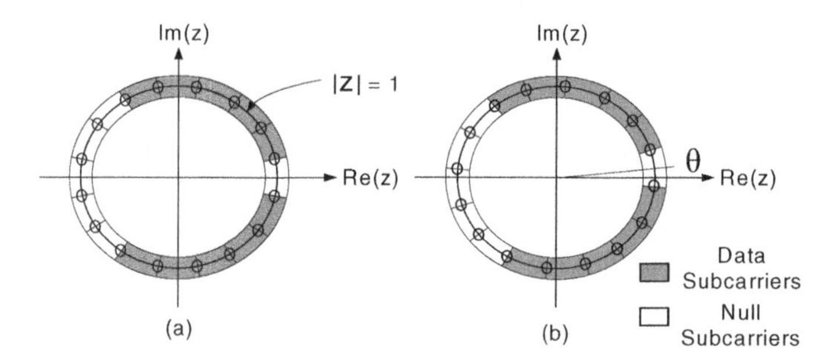

Figure 20.10 Placement of zeros of the comb filter $F(z)$: (a) nonrotated filter zeros ($\theta = 0$), and (b) rotated filter zeros.

These correspond to the transfer function

$$\begin{aligned} F(z) &= \prod_{k=0}^{N-1} \left[1 - e^{j(2\pi k/N + \pi/N)} z^{-1}\right] \\ &= 1 + z^{-N} \end{aligned} \tag{20.53}$$

Figure 20.10b, on the other hand, corresponds to the case where there is shift of θ in the spectrum of the OFDM signal. This corresponds to a carrier offset

$$\Delta f_c = \frac{\theta}{2\pi} \tag{20.54}$$

To align $F(z)$ with the shifted spectrum, its zeros should be modified as

$$z_k = e^{j(\theta+2\pi k/N+\pi/N)}, \quad \text{for } k = 0, 1, \ldots, N-1 \tag{20.55}$$

that is, the zeros of $F(z)$ should be rotated by the angle θ, over the unit-circle $|z| = 1$, as shown in Figure 20.10b. This leads to the transfer function

$$\begin{aligned} F(z) &= \prod_{i=0}^{N-1} \left[1 - e^{j(\theta+2\pi i/N+\pi/N)} z^{-1}\right] \\ &= 1 + e^{jN\theta} z^{-N} \end{aligned} \tag{20.56}$$

Recalling the relationship $\epsilon = \Delta f_c$, we obtain from Eq. (20.54),

$$\theta = 2\pi\epsilon \tag{20.57}$$

Using Eq. (20.57), we can describe the filter $F(z)$ as

$$F(z) = 1 + e^{j2\pi N\epsilon} z^{-N} \tag{20.58}$$

In Figure 20.10, one may note that the zeros are placed around null subcarriers as well as data subcarriers even though there is little signal power to track at locations near the null subcarriers. If zeros are placed only around data subcarriers then Eq. (20.58) will be more complicated. The uniform placement of zeros around the unit-circle is thus chosen to have the simple expression of Eq. (20.58).

Adaptation of the Comb Filter (Carrier Offset Tracking)

With $F(z)$ in place, it is now possible to derive update equations for carrier tracking. The transfer function $F(z)$ can be written as

$$F(z) = \frac{Y_f(z)}{Y(z)} = 1 + e^{j2\pi N\epsilon} z^{-N} \tag{20.59}$$

where $Y(z)$ and $Y_f(z)$ are the z-transforms of the received signal $y(n)$ (i.e., the received OFDM signal after demodulation, but before removing the carrier offset), and the comb filter output $y_f(n)$, respectively. If we rearrange Eq. (20.59) and take the inverse z-transform, we find that

$$y_f(n) = y(n) + e^{j2\pi N\epsilon} y(n-N) \tag{20.60}$$

On the other hand, for the development of an LMS-type algorithm, we remove the expectation from the right-hand side of Eq. (20.51) and accordingly define

$$\hat{\xi}(n, \epsilon) = |y_f(n)|^2 \tag{20.61}$$

Taking the derivative of the cost function $\hat{\xi}(n, \epsilon)$ with respect to ϵ gives

$$\begin{aligned} \frac{\partial \hat{\xi}(n, \epsilon)}{\partial \epsilon} &= y_f(n) \frac{\partial y_f^*(n)}{\partial \epsilon} + y_f^*(n) \frac{\partial y_f(n)}{\partial \epsilon} \\ &= 2\Re \left\{ y_f(n) \frac{\partial y_f^*(n)}{\partial \epsilon} \right\} \end{aligned} \tag{20.62}$$

where $\Re\{\cdot\}$ denotes the real part of the argument and $*$ denotes the complex conjugate. Using Eq. (20.62), the update equation for ϵ becomes

$$\epsilon(n+1) = \epsilon(n) + 2\frac{\mu}{4\pi N}\Re\left\{y_{\mathrm{f}}(n)\frac{\partial y_{\mathrm{f}}^{*}(n)}{\partial \epsilon}\right\} \tag{20.63}$$

where μ is the step-size parameter. We have divided the step-size μ by $4\pi N$ so as to simplify the final expression for the update equation.

The derivative of $y_{\mathrm{f}}(n)$ with respect to ϵ is

$$\frac{\partial y_{\mathrm{f}}(n)}{\partial \epsilon} = j2\pi N \mathrm{e}^{j2\pi N\epsilon} y(n-N) \tag{20.64}$$

Substituting Eqs. (20.60) and (20.64) in Eq. (20.63), we get the final form of the update equation as

$$\epsilon(n+1) = \epsilon(n) + \mu\Im\left\{\mathrm{e}^{-j2\pi N\epsilon(n)} y(n) y^{*}(n-N)\right\} \tag{20.65}$$

where $\Im\{\cdot\}$ denotes the imaginary part of the argument.

20.1.8 Channel-Tracking Methods

The ML and LMMSE channel estimation/equalization methods that were introduced in Section 20.1.5 for initial estimation of the channel and setting of the frequency domain equalizers can also be extended to tracking the channel variation and accordingly adjusting the frequency domain equalizer during the payload periods. For this purpose, pilot symbols are spread among data symbols. A few examples exemplifying different methods of insertion of pilots among data symbols is presented in Figure 20.11.

20.2 MIMO Communication Systems

Communication systems that use multiple antennas at both transmitter and receiver are known as MIMO. A MIMO system with N_{t} transmit antennas and N_{r} receive antennas connects the transmitter and receiver through $N_{\mathrm{t}}N_{\mathrm{r}}$ wireless links. One can benefit from these links by adopting two different approaches. The first approach transmits an independent stream of data from each antenna. This approach results in a significant increase in the system throughput, often called *space-multiplexing gain*. The second approach transmits correlated streams of data from two or more of the transmit antennas and uses proper combining techniques at the receiver to improve on the link reliability. This approach does not increase the system throughput, but improves the quality of the channel by avoiding deep fades in the combined channel. The result, thus, is a gain in quality of channel. This is known as *space-diversity gain*. The combination of the two approaches is also possible.

In this section, we present a brief review of the various signaling methods that are used to benefit from either or both the space-multiplexing and space-diversity gains. We also discuss the various MIMO detection methods that are used to extract the space-multiplexed data symbols that are mixed together after passing through the channel. We note that such detectors require an estimate of the channel response between all pairs of transmit and receive antennas. We, thus, also introduce and review a few approaches

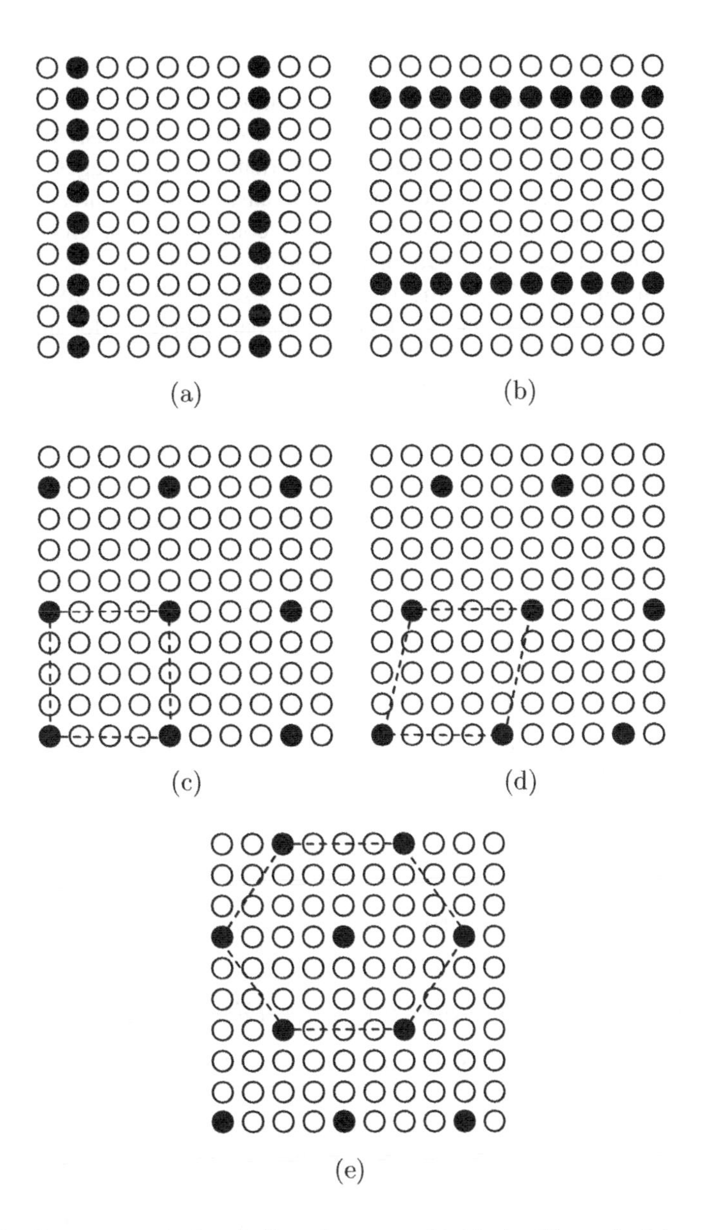

Figure 20.11 Common methods of pilot placement: (a) block, (b) comb, (c) rectangular, (d) parallelogram, and (e) hexagonal.

that may be used for channel estimation. All the developments in this section assume the channel gain between each pair of transmit and receiver antenna is a flat gain. This is a reasonable assumption, as OFDM with cyclic prefix can be (and is usually) used to flatten the gains over each subcarrier channel. The MIMO–OFDM systems are discussed in Section 20.3.

20.2.1 MIMO Channel Model

Figure 20.12 depicts an example of a MIMO communication system with N_t transmit antennas and N_r receive antennas. The reader may note that the basic blocks, including the transmit pulse-shaping, $p_T(t)$, modulation and demodulation stages, and the receive matched filtering, $p_R(t)$, are similar to those in the case of single antenna systems.

Here, we consider the case where the MIMO communication channel is flat fading, hence, the link between each pair of transmit and receive antenna is characterized by a complex-valued gain. Thus, the equivalent discrete-time baseband channel that relates the transmitted symbols and the received signal samples, after matched filtering, is expressed as

$$\mathbf{y}(n) = \mathbf{C}\mathbf{s}(n) + \mathbf{v}(n) \tag{20.66}$$

where $\mathbf{s}(n)$ is a column vector of the transmitted data symbols $s_0(n), s_1(n), \ldots, s_{N_t-1}(n)$ at time instant n, $\mathbf{y}(n) = \left[y_0(n) y_1(n) \cdots y_{N_r-1}(n)\right]^{\mathrm{T}}$ is the vector of the received signal samples, $\mathbf{v}(n)$ is the vector of noise samples after matched filtering, and

$$\mathbf{C} = \begin{bmatrix} c_{0,0} & c_{0,1} & \cdots & c_{0,N_t-1} \\ c_{1,0} & c_{1,1} & \cdots & c_{1,N_t-1} \\ \vdots & \vdots & \ddots & \vdots \\ c_{N_r-1,0} & c_{N_r-1,1} & \cdots & c_{N_r-1,N_t-1} \end{bmatrix} \tag{20.67}$$

is the channel gain matrix with its ijth element denoting the gain between the jth transmit antenna and the ith receive antenna. We assume that the elements of $\mathbf{v}(n)$ are a set of independent Gaussian random variables each with variance of σ_ν^2. Hence,

$$E[\mathbf{v}(n)\mathbf{v}^{\mathrm{H}}(n)] = \sigma_\nu^2 \mathbf{I} \tag{20.68}$$

where $\mathbf{I}$ is the $N_r \times N_r$ identity matrix.

20.2.2 Transmission Techniques for Space-Diversity Gain

In a communication system with single transmit and single receive antenna, Eq. (20.66) reduces to

$$y(n) = cs(n) + \nu(n) \tag{20.69}$$

where c is the channel gain (a scalar), $s(n)$ is the transmitted symbol, and $\nu(n)$ is the channel noise. To obtain an estimate of $s(n)$, one may premultiply Eq. (20.69) by $c^*/|c|^2$ to obtain

$$\begin{aligned} \tilde{s}(n) &= \frac{c^*}{|c|^2} y(n) \\ &= s(n) + \frac{c^*}{|c|^2} \nu(n) \end{aligned} \tag{20.70}$$

One may note that $\tilde{s}(n)$ is a noisy estimate of $s(n)$ with the additive noise component $\frac{c^*}{|c|^2}\nu(n)$. Noting that the latter has variance of $\sigma_\nu^2/|c|^2$, one will find that the SNR of the

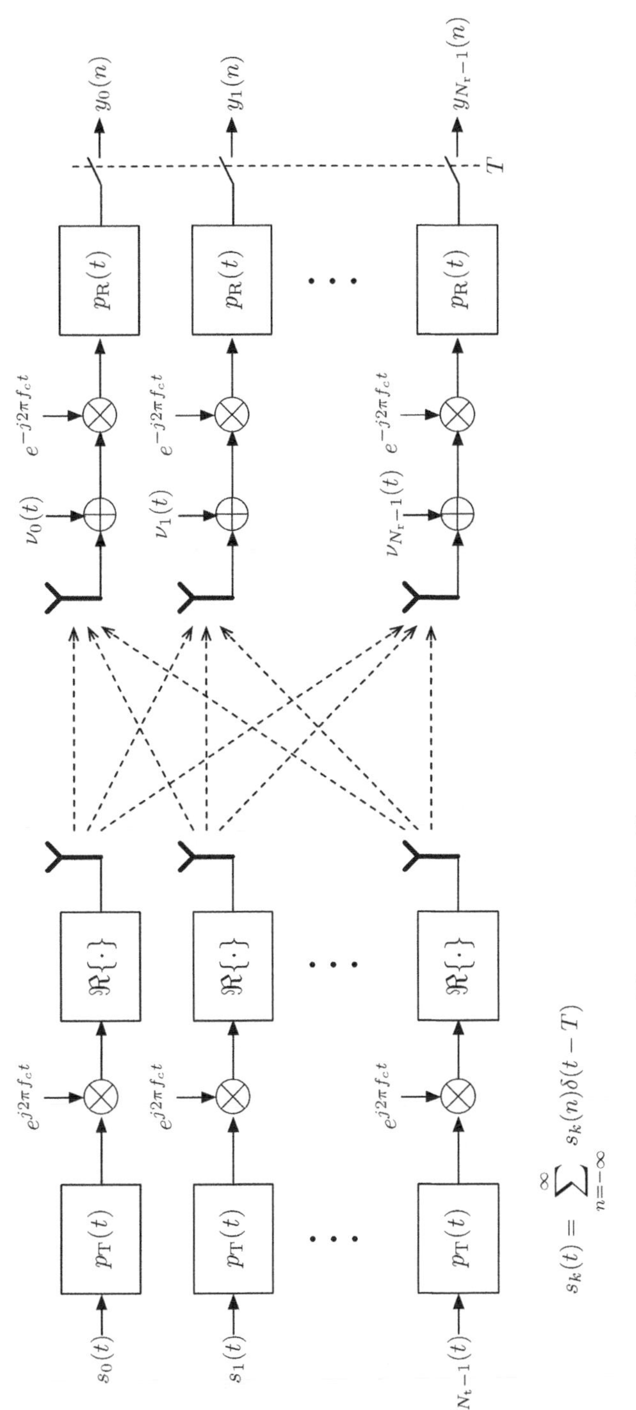

Figure 20.12 Channel model of a MIMO system.

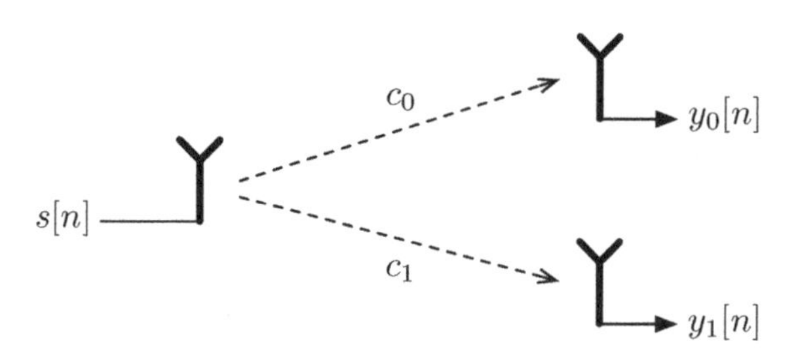

Figure 20.13 A communication channel with one-transmit antenna and two-receive antennas.

estimated symbol $s(n)$ as

$$\text{SNR} = |c|^2 \frac{\sigma_s^2}{\sigma_v^2} \tag{20.71}$$

where σ_s^2 and σ_v^2 are the variances of $s(n)$ and $v(n)$, respectively. When $|c|^2$ is small, that is, when channel is in a deep fade, SNR will be small and this may lead to a poor detection of the transmitted information.

Next consider a communication system with one-transmit antenna and two-receive antennas, as depicted in Figure 20.13. For this setup, Eq. (20.66) reduces to

$$\mathbf{y}(n) = \begin{bmatrix} y_0(n) \\ y_1(n) \end{bmatrix} = \begin{bmatrix} c_0 \\ c_1 \end{bmatrix} s(n) + \begin{bmatrix} v_0(n) \\ v_1(n) \end{bmatrix} \tag{20.72}$$

where $v_0(n)$ and $v_1(n)$ are additive noise samples at the two antennas. We assume that $v_0(n)$ and $v_1(n)$ are a pair of zero-mean, Gaussian, independent, and identically distributed variables. This implies that the vector

$$\mathbf{v}(n) = \begin{bmatrix} v_0(n) \\ v_1(n) \end{bmatrix} \tag{20.73}$$

has the covariance matrix

$$E[\mathbf{v}(n)\mathbf{v}^{\text{H}}(n)] = \sigma_v^2 \mathbf{I} \tag{20.74}$$

where σ_v^2 is the variance of each of the elements of $\mathbf{v}$ and $\mathbf{I}$ is the 2×2 identity matrix.

As the receiver task is to obtain the best estimate of $s(n)$ from the received signal samples of $y_0(n)$ and $y_1(n)$, if we limit this estimate to be based on a linear combination of $y_0(n)$ and $y_1(n)$ (i.e., be a linear estimator), one needs to solve the following problem.

Find the vector $\mathbf{w} = \begin{bmatrix} w_0 \\ w_1 \end{bmatrix}$ that results in the signal estimate

$$\tilde{s}(n) = \mathbf{w}^{\text{H}}\mathbf{y}(n) = s(n) + v(n) \tag{20.75}$$

with the maximum SNR.

The identity (20.75) implies that

$$\mathbf{w}^{\mathrm{H}}\mathbf{c} = 1 \tag{20.76}$$

where $\mathbf{c} = \begin{bmatrix} c_0 \\ c_1 \end{bmatrix}$. Equation (20.76) may be thought as a linear constraint on w_0 and w_1. Moreover, one may note that

$$\nu(n) = \mathbf{w}^{\mathrm{H}}\mathbf{v} \tag{20.77}$$

and to maximize SNR in Eq. (20.75), one should minimize the cost function

$$E[|\nu(n)|^2] = \mathbf{w}^{\mathrm{H}}E[\mathbf{v}(n)\mathbf{v}^{\mathrm{H}}(n)]\mathbf{w} = \sigma_\nu^2\mathbf{w}^{\mathrm{H}}\mathbf{w} \tag{20.78}$$

where the second identity follows from Eq. (20.74).

From the above results, we conclude that to maximize SNR in the estimator (20.75), one should choose to minimize the cost function

$$\xi = \mathbf{w}^{\mathrm{H}}\mathbf{w} \tag{20.79}$$

subject to the constraint (20.76). Obviously, such minimization can be performed using the method of Lagrange multipliers, whose solution leads to

$$\mathbf{w} = \frac{1}{|c_0|^2 + |c_1|^2}\begin{bmatrix} c_0 \\ c_1 \end{bmatrix} \tag{20.80}$$

Substituting Eqs. (20.80) and (20.72) in Eq. (20.75), we obtain

$$\tilde{s}(n) = s(n) + \frac{c_0^*\nu_0(n) + c_1^*\nu_1(n)}{|c_0|^2 + |c_1|^2} \tag{20.81}$$

Comparing Eqs. (20.70) and (20.81), one may find that while in the single-input single-output case, the system will suffer from the channel fade when the gain c is small, a system with a single transmit and double receive antennas will only suffer if both gains c_0 and c_1 are simultaneously in deep fades. Obviously, the chance of the latter is much smaller than the former. We, thus say, introduction of the diversity in the channel, by introducing an additional link, leads to a more reliable communication link. This added reliability, as mentioned earlier, is called *space-diversity gain.*

The setup presented in Figure 20.13 often happens when a mobile station transmits to a base station. The situation will be reverse when a base station transmits to a mobile station. This is presented in Figure 20.14. A question to ask here is how one can benefit from the diversity gain of this setup? In other words, can one design a scheme in which each transmit symbol goes through both the links c_0 and c_1 before reaching the receiver? It turns out that the answer to this question is positive and solution to it is provided by using code words that span across both time and space. Such code words are called *space–time block codes.*

An important subset of space–time block codes that allow simple decoding (separation of the transmitted symbols at the receiver) are *orthogonal* space–time block (OSTB) codes. A popular OSTB that has been widely used in practice is the so-called *Alamouti code* (Alamouti, 1998). It is designed for MIMO systems with two transmit antennas and is defined by the *code matrix*

$$\mathbf{S} = \begin{bmatrix} s_0 & s_1 \\ -s_1^* & s_0^* \end{bmatrix} \tag{20.82}$$

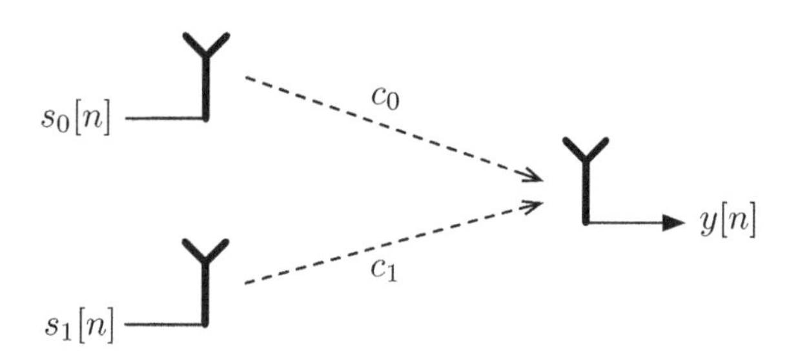

Figure 20.14 A communication channel with two transmit antennas and one receive antennas.

Each row of this matrix indicates the pair of symbols that are transmitted from the transmit antennas. The same pair of symbols are transmitted in the next time slot after conjugation and possibly a sign change. Note that in this arrangement two symbols are transmitted in two time slots and this, clearly, is equivalent of transmitting one symbol per time slot. We also note that the code matrix **S** satisfies

$$\mathbf{S}^{\mathrm{H}}\mathbf{S} = \left(|s_0|^2 + |s_1|^2\right)\mathbf{I} \tag{20.83}$$

where **I** is the identity matrix, hence the name OSTB code.

Assuming that the pair of data symbols $s(n)$ and $s(n+1)$ are transmitted according to the Alamouti code, the pair of the received signal samples received at the time slots n and $n+1$ are given by

$$\begin{bmatrix} y(n) \\ y(n+1) \end{bmatrix} = \begin{bmatrix} s(n) & s(n+1) \\ -s^*(n+1) & s^*(n) \end{bmatrix} \begin{bmatrix} c_0 \\ c_1 \end{bmatrix} + \begin{bmatrix} \nu(n) \\ \nu(n+1) \end{bmatrix} \tag{20.84}$$

To extract estimates of $s(n)$ and $s(n+1)$ from Eq. (20.84), we first note that Eq. (20.84) can be rearranged as

$$\begin{bmatrix} y(n) \\ y^*(n+1) \end{bmatrix} = \begin{bmatrix} c_0 & c_1 \\ -c_1^* & c_0^* \end{bmatrix} \begin{bmatrix} s(n) \\ s(n+1) \end{bmatrix} + \begin{bmatrix} \nu(n) \\ \nu^*(n+1) \end{bmatrix} \tag{20.85}$$

Defining

$$\mathbf{C} = \begin{bmatrix} c_0 & c_1 \\ -c_1^* & c_0^* \end{bmatrix} \tag{20.86}$$

and multiplying Eq. (20.84) through from left-hand side by $\dfrac{1}{|c_0|^2 + |c_1|^2}\mathbf{C}^{\mathrm{H}}$, we obtain

$$\begin{aligned} \begin{bmatrix} \tilde{s}(n) \\ \tilde{s}(n+1) \end{bmatrix} &= \frac{1}{|c_0|^2 + |c_1|^2}\mathbf{C}^{\mathrm{H}} \begin{bmatrix} y(n) \\ y^*(n+1) \end{bmatrix} \\ &= \begin{bmatrix} s(n) \\ s(n+1) \end{bmatrix} + \begin{bmatrix} \dfrac{c_0^*\nu(n) - c_1\nu^*(n+1)}{|c_0|^2 + |c_1|^2} \\[2ex] \dfrac{c_1^*\nu(n) + c_0\nu^*(n+1)}{|c_0|^2 + |c_1|^2} \end{bmatrix} \end{aligned} \tag{20.87}$$

Comparing the first line of this result with Eq. (20.81), one will find that apart from a change in the phase of the gains, which has no consequence on the SNR values, one finds that the estimates of $s(n)$ in both cases is the same. That is, the space-diversity gain achieved in the case of a one-transmit two-receive antennas systems is also achieved in the case where information is transmitted through two antennas and received by one antennas.

20.2.3 Transmission Techniques and MIMO Detectors for Space-Multiplexing Gain

Unlike, in the space-diversity, where correlated symbols are transmitted from all the transmit antennas, in space-multiplexing data streams transmitted from the transmitted antennas are in dependent of one another. When $N_r \geq N_t$, noise is absent, and the channel gain matrix $\mathbf{C}$ and the received signal $\mathbf{y}(n)$ are known, one can find the transmit symbol vector $\mathbf{s}(n)$ perfectly by solving the equation $\mathbf{C}\mathbf{s}(n) = \mathbf{y}(n)$, if the rank of $\mathbf{C}$ is equal to or greater than N_t. In presence of noise, a noisy estimate of $\mathbf{s}(n)$ will be obtained. The point to note here is that even though the set of symbol streams $s_0(n)$ through $s_{N_t-1}(n)$ are transmitted simultaneously and over the same band, they are detectable at the receiver, thanks to the use of multiple antennas at the receiver that will provide sufficient samples for the detection. In other words, the amount of information (considering that each symbol carries a certain amount of information) transmitted per unit of time and over a unit of bandwidth increases as N_t increases, as long as there are sufficient antennas at the receiver. Here, without involving ourselves in theoretical details, we note that the capacity of a MIMO channel, that is, given a total transmit power P and a bandwidth B, the amount of information that could be reliably transmitted per unit of time, increases with the minimum of N_t and N_r. This property of MIMO systems is often referred to as *capacity gain.*

In the rest of this section, we present a number of signal processing methods that attempt to estimate the transmitted symbols/information, given the received signal $\mathbf{y}(n)$ and the channel gain matrix $\mathbf{C}$. The first three methods obtain various estimates of the vector $\mathbf{s}(n)$, directly. The last two methods are to generate soft values of the detected bits following the same principles that the soft equalizers of Chapter 17 were based on.

Zero-Forcing Detector

In zero-forcing (ZF) detector an estimate of $\mathbf{s}(n)$ is obtained as

$$\tilde{\mathbf{s}}(n) = \mathbf{C}^{\dagger}\mathbf{y}(n) \tag{20.88}$$

where $\mathbf{C}^{\dagger}$ is the pseudoinverse of $\mathbf{C}$ given by

$$\mathbf{C}^{\dagger} = \left(\mathbf{C}^{\mathrm{H}}\mathbf{C}\right)^{-1}\mathbf{C}^{\mathrm{H}} \tag{20.89}$$

Substituting Eqs. (20.66) and (20.89) in Eq. (20.88), we obtain

$$\tilde{\mathbf{s}}(n) = \mathbf{s}(n) + \left(\mathbf{C}^{\mathrm{H}}\mathbf{C}\right)^{-1}\mathbf{C}^{\mathrm{H}}\mathbf{v}(n) \tag{20.90}$$

This is a noisy estimate of $\mathbf{s}(n)$ with a noise vector

$$\mathbf{v}'(n) = \left(\mathbf{C}^{\mathrm{H}}\mathbf{C}\right)^{-1}\mathbf{C}^{\mathrm{H}}\mathbf{v}(n) \tag{20.91}$$

Obviously, the performance of the ZF detector is determined by the variance of the elements of $\mathbf{v}'(n)$ that is quantified by the diagonal elements of its covariance matrix. Using Eqs. (20.91) and (20.68), the covariance matrix of $\mathbf{v}'(n)$ is obtained as

$$E[\mathbf{v}'(n)\mathbf{v}'^{\mathrm{H}}(n)] = \sigma_\nu^2(\mathbf{C}^{\mathrm{H}}\mathbf{C})^{-1} \tag{20.92}$$

This shows that the size of elements of $E[\mathbf{v}'(n)\mathbf{v}'^{\mathrm{H}}(n)]$ are related to the size of elements of the matrix $(\mathbf{C}^{\mathrm{H}}\mathbf{C})^{-1}$ which, in turn, are related to the condition number, respectively, the spread of eigenvalues of the matrix $\mathbf{C}^{\mathrm{H}}\mathbf{C}$. When one or more of eigenvalues of $\mathbf{C}^{\mathrm{H}}\mathbf{C}$ are small, the size of elements of $(\mathbf{C}^{\mathrm{H}}\mathbf{C})^{-1}$ will be large and, hence, ZF detector performs poorly. More specifically, one may quantify the performance of the ZF detector by considering the SNR values at the detector outputs. These SNR values are defined as

$$\mathrm{SNR}_k = \frac{\sigma_s^2}{\sigma_{\nu'_k}^2} \tag{20.93}$$

where $\sigma_s^2 = E[|s_k(n)|^2]$ and $\sigma_{\nu'_k}^2 = E[|\nu'_k(n)|^2]$. Note that while we assume σ_s^2 is the same for the data symbols transmitted from different antennas, $\sigma_{\nu'_k}^2$ differs for various choices of k and also it varies with the channel gain $\mathbf{C}$.

Numerical Examples: Part 1

Some of the properties of the ZF detector could be best understood through numerical examples. Let us consider the following two choices of $\mathbf{C}$.

$$\mathbf{C}_1 = \begin{bmatrix} 0.4209 + 0.0060i & -0.3859 + 0.1032i & 0.2284 - 0.0397i \\ -0.0327 + 0.1392i & 0.2426 - 0.0631i & 0.4007 - 0.1766i \\ -0.0280 + 0.1797i & -0.0037 - 0.4032i & -0.3683 - 0.0566i \end{bmatrix}$$

and

$$\mathbf{C}_2 = \begin{bmatrix} 0.0382 - 0.0424i & -0.0099 - 0.2285i & 0.1134 + 0.4595i \\ 0.4010 + 0.0661i & -0.0344 + 0.4154i & 0.3406 + 0.1319i \\ 0.2441 - 0.1635i & -0.1571 + 0.3103i & 0.0105 + 0.2018i \end{bmatrix}$$

The first choice is a poorly conditioned matrix; the eigenvalues of $\mathbf{C}_1^{\mathrm{H}}\mathbf{C}_1$ are 0.0081, 0.8843, and 1.1777. The second choice is a moderately conditioned matrix; the eigenvalues of $\mathbf{C}_2^{\mathrm{H}}\mathbf{C}_2$ are 0.0637, 0.2553, and 0.6810.

Letting $\sigma_\nu^2 = 0.01$, the SNR values at the detector outputs (calculated according to Eq. (20.93) for the two cases) are obtained as

$$\text{For } \mathbf{C}_1: \ \mathrm{SNR} = -5.46, -4.12, 1.18\mathrm{dB}$$

$$\text{For } \mathbf{C}_2: \ \mathrm{SNR} = 9.92, 11.37, 14.42\mathrm{dB}$$

Clearly, as expected, the poorly conditioned channel gain $\mathbf{C}_1$ leads to very low SNR values at the detector outputs.

Minimum Mean-Squared Error Detector

The minimum mean-squared error (MMSE) detector finds an estimate $\tilde{\mathbf{s}}(n)$ of $\mathbf{s}(n)$ according to the formula

$$\tilde{\mathbf{s}}(n) = \mathbf{G}^{\mathrm{H}}\mathbf{y}(n) \tag{20.94}$$

and set $\mathbf{G}$ by minimizing the cost function

$$\xi = E[\mathbf{e}^{\mathrm{H}}(\mathbf{n})\mathbf{e}(\mathbf{n})] \tag{20.95}$$

where

$$\mathbf{e}(n) = \mathbf{s}(n) - \tilde{\mathbf{s}}(n) \tag{20.96}$$

One, thus, may note that minimization of the cost function ξ leads to a MMSE solution, hence, the name MMSE detector.

Minimization of ξ leads to the solution

$$\mathbf{G} = \left(\mathbf{C}\mathbf{C}^{\mathrm{H}} + \frac{\sigma_v^2}{\sigma_s^2}\mathbf{I}\right)^{-1}\mathbf{C} \tag{20.97}$$

An alternative form of this result is obtained by applying the matrix inversion lemma to Eq. (20.97). This leads to

$$\mathbf{G}^{\mathrm{H}} = \left(\mathbf{C}^{\mathrm{H}}\mathbf{C} + \frac{\sigma_v^2}{\sigma_s^2}\mathbf{I}\right)^{-1}\mathbf{C}^{\mathrm{H}} \tag{20.98}$$

An interesting observation that can be made using Eq. (20.98) is that when $\sigma_v^2 = 0$, $\mathbf{G}^{\mathrm{H}} = \mathbf{C}^{\dagger}$. That is, when there is no noise, MMSE and ZF detectors are the same.

Substituting Eq. (20.98) in Eq. (20.94) and recalling Eq. (20.66), we obtain

$$\tilde{\mathbf{s}}(n) = \mathbf{H}\mathbf{s}(n) + \mathbf{v}'(n) \tag{20.99}$$

where

$$\mathbf{H} = \left(\mathbf{C}^{\mathrm{H}}\mathbf{C} + \frac{\sigma_v^2}{\sigma_s^2}\mathbf{I}\right)^{-1}\mathbf{C}^{\mathrm{H}}\mathbf{C} \tag{20.100}$$

and

$$\mathbf{v}'(n) = \mathbf{G}^{\mathrm{H}}\mathbf{v}(n) \tag{20.101}$$

Using Eqs. (20.101) and (20.68), the covariance matrix of $\mathbf{v}'(n)$ is obtained as

$$E[\mathbf{v}'(n)\mathbf{v}'^{\mathrm{H}}(n)] = \sigma_v^2\mathbf{G}^{\mathrm{H}}\mathbf{G} \tag{20.102}$$

Unlike ZF detector in which interference among symbols was completely removed, in MMSE there is usually some residual ISI. Hence, to measure the performance of MMSE detector, one should evaluate signal-to-interference plus noise ratio (SINR). Recalling Eq. (20.99), one finds that at the kth output of MMSE detector

$$\mathrm{SINR}_k = \frac{\sigma_s^2|h_{kk}|^2}{\sigma_s^2\sum_{l\neq k}|h_{kl}|^2 + \sigma_{v_k'}^2} \tag{20.103}$$

where h_{kl} denotes the klth element of $\mathbf{H}$ and $\sigma_{v_k'}^2$ is the kth diagonal element of the matrix $\sigma_v^2\mathbf{G}^{\mathrm{H}}\mathbf{G}$.

Numerical Examples: Part 2

Following the above equations, the SINR values of MMSE detector for the channels $\mathbf{C}_1$ and $\mathbf{C}_2$, of the "Numerical Example: Part 1," are obtained as

$$\text{For } \mathbf{C}_1 : \ \text{SINR} = 0.71, 2.89, 8.79 \ \ \text{dB}$$

$$\text{For } \mathbf{C}_2 : \ \text{SINR} = 10.10, 11.60, 14.51 \ \ \text{dB}$$

Comparing these numbers with their counterpart in ZF detector, one may conclude that while for moderately conditioned channels, the difference between ZF and MMSE detectors may not be significant, the two detectors perform significantly different when the channel is ill/poorly conditioned.

Successive Interference Cancellation Detector

From the numerical results presented above for channels $\mathbf{C}_1$ and $\mathbf{C}_2$, one may observe that the performance of both ZF and MMSE detectors for different symbols in a data vector $\mathbf{s}(n)$ may vary significantly. For instance, for channel $\mathbf{C}_1$, the MMSE detector detects $s_2(n)$ with an SINR of 8.79 dB, $s_1(n)$ with an SINR 2.89 dB, and $s_0(n)$ with an SINR of 0.71 dB. With these values of SINR, one may argue that while $s_2(n)$ may be detected correctly with a high probability, $s_1(n)$ and $s_0(n)$ cannot be detected reliably because they are corrupted by a significant amount of noise/interference. This observation has led to the design of a class of detectors that operate by taking the following steps:

1. Assuming that the channel gain matrix $\mathbf{C}$ and the noise variance σ_ν^2 are known, design a ZF/MMSE detector.
2. Detect the symbol $s_k(n)$ with the largest SNR/SINR.
3. Subtract the effect of the detected symbol $s_k(n)$ from the received signal $\mathbf{y}(n)$. This is equivalent of removing $s_k(n)$ from $\mathbf{s}(n)$ and also removing the kth column of $\mathbf{C}$. The new equation that relates the remaining elements of $\mathbf{s}(n)$ and the associated received signal will be

$$\mathbf{y}_{-k}(n) = \mathbf{C}_{-k}\mathbf{s}_{-k}(n) + \mathbf{v}(n) \tag{20.104}$$

   where the subscript "$-k$" on $\mathbf{s}(n)$ means the kth element of $\mathbf{s}(n)$ is removed, on $\mathbf{C}$ means the kth column of $\mathbf{C}$ is removed, and on $\mathbf{y}(n)$ indicates that the effect of $s_k(n)$ has been removed.
4. Consider Eq. (20.104) as the channel equation and repeat Steps 1, 2, and 3 until all the symbols are detected.

This detector is called *successive interference cancellation* (SIC) detector, because of obvious reasons. We note that, here, since after detection of each symbol, the number of transmit symbols in the channel equation (20.104) reduces by 1, but the number of receive antennas remain unchanged, one would expect to see improvement in the detection of the remaining symbols. Of course, this improvement is granted only if the detected symbols are correct. An error in detection will result in error propagation and thus may lead to a poor performance of the SIC detector.

Numerical Examples: Part 3

Following the above equations, the SINR values of SIC detector for the channels $\mathbf{C}_1$ and $\mathbf{C}_2$ of the above numerical examples are obtained as

$$\text{For } \mathbf{C}_1 : \text{ SINR} = 13.63, 10.82, 8.79 \text{ dB}$$

$$\text{For } \mathbf{C}_2 : \text{ SINR} = 14.06, 12.16, 14.51 \text{ dB}$$

Comparing these numbers with their counterpart in ZF/MMSE detector, one observes a significant performance gain, particularly for $\mathbf{C}_1$.

Soft MIMO Detectors

The soft equalization methods (i.e., the soft MMSE equalizer and the statistical soft equalizer) that were presented in Section 17.9 can be easily extended to MIMO detectors as well. These methods are widely studied by many authors in the literature. Examples of the relevant literature are: (i) The soft MMSE–MIMO detectors can be found in Abe and Matsumoto (2003) and Liu and Tian (2004). (ii) The MCMC–MIMO detectors that follow the statistical soft equalization method that was discussed in Chapter 17 can be found in Farhang-Boroujeny, Zhu, and Shi (2006) and Chen *et al.* (2010). Yet, another class of soft MIMO detectors has been proposed using the method of *sphere decoding*, see Hochwald and ten Brink (2003) and Vikalo, Hassibi, and Kailath (2004). A comparison of MCMC and sphere decoding MIMO detectors can be found in Zhu, Farhang-Boroujeny, and Chen (2005).

20.2.4 Channel Estimation Methods

As discussed in Chapter 17 and also earlier in this chapter, with regard to OFDM, channel estimation is usually performed through the use of a training sequence at the beginning of each packet or through training/pilot symbols that are inserted between data symbols. In a single-input single-output channel, when the channel is characterized by a flat gain, the minimum number of pilot symbols required to estimate the channel gain is 1. Additional pilot symbols may be used to average out the effect of channel noise and hence obtain a more accurate estimate of the channel.

In a MIMO channel with flat gain, one has to estimate the channel gain matrix $\mathbf{C}$. For this purpose, considering the channel equation (20.66), one may note that transmitting a single pilot vector, say, $\mathbf{s}_\text{p}$ may not be sufficient to estimate all the $N_\text{t}N_\text{r}$ elements of $\mathbf{C}$. Here, one needs to transmit a pilot matrix $\mathbf{S}_\text{p}$ of size $N_\text{t} \times L$, where $L \geq N_\text{t}$. Each column of $\mathbf{S}_\text{p}$ is a set of pilots transmitted from the N_t transmit antennas. Combining the received signals from the L pilot symbol vectors, the channel equation relating $\mathbf{S}_\text{p}$ with the matrix of received signal samples reads

$$\mathbf{Y}_\text{p} = \mathbf{C}\mathbf{S}_\text{p} + \mathbf{V} \tag{20.105}$$

where the matrix $\mathbf{V}$ contains channel noise samples. In Eq. (20.105), both $\mathbf{Y}_\text{p}$ and $\mathbf{V}$ are matrices of size $N_\text{r} \times L$. Assuming that $\mathbf{S}_\text{p}$ is of rank N_r, an estimate of $\mathbf{C}$ can be obtained by multiplying both sides of Eq. (20.105) from right by $\mathbf{S}_\mathrm{p}^\mathrm{H}(\mathbf{S}_\mathrm{p}\mathbf{S}_\mathrm{p}^\mathrm{H})^{-1}$, to obtain

$$\begin{aligned}\hat{\mathbf{C}} &= \mathbf{Y}_\mathrm{p}\mathbf{S}_\mathrm{p}^\mathrm{H}(\mathbf{S}_\mathrm{p}\mathbf{S}_\mathrm{p}^\mathrm{H})^{-1} \\ &= \mathbf{C} + \mathbf{V}\mathbf{S}_\mathrm{p}^\mathrm{H}(\mathbf{S}_\mathrm{p}\mathbf{S}_\mathrm{p}^\mathrm{H})^{-1}\end{aligned} \tag{20.106}$$

A convenient choice for $\mathbf{S}_\mathrm{p}$ is to be a square, that is, $L = N_\mathrm{t}$, and orthogonal matrix, that is, $\mathbf{S}_\mathrm{p}\mathbf{S}_\mathrm{p}^H = K\mathbf{I}$, where K is a constant, hence,

$$\begin{aligned}\hat{\mathbf{C}} &= \frac{1}{K}\mathbf{Y}_\mathrm{p}\mathbf{S}_\mathrm{p}^\mathrm{H} \\ &= \mathbf{C} + \frac{1}{K}\mathbf{V}\mathbf{S}_\mathrm{p}^\mathrm{H}\end{aligned} \tag{20.107}$$

A trivial choice of $\mathbf{S}_\mathrm{p}$ that satisfies the orthonormality condition $\mathbf{S}_\mathrm{p}\mathbf{S}_\mathrm{p}^H = K\mathbf{I}$ is when $\mathbf{S}_\mathrm{p} = \sqrt{K}\mathbf{I}$. This corresponds to the case where at each instant of time a pilot symbol is transmitted from one of the antennas, and the rest of the antennas stay quiet. In that case, the channel estimate is trivially given by

$$\hat{\mathbf{C}} = \frac{1}{\sqrt{K}}\mathbf{Y}_\mathrm{p} \tag{20.108}$$

This method is often referred to as *independent pilot pattern* – independent pilots are transmitted from different antennas. The independent pilot patterns may be problematic in cases where the amplifiers at transmitters are power limited. Therefore, injecting all the transmit power into one the transmitter amplifiers limits the amount of power used for channel estimation, hence, reduces the quality of the estimated channel.

An alternative pilot pattern, known as *orthogonal pilot pattern*, selects an $\mathbf{S}_\mathrm{p}$ in which all the elements are of the same level, and the orthogonality condition $\mathbf{S}_\mathrm{p}\mathbf{S}_\mathrm{p}^H = K\mathbf{I}$ holds. Examples of such designs for 2, 3, and 4 transmit antennas are, respectively,

- $N_\mathrm{t} = 2$

$$\mathbf{Y}_\mathrm{p} = \sqrt{\frac{K}{2}}\begin{bmatrix}1 & 1 \\ 1 & -1\end{bmatrix} \tag{20.109}$$

- $N_\mathrm{t} = 3$

$$\mathbf{Y}_\mathrm{p} = \sqrt{\frac{K}{3}}\begin{bmatrix}1 & 1 & 1 \\ 1 & \mathrm{e}^{j2\pi/3} & \mathrm{e}^{-j2\pi/3} \\ 1 & \mathrm{e}^{-j2\pi/3} & \mathrm{e}^{j2\pi/3}\end{bmatrix} \tag{20.110}$$

- $N_\mathrm{t} = 4$

$$\mathbf{Y}_\mathrm{p} = \sqrt{\frac{K}{4}}\begin{bmatrix}1 & 1 & 1 & 1 \\ 1 & -1 & -1 & 1 \\ 1 & 1 & -1 & -1 \\ 1 & -1 & 1 & -1\end{bmatrix} \tag{20.111}$$

20.3 MIMO–OFDM

Extension of OFDM to MIMO channel is straightforward. Figure 20.15 present a block diagram of a MIMO–OFDM system. For brevity, modulators/demodulators to/from RF are not included. Separate OFDM symbol/signal sequences are *synchronously* transmitted from different transmit antennas. At the receiver, a set of OFDM demodulators separate the received signals into their respective subcarriers. This includes the removal of CP from each OFDM symbol and applying an FFT. Recalling the discussion on OFDM in Section 20.1, one may note that OFDM partitions the channel into a set of flat-fading MIMO channels expressed by the set of equations

$$\mathbf{y}_k(n) = \mathbf{C}_k\mathbf{s}_k(n) + \mathbf{v}_k(n), \quad k \in \mathcal{K} \tag{20.112}$$

where $\mathcal{K}$ indicates the indices of the active subcarriers.

All the methods related to MIMO system that were discussed in Section 20.2 are obviously applicable to the set of channel models (20.112) as well. In particular, the space-diversity and/or space-multiplexing methods can be applied to each subcarrier channel separately. Moreover, the channel estimation methods that were discussed in Section 20.2.4 can be readily extended to each subcarrier channel. In addition, the pilot spreading and channel estimation methods that were developed for OFDM systems can extended to MIMO–OFDM. A few relevant literature that interested readers may refer to are Ghosh *et al.* (2011), Chiueh, Tsai, and Lai (2012), and Stüber *et al.* (2004).

Problems

P20.1 Study and present your observation on the effect of cyclic prefix extension of $\mathbf{x}(n)$ on the individual tones embedded in $\mathbf{x}(n)$. Specifically, confirm that the cyclic prefix extension does not introduced and discontinuity in the time index of the tones.

P20.2

(i) Starting with the MATLAB script "`OFDM.m`," available on the accompanying website, add the necessary line to form the autocorrelation coefficients $r_{yy}(n)$ and plot the result to confirm that you obtain a plot similar to the one in Figure 20.3.

(ii) Provide the necessary arguments to prove that the time indices n_1, n_2, n_3, and n_4 of Figure 20.3 are related accordingly as

$$n_1 + 192 = n_2 + 160 = n_3 + 32 = n_4$$

(iii) Check to see these relationships are applicable to the plot you obtained in (i).

(iv) Do the observations made above change in presence of a carrier frequency offset? Examine the developed code and also provide theoretical explanation.

P20.3 In the MATLAB script "`OFDM.m`," available on the accompanying website, let the carrier frequency offset be zero (Dfc = 0) and the noise variance be set equal to 10^{-8} (sigmav = 0.0001). Also, to begin assume that the channel is ideal ($c = 1$).

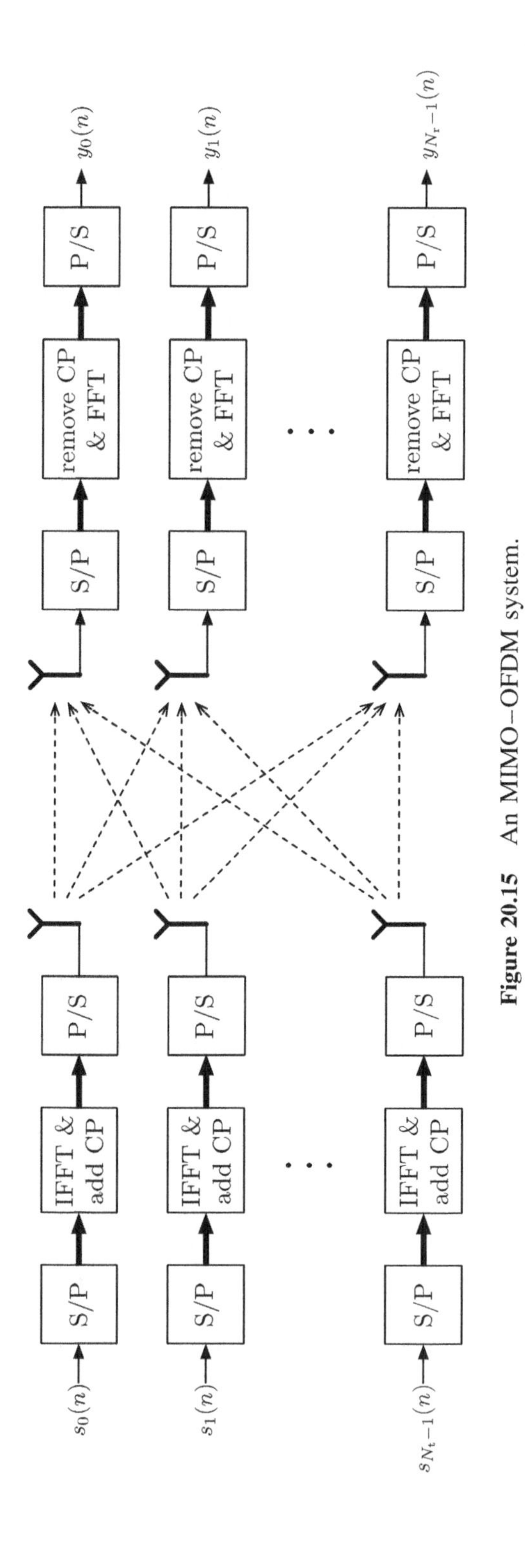

Figure 20.15 An MIMO–OFDM system.

(i) Complete the MATLAB script "`OFDM.m`" to perform the tasks of timing recovery/acquisition, equalizers setting, and data detection. Here, for setting the equalizers coefficients, training symbol T_2 can be used. Compare the recovered data symbols with the transmitted ones to confirm that your code works correctly. Also, by finding the difference between the recovered and transmitted data symbols obtain an estimate of the MSE at the receiver output.

(ii) Repeat (i) when the training symbol T_1 is used to adjust the equalizers coefficients, and compare your results with those of (i).

(iii) Repeat (i) when the average of the training symbols T_1 and T_2 is used to adjust the equalizers coefficients, and compare your results with those of (i).

(iv) Repeat the above part when the ideal channel $c = 1$ is replaced by $c = [1;$ zeros $(47, 1); 0.5;$ zeros $(28, 1); 0.8]$.

P20.4 Repeat problem P20.3 for the case where the carrier frequency offset Dfc is nonzero. Not that, in this case, your code should first perform carrier recovery and correct the received signal "y" according to the estimated carrier offset, before proceeding with setting the equalizer coefficients and data detection. Examine your code for values of Dfc = 1, 2, and 10.

P20.5 Recall the time window function $g(n)$ of Eq. (20.47). Also, recall from the results in Chapter 2 that if $f(n) = g(n) \star g(-n)$, then $F(\omega) = |G(\omega)|^2$.

(i) Show that

$$f(n) = \begin{cases} N + N_{\text{cp}} - |n|, & \text{for } |n| < N + N_{\text{cp}} \\ 0, & \text{otherwise} \end{cases}$$

(ii) Using the result of (i), show that $F(\omega)$ can be arranged to the form given on the right-hand side of Eq. (20.50).

P20.6 Develop and examine the necessary MATLAB codes to confirm the numerical results presented in "Numerical examples: Part 1 through 3" of Section 20.2.2.

P20.7 Show that the solution to the minimization of the cost function (20.95) is Eq. (20.97).
Hint: Note that the cost function ξ may be divided into the independent cost functions $\xi_k = E[|s_k(n) - \tilde{s}_k(n)|^2]$, for $k = 1, 2, \ldots, N_{\text{t}}$.

P20.8 By invoking the matrix inversion lemma, show that Eq. (20.97) may be rearranged as Eq. (20.98).

References

Abe, T. and Matsumoto, T. (2003) Space-time turbo equalization in frequency-selective MIMO channels. *IEEE Transactions on Vehicular Technology*, **52** (3), 469–475.

Ahmed, N., Hush, D., Elliott, G.R., and Fogler, R.J. (1984) Detection of multiple sinusoids using an adaptive cascaded structure. *Proceedings ICASSP'84*, pp. 21.3.1–21.3.4.

Alamouti, S.M. (1998) A simple transmit diversity technique for wireless communications. *IEEE Journal on Selected Areas in Communications*, **16** (8), 1451–1458.

Alexander, S.T. (1986a) *Adaptive Signal Processing: Theory and Applications*, Springer-verlag, New York.

Alexander, S.T. (1986b) Fast adaptive filters: a geometrical approach. *IEEE ASSP Magazine*, **3** (4), 18–28.

Amano, F., Meana, H.P., de Luca, A., and Duchen, G. (1995) A multirate acoustic echo canceller structure. *IEEE Transactions on Communications*, **43** (7), 2172–2176.

Anderson, B.D.O. and Moore, J.B. (1979) *Optimal Filtering*, Prentice Hall, Englewood Cliffs, NJ.

Ardalan, S.H. (1987) Floating-point error analysis of recursive least-squares and least-mean-square adaptive filters. *IEEE Transactions on Circuits and Systems*, **CAS-33**, 1192–1208.

Ardalan, S.H. and Alexander, S.T. (1987) Fixed-point roundoff error analysis of the exponentially windowed RLS algorithm for time-varying systems. *IEEE Transactions on Acoustics, Speech, & Signal Processing*, **ASSP-35** (6), 770–783.

Asharif, M.R., Amano, F., Unagami, S., and Muramo, K. (1986) A new structure of echo canceller based on frequency bin adaptive filtering (FBAF). *DSP Symp* (in Japan), *A3.1*. pp. 165–169.

Asharif, M.R., Takebayashi, T., Chugo, T., and Murano, K. (1986) Frequency domain noise canceller: frequency-bin adaptive filtering (FBAF). *Proceeding of ICASSP'86, April 7–11, Tokyo, Japan*, pp. 41.22.1–4.

Asharif, M.R. and Amano, F. (1994) Acoustic echo-canceller using the FBAF algorithm. *Transaction on Communications*, **42** (12), 3090–3094.

Ashihara, K., Nishikawa, K., and Kiya H. (1995) Improvement of convergence speed for subband adaptive digital filters using the multirate repeating method. *Proceedings IEEE ICASSP'95, Detroit, MI, May*, pp. 989–992.

Astrom, K.J., and Wittenmark, B. (1980) Self-tuning controllers based on pole-zero placement. *IEE Proceedings D, Control Theory and Applications*, **127** (3), 120–130.

Benallal, A. and Gilloire, A. (1988) A new method to stabilize fast RLS algorithms based on a first-order model of the propagation of numerical errors. *Proceedings ICASSP'88 Conference*, **vol. 3**, pp. 1373–1376.

Benesty, J., Morgan, D.R., and Sondhi M.M. (1998) A better understanding and an improved solution to the specific problems of stereophonic acoustic echo cancellation. *IEEE Transactions on Speech and Audio Processing*, **6** (2), 156–165.

Benveniste, A. (1987) Design of adaptive algorithms for the tracking of time-varying systems. *International Journal of Adaptive Control and Signal Processing*, **1** (1), 3–29.

Benveniste, A. and Ruget, G. (1982) A measure of the tracking capability of recursive stochastic algorithms with constant gains. *IEEE Transactions on Automatic Control*, **AC-27**, 639–649.

Adaptive Filters: Theory and Applications, Second Edition. Behrouz Farhang-Boroujeny.

Bergland, G.D. (1968) A fast Fourier transform algorithm for real-valued series. *Communications of ACM*, **11**, 703–710.

Bergmans, J.W.M. (1996) *Digital Baseband Transmission and Recording*. Boston: Kluwer.

Bershad, N.J. (1986) Analysis of the normalized LMS algorithm with Gaussian inputs. *IEEE Transactions on Acoustics Speech and Signal Processing*, **ASSP-34** (4), 793–806.

Bershad, N.J. and Macchi, O.M. (1991) Adaptive recovery of a chirped sinusoid in noise, Part 2: performance of the LMS algorithm. *IEEE Transactions on Signal Processing*, **39** (3), 595–602.

Botto, J.-L. and Moustakides, G.V. (1989) Stabilizing the fast Kalman algorithms. *IEEE Transactions on Acoustics Speech and Signal Processing*, **ASSP-37** (9), 1342–1348.

Brandwood, D.H. (1983) A complex gradient operator and its application in adaptive array theory. *IEE Proceedings*, **130** pts F and H, 11–16.

Buckley, K. and Griffiths, L. (1986) An adaptive generalized sidelobe canceller with derivative constraints. *IEEE Transactions on Antennas and Propagation*, **34** (3), 311–319.

Capon, J. (1969) High-resolution frequency-wavenumber spectrum analysis. *Proceedings of the IEEE*, **57**, 1408–1418.

Caraiscos, C. and Liu, B. (1984) A round-off analysis of the LMS adaptive algorithm. *IEEE Transactions on Acoustics Speech and Signal Processing*, **ASSP-32** (1), 34–41.

Chan, K.S. (2000) *Fast block LMS algorithms and analysis*. Ph.D. dissertation, National University of Singapore.

Chan, K.S. and Farhang-Boroujeny, B. (2001) Analysis of the partitioned frequency-domain block LMS (PFBLMS) algorithm. *IEEE Transactions on Signal Processing*, **49** (9), 1860–1874.

Chen, J., Bes, H., Vandewalle, J., and Janssens, P. (1988) A new structure for sub-band acoustic echo canceller. *Proceedings IEEE ICASSP'88, April, New York*, pp. 2574–2577.

Chen, R.-R., Peng, R.H., Ashikhmin, A., and Farhang-Boroujeny, B. (2010) Approaching MIMO capacity using bitwise Markov Chain Monte Carlo detection. *IEEE Transactions on Communications*, **58** (2), 423–428.

Chi, C.-Y., Chen, C.-Y., Chen, C.-H., and Feng, C.-C. (2003) Batch processing algorithms for blind equalization using higher-order statistics. *IEEE Signal Processing Magazine*, **20** (1), 25–49.

Chiueh, T.-D., Tsai, P.-Y., and Lai, I.-W. (2012) *Baseband receiver design for wireless MIMO-OFDM Communications*, 2nd edn, John Wiley & Sons, Singapore.

Cho, N.I. and Lee, S.U. (1993) On the adaptive lattice notch filter for the detection of sinusoids. *IEEE Transactions on Circuits and Systems – II: Analog and Digital Signal Processing*, **40** (7), 405–416.

Chu, D.C. (1972) Polyphase codes with good periodic correlation properties. *IEEE Transactions on Information Theory*, **18** (4), 531–532.

Cioffi, J.M. (1987) Limited-precision effects in adaptive filtering. *IEEE Transactions on Circuits and Systems*, **CAS-34**, 821–833.

Cioffi, J.M. (1987) A fast QR/frequency-domain RLS adaptive filter. *Proceedings ICASSP-87 Conference*, **vol. 1**, pp. 407–410.

Cioffi, J.M. (1990a) A fast echo canceller initialization method for the CCITT V.32 modem. *IEEE Transactions on Communications*, **38**, 629–638.

Cioffi, J.M. (1990b) The fast adaptive rotors RLS algorithm. *IEEE Transactions on Acoustics, Speech, & Signal Processing*, **ASSP-38**, 631–651.

Cioffi, J.M. and Kailath, T. (1984) Fast recursive least-squares transversal filters for adaptive filtering. *IEEE Transactions on Acoustics, Speech, & Signal Processing*, **ASSP-32**, 304–337.

Cioffi, J.M. and Kailath, T. (1985) Windowed fast transversal filters adaptive algorithms with normalization. *IEEE Transactions on Acoustics, Speech, & Signal Processing*, **ASSP-33**, 607–625.

Claasen, T.A.C.M. and Mecklenbr'auker, W.F.G. (1981) Comparison of the convergence of two algorithms for adaptive FIR digital filters. *IEEE Transactions on Acoustics, Speech, & Signal Processing*, **ASSP-29**, 670–678.

Clark, G.A., Mitra, S.K., and Parker, S.R. (1981) Block implementation of adaptive digital filters. *IEEE Transactions on Circuits and Systems*, **CAS-28** (6), 584–592.

Clark, G.A., Parker, S.R., and Mitra, S.K. (1983) A unified approach to time and frequency domain realization of FIR adaptive digital filters. *IEEE Transactions on Acoustics, Speech, & Signal Processing*, **ASSP-31**, 1073–1083.

Clark, A.P. and Hau, S.F. (1984) Adaptive adjustment of receiver for distorted digital signals. *IEE Proceedings*, **131** part F, 526–536.

Clark, A.P. and Hariharan, S. (1989) Adaptive channel estimator for an HF radio link. *IEEE Transactions on Communications*, **COM-37** (9), 918–926.

Crochiere, R.E. and Rabiner, L.R. (1983) *Multirate Digital Signal Processing*, Prentice-Hall, Englewood Cliffs, NJ.

Cupo, L.R. and Gitlin, R.D. (1989) Adaptive carrier recovery systems for digital data communications receivers. *IEEE Journal on Selected Areas in Communications*, **7**, 1328–1339.

David, R.D., Stearns, S.D., Elliott, G.R., and Etter, D.M. (1983) IIR algorithm for adaptive line enhancement. *Proceedings ICASSP'83*, pp. 17–20.

Davila, C.E. (1990) A stochastic Newton algorithm with data-adaptive step size. *IEEE Transactions on Acoustics, Speech, & Signal Processing*, **ASSP-38**, 1796–1798.

De Courville, M. and Duhamel, P. (1995) Adaptive filtering in subbands using a weighted criterion. *Proceedings IEEE ICASSP'95, May, Detroit, MI*, pp. 985–988.

de León, P.L. II and Etter, D.M. (1995) Experimental results with increased bandwidth analysis filters in oversampled, subband acoustic echo cancellers. *IEEE Signal Processing Letters*, **2** (1), 1–3.

Dembo, A. and Salz, J. (1990) On the least squares tap adjustment algorithm in adaptive digital echo cancellers. *IEEE Transactions on Communications*, **COM-38**, 622–628.

Demoment, G. and Reynaud, R. (1985) Fast minimum variance deconvolution. *IEEE Transactions on Acoustics, Speech, & Signal Processing*, **ASSP-33**, 1324–1326.

Douglas, S.C. (1997) Performance comparison of two implementations of the leaky LMS adaptive filter. *IEEE Transactions on Signal Processing*, **45** (8), 2125–2129.

Duttweiler, D.L. (1978) A twelve-channel digital echo canceler. *IEEE Transactions on Communications*, **COM-26** (5), 647–653.

Duttweiler, D.L. (1982) Adaptive filter performance with nonlinearities in the correlation multiplier. *IEEE Transactions on Acoustics Speech and Signal Processing*, **ASSP-30**, 578–586.

Egelmeers, G.P.M. and Sommen, P.C.W. (1996) A new method for efficient convolution in frequency domain by nonuniform partitioning for adaptive filtering. *IEEE Transactions on Signal Processing*, **44** (12), 3123–3129.

Eleftheriou, E. and Falconer, D.D. (1986) Tracking properties and steady-state performance of RLS adaptive filter algorithms. *IEEE Transactions on Acoustics, Speech, & Signal Processing*, **ASSP-34**, 1097–1109.

Eleftheriou, E. and Falconer, D.D. (1987) Adaptive equalization techniques for HF channels. *IEEE Journal on Selected Areas in Communications*, **SAC-5**, 238–247.

Elliott, S.J. and Nelson, P.A. (1993) Active noise control. *IEEE Signal Processing Letters*, **10** (4), 12–35.

Elliott, D.F. and Rao, K.R. (1982) *Fast Transforms: Algorithms, Analysis, Applications*, Academic Press, New York.

Elliott, S.J. and Rafaely, B. (1997) Rapid frequency-domain adaptation of causal FIR filters. *IEEE Signal Processing Letters*, **4** (12), 337–339.

Eneman, K. and Moonen, M. (1997) A Relation between subband and frequency-domain adaptive filtering. *IEEE International Conference on Digital Signal Processing*, **1**, 25–28.

Er, M. and Cantoni, A. (1983) Derivative constraints for broad-band element space antenna array processors. *IEEE Transactions on Acoustics, Speech, & Signal Processing*, **31** (6), 1378–1393.

Ersoy, O.K. (1997) *Fourier-Related Transforms, Fast Algorithms and Applications*, Prentice Hall PTR, Upper Saddle River, NJ, p. 07458.

Esterman, P. and Kaelin, A. (1994) Analysis of the transient behavior of unconstrained frequency domain adaptive filters. *IEEE International Symposium on Circuits and Systems*, **2**, 21–24.

Eweda, E. (1990a) Analysis and design of a signed regressor LMS algorithm for stationary and nonstationary adaptive filtering with correlated Gaussian data. *IEEE Transactions on Circuits and Systems*, **CAS-37**, 1367–1374.

Eweda, E. (1990b) Optimum step size of sign algorithm for nonstationary adaptive filtering. *IEEE Transactions on Acoustics, Speech, & Signal Processing*, **ASSP-38**, 1897–1901.

Eweda, E. (1994) Comparison of RLS, LMS, and sign algorithms for tracking randomly time-varying channels. *IEEE Transactions on Signal Processing*, **42** (11), 2937–2944.

Eweda, E. (1997) Tracking analysis of sign-sign algorithm for nonstationary adaptive filtering with Gaussian data. *IEEE Transactions on Signal Processing*, **45** (11), 1375–1378.

Eweda, E. and Macchi, O. (1985) Tracking error bounds of adaptive nonstationary filtering. *Automatica*, **21**, 293–302.

Eweda, E. and Macchi, O. (1987) Convergence of the RLS and LMS adaptive filters. *IEEE Transactions on Circuits and Systems*, **34** (7), 799–803.

Falconer, D., Ariyavisitakul, S.L., Benyamin-Seeyar, A., and Eidson, B. (2002) Frequency domain equalization for single-carrier broadband wireless systems. *IEEE Communications Magazine*, **40** (4), 58–66.

Falconer, D.D. and Ljung, L. (1978) Application of fast Kalman estimation to adaptive equalization. *IEEE Transactions on Communications*, **COM-26**, 1439–1446.

Falconer, D.D. and Magee, F.R. Jr. (1973) Adaptive channel memory truncation for maximum likelihood sequence estimation. *Bell System Technical Journal*, **52** (9), 1541–1562.

Farden, D.C. (1981) Tracking properties of adaptive signal processing algorithms. *IEEE Transactions on Acoustics, Speech, & Signal Processing*, **ASSP-29**, 439–446.

Farhang-Boroujeny, B. (1993) Application of orthonormal transforms to implementation of quasi-LMS/Newton algorithm. *IEEE Transactions on Signal Processing*, **41**, 1400–1405.

Farhang-Boroujeny, B. (1994a) Variable-step-size LMS algorithm: new developments and experiments. *IEE Proceedings - Vision, Image, and Signal Processing*, **141**, 311–317.

Farhang-Boroujeny, B. (1994b) Near optimum timing recovery for digitally implemented data receivers. *IEEE Transactions on Communications*, **38** (9), 1333–1336.

Farhang-Boroujeny, B. (1996a) Performance of LMS-based adaptive filters in tracking a time-varying plant. *IEEE Transactions on Signal Processing*, **44** (11), 2868–2871.

Farhang-Boroujeny, B. (1996b) Analysis and Efficient Implementation of Partitioned Block LMS adaptive filters. *IEEE Transactions on Signal Processing*, **44** (11), 2865–2868.

Farhang-Boroujeny, B. (1996c) Channel equalization via channel identification: algorithms and simulation results for rapidly fading channel. *IEEE Transactions on Communications*, **44** (11), 1409–1412.

Farhang-Boroujeny, B. (1997a) An IIR adaptive line enhancer with controlled bandwidth. *IEEE Transactions on Signal Processing*, **45** (2), 477–481.

Farhang-Boroujeny, B. (1997b) Fast LMS/Newton algorithms based on autoregressive modeling and their applications to acoustic echo cancellation. *IEEE Transactions on Signal Processing*, **45** (8), 1987–2000.

Farhang-Boroujeny, B. (2010) *Signal Processing for Software Radios*, 2nd edn, Lulu publishing house. Raleigh, NC.

Farhang-Boroujeny, B. (2011) OFDM versus filter bank multicarrier. *IEEE Signal Processing Magazine*, **28** (3), 92–112.

Farhang-Boroujeny, B. and Gazor, S. (1991) Performance analysis of transform domain normalized LMS Algorithm. *Proceedings ICASSP'91, Conference, Toronto, Canada*, pp. 2133–2136.

Farhang-Boroujeny, B. and Gazor, S. (1992) Selection of orthonormal transforms for improving the performance of the transform domain normalized LMS Algorithm. *IEE Proceedings - Part F: Radar & Signal Processing*, **139**, 327–335.

Farhang-Boroujeny, B. and Lim, Y.C. (1992) A comment on the computational complexity of sliding FFT. *IEEE Transactions on Circuits and Systems*, **39**, 875–876.

Farhang-Boroujeny, B. and Gazor, S. (1994) Generalized sliding FFT and its application to implementation of block LMS adaptive filters. *IEEE Transactions on Signal Processing*, **42**, 532–538.

Farhang-Boroujeny, B., and Wang, Z. (1995) A modified subband adaptive filtering for acoustic echo cancellation. *Proceedings of the International Conference Signal Processing Applications and Technology, Oct, Boston, MA*, pp. 74–78.

Farhang-Boroujeny, B., Lee, Y., and Ko, C.C. (1996) Sliding transforms for efficient implementation of transform domain adaptive filters. *Signal Processing, Elsevier*, **52**, 83–96.

Farhang-Boroujeny, B. and Wang, Z. (1997) Adaptive filtering in subbands: Design issues and experimental results for acoustic echo cancellation. *Signal Processing, Elsevier*, **61**, 213–223.

Farhang-Boroujeny, B., Zhu, H., and Shi, Z. (2006) Markov chain Monte Carlo algorithms for CDMA and MIMO communication systems. *IEEE Transactions on Signal Processing*, **54** (5), 1896–1909.

Fechtel, S.A. and Meyr, H. (1991) An investigation of channel estimation and equalization techniques for moderately rapid fading channels. *ICC'91 Conference Record, Sheraton-Denver Technological Center, June 23–26, Denver*, pp. 25.2.1–25.2.5.

Farhang-Boroujeny, B. and Chan, K.S. (2000) Analysis of the frequency-domain block LMS algorithm. *IEEE Transactions on Signal Processing*, **48** (8), 2332–2342.

Fechtel, S.A. and Meyr, H. (1992) Optimal feedforward estimation of frequency-selective fading radio channels using statistical channel information. *ICC'92 Conference Record, June 14–18, Chicago, IL*, pp. 677–681.

Feng, C.-C. and Chi, C.-Y. (1999) Performance of cumulant based inverse filters for blind deconvolution. *IEEE Transactions on Signal Processing*, **47** (7), 1922–1935.

Ferrara, E.R. (1980) Fast implementation of LMS adaptive filters. *IEEE Transactions on Acoustics, Speech, & Signal Processing*, **ASSP-28**, 474–475.

Ferrara, E.R. and Widrow, B. (1981) The time-sequenced adaptive filter. *IEEE Transactions on Circuits and Systems*, **CAS-28**, 519–523.

Fertner, A. (1997) Frequency-domain echo canceller with phase adjustment. *IEEE Transactions on Circuits and Systems Part II: Analog and Digital Signal Processing*, **44** (10), 835–841.

Feuer, A. and Weinstein, E. (1985) Convergence analysis of LMS filters with uncorrelated Gaussian data. *IEEE Transactions on Acoustics, Speech, & Signal Processing*, **ASSP-33**, 222–230.

Frost, O.L. III. (1972) An algorithm for linearly constrained adaptive array processing. *Proceedings of the IEEE*, **60**, 926–935.

Furukawa, I. (1984) A design of canceller of broad band acoustic echo. *International Teleconference Symposium, Tokyo*. pp. 1/8–8/8.

Gänsler, T., Hansson, M., Ivarsson, C.-J., and Salomonsson, G. (1996) A double-talk detector based on coherence. *IEEE Transactions on Communications*, **44** (11), 1421–1427.

Garayannis, G., Manolakis, D., and Kalouptsidis, N. (1983) A fast sequential algorithm for least squares filtering and prediction. *IEEE Transactions on Acoustics, Speech, & Signal Processing*, **ASSP-31**, 1394–1402.

Gardner, W.A. (1987) Nonstationary learning characteristics of the LMS algorithm. *IEEE Transactions on Circuits and Systems*, **CAS-34** (10), 1199–1207.

Gazor, S. and Farhang-Boroujeny, B. (1992) Quantization effects in transform domain normalized LMS algorithm. *IEEE Transactions on Circuits and Systems Part II: Analog and Digital Signal Processing*, **39**, 1–7.

Gilloire, A. and Vetterli, M. (1992) Adaptive filtering in subbands with critical sampling: analysis, experiments, and application to acoustic echo cancellation. *IEEE Transactions on Signal Processing*, **40** (8), 1862–1875.

Gitlin, R.D. and Weinstein, S.B. (1981) Fractionally-spaced equalization: An improved digital transversal equalizer. *Bell System Technical Journal*, **60**, 275–296.

Gitlin, R.D., Hayes, J.F., and Weinstein, S.B. (1992) *Data Communications Principles*, Plenum Press, New York.

Glendis, G.O. and Kalouptsids, N. (1992a) Efficient order recursive algorithms for multichannel least-squares filtering. *IEEE Transactions on Signal Processing*, **40** (7), 1354–1374.

Glendis, G.O. and Kalouptsids, N. (1992b) Fast adaptive algorithms for multichannel filtering and system identification. *IEEE Transactions on Signal Processing*, **40** (10), 2433–2459.

Glover, J.R. Jr. (1977) Adaptive noise canceling applied to sinusoidal interferences. *IEEE Transactions on Acoustics, Speech, & Signal Processing*, **ASSP-25** (6), 484–491.

Godara, L.C. (1997) Application of antenna arrays to mobile communications. II. Beam-forming and direction-of-arrival considerations. *Proceedings of the IEEE*, **85** (8), 1195–1245.

Golub, G.H. and Van Loan, C.F. (1989) *Matrix Computations*, 2nd edn, The John Hopkins University Press, Baltimore, MD.

Goodwin, G.C. and Sin, K.S. (1984) *Adaptive Filtering, Prediction and Control*, Prentice-Hall, Englewood Cliffs, NJ.

Ghosh, A., Zhang, J., Andrews, J.G., and Muhamed, R. (2011) *Fundamentals of LTE*, Prentice Hall, Upper Saddle River, NJ.

Gray, R.M. (1972) On the asymptotic eigenvalue distribution of toeplitz matrices. *IEEE Transactions on Information Theory*, **IT-18**, 725–730.

Griffiths, L.J. (1969) A simple adaptive algorithm for real-time processing in antenna arrays. *Proceedings ofthe IEEE*, **57**, 1696–1704.

Griffiths, L.J. (1978) *An adaptive lattice structure for noise-cancelling applications. ICASSP Conference, Tulsa, OK*, pp. 87–90.

Griffiths, L.J. and Jim, C.W. (1982) An alternative approach to linearly constrained adaptive beamforming. *IEEE Transactions on Antennas and Propagation*, **AP-30**, 27–34.

Hajivandi, M. and Gardner, W.A. (1990) Measures of tracking performance for the LMS algorithm. *IEEE Transactions on Acoustics, Speech, & Signal Processing*, **ASSP-38** (11), 309–316.

Harris, E.W., Chabries, D.M., and Bishop, F.A. (1986) A variable step (VS) adaptive filter algorithm. *IEEE Transactions on Acoustics, Speech, & Signal Processing*, **ASSP-34**, 309–316.

Hassibi, B., Sayed, A.H., and Kailath, T. (1996) H^{∞} optimality of the LMS algorithm. *IEEE Transactions on Signal Processing*, **44** (2), 267–280.

Hatzinakos, D. and Nikias, C.L. (1991) Blind equalization using a tricepstrum-based algorithm. *IEEE Transactions on Communications*, **39** (5), 669–682.

Haykin, S. (1991) *Adaptive Filter Theory*, 2nd edn, Prentice-Hall, Englewood Cliffs, NJ.

Haykin, S. (1996) *Adaptive Filter Theory*, 3rd edn, Prentice-Hall, Englewood Cliffs, NJ.

Hirsch, D. and Wolf, W.A. (1970) Simple adaptive equalizer for efficient data transmission. *IEEE Transactions on Communication Technology*, **18** (1), 5–12.

Hochwald, B.M. and ten Brink, S. (2003) Achieving near-capacity on a multiple-antenna channel. *IEEE Transactions on Communications*, **51** (3), 389–399.

Honig, M., Madhow, U., and Verdú, S. (1995) Blind adaptive multiuser detection. *IEEE Transactions on Information Theory*, **41** (4), 944–960.

Honig, M.L. and Messerschmitt, D.G. (1981) Convergence properties of an adaptive digital lattice filter. *IEEE Transactions on Acoustics, Speech, & Signal Processing*, **ASSP-29**, 642–653.

Honig, M.L. and Messerschmitt, D.G. (1984) *Adaptive Filters: Structures, Algorithms, and Applications*, Kluwer Academic, Boston, MA.

Horowitz, L.L. and Senne, K.D. (1981) Performance advantage of complex LMS for controlling narrow-band adaptive arrays. *IEEE Transactions on Acoustics, Speech, & Signal Processing*, **ASSP-34**, 722–736.

Houacine, A. (1991) Regularized fast recursive least squares algorithms for adaptive filtering. *IEEE Transactions on Signal Processing*, **39**, 860–871.

Householder, A.S. (1964) *Theory of Matrices in Numerical Analysis*, Blaisdell, New York.

Hoya, T., Loke, Y., Chambers, A., and Naylor, P.A. (1998) Application of the leaky extended LMS (XLMS) algorithm in stereophonic acoustic echo cancellation. *Signal Processing*, **64**, 87–91.

Hsu, F.M. (1982) Square root Kalman filtering for high-speed data received over fading dispersive HF channels. *IEEE Transactions on Information Theory*, **IT-28**, 753–763.

Hush, D.R., Ahmed, N., David, R., and Stearns, S.D. (1986) An adaptive IIR structure for sinusoidal enhancement, frequency estimation, and detection. *IEEE Transactions on Acoustics, Speech, & Signal Processing*, **ASSP-34**, 1380–1390.

Itakura, F. and Saito, S. (1971) Digital filtering techniques for speech analysis and synthesis. *Proceedings of the 7th International Conference on Acoustics*, paper 25C-1, **vol. 3**, pp. 261–264.

ITU-T Recommendation G.167 (03/93). (1993) *Acoustic echo controllers*. ITU.

Iyer, U., Nayeri, M., and Ochi, H. (1994) A poly-phase structure for system identification and adaptive filtering. *Proceedings IEEE ICASSP'94, Adelaide, South Australia*, **vol. III**, pp. 433–436.

Jaggi, S. and Martinez, A.B. (1990) Upper and lower bounds of the misadjustment in the LMS algorithm. *IEEE Transactions on Acoustics, Speech, & Signal Processing*, **ASSP-38**, 164–166.

Jain, A.K. (1976) A fast Karhunen-Loeve transform for a class of stochastic processes. *IEEE Transactions on Communications*, **COM-24**, 1023–1029.

Jayant, N.S. and Noll, P. (1984) *Digital Coding of Waveforms: Principles and Applications to Speech and Video*, Prentice-Hall, Englewood Cliffs, NJ.

Jeyendran, B. and Reddy, V.U. (1990) Recursive system identification in the presence of burst disturbance'. *Signal Processing, Elsevier*, **20**, 227–245.

Jim, C.W. (1977) A comparison of two LMS constrained optimal array structures. *Proceedings of the IEEE*, **65**, 1730–1731.

Johnson, D.H. and Dudgeon, D.E. (1993) *Array Signal Processing: Concepts and Techniques*, Prentice-Hall, Englewood Cliffs, NJ.

Kalouptsidis, N. and Theodoridis, S. (eds) (1993) *Adaptive System Identification and Signal Processing Algorithms*, Prentice-Hall, UK.

Karaboyas, S. and Kalouptsidis, N. (1991) Efficient adaptive algorithms for ARMA identification. *IEEE Transactions on Acoustics, Speech, & Signal Processing*, **ASSP-39**, 571–582.

Karaboyas, S., Kalouptsidis, N., and Caroubalos, C. (1989) Highly parallel transversal batch and block adaptive multichannel LS algorithms and application to decision feedback equalizers. *IEEE Transactions on Acoustics, Speech, & Signal Processing*, **ASSP-37**, 1380–1396.

Kay, S. (1988) *Advanced Topics in Signal Processing* (eds J.S. Lin and A.V. Oppenheim), Prentice-Hall, Englewood Cliffs, NJ. Chapter 2.

Kellermann, W. (1984) Kompensation akustischer echos in frequenzteilbändernin Aachener Kolloquium. *Proceedings Aachener Kolloquium*, FRG, pp. 322–325 (in German).

Kellermann, W. (1985) Kompensation akustischer echos in frequenzteil-bandern. *Frequenz*, **39** (7/8), 209–215 (in German).

Kellermann, W. (1988) Analysis and design of multirate systems for cancellation of acoustical echoes in Proc. *IEEE ICASSP'88, New York*, pp. 2570–2573.

Khong, A.W.H. and Naylor, P.A. (2005) Selective-tap adaptive algorithms in the solution of the nonuniqueness problem for stereophonic acoustic echo cancellation. *IEEE Signal Processing Letters*, **12** (4), 269–272.

Khong, A.W.H. and Naylor, P.A. (2007) Selective-tap adaptive filtering with performance analysis for identification of time-varying systems. *IEEE Transactions on Audio, Speech & Language Processing*, **15** (5), 1681–1695.

Kuo, S.M. and Morgan, D.R. (1996) *Active Noise Control Systems, Algorithms and DSP Implementations*, John Wiley & Sons, Inc., New York.

Kuo, S.M. and Morgan, D.R. (1999) Active noise control: a tutorial review. *The Proceedings of IEEE*, **87** (6), 943–973.

Kushner, H. (1984) *Approximation and Weak Convergence Methods for Random Processes with Applications to Stochastic Systems Theory*, MIT Press, Cambridge, MA.

Kwong, C.P. (1986) Dual sign algorithm for adaptive filtering. *IEEE Transactions on Communications*, **COM-34**, 1272–1275.

Lee, D.T., Morf, M., and Friedlander, B. (1981) Recursive least square ladder filter algorithms. *IEEE Transactions on Acoustics, Speech, & Signal Processing*, **ASSP-29**, 627–641.

Lee, E.A. and Messerschmitt, D.G. (1994) *Digital Communication*, 2nd edn, Kluwer, Boston, MA.

Lee, J.C. and Un, C.K. (1984) A reduced structure of frequency domain block LMS adaptive digital filter. *Proceedings of the IEEE*, **72**, 1816–1818.

Lee, J.C. and Un, C.K. (1986) Performance of transform-domain LMS adaptive digital filters. *IEEE Transactions on Acoustics, Speech, & Signal Processing*, **ASSP-34**, 499–510.

Lee, J.C. and Un, C.K. (1989) Performance analysis of frequency domain block LMS adaptive digital filters. *IEEE Transactions on Circuits and Systems*, **36**, 173–189.

Lev-Ari, H. (1987) Modular architecture for adaptive multichannel lattice algorithms. *IEEE Transactions on Acoustics, Speech, & Signal Processing*, **ASSP-87**, 543–552.

Lev-Ari, H., Kailath, T., and Cioffi, J. (1984) Least-squares adaptive lattice and transversal filters: a unified geometric theory. *IEEE Transactions on Information Theory*, **IT-30**, 222–236.

Levinson, N. (1974) The Wiener R.M.S. (root-mean-square) error criterion in filter design and prediction. *Journal of Mathematical Physics*, **25**, 261–278.

Lewis, P.S. (1990) QR-based algorithms for multichannel adaptive least square lattice filters. *IEEE Transactions on Acoustics, Speech, & Signal Processing*, **ASSP-38**, 421–432.

Li, G. (1997) A stable and efficient adaptive notch filter for direct frequency estimation. *IEEE Transactions on Signal Processing*, **45** (8), 2001–2009.

Li, J., Stoica, P., and Wang, Z. (2003) On robust Capon beamforming and diagonal loading. *IEEE Transactions on Signal Processing*, **51** (7), 1702–1715.

Li, X. and Jenkins, W.K. (1996) The comparison of the constrained and unconstrained frequency-domain block-LMS adaptive algorithms. *IEEE Transactions on Signal Processing*, **44** (7), 1813–1816.

Lim, Y.C. and Farhang-Boroujeny, B. (1992) Fast filter bank (FFB). *EEE Transactions on Circuits and Systems II: Analog and Digital Signal Processing*, **39**, 316–318.

Lim, T.J., Gong, Y., and Farhang-Boroujeny, B. (2000) Convergence analysis and chip- and fractionally-spaced LMS adaptive multiuser CDMA detectors. *IEEE Transactions on Signal Processing*, **48** (8), pp. 2219–2228.

Lin, J., Proakis, J.G., Ling, F., and Lev-Ari, H. (1995) Optimal tracking of time-varying channels: a frequency domain approach for known and new algorithms. *IEEE Transactions on Selected Areas in Communications*, **13**, 141–154.

Lin, S. and Costello, D.J. (2004). *Error Control Coding*, 2nd edn, Prentice Hall, Englewood Cliffs, NJ.

Ling, F. (1991) Givens rotation based least squares lattice and related algorithms. *IEEE Transactions on Signal Processing*, **39**, 1541–1552.

Ling, F. (1993) *Adaptive System Identification and Signal Processing Algorithms* (eds N. Kalouptsidis and S. Theodoridis), Prentice-Hall, UK, Chapter 6.

Ling, F. and Proakis, J.G. (1984a) Nonstationary learning characteristics of least squares adaptive estimation algorithms. *Proceedings ICASSP'84, San Diego, CA*, pp. 3.7.1–3.7.4.

Ling, F. and Proakis, J.G. (1984b) A generalized multichannel least squares lattice algorithm based on sequential processing stages. *IEEE Transactions on Acoustics, Speech, & Signal Processing*, **ASSP-32** (2), 381–389.

Ling, F. and Proakis, J.G. (1985) Adaptive lattice decision-feedback equalizers-their performance and application to time-variant multipath channels. *IEEE Transactions on Communications*, **COM-33** (4), 348–356.

Ling, F., Monolakis, D., and Proakis, J.G. (1986a) A flexible, numerically robust array processing algorithm and its relationship to the Givens transformation. *Proceedings of IEEE International Conference on ASSP, April, Tokyo, Japan.*

Ling, F., Monalakis, D.G., and Proakis, J.G. (1986b) A recursive modified Gram-Schmidt algorithm for least-squares estimation. *IEEE Transactions on Acoustics, Speech, & Signal Processing*, **ASSP-34**, 829–836.

Ling, F., Monalakis, D.G., and Proakis, J.G. (1986c) Numerically robust least-squares lattice ladder algorithms with direct updating of the reflection coefficients. *IEEE Transactions on Acoustics, Speech, & Signal Processing*, **ASSP-34**, 837–845.

Liu, J.C. and Lin, T.P. (1988) Running DHT and real-time DHT analyzer. *Electronics Letters*, **24**, 762–763.

Liu, S. and Tian, Z. (2004) Near-optimum soft decision equalization for frequency selective MIMO channels. *IEEE Transactions on Signal Processing*, **52** (3), 721–733.

Liu, H. and Xu, G. (1996) A subspace method for signature waveform estimation in synchronous CDMA systems. *IEEE Transactions on Communications*, **44** (10), 1346–1354.

Ljung, L. and Söderström, T. (1983) *Theory and Practice of Recursive Identification*, MIT Press, Cambridge.

Ljung, S. and Ljung, L. (1985) Error propagation properties of recursive least-squares adaptation algorithms. *Automatica*, **21**, 157–167.

Ljung, L., Morf, M., and Falconer, D. (1978) Fast calculation of gain matrices for recursive estimation schemes. *International Journal of Control*, **27**, 1–19.

Long, G., Shwed, D., and Falconer, D.D. (1987) Study of a pole-zero adaptive echo canceller. *IEEE Transactions on Circuits and Systems*, **34** (7), 765–769.

Long, G., Ling, F., and Proakis, J.G. (1989) The LMS algorithm with delayed coefficient adaptation. *IEEE Transactions on Acoustics, Speech, & Signal Processing*, **ASSP-37**, 1397–1405.

Long, G. and Ling, F. (1993) Fast initialization of data-deriven Nyquist in-band echo cancellers. *IEEE Transactions on Communications*, **41** (6), 893–904.

Macchi, O. (1986) Optimization of adaptive identification for time-varying filters. *IEEE Transactions on Automatic Control*, **AC-31**, 283–287.

Macchi, O. (1995) *Adaptive Processing, the Least Mean Squares Approach with Applications in Transmission*, John Wiley & Sons, Ltd., Chichester.

Macchi, O.M. and Bershad, N.J. (1991) Adaptive recovery of a chirped sinusoid in noise, Part 1: performance of the RLS algorithm. *IEEE Transactions on Signal Processing*, **ASSP-39**, 583–594.

Makhoul, J. (1975) Linear prediction: a tutorial review. *Proceedings of the IEEE*, **63**, 561–580.

Makhoul, J. (1978) A class of all-zero lattice digital filters: Properties and application. *IEEE Transactions on Acoustics, Speech, & Signal Processing*, **ASSP-26**, 304–314.

Makhoul, J. and Viswanathan, R. (1978) Adaptive lattice methods for linear prediction. *Proceedings ICASSP-78 Conference*, pp. 87–90.

Mavridis, P.P. and Moustakides, G.V. (1996) Simplified Newton-type adaptive estimation algorithms. *IEEE Transactions on Signal Processing*, **44** (8), 1932–1940.

Mansour, D. and Gray, A.H. Jr. (1982) Unconstrained frequency-domain adaptive filter. *IEEE Transactions on Acoustics, Speech, & Signal Processing*, **ASSP-30**, 726–734.

Markel, J.D. (1971) FFT pruning. *IEEE Transactions on Audio and Electroacoustics*, **AU-19**, 305–311.

Marshall, D.F., Jenkins, W.K., and Murphy, J.J. (1989) The use of orthogonal transforms for improving performance of adaptive filters. *IEEE Transactions on Circuits and Systems*, **CAS-36**, 474–483.

Marshall, D.F. and Jenkins, W.K. (1992) A fast quasi-Newton adaptive filtering algorithm. *IEEE Transactions on Signal Processing*, **40**, 1652–1662.

Mathew, G., Farhang-Boroujeny, B., and Wood, R. (1997) Design of multilevel decision feedback equalizers. *IEEE Transactions on Magnetics*, **33**, 4528–4542.

Mathew, G., Reddy, V.U., and Dasgupta, S. (1995) Adaptive estimation of eigensubspace. *IEEE Transactions on Signal Processing*, **43** (2), 401–411.

Mathews, V.J. and Cho, S.H. (1987) Improved convergence analysis of stochastic gradient adaptive filters using the sign algorithm. *IEEE Transactions on Acoustics, Speech, & Signal Processing*, **ASSP-35**, 450–454.

Mathews, V.J. and Xie, Z. (1990) Stochastic gradient adaptive filters with gradient adaptive step sizes. *Proceedings ICASSP'90, Conference, Albuquerque, NM*, pp. 1385–1388.

Mathews, V.J. and Xie, Z. (1993) A stochastic gradient adaptive filter with gradient adaptive step size. *IEEE Transactions on Acoustics, Speech, & Signal Processing*, **ASSP-41**, 2075–2087.

Mayyas, K. and Aboulnasr, T. (1997) Leaky LMS algorithm: MSE analysis for Gaussian data. *IEEE Transactions on Signal Processing*, **45** (4), 927–934.

Mazo, J.E. (1979) On the independence theory of equalizer convergence. *Bell System Technical Journal*, **58**, 963–993.

McLaughlin, H.J. (1996) System and method for an efficient constrained frequency-domain adaptive filter. US Patent, No. 5,526,426, Date of Patent: June 11, 1996.

Meyr, H., Moeneclaey, M., and Fechtel, S.A. (1998). *Digital communication receivers*. John Wiley and Sons, New York.

Mikhael, W.B., Wu, F.H., Kazovsky, L.G., Kang, G.S., and Fransen, L.J. (1986) Adaptive filters with individual adaptation of parameters. *IEEE Transactions on Circuits and Systems*, **CAS-33**, 677–685.

Montazeri, M. and Duhamel, P. (1995) A set of algorithms linking NLMS and block RLS algorithms. *IEEE Transactions on Signal Processing*, **43** (2), 444–453.

Moonen, M. and Vandewalle, J. (1990) Recursive least square with stabilized inverse factorization. *IEEE Transactions on Signal Processing*, **21** (1), 1–15.

Moose, P.H. (1994) A technique for orthogonal frequency division multiplexing frequency offset correction. *IEEE Transactions on Communications*, **42** (10), 2908–2914.

Morf, M. and Kailath, T. (1975) Square-root algorithms for least-squares estimation. *IEEE Transactions on Automatic Control*, **AC-20**, 487–497.

Morf, M., Dickinson, B., Kailath, T., and Vieira, A. (1977) Efficient solution of covariance equations for linear prediction. *IEEE Transactions on Acoustics, Speech, & Signal Processing*, **ASSP-25**, 423–433.

Morgan, D.R. (1995) Slow asymptotic convergence of LMS acoustic echo cancellers. *IEEE Transactions on Speech and Audio Processing*, **3**, 126–136.

Morgan, D.R. and Thi, J.C. (1995) A delayless subband adaptive filter architecture. *IEEE Transactions on Signal Processing*, **43** (8), 1819–1830.

Morgan, D.R. and Kratzer, S.G. (1996) On a class of computationally efficient, rapidly converging, generalized NLMS algorithms. *IEEE Signal Processing Letters*, **3**, 245–247.

Morgan, D.R., Hall, J.L., and Benesty, J. (2001) Investigation of several types of nonlinearities for use in stereo acoustic echo cancellation. *IEEE Transactions on Speech and Audio Processing*, **9** (6), 686–696.

Moulines, E., Amrane, O.A., and Grenier, Y. (1995) The generalized multidelay adaptive filter: structure and convergence analysis. *IEEE Transactions on Signal Processing*, **43** (1), 14–28.

Moulines, E., Duhamel, P., Cardoso, J., and Mayrargue, S. (1994) Subspace methods for the blind identification of multichannel FIR filters. *IEEE International Conference on Acoustics, Speech, and Signal Processing*, **4**, IV/57–IV/576.

Moustakides, G.V. (1997) Study of the transient phase of the forgetting factor RLS. *IEEE Transactions on Signal Processing*, **45** (10), 2468–2476.

Moustakides, G.V. and Theodoridis, S. (1991) Fast Newton transversal filters –A new class of adaptive estimation algorithms. *IEEE Transactions on Signal Processing*, **39**, 2184–2193.

Mueller, K.H. (1973) A new approach to optimum pulse shaping in sampled systems using time-domain filtering. *The Bell System Technical Journal*, **52** (5), 723–729.

Murthy, N.R. and Swamy, M.N.S. (1992) On the computation of running discrete cosine and sine transforms. *IEEE Transactions on Signal Processing*, **40**, 1430–1437.

Narayan, S.S. and Peterson, A.M. (1981) Frequency domain LMS algorithm. *Proceedings of the IEEE*, **69**, 124–126.

Narayan, S.S., Peterson, A.M., and Narasimha, M.J. (1983) Transform domain LMS algorithm. *IEEE Transactions on Acoustics, Speech, & Signal Processing*, **ASSP-31**, 609–615.

Nayebi, K., Barnwell, T.P., and Smith, M.J.T. (1994) Low delay FIR filter banks: design and evaluation. *IEEE Transactions on Signal Processing*, **42** (1), 24–31.

Nitzberg, R. (1985) Application of the normalized LMS algorithm to MSLC. *IEEE Transactions on Aerospace and Electronic Systems*, **AES-21**, 79–91.

Ogunfunmi, A.O. and Peterson, A.M. (1992) On the implementation of the frequency-domain LMS adaptive filter. *IEEE Transactions on Circuits and Systems II: Analog and Digital Signal Processing*, **39**, 318–322.

Oppenheim, A.V., Willsky, A.S., and Young, I.T. (1983) *Signals and Systems*, Prentice-Hall, Englewood Cliffs, NJ.

Oppenheim, A.V. and Schafer, R.W. (1975) *Digital Signal Processing*, Prentice-Hall, Englewood Cliffs, NJ.

Oppenheim, A.V. and Schafer, R.W. (1989) *Discrete-Time Signal Processing*, Prentice-Hall, Englewood Cliffs, NJ.

Paleologu, C., Benesty, J., and Ciochină, S. (2008) A variable step-size affine projection algorithm designed for acoustic echo cancellatin. *IEEE Transactions on Audio, Speech and Language Processing*, **16** (8), 1466–1478.

Papoulis, A. (1991) *Probability, Random Variables, and Stochastic Processes*, 3rd edn, McGraw Hill, Singapore.

Peng, R.-H., Chen, R.-R., and Farhang-Boroujeny, B. (2010) Markov chain Monte Carlo detectors for channels with intersymbol interference. *IEEE Transactions on Signal Processing*, **58** (4), 2206–2217.

Perkins, F.A. and McRae, D.D. (1982) A high performance HF modem (Harris Corp. Apr. 1982). *Presented at the International Defense Electron.Expo*, May, *Hanover, West Germany*.

Petraglia, M.R. and Mitra, S.K. (1991) Generalized fast convolution implementations of adaptive filters. *IEEE Symposium Circuits and Systems*, **5**, 2916–2919.

Petraglia, M.R. and Mitra, S.K. (1993) Adaptive FIR filter structure based on the generalized subband decomposition of FIR filters. *IEEE Transactions Circuits and Systems II: Analog and Digital Signal Processing*, **40**, 354–362.

Petillon, T., Gilloire, A., and Theodoridis, S. (1994) The fast Newton transversal filter: an efficient scheme for acoustic echo cancellation in mobile radio. *IEEE Transactions on Signal Processing*, **42**, 509–518.

Picchi, G. and Prati, G. (1994) Self-orthogonalizing adaptive equalization in the discrete frequency domain. *IEEE Transactions on Communications*, **COM-32**, 371–379.

Proakis, J.G. (1995). *Digital Communications*, 3rd edn, McGraw-Hill International Editions, New York.

Proakis, J.G., Rader, C., Ling, F., and Nikias, C. (1992) *Advanced Digital Signal Processing*, Macmillan, New York.

Proudler, I.K., McWhirter, J.G., and Shepherd, T.J. (1990) The QRD-based least square lattice algorithm: Some computer simulation using finite wordlenghths. *Proceedings IEEE International Symposium on Circuits and systems, New Orleans, LA, May*, pp. 258–261.

Proudler, I.K., McWhirter, J.G., and Shepherd, T.J. (1991) Computationally efficient QR decomposition approach to least squares adaptive filtering. *IEE Proceedings*, **138** (4) Pt. F, 341–353.

Qureshi, S.U.H. (1985) Adaptive equalization. *Proceedings of the IEEE*, **73**, 1349–1387.

Rabiner, L.R. and Gold, B. (1975) *Theory and Application of Digital Signal Processing*, Prentice-Hall, Englewood Cliffs, NJ.

Rabiner, L.R. and Schafer, R.W. (1978) *Digital Processing of Speech Signals*, Prentice-Hall, Englewood Cliffs, NJ.

Rao, H.I.K. and Farhang-Boroujeny, B. (2009) Fast LMS/Newton algorithms for stereophonic acoustic echo cancellation. *IEEE Transactions on Signal Processing*, **57** (8), 2919–2930.

Rao, H.I.K. and Farhang-Boroujeny, B. (2010) Analysis of the stereophonic LMS/Newton algorithm and impact of signal nonlinearity on its convergence behavior. *IEEE Transactions on Signal Processing*, **58** (12), 6080–6092.

Reddi, S.S. (1984) A time-domain adaptive algorithm for rapid convergence. *Proceedings of the IEEE*, **72**, 533–535.

Reddy, V.U., Egardt, B., and Kailath, T. (1981) Optimized lattice-form adaptive line enhancer for a sinusoidal signal in broad-band noise. *IEEE Transactions on Circuits and Systems*, **CAS-28** (6), 542–552.

Reddy, V.U., Shan, T.J., and Kailath, T. (1983) Application of modified least-squares algorithms to adaptive echo cancellation. *International Conference on Acoustic and Speech Signal Pprocess, March 14–16, Boston, MA*, pp. 53–56.

Reddy, V.U., Paulraj, A., and Kailath, T. (1987) Performance analysis of the optimum beamformer in the presence of correlated sources and its behavior under spatial smoothing. *IEEE Transactions on Acoustics, Speech, & Signal Processing*, **ASSP-35** (7), 927–936.

Regalia, P.A. (1991) An improved lattice-based adaptive IIR notch filter. *IEEE Transactions on Signal Processing*, **39**, 2124–2128.

Regalia, P.A. and Bellanger, M.G. (1991) On the duality between fast QR methods and lattice methods in least squares adaptive filtering. *IEEE Transactions on Signal Processing*, **39**, 879–892.

Rockafellar, R.T. (1970) *Convex Analysis*, Princeton Univ. Press, Princeton, NJ.

Samson, C. and Reddy, V.U. (1983) Fixed-point error analysis of normalized ladder algorithm. *IEEE Transactions on Acoustics, Speech, & Signal Processing*, **ASSP-31**, 1177–1191.

Santamaria, I., Pantaleon, C., Vielva, L., and Ibanez, J. (2004) Blind equalization of constant modulus signals using support vector machines. *IEEE Transactions on Signal Processing*, **52** (6), 1773–1782.

Satorius, E.H. and Alexander, S.T. (1979) Channel equalization using adaptive lattice algorithms. *IEEE Transactions on Communications*, **COM-27**, 899–905.

Satorius, E.H. and Park, J.D. (1981) Application of least-squares lattice algorithms to adaptive equalization. *IEEE Transactions on Communications*, **COM-29**, 136–142.

Sayed, A.H. (2003) *Fundamentals of Adaptive Filtering*, John Wiley & Sons, Inc., Hoboken, NJ.

Schobben, D.W.E., Egelmeers, G.P.M., and Sommen, P.C.W. (1997) Efficient realization of the block frequency domain adaptive filter. *Proceedings ICASSP'97, Conference*, **vol. 3**, pp. 2257–2260.

Sethares, W.A., Mareels, I.M.Y., Anderson, B.D.O., Johnson, C.R. Jr., and Bitmead, R.R. (1988) Excitation conditions for signed regressor least mean squares adaptation. *IEEE Transactions on Acoustics, Speech, & Signal Processing*, **ASSP-35**, 613–624.

Shahbazpanahi, S., Gershman, A.B., Luo, Z.-Q., and Wong, K.M. (2003) Robust adaptive beamforming for general-rank signal models. *IEEE Transactions on Signal Processing*, **51** (9), 2257–2269.

Sharma, R., Sethares, W.A., and Bucklew, J.A. (1996) Asymptotic analysis of stochastic gradient-based adaptive filtering algorithms with general cost functions. *IEEE Transactions on Signal Processing*, **44** (9), 2186–2194.

Shin, Y.K., and Lee, J.G. (1985) A study on the fast convergence algorithm for the LMS adaptive filter design. *Proc. KIEE*, **19** (5), 12–19.

Shin, H.-C. and Sayed, A.H. (2004) Mean-square performance of a family of affine projection algorithms. *IEEE Transactions on Signal Processing*, **52** (1), 90–102.

Shynk, J.J. (1989) Adaptive IIR filtering. *IEEE ASSP Magazine*, **5**, 4–21.

Shynk, J.J. (1993) Frequency-domain and multirate adaptive filtering. *IEEE Signal Processing Magazine*, **9**, 14–37.

Sidhu, G.S. and Kailath, T. (1974) Development of new estimation algorithms by innovations analysis and shift-invariance properties. *IEEE Transactions on Information Theory*, **IT-20**, 759–762.

Slock, D.T.M. (1991), Fractionally-spaced subband and multiresolution adaptive filters. *Proceedings IEEE ICASSP'91*, pp. 3693–3696.

Slock, D.T.M. (1993) On the convergence behaviour of the LMS and the normalized LMS algorithms. *IEEE Transactions on Signal Processing*, **41** (9), 2811–2825.

Slock, D.T.M. and Kailath, T. (1988) Numerically stable fast recursive least-squares transversal filters. *Proceedings ICASSP-88 Conference*, **vol. 3**, pp. 1365–1368.

Slock, D.T.M. and Kailath, T. (1991) Numerically stable fast recursive least-squares transversal filters. *IEEE Transactions on Signal Processing*, **39** (1), 92–114.

Slock, D.T.M. and kailath, T. (1992) A modular multichannel multiexperiment fast transversal filter RLS algorithm. *Signal Processing*, **28**, 25–45.

Slock, D.T.M., Chisci, L., Lev-Ari, H., and Kailath, T. (1992) Modular and numerically stable fast transversal filters for multichannel and multiexperiment RLS. *IEEE Transactions on Signal Processing*, **ASSP-40** (4), 784–802.

Slock, D.T.M. and Kailath, T. (1993) in *Adaptive System Identification and Signal Processing Algorithms* (eds N. Kalouptsidis and S. Theodoridis), Prentice-Hall, UK, Chapter 5.

Schmidl, T.M. and Cox, D.C. (1997) Robust frequency and timing synchronization for OFDM. *IEEE Transactions on Communications*, **45** (12), 1613–1621.

Solo, V. (1992) The error variance of LMS with time-varying weights. *IEEE Transactions on Signal Processing*, **40** (4), 803–813.

Somayazulu, V.S., Mitra, S.K., and Shynk, J.J. (1989) Adaptive line enhancement using multirate techniques. *Proceedings of IEEE ICASSP'89, Conference, May, Glasgow, Scotland*, pp. 928–931.

Sommen, P. (1988) On the convergence properties of a partitioned block frequency domain adaptive filter (PBFDAF). *Proceedings EUSIPCO*, pp. 1401–1404.

Sommen, P.C.W. (1989) Partitioned frequency domain adaptive filters. *Proceedings Asilomar Conference on Signals and Systems, Pacific Grove, CA*, pp. 676–681.

Sommen, P.C.W., Van Gerwen, P.J., Kotmans, H., and Janssen, A.J.E.M. (1987) Convergence analysis of a frequency domain adaptive filter with exponential power averaging and generalized window function. *IEEE Transactions on Circuits and Systems*, **34** (7), 788–798.

Sommen, P.C.W. and de Wilde, E. (1992) Equal convergence conditions for normal and partitioned-frequency domain adaptive filters *Proceedings ICASSP'92, Conference*, **vol. IV**, pp. 69–72.

Sondhi, M.M. and Kellermann, W. (1992) Adaptive echo cancellation for speech signals, in *Advances in Speech Signal Processing* (eds S. Furui and M.M. Sondhi), Marcel Dekker, New York, pp. 327–356, Chapter 11.

Sondhi, M.M., Morgan, D.R., and Hall, J.L. (1995). Stereophonic acoustic echo cancellation –an overview of the fundamental problem. *IEEE Signal Processing Letters*, **2** (8), 148–151.

Soo, J.S. and Pang, K.K. (1987) A new structure for block FIR adaptive digital filters. *IREECON Int. Dig. Papers, Sydney, Australia*, pp. 364–367.

Soo, J.S. and Pang, K.K. (1990) Multidelay block frequency domain adaptive filter. *IEEE Transactions on Acoustics, Speech, & Signal Processing*, **ASSP-38**, 373–376.

Soumekh, M. (1994) *Fourier Array Imaging*, Prentice-Hall, Englewood Cliffs, NJ.

South, C.R., Hoppitt, C.E., and Lewis, A.V. (1979) Adaptive filters to improve loudspeaker telephone. *Electronics Letters*, **15** (21), 673–674.

Stasinski, R. (1990) Adaptive Filters in Domains of Adaptive Transforms. *Proceedings of Singapore ICCS'90, Conference, 5–9 Nov Singapore*, pp. 18.2.1–18.2.5.

Stearns, S.D. (1981) Error surface of recursive adaptive filters. *IEEE Transactions on Circuits and Systems*, **CAS-28** (6), 603–606.

Stein, S. (1987) Fading channels issues in system engineering. *IEEE Journal on Selected Areas in Communications*, **SAC-5**, 68–89.

Stoica, P., Wang, Z., and Li, J. (2003) Robust Capon beamforming. *IEEE Signal Processing Letters*, **10** (6), 172–175.

Strang, G. (1980) *Linear Algebra and Its Applications*., 2nd edn, Academic Press, New York.

Stüber, G.L., Barry, J.R., McLaughlin, S.W., Li, Ye., Ingram, M.A., and Pratt, T.G. (2004) Broadband MIMO-OFDM wireless communications. *Proceedings of the IEEE*, **92** (2), 271–294.

Talbot, S.L. (2008) Carrier synchronization for mobile orthogonal frequency division multiplexing. PhD Thesis, Univ. of Utah.

Talbot, S.L., and Farhang-Boroujeny, B., (2008) Spectral method of blind carrier tracking for OFDM, *IEEE Transactions on Signal Processing*, **56** (7), Part 1, 2706–2717.

Tanrikulu, O., Baykal, B., Constantinides, A.G., and Chambers, J.A. (1997). Residual echo signal in critically sampled subband acoustic echo cancellers based on IIR and FIR filter banks *IEEE Transactions on Signal Processing*, **45** (4), 901–912.

Tarrab, M. and Feuer, A. (1988) Convergence and performance analysis of the normalized LMS algorithm with uncorrelated Gaussian data. *IEEE Transactions on Information Theory*, **IT-34**, 680–691.

Tong, L., Xu, G., and Kailath, T. (1994) Blind identification and equalization based on second-order statistics: a time domain approach. *IEEE Transactions on Information Theory*, **40** (2), 340–349.

Tourneret, J.-Y., Bershad, N.J., and Bermudez, J.C.M. (2009) Echo cancellation – the generalized likelihood ratio test for double-talk versus channel change. *IEEE Transactions on Signal Processing*, **57** (3), 916–926.

Tsatsanis, M.K. and Giannakis, G.B. (1997) Transmitter induced cyclostationarity for blind channel equalization. *IEEE Transactions on Signal Processing*, **45** (7), 1785–1794.

Tüchler, M., Singer, A.C., and Koetter, R. (2002) Minimum mean squared error equalization using a priori information. *IEEE Transactions on Signal Processing*, **50** (3), 673–683.

Tugnait, J.K. (1995) Blind equalization and estimation of digital communication FIR channels using cumulant matching. *IEEE Transactions on Communications*, **43** (2-4), 1240–1245.

Tummala, M. and Parker, S.R. (1987) A new efficient adaptive cascade lattice structure. *IEEE Transactions on Circuits and Systems*, **34** (7), 707–711.

Vaidyanathan, P.P. (1993) *Multirate Systems and Filter Banks*, Prentice-Hall, Englewood Cliffs, NJ.

Vaidyanathan, P.P. and Nguyen, T.Q. (1987) Eigenfilters: a new approach to least-squares FIR filter design and applications including Nyquist filters. *IEEE Transactions on Circuits and Systems*, **CAS-34**, 11–23.

Valin, J.-M. (2007) On adjusting the learning rate in frequency domain echo cancellation with double-talk. *IEEE Transactions Audio, and Language Processing*, **15** (3), 1030–1034.

van den Bos, A. (1994) Complex gradient and Hessian. *IEE Proceedings - Vision, Image, and Signal Processing*, **141** (1), 380–382.

van Waterschoot, T. and Moonen, M. (2011) Fifty years of acoustic feedback control: state of the art and future challenges. *Proceedings of the IEEE*, **99** (2), 288–327.

Verhaegen, M.H. (1989) Improved understanding of the loss-of-symmetry phenomenon in the conventional Kalman filter. *IEEE Transactions on Automatic Control*, **AC-34**, 331–333.

Verhaegen, M.H. (1989) Round-off error propagation in four generally-applicable, recursive least-squares estimation schemes. *Automatica*, **25** (3), 437–444.

Verhoeckx, N.A.M. and Claasen, T.A.C.M. (1984) Some considerations on the design of adaptive digital filters equipped with the sign algorithm. *IEEE Transactions Acoustic, Speech & Signal Processing*, **ASSP-32**, 258–266.

Vikalo, H., Hassibi, B., and Kailath, T. (2004) Iterative decoding for MIMO channels via modified sphere decoding. *IEEE Transactions on Wireless Communications*, **3** (6), 2299–2311.

von Zitzewitz, A. (1990), Considerations on acoustic echo cancelling based on realtime experiments *Proceedings IEEE EUSIPCO'90 V European Signal Process. Conference, Sept, Barcelona, Spain*, pp. 1987–1990.

Wackersreither, G. (1985) On the design of filters for ideal QMF and polyphase filter banks. *AEU – International Journal of Electronics and Communications, Elsevier*, 123–130.

Wang, X. and Poor, H.V. (1999) Iterative (turbo) soft interference cancellation and decoding for coded CDMA. *IEEE Transactions on Communications*, **47** (7), 1046–1061.

Wang, Z. (1996) Adaptive filtering in subband. M.Eng. thesis. Department of Electrical Engrg, National Univ. of Singapore.

Ward, C.R., Hargrave, P.J., and McWhirter, J.G. (1986) A novel algorithm and architecture for adaptive digital beamforming. *IEEE Transactions on Antennas and Propagation*, **AP-34**, 338–346.

Wei, P.C., Zeidler, J.R., and Ku, W.H. (1997) Adaptive recovery of a chirped signal using the RLS algorithm. *Ieee Transactions on Signal Processing*, **45** (2), 363–376.

Weiss, A. and Mitra, D. (1979) Digital adaptive filters: Conditions for convergence, rates of convergence, effects of noise and errors arising from the implementation. *IEEE Transactions on Information Theory*, **IT-25**, 637–652.

Widrow, B. and Hoff, M.E. Jr. (1960) Adaptive switching circuits. *IRE WESCON Conv. Rec.*, pt. 4 96–104.

Widrow, B., Mantey, P.E., Griffiths, L.J., and Goode, B.B. (1967). Adaptive antenna systems *Proceedings of the IEEE*, **55** (12), 2143–2159.

Widrow, B., McCool, J., and Ball, M. (1975) The complex LMS algorithm. *Proceedings of the IEEE*, **63**, 719–720.

Widrow, B., Glover, J.R., Jr., McCool, J.M., Kaunitz, J., Williams, C.S., Hearn, R.H., Zeidler, J.R., Dong, E. Jr., and Goodlin, R.C. (1975). Adaptive noise cancelling: principles and applications. *Proceedings of the IEEE*, **63** (12), 1692–1716.

Widrow, B., McCool, J.M., Larimore, M.G., and Johnson, C.R. Jr. (1976) Stationary and nonstationary learning characteristics of the LMS adaptive filter. *Proceedings of the IEEE*, **64** (8), 1151–1162.

Widrow, B. and Walach, E. (1984) On the statistical efficiency of the LMS algorithm with nonstationary inputs. *IEEE Transactions on Information Theory*, **IT-30**, 211–221.

Widrow, B. and Stearns, S.D. (1985) *Adaptive Signal Processing*, Prentice-Hall, Englewood Cliffs, NJ.

Widrow, B., Baudrenghien, P., Vetterli, M., and Titchener, P.F. (1987) Fundamental relations between the LMS algorithm and the DFT. *IEEE Transactions on Circuits and Systems*, **CAS-34**, 814–820.

Yang, B. (1994) A note on error propagation analysis of recursive least-squares algorithms. *IEEE Transactions on Signal Processing*, **42** (12), 3523–3525.

Yasukawa, H. and Shimada, S. (1987) Acoustic echo canceler with high speech quality. *Proceedings IEEE ICASSP'87, Dallas, TX*, pp. 2125–2128.

Yasukawa, H. and Shimada, S. (1993) An acoustic echo canceler using subband sampling and decorrelation methods. *IEEE Transactions on Signal Processing*, **41** (2), 926–930.

Yon, C.H. and Un, C.K. (1992) Normalised frequency-domain adaptive filter based on optimum block algorithm. *IEE Electronic letters*, **28** (1), 11–12.

Yon, C.H. and Un, C.K. (1994) Fast multidelay block transform-domain adaptive filters based on a two-dimensional optimum block algorithm. *IEEE Transactions on Circuits and Systems II: Analog and Digital Signal Processing*, **41** (5), 337–345.

Yuen, S., Abend, K., and Berkowitz, R.S. (1988) A recursive least-squares algorithm for multiple inputs and outputs and a cylindrical systolic implementation. *IEEE Transactions on Acoustics, Speech, & Signal Processing*, **ASSP-36**, 1917–1923.

Zhu, H., Farhang-Boroujeny, B., and Chen, R.-R. (2005) On performance of sphere decoding and Markov chain Monte Carlo detection methods. *IEEE Signal Processing Letters*, **12** (10), 669–672.

Index

Adaptive Filters: Theory and Applications, Second Edition. Behrouz Farhang-Boroujeny.

Printed and bound by CPI Group (UK) Ltd, Croydon, CR0 4YY

16/04/2025

14658398-0004